SPACE RESEARCH XII

Vol. 2

Organized by

THE COMMITTEE ON SPACE RESEARCH — COSPAR

and

THE UNITED STATES NATIONAL ACADEMY OF SCIENCES

* * *

THE SYMPOSIUM ON TOTAL SOLAR ECLIPSE OF 7 MARCH, 1970

Sponsored by
The Committee on Space Research — COSPAR
The International Union of Geodesy and Geophysics — IUGG
The International Astronomical Union — IAU
The International Union of Radio Science — URSI

THE SYMPOSIUM ON DYNAMICS OF THE THERMOSPHERE AND IONOSPHERE ABOVE 120 km

Sponsored by
The International Union of Radio Science — URSI
The Committee on Space Research — COSPAR

THE SYMPOSIUM ON HIGH ANGULAR RESOLUTION ASTRONOMICAL OBSERVATIONS FROM SPACE

Sponsored by
The Committee on Space Research — COSPAR
The International Astronomical Union — IAU

COSPAR

SPACE RESEARCH XII

Proceedings of Open Meetings of Working Groups of the Fourteenth Plenary Meeting of COSPAR

Seattle, Washington, USA, 21 June—2 July, 1971

and of

THE SYMPOSIUM ON TOTAL SOLAR ECLIPSE OF 7 MARCH, 1970

Seattle, Washington, USA, 18, 19 and 21 June, 1971

and of

THE SYMPOSIUM ON DYNAMICS OF THE THERMOSPHERE AND IONOSPHERE ABOVE 120 km

Seattle, Washington, USA, 24, 25 and 26 June, 1971

and of

THE SYMPOSIUM ON HIGH ANGULAR RESOLUTION ASTRONOMICAL OBSERVATIONS FROM SPACE

Seattle, Washington, USA, 28, 29, 30 June and 1 July, 1971

Edited by

S. A. BOWHILL
L. D. JAFFE
M. J. RYCROFT

Volume 2

AKADEMIE-VERLAG · BERLIN
1972

Executive Editor:
Dr. A. C. Stickland

Library of Congress Catalog Card Number 60–50878

202 · 100/466/72
Gesamtherstellung: VEB Druckerei „Thomas Müntzer“, 582 Bad Langensalza
Bestellnummer: 761 674 2 (3059/XII) · ES 18 D 1, 18 E 2
Printed in German Democratic Republic

CONTENTS LIST

Volume 1

Cosmic Dust

Lunar Results

Direct Observations

Optical Methods

The Earth observed from Space

Surface and Meteorological Surveys

Radio Interferometry

The Earth's Neutral Atmosphere

Tropospheric Refraction

Stratosphere and Mesosphere

Thermospheric Tides

Mid-latitude Ionosphere

Polar Ionosphere and Precipitation of Low Energy Charged Particles

Rocket Observations

Satellite Observations

Theory

Solar Terrestrial Relationships

Magnetosphere

Interplanetary Medium

Astronomy

Solar Physics

Interstellar Wind

Ultraviolet Radiation

Cosmic Rays

General Relativity

High Angular Resolution Astronomical Observations from Space

Ground-based Observations, Results, Limitations:
Astrophysical Expectations from Increased Angular Resolution

Dynamics of the Thermosphere and Ionosphere above 120 km

Space Research XII — Akademie-Verlag, Berlin 1972

STRUCTURE AND MOTION OF THE THERMOSPHERE DEDUCED FROM SATELLITE DRAG

M. YA. MAROV and A. M. ALPHEROV

Academy of Sciences of the USSR, Moscow, USSR

Results of the satellite drag measurements have provided significant progress in comprehensive understanding of the structure and dynamics of the earth's thermosphere. The data of temperature and density variations of the thermosphere due to solar and geomagnetic activity as well as semi-annual, seasonal, diurnal and latitudinal variations were deduced from the density and scale height distribution accounting for the data of the neutral composition measurements. The contemporary knowledge on the structure and principal variations of the thermosphere, the latter described by corresponding empirical relationships, are summarized in the new version of COSPAR Reference Atmosphere (to be adopted as CIRA-1971).

The analysis of secular change of satellite orbital inclinations leads to supposition of the existence of permanent westward winds above 180–200 km with speeds exceeding that of the earth's rotation. The speed of winds, revealed from this procedure, appears to increase depending on height up to 300–350 km.

The whole bulk of experimental data now available characterizes the thermosphere as a rather dynamic medium, its structure and motion being determined by solar heat input and by its interaction wtih lower regions of the atmosphere.

1. Introduction

The last few years when the new COSPAR Reference Atmospheres (CIRA-1971) were being developed were especially fruitful for the physics of the upper atmosphere. Rapid progress in the accumulation of experimental data has resulted in improved knowledge of the systematic variations of the structure parameters and in the discovery of new phenomena (for example, the winter helium bulge).

Determination of atmospheric density by satellite drag is still the most effective method of studying the global structure and upper atmosphere dynamics. This method provides a self-consistent picture of the distribution of the structure parameters, including their space–time variations, from the bottom of the thermosphere to the exosphere.

Important data on composition and temperature were obtained at heights between 100 and 300 km by means of direct *in situ* measurements with instruments installed on rockets and satellites and by the method of incoherent scattering. At the same time investigations associated with the development of upper atmosphere models were continued.

The measurements and theoretical research have led to the concept of a significant interconnection of the thermal regime, dynamics and composition of the upper atmosphere and the relation between the neutral and charged components. The full system of equations describing the upper atmosphere structure is rather complex and in spite of its formulation this system has not yet been completely solved.

Despite great progress in studying compound aeronomic and dynamic processes the theoretical models obtained so far do not adequately predict all phenomena observed in the upper atmosphere with the required accuracy. Therefore, in spite of the more physical character of the theoretical models, a new CIRA-1971 suggested by Jacchia [3] (Jacchia-1970) is based on empirical relations. The latter are derived in such a way as to provide the best fitting to the experimental data on the neutral upper atmosphere structure and variations deduced from analysis of the observed motion of satellites.

2. The Main Regularities

The satellite drag data cover almost the whole thermosphere and a significant part of the exosphere in the height interval from 130 to 1200 km. Since up to about 1000 km the ratio of the ion concentration to the neutral particle concentration does not exceed the value 10^{-1}, the dynamical behaviour of the thermosphere is essentially defined by the neutral gas. At the present time the height distribution of the upper atmosphere, its character and magnitude of variation are known reasonably accurately. All of the variations in one way or another are associated with intensity variations of solar energy absorbed by the atmosphere at a number of levels. The most prominent variations are those connected with different phases of solar activity and insolation periodicity due to diurnal and annual cycles. Some latitudinal dependence also exists, as well as short-term variations correlating mainly with fluctuations in the geomagnetic indices. Finally, there must be mentioned dynamical processes in the troposphere, on the surface and in the earth's crust (heating by infrasonic waves, cyclonic activity, seismicity etc.) that appear to exert an influence on the thermosphere.

Fig. 1 shows density and temperature variations with height in the same way as they are determined by Jacchia's model 1970. The profiles of density and temperature are taken for exospheric temperatures equal to $T_\infty = 700$ °K, 1000 °K and 1600 °K. Exospheric temperature variation from 700 °K to 1600 °K corresponds, respectively, to the mean variation of density and temperature with height from night conditions near minimum solar activity when the solar decimetric flux is equal to $F_{10.7} = 70 \times 10^{-22}$ W m^{-2} Hz^{-1} up to daily conditions near maximum solar activity when $F_{10.7} = 220 \times 10^{-22}$ W m^{-2} Hz^{-1}. Maximum amplitudes of possible variations of $F_{10.7}$ and T_∞ are even more. The exospheric temperature equal to 1000° is related to medium diurnal conditions when solar activity is not high. The temperature lapse rate is determined by the intensity of the solar short-wave radiation absorbed in the lower thermosphere.

As is seen in Fig. 1 the atmosphere is dynamic with height as well as depending on geophysical conditions. Density is especially subjected to dynamics because of its dependence on temperature variations and the distribution of volatile components in the gravitational field. Between 100 and 1000 km the value of the density decreases by about 8 orders of magnitude, and at a level of 500–600 km height it can vary by a factor of 100 or more. According to the steepness of the density gradient curve it is seen that the height range where the most variation occurs moves towards lower heights with lower solar activity. This is associated with molecular weight change. The defining role of atomic oxygen in the period of high solar activity in the height range from 300 to 900 km is replaced by a predominant abundance of helium and hydrogen at heights above 500 km. Fig. 1 (b) gives

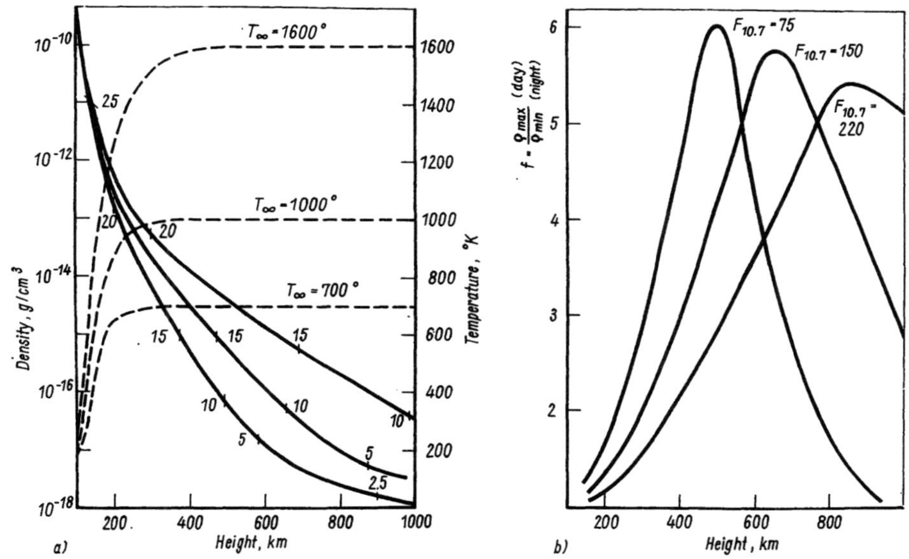

Fig. 1. (a) Variations of density, temperature and molecular weight with height according to Jacchia's 1970 models [3] (values of molecular weight are marked by numbers on the density profiles)

(b) The ratio of maximum daily density to minimum night density for low, high and medium solar activity.

values of diurnal variations expressed by the ratio of maximum density in daytime to minimum density in night-time for several levels of solar activity. As is seen maximum diurnal density variation with increased solar activity is displaced from the 500 km height to a height of about 800 km. Since a similar picture is observed for the other variations the Jacchia's model 1970 explains such a situation by stating that the distribution in the upper atmosphere of the regions of heating and the regions of extension is the same for several density variations. However, such a scheme underestimates the character of semi-annual variations [18, 41, 43] and those connected with geomagnetic activity [14].

The orbital motion analysis discovered secular variations in inclination of the satellite orbits found to be due to zonal winds at high heights. Thus, in specifying the structure of the upper atmosphere we can describe periodic and sporadic variations and stable motions.

Let us now describe the character and nature of the upper atmosphere parameter variations in more detail.

3. Variations Associated with Solar Activity

These variations comprise secular variations connected with the solar cycle (11-year cycle), recurrent variations connected with the rotational period of the sun about its axis (27-days) and irregular variations. The most significant

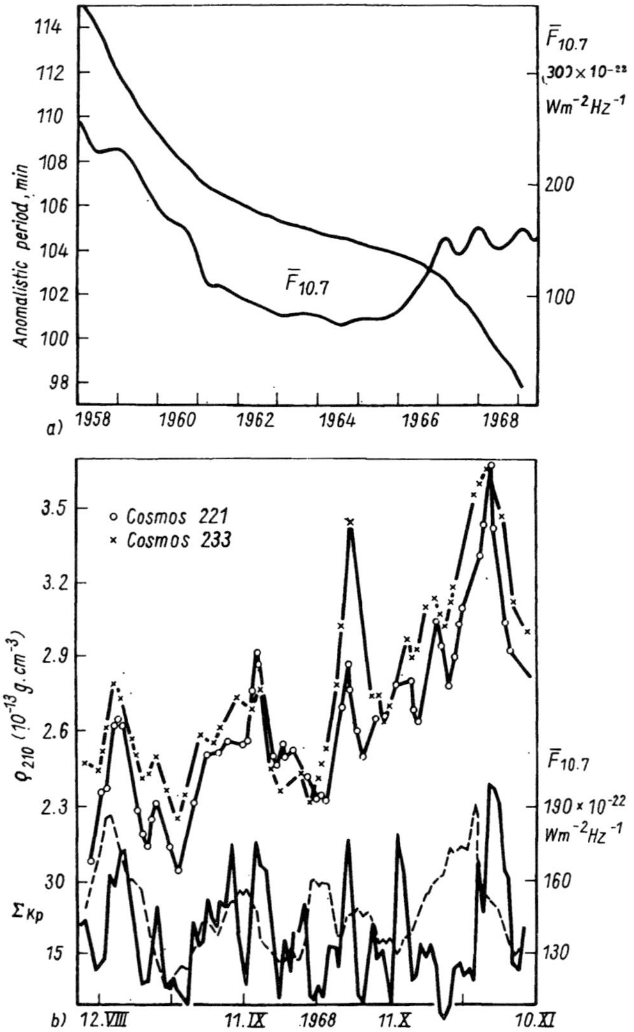

Fig. 2. Secular and short-period density variations with solar activity: (a) orbital period of Explorer 1 (1958 α) from 1958 to 1969; (b) current values of density at heights of 210 km obtained from analysis of Cosmos 221 and 233 satellite drag [7].

of these are the secular variations. As is seen in Fig. 1a, b during one half of the 19th solar cycle from the maximum in 1958 to the minimum in 1964 density decreased by a factor of approximately 20 at a height of 500 km. On the whole, long-period and short-period variations correlate satisfactorily with the index of solar activity for which the magnitude of the solar decimetric flux ($F_{10.7}$) is most often used. The connection between radio radiation and density of the upper atmosphere results from the fact that solar euv and X-radiation, which are effective in heating the atmosphere, are generated in the corona and the solar upper atmosphere — the same regions of the sun where radiation in centimetric and decimetric ranges is produced. The atmosphere reacts differently to long-period and short-period variations of $F_{10.7}$: the first is twice as large as the second. Both variations can be approximately expressed by a linear dependence on $F_{10.7}$. The correct relation for variations with solar activity can be established assuming an appropriate magnitude of the efficiency of conversion of short-wave radiation into heat and by systematic measurements of euv radiation because it does not correlate well enough with $F_{10.7}$ as is seen from direct measurements [4].

Explorer 1 drag data are a classic example illustrating secular variations. Fig. 2 (a) shows values of the anomalic period for this satellite in the period 1958–1969 [5]. The three solar-revolution average values of the decimetric flux $\overline{F}_{10.7}$ are given as well. As is seen from Fig. 2 (a) the rates of decreasing period in 1958–1960 and 1967–1969 are considerably higher than in 1962–1965. This means that density at a height of about 350 km was higher in periods of high solar activity than in a period of minimum solar activity. Night density values correlate well with values of $\overline{F}_{10.7}$ averaged for three solar rotations without any hysteresis effect [6].

Fig. 2 (b) gives an example of recurrent 27-day variations. Current values of density at heights of about 210 km in the August–November period of 1968 are obtained from analysis of Cosmos 221 and 233 satellite drag [7]. It is evident that the time-lag of the 27-day variations from the average diurnal values of $F_{10.7}$ is equal to 1.0 ± 0.12 days. This can be explained by the thermal inertia of the atmosphere [8, 9]. From Fig. 2 (b) it is seen that the 27-day density variations are intricately superimposed on variations correlated with the geomagnetic activity which is also characterized by a tendency to 27-day recurrence.

4. Density Variations during Periods of Geomagnetic Disturbance

Short-period fluctuations of density and temperature inferred from satellite drag data in the height range between 125 and 1000 km last from one to several days during strong or weak geomagnetic disturbances are observed. During intensive magnetic storms the density can increase several times for some hours and then sharply decrease.

Fig. 3 shows density variations at heights near 170 km obtained from the 1968-59A (OV1-15) satellite orbital drag with good time resolution and from the data of accelerometers installed on this satellite [10, 11]. We can see that each magnetic storm (the indicator of which is an increased 3-hour geomagnetic index Kp) is accompanied by a corresponding increase in density. The time delay is about 6 hours. Significant heating and "bulging" of the atmosphere occurs in the auroral zone. The most accurate statistical analysis of the geomagnetic effect was made by Roemer [13] who obtained a confirmation that the influence of geomagnetic activity at night-time and in the morning hours (from 20 h to 8 h LT) is 30% stronger than it is in daytime. He suggested for the 250–800 km height range the following empirical dependence of increasing exospheric temperature with increasing Kp index

$$\Delta T_\infty = (21^\circ\!.4 \sin \varphi + 17^\circ\!.9)\,\overline{Kp} + 0^\circ\!.03 \exp \overline{Kp}$$

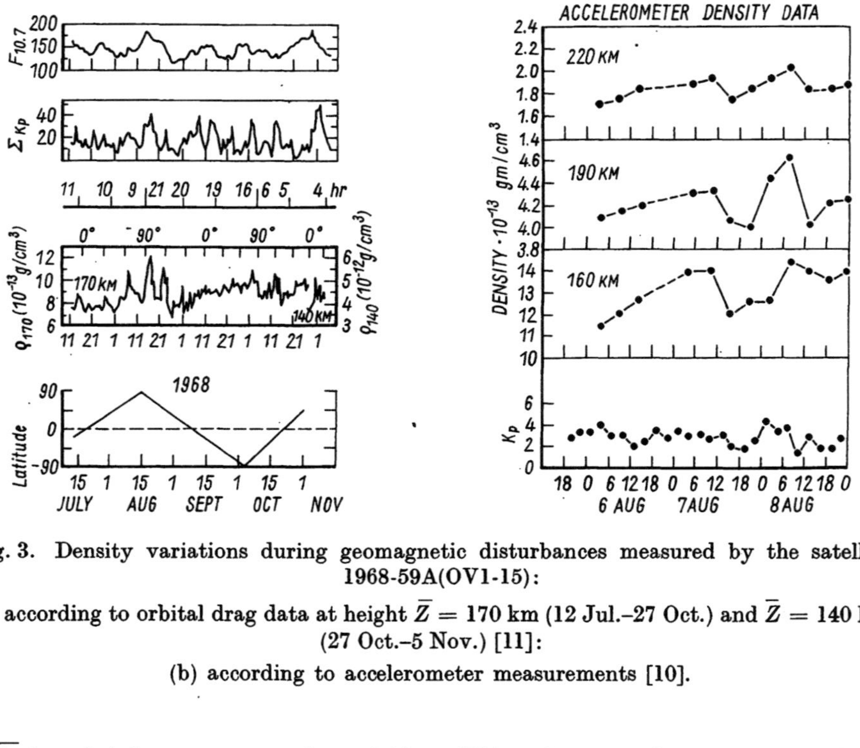

Fig. 3. Density variations during geomagnetic disturbances measured by the satellite 1968-59A(OV1-15):

(a) according to orbital drag data at height $\bar{Z} = 170$ km (12 Jul.–27 Oct.) and $\bar{Z} = 140$ km (27 Oct.–5 Nov.) [11]:

(b) according to accelerometer measurements [10].

($\overline{Kp}$ is a 0.4-day average value of Kp). This relation reflects the latitudinal dependence of the intensity of the geomagnetic effect and indicates that magnetic activity causes a considerable heating of the atmosphere in magnetically disturbed periods. The time delay of the influence on the atmosphere of increasing Kp is equal to $\Delta t = 5.5 \pm 0.3$ (h) $= 0.231 \pm 0.12$ (day); note the essential dependence of the time delay on latitude is not indicated. This conclusion does not agree with the results obtained [12], according to which the time delay at lower latitudes $|\varphi| < 55^\circ$, $\Delta t = 0.30 \pm 0.1$ (day) exceeds that at higher latitudes $|\varphi| > 55^\circ$, $\Delta t = 0.24 \pm 0.02$ (day). If additional heating of the atmosphere during a geomagnetic storm occurs mostly at high latitudes the explanation of this result requires very rapid heat transfer to lower latitudes.

Other indices of magnetic disturbances were used as well to relate density variations with geomagnetic activity. Yu. Truttse [14] suggested the indices *Dst* and *AE*. Although there is a good correlation between maximum values of ϱ and *Dst* for a storm and a time delay is practically absent at the beginning stage of a geomagnetic disturbance when the values of *Dst* are small, the correlation between the parameters is ambiguous. Probably it is due to the fact that *Dst* reflects not only the power of the equivalent current circuit but also other types of geomagnetic disturbances. The index *AE* is free of this disadvantage. The sum of the previous 10-hour values is proportional to *Dst* during almost all the time of the geomagnetic disturbance, excluding its initial stage. The linear dependence of density on the average *AE* index was found, but at the same time a non-linear dependence of density on the $F_{10.7}$ flux was determined. One can attempt to interpret this result as disproving a supposition about similar distributions with height of regions of heating associated with both geomagnetic and solar activity. However, the mechanism of creating wide regions of different height dependences of heat inflow into the thermosphere is still obscure. Such a conclusion results from a series of new results obtained by satellite drag in the lower thermosphere where a close connection between a sudden increase of density and geomagnetic disturbances was revealed. Thus, at a height of 170 km (Fig. 3) in the period of a magnetic storm a density increase up to 25% was observed and at a 140 km height a disturbance of about 35% took place. Two strong storms of $Ap > 100$ during the life of satellite ATS 2 were accompanied by density increases of about 30% and 70% [15]. It is interesting that these values considerably exceed estimates predicted by CIRA-1965 and Jacchia's 1970 models.

Unfortunately, it should be noted that the influence of a magnetic disturbance on the upper atmosphere is still obscure, although most investigators are of the opinion that it is associated with increases in the solar corpuscular flux. The indices *Kp* and *Ap* are used to characterize the interaction of the corpuscular fluxes with the earth's magnetosphere. It is known, however, that solar corpuscular fluxes are of two types — due to flares and due to slowly varying components — which are strongly different in both heliophysical properties and geophysical effects. These differences are mostly noticeable during the 11-year cycle [16, 17]. Therefore, while selecting geomagnetic data it is necessary to divide magnetic disturbances into two types, slowly varying or periodic and sporadic. Non-application of this can cause the empirical dependences obtained to the present time to comprise several controversial aspects and reflect actual disturbances in density with rather high dispersion.

5. Diurnal and Seasonal-Latitudinal Variations

These variations are explained by different insolation and heating of the thermosphere by short-wave solar radiation. As a consequence of the thermal inertia of the atmosphere the maxima in the diurnal density and temperature distributions lag about 40–50° behind the subsolar point. Since the temperature and density continuously decrease from day maximum to night minimum not only in longitude but in latitude as well, latitudinal variations of density and temperature must inevitably be present. If the maximum is replaced with latitude, together with the seasonal distribution of the subsolar point, one should speak about seasonal–latitudinal variations.

Fig. 4 gives an example of current and smoothed (for $\overline{F}_{10.7} = 140$) values of density observed at heights of about 270 km in the region of the perigee of the orbit for Cosmos 204 during 1968 [7]. As is seen diurnal, seasonal–

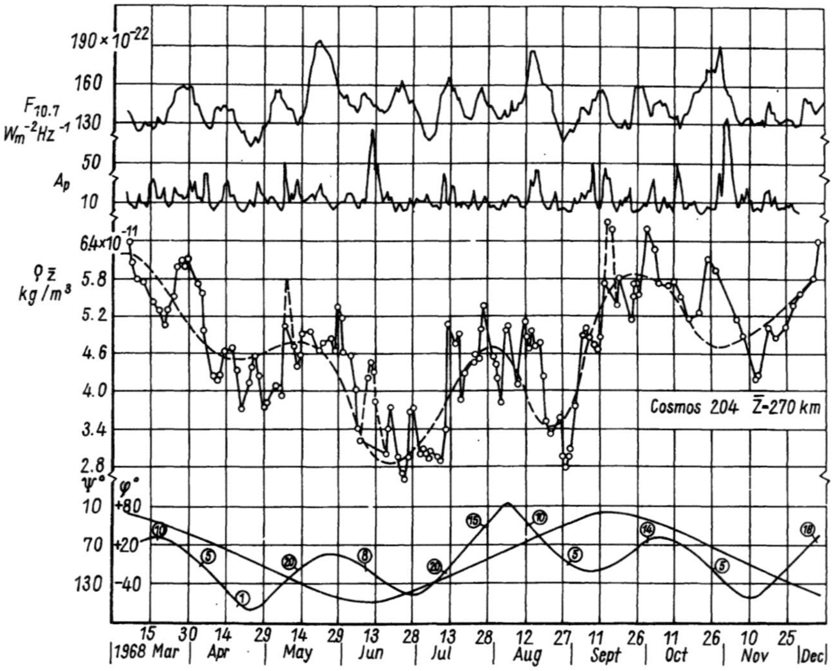

Fig. 4. Short-periodic and smoothed ($\overline{F}_{10.7} = 140$) diurnal, latitudinal, seasonal and semi-annual density variations at height $\overline{Z} = 270$ km calculated from data of Cosmos 204 satellite drag. Values of the geocentric angle between the direction towards the sun and the orbit perigee ψ and the perigee latitude φ are also given.

latitudinal and semi-annual variations are superimposed on each other. Therefore, the determination of an accurate form of a diurnal bulge seasonally migrating with latitude is difficult because of a masking by semiannual variations. One possible way for dividing these variations is an application of two satellite orbits which have similar heights of perigee. Fig. 5 gives an example of current and smoothed values of density obtained at a height of 270 km by Cosmos 204 and 222. In the July–August period of 1968 the difference in heights of perigee for these satellites did not exceed 10 km. From Fig. 5, without preliminary assumptions on the semi-annual variations, the

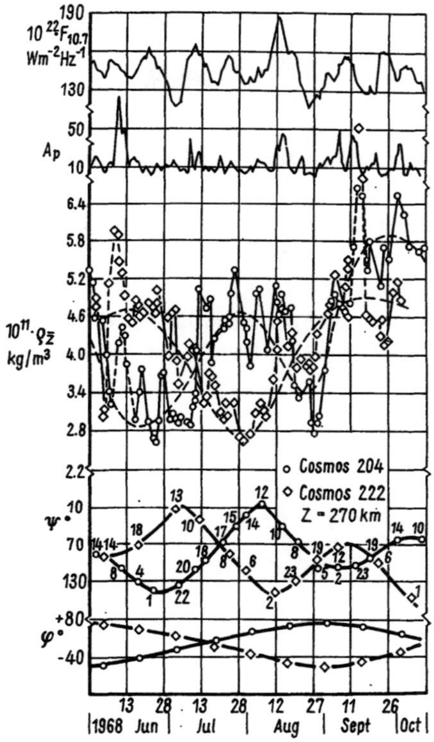

Fig. 5. Current and smoothed ($\bar{F}_{10.7} = 140$) values of density obtained for height $\bar{Z} = 270$ km from the Cosmos 204 and 222 satellites. Values of the geocentric angle between the direction towards the sun and the orbit perigee ψ and the value of perigee latitude φ are also given.

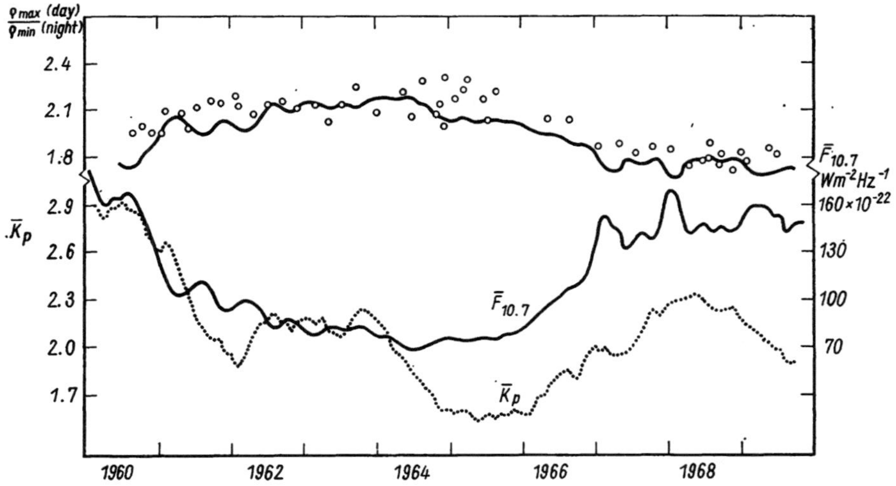

Fig. 6. Dependence of the diurnal variation at height $\bar{Z} = 270$ km obtained from drag data on 21 satellites [18] ○ and according to Jacchia's 1970 models [3] ——, compared with the decimetric flux (averaged over 3 rotations of the sun) and Kp index (a mid-annual running average).

conclusion can be reached that the combined contribution of diurnal and latitudinal effects at a height of 270 km in the summer of 1968 did not exceed 1.7. Several methods are proposed to describe the form of the diurnal bulge. The analytical form used in the model [3] is based on calculated variations of the exospheric temperature corresponding to the density variations observed. A confirmation of this is shown by the good agreement of the model with the actual magnitude of the diurnal density variations at a height of 270 km obtained from the drag data from the 21 Cosmos satellites [18] (Fig. 6). At the same time the accepted relations for describing a variety of forms of global exospheric temperature distribution hardly allow convincing considerations about the physical treatment of the phenomena. From this

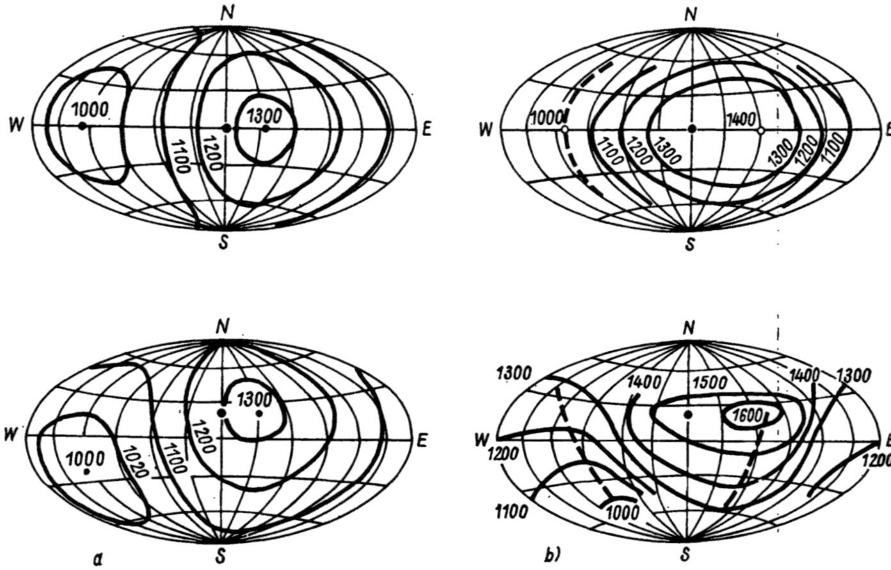

Fig. 7. Global distributions of neutral temperature (for magnetically undisturbed periods) for the conditions of equinox and summer solstice (a) obtained from satellite drag data [3] compared with (b) similar global distributions obtained by the method of incoherent scattering [24].

point of view it is worth while to consider parameter R introduced by Jacchia [13] and connecting the maximum day and minimum night temperatures by the relation $T_M/T_c = 1 + R$. Taking into account a satisfactory correlation of this parameter with a semiannual running average value of the geomagnetic index Kp ($R = 0.134 + 0.090\,Kp$) the conclusion was reached that the contribution of the solar wind is as important as (if not more than) that of short-wave solar radiation. This conclusion requires the most serious experimental test because of the essential modifications of ideas about the energy sources in the thermosphere [73].

The models usually assume that density and temperature variations coincide in phase. A more direct way of investigating temperature variations is measurement by the incoherent scattering method [19–26]. Waldteufel and McClure constructed maps of global temperature distribution for conditions of equinox and solstice that are compared in Fig. 7 with similar model distributions. We can see that the extreme values of density and temperature do not coincide in phase: maximum temperature in the thermosphere is at 16–17 hours, that is about 2 hours later than maximum density above 300 km according to satellite drag data. Theoretically such a situation is possible due to interaction of thermal and dynamical sources of energy in the thermosphere [27, 28]. Below 300 km a phase shift in the maximum of diurnal density variation and a tendency to follow more closely that of temperature has been revealed by analysis of the satellite drag data 29, 30, 7, 18.

In connection with this a peculiarity of global temperature distribution obtained by incoherent scattering should be noted. The observed diurnal bulge apparently cannot be described by the simple formula suggested in the model [3] (i.e. with the same number of parameters) according to which a contour of temperature intersection, in the direction of north–south for any season, is symmetrical relative to the subsolar point and is defined by a constant parameter η (that is a consequence of the totality of initial assumptions). Fig. 7 shows that the parameter η, first, can be changed with season, and secondly, at times different from the equinox an additional asymmetry appears relative to the subsolar point in the north–south direction. In other words, a parallel of latitude crossing a subsolar point at times different from the equinox is not an axis of symmetry for latitudinal variations.

The analysis of observations on satellites with high inclinations made it possible to define seasonal variations at large heights over polar areas. Keating and Prior [33, 34] concluded that annual variations at high altitude are associated with the relative content of helium. These variations were seen to change during a solar cycle. In the period of high solar activity the Explorer 19, 24 and 39 satellites detected that the maximum density at a height of 650 km shifted from high latitudes in the winter hemisphere to low latitudes in the summer one. Thus, a helium bulge at these heights in the period of increasing solar activity must decrease under the influence of a summer bulge of atomic oxygen. At the 900 km height in 1968 at the winter pole the density exceeded 3 times the density at the summer pole; at the same time at the 550 km height where oxygen was the major constituent the distribution was almost symmetrical. These results have been described empirically [35, 36] but have not yet been explained in terms of a theoretical model.

6. Semi-annual Variations

In spite of a great number of data obtained during the last few years one of the least understood phenomena in the dynamics of the upper atmosphere is the semi-annual variation. To illustrate the semi-annual effect in the density at heights of 470 and 600 km an example is given in Fig. 8 where the results are presented of the analysis of the circular orbit of the 1966-118A [44] and

Cosmos 144, 151, 156 and 184 satellites [7]. At large heights circular orbits are the most suitable for investigation of this effect because it is more global than the seasonal one and large diurnal variations are not to be excluded. Semi-annual variations were first noticed by Paetzold and Zschörner [37] in 1960 from the drag data of 111th Soviet satellite. The variations were further repeatedly confirmed in a wide range of heights from approximately 90 km in the mesosphere to about 1500 km in the exosphere. The dependence of semi-annual variations on height, latitude and phase of the solar activity cycle have been repeatedly described in the literature [18, 38–46]. It was shown that there are (within some days) peaks of density in early April and at the end of October and minima in January and July. The minimum in July is

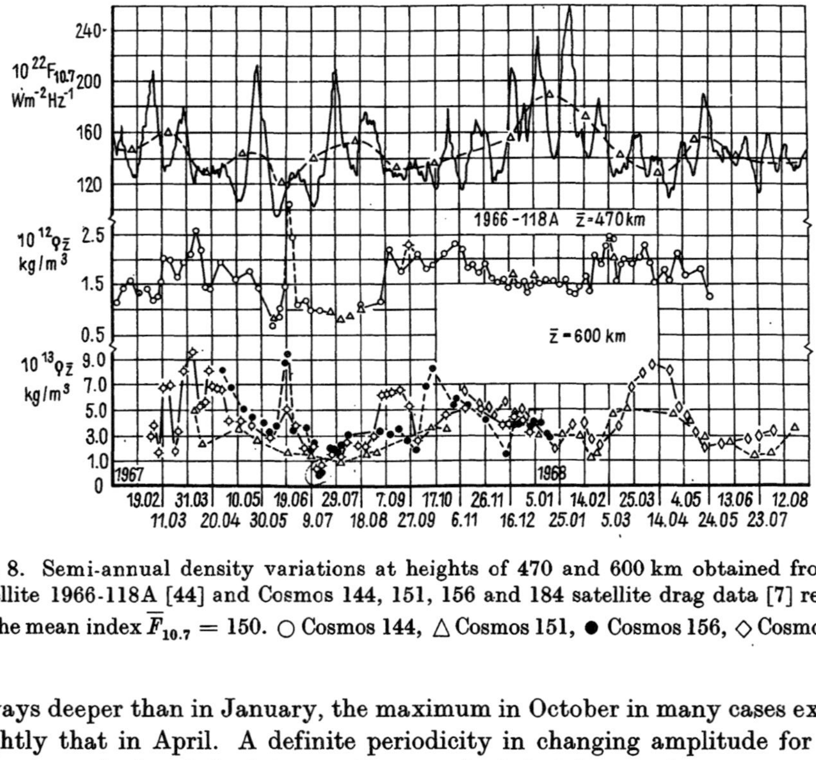

Fig. 8. Semi-annual density variations at heights of 470 and 600 km obtained from the satellite 1966-118A [44] and Cosmos 144, 151, 156 and 184 satellite drag data [7] reduced to the mean index $\bar{F}_{10.7} = 150$. ○ Cosmos 144, △ Cosmos 151, ● Cosmos 156, ◇ Cosmos 184.

always deeper than in January, the maximum in October in many cases exceeds slightly that in April. A definite periodicity in changing amplitude for some years is marked and the latter at thermospheric heights can be commensurable with diurnal density variations.

As for the character of semi-annual variations the mechanism responsible for their appearance has not yet been found. According to the analysis of satellite drag for the period 1958–1966 density variations observed were explained in terms of temperature variations in the thermosphere at the same levels as that specified for the effect of the solar activity [38]. This result is reflected in the model [3] in the form of a seasonal correction to the exospheric temperature. The magnitude of the correction is associated with the $F_{10.7}$ index by direct linear dependence. However, the data obtained during the last few years modified hypotheses proposing that semi-annual variations are only of thermal origin [7, 41–43]. It was shown that the height dependence observed over a period of several years cannot be explained by the proposed models either in form or in magnitude. While there is adequate agreement for heights near 600 km the large amplitudes in the thermosphere at heights below 300 km can only be explained by taking into account semi-annual variations in the structure parameters near the turbopause. It was found that the observed secular distribution of the intensity of the semi-annual variation significantly differs from the secular distribution of the $\bar{F}_{10.7}$ index. Such a conclusion is confirmed by the data of the drag on the Echo 2 satellite for the period

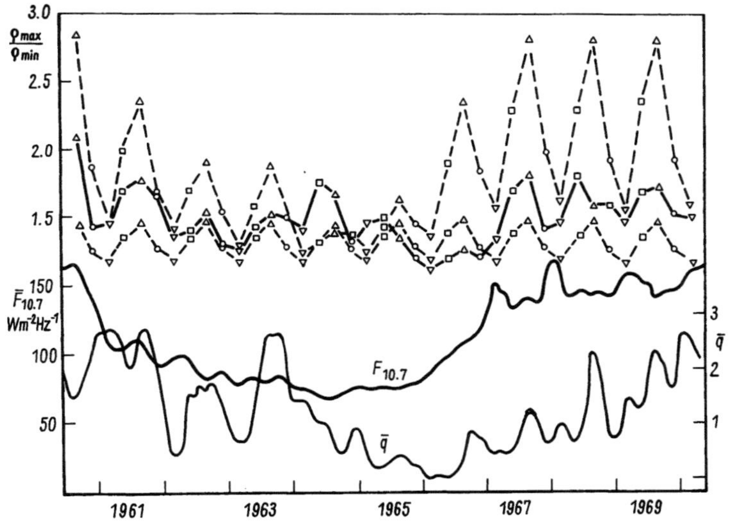

Fig. 9. Secular distribution of the amplitude of the semi-annual density variation at height $\bar{Z} = 270$ km for the period 1960–1970 ——— and the corresponding distribution from Jacchia's 1970 models — — — for heights of 270 (lower curve) and 600 km (upper curve), compared with the indexes of solar activity, $\bar{F}_{10.7}$ and $\bar{q}$ ($q = g^2/G$, where g is the number of groups of activity in a narrow central zone of the sun at the heliolatitude of the earth's projection and G is the number of groups of activity on the whole disk).

▽ January/April; □ April/July; △ July/October; ○ October/January.

1964–1968 at heights of about 1000 km and the results obtained for 1960–1970 from the data of Cosmos satellite drag at heights below 300 km that are given in Fig. 9. The results given in Fig. 9 show that the secular distribution at the 270 km height is considerably different from the model profiles and besides is not connected with the $\bar{F}_{10.7}$ index by an unambiguous relation. For heights below 300 km these results cannot be explained by a seasonal migration of the helium bulge (as was done by Jacchia's [3] treatment of Cook's results for large heights). One can suppose that this phenomenon is connected with solar corpuscular radiation. One possible index of the atmospheric heating efficiency, taking into account the narrowness and direction of the solar corpuscular flux, is given in Fig. 9. Comparison with this index shows that the secular distribution of the semi-annual variation at heights below 300 km depends on the influence of the solar wind rather than on the euv radiation. At the same time the agent connected with the solar wind efficiency cannot create a phase-stable effect in the upper atmosphere. Although a semi-annual variation takes place in averaged 11-year cycle magnetic indices, in several years these variations do not occur or strongly differ in phase from similar variations in the thermosphere.

A preferable mechanism causing semi-annual variations is a supposition about their connection with a major change in the meridional circulation system in the upper atmosphere that results in a large outflux of heat from the lower thermosphere in the periods of the solstices but not at the equinoxes [7]. By this hypothesis it is proposed that the lower atmosphere significantly

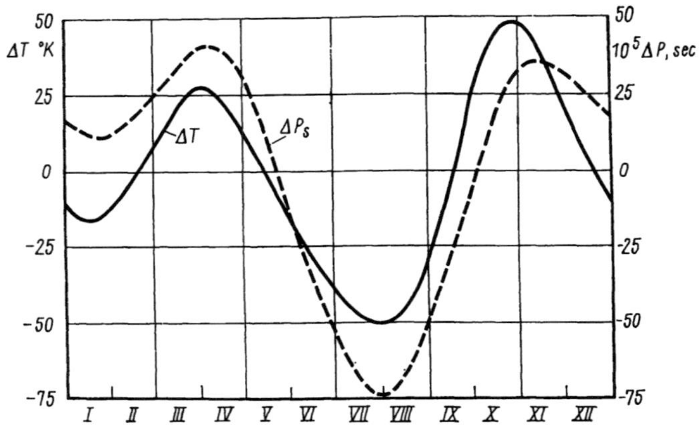

Fig. 10. Exospheric temperature variations due to semi-annual variations ($\bar{F}_{10.7} = 100$) [3] compared with diurnal duration variations (for the period 1962–1966) [52].

modulates effects of solar origin. Semi-annual variations in the winds in the lower atmosphere were shown by Cook [41]. Volland [50] associates a semi-annual effect with energy dissipation of a tidal wave coming from the homosphere. The results of observations of seasonal irregularities in the earth's rotation prove that the integrated effect of lower atmosphere circulations can generate highly stable oscillations of geophysical parameters. This irregularity is associated with momentum re-distribution between the earth and the atmosphere because of the seasonal circulation. Fig. 10 shows a comparison of semi-annual variations of the thermosphere with seasonal variations of the diurnal duration (according to data of the International Hours Bureau) for the period 1962–1966. A good agreement in form and phase of both phenomena is evident. The coupling of the semi-annual effect with the seasonal atmospheric circulation is confirmed by their interconnection with known 26-month changes of the velocity of the equatorial stratospheric wind [44, 52].

7. The Upper Atmosphere Rotation

If the earth and its atmosphere do not rotate, the aerodynamical force affecting a satellite should be a tangent to its orbit and should not change the orientation of the orbital plane. In fact, the atmosphere rotates and the drag force has a component normal to the orbital plane that causes small changes of the orbital inclination with time (for satellites moving eastward $\Delta i < 0$). Solving the problem by assuming that the angular velocity of the atmospheric rotation equals the angular speed of the earth's rotation shows that the calculated changes of inclination caused by the cross drag are different from the values observed.

If the estimates of the rotational velocity of the atmosphere from observations of the first satellite [54–56] included errors, the results obtained later are

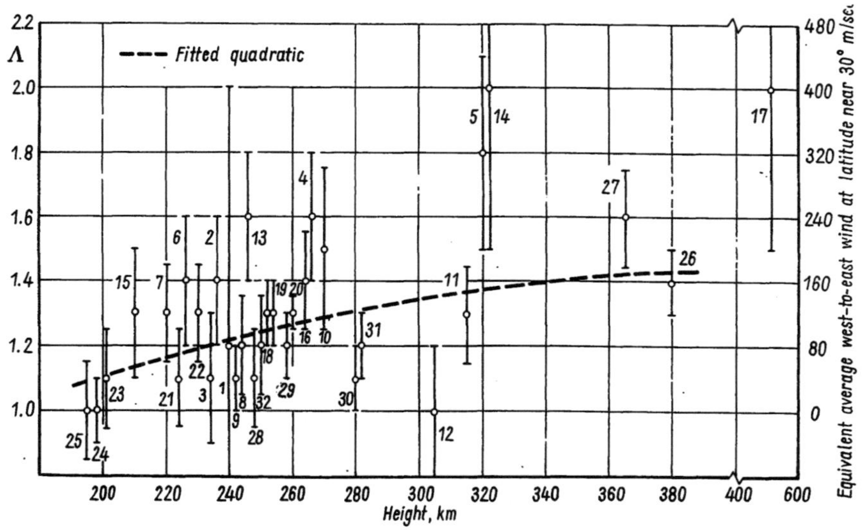

Fig. 11. Values of the Λ-parameter (equal to the ratio of the angular rotational velocity of the atmosphere to the angular rotational speed of the earth) obtained from the analysis of changing inclinations of 32 satellites [61].

stringent both in the accuracy of the data processing and in the mathematical treatment of the orbital theory [57–61]. These results give rise to the concept that the rotation of the upper atmosphere leaves the earth's rotation behind or, in other words, there is a systematic west wind component. Fig. 11 taken from paper [61] summarizes the results of observations of several years that are based on the change of inclination of 32 satellites. According to these data the Λ-parameter expressing the ratio of the angular rotational velocity of the atmosphere to the angular rotational speed of the earth at a given height increases from 1.1 at a height of 200 km to 1.25 at 250 km and 1.4 at 350 km. At a latitude near 30°, where the rotational speed of the earth is equal to about 400 m s^{-1}, these values correspond to average velocities of west-to-east wind of 40 ± 20, 100 ± 20 and 160 ± 40 m s^{-1}. The values of the rotational velocity of the atmosphere at heights below 180 km are uncertain. For the satellite 1968-59A (OV1-15) according to King-Hele's calculations [62] Λ is equal to 1.2 ± 0.07 at a height of 170 km and 1.1 ± 0.1 at 150 km. According to Ching [11] (using more accurate orbital data) Λ is equal to 1.32 ± 0.05 for a height $\bar{Z} = 175$ km.

The estimates of wind velocity obtained can be considered as the averages over latitudes up to $i/2$, where i is the inclination of the satellite orbit. Therefore, the curve given in Fig. 11 (the dashed line) shows an estimate of the average wind velocity only up to latitudes of about 30°. At mid-latitudes a secular dependence of variations from year to year or with latitude is not observed. Although the estimates obtained represent as a rule the average diurnal magnitudes several results give rise to the conclusion that the wind velocity is highest at night-time.

Other experimental data on the wind velocity in the lower thermosphere, in particular from observations of chemical releases [63], do not confirm the conclusions deduced from the satellite orbit inclination variation. At heights of about 170 km according to these experiments very strong winds with a 300 m s^{-1} velocity can be observed, but west and east components are observed with equal probability. Unfortunately, the problem of winds in the upper atmosphere encounters considerable difficulties and the bulk of data available is quite scarce. It is quite possible that estimates obtained from satellite drag data on the average are more representative than observations of luminescent clouds and ionospheric drifts since the wind determined is representative of the magnitude averaged over the extent of the near-perigee orbit segment and random local fluctuations are removed.

According to the model density distribution the thermosphere winds were calculated in some theoretical papers [64–72]. Disregarding several differences that are a consequence of the choice of the initial assumptions it can be expected that at 300–400 km heights the wind velocity is equal to 200–250 m s^{-1} at night-time and at daytime it is equal to 50–100 m s^{-1}. In addition, the zonal component is directed to the east from about 14 h to 4 h and to the west from 4 h to 14 h. The meridional component is directed towards a pole in the daytime and to the equator at night-time. Because of the horizontal velocity divergence [71, 72] a vertical velocity can arise which has a magnitude of several m s^{-1} at 300 km (i.e. the vertical velocity due to extending and compressing the atmosphere with heating and cooling). If the calculated results are correct, this velocity moving upward in daytime and downward at night-time, associated with a loss and a source of heat can contribute considerably to the heat balance of the thermosphere, as its magnitude is comparable with the magnitude of the radiant heat flow. The model of thermospheric winds can explain the results obtained from changing inclinations of satellite orbits. According to the model of Challinor [68–70] at latitudes $|\varphi| < 40°$ (which comprise estimates of $\Lambda > 1$) the zonal velocity along a parallel has an average component in the eastward direction.

8. Conclusion

An overall review of data on the structure and motion of the thermosphere from satellite drag gives an appreciation of the significance of this method for the physics of the upper atmosphere. This method was and still remains the only means of studying the global structure and space–time variations and establishing regularities in solar-terrestrial relations in the atmosphere at large heights smoothed of local and random effects. We can say that the main effects appearing in the form of variations of density and temperature can be investigated rather well. We know much less about their physical nature and the interconnection with solar and geomagnetic activity of dynamical processes in the lower atmosphere and other geophysical factors (periodicity of insolation, irregularity of the earth's rotation, tidal effects etc.). Taking into account the progress over the last few years in the realization of the accuracy of quantitative estimates of the atmospheric parameters it is now possible to consider investigating the meteorology of the thermosphere. Current satellite drag observations contribute systematic experimental material and as a result cause a continuous improvement of our knowledge about the structure and dynamics of the upper atmosphere.

References

[1] CIRA-1965 COSPAR International Reference Atmosphere 1965, North-Holland, Amsterdam (1965).
[2] L. G. Jacchia, Smithson. Contrib. Astrophys. 8, 215 (1965).
[3] L. G. Jacchia, Smithson. Astrophys. Obs. Spec. Rep. No. 313 (1970).
[4] L. A. Hall et al., J. Geophys. Res. **74**, 4181 (1969).
[5] D. G. King-Hele, R.A.E. Tech. Rep. 69102 (1969).
[6] L. G. Jacchia, J. Geophys. Res. **75**, 4347 (1970).
[7] M. Ya. Marov and A. M. Alpherov, Space Research X, 419 (1970).
[8] M. Roemer, Forschungsberichte der Astronomischen Institute 68-08 (1968).
[9] G. E. Thomas and B. K. Ching, J. Geophys. Res. **74**, 1796 (1969); Planet. Space Sci. **17**, 1737 (1969).
[10] F. A. Marcos, K. S. W. Champion and R. A. Schweinfurth, Space Research XI, 941 (1971).
[11] B. K. Ching, J. Geophys. Res. **76**, 197 (1971).
[12] L. G. Jacchia, J. Slowey and F. Verniani, J. Geophys. Res. **72**, 1423 (1967).
[13] M. Roemer, Space Research XI, 965 (1971).
[14] Yu. L. Truttse, Planet. Space Sci. **16**, No. 12 (1968); Space Research X, 467 (1970).
[15] D. G. King-Hele and D. M. C. Walker, paper presented at XIII COSPAR meeting, Leningrad 1970.
[16] E. R. Mustel, Nauchnie informatsii, No. 10, Astrosovet Akad. Nauk SSSR, M(1968).
[17] A. I. Ol, Trudy Gl. geophizich. observ. **245**, 35 (1969).
[18] M. Ya. Marov and A. M. Alpherov, paper presented at XIII COSPAR meeting, Leningrad 1970.
[19] H. Carru, M. Petit and P. Waldteufel, Planet. Space Sci. **15**, 944 (1967).
[20] H. Carru and P. Waldteufel, paper presented at XII COSPAR meeting, Prague 1969.
[21] J. S. Nisbet, J. Atmos. Sci. **24**, 586 (1967).
[22] K. K. Mahajan, J. Atmos. Terr. Phys. **31**, 93 (1969).
[23] J. P. McClure, J. Geophys. Res. **74**, 279 (1969).
[24] P. Waldteufel and J. P. McClure, Annls Géophys. **25**, 785 (1969).

[25] P. A. WALDTEUFEL, Planet. Space Sci. **18**, 741 (1970).
[26] W. E. SWARTZ and J. S. NISBET, J. Geophys. Res. **76**, 185 (1971).
[27] H. RISHBETH, Annls Géophys. **25**, 495 (1969).
[28] S. CHANDRA and P. STUBBE, Planet. Space Sci. **18**, 1021 (1970).
[29] M. YA. MAROV, Space Research VIII, 811 (1968).
[30] M. YA. MAROV and A. M. ALPHEROV, Space Research IX, 497 (1969).
[31] L. G. JACCHIA, Space Research X, 367 (1970).
[32] M. I. VOISKOVSKI et al., Space Research XI, 953 (1971).
[33] G. M. KEATING and E. J. PRIOR, Space Research VII, 1119 (1967).
[34] G. M. KEATING and E. J. PRIOR, Space Research VIII, 982 (1968).
[35] G. M. KEATING, J. A. MULLINS and E. J. PRIOR, Space Research X, 439 (1970).
[36] G. M. KEATING, J. A. MULLINS and E. J. PRIOR, Space Research XI, 987 (1971).
[37] H. K. PAETZOLD and H. ZSCHÖRNER, Space Research I, 24 (1960).
[38] L. G. JACCHIA, J. W. SLOWEY and I. G. CAMPBELL, Planet. Space Sci. **17**, 49 (1969).
[39] D. G. KING-HELE and J. HINGSTON, Planet. Space Sci. **16**, 675 (1968).
[40] D. G. KING-HELE and J. HINGSTON, Planet. Space Sci. **16**, 937 (1968).
[41] G. E. COOK, Annls Géophys. **25**, 451 (1968).
[42] G. E. COOK, Planet. Space Sci. **18**, 1573 (1970).
[43] G. E. COOK, Planet. Space Sci. **18**, 387 (1970).
[44] D. G. KING-HELE and D. M. C. WALKER, Planet. Space Sci. **17**, 197 (1969).
[45] D. G. KING-HELE and D. M. C. WALKER, Planet. Space Sci. **19**, 297 (1971).
[46] C. WULF-MATHIES, Space Research XI, 981 (1971).
[47] F. BARLIER et al., Annls Géophys. **25**, 439 (1969).
[48] J. VERCHEVAL, Annls Géophys. **25**, 405 (1969).
[49] G. E. COOK, Nature, Lond. **222**, 969 (1969).
[50] H. VOLLAND, Planet. Space Sci. **17**, 1709 (1969).
[51] V. P. BHATNAGAR, Planet. Space Sci. **18**, 1051 (1970).
[52] M. YA. MAROV and A. M. ALPHEROV, Space Research XII, 803 (1972).
[53] N. S. SIDORENKOV, Astron. J. **45**, 892 (1968).
[54] D. G. KING-HELE, Satellites and Scientific Research, London (1960).
[55] G. E. COOK and R. N. A. PLIMMER, Proc. Roy. Soc. A **258**, 516 (1960).
[56] B. N. TRUBNIKOV, in: Issled. Kosm. prostranstva, Nauka, Moscow 1965.
[57] D. G. KING-HELE, Planet. Space Sci. **12**, No. 7 (1964).
[58] D. G. KING-HELE and D. W. SCOTT, Planet. Space Sci. **14**, 1339 (1966).
[59] D. G. KING-HELE and D. W. SCOTT, Planet. Space Sci. **17**, 217 (1969).
[60] D. G. KING-HELE and D. W. SCOTT, R.A.E. Tech. Rep. 70208 (1970).
[61] D. G. KING-HELE, D. W. SCOTT and D. M. C. WALKER, Planet. Space Sci. **18**, 1433 (1970).
[62] D. G. KING-HELE, Planet. Space Sci. **18**, 1585 (1970).
[63] J. E. BLAMONT et al., Dokl. Akad. Nauk SSSR **193**, 1031 (1970).
[64] J. E. GEISLER, J. Atmos. Terr. Phys. **28**, 703 (1966).
[65] J. E. GEISLER, J. Atmos. Terr. Phys. **29**, 1469 (1967).
[66] H. KOHL and J. W. KING, J. Atmos. Terr. Phys. **29**, 1045 (1967).
[67] R. S. LINDZEN, J. Geophys. Res. **72**, 1591 (1967).
[68] R. A. CHALLINOR, Planet. Space Sci. **16**, 552 (1968).
[69] R. A. CHALLINOR, Planet. Space Sci. **17**, 1097 (1969).
[70] R. A. CHALLINOR, Planet. Space. Sci. **18**, 1486 (1970).
[71] R. E. DICKINSON and J. E. GEISLER, Mon. Weather Rev. **96**, 606 (1968).
[72] H. RISHBETH, R. J. MOFFETT and G. J. BAILEY, J. Atmos. Terr. Phys. **31**, 1035 (1969).
[73] YU. I. GALPERIN, Izv. Akad. Nauk SSSR, Ser. Geofiz. No. 2 (1962).

Space Research XII — Akademie-Verlag, Berlin 1972

DENSITY VARIATIONS AND ATMOSPHERIC ROTATION BELOW 200 KM FROM THE DRAG ON THE SATELLITE OV1-15

B. K. Ching

The Aerospace Corporation, El Segundo, California, USA

Density variations below 200 km from the drag on the satellite OV1-15 have been examined in detail. The observational data were obtained over a period of nearly 4 months at all latitudes and at various local-time periods. Correlation between short-term density changes and geomagnetic activity was evident, but the relationship between the magnitude of the density perturbation and the geomagnetic activity index was not well defined. The importance of the semiannual density effect at low altitudes was established by the observations, but not the diurnal bulge effect. The data indicated that density enhancement toward the polar regions may be more important at low altitudes than enhancement in the subsolar region of the atmosphere. From the change in the orbital inclination of the satellite the atmospheric rotation rate near 175 km was inferred to be a factor of about 1.3 greater than that of the earth.

1. Introduction

The satellite OV1-15 (1968-059A) was launched by the U.S. Air Force on 11 July 1968 into a low-perigee polar orbit for the express purpose of obtaining data over an extended period of time on the density and other atmospheric parameters. Information on the spacecraft, the orbit, and the complement of experiments are given in [1]. The density and atmospheric rotation rate below 200 km deduced from changes in the orbit are presented in [2]. Additional results are presented in this paper.

2. Density

Analysis of the drag on OV1-15 [2] yielded densities at heights between 135 and 195 km in the period from 11 July to 6 November 1968. Densities were obtained at all latitudes in both hemispheres and at local-time periods 11–09 h, 21–18 h, and 06–04 h.

Fig. 1 shows the measured density, transformed to fixed altitudes, and daily indices of solar ($F_{10.7}$) and geomagnetic (ΣKp) activity. Altitude transformation of the data was accomplished using the Jacchia-1970 model [3] with exospheric temperature of 1050 °K, the average temperature corresponding to the measurements. The most noticeable features of the density were lively day-to-day fluctuations, usually associated with geomagnetic activity, and an increase in the average density after September in phase with the semiannual effect. Exceptionally large short-term enhancements were also observed in the northern auroral region, but these anomalies will not be discussed in the present paper.

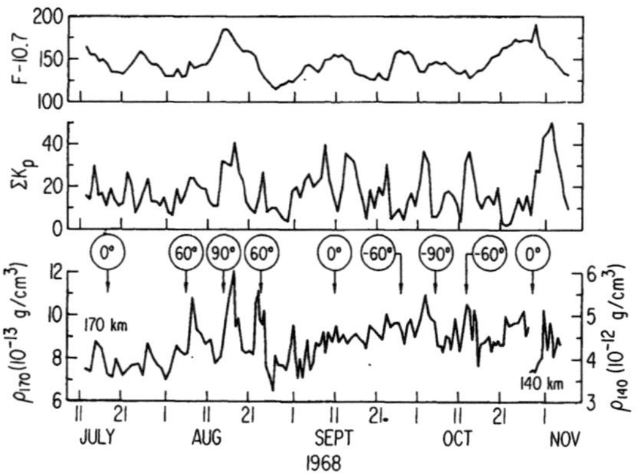

Fig. 1. Drag-deduced density data from OV1-15 in the period 11 July–6 November 1968 transformed to altitudes of 170 km and 140 km by means of the Jacchia-1970 model with exospheric temperature 1050 °K, daily indices of solar ($F_{10.7}$) and geomagnetic, (ΣKp) activity. Circled numbers are the perigee latitude.

2.1. Geomagnetically Correlated Density Changes

The transient density changes could often be identified with changes in geomagnetic activity, but the relationship between the magnitude of the density perturbation and the geomagnetic activity index was not well defined. This is in agreement with other observations (e.g. [4, 5]). On the average, however, the density perturbation did increase with geomagnetic activity, as shown in

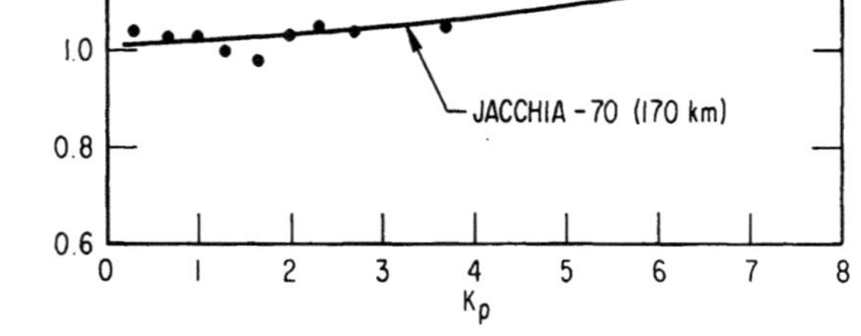

Fig. 2. Average ratio of the OV1-15 drag-deduced density to the Jacchia-1970 model density without the geomagnetic effect as a function of geomagnetic activity index Kp. The ratio of the model density including the geomagnetic effect to the model density without the geomagnetic effect is indicated by the continuous curve.

Fig. 2. The value of Kp assigned to an observation was taken to be the average of the 3-hourly values during an interval of time commensurate with the resolution of the data but preceding the measurement by 6 hours. The results shown in Fig. 2 indicate that the average density response for $Kp > 3$ was slightly larger than the prediction of the model.

2.2. Long-term Density Change

The ratio between the measured and the Jacchia-1970 model density is shown in Fig. 3. The observations and the model are seen to agree to within $\pm 20\%$ with few exceptions. In view of the fact that very little low-altitude drag data have been available for use in modelling, the agreement on the whole is regarded as quite good. Close examination of Fig. 3 reveals a long-term systematic difference between the data and the model. In fact, the long-term

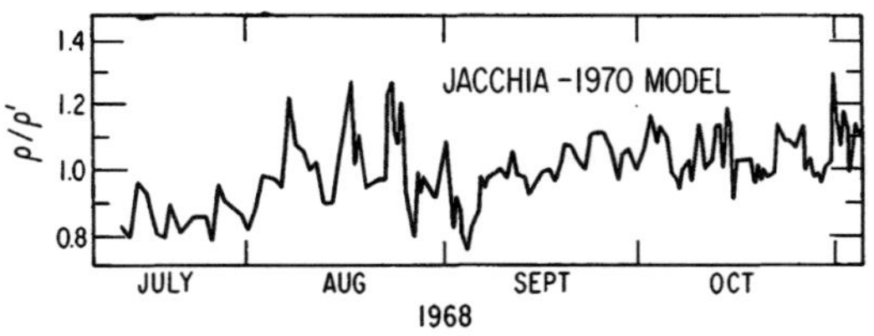

Fig. 3. Ratio of the OV1-15 drag-deduced density to the Jacchia-1970 model density in the period 11 July–6 November 1968.

trend of the drag-to-model ratio shown in Fig. 3 is strikingly similar to that of the measured density of Fig. 1, indicating that the long-term model density variation was extremely small. Now, the model density variations between July and October corresponding to the pure semiannual and diurnal bulge effects were each of the order of 10–15%. By coincidence, however, the motion of perigee was such that the effects were opposed, resulting in no significant long-term variation. Since the observed density did exhibit a long-term variation, the model either overestimated the diurnal bulge effect or underestimated the semiannual effect. We believe that the former condition was responsible for the major part of the discrepancy between the data and the model.

In [2] it was pointed out that the long-term density increase measured by OV1-15 gave only the lower limit for the pure semiannual effect, since the diurnal effect, if it existed, was expected to be in opposition. The local time of perigee for OV1-15 was about 11 h in July and 05 h in October, and at both times the latitude was near-equatorial. In orbit at the same time as OV1-15, but at a slightly higher altitude, was the satellite ATS-2 [6]. This satellite was in a low inclination orbit and the local times of perigee in July and October were approximately 06 h and 11 h respectively. Ratios of the densities from OV1-15 and ATS-2 [6] were approximately the same in July and October, implying that the local-time differences did not affect the density significantly. Consequently, it appears that the difference in the average drag-to-model ratio before and after September seen in Fig. 3 results from the model's overestimation of the bulge effect rather than underestimation of the semiannual effect.

2.3. Modification of the Bulge Shape

Simple changes were made in the shape and orientation of the bulge described in the model in order to determine whether the agreement between the data and the model density could be improved. Specifically, the revisions involved the reduction of the amplitude of the bulge, a shift in the local time of the maximum (from 14 h to 18 h), and a change in the latitude of the maximum (from subsolar to polar). The effect of eliminating the bulge entirely was also investigated.

Table 1

Correlation coefficients between OV1-15 drag-deduced density transformed to fixed altitude of 170 km and Jacchia-1970 model modified with different bulge shapes

Case	Bulge description*	Correln coeff.
1	$R = 0.35$; $\tau = 14$ h	0.52
2	$R = 0.35$; $\tau = 18$ h	0.43
3	$R = 0.2$; $\tau = 14$ h	0.63
4	$R = 0.2$; $\tau = 18$ h	0.57
5	$R = 0$ (no bulge)	0.61
6	$R = 0.35$ (polar bulge)	0.71

* R is the bulge amplitude parameter defined in [3]; τ is the local time of the bulge maximum.

Correlation coefficients between the data and the model densities are given in Table 1 for different bulge shapes. Case 1 represents the unrevised bulge of the Jacchia-1970 model; Cases 2–6 are the revisions. R is the bulge-amplitude parameter associated with the global exospheric temperature variation defined in [3] and τ is the local time of maximum. The latitude of the maximum is subsolar in Cases 1–4 and polar in Case 6. Case 5 is that of no bulge effect. The global variation of the exospheric temperature for Cases 2–4 was taken to be that given by Jacchia in Eq. (17) of [3] with appropriate changes in R and in the phase of the local-time dependent term. In Case 6, the term containing the local-time dependent function was set to zero. It is seen that four of the five cases of bulge modification resulted in better correlation than the original unrevised model. The common feature of these four cases is that the local-time variation of the density is less than that of the unrevised bulge model. The best correlation was achieved with the use of the polar bulge model. The improvement of this model over the unrevised subsolar model was statistically significant at better than the 95% level.

Fig. 4 shows the drag-to-model density ratios resulting from the use of the polar bulge. The difference in the average ratio before and after September which was seen in Fig. 3 does not appear in Fig. 4. It is important to note that the anomalous northern auroral zone fluctuations are still improperly modelled in Fig. 4, and hence the improved correlation coefficient achieved with the polar bulge model did not result from these high-latitude data. The polar bulge model was intended to represent a high-latitude effect in the simplest possible way. No attempt was made to optimize the magnitude or width of the bulge. An auroral-region bulge would probably have been more realistic, but the resolution of the data was not sufficiently good to warrant

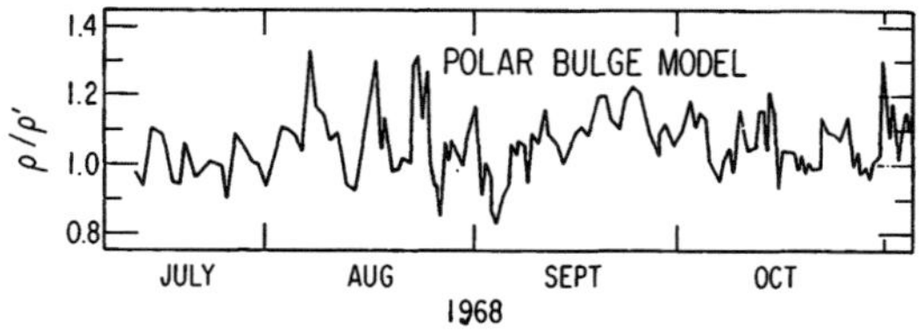

Fig. 4. Ratio of the OV1-15 drag-deduced density to the Jacchia-1970 model modified with a north and south polar bulge instead of a subsolar bulge.

the refinement. The success of the polar bulge model, despite its simplicity, is suggestive of high latitude heating of a permanent, or at least quasi-permanent nature. The source of the heating could be the auroral electrojet or particle precipitation. It should be noted that observations of a high-latitude density effect at low altitudes have been reported previously, e.g. [4, 7, 8].

3. Atmospheric Rotation Rate

The decrease in the orbital inclination of OV1-15 over its lifetime was used to determine the average atmospheric rotation rate. The ratio of the atmospheric rotation rate to that of the earth was found to be 1.3 with an estimated uncertainty of about ± 0.2. This value represents an average over middle and low latitudes at a height of about 175 km during the local-time periods sampled by the satellite. However, the local-time distribution of the data was such that the inferred rate is weighted toward evening hours (19–20 h). Although the inferred rotation rate is higher than expected on the basis of other observations made near 200 km [9], the result is in agreement with that obtained independently by King-Hele [10] using orbital data for the same satellite but from different sources. An attempt was made to determine a possible local-time influence on the rotation rate by using subsets of the original data. No difference was detected that could be considered significant.

4. Summary

Time variations of the density below 200 km have been examined using drag data from the satellite OV1-15. An association between transient density changes and geomagnetic activity exists, but the relationship between the magnitude of the density perturbation and the geomagnetic activity index Kp is not well defined. On the average the density increase for $Kp > 3$ is larger than predicted by the Jacchia-1970 model.

The long-term density change between July and the end of October in 1968 which was measured by OV1-15 has been interpreted to represent the semi-annual effect with little or no contribution from the diurnal bulge effect. Comparisons of the drag data with the Jacchia-1970 model with modified bulge parameters indicate that at heights near 170 km the global density variation may be different from what is currently believed. The results of the tests of different bulge shapes suggest that density enhancement toward the polar regions may be more important at low altitudes than enhancement in the subsolar region. This implied high-latitude effect must of course be considered tentative pending further observation from other satellites.

The atmospheric rotation rate near 175 km was found to be a factor of about 1.3 greater than the earth's rotation rate. No local-time dependence in the rotation rate could be discerned with the data.

References

[1] V. L. CARTER et al., J. Geophys. Res. **74**, 5083 (1969).
[2] B. K. CHING, J. Geophys. Res. **76**, 197 (1971).
[3] L. G. JACCHIA, Smithsonian Astrophysical Observatory Special Report No. 313 (1970).
[4] F. A. MARCOS et al., Space Research XI, 941 (1971).
[5] B. R. MAY and D. E. MILLER, Planet. Space Sci. **19**, 39 (1971).
[6] D. G. KING-HELE and D. M. C. WALKER, Royal Aircraft Establishment Technical Report No. 70084 (1970).
[7] R. L. JACOBS, J. Geophys. Res. **72**, 1571 (1967).
[8] U. VON ZAHN and J. GROSS, J. Geophys. Res. **74**, 4055 (1969).
[9] D. G. KING-HELE and D. W. SCOTT, Planet. Space Sci. **15**, 1913 (1967).
[10] D. G. KING-HELE, Planet. Space Sci. **18**, 1585 (1970).

Space Research XII – Akademie-Verlag, Berlin 1972

MEASUREMENTS OF UPPER-ATMOSPHERE ROTATIONAL SPEED FROM CHANGES IN SATELLITE ORBITS

D. G. King-Hele

Royal Aircraft Establishment, Farnborough, Hants, England

The rotation of the upper atmosphere subjects a satellite to an aerodynamic force normal to the orbit, which has the effect of slightly reducing the inclination of the orbit to the equator. The average rotational speed of the upper atmosphere at heights a little above that of perigee can therefore be evaluated from the observed changes in orbital inclination. Since the change in inclination is small (less than 0.1°), the values generally have to be averaged over several months, and they can also be regarded as applying over latitudes up to about half the inclination, the effects being strongest at the equator.

Recent results reviewed in this paper confirm the previous finding that the upper atmosphere at heights of 200 to 350 km rotates on average faster than the earth, and that the rate of rotation increases with height from about 1.1 rev/day at 200 km to 1.4 rev/day at 350 km. However, the newest results suggest that the rotation rate decreases above 350 km, to about 1.0 rev/day at 450 km and 0.7 rev/day at 500 km. In addition, new studies at heights of 120–230 km indicate wide variations in the rotation rate over short time intervals.

1. Introduction

The rotation of the upper atmosphere subjects a satellite to an aerodynamic force normal to the orbit, and this force continuously but very slowly reduces the inclination of the orbit to the equator. The inclination can decrease by as much as 0.1° for a satellite in an orbit of low perigee and high inclination, with an initial orbital period greater than 100 minutes. By measuring the observed decrease in inclination, we can determine the average rotational speed of the upper atmosphere at heights a little above that of perigee. Since the change of inclination per month is usually very small, the values of rotational speed so far obtained have usually been averaged over several months; this generally means that the local time at perigee is an average over at least 6 hours. The values can also usually be regarded as being averaged in latitude over latitudes up to about half the inclination, the effects being strongest at the equator. So far, most results have been confined to the height range 170–380 km, where the influence of air drag is strongest; and the errors (sd) have been about 3% at best and 10% at worst.

Results from 27 orbits were available in 1969, and were reviewed at the meeting of COSPAR in Prague [1]. These results showed that the upper atmosphere at heights between 200 and 360 km was on average rotating faster than the earth, and the speed increased with height from about 1.1 rev/day at

200 km to about 1.4 rev/day at 350 km. This would correspond to average west-to-east winds at 30° latitude increasing from 40 m s^{-1} at 200 km to 160 m s^{-1} at 350 km height. Since 1969 great improvements have been made in the accuracy of orbit determination of high-drag satellites, largely as a result of a campaign of observations of low-perigee satellites by the Hewitt cameras at Malvern and Edinburgh, and the 200-mm camera at Meudon. The first results from these accurate orbits are now becoming available and are reviewed in this paper.

2. Results from Individual Satellites

2.1. Explorer 1 (1958α)

A good example of the use of the method is shown in Fig. 1, which gives the observational values of orbital inclination for Explorer 1, as determined by the Smithsonian Astrophysical Observatory [2] over $11^1/_2$ years from its launch

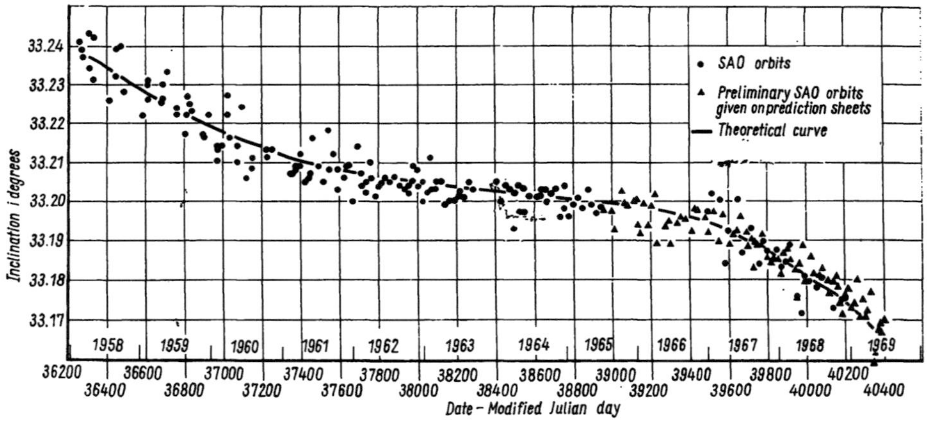

Fig. 1. Orbital inclination of Explorer 1 (1958 α), from 1958 to 1969. ● SAO orbits; ▲ preliminary SAO orbits given on prediction sheets; ——— theoretical curve ($\Lambda = 1.4$).

in 1958 to the end of 1969. The inclination decreased from 33.24° initially to 33.17° at the end of 1969. Fig. 1 also shows the theoretical curve for an atmospheric rotational speed 1.4 times that of the earth ($\Lambda = 1.4$)*. The agreement between theory and observation is excellent, allowing for errors of about 0.005° in the observational values, and shows [3] that, at latitudes up to about 20°, the upper atmosphere at heights near 380 km was rotating on average at 1.4 ± 0.1 rev/day over the complete cycle of solar activity between 1958 and 1969.

*) Λ is defined as the ratio of atmospheric angular velocity to the earth's angular velocity; or, in other words, the atmospheric rotation rate in revolutions per day.

2.2. 1968-59A

The satellite 1968-59A was launched in July 1968 into a polar orbit with an initial perigee height of 160 km and an initial period of 104.8 minutes: it offered the first good opportunity for evaluating the atmospheric rotational speed below 200 km. The orbit of the satellite was determined at the Royal Aircraft Establishment (RAE) from Baker-Nunn, visual and radar observations [4], and Fig. 2 shows the nine RAE observational values of inclination, with standard

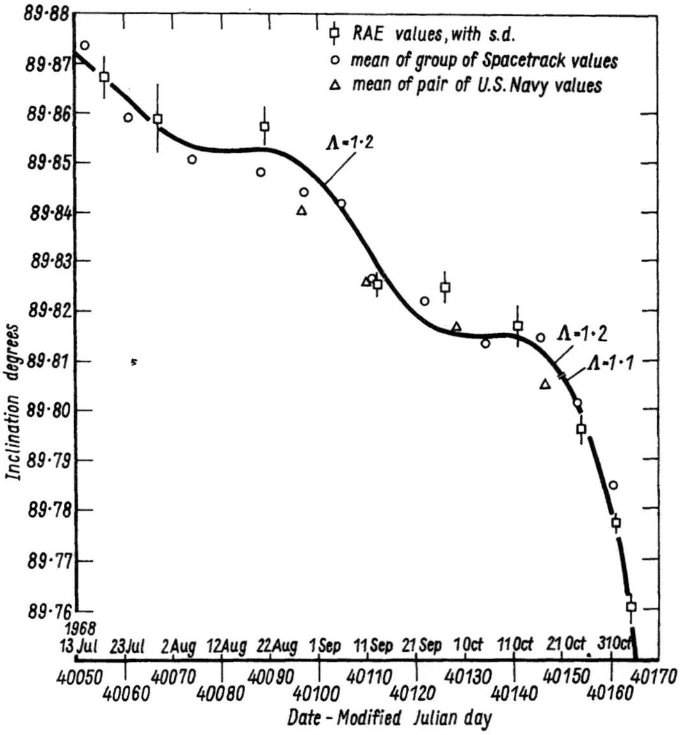

Fig. 2. Values of inclination for 1968-59A, corrected for luni-solar perturbations, with theoretical curve for $\Lambda = 1.2/1.1$. RAE values, with s.d.; ○ mean of group of Spacetrack values; △ mean of pair of US Navy values.

deviations, together with values from USAF Spacetrack and US Navy orbital elements, all after correction for luni-solar and other relevant perturbations. The theoretical curve fitted is divided into two parts, with $\Lambda = 1.20 \pm 0.07$ up to MJD 40150, applicable for a height of 175 km, and $\Lambda = 1.1 \pm 0.1$ after MJD 40150, applicable for a height of 150 km.

The orbit of this satellite has also been determined, quite independently and from different observations, at the Aerospace Corporation. By fitting these data, Mrs. Ching [5] obtained $\Lambda = 1.32$ with nominal standard deviation of 0.05. However, other statistical tests applied by Mrs. Ching suggested that a realistic standard deviation would be somewhat larger, and it is probably fair to quote $\Lambda = 1.32 \pm 0.1$ at a height of 175 km as a value comparable with the $\Lambda = 1.20 \pm 0.07$ from the results in Fig. 2. So the difference between the two independent determinations is not significant.

2.3. Cosmos 307 Rocket, 1969-94B

Cosmos 307 rocket entered orbit on 24 October 1969 and decayed on 20 July 1970. The initial perigee height was 210 km, the inclination 48.4°, and the initial period 108.8 minutes. So the object was very suitable for studies of atmospheric rotational speed, and was intensively observed by the British

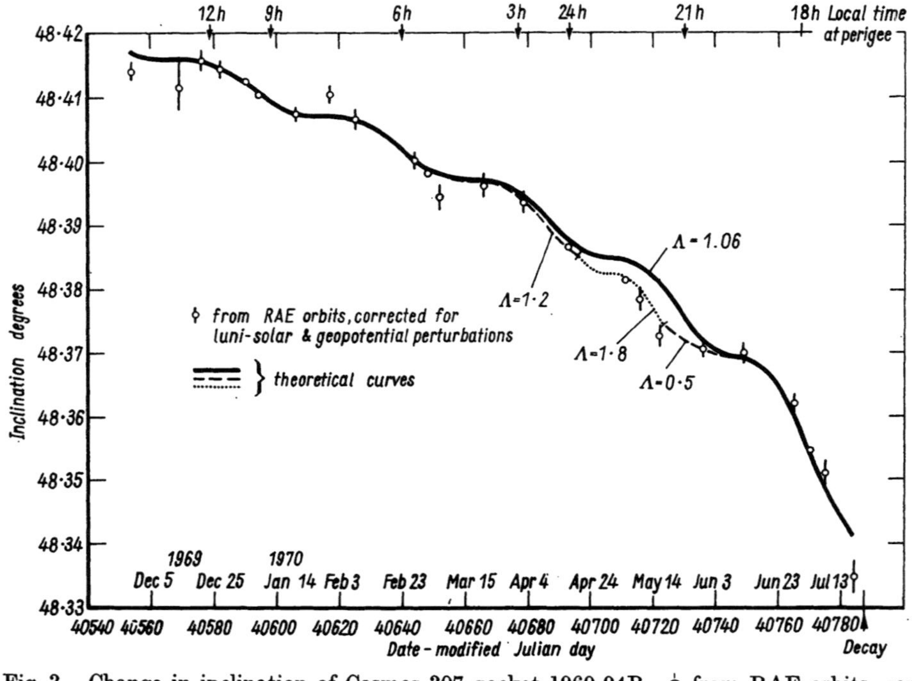

Fig. 3. Change in inclination of Cosmos 307 rocket 1969-94B. ♁ from RAE orbits, corrected for luni-solar and geopotential perturbations; ———, - - - - -, · · · · · theoretical curves.

optical tracking network, including the Hewitt cameras, by the camera at Meudon, and also by radar. From the observations the orbit has been determined by Hiller [6] at 25 epochs, and the values of inclination obtained are plotted in Fig. 3, with standard deviations, after correction for luni-solar, odd zonal harmonic, and tesseral harmonic perturbations. Six of the standard deviations are too small to show, being less than 0.0004°. The solid line in Fig. 3 is the theoretical curve for $\Lambda = 1.06$, which gives the best fit over the complete lifetime. The standard deviation of this mean value of Λ is estimated as 0.03, and the height at which it applies is about 230 km.

This is the most accurate set of values of inclination yet produced for a high-drag satellite with an initial period of more than 100 minutes, and some of the observational values seem to differ significantly from the values given by the theoretical curve: in particular, there is a run of nine values below the curve between 1 March and 1 June 1970. To fit all these points satisfactorily requires at least three different values of Λ during the three months. The broken and dotted curves in Fig. 3 show the values given by taking $\Lambda = 1.2$ between MJD 40653 and 40696, $\Lambda = 1.8$ between MJD 40696 and 40723, and $\Lambda = 0.5$ from MJD 40723 to 40747. The fit is of course much better, but the numerical

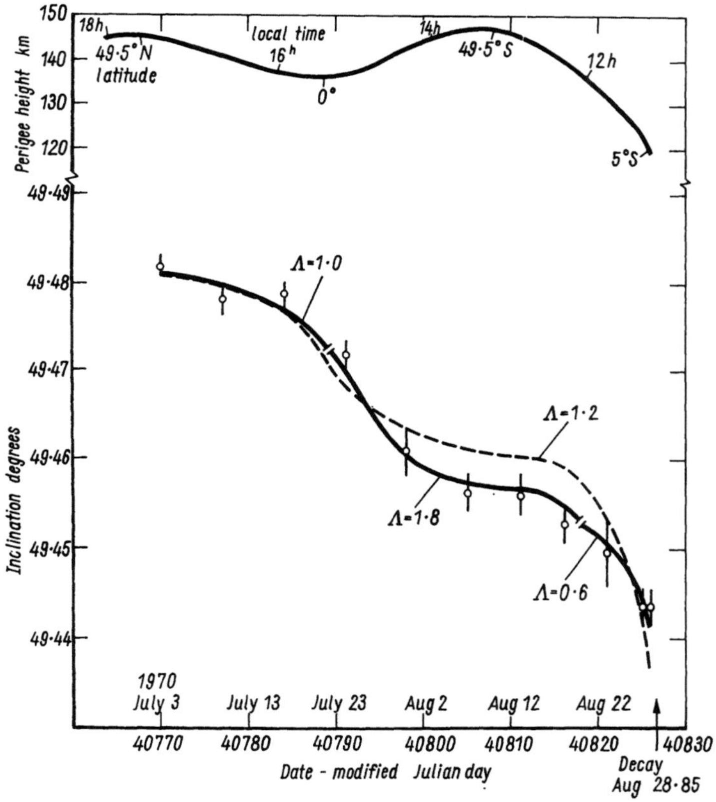

Fig. 4. Values of inclination for 1969-108A, with theoretical curves. ϕ values from RAE orbits, with s.d., corrected for luni-solar, zonal and tesseral harmonic perturbations.

values of Λ may be unreliable, because a different choice of values might be made if one of the orbits was in error.

At first sight the value $\Lambda = 1.8$ may seem rather outrageous, since it implies west-to-east winds of order 300 m s^{-1} at low latitudes, while $\Lambda = 0.5$ implies east-to-west winds of order 200 m s^{-1}. It seems possible, however, that upper-atmosphere rotation may sometimes depart greatly from normality: Rees [7] has measured local east-to-west neutral air winds of up to 500 m s^{-1} in the auroral region at times of geomagnetic disturbance (corresponding to an even more outrageous local value, $\Lambda = -2$); and Feess [8] reports even higher neutral wind speeds, of order 1000 m s^{-1}. It may be relevant that a geomagnetic storm, with $Ap = 90$, occurred at MJD 40697, at the beginning of the section with $\Lambda = 1.8$ in Fig. 3.

Even though the values of Λ for the broken and dotted curves in Fig. 3 are probably inaccurate, the rotation is presumably most rapid, i.e. west-to-east winds are strongest, at the time when $\Lambda = 1.8$; and the scale at the top of Fig. 3 indicates that if local time is the most important controlling factor, west-to-east winds are strongest between 21 h and 24 h local time, in conformity with theoretical studies [9, 10] and observational results from the orbit of 1966-51C and rocket firings [11, 12, 13].

2.4. Cosmos 316, 1969-108A

Cosmos 316 was launched on 23 December 1969 into an orbit with initial perigee height 150 km, inclination 49.5° and initial period 102.8 minutes. The satellite was very massive, and despite its low initial perigee remained in orbit until 28 August 1970. Because of its obvious interest for studies of upper-atmosphere density and rotational speed, Cosmos 316 was intensively observed, and its orbit is now being determined [14] at RAE from visual, radar and photographic observations, including observations from the Hewitt camera at Malvern. Computations are now complete on the last 11 orbit determinations, between 3 July and 28 August 1970, and the resulting values of inclination with standard deviations, are shown in Fig. 4 after correction for luni-solar, zonal harmonic, and tesseral harmonic perturbations. The curve at the top of Fig. 4 shows the perigee height, which ranges between 147 and 119 km, and the local time and latitude at perigee are marked on this curve. In Fig. 4 the theoretical curve for $\Lambda = 1.2$ is drawn as a broken line. Although 1.2 may be approximately the correct average value, it is clear that the points cannot be properly fitted if Λ is kept constant. The curve has therefore been split into three parts with $\Lambda = 1.0$, 1.8 and 0.6, shown by the three unbroken curves in Fig. 4. The values of inclination are now fitted to within one standard deviation, though it should be remembered that the numerical values of Λ may not be reliable, because the observational values of i are not adequate either in number or accuracy to define Λ precisely over such short time intervals. However, there seems little doubt that Λ is exceptionally large between 24 July and 31 July, when the local time at perigee was about 15 h, perigee height was 140 km, and perigee latitude was between 0 and 20° S. It would be rash to guess whether date, height, local time or latitude was the most important controlling factor; a clearer picture may emerge when the orbit determinations are completed.

2.5. Ariel 3, 1967-42A

In a careful analysis of the changes in orbital inclination of Ariel 3 during its first two years in orbit, Gooding [15] has determined lumped fifteenth-order tesseral harmonics in the gravitational field, and as a by-product has obtained a value of 0.7 ± 0.1 rev/day for the average atmospheric rotational speed at a height of 500 km in the time interval between May 1967 and August 1969.

At first sight, this result is unexpected, since previous determinations of the atmospheric rotational speed have indicated that Λ increases from 1.1 at 200 km to 1.4 at 350 km. However, the previous determinations only extend to a height of 380 km, and it is possible that Λ reaches a maximum near this height and then decreases.

2.6. 1968-86A

The satellite 1968-86A was in a circular orbit of inclination 75°, at heights near 480 km initially. The orbit is about to be determined from observations at RAE, but a preliminary determination of Λ has already been made using US Navy elements. This gives $\Lambda = 1.0 \pm 0.1$ at an average height of 430 km between mid 1969 and the satellite's decay in March 1971, thus confirming that the rotation rate probably does decrease at heights above 400 km.

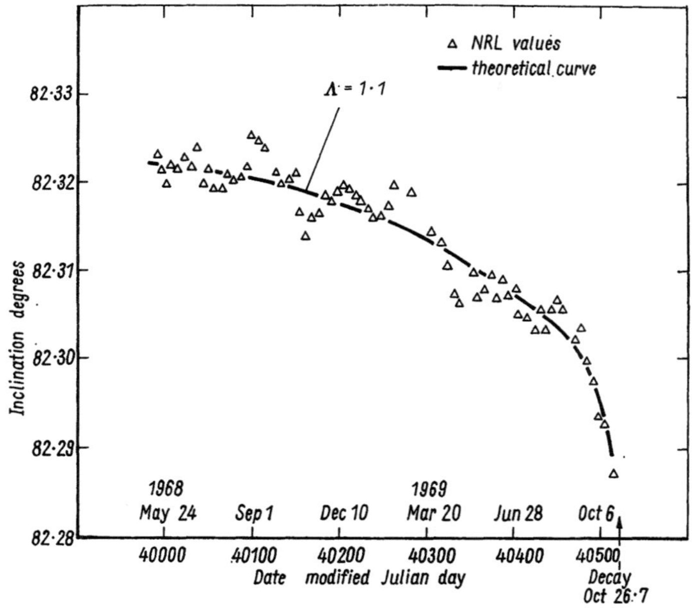

Fig. 5. NRL values of inclination for 1963-27A, with theoretical curve.

2.7. 1963-27A

US Navy values of inclination for 1963-27A between May 1968 and decay are plotted in Fig. 5, with the theoretical curve for $\Lambda = 1.1$. The value applies at a mean height of 420 km and the estimated sd is 0.15.

3. Conclusions

3.1. Mean values of Λ

The 27 most accurate mean values of Λ so far obtained, including the new values from Ariel 3, 1968-86A and 1963-27A, are plotted against height in Fig. 6. No value from Cosmos 316 is given, in view of the difficulty of assigning a mean value with the data so far available, and the values for 1968-59A are those from Fig. 2 rather than that of [5].

The broken line in Fig. 6, which consists of two straight lines with their intersection rounded off, is intended to indicate the trend of the values. Fig. 6 confirms that the mean rotational speed of the upper atmosphere increases

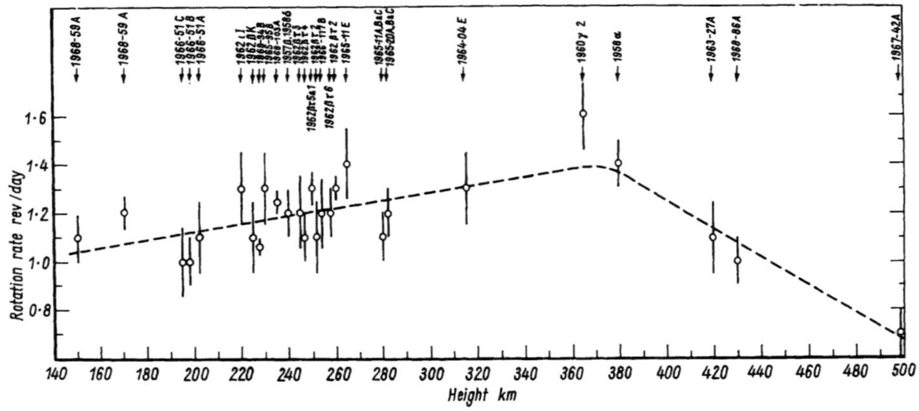

Fig. 6. Mean upper-atmosphere rotation rate, derived from analysis of 31 satellite orbits.

with height between 150 and 350 km, from about 1.1 rev/day at 200 km height to about 1.4 rev/day at 350 km.

Above 400 km the evidence is at present decidedly sparse; but it seems likely that the mean rotation rate of the upper atmosphere decreases to approximately 1.0 rev/day at 450 km height and approximately 0.7 rev/day at 500 km. This decrease is not particularly surprising, since Λ may possibly decrease very much below 1 at great heights. The maximum value of Λ occurs at a height near 350 km, and it may be significant that this is quite near the height of peak electron concentration at night in the F layer of the ionosphere.

3.2. Variations in Λ

The other main conclusion concerns the degree of detail revealed by these new orbital studies. It is clear from Figs. 3 and 4 that the accurate orbits of low-perigee satellites now being determined, particularly when Hewitt camera observations are available, are not adequately fitted by choosing a single mean value of Λ. The fitted curve must be split into several sections, with differing values of Λ. The parameters controlling the variations in the rotation rate cannot yet be certainly identified; but local time is likely to be an important factor, and, if so, the results indicate a rapid rotation rate, i.e. strong west-to-east winds, between 20 h and 24 h local time at a height near 230 km.

The new results also strongly suggest that the upper-atmosphere rotation rate may sometimes depart greatly from normality. In both Fig. 3 and Fig. 4 values of Λ near 1.8 are sometimes required in fitting the observational points, and although the accuracy of these numerical values is questionable, the indication of strong departures from the average is probably valid, especially in view of recent measurements of very strong local winds [7, 8].

References

[1] D. G. KING-HELE, Space Research X, 537 (1970).
[2] I. G. IZSAK, B. MILLER et al., Smithson. Astrophys. Obs. Spec. Rep. Nos. 50, 78, 86, 93, 113, 117, 120, 126, 142, 159, 160, 169, 208, 209, 225 and 287 (1960–1968).
[3] D. G. KING-HELE, D. W. SCOTT and D. M. C. WALKER, Planet. Space Sci. **18**, 1433 (1970).
[4] D. G. KING-HELE, Planet. Space Sci. **18**, 1585 (1970).
[5] B. K. CHING, J. Geophys. Res. **76**, 197 (1971).
[6] H. HILLER, RAE Technical Report 71151 (1971).
[7] D. REES, J. Brit. Interplan. Soc. **24**, 233 (1972).
[8] W. A. FEESS, paper presented at COSPAR Symposium on the Dynamics of the Thermosphere and Ionosphere, Seattle 1971.
[9] R. A. CHALLINOR, Planet. Space Sci. **17**, 1097 (1969); **18**, 1485 (1970).
[10] G. J. BAILEY, R. J. MOFFETT and H. RISHBETH, J. Atmos. Terr. Phys. **31**, 253 (1969).
[11] D. G. KING-HELE and D. W. SCOTT, Nature, Lond. **213**, 1110 (1967).
[12] D. REES, paper presented at COSPAR Meeting, Prague 1969.
[13] J. F. BEDINGER, J. Geophys. Res. **75**, 683 (1970).
[14] D. G. KING-HELE, The Orbit of Cosmos 316 (to be published).
[15] R. H. GOODING, Nature Physical Science **231**, 168 (1971).

Space Research XII — Akademie-Verlag, Berlin 1972

UPPER ATMOSPHERE HEATING AT HIGH LATITUDES

R. R. ALLAN

Royal Aircraft Establishment, Farnborough, Hants, England

The orbit of Molniya 1K (1968-85A) from November 1968 to November 1970 has been analysed to study variations in air density near 65° S. This object is unique among the Molniya satellites in that its perigee height, which varies with a 7-year period due to luni-solar forces, reached a minimum near 400 km about a year after launch. The air density is sampled over very short arcs because of the high eccentricity (0.74), and the sampling is confined to two regions of the earth since the orbit is 2:1 commensurable with the earth's rotation.

The air density at 400 km shows great variability, apparently not related to *Kp*, with maxima which exceed the corresponding values at the equator by a factor of at least 6, and the highest densities occur near 18 h local time. The heating presumably occurs during auroral substorms, and might be due either to ohmic losses in the auroral electrojet or to absorption of precipitating electrons. There is some evidence that the source might be enhanced near the equinoxes, and, if so, might be responsible for the semiannual variation in density.

1. Introduction

It has long been supposed [1–4] that there is a significant heat input into the upper atmosphere at high latitudes whether by Joule heating or by absorption of charged particles. Cole [2, 3], in particular, has estimated that Joule heating by the auroral electrojet at times of severe disturbance could lead to temperature increases of the order of 1000 °K above 150 km. Likewise the average flux of electrons precipitating into the auroral zones [5] is 4 erg cm^{-2} s^{-1}, and can occasionally rise to 2000 erg cm^{-2} s^{-1}; by comparison the energy requirement of the thermosphere is equivalent to [6] a steady heat flux of $\sim$0.4 erg cm^{-2} s^{-1} at 120 km over the whole earth.

Certainly the effects associated with geomagnetic activity are evident in the density of the upper atmosphere [7, 6], and appear to be enhanced [8, 9] at high latitudes. Similarly the time delay between the peak geomagnetic disturbance and the maximum density is smaller at higher latitudes [9]. Thus it seems likely that the geomagnetic activity effect is produced by energy injected in or near the auroral zones, and that the density perturbation propagates from there. The recent experimental results of DeVries [10] confirm this view.

2. The Molniya Satellites

A unique opportunity of studying the density of the upper atmosphere above about 400 km in the auroral zones is presented by the Russian communications satellite Molniya 1K (1968-85A). Seventeen Molniya satellites have now been launched; their orbits are inclined at $\sim 65°$ and are highly eccentric, with the initial eccentricity consistently 0.74, corresponding to a perigee height of ~ 500 km. The semi-major axis is $\sim 26\,550$ km giving a period of approximately 12 h, and is moreover adjusted after injection so that the ground track is nearly fixed relative to the earth. Also the argument of perigee is near 270°; in fact the initial value is consistently 285° but is decreasing slowly at a rate between 5°/yr and 10°/yr depending on the particular object. Apogee therefore occurs in the northern hemisphere near the apex of the orbit.

Clearly, so long as perigee remains at a suitable height, a Molniya-type satellite would provide a fairly unique probe of upper-atmosphere density. Because of the high eccentricity, the atmosphere is sampled over only a short arc; assuming a scale height of 60 km appropriate to perigee at ~ 400 km, the satellite is within one scale height of perigee only over an arc of about $\pm 12°$. Moreover perigee remains near 65° S and each arc is at virtually a fixed latitude with the extension purely in longitude. Furthermore perigee occurs at two (slowly moving) longitudes since the orbit is nearly 2:1 commensurable with the earth's rotation. The perigee height of the Molniya satellites is, however, severely perturbed by luni-solar forces [11] owing to the high eccentricity. Essentially the height of perigee varies sinusoidally with the orientation of the orbital plane in space. Since the plane is regressing at a rate $\dot{\Omega} \sim -50°$/yr, the period of the oscillation is about seven years; the amplitude is of the order of ± 500 km. Usually perigee either ascends fairly rapidly so that the object will survive for at least about seven years, or else descends into the denser atmosphere and the object decays within about two years. Molniya 1K is unique, however, in that perigee fell from an initial value near 500 km to a minimum of about 380 km a year later and then rose again, being now above 600 km. The variation of perigee height from Spacetrack bulletins between November 1968 and November 1970 is given in the lower part of Fig. 1; there is an obvious oscillation with a period of about 150 days which is a short-period term due to the sun.

The fact that the ground track is nearly stationary relative to the earth brings the attendant difficulty that the satellite is in resonance [12, 13] with the longitude-dependent part of the earth's gravitation. It can be shown [14], however, that the effect on Molniya-type orbits can be described in terms of the resonant variable

$$\Pi_{1,2} = M + 2\,(\Omega - \nu)\,, \tag{1}$$

where Ω and M are two of the usual set of osculating elliptic elements, a, e, I, Ω, ω, M, and ν is the sidereal hour angle. By far the largest contribution arises from the gravity coefficient $J_{2,2}$ which corresponds to the 'ellipticity of the

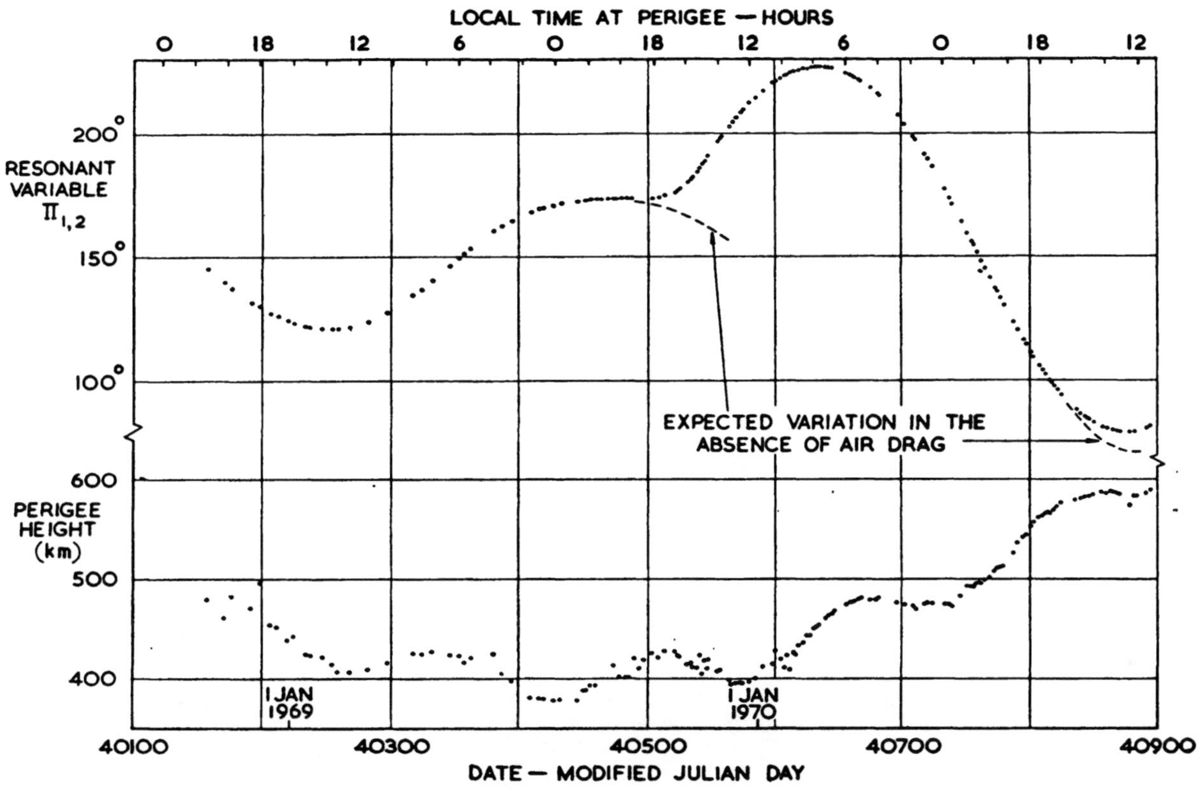

Fig. 1. The behaviour of the resonant variable and the perigee height for Molniya 1 K (1968-85A) from November 1968 to November 1970.

equator', and the resonant behaviour can be described by the equation

$$\frac{d^2 \Pi_{1,2}}{dt^2} = -3\, n^2 J_{2,2} \left(\frac{R}{a}\right)^2 A \sin\left(\Pi_{1,2} + \delta - 2\,\varphi_{2,2} - \pi\right) \tag{2}$$

$$+ \text{ effects of higher order gravity coefficients, } J_{3,2} \text{ etc.} \ldots,$$

where [14] the constants A and δ depend mainly on ω and only slightly on e and I. The phase δ cannot, however, exceed about 10°.

In the absence of other perturbing forces the variable $\Pi_{1,2}$ will oscillate about the stable value $2\,\varphi_{2,2} + \pi - \delta$; in practice the stable position is changing slowly (and linearly) due to the dependence of δ on ω. For other Molniya satellites the variable $\Pi_{1,2}$ does indeed behave according to (2) unless and until perigee descends deeply into the atmosphere. The simple approximation of including only the $J_{2,2}$ gravity coefficient, and taking the values given by Kaula's 'combined-adjusted' solution [15], namely $J_{2,2} = 1.79 \times 10^{-6}$ and $\varphi_{2,2} = -15°$, reproduces the actual libration period and gives the stable position to within about 1°. For Molniya 1K on the other hand, the resonant variable, shown in Fig. 1, appears to behave normally until about MJD 40450, but thereafter the object suffers a large and varying perturbing force which apparently vanishes again about MJD 40550 so that $\Pi_{1,2}$ resumes a simple libration but with greatly increased amplitude. The expected behaviour after MJD 40450 is shown by the broken line in Fig. 1.

Although it may not be obvious from Fig. 1, the libration curve also displays a slightly increased curvature somewhere near the minimum about MJD 40250. Moreover Molniya 1K again shows an anomalous acceleration in August–September 1970. The libration curve is in accord with (2) from MJD 40550 up to about MJD 40820. After this the curve should follow the broken line to continue in agreement, but instead turns up much more sharply despite the fact that perigee is at nearly 600 km.

3. Analysis of the Observations

We shall first consider the period around October–November 1969, when the anomalous acceleration is highest so that its variation can be determined with some accuracy. Fig. 2 shows the drag acceleration from MJD 40450 to 40600 deduced from Spacetrack data. The points, joined by slant lines, are derived by taking double differences of the variable $\Pi_{1,2}$ and the data are smoothed to the extent of omitting values which give rise to absurd swings. The drag acceleration derived by single-differencing the values of the mean motion n is also shown in Fig. 2 in the form of a histogram, with some parts of the lines omitted for clarity; here no element sets have been rejected since the effect of a 'wild' value is less disastrous in forming single differences. In both cases the effect of the resonance forces is calculated according to (2) and duly subtracted. It is at least reassuring that the points and the histogram are in reasonable agreement.

It must be emphasized that we are dealing with drag acceleration, not air density. The Molniya satellites are stabilized so that their large area of solar cells points at the sun. From press releases, the maximum cross-sectional area appears to be about (20 ± 4) m^2 and the minimum about (7 ± 1) m^2. With ω near 270°, the satellite at perigee presents nearly its maximum cross-sectional area to the atmosphere when perigee is at 06 h or 18 h local time, and nearly its minimum cross-sectional area at 12 h and 24 h; the drag experienced will be weighted accordingly.

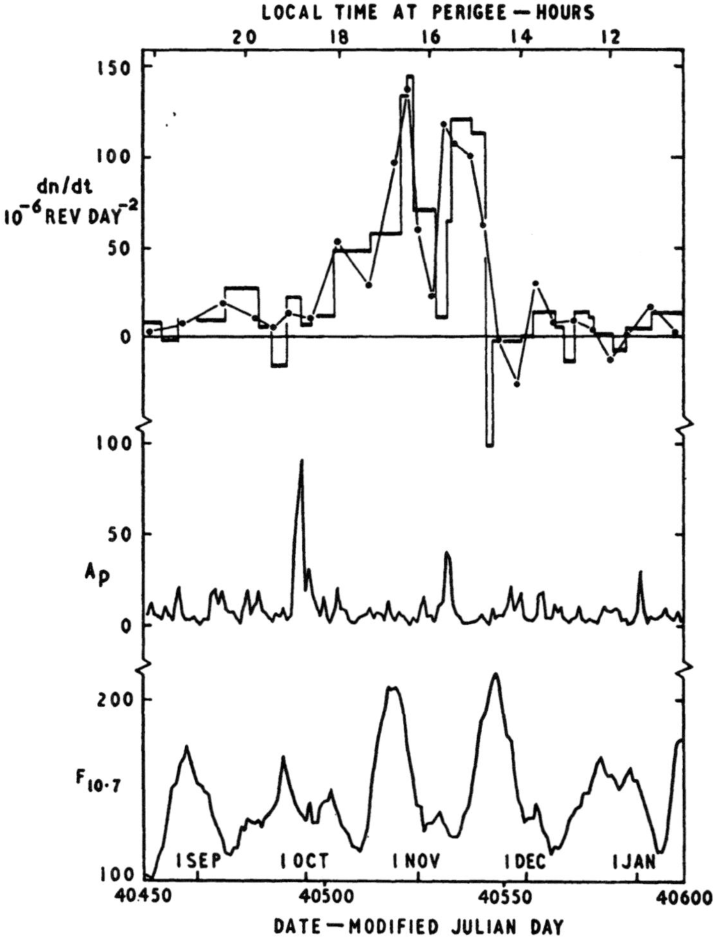

Fig. 2. The drag acceleration of Molniya 1K from August 1969 to January 1970 with geomagnetic and solar indices. The points connected by slant lines are derived from double differences of the resonant variable $\Pi_{1,2}$, and the histogram from single differences of n.

Both the points and the histogram in Fig. 2 show a double-peaked structure with the first narrower peak just before 1 November and the second broader peak from 10 to 20 November. The local time of perigee, shown at the top of Fig. 2, is decreasing slowly (by 2 h about every 25.3 days) mainly due to the annual motion of the earth, and the two peaks occur just after and just before 16 h local time. The great variability makes it clear, however, that we are not dealing simply with a diurnal maximum due to solar euv heating. There does not seem to be any correlation with the $F_{10.7}$ index of solar activity. Nor is there any correlation with global magnetic activity as measured by the index Ap, which was high at the end of September 1969 but below average throughout October and November; the only significant events are $Kp = 5+$ on 2 October, and $Kp = 6-$ on 9 and 10 November.

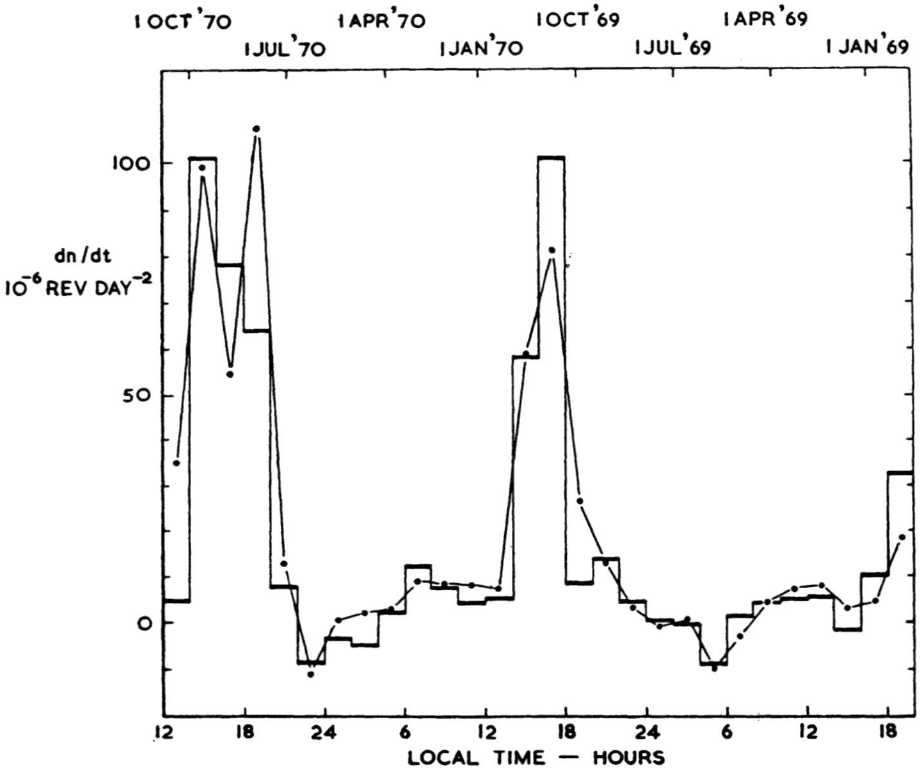

Fig. 3. The drag acceleration (two-hourly means) against local time at perigee adjusted to a standard height of 400 km assuming a scale height of 60 km. The significance of the points and histogram is as in Fig. 2.

A detailed record of the drag acceleration similar to Fig. 2 for the whole two years is not very illuminating since the noise level is high; it is just possible to discern the other periods with a significant positive acceleration. Accordingly we have chosen to average the drag acceleration by combining all the individual values in each two-hour interval of local time. The result is shown in Fig. 3, plotted with local time increasing to the right so that the real date increases to the left. All values have been adjusted to a standard height of 400 km assuming a scale height of 60 km, and the significance of the points and the histogram is as in Fig. 2. The period covered by Fig. 2 now appears as the portion from 12 h to 22 h in the centre of Fig. 3. The detail, in particular the double peak apparent in Fig. 2, has been averaged out, and there now appears to be a broad peak from 14 h to 18 h.

The anomalous acceleration from 40800 onwards (August–September 1970) produces the broad peak from 14 h to 20 h on the left of Fig. 3; the points show a double-peaked structure while the histogram does not. The important point, however, is that the drag is consistent with October–November 1969 (assuming a scale height of 60 km); the actual acceleration is much smaller since perigee is between 550 and 600 km. Taking $F = 0.96$, $S = 20\ \mathrm{m}^2$, $C_D = 2.2$, $m = 1000$ kg, and $H = 60$ km, these peak accelerations $dn/dt \simeq \simeq 10^{-4}$ rev day^{-2} imply an air density of $\sim 2.4 \times 10^{-14}$ g cm^{-3} at 400 km. The peaks in Fig. 2 would give densities, over shorter intervals, roughly twice as large; thus the first peak in Fig. 2 gives $\sim 5.0 \times 10^{-14}$ g cm^{-3} when reduced to 400 km. These values compare with a maximum equatorial density (at $\sim$15 h) of 0.76×10^{-14} g cm^{-3} at 400 km for $\bar{F}_{10.7} = 150$ given by the CIRA 1965 model atmosphere. These maximum densities should not be interpreted too literally; in the writer's view, they probably arise from even greater increases in density which occur only on some passages through perigee.

On the other hand the anomalous acceleration around MJD 40200 (December 1968), when perigee was at approximately 450 km, is comparatively small, and produces the rather insignificant peak on the right of Fig. 3, implying densities smaller by a factor of three or four. Admittedly the earlier observations are not so accurate or numerous, but an increase in density in line with the other two peaks would have produced an obvious effect on the libration curve. The common feature, however, is that the peaks in density are centred on the local time interval 16 h–18 h.

Can the disparity between the peak densities be due to the different locations of perigee relative to the earth? In Table 1, λ and φ are the geographic (and Λ and Φ the geomagnetic) latitude and longitude, and the corresponding values of L ($= \sec^2 \Lambda$) and geomagnetic local time (MLT) are also given.

Table 1

Locations of perigee at density peaks

Local time	~18 h	20 h–14 h	20 h–14 h
Date	MJD 40200 Dec 1968	MJD 40480–40550 Oct, Nov 1969	MJD 40780–40860 Aug, Sept 1970
Perigee height	~450 km	~400 km	550–600 km
λ	62° S	65° S	65° S
φ	0° E, 180° E	15° E, 195° E	30–50° W, 150°–130° E
Λ	57° S, 68° S	60° S, 65° S	52° S, 80° S
Φ	50° E, 285 °E	60° E, 300° E	30°–20° E, 250°–210° E
L	3.4, 7.1	4.0, 5.6	2.6, 33
MLT	16 h, 21 h	18 h–12h, 22 h–16 h	19 h–14 h, 22 h–15 h

It is clear from Table 1, however, that the disparity between December 1968 and October–November 1969 can scarcely be due to differences in geomagnetic latitude or geomagnetic local time. Indeed the peaks in October–November 1969 and August–September 1970 differ much more in geomagnetic latitude; in the latter case one perigee is at $L = 2.6$ and the other at $L > 30$. Possibly the difference is that the two large peaks occurred near the equinoxes while the other occurred in mid-summer. Indeed there seem to be minor peaks at the remaining equinoxes, namely between 10 h and 14 h in February–March 1969 and between 6 h and 10 h in February–March 1970, although this latter will be partly due to the variation of the cross-sectional area.

4. Physical Mechanisms

The present findings strongly suggest that the heating takes place during auroral substorms which appear to be produced [16–18] by a contraction or collapse of the magnetospheric tail. This is possibly caused [16] by a reversal of the vertical component of the interplanetary magnetic field from north to south, which strips flux tubes from the dayside magnetosphere and transports them to the tail without compressing the magnetosphere. A substorm occurs [19] every few hours and the active phase typically lasts for $^1/_2$ to 1 h. Two related effects are involved, either of which might account for the heating: (1) large electric currents, the auroral electrojets, flow round parts of the auroral oval ($\Lambda \simeq 70°$), and (2) intense fluxes of low energy electrons precipitate into the auroral zones.

An analysis [19] of magnetograms from auroral observatories shows that an eastwards current centred on 18 h MLT flows in the evening sector, and a westwards current centred on 04 h MLT in the morning sector; each current region might be up to a few thousand km long, a few hundred km in north–south extent, and a few tens of km in vertical extent. Cole [20] estimates that the ohmic loss peaks at roughly 150 km, and that the energy input to the auroral thermosphere is of order 1 erg cm^{-2} s^{-1} at quiet times rising to ~100 erg cm^{-2} s^{-1} during strong disturbances.

Turning to electron precipitation, there appear [21] to be two separate regions with quite different characteristics. One region, centred near the auroral oval at ~22 h MLT, is characterized by a highly variable flux with a very soft spectrum consistent with the tail population. O'Brien and Taylor [5] suggest that the energy spectrum peaks near 1 keV; an electron of this energy will mainly dissipate its energy at ~150 km, and lower energy electrons at even greater heights. Further, the average energy flux of electrons precipitating into the auroral zones is ~4 erg cm^{-2} s^{-1}, rising occasionally to as much as 2000 erg cm^{-2} s^{-1}. The other type [21] of particle precipitation is from the trapped population and occurs at slightly lower latitudes mainly in the morning sector around 08 h; the flux is roughly constant and the spectrum is flat.

It would appear that either Joule heating in the auroral electrojet, or the absorption of precipitating electrons, could easily supply the energy required to account for the behaviour of Molniya 1K. It is not clear, however, why the effect on air density appears to be confined to near 18 h local time.

5. Discussion

The only comparable results [22, 23] known to the writer are based on densities measured by gauges on Explorer 32, and cover the period May to October 1966, when perigee was in the northern hemisphere. Newton and Pelz [22] find that the density tends to increase with latitude above 45° N and shows a significant structure even during geomagnetically quiet times. The effects are most marked in the highest latitude band, 55°–65° N, and for the local time intervals 20–24 h, and 00–04 h. Moreover there is evidence of great variability.

Two other recent investigations are also highly relevant. DeVries [10] has analysed air densities measured by a satellite-borne accelerometer during a period of very high geomagnetic activity ($a_p \sim 400$) on 25–26 May 1967. The increase in density in the night side auroral zone (crossed at ~22.30 h local time at 190 km height) coincided with the onset of activity, and the increase propagated to mid-latitudes after a delay of some hours. Rees [24], using rocket-released clouds at Kiruna, found that the neutral temperature around 150 km was strongly enhanced during local geomagnetic activity, and that the increase correlated well with the C9 index rather than with *Kp*.

6. Conclusions

The results given here imply that a heat source in the sector around 18 h local time is present at high latitudes even during geomagnetically quiet times, at least as measured by the indices *Kp* or *Ap*. The heating is presumably associated with auroral substorms, and might be due either to ohmic losses in the auroral electrojet or to absorption of precipitating electrons. The source is probably sufficiently strong to be significant in the global heat balance of the thermosphere. If, moreover, the suggestion that the source is enhanced near the equinoxes should be correct, then it might well be responsible for the semiannual variation in density.

References

[1] V. I. Krassovsky, Planet. Space Sci. **1**, 14 (1959).
[2] K. D. Cole, Nature **194**, 761 (1962).
[3] K. D. Cole, Aust. J. Phys. **15**, 223 (1962).
[4] C. O. Hines, J. Geophys. Res. **70**, 177 (1965).
[5] B. J. O'Brien and H. Taylor, J. Geophys. Res. **69**, 45 (1964).
[6] W. Priester, M. Roemer and H. Volland, Space Sci. Rev. **6**, 707 (1967).
[7] L. G. Jacchia, Nature **183**, 1662 (1959).
[8] L. G. Jacchia and J. Slowey, J. Geophys. Res. **69**, 905 (1964).
[9] L. G. Jacchia, J. Slowey and F. Verniani, J. Geophys. Res. **72**, 1423 (1967).
[10] L. L. DeVries, NASA TM X-64568 (1971).
[11] G. E. Cook and D. W. Scott, Planet. Space Sci. **15**, 1549 (1967).

[12] R. R. ALLAN, Planet. Space Sci. **15**, 53 (1967).
[13] R. R. ALLAN, Planet. Space Sci. **15**, 1829 (1967).
[14] R. R. ALLAN, Cel. Mech. **3**, 320 (1971).
[15] W. M. KAULA, J. Geophys. Res. **71**, 5303 (1966).
[16] M. P. AUBRY, C. T. RUSSELL and M. G. KIVELSON, J. Geophys. Res. **75**, 7018 (1970).
[17] D. H. FAIRFIELD and N. F. NESS, J. Geophys. Res. **75**, 7032 (1970).
[18] E. W. HONES, JR. et al., J. Geophys. Res. **75**, 7060 (1970).
[19] T. N. DAVIS and M. SUGIURA, J. Geophys. Res. **71**, 785 (1966).
[20] K. D. COLE, Space Sci. Rev. **5**, 699 (1966).
[21] T. R. HARTZ and N. M. BRICE, Planet. Space Sci. **15**, 301 (1967).
[22] G. P. NEWTON and D. T. PELZ, J. Geophys. Res. **74**, 4169 (1969).
[23] G. P. NEWTON, J. Geophys. Res. **75**, 5510 (1970).
[24] D. REES, Planet. Space Sci. **19**, 233 (1971).

Space Research XII – Akademie-Verlag, Berlin 1972

STRUCTURE AND MOTION OF THE THERMOSPHERE SHOWN BY DENSITY DATA FROM THE LOW-G ACCELEROMETER CALIBRATION SYSTEM (LOGACS)

L. L. DeVries

Aero-Astrodynamics Laboratory, Aerospace Environment Division, NASA-George C. Marshall Space Flight Center, Alabama, USA

Neutral atmospheric density data obtained from a Low-G Calibration System (LOGACS) flown on a polar-orbiting satellite during a period of high geomagnetic activity, 22–26 May 1967, provide new information concerning the dynamics of the thermosphere. This experiment produced density data from more than 40 orbits with a time resolution of one second and with an accuracy to within 10%. The unique design of the experiment also permitted the measurement of lateral wind components. Analyses of these data show that density increases in the auroral zone almost simultaneously with enhanced geomagnetic activity. The largest density increases occurred in the region of the maximum current of the auroral electrojet with no significant increases in the equatorial region until several hours later. Atmospheric waves, apparently associated with joule heating, appear to originate in the auroral region at altitudes lower than 150 km and propagate upward and toward the equator. The enormous wind velocities calculated from the satellite lateral forces and the steep density gradients observed in the auroral zone may also result from joule dissipation of the electrical fields. Fine-scale details of the structure of the polar thermosphere are shown in these data.

1. Introduction

The response of the neutral atmosphere to heating associated with geomagnetic activity was discovered by Jacchia [1] in 1959. Studies of satellite-drag-deduced density data during the past 10 years by Jacchia et al. [2] and Roemer [3] have shown the following general details concerning this phenomenon: (i) even the smallest variations in geomagnetic activity are reflected in atmospheric variations; (ii) atmospheric variations lag behind geomagnetic perturbations by several hours; (iii) the temperature increase during a geomagnetic disturbance is enhanced at high latitudes; (iv) a diurnal dependence on specific heating is apparent, with group means for night conditions (20 h to 08 h local time) about 30% higher than the daytime values. Delay times of atmospheric response to geomagnetic activity are especially difficult to determine from orbital drag data and there is disagreement concerning whether or not they are dependent on latitude. DeVries et al. [4], in analyzing high-resolution density data deduced from precise radar tracking data from eleven low-altitude, attitude-stabilized Agena satellites, found that delay times of density increases associated with geomagnetic activity were latitude dependent, varying from near-instantaneous in the auroral regions to about 17 hours at low latitudes. On the other hand, in a very comprehensive study of density data from orbital drag measurements, Roemer [3], in analyzing the delay time between peak geomagnetic disturbances and maximum density for more than 200 cases, concluded that the peak density is reached world-wide with the same delay after the maximum of geomagnetic disturbances. Studies of orbital drag density data have not revealed sufficient details concerning this phenomenon to identify the energy source or the mechanism responsible for atmospheric heating following geomagnetic disturbances. It is recognized that short-period, transient density variations are difficult to analyze from orbital drag data, because the time and spatial resolutions (about three hours, or two orbits) are so large.

A few direct measurements of thermospheric composition, temperature and density have become available during the past few years. Taeusch et al. [5] analyzed *in situ* atomic oxygen measurements from a quadrupole mass spectrometer on the OGO 6 satellite, and found that the time response of this constituent in the sunlit sector of the auroral region to heating associated with geomagnetic activity was very short, apparently less than one hour. The OGO 6 neutral mass analyzer data also indicated that variations in atmospheric composition, density, and probably temperature, are most prominent in the polar regions; the density enhancements seemed to increase with latitude but, in some cases, seemed to dip at the pole. The orbit of the OGO 6 satellite during the period reported by Taeusch et al. [5] did not provide sufficient data over the South Pole or on the night side of either pole to analyze the atmospheric response to geomagnetic activity in these regions.

Blamont et al. [6] analyzed thermospheric temperature measurements from data obtained by a spherical Fabry–Perot interferometer on the OGO 6 spacecraft in a period during which two large geomagnetic disturbances occurred. These measurements showed a temperature increase of 300 °K in the polar regions, while the temperature near the equator increased by only 90 °K.

Two research satellites, Cannon Ball 1 (1968-59B) and SPADES (1968-59A), were orbited to measure atmospheric density and related properties in the altitude region 120–200 km. Both satellites were placed in near-polar orbits and carried accelerometers. Although considerable excellent density data were obtained during the 39-day lifetime of the Cannon Ball 1 satellite, the absence of highly disturbed geomagnetic activity and the lack of an on-board tape recorder restricted the data from which to study density variations associated with geomagnetic activity. Analyses of density data from the SPADES satellite [7] indicated that the density enhancement subsequent to geomagnetic activity was greater in the high northern latitudes than in the equatorial regions, with an apparent six-hour time lag at all latitudes. A density bulge was found in the auroral region, followed by a trough near the North Pole. Marcos et al. [7] concluded that, because of the wave-like nature of the density variations, this type of disturbance may be related to varying electric currents in the ionospheric E region, probably in the auroral electrojet.

A report by the United States Space Science Board of the National Academy of Sciences [8] summarizes some of the consequences of the modification of the global atmospheric circulation if enhanced heating occurs at high latitudes during periods of high solar activity: (i) the observed increase in density and temperature; (ii) the changes in the F region of the ionosphere: either wind-induced vertical ionization drifts or changes in neutral composition would be effective in this regard; and (iii) ionospheric electric currents which perturb the magnetic field observed on the ground. The authors of this study concluded that the response of the global circulation to high-latitude heating is one of the most provocative questions facing us at present.

One of the most serious difficulties in studying this very important problem is the lack of atmospheric data with sufficient resolution, accuracy, and geographic coverage. The LOGACS experiment, conducted in May 1967, produced some excellent density data which provide an improved understanding of thermospheric response to heating associated with geomagnetic activity.

2. The LOGACS Experiment

The United States Air Force placed an Agena satellite in a near-polar orbit during the period 22–26 May 1967, with an experiment designed to calibrate an extremely sensitive accelerometer in the orbital environment. This experiment was designated the **Low-G** **Accelerometer** **Calibration** **System** (LOGACS). The Bell **Miniature** **Electrostatic** **Accelerometer** (MESA) was selected. The MESA is a single-axis, electrostatically pulse-rebalanced accelerometer designed so that both the suspension and rebalance forces can be scaled by changing the voltage levels. The accelerometer was placed on a turntable device so that it could be rotated at two precisely known rates, or held in two fixed positions. The Agena vehicle was attitude-controlled so that the plane of the turntable was always held in the plane of the local horizon. In its fixed positions the sensitive axis of the MESA was aligned in the orbital plane (one position in the direction of motion and the other position opposite to the direction of motion). The two distinct rotating modes of the accelerometer were approximately $^1/_2$ and 1 rev min^{-1}. Fortuitously, this permitted measurements of accelerations normal to the plane of the orbit during the rotating modes of the accelerometer. The sampling rate was about one observation per second. Bruce [9] estimated the actual errors in the accelerations measured during the LOGACS experiment to be less than one % and estimated the absolute range of uncertainty in the density values to be within $\pm 10\%$. The primary source of error is the uncertainty in the value of the drag coefficient. The LOGACS exepriment was placed in orbit on 22 May 1967. The orbit has an inclination of 91.5°, a perigee altitude of 147 km, an apogee altitude of 403 km, and a perigee latitude of 40.7° N. During the following three days the apogee decayed to 296 km, the perigee decayed to about 141 km, and the latitude of perigee moved to 59° N. Solar and geomagnetic activity in this period was extremely high, with two class 2B and one class 3B solar flares occurring on 23 May, followed during the next two days by several large geomagnetic disturbances in which values of the geomagnetic planetary index, *ap*, reached 400 units for two successive three-hourly periods. The values of the 10.7 cm solar radio flux and the *ap* index during the LOGACS experiment are shown in Fig. 1, along with the LOGACS revolution numbers during which useful density data were obtained.

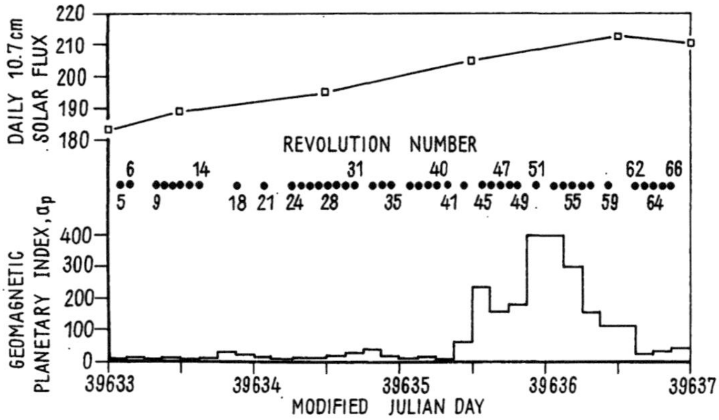

Fig. 1. Solar and geomagnetic data during LOGACS experiment, with corresponding revolutions during which data were obtained.

3. Latitudinal Response of the Atmosphere to Heating associated with Geomagnetic Activity

The orbit geometry of the LOGACS experiment provided an excellent opportunity to study the atmospheric response associated with geomagnetic activity as a function of geomagnetic latitude. During a typical orbit, the satellite was at an altitude of about 300 km at the ascending node, descended to about 200 km over the North Pole and to about 145 km at perigee, at about 45° N, then ascended to about 170 km over the equator and to about 300 km over the South Pole. Accelerometer data at altitudes above 300 km were not included in this analysis because of the relative insensitivity of the instrument at the higher altitudes.

During the first 41 revolutions of the LOGACS experiment, the geomagnetic activity was relatively low, with values of the *ap* index not exceeding 32 units. For nine hours preceding revolution 41, the *ap* index never exceeded 7 units. Density values during revolution 41 were therefore selected as being representative of those existing during geomagnetically quiet periods. For the analyses of latitudinal atmospheric response, all of the density data subsequent to revolution 41 were normalized by computing the ratios of density values at discrete altitudes to those occurring during revolution 41. These ratios were then plotted as a function of geomagnetic latitude, since analyses indicated that the density variations in the LOGACS data are more dependent on geomagnetic than on geocentric latitude.

The normalized density values during revolutions 43, 46, and 51 are shown in Fig. 2. During revolution 43 the value of the *ap* index had increased from four units during revolution 41 to 56 units. In general, the normalized density values had not changed more than a few per cent at geomagnetic latitudes less than 50°. However, the density ratios at a geomagnetic latitude of 75° (on the night-time side of the hemisphere) had increased about 30%, while those at the sunlit auroral region had decreased about 20%.

The first large geomagnetic disturbance during the LOGACS experiment occurred during revolution 46, about $4^1/_2$ hours after revolution 43. The *ap* index at this time had increased to 236 units. The highest density ratios again occurred in the night-time auroral zone, with a peak increase of about 65% over that during revolution 41 at an altitude of about 185 km. A small (about

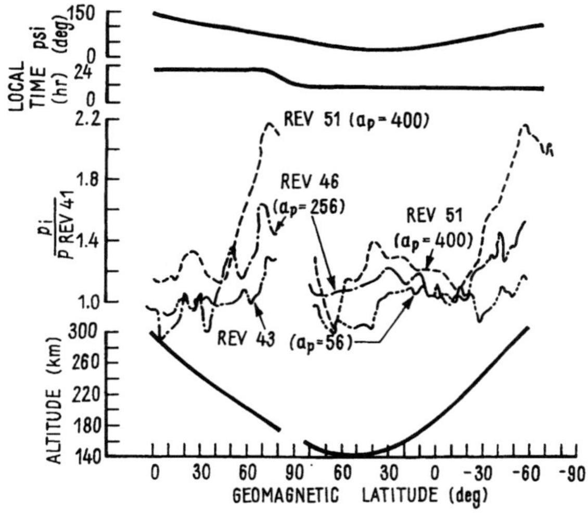

Fig. 2. Normalized density values versus geomagnetic latitudes during revolutions 43, 46 and 51.

5%) density increase occurred in the sunlit auroral region. Density values in the sunlit equatorial region had increased, while those in the night-time equatorial region had decreased.

The initial phase of one of the largest geomagnetic disturbances of solar cycle 19 occurred during revolution 51, with the *ap* index reaching 400 units. As was evident during revolutions 43 and 46, the peak density increases over the northern hemisphere again occurred in the vicinity of the night-time auroral region, with a density increase of about 120% over that during revolution 41 at an altitude of 190 km. The density in the sunlit auroral region had decreased to about 75% of that during geomagnetically quiet conditions. The density in the equatorial regions had increased about 30%, probably as a result of heating associated with the geomagnetic activity of the past 12 hours. A large density increase can be seen over the southern hemisphere auroral region. This increase does not appear to be consistent with the decreased density values observed over the northern hemisphere sunlit auroral region.

Although one recognized shortcoming of the LOGACS data is that nearly all measurements are at only two local times (about 2230 and 1030), the data shown in Fig. 2 indicate that neutral density increases associated with geomagnetic activity occur in the night-time auroral region almost simultaneously with the associated geomagnetic disturbances, and the peak density values are reached in this region. This behavior suggests a possible relationship of this heating to ionospheric current systems. A proposed current system model for an intense magnetic substorm, constructed by Akasofu et al. [10] and the corresponding traverse of the LOGACS experiment through such a system are

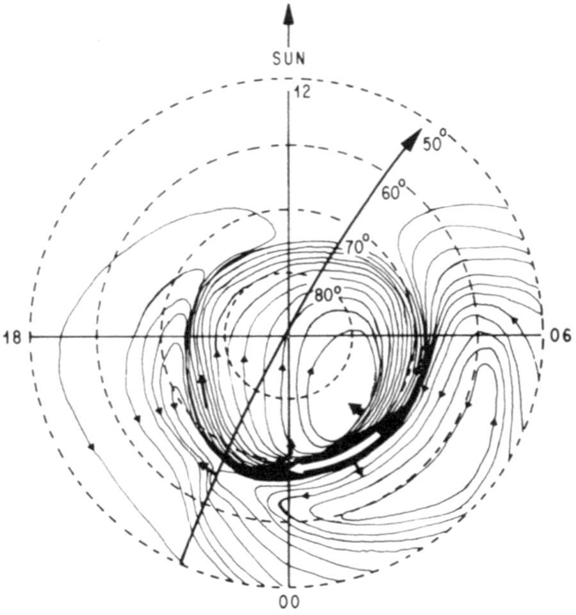

Fig. 3. Proposed model current system for intense polar magnetic storm, with the corresponding LOGACS track (after Akasofu [10]).

shown in Fig. 3. The portions of the orbits in Fig. 2 which correspond to the track in Fig. 3 include the region from geomagnetic latitudes of 50° N on the night-time side of the hemisphere over the North Pole, and then south to 50° N on the sunlit side of the hemisphere. The most intense enhancements of density for revolutions 43, 46, and 51 occur almost coincidentally with the most intense portion of the auroral electrojet current system penetrated by the satellite. The density values in the sunlit auroral region show no significant enhancements; Akasofu's current system model also shows relatively small currents in that area. We are planning to construct the current system which existed during the LOGACS flight.

4. Indications of Thermosphere Circulation Changes during Geomagnetically Disturbed Conditions

Atmospheric density data calculated from the LOGACS accelerometer measurements during revolutions 41 and 51 from an altitude of about 300 km down to the perigee altitude of about 140 km are shown in Figs. 4 and 5. These data were selected to show comparisons of density anomalies during geomagnetically quiet and geomagnetically disturbed conditions. The individual points in the graphs in the upper part of the figures are the calculated density values. The solid lines are least-squares fits to the data of a second-degree

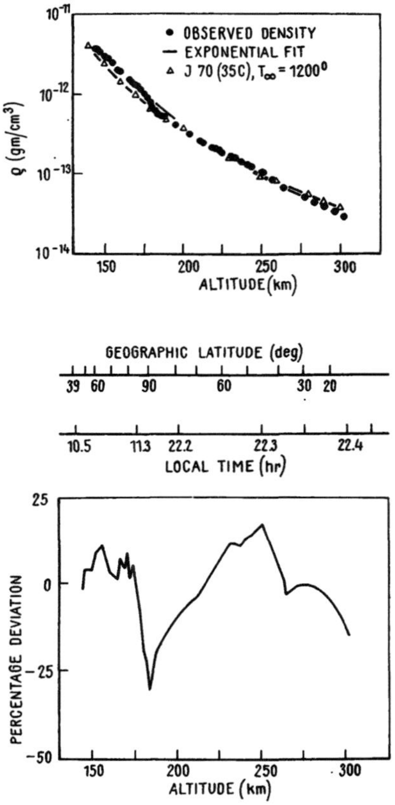

Fig. 4

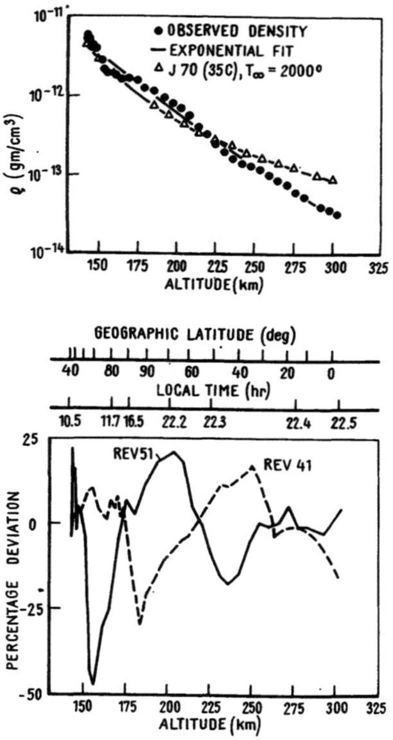

Fig. 5

Fig. 4. Atmospheric densities and the percentage differences between the data points and the smoothed exponential curve for revolution 41.

Fig. 5. Atmospheric densities and the percentage differences between the data points and the smoothed exponential curve for revolution 51.

exponential dependence of density on altitude. For comparison purposes, the Jacchia 1970 [11] model for an exospheric temperature of 1200° is shown for revolution 41 and the model for an exospheric temperature of 2000° is shown for revolution 51. The percentage differences between each data point and the exponential fits are shown in the lower part of Figs. 4 and 5. During each revolution the satellite was simultaneously changing altitude, local time, and latitude. These values are also shown in the figures.

Density observations during revolution 41 are shown in Fig. 4 to illustrate latitudinal density anomalies during a geomagnetically quiet period. From the perigee latitude of about 40° to the northern edge of the sunlit auroral region, the density values were about 5–10% higher than the exponential fit, indicating the existence of a density bulge in the sunlit auroral region. Analyses of other LOGACS revolutions during geomagnetically quiet conditions indicate this bulge to be a feature of this region during geomagnetically quiet periods. From the northern edge of the sunlit auroral region over the polar cap and south to about 65° latitude on the night-time side of the hemisphere, negative departures from the exponential fit are evident. The largest departure occurred directly over the North Pole, with a density value of about 30% less than the exponential fit. Another density "bulge" can be seen between latitudes of about 38° and 65°, with a maximum density increase of about 15%. From about 38° to the equator all density values are less than the exponential fit, possibly because of the relative insensitivity of the accelerometer or increased drag coefficient errors at the higher altitudes (265–300 km). The apparent density ridges and troughs in these data indicate the existence of cellular thermospheric circulation patterns, with a scale somewhat similar to that found in the tropospheric circulation patterns.

Fig. 5 shows the density values during revolution 51. The percentage differences from the exponential fit of density values during revolution 41 are shown to illustrate the dramatic difference in the trend of the latitudinal density anomalies between geomagnetically quiet and geomagnetically disturbed conditions. In the sunlit auroral region, the density values decreased from about 10% above to more than 45% below the exponential fit, indicating a rather surprising phenomenon: a significant density depletion in the sunlit auroral region during a period of highly disturbed geomagnetic activity. This feature persisted for at least 15 hours after peak geomagnetic values were reached. The extremely strong gradients of density in the auroral regions indicate large pressure gradients and associated wind velocities. Over the polar cap and through the night-time auroral region, the density values had increased from about 25% below to more than 25% above the exponential fit, indicating that a significant amount of the energy which causes atmospheric heating during periods of high geomagnetic activity enters the atmosphere near the night-time auroral region. From the southern edge of the night-time auroral region to about 30° latitude, the apparent density bulge had been replaced by a density trough, with the density at 45° latitude decreasing from about 15% above to about 15% below the exponential fit. From about 30° latitude to the equator, the density values had increased slightly, probably because of residual heating associated with geomagnetic disturbances during the previous 15 hours. The changes in the atmospheric profiles during these two revolutions indicate that the thermospheric circulation during geomagnetic disturbances is significantly different from the thermospheric circulation during periods of low geomagnetic activity.

To gain insight into the behavior of the atmospheric density anomalies seen in the LOGACS data, ratios of LOGACS density values to those depicted by the Jacchia 1965 model atmosphere [12] modified by Bruce [13] were computed and plotted as a function of geomagnetic latitude. The ratios for revolutions 41 and 43 are shown in Fig. 6. The altitude range includes all LOGACS data below 300 km, and the portion of the orbit included extends over the entire northern hemisphere and to about $-75°$ in the southern hemisphere. At altitudes greater than about 250 km, the observed density values are less than those depicted by the model, indicating that the accelerometer may have a systematic error at high altitudes. The geomagnetic *ap* value during revolution 43 had increased from 4 units during revolution 41 to 56 units. During revolution 41, a density bulge, with a density of about 35% higher than shown by the model, existed in the sunlit auroral region. In the night-time auroral

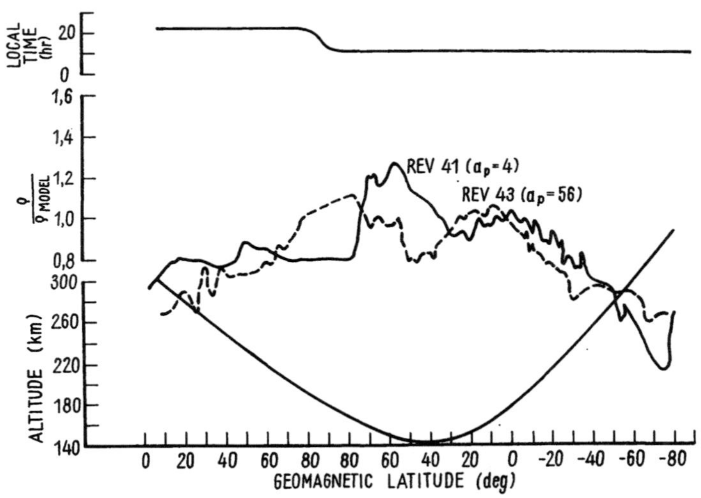

Fig. 6. Normalized density values versus geomagnetic latitude for revolutions 41 and 43.

region, the observed density was about 10% lower than the model density. During revolution 43, three hours later, the density in the sunlit auroral region had decreased from 35% higher to 10% lower than the model, and the density in the night-time auroral region had increased from 10% lower to about 20% higher than the model. These data indicate that a density bulge in the auroral region exists on the sunlit side of the hemisphere during geomagnetically quiet conditions, but appears on the night-time side during periods of increased geomagnetic activity. There would obviously also be changes in the thermospheric pressure and winds during such periods.

If one accepts the hypothesis that atmospheric heating associated with geomagnetic activity is caused by some mechanism or energy source in the auroral regions, some means of transporting this energy to the equatorial regions must be present. Various mechanisms that have been suggested include heat conduction, winds, and wave propagation. The rate of molecular heat conduction is obviously too slow to account for the observed time delay, at the equator, of about six hours between peak geomagnetic activity and peak atmospheric response. On the other hand, wave propagation (assuming wave speed is based on a group velocity about one third the speed of sound), or bulk motion (assuming the existence of sufficiently high wind speeds) could transport the energy from the auroral regions to the equatorial region within time limits consistent with observations.

The ratios of observed density values to those depicted by the Bruce model during revolution 54 are shown in Fig. 7. During this revolution, the *ap* value

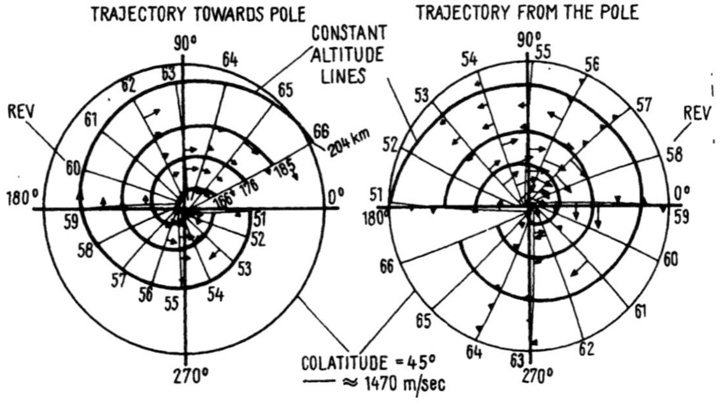

7a

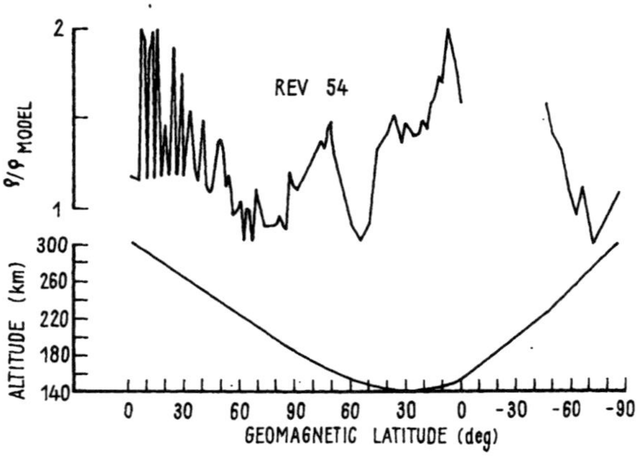

7b

Fig. 7. Normalized density values versus geomagnetic latitude for revolution 54 and computed lateral winds.

had been 400 units for about $4^1/_2$ hours. Density values lower than the model can be seen in both northern auroral zones and in the southern sunlit auroral region.

The high-latitude density bulge during this revolution appears to be directly over the northern polar cap. Density values in the equatorial regions had about doubled. These data indicate that the atmospheric heating associated with this highly disturbed geomagnetic activity was propagating to equatorial regions and show a rather surprising phenomenon of lower than normal auroral region density values during a period in which the *ap* index was 400 units.

Lateral wind values during this period, computed by Feess [14], are shown at the top of Fig. 7. During the portion of orbit 54 in which the satellite trajectory was toward the pole (left portion of Fig. 7), wind data are not available; however, the lateral direction reversed between revolutions 53 and 55, indicating that these high winds could be associated with thermospheric electric current systems. In the portion of the trajectory from the North Pole during revolution 54, a reversal in the lateral wind component occurs near the sunlit auroral region, indicating that the circulation north of the auroral region was in a different direction from that south of this region. This feature persisted for at least six hours following revolution 54.

5. Comparison of LOGACS Observations with Proposed Mechanisms for Atmospheric Heating associated with Geomagnetic Activity

Several mechanisms by which energy associated with geomagnetic disturbances could be transformed into the thermal energy required to heat the atmosphere as observed from satellite drag observations have been proposed. Dessler [15] advanced the theory that such heating results from deposition of hydrodynamic waves created on the magnetosphere from interaction with the solar wind. Such a mechanism would cause thermospheric heating from above. LOGACS data and satellite decay data do not appear to support this theory. McIlwain [16] suggested that energetic particles precipitating into the auroral zones during geomagnetic disturbances would heat the thermosphere directly by absorption of their energy. LOGACS observations in the sunlit auroral region do not support this theory. Volland [17] considered the geomagnetic effect by a one-dimensional model in which horizontal exchanges of mass, momentum and energy were neglected. He considered three possible means of energy input to the neutral thermosphere: (i) heating at the bottom by joule or wave energy, (ii) direct energy input within the entire thermosphere, and (iii) heating at the top of the thermosphere. From this study, the origin of the geomagnetic effect was interpreted as being from corpuscular radiation into the auroral zone in the general form of heat source (ii) and a subsequent heat distribution by heat conduction and convection. This type of input seemed to explain the observed height distribution of the density variations, as well as the time lag of about six hours between maximum geomagnetic activity and maximum density observed by Roemer [3] and Jacchia et al. [2]. LOGACS data indicate the input of energy as being Volland's source (i), i.e. heating at the bottom of the thermosphere by joule heating or wave energy.

In an examination of gravity waves as a candidate for a mode of rapid energy transport as required by an auroral heat deposition from particle deposition and/or joule dissipation of electrojets and currents, Chiu et al. [19] developed a one-dimensional model of upper atmospheric response to transient heating which includes the effect of energy exchange between the ambient atmosphere and propagating gravity waves. This model predicted density time lags from 6 ± 2 hours at altitudes directly over the heating source. The primary disagreement between these calculations and the LOGACS data is Chiu et al.'s [18] 6 ± 2 hour time lag for heating directly over the energy source. The LOGACS data indicate an almost instantaneous response in this region.

In a recent study, Volland and Mayr [19] calculated the response of thermospheric density during geomagnetic disturbance under the assumption that an impulse type of heat input is injected into a small band of latitude within the auroral oval during local night-time. He determined a "system transfer function" of the thermospheric density and calculated, from that function and from the Fourier spectrum of the heat input, the temporal response of the thermospheric density to the heat input. This theory indicates a dependence of the atmospheric density variations on latitude and longitude. The LOGACS data support this type of response, and the general trend of the latitudinal and longitudinal density variations proposed by Volland and Mayr [19] is in agreement with the trend seen in the LOGACS observations.

Cole [20] applied the results of earlier work by Cowling and Piddington on joule heating and motion of uniform ionized gas to the ionosphere during geomagnetic disturbances to explain the source of energy available to the atmosphere by this process. The joule heating rate $Q(J)$ by this mechanism is given by j^2/σ_3, where j is the electric current and σ_3 is the Cowling conductivity. The peak heating would thus occur at the peak current of the auroral electrojet or near the night-time auroral zone. The LOGACS data support this hypothesis. Cole [21] has enumerated several atmospheric phenomena which would be associated with joule heating. These phenomena and comparisons with LOGACS data are as follows: (i) Velocities of the neutral atmospheric wind in direct proportion to the electric current, with wind velocities exceeding 1500 m s^{-1} possible. Wind velocities calculated from LOGACS data increase significantly with an increase in geomagnetic activity and exceeded 1500 m s^{-1}. High wind velocities also seem to occur in the sunlit auroral region, where a relatively small electric current would be expected. However, these winds may result from the circulation pattern established by the high winds in the night-time auroral region. (ii) A significant pressure gradient over the auroral regions. LOGACS density data indicate extremely high density gradients associated with up to a twofold density increase over a relatively short distance in the auroral regions. (iii) Dissipation of energy by generation of winds, lifting up of the atmosphere, and the formation of cells on either side of the auroral region as a result of air rising over the heated region and coming in from the sides. The LOGACS wind data indicate the existence of the high winds, and indicate the convective cells by the apparent reversal of lateral wind components at the edge of the auroral region. The lifting up of the atmosphere or vertical air velocities has not been calculated from the LOGACS data. Such calculations may be possible by examining the pitch control jet data.

6. Conclusions

Analyses of accelerometer measurements from the LOGACS experiment during a period in which the geomagnetic activity varied from extremely low to very high indicate the following phenomena which have a significant effect on the dynamics and circulation of the thermosphere: 1. Density increases about proportional to the level of geomagnetic activity occur in the night-time auroral zone almost simultaneously with geomagnetic disturbances. 2. During high geomagnetic activity, the density in the sunlit auroral zone decreased to about 70% of that existing during low geomagnetic activity. 3. Density values over the polar cap that are abnormally low during low geomagnetic activity increase significantly during high geomagnetic activity. 4. Extremely high pressure gradients exist in the auroral regions. 5. Atmosphere waves, with half-amplitude of density of up to 35%, were evident a few hours subsequent to disturbed geomagnetic conditions. 6. Lateral wind speeds in excess of 1500 m s^{-1} were calculated from the lateral accelerations detected by the LOGACS experiment. 7. The reversal of the lateral wind direction at the edge of the auroral zone indicates the existence of a cellular type of thermospheric circulation pattern. The LOGACS experiment has demonstrated the usefulness of a limited amount of density data obtained during a period of only a few days at only two local times. Future *in situ* atmospheric measurements spanning long time periods and all local times have the potential to furnish answers to many of the perplexing questions regarding the complex nature of the dynamics and circulation of the thermosphere.

References

[1] L. G. Jacchia, Nature, Lond. **183**, 1662 (1959).
[2] L. G. Jacchia et al., J. Geophys. Res. **72**, 1423 (1967).
[3] M. Roemer, Space Research XI, 965 (1971).
[4] L. L. DeVries et al., Space Research VII, 1174 (1967).
[5] D. R. Taeusch et al., Space Research XI, 995 (1971).
[6] J. E. Blamont et al., paper presented at XIII COSPAR Meeting, Leningrad 1970.
[7] F. A. Marcos, K. S. W. Champion and R. A. Schweinfurth, Space Research XI, 941 (1971).
[8] National Academy of Sciences, National Research Council, A Program of Research 1968–1975, August 1968 (p. 96).
[9] R. W. Bruce, Aerospace Corp. Rep. No. TOR-0158 (3110-01)-16 (1968).
[10] S. I. Akasofu et al., J. Atmos. Terr. Phys. **27**, 1275 (1965).
[11] L. G. Jacchia, Smithson. Astrophys. Obs. Spec. Rep. No. 313 (1970).
[12] L. G. Jacchia, Smithson. Contr. Astrophys., **8**, No. 9 (1965).
[13] R. W. Bruce, Aerospace Corp. Rep. No. TOR-1001 (2110-08)-8 (1966).
[14] W. A. Feess, Aerospace Corp. Rep. No. TOR-0200 (9990)-1 (1968).
[15] A. J. Dessler, J. Geophys. Res. **64**, 397 (1959).
[16] C. E. McIlwain, J. Geophys. Res. **65**, 2767 (1960).
[17] H. Volland, Planet. Space Sci. **17**, 1709 (1969).
[18] Y. T. Chiu et al., Aerospace Corp. Rep. No. SPL-0101 (1970).
[19] H. Volland and H. G. Mayr, Goddard Space Flight Center Rep. No. X-621-71-57 (1971).
[20] K. D. Cole, Austr. J. Phys. **15**, 223 (1962).
[21] K. D. Cole, La Trobe University Sci. Rep. SP 20-1-71 (1971).

Space Research XII – Akademie-Verlag, Berlin 1972

MEASUREMENT OF THERMOSPHERIC COMPOSITION

A. O. NIER

University of Minnesota, Minneapolis, Minnesota, USA

A comparison is made between atmospheric densities determined from the low-altitude satellites OV1-15 and OV1-16 and those computed from mass spectroscopically measured composition data. Although the absolute values for mass spectroscopic densities are lower by a factor of approximately 1.4 than those found from satellite drag, the best agreement between relative densities over an altitude range is obtained by assuming that atomic oxygen abundances relative to N_2 and O_2 as reported are correct and should not be increased by large factors as has been suggested by several investigators.

1. Introduction

Atmospheric densities computed from satellite drag measurements have generally been higher than those determined by instruments such as ionization gauges or mass spectrometers. Until recently, comparison have been difficult to make since virtually all of the satellite data were obtained at altitudes above 250 km whereas the bulk of the instrument measurements were made on sounding rockets in the 120–200 km altitude range and various models had to be invoked to extrapolate the satellite determinations down to the sounding rocket region. The availability of data from low-altitude satellites such as OV1-15 [1] and OV1-16 [2] makes possible direct comparisons without the use of models. Jacchia [3] has given an excellent summary of the state of the field at the time of the 1969 COSPAR meeting.

More recently, von Zahn [4] compared various composition measurements available at 150 km with drag measurements and with models, and concluded that if one assumed drag measurements were high by 10% whereas mass spectrometric determinations of atomic oxygen were low by a substantial factor, perhaps 3 to 5, or even more, agreement could be obtained. Indeed, in a 1967 review paper he [5] suggested that the O/O_2 number density ratio at 120 km might be as high as 4 compared with the average value of approximately 1 obtained from the mass spectroscopic data then available. More recer t Minnesota values are even lower and lie in the range 0.4–0.7. Very recently Offermann and von Zahn [6] made a new mass spectrometric measurement at 120 km and obtained a tentative value of 3.5. This determination was made with an ingenious new helium cooled mass spectrometer which presumably solves the selective atomic oxygen loss problem.

The view that mass spectrometrically determined atomic oxygen values are low by substantial factors seems to have been adopted by others as well [3, 7, 8]. Since most of the "low" atomic oxygen values appear to have come from University of Minnesota rocket experiments, it is appropriate to review the Minnesota results, making comparisons with the low-altitude satellite data. It is concluded that it is highly unlikely that the Minnesota values are low by the large factors proposed and that it is not necessary drastically to modify mass spectrometric composition determinations in order to have agreement between mass spectrometric and drag determinations. It is necessary, however, to postulate a somewhat larger systematic difference between satellite drag and mass spectrometer-determined densities than is currently accepted.

2. The Experimental Procedure

In the various rocket flights in question [9, 10, 11] the standard three-mass spectrometer payload configuration shown in Fig. 1 was flown. Mass spectrometer 1 had its ion source exposed directly to the ambient atmosphere in order to minimize the number of wall collisions a particle made before escaping from the source or passing through the electron beam to be ionized. The openness leads to some uncertainty in the relation between the number density of particles in the path of the ionizing electron beam and the ambient atmosphere. To remove this uncertainty and also provide redundancy, a pair of mass spectrometers, 2 and 3, are attached to a cavity in a "V-block", which in turn connects to the ambient atmosphere through a gas inlet hole having a knife edge. Instruments 1 and 2 scan mass sepctra with periods of approximately 2 seconds. Mass spectrometer 3 is tuned permanently to mass 28 (N_2). The closed instrument 2 presumably reads all the chemically inactive gases accurately while instrument 1 reads active gases as well as inactive gases. In the case of inactive gases the agreement between instruments Nos. 2 and 3 is remarkably good [9]. It has always been recognized that the atomic oxygen read by the No. 1 instrument might be low, but by how much? The various Minnesota publications have given the results of the computations on the basis of no atomic oxygen loss with a precautionary statement warning that the atomic oxygen values may have to be increased by a modest factor, unknown at the time of publication.

3. Experimental Results

In the Minnesota experiments, the rocket spin rate lies in the approximate range 0.2–0.3 revolution/sec. This, together with the spectral scan period of 2 seconds, results in apparently complicated raw telemetry data [12], which, however, readily lends itself to analysis [13] and which, in addition, provides some highly desirable cross-checks on the internal consistency of the results. Consider, for example, the unprocessed O and N_2 data from flight 3.115F [12] shown in Fig. 2. The modulations are caused by the rocket velocity, which is comparable with average molecular velocities. Thus if the opening to the mass spectrometer looks "forward" at the time a peak is read, one obtains a high reading because of the ram effect. On the other hand, if the opening looks "backward", only the molecules in the fast tail of the Maxwellian distribution

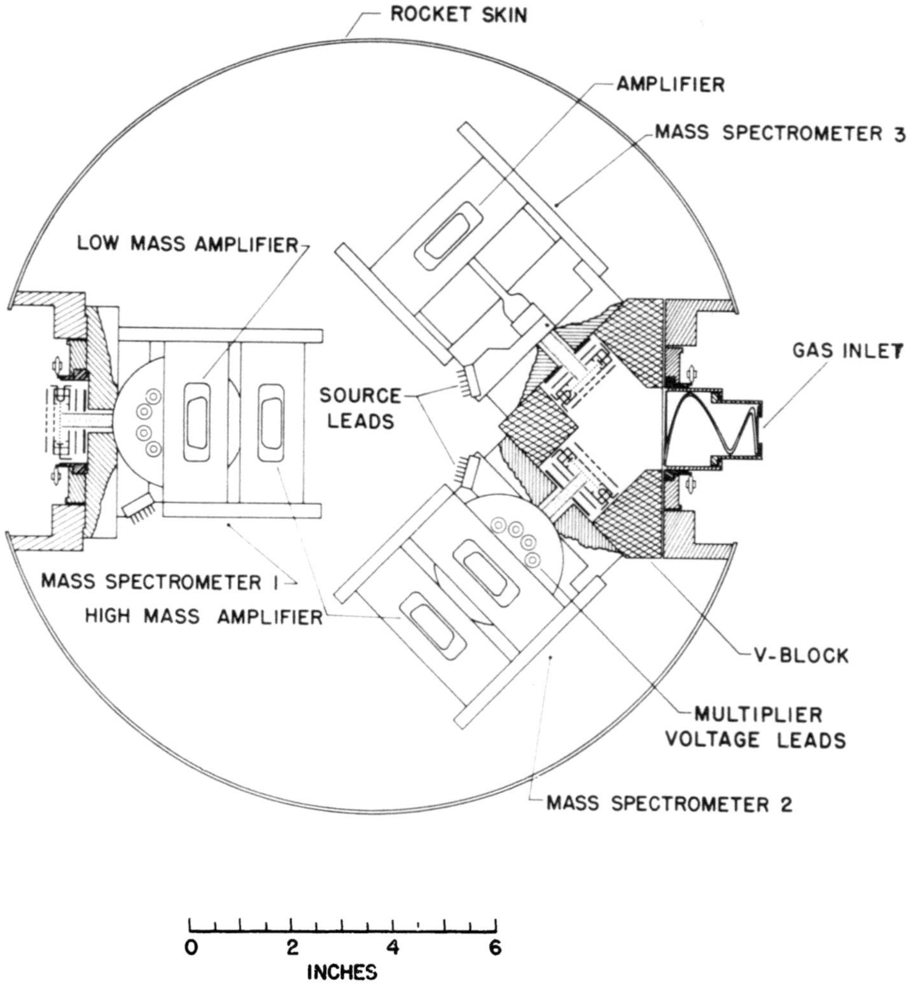

Fig. 1. Cross section view of rocket showing three magnetic deflection mass spectrometers as employed in a typical payload.

have sufficient velocity to overcome the rocket velocity and enter the instrument. One thus obtains a very low reading.

Through a fortuitous coincidence in this particular flight, the spin period of the rocket was almost, but not exactly, twice that of the spectral scan period. This resulted in the interesting patterns shown. The nodes in the curve correspond to points where readings are obtained when the instrument opening is at 90° to the forward direction, and a reasonably accurate density curve can be obtained merely by drawing a smooth curve through the nodal points.

As shown by Hedin et al. [13], by making use of rocket velocity, rocket aspect and spin data, one may reduce the original data to a smooth curve giving particle density as a function of altitude. Atmospheric temperature is obtained as a by-product of the analysis. Even without a detailed analysis, Fig. 2 exhibits some interesting features bearing on the question of atomic

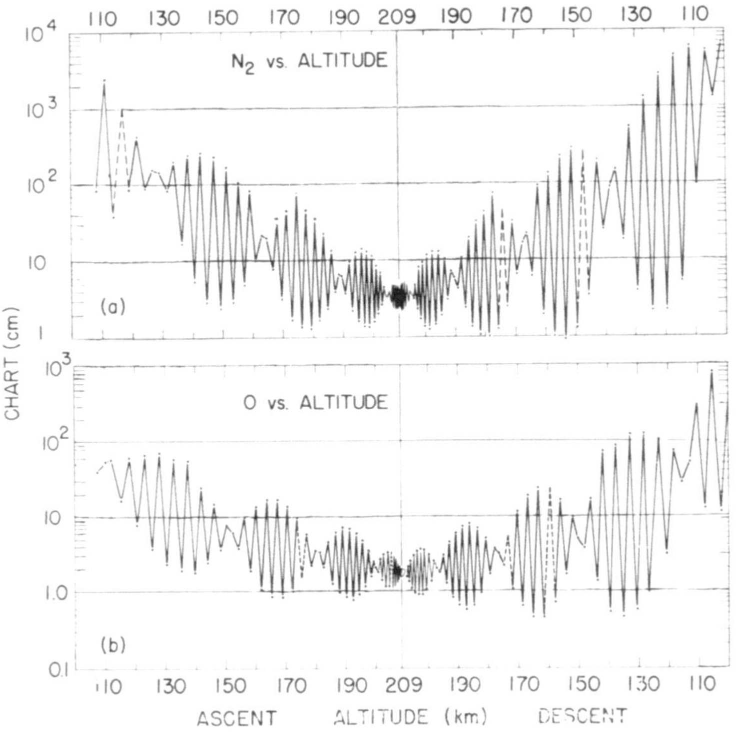

Fig. 2. Variation of N_2 and O signals with altitude in flight 3.115F. Spectra are scanned continuously with repetition rate of 2 seconds. Period of rocket rotation is approximately double the scan period. High readings correspond to mass spectrometer being in ram position, low readings to wake position. Readings for O at low altitudes as shown are slightly high due to uncorrected amount of O arising from dissociation of O_2 by electron impact in mass spectrometer ion source.

oxygen measurement. First of all, we note that the nodal points for the up and down legs of the flight are similar for the O as well as for the N_2 data. Also, we see that the relative scale heights for O and N_2 appear reasonable for an atmosphere in diffusive equlibrium, i.e. N_2 drops off faster with altitude than does O. Moreover, the amplitude of the modulations for N_2 are greater than for O, as one excepts, since the thermal velocity of an oxygen atom in the atmosphere is greater than that of a nitrogen molecule. The rocket velocity which causes the modulation thus is less important for O than for N_2. Finally, the symmetry of the up and down legs, together with the rapid time response of the signal to rotation of rocket (the signal faithfully changes amplitude by a factor of over 100 in 2 seconds), eliminates the need to invoke complicated time-dependent surface loss mechanisms for atomic oxygen. In short, atomic oxygen behaves in these experiments much as any other gas, except that the absolute ambient density may be higher than the instrument indicates by some constant factor, independent of altitude or time. This qualitative observation is sub-

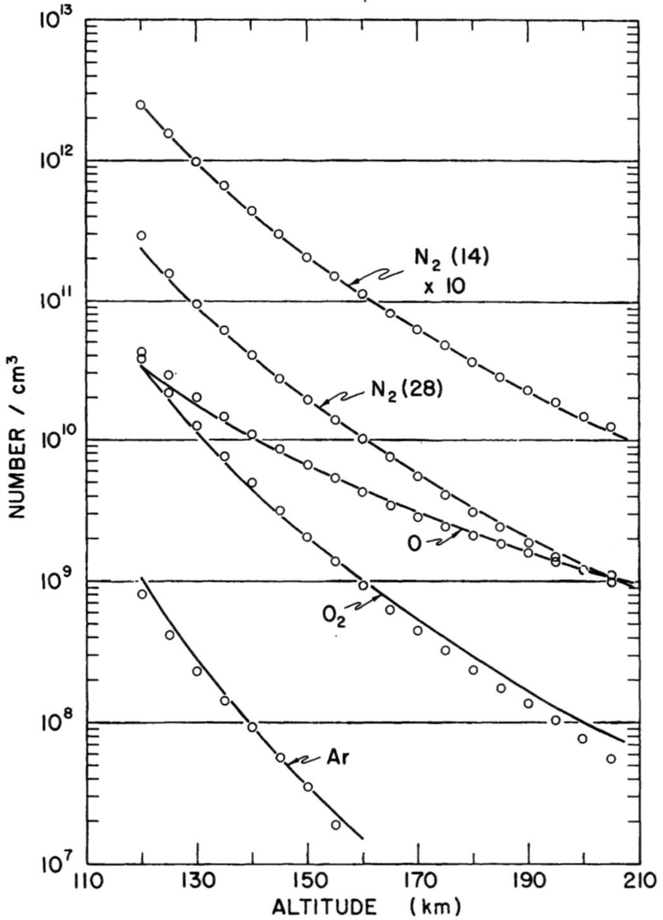

Fig. 3. Variation of number densities with altitude in rocket flight 3.115F (6 June 1963) showing existence of diffusive equilibrium. Circles are experimental data, solid curves give the predicted variations based on temperature profile. Data are normalized at 150 km for convenience. Results also show redundancy of N_2 data as a result of reading both the molecular ion (mass 28) and the ionized fragment of N_2 (mass 14).

stantiated by the detailed analysis made by Hedin and Nier [14], who showed that in the flight in question, the various constituents, including O, were in diffusive equilibrium in the altitude range 120–200 km. Their result is reproduced here as Fig. 3.

Subsequent rocket flights [9, 10, 11] have consistently led to the same conclusions. There is thus every reason to conclude that in over eight flights to date, carrying similar mass spectrometers, the atomic oxygen correction factor, if indeed one exists, is constant through the flight. Whether the factor is the same for all flights remains an open question. Kasprzak at al. [10] showed that even the closed source instrument 2 of Fig. 1 indicated some atomic oxygen. A quantitative analysis making use of this fact led to the conclusion that the atomic oxygen values may have been low by only 20%. Subsequent flights qualitatively confirmed the magnitude of the correction [11].

4. Density Comparisons with Low Altitude Satellites

From absolute composition measurements as determined by rocket-borne mass spectrometers one can compute absolute densities as a function of altitude and compare the results with values found from satellite drag. This comparison can be made directly without resorting to models, since the altitude ranges coincide. In the previous section it was pointed out that all the evidence showed that for a particular rocket flight the atomic oxygen would be off at most by a constant factor, independent of altitude. This being the case, it is instructive to compute densities from mass spectrometer measurements using as a parameter the factor f, where f represents the amount by which the "measured" atomic oxygen determinations should be multiplied to give "true" values.

Probably the greatest weakness in the drag determination of absolute density is the uncertainty in the drag coefficient C_D. On the other hand, relative densities as found by a single satellite are quite accurate. The greatest error in mass spectrometer measurements is almost always the absolute calibration. Because of the linearity of the instruments, relative measurements are generally quite reliable. Hence a comparison of densities over an altitude range by the two methods employs the most desirable features of the measurements, their respective linearities. A comparison at one altitude as employed by von Zahn [4] puts all the burden on the much less certain absolute measurements.

Fig. 4 shows a plot of densities against altitude derived from the OV1-15 and OV1-16 data [15–17]; also shown is a plot for the mass spectrometer rocket flight NASA-4.211 UA [11] which took place at White Sands, New Mexico, at 1224 MST on 20 July 1967. The densities determined by mass spectrometer have all been increased by a factor of 1.5 in order to bring the absolute densities into reasonable agreement in the low-altitude region. Whether the discrepancy in absolute values is due to an error in absolute mass spectrometer calibration, an underestimate of the drag coefficient, or a combination of the two is immaterial in the present comparison. What is of interest is the relative shape of the curves. Four curves are shown for the mass spectrometer data: $O \times 1$ gives the density computed from mass spectra assuming *no* correction need be applied to the atomic oxygen raw data (we believe a factor of 1.25 may be applicable [10]); $O \times 2$ gives the density assuming the atomic oxygen values measured by mass spectrometer should be multiplied by a factor of 2, etc. Increasing the proportion of atomic oxygen has a relatively larger effect on the densities at high altitudes compared with low, as can be seen from the curves in Fig. 4. In spite of the fact that the exospheric temperature was no higher in summer 1967 than in 1968, the time of the satellite flights, even the lowest mass spectrometer density curve ($O \times 1$) fell off with altitude less rapidly than the satellite drag-determined density curve. Increasing the atomic oxygen abundance would only make the fit worse.

Fig. 5 shows a similar comparison for rocket flight NASA-4.180 UA [10] which took place at White Sands, New Mexico, at 1409 MST on 2 December

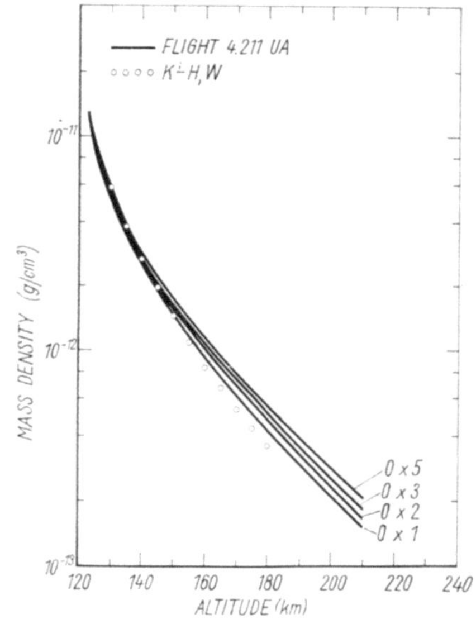

Fig. 4

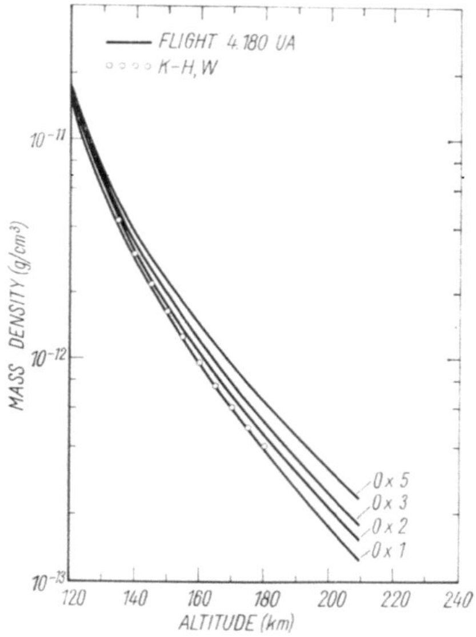

Fig. 5

Fig. 4. Comparison between mass density variation with altitude as given for satellites OV1-15 and OV1-16 [15] and as computed from composition found in rocket flight 4.211 UA (20 July 1967). The mass spectrometer values have been arbitrarily increased by a factor 1.5 in order to produce agreement at the lower altitudes (this normalization could also have been made at the higher altitudes without affecting the validity of the conclusions drawn from the comparison).

Fig. 5. Comparison similar to that of Fig. 4 for rocket flight 4.180 UA (2 Dec. 1966).

1966. In this case the mass spectrometer-determined densities are multiplied by 1.4 in order to bring the absolute mass spectrometer and drag-determined densities into agreement at the low altitudes. Again, the best agreement is obtained for $O \times 1$, i.e. the unaltered atomic oxygen curve. Similar comparisons were made for flights 4.181 UA, 4.179 UA, and 4.212 UA (all referred to by von Zahn [4] in his comparisons). In each case the King-Hele–Walker [15] values fell off with altitude at least as rapidly as the lowest of the four oxygen curves shown. In other words, the best agreement between drag and mass spectrometer measurement over the altitude range 130–180 km is obtained if it is assumed that atomic oxygen values relative to N_2 and O_2 are *not* increased above the values originally reported in the publications.

The total density multiplying factors applied to the flights 4.181 UA, 4.179 UA and 4.212 UA corresponding to the 1.5 and 1.4 for 4.211 UA and 4.180 UA respectively, were 1.4, 1.25 and 2.5 respectively. The last number deviates so far from the others, as do the absolute abundances of the several constituents, that it must be assumed that the mass spectrometer on which the data were primarily based suffered a serious drop in sensitivity between calibration and flight. We now believe the effect could be due either to a mechanical shift in electrodes in the mass spectrometer ion source during powered flight or a change in the value of the input resistance of the amplifier measuring the ion currents. In any case, the relative values of the several constituents appear reliable and there is no evidence for selectively low atomic oxygen values.

Von Zahn [4], in his comparisons of densities at 150 km, included data from our 6 June 1963, 0730 MST flight NC 3.115F [12] and 15 April 1965, 0345 MST flight 4.127 UA [9]. In the former case the King-Hele–Walker curve is definitely less steep than the one based on the mass spectrometer data although this flight showed one of our highest atomic oxygen abundances to date. The flight occurred at a time when the atmosphere was much cooler than in 1968, so a comparison is not appropriate. In the case of the 1965 flight the agreement between the King-Hele–Walker curve and our lowest assumed atomic oxygen concentration (corresponding to $O\times 1$ in Figs. 4 and 5) is quite good in spite of the fact that this flight gave one of the lowest atomic oxygen abundances yet measured! While the results support the conclusions reached in this paper, the agreement may only be fortuitous. The flight took place over two years before the satellite flights used in the comparison and probably should not be included in the present discussion.

In conclusion, a comparison between five 1966 and 1967 mass spectroscopically determined densities over an altitude range 130–180 km and the 1968 satellite drag-determined densities for the same altitude range shows no reason for assuming that mass spectroscopically determined atomic oxygen values are selectively in error. It is necessary to assume, however, that mass spectroscopically determined densities are on the average only about 70% of those found from drag measurements. While part of the discrepancy may be due to mass spectroscopic errors, the almost constant value for the difference suggests that some, if not most, of it probably should be attributed to the value assumed for the drag coefficient C_D. To bring the results into complete agreement could require an increase of more than the 10% suggested by von Zahn [4]. Differences which remain should probably be attributed to actual variations in the atmosphere. With growing evidence for variations in the atmosphere at even these low altitudes, the need for coordinated rocket and satellite experiments is clearly indicated.

Acknowledgments

This work was supported in part by grants NGL-24-005-009 and NGL-24-005-196 and contract NASA-248/24-005-025 from the National Aeronautics and Space Administration.

References

[1] K. S. W. Champion, F. A. Marcos and J. P. McIsaac, Space Research X, 450 (1970).
[2] K. S. W. Champion, F. A. Marcos and R. A. Schweinfurth, Space Research X, 459 (1970).
[3] L. G. Jacchia, Space Research X, 367 (1970).
[4] U. von Zahn, J. Geophys. Res. **75**, 5517 (1970).
[5] U. von Zahn, J. Geophys. Res. **72**, 5933 (1967).
[6] D. Offermann and U. von Zahn, J. Geophys. Res. **76** (1971) 2520.
[7] K. S. W. Champion, Space Research IX, 459 (1969).
[8] K. S. W. Champion, Space Research X, 347 (1970).
[9] A. E. Hedin and A. O. Nier, J. Geophys. Res. **71**, 4121 (1966).
[10] W. T. Kasprzak, D. Krankowsky and A. O. Nier, J. Geophys. Res. **73**, 6765 (1968).
[11] D. Krankowsky, W. T. Kasprzak and A. O. Nier, J. Geophys. Res. **73**, 7291 (1968).
[12] A. O. Nier et al., J. Geophys. Res. **69**, 979 (1964).
[13] A. E. Hedin, C. P. Avery and C. D. Tschetter, J. Geophys. Res. **69**, 4637 (1964).
[14] A. E. Hedin and A. O. Nier, J. Geophys. Res. **70**, 1273 (1965).
[15] D. G. King-Hele and D. M. C. Walker, Planet. Space Sci. **17**, 2027 (1969).
[16] B. K. Ching, J. Geophys. Res. **76**, 197 (1971).
[17] B. K. Ching, Aerospace Corp., El Segundo, California Rep. SPL-1108 (1971).

Space Research XII — Akademie-Verlag, Berlin 1972

DIURNAL VARIATIONS OF ATMOSPHERIC NEUTRAL COMPOSITION AT ALTITUDES OF 130–200 KM

A. D. Danilov

Institute of Applied Geophysics, Hydrometeorological Service, Moscow, USSR

The analysis of the published results of about 50 flights on neutral composition measurements allow one to draw some conclusions on the behaviour of atmospheric components at 130–200 km. The main conclusion is that the concentration ratio [O]/[N_2] at these height decreases from day to night in agreement with conclusions based on ion photochemistry.

The question of the variations of neutral composition above 100 km is now one of the main problems of aeronomy. The results of the first measurements of the main constituents were summarized in several works (eg. [1–3]) to obtain average curves of distribution of the main components. Later [4–8, 36] some conclusions were obtained about the behaviour of absolute and relative concentrations of O, N_2 and O_2 on the basis of separated groups of experiments.

An attempt is made in this report to investigate the variation of neutral composition at 130–200 km as a function of different parameters (altitude, local time, solar activity). For that purpose the results of 58 experiments (see Table 1) carried out by different methods mainly in middle latitudes (except two flights in the equatorial zone and several flights at Fort Churchill) were used.

The most often measured component is molecular nitrogen. The values [N_2] as a function of local time at altitudes of 180 and 200 km are shown in Fig. 1. Crossed and uncrossed points show results of measurements by optical and mass-spectrometric methods respectively. It is seen that open circles, corresponding to the results obtained at low solar activity ($P_{\odot}^{10} < 100$), are situated below closed circles (moderate activity, $100 < P_{\odot}^{10} < 200$). It shows, too, the existence of [N_2] variations, related to solar activity (P variations). The dependence of [N_2] on solar radio-emission flux at 10.7 cm for chosen heights is presented in Fig. 2. It has to be pointed out, that in the figure only the results of mass-spectrometric daytime mid-latitude measurements are presented to avoid scattering of the data due to diurnal and latitudinal variations and differences in method. It may be seen from Fig. 2 that P variations do exist and that their amplitude increases with height and at 200 km reaches a factor of 2–2.5. Moreover, the molecular nitrogen concentration, under both low and moderate solar activity, shows well pronounced dependence on time of day. The ratio $[N_2]_{day}/[N_2]_{night}$ increases from 1.4 at 140 km to 2 at 200 km. One can see that the scattering of the points in Fig. 1

Table 1

No.	Date	Local time	Author	Place	Method	Measured components	$P_{\odot}^{10.7}$	Kp	Ap
1.	14 Jul. 1959	0740	Pokhunkov [9–11]	Mid. lat. USSR	r.f. mass-spectr.	O_2	140	4_-	16
2.	22 Jul. 1959	0515	Pokhunkov [12]	Mid. lat. USSR	r.f. mass-spectr.	N_2	160	2_0	13
3.	23 Sep. 1960	0056	Pokhunkov [13, 14]	Mid. lat. USSR	r.f. mass-spectr.	O, O_2, N_2	175	3_+	14
4.	23 Aug. 1961	1003	Hall et al. [15]	White Sands	optical	O, O_2, N_2	98	0_+	3
5.	15 Nov. 1961	1600	Pokhunkov [14]	Mid. lat. USSR	r. f. mass-spectr.	N_2	100	0_+	1
6.	18 May 1962	1302	Shaefer [16, 17]	Wallops Is.	mass-filter	O, O_2, N_2	95	1_0	2
7.	05 Jun. 1962	0545	Hall et al. [18]	White Sands	optical	O, O_2, N_2	85	1_+	10
8.	25 Oct. 1962	1630	Hall et al. [18]	White Sands	optical	O, O_2, N_2	87	5_-	34
9.	06 Mar. 1962		Jursa et al. [19]	White Sands	optical	O_2	81	—	31
10.	20 Nov. 1962	1641	Spencer et al. [16]	Wallops Is.	omegatron	N_2	89	1_0	4
11.	28 Mar. 1963	0255	Shaefer et al. [20, 21]	Wallops Is.	quadrupole mass-spectr.	O/O_2; O_2/N_2	73	0_0	4
12.	18 Apr. 1963	1604	Spencer et al. [6]	Wallops Is.	omegatron	N_2	88	4_-	27–38
13.	06 Jun. 1963	0730	Nier et al. [22, 23, 25]	White Sands	2 magnetic mass-spectr.	O, O_2, N_2	77	0_+	16
14.	10 Jul. 1963	1004	Hall et al. [24]	White Sands	optical	O, O_2, N_2	76	2_-	10
15.	20 Jul. 1963	1654	Spencer et al. [6]	Wallops Is.	omegatron	N_2	74	3_-	6
16.	26 Nov. 1963	1316	Shaefer et al. [4, 26]	Wallops Is.	quadrupole mass-spectr.	O/O_2; O_2/N_2	82	1_-	2
17.	29 Jan. 1964	2209	Spencer et al. [6]	Wallops Is.	omegatron	N_2	78	3_-	27
18.	18 Feb. 1965	1409	Shaefer [26, 27]	F. Churchill	quadrupole mass-spectr.	O/O_2; O_2/N_2	72	1_+	4
19.	19 Feb. 1965	0317	Shaefer [26, 27]	F. Churchill	quadrupole mass-spectr.	O/O_2; O_2/N_2	72	2_-	4
20.	02 Mar. 1965	1713	Hall et al. [28]	White Sands	optical	O, N_2	75	3_+	8
21.	03 Mar. 1965	0719	Hall et al. [28]	White Sands	optical	O, N_2	75	4_0	26
22.	04 Mar. 1965	1715	Hall et al. [28]	White Sands	optical	O, N_2	75	1_+	24
23.	11 Mar. 1965	0435	Shaefer [4, 29]	$\varphi = 9°27'$ S	quadrupole mass-spectr.	O/O_2; O_2/N_2	72	1_+	4
24.	11 Mar. 1965	1507	Shaefer [4, 29]	$\varphi = 9°31'$ S	quadrupole mass-spectr.	O/O_2; O_2/N_2	72	1_0	4
25.	15 Apr. 1965	0345	Nier et al. [30]	White Sands	3 magnetic mass-spectr.	O, O_2, N_2	75	2_0	4
26.	09 Nov. 1965	1316	Spencer et al. [7]	F. Churchill	omegatron	N_2	80	1_0	5
27.	10 Nov. 1965	0100	Spencer et al. [7]	F. Churchill	omegatron	N_2	80	0_0	0
28.	11 Dec. 1965	0443	Von Zahn et al. [31]	Sardinia	quadrupole mass-spectr.	O, N_2, O_2	73	3_+	10
29.	20 Mar. 1965	0042	Spencer et al. [7]	Wallops Is.	omegatron	N_2	74	2_0	5

30.	20 Mar. 1965	1309	Spencer et al. [7]	Wallops Is.	omegatron	N_2	76	2_+	5
31.	02 Mar. 1966	1259	Hall et al. [28]	White Sands	optical	O, N_2	78	0_+	3
32.	11 Jul. 1966	1318	Shaefer et al. [27]	F. Churchill	quadrupole mass-spectr.	O, O_2	105	1_-	8
33.	26 Aug. 1966	1331	Spencer et al. [7]	Wallops Is.	omegatron	N_2	128	1_-	6
34.	28 Aug. 1966	2303	Spencer et al. [7]	Wallops Is.	omegatron	N_2	130	2_0	4
35.	30 Nov. 1966	0445	Nier et al. [32]	White Sands	3 magnetic mass-spectr.	O, O_2, N_2	97	3_0	28
36.	02 Dec. 1966	1409	Nier et al. [32]	White Sands	3 magnetic mass-spectr.	O, O_2, N_2	98	1_+	6
37.	12 Dec. 1966	1320	Muller et al. [33, 34]	F. Churchill	quadrupole mass-spectr.	O, O_2, N_2	168	0_+	2
38.	24 Jan. 1967	0326	Taeusch et al. [35]	Cape Kennedy	omegatron	N_2	147	0_+	2
39.	24 Jan. 1967	0818	Taeusch et al. [35]	Cape Kennedy	omegatron	N_2	147	0_0	2
40.	24 Jan. 1967	0935	Taeusch et al. [35]	Cape Kennedy	omegatron	N_2	147	0_0	2
41.	24 Jan. 1967	1400	Taeusch et al. [35]	Cape Kennedy	omegatron	N_2	147	0_+	2
42.	24 Jan. 1967	1712	Taeusch et al. [35]	Cape Kennedy	omegatron	N_2	147	0_+	2
43.	24 Jan. 1967	2126	Taeusch et al. [35]	Cape Kennedy	omegatron	N_2	147	1_+	2
44.	21 Jun. 1967	1249	Nier et al. [8]	White Sands	3 magnetic mass-spectr.	O, O_2, N_2	123	1_0	4
45.	20 Jul. 1967	0200	Nier et al. [8]	White Sands	3 magnetic mass-spectr.	O, O_2, N_2	136	2_-	5
46.	20 Jul. 1967	1224	Nier et al. [8]	White Sands	3 magnetic mass-spectr.	O, O_2, N_2	135	2_0	5
47.	18 Sep. 1967	1410	Taeusch et al. [4]	Wallops Is.	2 omegatrons	N_2	129	4_-	12
48.	04 Oct. 1967	1401	Bitterberg et al. [42]	Sardinia		O, O_2, N_2	130		4
49.	10 Oct. 1967	1511	Bitterberg et al. [42]	Sardinia		N_2	127		33
50.	01 Nov. 1967	1430	Wildman et al. [43]	Woomera	optical	O_2	140		3
51.	17 Mar. 1968	0253	Spencer et al. [44]	$\varphi = 18°$ N	omegatron	N_2	128		12
52.	17 Mar. 1968	0452	Spencer et al. [44]	$\varphi = 18°$ N	omegatron	N_2	128		12
53.	08 Aug. 1968	1403	Spencer et al. [44]	Wallops Is.	omegatron	N_2	140		14
54.	09 Aug. 1968	0155	Spencer et al. [44]	Wallops Is.	omegatron	N_2	142		15
55.	04 Dec. 1968	1251	von Zahn et al. [45]	F. Churchill	monopole mass-spectr.	O, O_2, N_2	148		21
56.	04 Feb. 1969	0835	Nier et al. [46]	F. Churchill		O, O_2, N_2	138		13
57.	06 Feb. 1969	1345	Giraud et al. [47]	Landes	r.f. mass-spectr.	O, O_2, N_2	139		16
58.	07 Feb. 1969	0122	Giraud et al. [47]	Landes	r.f. mass-spectr.	O, O_2, N_2	140		9

is small — not more than a factor of 1.2 for medium activity and 1.4 for low activity. The same diurnal variations were obtained in [36, 5], where it was found that $[N_2]$ increases from night to day by a factor of 1.5. We failed to find in Fig. 1 any systematic $[N_2]$ variations with season or geomagnetic activity. The comparison between experiments 45 and 38 (see Table 1) and

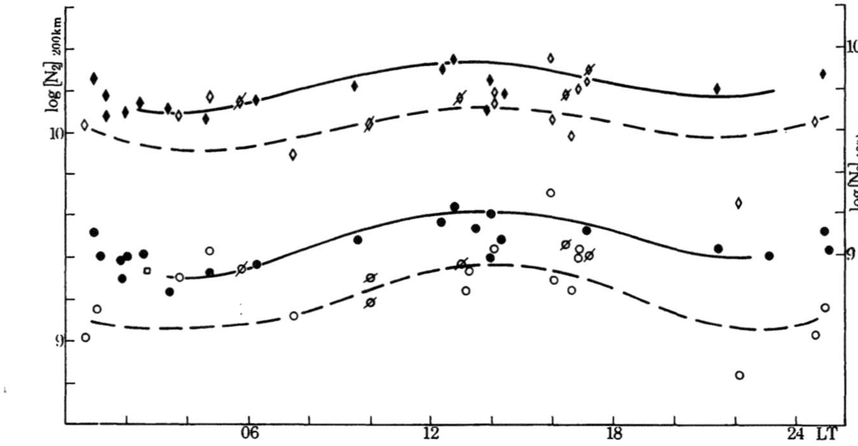

Fig. 1. Values of $[N_2]$ as a function of local time at altitudes of 180 km ◆ and 200 km ●. Open symbols, low solar activity; filled symbols, moderate solar activity. Crossed symbols, optical method; uncrossed symbols, mass-spectrometric method.

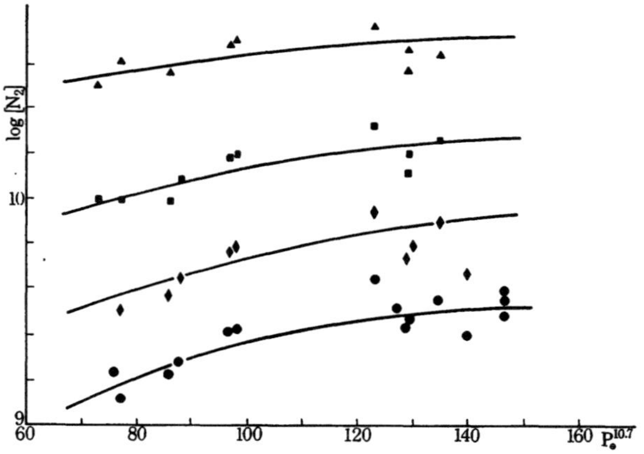

Fig. 2. Dependence of $[N_2]$ on the solar radio-emission flux at 10.7 cm for altitudes 140 km ▲, 160 km ■, 180 km ◆, 200 km ●.

also between 10 and 15 shows, that $[N_2]$ is lower in winter than in summer by a factor of about 1.5; but because the A_p index is higher for summer experiments, the difference may be due to a geomagnetic effect.

Measurements of atomic oxygen are very difficult to carry out, so there are less data on [O] distribution than on $[N_2]$. Diurnal variations of [O] at 200 km are shown in Fig. 3 (legends are the same as in Fig. 1). Squares denote the data, obtained on Explorer 17 and extrapolated down to 200 km in [36]. One can see from Fig. 3 that all the points form two distinct groups: 1) data, obtained under low solar activity (open circles); 2) data, obtained in Nier's experiments under moderate solar activity (filled circles); the open point in the lower curve corresponds to Nier's experiment under high geomagnetic activity (see Table 1) so we have attributed it to second group. Though the fact that all the filled points lie below the open points to the existence of [O] dependence on solar activity, it is necessary to point out that most of the open

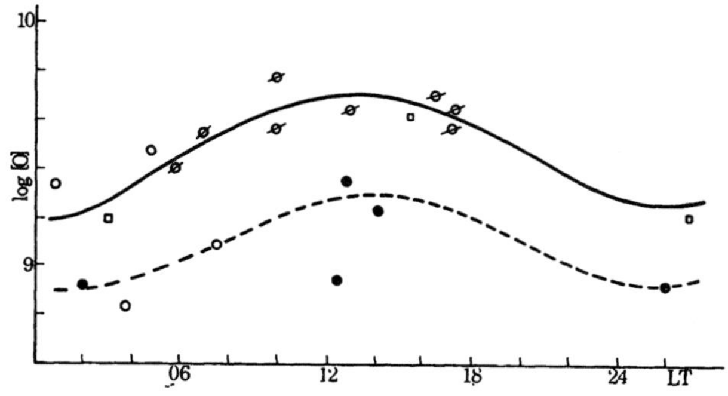

Fig. 3. Values of [O] at 200 km as a function of local time. Symbols as in Fig. 1. Squares, Explorer 17 data extrapolated to 200 km.

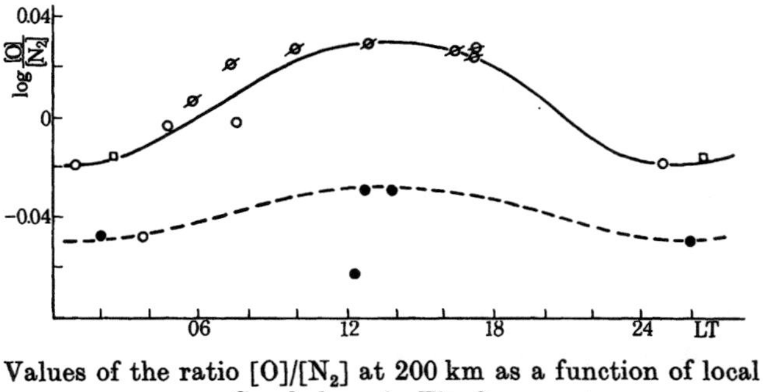

Fig. 4. Values of the ratio $[O]/[N_2]$ at 200 km as a function of local time. Symbols as in Fig. 3.

points correspond to optical experiments and all the closed to mass-spectrometric, so there is some danger that the difference in question reflects only the difference in method. One can see from Fig. 3, that the concentration of atomic oxygen increases from night to day with maximum at about 13 h LT. The amplitude of the variation achieves a factor of 2.3 for moderate activity and a factor of 3.6 for low activity.

The question of the behaviour of relative concentrations $[O]/[N_2]$ and $[O]/[O_2]$ is of great importance. The character of the variations of the atomic-to-molecular ratio is still unknown. It follows from the barometric law that the ratio has to be higher at night, because it has to be inversely proportional to the temperature. This is the general idea of all atmospheric models available [37, 39–41]. The Nier experiments 45 and 46 (see Table 1) give $([O]/[N_2])_{night}$

$>$ ([O]/[N_2])$_{day}$ but the comparison of experiments 44 and 45 gives the reverse effect. Shaefer [4, 29] has found that at 180 km ([O]/[N_2])$_{day}$ $>$ ([O]/[N_2])]$_{night}$, but below 160 km there exists the reverse dependence of the ratio. The behaviour of the [O]/[N_2] ratio, which is based on all the experiments in Table 1, is shown in Fig. 4. One can see, that [O]/[N_2] increases from night to day with a maximum at about 13 h LT. The amplitude of variation is about 2.5 for low solar activity and 1.5 for moderate activity. Decrease of the ratio with solar activity follows from the above mentioned fact, that [O] decreases and [N_2] increases with the growth of $P_{\odot}^{10}$. Essential confirmation of the conclusion about the increase of atomic-to-molecular ratio from day to night was obtained in the recent results of Pokhunkov [48]. For two flights, carried out within two days interval in closely situated points he obtained at 150 km the following values for [O]/[O_2]: 5 at 1400 LT; 2.8 at 0050 LT. Moreover, the same data point to strong variations of neutral composition with geographic latitude. For three flights carried out at times close to local noon, the following values of [O]/[O_2] were obtained:

12 May 68	1415 LT	10° N	1.8
21 May 68	1400 LT	22° N	5
30 March 68	1057 LT	45° N	8

Some conclusions about the character of [O]/[N_2] and [O]/[O_2] diurnal variations may be drawn on the basis of the behaviour of positive ion concentrations. Thus photochemical theory shows that [O^+] and [N_2^+] concentrations are directly dependent on the ratio [O]/[N_2], and the ratio [N^+]/[N_2^+] is proportional to the ratio [O]/[O_2]. It was shown [42] that the analysis of diurnal variations of ion concentrations leads to the same conclusion, that the proportion of atomic oxygen decreases from noon to midnight. It has to be pointed out that this fact, based both on the analysis of experimental data on neutral composition and on the data on ion composition, seems to be the vital point for the question of constructing modern atmospheric models.

It is difficult on the ground of the data in Table 1 to make any definite conclusions about the existence or non-existence of variations of the atomic-to-molecular ratio with season. However the comparison of the Nier experiments Nos. 36–44 (White Sands) and Nos. 18, 32 (Fort Churchill) shows that probably, other conditions being equal, this ratio is higher in summer than in winter, the amplitude of the variation achieving a factor of 1.5–2.

Table 2

[N_2] moderate solar activity ($100 < P_{\odot}^{10.7} < 200$)

h(km)	LT 02	04	06	08	10	12	14	16	18	20	22	24	
130	7.9	8.2	8.7	9.4	10.1	10.2	10.6	10.4	9.8	9.1	8.4	7.9	10^{10}
140	3.9	3.9	3.9	4.4	5.0	5.5	5.5	5.4	4.8	4.3	4.0	3.9	10^{10}
150	2.4	2.2	2.3	2.5	3.0	3.3	3.3	3.2	2.7	2.3	2.2	2.4	10^{10}
160	1.4	1.3	1.4	1.5	1.8	2.0	2.2	1.9	1.6	1.4	1.3	1.4	10^{10}
170	0.89	0.77	0.88	0.98	1.1	1.3	1.3	1.2	1.0	0.87	0.80	0.89	10^{10}
180	5.6	4.8	5.5	6.3	7.6	8.5	8.7	7.7	6.5	5.5	5.3	5.6	10^{9}
190	3.5	3.1	3.5	4.1	5.0	5.6	5.9	5.4	4.4	3.7	3.4	3.5	10^{9}
200	2.5	2.0	2.2	2.7	3.5	3.9	4.0	3.7	3.0	2.4	2.5	2.5	10^{9}

Table 2 (Cont.)

$[N_2]$ low solar activity ($P^{\odot}_{10.7} \leqq 100$)

h(km)	LT 02	04	06	08	10	12	14	16	18	20	22	24	
130	7.9	8.2	8.7	9.4	10.1	10.2	10.6	10.4	9.8	9.1	8.4	7.9	10^{10}
140	3.9	3.9	3.9	4.4	5.0	5.5	5.5	5.4	4.8	4.3	4.0	3.9	10^{10}
150	2.4	2.2	2.3	2.5	3.0	3.3	3.3	3.2	2.7	2.3	2.2	2.4	10^{10}
160	1.4	1.3	1.4	1.5	1.8	2.0	2.1	1.9	1.6	1.4	1.3	1.4	10^{10}
170	6.3	6.3	6.4	7.1	7.9	8.8	9.0	8.3	7.2	6.6	6.3	6.3	10^{9}
180	4.1	4.0	4.1	4.4	5.0	5.6	5.8	5.4	4.7	4.2	4.0	4.1	10^{9}
190	2.5	2.5	2.5	2.7	3.1	3.6	3.7	3.5	3.0	2.6	2.5	2.5	10^{9}
200	1.6	1.5	1.6	1.7	1.9	2.2	2.3	2.2	1.9	1.6	1.6	1.6	10^{9}

[O] moderate solar activity ($100 < P^{\odot}_{10.7} < 200$)

h(km)	LT 02	04	06	08	10	12	14	16	18	20	22	24	
130	0.56	0.60	0.70	0.87	1.1	1.2	1.2	1.15	0.98	0.75	0.63	0.56	10^{10}
140	3.9	4.4	4.9	6.3	7.6	8.1	8.5	7.8	6.6	5.0	4.2	3.9	10^{9}
150	2.8	3.1	3.4	4.7	5.7	5.9	6.1	5.8	4.8	3.6	3.0	2.8	10^{9}
160	2.15	2.3	2.5	3.5	4.4	4.5	4.7	4.3	3.5	2.7	2.2	2.15	10^{9}
170	1.6	1.8	1.9	2.7	3.3	3.5	3.6	3.3	2.7	2.0	1.7	1.6	10^{9}
180	1.24	1.3	1.5	2.0	2.5	2.7	2.8	2.6	2.1	1.6	1.3	1.24	10^{9}
190	1.0	1.0	1.15	1.5	2.0	2.2	2.3	2.0	1.7	1.3	1.0	1.0	10^{9}
200	0.77	0.81	0.91	1.2	1.5	1.8	1.8	1.66	1.3	1.05	0.85	0.77	10^{9}

[O] low solar activity ($P^{\odot}_{10.7} \leqq 100$)

h(km)	LT 02	04	06	08	10	12	14	16	18	20	22	24	
130	1.4	1.6	1.9	2.5	3.0	3.15	3.15	3.0	2.5	2.0	1.6	1.4	10^{10}
140	0.98	1.1	1.3	1.7	2.1	2.3	2.4	2.2	1.8	1.4	1.1	0.98	10^{10}
150	0.67	0.74	0.98	1.25	1.5	1.7	1.8	1.6	1.3	1.0	0.77	0.67	10^{10}
160	0.47	0.54	0.71	0.95	1.15	1.35	1.35	1.25	1.0	0.74	0.55	0.47	10^{10}
170	3.4	4.0	5.1	6.9	8.7	10.0	10.0	9.6	7.4	5.1	4.0	3.4	10^{9}
180	2.5	3.0	3.8	5.2	6.8	7.8	7.8	7.3	5.5	3.9	2.9	2.5	10^{9}
190	1.8	2.2	2.8	4.0	5.5	6.0	6.0	5.7	4.2	2.9	2.1	1.8	10^{9}
200	1.3	1.7	2.1	3.0	4.2	4.7	4.7	4.5	3.2	2.2	1.5	1.3	10^{9}

$[O_2]$

h(km)	LT 02	04	06	08	10	12	14	16	18	20	22	24	
130	1.0	1.1	1.2	1.25	1.3	1.35	1.35	1.3	1.25	1.2	1.1	1.0	10^{10}
140	4.0	4.35	4.7	5.15	5.4	5.5	5.5	5.4	5.15	4.5	4.4	4.0	10^{9}
150	2.0	2.1	2.3	2.5	2.7	2.8	2.8	2.7	2.5	2.2	2.1	2.0	10^{9}
160	1.05	1.1	1.2	1.3	1.4	1.5	1.5	1.4	1.3	1.2	1.1	1.05	10^{9}
170	6.0	6.3	6.8	7.7	8.3	8.7	8.7	8.3	7.6	7.1	6.3	6.0	10^{8}
180	3.5	3.8	4.3	4.7	4.9	5.2	5.2	5.15	4.7	4.3	3.9	3.5	10^{8}
190	2.2	2.4	2.8	3.0	3.15	3.3	3.3	3.15	3.0	2.7	2.4	2.2	10^{8}
200	1.4	1.5	1.8	1.9	2.1	2.2	2.2	2.15	2.0	1.8	1.6	1.4	10^{8}

On the basis of the data described above an attempt is made to construct an empirical atmospheric model at altitudes of 130–200 km (see Table 2). This model reflects the details of the diurnal behaviour of the concentrations and observed variations of [O] and $[N_2]$ with solar activity. Because we did not succeed in obtaining the $[O_2]$ variations with $P^{10}_{\odot}$, the average distribution of $[O_2]$ as a function of height and LT is presented in Table 2.

References

[1] A. D. Danilov, Kosm. Issled. **4**, 47 (1966).
[2] A. A. Pokhunkov, Annls Géophys. **22**, 92 (1966).
[3] A. O. Nier, Annls Géophys. **22**, 102 (1966).
[4] K. S. W. Champion, Space Research X, 347 (1970).
[5] U. von Zahn and J. Gross, J. Geophys. Res. **74**, 4055 (1969).
[6] N. W. Spencer and L. H. Brace, J. Geophys. Res. **70**, 2665 (1965).
[7] N. W. Spencer et al., Space Research VIII, 926 (1968).
[8] W. T. Kasprzak et al., J. Geophys. Res. **73**, 7291 (1968).
[9] A. A. Pokhunkov, Space Research I, 104 (1960).
[10] A. A. Pokhunkov, Izv. Akad. Nauk SSSR, Ser. Geophys. No. 11, 1649 (1960).
[11] A. A. Pokhunkov, Iskusstv. sputniki zemli **12**, 133 (1962).
[12] A. A. Pokhunkov, Iskusstv. sputniki zemli **7**, 89 (1961).
[13] A. A. Pokhunkov, Iskusstv. sputniki zemli **13**, 110 (1962).
[14] A. A. Pokhunkov, Space Research III, 134 (1963).
[15] H. E. Hinteregger, J. Atmos. Sci. **19**, 351 (1962).
[16] E. J. Schaefer, J. Geophys. Res. **68**, 1173 (1963).
[17] E. J. Schaefer, J. Geophys. Res. **69**, 4649 (1964).
[18] L. A. Hall et al., J. Geophys. Res. **68**, 6413 (1963).
[19] A. S. Jursa et al., J. Geophys. Res. **68**, 6145 (1963).
[20] E. J. Schaefer, Aeronomical Rep. **1**, 129 (1963).
[21] E. J. Schaefer and M. N. Nicols, Space Research IV, 205 (1964).
[22] A. O. Nier et al., J. Geophys. Res. **69**, 979 (1964).
[23] A. O. Nier, Aeronomical Rep. **1**, 120 (1963).
[24] L. A. Hall and H. E. Hinteregger, J. Geophys. Res. **70**, 105 (1965).
[25] A. O. Nier et al., J. Geophys. Res. **69**, 4637 (1964).
[26] E. J. Schaefer, Aeronomical Rep. **10**, 219 (1965).
[27] E. J. Schaefer, Space Research VIII, 959 (1968).
[28] L. A. Hall et al., J. Geophys. Res. **72**, 3425 (1967).
[29] E. J. Schaefer, J. Geophys. Res. **74**, 3488 (1969).
[30] E. A. Hedin and A. O. Nier, J. Geophys. Res. **71**, 4121 (1966).
[31] U. von Zahn et al., J. Geophys. Res. **73**, 1071 (1968).
[32] W. T. Kasprzak et al., J. Geophys. Res. **73**, 6725 (1968).
[33] D. Müller and G. Hartmann, J. Geophys. Res. **74**, 1287 (1969).
[34] J. Gross et al., Space Research VIII, 920 (1968).
[35] D. R. Taeusch et al., Space Research VIII, 930 (1968).
[36] L. A. Antonova et al., Geomagn. i Aeron. **8**, 348 (1968).
[37] K. B. Norton and T. E. Van Zandt, J. Atmos. Terr. Phys. **26**, 1047 (1967).
[38] Solar-Geophysical Data, 1959–1968.
[39] M. Nicolet, Aeronomy, Mir, Moscow 1964.
[40] L. G. Jacchia, Smithson. Astrophys. Obs. Spec. Rep., No. 170 (1964).
[41] COSPAR International Reference Atmosphere, 1965.
[42] W. Bitterberg et al., J. Geophys. Res. **75**, 5528 (1970).
[43] P. J. U. Wildman et al., J. Atmos. Terr. Phys. **31**, 951 (1969).
[44] N. W. Spencer et al., Space Research X, 389 (1970).
[45] J. Gross and U. von Zahn, Space Research XI, 875 (1971).
[46] D. R. Hickman and A. O. Nier, EOS **51**, 239 (1970).
[47] A. Giraud et al., Space Research X, 751 (1970).
[48] A. A. Pokhunkov, Space Research XII, 657 (1972).

Space Research XII – Akademie-Verlag, Berlin 1972

STRUCTURE OF THE THERMOSPHERE AS INFERRED FROM INCOHERENT SCATTER MEASUREMENTS

P. Waldteufel[a], P. Bauer[a] and J. P. McClure[b]

[a] Centre National d'Etudes des Telecommunications, R.S.R., Issy-les-Moulineaux, France
[b] The University of Texas at Dallas, Dallas, Texas, USA

In this review we summarize incoherent scatter (i.s.) results on the temperature, composition and density of the thermosphere. Many ionospheric results can be interpreted in terms of thermospheric parameters, and the excellent time and altitude resolution of i. s. has allowed a significant improvement in our understanding of short-term as well as long-term thermospheric variations. Near 100 km altitude the "collision dominated" i. s. spectrum yields the neutral temperature and density. Near 200 km altitude neutral composition variations can be inferred from i. s. measurements. The exospheric temperature and the atomic oxygen concentration near 400 km have also been obtained from i. s. These and other results are discussed and interpreted in the light of information obtained from satellite orbital analysis and from other sources. This leads to a unified picture of the thermosphere: although some i. s. results appeared at first to contradict earlier findings, many of the apparent discrepancies have tended to disappear when various types of valid results were properly combined.

1. Introduction

Since the ionosphere and the thermosphere are intimately related by a number of physical processes, each study of the ionized medium leads in principle to information about the thermosphere. Such a point of view was not widely adopted, however, until about twelve years ago. At this time interest in the thermosphere was considerably enhanced by the dawn of the space age, and new and more refined techniques for investigating the ionosphere were developed, allowing the creation of detailed theoretical ionospheric models which incorporate many thermospheric parameters.

In this respect, the incoherent scatter (i. s.) technique has been particularly successful, because it yields many ionospheric parameters with a good accuracy. Although i. s. observations were reported as early as 1958 [1], an attempt to study the thermosphere was not made until many years later. Indeed, the 1967 Thomson Scatter Conference, organized by Bowhill and Evans at the University of Illinois, was probably the first time when the community of scientists interested in i. s. formally acknowledged the promising future of the technique for thermospheric studies and discussed experimental results. Shortly afterwards, several publications appeared [2–5] dealing with the diurnal variation of the exospheric temperature T_∞; in the following years, incoherent scatter studies of thermospheric properties have proliferated and become more elaborate, treating such subjects as the shape of the profile of neutral temperature T_n

and the variations of neutral concentrations and motions at various altitudes.

We attempt here to review the results of i.s. concerning neutral temperature and density variations in the thermosphere. The question of winds and of waves are dealt with in other papers. No presentation of the measurement technique itself will be given; a comprehensive review of the theory and practice of i. s. was published recently [6].

2. Methods of Analysis of Incoherent Scatter Data and their Limitations

An i. s. radar is in principle capable of measuring six parameters: the electron concentration and temperature N_e and T_e, the ion temperature T_i, an ionic composition parameter p, the ion drift velocity $\boldsymbol{V}$ (or at least one component of $\boldsymbol{V}$) and the ion–neutral collision frequency ν_{in}. These parameters are not, however, obtained simultaneously over the entire range of altitudes covered by the i.s. systems (approximately 90 km to over 1000 km), and do not all yield the same amount of insight regarding the neutral atmosphere.

Table 1

Incoherent scatter probing of the thermosphere

Altitude (km)	Region	Measured parameters	Method of analysis		Resulting parameters
550–230	F2	N_e, T_e, T_i, V_z	$L_{ei} = L_{in}$	⟶	daytime: $[O], s, T_\infty$
			continuity equation	⟶	$q \propto [O]$; $\beta \propto [N_2]$
230–130	F1	$N_e, p = \frac{O^+}{N_e}$ (daytime only)	continuity equations	⟶	$[O]$; $[N_2]$
130–110	E	T_i	$T_n = T_i$	⟶	T_n
110–90	E	T_i, ν_{in}	$[N_2] \propto \nu_{in}$	⟶	$[N_2]$
			$T_n = T_i$	⟶	T_n
			heating experiment	⟶	$[N_2]$

The conventional division of the ionosphere into various regions provides a practical way of classifying the coupling of the ionosphere to the neutral atmosphere; such a procedure is followed here. This division is summarized in Table 1 and discussed in detail below.

2.1. F2 Region (230–550 km)

The F2 region is characterized by a single dominant ion (O^+), by a large daytime temperature separation between electrons and ions and by the increasing importance of diffusion processes with height. The measured parameters are T_e, T_i, N_e, and V. The ionosphere–thermosphere coupling can be described by two basic equations which are reduced to two relatively simple forms: the continuity and energy equations for the ion gas.

(a) The energy equation for the ions reduces to a local one, assuming no thermal conduction by the ions. The energy yielded by the electrons to the ions is balanced by the energy given from the ions to the neutrals [7–9,3]:

$$L_{ei} = L_{in} \tag{1}$$

L_{ei} depends upon N_e, T_e and T_i which are determined by i.s. probing. L_{in} depends on the neutral temperature and concentration, and the measured parameters N_e and T_i. By making several assumptions about the temperature profile it is possible to deduce from Eq. (1), considered over the F2 region, the values of T_∞, the atomic oxygen concentration [O], and a temperature shape parameter s [2–5, 10–12].

Possible limitations of the method can arise from: (i) Relative drifts of the ions with respect to the neutral gas leading to additional frictional terms in the energy equation [13, 14]. Neglect of such drifts would lead to an overestimate of the exospheric temperature. However, the behavior of i.s. data indicates that frictional heating is often quite small ($T_i - T_n < 20$ °K), although at other times this temperature difference appears to be as large as 50 °K or perhaps 100 °K [15]. The larger temperature differences (up to 200 °K) predicted for night-time by Stubbe and Chandra [14], which correspond to drift velocities up to more than 500 m s^{-1}, seem to be ruled out by i.s. observations. (ii) Errors in the neutral composition: under or overestimation of the molecular nitrogen concentrations would slightly affect the determination of the temperature shape parameters. (iii) Errors in the ionic composition: the method is based on the assumption of a single ionic species [O^+], yet, the range of altitudes where this assumption is valid can vary from day to night and over the solar cycle. At low altitudes molecular ions can participate in the cooling of the plasma, but considering the near equality of the neutral and ion temperatures at low altitudes, this should not greatly affect the measurements. This is not so, however, for the upper F region where it is sometimes necessary to consider also the high electron to light ion cooling rate, which complicates the calculations and introduces an additional uncertainty, the estimation of the light ion concentrations. (iv) The assumption that the neutral temperature profile can be approximated by the form proposed by Bates [16]. (v) Errors in the collision coefficients: the electron to ion cooling rate, and the ion to neutral cooling rate. These are not expected to be in error by more than $\pm 10\%$ [9]. None of the above limitations appear to be very serious, and the resulting errors should not be larger than a few per cent for the temperature and 10–20% for the oxygen concentration [O].

(b) The daytime continuity equation is dominated by the production, loss and divergence terms and to a lesser extent by the time dependent term:

$$\frac{\partial N_e}{\partial t} = q - L - \operatorname{div}(n\, \boldsymbol{V}_i)\,. \tag{2}$$

The divergence and time dependent terms contain only ionospheric parameters determined by incoherent scatter. The production term depends upon the oxygen concentration [O] and the solar flux; the loss term depends upon N_e and the molecular species $[N_2]$ and $[O_2]$.

Use of the fact that [O], $[N_2]$ and $[O_2]$ are in diffusive equilibrium in the F2 region leads to a determination of these quantities (or quantities proportional to them) by considering the continuity equation over the F2 region [17–19].

The importance of ionospheric F region transport effects due to neutral winds and/or electromagnetic drift has been established in many studies [20–26]. It is clear that for the purpose of determining the parameters of the thermosphere via the continuity equation, incoherent scatter measurements of plasma drift are a great asset.

The limitations of the method are as follows: (i) Errors in the optical depth. However, these errors should be quite small in the lower F2 region and have no effect at all at higher altitudes on the estimate of the neutral densities. (ii) Uncertainty about the relative importance of the molecular species O_2 and N_2 in the recombination process. Although O_2 is a minor constituent, the recombination rate coefficient for O_2 is larger than that for N_2, and it is believed that O_2 and N_2 are almost equally responsible for recombination. Thus changes in the loss term might reflect either changes in $[N_2]$ or changes in $[O_2]$ or even changes in both species. The effect of temperature on the recombination rate could also affect the results. (iii) Uncertainties in the wind measurements. For middle and high latitude stations the information needed to compute the divergence term is the vertical component of the ion drift. Monostatic systems can provide this information, but technical problems have thus far prevented complete determination of the vertical ion wind. The St. Santin bistatic system yields accurate measurements of one component of the ion drift (in a direction close to the magnetic field). In the absence of electric fields this component would correspond to the total ion drift in the F2 region. Therefore the validity of the St. Santin i. s. determination of the divergence term relies upon the non-existence of electric fields. Such difficulties should be overcome in the near future when the new quadristatic radar systems under construction at Malvern and St. Santin are complete. (iv) Presence of molecular and light ions in the lower and upper F2 region. This should not be a serious limitation, since incoherent scatter spectra provide information about the ion composition.

2.2. F1 Region (130–230 km)

This region is characterized by a gradual equalization of the electron, ion and neutral temperatures as the altitude decreases and by a change in the ionic composition from predominantly O^+ to molecular.

The measured parameters are:

$$N_e,\ T_i,\ T_e,\ p = \frac{O^+}{N_e}, \quad \text{and} \quad \boldsymbol{V}.$$

The basic equations governing the coupling of the ionosphere to the neutral atmosphere are the motion and continuity equations for the various ions. The dynamic coupling between the ions and the neutral particles in the F1 region has not yet been treated extensively, and this aspect will not be reviewed here.

The dominant terms in the continuity equations are the production and loss terms which depend upon the ionospheric as well as the atmospheric concentrations. Applying a procedure similar to the one used in the F2 region, that is, by inspection of the different continuity equations at various heights, it is possible to determine values related to the neutral densities [O], [O_2] and [N_2] [27–30, 18].

Limitations to this technique arise from: (i) the choice of the chemical model; (ii) the uncertainties in the reaction rates; (iii) the uncertainties in the recombination rates; (iv) the determination of the solar flux; (v) the practical determination of the composition parameter p which is not completely unambiguous.

2.3. E Region (upper part, 110–130 km)

This region is characterized by thermal equilibrium between the electrons, the ions and the neutral particles and by a molecular dominated ion composition. The incoherent scatter data produced in this altitude range include T_i ($= T_e$), N_e, $\boldsymbol{V}$. The thermal equilibrium and the dynamic coupling between the ionosphere and the atmosphere allow for determinations of the neutral winds and temperatures $\boldsymbol{V}_n$ and T_n [31–33, 18, 29].

Limitations result from the usual statistical uncertainties and from the possibility that thermal equilibrium between electrons, ions and neutral particles does not prevail. While there is no evidence for a temperature separation between the electrons and the ions in the incoherent scatter data, some rocket data (see e.g. [34]) give a different description. It has been suggested that differences could possibly arise because of the different temporal and spatial scales of the measurements.

2.4. E Region (lower part, 90–110 km)

The physical characteristics of this region are not different from those of the upper E region. There is, however, a difference in the information obtained from the i.s. spectrum which, due to increased neutral concentration, is directly affected by the ion–neutral particle collision frequency ν_{in}. The parameters obtained are T_i, N_e, $\boldsymbol{V}$, ν_{in}. As previously, T_n and $\boldsymbol{V}_n$ can be derived in this region from T_i and $\boldsymbol{V}$. The ion–neutral particle collision frequency is proportional to the neutral concentration which is dominated by molecular nitrogen. This gives a method for determining [N_2] [35–38, 18, 29].

A new opportunity for studying the atmosphere near 100 km has recently begun to be explored. It consists of electromagnetic heating of the electron gas near 100 km. Diagnostics obtained by incoherent scatter techniques obtained under such conditions permit an investigation of the electron–neutral particle collision processes [39]. Limitations similar to those found in the upper E region are also present. The determination of [N_2] is limited by the state of the theory of the effect of neutral collisions on the incoherent scatter signals. Some of the limitations in the E and higher altitude regions could be minimized by comparisons with data gathered by other techniques, leading to a better knowledge of the physical processes involved. Other limitations, like the presence of light ions in the upper F2 region [40–45], should offer new possibilities of determination of neutral parameters (such as [O]/[H]).

3. Incoherent Scatter Results

3.1. Comparisons with Other Experimental Data

Direct comparisons between independent measurements are quite difficult when i.s. is involved. In many cases the measured quantities are not strictly identical, and furthermore, i.s. facilities are not mobile and their operating time is severely restricted; perfect coincidence in time and space is therefore difficult to achieve. Thus it is often necessary to infer one parameter from

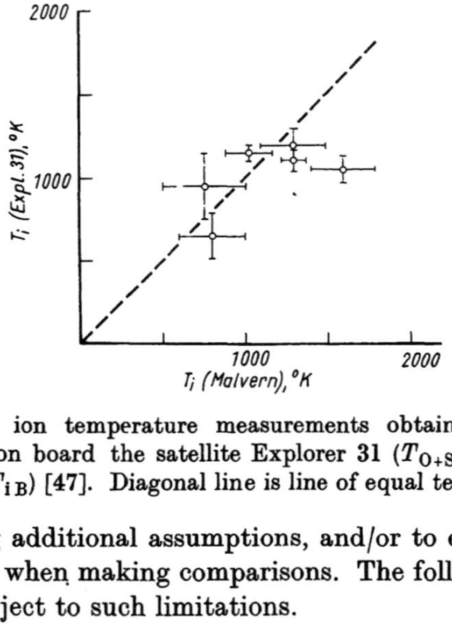

Fig. 1. Comparison of ion temperature measurements obtained simultaneously from a spherical ion probe on board the satellite Explorer 31 (T_{O+S}) and from radar i. s. at Malvern (T_{iB}) [47]. Diagonal line is line of equal temperatures.

another, introducing additional assumptions, and/or to extrapolate the results in time and location when making comparisons. The following examples are to a various extent subject to such limitations.

3.1.1. Parameters of the Ionosphere

Most of the first comparisons involving i.s. measurements concerned electron concentrations and temperatures [46]. These parameters are in general remote from thermospheric parameters. However, in equatorial latitudes T_e is nearly equal to T_n at some altitudes both night and day [4]. Also, in middle latitudes at night T_e is often equal to T_i and T_n. Comparisons of such T_e measurements are relative to studies of the thermosphere and are included here. Ion temperatures observed by i. s. have been compared with several *in situ* measurements (see e.g. [34]). The results are summarized in Table 2 and in Figs. 1–5.

Ion composition in the F1 region was measured simultaneously at St. Santin and by a rocket-borne ion mass spectrometer. The results [51] are summarized in Fig. 6; the relative concentrations measured are in reasonable agreement.

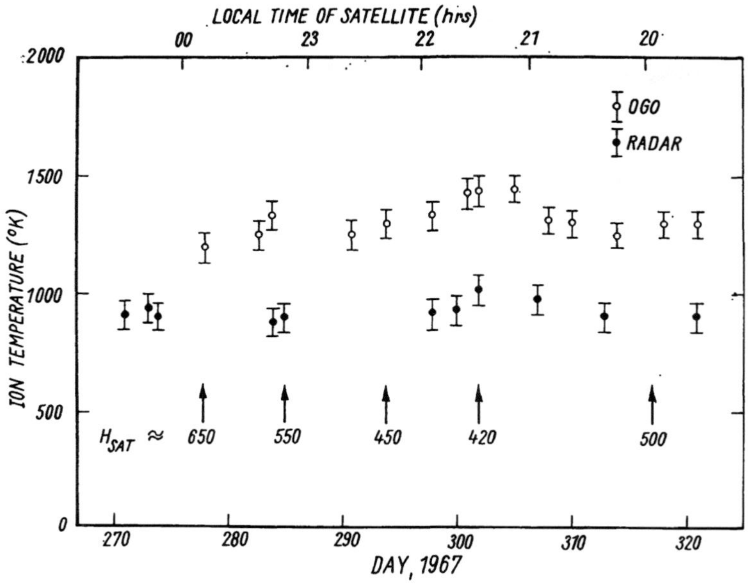

Fig. 2. Conflicting Jicamarca i.s. and OGO 4 retarding potential analyzer values of ion temperature [43].

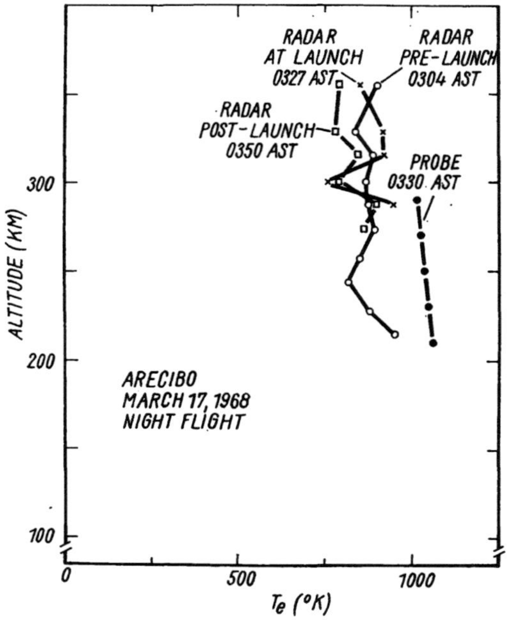

Fig. 3. Comparison of probe and backscatter T_e measurements made at night, when $T_e \doteq T_i = T_n$ at Arecibo. The radar values taken 'at launch' (0327 AST) are 6% lower than the probe values at 290 km. Radar values before and after the launch are systematically lower by about 15%, an amount that approaches the temperature variation between radar runs [50].

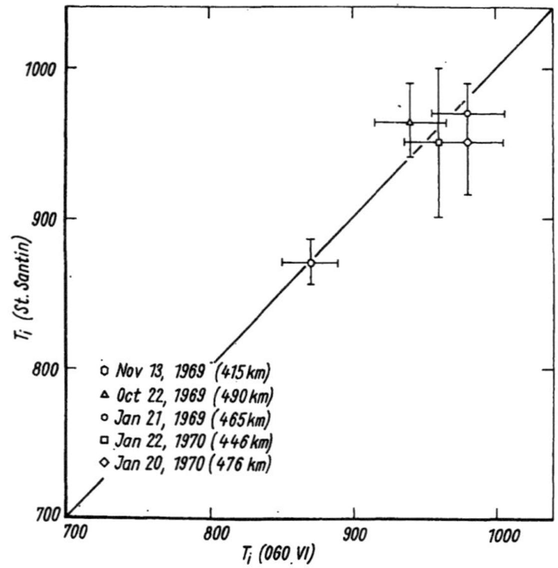

Fig. 4. Simultaneous ion temperature measurements obtained from the OGO 6 retarding potential analyzer and from the St. Santin i.s. radar.

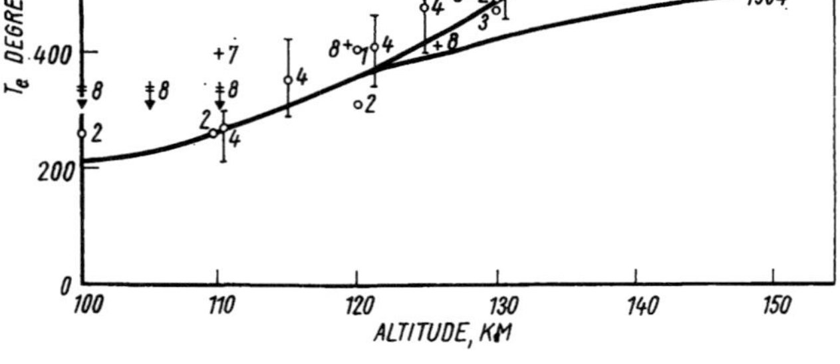

Fig. 5. Measurements of T_e in the E region by Langmuir probes and incoherent scatter compared with gas temperature from CIRA 1965 (solid lines). 6 and 7 refer to similar rocket flights, 6 near midnight and 7 near noon. 8 and 9 were also similar, the two flights occurring at approximately the same local time, two days apart [34]. o, incoherent scatter: 1, Carru, Petit and Waldteufel [67]; 2, Wand and Perkins [37]; 3, Evans [27]; 4, Carru, Petit and Waldteufel [2]; +, Langmuir probe: 6, 7, Brace, Mayr and Carignan (1969); 8, 9, Aubry et al. (1966).

Table 2

T_e and T_i comparison relevant to T_n studies

Ref.	Data compared		Date/Time	Altitude	Comments	Fig.
[47]	Malvern	Explorer 31	Daytime and night-time fall 1968	F2	Agreement	1
[34]	St. Santin	Explorer 31	Daytime — Spring 1968	F2	Altitude extrapolation of incoherent scatter measurements seems to indicate a good agreement	—
[48]	Jicamarca	Explorer 32	Daytime and night-time Aug.–Nov. 1966	F2	T_e (sat.)/T_e (i.s.) $\sim$ 1.7; T_e (i.s.) = T_i (i.s.)	—
[49]	Jicamarca	OGO 4	Night-time — Oct. 1967	F2	T_i (sat.)/T_i (i.s.) $\sim$ 1.4	2
[50]	Arecibo	Rocket	Night-time — Spring 1968	F2	T_e (rocket)/T_e (i.s.) $\sim$ 1.15; T_e (i.s.) = T_i (i.s.)	3
(a)	St. Santin	OGO 6	Night-time — Fall and Winter 1969–70	F2	Good agreement	4
[34]	Arecibo, Millstone, St. Santin	Various rockets	Daytime	E	T_e (i.s.) = T_i (i.s.) $\sim T_n$; R_e (rocket)/T_n = 1–3 (indirect comparisons)	5

(a) unpublished

3.1.2. Neutral Atmosphere Parameters

A comparison of T_∞ obtained at two incoherent scatter facilities, St. Santin and Jicamarca, was made by Waldteufel and McClure [52]. These radars are widely separated; however the measurements were made in northern summer,

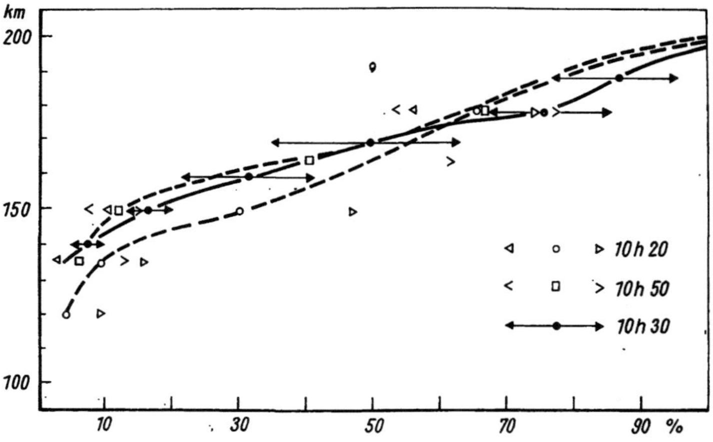

Fig. 6. Simultaneous ion composition measurements obtained from a rocket-borne mass spectrometer and from the St. Santin i.s. radar [51]. Open symbols: Thomson sounder; solid symbols: mass spectrometer.

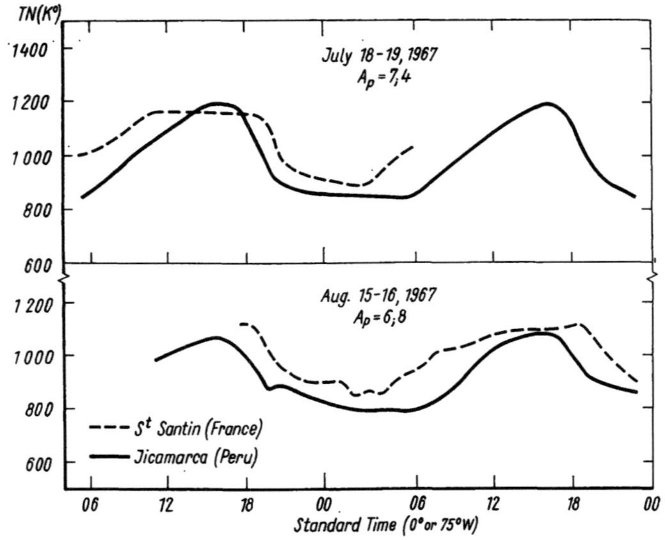

Fig. 7. Diurnal variation of T_∞ obtained from two i. s. facilities at times when the declination of the sun was approximately midway between the two stations [52].

when the declination of the sun was approximately midway between the two stations and therefore some similarity could be expected. Indeed, the temperatures at the two sites always agreed very well, particularly for the mean values (see Fig. 7). Recently it has become possible to determine T_∞ from the interferometric measurement of the width of the 6300 Å oxygen line. Such estimates have been compared with data obtained at Millstone Hill [53] and Arecibo [54]. The results of these comparisons are shown in Figs. 8 and 9. On the whole, they show that both sets of measurements agree within the experi-

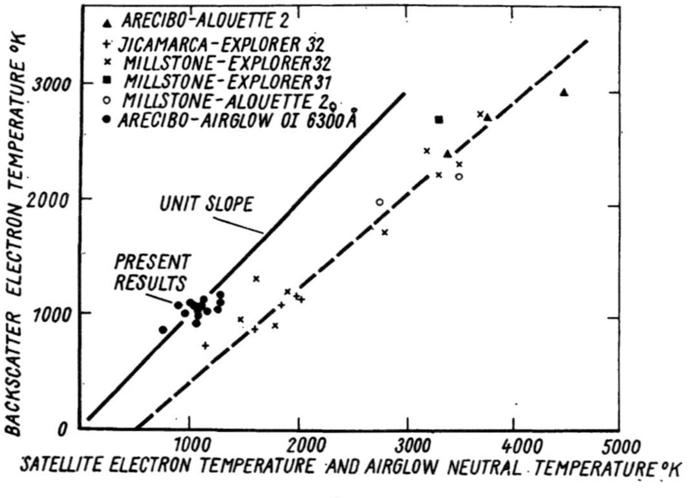

Fig. 8. Comparison of the airglow [OI] 6300 Å neutral temperature with i.s. electron temperature values from Arecibo (solid line) at night when $T_e = T_i = T_\infty$. Also shown are some results of satellite–radar backscatter electron temperature comparisons (dashed line) [54].

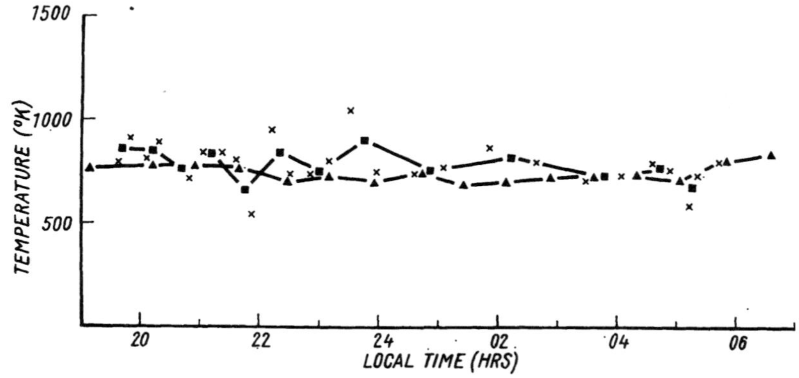

Fig. 9. A comparison of optical and i.s. neutral temperature measurements made above the Millstone Hill Radar on 14–15 October 1969 [53]. (Symbols as in Fig. 8.)

mental uncertainties. Also, data from an optical interferometer on board OGO 6 (J. Luton and J. Blamont, private communication) have been compared with daytime St. Santin observations of T_∞, and good agreement has been found.

A global T_∞ model based upon T_∞ measurements at St. Santin and Arecibo has been constructed [55], and values predicted using this model have been compared with a series of thermospheric temperatures inferred from the scale height of *in situ* nitrogen density measurements [56] obtained between 1962 and 1968. This comparison is shown in Fig. 10. Again, one can say that the differences are easily accommodated within the experimental error bars. (The same holds true for the 1970 model of Jacchia [57], as illustrated in Fig. 10.) Finally, atomic oxygen concentration values at 400 km altitude were measured at St. Santin during a 5-day period in February 1969 by a technique developed by Bauer et al. [11]. These values are compared with density values deduced from orbital analysis in Fig. 11 [58]. Although there is a relatively large uncer-

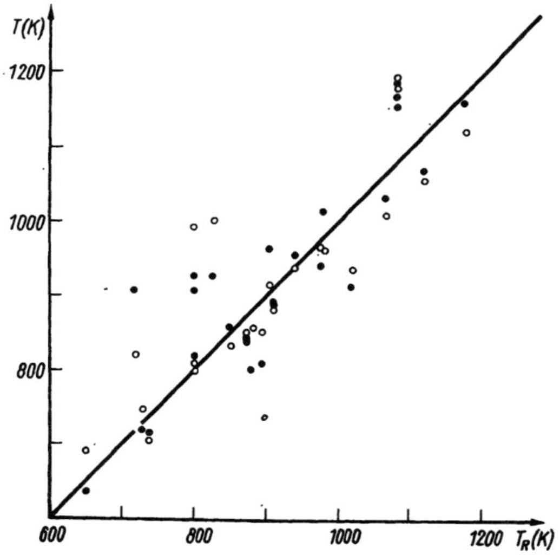

Fig. 10. A comparison of MUMP rockets temperature measurements [56] with predictions from: ○ Jacchia model [57] and ● i.s. model [63].

tainty in the [O] measurements and a number of corrections are necessary in order to make the comparison, the agreement is most satisfactory (see also [59]).

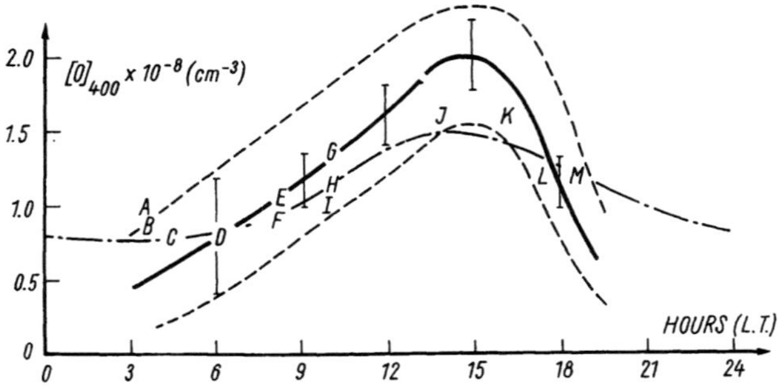

Fig. 11. A comparison of measured and model values of atomic oxygen number density at 400 km [58]. The dashed line gives the range of $[O]_{400}$ inferred from i.s. for 6–10 February 1969; the solid lines gives the mean $[O]_{400}$, and typical error bars are shown. Mean satellite drag values for the 5-day period and the inclination and perigee heights of the 13 satellites are presented. The quasi-sinusoidal curve is from the 1970 model of Jacchia [57].

	Satellites	Incl. (°)	Perigee (km)		Satellites	Incl.(°)	Perigee(km)
A	68 41 A	28	330	I	68 43 B	10	225
C	68 38 B		195	K	66 44 C	18	305
B	66 34 B	− 2.5	355	E	66 44 B	2	250
D	61 15 94	51.8	290	L	65 11 D	18	250
M	66 97 B	−27.5	305	H	68 84 A	58	265
G	68 119 A	−40	250	J	64 06 C	44	405
F	68 119 B	−40	250				

3.1.3. Conclusion

Clearly, the overall conclusion is that agreement is found between incoherent scatter measurements and other results. There is one exception in the case of ion temperatures from *in situ* equipment, where some results agree well while others disagree widely; and another exception in the case of discrepant electron temperatures at times and places where incoherent scatter indicates $T_e = T_i = T_n$ approximately. Difficulties connected with the actual measuring technique are suspected in these cases.

The main limitations of these comparisons, in addition to those previously mentioned, are their small number. Although in some cases (e.g. rocket comparison) the cost of the experiment is such that it can hardly be repeated, in those cases where satellites are involved there are frequently many opportunities for comparison. It must be recognized that so far only moderate use has been made of this possibility. Efforts are presently being made in this direction.

Nevertheless, the limited number of results reviewed leads us to suggest that (except for the cases already mentioned) the agreement is satisfactory, which means that each technique does indeed measure what it claims to, and that the results can be accepted at their face value.

3.2. Incoherent Scatter Results for the Upper Thermosphere

3.2.1. Introductory Remarks

In the morphological study of the upper atmosphere, most of the approaches used thus far have a strong similarity to the one used to interpret satellite drag data, i.e. the observed parameters are analyzed in terms of periodic components (annual, semi-annual, diurnal) and their possible correlation with geophysical indices (solar flux, magnetic activity). The reason for this is twofold: to permit direct comparison of the i. s. and satellite drag results, and because of the high degree of suc·ess demonstrated for this approach, which has been used to construct very successful models of the upper thermosphere [60, 61, 57]. As will be seen, this approach has been quite successful for i.s. data too. It must be remembered, nevertheless, that: (i) due to the discontinuous character of the i. s. observations at each site, the statistical analysis is based on a relatively small number of data points (of the order of 100 at most) and therefore some variations are probably not as firmly established as desirable, and (ii) owing to the very limited geographical coverage of i.s. (Jicamarca 12° S, Arecibo 18° N, St. Santin 45° N), some phenomena (for example, high latitude phenomena) are outside the view of incoherent scatter. For example, the winter helium bulge [62] is not only outside the view of i.s. but, in addition, no analysis of high altitude i.s. data to date has been sufficiently refined to yield statistical information about neutral helium. In summary, the i.s. study of the higher thermosphere is limited in that it is mainly a study of the atomic oxygen, mid-latitude thermosphere during the rising part and the maximum of the most recent solar cycle. On the other hand, the good time resolution of i.s. data has already allowed a significant improvement in the treatment of relatively short term (24 hr and less) variations.

3.2.2. Exospheric Temperature Results (excluding the Diurnal Variation)

The exospheric temperature has been studied in detail at St. Santin [18], Jicamarca [49] and Arecibo [55]. Numerical results will not in most cases be reviewed here, as they are available in the publications cited above (see also [63]).

In the long run, the dominant influence on T_∞ is the solar activity, the effect of which can be clearly identified between 1965 and 1970. If one uses the 10.7-cm solar flux φ to characterize this activity, it is useful to split this index into a slowly varying component $\bar{\varphi}$ (e.g. the 5-month running average) and the difference $\varphi - \bar{\varphi}$. The linear heating coefficient for the $\bar{\varphi}$ term is about twice that for the $\varphi - \bar{\varphi}$ term.

There is firm evidence that magnetic activity heats the thermosphere. If the index of such activity is chosen to be the planetary Kp index, it is found that the magnetic heating coefficient C_M (C_M = the increase of T_∞ in °K per unit increase in Kp) is larger at 45° N ($C_M \sim 17$) than at low latitude ($C_M \lesssim 12$). Also, the time delay τ_M between the Kp increase and the T_∞ increase is larger at Arecibo ($\tau_M \sim 9$ hr) than at St. Santin ($\tau_M \sim 4.5$ hr). Actually this is a very crude description, as magnetic activity is strongly correlated with the appearance of oscillatory phenomena. Also, the response of the thermosphere is a complex function of the structure of the magnetic events themselves, and a linear description tends to fail for high levels of activity.

There is a large seasonal variation in T_∞ at 45° N with a summer maximum. The seasonal variation becomes somewhat smaller at Arecibo and it is reversed in phase at Jicamarca, i.e. the maximum always occurs in local summer (cf.

Fig. 16). A semi-annual variation of T_∞ with maxima at the equinoxes is observed at Arecibo and St. Santin. These effects are all, on the whole, in good agreement with the exospheric temperature which is used as a parameter in static diffusion models such as that of Jacchia [57], abbreviated here as J70. A number of differences or additions should, however, be noted: (a) The heating coefficients for φ, $\varphi - \bar{\varphi}$ and Kp are lower than those given by J70. Also, the dependence of T_∞ on φ is clearly nonlinear, with T_∞ apparently reaching a saturation value at the higher φ values. (b) There is at 45° N a considerable semi-annual component proportional to $\varphi - \bar{\varphi}$, which is absent in the models. (c) The semi-annual variation of T_∞ at 45° N deduced from incoherent scatter is nearly the same as that given by J70, but the seasonal effect at 45° N is twice as large as that given by the model, and at 12° S it is over three times as large [15].

3.2.3. Diurnal Variations in the Upper Thermosphere

The i.s. observations of T_∞ show a strong diurnal effect, with a night-time minimum and a daytime maximum. Fig. 12 [52] shows typical diurnal variations obtained from i.s. observations in local summer and winter at 12° S and 45° N.

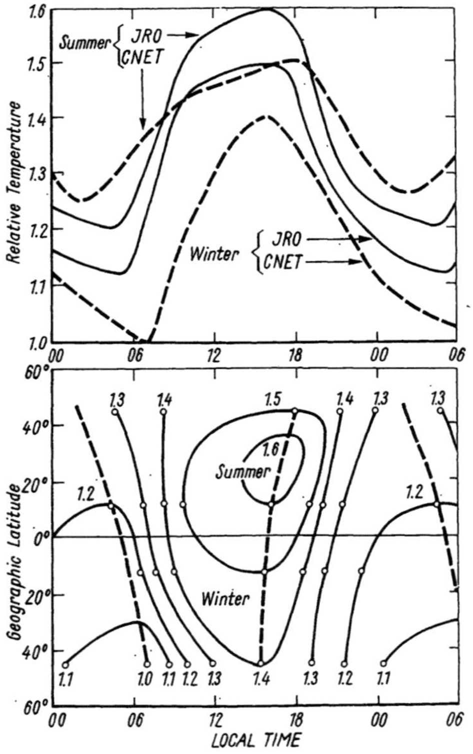

Fig. 12. Typical quiet day summer and winter exospheric temperature diurnal variations for France (CNET) and Peru (JRO), and a contour map giving the global temperature variations derived from them [52].

The resulting contour map of T_∞ shows a number of features that do not correspond to the model. For example: (a) The time of the extrema depends on latitude and season in a manner which clearly indicates that the diurnal variation is strongly dependent on the length of the day. At 45° N the time of maximum varies between 1430 hours in winter and 19 hours in summer (instead of occurring at a constant local time near 14 hours as in the J70 model). (b) Similarly, the amplitude $R = T_{max}/T_{min}$ of the diurnal variation varies between 1.2 and 1.45, being larger in winter, while no dependence of R on solar activity has been detected. The ratio of the global extrema appears to exceed 1.55, which is much higher than the typical J70 value of 1.33. Actually, the fact that the diurnal maxima of T_∞ occurred during the late afternoon, which was the first important result provided by i.s. relevant to the thermosphere, was difficult to accept because it apparently pointed to a disagreement with the occurrence near 14 h of the density maximum, supported and confirmed by many thousands of satellite orbital drag observations. Therefore measurements of the atomic oxygen number density by i.s. techniques [11] were of particular interest.

Such measurements are illustrated in Fig. 11, which shows the oxygen concentrations at 400 km $[O]_{400}$. The diurnal variation of this quantity seems quite different from the T_∞ variation; the diurnal maximum occurs near 14 h, and the rate of decrease of $[O]_{400}$ in the afternoon appears to be very steep. This behavior of $[O]_{400}$ and T_∞ are perfectly compatible provided that either the assumption of diffusive equilibrium is abandoned or appropriate changes are introduced in the lower thermosphere. It has been established by Swartz and Nisbet [12] that merely accounting for the observed diurnal changes in the shape of the temperature profile was sufficient to introduce a phase difference between $[O]_{400}$ and T_n; the parameter s, which characterizes the steepness of the T_n profile, exhibits a maximum during the morning. Thus observed changes in the thermal structure of the lower thermosphere can account for a dephasing of the density and temperature maxima in the upper thermosphere.

We should note here that these i.s. observations concern atomic oxygen only and that the diurnal behaviour just described may be different for the other neutral constituents.

3.2.4. Lower Thermosphere

Several results have been obtained on the lower thermosphere by assuming that the atomic oxygen densities and exospheric temperatures are given respectively by satellite drag and i.s. data, and inferring from this the variations in the E and F 1 region which were necessary to make the i.s. measurements compatible with atmospheric models. It is convenient to assume that diffusive equilibrium prevails above 120 km and to present the relevant results as variations at this altitude. It is then found that $[O]_{120}$ exhibits a solar cycle variation and that it increases by a factor 1.7 from solar minimum to solar maximum [55, 64, 59], and that $[O]_{120}$ also varies seasonally by a factor of 1.7 at 45° N, being larger in the winter [18, 65]. The first of these effects is corroborated by other experimental evidence, and in particular by ionospheric observations [66].

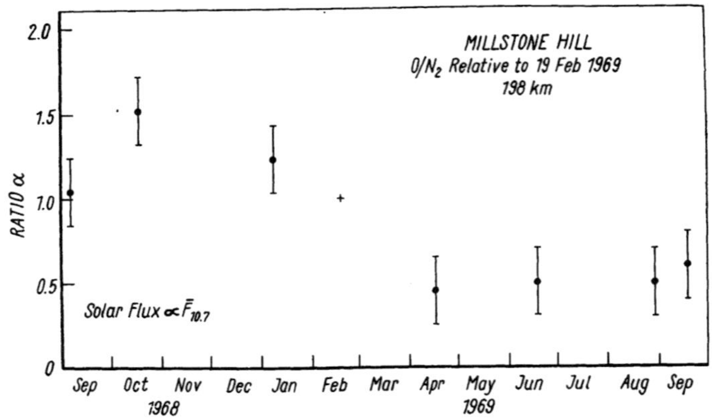

Fig. 13. The ratio O/N_2 at 198 km derived from i.s. measurements and normalized to the value derived on 19 February 1969 [28].

The same is true for the seasonal variation, which has the characteristics required to explain the F region winter anomaly in terms of variations of the neutral concentration. Moreover, i.s. observations of $p = [O^+]/N_e$ in the F2 region exhibit a strong seasonal variation [27, 29, 30]: at a given altitude, p is higher in winter than in summer. Under certain hypotheses, this variation leads to estimates of changes in [O] or $[O]/[N_2]$ in the F 1 region [18, 28] and it is found that $[O]/[N_2]$ increases considerably from summer to winter, as shown by Fig. 13.

One should, however, be cautious about interpreting these observations in terms of seasonal variations, because no experimental data are available from the southern hemisphere. Indeed, the thermosphere undergoes an annual effect

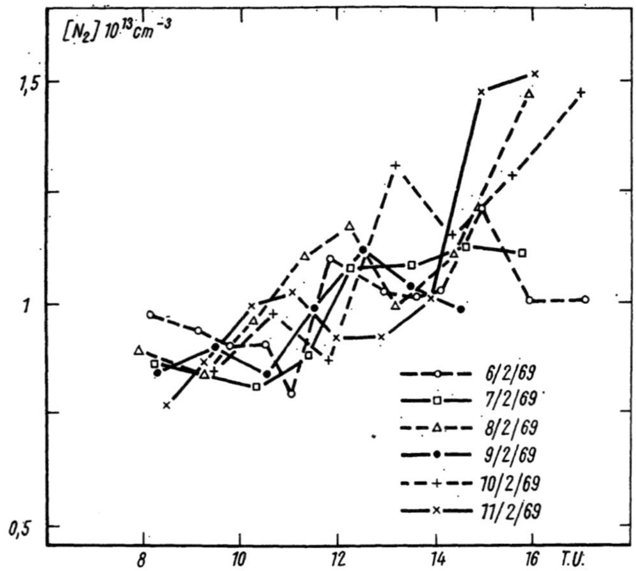

Fig. 14. Diurnal variation of $[N_2]$ at 100 km measured by i. s. [18]–St. Santin, Feb. 1969.

with a minimum in July; this effect is not completely understood but it is experimentally well confirmed. (It corresponds to the annual component in the expression for the semi-annual variation in Jacchia's models.) Thus, part at least of the apparent seasonal effect on [O] or [O]/[N_2] must in fact be an annual effect.

Systematic variations have also been detected by i.s. in [N_2], which is measured directly around 100 km. At 45° N, there is a marked seasonal variation of [N_2] with a minimum in summer [29]. Furthermore, the [N_2] values tend to increase linearly throughout the day [18] as shown in Fig. 14. There is an apparent

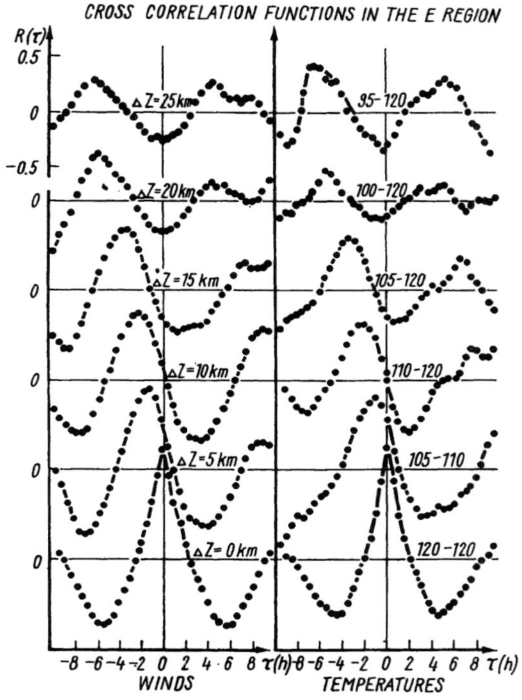

Fig. 15. The auto- and cross-correlation of the winds and the temperatures measured by i.s. at 95 to 120 km altitude between 1966 and 1969 [33] — St. Santin. These statistical averages clearly show that the semidiurnal tidal mode is dominant.

long-term variation in T_n (120 km), with an increase throughout the solar cycle, and a weak seasonal effect, with a 20° K increase from summer to winter at 120 km.

However, the long-term variations are difficult to distinguish unambiguously because they are obscured by the much larger short-term variations. Various types of short-term changes have been identified: Carru et al. [67] discussed variations of the neutral temperatures and winds; Perkins and Wand [31] and Wand [32] discussed variations of the neutral temperatures and densities. The change in phase with altitude of these variations indicated that wave propagation, which is not completely identical from day to day, is present. Using i.s.

data, Bernard and Spizzichino [33] have shown through a correlation analysis illustrated in Fig. 15, that a semidiurnal tidal mode is dominant in the E region.

4. Conclusions

We have tried to review as thoroughly as possible the thermospheric results which have been obtained using i. s., and to discuss the variety of important thermospheric phenomena that are accessible to incoherent scatter. Looking backwards, it is clearly the results concerning the upper thermosphere which are the most impressive. With the inclusion of incoherent scatter oxygen density data, the picture that is being built begins to approach, from the point of view of completeness, the picture based on satellite drag data.

It should be stressed here again that, although i.s. results appeared at first to contradict earlier findings, the apparent discrepancies tend to disappear when various types of valid results are combined. Indeed, the results from various observational techniques are converging as shown for example in Fig. 10 [63].

While i. s. results have been valuable in a worldwide description of the thermosphere, we are aware that the lack of a sufficient number of i.s. observatories, and mainly of a sufficient latitude coverage by i.s., severely limits the possibilities in this direction. For example, there are still no i.s. data from the auroral zone, although we know that the auroral regions have a critical importance at least during disturbed periods and also possibly for the semi-annual variation. Similarly, there is no observatory in the southern mid-latitudes. Data from that region could unambiguously resolve the question of seasonal versus annual variations in the thermosphere.

In the future, the situation will probably be improved by the construction of new i.s. facilities, especially in the auroral zone. However, the cost of i.s. radars is such that building a network of such stations, in any sense of the word, is completely impossible. The rational way to exploit the i.s. data, then, will be in conjunction with satellite observations, by achieving as many rendezvous as possible and by uniting the worldwide picture provided by satellites (e.g. [68]) with the profile and diurnal variation information supplied by the i.s. data. Such studies, as indicated previously, are already under way.

We believe that, in the long run, information about the lower thermosphere will become at least comparable in importance with the results reviewed here relevant to the thermopause level. Actually, i.s. data already available probably contain more information about this region than those which were previously available from other sources, but this fact is only being realized and the corpus of data is just now being organized and examined. The demonstration of the propagation of a semidiurnal tidal mode (see Fig. 15) is an example of what the i.s. method can bring. We believe that incoherent scatter will continue to help solve the many problems that still exist in understanding the behavior of the thermosphere.

Acknowledgments

We thank Professor W. B. Hanson for permission to include unpublished data from the OGO 6 satellite. The C.N.E.T. incoherent scatter facility was constructed with financial support of the C.N.E.S. and is operated with financial support of the C.N.R.S. Part of the work on this project was done while one of us (P.B.) was visiting at The University of Texas at Dallas, where this project was supported by NASA Research Grant NGL 44-004-001 and by NOAA Research Grant E22-92-71(G).

References

[1] K. L. Bowles, Phys. Rev. Lett. **1**, 454 (1958).
[2] H. Carru, M. Petit and P. Waldteufel, Planet. Space Sci. **15**, 944 (1967).
[3] J. S. Nisbet, J. Atmos. Sci. **24**, 586 (1967).
[4] J. P. McClure, J. Geophys. Res. **74**, 279 (1969).
[5] K. K. Mahajan, J. Atmos. Terr. Phys. **31**, 93 (1969).
[6] J. V. Evans, Proc. Inst. Elect. Electron. Engrs **57**, 496 (1969).
[7] W. B. Hanson, Space Research III, 282 (1963).
[8] P. M. Banks, Earth and Planet. Sci. Lett. **1**, 270 (1966).
[9] P. M. Banks, J. Geophys. Res. **72**, 3365 (1967).
[10] H. Carru and P. Waldteufel, Annls Géophys. **25**, 485 (1969).
[11] P. Bauer, P. Waldteufel and D. Alcayde, J. Geophys. Res. **75**, 4825 (1970).
[12] W. E. Swartz and J. S. Nisbet, J. Geophys. Res. **76**, 185 (1971).
[13] W. B. Hanson and S. Sanatani, J. Geophys. Res. **75**, 5503 (1970).
[14] P. Stubbe and S. Chandra, Planet. Space Sci. **19**, 731 (1971).
[15] J. P. McClure, J. Geophys. Res. **76**, 3106 (1971).
[16] D. R. Bates, Proc. Roy. Soc. London, A **253**, 451 (1959).
[17] G. Vasseur and P. Waldteufel, J. Atmos. Terr. Phys. **30**, 779 (1968).
[18] P. Waldteufel, Thèse de Doctorat ès Sciences Physiques (1970).
[19] G. Vasseur, Thèse de Doctorat ès Sciences Physiques (1971).
[20] H. Kohl and J. W. King, J. Atmos. Terr. Phys. **29**, 1045 (1967).
[21] H. Kohl, J. W. King and D. Eccles, J. Atmos. Terr. Phys. **30**, 1735 (1968).
[22] G. J. Bailey, R. J. Moffett and H. Rishbeth, J. Atmos. Terr. Phys. **31**, 253 (1969).
[23] E. N. Bramley, J. Atmos. Terr. Phys. **31**, 1223 (1969).
[24] P. Stubbe and S. Chandra, J. Atmos. Terr. Phys. **32**, 1909 (1970).
[25] H. Rishbeth, J. Atmos. Terr. Phys. **29**, 225 (1967).
[26] H. Rishbeth, J. Atmos. Terr. Phys. **30**, 63 (1968).
[27] J. V. Evans, J. Geophys. Res. **72**, 3343 (1967).
[28] L. P. Cox and J. V. Evans, J. Geophys. Res. **75**, 6271 (1970).
[29] P. Waldteufel, Planet. Space Sci. **18**, 741 (1970).
[30] J. V. Evans and L. P. Cox, J. Geophys. Res. **75**, 1591 (1970).
[31] F. W. Perkins and R. H. Wand, Univ. of Illinois Aeron. Rep. No. 19, 125 (1967).
[32] R. H. Wand, J. Geophys. Res. **74**, 5688 (1969).
[33] R. Bernard and A. Spizzichino, J. Atmos. Terr. Phys. **33**, 1345 (1971).
[34] A. P. Willmore, Space Sci. Rev. **11**, 607 (1970).
[35] J. P. Dougherty and D. T. Farley, J. Geophys. Res. **68**, 5473 (1963).
[36] M. Petit and P. Waldteufel, Space Research VII, 241 (1967).
[37] R. H. Wand and F. W. Perkins, J. Geophys. Res. **73**, 6370 (1968).
[38] P. Waldteufel, Planet. Space Sci. **17**, 725 (1969).
[39] R. L. Showen, to be published in J. Geophys. Res. (1972).
[40] D. R. Moorcroft, J. Geophys. Res. **69**, 955 (1964).

[41] H. C. CARLSON and W. E. GORDON, J. Geophys. Res. **71**, 5573 (1966).
[42] D. T. FARLEY, Annls Géophys. **22**, 448 (1966).
[43] D. T. FARLEY et al., J. Geophys. Res. **72**, 5837 (1967).
[44] S. S. PRASAD, J. Geophys. Res. **73**, 6795 (1968).
[45] D. R. MOORCROFT, J. Geophys. Res. **74**, 315 (1969).
[46] H. G. BOOKER and E. K. SMITH, J. Atmos. Terr. Phys. **32**, 467 (1970).
[47] G. N. TAYLOR and G. L. WRENN, Planet. Space Sci. **18**, 1663 (1970).
[48] W. B. HANSON et al., J. Geophys. Res. **74**, 400 (1969).
[49] J. P. McCLURE and B. E. TROY, JR., J. Geophys. Res. **76**, 4534 (1971).
[50] L. H. BRACE, H. C. CARLSON and K. K. MAHAJAN, J. Geophys. Res. **74**, 1883 (1969).
[51] A. GIRAUD, Space Research XI, 1057 (1971).
[52] P. WALDTEUFEL and J. P. McCLURE, Annls Géophys. **25**, 785 (1969).
[53] P. B. HAYS et al., J. Geophys. Res. **75**, 4881 (1970).
[54] L. L. COGGER et al., J. Geophys. Res. **75**, 4887 (1970).
[55] P. WALDTEUFEL and L. COGGER, J. Geophys. Res. **76**, 5322 (1971).
[56] N. W. SPENCER et al., Space Research X, 389 (1970).
[57] L. G. JACCHIA, Smithson. Astrophys. Obs. Spec. Rep. No. 313 (1970).
[58] D. ALCAYDE et al., J. Geophys. Res. 76, 7814 (1971).
[59] A. GIRAUD, G. SCIALOM and A. A. POKHUNKOV, J. Geophys. Res. (1972), in press.
[60] CIRA 1965, North Holland Publ. Co., Amsterdam 1965.
[61] L. G. JACCHIA, Smithson Contrib. Astrophys. 8, 215 (1965).
[62] G. M. KEATING, J. A. MULLINS and E. J. PRIOR, Space Research X, 439 (1970).
[63] P. WALDTEUFEL, J. Geophys. Res. 76, 6990 (1971).
[64] K. K. MAHAJAN, J. Geophys. Res. **76**, 4621 (1971).
[65] D. PERRET, Thesis 1971 Faculté des Sciences de Paris.
[66] J. R. HERMAN and S. CHANDRA, Planet. Space Sci. **17**, 815 (1969).
[67] H. CARRU, Annls Géophys. **23**, 455 (1967).
[68] J. M. LUTON and J. E. BLAMONT, to be published in J. Geophys. Res. (1972).

Space Research XII — Akademie-Verlag, Berlin 1972

THERMOSPHERIC MOTIONS MEASURED BY CHEMICAL RELEASES

J. F. Bedinger

GCA Technology Division, Bedford, Mass., USA

The motion of controlled chemical releases is observed from the ground and interpreted as direct measurements of thermospheric winds. Observations have been made from numerous locations using trails, point releases and multiple releases separated in both space and time. Trail releases are a unique method of obtaining a highly accurate and detailed vertical profile of the horizontal wind over a large altitude range. Chemiluminescent trails allow measurements from 90 to 150 km during the night; sunlit alkali vapor trails during twilight have been used from 80 to 260 km. Precise height resolution of 0.1 km at 100 km and 1 km at 130 km are reported. The reported accuracy of wind speed determination is 3 m s^{-1} at 100 km, decreasing to 10 m s^{-1} at 130 km. Below 130 km, the observed winds are highly variable in both speed and direction. Speeds of over 100 m s^{-1} and shear of 0.3 s^{-1} are generally present in this region. Vertical winds are usually considered to be small compared with the horizontal component and are difficult to measure. The major features of the wind profile have a horizontal extent of hundreds to thousands of kilometers and persist for hours. Smaller features which may have a vertical extent of less than a kilometer are also present. Measurements above 150 km can be made only during twilight and also show wind speeds exceeding 100 m s^{-1} but the direction is less variable and the shears are small compared with the lower region. This paper discusses the techniques associated with the measurement of winds by chemical releases and summarizes the results obtained at various locations.

1. Introduction

The vapor trail method of probing the earth's upper atmosphere has been used successfully by many different countries at numerous launching sites around the world. The relative simplicity and low cost of the method compared with most sounding rocket programs, as well as the visual display, have contributed to its initial widespread use. The continued use of the method is due to the fact that it has proven to be a very useful research tool. The simplicity of the method results from the fact that the rocket is used as a carrier vehicle only in order to release or eject a vapor into the atmosphere in a prescribed manner. Thus, expensive and complicated rocket-borne instruments are not required. All observations are made with ground-based equipment which need not be expensive in order to obtain useful data.

The first chemical release in the upper atmosphere, resulting from a suggestion by Bates [1], was a trail of sodium vapor ejected from an Aerobee rocket over

White Sands, New Mexico, during evening twilight on 12 October 1955 [2]. Since that time, hundreds of chemical releases have been made in the earth's upper atmosphere. Many different materials have been used. These include sodium, lithium, potassium, cesium, nitric oxide, trimethylaluminium, tetraethyl lead, barium, various high explosives, and hydrocarbons. The major requirements for a release material are that it participate in an illumination process which is intense enough for detection and long lived enough for observation. The releases have been made in single puffs, multiple puffs and continuous trails. Objectives have included: (i) the measurement of winds and diffusion from the motion and growth of the cloud; (ii) determination of temperature, composition, and chemical processes from emitted or scattered radiation from the cloud; (iii) the relation of neutral and ionized species and study of magnetic and electric fields from neutral and ionized clouds. This discussion is restricted to those measurements of motion of the released gas which are equated with motions of the neutral atmosphere and are termed winds.

A very accurate and detailed vertical profile of the horizontal wind over a large altitude range may be determined from a single vapor trail. A sunlit trail of alkali vapor during twilight may allow a continuous measurement of winds from 50 to 260 km. During the night, the altitude range is restricted to that region in which a sufficiently bright and persistent chemiluminescent process occurs. Trimethylaluminium is the most used material at night and measurements can usually be obtained from 90 km to between 140 and 160 km. At the present time, there have been no vapor trail measurements during the daytime. However, it has been demonstrated by Bedinger [3] and independently by Best [4] that a lithium vapor trail may be tracked during the daytime with a differential radiometer employing a narrow-band interference filter. Thus, vapor trail measurements during the day are possible and may soon be used.

A limitation of the vapor trail method is that observations from at least two locations are required in order to determine position of the release by triangulation. This requirement of cloud-free skies over a large area has been particularly restricting during times when the objective was coordinated observations from many sites or many consecutive observations from one site. However, such series of observations have been successfully completed and yielded good data. International coordinated programs were successfully conducted at least twice. Multiple launches throughout the night have been reported from many sites.

Because of the large quantity of data that have been reported from many locations, this report only summarizes the methods and general results: special emphasis is placed on the precision of the measurements and the special character of the method as it relates to other measuring techniques. The observed characteristics of the winds are described, but discussion of the various interpretations of the observations is avoided. Analytical results are included only when they are primarily descriptive rather than interpretive. The list of references was chosen as being representative of the reported observations and is not intended to be complete.

2. Methods and Accuracy

2.1. Methods

The method employed in the release of chemicals is dependent upon the nature of the substance involved. The criteria for the release technique are that the substance be in a form to produce the desired reaction and that it be as efficient as possible due to the high cost of delivery. Metallic vapors are usually produced by explosions or are boiled off in very hot thermite reactions. Liquids and gases are usually released from pressurized containers. The releases have been made from sounding rockets and gun-launched probes. These techniques have been described in detail by various investigators [2, 5, 6].

The sole purpose of the rocket or gun-launched probe is to deliver the chemical vapor. Usually all data from which winds are to be determined are collected on the ground in the form of simultaneous, sequential photographs. The equipment and methods used to acquire the photographs vary widely depending on the material and the investigation. Highly specialized equipment such as high speed cameras, long focal length lenses and numerous combinations of film and filters have been employed to observe releases. However, only medium speed lenses and simple cameras with modern film, are required for determination of winds.

These techniques and equipment are described in [7–11]. The winds are determined from the motion or distortion of the released gas which requires accurate determination of the position of the release at separate times. The release position is determined using triangulation techniques from carefully oriented and timed synchronous photographs which are obtained at widely separated and accurately located sites. In principle, only two simultaneous photographs from separate locations are required for a position determination, but more are usually desired to furnish redundant data for improving confidence in the results, in the event of poor "seeing" from some sites, or because some sites may be unfavorably oriented to observe certain release motions. After determining the release position at two separate times, the wind velocity at a given height is obtained by dividing the displacement appropriate to that height by the time interval separating the two position measurements. The method is described mathematically in [12].

Two distinctively different methods have been used to perform the triangulation necessary for determination of release position. The first wind measurements from vapor trails were obtained with an analog system in which vapor trail photographs were projected on screens and positions were measured on a scale model of the launch and observing site [13]. Star images were used to insure that the magnification and direction were exactly the same as at the time of the observation. Thin nylon filaments were stretched from each projector to the corresponding projected image on the screen. When the filaments from each projector cross at the same point, the end points of each filament in the projected image are known to correspond and the crossing point of the filaments is the spatial position of the release with the appropriate scaling factor.

The currently popular method of determining release position uses a digital computer to perform the numerous calculations necessary for triangulation [14, 15]. Procedural details used by various groups differ, but the general method is the same. Star photographs recorded either on the same frames as the release or on a different frame which has effective fiducial markings are used to determine a set of "plate constants" or standard coordinates so that real angular positions on the photograph may be determined. A series of points is chosen on the simultaneous photographs from the triangulation sites and two angular coordinates such as azimuth and elevation angles are determined for each point. This information together with the geometry of the observing sites allows the properly programmed computer to complete the triangulation.

2.2. Accuracy

Both of these methods have been used extensively at the GCA Corporation. On several occasions, the two systems have been employed to perform measurements on the same trail. The comparison of these measurements reveals that the two very different systems yield essentially identical results, with the same resolution and accuracy. Similar comparisons were made and the same conclusions were reached by Jarrett et al. [16]. This finding indicates that the limitations on accuracy and resolution are determined by physical considerations pertaining to the trail and not by the method of data reduction. The comparison measurements at GCA have led this author to believe that the only accurate assessment of resolution and accuracy of these measurements is obtained by comparison of redundant data.

We use data from all available sites and in all combinations which includes matching trails starting from the top and from the bottom. For N observing sites, one may obtain $2\,N\,(N-1)$ different triangulation combinations. We usually have four sites and all twenty-four profiles are computed. Often the orientation of the trail with some sites is not good and incorrect profiles may be generated, but these are easily detected and eliminated through comparison with other triangulation combinations. Some of the factors which control the accuracy and resolution of vapor trail measurements are summarized below.

Exposure times of from 1 to 10 seconds are generally employed in vapor trail photography. Horizontal wind speeds exceeding 100 m s^{-1} are usually present at some height in the thermosphere. This causes a spreading or smearing of trail features which may produce uncertainties of 0.1–1.0 km in position.

Cloud dimensions are often a limiting factor in determining position. Usually, the cloud center is the measured position and uncertainties result from several causes. At 100 km, a trail release reaches a diameter of about 2 km within 200 seconds after ejection. Uncertainties in the location of the trail center are caused by distortion due to wind shear, uneven illumination by solar rays if it is a sunlit trail, irregular trail structure or unfortunate camera orientation. Analyses at GCA using redundant data and cameras with varying focal lengths have determined that with good data, absolute accuracies to 0.1 km are obtained at 100 km and to 1 km at 130 km. Relative measurements such as the separation of clearly identified features or the extent of a small feature may be made with greater accuracy.

The wind velocities are determined by dividing a measured distance by the appropriate time interval. The measured distance is determined by two absolute position measurements and thus at 100 km is usually uncertain by a length of at least 200 meters (2×100). If trail photographs separated by 200 seconds in time are used to obtain wind velocity, the uncertainty in the derived speeds is then $2 \times 100/200 = 1$ m s^{-1}. If, however, the time interval employed is only 20 seconds, the uncertainty in the wind measurement is 10 m s^{-1}. The uncertainty increases at higher altitudes where releases expand more rapidly due to diffusion. Some estimates of accuracy which have been reported are listed in Table 1.

Table 1
Reported accuracy of wind measurements

1 m s^{-1} at 100 km; 10 m s^{-1} at 130 km*	[17]
± 3 m s^{-1} below 150 km	[18]
10 m s^{-1} between 100 and 200 km	[19]
± 3 m s^{-1} below 150 km	[20]
± 5 m s^{-1} below 150 km	[21]

* The 1 m s^{-1} value is an optimum value applicable to regions of low wind shear. The height uncertainty of 0.1 km causes a greater effective error in high shear regions.

3. Observed Characteristics of Thermospheric Winds

3.1. General Characteristics

When discussing the winds observed from vapor trails, it is convenient to consider separately the lower and the upper thermosphere. Below about 150 km the winds are usually highly variable in both speed and direction with regions of high shear. Above that altitude the variability is much less. Also, below 150 km measurements may be made both at twilight and at night and thus there are a larger number of observations for statistical studies and studies of temporal variations. Above 150 km measurements may be made in twilight only and are thus much fewer in number as well as restricted to the short, twilight interval.

The observed characteristics have been reported often [8, 12, 20, 22, 23]. Wind profiles from nearly 100 vapor trails from 13 locations were collected by Bedinger [24]. Most of the observations are from northern mid-latitudes and below 150 km. The description of winds in this report will pertain to that region unless otherwise stated.

Although observed winds vary greatly from day to day at all heights, certain characteristic profiles as well as average and limiting values are observed. Many of the characteristic features observed from vapor trails are illustrated in Fig. 1 which is a photograph of a sodium–lithium vapor trail released over Wallops Island, Virginia, during evening twilight 13 February 1969. Wind velocities between the altitudes 90 and 260 km were determined from the trail. The characteristics of the upper and lower regions are clearly shown. The lower region has variable winds with rapidly changing speed and direction. The variability decreases with height up to at about 160 km, the winds remain constant at 100 m s^{-1} up to 260 km, the peak of the rocket trajectory. The winds in the lower region reach maximum velocities of about 70 m s^{-1} which is the average observed value. Wind speeds of 150 m s^{-1} are often observed, however. Wind shears in the lower region are found to be about 0.03 s^{-1} on the average, but values up to 0.1 s^{-1} have been observed.

The relatively small diameter of the trail at the lower altitudes as well as the clearly identifiable small features allow good accuracy of position determination by triangulations. It is readily apparent that the precision of measurement decreases with altitude due to increasing trail diameter and absence of identifiable structure.

3.2. Spiral Structure and Stratification

There are two specific characteristics of the vertical profiles of the horizontal winds which have been noted and discussed in the literature. The observed

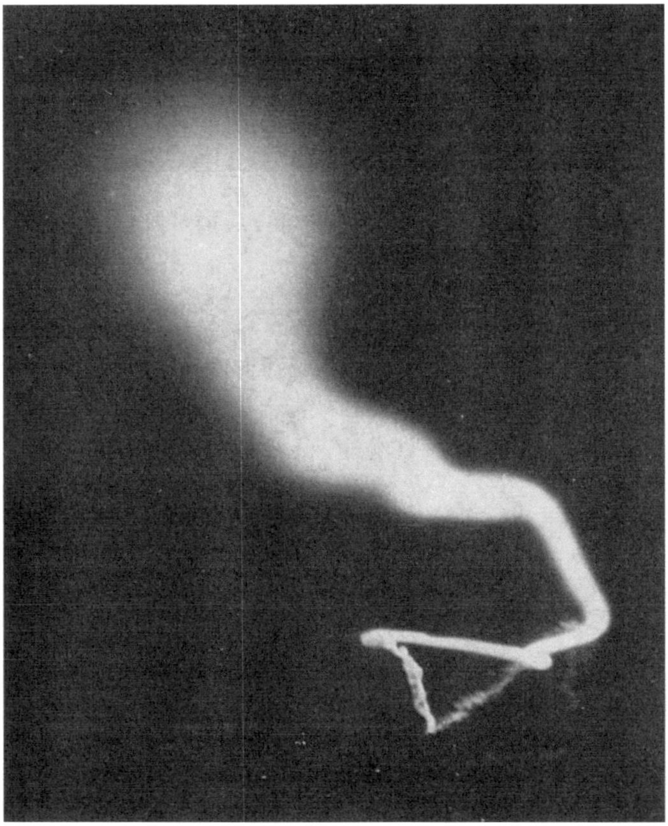

Fig. 1. Trail of sodium and lithium vapor over Wallops Island, Virginia, during evening twilight on 13 February 1969. Photograph is from Dover, Delaware.

profiles often have the form of a large spiral in which the sense of rotation when viewed from above is clockwise with increasing altitude in the northern hemisphere. The opposite direction of rotation is present in the southern hemisphere [19, 24]. The few equatorial results [25] are variable, usually exhibiting many reversals in direction. The spiral feature is evident in Fig. 1, but it is sometimes much more apparent, as in Fig. 2. Although the presence

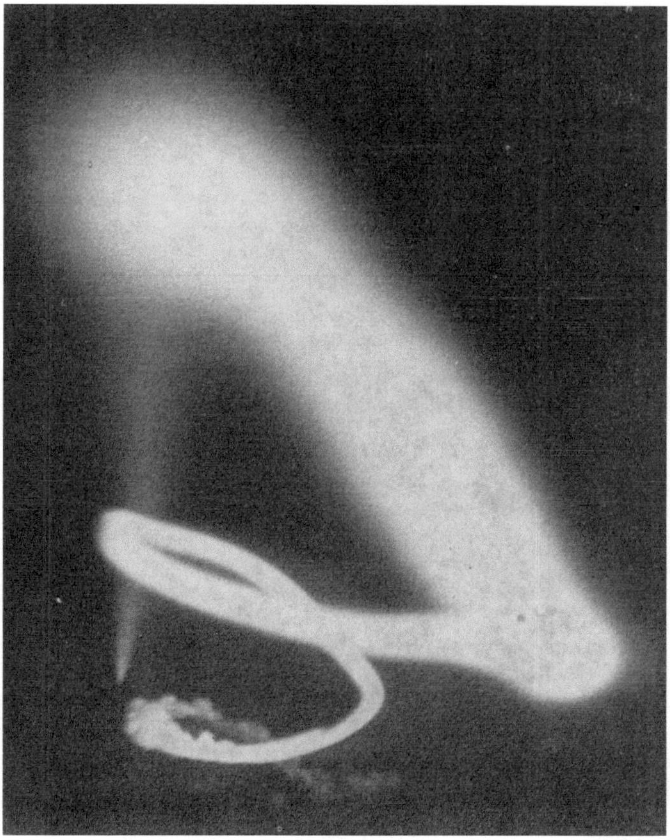

Fig. 2. Trail of sodium and lithium vapor over Wallops Island, Virginia, during evening twilight on 12 December 1969. Photograph is from Andrews AFB, Maryland.

of the spiral is noted at most sites, Rosenberg and Justus have emphasized this feature in the data obtained at Eglin, Florida. They described the variations of the winds throughout the night at heights above 105 km in terms of a well-organized sinusoidal pattern with a dominant wavelength from 2 to 4 scale heights which progressed downward at a rate of 1 wavelength per night. It was found that the north–south component of the dominant wave appeared to be vertically displaced in simultaneous, spatially separated observations. This slight variation was associated with a horizontal wavelength of 1000 to 4000 km. No displacement was found in the east–west component and a substantially larger horizontal wavelength was inferred. Similar observations have been reported from Yuma, Arizona [20] and by inference from Wallops Island, Virginia [12] where the difference between winds obtained from up and down trails is considered to be within the observational uncertainty. Blamont [26] reported also nearly identical profiles taken during the same twilight and separated by 650 km.

Usually, the spiral is not smooth and continuous with height. Often reversals in direction are present and the profile appears to be stratified so that directional

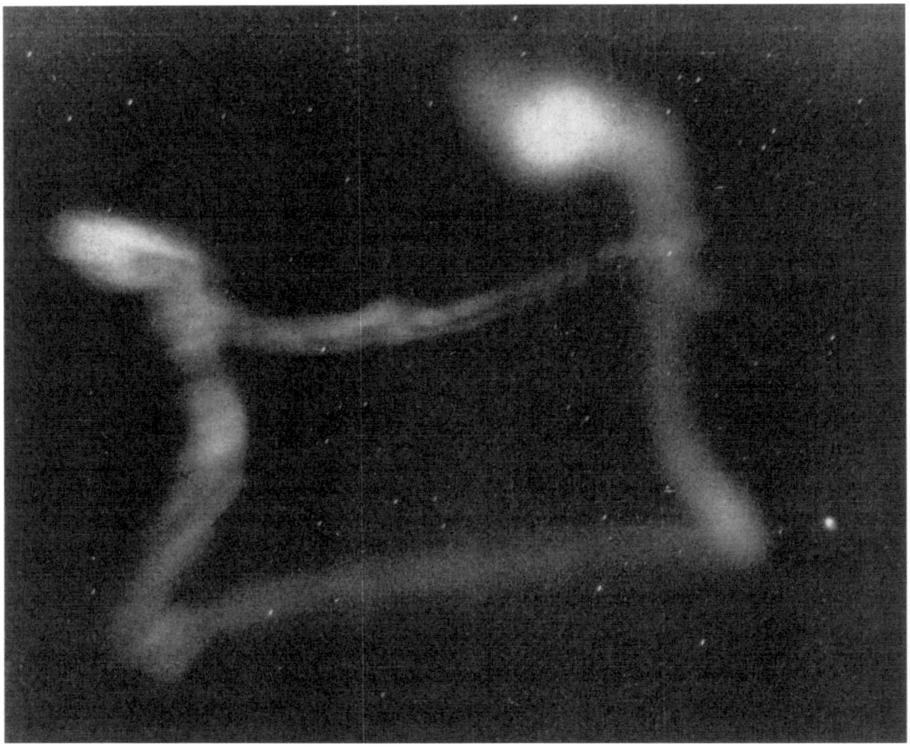

Fig. 3. Trail of TMA over Wallops Island, Virginia, at 2208 EST on 14 December 1970. Photograph is by NASA from Wallops Island, Virginia.

changes occur in small height intervals which are separated by regions of little change. This effect is present in the trail in Fig. 1 and dominates the entire profile in Fig. 3. Bedinger et al. [12] emphasized this property of the Wallops Island observations. They observed that the spiral pattern was associated with large wind speeds, whereas irregular patterns tended to contain low wind speeds. Murphy reports the presence of the spiral pattern and stratification at both Yuma, Arizona [20] and Barbados, West Indies [27]. At Yuma (33° N 114° W) the spiral pattern is observed to be dominant, whereas stratification is much more often observed at Barbados (13° N 58° W). The winds speeds at Yuma were also larger than at Barbados.

Examples of the large spiral pattern and a small irregular profile are shown in Fig. 4 and 5 respectively. The large spiral is from Wallops Island [28]. Unusually high winds of 200 m s^{-1} are present. The irregular profile is from Barbados, West Indies [29]. Most of the profiles observed at Wallops Island show evidence of both the spiral and stratification. Often the irregular profiles at Wallops have smaller wind speeds [12]. Fig. 6 is an example of a trail which exhibits a very smooth spiral structure in the upper portion but is highly stratified in the lower portion. This trail was formed on the downward portion

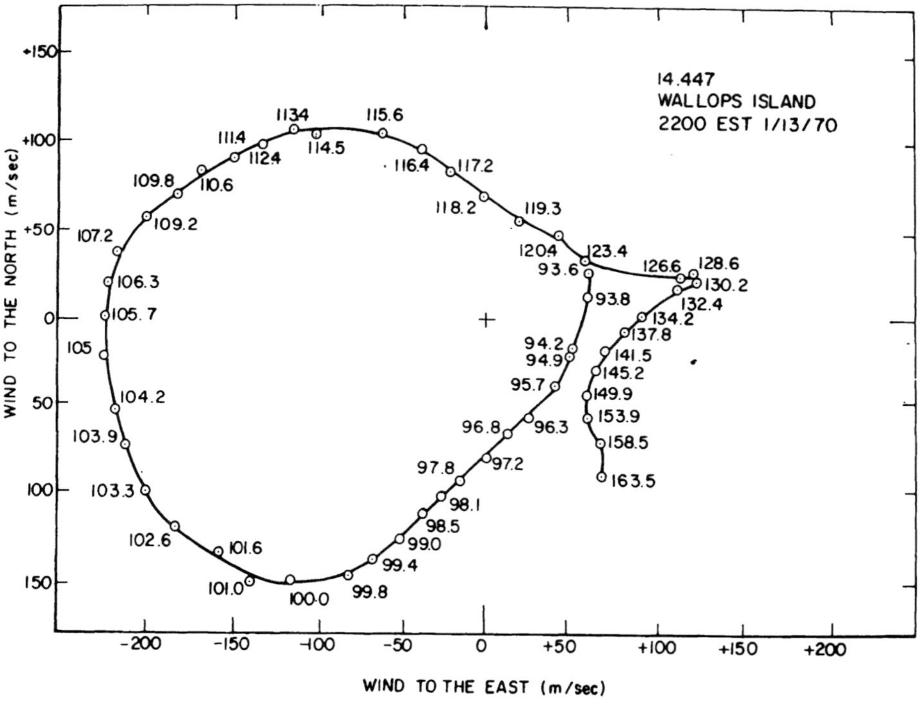

Fig. 4. Hodograph of winds obtained from trail of TMA over Wallops Island, Virginia, at 2200 EST, 13 January 1970.

of the rocket trajectory whereas Fig. 1, 2, and 3 were formed from an ascending rocket.

3.3. Small Scale Structure

There are two kinds of small scale structure which are visually apparent on vapor trails. One kind is associated with the stratification phenomenon when rapid changes in wind direction or speed occur in small height intervals. Several such regions are apparent in Figs. 1, 3 and 6. At Wallops Island, Virginia [12] these regions were found to be most evident on small wind profiles below 130 km. At Barbados, directional changes of greater than 145° within a 3 km height interval were observed on nine of 23 profiles [9]. Examples of the reversals in a Barbados profile are shown in Fig. 5.

The other type of small structure is evident in the lower portion of Figs. 1 and 2. It has been described as "globules" which suddenly appear on alkali vapor trails and grow at a rapid rate. At Wallops Island the "globules" appear only on trails released during the upward part of the trajectory and not on those released during the downward part. Compare Fig. 6 with Figs. 1 and 2. The Wallops Island results are completely discussed in [30]. Other investigators

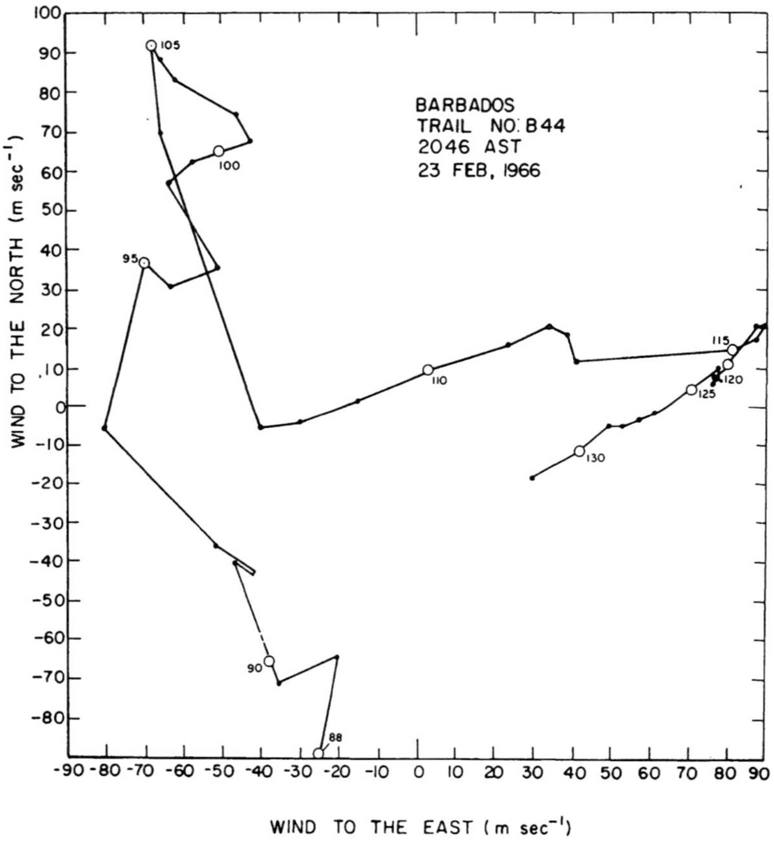

Fig. 5. Hodograph of winds obtained from trail of TMA over Barbados at 2046 AST, 23 February 1966 [29].

have reported the structure on both up and down trails, but photographs of a similar disturbance on both up and down trails have been published only by Blamont [31]. The comparison of TMA trails in the up and down part of the trajectory is complicated by the fact that the down trail is disturbed by the re-entry and vaporization of the fluid which was frozen at the time of ejection. The differences in structure of up and down trails is important in distinguishing between those features which are the result of motions pre-existing in the ambient atmosphere and effects associated with the release.

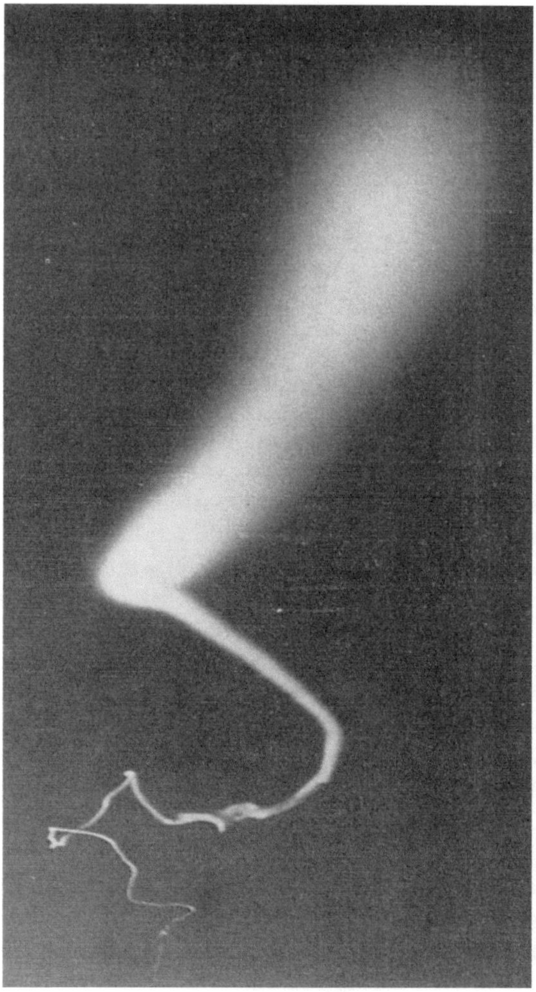

Fig. 6. Trail of sodium and lithium vapor over Wallops Island during morning twilight on 22 February 1968. Vapor was released on the downward part of the rocket trajectory and photographed from Dover, Delaware.

3.4. Temporal Variations

The measurement of temporal variations of the winds is severely restricted by the inability to observe during the daytime. However, several series of observations have been obtained throughout the night. At many observing sites, recognizable features have been observed to descend slowly throughout the night. The rate of descent infers a scale and wavelength as described in the previous section, i.e. several thousand km and periods of 24 hours and its major harmonics. Reports based on the observations from Eglin, Florida, indicate that the spiral pattern is the major feature throughout the night. Murphy reports similar results at Yuma, but also that the spiral tends to build up during the night. At Wallops Island, the spiral has been observed to persist through the night and also to collapse into a small irregular profile around midnight and begin to reform by morning. At Barbados, the spiral pattern has been seldom observed, but the average winds between 90 and 120 km appear to exhibit seasonal variations.

It should be noted that although the number of observations at several launch sites are numerous enough to provide a meaningful insight into the general nature of the winds, the number of closely spaced observations is small and interpretation is severely limited by the restriction to night-time observations. Thus, the temporal variations are not adequately observed.

3.5. Vertical Winds

It is generally concluded that vertical winds are small compared with the horizontal winds and indeed are usually neglected during data reduction. However, some measurements have been reported. Manring et al. [32] reported that a downward velocity of 3.67 m s^{-1} could explain the observed variations in a shear region around 109 km. An actual upward motion of 10 m s^{-1} was observed from some small filaments or plumes which separated from a vapor trail at 103 km. In a later analysis [30] this motion is not interpreted as due to vertical winds, but rather as a peculiarity of the trail expansion. Observations of vertical winds were reported by Edwards et al. [8] from various releases at Eglin, Florida. Vertical velocities up to 15 m s^{-1} in the region 97–111 km were reported from motion of identifiable "globular" structures. More recently, investigators at AFCRL have used multiple "puff" releases in order to measure vertical winds. Initial results were reported by MacLeod at the meeting of AGU in Washington, D.C., April 1971. A complicated small-scale motion which includes vertical motion was described. Final results have not yet been published. The greatest vertical velocities reported to date are from Ackerman et al. [33]. Upward speeds of 37 and 22 m s^{-1} were obtained from observations of solar illuminated AlO clouds at 160 km.

Several investigators have reported a downward drift of persistent features of wind profiles from series of observations spaced through the night. These motions are usually considered to be a phase propagation and not mass motion. It appears that at best, vertical motions are difficult to measure with the vapor release method.

3.6. High Latitudes

Most of the reported winds at high latitudes are from Ft. Churchill, Canada (59° N 94° W). The profiles are highly variable and the spiral pattern appeared strongly in only one of five observations during 1963–1965 [24]. The wind speeds were often high with one observation of a speed of 300 m s^{-1} between 110 and 115 km. Four more profiles were reported during September 1966 [34]. Speeds were not extremely large and the spiral was more in evidence as were high shear regions. The very irregular structure was again found on six trails during January and February 1967 [35]. The wind patterns appear to be very complicated and more observations will be required before they can be reliably described.

The most interesting recent results at high latitudes have involved ionized as well as neutral releases. However, these results are beyond the scope of this review.

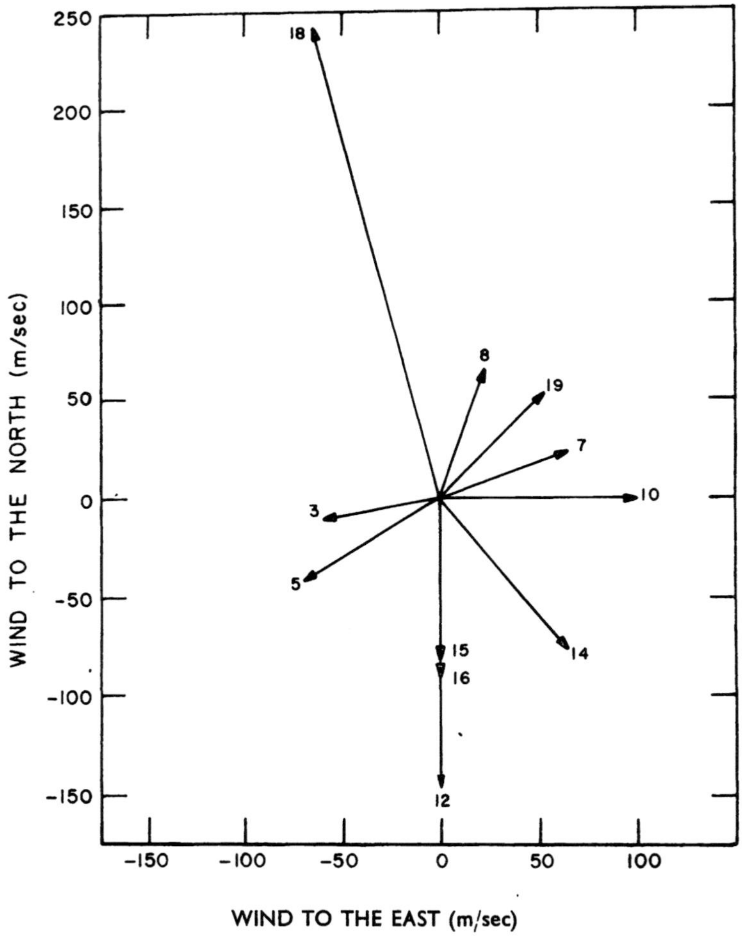

Fig. 7. Winds above 180 km during evening twilight. The arrows indicate the wind speed and the direction. The numbers on the arrows refer to a table in [36].

3.7. Winds above 150 km

Above 150 km winds have been observed with vapor releases only during twilight. Often the apogee of the release vehicle is not much above that height so that even at twilight the reported observations are few. The wind speeds are variable as in the lower altitude, ranging from 40 to 250 m s^{-1}. However, the wind shears are small and directional changes are nearly absent. The spiral pattern is not observed above 150 km and often the direction is nearly constant over a large height range. A summary of reported observations above 180 km was given by Bedinger [36]. A similar summary was published by King-Hele and Allan for results prior to 1965 [37]. The results for evening and morning twilight are shown in Figs. 7 and 8. Some reports not included are two firings from Australia in 1961–1962 for which the firing times are uncertain. They are thought to have been evening firings and the winds are 130 m s^{-1} at 300° and 110 m s^{-1} at 345°. More recent observations are reported by Rees et al. [19]

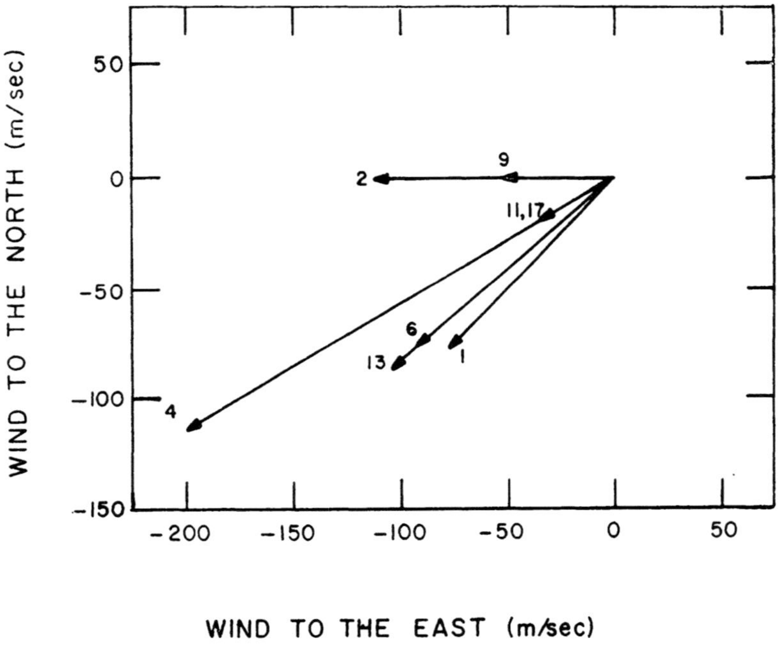

Fig. 8. Winds above 180 km during morning twilight. The arrows indicate the wind speed and the direction. The numbers on the arrows refer to a table in [36].

in Australia (31° S 137° E) contain one evening profile with a large eastward component and a smaller southward component and one morning profile with a large northward component and a small eastward component. These are about opposite to those shown in Figs. 1 and 2 and thus may reflect the difference in northern and southern latitudes. Two other recent measurements are reported by Ackerman et al. [33]. Both of these measurements were during evening twilight from Sardinia and large eastward and relatively small northward components were observed. These may be compared with numbers 12, 14, 15 and 16 in Fig. 7 which were also observed in Sardinia and which show primarily large southerly components. Many more observations are required before one can adequately describe the motion in this region.

4. Conclusions

(i) It appears that more measurements are required at high altitudes, high latitudes, for both spatially and temporally separated trails and for vertical wind components. Measurements in the daytime should begin.

(ii) During twilight a vapor trail allows a direct, detailed and continuous measurement of winds from 50 to 250 km and from 90 to 150 km at night. This range and precision is especially valuable in relation to measurements by other methods which are effective over lesser height intervals such as radar observations of meteor trails, ionospheric drifts and satellite drag measurements.

(iii) The accuracy of the wind measurements is reported to be the same at all observing sites. The accuracy in wind speed is 3 to 5 m s^{-1} for winds which exceed 100 m s^{-1}. It thus appears that the precision of the measurements is adequate and that different conclusions concerning the origin and structure of the thermospheric winds are due to differing interpretations of the data.

References

[1] D. R. Bates, J. Geophys. Res. **55**, 347 (1950).
[2] H. D. Edwards et al., in: Aurora and Airglow, Pergamon Press, New York 1955 (p. 122).
[3] J. F. Bedinger, Rev. Sci. Instrum. **41**, 1234 (1970).
[4] G. T. Best, J. Atmos. Sci. **27**, 979 (1970).
[5] N. W. Rosenberg et al., J. Geophys. Res. **68**, 5895 (1963).
[6] E. R. Manring, Aerospace Engineering **20**, No. 9, 8 (1961).
[7] J. F. Bedinger et al., J. Geophys. Res. **63**, 19 (1958).
[8] H. D. Edwards et al., J. Geophys. Res. **68**, 3021 (1963).
[9] C. H. Murphy et al., J. Geophys. Res. **71**, 4535 (1966).
[10] G. Groves, Nature, Lond. **187**, 1001 (1960).
[11] J. E. Blamont and M. L. Lory, Nature, Lond. **201**, 593 (1964).
[12] J. F. Bedinger et al., Planet. Space Sci. **16**, 159 (1968).
[13] E. R. Manring et al., J. Geophys. Res. **64**, 587 (1959).
[14] D. L. Albritton et al., Photogram. Eng., Sept. 1962.
[15] F. J. Smith, Planet. Space Sci. **11**, 1311 (1963).
[16] A. H. Jarrett et al., Planet. Space Sci. **11**, 1319 (1963).
[17] E. R. Manring et al., Meteor. Monog. **8**, No. 31, 196 (1968).
[18] N. W. Rosenberg et al., Rad. Sci. **1**, 149 (1966).
[19] D. Rees et al., Sci. Report No. 1, USAF Contract F61052-68-C-0057, AFCRL.
[20] C. H. Murphy et al., J. Geophys. Res. **73**, 3005 (1968).
[21] M. Huruhata et al., Sixth Internat. Symp. on Space Technology and Science, Tokyo 1965.
[22] A. Kochanski, J. Geophys. Res. **69**, 3651 (1964).
[23] N. W. Rosenbbrg, J. Atmos. Terr. Phys. **30**, 907 (1968).
[24] J. F. Bedinger, CGA Tech. Rept. 66-7-N NASA Contract NASW-1083 (March 1966).
[25] P. D. Bhavsar et al., Space Research V, 986 (1965).
[26] J. E. Blamont, Planet. Space Sci. **10**, 89 (1963).
[27] C. H. Murphy, J. Geophys. Res. **74**, 339 (1969).

[28] GCA Tech. Rep. 70-10-N, Final Report NASA Contract NAS5-11591 (October 1970).
[29] BRL Contract 169 Rep. No. 5, U.S. Army Contract No. DA01-009-AMC-169 (X), January 1967.
[30] D. LAYZER et al., Planet. Space Sci. **17**, 1891 (1969).
[31] J. E. BLAMONT, Annls Géophys. **17**, 134 (1961).
[32] E. R. MANRING et al., Space Research II, 1107 (1961).
[33] M. ACKERMAN et al., J. Geophys. Res. **76**, 3162 (1971).
[34] GCA Tech. Rep. 67-11-N, Final Report on NASA Contract NASW-1083 (June 1967).
[35] GCA Tech. Rep. 68-11-N, Final Report NASA Contract NAS5-3970 (October 1968).
[36] J. F. BEDINGER, J. Geophys. Res. **75**, 683 (1970).
[37] D. G. KING-HELE et al., Space Sci. Rev. **6**, 248 (1966).

Space Research XII — Akademie-Verlag, Berlin 1972

OBSERVATIONS OF IONOSPHERIC MOTIONS USING BARIUM ION CLOUDS

T. N. Davis and D. D. Wallis

Geophysical Institute, University of Alaska, College, Alaska, USA

Observations of small barium ion clouds in the low- and middle-latitude ionosphere show eastward and equatorward movement in evening and westward in morning at speeds of order 10 m s^{-1}. At the auroral zone a definite pattern of westward motion in evening and eastward in morning persists with speeds of several hundred m s^{-1}. Over the polar cap the motion is in the direction away from the sun. The auroral zone and polar cap motions appear to be chiefly determined by the electric field originating in the outer magnetosphere and carried to the ionosphere along highly conducting magnetic field lines. On the other hand, the motions at middle and low latitudes, at least during quiet times, appear to be caused by polarization fields established by dynamo action in the ionosphere. The electric field so generated is transmitted along the magnetic field to the inner magnetosphere to affect plasma motions there.

1. Introduction

The development of the barium vapor release technique during the past few years has provided a significant new tool for measurement of motions of magnetospheric plasma and the ionized component of the atmosphere at altitudes above 120 km [1, 2]. Furthermore, this new method serves as an aid in the calibration of other types of measurement so that they can be more readily interpreted. In particular, observations of auroral alignments and motions, and to a limited extent ground-based measurement of magnetic variations, are now seen to be useful indicators of the directions and speeds of the ionospheric plasma. Other tools are balloon [3], rocket [4] and satellite borne [5] measurements of electric fields and ground-based Doppler and pulse radar measurements [6].

2. The Barium Release Technique

The release technique most commonly used makes use of a cannister filled with a finely divided mixture of CuO and Ba metal with an excess of Ba so that the exothermic reaction

$$\mathrm{Ba} + \mathrm{CuO} \rightarrow \mathrm{BaO} + \mathrm{Cu} \tag{1}$$

provides heat to vaporize the excess Ba [1]. At a chosen altitude the mixture is ignited and the vapor released rapidly expands to form a sphere in equilibrium with the ambient atmosphere. Following this initial phase both an ion and a neutral cloud become visible. The initial spherical shape of the neutral cloud changes by diffusion and may become distorted by wind shear. Superposed within it initially is the ion cloud which is elongated along the direction of the local magnetic field.

Numerous resonance or other emission lines of neutral and ionized Ba are excited. In the visible spectrum strong neutral Ba lines include those at wavelengths 5535, 5826, 5971, 5997, 6019, 6063, 6111, 6595, and 6527 Å. Strong resonance Ba ion lines are at 4554, 4934, 5853, 6141 and 6497 Å. Also BaO bands underlie these emissions [7]. The neutral and ion clouds are easily separable since the neutral cloud appears green and the ion cloud appears red-violet. The clouds remain visible as long as they are in sunlight.

3. Motion of the Barium Neutral and Ion Clouds

The motion of the Ba neutral cloud follows the motion of the ambient neutral atmosphere and so provides a direct measure of high altitude wind. Motion of the Ba ion cloud normal to the direction of the magnetic field is related to the neutral wind $\boldsymbol{V}_{\mathrm{n}}$, the perpendicular electric field $\boldsymbol{E}_{\perp}$ and the ambient magnetic field $\boldsymbol{B}$ by a relation given by Haerendel et al. [8] and by Hašer [1].

$$\boldsymbol{V}_{\perp} = \frac{1}{K}\frac{2}{1+\lambda^*}\left[\frac{\boldsymbol{E}_{\perp}}{|\boldsymbol{B}|} + \boldsymbol{V}_{\mathrm{n}} \times \frac{\boldsymbol{B}}{|\boldsymbol{B}|}\right] + \frac{2}{1+\lambda^*}\frac{\boldsymbol{E}_{\perp}\times\boldsymbol{B}}{|\boldsymbol{B}|^2} + \left[1 - \frac{2}{1+\lambda^*}\right]\boldsymbol{V}_{\mathrm{n}} \tag{2}$$

where $K = \omega_{\mathrm{i}}/\nu_{\mathrm{in}}$ is the ratio of the ion gyrofrequency to the ion–neutral collision frequency and λ^* is the ratio of the height-integrated Pederson conductivity inside the ion cloud to that outside. If, as may be the case at middle and low latitude, the electric field is very small or zero, Eq. (2) reduces to

$$\boldsymbol{V}_{\perp} = \frac{1}{K}\frac{2}{1+\lambda^*}\boldsymbol{V}_{\mathrm{n}} \times \frac{\boldsymbol{B}}{|\boldsymbol{B}|} + \left(1 - \frac{2}{1+\lambda^*}\right)\boldsymbol{V}_{\mathrm{n}}\,. \tag{3}$$

Alternatively, if $\boldsymbol{E}_{\perp}$ is large, as is often the case at auroral latitudes, then Eq. (2) reduces to

$$\boldsymbol{V}_{\perp} = \frac{1}{K}\frac{2}{1+\lambda^*}\frac{\boldsymbol{E}_{\perp}}{|\boldsymbol{B}|} + \frac{2}{1+\lambda^*}\frac{\boldsymbol{E}_{\perp}\times\boldsymbol{B}}{B^2}\,. \tag{4}$$

In order to measure $\boldsymbol{E}_{\perp}$, it is common to use small releases (1–3 kg) at altitudes above 180 km. Then $1/K \to 0$ and $\lambda^* \to 1$. Eq. (2) reduces to

$$\boldsymbol{V}_{\perp} = \frac{\boldsymbol{E}_{\perp}\times\boldsymbol{B}}{|B|^2}\,. \tag{5}$$

In practice, this is the equation used for nearly all determinations of $\boldsymbol{E}_{\perp}$ by the barium release technique. Even in those cases where large ($\geqq$20 kg) releases are made at altitudes below 200 km it is possible to measure $\boldsymbol{E}_{\perp}$ approximately by observing the motion of a specific part of the ion cloud wherein the density is low and hence $\lambda^* \cong 1$. K rapidly increases with altitude above the E region and has values >1 at altitudes above ~120 km [8]. If K is sufficiently large or if its value can be assumed without serious error, Eq. (2) can be used to determine $\boldsymbol{E}_{\perp}$ from observations of a large barium release even in a case where

$E_\perp$ is quite small. In such a case it is found empirically that the ion cloud tends to spread horizontally from the initial point of release toward the direction in which the neutral Ba cloud is moving under the influence of the ambient wind. That portion of the cloud remaining near the initial point of release is observed to have low ion density, whereas the part moving off in the direction of the neutral Ba cloud has a higher ion density that may actually increase with time. Thus one interprets the spreading of the ion cloud to be the result of non-uniform λ^* within the cloud.

4. Internal Morphology of Barium Clouds

Rather spectacular developments occur within Ba ion clouds causing many of them to develop into configurations quite similar to auroral draperies [9].

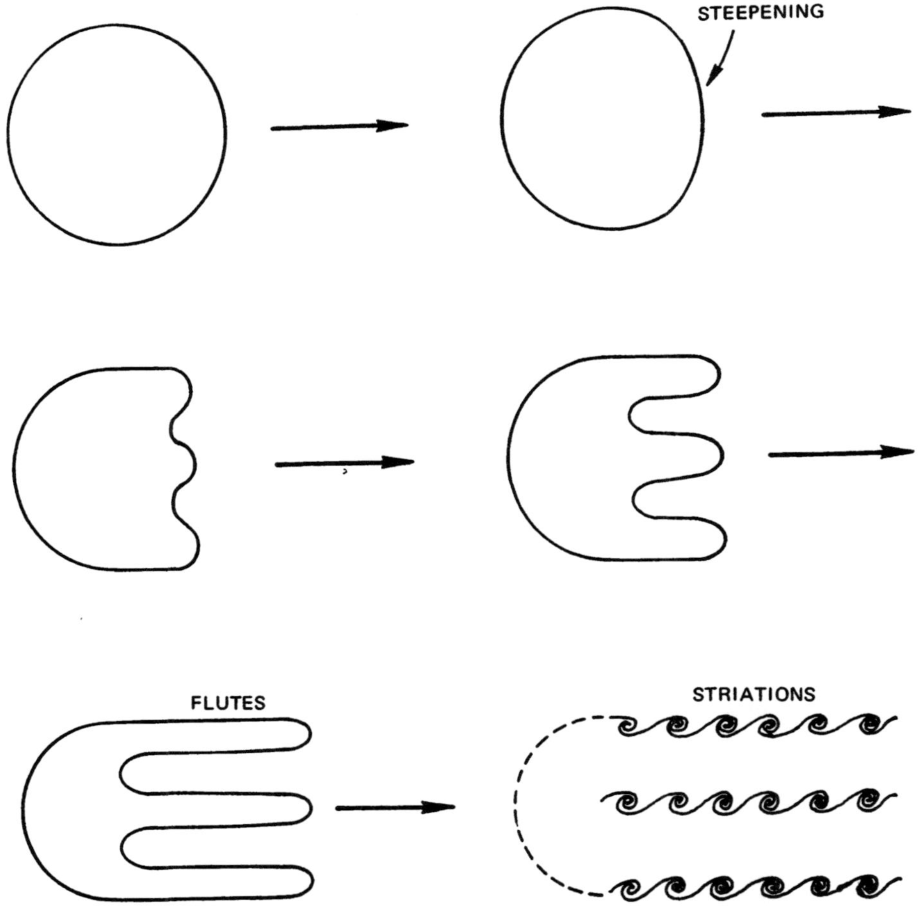

Fig. 1. Schematic illustration of the development of a large barium ion cloud as seen looking along the magnetic field. The development is from a field-aligned cigar shape to striated sheets. After Davis et al. [9].

Fig. 1 illustrates our present understanding of the typical morphological development of a Ba ion cloud. The development depicted in Fig. 1 is independent of the relative effect of $\boldsymbol{E}_\perp$ and $\boldsymbol{V}_n$ upon the cloud motion. That is, there appears to be no difference between the extreme cases (i) $\boldsymbol{E}_\perp \cong 0$ wherein one edge of the ion cloud remains near the release point and the opposite moves approximately in the direction of $\boldsymbol{V}_n$ and (ii) $\boldsymbol{E}_\perp > 0$, $\boldsymbol{V}_n \cong 0$ wherein the neutral cloud remain at the release point and the Ba ion cloud moves away under the influence of $\boldsymbol{E}_\perp$.

Starting from an initial cigar shape the ion cloud first becomes horizontally elongated in the direction of $\boldsymbol{V}_n$ if $\boldsymbol{V}_n \gg \boldsymbol{V}_\perp$, or in the direction of $\boldsymbol{V}_\perp$ if $\boldsymbol{V}_\perp \gg \boldsymbol{V}_n$. On the edge of the cloud nearest the Ba neutral cloud (the trailing edge in a frame fixed to the neutral wind) the ion density gradient steepens. Shortly thereafter sheets (or flutes) emerge from the trailing edge. These sheets extend in the plane defined approximately by the direction of the magnetic field $\boldsymbol{B}$ and the direction of ion motion $\boldsymbol{V}_\perp$. As many as 14 such sheets have been observed in large releases. Each is of order 100 m thick and they are spaced approximately 500 m to 1 km apart. These sheets often subsequently become convoluted into quasi-sinusoidal forms or develop rod-like structures within the individual sheets and particularly at their ends nearest the neutral cloud [10]. The onset of this "striation" can be very abrupt; in one case the onset appeared to develop fully in a time of order 1 second. Thus it seems likely that at least two types of instabilities occur to create the complex internal morphology of the barium ion cloud.

From study of the internal morphology and particularly of the distribution of λ^* within the cloud, one concludes that the velocity $\boldsymbol{V}_\perp$ of the leading edge (in the reference frame fixed to the neutral wind) is the proper portion of the cloud to observe to determine the ambient electric field $\boldsymbol{E}_\perp$.

5. Observed Motions of Barium Ion Clouds

5.1. Reference Coordinates

We attempt here to present observations of barium ion cloud motions made at various locations in the ionosphere on an orthographic projection using geomagnetic latitude and geomagnetic local time as coordinates. From one point of view, namely that of an observer on the rotating earth, this is somewhat awkward because of the transformation between co-rotating and non-rotating coordinate systems. A barium cloud stationary over the equator (seen from the ground) moves eastward at 460 m s^{-1} relative to the non-rotating coordinate system; at the auroral zone (dip latitude 67°) the difference is 180 m s^{-1}.

For phenomena strongly influenced by processes related primarily to the rotating earth (such as motions of the neutral atmosphere) the rotating reference frame is most useful. Similarly, phenomena strongly influenced by processes in the outer magnetosphere are most usefully represented in the fixed reference frame. We find that the motions of ionospheric plasma, as observed using barium clouds and other techniques, fall into both categories and therefore we present the barium ion motion data in two formats, Figs. 2 and 3. Fig. 2 has a fixed coordinate system, but the barium ion motions are shown with the co-rotation removed as though the motions were relative to the earth's surface. This presentation is somewhat improper in a strict sense. The motions are properly given in the fixed frame in Fig. 3.

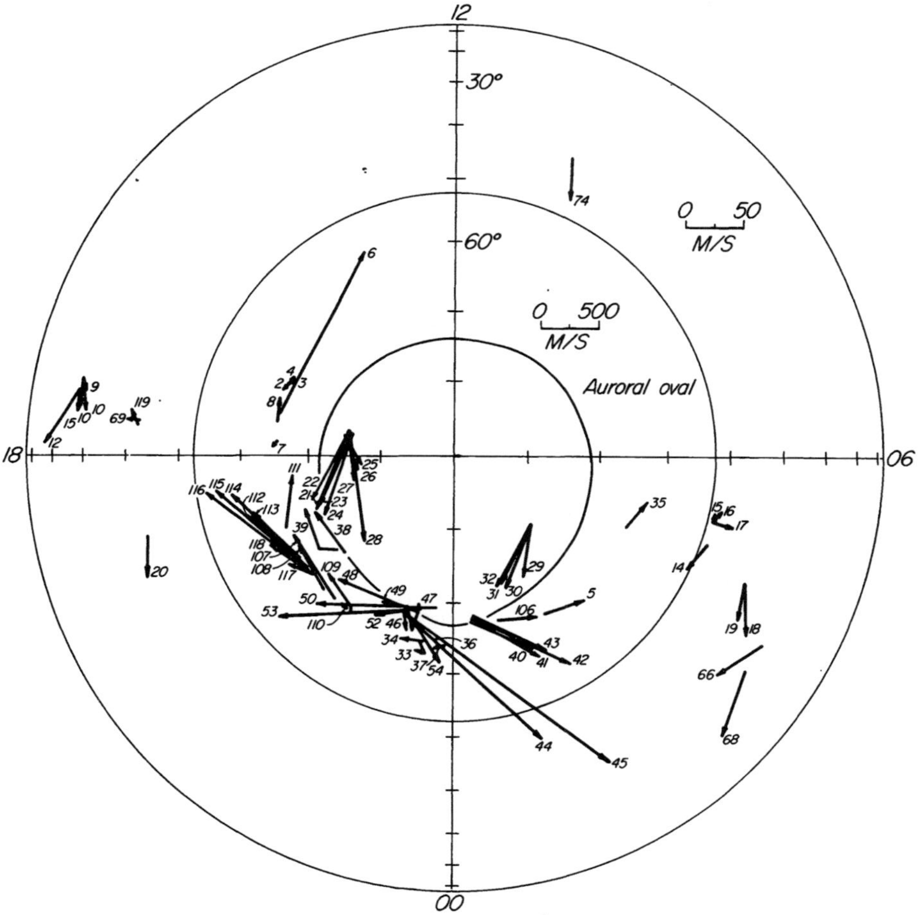

Fig. 2. Average ion cloud velocity vectors plotted relative to the rotating earth, in geomagnetic local time and geomagnetic latitude on an orthographic projection. The magnitude scale changes at low latitudes. Directions are measured from north increasing clock wise. The numbers near the ends of vectors refer to listings in Table 1.

5.2. Observed Ion Cloud Motions

Figs. 2 and 3 represent the majority, but not all, of motions measured using the barium release technique. Not all observations shown are reported in the literature. The available data on barium ion motions are primarily restricted to the twilight regions since sunlight is required for ionization and excitation and a relatively dark sky background is required for observation; thus the observations are as yet primarily limited to the morning and evening sectors at low and middle latitude, to the night side of the auroral zone and to the polar cap. By proper choice of season and release location it is possible to accumulate data in all local time sectors of the auroral zone and polar cap regions.

Speeds of barium ion clouds observed from the earth at the auroral zone range up to approximately 2000 m s^{-1} when the level of auroral activity is

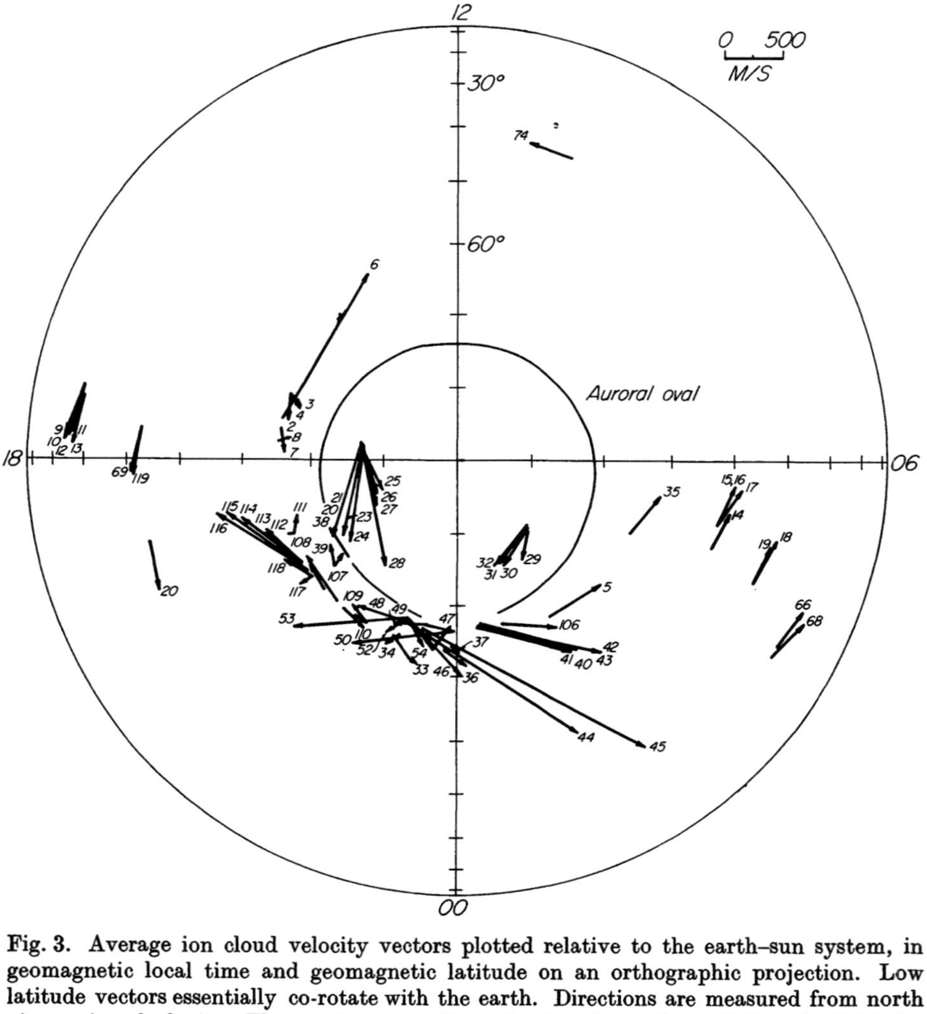

Fig. 3. Average ion cloud velocity vectors plotted relative to the earth–sun system, in geomagnetic local time and geomagnetic latitude on an orthographic projection. Low latitude vectors essentially co-rotate with the earth. Directions are measured from north increasing clockwise. The numbers near the ends of vectors refer to listings in Table 1.

high. At times of moderate activity speeds of several hundred m s^{-1} are common, and if the level of activity is very low speeds of several tens of m s^{-1} occur. At low and middle latitude the speed is of order 10 m s^{-1} indicating that the ions are essentially co-rotating with the earth. Ion motions are presented in Figs. 2 and 3 as vectors, the directions of which are measured from geomagnetic north as though the orthographic projection used preserves angles. Below geomagnetic latitude 52° on Fig. 2 the vector magnitude is plotted as being ten times the scale used at higher latitude. In cases where a cloud radically changed or reversed its direction of motion, bent vectors are shown; each leg of such a vector can itself be interpreted as a vector. From the published data available for some cloud motions we have been able to determine only approximate motion vectors. See also Table 1. At low and middle latitudes barium ion clouds move eastward

Table 1

List of barium releases giving the velocity of ion cloud motions plotted in Figs. 2 and 3 Velocities are relative to the rotating earth; directions are measured eastward from geomagnetic north.

Release number (Ref)	UT date	Time (UT)	Location	Weight (kg)	Alt. (km)	Velocity Azim. (deg)	Velocity Mag. ($m\,s^{-1}$)
1 [a]	5 Mar. 69	0430	Fairbanks	6	140	—	—
2 [a]	5 Mar. 69	0430	Fairbanks	2.4	170	109	61.6
3 [a]	6 Mar. 69	0430	Fairbanks	12	140	295	95.4
4 [a]	6 Mar. 69	0430	Fairbanks	12	170	289	90.3
5 [a]	11 Mar. 69	1409	Fairbanks	12	186	103	375
6 [a]	15 Mar. 69	0506	Fairbanks	Large	165	284	1604
7 [a]	19 Mar. 69	0517	Fairbanks	Large	168	104	52.8
8 [a]	20 Mar. 69	0537	Fairbanks	Large	176	270	216
9 [b]	27 Nov. 64	1740	Hammaguir	Small	152	67	14
10 [b]	30 Nov. 64	1740	Hammaguir	Small	160	72	28
11 [b]	30 Nov. 64	1740	Hammaguir	Small	200	85	30
12 [b]	18 Nov. 65	1745	Hammaguir	Small	135	113	58
13 [b]	19 Nov. 65	1745	Hammaguir	Small	195	85	16
14 [b]	24 Sep. 66	0930	Wallops	Small	200	290	26
15 [b]	25 Sep. 66	0951	Wallops	Small	216	99	6
16 [b]	25 Sep. 66	0951	Wallops	Small	270	115	12
17 [b]	25 Sep. 66	0951	Wallops	Small	?	182	18
18 [b]	16 Oct. 69	1916	Woomera	Small	240	247	46
19 [b]	16 Oct. 69	1916	Woomera	Small	290	261	34
20 [b]	18 Oct. 69	0945	Woomera	Small	240–300	105	34
21 [c]	8 Mar. 69	0320	Cape Parry	3	218	104	773
22 [c]	8 Mar. 69	0322	Cape Parry	3	274	105	696
23 [c]	8 Mar. 69	0323	Cape Parry	3	276	100	720
24 [c]	8 Mar. 69	0325	Cape Parry	3	189	96	747
25 [d]	9 Mar. 69	0332	Cape Parry	3	210	54	307
26 [d]	9 Mar. 69	0333	Cape Parry	3	263	64	305
27 [d]	9 Mar. 69	0335	Cape Parry	3	257	72	406
28 [d]	9 Mar. 69	0336	Cape Parry	3	168	72	915
29 [d]	7 Mar. 69	1314	Cape Parry	3	221	235	403
30 [d]	7 Mar. 69	1316	Cape Parry	3	286	250	525
31 [d]	7 Mar. 69	1317	Cape Parry	3	286	255	545
32 [d]	7 Mar. 69	1319	Cape Parry	3	210	257	297
33 [e]	7 Apr. 67	1957	Kiruna	2.6	230	146–280	115–90
34 [e]	8 Apr. 67	2003	Kiruna	2.6	230	266	200
35 [e]	9 Apr. 67	0100	Kiruna	2.6	230	108	290
36 [e]	10 Apr. 67	2014	Kiruna	2.6	230	142	120
37 [e]	11 Apr. 67	2021	Kiruna	2.6	230	255	74
38 [f]	17 Aug. 66	0345	Ft. Churchill	Small	258	~274	~445

Table 1 (con't)

Release number (Ref)	UT date	Time (UT)	Location	Weight (kg)	Alt. (km)	Velocity Azim. (deg)	Velocity Mag. (m s^{-1})
39 [f]	17 Aug. 66	0346	Ft. Churchill	Small	376–230	~220–295	~200–380
40 [g]	31 Aug. 67	2208	Andenes	3	230	120	585
41 [g]	31 Aug. 67	2209	Andenes	3	278	123	690
42 [g]	31 Aug. 67	2211	Andenes	3	310	120	950
43 [g]	31 Aug. 67	2213	Andenes	3	225	122	720
44 [g]	2 Sep. 67	2033	Andenes	3	230	113	1625
45 [g]	2 Sep. 67	2034	Andenes	3	290	108	2205
46 [g]	2 Sep. 67	2036	Andenes	3	320	132	535
47 [g]	2 Sep. 67	2039	Andenes	3	230	152–358	195–220
48 [g]	12 Sep. 67	2059	Andenes	3	230	278	760
49 [g]	12 Sep. 67	2100	Andenes	3	290	271	350
50 [g]	12 Sep. 67	2102	Andenes	3	320	78–266	129–1060
51 [g]	12 Sep. 67	2104	Andenes	3	—	No data	—
52 [c]	19 Sep. 68	1933	Andenes	3	207	249	250
53 [c]	19 Sep. 68	1934	Andenes	3	251	251	1110
54 [c]	19 Sep. 68	1936	Andenes	3	257	157	158
66 [h]	2 May 68	0918	Puerto Rico	Large	196	297	57
68 [h]	7 May 68	0911	Puerto Rico	Large	191	255	60
69 [i]	16 Jan. 71	2335	Eglin, Fla.	Large	150	187	10
74 [l]	2 Feb. 71	1634	Eglin, Fla.	Large	185	341	36
106 [f]	17 Apr. 69	2155	Andenes	Small	250	~100	~350
107 [f]	17 Mar. 69	1814	Kiruna	Small	255	~286	~650
108 [f]	15 Mar. 69	1759	Kiruna	Small	211	~284	~430
109 [f]	23 Mar. 68	1839	Kiruna	Small	200–210	~326–~290	~ 40–~360
110 [f]	20 Mar. 69	1827	Kiruna	Small	215	~ 89	~105
111 [f]	23 Oct. 67	1609	Kiruna	Small	217	~304–~295	~160–~330
112 [j]	15 Mar. 69	1755	Kiruna	Small	195	254	582
113 [j]	15 Feb. 69	1755	Kiruna	Small	235	257	574
114 [j]	17 Mar. 69	1810	Kiruna	Small	150	262	928
115 [j]	17 Mar. 69	1810	Kiruna	Small	197	258	1054
116 [j]	17 Mar. 69	1810	Kiruna	Small	235	254	1111
117 [j]	18 Mar. 69	1816	Kiruna	Small	160	240	197
118 [j]	18 Mar. 69	1816	Kiruna	Small	165	262	444
119 [k]	16 Jan. 67	2336	Eglin, Fla.	2.4	194	237	14.4

[a] H. C. Olsen, thesis, Univ. Alaska; [b] [27]; [c] E. M. Wescott et al., Particles and Fields in the Magnetosphere, Reidel, Dordrecht 1970 (p. 229); [d] [15]; [e] [11]; [f] [32]; [g] [13]; [h] T. Lumenello et al., Tech. Rep. B-4088, E. G. and G. Co., Bedford, Mass. 1969; [i] T. N. Davis et al., in preparation; [j] [30]; [k] [10]; [l] [20].

relative to the fixed reference frame (Fig. 3) at approximately the co-rotation velocity. Relative to the earth's surface (Fig. 2) the observations indicate a trend for the clouds to drift eastward and southward in evening and westward in morning. In the morning hours, no significant trend appears in the meridional component, but the observations are as yet quite sparse.

In the auroral zone the barium ion clouds tend to move primarily in the east–west direction approximately parallel to auroral forms. The motion is generally westward before midnight and eastward after. Near midnight both east and west motions are observed depending upon the stage within an auroral substorm if one is in progress when the barium release is made. At least a crude relationship exists between the direction of motion and the sign of horizontal magnetic disturbance ΔH [11]. If ΔH is positive, the ion cloud typically moves west, and if ΔH is negative, the ion cloud moves east. Although the motion is primarily longitudinal, ion clouds have been observed to move slowly across magnetic field lines occupied by auroral forms [12]. The speed of the ion cloud is generally less if the ion cloud is within or very near an auroral form than it is if the cloud is some tens of kilometers away [13, 14]. Owing to both temporal and spatial variations, barium ion motions tend to be considerably more complex at the auroral zone than in the middle latitudes and polar regions.

The few barium releases performed in the outer part of the polar cap produced ion clouds that move toward the midnight meridian (relative to the fixed coordinate system, Fig. 3), that is, in direction opposite to the motions observed at the auroral zone [15].

Plotted in Figs. 2 and 3 are observed motions of several ion clouds produced by large barium releases at altitudes ranging from 140 to 320 km. With the exception of vector numbers 66, 68 and 74, the motion vectors refer to the leading edge of the clouds, that is, the portion of the ion cloud which appears to have the lowest value of λ^* and hence the part most likely to move under the influence of an electric field rather than under the influence of the neutral wind: see Eq. (2)–(5) and the related discussion in § 3. The more dense portions of these large ion clouds move at higher speeds approximately in the direction of the neutral wind $\boldsymbol{V}_n$ with speeds approximately 0.5 V_n to 0.8 V_n. Vector No. 74 is included as it represents the only daytime observation using a technique involving a large barium cloud containing strontium to allow optical tracking of the neutral cloud and rf tracking of the ion cloud. Here the plotted vector refers to the most dense part of the ion cloud; it moved in the same direction as the neutral cloud at approximately half its speed.

6. Comparison with Other Measurements

At least on a statistical basis, the motions of barium ion clouds show general agreement with other types of observations allowing direct measurement or inference of the plasma motion or of the electric field to which it is related through the equation $\boldsymbol{V} = \boldsymbol{E} \times \boldsymbol{B}/B^2$.

Although Doppler shift and pulse-radar methods to measure the motions of radio auroras do not necessarily give the same results [6] it is generally found by these methods that primarily westward motions occur at the auroral zone in evening and eastward motions occur in morning. Using data obtained at Jodrell Bank, Bullough and Kaiser [16] found mean westward speeds of 600 m s^{-1} in evening and similar mean speeds in morning. Unwin [17] using data from Invercargill, New Zealand, obtained similar results. He concluded, as did Nichols [18] and Bullough et al. [19] that the observed motions correlated sufficiently well with auroral zone magnetic variations to suggest that the obser-ed motions were of the ionization that moves to create the auroral electrojets. Fig. 4, compiled from data given by Bullough [20] shows individual observations which show general agreement with Figs. 2 and 3 and also serve to give additional data in the polar cap and auroral zone regions.

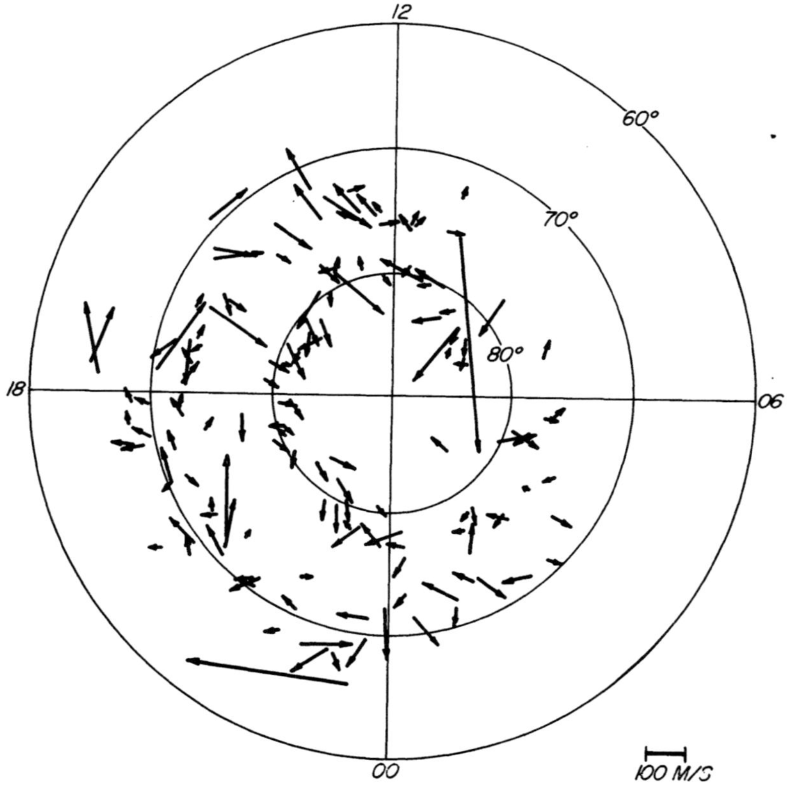

Fig. 4. Motions of radio aurora measured by Bullough [20] using the radio echo technique at 73 MHz. This diagram is compiled from diagrams representing data for the periods May 1957 to January 1958.

Through use of the assumption that the auroral electrojet currents are primarily Hall currents, Heppner [21] and Davis [22] inferred the directions of $\boldsymbol{E}$ in the ionosphere and mapped these to the equatorial plane to indicate the magnetospheric convection. Davis also mapped the directions of auroral motions for the same purpose using the assumption that the longitudinal auroral motions (motions parallel to auroral orientations) are given by $\boldsymbol{V} = \boldsymbol{E} \times \boldsymbol{B}/B^2$. Their results are compatible with the observed motions of barium ion clouds. In fact, the barium ion cloud motions observed in the vicinity of auroras helped provide confidence that the longitudinal auroral motions used by Davis are approximately expressed by the relation $\boldsymbol{V} = \boldsymbol{E} \times \boldsymbol{B}/B^2$. A statistical summary of auroral motions observed along and just inside the auroral zone is given in Fig. 5.

Balloon-borne probes have been employed by Mozer and co-workers [3] to measure ionospheric electric fields. This method has advantage in that it allows relatively long-term observations in one area and is not restricted by sunlight considerations. Rocket-borne probes allow measurement of $\boldsymbol{E}$ at ionospheric altitude but, owing to the short duration of flight, provide measurements at only one point in time. The balloon and rocket methods have yielded

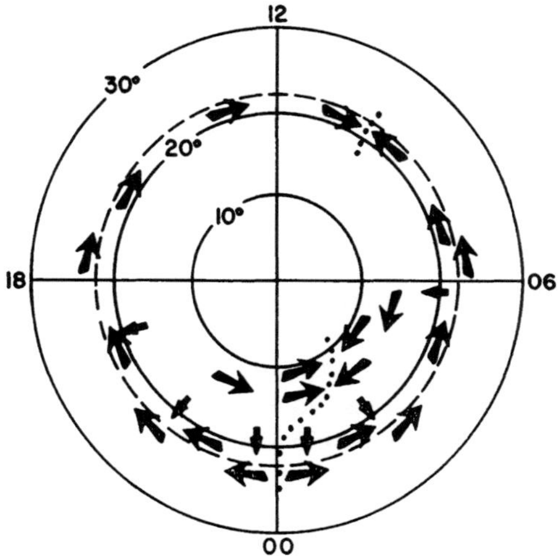

Fig. 5. Summary diagram of auroral motions in geomagnetic time determined from all-sky camera observations. After Davis [23].

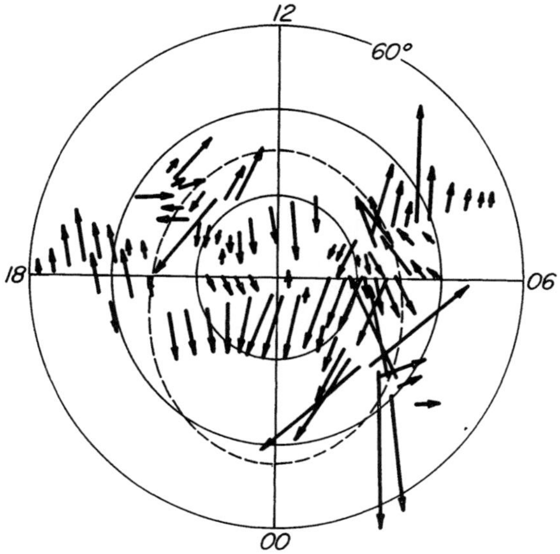

Fig. 6. Sketch drawing compiled from Injun 5 data [5] showing components of convection velocity sensed (not full vector direction) obtained on several polar passes of the satellite.

results generally compatible with those obtained from barium releases [3, 4].

Polar-orbiting satellite measurement of electric fields has the special advantage that the field can be measured along a path traversing the instantaneous auroral oval in a very short time, in fact, within a period of about 10 minutes the probe can traverse the auroral oval, cross the polar cap and traverse the oval on the other side. Of particular interest are the recent results of Cauffman and Gurnett [5] from the Injun 5 satellite. These show sharp reversals in the direction of electric field directions as the satellite traverses the auroral zone on individual passes, a finding in agreement with earlier observations of auroral motions by Stoffregen [24] and others [25, 26]. Though exceptions occur, the Injun 5 data show a plasma flow away from the sun in the polar cap (see Fig. 6). This result agrees with and extends the barium cloud observations of Heppner et al. [15]; also, compare Fig. 6 with Fig. 4.

7. Observations of Neutral Wind using Barium Releases

An important attribute of the barium release method is the ability to measure both ion and neutral motions. At low and middle latitudes, dynamo action of the neutral wind near altitude 115 km is considered to be the main agent causing transport of the ionospheric plasma which in turn causes the Sq current system [27]. Owing to the high conductivity along magnetic field lines, the polarization electric field set up by the dynamo pervades the inner magnetosphere and there partly determines the motion of the magnetospheric plasma [28]. Especially at low latitude, where $E_\perp$ is small, one can choose the altitude and quantity of barium released to alter the parameters λ^* and K in Eq. (2) and thereby experimentally measure each of the terms in that equation. In essence, one can use a small barium release to obtain $\lambda^* \sim 1$ and thereby perform observation on the ambient conditions. Alternatively increase of λ^* allows study of perturbation of the ambient environment.

At auroral and polar latitudes, the magnetospheric electric field rather than the ionospheric wind appears to be the dominating factor in determining the motion of the ionospheric plasma. In fact, new results obtained using the barium release technique demonstrate that the high altitude neutral wind is considerably affected at times of geomagnetic activity by ion drag and perhaps direct heating. Barium releases in Alaska [29] and Sweden [30] at altitudes 140 km and above show the neutral wind velocity to be strongly correlated to ion cloud velocity and to the direction and magnitude of magnetic disturbance. Rees [30] obtained the relationship $V_{ny} = 2.5\,\Delta X$ where V_{ny} is the east component of the neutral wind velocity in m s^{-1} and ΔX is the 2-hour mean of the north component of the preceding magnetic disturbance in gammas. Associated with Release No. 6 (Table 1) Romick et al. [29] observed a remarkable shearing of the barium neutral cloud. Extending from 160 km to 175 km in altitude, the neutral cloud evolved into a cigar shape, approximately 300 km long. It was aligned in the magnetic east–west direction, along which the barium ion cloud moved westward at a maximum speed exceeding 2 km s^{-1}. In another case a neutral wind speed of 800 m s^{-1} was observed during an auroral disturbance [31].

8. Summary and Discussion

The introduction of the barium release technique by Professor Lüst's group at the Max-Planck-Institut für Physik und Astrophysik is the most significant experimental development of the past few years in the area of ionospheric and magnetospheric physics. Of the various methods available to measure plasma motions and ionospheric and magnetospheric electric fields, the barium release method is the most sensitive and direct. Also, the application of this method has allowed greater confidence in the interpretation of such other indicators of plasma motion and electric fields as auroral motions, magnetic disturbance variations, motions of radio aurora and direct probe measurements. Well over a hundred barium releases have been made in the ionosphere as has one in the distant magnetosphere [32]. As more of the results from these releases become available and as new releases are performed, the spatial and temporal patterns of electric fields and of the ionized and neutral atmosphere motions will become better known.

With the data now available from barium releases and other observations it is possible to be fairly confident of speculations or interpretations regarding ionospheric and magnetospheric motions. Based upon the data of the type reviewed here and the ideas of the authors referenced, we attempt to summarize and interpret as follows.

At latitudes below the auroral ovals, dynamo action (Sq and to a much lesser extent, lunar variations) is the primary cause of horizontal plasma motion in the ionosphere, at least during geomagnetically undisturbed periods. Polarization electric fields generated by the dynamo are transmitted on highly conducting magnetic field lines to the inner magnetosphere to affect plasma motion there. This portion of the ionosphere–magnetosphere system co-rotates with the earth in the sense that the important motions are best described relative to the rotating earth. Owing to the relatively low speed of these motions, they transform into the fixed frame (Fig. 3) to give eastward motion essentially at the speed of co-rotation. Relative to the observer, the mid-latitude ion motion of small, high altitude barium releases is near 10–20 m s^{-1}, indicating electric field strengths of 0.5–1.0 mV m^{-1}. Results obtained thus far indicate a pattern of eastward and southward motion in evening and, in morning, westward motion. The few results available indicate considerable variability, evidently as a result of magnetic activity, and so interpretation is not simple. A factor complicating interpretation is perturbation by the injection of the barium plasma, especially if other than the very smallest release is employed.

Barium ion cloud speeds at the auroral zone during quiet times are similar to those at mid-latitude; most auroral zone releases are made under active conditions. Then, speeds of several hundred m s^{-1} are typical; maximum speeds are in excess of 2000 m s^{-1}. At the auroral oval and in the polar cap, the ion cloud motions are governed by the magnetospheric electric field mapped down to the ionosphere along the highly conducting geomagnetic field. Owing to the generally high speeds of plasma motions only slight differences in detail appear between the representations given in the co-rotating (Fig. 2) and the fixed (Fig. 3) reference frames. However, the fixed frame clearly is the more useful representation. In this sense we can think of this portion of the ionosphere–magnetosphere system as the non-co-rotating part. The boundary between the co-rotating and fixed portions of the system probably corresponds approximately to the equatorward boundary of the instantaneous auroral oval and hence its location depends upon the level of geomagnetic activity.

The statistical pattern of motion at the auroral zone is eastward between geomagnetic local hour 23 and approximately 10 hours and westward at other local hours. As a means to refer to individual cases and to relate to the magnetic substorm it is also correct to state that at the auroral zone the motion is westward when ΔH is positive and is eastward when ΔH is negative. Injun 5 passes, primarily approximately along the dawn–dusk meridian, show a sharp reversal in the direction of the plasma motion near the auroral oval [5], and this observation is in agreement with observed auroral motions [24]. Hence, in the pre-midnight sector the motion is eastward poleward of the auroral zone and in the post-midnight sector it is westward. The few barium releases in the polar cap show the motion to be away from the sun. That this pattern is general over the whole polar cap is indicated by the radar aurora motions shown in Fig. 4 and the Injun 5 results shown in Fig. 6.

The overall pattern of motions in the auroral and polar regions as outlined above is essentially that deduced a decade ago by Axford and Hines [33] on the basis of magnetic disturbance and auroral motion observations available at the time. The barium release technique and other new methods of measurement have placed the interpretations of Axford and Hines and others on a much firmer basis.

The barium release technique, as well as being the most sensitive means to measure plasma motions, is extremely versatile. Neutral gas motion as well as plasma motion can be measured. The barium plasma can be used as a tracer or as a perturbing agent in the ionosphere and in the magnetosphere. Beyond the plasmapause, the barium release method is the most promising means of measuring magnetospheric plasma motions. Its use at the auroral zone has shown conclusively that the Hall conductivity dominates. As a tracer and a perturbing agent, the barium release method allows direct exploration of instabilities and irregularities in wall-free plasmas. Currently, methods are being explored to generate barium plasmas with velocities near 10 km s^{-1} to allow the plotting of geomagnetic field lines over their whole length and thereby provide better knowledge of the geomagnetic field configuration [34, 35]. Brice [36] has suggested that a cold plasma injection (using barium or other material) in the equatorial plane near $L = 6$ may trigger auroral acceleration processes.

Acknowledgments

This manuscript was prepared utilizing support from Air Force Rome Contract F-30602-70-C-0179 and Atmospheric Sciences Section, National Science Foundation NSF Grant GA-28079.

References

[1] L. Haser in: Aurora and Airglow, Reinhold Publ. Corp., New York 1967 (p. 391).
[2] G. Haerendel and R. Lüst, Scient. Amer. **81**, 80 (1968).
[3] F. S. Mozer and R. Serlin, J. Geophys. Res. **74**, 4739 (1969).
[4] T. L. Aggson in: Atmospheric Emissions, Van Nostrand-Reinhold, Co., New York 1969 (p. 305).
[5] D. P. Cauffman and D. A. Gurnett, University of Iowa report 71:4, Iowa City, Iowa (1971).
[6] B. Hultqvist and A. Egeland, Space Sci. Rev. **3**, 27 (1964).
[7] V. Degen, N. B. Brown and G. J. Romick, Planet. Space Sci. **19**, December 1971.
[8] G. Haerendel, R. Lüst and E. Rieger, Planet. Space Sci. **15**, 1 (1967).
[9] T. N. Davis, G. J. Romick and E. M. Wescott, RADC TR 70-216, Stanford Research Institute **4**, 141 (1970).
[10] N. W. Rosenberg, J. Geophys. Res. **76**, 6976 (1971).
[11] H. Föppl et al., J. Geophys. Res. **73**, 21 (1968).
[12] E. M. Wescott et al., report in preparation.
[13] E. M. Wescott, J. D. Stolarik and J. P. Heppner, J. Geophys. Res. **74**, 3469 (1969).
[14] E. M. Wescott (private communication).
[15] J. P. Heppner, J. D. Stolarik and E. M. Wescott, J. Geophys. Res. **76**, 6028 (1971).
[16] K. Bullough and T. R. Kaiser, J. Atmos. Terr. Phys. **6**, 198 (1955).
[17] R. S. Unwin, Nature, Lond. **183**, 1044 (1959).
[18] B. Nichols, J. Atmos. Terr. Phys. **11**, 292 (1957).
[19] K. Bullough et al., J. Atmos. Terr. Phys. **11**, 237 (1957).
[20] K. Bullough, Annls Géophys. **17**, 195 (1961).
[21] J. P. Heppner, in: Atmospheric Emissions, Van Nostrand-Reinhold Co., New York 1969.
[22] T. N. Davis, J. Geophys. Res. **76**, 5978 (1971).
[23] T. N. Davis, in: Introduction to Space Science, Gordon and Breach, New York 1965 (p. 205).
[24] W. Stoffregen, J. Atmos. Terr. Phys. **21**, 257 (1961).
[25] T. N. Davis, J. Geophys. Res. **67**, 75 (1962).
[26] T. N. Davis and R. N. DeWitt, J. Geophys. Res. **68**, 6237 (1963).
[27] E. Rieger, Max-Planck-Inst. Rep., Sept. 1970.
[28] S. Matsushita and J. D. Tarpley, J. Geophys. Res. **75**, 5433 (1970).
[29] G. J. Romick et al., Final Rep. DASA Contract 01-69-C-0158, Oct. 1970.
[30] D. Rees, J. Brit. Interplan. Soc. **24**, 233 (1971).
[31] J. W. Meriwether, Jr. et al., presented at 52nd Annual meeting of the AGU, Washington, D.C., April 1971.
[32] G. Haerendel and R. Lüst, in: Particles and Fields in the Magnetosphere, D. Reidel Publ. Co., Dordrecht 1970 (p. 213).
[33] W. I. Axford and C. O. Hines, Canad. J. Phys. **39**, 1433 (1961).
[34] K. W. Michel, presented at 52nd Annual meeting of the AGU, Washington, D.C., April 1971.
[35] E. M. Wescott, (presented at 52nd Annual meeting of the AGU, Washington, D.C., April 1971.
[36] N. Brice, J. Geophys. Res. **75**, 4890 (1970).

Space Research XII — Akademie-Verlag, Berlin 1972

INCOHERENT SCATTER OBSERVATIONS OF MERIDIONAL WINDS IN THE 150–225 KM REGION

C. A. REDDY* and G. VASSEUR

C. N. E. T. — R. S. R., Issy-les-Moulineaux, France

The incoherent scatter facility at Saint-Santin-Nançay is capable of measuring the ionization drifts in a direction nearly parallel to the geomagnetic field. The measured drifts in the 150–225 km region can be interpreted mainly in terms of the meridional neutral winds, since the diffusion velocities in this region are very small compared with the observed velocities (which are a few tens of metres per second). The analysis of the daytime drifts reveals the following features: (i) the semidiurnal wind component, which is the predominant one below ~135 km, becomes less important above 150 km compared with the diurnal and other components; (ii) there is a significant seasonal variation in the magnitude of the observed velocities; (iii) quasi-periodic variations of 2–3 hours period are persistently present on days of moderate or high magnetic activity; (iv) apart from the quasi-periodic fluctuations, the average daytime velocities are appreciably altered on magnetically disturbed days compared with those on quiet days.

1. Introduction

The observational knowledge of the daytime thermospheric winds above 120 km is very limited. The incoherent scatter technique has recently emerged as an important tool for deducing the neutral winds in the upper atmosphere from the measured ion drift velocities. Using the incoherent scatter facility of Saint-Santin-Nançay it is possible to measure one component of the ion velocity [1]. The geometry of the equipment is such that in the 150–225 km region the direction of this measured component is within 6° of the geomagnetic field direction.

As discussed earlier [2], the ion velocities parallel to the magnetic field are caused, in the F region, mainly by the neutral wind component in the magnetic meridian plane and by the ambipolar diffusion. The diffusion velocities become increasingly important at altitudes above 200 km, and these must be evaluated for obtaining reliable information about the neutral winds. However, in the 150–200 km region, diffusion velocities are small and negligible in comparison with the wind-induced velocities. Therefore the observed ion velocities in this region can be interpreted directly in terms of neutral winds in the magnetic meridian plane.

In this paper we present the most pertinent features of the observed ionization drifts in the 150–225 km height range. More detailed investigations of

* On leave of absence from Space Physics Division, S. S. T. C., Trivandrum, India.

some of the features presented here are in progress and the results will be reported elsewhere.

2. Results

The results presented in this section show the time variations of $V_{\parallel z}$ which is the projection upon the vertical of the observed ion drift $V_{\parallel}$ along the direction of the magnetic field. Thus,

$$V_{\parallel z} = V_{\parallel} \sin I \, .$$

I is the magnetic dip angle ($\sim 61°$ at Saint-Santin). It is to be noted specifically that $V_{\parallel z}$ is *not* a measured vertical velocity.

2.1. Ion Drifts on Quiet Days

The daytime variations of the ion drifts are characterized by three important features: (a) the frequent presence of short period fluctuations with quasi-

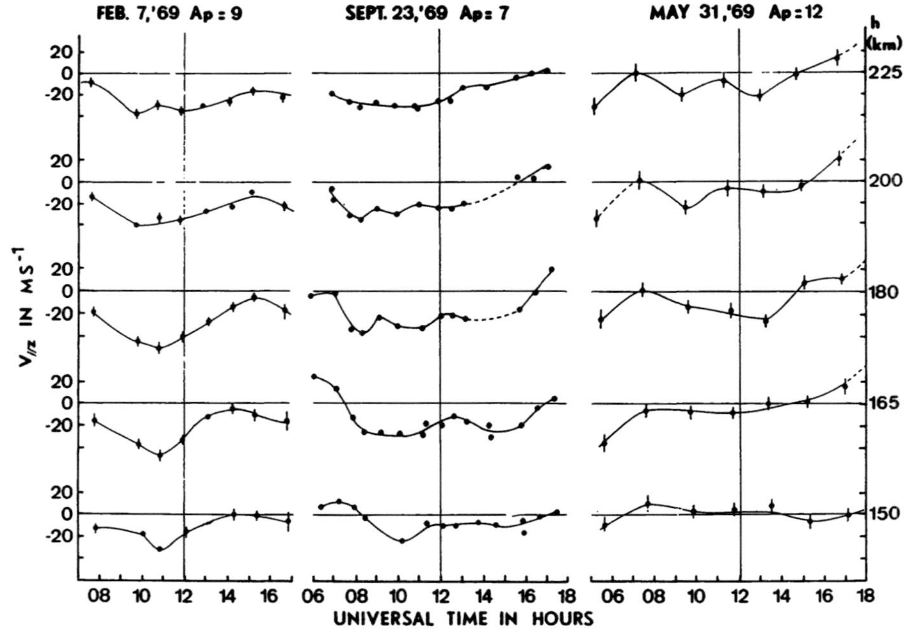

Fig. 1. Typical quiet day variations of the vertical component ($V_{\parallel z}$) of the observed ion velocities on a winter day, an equinoxial day and a summer day.

periods of 1–3 hours, (b) a general decrease of amplitude from winter to summer, (c) the consistency of the general daytime variation on different days.

The occurrence of short period fluctuations even on quiet days can be seen in Figs. 1 and 2. These fluctuations do not occur as a train of waves with a single dominant period; rather they occur commonly as a single disturbance persisting for the duration of one or two quasi-periods only. Any quantitative analysis of these disturbances is made extremely difficult by two factors: insufficient time resolution, and more important, the unknown nature of the longer period (24 hr, 12 hr, etc.) variations which have to be subtracted to obtain these shorter period fluctuations.

The decrease of the overall magnitude of the drifts from winter to summer is illustrated in Fig. 1. The change from winter to equinoxial months is rather small. The semidiurnal tide, which is the predominant wind component up to about 135 km [3], is still clearly evident in the 150–200 km region in the winter months only. In other seasons it is obscured by other wind components of

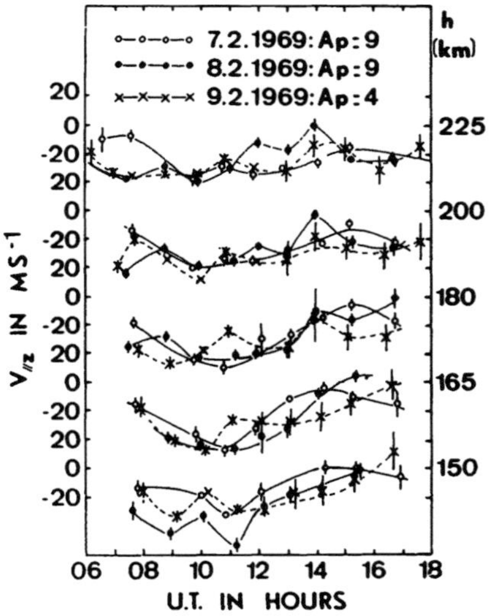

Fig. 2. Ionization drifts on three consecutive quiet days showing a consistency of the general pattern of daytime variations and some differences over short periods.

comparable or greater amplitude. The diurnal component seems to increase with altitude to become the dominant wind component above 225 km.

The repetition of the general pattern of the daytime variation on consecutive quiet days is illustrated in Fig. 2. The small but significant day to day changes over a limited duration are obviously caused by the short-period fluctuations described earlier. Large day to day variability in the general pattern of daytime variations is not uncommon at altitudes below 135 km, but it seems to be rare (on quiet days) for the 150–200 km region.

2.2. Ion Drifts on Disturbed Days

When the magnetic activity reaches even a moderate level, as indicated by *Ap* values of about 13–20, wave-like perturbations show up invariably in this altitude range (150–225 km). The apparent periods are mostly in the 1–3 hours range. However, shorter period perturbations can be missed in the observations which have a typical time resolution of about one hour. The

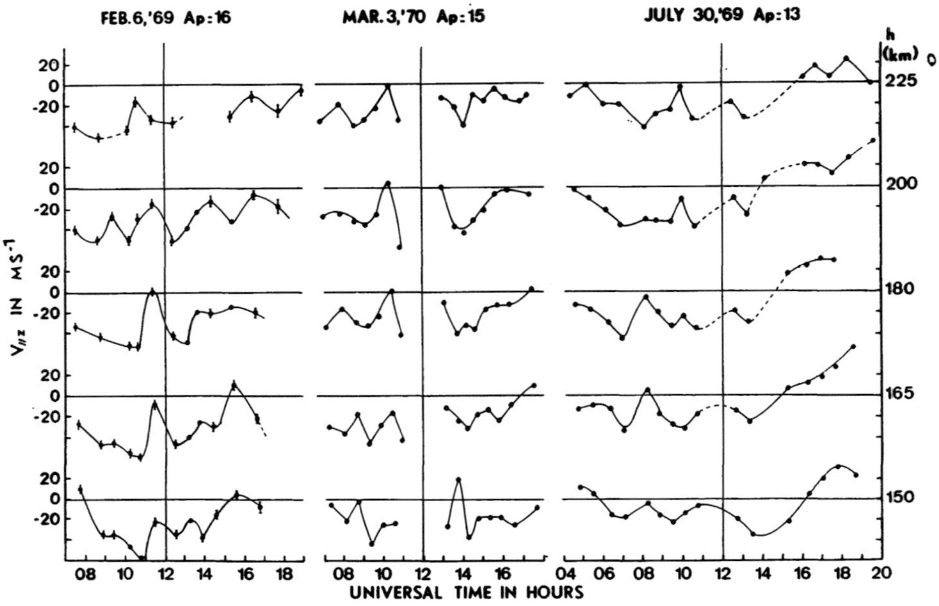

Fig. 3. Three cases illustrating the persistence of short period disturbances on moderately disturbed days.

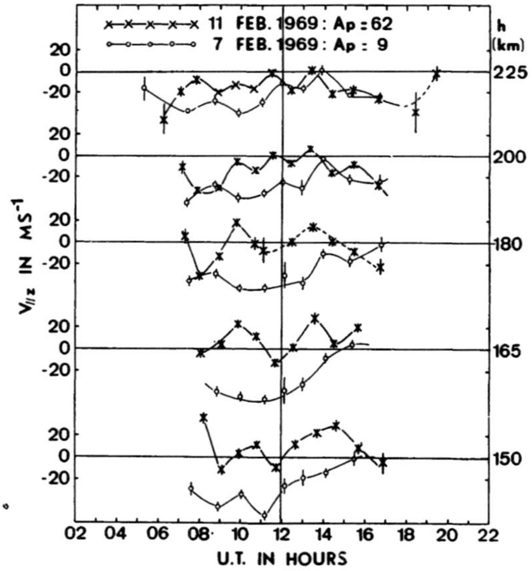

Fig. 4. Daytime ionization drifts on a quiet day and a disturbed day.

perturbations have a larger amplitude in general and they persist throughout daytime on these moderately disturbed days in contrast to those on quiet days. Fig. 3 shows a few cases of the wave activity on moderately disturbed days. Apart from these shorter period disturbances, the general pattern of daytime variation itself seems to undergo some change on these moderately disturbed days, at least in a few cases. But the evidence is not unambiguous and no conclusion can be drawn without further study.

What is remarkably clear even in the few cases studied is that large changes do occur in the general daytime pattern on disturbed days with *Ap* values of 50 or more; one example is shown in Fig. 4. The generally poleward daytime drift observed on quiet days (with an amplitude of 25–35 m s^{-1}) is reduced to near zero value or even reverses to a small equatorward drift on disturbed days. One interesting feature of this magnetic activity effect is that the change seems to be largest in the 150–165 km height range, with a decrease at higher and lower altitudes. A detailed analysis of this magnetic activity effect is in progress and the results will be reported elsewhere.

3. Discussion

As stated in § 1, the observed ionization drifts in the 150–200 km region may be attributed to the neutral wind component in the magnetic meridian plane. During daytime the ambipolar diffusion velocities do not exceed a few m s^{-1}, whereas the observed $V_{\parallel z}$ is in the range of some tens of m s^{-1}. Using this interpretation, the values of $V_{\parallel z}$ can be immediately converted to horizontal meridional wind U_n by the relation:

$$U_n = \frac{V_{\parallel z}}{\sin I \cos I} \sim 2.2\ V_{\parallel z}\,.$$

U_n is taken as positive when the wind is directed equatorward. One important feature of the observed winds is the frequent presence of short-period waves in addition to the winds of tidal periods. The short-period fluctuations are irregular in their nature and are probably due to the propagation of gravity waves; on the contrary, the longer period daytime variation appears to be more regular. This regular part of the wind apparently consists of the diurnal and semidiurnal tides and probably of other shorter period tidal components (such as the 8 hours component).

The observed presence of the shorter period gravity waves has an important implication for the interpretation of vertical wind profiles as obtained by rocket measurements. Even on moderately disturbed days the amplitude of these waves reaches a value comparable with that of the tidal winds. The vertical wavelengths of the gravity waves and of some tidal components can be of comparable magnitude. Therefore care is needed in interpreting rocket measured wind profiles which are spaced a few hours apart. In the lower region of the thermosphere (i.e. below 135 km) the observed winds mainly consist of semidiurnal tidal components [3–5]. In the upper thermosphere (above 250 km) the diurnally varying wind is the dominant one [2–6]. As is to be expected, in this transition region of 150–225 km the relative importance of the semidiurnal component in relation to the diurnal appears to decrease with altitude. The quantitative estimate of the various tidal winds is made very difficult by the limited duration of the observations (daytime only) and by the presence of the shorter period disturbances. An attempt is being made, however, to extract the semidiurnal wind component and its height variation.

Acknowledgments

We are indebted to Dr. P. Waldteufel, for helpful discussions and comments, and to Drs. A. Spizzichino, I. Revah, P. Amayenc and J. Testud for the benefit of useful discussions.

References

[1] H. Carru et al., Annls Géophys. **23**, 455 (1967).
[2] G. Vasseur, J. Atmos. Terr. Phys. **31**, 397 (1969).
[3] R. Bernard and A. Spizzichino, J. Atmos. Terr. Phys. **33**, 1345 (1971).
[4] J. S. Greenhow and E. L. Neufeld, Phil. Mag. **1**, 1157 (1956).
[5] A. Spizzichino, Annls Géophys. **25**, 693 (1969).
[6] H. Kohl, Space Research X, 550 (1970).

Space Research XII — Akademie-Verlag, Berlin 1972

OBSERVATION AND INTERPRETATION OF IONIZATION DRIFT MEASUREMENTS IN THE F REGION AT ST-SANTIN-NANÇAY

P. Amayenc and G. Vasseur

C. N. E. T. — R. S. R., Issy-les-Moulineaux, France

Ionization drifts observed at St-Santin-Nançay are presented. Altitude variation, diurnal and seasonal behaviour are obtained, and it is shown that neutral air winds may be obtained from the observations. These neutral air winds are compared with a theoretical model of thermospheric circulation due to horizontal pressure gradients. A general agreement is found; however, electric field induced motions, not observed with the existing equipment, have to be introduced during various circumstances.

1. Introduction

Whereas the concentrations and temperatures of the atmosphere and of the ionosphere have been extensively studied using a great variety of observational techniques, the dynamics of the thermosphere and ionosphere is still poorly known. Most of the earlier experimental techniques used are either difficult to interpret (e.g. the "fading method") or very limited in time coverage (for example, the artificial cloud release by rockets). On the contrary, the incoherent scatter technique gives a continuous and unambiguous determination of the ion drift velocity as a function of altitude and time. The measurement of the ion drift is deduced from a slight change in the frequency of the returned signal. The direction of the measured velocity component depends upon the geometry of the equipment.

From 1966, the incoherent scatter facility of Saint-Santin-Nançay (France) has been used to measure one component of the ion drift in the ionosphere. The geometry of the equipment is such that the measured component is very close to the direction of the magnetic field. This paper is concerned with the description and interpretation of the ion drifts observed in the F region above 200 km. In previous papers other observations have been presented and discussed in the light of the theory of thermospheric tides [1, 2], the theory of the formation of the F layer [3, 4] and the theory of atmospheric gravity waves for short period phenomena [5, 6].

In § 2 of this paper the results of ion drift measurement, over a three-year period, are presented and the salient features are pointed out. Some new features are found and the mean behaviour of the observed drift is traced. § 3 is concerned with the interpretation of the observed phenomena. The influence of neutral winds, electric fields and ambipolar diffusion is discussed.

2. Experimental Results

As stated before, the measured component lies nearly parallel to the earth's magnetic field. However, the velocity V considered here is the projection upon the vertical of the observed velocity. The altitude and time evolution of V is now given using various presentations of the results.

2.1. Some Examples of Diurnal Variation

In Fig. 1 diurnal variations for three particular observational periods are shown. Such diurnal variations are obtained from a running average of V values observed in a given altitude range (225–350 km) and a given time interval

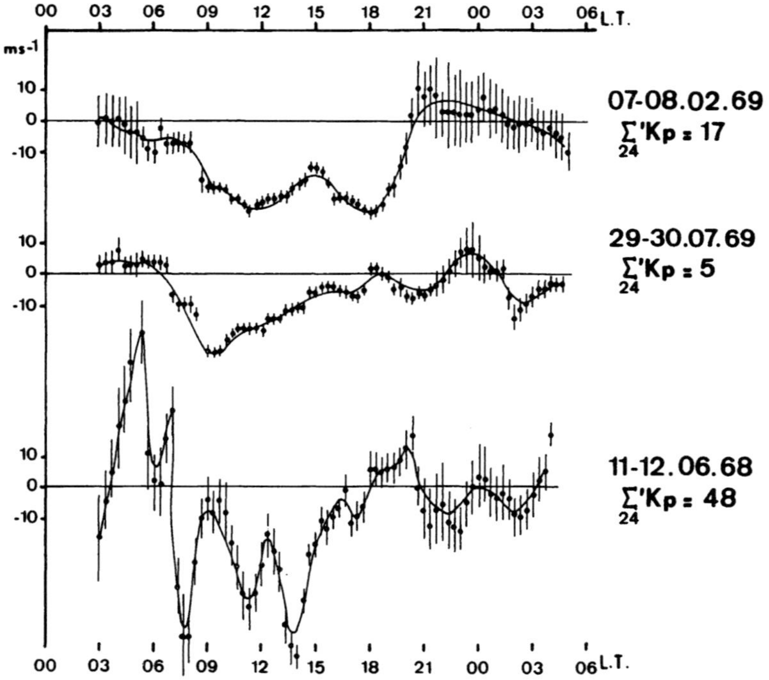

Fig. 1. Diurnal variation of smoothed velocities in the F region. Each point and its error bar represents the running mean obtained between 225 and 350 km altitude during one hour period around the printed time. The two top curves correspond to quiet days in winter and summer and the third curve is an example of results obtained for a disturbed day.

(1 hour). The two upper diurnal variations (7–8 Feb. 1969 and 29–30 July 1969) correspond to moderate low magnetic activity; the wind behaviour is then quite smooth. During the day, V is negative (downwards) reaching values of -20 to -30 m s^{-1} whereas during the night V lies around zero. There is a clear amplitude and phase difference between these two curves which will be discussed later (seasonal effect). The bottom curve corresponds to a day with high

magnetic activity. The time variation exhibits strong oscillations with a few hours period. The presence of such oscillations appears to be very closely related to magnetic activity; such phenomena will not be discussed further here.

2.2. Sunset and Sunrise Effect

Strong vertical fluxes of ionization at sunrise and sunset were detected in the upper part of the F2 layer at Millstone Hill [7]. Similar phenomena can be obtained using the Saint-Santin-Nançay observations (Fig. 2). Above 400 km a large upward velocity is noted at sunrise whereas there is a strong downward velocity peak after sunset. The time variation of V at various altitudes shows that the amplitude of the peaks increases strongly with altitude in agreement with the results of Evans et al. [7].

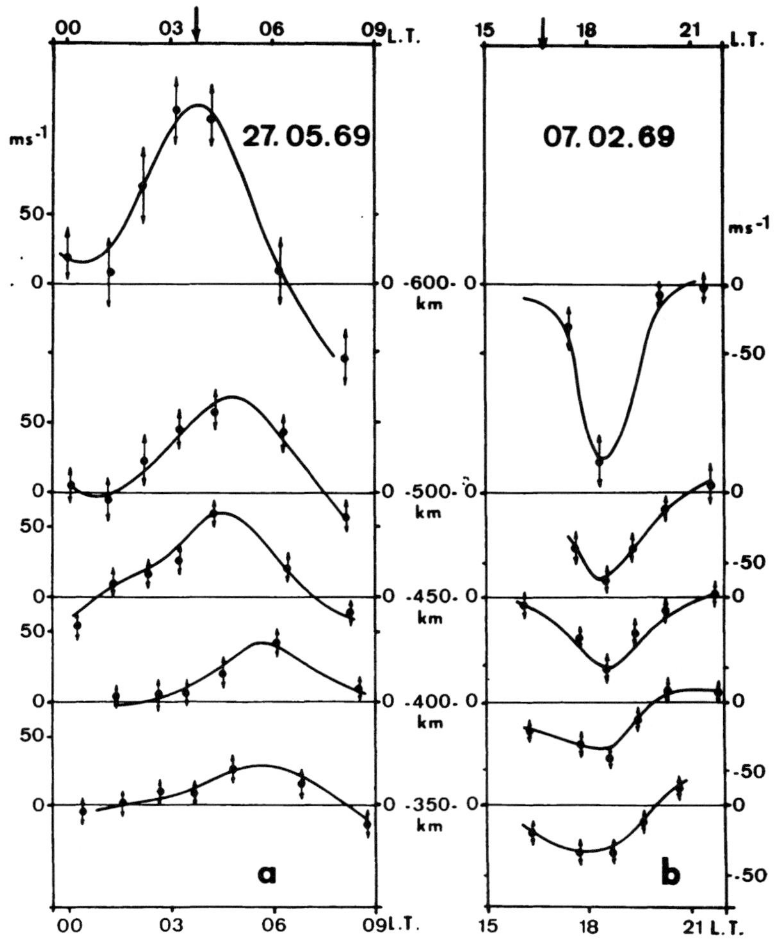

Fig. 2. Two examples of temporal variation of V for different altitudes: a, 27 May 1969 at sunrise: the sunrise time at the ground level is indicated by an arrow; b, 7 February 1969 at sunset: the sunset time at the ground level is indicated by an arrow.

2.3. Altitude Variation

Except during sunrise and sunset, the altitude variation of the observed ion drift is quite small. Vertical gradients have however been computed on a statistical basis. During the night it was not possible to draw any significant conclusion about vertical gradients because the error bars are then too strong. During the day (0900 to 1500 LT) significant values of the vertical gradient of V are obtained between 200 and 400 km. The value of $\partial V/\partial z$ for two periods of the year are shown in Table 1.

Table 1
Results of 1967–1969

Epoch of the year	$\partial V/\partial z$ (m s^{-1} km^{-1})
April–September	$+0.028 \pm 0.008$
October–March	$+0.052 \pm 0.005$
Whole year	$+0.043 \pm 0.004$

Since V is negative during daytime, such a positive gradient would indicate that V reverses at some particular level. When extrapolating the values observed, it is found that above 650 km V becomes positive. This result has to be compared with measurements at Millstone Hill where such a change was found to occur lower, around 450 km.

2.4. Seasonal Effect

In Fig. 3 the mean diurnal variations of V over a three-year period are obtained for five altitude ranges in the F region and for each season. On these curves it is possible to see the main features previously observed. The comparison of results for several reasons shows a strong seasonal variation. During winter, V remains low during the whole day whereas during summer a minimum appears before noon and V increases very early in the afternoon. Spring and autumn behaviours appear to be intermediate between summer and winter conditions.

2.5. Magnetic Activity Effects

On some occasions, observations were made for successive days with different magnetic activities. It is interesting to look then for the difference of the behaviour of V. This difference may be obtained from the steady component, i.e. the 24-hour mean value of V computed in order to smooth out the dominant 24 hours component and the smaller period fluctuations. In Fig. 4 it can be seen that this steady component is related to the magnetic activity: the algebraic value increases with increasing magnetic activity.

3. Interpretation

The previous results are now discussed and interpreted. It is shown that the observed drifts result from ambipolar diffusion effects and neutral wind drag, and these two effects are separated; the values of neutral winds are obtained and compared with a theoretical model.

3.1. Theoretical Expression for V

Using the hypothesis of ambipolar diffusion it can be shown [1] that the observed ion drift velocity may be written as:

$$V = W_n + W_d\,; \tag{1}$$

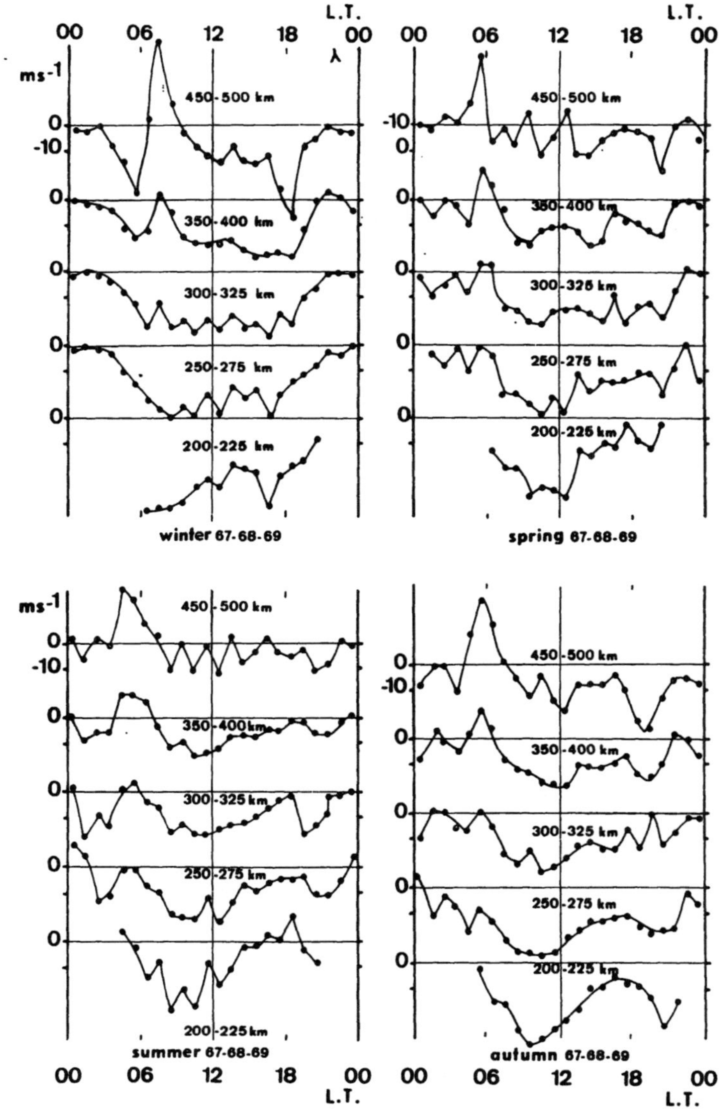

Fig. 3. Mean value of winds obtained from 1967 to 1969. For each of the four seasons five diurnal variations are drawn corresponding to five altitude ranges in the F region. Each point is a mean of measured V in the altitude range and one hour interval around the point considered.

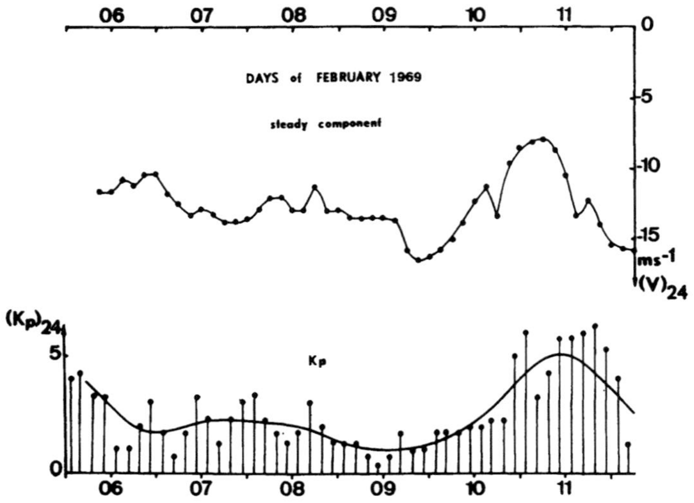

Fig. 4. Plots against time, for some days in February 1969, of: top, the prevailing component of V calculated as:

$$\frac{1}{24}\int_{t-12}^{t+12} V\,dt$$

bottom, the corresponding value of Kp (from "Solar and Geophysical Data"). The continuous curve is a 24-hour running mean, determined as for V.

W_n is the drag of ionization by neutral air wind and W_d is the ambipolar diffusion velocity. Supposing this neutral wind to be horizontal, then W_n is related to V_n by

$$W_n = V_n \sin I \cos I \tag{2}$$

where V_n is the meridional component of neutral wind and I the magnetic dip angle; W_d is given by

$$W_d = -D_a \sin^2 I \left[\frac{\partial[N(T_e + T_i)]}{2NT_i\,\partial z} + \frac{m_i g}{2kT_i}\right] \tag{3}$$

where D_a is the ambipolar diffusion coefficient, N the electron concentration, $T_e(T_i)$ the electron (ion) temperature, m the ion mass and k is Boltzmann's constant. Some authors [8] have chosen a slightly different expression for the temperature gradient effect in order to account for thermal diffusion. Below 400 km however, this difference has very small effect.

It must be pointed out here that V is not the vertical ion velocity but the projection upon the vertical of the velocity parallel to the magnetic field. In the presence of an east–west electric field, the vertical velocity of ionization V_z becomes:

$$V_z = V + W_E = W_d + W_n + W_E \tag{4}$$

where W_E is the vertical component of the ion velocity generated by this electric field.

At middle latitudes around the F2 peak vertical transport of ionization plays an important role in the formation of the F layer. Rishbeth [9] has shown the existence of an approximate relation between the value of V_z and the value of the recombination coefficient β at the F layer peak altitude z_m. This relation is valid only under stationary conditions and may be written

$$V_z \sim -\frac{\beta(z_m)\, H}{\alpha}$$

where $\beta(z_m)$ is the recombination coefficient at z_m, H is the scale height of atomic oxygen, α is a coefficient close to 1. Using these results, it is now possible to interpret the observed values of V.

3.2. Diurnal Variation

If we restrict this interpretation to the vicinity of the peak altitude z_m where the gradients of ionization density and temperature are small it is possible to explain easily the day to night variation: During the day, as the layer is low, diffusion effects are small so that V is very close to W_n. W_n is then negative, so that V_n is directed polewards. During the night, on the contrary, diffusion effects become very important, corresponding to a strong negative value of W_d ($\sim$100 m s^{-1}). As the mean value of V is near zero, it follows from Eq. (1) that W_n is positive corresponding to V_n directed equatorwards. The reason for the fact that W_n counterbalances W_d may be found immediately from Eq. (5) which can be written:

$$V + W_E + \frac{\beta(z_m)\, H}{\alpha} \sim 0\,. \qquad (6)$$

Since during night time the height of the peak z_m is very high, the recombination coefficient $\beta(z_m)$ is very small ($\beta(z_m)\, H/\alpha$ is of the order of a few m s^{-1}); therefore the sum of $W + W_E$ is also near zero. As was shown previously the mean observed value of V is near zero, so the conclusion is that, during the night, W_E is also near zero and east–west electric fields are negligible.

For daytime electric fields, it is not possible to derive a definite conclusion because then the three terms in Eq. (6) have the same order of magnitude. However, we know that $\beta(z_m)\, H/\alpha$ is always positive. Thus, when V is positive as happens sometimes (in summer afternoon) this indicates a negative value of W_E.

3.3. Altitude Variation

The altitude variation of W_d may be obtained from incoherent scatter observation of temperature and ionization density as a function of altitude. Such a computation is, however, inaccurate, since derivatives of these various quantities are involved. Further study on this point is in progress. The change of sign of V at high altitude during daytime is related to the replenishment of the protonosphere. The sunrise and sunset effects have been studied in detail by Evans ([7] and private communication); they are attributed to changes in the upper electron content related to rapid variation of electron density and temperature at these moments.

3.4. Deduced Neutral Winds

Using Eq. (1)–(3) and a model for the ambipolar diffusion coefficient, it is possible to deduce from incoherent scatter measurements of N, V, T_e and T_i the value of the meridional component V_n of the neutral wind [1, 2]. In Fig. 5, two examples of such a determination are shown. Only values at the peak of the layer z_m are presented because the expression for W_d is then simplified. The

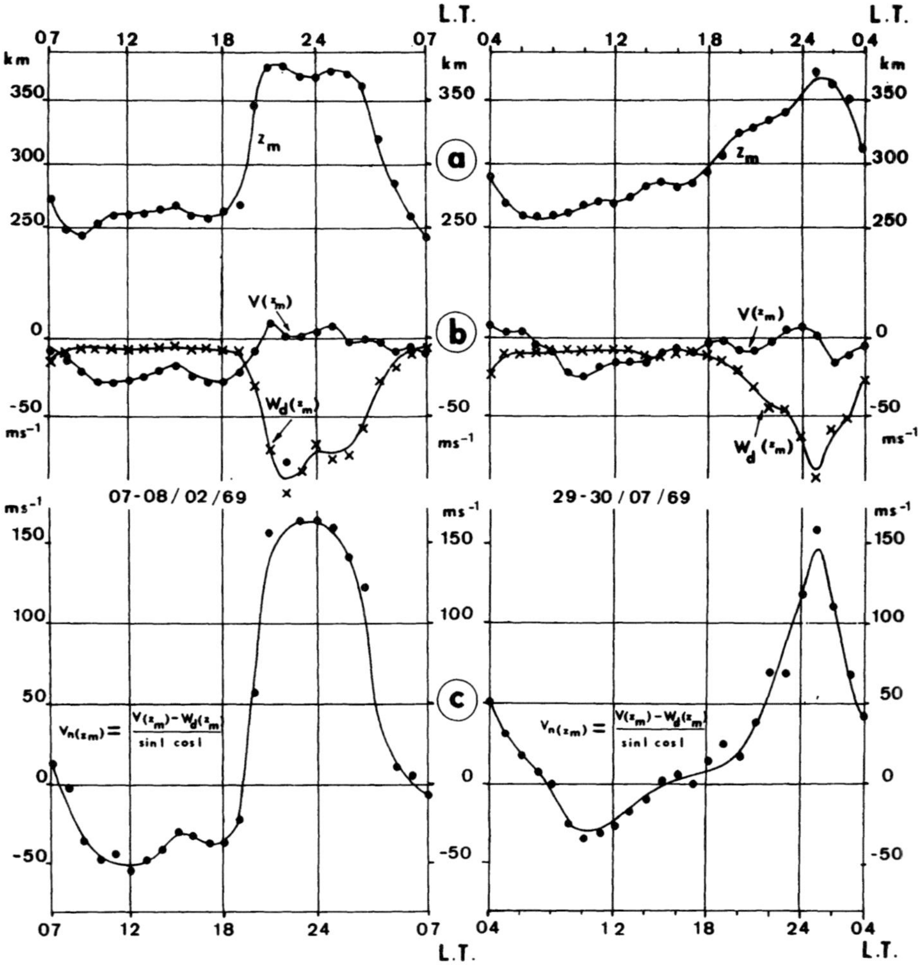

Fig. 5. Evaluation of the neutral atmosphere meridional velocity from experimental results: a, diurnal variation of z_m; b, diurnal variation of $V(z_m)$, observed velocity, and $W_d(z_m)$, evaluated diffusion velocity, from density and temperature measurements; c, diurnal variation of horizontal neutral velocity V_n. These results are obtained for a summer case (29–30 July 1969) and a winter case (7–8 Feb. 1969).

values of V_n calculated exhibit a strong diurnal variation, V_n is negative (i.e. directed polewards) during the day and positive (i.e. directed equatorwards) during the night. Comparing the diurnal variations for these two particular days, a strong difference can be noticed. In the summer case (29–30 July 1969) V_n becomes positive as early as 1500 LT whereas for the winter case (7–8 Feb. 1969) V_n remains negative with a value around -50 m s^{-1} during the whole day. These results are typical of the results obtained for about 25 cases from 1967 to 1969.

Theoretical values of the neutral wind velocity have been obtained by many authors using various assumptions [10–14]. It is interesting to compare the values of V_n obtained here with a theoretical model.

3.5. Theoretieal Model of Neutral Winds

The equation of horizontal motion for a neutral atmosphere can be written

$$\frac{d\boldsymbol{V}_n}{dt} = -2\boldsymbol{\Omega} \times \boldsymbol{V}_n - \nabla p_n + \mu \nabla^2 V_n + \frac{\varrho_i \nu_{in}}{\varrho_n}(\boldsymbol{V}_i - \boldsymbol{V}_n) \qquad (7)$$

where $V_n(V_i)$ is the horizontal component of the neutral (ionization) velocity vector, $\varrho_n(\varrho_i)$ is the density of neutral (ionized) air, p_n is the neutral pressure,

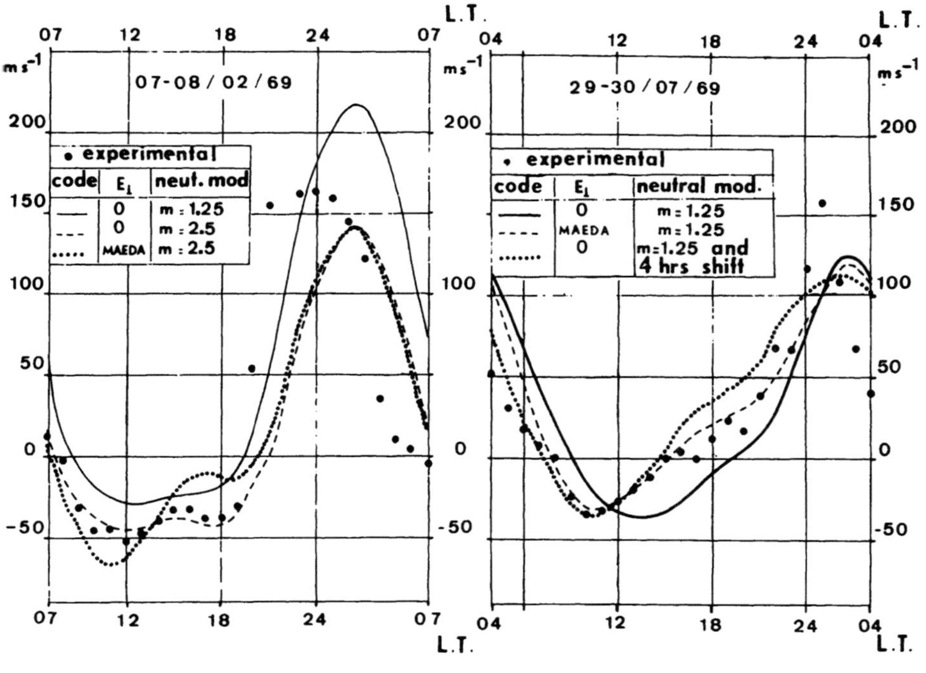

Fig. 6. Comparison between "experimental neutral velocities" and "theoretical neutral velocities" deduced from equation of neutral atmosphere movement for summer and winter cases. Different hypotheses have been made on the value of horizontal pressure gradient, fixed by the value of m. Perpendicular electric fields are considered to be present or not.

μ is the viscosity coefficient. $\varrho_i \nu_{in}/\varrho_n$ represents the momentum transfer from neutral atmosphere to the ions; this coefficient is therefore proportional to N.

This partial differential equation was solved for V_n using (a) a neutral atmospheric model for ϱ_n and p_n deduced from satellite drag data measurements [15]. Several values of the parameter m fixing the shape of the diurnal pressure bulge were tried; (b) the observed electron density to compute the ion drag term $\varrho_i \nu_{in}/\varrho_n$; (c) the observed ion velocity parallel to the magnetic field $\boldsymbol{B}$ in order to obtain $\boldsymbol{V}_i$. In fact, only one component is measured so that two different assumptions were used: (i) there is no electric field perpendicular to $\boldsymbol{B}$ so that the ion velocity is parallel to $\boldsymbol{B}$; (ii) there is such an electric field $E_\perp$ introducing a motion perpendicular to $\boldsymbol{B}$; the value of this ion velocity component was taken from the results of Maeda [16]. The "theoretical value of V_n" resulting from this computation is shown in Fig. 6 and compared with the values of V_n previously obtained ("experimental V_n") by subtracting diffusion effects from the observed value.

For the winter case the best fit between experimental and theoretical V_n in daytime is obtained from the computation using no electric field ($E_\perp = 0$) and with $m = 2.5$. For the summer case the best fit can be achieved using the electric field deduced from Maeda and with $m = 1.25$. For this case good agreement is also obtained without electric field ($E_\perp = 0$) when shifting the phase of the atmospheric diurnal bulge by four hours towards morning hours. This last hypothesis is however in contradiction with the satellite drag data and leads us to prefer the first explanation. This comparison gives evidence for the presence of electric field during summer and presumably its absence during winter which could explain the observed seasonal variation.

The variation of the steady component of the observed velocity with geomagnetic activity can be interpreted easily. Satellite measurements of neutral atmospheric temperature [17] have shown that, during magnetic storms, the atmospheric temperature increases in the auroral zone. Therefore the horizontal north–south pressure gradients are modified, resulting in an increase of the equatorward component of the neutral wind in agreement with our observations.

4. Discussion

Results of three years' measurements of ion drifts parallel to the magnetic lines can be satisfactorily interpreted in terms of horizontal neutral winds. The evaluation of the neutral winds was limited to the altitude of the electron density maximum; further work to study the altitude behaviour of these winds is in progress. The meridional winds obtained from the data were compared with theoretical models of neutral winds induced by horizontal pressure gradients and a good agreement was obtained. However it was necessary to include, only during summer, an electric field perpendicular to the magnetic field, similar to that deduced by Maeda [16]. Another evidence for the presence of such electric fields during summer was also obtained from incoherent scatter results using the ionization continuity equation [4]. This variation of electric field from summer to winter needs further explanation. If magnetic lines of force are supposed to be electric equipotentials, the electric field variation which has been obtained here is an annual effect. On the contrary, if this is related to a seasonal effect, we must admit the existence of electric currents along field lines in order to cancel the electric fields produced in one of the two hemispheres. Experimental measurements of electric fields will be possible in the near future using a quadristatic equipment. The three components of the ion velocity vector will then be measured.

Acknowledgments

We wish to thank all the members of the group concerned with C.N.E.T. incoherent scatter facility. We are particularly indebted to Drs. Petit, Waldteufel, Bauer, Testud for helpful suggestions. One of the authors is grateful to Dr. Rishbeth from R.S.R.S. Slough for stimulating discussions and fruitful advice. The Saint-Santin-Nançay facility was built by C.N.E.T. with funds of C.N.E.S. It is now supported by C.N.R.S.

References

[1] G. Vasseur, J. Atmos. Terr. Phys. **31**, 397 (1969).
[2] G. Vasseur, Annls Géophys. **25**, 517 (1969).
[3] G. Vasseur and P. Waldteufel, J. Atmos. Terr. Phys. **31**, 885 (1969).
[4] G. Vasseur, Thèse de Doctorat ès Sciences phys., Fac. Sciences, Paris (13 janvier 1971).
[5] J. Testud and G. Vasseur, Annls Géophys. **25**, 525 (1969).
[6] J. Testud, J. Atmos. Terr. Phys. **32**, 1793 (1969).
[7] J. V. Evans et al., Radio Sci. **5**, 27 (1970).
[8] L. A. Carpenter and S. A. Bowhill, Proc. on Upper Atmospheric Currents and Electric Fields, Boulder 1970.
[9] H. Rishbeth, J. Atmos. Terr. Phys. **32**, 413 (1970).
[10] J. E. Geisler, J. Atmos. Terr. Phys. **28**, 703 (1966).
[11] H. Kohl and J. W. King, J. Atmos. Terr. Phys. **29**, 1045 (1967).
[12] G. J. Bailey, R. J. Moffett and H. Rishbeth, J. Atmos. Terr. Phys. **31**, 253 (1969).
[13] H. R. Cho and K. C. Yeh, Annls Géophys. **26**, 801 (1970).
[14] P. Stubbe, J. Atmos. Terr. Phys. **32**, 865 (1970).
[15] L. G. Jacchia and J. Slowey, Space Research VII, 1077 (1967).
[16] H. Maeda, in: Proc. Conf. Ionosphere, London, Inst. of Phys. and Phys. Soc. London 1963 (p. 187).
[17] J. E. Blamont and J. M. Luton, to be published in J. Geophys. Res. (1972).

Space Research XII — Akademie-Verlag, Berlin 1972

EAST-WEST IONOSPHERIC DRIFTS AT THE MAGNETIC EQUATOR

R. F. Woodman

Jicamarca Radar Observatory, Instituto Geofísico del Perú, Lima, Perú

A technique has been developed to measure the electromagnetic east–west drift velocity of the F region ionosphere by means of the Jicamarca incoherent scatter radar. Results are presented. They show a fairly consistent behaviour from day to day even during a magnetically disturbed day. Velocities are westward with a maximum of the order of 50 m s^{-1} during the day, and eastward with a maximum of the order of 135 m s^{-1} during the night. They are shown as experimental evidence for the super-rotation of the neutral atmosphere at equatorial latitudes, but with a velocity smaller than the values inferred from satellite drag.

1. Introduction

East–west velocities of the earth ionosphere are important not only for a good understanding of the dynamics of the ionosphere but for their effect on the dynamics of the neutral atmosphere. At F region heights, in equatorial latitudes, the velocity of the neutral atmosphere is determined mainly by the balance of pressure gradient forces and frictional forces with the ions. This force depends on the differential velocity between the neutral and ion gas and, therefore, depends on the velocity of the latter.

The purpose of this note is to report on recent measurements of the eastward component of the ionospheric drift. The measurements were made at the Jicamarca Radar Observatory (2° dip) using an incoherent scatter technique in the height range 200–600 km. At these heights drift velocities are a direct measurement of the electric field in a direction orthogonal to drift and to the direction of the magnetic field ($\boldsymbol{v} = \boldsymbol{E} \times \boldsymbol{B}/B^2$). At the latitude of the observatory an eastward drift corresponds to an almost vertical electric field (40 m s^{-1} corresponds to approximately 1 mV m^{-1}). Measurements of the vertical ionospheric drifts (eastward electric field) have been reported [1] using a technique described in a previous paper [2]. The same technique, differing only in the pointing direction of the antenna, has been used to obtain the horizontal component.

2. Description of the Technique

As described in [1, 2] "vertical" velocities are obtained by pointing the antenna to the north about two degrees from vertical, so that the "probing wave vector $\boldsymbol{k}$" is perpendicular to the magnetic field. When perpendicular to the magnetic field the spectrum of the incoherently scattered signal narrows considerably and allows the measuring of any Doppler shift imposed on the signal, with sufficient accuracy to determine the velocities along $\boldsymbol{k}$ with random errors of a few metres per second, with just 5–10 minutes of integration time. Although the antenna at Jicamarca is physically fixed, it is possible to tilt it slightly from its on-axis (−1.5° to the south-west) position by properly phasing each of the 64 sections of the antenna. There are actually two antennas, one for each linear polarization. In order to obtain a projection from the eastward component of the drift we have pointed one polarization as far to the NE, and the other as far as to NW, as possible (we were limited by the beam pattern of the 64 sections). This allows us to point the antenna perpendicular to the magnetic field, but in two directions, one 2.45° to the east, the other 4.33° to the west giving a split of 6.78°. Let $\boldsymbol{\varepsilon}_1$ and $\boldsymbol{\varepsilon}_2$ be two unit vectors pointing in the two directions mentioned above. Let $\boldsymbol{v}$ be the drift velocity of the ionosphere, then from the Doppler shifts measured from each antenna we can obtain the two projected velocities $\boldsymbol{v}.\ \boldsymbol{\varepsilon}_1$, and $\boldsymbol{v}.\ \boldsymbol{\varepsilon}_2$. The linear combinations

$$v_z = \frac{\boldsymbol{v} \cdot (\hat{\varepsilon}_1 + \hat{\varepsilon}_2)}{2 \cos \alpha} \tag{1}$$

and

$$v_x = \frac{\boldsymbol{v} \cdot (\hat{\varepsilon}_1 - \hat{\varepsilon}_2)}{2 \sin \alpha} \tag{2}$$

give the velocities in the direction of two unitary orthogonal vectors $\hat{\varepsilon}_z$ and $\hat{\varepsilon}_x$ given by

$$\boldsymbol{\varepsilon}_z = \frac{\boldsymbol{\varepsilon}_1 + \boldsymbol{\varepsilon}_2}{2 \cos \alpha} \tag{3}$$

$$\boldsymbol{\varepsilon}_x = \frac{\boldsymbol{\varepsilon}_1 - \boldsymbol{\varepsilon}_2}{2 \sin \alpha} \tag{4}$$

where $\alpha = 3.39°$ is the angle between $\boldsymbol{\varepsilon}_z$ and either $\boldsymbol{\varepsilon}_1$ or $\boldsymbol{\varepsilon}_2$. Table 1 gives the directions of these vectors for the positions used in the experiment hereby described. For all practical purposes $\boldsymbol{\varepsilon}_z$ points in a direction perpendicular to $\boldsymbol{B}$ and in a vertical plane, and $\boldsymbol{\varepsilon}_x$ in the magnetic east direction and on a horizontal plane. The directions deviate from a truly vertical or horizontal plane by one degree which we shall neglect in what follows.

Table 1

Directional cosines with respect to geomagnetie coordinates*

	Magnetic North Projection	Magnetic East Projection	Vertical Projection
$\boldsymbol{\varepsilon}_1$	0.0427	0.0427	0.9984
$\boldsymbol{\varepsilon}_2$	0.0427	−0.0755	0.9964
$\boldsymbol{\varepsilon}_x$	0.0000	0.999	0.0164
$\boldsymbol{\varepsilon}_z$	0.0428	−0.0164	0.999

* Declination 6° E, dip 2° N.

An error in the determination of the velocity along the antenna direction is translated into an error six times larger in the horizontal direction ($\sqrt{2}/4 \sin \alpha = 6$)

Fortunately, the magnitudes in the eastward direction are of the order of 100 m s^{-1}, so that an error of 2 m s^{-1} in the determination of $\boldsymbol{v}.\boldsymbol{\varepsilon}_1$ or $\boldsymbol{v}.\boldsymbol{\varepsilon}_2$ which correspond to the actual measured r.m.s. errors, gives errors of the order of 12 m s^{-1} for the horizontal component, which is an acceptable fraction of the magnitudes we want to measure.

3. Results and Discussion

Drift velocities have been obtained from 275 to 500 km, or to 700 km depending on the ionospheric conditions. The most accurate measurements are taken within the range 300–400 km. The average velocity within this reduced range of heights has been scaled, but it can be taken as indicative of the velocities throughout the observed region since no height dependence is apparent on the data. At times, when the F region was too high or when there was spread F, the lower heights were either too noisy or contaminated with spread F, and a higher 100 km range was taken.

Fig. 1(a)–(c) show vertical and east–west velocities for three representative days around the (September) equinox; they include the corresponding K indexes scaled from the Huancayo magnetograms. Note that 16 October 1970 was a magnetically active day. The two September days are relatively quiet consecutive days. In contrast with the very variable behaviour of the vertical velocities shown in Fig. 1 (see also [1]) the large time scale features (diurnal and first harmonics) of the east–west velocity show a very consistent behaviour from day to day. This includes the magnetically disturbed day in October. The effects of the magnetic activity are evident on the unusual negative excursion of the vertical drift during the day. The short time scale fluctuations will not be discussed here, since they may be due partially to a contamination of vertical fluctuations. At present, due to computer limitations the two projected velocities $\boldsymbol{v}.\boldsymbol{\varepsilon}_1$ and $\boldsymbol{v}.\boldsymbol{\varepsilon}_2$ are taken consecutively at 10-minute intervals and their difference (Eq. 2) is sensitive to fluctuations in the vertical component.

We have at present twelve days of east–west drift observations randomly distributed in the period August 1970–May 1971. Fig. 2 shows the average of these data. The results are preliminary, in the sense that they represent a free-hand smoothing drawn on a composite picture of all available data. The data are being digitally processed to obtain the seasonal dependence of the first few (diurnal) Fourier components, but we can anticipate that the results will not deviate appreciably from the curve shown in Fig. 2. A feature of importance in the E–W drifts is the asymmetric behaviour of the eastward drifts from day to night. The velocities at night have a maximum of the order of 135 m s^{-1} compared with a maximum of 50 m s^{-1} during the day.

Rishbeth [3, 4] has recently proposed a mechanism to explain the super-rotation of the earth's neutral atmosphere [5] based on the different ion drag which the ionosphere should present to neutral particles from day to night. This difference in drag is produced not so much by the difference in electron density as by a difference in the velocities of the ionized medium as a consequence

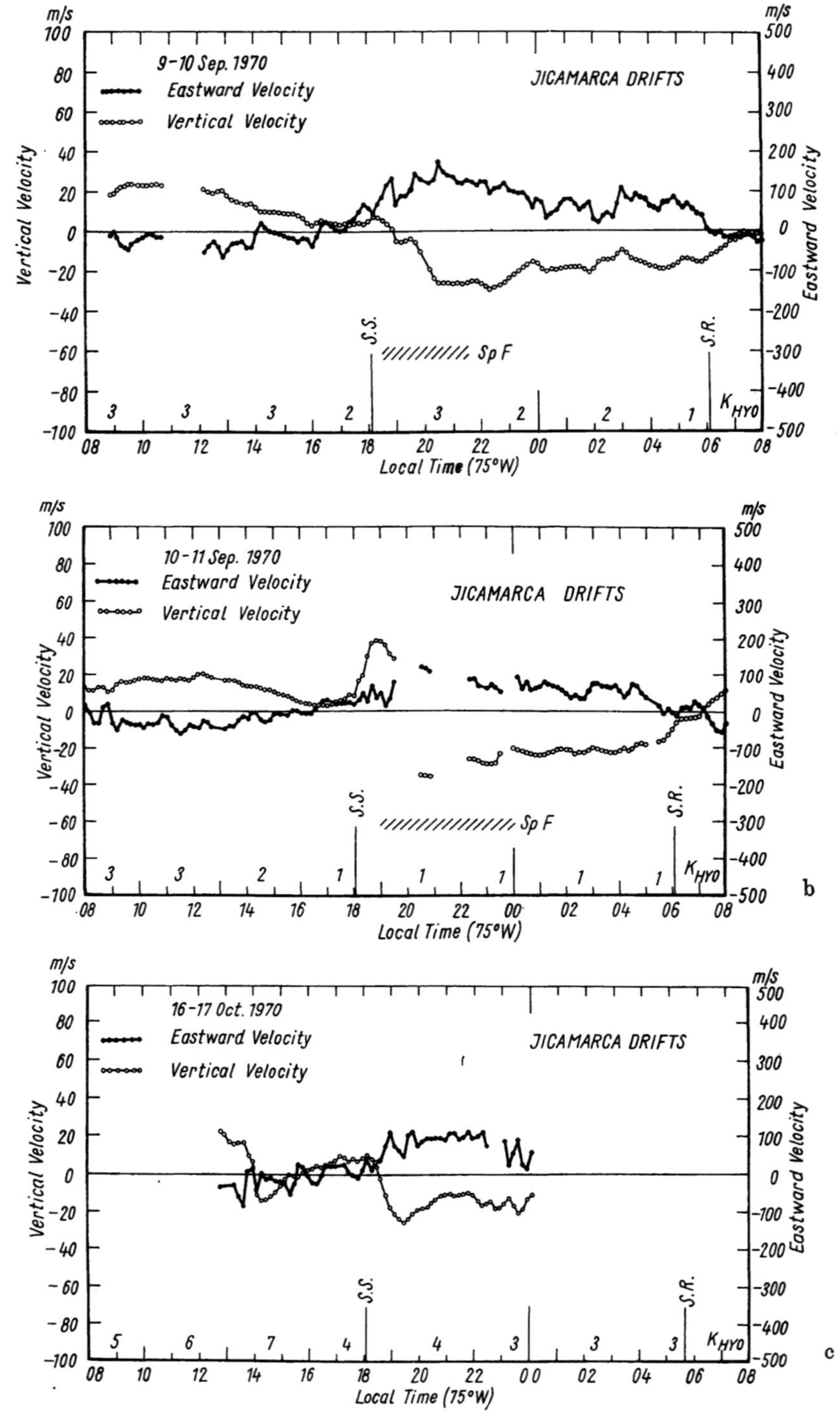

Fig. 1(a)–(c). Vertical and eastward velocities for three different days as shown. Numbers above time scale are Huancayo magnetic indexes. S.S. = sunset, S.R. = sunrise.

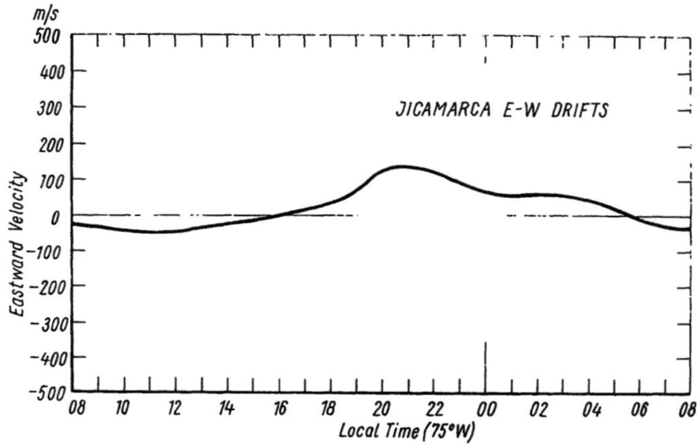

Fig. 2. Smoothed average of 12 days of east–west drift observations.

of different degrees of electromagnetic coupling between the E and F regions. Our observations support the mechanism he proposes. One can explain the observed super-rotation of the ionized medium as follows. During the day the F region is strongly coupled to the E region owing to the high conductivity along the magnetic field lines. The E region itself presents high conductivity so the F region responds to the electric field imposed by the E region. The observed eastward velocities at F region heights are then an indication of the eastward velocities of the E region neutral winds which generate this field. During the night the conductivity at E region heights decreases with the ionization as does the conductivity along the field lines at regions immediately above the E region, decoupling the F from the E region. This allows the F region to be "blown along" by the neutral particles. This "blowing along" takes place indirectly. Collisional forces make ions and electrons cross-drift in opposite directions and, because of vertical density gradients, a polarization field is built up making the ionization drift with a velocity equal (if the decoupling were perfect) to that of the neutral particles. (This polarization could not exist if there was good conductivity down to a conducting E region.) Consequently, the eastward velocities for the F region are low and of the order of eastward E region neutral velocities during the day, and high during the night with velocities in the order of F region neutral velocities with small ion drag.

Although we only observe the velocity of the ionized medium we can infer from our measurements the velocity of the neutral atmosphere at our latitudes since its velocity is determined mainly by ion-drag frictional forces balancing the pressure gradient forces. We can roughly state that during the day the neutral velocity should not differ more than a few tens of metres per second from the observed F region values. Lagos [6] has computed neutral velocities at F-region heights including ion-drag for different conditions of the ionosphere. He assumes that there is no E field (stationary ions) and obtains velocities of the order of 20–30 m s^{-1} for typical day conditions. The neutral velocities should be higher than our measurements by this amount in order to have a frictional ion-drag force of the right magnitude to balance the driving pressure gradient. Rishbeth [3, 4] estimates that during the night the ion velocities should be of the order of 80% of that of the neutral particles. If we allow for these differences we conclude that the neutral atmosphere velocities which correspond to our observed ionospheric velocities must super-rotate with excess velocities close to 50 m s^{-1}, a velocity about half as large as the velocities inferred from satellite drag [5].

4. Conclusions

We have described a technique to measure E–W ionization drifts at F region heights using the Jicamarca incoherent scatter radar. We have presented some observations taken with this technique; although few in number, they are sufficiently consistent as to be indicative of the daily behaviour of the E–W drift at equatorial latiudes. We have included one magnetically disturbed day showing the same general behaviour as the quiet days. The data show that there is a super-rotation of the F region ionization which can be interpreted as experimental evidence of the super-rotation of the neutral atmosphere as well as of the mechanism proposed by Rishbeth to produce it.

Acknowledgments

The author gratefully acknowledges the staff of the Jicamarca Radar Observatory for their assistance in this work. Thanks are given to the National Science Foundation, the National Aeronautics and Space Administration (NGR-52-158-001) and the Organization of the American States for their economic support.

References

[1] R. F. WOODMAN, J. Geophys. Res. **75**, 6249 (1970).
[2] R. F. WOODMAN and T. HAGFORS, J. Geophys. Res. **74**, 1205 (1969).
[3] H. RISHBETH, Planet. Space Sci. **19**, 357 (1971).
[4] H. RISHBETH, Nature, Lond. **229**, 333 (1971).
[5] D. G. KING-HELE, Space Research XII, 847 (1972).
[6] C. P. LAGOS, private communication.

Space Research XII — Akademie-Verlag, Berlin 1972

A MODERNIZED TECHNIQUE FOR IONOSPHERIC DRIFTS WITH SPECTRAL ANALYSIS

W. PFISTER[a] and K. BIBL[b]

[a] Air Force Cambridge Research Laboratories, Bedford, Mass., USA
[b] Lowell Technological Institute Research Foundation, Lowell, Mass., USA

The new drift technique measures fluctuations of the complex signal reflected from the ionosphere. A spectral analysis reveals the Doppler spectrum of the echos returned from a range given by the pulse width. With the help of the phase spectra between pairs of antennas, the angular direction of arrival for individual Doppler shifted lines can be determined. Samples taken from E layer reflections show wavelike structures and motions. The latest development is a so-called digital drift system which records sine and cosine samplings directly on computer tape. The instrumentation becomes especially simple when combined with a digital ionospheric sounder, and allows recording on a number of arbitrarily chosen frequencies at the same time.

1. Introduction

Drift measurements using reflected pulses and the closely spaced receiver technique are possible because the reflecting medium in the ionosphere contains a more or less irregular structure which exhibits a common velocity component. Usually it is assumed according to Briggs et al. [1] that the reflected signal comes from a large number of scattering centers in the reception cone with an approximately random distribution of position and velocity superimposed on the drift. It is important to examine this assumption more closely.

2. Principles of Spectral Analysis

This can be done with a new technique of spectral analysis which will be described briefly. Let us assume for the moment the signal on the ground is produced by the interference of a limited number of rays specularly reflected from different parts in the reflecting medium. A simple case would be a wave structure in the ionosphere with sufficient amplitude, as for instance produced by an internal gravity wave. Each of the reflecting areas at a given time interval can be identified by a Doppler shift and an angular direction. The aim of the spectral analysis is to measure the various Doppler shifts and the corresponding angles of arrival, which can then be used to study the structure and the motions in the ionosphere. The basic measurements are the amplitude and phase of the pulsed signal at three closely spaced antennas on the ground. The complex signal is then Fourier analyzed over an appropriate sampling period and yields a complex spectrum or two spectra in quadrature for each antenna:

$$\boldsymbol{S}(\Omega) = \int \boldsymbol{f}(t) \exp \{j\, \Omega\, t\}\, dt\,.$$

Ω is the Doppler shift between transmitted and received signal and covers both the positive and negative range. The power spectrum is obtained by multiplying the spectrum with its complex conjugate value

$$\boldsymbol{W}(\Omega) = \boldsymbol{S}^*(\Omega) \cdot \boldsymbol{S}(\Omega)\,.$$

It is a real quantity. Similarly a cross spectrum is obtained by multiplying the complex conjugate spectrum of one antenna with the complex spectrum of a second antenna. The result is again a complex spectrum.

$$\boldsymbol{W}_{12}(\Omega) = \boldsymbol{S}_1{}^*(\Omega) \cdot \boldsymbol{S}_2(\Omega)\,.$$

If band averaging is applied it may be represented by a combination of a power spectrum, a coherence (under the square root) and a phase spectrum $\Psi_{12}(\Omega)$

$$\boldsymbol{W}_{12}(\Omega) = W(\Omega) \left(\frac{|W_{12}(\Omega)|^2}{W_1 \cdot W_2}\right)^{1/2} \exp \{j\, \Psi_{12}(\Omega)\}\,.$$

The angular direction of a particular reflecting area can be determined whenever it can be resolved in the Doppler spectrum. Azimuth and elevation are computed from the formula

$$\Psi_{12}(\Omega) = \frac{2\,\pi}{\lambda}\, l \cos \theta \;\sin \delta$$

where θ is the azimuth with respect to the antenna base line, l is the antenna distance, λ the wavelength of the probing frequency and δ is the deviation from the vertical. $\Psi_{12}(\Omega)$ of course is the measured phase at the Doppler frequency Ω.

More details about the internal structure of a particular reflection area can be found by studying the shape of the individual Doppler line in the power spectrum and the coherence for the antenna separation at the same line. These characteristics are a function of the sampling length, the duration of the reflection and change of Doppler shift during a sampling period, and the parameters of the internal structure. The larger structure like the effect of internal gravity waves can be studied by looking at the relationship between Doppler shift and angles of arrival over the entire spectrum and at variations of this relationship with time.

3. E Region Data

So far only limited experience has been gained with this technique of complex spectral analysis. A preliminary version has been applied by Pfister [2] to data from E layer reflections. The results are interesting and can illustrate to a certain extent what could be expected also from reflections higher up in the ionosphere.

Measurements have been made by LTI Research Foundation during daytime at the field site in Billerica, Massachusetts, with a frequency of 2.89 MHz. Since the instrumentation and the technique of analysis is outdated by now,

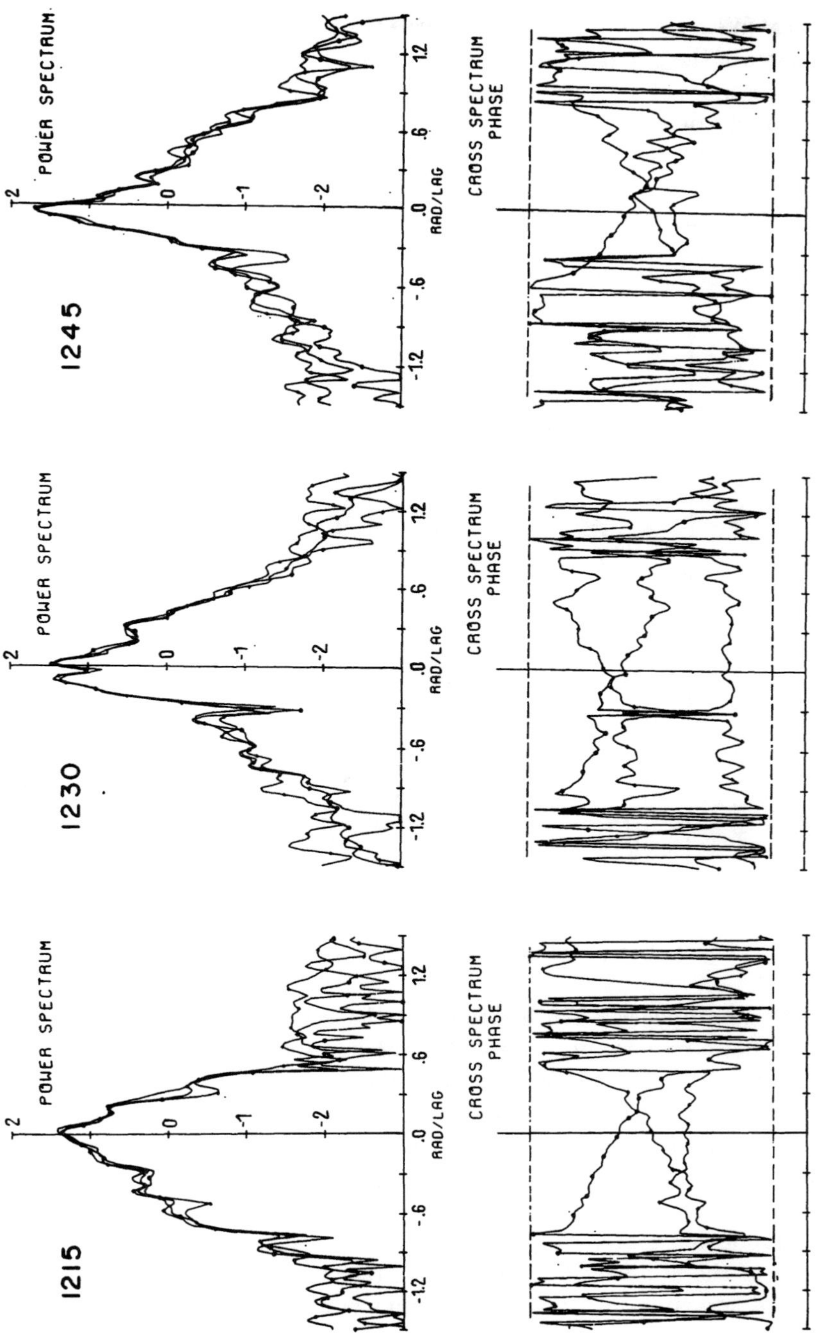

Fig. 1. Power spectra and phase spectra taken at 15 minute intervals on date 690826.

it is sufficient to show some of the results for the power spectrum and the phase spectrum of $2^1/_2$ minute samples. In Fig. 1 the Doppler frequency is plotted in radians/lag. With a unit of lag of 0.3 records the total range is roughly ± 0.8 Hz. The ordinates of the power spectra cover five orders of magnitude in logarithmic scale. The samples show a superposition of a number of individual lines whose width is largely determined by the length of the sample. In other words the fluctuations on the ground are produced essentially by an interference effect of a limited number of Doppler shifted rays. The second point of interest in the results is the form of the phase spectrum which is plotted in the range $-\pi$

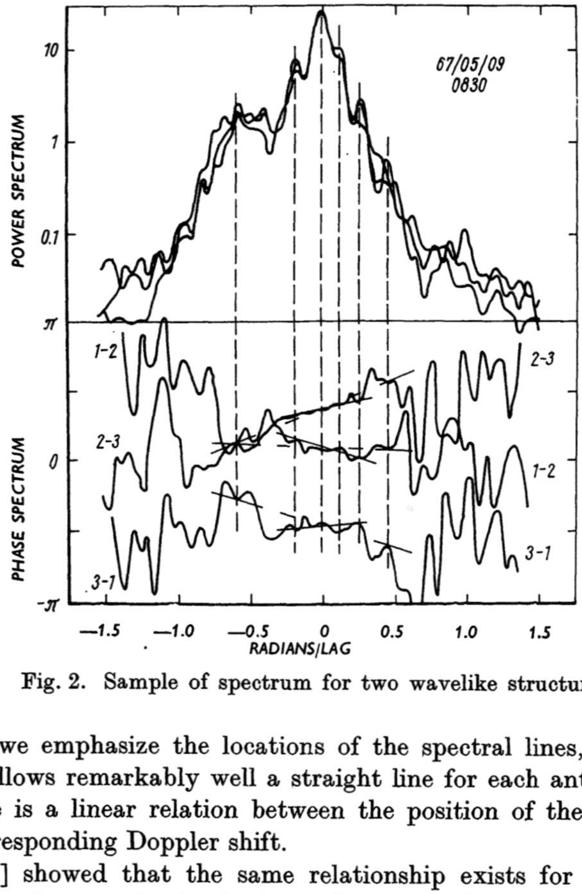

Fig. 2. Sample of spectrum for two wavelike structures.

to $+\pi$. If we emphasize the locations of the spectral lines, the cross-phase spectrum follows remarkably well a straight line for each antenna pair. This means there is a linear relation between the position of the reflecting areas and the corresponding Doppler shift.

Pfister [2] showed that the same relationship exists for reflections from a wavelike surface moving in the ionosphere. On the other hand the position of the Doppler-shifted lines and their appearance indicates that the structure is more complicated than a single plane wave. We therefore talk about a wave-like nature of the inhomogeneities in the E region and favor an interpretation in terms of a large internal gravity wave and a superposition of relatively short-lived ripples moving essentially with the wave. Occasionally two wave-like structures have been observed to occur during the same sampling period as for instance in the case of Fig. 2. Six Doppler-shifted lines have been identified in the power spectrum. As indicated by the slopes that can be drawn through the intersection points in the phase spectrum, four inner lines belong to one wavelike motion and two outer ones to a different wavelike motion. The fact that corresponding pairs of slopes intersect at zero Doppler shift indicates that the motions occur in parallel horizontal planes.

4. Features of the Modernized Technique

A test run of the new modernized technique has been carried out in May 1971 at the Eglin rocket launching site in Florida. The principle of the technique is that of a spectral analysis as outlined earlier in the paper. Sampling occurs every quarter of a second for six frequencies and three antennas requiring 18 pulses 0.01 second apart. Each pulse creates a cosine sample $A_n \cos \varphi_n$ and a sine sample $A_n \sin \varphi_n$. In the recording technique, two fast variable-step forward–reverse counters follow the if. signal excursions in two consecutive quarter periods at a selected height range. This technique is better than using two devices, one for amplitude and one for phase, and it saves a step in the performance of the complex Fourier transform

$$R[S(\Omega_m)] = \sum_{n=1}^{512} A_n \cos \varphi_n \cos \Omega_m t_n - A_n \sin \varphi_n \sin \Omega_m t_n ,$$

$$I[S(\Omega_m)] = \sum_{u=1}^{512} A_n \cos \varphi_n \sin \Omega_m t_n + A_n \sin \varphi_n \cos \Omega_m t_n .$$

For the time being the original complex function is recorded in digital form while the Fourier spectrum is generated in the computer by fast Fourier transform. We hope to have in the near future a digital device which performs the Fourier transform in real time at the recording site. Then the real and imaginary component of the Fourier spectrum will be recorded instead of the original function, resulting in savings of tape.

Another feature of the modernized technique is a digitized print-out of the information that is recorded on tape. This gives the operator a quality control for the data taken and is especially valuable when Fourier spectra are recorded. An idea of what can be accomplished with this particular technique of digitized print-out is illustrated in Fig. 3. In this case data recorded in the older style [2] have been analyzed on a computer to show the variation of the power spectrum in radians s^{-1} as a function of diurnal time. Each individual recorded data point consists of a number between 0 and 15 representing a measure of the spectral intensity. The higher the number the darker it appears. There are 127 numbers in a vertical row representing the entire spectrum. The black vertical rows are time marks. The open space on the bottom is used to print the preface, containing information such as date, time, frequency, and range. Here the preface has been omitted. Three successive rows show similar spectra measured at the three different antennas. The spectral variation is not a continuous function of time because the samples have been taken only every 15 minutes and in reflections from the E region the variation occurs obviously more rapidly. The spectrum can be quite broad or rather narrow. Quite frequently several spectral lines can be distinguished and most of the time the spectrum is Doppler-shifted and not symmetrical. The first record is a calibration and shows the resolution that can be achieved.

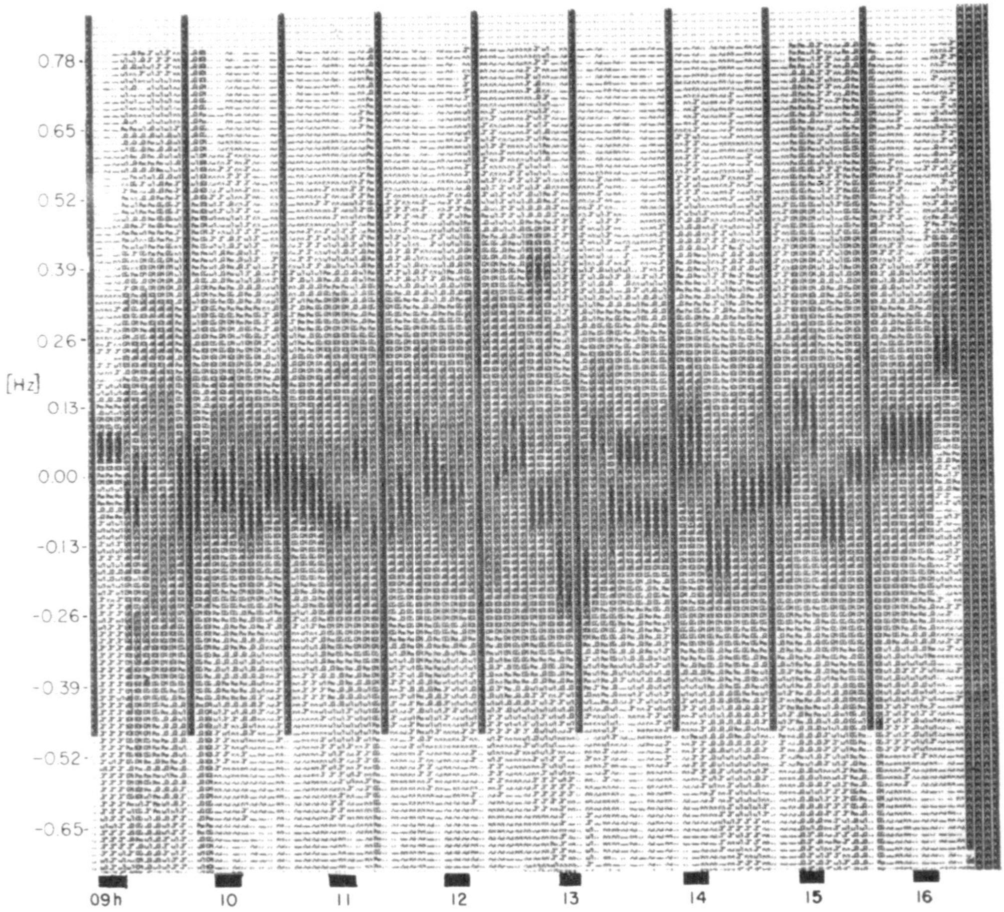

Fig. 3. Digitized printout of power spectrum for 3 antennas taken on date 700121.

A similar digitized printout for the variation of the three-phase spectra is shown in Fig. 4. The pattern is less easily recognized because of the interlacing between various antenna combinations. Evident are the rapid fluctuations in the sidebands, where the phase spectrum is meaningless, and the smooth variations in the center of the spectrum, where only a closer examination reveals the slopes. This type of three-dimensional data representation can easily be optimized for specific purposes by additional data processing. What comes to mind is a presentation of Doppler spectra and angular directions of arrival as a function of time.

5. Concluding Remarks

The modernized technique of spectral analysis has been developed primarily as a research tool for the study of inhomogeneities in the ionosphere and their

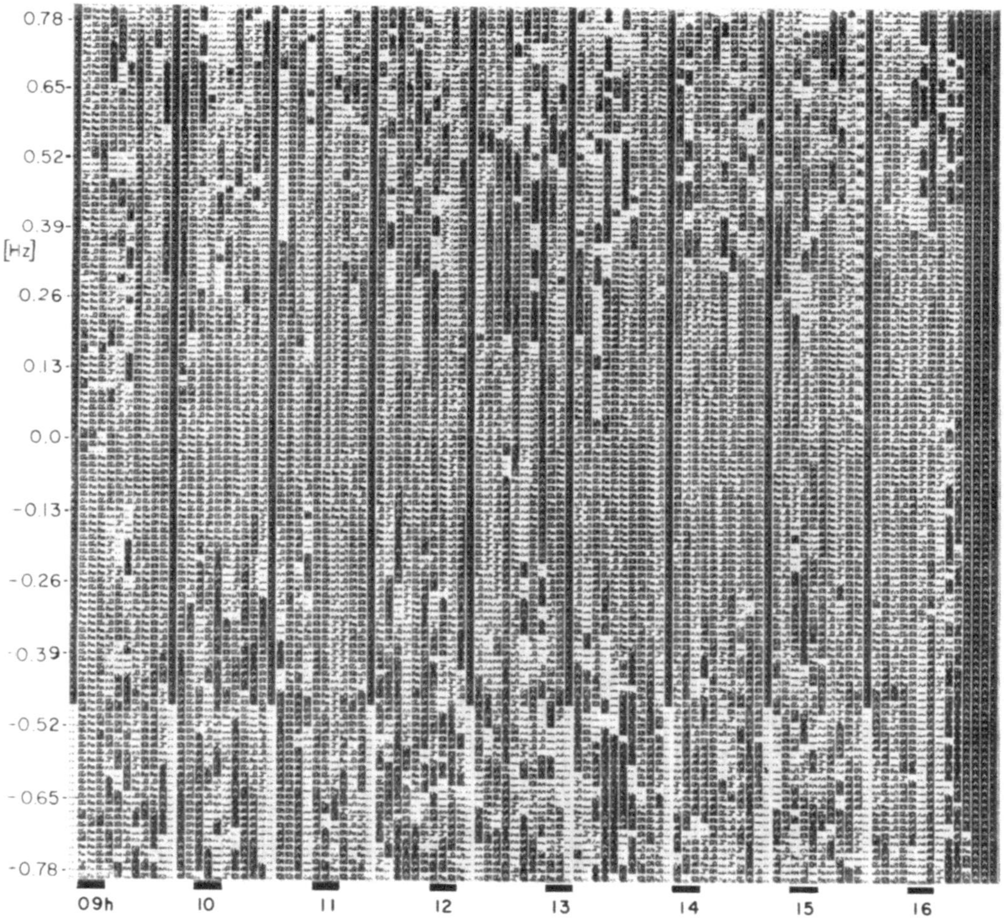

Fig. 4. Digitized printout of phase spectrum for 3 antennas taken on date 700121.

motions. However its application is advantageous in many other fields. In the area of drift measurements it provides more parameters and presents them in a form more easily visualized. It is therefore valuable for the proper design of drift experiments, for the correct interpretation of more simple drift experiments, or even for a clarification of the meaning of "drift". The technique can also be used with advantage in connection with ionosondes, to provide the necessary information for the interpretation of complicated ionograms. Finally we should not forget oblique propagation through the inhomogeneous ionosphere, where a knowledge of the spread in Doppler shift and angular deviation and the interdependence between the two can be brought to practical use in various applications.

References

[1] B. H. Briggs et al., Proc. Phys. Soc. B **63**, 106 (1950).
[2] W. Pfister, J. Atmos. Terr. Phys. **33**, 999 (1971).

Space Research XII — Akademie-Verlag, Berlin 1972

FEATURES OF THE IONOSPHERIC DRIFT OVER THE MAGNETIC EQUATOR

R. G. Rastogi, H. Chandra and R. K. Misra

Physical Research Laboratory, Ahmedabad, India

The horizontal drift measurements in the ionospheric E and F regions over Thumba at magnetic equator show close association of the drift characteristics with the electrojet both on a long-term and on a short-term basis. The drift speeds increase linearly with the electrojet current strength as estimated from the variation of the H component of the geomagnetic field. On magnetically disturbed days the drift speeds are reduced. This can be explained in terms of an opposing field to the normal Sq field which originates in the magnetosphere. Solar activity does not have any effect on apparent drift but true drift speed decreases with solar activity. This can also be interpreted in terms of a field due to magnetospheric currents opposite to the Sq field which will have higher magnitude when solar activity is high.

The drift speed calculated by second order echoes is shown to be about two thirds of the speed calculated from first order echoes. Thus the division by two to get the drift of irregularities in ionosphere looks doubtful at Thumba. Spectral analysis shows increasing drift speed for higher fading frequency components. Apparent drift seems to be the drift speed at the highest significant Fourier frequency while V_c (characteristic velocity) is caused by the spectral variation of drift speed.

1. Introduction

Ionospheric drift measurements at Thumba (dip 0.6° S, a station very close to the magnetic equator) using the method of closely spaced receivers [1, 2] were started in January 1964. In this paper we present results of these measurements in the period 1964–1969 which covers half a solar cycle. The observations were restricted to daytime only in the years 1964–1965, but from 1967 regular night-time F region drift measurements were started and with an additional independent unit having a high power transmitter installed in January 1968, it was possible to record night-time Es drifts on certain occasions during the years 1968–1969.

The earliest results at Thumba [3] showed an entirely westward daytime drift at Thumba, there being no measurable time shift between the N–S pair of aerials. In 1966 the N–S separation was doubled (240 m). Further aerials in the N–S and E–W directions were added in 1967 up to a separation of 480 m each. Even these measurements have not indicated any significant time shift along the N–S direction.

The observations in the period 1968–1969 were obtained using an antenna separation of 60 m along the E–W direction and a separation of 480 m along the N–S direction. With the availability of different aerials and two independent drift units linked with a common time marker it was possible to study the simultaneous fading recordings at two different aerial systems, two different frequencies or two different multiples.

The entire data of the period 1964–1969 have been analysed by a similar fade method, and a fairly large number of records have been analysed by the correlation technique developed by Briggs et al. [4], Phillips and Spencer [5] and Fooks [6].

2. Results at Thumba

The 24-hourly E and F region apparent drifts in the period 1968–1969 are shown in Fig. 1 as the drift vectors. The steady component is very small (about 20 m s^{-1}) for both E and F regions and is directed towards the west. The diurnal

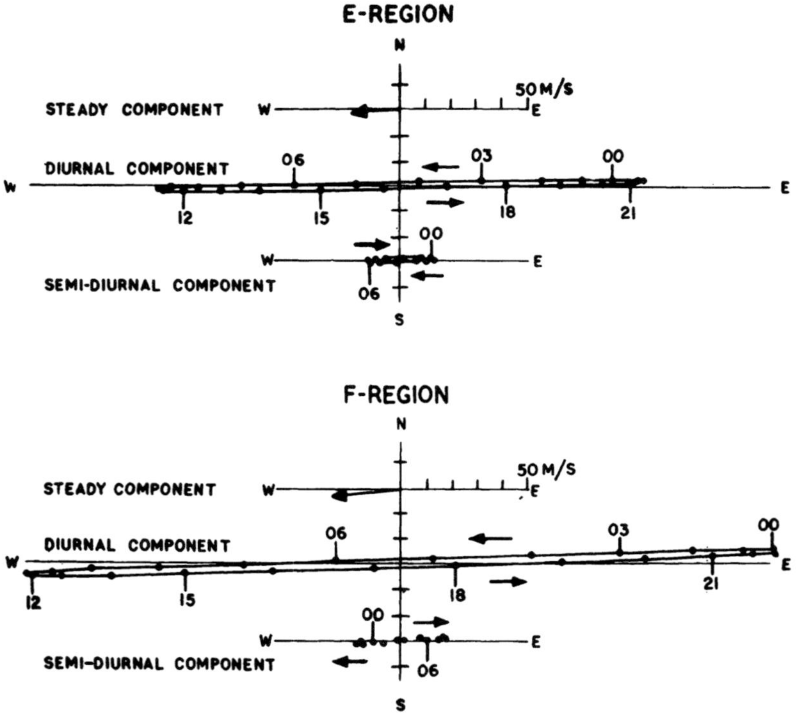

Fig. 1. Steady, diurnal and semidiurnal apparent drift vectors for thr E and F regions over Thumba averaged for the period 1968–1969.

component is the major one and has an amplitude of about 100 m s^{-1} for the E region and about 150 m s^{-1} for the F region, vector rotation being anticlockwise. The N–S component is insignificant hence the vector rotates along a highly elongated ellipse in the E–W direction. The semidiurnal vector again is very small in amplitude (about 20 m s^{-1}) and rotates clockwise.

The diurnal variations of the E–W component of the F region drifts during different seasons in the period 1967–1969 are shown in Fig. 2. The smooth curve drawn through the mean points is built up from the first four harmonics. The drift is westward during the day and eastward during the night. The reversals occur between 06 and 07 hr in the morning and between 19 and 20 hr in the evening. Seasonally the time of morning reversal does not show much

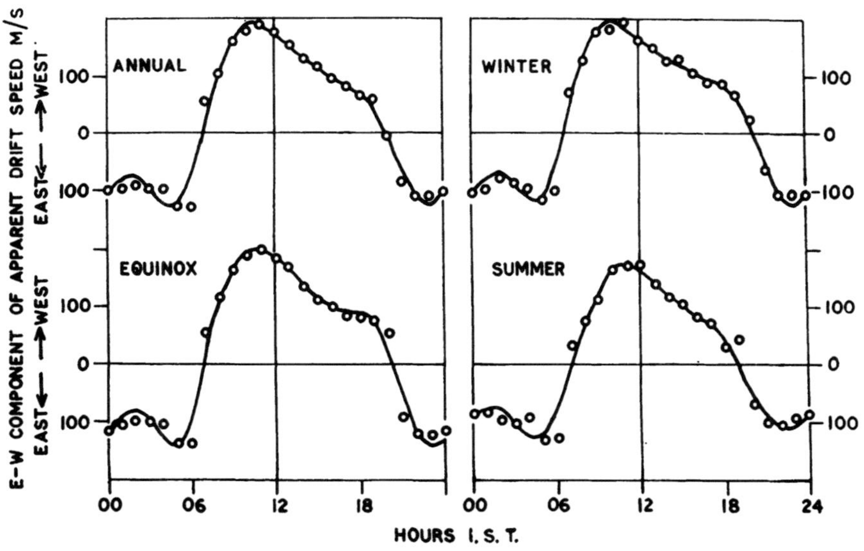

Fig. 2. Diurnal variation of the E–W component of the F region apparent drift during different seasons averaged for the period 1967–1969, Thumba.

variation, but the evening reversal occurs about an hour earlier during summer. There is a tendency for the drift speed to increase just before the morning reversal and a similar suggestion before the evening reversal (clearer during the equinoxes). The maximum daytime westward speed occurs at about 10–11 hr. The harmonic coefficients as well as the time of maxima are collected in Table 1.

Table 1

Harmonic coefficients of the daily variation of the E–W component of apparent drift speed in the F region at Thumba for the period 1967–1969

Period	Amplitude (m s^{-1})					Time of maxima (LT)				Morning reversal (LT)	Evening reversal (LT)
	a_0	r_1	r_2	r_3	r_4	φ_1	φ_2	φ_3	φ_4		
Winter	−30	150	26	40	31	0.8	3.5	5.7	5.0	6.6	20.2
Equinoxes	−25	160	30	44	24	1.0	3.8	6.0	4.8	6.8	20.2
Summer	−10	140	30	35	7	0.8	5.2	5.6	4.4	7.2	19.0
Annual	−22	152	27	40	16	0.8	4.3	5.7	4.9	6.8	19.8

3. Latitudinal Variation of Drift and Anisotropy Parameters

The drift speeds obtained at Thumba [3] were the highest reported for any station. From all the available results a latitudinal survey of the daytime drift speed and other features such as the axial ratio r, orientation of the major axis ψ and semi-minor axis b which define the ground diffraction pattern, was undertaken and the results are shown in Fig. 3. The drift speed is around 80 m s^{-1} at all stations except very close to the magnetic equator ($\pm 5°$ magnetic latitude) where there is a sharp increase in drift speed. At the equator the drift speed

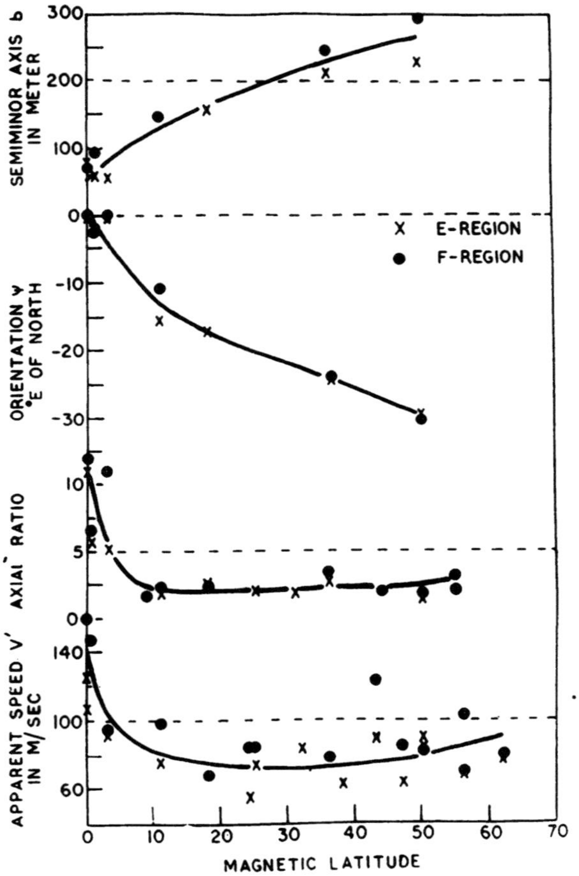

Fig. 3. Latitudinal variation of apparent drift speed V' axial ratio r, orientation of major axis ψ and semi-minor axis b of the characteristic ellipse representing the average size of the diffraction pattern at ground.

is about 150 m s^{-1}. Similar features near the equator can be noticed in the latitudinal variations of other parameters also. The axial ratio r is about 2 at all stations but increases to more than 10 at the equator. The orientation of the ground diffraction patterns is N–S at the equator but departs slowly from it. At latitudes of about 50° the orientation is along the NW–SE quadrant. The transverse size of these patterns is less than 100 m near the equator and increases steadily; it is about 300 m at high latitudes.

4. Drift at Thumba and the Electrojet Currents

These results indicate a close association between the drift at the equator and the electrojet. To test the relation between the equatorial electrojet currents and drifts on a day-to-day basis the E region midday drift was compared with the range in H at Trivandrum (ΔH_T) which can be taken as a measure of the currents at E region height. To avoid any non-ionospheric contribution in the magnetic field variation, e.g. magnetospheric contributions, the difference between the ranges at Trivandrum and Alibag (a station away from the electrojet

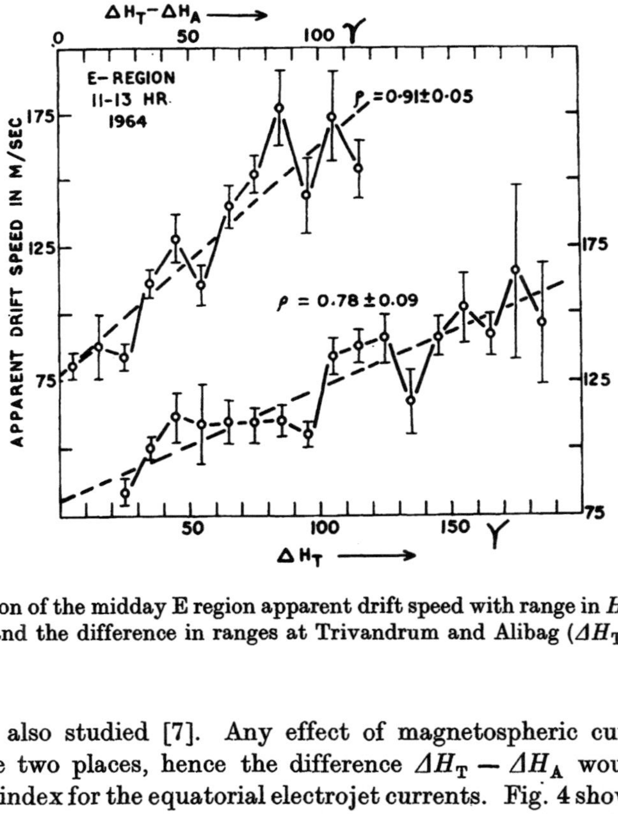

Fig. 4. Relation of the midday E region apparent drift speed with range in H at Trivandrum (ΔH_T) and the difference in ranges at Trivandrum and Alibag ($\Delta H_T - \Delta H_A$).

region) was also studied [7]. Any effect of magnetospheric currents will be equal at the two places, hence the difference $\Delta H_T - \Delta H_A$ would be a more appropriate index for the equatorial electrojet currents. Fig. 4 shows the relation between midday E region drift speed and the parameters ΔH_T and $\Delta H_T - \Delta H_A$. A linear increase of drift speed is clear with both ΔH_T and $\Delta H_T - \Delta H_A$. However the correlation is better with $\Delta H_T - \Delta H_A$ than with ΔH_T. Thus the drifts at equator are an index of the equatorial electrojet currents, measured as the increase of magnetic field during daytime from the constant night level. The non-ionospheric currents do have some contribution in the magnetic field variations which can be removed by taking the differences between the range H at an equatorial station and another station away from the equator such as Alibag.

5. Drift at Thumba and Magnetic Disturbance

Drifts at high latitudes are found to increase with magnetic activity [8]; a similar study has not been done in detail so far for the equator. We have compared the drifts at Thumba for low sunspot period (1964) and high sunspot period (1967–1968) on five international quiet and five international disturbed days; Fig. 5 shows the results obtained. The daytime results for the E and F region during 1964 show the drift speed to be reduced on disturbed days in each

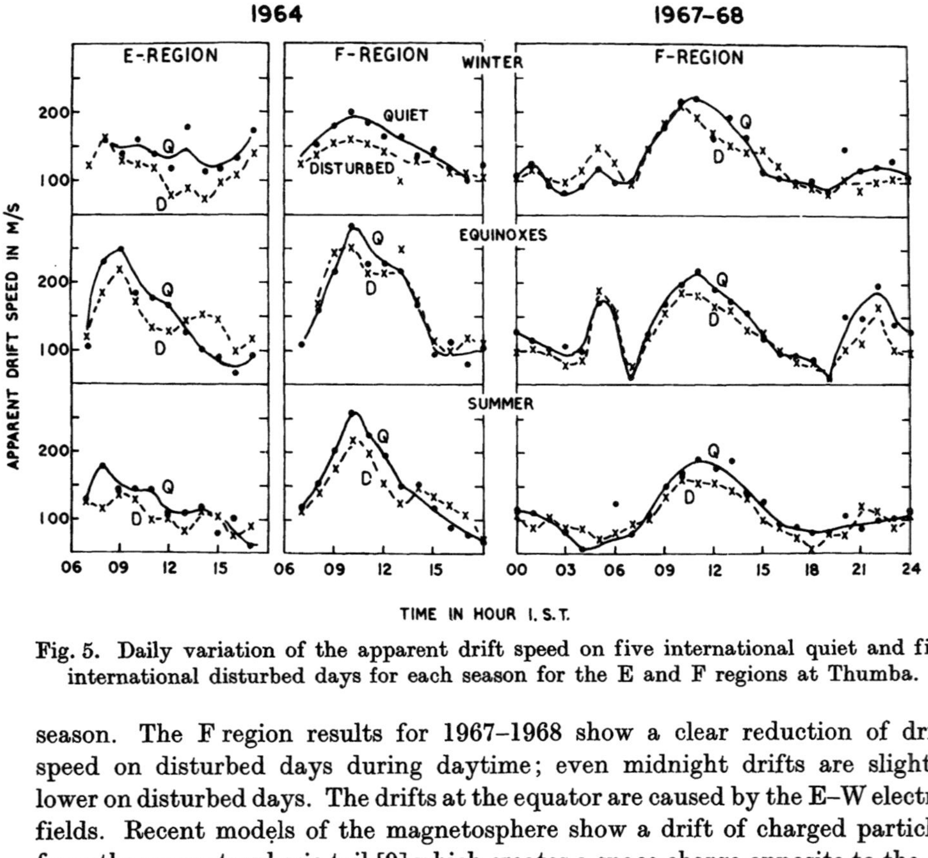

Fig. 5. Daily variation of the apparent drift speed on five international quiet and five international disturbed days for each season for the E and F regions at Thumba.

season. The F region results for 1967–1968 show a clear reduction of drift speed on disturbed days during daytime; even midnight drifts are slightly lower on disturbed days. The drifts at the equator are caused by the E–W electric fields. Recent models of the magnetosphere show a drift of charged particles from the magnetospheric tail [9] which creates a space charge opposite to the *Sq* field. At times of magnetic storms more particles enter the tail, hence this space charge would increase. Thus the reduction of drift speed at the equator can be explained in terms of the reduction of the E–W *Sq* electric field.

6. Short-term Fluctuations in Drift and Magnetic Field at Thumba

Though the daytime drifts at Thumba are mostly westward, on certain days it has been noticed that there is a reversal of drift direction towards the east for a short time, after which it reverts to the normal westward direction. The magnetograms on these days show sudden decreases in H in the same time sequence, while the ionograms show a disappearance of Es [10]. In Fig. 6 are shown the daily variations of the magnetic H component at Kodaikanal, f_0Es at Kodaikanal and the E–W E region drift at Thumba on one abnormal day (23 November 1964) and one normal day in the same month (20 November 1964). A sudden depression in H at 11 hours is followed by the disappearance of Es at the same time and a reversal of drift. The drift, f_0Es and H all regain normal values after 1430–1600.

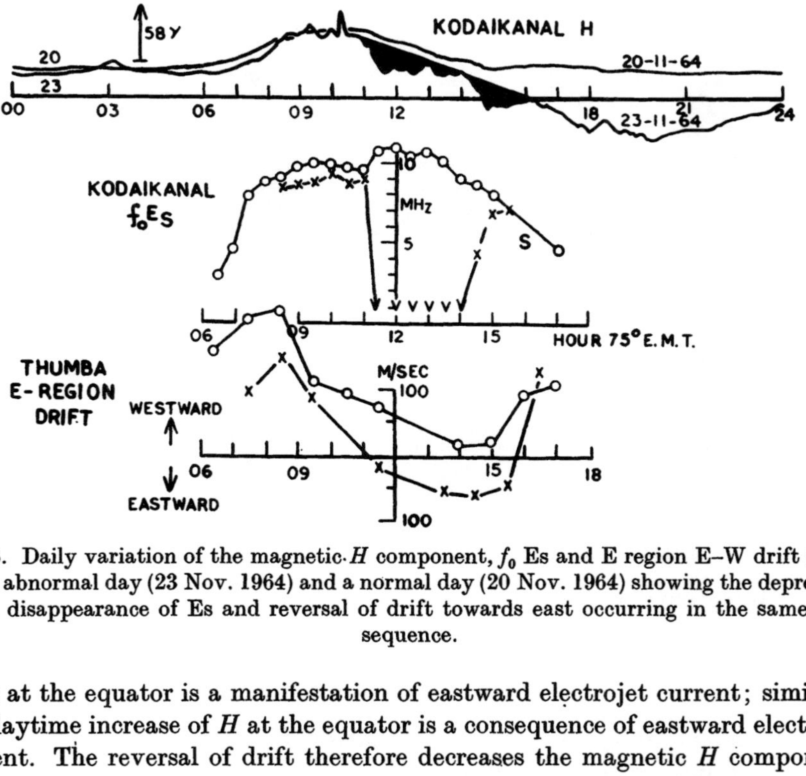

Fig. 6. Daily variation of the magnetic H component, f_0 Es and E region E–W drift speed on an abnormal day (23 Nov. 1964) and a normal day (20 Nov. 1964) showing the depression in H, disappearance of Es and reversal of drift towards east occurring in the same time sequence.

Es at the equator is a manifestation of eastward electrojet current; similarly the daytime increase of H at the equator is a consequence of eastward electrojet current. The reversal of drift therefore decreases the magnetic H component. Further it shows that the Es condition is not created when drift is eastward (westward current). Thus, even in a short time scale, the drifts at the equator are associated with the electrojet currents.

7. Solar Cycle and Drift at Thumba

The entire data at Thumba have been grouped into two periods: 1964–1966 which is a low sunspot period, and 1967–1969 which is a high sunspot period. The apparent drift speed V' and the true drift speed V in these periods are shown in Fig. 7, as histograms of percentage occurrences. Only daytime (08–16 hr) drifts are compared here. The median values are marked in the histograms by arrows. The apparent drift speeds are not changed in this period. However the true drift speeds are lower for both E and F regions during high sunspot years. These results do not tally with those of Morris and Lyon [11] for Ibadan where drift was found to be greater during high sunspot years. As the electron density N increases with solar activity, the reduction found by us can be interpreted only in terms of increase in electric fields in the magnetospheric region opposing the dynamo field.

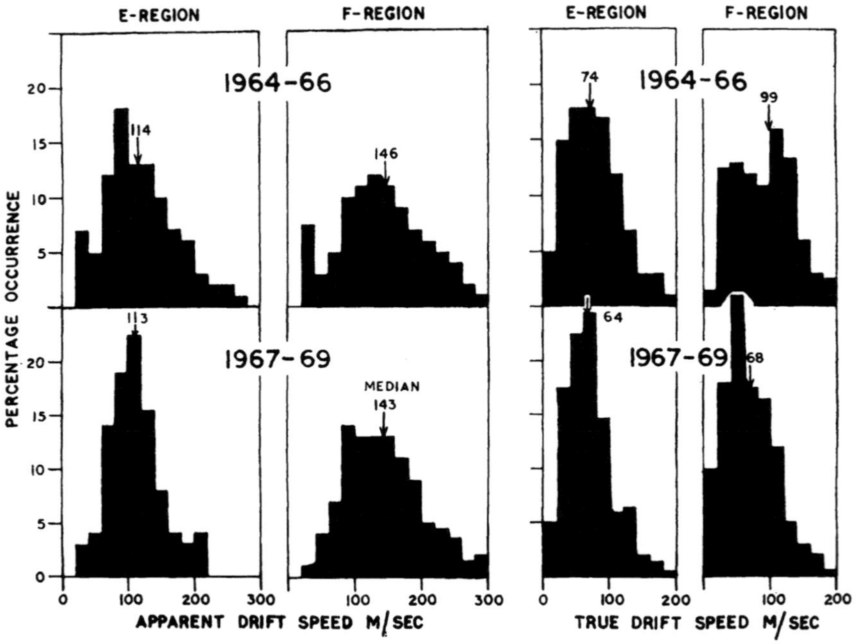

Fig. 7. Histograms of drift speed for the E and F region over Thumba for a low sunspot period (1964–1966) and a high sunspot period (1967–1969).

8. Drifts by Different Order Echoes

The apparent drift parameters were calculated from the closely spaced fadings of the first and second order echoes. The drift direction calculated from first or second reflection were almost same and are not shown in the figure. The apparent drift speeds calculated from the fadings of two echoes are shown in Fig. 8. The drift speeds as calculated from the second order echoes are always smaller than the speeds calculated by the first order echoes being about two thirds of the speed calculated by the first order echoes. This shows that the process of reflection is neither purely specular nor completely diffusive, hence division by a factor of two in the drift calculation may not be correct.

9. Spectral Variation of the Drift Speed

Some of the long fading records were subjected to cross spectral analysis following Jones and Maude [12] to find the cross spectral estimates and to study the drift speed dependence on fading frequency. Results of the few daytime cases are shown in Fig. 9. The peak power is obtained at the optimum fading frequency. The apparent drift speed increases with the fading frequency in all the cases studied, the increase being more for the F region. The drift speed at the highest significant Fourier frequency is found to be nearly equal to the apparent drift speed, while the total variation in $V'(f)$ is found to be nearly equal to the characteristic drift speed V_c. Thus the apparent drift speed cal-

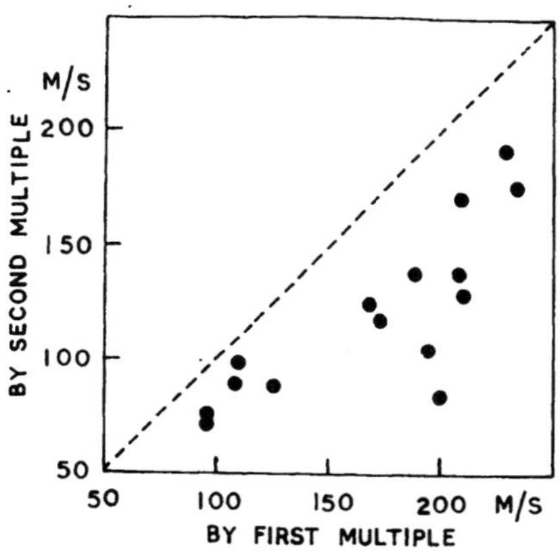

Fig. 8. Mass plot of the apparent drift speeds calculated from first and second order echoes ('multiples' in the figure) recorded simultaneously. Thumba 1968–1969.

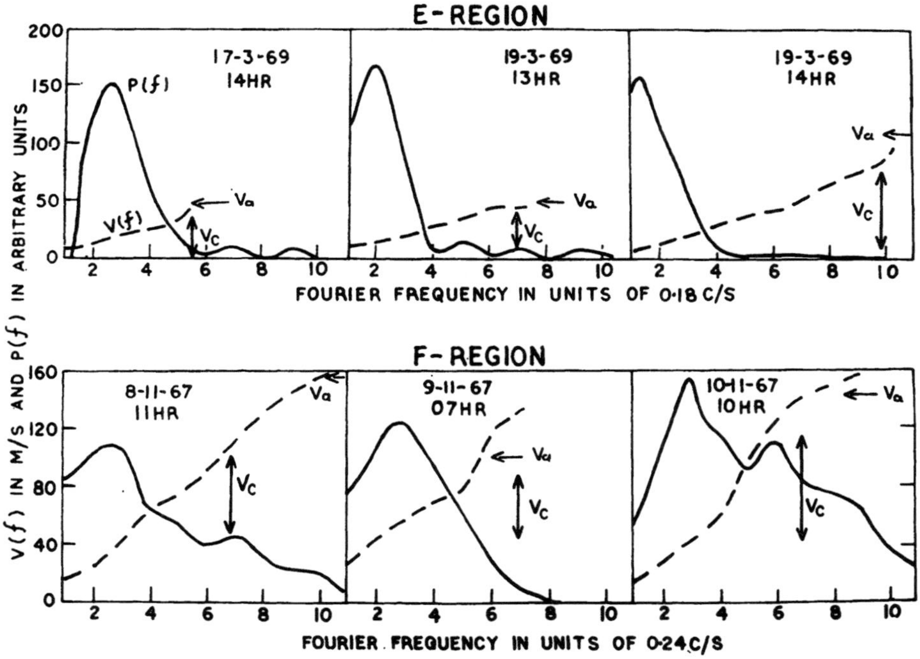

Fig. 9. Spectral variation of the apparent drift speeds for daytime fading records of the E and F region over Thumba.

culated by the correlograms is the drift speed at the highest fading frequency while the random drift V_c is caused by the spectral variation of drift speed.

10. Conclusion

The drifts measured at Thumba are a good index of the electrojet currents. The drifts are caused by the E–W electric field at the dynamo level. Electric fields of magnetospheric origin are opposite to the dynamo fields. Thus with increasing solar activity or at times of magnetic activity, the E–W electric field in the dynamo region is reduced, resulting in reduced drift speeds.

The mechanism of reflection operative at Thumba is neither completely diffusive nor purely specular. The division of the drift speed measured at ground by a factor of two is therefore doubtful.

Acknowledgments

Thanks are due to Prof. K. R. Ramanathan and Prof. V. A. Sarabhai for their keen interest in the establishment of the Ionospheric Drift Station at Thumba and to Dr. M. R. Deshpande for supervising the station during the initial stage.

References

[1] S. N. MITRA, Proc. Instn Elect. Engrs, Part III, **96**, 441 (1949).
[2] J. KRAUTKRAMER, Arch. Elek. Übertr. **4**, 153 (1950).
[3] R. G. RASTOGI, M. R. DESHPANDE and N. D. KAUSHIKA, J. Atmos. Terr. Phys. **28**, 137 (1966).
[4] B. H. BRIGGS, G. J. PHILLIPS and D. H. SHINN, Proc. Phys. Soc. B **63**, 106 (1950).
[5] G. J. PHILLIPS and M. SPENCER, Proc. Phys. Soc. B **68**, 481 (1955).
[6] G. F. FOOKS, J. Atmos. Terr. Phys. **27**, 979 (1965).
[7] K. N. NAIR, R. G. RASTOGI and V. SARABHAI, Nature, Lond. **226**, 740 (1970).
[8] J. H. CHAPMAN, Canad. J. Phys. **31**, 120 (1953).
[9] L. D. KAVANAGH, J. W. FREEMAN and A. J. CHEN, J. Geophys. Res. **73**, 5511 (1968).
[10] R. G. RASTOGI, H. CHANDRA and S. C. CHAKRAVARTY, Proc. Ind. Acad. Sci. **74**, 62 (1971).
[11] R. W. MORRIS and A. J. LYON, Nature, Lond. **310**, 617 (1966).
[12] D. JONES and A. D. MAUDE, J. Atmos. Terr. Phys. **30**, 1487 (1968).

Space Research XII – Akademie-Verlag, Berlin 1972

THE GLOBAL ENERGY BUDGET OF THE THERMOSPHERE

W. E. Swartz, J. L. Rohrbaugh and J. S. Nisbet

The Ionosphere Research Laboratory, The Pennsylvania State University,
University Park, Penn., USA

Experimental neutral atmospheric density and temperature profiles can be used to calculate the energy content variations of a given region as a function of time, season and location for which the data are available. Such calculations may then be compared with estimated energy inputs and losses to determine what other mechanisms may be important. This process is much simpler than the construction of a completely self-consistent theoretical model based on all energy sources, losses and motions. In the present paper a model based on both incoherent scatter data from Arecibo, Nançay, and Jicamarca and satellite density measurements is used to derive the global distributions of thermal energy content and losses in the thermosphere. The energy sources and transport mechanisms required for global thermal balance are then discussed.

1. Introduction

The upper atmosphere must simultaneously satisfy the mass number, momentum, and energy continuity equations. Harris and Priester [1] showed that this could not be done with a two-dimensional model using realistic parameters. This led Kohl and King [2] and others to speculate that horizontal transport provides the compressional heating at night and expansive cooling during the day necessary to satisfy the equations.

The coupled three-dimensional equations are extremely difficult to solve. However, estimates of the wind systems have been made by Kohl and King [2], and Geisler [3] using the momentum conservation equation starting with an assumed density model. If the winds do play a major role in controlling the energy budget, then the energy conservation equation should also provide a method of calculating the divergent parts of the wind system which should be consistent with the winds obtained from the momentum equation for a correct neutral model. Calculation of the wind system in this manner should also provide acheck on the need and location of any additional heat sources.

In this paper an attempt is made to calculate wind systems from the energy continuity equations for an empirical model based on incoherent scatter temperature measurements and satellite orbital decay density measurements.

2. The Neutral Model

Although the Jacchia models [4–6] reproduce the density at satellite altitudes rather well, the temperatures derived from the density scale heights have been in poor agreement with the incoherent scatter temperature measurements. The Jacchia [6] model has temperatures which on the average are closest to the incoherent scatter data except for the shape of the diurnal variation of the exospheric temperature and except for the one-to-one correspondence of exospheric temperature and temperature profile shape assumed in the model. Since the Jacchia model is basically a density model while most of the incoherent scatter data have been reduced only for the temperatures, Nisbet et al. [7] developed the Penn State MK2 to link the best parameters of these two data sources.

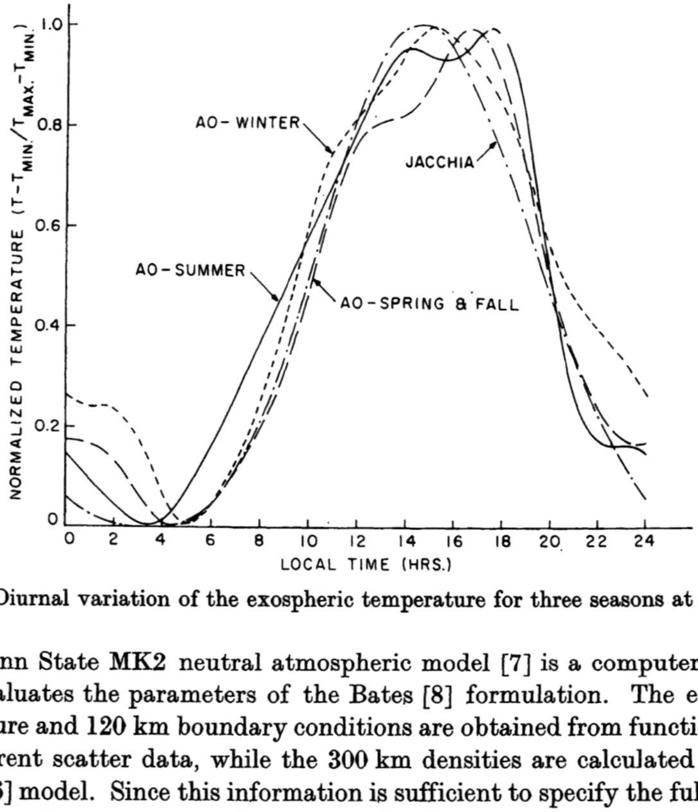

Fig. 1. Diurnal variation of the exospheric temperature for three seasons at Arecibo.

The Penn State MK2 neutral atmospheric model [7] is a computer program which evaluates the parameters of the Bates [8] formulation. The exospheric temperature and 120 km boundary conditions are obtained from functions fitted to incoherent scatter data, while the 300 km densities are calculated from the Jacchia [6] model. Since this information is sufficient to specify the full altitude distribution of temperature and density under the assumed diffusive equilibrium conditions, the program then calculates the unknown shape parameter. Using measurements from the Arecibo Observatory, Swartz and Nisbet [9] showed that changes in the temperature gradient in the lower thermosphere are consistent with the phase difference between the temperature and density variations at higher altitudes. This allows specification of a complete analytic model in terms of three separate simple models of exospheric temperature, 300 km densities, and of the temperatures and densities at some lower boundary.

The diurnal variation of the exospheric temperature has a pronounced seasonal effect as can be seen in the normalized Arecibo data presented in Fig. 1. Note that the timings of the maxima and minima shift to later and earlier hours as the day length increases from winter to summer. This may be compared with the constant shape of the Jacchia [6] model also shown in the figure. This behavior was incorporated in the model through functions of day length giving the Fourier components of the normalized variation. Incoherent scatter temperatures from Nançay and Jicamarca [10–12] were also used in developing these functions. The latitude dependence of the diurnal maximum and minimum temperatures followed a simple cosine function with its maximum at the latitude of the subsolar point as in the Jacchia model.

3. Energy Input and Loss of the Thermosphere

The two energy loss mechanisms considered here are conduction down through the lower boundary and infrared radiation by atomic oxygen. Based on the model described above the average global loss was about 0.81 erg cm^{-2} s^{-1} above 120 km, and 0.059 erg cm^{-2} s^{-1} above 250 km. The infrared radiation loss varied between 15 and 35% of the total. The energy input due to the absorption of solar euv was calculated using the Hinteregger [13] fluxes for wavelengths between 30 and 1300 Å, and the longer wavelength fluxes of Hinteregger et al. (14). Hinteregger in [13] suggested that the earlier data for wavelengths greater than 1300 Å may have been over-estimated by about a factor of three. Recently, however, he reported [15] that some new data are more in line with the larger fluxes. It was also suggested that these fluxes may have a greater variability than was previously thought possible and thus this uncertainty is reflected in the energy input calculations.

Since the total global energy input and loss must balance, the uncertain euv heating efficiency and effects of auroral heat sources may be estimated. Although there is evidence that large amounts of energy propagate out of the auroral region during magnetic disturbances, most of this energy is dissipated before reaching the equator. Thus, even if auroral heating is significant during quiet times, its influence in the lower latitudes should be small, and the boundary conditions of the divergence term of the energy conservation equation then suggest that the euv input alone should balance the losses over a 24 hour period at the equator during equinox. This equatorial balance for the given model and euv flux data required an euv heating efficiency of 38% at 120 km and 43% at 250 km. For both cases these efficiencies gave total global euv inputs that were about 10% less than the total losses. An examination of the energy equation on a global scale should show whether this difference results simply from model uncertainties or whether it is also consistent with an auroral heat source.

4. Energy Transport

By integrating the thermal balance equation given in the form used by Volland [16], one may obtain:

$$\int v\, dx_{\perp} = \frac{\iint \left[Q_{euv} - Q_{loss} + \left(\varkappa \frac{\partial T}{\partial z} \right) - \int c_p\, \varrho \frac{dT}{dt}\, dz \right] dx_{\perp}\, dx_{\parallel}}{\int p\, dz} \tag{1}$$

where $\boldsymbol{v}$ = effective velocity of energy flux (cm s^{-1}); $x_{||}$ = horizontal coordinate in the direction of $\boldsymbol{v}$ (cm); $x_{\perp}$ = horizontal coordinate perpendicular to $\boldsymbol{v}$ (cm); z = vertical coordinate (cm); Q_{euv} = column rate of euv input (erg $cm^{-2}\,s^{-1}$); Q_{loss} = column rate of loss due to the infrared radiation of atomic oxygen and viscous heating (erg $cm^{-2}\,s^{-1}$); H = scale height (cm); c_p = specific heat at constant pressure (erg $deg^{-1}\,g^{-1}$); ϱ = density (g cm^{-3}); $\varkappa$ = conductivity (erg $cm^{-1}\,s^{-1}\,deg^{-1}$); T = temperature (°K); p = pressure (erg cm^{-3}).

The right-hand side of this equation may be evaluated from the model parameters and the energy input and loss calculations. By selecting a surface geometry such that the velocity is perpendicular to its boundary as implied by the form of the integral on the left-hand side of the equation, one may perform a numerical integration along a flow line to evaluate the horizontal velocity of the energy flux.

The integration begins at the apex of a spherical triangle with its sides along the flow lines originating at a zero velocity point. Successive integrations follow two closely spaced flow lines to the second zero velocity point in the opposite hemisphere. Tesseral harmonics of the terms on the right-hand side of Eq. (1) were generated from the grid of points obtained from the calculations described in § 3. The largest uncertainties in the velocity calculations for the results to be presented here come from the assumptions about the wind directions. A wind system could be found following the methods used by Kohl and King [2] or Geisler [3] and then, with the direction known, checking the consistency of the energy balance through Eq. (1). Kohl and King found, however, that a wind system based on the momentum conservation equations follows the direction of the pressure gradient rather closely. Thus, tesseral harmonics were also generated for the pressure at 300 km, and the gradient of these harmonics provided a first approximation to the wind direction.

5. Results

Figs. 2 and 3 compare the energy inputs and losses above the 120 km and 250 km levels respectively. Although the energy inputs are essentially the same as those obtained from earlier density models, marked differences occur in the column loss rates, where the lack of symmetry may be contrasted with the perfect symmetry which would result from the Jacchia model with its one-to-one correspondence of the temperature profile shape and the exospheric temperature. Examination of the difference between the energy inputs and losses shown in Fig. 4 reveals the effect of the term dependent on the divergence of the wind velocities. It differs from the "second heat source" of Harris and Priester [1] somewhat in form and because it has a zero mean. The global wind system developed from Eq. (1) for the equinox case and medium solar flux levels is shown in Fig. 5. It is apparent that general agreement with the form of the wind system of Kohl and King has been obtained.

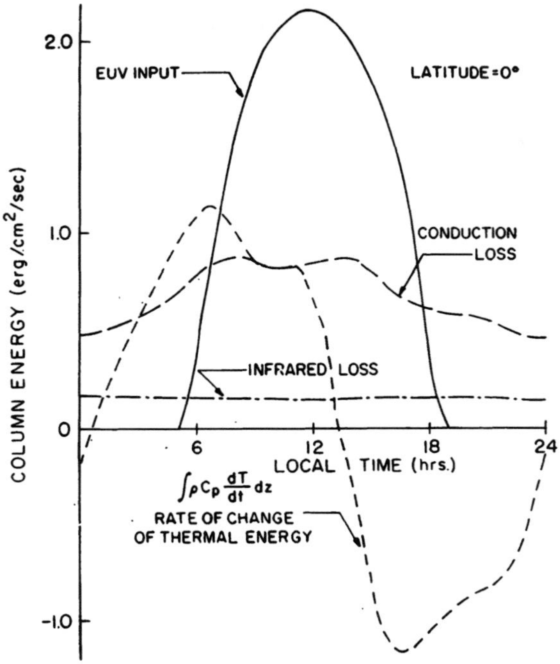

Fig. 2. Energy input and losses above 120 km at the equator.

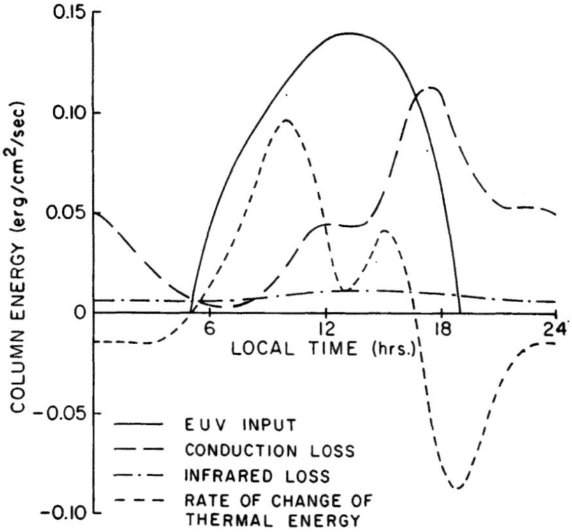

Fig. 3. Energy input and losses above 250 km at the equator.

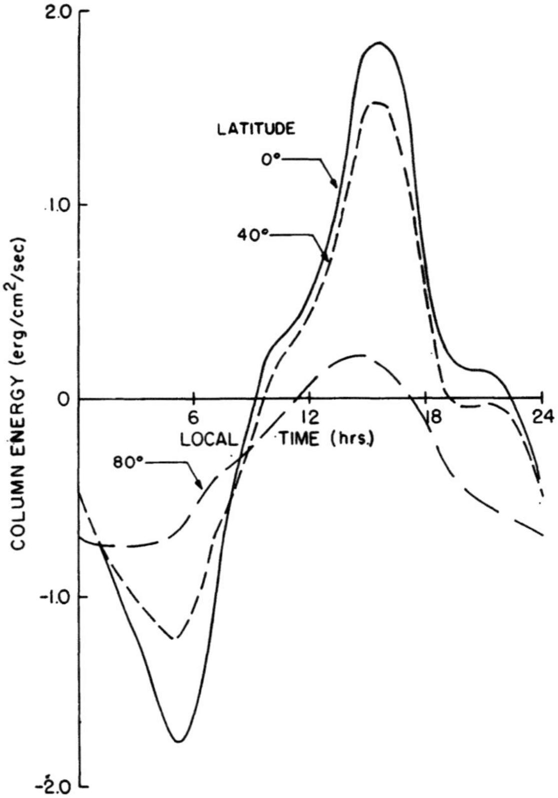

Fig. 4. Diurnal variation in the compressional heating at three latitudes.

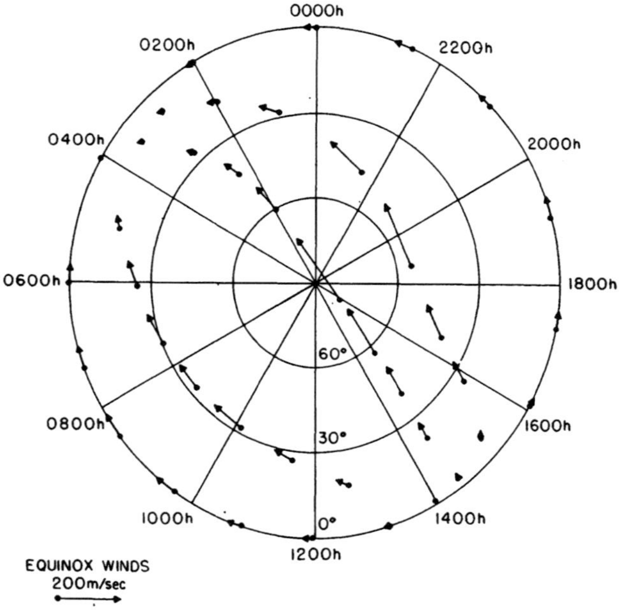

Fig. 5. Global wind map for equinox derived from the energy continuity equation.

6. Conclusions

Much work remains to be done. The temperature models in particular are based on a very limited sample of incoherent scatter sounding stations and results at higher latitudes and in the southern hemisphere are urgently required. The boundary conditions at the lower altitudes are particularly important and here incoherent scatter sounding and low altitude satellite accelerometer measurements are required. The solar flux measurements are of vital importance as is the estimation of the heating efficiency. In the present work the heat sources in the auroral region have been completely neglected. It should also be pointed out that the wind system determined here is that responsible for the divergent component and that any non-divergent steady wind component such as has apparently been measured by satellites would not be seen.

The comparison of the wind systems developed from the momentum and energy conservation equations is, however, most important in arriving at physically realistic neutral atmospheric models.

Acknowledgments

This work was supported by NASA Grant NGL 39-009-003. Mr. Swartz was a NASA trainee. The Arecibo Observatory of the National Astronomy and Ionosphere Center is operated by Cornell University under contract to the National Science Foundation.

References

[1] I. Harris and W. Priester. NASA Tech. Note D-1443, 1-71 (1962).
[2] H. Kohl and J. W. King, paper presented at XV URSI General Assembly, Munich 1966.
[3] J. E. Geisler, J. Atmos. Terr. Phys., **28**, 703 (1966).
[4] L. G. Jacchia, Smithson. Astrophys. Obs. Spec. Rep. No. 313 (1970).
[5] L. G. Jacchia, private communication, revised version of SAO Spec. Rep. No. 313 (1970).
[6] L. G. Jacchia, Smithson. Astrophys. Obs. Spec. Rep. No. 332 (1971).
[7] J. S. Nisbet, J. L. Rohrbaugh and W. E. Swartz, Trans. Am. Geophys. Un. **51**, 788 (1970).
[8] D. R. Bates, Proc. Roy. Soc. A **253**, 451 (1959).
[9] W. E. Swartz and J. S. Nisbet, J. Geophys. Res. **76**, 185 (1971).
[10] H. Carru et al., Annls Géophys. **23**, 455 (1967).
[11] H. Carru and P. Waldteufel, Annls Géophys. **25**, 485 (1969).
[12] J. P. McClure, J. Geophys. Res. **74**, 279 (1969).
[13] H. E. Hinteregger, Annls. Géophys. **26**, 547 (1970).
[14] H. E. Hinteregger et al., Space Research V, 1175 (1965).
[15] H. E. Hintergeger, Conf. on Theoretical Ionospheric Models at Pennsylvania State University (1971).
[16] H. Volland, J. Geophys. Res. **72**, 2831 (1967).

Space Research XII — Akademie-Verlag, Berlin 1972

PROPERTIES OF THE THREE-DIMENSIONAL THERMOSPHERE DYNAMICS

H. VOLLAND[a] and H. G. MAYR[b]

[a] Astronomical Institutes of the University, Bonn, FRG
[b] NASA/Goddard Space Flight Center, Greenbelt, Md, USA

Assuming spherical harmonics P_n^m as the eigenfunctions of the atmosphere dynamics above 200 km altitude, a transfer function $G_n = \varrho_n^m/Q_n^m$ can be derived which gives the ratio between density component ϱ_n^m and the corresponding heat input component Q_n^m in a development into series of spherical harmonics. G_n has the properties of a filter suppressing low periods (smaller than about 10 hours) and waves with small horizontal scales ($n > 2$). The implications of this filtering effect are discussed to show that it can explain a number of general properties in the thermosphere dynamics associated with tides, planetary waves and magnetic disturbances.

The objective of this paper is to present a very simplified three-dimensional model. It was designed to aid in the interpretation of rather general phenomena in thermosphere dynamics such as the global nature of geomagnetic storm effects or the amplitude and time delay of the tidal and planetary variations. For that purpose it was desirable to have a quasi-analytic model to enable the separation of the various energy components and their associated thermospheric structures. Therefore we started from the assumption that the following approximations be valid: (a) perturbation approximation, (b) isothermal dissipative thermosphere, (c) spherical harmonics P_n^m as the eigenfunctions of the thermo-

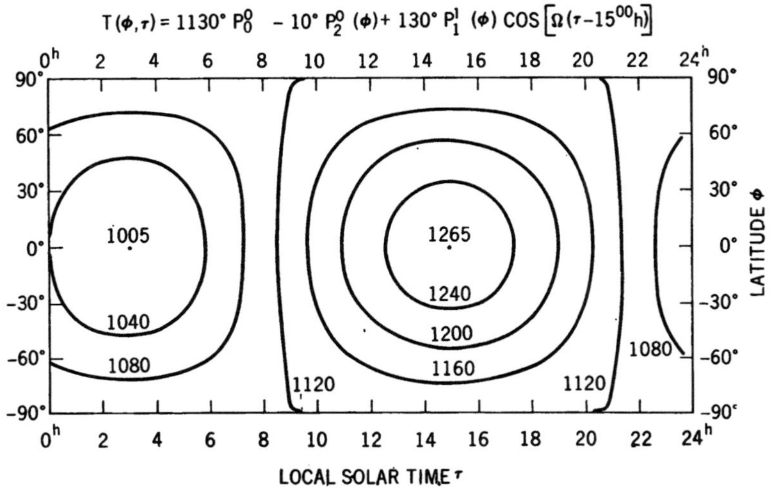

Fig. 1. Exospheric temperature distribution according to Jacchia and Slowey [1] approximated by three spherical harmonics.

sphere. These assumptions are fairly good approximations at thermospheric heights above 200 km. Non-linear effects in general do not dominate the thermosphere dynamics, the thermosphere is nearly isothermal above 200 km, and the thermospheric structure above this height justifies the spherical analysis.

This last point is illustrated in Fig. 1, which gives the exospheric temperature distribution with local time and with latitude in terms of three spherical harmonics of lowest degree. This distribution which is valid during equinox approximates within 5% the exospheric temperature derived by Jacchia and Slowey [1] from satellite drag density data. In Fig. 1 the dominant term is the term with $P_0^0 = 1$ which describes the spatial and temporal average of the temperature. The term with $P_1^1 = \sin \vartheta$ (ϑ is the co-latitude) depends on local time and is responsible for the pressure and temperature bulge during the afternoon at the equator. It has a peak at 15 hours local time. A less significant zonal term with P_2^0 depending only on latitude is added which gives rise to a zonally averaged temperature gradient directed from the equator to the poles. Its existence can be explained as resulting from the mean surplus of solar energy deposited at lower latitudes when compared with the energy input at higher latitudes. Several other quite interesting terms have been neglected for simplicity in Fig. 1, e.g. a semidiurnal term which shifts the maximum of the pressure bulge from 15 hours to the observed 14 hours local time. Local time dependent terms with wave domain numbers $m > 0$ can be considered as tidal waves. The term with P_1^1 is the fundamental symmetric diurnal tidal wave mode. The other terms with number $m = 0$ are called planetary wave modes. Although Jacchia's exospheric temperature is only a measure for the observed density, Fig. 1 nevertheless shows that the large scale features of thermosphere dynamics appear to be rather simple and can indeed be described in terms of few spherical harmonics of lowest degree (n, m).

It can be shown from an analysis of three-dimensional thermosphere dynamics that the spherical harmonics become, to a fair degree of accuracy [2], the eigenfunctions of the thermosphere above 200 km. If one assumes the heat input Q developed into a series of spherical harmonics and if one assumes a height profile of the coefficients Q_n^m in this series proportional to the mean pressure p_0, then one finds an asymptotic relationship between the heat input coefficient Q_n^m and the corresponding density coefficient ϱ_n^m:

$$\varrho_n^m = G_n(\omega)\, Q_n^m \,.$$

The function $G_n(\omega)$, depending only on the zonal wave domain number in our simplified theory, is a function of the angular frequency ω of the respective wave. It can be considered as a system transfer function which essentially represents the efficiency for the excitation of the density amplitudes by the respective heat input component. Numerical calculations, using plausible data within the height range between about 300 and 400 km, were made to derive magnitude and phase of the transfer function G_n. It is shown in Fig. 2 as function of frequency ω or of period t (the upper scale in Fig. 2), the parameter n being the zonal wave domain number.

From Fig. 2a we notice a low-pass filtering effect which is due to the long thermospheric response time. Waves with periods smaller than about 10 hours are suppressed except for large numbers n in a window near periods of about 2 hours which is just the range of internal gravity waves. This results from resonance effects within the thermospheric cavity.

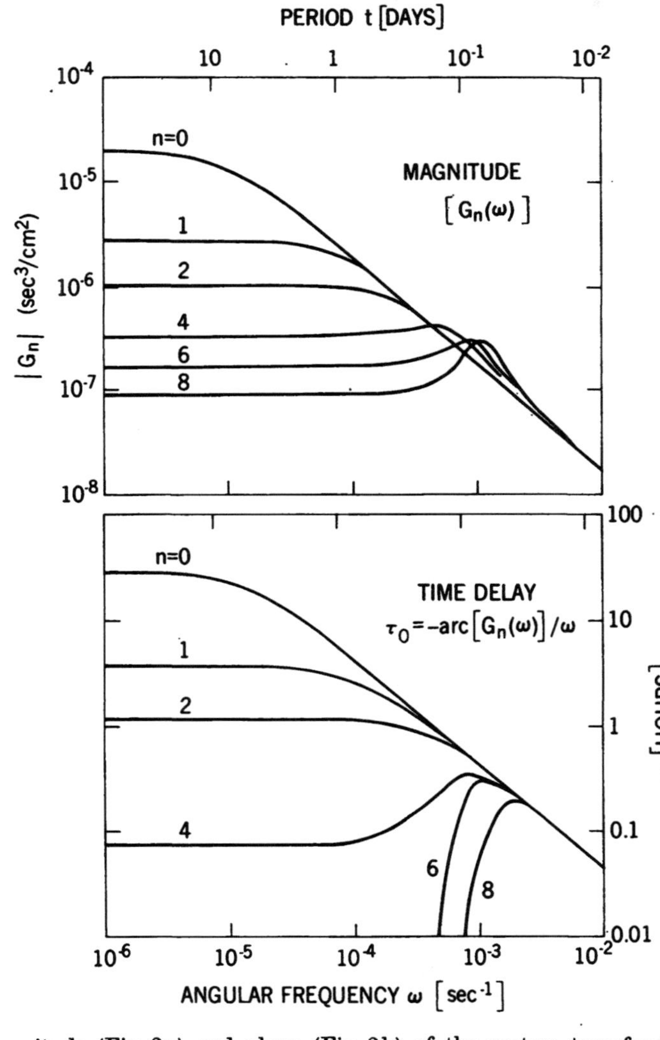

Fig. 2. Magnitude (Fig. 2a) and phase (Fig. 2b) of the system transfer function $G_n(\omega)$ versus angular frequency ω. $G_n(\omega) = |G_n(\omega)| \exp(-j\,\omega\,\tau_0)$. The parameter is the zonal wave domain number n. The time delay has been determined from the phase by the relation $\tau_0 = -\varphi/\omega$.

A second filtering effect is apparent in connection with the increasing zonal wave domain number n. Since n is a measure for the meridional scale length of the respective spherical harmonics P_n^m, we can interpret this filtering effect as due to the increasing number of circulation cells of the wind field associated with the density distribution ϱ_n^m. Both density ϱ_n^m and the associated winds are generated by the heat input term Q_n^m. If the horizontal wind structure is simple or even non-existent as in the case of $n = 0$, most of this heat input is transferred into internal energy of the gas, that means, into temperature and density amplitudes. With increasing number n, an increasing fraction of the energy is needed to drive the wind field, and a decreasing amount of heat input remains to increase the corresponding density amplitudes.

This can be seen most drastically in the case of zonal wave domain number $n = 0$ which describes the world-wide mean increase of the density. Here nearly the entire solar heat input is transferred into internal energy of the gas, increasing in a most efficient way the density and temperature of planetary waves with periods larger than about 1 day. However, in the case of $n = 1$ which includes the fundamental diurnal tidal wave of Fig. 1, a significant fraction of the corresponding solar heat input component is transferred into kinetic energy to drive the Kohl–King–Geisler wind [3, 4] which reduces the available fraction of the heat input to increase the density. The same is true for the fundamental symmetric semidiurnal wave with wave domain numbers (2,2). Although the corresponding solar heat input coefficient Q_2^2 is about half the value of the fundamental diurnal component Q_1^1, the filtering effect of the thermosphere furthermore reduces the density amplitude by a factor of about 3, so that the ratio between diurnal and semidiurnal density component is about 6:1 in rough agreement with the observations. From Fig. 2b we note a time delay of 3 hours in the case of the diurnal wave ϱ_1^1 and a delay of 1 hour in the case of the semidiurnal wave, both in agreement with Jacchia's observations [1].

Fig. 2a and 2b allow us also to explain the results of the Harris–Priester model [5] and the meaning of the second heat source. Harris and Priester, using a one-dimensional model, treated the ϱ_0^0 component with a period of one solar day. According to Fig. 2, this leads to a large density amplitude and a time delay of about 6 hours. However, the correct density component of the diurnal tide is the ϱ_1^1 which has the right amplitude and phase in agreement with the observations. The second heat source of Harris and Priester was therefore an ingenious trick to simulate the effect of the kinetic energy in the (1,1) component. Moreover, this simple model can describe general features of the geomagnetic activity effect [6]. During a geomagnetic storm, energy stored within the magnetosphere is deposited within a small strip within the auroral ovals in the night-time hemisphere. If the geomagnetic storm is of the substorm type with a duration of a few hours, it generates predominantly short periodic gravity waves and does not affect significantly the large scale density features. If, however, a big storm occurs with an impulse width of about one day, then a sufficient fraction of its spectral energy is transferred into the ϱ_0^0 and the ϱ_2^0 components of the density. The time response of the ϱ_0^0 component is about 6 hours while that of the ϱ_2^0 component decreases to 1 hour. The sum of the ϱ_0^0 and the ϱ_2^0 components gives then a larger density amplitude and a smaller time delay at the poles than at the equator which is in general agreement with the observations [7, 8].

It is obvious that such a simplified model as presented here cannot describe details of the thermosphere dynamics, and it cannot replace in any way empirical models like the Jacchia model [9]. However, it seems to be helpful in the understanding of rather fundamental dynamic properties of the thermosphere in which temporal and spatial distributions of heat sources are involved.

References

[1] L. G. JACCHIA and J. SLOWEY, Space Research VII, 1077 (1967).
[2] H. VOLLAND and H. G. MAYR, NASA-Document X-621-71-240, GSFC, Greenbelt, Md. USA 1971.
[3] H. KOHL and J. W. KING, J. Atmos. Terr. Phys. **29**, 1045 (1967).
[4] J. E. GEISLER, J. Atmos. Terr. Phys. **29**, 1469 (1967).
[5] I. HARRIS and W. PRIESTER, J. Atmos. Sci. **19**, 286 (1962).
[6] H. VOLLAND and H. G. MAYR, J. Geophys. Res. **76**, 3764 (1971).
[7] M. ROEMER, Space Research XII, 797 (1972).
[8] L. L. DE VRIES, Space Research XII, 777 (1972).
[9] L. G. JACCHIA, Smithson. Astrophys. Obs. Spec. Rep. No. 332 (1971).

Space Research XII — Akademie-Verlag, Berlin 1972

CORPUSCULAR RADIATION AS AN UPPER ATMOSPHERIC ENERGY SOURCE

W. P. Olson

Space Sciences Department, McDonnell Douglas Astronautics Company, Huntington Beach, Calif., USA

The deposition of energy into the upper atmosphere by low energy charged particles is examined quantitatively. Most of this corpuscular energy is carried by (<10 keV) protons originating in the solar wind. Recent satellite observations show that these particles enter the magnetosphere directly in the high latitude dayside regions and that about 10% travel along magnetic field lines toward the dayside auroral atmosphere. Programs were developed to study the guiding center motion of charged particles through the magnetosphere and their interaction with the atmosphere. It is found that enough energy is placed in the atmosphere by these particles that they must be considered an important atmospheric energy source. The largest amount of energy deposited by protons is 8×10^5 eV cm^{-3} s^{-1} at a height of 168 km. The continuous precipitation of low energy protons should heat the atmosphere around 75° geomagnetic latitude near noon and influence dayside auroras and the conductivity of the high latitude ionosphere. The remaining particles move back into the tail of the magnetosphere. Their trajectories are very complex and are controlled by several magnetospheric processes which at present are not well understood.

1. Introduction

The ways in which energy is deposited in the atmosphere by precipitating charged particles are examined. The source of most of these particles is a neutral plasma (the solar wind) which moves continuously outward from the sun and past the earth. The path of corpuscular energy (from its source) in the solar wind to where it is deposited in the atmosphere is calculated and the amount of heating in the atmosphere is determined as a function of position and source strength.

Indirect evidence that precipitating charged particles are an important energy source for the upper atmosphere is provided by satellite drag which is noticeably enhanced at auroral latitudes [1, 2, 3]. The latitude dependence cannot be explained in terms of a source that deposits energy at all latitudes (e.g. solar electromagnetic radiation). It can however, be understood in terms of charged particles constrained by the magnetic field to precipitate only at high latitudes. Additional indirect evidence is provided by the observation that increases in the density of the upper atmosphere are associated with large values in the planetary magnetic disturbance index, Kp. The magnitude of Kp is known to correlate well with the intensity of auroras, solar wind parameters, and other magnetospheric phenomena associated with charged particles. Recent observations of charged particle fluxes in the high latitude, outer magnetosphere provide *direct* evidence of the importance of corpuscular radiation as an atmospheric energy source. Particle detectors on board ISIS 1 and IMP 5 have observed solar wind (or magnetosheath) particles to move directly into the magnetosphere (in the "dayside cusp" region, see Fig. 1) and travel along magnetic field lines down to the high latitude upper atmosphere [4, 5, 6]. These data help to explain the observed dependence of atmospheric heating on latitude. About one tenth of the entering plasma moves along field lines toward the dayside atmosphere at auroral latitudes. The remainder flows back into the antisolar region of the magnetosphere. The amount of energy entering the magnetosphere during quiet times is about 5×10^{18} ergs s^{-1} [5]. This is sufficient to drive the magnetospheric phenomena which depend on the presence of charged particles.

2. Satellite Observations of Plasma Entry into the Magnetosphere

Data on the geometry of the dayside cusps (see Fig. 1) and the characteristics of the particle fluxes observed there are given by [4, 5, 6]. Data from the cusp region near the magnetopause are not used since most of the particles present there propagate back into the tail of the magnetosphere. The cusp at

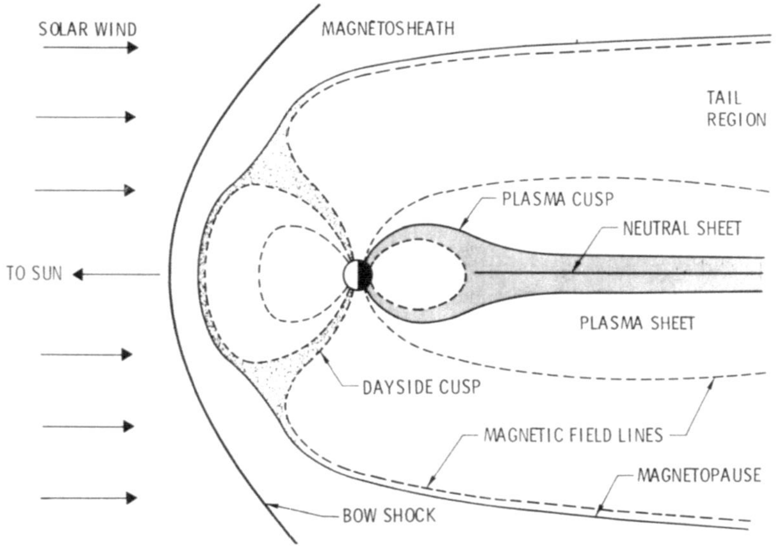

Fig. 1. The dayside cusp connects the magnetopause with the dayside auroral atmosphere. The "mid-altitude" region of the dayside cusp has a geocentric distance of 5–7 R_e.

"mid-altitude" (at a geocentric distance of 5–7 earth radii) has an oval cross section and is centered on the magnetic noon meridian plane. It is about 1 earth radius (R_e) in height along the noon meridian plane and extends about 8 R_e in longitude (about 70°). In this region the peak proton intensity (2×10^4 particles cm^{-2} sr^{-1} s^{-1} eV^{-1}) occurs at about 1 keV while the peak electron intensity (about 5×10^4 particles $cm^{-2}\,sr^{-1}\,s^{-1}\,eV^{-1}$) is at 100 eV. The protons carry the bulk of the energy. To determine the fraction of the mid-altitude particles that will enter the atmosphere, the pitch angle distribution must be known. Only particles with angles 6° or less at mid-altitude will have mirror points within the atmosphere. The pitch angle distribution is peaked toward $\alpha = 0°$ but only the range $150° < \alpha < 30°$ is observed [5]. However, other recent satellite observations of ions at auroral altitude suggest a large anisotropy in pitch angles. The intensity of 6 keV protons with $\alpha = 10°$ is frequently observed to be 50 times the intensity of the same energy particles with $\alpha = 80°$ [7]. Here it is assumed that 10% of the mid-altitude particles have pitch angles less than 6°. This means that about 1% (or 5×10^{16} ergs s^{-1}) of the energy carried by the low energy particles when they enter the magnetosphere is deposited in the dayside high latitude atmosphere. These data from the dayside cusps are used as inputs to study the trajectories of low energy particles and their interaction with the atmosphere.

3. Particle Trajectories

Because of the low particle densities throughout the magnetosphere, it can be assumed that charged particles moving through the magnetosphere do not suffer collisions except when they enter the upper atmosphere. The maximum cross section for all processes and all particle species is about 2×10^{-16} cm^2 and occurs for incident particle energies of 1–10 keV. The particle number density in the magnetosphere is almost everywhere less than 100 cm^{-3}. Thus, the average mean free path is about 10^9 km which is larger than the scale length of the magnetosphere.

Charged particles in the magnetosphere are therefore influenced only by their electromagnetic environment. The magnetospheric magnetic field model used here consists of three sources, a dipole field representing the main (internal) field of the earth, and scalar potential representations of the fields produced by the magnetopause and tail current systems. This quantitative magnetospheric magnetic field model was developed by the author [8, 9] and is available from the National Space Science Data Center. A computer code was developed to trace the motion of the guiding center of particles in this field. This motion averages out the particle gyrations about the magnetic field [9, 10]. The guiding center program was used to determine the extent in latitude and longitude of the intersection of the polar cusp with the atmosphere. The extent in latitude is found to be very narrow, about 0.5° (or 50 km) while the extent in longitude is larger, about 100° (3000 km). Note that this is very similar to the extent in longitude of the mid-altitude dayside cusp while the extent in latitude for the cusp is much larger (about 10°). The guiding center program was also used to provide a cursory examination of propagation of particles from the distant dayside cusp (near the magnetopause) to the tail of the magnetosphere. It is found (as expected) that the particles do not appreciably cross field lines but propagate through a region close to the magnetopause along field lines back to the outer regions of the tail. How the particles subsequently penetrate to the plasma sheet and then precipitate was not investigated in detail. However, the same mechanisms that are responsible for the formation of the dayside cusp may also be present in the tail and permit the entry of particles into the (plasma sheet) region where the total magnetic field is small.

4. Interaction with the Atmopshere

The mean free path λ of a particle moving in the atmosphere is given by

$$\lambda = \frac{1}{\sigma N}$$

where σ is the total cross section for energy loss to the atmosphere and N is the number density of the atmospheric gas. N is determined from the total atmospheric density and mean molecular weights obtained from models. Below

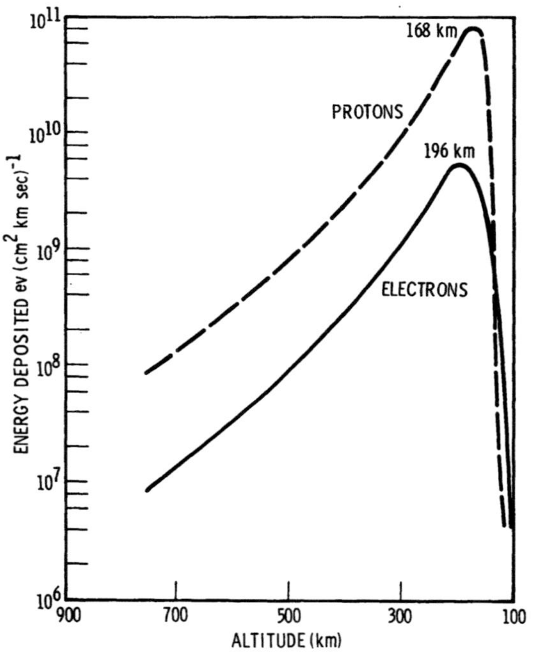

Fig. 2. The energy deposited is given in eV $cm^{-2}\,km^{-1}\,s^{-1}$. The electron and proton contributions are equal at 133 km. The data are from periods when $Kp \leqq 3.0$. The intensities shown are about 2.5 larger than those present when $Kp \leqq 1.0$.

120 km the USSAS 1966 model was used and from 120 to 800 km the CIRA 1965 model 1 was used. (CIRA 1965 models 5 and 9 were also examined and led to results almost identical with model 1.) The total cross sections are different for electrons and ions. For electrons in the energy range 100 eV to

10 keV, collisional ionization with O, O_2, and N_2 is the dominant loss process. For protons above 5 keV, collisional ionization is the most important process while below 1 keV charge exchange dominates. It is assumed that the total cross sections are constant. For protons $\delta = 1.9 \times 10^{-16}$ cm² and for electrons $\delta = ^2 0.9 \times 10^{-16}$ cm². This is justified for both protons and electrons because of their fairly flat cross sections over the range of energies considered here (less than several keV) and because the cross sections for O_2, N_2 and O are very similar. In one mean free path the particle loses about 35 eV.

Because of their larger cross sections, protons do not penetrate as far into the atmosphere as electrons. The particle energy loss per unit altitude increases monotonically with pitch angle since particles with large pitch angles move a larger distance through the atmosphere for a given decrease in altitude. Throughout the region where the dayside cusp particles interact with the atmosphere, the angle between the vertical direction and $\boldsymbol{B}$ is about 7.5°. For protons (or electrons) the energy loss per distance traveled through a given region of the atmosphere is the same for all incident particle energies since the proton (or electron) cross section is assumed to be independent of particle energy.

For constant cross sections, all particles penetrating to a given altitude deposit the same amount of energy. At that altitude, the amount of energy deposited per kilometer is multiplied by the total number of particles that have not yet been thermalized to obtain total energy loss (see Fig. 2). The protons deposit an order of magnitude more energy than the electrons for all altitudes above 150 km. The altitude at which the electrons deposit the maximum amount of energy is higher than that for the protons even though the electron cross section is smaller than the proton cross section because the electron energy spectrum is peaked toward low energies (~100 eV). At 133 kms the proton and electron energy fluxes are equal. Below that level the electrone rapidly become the primary energy source. However, at these altitudes, the total amount of energy deposited is down more than $2\frac{1}{2}$ orders of magnitudn from the peak values in the 150–200 km range. At the peak of the proto. intensity (168 km), the amount of energy deposited is 0.136 erg cm^{-2} s^{-1} km^{-1} The total amount of energy deposited in the atmosphere is about 1.5 ergs cm^{-2} s^{-1} and agrees with observations [12].

The rate of heating of the atmosphere is given by

$$\frac{dT}{dt} \cong \frac{dQ}{dt}\left(\frac{1}{m\, c_v}\right)$$

where Q is the energy input, T is the temperature and c_v is the specific heat at constant volume ($c_v \cong 0.1715$ cal g^{-1} deg^{-1}). The mass m is determined from the atmospheric model. The proton heating rate within the cusp varies from 0.0001 degK s^{-1} at 110 km to 1.82 degK s^{-1} at 800 km. Above 200 km protons with initial energies below 2 keV produce over 90% of the heating. At atmospheric heights the cusp region extends over 100° in longitude so a given point (in geographical coordinates) will be heated for over six hours. The heating rates have been used to calculate the *total* change in temperature,

ΔT, as a function of the unperturbed temperature $\bar{T}$. Above 175 km $\Delta T/\bar{T}$ is greater than 1 and at 700 km it is about 19. These values consider only the corpuscular energy as a source. Certainly, there are several ways in which the energy is dissipated and transported out of the cusp region.

The maximum temperature increase should occur in the region of the atmosphere that has just passed through the cusp. The local time of maximum heating should occur about two hours after passage through the center of the cusp (on the noon meridian in solar-magnetic coordinates). The position of the center of the cusp at atmospheric heights was calculated and found to vary from 0835 to 1525 geographic LT in hemispheric winter and from 0858 to 1502 LT in the summer. It ranges in geographic latitude from 86.5° at 0500 UT to 63.5° at 1700 UT throughout the year. This suggests a density bulge in the upper atmosphere whose geographic position ranges 23° in latitude and over six hours in local time (centered on ~1500 LT). The present analysis therefore does not agree with the geographically fixed atmospheric density bulges suggested previously.

The relative contributions of solar electromagnetic and corpuscular radiation to auroras, atmospheric heating, and ionization can be determined from their latitudinal (discussed above) and seasonal dependences. To study the seasonal dependence, the zenith angle of the center of the cusp was calculated as a function of time of day and year. For most of the day near hemispheric winter solstice, the center of the cusp at 200 km altitude is in total darkness. At that time particle precipitation should be the main direct source of heating and ionization. Precipitating low energy particles also influence the near earth environment by enhancing the conductivity of the high latitude F-region ionosphere. A large portion of the incident energy is transferred to the atmosphere by collisional ionization. For example, at 168 km collisions of protons with neutral atmospheric particles produce about 5×10^3 ions $cm^{-2}\,s^{-1}$. Typical electron densities at that altitude range from 10^5 to 10^6 cm^{-3}. Thus, for the corpuscular radiation to be considered an important ionization source, the electron lifetimes at that position must be of the order of 100 seconds. At 168 km the formation of negative ions is negligible and the dominant loss mechanism is the dissociative recombination of molecular ions. For that process the lifetime of electrons is about 50 seconds. This suggests that the cusp particles may be a major source for maintaining the dayside auroral ionosphere. Indeed, charged particles are also responsible for the ionization of the polar D region [13]. However, those particles are of higher energy (~50 keV) than the ones studied here. The same higher energy particles (~40 keV protons and ~6 keV electrons) may contribute significantly to dayside auroras. The geomagnetic latitude of maximum auroral intensity at noon is near 75° and coincides with the region of precipitation of the dayside cusp particles.

5. Conclusions

In this study, dayside cusp data were used to determine the extent of atmospheric heating by low energy charged particles. The dayside cusp particles are found to produce an atmospheric density bulge in mid afternoon around 75° geomagnetic latitude. In addition, they help maintain the high latitude ionosphere and contribute to dayside auroras. Quiet time mid-altitude cusp data ($Kp \leqq 1.0$) were used throughout this study. These charged particles will exert an even greater influence on the atmosphere during periods of increased magnetic activity.

References

[1] L. L. De Vries, Space Research XII, 777 (1972).
[2] R. R. Allan, Space Research XII, 857 (1972).
[3] R. W. Bruce, Paper presented at COSPAR Meeting, Seattle 1971.
[4] W. J. Heikkila, Nature, Lond. **225**, 369 (1970).
[5] L. A. Frank, University of Iowa Report No. 70:55 (1970).
[6] W. J. Heikkila and J. D. Winningham, J. Geophys. Res. **76**, 883 (1971).
[7] B. Hultqvist, H. Borg, W. Reidler and P. Christophersen, Planet. and Space Sci. **19**, 279 (1971).
[8] W. P. Olson, J. Geophys. Res. **74**, 5642 (1969).
[9] W. P. Olson, McDonnell Douglas Astronautics Co. Rep. No. WD-1332 (1970).
[10] H. Alfvén, Cosmical Electrodynamics, Clarendon Press, Oxford 1950.
[11] T. G. Northrop, The Adiabatic Motion of Charged Particles, Interscience, New York 1963.
[12] L. A. Frank and K. L. Ackerson, J. Geophys. Res. **76**, 3612 (1971).
[13] W. J. Helms and H. M. Sworm, J. Geophys. Res. **74**, 6341 (1969).

Space Research XII — Akademie-Verlag, Berlin 1972

DYNAMICS OF THE THERMOSPHERE

R. E. Dickinson

National Center for Atmospheric Research, Boulder, Col., USA

Dynamics of the thermosphere in the past has referred to either (a) determination of the distribution of the major constituents of the neutral thermospheric gas by vertical diffusion calculations, or (b) calculation of horizontal winds which are driven by pressure forces coupled to the thermodynamics by the hydrostatic and continuity equations for the gas as a whole. Past studies along these lines are reviewed. It is now time to recognize that bulk transport is probably as important as vertical diffusion in determining the global compositional structure of the thermosphere. A knowledge of thermospheric composition, in turn, is required to determine realistically the response of the winds and temperatures to solar radiation. The two viewpoints referred to above are thus actually closely coupled parts of the same problem. The intent of this paper is to provide a theoretical framework for thermospheric dynamics that recognizes this coupling of composition and mass diffusion on the one hand with motions and temperature on the other hand.

1. Introduction

The problem of the dynamics of the thermosphere is defined here as the derivation of time and spatial dependence of the macroscopic variables of the neutral gas, starting with incoming solar radiation and other basic physical parameters. In other words, dynamics refers to the deduction of the bulk motions, density, composition, and temperature. Historically, this designation has been applied to widely different theoretical approaches emphasizing the calculation of one or another of the above observables, assuming the remaining are known. It perhaps has not been obvious why vertical diffusion calculations and calculations of horizontal winds should both be referred to as "dynamics". One reason, of course, is that both calculations involve solution of an equation of motion. It should be clear from the discussion to follow that a unified, self-consistent theory of thermospheric dynamics involves the joint deduction of all the neutral gas parameters. To limit somewhat the subject, I assume the existence of the observed ionosphere and knowledge of relevant atmospheric processes at the mesopause.

First, the different approaches which have been developed in the past to calculate one or another of the dynamic variables are briefly summarized, some of the difficulties that have indicated the necessity for a more unified formulation are pointed out and some indication given of progress that has been made in joint calculations of wind and temperature. Second, the theoretical formulation of thermospheric dynamics for fixed composition is reviewed in a simple context, and finally, it is suggested on the basis of scaling arguments that the coupling between winds and composition is also an important aspect of thermospheric dynamics.

2. Past Approaches

One past approach has been that of Kohl and King [1] and Geisler [2], among others, who have derived horizontal wind systems in the thermosphere from Jacchia's [3] empirical models for global density. The dependence of these winds on horizontal gradients of density below 200 km where drag data are sparse appears to be one of the weakest links in this approach. Further low-level drag data [4, 5], mass spectrometry [6], and improved description of the temperature structure in the lower thermosphere through incoherent scatter measurements [7, 8] should decrease the uncertainty of empirically derived pressure forces. Thermospheric winds also depend on the concentration and possible drift of the ionization. Rishbeth [9], for example, has recently discussed equatorial F layer ionization drift as a possible eastward momentum source.

A second approach to thermospheric dynamics has been to emphasize the sources, sinks, and vertical transport of the major neutral constituents of the thermosphere in determining thermospheric density and composition [10–12]. The most significant physical process is the dissociation and subsequent downward transport of O to a region above the mesopause where it recombines into O_2. This O_2 must then eventually return to higher levels to compensate for the O_2 lost by dissociation. One-dimensional calculations may provide an adequate description of the global mean distribution of composition. However, the deviation from the global mean of the O_2 photodissociation rate is comparable in magnitude with its mean value. Indeed, there is no photoproduction whatsoever of O in the winter hemisphere polar night region where observations indicate no great deficit of O and there is no reason for downward transport to cease. It is thus clear that latitudinal and also vertical transport by large-scale circulations [13, 14] must be as significant as photoproduction and vertical diffusion in determining the latitudinal distribution of O.

A third approach followed by Harris and Priester [15] and others [16, 17] has been to attempt to derive the diurnal and spatial variation of the thermospheric temperature field through the solution of a one-dimensional heat conduction equation assuming a knowledge of thermospheric heat sources. Unfortunately, the adiabatic redistribution of the heat by atmospheric motions is comparable in magnitude with the direct solar heating. This point is readily established assuming that the horizontal wind systems derived from Jacchia's model densities give at least a crude description of the actual thermospheric winds [18, 19]. For nearly isothermal conditions, the dominant heating term is largely given by the product of the gravitational acceleration g with that part of vertical motion due to horizontal mass convergence.

One-dimensional heat conduction models are appropriate for discussion of global mean transients in the temperature and density.

When it was realized that the theoretical deduction of temperature was closely linked with the calculation of winds, theory was developed for the coupling between the winds and temperatures and simple models were integrated to illustrate the importance of these effects [20, 21]. It was recognized [21] that appropriate approximations introduced on the basis of scaling arguments lead to a system of equations physically and mathematically equivalent to those considered in the theory of atmospheric tides [22] with the addition of thermal conduction in the thermodynamic equation and viscous and ion drag terms in the equations of motion. Volland [23, 24] discussed the possible role of upward tidal propagation on the basis of a simple, heuristic, two-dimensional model neglecting viscosity.

It is not possible to employ the usual separation of variables procedure of tidal theory when the drag terms and viscosity are present. Lindzen has recently developed a formalism to avoid this difficulty in evaluating the propagation of tidal oscillations from the lower atmosphere into the thermosphere [25, 26]. The procedure is to solve for the upward propagation into the thermosphere of gravity waves with the same vertical wavelength as the various tidal modes. Lindzen's theory indicates that tides from the lower atmosphere have only a small effect on the diurnal component of thermospheric winds, temperature, and density, but produce large semidiurnal components, indeed considerably larger than indicated by current observational description. More attention should be paid to the resolution of observations into diurnal and semidiurnal components. As Volland [27] has pointed out, the difference between the phase of satellite drag densities and incoherent scatter is greatly reduced if the phases of the diurnal components (wavenumber one) are compared rather than the times at which the maximum values are achieved.

At low-enough levels most of the radiative heating is balanced by adiabatic cooling or storage of heat rather than heat conduction. For long-period motions where the thermal storage is also negligible, net radiative heating and adiabatic cooling should nearly balance. Johnson and Gottlieb [28] have used this balance to estimate the mean meridional circulation in the lower thermosphere.

3. Theoretical Formulation of Thermospheric Dynamics

Those unfamiliar with planetary-scale atmospheric hydrodynamics experience some difficulty in following presentations of the various dynamic models for thermospheric motions. This difficulty is due in part to the great complexity of "exact" equations, so it is necessary, at the start of an analysis, for a large number of more or less acceptable approximations to be introduced. Furthermore, unrealistic assumptions are usually also made to derive models intended to demonstrate important physical processes in a sufficiently simple framework. A "realistic" model of planetary thermospheric dynamics is necessarily a three-dimensional time-dependent numerical solution of the coupled dynamic and thermodynamic equations [29]. Such a calculation has not yet been successfully carried out and would be impossible without the prior development and thorough understanding of a hierarchy of simpler, i.e. more or less unrealistic, models. Our first example of an "unrealistic" assumption is to assume a rotating Cartesian (x, y, z) geometry where z is in the upward direction. This Cartesian geometry does not change the physics of the equations, but a spherical geometry must be used for realistic calculations. First, we consider the equation for neutral horizontal motion to calculate the horizontal vector velocity $\boldsymbol{v}$. The substantial derivative, i.e. time change following a fluid parcel, is written

$$d/dt = \partial/\partial t + \boldsymbol{v} \cdot \nabla + w\,\partial/\partial z$$

and the motion equation is

$$\frac{d\boldsymbol{v}}{dt} + \lambda\,(\boldsymbol{v} - \boldsymbol{v}_{\mathrm{i}}) + 2\,\Omega \times \boldsymbol{v} - \frac{1}{\varrho}\frac{\partial}{\partial z}\mu\frac{\partial \boldsymbol{v}}{\partial z} = -\frac{1}{\varrho}\nabla p\,. \tag{1}$$

Here ϱ is the neutral gas density.

The equation merely states that momentum added by gradients of pressure p is balanced by the following terms: (i) the fluid acceleration $d\boldsymbol{v}/dt$ in the rotating frame; (ii) an exchange of momentum through collisions with ions moving with velocity $\boldsymbol{v}_{\mathrm{i}}$ (λ is an "ion drag" coefficient proportional to the ion number density); (iii) a Coriolis acceleration (Ω is the rotation vector); (iv) a viscous drag (μ is viscosity coefficient). If p is known, the left-hand side is readily integrated at a given point (x, y) as a two-dependent variable partial differential equation in height and time. This is the Kohl–King [1], Geisler [2] type of calculation. See also [30–35].

The $\boldsymbol{v}_{\mathrm{i}}$ must be determined through solution of the ion equations of motion. At F layer heights, the plasma is locked to the lines of the earth's magnetic field which do not move in the absence of electric fields. The further assumption has sometimes been made that the horizontal component of plasma drift along field lines is small so that $\boldsymbol{v}_{\mathrm{i}} = 0$, eliminating coupling with the plasma dynamics. In the E layer, the ions driven by collisions move as the Hall current at right angles to the neutral velocities, accelerating the neutral particles in the opposite direction to the Coriolis force.

If polarization electric fields are neglected, there is no difficulty in calculating the ion densities and velocities. A realistic three-dimensional dynamo model has not yet been integrated for given winds, so presumably solution of a neutral wind model which includes realistic calculation of ion drifts due to polarization electric fields would be quite difficult. The effects of ion drifts given by polarization electric fields on neutral winds are apparently of second-order importance with the possible exception of equatorial latitudes [9]. See, however, [36].

Scale analysis indicates that the total acceleration $d\boldsymbol{v}/dt$ can be approximated by the local acceleration $\partial v/\partial t$ with relative error less than unity for planetary-scale motions having velocities up to several hundred m s^{-1}. This approximation has been commonly used since the resulting approximate equation is linear in the velocities and so easier to solve. The viscous drag term has sometimes also been discarded, further reducing Eq. (1) to an ordinary differential equation in time. The omission of vertical diffusion of momentum by viscosity is of doubtful validity, especially during night and above the F2 layer peak. Horizontal diffusion of momentum by viscosity, on the other hand, is usually negligible. For planetary-scale motions, the ratio of horizontal to vertical diffusion terms, in general, is order of $(H/r)^2 \sim 10^{-4}$ where H is neutral scale height, r is radius of the earth. The only significant horizontal diffusion process for planetary-scale dynamics is the diffusion of neutral particles through ions as described by the ion drag term. Other horizontal diffusion processes will be omitted without further discussion in the remainder of this paper.

The pressure temperature and density are simply related by the equation of state for an ideal gas

$$p = \varrho\, R^*\, T/m \tag{2}$$

where R^* is the universal gas constant, T is the kinetic temperature of the neutral gas, and m is the average mass of the gas mixture in atomic units. The current empirical models of thermospheric structure which have been used for wind calculations are constructed to fit satellite drag observations of density ϱ and the still rather crude description of thermospheric composition which, together with assumed hydrostatic balance of each constituent, gives m as a function of height.

It is clear that the actual densities, temperature, composition, and hence pressures are ultimately determined by solar heating, so the equation for the neutral response to radiative heating is a very important constraint on the thermospheric structure. An understanding of how the thermodynamic equation is satisfied is thus basic to understanding thermospheric dynamics. This equation is written here

$$c_p \frac{dT}{dt} - \frac{1}{\varrho}\frac{\partial}{\partial z} k \frac{\partial T}{\partial z} - \frac{1}{\varrho}\frac{dp}{dt} = Q \tag{3}$$

where k is the average heat conduction coefficient, c_p is the average specific heat at constant pressure for given composition, and Q is the rate of radiative heat addition per unit mass. Heat addition is balanced by the three terms from left to right: (i) $c_p\, dT/dt$ which is the rate of change of internal energy following a fluid parcel; (ii) the loss of heat by vertical conduction; and (iii) $\varrho^{-1}\, dp/dt$ the warming by adiabatic compression. We can write

$$\frac{dT}{dt} = \frac{\partial T'}{\partial t} + w\frac{\partial T_0}{\partial z} + \left[u\frac{\partial T'}{\partial x} + w\frac{\partial T'}{\partial t}\right]$$

where T_0 is a specified standard atmosphere depending only on z, $T = T_0 + T'$. It is usually possible for a crude first approximation to omit the nonlinear terms in brackets as is done in tidal theory [22].

The compression term dp/dt can be related to that part of the downward vertical motion determined by horizontal mass divergence, so to lowest order the adiabatic cooling depends only on upward vertical motion. This does not mean, however, that horizontal motions are not also required. For example, in considering conservation of total energy, we can relate the compressional heating term to horizontal transport of potential energy. It is essential in calculating temperature to include at least the linear part of the adiabatic cooling which is always comparable in magnitude with the radiative heating.

An especially simple formulation of the theory is obtained if we transform our vertical coordinate to $Z = \log (p_0/p)$ where p_0 is some reference pressure. The compressional heating is then proportional to

$$\frac{dp}{dt} = -p\frac{dZ}{dt} = -p\,W \tag{4}$$

where W is the vertical motion associated with the new coordinate. The equation for mass continuity is then simply written

$$\nabla \cdot \boldsymbol{v} + \frac{\partial W}{\partial Z} - W = 0 \tag{5}$$

which gives us the vertical motion determining the adiabatic cooling or warming in terms of horizontal velocities obtained by integrating Eq. (1). The description of the coupling between the motions and temperature is now completed by prescribing how the pressure force is derived from the temperature field through the hydrostatic equation

$$-\frac{1}{p}\frac{\partial p}{\partial z} = \frac{g\,m}{R^*\,T}. \tag{6}$$

For using the pressure coordinates, we regard z in Eq. (6) as the dependent variable and pressure (or Z) as the independent variable and we replace the pressure force in the equations of motion with the gradient of height z on a pressure surface

$$\frac{1}{\varrho}(\nabla p)_z = g(\nabla z)_p. \tag{7}$$

The equations summarized above form a complete set for the calculation of the neutral gas macroscopic variables provided we know the composition and thus m and c_p. The system of equations has been closed in the past by assuming a "standard atmosphere" description of the composition. The idea that will be stressed in the next section is that this assumption is probably invalid from the viewpoint of realistic dynamic calculations.

4. Coupling of Dynamics with Composition

The role of motions in determining the atomic oxygen distribution is apparent when we compare time scales for photoproduction, bulk transport, and diffusion as indicated in Table 1 at three levels. The photoproduction time is the time required to convert O_2 to O at a level in the absence of transport. I have used the recently suggested low Schumann–Runge photon fluxes [37]. The bulk transport times are defined as the time required for planetary-scale motions to transport mass 5000 km assuming typical velocities of 5, 15 and 50 m s^{-1} at 100, 120 and 160 km, respectively [28]. Vertical motions determined by continuity must transport mass vertically through a distance of the order of a scale height in these times. The diffusion time is a lower limit defined as the time scale on which the O in a fully mixed mixture is restored to diffusive equilibrium. Molecular diffusion is capable in the mean of transporting O to around the

Table 1
Time scale in days of atomic oxygen sources and sinks

Level (km)	160	120	100
Photoproduction time (day/night average)	10	20	100
Bulk transport time	1	3	10
Diffusion transport time	0.1	1.0	10.0

100 km level. Below this level, the combined effects of eddy mixing and photochemistry sharply deplete the O concentration.

The departure of partial pressure gradients from diffusive equilibrium can be estimated by comparing the source time scales with the diffusion time scales. According to this argument, photoproduction is sufficiently rapid to produce departures from diffusive equilibrium which are of the order of 1% at 160 km to the order of 10% at 100 km. The scale analysis indicates that bulk transport is as much as an order of magnitude more effective in determining departures from diffusive equilibrium at a given location than is photoproduction and thus may largely control the latitudinal distribution of atomic oxygen. The inverse of the photoproduction time gives the amplitude of daily variation of atomic oxygen mixing ratios in the absence of transport. These values range from 0.01 at 100 km to 0.1 at 160 km. Diffusion should limit the maximum variation to around 0.05. Diurnal vertical velocities with amplitudes ranging from 10 cm s^{-1} at 120 km to 50 cm s^{-1} at 160 km will produce variations of the O mixing ratio of about 0.1.

The above scaling arguments indicate that transport by large-scale motions is as significant a factor as photoproduction in determining the latitudinal and diurnal variability of atomic oxygen. We now discuss how compositional variations, in turn, influence the motions.

Below 100 km, variations in the atomic oxygen concentration are largely significant in producing a fluctuating heat source [13, 38, 39]. At higher levels, compositional variations influence the distribution of mass and thus pressure in a fashion similar to temperature variations, as is seen from the hydrostatic equation Eq. (6). It is convenient to introduce a virtual temperature T_v defined as the temperature obtained if the actual mean molecular weight $m(x, y, Z, t)$ is replaced by some standard atmosphere value $m_0(Z)$ depending only on pressure. That is, the virtual temperature is defined by

$$T_v(x, y, Z, t) = T(x, y, Z, t)\, m_0(Z)/m(x, y, Z, t)\,. \tag{8}$$

The hydrostatic law giving the height of a constant pressure surface is then written

$$\frac{\partial h}{\partial Z} = R\, T_v \tag{9}$$

where the "gas constant" $R = R^*/m_0(Z)$.

Diffusive equilibrium model atmospheres are usually calculated from fixed constituent number densities at a lower boundary. The resulting average molecular weight depends essentially only on pressure. Thus, for example, Jacchia's models generate virtual temperatures whereas, on the other hand, incoherent scatter observations give actual temperatures. The determination of pressure forces for motion calculations requires directly the virtual temperature. For this reason, the virtual temperature may be a more useful dependent variable than the actual temperature. When the thermodynamic equation is rewritten in terms of virtual temperature, it acquires an additional source term, essentially proportional to the substantial derivative of the mass mixing ratio of atomic oxygen. In other words, the virtual temperature can be increased following a parcel either by addition of heat or by increasing the relative amount of O. Horizontally varying sources and sinks of atomic oxygen can drive motions by changing the virtual temperature. For example, diurnal production of O which provides 5% diurnal oscillation in the mean molecular weight at a given pressure level where the temperature is order of 500 °K gives the same change of virtual temperature as the compressional warming provided by a diurnal vertical motion of amplitude 10 cm s^{-1}.

In the absence of mass sources, the consideration of possible compositional variation for a given motion time scale suggests three regimes. At sufficiently high levels for diffusion to be fast relative to the motion–time scale, the deviations of mean molecular weight produced by the motions will be unimportant. On the other hand, at levels low enough for diffusion to be slow relative to the motion–time scale, the average molecular weight will be conserved following a fluid parcel. The restoring buoyancy force acting on the parcel above 100 km is increased as a consequence. In the thermodynamic equation written for T_v, the mass source term will be negligible.

Finally, the most interesting situation occurs where the diffusion and motion time scales are comparable. Diffusion partially relaxes to a basic state the value of the mass mixing ratio following a parcel of air. This process damps the temperature and thus dissipates the motions. For motions with vertical scales small compared with the O scale height, this dissipation can be comparable in magnitude with dissipation by viscosity and thermal conduction. For motions with very large vertical scales, this partial compositional relaxation could be more effective than viscosity and heat conduction in dissipating the motions. For example, the upward propagating semidiurnal tide [26] should be significantly damped by this process between 140 and 200 km.

Acknowledgments

The National Centre for Atmospheric Research is sponsored by the National Science Foundation.

References

[1] H. KOHL and J. W. KING, J. Atmos. Terr. Phys. **29**, 1045 (1967).
[2] J. E. GEISLER, J. Atmos. Terr. Phys. **29**, 1469 (1967).
[3] L. G. JACCHIA, Smithson. Astrophys. Obs. Spec. Rep. No. **313** (1970).
[4] D. G. KING-HELE and D. M. C. WALKER, Planet. Space Sci. **17**, 985 (1969).

[5] B. K. CHING, J. Geophys. Res. **76**, 197 (1971).
[6] G. SWENSON, J. Geophys. Res. **74**, 4074 (1969).
[7] P. BAUER, P. WALDTEUFEL and D. ALCAYDE, J. Geophys. Res. **75**, 4825 (1970).
[8] W. E. SWARTZ and J. S. NISBET, J. Geophys. Res. **76**, 185 (1971).
[9] H. RISHBETH, Planet. Space Sci. **19**, 357 (1971).
[10] F. D. COLEGROVE, W. B. HANSON and F. S. JOHNSON, J. Geophys. Res. **70**, 4931 (1965).
[11] T. SHIMAZAKI, J. Atmos. Terr. Phys. **30**, 1279 (1968).
[12] T. J. KENESHEA and S. P. ZIMMERMAN, J. Atmos. Sci. **27**, 831 (1970).
[13] W. W. KELLOGG, J. Met. **18**, 373 (1961).
[14] G. A. M. KING, J. Atmos. Sci. **21**, 231 (1964).
[15] I. HARRIS and W. PRIESTER, J. Atmos. Sci. **22**, 3 (1965).
[16] C. P. LAGOS and J. R. MAHONEY, J. Atmos. Sci. **24**, 88 (1967).
[17] D. R. BATES and R. J. MOFFETT, Planet. Space Sci. **16**, 1531 (1968).
[18] R. E. DICKINSON and J. E. GEISLER, Mon. Weath. Rev. **96**, 606 (1968).
[19] H. RISHBETH, R. J. MOFFETT and G. J. BAILEY, J. Atmos. Terr. Phys. **31**, 1035 (1969).
[20] H. VOLLAND, Space Research VII, 1093 (1967).
[21] R. E. DICKINSON, C. P. LAGOS and R. F. NEWELL, J. Geophys. Res. **73**, 4299 and 6867 (1968).
[22] S. CHAPMAN and R. S. LINDZEN, Atmospheric Tides, D. Reidel, Dordrecht-Holland 1970.
[23] H. VOLLAND, Planet. Space Sci. **17**, 1581 (1969).
[24] H. VOLLAND and H. G. MAYR, Annls Géophys. **26**, 907 (1970).
[25] R. S. LINDZEN, Geophys. Fluid Dynamics **1**, 303 (1970).
[26] R. S. LINDZEN, Geophys. Fluid Dynamics **2**, 89 (1971).
[27] H. VOLLAND, J. Geophys. Res. **75**, 5618 (1970).
[28] F. S. JOHNSON and B. GOTTLIEB, Planet. Space Sci. **18**, 1707 (1970).
[29] M. P. FRIEDMAN, Smithson. Astrophys. Obs. Spec. Rep. No. 250 (1967).
[30] P. STUBBE, J. Atmos. Terr. Phys. **30**, 1965 (1968).
[31] G. J. BAILEY, R. J. MOFFETT and H. RISHBETH, J. Atmos. Terr. Phys. **31**, 253 (1969).
[32] G. VASSEUR, J. Atmos. Terr. Phys. **31**, 397 (1969).
[33] R. A. CHALLINOR, Planet. Space Sci. **17**, 1097 (1969) and **18**, 1485 (1970).
[34] H. R. CHO and K. C. YEH, Radio Sci. **5**, 881 (1970).
[35] D. F. STROBEL and M. B. MCELROY, Planet. Space Sci. **18**, 1181 (1970).
[36] N. MATUURA, J. Atmos. Terr. Phys. **30**, 763 (1968).
[37] H. E. HINTEREGGER, Annls Géophys. **26**, 547 (1970).
[38] C. YOUNG and E. S. EPSTEIN, J. Atmos. Sci. **19**, 435 (1962).
[39] C. B. LEOVY, J. Atmos. Sci. **23**, 223 (1966).

Space Research XII — Akademie-Verlag, Berlin 1972

THERMOSPHERIC CONVECTIVE INSTABILITY

Y. T. Chiu

Space Physics Laboratory, The Aerospace Corporation, El Segundo, Calif., USA

It is shown that the high speed winds observed by the 1967-50B (LOGACS) satellite near 200 km altitude at the north polar cap during the historic magnetic storm of 26 May 1967 may be interpreted as convective motion due to excessive heating of the neutral polar thermosphere. The observed persistent wind pattern implies a convective solution while the high wind speed necessitates an instability theory that allows small stable oscillations to grow to anomalously high amplitudes. It is shown that an unstable temperature profile can be set up by energy deposition equivalent to a fraction of the measured storm ring current loss. The thermohydrodynamic instability of an exponential atmosphere heated from below allows calculation of the horizontal wavelength and the ratio between vertical and horizontal wind speeds. It is pointed out that there is evidence of an unstable density profile existing concurrently with the winds.

The thermosphere, having a temperature gradient increasing with altitude, is generally stable against thermal convection. Here we wish to offer arguments that the stability may be destroyed if excessive heating takes place in a thin layer during a large magnetic storm. The characteristics of such convective flow turn out to be in fairly good agreement with the cross-track winds observed by the LOGACS satellite on 26 May 1967 [1].

The winds were recorded when the satellite passed in near circular orbit, within approximately 30° of the north pole at an altitude of 160–220 km. Approximately 8 hours after the peak of the magnetic storm at the polar regions, wind patterns were seen to persist for 15 hours with very little change. The horizontal components of wind velocity exhibit well organized direction changes with a characteristic length of 2000–4000 km. Since the horizontal component changes very smoothly within one horizontal scale length, the altitude change of the satellite within that horizontal length allows us to determine the minimal vertical characteristic length, which turns out to be 30–60 km. The maximal horizontal speed observed was approximately 1 km s^{-1}, while the maximal vertical speed was 100 $\pm$ 100 m s^{-1}. There seems to be an increase of horizontal speed with altitude of less than one e-fold. The growth phase of the wind pattern seems to last one or two orbits.

To attempt a physical interpretation of these observations, we first point out that a gravity wave interpretation is unlikely because it would require a phase coherence between wave and satellite orbit in order to account for the persistence of the wind pattern. Secondly, although ions can be made to move with such high speeds by appropriate electric and magnetic fields, the low altitude occurrence of the event would require the ions to overcome the inertia of the predominant neutral particles. We are inclined to a third possibility, namely, that the wind pattern is the result of a fully developed convective instability due to excessive heating of a thin layer. Stationary convective flow is certainly in agreement with the observed long duration of wind patterns. Furthermore, the density observations of the same experiment indicate that a density inversion, a necessary and sufficient condition for convective instability, seems to exist at the same spatial and temporal locations.

The primary pre-condition for convective instability of a fluid layer heated from below is the setting up of an adverse temperature gradient, i.e. decreasing temperature with altitude in the fluid layer. Thomas and Ching [2] have

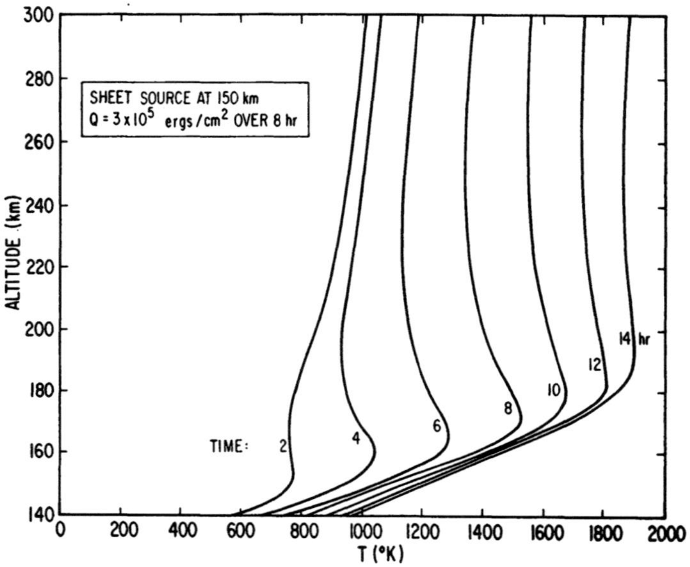

Fig. 1. Thomas–Ching model temperature profile of a CIRA 2 atmosphere heated by a sheet source at 150 km depositing 3×10^5 ergs cm^2 spread over the polar caps, equivalent to 7×10^{18} ergs s^{-1} energy deposition rate over 8 hr.

studied models of atmospheric response to heating at 150 km. These models, when subjected to a moderate amount of heating to account for global density changes, show no tendency to form unstable inverse temperature gradients. Our central question is: what must the energy input be in order to set up an adverse temperature gradient for a thin layer of the atmosphere; and, more important, can the energy input required be obtained from the energy loss of the storm? Fig. 1 shows the temperature profile calculated from the Thomas–Ching model in which a total energy of 3×10^5 ergs cm^{-2} is deposited over eight hours as a sheet source at 150 km. The above amount of energy deposited over an area within 30° of both poles is equivalent to a total energy dissipation rate of 7×10^{18} ergs s^{-1} over eight hours. This amount, in turn, is approximately one tenth of the total ring current loss of the storm. From Fig. 1 we see that a temperature inversion occurs in a layer of approximately 60 km for several hours with a time delay of approximately four hours. Thus, a layer with an adverse temperature gradient can be set up locally if energy deposition during a large magnetic storm is increased by one order of magnitude over the global density response calculations.

Convective instability of a compressible fluid layer heated from below has been studied extensively by many authors (see [3] for further references). Their conclusions are that if the layer thickness is smaller than or comparable with the scale height, the convective flow of the compressible case is not substantially different from the incompressible case studied by Pellew and Southwell

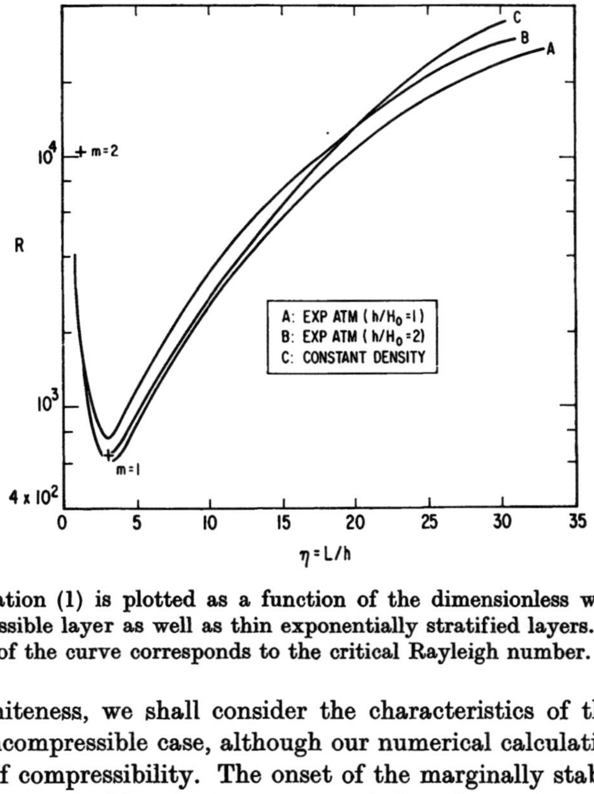

Fig. 2. The relation (1) is plotted as a function of the dimensionless wavelength η for a thin incompressible layer as well as thin exponentially stratified layers. The minimum of the curve corresponds to the critical Rayleigh number.

[4]. For definiteness, we shall consider the characteristics of the convective flow for the incompressible case, although our numerical calculations will show some effects of compressibility. The onset of the marginally stable convective normal mode, identified by the integer m, and the relation between horizontal scale length L, layer thickness h and the parameters of the fluid can be distilled into the solution of the following eigenvalue equation:

$$R = 16\,\pi^4 \frac{1}{\eta^4}\left(1 + \frac{m^2\,\eta^2}{4}\right)^3, \qquad (1)$$

in which $\eta = L/h$ and the Rayleigh number R is defined by

$$R = g\,c_p\,\varrho_0\,h^3\,\Delta T/\varkappa\,\nu\,T_0 \qquad (2)$$

and g is the gravitational acceleration, c_p is the specific heat at constant pressure, ϱ_0 is the density at the middle of the compressible layer [3], ΔT is the temperature difference across the layer, T_0 is the temperature at the middle of the layer, $\varkappa$ is the heat conductivity and ν is the kinematic viscosity. Putting the middle of the inversion layer at 190 km, we have $c_p \sim 10^7$ ergs g^{-1} deg^{-1}, $\varkappa \sim 10^4$ ergs cm^{-1} s^{-1} deg^{-1} [5], $\nu \sim 10^9$ cm^2 s^{-1} [6]. From Fig. 1, we have $h \sim 60$ km and $\Delta T/T_0 \sim 10^{-1}$, while $\varrho_0 \sim 5 \times 10^{-13}$ g cm^{-3}. For these values we obtain $R \sim 9700$. In Fig. 2, we show the relation (1) between the Rayleigh number R and the scale ratio η. In the linear analysis, (1) is valid only near the minimal value of R, marked by crosses for $m = 1$ and 2 modes. For the nonlinear case, all modes up to the maximum size determined by (1) are allowed. For our value of R, the $m = 1$ mode is excited while the $m = 2$ mode may be marginally excited. For the $m = 1$ mode solution, we obtain the maximal value $L/h \sim 20$, which implies that the horizontal scale length is 1200 km. Since $R \propto h^3$, the value of L determined is sensitive to h; however, our predicted value of L is in fair agreement with observations.

The ratio between horizontal and vertical flow speed maxima for the $m = 1$ mode can also be obtained from the normal mode eigenfunction via the equation of continuity

$$v_{x,\max}/v_{z,\max} \simeq \tfrac{1}{2}\eta \sim 10 . \tag{3}$$

Although the vertical wind component is difficult to determine from the observations due to the large error, it is fairly certain that the horizontal components are much larger than the vertical components as indicated by (3).

Thus, we see that the characteristics of the convective flow at onset show substantial agreement with the characteristics of the observed wind patterns. The only assumptions that we have made are the magnitude of the heat input and the validity of the Thomas–Ching model of atmospheric heating. Furthermore, we have shown that the energy input required is but a very small fraction of the total ring current loss to the atmosphere. Although there is observational evidence that a density inversion may exist concurrently with the winds, we must emphasize that the present consideration deals with the characteristics rather than the magnitude of the winds. For finite amplitude convection, the extremely anisotropic modes considered here are susceptible to finite amplitude instability.

Acknowledgment

This work was conducted under US Air Force Space and Missile Systems Organization (SAMSO) Contract No. F04701-70-C-0059.

References

[1] W. A. FEESS, Rep. TOR-0200(9990)-1, The Aerospace Corp., El Segundo, Calif. (1968).
[2] G. E. THOMAS and B. K. CHING, J. Geophys. Res. **74**, 1796 (1969).
[3] E. A. SPIEGEL, Astrophys. J. **141**, 1068 (1965).
[4] A. PELLEW and R. V. SOUTHWELL, Proc. Roy. Soc. A **176**, 312 (1940).
[5] I. HARRIS and W. PRIESTER, J. Atmos. Sci. **19**, 286 (1962).
[6] H. KOHL and J. W. KING, J. Atmos. Terr. Phys. **29**, 1045 (1967).

Space Research XII – Akademie-Verlag, Berlin 1972

THEORETICAL MODEL OF DIURNAL VARIATIONS OF THE EQUATORIAL THERMOSPHERE AT EQUINOX

M. N. Izakov, S. K. Morozov and E. E. Shnoll

Institute for Space Research, Moscow, USSR

Within the limits of the spatially two-dimensional non-stationary model, the diurnal variations of temperature, density and winds in the equatorial thermosphere during the equinox at the altitudes 120–320 km are calculated.

A comparison between one- and two-dimensional models shows the importance of interconnection of thermal regime and dynamics, which may lead to a variable shape of diurnal density and temperature variations, and in some cases to supersonic motion of the atmospheric gas relative to the heat source, and to non-linear effects in wind distribution.

Calculations of theoretical models of the thermosphere help us not only to improve the picture of processes which take place there, but also to clear up the physical mechanisms of the dependence of thermosphere parameters on determining factors.

Such calculations are rather urgent because of some discrepancies between the data obtained with different methods (e.g. [1, 2]), because the mechanisms of some variations (for example, semi-annual variations) are not yet clear. A suggestion has been made that the solar wind can make a contribution to the thermal regime of the thermosphere comparable with that of solar xuv radiation, even during magnetically quiet periods [3].

1. Previously it has been shown [4–6] that, though a full description of the thermosphere in terms of generalized Boltzmann equations and radiation transfer equation is needed, at least up to the altitudes of 300–400 km (where the ratio of molecular mean free path to scale height remains considerably less than unity), the system of Navier–Stokes hydrodynamic equations, with heat source due to absorbed xuv solar radiation in the right-hand side of the energy balance equation is sufficient. However, this system is still very complicated and its solution is a matter for the future. Let us limit ourselves to the problem of consideration of diurnal variations of thermosphere structure above the equator at equinox. Then structure parameter distributions are evidently symmetrical relative to the equator, and we can assume in some approximation, neglecting the meridional component, that motion near equator is zonal. This allows us to consider a spatially two-dimensional problem with coordinates r along the vertical and φ the longitudinal angle from west to east (the coordinate system is stationary, $\varphi = 0$ corresponds to the earth–sun direction) and t is the time. Then, neglecting some second-order terms [6], we obtain the following

system of equations:
continuity equation:

$$\frac{\partial \varrho}{\partial t} + \frac{1}{r^2}\frac{\partial}{\partial r}(r^2 \varrho v_r) + \frac{1}{r}\frac{\partial}{\partial \varphi}(\varrho v_\varphi) = 0 \tag{1}$$

equation of motion (horizontal component):

$$\varrho\left(\frac{\partial v_\varphi}{\partial t} + v_r \frac{\partial v_\varphi}{\partial r} + + \frac{v_\varphi}{r}\frac{\partial v_\varphi}{\partial \varphi}\right) = -\frac{1}{r}\frac{\partial p}{\partial \varphi} + \frac{\partial}{\partial r}\left(\eta \frac{\partial v_\varphi}{\partial r}\right) + \frac{1}{2}\varrho_i \nu_{in}(v_{i\varphi} - v_\varphi) \tag{2}$$

barometric low (obtained from vertical component of equation of motion [5, 6]):

$$\frac{\partial p}{\partial r} = -\frac{M g}{R_0 T} \tag{3}$$

energy balance equation:

$$\left.\begin{aligned} &\varrho c_v\left(\frac{\partial T}{\partial t} + v_r \frac{\partial T}{\partial r} + \frac{v_\varphi}{r}\frac{\partial T}{\partial \varphi}\right) + \frac{1}{r^2}\frac{\partial}{\partial r}\left(r^2 \lambda \frac{\partial T}{\partial r}\right) + \eta\left(\frac{\partial v_\varphi}{\partial r}\right)^2 + \\ &\quad + p\left(\frac{1}{r^2}\frac{\partial (r^2 v_r)}{\partial r} + \frac{1}{r}\frac{\partial v_\varphi}{\partial \varphi}\right) = Q_{ST} - Q_{IR} + \frac{1}{2}\varrho_i \nu_{in}(v_i - v)^2 \end{aligned}\right\} \tag{4}$$

equation of state:

$$p = \varrho \frac{R_0}{M} T . \tag{5}$$

Here, as usual, ϱ, p, T, M are density, pressure, temperature and molecular weight of gas, R_0 is universal gas constant, η is a coefficient of viscosity, λ is a coefficient of thermal conductivity; ϱ_i, v_i are density and velocity of ions, ν_{in} is collision frequency for ions and neutrals ; Q_{ST}, Q_{IR} are heat source and sink that will be defined below.

This system does not contain continuity equations for separate components, because, as one can see from Eqs. (1)–(5) the variations of composition influence the thermal regime and dynamics only through the variations of molecular weight M, and evaluations (due to available model data [1, 7]) have shown that diurnal variations of M are less than $\pm 5\%$. Therefore, at the beginning we computed $M = \sum M_\alpha n_\alpha (\sum n_\alpha)^{-1}$ from the model values of n_α (as well as coefficients of viscosity η and thermal conductivity λ that depend on n_α), and then, after integration of equations (1)–(5), $M(r)$ was improved with the use of $n_\alpha(r)$ computed by the formula

$$n_\alpha(r) = n_\alpha(r_0)\frac{T(r_0)}{T(r)}\exp\left(-\int_{r_0}^{r}\frac{M_\alpha g}{R_0 T}\,dr\right). \tag{6}$$

Coefficients of viscosity η and thermal conductivity λ for the whole gas were computed from simplified formulae suitable for major atmospheric components [8]:

$$\lambda = \sum \lambda_\alpha n_\alpha (\sum n_\alpha)^{-1}, \qquad \eta = \sum \eta_\alpha n_\alpha (\sum n_\alpha)^{-1}$$

where the values of λ_α and η_α for each component were taken from the experimental (and for atomic oxygen, from theoretical) data available.

The heat source due to euv solar radiation absorption may be represented in the form [9]:

$$Q_{ST}(r, \chi) = \int_{1300\,Å}^{1800\,Å} \varepsilon_I(\lambda, r)\, \sigma_{02}\, n_{02}\, F_{\lambda 0} \exp\left(-\int_r^\infty \sigma_{02}\, n_{02} \sec\chi\, dr\right) d\lambda +$$

$$+ \bar{\varepsilon}_{II} \int_{30\,Å}^{1050\,Å} \Sigma\, \sigma_{\alpha\lambda}\, n_\alpha\, F_{\lambda 0} \exp\left(-\sum_\alpha \sigma_{\alpha\lambda} \int_r^\infty n_\alpha \sec\chi\, dr\right) d\lambda\,. \qquad (7)$$

Values of spectral solar flux $F_{\lambda 0}$ were taken from the experimental data [10] corresponding to low solar activity ($F_{10.7} \approx 75 \times 10^{-22}$ W m^{-2} Hz^{-1}). Values of cross sections $\sigma_{\alpha\lambda}$ were taken from laboratory measurements [11]. Values of thermal efficiency $\varepsilon_I(\lambda, r)$ were calculated in [12]. In the present calculations the minimal curve from [12] was used, which yields $\bar{\varepsilon}_I \approx 0.1$, because there are some indications [13] that the excitation energy of the O(^{1}D) level of atomic oxygen transfers mainly into the energy of vibrational degrees of freedom of molecular nitrogen N_2 rather than into the energy of translational degrees of freedom. The value $\bar{\varepsilon}_{II} \approx 0.5$ was taken according to approximate estimations.

Values of Q_{ST} were calculated beforehand by formula (7) (with data n_α taken from models), and because of the complexity of this formula, the dependence obtained was approximated by expression

$$Q_{ST} = A_1\, \varrho \exp(-B_1\, \varrho\, T \sec\chi) \qquad (7\,a)$$

(where A_1 and B_1 are constants). The latter expression was used in integration of equations (1)–(5). The dependence of $q = (Q_{ST} - Q_{IR})/\varrho$ on r and t is shown in Fig. 1.

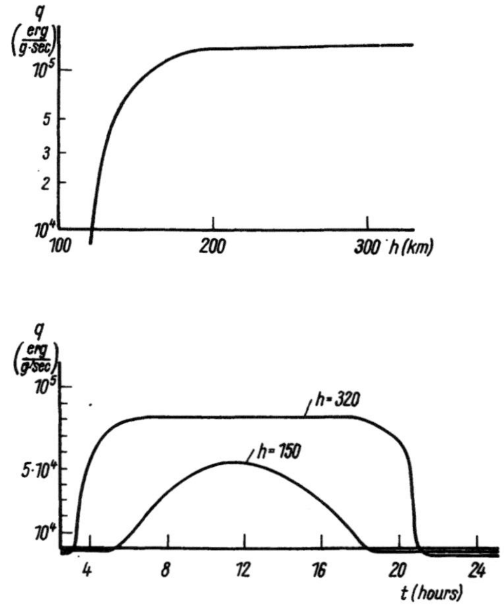

Fig. 1. Heat source due to solar xuv radiation $q = (Q_{ST} - Q_{IR})\,\varrho^{-1}$ depending on altitude h and time t.

The heat sink due to infrared radiation at $\lambda = 63$ µm was taken in accordance with [14]:

$$Q_{\mathrm{IR}} = n_0\, w_1\, A_{12} \frac{\omega_1 \exp(-w_1/k\,T)\,\xi(r)}{\omega_2 + \omega_1 \exp(-w_1/k\,T) + \omega_0 \exp(-w_0/k\,T)}, \tag{8}$$

(where w_α, ω_α are energies and statistical weights of atomic levels, A_{12} is the Einstein coefficient), with the simplification $\xi = 0.8$, that is close to results of [15, 16].

Boundary conditions for system (1)–(5) were specified in the form:

$$\left.\begin{aligned} r_0 = R_e + 120\ \text{km}:\quad & \text{a) } v_r(r_0) = 0; \qquad \text{b) } v_\varphi(r_0) = \Omega\,(R_e + r_0) \\ & \text{c) } T(r_0) = 350\ {}^\circ\text{K}; \\ r_1 = R_e + 320\ \text{km}:\quad & \text{d) } \left.\frac{\partial^2 v_r}{\partial r^2}\right|_{r_1} = 0\,; \qquad \text{e) } \left.\frac{\partial v_\varphi}{\partial r}\right|_{r_1} = 0; \\ & \text{f) } \left.\frac{\partial T}{\partial r}\right|_{r_1} = 0 \end{aligned}\right\} \tag{9}$$

where Ω is the angular velocity of the earth's rotation, and R_e is the earth's radius.

The conditions on the lower boundary mean that we neglect the influence of the mesosphere on the thermosphere, which seems to be reasonable at the first stage of investigation, though in reality this influence may be considerable [17]. The conditions on the upper boundary are consequences of equating the heat and mass fluxes to zero at infinity and of the fast growth of the kinematic coefficient of viscosity with height. In deriving condition (9d) from the condition $\varrho\, v_r|_\infty = 0$ Eqs. (1) and (3) were used. The condition for the φ coordinate is that the solution is periodic. Initial conditions were specified according to the models available, and the problem was solved by the stabilization method (i.e. integration with respect to time continued till the stationary periodic regime was established). Proper choice of initial conditions decreases the time of establishment.

2. The following computational scheme was used: The initial non-linear differential equations were transformed according to common rules into non-linear difference equations. A "two-layer" difference scheme was chosen, i.e. difference equations contain unknowns for two moments close in time: t_n and t_{n+1}. The difference scheme was chosen implicit in r and explicit in φ. Approximation in r is the simplest implicit one. Approximation in φ was obtained by the process of iteration toward the implicit difference scheme, so that the structure of approximation in φ depends on the number of iterations. For accurate solution of non-linear difference equations iterations in r were also used.

Let us consider some peculiarities of the algorithm. Variable $\ln \varrho$ instead of ϱ was included in Eqs. (1)–(5) (because of the fast variation of ϱ with r by several orders of magnitude). During the solution at a given iteration of the linear boundary-value problem for a system of one-dimensional difference equations in variable r ($t = t_{n+1}$ and $\varphi = \varphi_k$ are fixed) three of the four equations (1), (3), (4) are solved simultaneously (by some special variant of the exclusion method) and the values obtained are used then for the solution of Eq. (2), which is performed by means of the standard exclusion method.

Computations were performed on computer BESM-4 (2×10^4 counting operations per second, operative memory capacity 8000). Twenty counting points were chosen in variable r (at distances between them increasing with r), 15 or 30 radii in variable φ, and $\Delta t = t_{n+1} - t_n = 0.01$ day. After suitable choice of initial conditions computation was performed until the stabilization of the periodic regime was reached or until discontinuities in the solution appeared (see later). Comparison of solutions for 30 and 15 radii has shown that they are sufficiently close.

3. Results of calculations for low solar activity conditions (according to solar radio flux at wavelength 10.7 cm, $F_{10.7} = 75 \times 10^{-22}$ W m^{-2} Hz^{-1}) are presented in Figs. 2b, c (altitude dependence) and 2a (time dependence). In the same figure the results of computation are presented (dotted curves), neglecting horizontal transfer (i.e. $v_\varphi = \Omega r$), when the problem becomes spatially one-dimensional.

One can see that the results obtained are fairly close in some respects to the data of the other models and experiments, but there are some differences.

The main conclusion follows from the comparison of the variants with and without taking into account horizontal transfer. It is evident that interconnection of the thermal regime and dynamics plays an important part in the diurnal variations of the thermosphere. The consideration of wind effects leads to: a) the shift of ϱ and T maxima to earlier time; b) some differences in time location of maxima of ϱ and T. In the second variant of computation (see below) the amplitudes of the diurnal variations of ϱ and T also decrease.

The comparison of this model with experimental data shows the possibility of some overestimation of horizontal transfer influence, because the shift of temperature maximum seems to be too large. Indeed, there are cases in incoherent scatter data where the maximum of diurnal variation of temperature lies near 12 hours, whereas in the model a broad maximum lies at an altitude of 320 km at 7–10 hours (at altitude 150 km it lies near 11–13 hours). It is possible that real diurnal variations are intermediate between the variants with and without taking into account horizontal transfer.

This may be a consequence of the long period of the stabilization process, for, during the computation from initial conditions in which the atmosphere rotates as a whole with the earth ($v_\varphi(t_0) = \Omega r$) the stabilization process lasts more than 10 days; in other variants it lasts several days. This is a consequence of the fact that all disturbances that arise here do not flow away, but must fade out due to dissipative processes. Dissipative processes are much slower than dynamical processes; thus, computing thermal relaxation time (time required for fading out temperature disturbances due to thermal conductivity) $\tau_t = \varrho c_p H^2 \lambda^{-1}$ we can see that near 120 km $\tau_t \sim 10^5$ s, and near 300 km it lies within the limits $\tau_t \approx 10^2$–10^3 s, whereas the dynamical relaxation time (with velocity scale $V \sim 10^2$ m s^{-1} and scale length $L \sim R_e$) is of the order of $\tau_d \sim 10^3$ at all heights, so that τ_t decreases up to the order of τ_d only near the upper boundary of the height range under consideration.

Investigation of the stabilization process shows that the distribution of winds adapts quickly to the given distributions of ϱ and T (during several hours) and the reverse influence of winds on the thermal regime manifests itself much more slowly.

As a consequence of this fact, during the time variations of source Q_{ST} intermediate regimes can be realized, some of which may be close to the model without

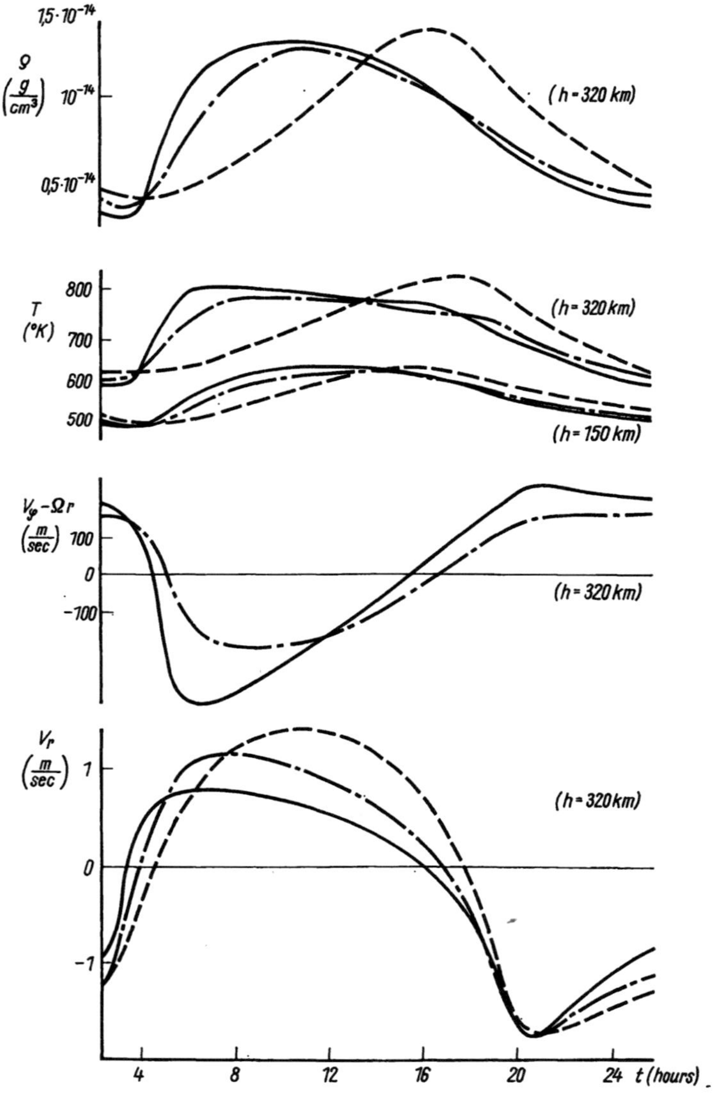

Fig. 2a

Fig. 2. Distributions of temperature T, density ϱ, vertical velocity v_r, horizontal velocity v_φ, and velocity of sound a for conditions corresponding to low solar activity ($F_{10.7} = 75 \times 10^{-22}$ W m^{-2} Hz^{-1}); 2a, dependence on time t; 2b, c, dependence on altitude h. Full curves, spatially two-dimensional model; dotted curves, spatially one-dimensional model; chain curves, intermediate regime.

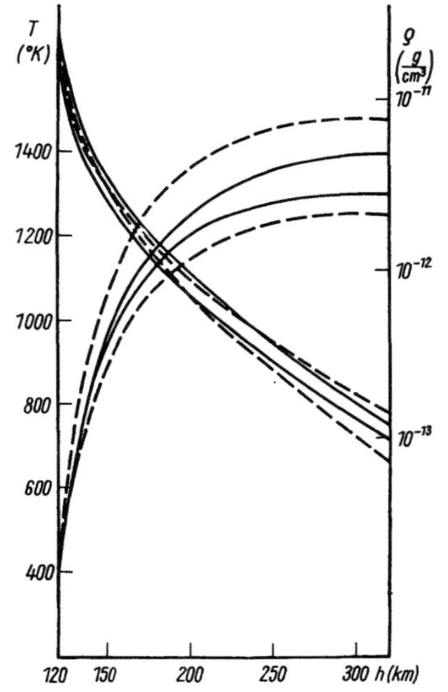

Fig. 2 b

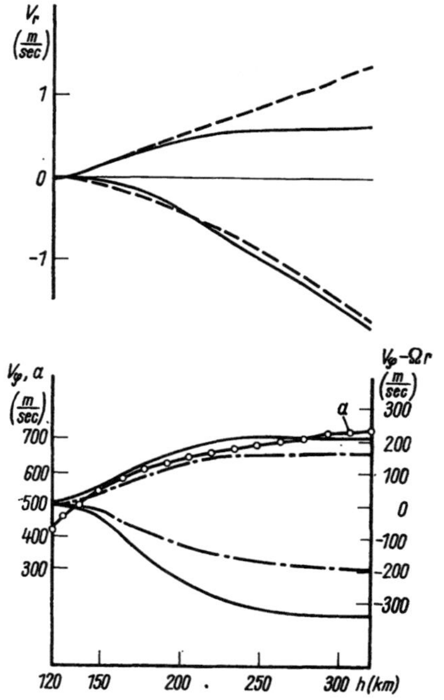

Fig. 2 c

horizontal transfer. In Fig. 2 the distributions of ϱ, T, v are shown as the chain lines which correspond to an intermediate time moment (9 hours, real time) for calculation from the initial conditions according to the one-dimensional model.

The influence of horizontal transfer in our model may be overestimated also due to the following two reasons. First: due to not taking into account meridional motions whose existence must lead to some decrease of a zonal component. Second: high Reynolds numbers for horizontal motions (which in the lower part of the height interval under consideration reach the value $Re \sim 10^7$ [5]) indicate the possibility of the existence in the thermosphere of large-scale horizontal (two-dimensional) turbulent vortices, that may lead to increase of viscosity coefficient in comparison with its molecular value, and therefore to decrease of horizontal velocity. The possibility of such a phenomenon follows also from the great variability in the location of temperature maximum which has been noticed from the incoherent scattering data (location of a maximum of diurnal temperature variation at 250–300 km varies from 12 to 19 hours, though mainly it lies near 16–17 hours). This may result from the inherent variability of a turbulent dynamical regime in the presence of its close interconnection with a thermal regime. On the other hand, this may also be caused by the different degrees of stabilization in the presence of irregular variations of the xuv source Q_{ST}.

All the above mentioned considerations point to the necessity of calculating a three-dimensional model for the complete understanding of thermosphere structure and dynamics. It should be noted that in three-dimensional Friedman's model [18] interconnection of thermal regime and dynamics seems to be taken into account incompletely due to insufficient consideration of vertical velocities and due to inaccurate upper boundary conditions for velocity.

A very interesting phenomenon exists (within the limits of our two-dimensional model) in the distribution of winds: a braking region arises in the morning in which the air, initially outstripping the earth's rotation, then lags behind. In the process of stabilization this region get narrower and transforms into a shock wave. From the point of view of the ground observer, a strong eastward wind (with velocity of the order of 100 m s^{-1}) changes into a similar westward wind. A physical explanation of this phenomenon may be the following: the atmospheric gas, rotating in essence together with the earth (due to inertial and viscous forces), in the morning begins to be heated by source Q_{ST}, and under some conditions the motion of gas relative to the source may become supersonic

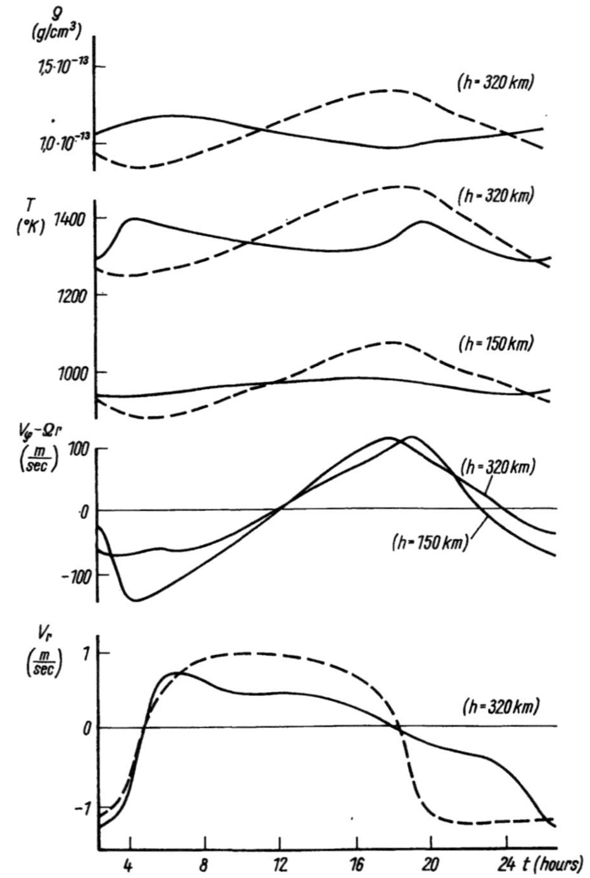

Fig. 3a

Fig. 3. As Fig. 2 for larger Q_{ST}.

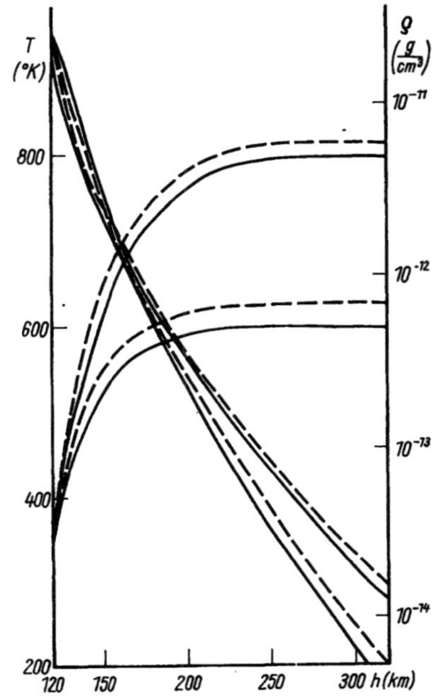

Fig. 3b

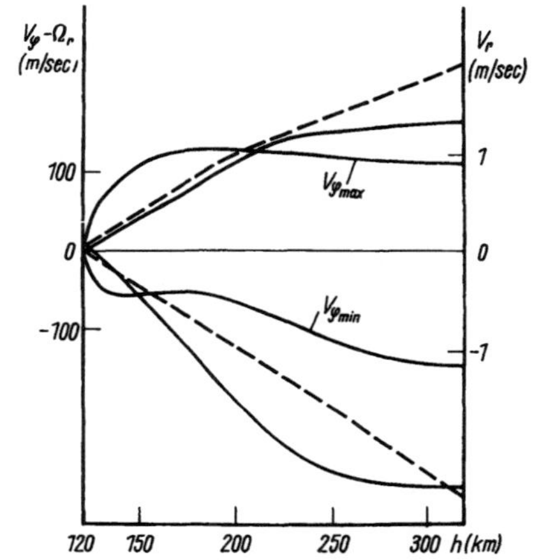

Fig. 3c

(indeed, as is seen from Fig. 2c, the velocity of the gas is larger than the local velocity of sound a almost within the whole height range). And we know from gas-dynamics (see, e.g., [19]), that supersonic flow brakes when heating (in contrast to subsonic flow). This explanation is confirmed by the fact that in the second variant with somewhat larger source Q_{ST} and larger T the velocities became subsonic and a shock wave did not arise (see Fig. 3).

In calculation of a variant with shock wave we could not obtain full stabilization of the periodic regime (because our computational scheme is not suitable for shock wave calculation) and the computation was performed until the braking region occupied more than two computational points. Now we can affirm only that a phenomenon of morning shock wave *can* exist in nature. For final conclusion a three-dimensional model must be calculated. We have not found direct confirmation of this phenomenon in the experimental data. This phenomenon (if it exists in reality) is low-latitudinal, because evaluation shows that even at latitudes $\theta \gtrsim 40°$ the gas flow is always subsonic.

In Fig. 3 a second variant of the calculations is presented which corresponds approximately to somewhat higher solar activity compared with the first variant (for this purpose the source Q_{ST} used in the first variant was about doubled). In this case a practically stabilized periodic regime was achieved, because the flow is everythere subsonic and a shock wave is absent. Here distortion of diurnal variation of ϱ and T due to the winds was shown to be more restricted than in the first case and their amplitudes considerably decreased. Probably, in this case real variations are also intermediate between those with and without taking into account horizontal motions. Additional calculations are being performed for analysis and improvement of the model described.

Thus, the main conclusion of this work is that a close interconnection exists between the thermal regime and the dynamics of thermosphere which may lead to variable shapes of the diurnal variations of temperature and density, and that for the complete understanding of this phenomenon the calculation of a three-dimensional model is necessary.

References

[1] L. G. JACCHIA, Smithson. Astrophys. Observ. Spec. Rep. 313 (1970).
[2] P. WALDTEUFEL and J. P. MCCLURE, Annls Géophys. **25**, 785 (1969).
[3] L. G. JACCHIA, J. Geophys. Res. **75**, 4347 (1970).
[4] M. N. IZAKOV, Space Sci. Rev. **7**, 579 (1967).
[5] M. N. IZAKOV, Dokl. Akad. Nauk SSSR **177**, 1342 (1967).
[6] M. N. IZAKOV, Space Sci. Rev. **12**, 261 (1971).
[7] CIRA-65, COSPAR International Reference Atmosphere (1965).
[8] M. NICOLET, in: Physics of the Upper Atmosphere (ed. by J. A. Ratcliffe) (1960).
[9] M. N. IZAKOV, Geomagn. i Aeron. **10**, 283 (1970).
[10] H. E. HINTEREGGER, paper presented at IQSY–COSPAR Symposium, London 1967.
[11] R. E. HUFFMAN, Canad. J. Chem. **10**, 1823 (1969).
[12] M. N. IZAKOV and S. K. MOROZOV, Geomagn. i Aeron. **10**, 630 (1970).
[13] V. B. LEONAS, private communication, 1970.
[14] D. R. BATES, Proc. Phys. Soc. B **64**, 805 (1951).
[15] G. M. SHVED and A. F. NERUSHEV, Izv. Akad. Nauk SSSR, Fiz. Atmos. Okeana **5**, 473 (1969).
[16] G. KOCKARTS and W. PEETERMANS, Planet. Space Sci. **18**, 271 (1970).
[17] R. LINDZEN, J. Geophys. Res. **75**, 6868 (1970).
[18] H. FRIEDMAN, Smithson. Astrophys. Observ. Spec. Rep. 316 (1970).
[19] L. D. LOJKANSKY, Mekhanika zhidkosti i gasa, Moscow (1957).

Space Research XII — Akademie-Verlag, Berlin 1972

EFFECTS OF VERTICAL MASS MOTIONS ON THE COMPOSITION STRUCTURE IN THE THERMOSPHERE

T. SHIMAZAKI *

Aeronomy Laboratory, Department of Electrical Engineering, University of Illinois, Urbana, Ill., USA

Vertical mass motions could occur in the thermosphere by meridional air circulations caused by unequal uv heating between summer and winter hemispheres and by auroral heating and the return currents at lower latitudes during magnetic disturbances. It has been suggested that these vertical motions affect markedly the composition structure in the thermosphere, and changes in [O]/[N_2] and [O]/[O_2] ratio may explain the seasonal and stormtime variations in the F region electron density. These effects are examined quantitatively by solving the time-dependent continuity equations for [O], [O_2] and [N_2], including the transport terms by molecular and eddy diffusion and vertical mass motions as well as the photodissociation of O_2 and the three-body recombination of O. The result indicates that the effects of vertical motions are more favorable for explaining the seasonal and stormtime variations in the F region electron density than the effects of turbopause changes, which have often been suggested as explanation of the required composition changes in the thermosphere.

It is well known that the daytime F region electron density is larger in winter than in summer, contrary to the expectation from the seasonal variation in ionization intensity (the solar zenith angle). This seasonal variation in the F region electron density is ordinarily attributed to the seasonal variations in the ratio of [O]/[O_2] or [O]/[N_2]. Similar composition changes may also explain the general features of the stormtime variations in the F region electron density. What causes the seasonal and stormtime variations in the thermospheric composition is still, however, a matter of controversy. The change in the turbopause has often been assumed to explain the required composition changes, but no theory has explained favorably what caused such a change in the turbopause. Moreover, the required seasonal variation in the turbopause seems to be too large and to be opposite from the theoretical expectation. We need an eddy diffusion coefficient smaller by more than an order of magnitude or a turbopause lower by more than 10 km for winter than for summer [1]. This seems to contradict our knowledge that the winter upper atmosphere is generally more turbulent than the summer upper atmosphere.

A large scale meridional circulation transporting O atoms from summer polar regions to winter polar regions has been suggested as a cause of the excess O over the winter hemisphere [2, 3]. The general circulation below the turbopause could produce composition changes at the turbopause which will transmit to higher regions by diffusion. A large scale meridional circulation

* Visiting Professor 1970–1971: permanent address National Oceanic and Atmospheric Administration, Boulder, Colorado 80302, USA.

could also be produced within the thermosphere, because a large inequality of heating by the solar uv radiation between the summer and winter hemispheres would cause horizontal mass transports from summer to winter hemisphere, which eventually produce an upward velocity over the summer hemisphere and a downward velocity over the winter hemisphere. In this paper it is shown that the vertical mass motions within the thermosphere can produce appreciable changes in the composition, which may favorably explain the seasonal variations and perhaps also the stormtime variations in the F region electron density.

The winter-to-summer ratio of the observed electron density at $\sim$200 km in middle latitudes is $\sim$1.03–1.75 depending upon the time in the solar cycle [4]: the ratio is largest at solar maximum and the difference almost disappears at solar minimum. Similar values have been obtained from incoherent back scatter experiments at Millstone [5]. Considering the seasonal effect of the solar zenith angle on the ionization intensity, we may say that for the same solar zenith angle there is an average seasonal difference of the F region electron density of about 2.0 that must be explained. The seasonal temperature variation can produce changes in the $[O]/[N_2]$ ratio in the right direction for explaining the seasonal variations in the electron density, but we need an exospheric temperature variation between 800–1800 °K for this purpose [1]. This seasonal temperature variation is unbelievably large since it has been well established by observations that the winter thermosphere is about as warm as the summer thermosphere [6].

In order to calculate the quantitative effects of vertical mass motions on the composition structure in the thermosphere, the existing computer programs by the author [1, 7] have been modified. We have solved the time-dependent continuity equations for $[O]$, $[O_2]$ and $[N_2]$, including transport effects by molecular and eddy diffusion and vertical mass motions as well as various chemical reactions. It is well known that an eddy diffusion coefficient of the order of 10^6–10^7 cm^2 s^{-1} is needed to obtain reasonable values for $[O]/[O_2]$ ratio at 120 km [1, 8].

For the eddy diffusion coefficient in the present calculations, we have used a profile developed by the author [7] which is defined mainly by the maximum eddy diffusion coefficient E_M and the height where that maximum occurs, z_0. Since we do not know the exact seasonal variations in these parameters, we now calculate the effects of changing mass motion velocity rather than changing z_0 and E_M. It may be justifiable to assume the same temperature profile for summer and winter for our present purpose; there are small seasonal temperature variations and the thermal balance is almost maintained for each season. We have taken the temperature profile from the model in [7] with $T_\infty = 1200$ °K. The solar radiation intensity, absorption cross sections of O_2, and recombination coefficient of O are all assumed the same as in [7].

Fig. 1 shows the effects of z_0 and E_M on the $[O]/[N_2]$ and $[O]/[O_2]$ ratios at 200 km based on the result of calculations reached after 20 days. In these calculations the effect of vertical mass motions is not included. It is interesting to note that these ratios have minimum values when z_0 is $\sim$100–102 km, where the turbopause is usually observed. The ratios are smaller for larger E_M, and since it is expected from rocket released vapor trail experiments [9] that E_M is larger in winter than in summer, winter values for the ratio should have been smaller than the summer values. For instance, if we take the summer E_M as 5×10^6 cm² s⁻¹ and the winter E_M as 5×10^7 cm² s⁻¹, the winter [O]/[N_2] should be smaller by a factor of ~0.45 than the summer value for the same z_0. This is opposite to the required seasonal variation in [O]/[N_2].

Vertical mass motions can reverse the relative magnitude of [O]/[N_2] between summer and winter, if upward velocity is assumed for summer and downward

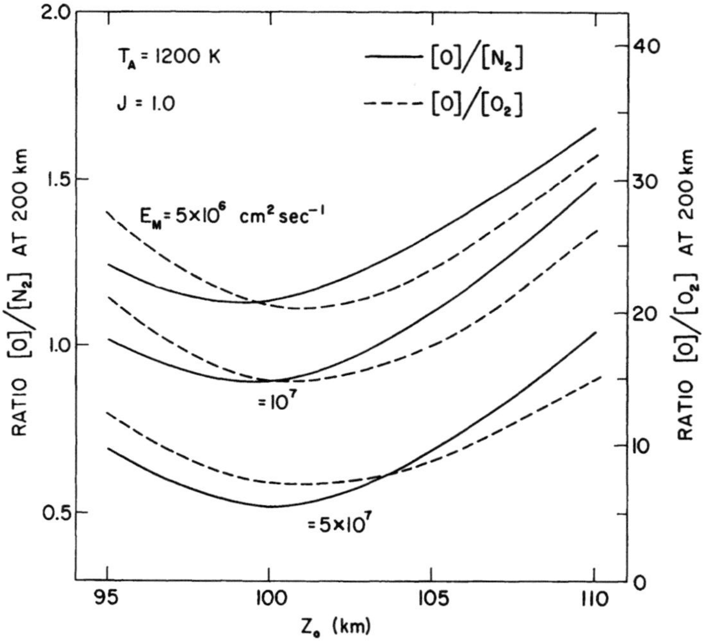

Fig. 1. Effects of E_M and z_0 on [O]/[N_2] and [O]/[O_2] ratios at 200 km.

velocity for winter and also if we assume proper magnitude and profiles for the vertical velocities. To examine this effect model calculations have been continued for an additional five days by introducing the vertical mass velocities. The initial conditions for these calculations are taken from the results shown in Fig. 1. Three representative cases of E_M have been considered: 5×10^6 cm² s⁻¹ for summer, 10^7 cm² s⁻¹ for an average and 5×10^7 cm² s⁻¹ for winter.

There are no direct observations of the vertical velocities in the thermosphere; therefore we have assumed different profiles for the vertical velocity as is illustrated in Fig. 2. Three sets of profiles are considered; model A assumes constant momentum ($V \propto \varrho^{-1}$), model B constant energy ($V \propto \varrho^{-1/2}$), and model C constant velocity with height. Fig. 2 also includes theoretical curves [10], calculated according to the condition that effects of adiabatic cooling by upward motion over the summer hemisphere and of compressional heating by downward motion over the winter hemisphere should compensate the unequal heat input between the two hemispheres. The profiles of these curves resemble most closely our model B, since both were based on a principle of energy conservation. Fig. 3 shows the result of calculations for the case of an average E_M ($= 10^7$ cm^2 s^{-1}) and $z_0 = 100$ km. The upper portion indicates the effects of vertical velocities on [O]/[N_2] at 200 km, and the lower portion illustrates the changes in ϱ_{90} as function of V_0, the velocity at 90 km. Solid curves indicate effects of upward velocities, and broken curves effects of downward velocities. It is generally seen in Fig. 3 that the upward velocities reduce the [O]/[N_2] ratio, whereas the downward velocities increase the ratio.

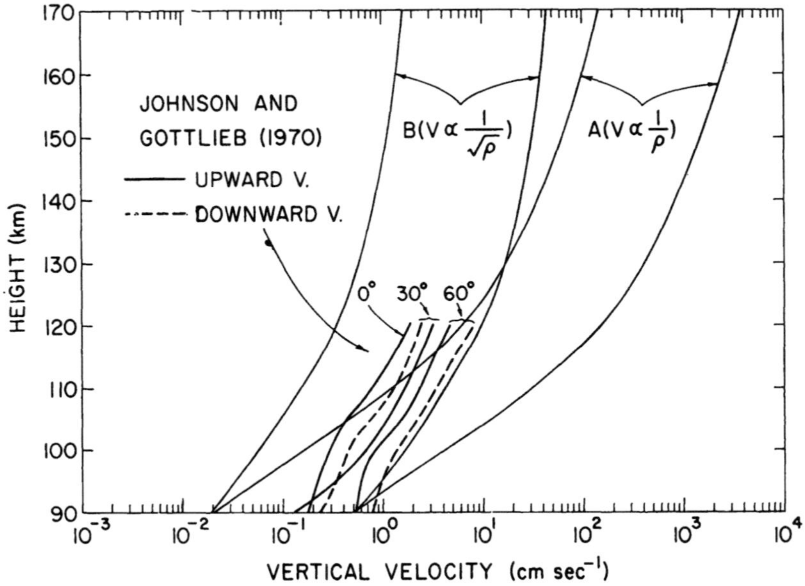

Fig. 2. Assumed profiles of vertical velocities. Theoretical curves [10] are calculated so that these velocities may compensate an unequal uv heating between summer and winter hemispheres.

Model A gives the largest effects on the composition ratio; the effect of downward velocities are particularly large. Very small velocities of $V_0 \sim 0.015$ cm s^{-1} are sufficient to produce the required magnitude of the seasonal variation in [O]/[N_2]. Model A produces the least change in ϱ_{90}; it does not change appreciably for $|V_0| < \sim 0.03$ cm s^{-1}. A sudden change in ϱ_{90} between $V_0 = -0.025$ and -0.05 cm s^{-1} was caused by a very large increase in [O]. Atomic O could be the most dominant constituent even at 90 km if V_0 is negative and its magnitude exceeds $\sim$0.05 cm s^{-1}. This is of course against the observations. Model C produces the least effects on the composition but the largest effects on ϱ_{90}. We need very large V_0 to explain the required seasonal variation in [O]/[N_2], but such large V_0 produces unbelievably large changes in ϱ_{90}. The observational evidence [11] indicates that the mass density is kept stable within $\sim$10% at 90 km (isopycnic level). Thus the profiles of model C do not satisfy this condition for ϱ_{90}. Model B has moderate effects on both the composition ratio at 200 km and ϱ_{90}. The effects become appreciable when $|V_0| \geq 0.1$ cm s^{-1}, and cases of $V_0 = \pm 0.2$–0.3 cm s^{-1} would explain the required seasonal variations in [O]/[N_2]. These V_0's may be the maximum allowable values in order to prevent large changes in ϱ_{90} from occurring.

In Fig. 4 we illustrate how the relative magnitude of the [O]/[N_2] ratio (summer value > winter value) produced by a possible seasonal change in

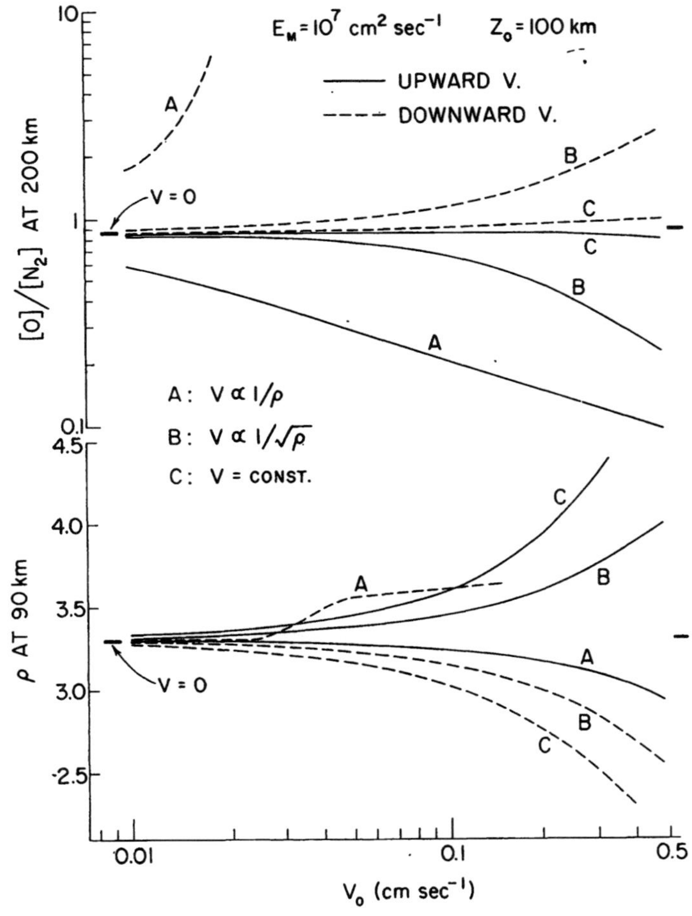

Fig. 3. Effects of vertical mass velocities on the [O]/[N_2] ratio at 200 km and on the mass density at 90 km for the case of an average E_M.

E_M could be reversed by introducing upward velocities for summer and downward velocities for winter. In order to explain the winter-to-summer ratio of 2.0 for [O]/[N_2], we need $V_0 \sim \pm 0.013$ cm s^{-1} for model A and $V_0 \sim \pm$ ± 0.3 cm s^{-1} for model B. But, because of the difference in the shape of velocity profile, vertical velocities at higher regions could be larger for model A than for model B; for instance, corresponding to the above values of V_0, model A gives $\sim$97 cm s^{-1}, while model B gives $\sim$25 cm s^{-1} for the vertical velocity at 170 km. Since model B produces larger deviation for ϱ_{90} from the isopycnic value than model A, and since there is a possibility that for model B, but not for model A, the deviation becomes even much larger if time-dependent model calculations are continued, model A seems to give the best profile for the vertical velocity for our present purpose.

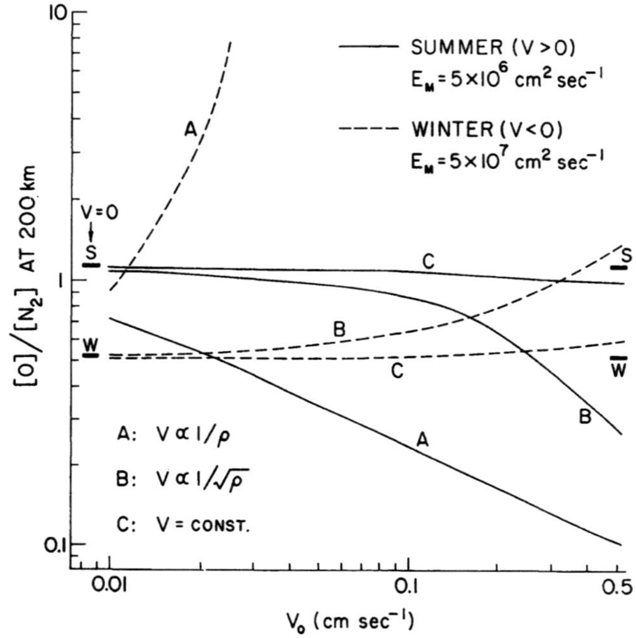

Fig. 4. Effects of vertical mass velocities on the [O]/[N_2] ratio at 200 km for representative values of E_M for summer and winter.

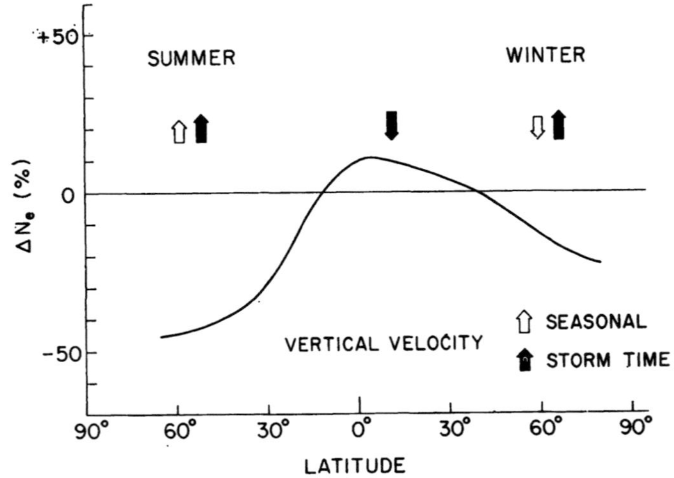

Fig. 5. Schematic illustration of vertical motion and corresponding changes in the F region electron density during magnetic disturbances over summer and winter hemispheres.

The effects of vertical motions could also explain the changes in F region electron density during magnetic storms. It has been established [12, 13] for the bottomside F region that the electron density generally decreases at high and middle latitudes whereas it increases at low latitudes in all seasons and at some middle latitude stations in winter. During magnetic disturbances, heating by precipitating particles should produce the upward motions at high latitudes in both hemispheres, as is shown schematically in Fig. 5. Over the summer hemisphere, this enhances an already existing upward motion, while over the winter hemisphere the seasonal downward motion is reduced and may be reversed at most stations at high latitudes. These net upward motions explain the decrease in the electron density observed at high latitudes in all seasons and at middle latitudes except for winter during magnetic storms.

Since the net air motion is upward at high latitudes, there must be downward motion at low latitudes in order to close the meridional circulation, and the electron density should increase at low latitudes in all seasons and at middle latitudes in winter where the seasonal downward motion still dominates. Thus, the combination of the seasonal and the stormtime variations in vertical velocities may explain well the seasonal and latitudinal differences in the complicated storm effects on the F region electron density.

Acknowledgment

This work was supported by Grant No. NRG 013 from the National Aeronautics and Space Administration.

References

[1] T. Shimazaki, J. Atmos. Terr. Phys. **29**, 723 (1967).
[2] W. W. Kellogg, J. Met. **18**, 373 (1961).
[3] G. A. M. King, J. Atmos. Sci. **21**, 231 (1964).
[4] J. W. Wright, Nature, Lond. **194**, 461 (1962).
[5] L. P. Cox and J. V. Evans, J. Geophys. Res. **75**, 6271 (1970).
[6] L. G. Jacchia, Space Research V, 1153 (1965).
[7] T. Shimazaki, J. Atmos. Terr. Phys. **33**, 1383 (1971).
[8] F. D. Colegrove, W. B. Hanson and F. S. Johnson, J. Geophys. Res. **70**, 4931 (1965).
[9] J. E. Blamont and C. de Jager, Annls Géophys. **17**, 134 (1961).
[10] F. S. Johnson and B. Gottlieb, Planet. Space Sci. **18**, 1707 (1970).
[11] US Standard Atmosphere, Supplement (1966).
[12] T. Obayashi, J. Geomag. Geoelect. **16**, 1 (1964).
[13] S. Matsushita, J. Geophys. Res. **64**, 305 (1959).

Space Research XII — Akademie-Verlag, Berlin 1972

DEUTERIUM DISTRIBUTIONS IN THE EARTH'S UPPER ATMOSPHERE

G. Kockarts

Institut d'Aéronomie Spatiale de Belgique, Bruxelles, Belgium

The physical conditions leading to a deuterium enhancement in the earth's upper atmosphere are discussed as a function of the thermopause temperature and of the eddy diffusion coefficient. It appears that changes in the eddy diffusion can induce large variations in the deuterium concentration for temperatures less than 1500 °K. Under those conditions, the inefficiency of thermal escape should be compensated by the polar wind process. As the deuterium total content is relatively small compared with that of atomic hydrogen, the best observational conditions occur for thermopause temperature around 1250 °K.

Deuterium ions were detected [1, 2] in the earth's upper atmosphere by a magnetic mass spectrometer flown in 1965 on the Explorer 31 satellite. The ratio of the concentrations $n(D^+)/n(H^+)$ was estimated to be of the order of 1.1×10^{-3} below 1000 km, but it decreased by a factor of 10 above 2000 km [2]. High-resolution profiles of the solar H Lyman α line [3, 4] seem to indicate an atmospheric absorption effect due to deuterium at 1215.346 Å.

The presence of deuterium above 100 km probably results from a complex set of photochemical reactions in the chemosphere and an approach to the problem should be made in a way similar to Nicolet's method for atomic hydrogen [5], although unknown isotopic effects could be important in determining the reaction rates. In the present analysis, a ratio $n(D)/n(H) = 1.6 \times 10^{-4}$ is adopted at 100 km altitude, without trying to explain it by photochemical and transport processes in the stratosphere and mesosphere. Such a value corresponds to the deuterium content of standard mean ocean water [6, 7]. With an atomic hydrogen concentration $n(H) = 3 \times 10^7$ cm^{-3} at 100 km, which is necessary to explain the OGO 4 Lyman-α observations [8], the lower boundary condition adopted for the concentration of deuterium is 4.8×10^3 cm^{-3}. This value could be decreased by a factor of 2, since a recent analysis [9] of several Lyman-α observations shows that the hydrogen concentration could be between 1.25 and 2×10^7 cm^{-3}.

In a previous theoretical analysis, Kockarts and Nicolet [10] showed that the ratio $n(D)/n(H)$ can increase by a factor 20 between 100 and 500 km for a large range of thermopause temperatures when molecular diffusion is introduced in the thermosphere. In the present work, a general diffusion equation, including molecular and eddy diffusion transport, is used in the way described by Kockarts [11]. Similar equations were applied by Donahue [12] and by

McElroy and Hunten [13] in different studies on deuterium in the atmosphere of Venus.

Above 100 km in the earth's thermosphere, the vertical distribution of deuterium depends mainly on the temperature, on the escape flux and on the value adopted for the eddy diffusion coefficient. The relative importance of these factors is determined by the atmospheric structure. In Fig. 1 the variation of the deuterium concentration is shown as a function of the thermopause temperature when the eddy diffusion coefficient K is 10^7 cm^2 s^{-1}. For atomic hydrogen it has been shown [10] that the diffusion flow leads to an

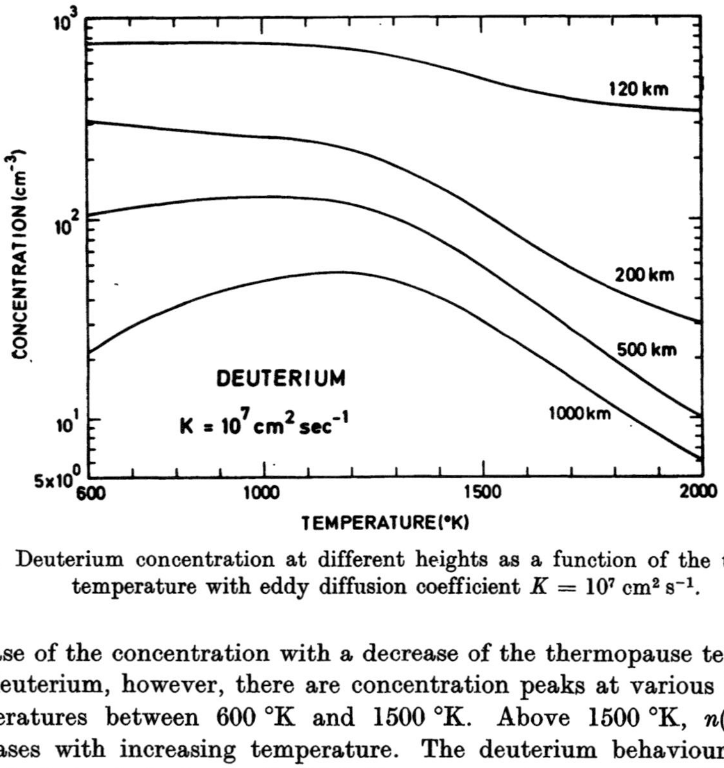

Fig. 1. Deuterium concentration at different heights as a function of the thermopause temperature with eddy diffusion coefficient $K = 10^7$ cm^2 s^{-1}.

increase of the concentration with a decrease of the thermopause temperature. For deuterium, however, there are concentration peaks at various heights for temperatures between 600 °K and 1500 °K. Above 1500 °K, n(D) always decreases with increasing temperature. The deuterium behaviour is similar to that of atomic hydrogen for high thermopause temperatures. Nevertheless, if $T < 1500$ °K, the vertical distribution of deuterium has a tendency to be similar to that of helium, since the concentration at 500 km hardly varies with changes in thermopause temperature. Moreover, the greatest variation of n(D) at 1000 km does not reach a factor of 10 whereas, under steady state conditions, the atomic hydrogen concentration increases by a factor of 200 between 2000 °K and 600 °K. Furthermore, the deuterium escape flux cannot be constant over a wide temperature range, since the effusion velocity decreases by a factor 2×10^6 from $T = 2000$ °K to $T = 600$ °K. For K $= 10^7$ cm^2 s^{-1}, the deuterium escape flux decreases actually from 8×10^3 cm^{-2} s^{-1} ($T = 2000$ °K) to 1.4 cm^{-2} s^{-1} ($T = 750$ °K). If the escape flux corresponds to the possible flow of 10^4 cm^{-2} s^{-1} at 100 km, it is necessary to consider a charge exchange process with atomic oxygen ions in order to obtain an equivalent ionic escape flux through the polar wind.

The vertical distribution of a minor constituent in the thermosphere depends strongly on the adopted values for the eddy diffusion coefficient, when the

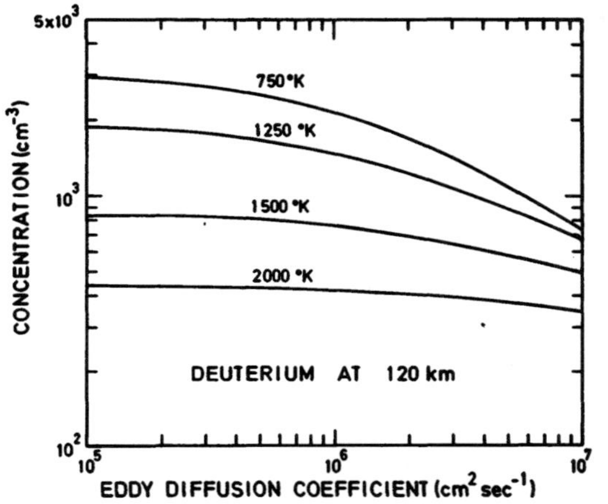

Fig. 2. Variation of the deuterium concentration with eddy diffusion coefficient for thermopause temperatures between 750 °K and 2000 °K.

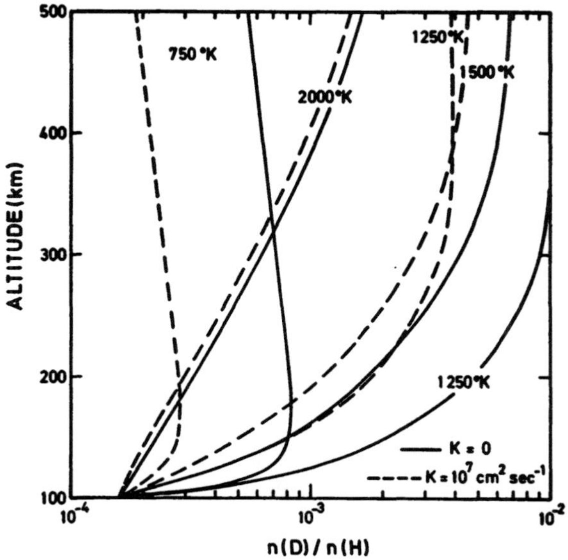

Fig. 3. Height dependence of the ratio $n(\mathrm{D})/n(\mathrm{H})$ for different thermopause temperatures. The adopted eddy diffusion coefficients are respectively $K = 0$ and $10^7\ \mathrm{cm^2\,s^{-1}}$.

escape of neutral particles is negligible compared with the maximum diffusion flow around 100 km [11]. Fig. 2 shows the deuterium concentration at 120 km as a function of K for different thermopause temperatures. The most important variations appear for temperatures less than 1500 °K. Actually, for $T = 1500$ °K the escape flux is respectively 9.8×10^3 cm^{-2} s^{-1} and 5.2×10^3 cm^{-2} s^{-1} with $K = 0$ and $K = 10^7$ cm^2 s^{-1}, whereas the maximum diffusion flow at 100 km is 1.3×10^4 cm^2 s^{-1}. Below 1500 °K, the vertical deuterium distribution depends both on the temperature and on the adopted value for K. For example, at 1000 km, $n(D) \simeq 3 \times 10^1$ cm^{-3}, when $T = 1500$ °K or 750 °K and with $K = 10^7$ cm^2 s^{-1}; however at the same height, $n(D) \simeq 1.6 \times 10^2$ cm^{-3}, when $T = 1250$ °K or 750 °K and with $K = 0$.

The effect of the escape flux for $T > 1500$ °K and the effect of eddy diffusion for $T < 1500$ °K together influence the vertical distribution of the ratio $n(D)/n(H)$ given in Fig. 3. This ratio actually goes through a maximum which occurs at an altitude depending on the thermopause temperature. In Fig. 3 it is shown that the maximum enhancement of the ratio $n(D)/n(H)$ occurs around 400 km for $T = 1250$ °K; it corresponds to almost 60 times the adopted natural abundance of 1.6×10^{-4} when $K = 0$ above 100 km. This enhancement is about a factor of 3 less than that deduced by Bruner and Wilson [3] at 160 km altitude. Moreover, the theoretical ratios given in Fig. 3 are of the same order of magnitude as the ratios $n(D^+)/n(H^+)$ measured by Hoffman et al. [2]. It should be noted that $n(D)/n(H)$ begins to decrease above a certain height and can even reach values less than the assumed abundance at 100 km. This means that the diffusion transport can lead to an increase of the ratio $n(D)/n(H)$ only over a certain height range.

Finally, it should be pointed out that optical observations of deuterium in the earth's atmosphere are extremely difficult, since the total content is such that the optical depth at the centre of the deuterium Lyman-α line is very small. For example, at 100 km, the vertical content of deuterium ranges from 6.7×10^{10} cm^{-2} to 5.6×10^9 cm^{-2}, depending both on the adopted eddy diffusion coefficient and on the thermopause temperatures. In any case observations above 200 km should be performed when the deuterium enhancement reaches its maximum, i.e. for a thermopause temperature of the order of 1250 °K.

References

[1] J. H. HOFFMAN, Science **155**, 322 (1967).
[2] J. H. HOFFMAN et al., J. Geophys. Res. **74**, 6281 (1969).
[3] E. C. BRUNER, JR. and T. E. WILSON, J. Geophys. Res. **74**, 6491 (1969).
[4] R. A. JONES, E. C. BRUNER, JR., and W. E. RENSE, J. Geophys. Res. **75**, 1849 (1970).
[5] M. NICOLET, Aeronomica Acta A 79, 100 pp. (1970).
[6] E. ROTH, J. Chim. Phys. Paris **60**, 339 (1963).
[7] R. HAGEMANN, G. NIEF and E. ROTH, Tellus **22**, 712 (1970).
[8] R. R. MEIER and P. MANGE, Planet. Space Sci. **18**, 803 (1970).
[9] R. R. MEIER and D. K. PRINZ, J. Geophys. Res. **75**, 6969 (1970).
[10] G. KOCKARTS and M. NICOLET, Annls Géophys. **18**, 269 (1962).
[11] G. KOCKARTS, Aeronomica Acta A 76, 24 pp. (1970).
[12] T. M. DONAHUE, J. Geophys. Res. **74**, 1128 (1969).
[13] M. B. MCELROY and D. M. HUNTEN, J. Geophys. Res. **74**, 1720 (1969).

Space Research XII — Akademie-Verlag, Berlin 1972

MAGNETOSPHERIC PROCESSES AND THE BEHAVIOR OF THE NEUTRAL ATMOSPHERE

P. M. BANKS

Department of Applied Physics and Information Science, University of California, La Jolla, Calif., USA

A review is given of the quiet time couplings which exist between the magnetosphere and the thermosphere. Joule heating arising from aurorae and the combination of Joule heating and ion drag arising from magnetospheric convection normally provide as much energy to the thermosphere as solar ultraviolet radiation. Consequently, it appears that theoretical atmospheric models which ignore such contributions cannot be correct, especially at high latitudes.

Ion drag is especially important for establishing wind systems in the thermosphere. Using the qualitative ideas of convective electric fields, a description of a possible model of the global wind system is given. It is pointed out that the existence of a convectively driven wind source at high latitudes consistent with magnetospheric convection leads to the possibility of a mean easterly wind, i.e. super-rotation.

1. Introduction

Satellite and rocket measurements provide ample evidence for the existence of an appreciable interaction between the magnetosphere and the upper atmosphere [1, 2]. The most dramatic effects occur during large scale geomagnetic disturbances when rapid increases in neutral gas temperature lead to complex changes in atmospheric composition [2–5]. More recently, it has also been found that enhancements in neutral winds occur at the same time, speeds greater than 1 km s^{-1} having been observed in the middle thermosphere [6]. While there is some reason to believe that the source for such disturbances lies in the polar regions, the atmospheric effects are observed on a global scale within a matter of hours after the initial onset seen in magnetic records [7].

Because the stormtime variations are large and rapid, there is a natural tendency to consider them as the principal means of coupling between the magnetosphere and neutral atmosphere. In this review, however, the quiet-time couplings are emphasized to show that there is a substantial influence even during geomagnetically undisturbed periods, i.e. when the usual processes of magnetospheric convection and auroral displays are in progress. It appears that these two features play an important part in determining the atmospheric temperature and density at high latitudes and act to establish global wind patterns which must influence the entire upper atmosphere.

The various energy sources of the thermosphere can be listed as: (i) solar euv radiation, (ii) tidal dissipation including gravity waves, (iii) energetic particles, (iv) Joule heating, (v) viscous dissipation, (vi) other effects. All existing theoretical models of the upper atmosphere are based on process (i) alone, although special studies have been made to assess the importance of alternate sources [8, 9].

It is important to note that the magnetospheric interactions (iii), (iv), (v) affect the neutral atmosphere in two fundamentally different ways. Effects which produce thermal energy directly alter the pressure of the neutral gas through the energy balance equation. Such sources can affect the lateral transport only through the pressure gradient term in the neutral gas momentum equation. Thus, Joule heating, particle precipitation and euv absorption introduce lateral flow winds only when these sources of thermal energy are not uniformly distributed over the surface of the earth. In a similar manner, the vertical composition is changed by a vertical flow under the influence of the vertical pressure gradient. Although heat sources are important, global wind patterns can also be greatly affected by viscosity and ion drag (resulting from magnetospheric convection, for example) since these terms occur directly in the momentum equation. As a result, the existence of ion drift motions in any region of the atmosphere, but especially those regions outside the plasmapause, can be expected to lead to neutral winds of substantial magnitude. This topic will be discussed in § 4.

The emphasis of this review is placed on the normal magnetospheric interactions ignoring energetic particles. In essence, it is the ionospheric ions, driven by a variety of electric fields, which can heat and set the neutral atmosphere in motion. Although the sources and variations of these fields are thought to be understood in principle, in fact, their global distribution and magnitudes remain a topic of intense study.

2. Structure of the Magnetosphere

The major features of the magnetosphere are shown in Fig. 1. The distribution of ionization in the ionosphere is strongly controlled by the geomagnetic field and the structure of the magnetosphere. As indicated in Fig. 1, there are three essentially different regions where the behavior of thermal plasma is determined by different effects. At low geomagnetic latitudes the *plasmasphere* is a region of relatively high plasma density contained within closed lines of magnetic force. The ionosphere in these regions is relatively quiescent and corotates with the earth, subject to small diurnal and sporadic electric fields which cause some lateral ion flow with associated neutral gas effects [10]. The termination of the plasmasphere occurs at the *plasmapause*, a rather uneven surface defining the shape of the plasmasphere as an asymmetrical torus. The behavior of the ionosphere outside the plasmapause is very complex as a result of magnetic convection. This process arises from the interaction of the solar wind with the geomagnetic field. Following the ideas of Dungey [11] and Axford and Hines [12] it is thought that the geomagnetic field is able to merge with the southward component of the interplanetary field. As a result, a large scale electric field is established within the magnetosphere which is directed roughly parallel to the dawn to dusk line as seen in the equatorial plane [13]. This electric field is present throughout the magnetosphere and drives thermal plasma with an $\boldsymbol{E} \times \boldsymbol{B}$ drift towards the front of the magnetosphere in regions where field lines are closed, and towards the tail of the magnetosphere in regions where field lines are open to the interplanetary field. (For a comprehensive review see [14].)

Following the concepts of magnetohydrodynamics, it is possible to visualize the behavior of the moving thermal plasma in terms of motions of the magnetic field lines themselves. Thus, in the initial step of the convection cycle, the dipole-like field lines from the earth become connected to the interplanetary magnetic field lines at the front of the magnetosphere. These field lines are then swept back over the polar cap into the tail of the magnetosphere where

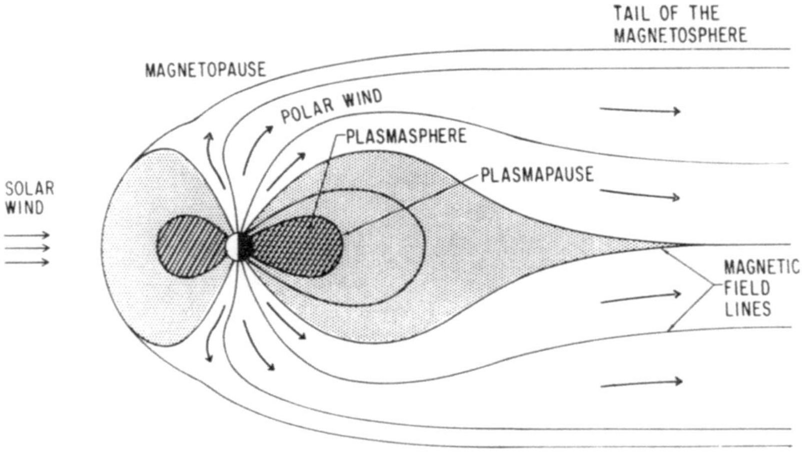

Fig. 1. Schematic view of the ionosphere and magnetosphere.

they eventually reconnect and shorten (to relieve magnetic tension). The field lines then migrate around the earth at relatively low geomagnetic latitudes. As the field lines approach the front of the magnetosphere they begin to move polewards towards higher magnetic latitudes and eventually begin the convection cycle anew. During the convection cycle thermal plasma is free to flow along the field tubes and, when the field lines are open, escape to interplanetary space. This thermal ion flow, called the polar wind [15], is present on both open and closed field lines as a dynamic process which greatly affects the topside ionosphere. Due to the polar wind flow, electron densities are very low on convecting field tubes. The plasmapause itself reflects the change in density between convecting and quasi-corotating plasma [16].

The motions of the magnetic field tubes as seen in their intercepts with the equatorial plane are shown in Fig. 2. Arrows indicate the direction of motion up to the magnetopause where disconnection occurs and the field lines no longer intersect the equatorial plane. Also shown is the plasmapause boundary where approximate corotation of thermal plasma begins under the influence of the corotation electric field. The projection of the field tube motion on to the northern hemisphere is shown schematically in Fig. 3. The region of open field tubes is indicated by cross-hatching and the complete convection cycle is present as closed loops. The dark line near 75° Λ represents the *polar cusp* region where there is direct access of magnetosheath plasma to the upper atmosphere [17–19]. It is just in these regions that the mid-day aurorae are seen, along with numerous ionospheric phenomena. The night-time auroral

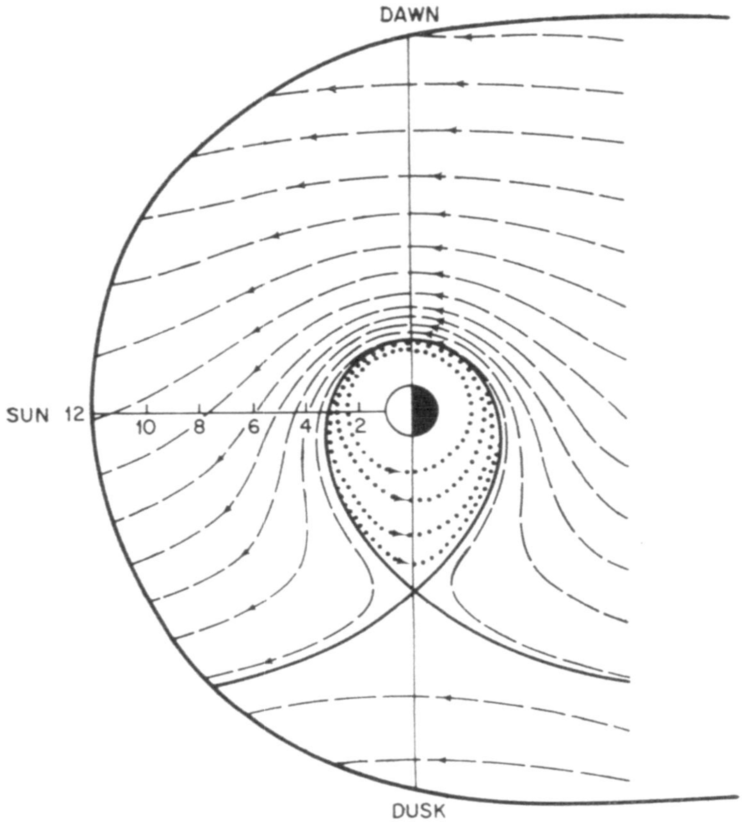

Fig. 2. View of field tube trajectories in the equatorial plane.

zones are normally found in the areas separating the two directions of convective motion, i.e. polewards of the auroral zone convection is in the anti-solar direction and equatorwards it is towards the sun [20]. As discussed later, the convective motions of the ions are of great importance to the establishment of wind systems in the polar regions which are substantially different from those predicted from solar heating alone.

The model of magnetospheric convection described above is not complete in one important aspect. From polar electric field measurements [20, 21] it is found that the convection electric field is not steady but continually fluctuates.

The power spectrum of this variation has been measured below the ionosphere with rapid changes in amplitude not being uncommon [21]. Typical steady values of the electric field seem to be in the range 10–40 mV m^{-1} [20, 22, 23]. These correspond to ion velocities in the range 0.36–0.72 km s^{-1} according to the equation

$$v_\mathrm{i} = \left(\frac{E_\perp}{55\ \mathrm{mV\ m^{-1}}}\right) \mathrm{km\ s^{-1}} \tag{1}$$

which applies in regions where ion–neutral particle collisions are very infrequent. Other electric fields act within the magnetosphere to move thermal plasma and cause Joule heating [24, 25]. At low latitudes the equatorial electrojet

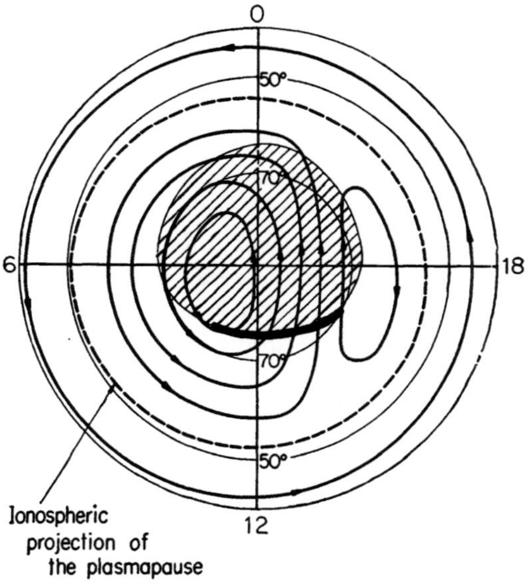

Fig. 3. Projection of field line motions on to the north polar region. The dotted line represents the plasmapause while the thick line is located at the polar cusp intersection with the surface of the earth.

involves a substantial Joule dissipation arising from neutral winds driving thermal plasma in response to diurnal heating. At mid-latitudes similar effects occur [26, 27] on a daily basis resulting in some motion of the ionospheric plasma. Probably the most important source of Joule heating, however, results from auroral electric fields which drive the auroral electrojet [8, 28–30]. In this instance the north to south electric field establishes a large Pederson current which produces large amounts of thermal energy directly in the neutral gas between 110 and 150 km altitude. This energy input is able to heat the auroral atmosphere and drive a substantial neutral wind system. The westward electrojet current, in contrast, does not involve a large energy dissipation since it results from a Hall drift of electrons.

3. Effects of Joule Heating

The ultimate source for the auroral electric field is not known. Cole [31] was the first to investigate in detail the release of thermal energy near aurorae and to point out the relevance of such heating to behavior of the thermosphere. In essence, it is known that large current sheets exist outside auroral forms with current flow parallel to the lines of magnetic force. Within the ionosphere Pederson current flows horizontally through the ionospheric E region establishing strong Joule heating through ion–neutral particle collisions. Since the electric field is low inside the auroral forms, most of the energy dissipation occurs on the outer regions of the actual visible auroral display [30]. In a series of computations, Cole [30] has estimated that a maximum value of 50 ergs $cm^{-2}\,s^{-1}$ can be released in the dayside auroral region due to Joule heating. For the night-time aurorae, the ionospheric conductivity is very irregular but with some estimates, Cole suggests that during very active periods the energy dissipation could be as large as 500–5000 ergs $cm^{-2}\,s^{-1}$. Since the total thermal energy of the thermosphere above 100 km is of the order 5×10^5 ergs cm^{-2}, such results indicate that a local auroral atmosphere could be rapidly heated.

The global implications of the auroral zone heat input cannot be precisely determined. The solar euv heat input averages about 1–5 ergs $cm^{-2}\,s^{-1}$ in the thermosphere [8]. Thus, over the surface of the earth an average of about 3×10^{18} ergs s^{-1} is absorbed per day. Joule heating, in contrast, can be related to the intensity of magnetic fluctuations at the earth's surface arising from the auroral electrojets [30]. If ΔH represents the magnetic fluctuation, measured in gammas, the global Joule heat input is $\sim10^{13}(\Delta H)^2$ ergs s^{-1}. Thus, if ΔH_{av} is greater than several hundred gamma, Joule heating is as important globally as solar euv. Furthermore, the Joule heating occurs principally in the dark sectors where euv heating is absent.

One direct consequence of auroral zone heating should be the presence of substantial equatorward winds in the thermosphere [32, 33]. In fact, observations of this effect appear to have been made. In a recent paper Stoffregen [34] has compiled data taken from the drift of 25 neutral barium clouds observed at twilight from high latitudes in northern Sweden. On the basis of the theory of Kohl and King [35] it was expected that the drift direction of these clouds would show a continual variation throughout the day (see Fig. 4) with a total wind rotation of 360°. The theoretical variation, derived from four different models of atmospheric winds [35, 36] is given in Fig. 5 as a function of local time. When the experimental observations are plotted on the same figure, however, it is found that two basic groups emerge. The first corresponds to observations made during quiet magnetic conditions in the absence of aurorae. For this case the cloud speed is relatively low and the direction is near that predicted by the standard models. More interesting, however, are the data clustered near a neutral wind direction of 270°, all of which have high speeds. These data were gathered during disturbed conditions and correspond to a 180° shift from the expected direction.

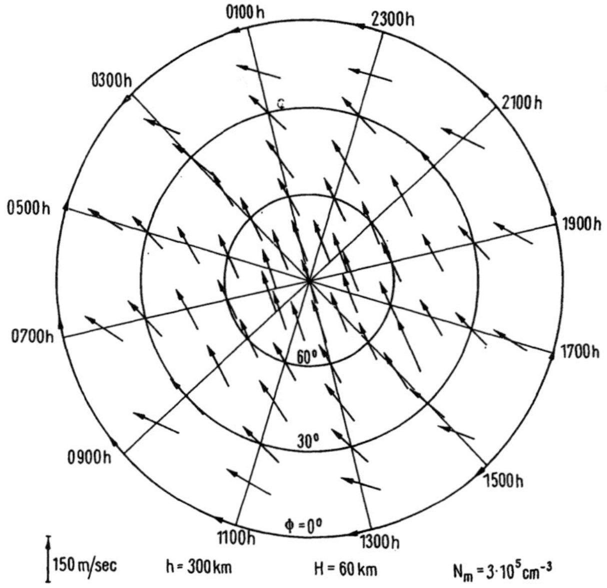

Fig. 4. Neutral gas winds at 300 km based on euv heating and ion drag with corotating ionosphere. (From Kohl and King [26]).

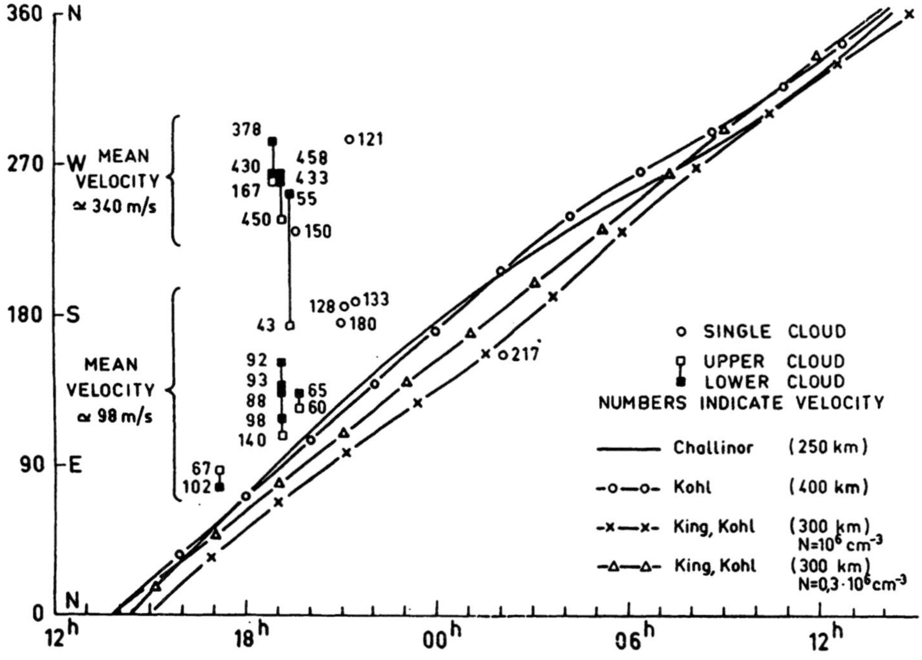

Fig. 5. Neutral wind direction showing theoretical values and results obtained by Stoffregen in northern Sweden (68° N) from barium cloud observations. (From Stoffregen [25].)

To explain the large shifts in wind direction, Stoffregen has assumed the existence of an auroral heat source which is sufficiently strong to alter the wind direction from that found for euv heating alone. In Fig. 6, prepared by Stoffregen, the standard Jacchia global temperature diagram has been changed

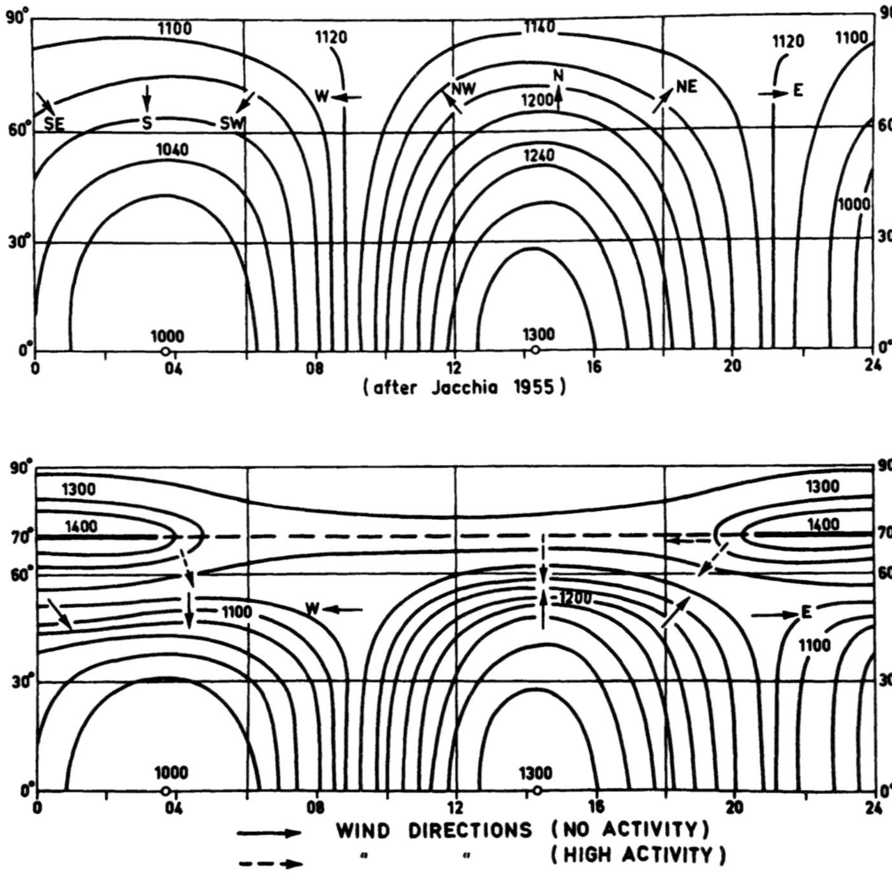

Fig. 6. Qualitative effects of auroral zone heating. Top, taken from standard model; bottom, effects of high pressure in the auroral zone. (From Stoffregen [25].)

to include an auroral zone temperature enhancement. Gradients in temperature are shown by arrows to indicate possible wind directions arising from the new temperature distribution. The presence of high temperature regions in the auroral zones can explain, at least qualitatively, the observed wind variations.

4. Effects of Magnetospheric Convection

In the regions outside the plasmasphere there is a continual convection of ionospheric plasma. This motion, which has its known source in the cross-tail magnetospheric electric field, drives the plasma through the neutral atmosphere.

The momentum thus transferred by ion–neutral particle collisions acts to set the neutral atmosphere in motion at all altitudes. Because of the variation of mass density with altitude, however, the neutral gas is not free to respond equally at all altitudes and a velocity shear is established which leads to a downward viscous transfer of momentum. In addition, there is substantial Joule heating within the E region where the ions, fixed by collisions with the neutral gas, cannot respond to the imposed electric field in the same way as electrons. As a consequence of these different processes, the neutral atmosphere is both heated from below and encouraged to share in the plasma motion at all altitudes.

Initial computations of the rate at which energy would be supplied to the neutral atmosphere for typical convection electric fields have been made by Walbridge [37]. It was found that about 3×10^{18} ergs s^{-1} would normally be released over the polar caps during undisturbed periods, i.e. the energy released by Joule heating and viscous dissipation in polar regions is of the same order of magnitude as that released globally from the absorption of euv radiation.

To investigate these processes further, new computations of the effects of a convection electric field upon the neutral atmosphere have been made. The equations of momentum and energy balance have been solved for typical summer-time polar cap conditions to find the time-dependent ion and neutral flow velocities and ion temperature following the onset of various electric fields.

To illustrate these results a standard coordinate system is adopted where X is the anti-solar direction and Z is pointed vertically. An electric field $E_\perp(t)$ is applied which drives the high altitude plasma in the X direction. Two forms for $E_\perp(t)$ were chosen

A. A quasi-step function

$$E_\perp(t) = 0 \qquad t < 0$$
$$E_\perp(t) = -20\, \hat{Y}\, [1 - \cos(t/3600)]\ \mathrm{mV\ m^{-1}} \qquad 0 \leqq t \leqq 1800\ \mathrm{s}$$
$$E_\perp(t) = -40\, \hat{Y}\ \mathrm{mV\ m^{-1}} \qquad t > 1800\ \mathrm{s}$$

B. A periodically varying function

$$E_\perp(t) = -20\, \hat{Y}\, [1 - \cos(t/3600)]\ \mathrm{mV\ m^{-1}}\,.$$

For the results discussed here a standard model atmosphere with $T_\infty = 1100$ °K was used. The electron density profile, listed in Table 1, is typical of daytime polar conditions. Complete details of the equations, boundary conditions and method of solution will be given elsewhere.

Table 1

Electron densities

z (km)	n_e (cm^{-3})
100	1.00×10^5
150	2.07×10^5
200	3.17×10^5
250	4.26×10^5
300	5.35×10^5
350	5.59×10^5
400	4.56×10^5

Results for the quasi-step function are shown in Figs. 7–9. Fig. 7 presents the Y component of neutral gas velocity normalized to $-800\ \mathrm{m\ s^{-1}}$ following initiation of the electric field. This figure shows that in the initial surge of Pederson current, ion momentum is transferred to the neutral gas, which begins to move the initially stationary gas in the $-Y$ direction. As time goes on this flow ceases progressively from high altitudes to lower due to the increase in the X component of neutral gas velocity which eliminates the collision

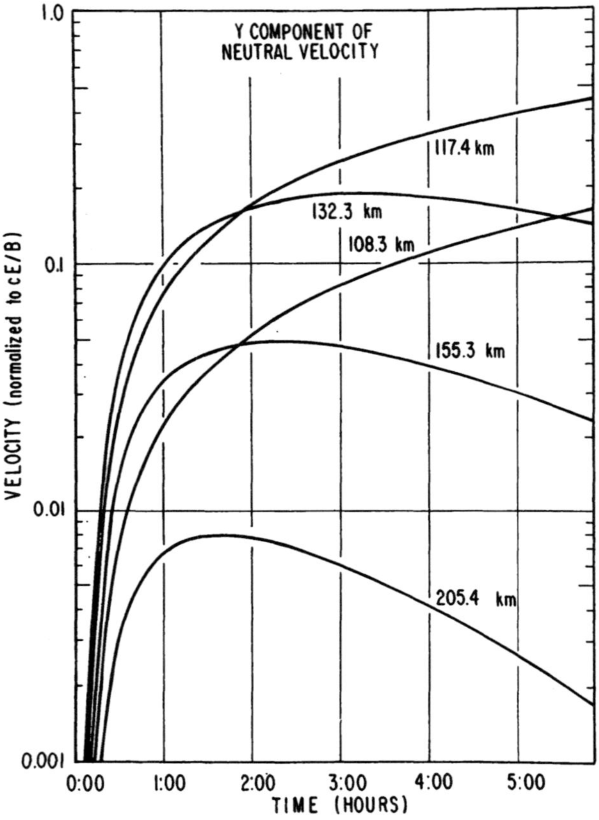

Fig. 7. Y component of neutral wind arising from the quasi-step function electric field. (Normalized velocity equals $-800\ \mathrm{m\ s^{-1}}$ for all figures.)

effect retarding the ion motions. The variation of the X component of neutral velocity is shown in Fig. 8. Here there is a progressive increase in neutral wind speed with time at all heights. It is interesting to note that one hour after onset, the neutral gas speed is about $160\ \mathrm{m\ s^{-1}}$ even at an altitude of 132 km. Such a rapid response would not be expected through the ion drag term alone: the viscous dissipation term plays an important part in the time variation of the velocities in the E region.

When the external electric field is applied, thermal ions are driven through the neutral gases. This results in appreciable heating of the ions as long as there is a relative velocity between the ions and neutrals. For a stationary neutral gas the expression

$$T_i = T_n + 1217\,(E_\perp/55)^2\ \text{°K} \tag{2}$$

with E measured in millivolts per meter gives the rise in ion temperature at polar latitudes. In practice the neutral gas is not stationary and $T_i - T_n$ is

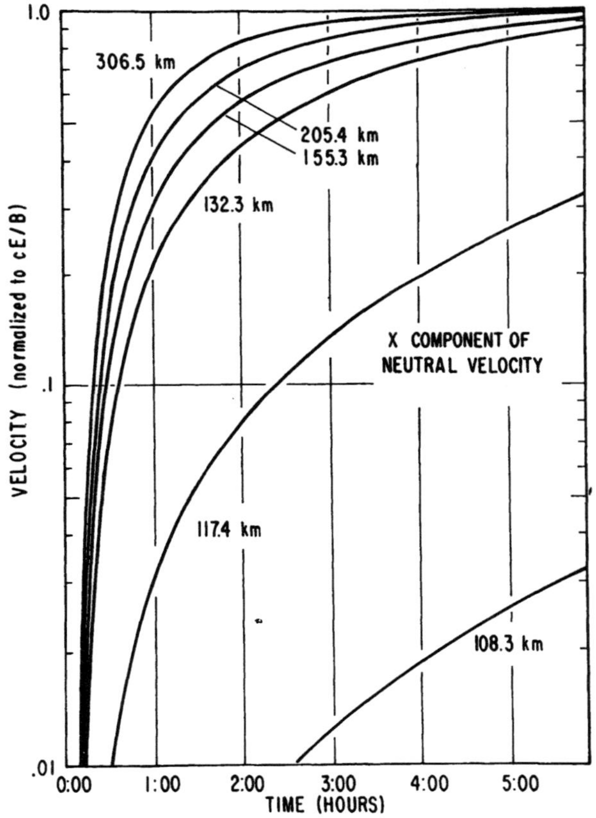

Fig. 8. X component of neutral wind arising from the quasi-step function electric field.

less than predicted by Eq. (2). For the present example computations of T_i yield the values shown in Fig. 9. There is an initial rise in T_i at all altitudes but, as the neutral gases reach the plasma drift speed, the heating becomes small. In this model ions transfer energy to the neutral gas in two ways: Joule heating and viscous dissipation. Computations of these heat sources are shown in Fig. 10. In the initial onset, the integrated Joule heat input rises rapidly to 15 ergs cm^{-2} s^{-1} with the principal heat input in the range 110–180 km. As the neutral gas speeds increase there is a gradual decline. In contrast, viscous dissipation never exceeds 2 ergs cm^{-2} s^{-1} and its behavior is to increase with time. Nevertheless, the combination of the two sources again indicates the

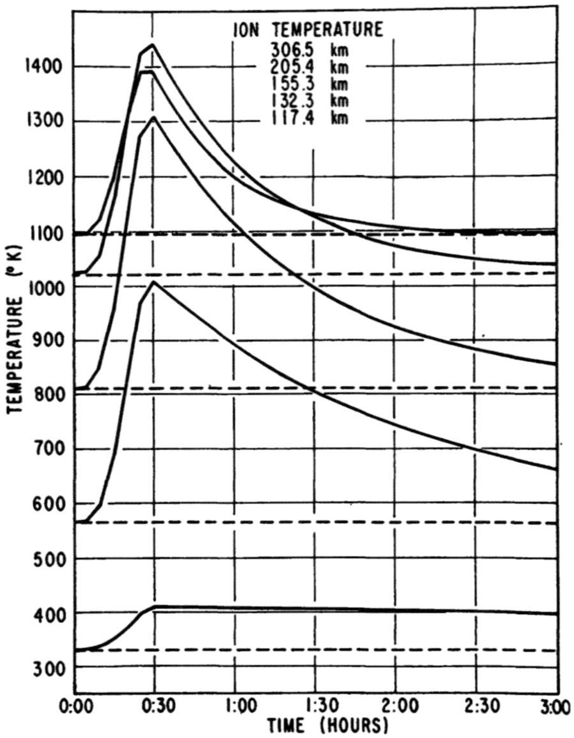

Fig. 9. Ion gas temperatures computed for the quasi-step function electric field of $E_{\perp} = 40 \text{ mV m}^{-1}$.

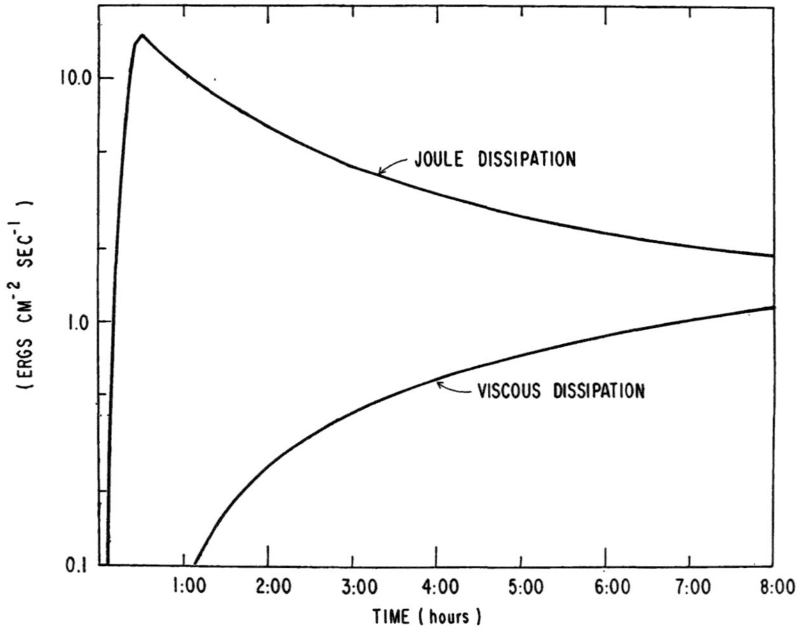

Fig. 10. Height integrated energy input to the neutral gas for the quasi-step function electric field.

importance of the convection electric field to the structure of the thermosphere. As a final point, from the data of Figs. 7 and 8 it can be deduced that there is a rotation of wind direction with altitude. At high altitudes the wind blows in the X direction, but swings progressively towards $-Y$ for lower heights. This rotation also depends on time, with a progression from $-Y$ to X as time goes on.

Experiments [20–23] show that the convection electric field is continually fluctuating. Thus, the periodic E defined by B above, is useful to indicate

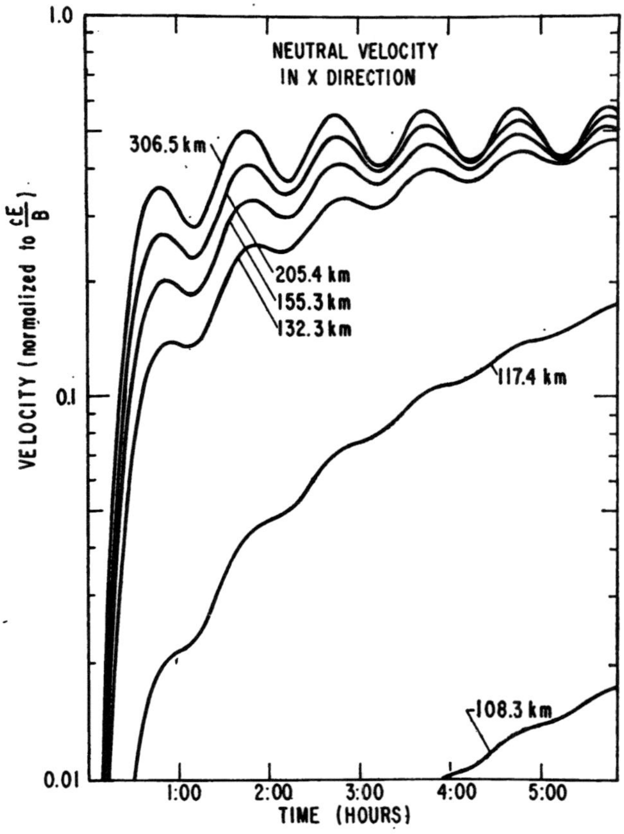

Fig. 11. X component of neutral gas velocity for periodic electric field.

the physical processes which may occur. Fig. 11 shows the X component of neutral velocity to indicate the rapid approach to steady state conditions. Again, the viscosity of the neutral gas is important in driving the neutral gas into motion at low altitudes. The corresponding variations in ion temperature are shown in Fig. 12. The double frequency behavior of the temperature follows from the inertia of the neutral gas. As $E_{\perp}$ varies in time the ion gas is alternately faster or slower than the neutral gas. Thus the relative velocity has two peaks per cycle as shown. An important aspect of this process relates to the behavior of the ion–neutral gas system when the magnetospheric source field fluctuates rapidly or goes to zero. In these cases the inertia of the moving neutral atmosphere will act as a dynamo to continue to drive the ionosphere and magnetospheric plasma at a speed similar to that before the sudden decrease in $E_\perp$. Thus, the neutral winds can act as an enormous flywheel, smoothing out plasma motions even when the fluctuations of the magnetospheric source field are large.

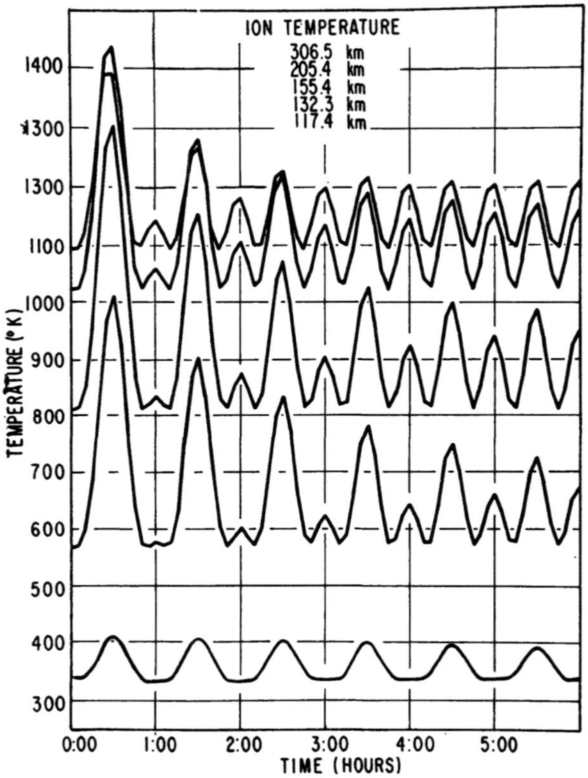

Fig. 12. Ion temperature for periodic electric field showing double frequency behavior.

The results given here deal with a specific electric field and ion density profile. Smaller values of $E_\perp$ or ion densities lead to smaller amounts of energy dissipation and larger time constants for motion. More complete details will be given in a later paper, but it is noted that the parameters used here tend to be typical of average rather than maximal conditions.

5. Global Winds including Polar Convection

A comprehensive model of global winds including the effects of convection and Joule heating can be qualitatively described. At low latitudes solar euv absorption heats the atmosphere in the usual fashion. In response to the daytime pressure maximum winds will blow towards the dark side of the earth, subject to Coriolis effects and ion drag. At high latitudes ion drag (convection) will assist this flow by accelerating the neutral gases over the polar cap into the night-time regions. High pressure regions will also be formed in the auroral oval as a result of energetic particle impact and Joule heating. Below the auroral zones, but outside the plasmapause, ion drag will oppose the normal day to night flow of the neutral atmosphere and may be strong enough to establish westward winds and a separate circulation pattern as shown in Fig. 13. Actually, due to Coriolis effects a further aspect may be that the westward winds in the dusk sector are enhanced by the south winds arising from auroral heating. Likewise, polewards of the dusk sector auroral

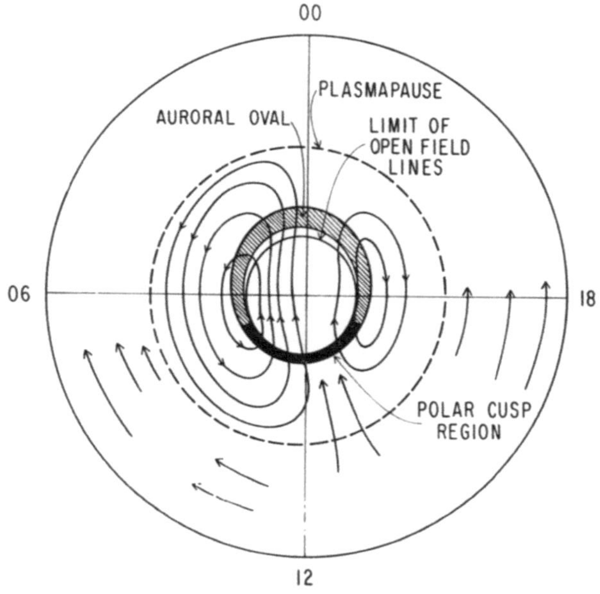

Fig. 13. Qualitative view of global winds including the effects of ion convection and auroral zone heating.

zone the winds generated by the aurorae will be deviated to the east, assisting the convection flow. Just the opposite effect occurs on the dawn side. Here the winds from the auroral zone will be in opposition to those generated from ion drag both polewards and equatorwards of the auroral zone. Since auroral activity is more frequent before midnight than after, the net effect of these considerations is difficult to assess. However, it is entirely possible that the dawn sector westward motion may predominate since most ion convection paths pass over this side of the earth (a corotation effect [38]). In such a situation the combination of westward motion outside the plasmapause, when coupled with the corotating atmospheric regions by horizontal viscosity, may explain the tendency for there to be an average movement of the upper atmosphere which is more rapid than corotation, i.e. a mean easterly wind (see also [39]).

There exist recent satellite data [6] which support the idea of a wind reversal occurring at the auroral zone. Fig. 14, presented by Feess [6], results from a precise measurement of the drag experienced by a low altitude satellite (200 km) during disturbed magnetic conditions on 26 May 1967. Three separate

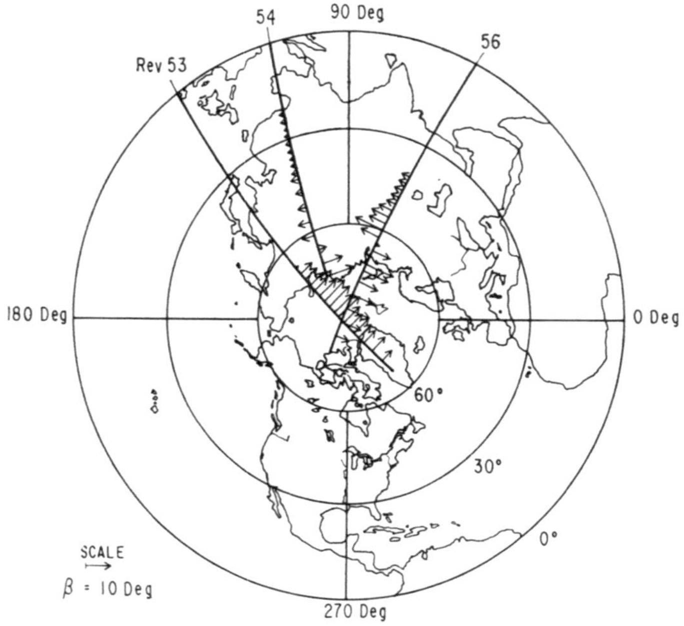

Fig. 14. LOGACS satellite data showing direction of neutral gas flow for three polar passes in May 1967. The local time is 1000 LT for all passes. (From Feess [6].)

polar passes were made at a constant local time (~1000 LT). The parameter β, plotted for each track, is proportional to the wind speed. On two of the three passes, wind reversals were noted. Furthermore, it can be seen that these changes occurred at approximate auroral zone boundaries with the wind blowing in an anti-solar sense over the polar cap and westward south of the auroral zone. Careful examination of data such as these should provide valuable clues as to the true behavior of neutral winds at high latitudes.

6. Summary

There is substantial reason to believe that convection and auroral heating play important parts in the energetics and dynamics of the thermosphere. The variability of these phenomena makes it difficult to make precise models of the global thermospheric behavior. During periods of exceptionally low magnetic activity euv may well be the principal heat source, but when convection is present, i.e. a plasmapause exists, the high latitude contributions cannot be ignored. It is found that the polar atmosphere may also affect magnetospheric motions, especially with regard to time variations in the cross-tail electric field.

Acknowledgments

This work was supported through NASA Grant NGR-05-009-075. The calculations of convective motions and energy dissipation were made by Mr. Joel Fedder whose assistance and comments are greatly appreciated.

References

[1] L. Broglio, Space Research X, 493 (1970).
[2] N. W. Spencer et al., Space Research X, 389 (1970).
[3] Y. L. Truttse, Planet. Space Sci. **17**, 181 (1969).
[4] L. G. Jacchia, J. Slowey and F. Verniani, J. Geophys. Res. **72**, **1423** (1967).
[5] B. R. May and D. E. Miller, Planet. Space Sci. **19**, 39 (1971).
[6] W. A. Feess, paper presented at XIVth COSPAR Meeting, Seattle 1971.
[7] L. G. Jacchia, Space Research X, 367 (1970).
[8] H. Volland and H. G. Mayr, J. Geophys. Res. **76**, 3764 (1971).
[9] S. Chandra and J. R. Herman, Planet. Space Sci. **17**, 841 (1969).
[10] M. F. K. Abub-Robb, Planet. Space Sci. **17**, 1269 (1969).
[11] J. W. Dungey, Phys. Rev. Lett. **6**, 47 (1961).
[12] W. I. Axford and C. O. Hines, Canad. J. Phys. **39**, 1433 (1961).
[13] T. Obayashi and N. Nishida, Space Sci. Rev. **8**, 3 (1968).
[14] W. I. Axford, Rev. Geophys. **7**, 421 (1969).
[15] P. M. Banks and T. E. Holzer, J. Geophys. Res. **74**, 6317 (1969).
[16] A. Nishida, J. Geophys. Res. **71**, 5669 (1966).
[17] D. M. Willis, Planet. Space Sci. **17**, 339 (1969).
[18] W. J. Heikkila and J. D. Winningham, J. Geophys. Res. **76**, 883 (1971).
[19] L. Frank, J. Geophys. Res. **76**, 5202 (1971).
[20] D. P. Cauffman and D. A. Gurnett, J. Geophys. Res. **76**, 6014 (1971).
[21] F. Mozer, J. Geophys. Res. **76**, 3651 (1971).
[22] E. M. Wescott, J. D. Stolarik and J. P. Heppner, J. Geophys. Res. **74**, 3469 (1969).
[23] F. S. Mozer and R. H. Manka, J. Geophys. Res. **76**, 1697 (1971).
[24] J. D. Tarpley, Planet. Space Sci. **18**, 1025 (1970).
[25] R. J. Slening, Planet. Space Sci. **18**, 423 (1970).
[26] S. Matsushita, Radio Sci. **6**, 279 (1971).
[27] R. J. Slening, Planet. Space Sci. **18**, 121 (1970).
[28] G. E. Thomas and B. K. Ching, J. Geophys. Res. **74**, 1796 (1969).
[29] H. Volland, Planet. Space Sci. **17**, 1709 (1969).
[30] K. D. Cole, Planet. Space Sci. **19**, 59 (1971).
[31] K. D. Cole, Aust. J. Phys. **15**, 223 (1962).
[32] J. H. Piddington, Geophys. J. Roy. Astron. Soc. **7**, 415 (1963).
[33] G. Chimonas and C. O. Hines, Planet. Space Sci. **18**, 565 (1970).
[34] W. Stoffregen, in: Magnetosphere–Ionosphere Interactions, Univ. of Oslo Press, Oslo (in press).
[35] H. Kohl and J. W. King, J. Atmos. Terr. Phys. **29**, 1045 (1967).
[36] R. A. Challinor, Planet. Space Sci. **18**, 1485 (1970).
[37] E. Walbridge, J. Geophys. Res. **72**, 5213 (1967).
[38] D. G. King-Hele, D. W. Scott and D. M. C. Walker, Planet. Space Sci. **18**, 1433 (1970).
[39] C. O. Hines, Planet. Space Sci. **13**, 169 (1965).

Space Research XII — Akademie-Verlag, Berlin 1972

THE CALCULATION OF WINDS FROM THERMOSPHERIC MODELS

H. KOHL

Max-Planck-Institut für Aeronomie, Lindau/Harz, FRG

Possible sources of inaccuracies are considered, which affect the calculation of winds from thermospheric models. Firstly, the variation of the pressure gradient derived from models may be in error, because of errors in the density, temperature and composition. A statistical test shows that the reliability of density gradients is quite satisfactory. Several authors have pointed out, however, that temperature and composition seem to be sometimes appreciably different from what models predict. The effect of these discrepancies on the calculated winds is discussed. Secondly, there are also inaccuracies in the methods of calculation. Most authors in this field neglect the non-linear term in the equation of motion, which is of some importance during the early morning. Furthermore, there is no realistic upper boundary condition. All boundary conditions used so far are more or less problematic. In spite of all these uncertainties there is ionospheric evidence that at least at medium latitudes the calculated winds are fairly correct, as many F layer phenomena can be fairly well interpreted as an effect of such winds.

1. Introduction

Neutral air motions in the thermosphere occur for different reasons. For instance, motions of ionization may exert a drag force on the neutral air which will be set into motion, gravity waves will move the neutral air in a wavelike manner, etc. In this paper only those neutral air motions will be considered that are produced by more or less regularly varying global pressure variations in the thermosphere. There are two procedures for calculating such winds. In the first the pressure gradients that produce the wind are taken from observations or from atmospheric models that are based on observations. The second approach starts from more general equations including the thermal balance of the atmosphere and may even result in atmospheric models rather than being based on them. Here only calculations belonging to the first category are discussed.

The question is, how reliable are those calculations? This question may be split in two: (i) What is the accuracy of atmospheric models used for the calculations? Are they systematically in error or are there occasional deviations between models and observations, for instance during disturbed periods? (ii) All calculations performed so far make use of simplifying assumptions and neglects. Some authors neglect viscosity or the non-linear term in the equation of motion. Also the choice of suitable boundary conditions is problematic. These limitations will be discussed.

2. The Usability of Jacchia's Atmospheric Models

The full equation of motion, which has to be solved in order to calculate the wind velocity $\boldsymbol{v}_n$ is

$$\frac{\partial \boldsymbol{v}_n}{\partial t} + (\boldsymbol{v}_n \nabla)\boldsymbol{v}_n + 2\,(\omega \times \boldsymbol{v}_n) - \frac{1}{\varrho_n}\nabla(\eta\,\nabla \boldsymbol{v}_n) + \alpha\,(\boldsymbol{v}_n - \boldsymbol{v}_i) = -\frac{1}{\varrho_n}\nabla p_n + \varrho_n\, g\,, \tag{1}$$

where α is proportional to the ion density and describes the friction between ions and neutral particles, and the other symbols have their usual meanings. The driving force on the right-hand side of the equation which produces the wind is

$$-\frac{1}{\varrho_n}\nabla p_n = -\frac{1}{\varrho_n}(k\,T \cdot \nabla n + k\,n \cdot \nabla T)\,. \tag{2}$$

As Jacchia's atmospheric models give information on the space and time variations of both n and T, it is possible to derive also ∇n and ∇T and consequently the driving force. In order to discuss the accuracy of the driving force determined in this way the various parameters ϱ_n, T, ∇T and ∇n have to be considered.

2.1. The Accuracy of the Density Gradient

The original information on which Jacchia's models (subsequently denoted J models) are based is the neutral air density ϱ_n derived from satellite drag measurements. Many ϱ_n data have been approximated by more or less empirical formulae making use of physical evidence as, for example, the validity of the barometric law. These formulae together form the atmospheric model. It has been shown that the average deviation of the J 1965 model from the observations is $\pm 15\%$ for a height of 325 km [1]. Thus, the model seems to approximate the measurements quite well. It cannot, however, be concluded from this that $\nabla \varrho_n$ derived from the J-models has the same accuracy. It is evident that when a function fits a number of points the gradients may not fit. In order to check the reliability of $\nabla \varrho_n$ inferred from the J-models a test was performed using daily mean values of ϱ_n as deduced from Explorer 1 and Explorer 8 for the period November 1960 to August 1961 [2]. The ϱ_n values were converted into exospheric temperature T_∞ using Jacchia's formula. For every day the ratio T_∞ (Explorer 1)/T_∞ (Explorer 8) was calculated and drawn in the ordinate of Fig. 1. This represents, in other words, the observed ratio of the densities at the two perigees on a non-linear scale. On the other hand the exospheric temperature is given by Jacchia's formula as

$$T_\infty = T_0 f(\varphi, \delta, \tau)\,,$$

where f is a given function of latitude φ, solar declination δ and the hour angle of the sun τ. Hence

$$(T_{\infty 1}/T_{\infty 2})\ \text{model} = f(\varphi_1, \delta, \tau_1)/f(\varphi_2, \delta, \tau_2)$$

and is only a function of the position of the sun and of the two perigees. This parameter is drawn in the abscissa of Fig. 1. It is obvious that there is a very good correlation between these two parameters, the correlation coefficient being 0.96. This means that the model predicts not only the densities but also the density ratios at different locations quite accurately, and it may be concluded with some caution, therefore, that ∇p_n values can also be derived from the J-models with sufficient reliability. It should be noted that in Eq. (2) ∇p_n depends on ∇n and not on $\nabla \varrho_n$, so that the composition of the atmosphere

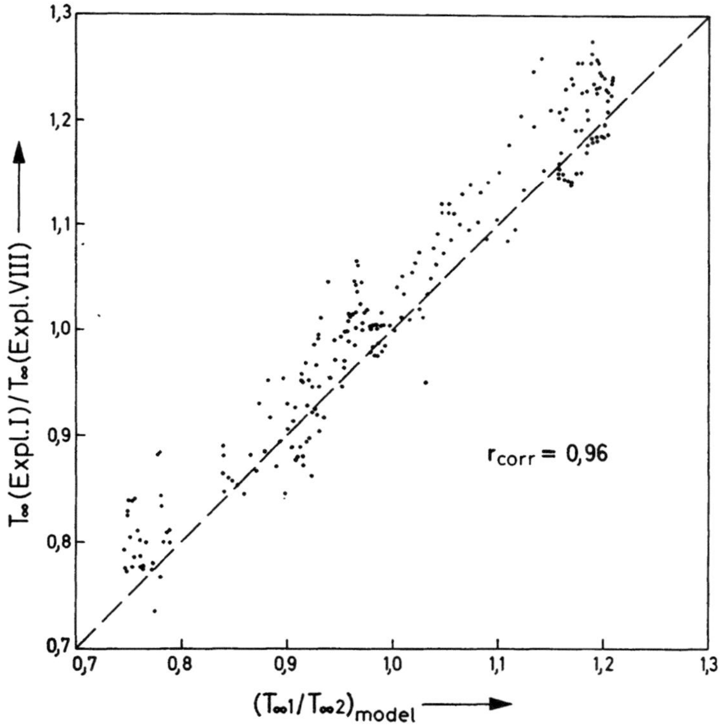

Fig. 1. Ratio of exospheric temperatures at the perigees of two satellites (Explorer 1 and Explorer 8) as predicted by atmospheric model compared with the same parameter as deduced from satellite drag measurements.

becomes involved. However, the heights of the two perigees were 350 km and 420 km, where atomic oxygen is dominant.

This test applies only to low and medium latitudes as the orbital inclination of the satellites was $<50°$. It is well known that during geomagnetic disturbance the density increases, especially at high latitudes [3]. The present models may be in error to an unknown extent during such periods, and there is evidence that then the wind velocities, indeed, are different from what is expected during quiet conditions [4, 5].

2.2. The Accuracy of the Temperature and the Temperature Gradient

The temperature given by the J-models is not an observed parameter, but is introduced into the models using different assumptions. It is interesting, therefore, to compare model temperatures with temperatures derived from measurements, e.g. from incoherent scatter observations. Fig. 2 shows the variation of T for three days as observed at St. Santin [6] together with temper-

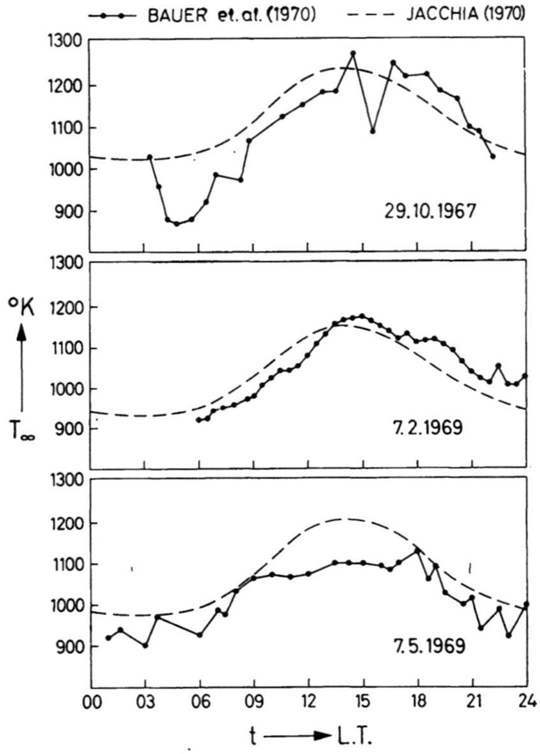

Fig. 2. Diurnal variations of exospheric temperature for three individual days according to incoherent scatter observations compared with Jacchia's model.

atures according to the J 1971 model for the same period. It turns out that on the average measured and observed temperatures agree within 10%. It should be noted that in [6] a comparison with the J 1965 model was made and the model temperatures appeared to be too high. It is difficult to estimate the accuracy of the temperature gradient because of lack of direct measurements. It has been concluded from incoherent scatter observations that the diurnal temperature maximum occurs between 1600 and 1700 LT instead of 1400 LT as assumed in the J-models. This means that at least in the afternoon ∇T as it is deduced from the models may be in error. It cannot be easily seen, however, how large the error is.

From simultaneous incoherent scatter measurements at Jicamarca and St. Santin [7] it was concluded that during solstice conditions the temperature variation over the globe is much higher than was expected. While the J-models assume that the temperatures vary from the coldest to the hottest point over the globe by a ratio of about 1:1.3 it is suggested in [7] that this ratio may be as high as 1:1.6 during solstice. If this is true, then ∇T can be about twice as large as was formerly assumed. In order to estimate the influence of this deviation on the driving force the relative magnitude of the two terms on the right-hand side of Eq. (2) was calculated. Fig. 3 shows the results for two exospheric temperatures. It appears that at high altitudes the ∇T term contributes about one quarter to the total pressure gradient; at low altitudes its

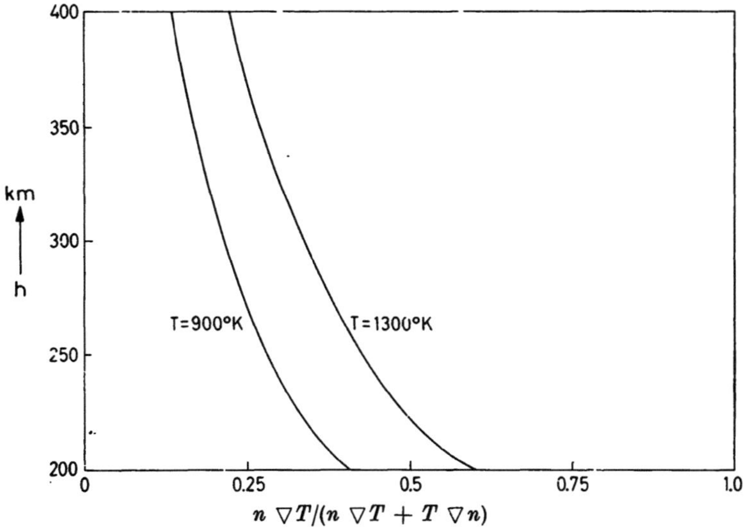

Fig. 3. Relative influence of the temperature gradient on the total driving force. $\nabla p_n = k\,(n \,\nabla T + T\, \nabla n)$ according to Jacchia's model.

influence increases to about one half. If the above-mentioned temperature ratio is really so large, this would result in an increase of the wind velocity during solstice of about 25%–50% compared with former calculations.

3. The Accuracy of the Calculations

As well as by limitations in the atmospheric models, inaccuracies are also introduced by simplifications and neglects in the procedure of calculation itself. In all but one [8] calculations performed so far the non-linear term in Eq. (1) was neglected. This is justified for velocities smaller than $100\ \mathrm{m\,s^{-1}}$ [9]. Recently it was demonstrated [10] that during the early morning hours, when the wind velocities exceed this limit, the non-linear term becomes important and changes the phase of the diurnally rotating wind vector by one or two hours.

In Eq. (1) the frictional term is very important. It is proportional to the ion density N_i and thus the reliability of the wind calculations depends on how accurately N_i is known. Generally, there are two ways of obtaining N_i: it may be calculated from the continuity equation of the ionization, or observational material can be used to construct a model of the N_i distribution. Both methods are subject to errors, and the resulting wind velocities cannot be more accurate than the assumed N_i values.

Some authors have omitted the viscosity term in the equation of motion, Eq. (1). It can be shown that this may change the results of wind calculations

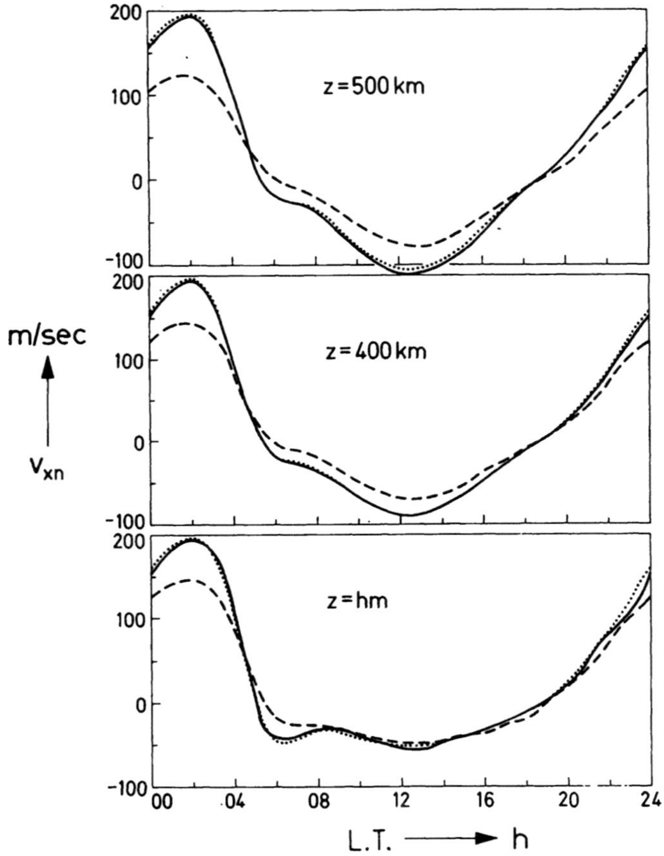

Fig. 4. Calculated meridional wind velocity at Ft. Monmouth, Sept. 1960, for three altitudes $z = 500$ km, 400 km and h_m (h_m = height of F2 layer maximum) and for three different boundary conditions: ——— $\partial v_n/\partial z = 0$ at 630 km altitude, ·········· $\partial v_n/\partial z = 0$ at 800 km altitude, - - - - $v_n = 0$ at 1100 km altitude.

by up to a factor of two. If, however, the viscosity term is included in the calculations, the problem of applying suitable boundary conditions arises, because by viscosity the wind velocities at different points in space are coupled. As in the viscosity term the derivatives are of second order in each of the three coordinates, six boundary conditions are necessary. In calculations performed so far it was assumed that $v_n = 0$ at the lower boundary, say at 120 km altitude, because there the driving force vanishes according to the J-models. This condition is unrealistic, since we know from observations that quite large wind velocities exist at this height level. However, the wind velocities at greater altitudes are largely independent of what is assumed below, so no complications arise from this point. A more serious problem is the choice of the upper boundary conditions. Without detailed discussion it may be stated here that above an altitude of 600–700 km the concept of hydromechanics breaks down more or less, and below this altitude no convincing boundary condition can be found. In current calculations of the wind velocity it is usually assumed that all horizontal derivatives can be neglected and that $\partial \boldsymbol{v}_n / \partial z = 0$ at a certain altitude around 600–700 km. This boundary condition is a compromise between conflicting arguments, which cannot been discussed here. Instead, the influence of different boundary conditions on the results of the calculations may be demonstrated. Fig. 4 shows the diurnal variation of the north–south component of the wind calculated for three different altitudes and three different conditions:

$$(a)\ \frac{\partial \boldsymbol{v}_n}{\partial z} = 0 \text{ at } z = 630 \text{ km}, \quad (b)\ \frac{\partial \boldsymbol{v}_n}{\partial z} = 0 \text{ at } z = 800 \text{ km},$$

$$(c)\ \boldsymbol{v}_n = 0 \text{ at } z = 1100 \text{ km}.$$

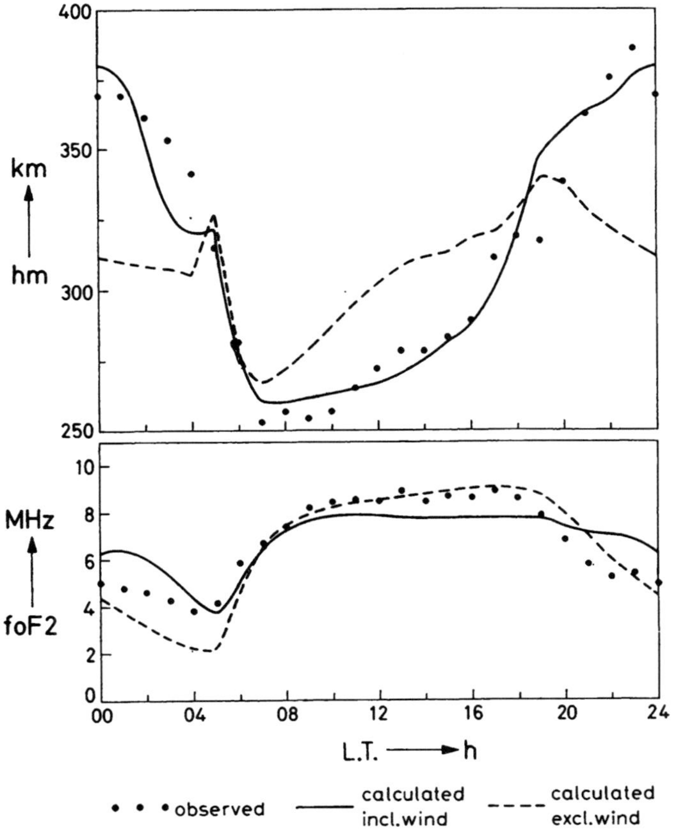

Fig. 5. Observed and calculated variations of h_m F2 and f_o F2. St. John's (48° N 53° W), Sept. 1960.

In all these calculations the horizontal derivatives were neglected. It turns out that during most of the day the three results agree very well. Only for the time from 00 LT to 04 LT appreciable differences exist between condition (c) and the two others. Thus, it seems that the results for $z \leqq 500$ km are not very sensitive to the upper boundary condition.

4. Ionospheric Evidence for Thermospheric Winds

Neutral air winds will drive the F region ionization up or down the magnetic lines of force, and it can be expected that information on thermospheric winds

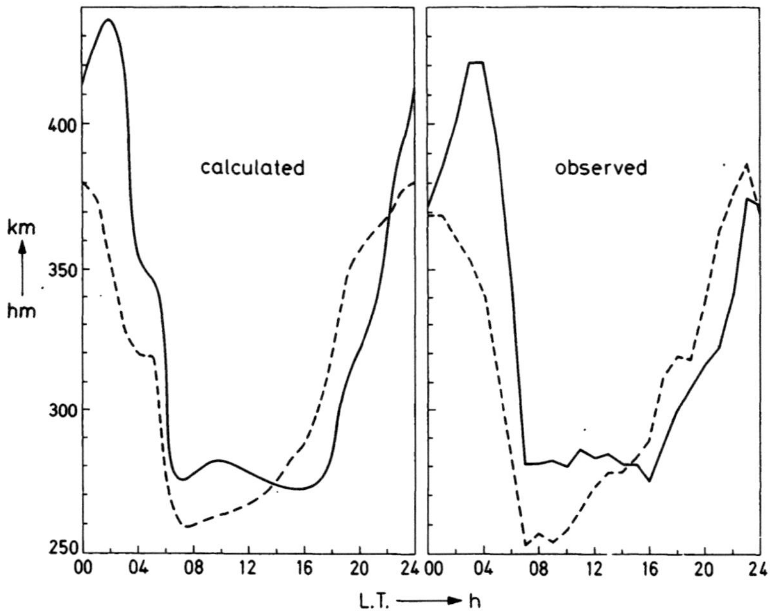

Fig. 6. Observed and calculated variation of h_m F2 for two locations demonstrationg the effect of magnetic declination. Sept. 1960. ——— Adak (52° N, 177° W), $I = 63°$, $D = 8°$. - - - St. John's (48° N 53° W), $I = 71°$, $D = -27°$.

can be obtained from ionospheric observations. Indeed, the meridional component of winds could be deduced from Doppler-shift measurements of incoherent scatter spectra [11], and it was shown that it agrees with calculated winds within the error limits of the observations [12].

It may also be expected that height variations of the F layer are caused by atmospheric winds. In Fig. 5 observed diurnal variations of h_m F2 are shown and compared with h_m F2 variations calculated including and excluding winds. Fig. 6 represents height changes for two stations which have different magnetic declination, the other coordinates being approximately equal. The calculations on the left-hand side of Fig. 6 are performed including winds. It can be seen in both figures that observation and calculation are in good agreement provided that winds are taken into account.

In the preceding sections some possible sources of error were enumerated. It is surprising that, nevertheless, the calculated wind velocities seem to be precise enough to enable a nearly quantitative explanation to be given of the h_m F2 behaviour and even of its declination effect. It should be noted, however, that the height of the F layer peak does not depend very sensitively on the wind velocity. A change in v_n by a factor of e will change h_m F2 by about one scale height. Thus, if v_n is in error by 40%, this will result in an error in h_m F2 of only about 20 km. Moreover, the agreement is not in all cases as good as in Figs. 5 and 6. Fig. 7 shows an example where the calculations do not fit the observations. This happens particularly at low latitudes.

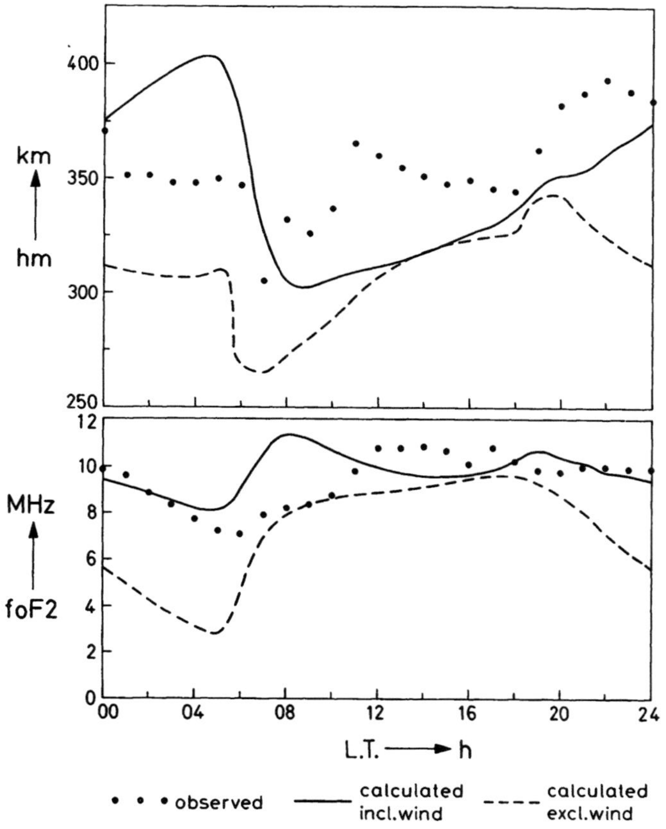

Fig. 7. Observed and calculated variations of h_m F2 and f_0 F2. Puerto Rico (19° N 66° W), June 1960.

5. Summary

Many sources of errors may exist that affect the calculations of wind velocities in the thermosphere from atmospheric models: inaccuracies may be inherent in the models or may be introduced by simplifications in the calculations, such as neglect of the viscosity or of the non-linear term in the equation of motion, and by the assumption of problematic boundary conditions. The accuracy achieved seems to be adequate for the explanation of ionospheric phenomena.

Acknowledgments

The author is indebted to the World Data Center, Boulder, for making available h_m F2 values and to Drs. Rüster and Stoffregen for information in advance of publication.

References

[1] L. G. JACCHIA, Smithson. Contrib. Astrophys. **8**, 215 (1965).
[2] L. G. JACCHIA, Smithson. Astroph. Obs. Spec. Rep. No. 171 (1965).
[3] L. G. JACCHIA and J. SLOWEY, J. Geophys. Res. **69**, 905 (1964).
[4] H. KOHL, Space Research X, 550 (1970).
[5] W. STOFFREGEN, private communication.
[6] P. BAUER, P. WALDTEUFEL and D. ALCAYDE, J. Geophys. Res. **75**, 4825 (1970).
[7] P. WALDTEUFEL and J. P. MCCLURE, Annls Géophys. **25**, 785 (1969).
[8] R. A. CHALLINOR, Planet. Space Sci. **16**, 557 (1968).
[9] H. KOHL and J. W. KING, J. Atmos. Terr. Phys. **29**, 1045 (1967).
[10] R. RÜSTER and J. R. DUDENEY, J. Atmos. Terr. Phys., in press.
[11] G. VASSEUR, J. Atmos. Terr. Phys. **31**, 397 (1969).
[12] G. J. BAILEY and R. J. MOFFETT, J. Atmos. Terr. Phys. **32**, 417 (1970).

Space Research XII — Akademie-Verlag, Berlin 1972

THERMOSPHERIC MOTION AND TEMPERATURE PERTURBATIONS FROM GLOBAL-SCALE WINDS FLOWING THROUGH F REGION IONIZATION ANOMALIES

R. E. Dickinson and R. G. Roble

National Center for Atmospheric Research, Boulder, Col., USA

There exist in the F region departures from a mean global-scale ionization distribution. Global-scale winds, forced by global-scale pressure gradients, flowing past these anomalies are accelerated where the ion drag is locally less than its global-scale smooth value and, likewise, are decelerated where the ion drag is locally greater. This interaction sets up a perturbation in the horizontal flow.

The temperature perturbation and circulation pattern caused by the ion drag anomaly are calculated using a time-independent model. Perturbation back-pressures almost completely balance momentum sources having a horizontal scale of a few hundred kilometers, but not sources having a horizontal scale of a few thousand kilometers or greater. We discuss the circulation and temperatures due to the interaction of the global-scale wind with the electron density variation at sunrise and with the electron density depression within a stable auroral red arc.

1. Introduction

Global-scale winds with amplitudes up to several hundred meters per second at the F layer heights are driven by global-scale pressure forces in the thermosphere [1, 2].

Ionization anomalies, defined as departures from a mean global-scale ionization distribution, are commonly found at F layer heights where the ion drag on the neutral motions is linear in the ion concentration. Global-scale winds flowing past these ionization anomalies experience an acceleration where the ion drag is locally less than its global-scale smooth value and, likewise, they experience deceleration where the ion drag is locally greater. A perturbation in the horizontal flow therefore results, and vertical velocities must also occur which give adiabatic warming and cooling. The consequent perturbation temperatures determine pressures whose horizontal gradients partially balance the perturbation ion drag. Hence, the magnitude of the anomaly in the winds should be less than that which would be inferred from the ion drag anomaly alone.

We discuss here the circulations that are forced by ionization anomalies interacting with the neutral wind in terms of a self-consistent dynamic model in which the resulting back-pressure force is included in the calculations [3]. For comparison, the circulation obtained from a dynamic model without the generated back-pressure is also presented.

2. Dynamic Model

The forcing for the dynamic model is given by a momentum source which depends on the interaction of a basic state neutral wind with the anomaly. The nature of the momentum source in the equation of motion is as follows (viscous forces are neglected for illustration, but are included in the calculations).

Equation of motion:

$$\varrho_0 (\bar{\lambda}\,\bar{u} + \bar{\lambda}\,u' + \lambda'\,\bar{u}) + \frac{\partial \bar{p}}{\partial x} + \frac{\partial p'}{\partial x} = 0$$

Basic wind:

$$\varrho_0 \bar{\lambda}\,\bar{u} + \frac{\partial \bar{p}}{\partial x} = 0$$

Perturbation wind:

$$\varrho_0 (\bar{\lambda}\,u' + \underbrace{\lambda'\,\bar{u}}_{\text{Momentum source}}) + \frac{\partial p'}{\partial x} = 0$$

where $\bar{u} = \bar{u}(z)$, $\bar{\lambda} = \bar{\lambda}(z)$, $u' = u'(x, z)$, $\lambda' = \lambda'(x, z)$. The basic state wind $\bar{u}(z)$ is determined from a balance between the basic state pressure force $\partial\bar{p}/\partial x$ and the global mean value of the ion drag force $\varrho_0 \bar{\lambda}\,\bar{u}$.

The perturbation wind is determined from the balance between the perturbation pressure force $\partial p'/\partial x$ and the forces resulting from the product of the basic wind with the ionization anomaly $\varrho_0 \lambda'\,\bar{u}$ and the product of the perturbation wind with the global mean value of ion drag $\varrho_0 \bar{\lambda}\,u'$. The term $\varrho_0 \lambda'\,\bar{u}$ is the momentum source determining the perturbation circulation. The perturbation parameters are functions of height and the horizontal distance from the ionization anomaly.

We examine the response of the neutral thermosphere to the momentum source generated by the interactions of the global wind with two ionization anomalies with different horizontal scale. These are: (i) the day–night electron density variation in the F region with a horizontal scale of about 3000 km, and (ii) the electron density depression within the stable auroral red arc with a horizontal scale of about 100–200 km.

The illustrations presented here are based on the detailed numerical integrations and scale analysis of the dynamic model described by Dickinson et al. [3].

3. Atmospheric Response

A diagram of an ion drag momentum source in cm s^{-2} and corresponding to a model of a night-time depression of ionization is shown in Fig. 1. The ordinate in the figure is measured from the 10^{-4} μbar level in units of scale height. The characteristic ionization anomaly scale size is a.

3.1. Atmospheric Response to the Day–Night Electron Density Variation

During sunrise and sunset, the electron density in the F region changes between day and night values in a period of about two hours [4]. In Fig. 2

we illustrate the sunrise variation between the night-time and daytime values of the electron density by a gradual increase in ionization over a horizontal distance of several thousand kilometers. The perturbation momentum source in this example also has much longer scale components than the 2000–3000 km since the "perturbation" ionization distribution does not go to zero at great distances. A wind driven by a constant pressure force and flowing into a region of increasing ionization will experience a deceleration due to an increase in the ion drag. We consider the problem with and without thermodynamics to illustrate the importance of the perturbation pressure in limiting the magnitude of the winds. In Fig. 2, upper part, a basic wind of 200 m s^{-1}, determined using the day–night average of ion drag, flows from a region of low electron

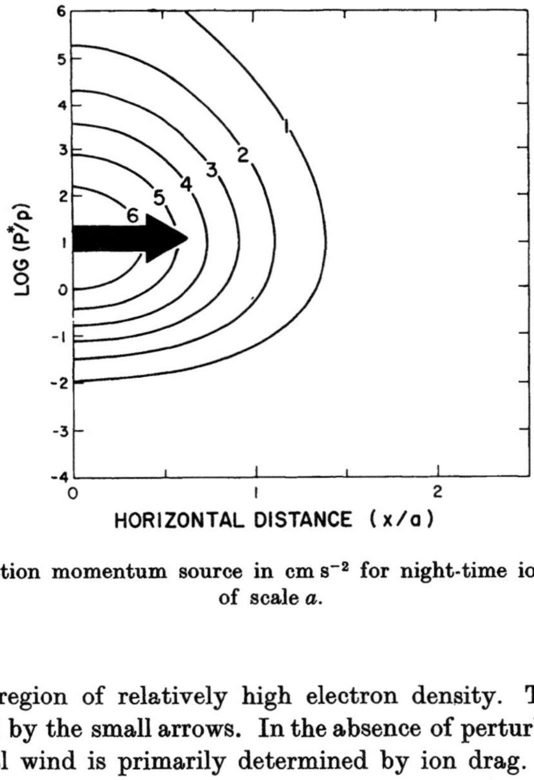

Fig. 1. Perturbation momentum source in cm s^{-2} for night-time ionization anomaly of scale a.

density into a region of relatively high electron density. The perturbation flow is indicated by the small arrows. In the absence of perturbation pressures, the total neutral wind is primarily determined by ion drag. Thus, the wind velocity given by the basic wind and a positive perturbation flow remains constant until it experiences a variation in ion drag. As the ionization increases, the horizontal wind decelerates. The horizontal convergence of the neutral wind forces a downward vertical wind in the immediate region of the ionization anomaly. The neutral wind continues to slow down until it once again reaches a constant ion drag on the opposite side of the ionization ledge. The increase in ionization causes a decrease in the wind of about 100 m s^{-1}. At the inflection point of the electron density curve, a maximum downward vertical velocity of 2 m s^{-1} is caused by convergence of the horizontal wind.

In the same case, with the perturbation pressure included (Fig. 2, lower part), the wind is slowed down considerably by the perturbation pressure long before it reaches the ionization anomaly. This reduces the magnitude of the pertur-

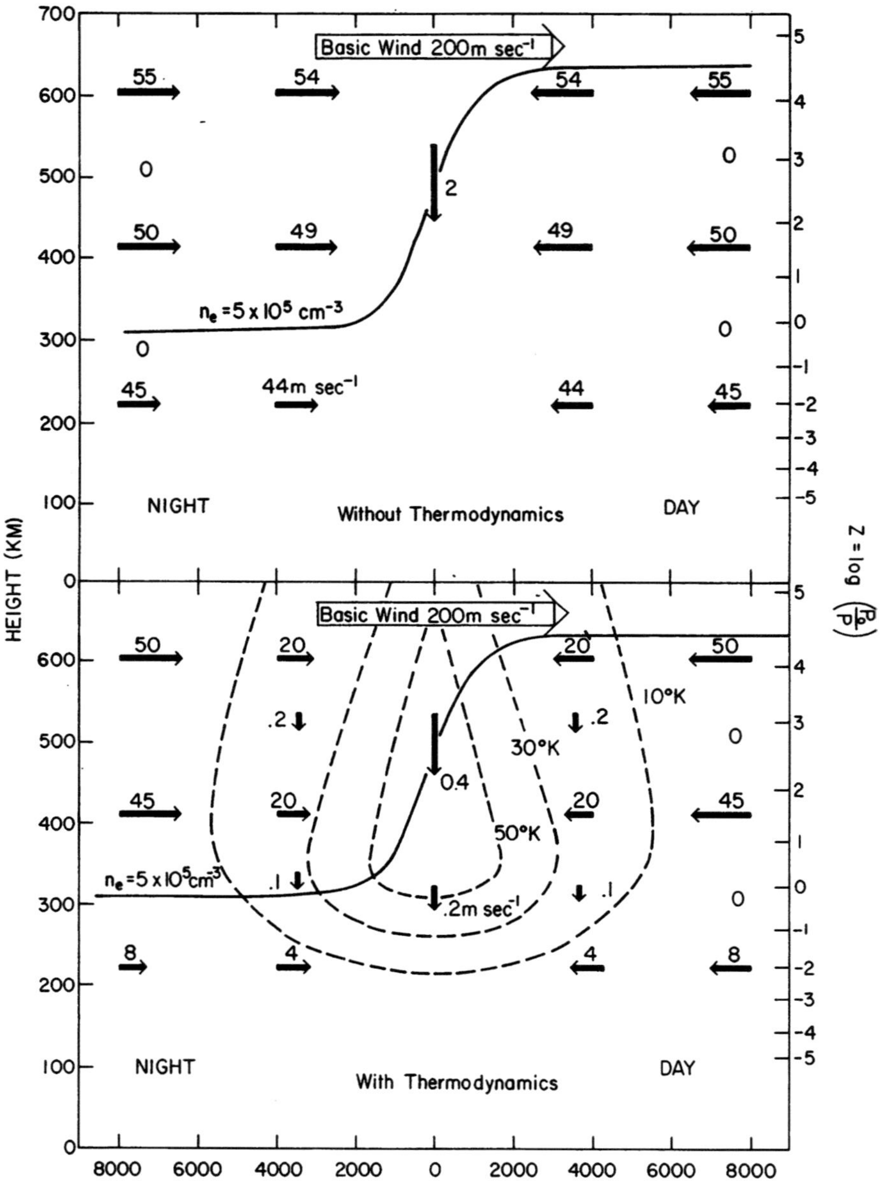

Fig. 2. Schematic diagram illustrating the atmospheric response to day–night ionization anomaly: above, response without perturbation pressure; below, response with perturbation pressure. Arrows represent perturbation wind (m s^{-1}); dashed contours are perturbation temperature (°K).

bation wind in the immediate region of the ionization anomaly, resulting in a weaker but much more extended circulation than in the previous case. The temperature perturbation caused by adiabatic heating is shown by the dotted lines in the figure.

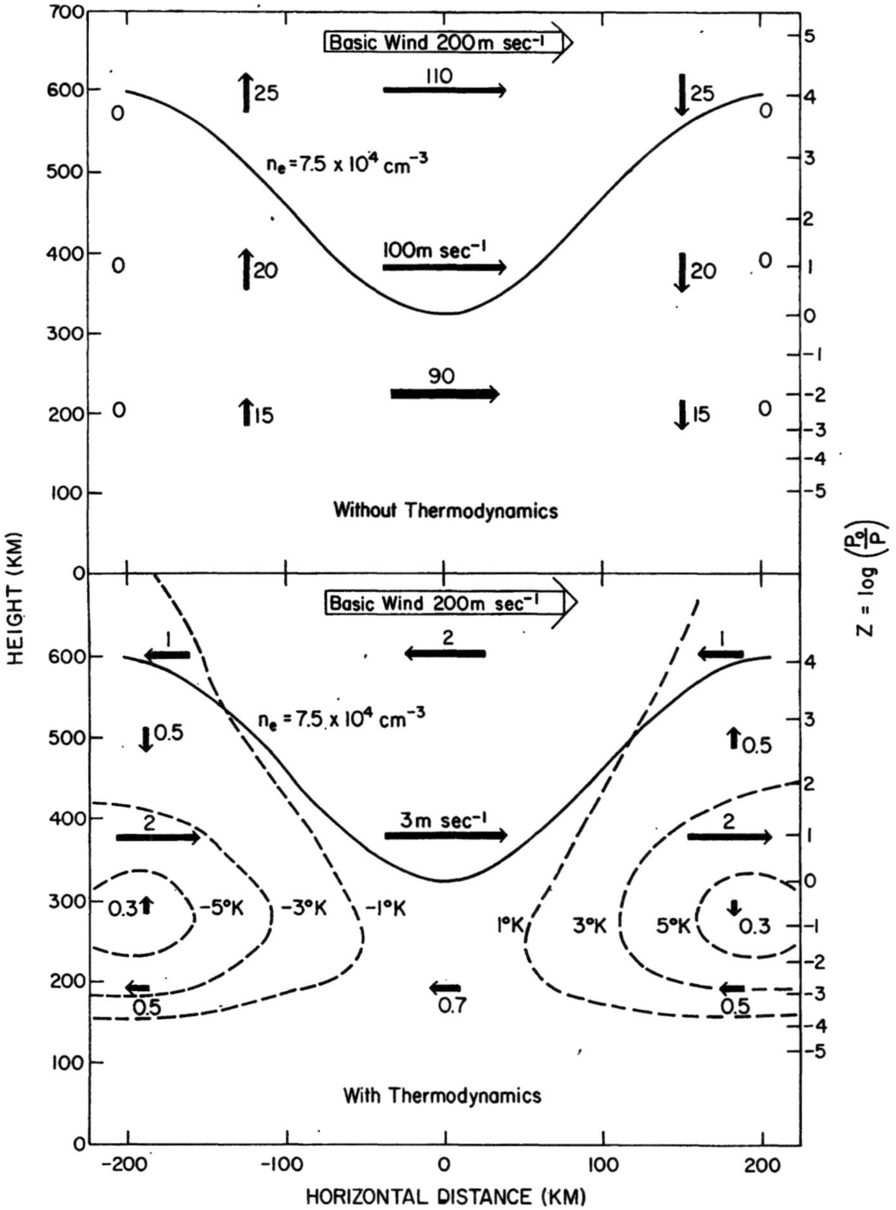

Fig. 3. Schematic diagram illustrating the atmospheric response to an electron density depression within a stable auroral red arc, 28–29 Sept. 1967; details as in Fig. 2.

3.2. Atmospheric Response to Electron Density Depression in a Stable Auroral Red Arc

An electron density depression within the stable auroral red arc has been observed as a common feature of all the arcs examined thus far [5]. The north–south horizontal scale of the depression is approximately 100–200 km and the electron density may vary by an order of magnitude within the arc compared with the surrounding region. Hays and Roble [6] have reported winds of 250–300 m s^{-1} observed during an arc. This situation is illustrated schematically in Fig. 3 where a basic wind profile, with an exospheric velocity of 200 m s^{-1} interacts with an electron density depression, represented as the solid line. In the upper part of Fig. 3, the perturbation wind without the perturbation pressure is controlled by ion drag within the depression. On the windward edge of the depression, the perturbation wind velocity is zero. As the basic wind penetrates the depression, the ion drag is reduced and the perturbation wind accelerates to a maximum in the center of the depression. The horizontal divergence of the wind gives rise, by continuity, to an upward vertical wind reaching a magnitude of the order of 20 m s^{-1}. On the leeward side of the depression, the perturbation wind slows with increase of ion drag until it reaches zero at the edge of the depression. The convergence of the horizontal wind results in a downward vertical wind of the same magnitude as the upward velocity on the windward side of the depression. The vertical winds would cause a large temperature perturbation and thus a perturbation back-pressure to oppose the momentum addition. Therefore the lower part of Fig. 3 shows the same case with the perturbation pressure included in the dynamic model. The perturbation pressure acting in response to the momentum addition extends well beyond the ionization anomaly to slow the basic wind. For the small horizontal-scale anomaly, the momentum source due to perturbation ion drag is almost completely balanced by the perturbation pressure force. The perturbation temperature and circulation are small with the magnitudes indicated.

4. Conclusions

These examples show the importance of including the perturbation pressure in the steady-state dynamic model to calculate the neutral response to momentum sources. For ionization anomalies with horizontal scales of the order of a few hundred kilometers, the perturbation temperature and circulation response are small and the mean wind is not greatly disturbed by the ionization anomaly. On the other hand, for horizontal scales of the order of a few thousand kilometers, significant horizontal velocity, vertical motion, and temperature fields will be superimposed upon the basic-state wind.

It appears that only variations of ion drag on a global scale can influence steady flows. For calculations of winds from empirically determined global-scale pressure distributions, less-than-global-scale variations in ion drag should be removed by smoothing.

Acknowledgment

The National Center for Atmospheric Research is sponsored by the National Science Foundation.

References

[1] H. Kohl and J. W. King, J. Atmos. Terr. Phys. **29**, 1045 (1967).
[2] J. E. Geisler, J. Atmos. Terr. Phys. **29**, 1469 (1967).
[3] R. E. Dickinson, R. G. Roble and E. C. Ridley, J. Atmos. Sci. **28**, 1280 (1971).
[4] J. V. Evans, J. Geophys. Res. **73**, 3489 (1968).
[5] R. G. Roble, P. B. Hays and A. F. Nagy, J. Geophys. Res. **75**, 4261 (1970).
[6] P. B. Hays and R. G. Roble, J. Geophys. Res. **76**, 5316 (1971).

Space Research XII – Akademie-Verlag, Berlin 1972

IONOSPHERIC F2 MOTIONS INTERACTING WITH THE *Sq* AND *L* FIELDS

S. Matsushita

High Altitude Observatory, National Center for Atmospheric Research, Boulder, Colo., USA

Based on reasonable models of anisotropic electrical conductivities and solar and lunar tidal wind systems in the dynamo region (90–150 km altitude), distributions of electrostatic fields for *Sq* and *L* geomagnetic variations on quiet days in equinoctial months during moderate sunspot periods have previously been estimated. These dynamo-region electrostatic fields are mapped to the ionospheric F2 region (300 km altitude) along the earth's magnetic lines of force, and horizontal and vertical drift motions in the F2 region caused by the coupling of the electrostatic and earth's magnetic fields are computed. Since a possibility of producing *Sq*-type electric currents in the dynamo region by the plasmaspheric electrostatic field due to magnetospheric motions has previously been examined, F2 drift motions caused by this electrostatic field are also computed. Discussions are given on which of the two mechanisms is better, taking into consideration observed results of F2 motions and estimated results of F2 density variations caused by drifts and neutral winds.

1. F2 Motions caused by Dynamo Region Electrostatic Fields

Based on reasonably assumed models of anisotropic electrical conductivities and solar thermal and lunar gravitational tidal wind systems in the dynamo region (90–150 km altitude) on quiet days in equinoctial months during moderate sunspot periods [1–3], distributions of electrostatic fields for solar-quiet (*Sq*) and lunar (*L*) geomagnetic variations produced in the dynamo region were estimated by Matsushita and Tarpley [1, 4, 5]. Although those computations were made for every 5 km altitude, each parameter in the height-integrated form (from 90 to 150 km) for *Sq* is shown in Fig. 1 to present an overall picture and for reference in later discussions. The top left diagram of Fig. 1 shows contours of electrical conductivities Σ_{xx}, Σ_{yy}, and Σ_{xy} in emu units for the forenoon (symmetric with respect to the noon meridian), and the bottom left diagram presents wind vectors (plotted for every 15° longitude and every 10° latitude) of the lowest order solar diurnal tide designated as the (1, −1) mode with a maximum speed 130 m s^{-1} (5°-latitude length corresponds to 50 m s^{-1} wind speed). These conductivity and wind models give a counterclockwise electric current system shown at the top right (current intensity between two consecutive lines is 10^4 amp) that is very similar to the *Sq* current pattern estimated from geomagnetic variations. The electrostatic field distribution plotted in 10° steps of latitude and longitude obtained from this model is shown in the bottom right of Fig. 1.

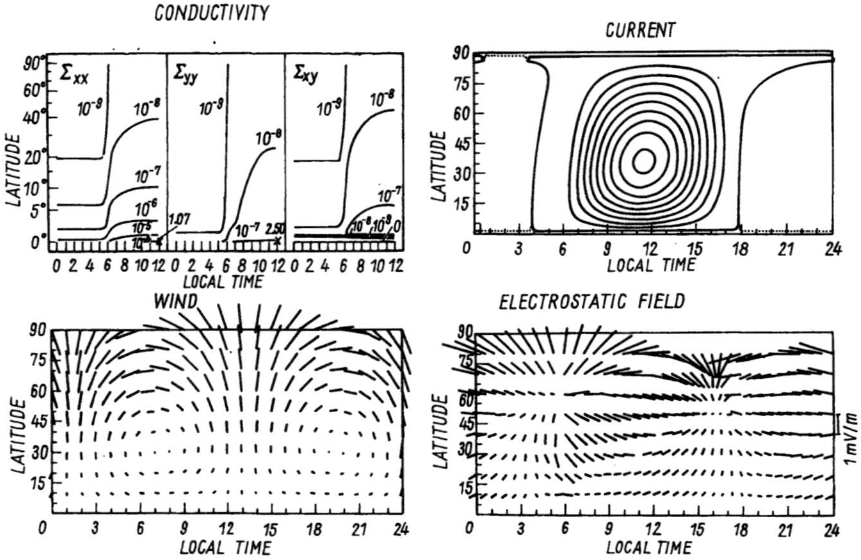

Fig. 1. Models of height-integrated electrical conductivities and a wind system (shown at the left) produce an *Sq*-type current system and electrostatic field distribution (shown at the right) (from Matsushita [5]).

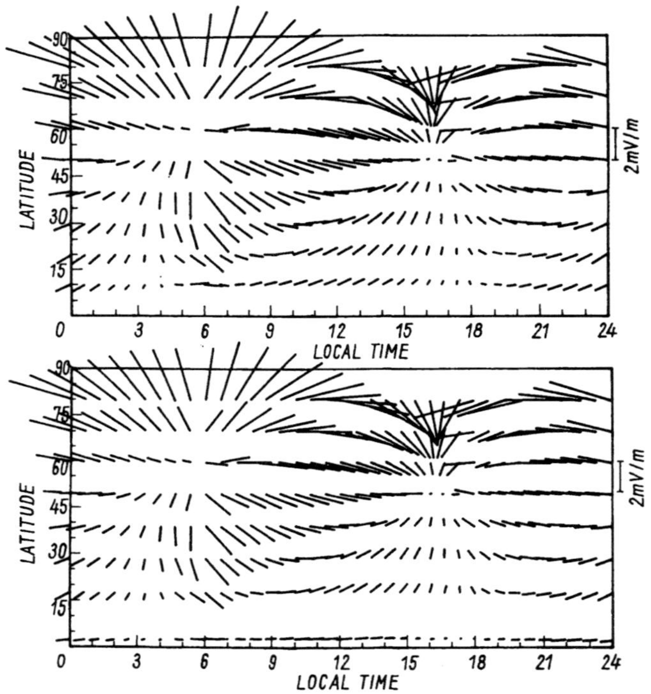

Fig. 2. Dynamo-region electrostatic field for the *Sq* (top) and its mapping to 300 km altitude (bottom).

This electrostatic field may be mapped to higher altitudes along the earth's magnetic lines of force in the way discussed by Matsushita and Tarpley [4]. The top diagram of Fig. 2 is the electrostatic field shown in Fig. 1 (assumed to be at 110 km altitude), and the bottom diagram is that same field mapped to 300 km altitude. Although the difference between the two diagrams is small at high latitudes, slight vector changes and equatorward shifts of locations are noticeable in low latitudes. Horizontal and vertical components of $\boldsymbol{E} \times \boldsymbol{B}$ drifts caused by this $\boldsymbol{E}$ at 300 km altitude are shown in Fig. 3. The vertical-drift vectors are presented by 45° slant lines to avoid superposition. It is

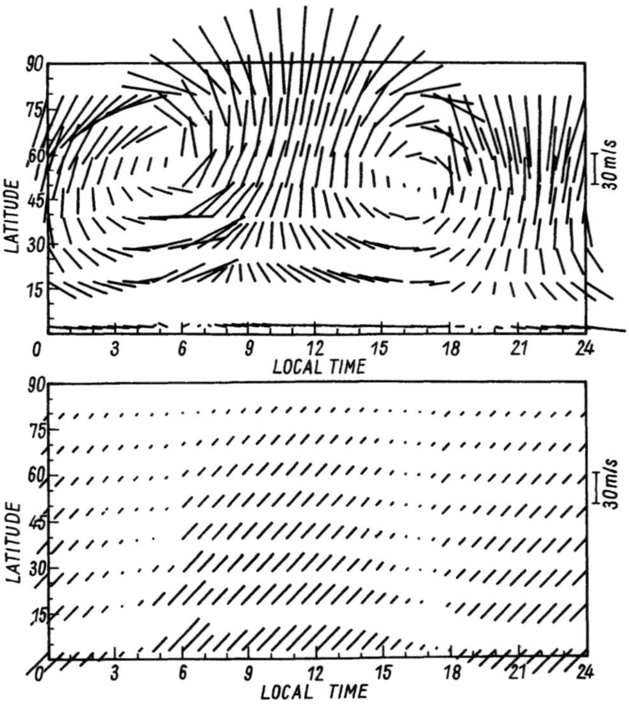

Fig. 3. Horizontal (top) and vertical (bottom) components of $\boldsymbol{E} \times \boldsymbol{B}$ drifts due to the *Sq* at 300 km altitude.

interesting to note that the general pattern of the horizontal drifts is fairly similar to the (1, −1) tidal mode shown in Fig. 1. The vertical upward drifts occur in the daytime, and the downward drifts occur at night; the changeover of the direction is around dawn and dusk.

Assuming the same electrical conductivity model, the lunar (2, 2) mode with the maximum speed of 7.5 m s^{-1} and a vertical wavelength of 85 km gives an electric current system that is very similar to the current pattern deduced from geomagnetic lunar variations [1, 2]. The dynamo-region electrostatic fields produced by this model when the moon is over the 0- or 12-hour meridian (new or full moon) and over the 3- or 15-hour meridian were obtained by

Matsushita and Tarpley [4]. These electrostatic fields are mapped to 300 km altitude, and $E \times B$ drifts are computed. Fig. 4 shows horizontal and vertical components of the drifts for a new or full moon day. Although the drift speeds

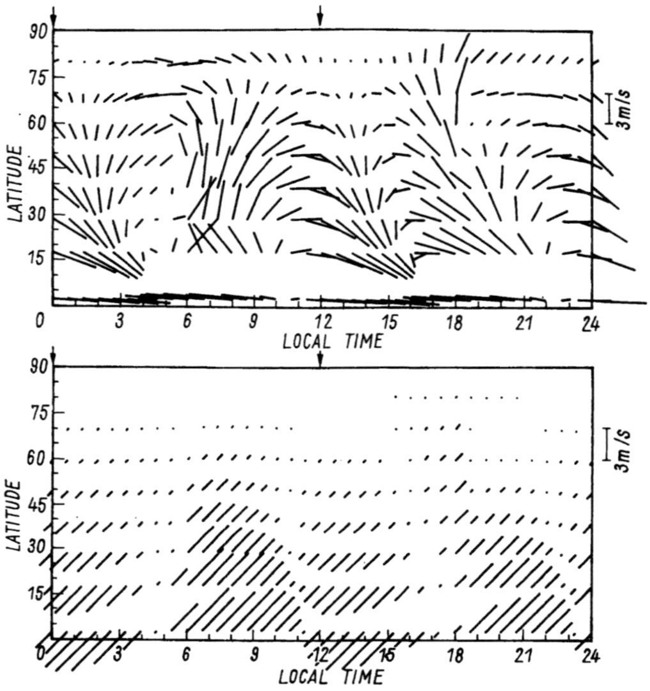

Fig. 4. Horizontal (top) and vertical (bottom) components of $E \times B$ drifts due to the L field at 300 km altitude when the moon is over the 0- or 12-hour meridian (new or full moon).

for L are roughly 10% of those for Sq, the changeover time of the Sq-drift directions in low latitudes may be shifted about one hour by L drifts depending upon the lunar phase.

2. F2 Motions caused by Plasmaspheric Electrostatic Fields

As discussed by Matsushita [5], Sq-type electric currents in the dynamo region may be produced by the plasmaspheric electrostatic field due to magnetospheric motions. When the electrical conductivities are the same as for preceding discussions, the electrostatic field in the dynamo region required for this mechanism is shown at the top of Fig. 5, and its mapping to 300 km altitude in the bottom diagram. Horizontal and vertical components of $E \times B$ drifts in this case are exhibited in Fig. 6. When Fig. 6 is compared with Fig. 3, differences of drift directions are obvious although the speeds are roughly similar. It is interesting that the changeover time of vertical drift directions shifts greatly to an earlier time with increasing latitudes from the equator to about 50° in Fig. 6, whereas the shift is slight in Fig. 3. Also, the equatorward drift is predominant in middle and high latitudes in the afternoon (Fig. 6), while the poleward drift is predominant in the morning through the early afternoon (Fig. 3).

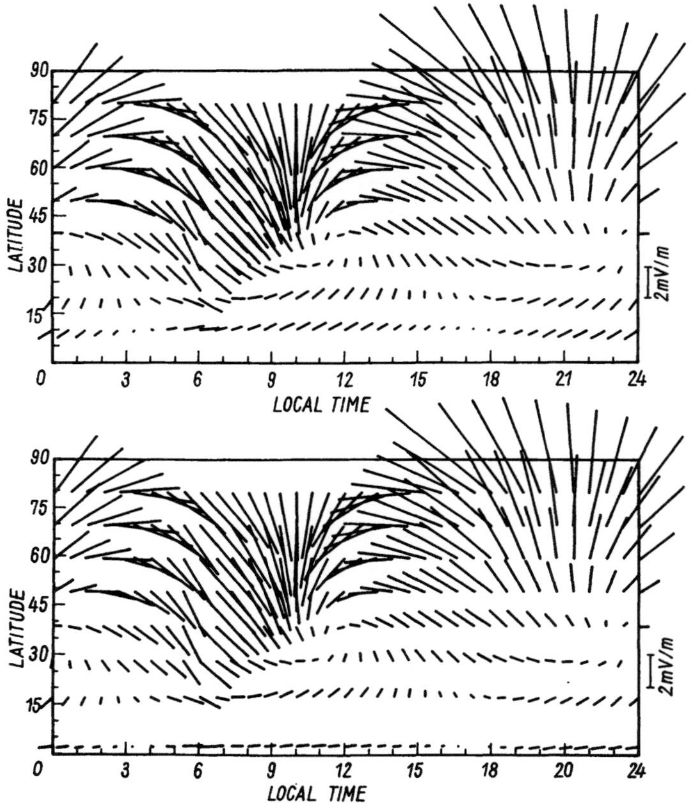

Fig. 5. Dynamo-region electrostatic field for the *Sq* (top) and its mapping to 300 km altitude (bottom), when the plasmaspheric electrostatic field due to magnetospheric motions may produce *Sq*-type electric currents in the dynamo region.

In the same way as for *Sq*, drifts for *L* caused by plasmaspheric electrostatic fields have also been obtained. However, since the ionospheric dynamo action seems to be the main cause of *L* variations, results shown in Fig. 4 may be sufficient.

3. Discussion

Based on different mechanisms for producing the electrostatic field responsible for *Sq*, two probable models of $\boldsymbol{E} \times \boldsymbol{B}$ drifts in the F region are presented in this report. Since barium cloud observations are limited to the twilight hours

[6], incoherent scatter observations may be the best means of judging which of the two models is better. Drift observations at Millstone Hill, e.g. [7, 8], Arecibo (R. Harper, private communication) and Saint Santin-Nançay [9, 10] provide middle-latitude data, and those at Jicamarca [11–13] provide magnetic equatorial data. The two models are very similar in the equatorial region, and they agree well with observations at Jicamarca (except for a sharp upward drift over that station at 18–19 hours). However, it is not easy to conclude which model is better in middle latitudes, because theoretically estimated drift vectors are for quiet time in equinoctial months during moderate sunspot

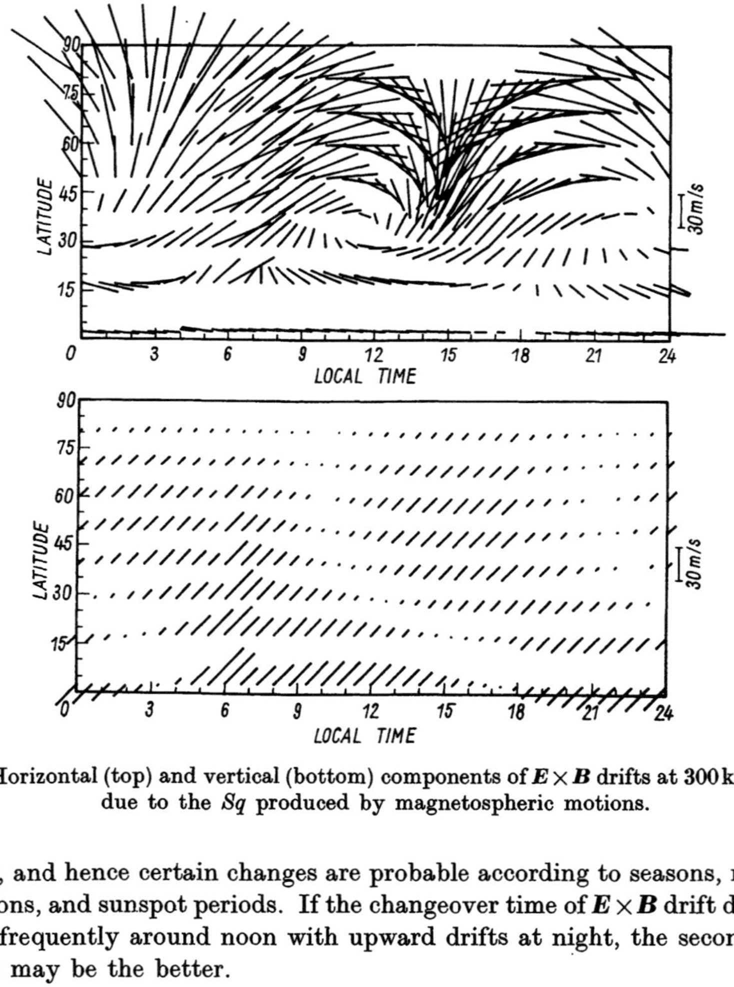

Fig. 6. Horizontal (top) and vertical (bottom) components of $\boldsymbol{E}\times\boldsymbol{B}$ drifts at 300 km altitude due to the Sq produced by magnetospheric motions.

periods, and hence certain changes are probable according to seasons, magnetic conditions, and sunspot periods. If the changeover time of $\boldsymbol{E}\times\boldsymbol{B}$ drift directions occurs frequently around noon with upward drifts at night, the second model (Fig. 6) may be the better.

To discuss ionization motions in the F region, however, neutral wind effects as well as $\boldsymbol{E}\times\boldsymbol{B}$ effects should be considered. Many scientists, e.g. [14–18] have discussed motions at middle latitudes, where neutral winds seem to produce stronger effects than $\boldsymbol{E}\times\boldsymbol{B}$, and several scientists, e.g. [19, 20] have investigated ionization motions at low latitudes, where $\boldsymbol{E}\times\boldsymbol{B}$ effects may be more important than neutral winds. Also, as Rishbeth [21, 22] has recently discussed, F region dynamo action in middle latitudes and polarization fields produced by winds in the equatorial F region should be taken into consideration. Future direct measurements of electrostatic fields and neutral winds as well as further incoherent scatter observations are desirable for examining the two suggested models.

Acknowledgments

I thank Dr. J. D. Tarpley for his helpful discussion and Mrs. Elaine L. Miller for her great asisstance in the computation and production of diagrams of the present paper. The National Center for Atmospheric Research is sponsored by the National Science Foundation.

References

[1] S. Matsushita, Radio Sci. **4**, 771 (1969).
[2] J. D. Tarpley, Planet. Space Sci. **18**, 1075 (1970).
[3] J. D. Tarpley, Planet. Space Sci. **18**, 1091 (1970).
[4] S. Matsushita and J. D. Tarpley, J. Geophys. Res. **75**, 5433 (1970).
[5] S. Matsushita, Radio Sci. **6**, 279 (1971).
[6] G. Haerendel and R. Lüst, in: Earth's Particles and Fields, Reinhold, New York 1968 (p. 271).
[7] J. V. Evans et al., Radio Sci. **5**, 27 (1970).
[8] L. A. Carpenter and S. A. Bowhill, Radio Sci. **6**, 203 (1971).
[9] G. Vasseur, J. Atmos. Terr. Phys. **31**, 397 (1969).
[10] G. Vasseur, J. Atmos. Terr. Phys. **32**, 775 (1970).
[11] B. B. Balsley, J. Geophys. Res. **74**, 1213 (1969).
[12] B. B. Balsley and R. F. Woodman, J. Atmos. Terr. Phys. **31**, 865 (1969).
[13] R. F. Woodman, J. Geophys. Res. **75**, 6249 (1970).
[14] P. Stubbe and S. Chandra, J. Atmos. Terr. Phys. **32**, 1909 (1970).
[15] E. N. Bramley and R. Rüster, J. Atmos. Terr. Phys. **33**, 269 (1971).
[16] R. Rüster, J. Atmos. Terr. Phys. **33**, 275 (1971).
[17] R. A. Challinor and D. Eccles, J. Atmos. Terr. Phys. **33**, 363 (1971).
[18] D. Eccles, J. W. King and P. Rothwell, J. Atmos. Terr. Phys. **33**, 371 (1971).
[19] D. L. Sterling et al., Radio Sci. **4**, 1005 (1969).
[20] D. Anderson, Ph. D. thesis, Astro-Geophysics Department, University of Colorado, Boulder (1971).
[21] H. Rishbeth, Planet. Space Sci. **19**, 263 (1971).
[22] H. Rishbeth, Planet. Space Sci. **19**, 357 (1971).

Space Research XII – Akademie-Verlag, Berlin 1972

THE DEGENERATION OF THE HOUGH FUNCTIONS WITHIN THE THERMOSPHERE

H. VOLLAND [a] and H. G. MAYR [b]

[a] Astronomical Institutes of the University, Bonn, FRG
[b] NASA/Goddard Space Flight Center, Greenbelt, Md, USA

Within the lower non-dissipative atmosphere, the eigenfunctions of the atmosphere are the Hough functions which describe the tidal wave motions. Under the influence of dissipation mechanisms like heat conduction, viscosity and neutral–ion drag these Hough functions become modified at thermospheric heights and are shown to degenerate ultimately into spherical harmonics at heights above 200 km. The implications of this degeneration on the latitudinal structure of the tidal modes and on the vertical wave propagation are discussed. It is suggested in particular that the semidiurnal mode becomes quasi-trapped at altitudes above 160 km thus explaining the apparent absence of semidiurnal variations in the dynamics of the upper thermosphere.

A common way to treat thermosphere dynamics is to separate the variables and this is only possible under the assumption that perturbation theory is a good approximation. One asks for solutions of the form

$$F(z)\,\Theta(\delta)\,\Lambda(\lambda, t)\,, \tag{1}$$

where F is a function of height z, Θ is a function of co-latitude δ and Λ is a function of longitude λ and universal time t. In the case of regular tidal and planetary waves, Λ is a simple exponential function

$$\Lambda = \exp\,[j\,m\,\lambda + j\,\omega\,t]\,, \tag{2}$$

where m is an integer and ω is the angular frequency of the wave. It remains then to determine the two functions F and Θ in Eq. (1).

Since the earth's atmosphere is a cavity, it behaves like a spherical waveguide in which only discrete geometrical structures of the waves can exist. The eigenfunctions that describe these wave modes are the Hough functions within the lower atmosphere [1] and are identical with the function Θ in (1) in this height region. Hough functions can be represented by a series of spherical harmonics and generally have a rather complicated latitudinal structure due to the strong influence of the Coriolis force on the wave propagation. However, within the thermosphere, dissipation effects like heat conduction, viscosity and neutral–ion drag become increasingly important with increasing heights and reduce the influence of the Coriolis force. The Hough functions therefore change their latitudial structures with altitude between about 100 and 200 km and ultimately degenerate into spherical harmonics above about 200 km [2]:

$$\Theta_n^{m,f} \to \mathrm{P}_n^m \qquad \text{above 200 km} \tag{3}$$

The indices n and m of the function Θ are identical with the wave domain numbers n and m of the corresponding spherical function P_n^m, and

$$f = \omega/\Omega \tag{4}$$

is the Coriolis parameter (Ω is the angular frequency of the earth's rotation).

Fig. 1 shows as an example the pressure and the horizontal wind components of the fundamental symmetric diurnal tidal wave mode $\Theta_1^{1,1}$ of gravity wave type. In Figure 1a the pressure function of this mode is plotted against co-latitude for four different heights between 100 and 300 km. At 100 km altitude a non-dissipative atmosphere has been assumed. Here the Hough function is

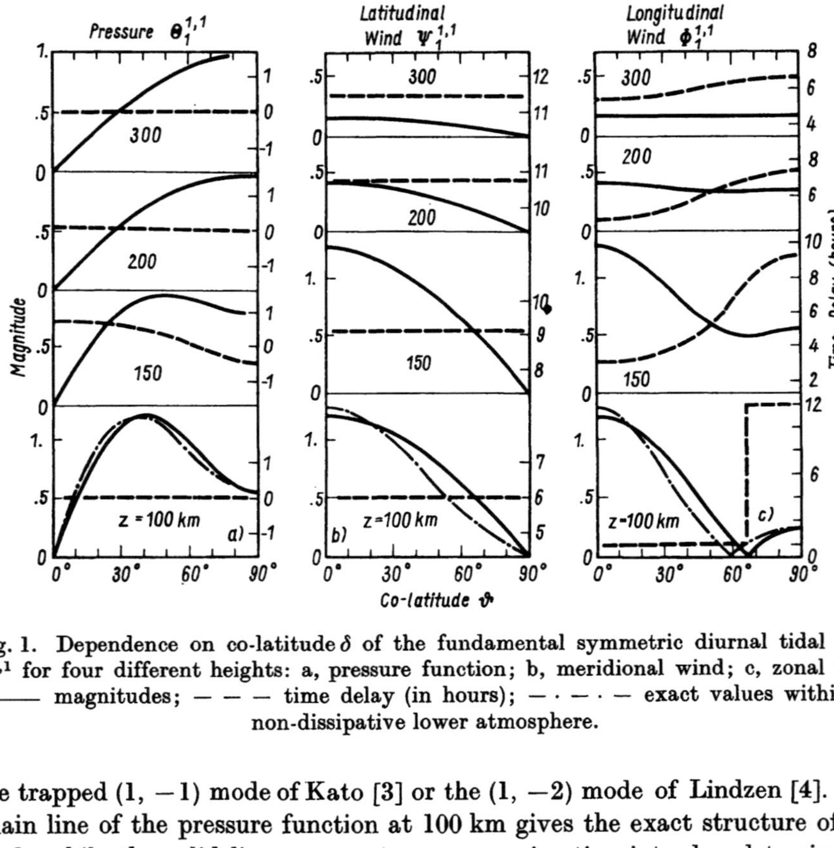

Fig. 1. Dependence on co-latitude δ of the fundamental symmetric diurnal tidal mode $\Theta_1^{1,1}$ for four different heights: a, pressure function; b, meridional wind; c, zonal wind. ——— magnitudes; — — — time delay (in hours); — · — · — exact values within the non-dissipative lower atmosphere.

the trapped (1, −1) mode of Kato [3] or the (1, −2) mode of Lindzen [4]. The chain line of the pressure function at 100 km gives the exact structure of this mode while the solid line represents an approximation introduced to simplify the analytic solutions. The dashed lines give the phase of the (generally complex) pressure function which is zero at 100 km indicating that Θ is a real function there. Between 100 and 200 km, the structure of the pressure function changes until finally it is transformed into the spherical harmonic P_1^1 which describes there the thermospheric diurnal pressure bulge of the Jacchia model [5].

In Figs. 1b and 1c the meridional and zonal wind components (blowing from the north and from the west, respectively) of the diurnal mode $\Theta_1^{1,1}$ are plotted against co-latitude in the same manner as in the case of the pressure function in Fig. 1a. Here the zonal wind at 100 km altitude and below changes its direction at about 30° latitude which is indicated by a phase jump of 12 hours. This wind field is mainly responsible for the generation of the geomagnetic *Sq* current at E layer heights [3, 6, 7]: During its transition within the height range between 100 and 200 km, this reversal of the zonal wind direction disappears, and the horizontal wind field above 200 km is essentially identical with the wind field derived by Geisler [8] and by Kohl and King [9] from the Jacchia model [5]. The transition region between 100 and 200 km of this wave mode and likewise of all the other wave modes, too, imposes a problem in connection with our basic assumption that the variables in Eq. (1) can be separated. As we note from Fig. 1, within this height range the latitude functions Θ depend not only on co-latitude but also on height, which contradicts this assumption. Evideently, a coupling between the various wave modes is to be expected in that height range. The degree of coupling can only be determined by an exact solution of the system of partial differential equations without separation of the variables. If we nevertheless do such a separation as in Fig. 1 and consider the different wave modes as if they were decoupled from each other we in fact make a kind of ray optics approximation well known from the electromagnetic wave propagation within the ionosphere. This coupling between the various wave modes takes place although perturbation approximation has been assumed. Taking into account the non-linear terms in the hydrodynamic equations gives rise to an additional coupling between the modes and complicates the problem further [10, 11].

Assuming the separation of the variables as a sufficient approximation, one obtains separation constants for the various wave modes called the equivalent depths. These equivalent depths are a measure for the eigenvalues of the height function F in Eq. (1). Taking into account heat conduction, viscosity and neutral–ion drag, gives eight complex eigenvalues corresponding to eight characteristic solutions or characteristic waves of the system of the eight complex coupled ordinary differential equations of first order for the height function F of each wave mode. Within a thin strip at a certain height in which the physical parameters of the thermosphere can be considered as constant, the characteristic waves are decoupled from each other, and each characteristic wave has the height structure

$$\exp\left[-\left(j\,\alpha_i + \beta_i - 1\right) z/2H_0\right] \tag{5}$$

where α_i and β_i are the real and imaginary parts respectively of the eigenvalues of the ith characteristic wave of the wave mode (n, m, f), and H_0 is the pressure scale height within this strip. Within the lower atmosphere, only characteristic waves of gravity wave type can exist. They are either propagation modes ($\alpha_i \neq 0$, $\beta_i = 0$) or trapped modes ($\alpha_i = 0$, $\beta_i \neq 0$). The (1, −1) mode is such a trapped mode with $|\beta_i| > 1$. Within the thermosphere, heat conduction waves and two kinds of viscosity waves can exist besides the gravity waves. Each wave type consists of an upgoing wave ($\beta_i > 0$) and a downgoing wave ($\beta_i < 0$). Above about 200 km, all characteristic waves are quasi-trapped ($\alpha_i \neq 0$, $|\beta_i| > 1$).

In Fig. 2, we plotted against altitude the attenuation factors $\beta - 1$ and the propagation factors α of the upgoing gravity wave type of three tidal wave modes: the fundamental symmetric diurnal mode $\Theta_1^{1,1}$ or the (1, −1) mode of Kato [3], the fundamental symmetric semidiurnal mode $\Theta_2^{2,2}$ or the (2, 2) mode according to Chapman and Lindzen [1] and the fundamental antisymmetric diurnal mode $\Theta_2^{1,1}$. The $\Theta_1^{1,1}$ mode, being a trapped mode within the lower atmosphere, remains a quasi-trapped mode at thermospheric heights with decreasing attenuation factor above 140 km. The semidiurnal $\Theta_2^{2,1}$ mode is a propagation mode at 100 km ($\beta = 0$). However, with increasing height, the attenuation factor β decreases rapidly (the dashed line in Fig. 2a which indicates negative values of $\beta - 1$) and becomes greater than one at 160 km. Above this

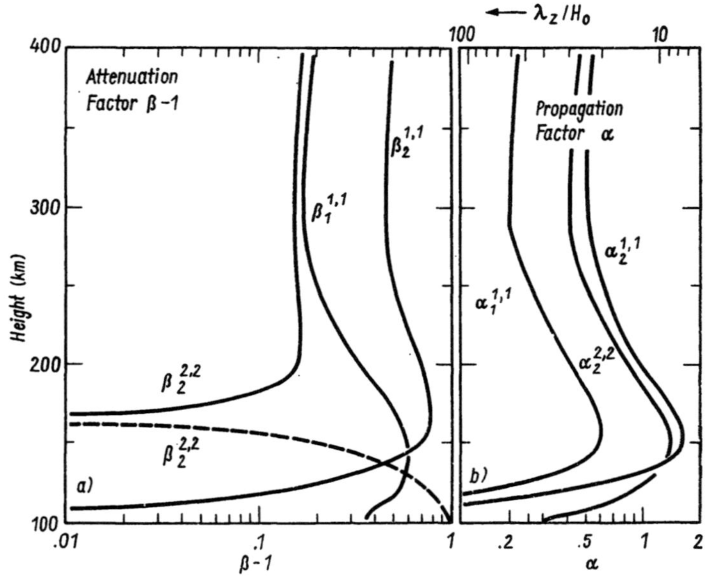

Fig. 2. Normalized eigenvalues of three tidal wave modes plotted against altitude. a, attenuation factors $\beta - 1$ (β is the imaginary part of the eigenvalue); $\beta_2^{2,2}$ is negative below 160 km (dashed line); b, propagation factors α (α is the real part of the eigenvalue).

height, the $\Theta_2^{2,2}$ mode behaves like a quasi-trapped mode. The antisymmetric $\Theta_2^{1,1}$ mode which describes the annual modulation of the diurnal variation of pressure and winds has the eigenvalue $\beta = 1$, $\alpha = 0$ within the lower atmosphere. This mode too changes its character into a quasi-trapped mode above 100 km. The propagation factors α which are a measure for the vertical phase velocities increase with altitude between 100 and 150 km and then decrease reaching asymptotic values above about 300 km. At each height one can define a vertical wavelength according to

$$\lambda_z = 4\pi H_0/\alpha \,. \tag{6}$$

The upper abscissa in Fig. 2b is scaled according to λ_z/H_0 and shows that at 150 km the tidal waves have minimum vertical wavelengths.

Fig. 2 shows only the one most important upgoing characteristic gravity wave for the three respective tidal modes. This characteristic wave together with the seven others can be excited in any height by a heat source component corresponding to the respective tidal mode. The eight characteristic waves propagate up and down and become coupled with each other at any height within a realistic thermosphere.

If one considers a heat input limited within a certain height region between z_0 and z_1, the boundary conditions are the radiation conditions which imply zero wave energy flux from outside through the boundaries (z_0, z_1) of the model. These conditions are fulfilled by setting the incoming characteristic waves from above (i.e. waves with $\beta_i < 0$) equal to zero at the upper boundary z_1 and by setting waves from below ($\beta_i > 0$) equal to zero at the lower boundary z_0. Such procedure can be handled without great mathematical difficulty in numerical calculations if one neglects the viscosity waves [12]. Otherwise, one must use the conditions of zero flux of mass, momentum and energy through the upper boundary as an alternative upper boundary condition [13–15]; one then loses, however, the advantage of discriminating the various characteristic waves.

References

[1] S. Chapman and R. S. Lindzen, Atmospheric Tides, D. Reidel Publ. Co., Dordrecht-Holland 1970.

[2] H. Volland and H. G. Mayr, NASA-Document X-621-71-240, GSFC, Greenbelt, Md.

[3] S. Kato, J. Geophys. Res. **71**, 3201 (1966).

[4] R. S. Lindzen, Mon. Weath. Rev. **94**, 295 (1966).

[5] L. G. Jacchia and J. Slowey, Space Research VII, 1077 (1967).

[6] R. J. Stening, Planet. Space Sci. **17**, 889 (1969).

[7] J. D. Tarpley, Planet. Space Sci. **18**, 1091 (1970).

[8] J. E. Geisler, J. Atmos. Terr. Phys. **29**, 1469 (1967).

[9] H. Kohl and J. W. King, J. Atmos. Terr. Phys. **29**, 1045 (1967).

[10] A. Spizzichino, Annls Géophys. **25**, 693 and 755 (1969).

[11] A. Spizzichino, Annls Géophys. **26**, 9 (1970).

[12] H. Volland and H. G. Mayr, Annls Géophys. **26**, 907 (1970).

[13] R. E. Dickinson, C. P. Lagos and R. E. Newell, J. Geophys. Res. **73**, 4299 (1968).

[14] R. S. Lindzen, Geophys. Fluid Dynamics **1**, 303 (1970); **2**, 89 (1971).

[15] R. S. Lindzen and D. Blake, Geophys. Fluid Dynamics **2**, 31 (1971).

Space Research XII — Akademie-Verlag, Berlin 1972

A COMPARISON OF TIDAL THEORY WITH LOWER THERMOSPHERIC WIND OBSERVATIONS

E. S. Batten *

The Rand Corporation, Santa Monica, Calif., USA

The wind in the lower thermosphere is generally considered to be composed of prevailing, tidal and gravity wave components. However, the observations of the wind, primarily from meteor trail and vapor trail drifts, are available at only a few locations and in the case of vapor trail drifts, only during the night hours. Because of the heterogeneous nature of the data, it has not been possible to specify definitely the contribution of each of these components to the total wind. In an attempt to improve the interpretation of the observations, tidal winds are computed from theory. To do this, the effects of dissipation and excitation in the thermosphere must be considered. In this work, the effects of heat conductivity and the absorption of solar radiation in the thermosphere are added to the tidal theory. The results of comparison of theory with observations indicate that major features of the vapor trail drifts (for example, the wind speed maximum near 105 km, the velocity oscillations below the maximum and the quiescent region above the maximum) can be explained in terms of interference between tidal modes propagating in an atmosphere with dissipation. The results also indicate that variations in water vapor excitation may be a source of the day-to-day changes observed in the *Sq* current system.

1. Introduction

As a result of new wind observations and improvements in the theory of atmospheric tides [1–3] there has been a renewed interest in the effects of tides on atmospheric phenomena in the lower ionosphere. For example, Stening [4], Matsushita [5] and Tarpley [6] have used the tidal theory to estimate the effects of various tidal modes on the production of *Sq* current systems. But, while it appears from their work that the (1, —2) mode is the most important contributor to the diurnal variation of the *Sq* current system, there is still a discrepancy between the amplitude of the (1, —2) mode required by the dynamo theory and the amplitude as calculated from tidal theory. Attempts to determine the relative contribution of the tidal modes using observed winds have not been too successful. Meteor-trail measurements can be used to separate tidal components with different periods. However, the vertical resolution of all but the most recent meteor measurements is poor. Furthermore, the vertical structure of the tides is expected to vary with height because of

* Any views expressed in this paper are those of the author. They should not be interpreted as reflecting the views of The Rand Corporation or the official opinion or policy of any of its governmental or private research sponsors. Papers are reproduced by The Rand Corporation as a courtesy to members of its staff.

vertical variations in the background temperature and the effects of dissipation [7, 8]. These factors will impede the identification of the component modes. Even the excellent vertical resolution obtained in the vapor-trail data will not aid separation of the modes unless one has prior knowledge of their vertical wavelengths. Despite these problems, or perhaps because of them, it is desirable to compare qualitatively theoretical calculations of the tides with lower thermospheric wind observations. In the following sections, the results of recent tidal wind calculations [7] are compared with observations in order to determine the extent to which tidal theory can explain observed features of the lower thermospheric winds.

2. Theory

A complete discussion of the theory together with an extensive bibliography is given in [7]. For present purposes it is sufficient to outline the excitation and dissipative mechanisms that have been included in the tidal calculations used here. Recent developments of the tidal theory have shown that solar tides in the atmosphere are almost wholly thermally induced. Of the several heat sources available to the atmosphere that may contribute to the excitation of the tides, we consider only the heating due to the absorption of solar radiation by water vapor in the troposphere, the absorption of solar radiation by ozone in the stratosphere and mesosphere, and the absorption of solar radiation by O, O_2 and N_2 in the thermosphere. It is assumed that the thermal excitation is symmetric about the equator. Thus the results of the calculations are biased towards equinoctial conditions.

In the thermosphere both heat conduction and viscosity become important. To avoid mathematical complications imposed by including viscosity, heat conduction is the only dissipative mechanism considered here. Recently Lindzen [8] has used an approximation to the tidal equations to study the effects of both viscosity and conductivity. A comparison of his results with those presented here shows that the gross features of the tides in the lower thermosphere remain unchanged when viscosity is added.

3. Observations

Three principal techniques have been used to measure neutral atmospheric motions in the ionosphere: ionization drifts, meteor-trail drifts, and vapor-trail drifts (see [7] for references to the original studies). Only a summary of the principal features of the observed wind field will be given here.

From meteor-trail drifts measured throughout a 24 hour period, the following features of the diurnal and semidiurnal components have been found: (i) the semidiurnal component dominates at high latitudes; (ii) the semidiurnal component has marked seasonal changes; (iii) the diurnal component is very irregular at high latitudes; (iv) at low latitudes the diurnal component is the larger but the semidiurnal increases relative to the diurnal as the altitude increases.

Studies of vapor trails [9] have revealed the following persistent features of the vertical wind profile: (i) velocity oscillations in the altitude range 70–120 km occur at vertical intervals (maximum to minimum) of approximately 6–8 km; (ii) a prominent velocity maximum nearly always occurs at heights between 100 and 111 km; the average height of this maximum is 105 km and the mean velocity attained is 118 m s^{-1}; (iii) a quiescent region from 140 km extends to at least 200 km; the oscillations that rarely occur in this region exist over vertical intervals greater than 40–50 km; (iv) in general, there is a tendency for the wind to rotate clockwise: in about a third of the data the winds spiral in a clockwise direction throughout the profile, below 130 km the direction of the wind can change by as much as 360° in about 29 km but at higher altitudes it changes slowly with height.

From dawn and dusk vapor-trail profiles it is possible to obtain the vertical structure of the diurnal component [7, 9–12]. Hines [10], for example, used averaged dawn and dusk profiles and found a clockwise rotating diurnal component with a wavelength of 20 $\pm$ 3 km. The amplitude increased to 40 m s^{-1} at 105 km and decreased above this level [10]. However, by using averaged dawn and dusk data, the amplitude may have been underestimated. With observations from a single day the diurnal component can exceed 100 m s^{-1} [7]. In the next section it will be shown that many of these features can be explained by interference between tidal modes that are propagating through an atmosphere with dissipation.

4. Comparison of Theory with Observations

For the following discussion it is important to recall the behavior of the tides in an atmosphere without dissipation. The vertical wavelengths of vertically propagating modes will not change in an isothermal atmosphere. As a point of reference, the wavelengths are approximately 28, 11 and 7 km for the (1, 1), (1, 3) and (1, 5) diurnal propagating modes. For the (2, 2) and (2, 4) semidiurnal modes, typical values of the vertical wavelengths are 180 and 50 km. Further, the amplitudes of the propagating modes will increase like $\varrho^{-1/2}$. The phases of the non-propagating modes (for example, the (1, -2) mode) are constant with height and, outside the region of heat input, the amplitudes decrease with height. In a non-isothermal atmosphere without dissipation, the vertical wavelength will change somewhat due to the vertical variation of the basic state temperature.

When dissipation is included in the tidal theory, the vertical structure of the tidal modes is significantly changed. An example of the vertical structure of the most important tidal modes at 38° N and 1945 LT is given in Fig. 1 which shows several of the features observed; for example, the amplitudes of the diurnal propagating modes, (1, 1) and (1, 3), decay above the 100–110 km region; for the (1, 3) mode the decay starts at 98 km and for the (1, 1) mode it starts at 110 km. The decay of the semidiurnal (2, 2) mode is not so obvious: since it decays less, the (2, 2) mode increases in importance as the altitude increases.

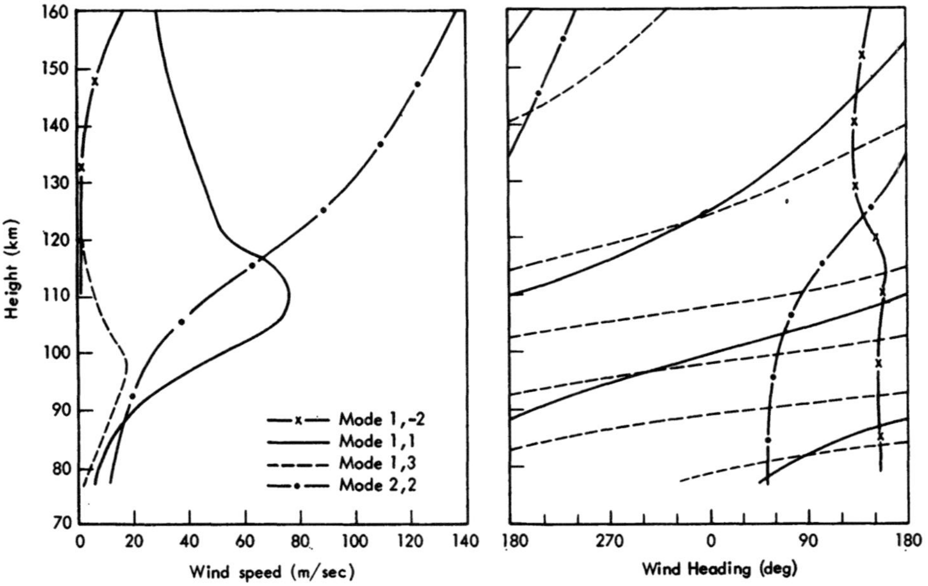

Fig. 1. An example of the vertical structure of the principal modes. The results are computed for 38° N and 1945 LT.

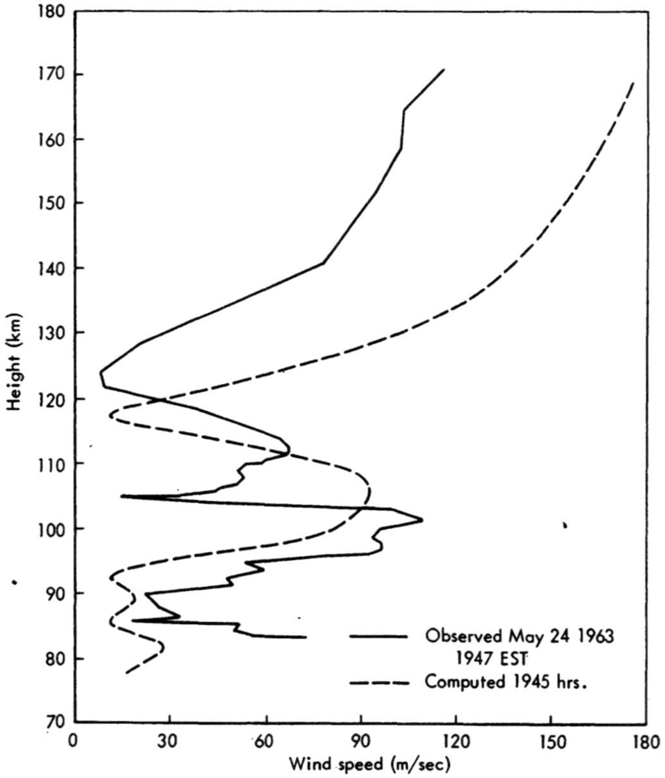

Fig. 2. Comparison of the theoretical wind speeds at 1945 LT (dashed line) with wind speeds observed (solid line) from Wallops Island at 1947 EST, 24 May 1963.

From the wind heading curves (right-hand side of Fig. 1), we note that the vertical wavelength of the diurnal propagating modes increases with height. For example, the (1, 1) mode has a vertical wavelength of about 22 km near 90 km and 45 km near 130 km. Below 100 km, where the (1, 1), (1, 3) and (2, 2) modes have similar amplitudes, it is easy to see how the (1, 3) mode, with a rapid change of direction and height, can add to and subtract from the more gradual changing wind vectors of the (1, 1) and (2, 2) modes. This interference between modes will produce wind speed maxima and minima in the total speed profile. Above 110 km the (1, 3) mode has almost completely

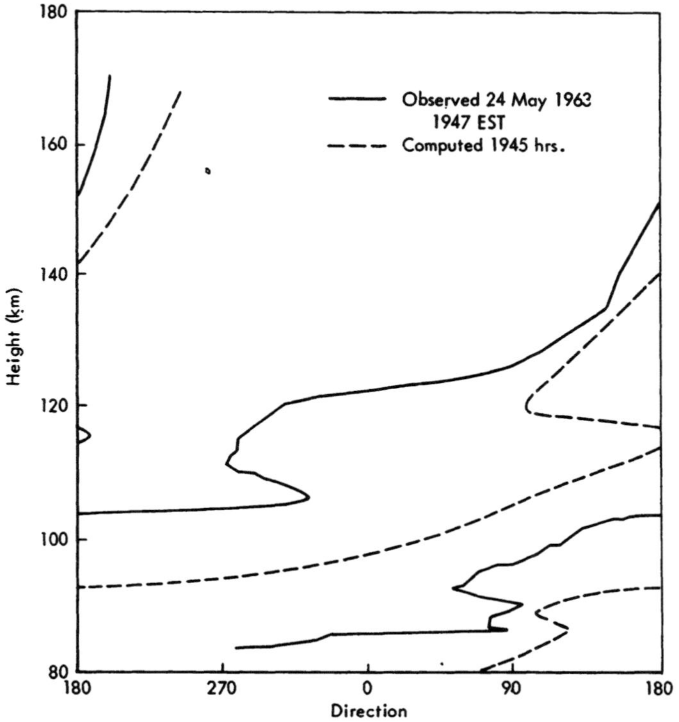

Fig. 3. Comparison of the theoretical wind directions at 1945 LT (dashed line) with wind directions observed from Wallops Island at 1947 EST, 24 May 1963.

decayed, the vertical wavelength of the (1, 1) mode has increased, and the (2, 2) mode with a very long wavelength becomes the largest. The combination of these factors will produce a total wind vector that slowly changes direction with height.

In Figs. 2 and 3 the tidal components shown in Fig. 1 are summed and compared with the wind observed from Wallops Island at 1947 EST, 24 May 1963. In Fig. 2 the theoretical and observed wind speeds are compared: the gross features of the observed wind speeds are reproduced by theory. However, Fig. 3 shows that the theoretical wind vector rotates faster than is observed. It is anticipated that this discrepancy will be removed when viscosity is included in the calculations. The theoretical calculations also compare favorably with meteor-trail data from Jodrell Bank (53° N). Greenhow and Neufeld [13] found the magnitude of the northerly semidiurnal component to range from 10 m s^{-1} in summer to 30 m s^{-1} in winter and the time of maximum north wind to range from 0500 LT to 0700 LT. Their theoretical computations show at 80 and 105 km a maximum northerly wind of 10 m s^{-1} at 0612 LT and 25 m s^{-1} at 0636 LT. They also found a smaller and more irregular diurnal component with an amplitude of approximately 5–10 m s^{-1}. A small diurnal component at high latitudes is easily explained by the tidal theory. The propagating diurnal modes have latitudinal structures with small amplitudes at high latitudes, and the non-propagating modes, which are large at high latitudes, decay in the meteor-trail region where there is very little heating to excite them.

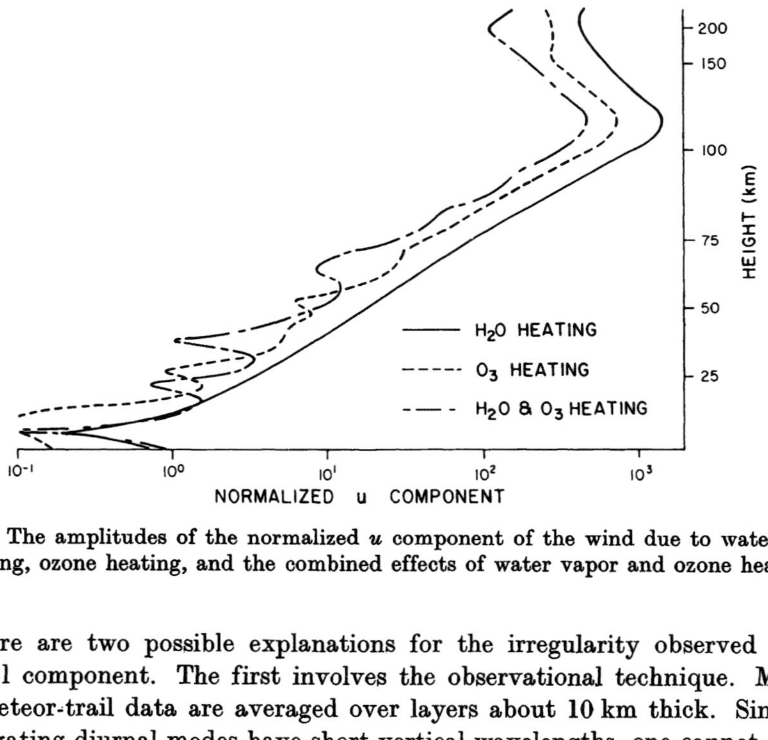

Fig. 4. The amplitudes of the normalized u component of the wind due to water vapor heating, ozone heating, and the combined effects of water vapor and ozone heating.

There are two possible explanations for the irregularity observed in the diurnal component. The first involves the observational technique. Most of the meteor-trail data are averaged over layers about 10 km thick. Since the propagating diurnal modes have short vertical wavelengths, one cannot expect the observations to have stable values. The second explanation involves the nature of the tidal excitation. Results of calculations with and without thermospheric heating show that the most important sources of excitation for the tides up to 160 km are water vapor and ozone heating [7]. Further, the calculations show that there is interference between the (1, 1) mode excited by water vapor heating in the troposphere and by ozone heating in the mesosphere. In Fig. 4 the amplitudes of the normalized u component of the (1, 1) mode are plotted for three cases: water vapor excitation, ozone excitation, and water vapor plus ozone excitation. The amplitudes excited by water vapor are in general a factor of 2 greater than those produced by ozone excitation. The amplitudes due to the combined ozone and water vapor excitation are, however, in general smaller than the amplitudes produced by either excitation mechanism acting alone. In other words, the two waves are out of phase. They are in fact almost 180° out of phase throughout the atmosphere above 20 km. The implication of this lack of matching is very important. It means that any significant variation in the heating rates of either source can produce large variations in the upper atmospheric diurnal tidal fields. The water vapor heating is the more likely to show appreciable variations: the amount of water vapor in the troposphere varies significantly with latitude, longitude, and time, while ozone is controlled more directly by the position of the sun at the altitudes considered (~50 km).

If the meteor-trail data were averaged in time and over the layer from 80 to 105 km, we might expect that the propagating modes, variable as they are, would not affect the yearly means. On the other hand, the (1, −2) mode which does not change phase with height should still be evident in the yearly means. Observations indicate that the yearly values of the N–S wind are more regular [13]. Data for the years 1953–1957 showed a yearly average amplitude of the N–S component that varies from 3.5 to 6 m s^{-1} and a time of maximum north wind that varied from 1000 to 1300 LT. In the two-year period 1954–1956 the amplitude was 4 m s^{-1} and the maximum north wind occurred near 1200 LT. The amplitudes and time of maximum of the north wind for the (1, −2) mode at 54° N latitude given in Table 1 show excellent agreement with this observation.

Table 1

The N–S component of the (1, −2) mode computed for 54° N latitude

Amplitude (km)	80	85	90	95	100
Amplitude (m s^{-1})	5.1	4.7	4.3	3.9	3.8
Time of max. north wind (LT)	1200	1200	1200	1200	1200

5. Discussion

The general agreement of the theory with observations and the excellent agreement of the calculated (1, −2) mode with the observed yearly averaged values of the diurnal tide at Jodrell Bank would seem to indicate that the tidal model used here could be a useful tool to study ionospheric plenomena. In particular, two of the results presented here have interesting implications to the *Sq* current systems. While it appears that the calculations correctly predict the amplitude of the (1, −2) mode at high latitudes and below 100 km, the calculated amplitudes above 100 km are still a factor of 2 or 3 smaller than those required by the dynamo theory. This implies that another source of tidal excitation exists above 100 km in addition to the thermospheric heating used here.

The results also indicate that the role of the (1, 1) mode as a contributor to the *Sq* current system needs to be re-evaluated. Prevously the (1, 1) mode has been rejected because of the short vertical wavelength predicted by a tidal theory neglecting dissipation. However, in an atmosphere with dissipation the vertical wavelength increases significantly. By increasing the vertical wavelength, it may turn out that the contribution of (1, 1) mode to the *Sq* currents also increases. It is then possible that the variability of this mode due to variations in water vapor heating is the source of the day-to-day variability of the *Sq* current systems during geomagnetically quiet periods.

References

[1] S. T. BUTLER and K. A. SMALL, Proc. Roy. Soc. A **274**, 91 (1963).
[2] R. S. LINDZEN, Mon. Weath. Rev. **94**, 295 (1966).
[3] S. KATO, J. Geophys. Res. **71**, 3201 (1966).
[4] R. J. STENING, Planet. Space Sci. **17**, 889 (1969).
[5] S. MATSUSHITA, Radio Sci. **4**, 771 (1969).
[6] J. D. TARPLEY, Planet. Space Sci. **18**, 1091 (1970).
[7] E. S. BATTEN, Ph. D. Thesis, Department of Meteorology, UCLA (1970); also published by The Rand Corporation, R-585-PR (1970).
[8] R. S. LINDZEN, Geophys. Fluid Dynamics **1**, 303 (1970).
[9] A. KOCHANSKI, Mon. Weath. Rev. **94**, 199 (1966).
[10] C. O. HINES, J. Geophys. Res. **71**, 1453 (1966).
[11] A. WOODRUM and C. G. JUSTUS, J. Geophys. Res. **73**, 467 (1968).
[12] A. WOODRUM et al., J. Geophys. Res. **74**, 4099 (1969).
[13] J. S. GREENHOW and E. L. NEUFELD, Quart. J. Roy. Met. Soc. **87**, 472 (1961).

Space Research XII — Akademie-Verlag, Berlin 1972

OBSERVATIONS OF WAVES AND TRAVELLING DISTURBANCES

G. Vasseur[a], C. A. Reddy[a,b] and J. Testud[a]

[a] C.N.E.T.-R.S.R., Issy-les-Moulineaux, France

[b] S.S.T.C., Trivandrum, India

Travelling ionospheric disturbances have been studied extensively during the past 25 years using various ground-based techniques. Earlier techniques, which are able to detect perturbations in the electron density at various heights in the ionosphere, are reviewed and the results obtained are discussed. More recently, new techniques such as incoherent scatter and *in situ* measurements by rockets and satellites have made possible the extension of the study of perturbations to several other parameters of the medium. These new results are reviewed in detail and it is shown that a far better understanding of the nature and origin of these disturbances can be achieved on the basis of these observations. The large scale, medium scale and very short period oscillations are discussed separately with regard to their characteristics and origin. The large scale disturbances seem to be gravity waves originating in the auroral region while the medium scale disturbances appear to be caused by gravity waves the origin of which is yet to be clearly identified. One category of very short period oscillations is associated with tropospheric weather disturbances but the origin of other categories of waves with infrasonic periods is not yet established.

1. Introduction

The upper thermosphere as a whole is dominated by the diurnal cycle of variations in its particle concentrations, temperatures and bulk velocity. The wind field in the upper thermosphere is mainly controlled by the thermally driven diurnal tide generated by solar radiation heating. The lower thermosphere (i.e. below about 180 km) is moreover subject to wind fields of upward propagating tidal waves. In fact, in the lowest region (90–120 km), the predominant winds are those due to propagating diurnal and semidiurnal modes, the relative amplitude of these modes changing with latitude. The theory and observation of these tidal waves have been reviewed elsewhere [1–5] and will not be included in this article.

The existence in the upper atmosphere from 75 to 110 km altitude of quasi-periodic winds of shorter periods than those of tidal nature were clearly demonstrated by meteor radar observations of winds. The frequent presence of wave-like perturbations in electron concentration at all levels up to the F2 layer peak (and above) was also revealed by ground-based observations. These irregularities have been studied with various experimental techniques and their observed magnitude and properties depend very much on the sensitivity and other characteristics of the particular observation method. It is clear however that a large variety of perturbations exists in the ionosphere; their sizes range from a few metres to several hundreds of kilometres. The structure and the origin of the smaller scale irregularities (scale less than 10 km) is as yet poorly known since the observational methods generally do not fulfil the required time and space resolution. This is not the case for the larger scale phenomena such as, for example, the so-called Travelling Ionospheric Disturbance (TID) for which a vast number of data have been gathered. Although the nature of the small scale irregularities may not differ essentially from the larger scale one this study will be restricted to mesoscale perturbations, i.e. those with a scale larger than a few tens of kilometres.

The observations of these "mesoscale perturbations" in electron density provide indirect but quite clear evidence of similar perturbations in the neutral atmosphere at these heights. Recent measurement by more sophisticated techniques like direct *in situ* measurements by rockets and satellites and the incoherent scatter observations have confirmed earlier deductions and added considerably to our knowledge of the nature of these perturbations. The interpretation of these phenomena received a new impetus with the proposal of a unified theory [6] which could successfully explain the characteristics of the observed disturbances in terms of atmospheric gravity waves. This interpretation seems to be now well accepted; this article uses the view point that most of the mesoscale perturbations can be interpreted in terms of acoustic gravity waves.

In § 2 of this article various observational techniques are mentioned and the main results obtained are summarized. Emphasis is given to the new observations which give a better understanding of the structure of the disturbances. In § 3, a unified view of the observations is presented. The results are discussed in terms of the current theoretical knowledge. It is pointed out that future experiments must be designed and directed towards understanding specific aspects of the disturbances and mainly their origin.

2. Experimental Results

In this section, the experimental techniques have been divided into six categories. The earlier sub-sections are concerned with earlier techniques such as ionosonde and backscatter measurements whereas the latter deal with the relatively new techniques such as satellite measurements and incoherent scatter observations.

2.1. Virtual Height and Ionosonde Observations

2.1.1. Virtual height vs. time records (h', t)

In Australia, Munro [7–10] made an extensive series of early measurements on TIDs. In addition to analysing the ionograms, he recorded the time variations of the virtual heights h' for fixed-frequency pulsed transmissions from three spaced transmitters. The perturbations of electron density at the reflection heights were indicated by complexities in the (h', t) traces. The time delays in the appearance of the same complexity on all the three traces were used to deduce the horizontal velocities of disturbances. The most pertinent features of the disturbances which emerged out of these (h', t) recordings over several years are: (i) The perturbations show quasi-periods in the 10–60 min range; the most frequently occurring periods are in the 20–30 min. range. (ii) The speeds of the disturbances are mostly in the 80–170 m s^{-1} range with a median value of about 130 m s^{-1}. (iii) The direction of movement is 30° E of N in local winter (May–Aug.), and 130° E of N in local summer (Nov.–Feb.). Very abrupt change takes place in Sept.–Oct., and a more gradual one in March–April. Both transitions are through east only. (iv) A small seasonal variation in speeds in indicated with largest mean speeds in local winter and with secondary maximum in local summer. (v) The mean occurrence rate is six disturbances per day. This rate shows a large seasonal variation with two broad maxima (8 per day) in summer and winter and sharp minima (4 per day) in April and September. (vi) The speeds, directions and occurrence rates all show a large day-to-day variability. (vii) The moving disturbances have their fronts tilted from the horizontal by about 45°. Munro [10] also discusses the changes in the speeds and directions with the time of the day but the regularity or statistical significance of these daytime variations is not clearly established.

Munro's experimental set-up was most effective in detecting medium scale disturbances with horizontal scale sizes of a few tens to a few hundreds of kilometres and with horizontal speeds of 50–200 m s^{-1}. The small separation of the reflection points ($\sim$20 km) and the time resolution ($\sim$45 s) preclude the reliable detection of very large scale (thousands of kilometres in extent) or very fast moving ($>$300 m s^{-1}) disturbances. Munro [8, 9] interpreted the observed disturbance in terms of quasi-sinusoidal perturbations in the ionospheric plasma. The observations in western Australia by Price [11] gave results similar to those of Munro. Using a technique similar to Munro's, Toman [12] observed TIDs in USA. The observed speeds ranged from 70–170 m s^{-1} and direction was about 120° E of N.

2.1.2. Ionosonde observations

Ionograms at middle and high-latitude stations often show unusual features in the F region traces: a kink, a cusp, an extra trace or a break, none of which is expected for reflections from normal F region. Even in the absence of such marked abnormalities, rapid sequence ionograms show the occurrence of wave-like perturbations in foF2 and h'F2. Careful (N_e, h) analysis of the ionograms would give the height structure of the perturbation, and spaced ionosonde observations would give the velocity of movement. A large amount of useful information on TIDs has been gathered through such studies of ionograms [13–21]. A typical case of a large TID observed with ionosonde is shown in Fig. 1.

McNicol, Webster and Bowman [22] studied the night-time occurrence of disturbances on ionograms (see also [23–26]). The important features of the disturbances as revealed by these ionograms studies are: (i) The quasi-periods of the disturbances are mostly centred around a 30 min period. (ii) Electron concentration perturbations are usually in the range of 10–30% of the ambient but much larger perturbations are observed in a few cases. (iii) Horizontal velocities of the disturbancees are mostly in the 150–200 m s^{-1} range, and typical horizontal scales seem to be several hundred kilometres. (iv) Some of the disturbances have fronts broader than 1000 km and travel distances of 3000 km or more with little change of shape or amplitude. (v) The directions of movement generally agree with those deduced by Munro from his (h', t)records.

According to Heisler [15] the disturbances observed on ionograms are necessarily greater in amplitude than those detected in (h', t) recordings. He also points out that the routine 15-min interval ionograms may not show the presence of the more frequently occurring shorter period ($\sim$30 min) perturbations. Heisler could observe [15] the interesting case of two disturbances travelling

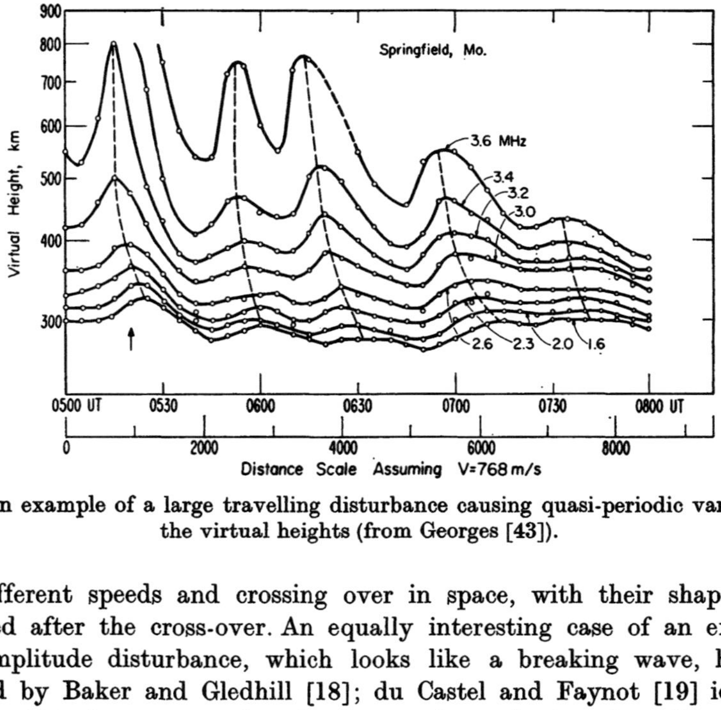

Fig. 1. An example of a large travelling disturbance causing quasi-periodic variations in the virtual heights (from Georges [43]).

with different speeds and crossing over in space, with their shapes being preserved after the cross-over. An equally interesting case of an extremely large amplitude disturbance, which looks like a breaking wave, has been observed by Baker and Gledhill [18]; du Castel and Faynot [19] identified a moving disturbance appearing simultaneously on the topside and bottomside ionograms and deduced characteristic dimensions and speeds which are in the same range as observed by other methods.

From (N_e, h) analysis of ionograms, Bowman [27] found the disturbances in ionization to have sloping fronts. He also found an enhanced sporadic E activity at the positions where the extension of the sloping front meets the E layer level. Bowman [28] further argues with some evidence that the "spread F" ionograms at mid-latitudes may result from the overlap of many satellite traces generated by spatially periodic structures (of about 35 km scale) in the electron concentration. From a careful true height analysis of night-time ionograms at three stations, Klostermeyer [26] could obtain the period, the phase velocity and the wave tilt of the disturbance. A study of disturbances at two magnetically conjugate stations (one in Kerguelen Island and another in USSR) was undertaken by Benkova et al. [29] using rapid sequence analysis. They found perturbations of the same periods at the two stations (about 2 h 20 min). Also a downward phase propagation was observed.

2.2. Ground Backscatter Observations

Range vs time records of hf radiowaves propagated through the ionosphere and backscattered from ground clearly reveal the presence and movement of "focusing" irregularities in the ionosphere. Many interesting observations of the sizes and velocities of moving disturbances were made by this technique, notably by Valverde [30], Tveten [31] and Hunsucker and Tveten [32]. Fairly large scale disturbances were observed by Valvarde [30] and these moved with speeds of 200–600 m s^{-1} in the direction 180°–225° E of N. Tveten's [31] observations (confirmed to Dec. 1952) showed the frequent occurrence of medium scale disturbances with typical scale sizes of 150–250 km. The speeds were mostly in the 50–250 m s^{-1} range, with a median value of 150 m s^{-1}, while the directions were mostly in the range of 120°–200° E of N. Even in the short observational period of one month, many large scale disturbances were observed; for these, speeds ranged from 250–750 m s^{-1} and directions from 140°–200° E of N. Tveten [31] points out that the perturbed isoionic surfaces have to be tilted by about 2° from the horizontal to produce the observed focusing of signals, whatever the precise shape of the irregularity and the nature of the focusing mechanism. Much earlier, Bramley and Ross [33] deduced tilts of the same order from their direction of arrival measurements. Bramley [34] also deduced vertical depths of about 4 km for the perturbation in the isoionic surface.

In one series of observations, Hunsucker and Tveten [32] deduced the velocities of 37 TIDs. The speeds were mostly in the 100–200 m s^{-1} range, but one disturbance moved with a speed of 325 m s^{-1}. In 22 out of the 37 cases, they found the TIDs to have an association (indicating a possible origin) in an "auroral zone event" which was defined by them as "a large peak in the measured values of telluric currents, magnetic H vector, ionospheric absorption or a start of a large disturbance in two or more of these parameters".

De Maistre et al. [35] observed medium scale disturbances in France; the speeds were mostly in the range of 140–250 m s^{-1}. The disturbances occurred mostly during daytime in local winter and moved equatorward, while they occurred mostly at night-time during local summer and moved poleward.

2.3. HF Doppler and Phase Path Measurements

The hf Doppler technique consists of measuring the changes in the path P of a radiowave reflected from the ionosphere. The phase path is defined as:

$$P = \int_{P'} \mu \, ds$$

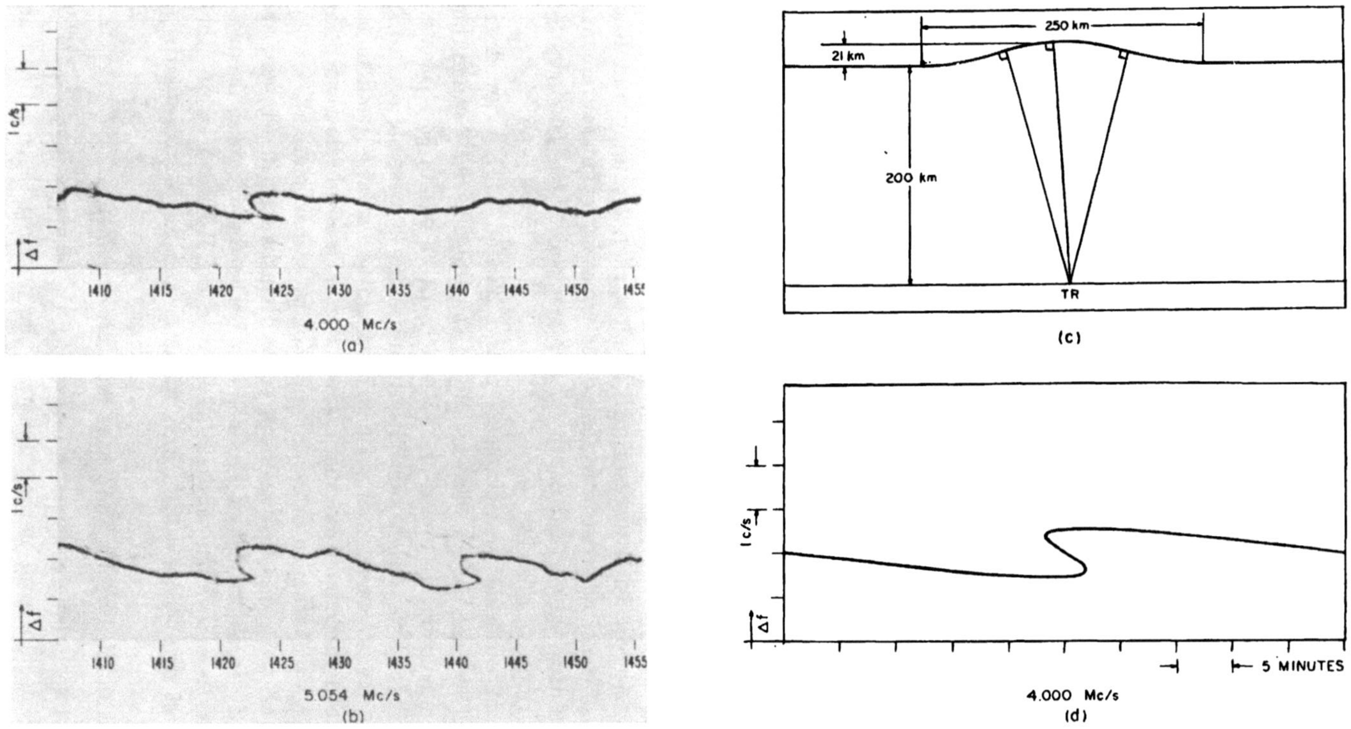

Fig. 2. Observed variations of Doppler frequency shift due to a TID; and a theoretical construction of the Doppler frequency variation from an assumed perturbation model (from Davies and Baker [42]).

a, b vertical incidence 220 ct. 1963, Boulder, Col.; c, illustrating the possibility of simultaneously receiving signals along 3 separate ray paths from a concave ionospheric disturbance; d, vertical incidence theoretical Doppler shift due to above disturbance travelling horizontally at 83.3 m s^{-1}; mirror-like reflection is assumed.

where μ is the phase refractive index and P' the ray path. The derivative of P with respect to time gives the Doppler frequency shift. Two different techniques exist for making phase path measurements. The first one, proposed by Findlay [36] uses pulsed transmission and can measure simultaneously the changes in phase path and group delay. This technique is capable of detecting very small perturbations though the analysis is somewhat laborious [37, 38]. A second technique, which came into more common use recently, makes use of continuous waves transmitted from a distant or nearby stable frequency transmitter [39, 40].

The Doppler frequency shift of the returned signal is caused by a phase change anywhere along the path. So it may arise due to vertical motions of the reflection surface as well as fluctuations in the refractive index (electron density) anywhere along the ray path. Under simplifying assumptions, it was shown [41–43] that, for changes in the reflection level, the Doppler shift is proportional to the frequency used, whereas for changes of the electron density below the reflection level, the Doppler shift is inversely proportional to the frequency. This property allows a discrimination of various phenomena causing the Doppler shift like solar flare effects which are not relevant here. Using this technique, which was mainly developed for the study of disturbances, many experiments were performed with various transmitted frequencies and various distances between transmitter and receiver. An example of a typical Doppler shift record showing the mechanism of perturbation is given in Fig. 2.

Earlier experiments, using the WWV transmitter at a long distance from the receiving station showed the occurrence of Doppler frequency variations with apparent long period and these were interpreted as due to ripples in the electron density at the reflection level [40, 41]. A horizontal propagation predominantly in the north–south direction was found [40] and the observed phenomenon was identified with large scale high speed travelling disturbances in the ionosphere; these were thought to be related to magnetic activity.

Later, continuous Doppler shift records have shown a large variety of quasi-periodic perturbations in the F region of the ionosphere. Based on the study of many hf Doppler records of TIDs, Georges [44] (see also [43]) proposed a classification of these disturbances: (i) Very large travelling disturbances, with periods of 20 min to 2 hours, horizontal velocities mostly in the range 200–400 m s^{-1}. Their effects are also observed on the ionograms which give the height structure of these disturbances. Their occurrence is strongly related to high magnetic activity. (ii) Medium scale disturbances, observed more often during the day with periods of 10–30 min and horizontal velocity in the range 100–200 m s^{-1}. These disturbances did not preserve their shape when the separation between observation points was of the order of 200 km [43]. In a recent publication [45] it was shown that the vertical propagation velocity of such disturbances could be obtained using multifrequency Doppler soundings. This study was however restricted to the sunset period. In the direction of the wave vector, the total speed had a mean value of 143 m s^{-1} (with a small seasonal variation) and an inclination of 45° downwards with respect to the vertical for the disturbance front; this corresponds to a horizontal velocity of

200 m s^{-1}. The observed periods and the horizontal and vertical wavelengths agree with the theoretical dispersion relation for internal gravity waves. The velocity vector has a north–south component directed towards the north in summer and towards the south in winter changing direction via the west. (iii) Short period oscillations [44, 46–48] with periods from 2 to 5 minutes. These oscillations were found to propagate upward [48] and to be related to severe weather activity [46–48]; strong storms were generally found less than 250 km away from the observation point. In addition to the well defined individual disturbances a spectrum analysis of the apparently random frequency fluctuations during otherwise quiet periods [49] has shown that at night fluctuations with periods in the range 15–90 min dominate, whereas during the day fluctuations with periods of 5–15 min are dominant. It appears that the Doppler sounding technique provides a powerful and relatively inexpensive method for monitoring the properties of travelling disturbances.

2.4. Electron Content Measurements

In these experiments, the total electron content along the ray path between an emitting satellite and a ground-based receiver can be derived through the measurements of Faraday rotation or Doppler frequency shifts. The observations of ionospheric irregularities using electron content measurements can be classified into two types: measurements using a low orbiting satellite moving with respect to the ground-based station, and measurements using a geostationary satellite.

2.4.1. Measurements using a low orbiting satellite

The earlier experiments have used a single receiver [50]. The speed of the satellite being very high (6 km s^{-1}), if we assume that the ionospheric irregularities move with speeds which are much smaller, these measurements provide a quasi-instantaneous cross section of the irregularity along the satellite trajectory. It was generally assumed that the irregularities contributing significantly to the observed variations of the electron content were localized in a small height interval around the peak of the F2 layer. Using these assumptions, two parameters of the irregularities can be derived from the measurements: the amplitude of the electron content fluctuation and the horizontal size of the irregularity.

A large number of wave-like irregularities were observed by Titheridge [50]. A statistical analysis gave the following results:

The occurrence rate of disturbances does not show any seasonal or diurnal variation: their horizontal sizes range from 10 to 500 km; the amplitude of the electron content fluctuation is an increasing function of the size. However, several limitations in the interpretation of these observations are to be recognized since we know of the presence in the ionosphere of wave-like irregularities propagating with horizontal phase velocity of several hundreds, or even several thousands of m s^{-1}. The actual speed of the ray path in the height region where the disturbances occur may be much smaller that the speed of the satellite, which is generally at a much higher altitude (like 1000 km). In such a situation the speed of the disturbance may be comparable with the speed of the ray path in the relevant region; the horizontal scales will be seriously in error if they are deduced on the assumption of an instantaneous cross section. Moreover, when the angle between the irregularity fronts and the trajectory of the satellite is very acute, the apparent phase velocity in the direction of this trajectory may be much higher than the actual horizontal speed of the fronts, so that the observed "horizontal size" of the irregularity may be very different from the actual wavelength between two adjacent fronts.

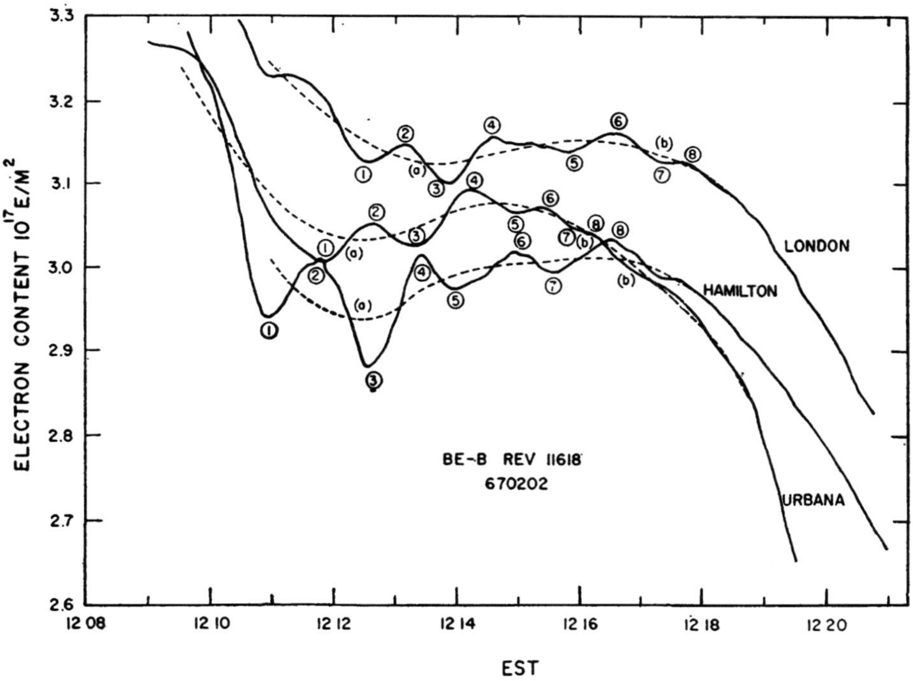

Fig. 3. Wave-like perturbations in the total electron content at spaced stations (from Rao et al. [51]).

For this reason, in the past few years, observational methods using spaced receiving stations have been developed [51]. Then the reconstitution of the fronts of the irregularities becomes possible (see Fig. 3) and the analysis of the data permits the determination of the following parameters: the wavelength of the irregularities, their horizontal speed of propagation and their direction of propagation. Following this method, Rao et al. [51] have observed propagating waves with the characteristics of acoustic neutral waves, i.e. with periods in the range of 20 s–3.5 min. Comparison between the apparent horizontal phase velocity ($\sim$4 km s^{-1}), and the theoretical speed of the wave front ($\sim$700 m s^{-1}) led them to consider that the wave fronts were propagating almost vertically. The possible generation of these acoustic waves at low heights by meteorological phenomena like thunderstorms is pointed out by the authors.

2.4.2. Measurements using a geostationary satellite

In the experiments with a geostationary satellite, the satellite has no apparent motion relative to the observing station; this method permits the continuous recording of the time variations of the electron content. The observation of travelling ionospheric disturbances using several ground-based receivers has been made by Davis and Da Rosa [52] with a spacing of several hundred kilometres between the receivers. Cowling et al. [53] made similar observations with a 50 km spacing between the stations. The parameters of the wavelike disturbances which were derived from these observations are the period and the horizontal speed of the disturbance. Davis and Da Rosa [52] have shown that the large scale TIDs observed with their equipment were the manifestation of gravity waves with periods ranging from 1 to 3 hours and horizontal speeds of about 500 m s^{-1}. The occurrence of these waves was highly correlated with magnetic activity, and their equatorward direction of propagation suggests an auroral origin. The observed diurnal variation in the direction of propagation leads to the conclusion that there is a preferential location of generation on the auroral oval. This location is near the longitude corresponding to 1900 geomagnetic time.

Using three 50 km-spaced receiving stations, Cowling et al. [53] have observed medium scale disturbances. They found waves with periods ranging 20–100 minutes and horizontal speeds 50–300 m s^{-1}. All azimuths of horizontal propagation have been observed but the authors concluded that there was a preferential azimuth of propagation depending on the hour of the day. This conclusion however was reached using a limited number of samples. The interpretation of Cowling et al. [53] is that the system of the thermospheric neutral winds acts like a selective filter of medium scale gravity waves, which favours certain directions of propagation. Cowling et al. have also looked for the possible origin of these waves; a backward ray tracing was used in order to show that either the jet stream or weather disturbances may be the source for most of the waves observed.

In conclusion, the electron content measurements using several spaced receivers appear to be a very fruitful means of observing wave-like ionospheric disturbances. However, when interpreting them in terms of atmospheric waves, it must be considered that the response of the F2 ionospheric layer to atmospheric waves is largely anisotropic and that the observation of waves by electron content measurements may introduce serious biases [54].

2.5. *In situ* Measurements by Rockets and Satellites

Some parameters, not measurable by ground-based apparatus, can be obtained by direct *in situ* observations from rockets and satellites. Rockets give mainly the vertical structure of the medium, whereas satellites can measure the structure along the satellite path. The satellite orbit is generally elliptic so that the variation of the measured parameter has a horizontal as well as a vertical component. Generally the velocity of the wave perpendicular to the phase plane (<500 m s^{-1}) is small compared with the satellite velocity ($\gtrsim 6$ km s^{-1}). However, the measured spatial scales of the perturbation may be in error when the satellite path is nearly parallel to a phase plane (cf § 2.4.1). Optical observations of rocket released sodium and TMA vapour trails have been used to measure horizontal neutral winds in the 90–180 km region with an altitude resolution of a few kilometres. The time evolution of the winds has also been studied from a series of rocket firings spaced a few hours apart [55], this time resolution being inadequate for an unambiguous detection of shorter period waves. In spite of this limitation, Kochanski [56] attempted, from a sizable sample of vapour trail data, to separate out the contributions

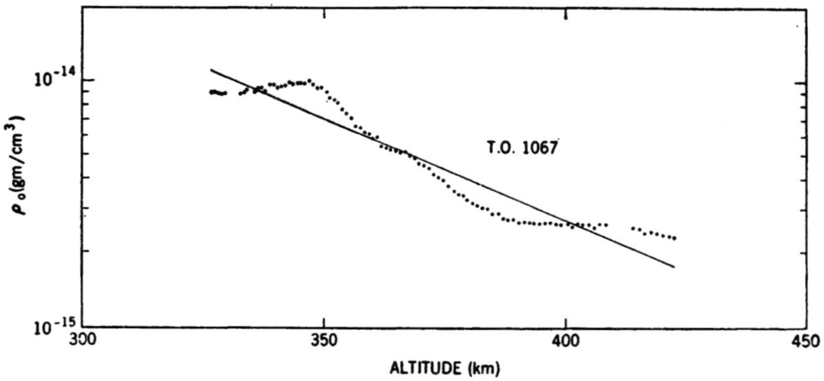

Fig. 4. An example of a perturbation in neutral density observed by satellite-borne density gauge. The straight line represents the altitude profile of the undisturbed atmosphere (from Newton et al. [62]).

of "gravity waves" (i.e. the smaller periods and spatial length) and the tidal (and prevailing) winds. Since the gravity waves of shorter periods and propagating tidal waves (especially the diurnal mode) have comparable vertical wavelengths in this region, such a separation must contain a large measure of uncertainty [57]. There is other direct evidence however for the frequent presence of gravity waves in this region [58].

A more direct evidence for the presence of gravity waves in the lower thermosphere was provided by rocket measurements of the ion temperature from 100 to 160 km with good height definition [59]. A vertical structure with ion temperature oscillations as high as 100 °K and a wavelength around 12 km was found. Such wave-like perturbations in T_i can only be due to short period waves, since otherwise conduction will smooth out the temperature gradients over long periods like those of tidal oscillations. Some wave structure was also observed in ionic temperature measurements made from a low altitude satellite around 200 km [60]. The oscillations occur mainly at geographic latitudes above 60°. Their amplitude reached some ±10% from the mean and the apparent horizontal wavelength was of the order of 200 km. Higher in the thermosphere, in the height range from 290 to about 500 km, some wave structure of the neutral density has been detected from the density gauge measurements on the Explorer 32 satellite. Oscillations of ϱ, with an amplitude of ±10 to 20%, were observed mainly at auroral latitudes during the late evening and the early morning hours [61]. An example of such an observation is given in Fig. 4. The apparent vertical half wavelength ($\lambda_z/2$) of the oscillation was obtained from the altitude difference between adjacent maxima and minima: it was found to vary consistently from 1 km at the perigee to 70 km around 510 km. It was recognized that this altitude variation of λ_z' is primarily due to the variation of the satellite vertical velocity [62]. These oscillations were interpreted in terms of neutral atmospheric gravity waves, propagating predominantly horizontally and having horizontal wavelengths less than 520 km [62]. Two other parameters, N_e and T_e, measured simultaneously on the same satellite were then found to exhibit the same kind of variation as ϱ. The amplitude of the electron density oscillations (±10–30%) was somewhat larger than the amplitude of neutral density oscillations [63]. An antiphase was found between ΔN_e and ΔT_e; the phase relation between ΔN_e and $\Delta\varrho$ was not always clearly discernible.

Such satellite measurements are clearly very important as they provide the direct detection of wave phenomena in the neutral atmosphere. The various parameters of the wave are however difficult to obtain unambiguously from the single satellite results.

2.6. Incoherent Scatter Observations

The incoherent scatter technique has been intensively developed in recent years. Using this technique it is possible to measure from the ground a large number of parameters in the ionosphere. Concerning travelling disturbances the more interesting features of this technique are: (i) its ability of making measurements over a wide range of altitudes both below and above the F layer peak; (ii) the possibility of measuring simultaneously several parameters such as electron concentration N_e, electron and ion temperatures T_e and T_i and ion velocity V_i; the last two parameters are related to the neutral atmosphere temperature and velocity. The height and time resolution and the ability to measure several parameters simultaneously depends on the type of equipment used.

Incoherent scatter observations of travelling ionospheric disturbances were first reported by Thome [64] using the radar at the Arecibo Ionospheric Observatory (Puerto Rico). Disturbances in the electron concentration, with periods ranging from about 20 minutes to several hours were detected from the lower F region up to about 700 km [64]. An attractive feature of the Arecibo facility consists in the possibility of tilting the antenna beam 20° off the vertical and rotating it continuously in azimuth. The horizontal scan of the ionosphere by

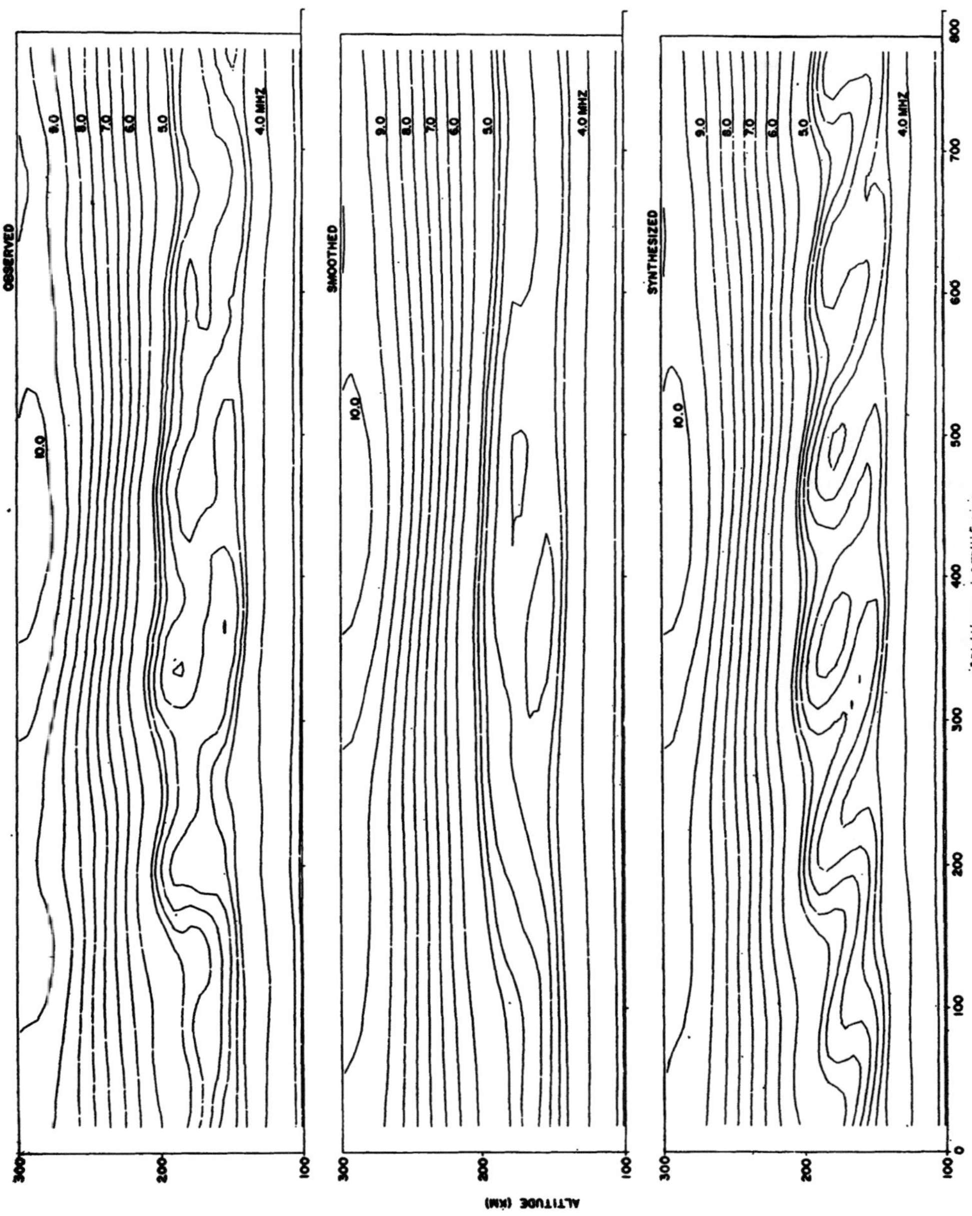

Fig. 5. Comparison of the electron density perturbations observed at Arecibo 12 June 1967 and the theoretically deduced perturbations using a gravity wave model (from Thome and Rao [66]).

the tilting and rotation of the beam leads to the detection of horizontal gradients of electron concentration. Therefore the horizontal wavelength as well as the period of the wave can be obtained [65] in addition to vertical structure. However this technique can be used only for disturbances with long wavelength and low horizontal velocity.

Thome [65] discussed in great detail some cases of long period waves observed at Arecibo. A horizontal velocity of about 700 m s^{-1} towards the south could be determined from a network of ionosondes. This phenomenon was interpreted in terms of a surface wave. The occurrence of such waves was found to be related to geomagnetic disturbances. Some other observations were performed at Arecibo using the incoherent scatter facility in conjunction with three spaced hf Doppler sounders [66]. These equipments were used to measure the horizontal velocity (100–200 m s^{-1}) of perturbation in electron density with periods of 20 minutes to 1 hour. The data were analysed in order to test the theory of gravity wave interaction with the ionosphere. Good agreement was found between the observed perturbation and the gravity wave theoretical model (Fig. 5).

At the Jicamarca Observatory, Peru, near the magnetic equator, weak wave-like disturbances in the F region electron concentration with a period of about 20 minutes and vertical wavelength around 300 km at the F layer peak have been observed during magnetic quiet periods [67–68]. Using the incoherent scatter equipment of Saint-Santin-Nançay (France) only long period disturbances (about 2 hours or more) could be detected. Simultaneously oscillations of the various parameters measured (N_e, T_e, T_i and V_i the ion velocity component parallel to the magnetic field) were observed in the F2 region [69–70] and also in the F1 region from 120 to 200 km [58]. Variations of T_i and V could be interpreted immediately as variations of the neutral temperature and neutral velocity. In some cases a vertical propagation was noticed; also a horizontal propagation toward the south was deduced from ionosonde data. The horizontal wavelength (4000 km), the vertical wavelength (500 km) and the observed period (2 hours) were used in order to check the dispersion relation of free gravity waves. The amplitudes and phases of disturbances of the various parameters measured were also successfully compared with the theory of gravity waves [70]. Such long period oscillations appear to be closely related to magnetic substorms [71] suggesting an origin in the auroral region. An example of oscillation of T_e and V_z is shown in Fig. 6.

Finally the Millstone Hill scatter facility (Mass., USA) also gave evidence of the presence of wave motions of the vertical ionic velocity in the upper F region (above 450 km) [72–73], with periods about 1 hr. An increase of the amplitude with altitude was found. Evans noted that, for this altitude range, the ion velocity is more controlled by ionization diffusion than by the neutral atmosphere wind. Nevertheless this observation shows the existence of wave-like phenomena in the topside ionosphere.

Thus the incoherent scatter technique provides a very powerful tool for the observation of travelling disturbances as it can measure continuously as a function of time and altitude several simultaneous ionospheric parameters.

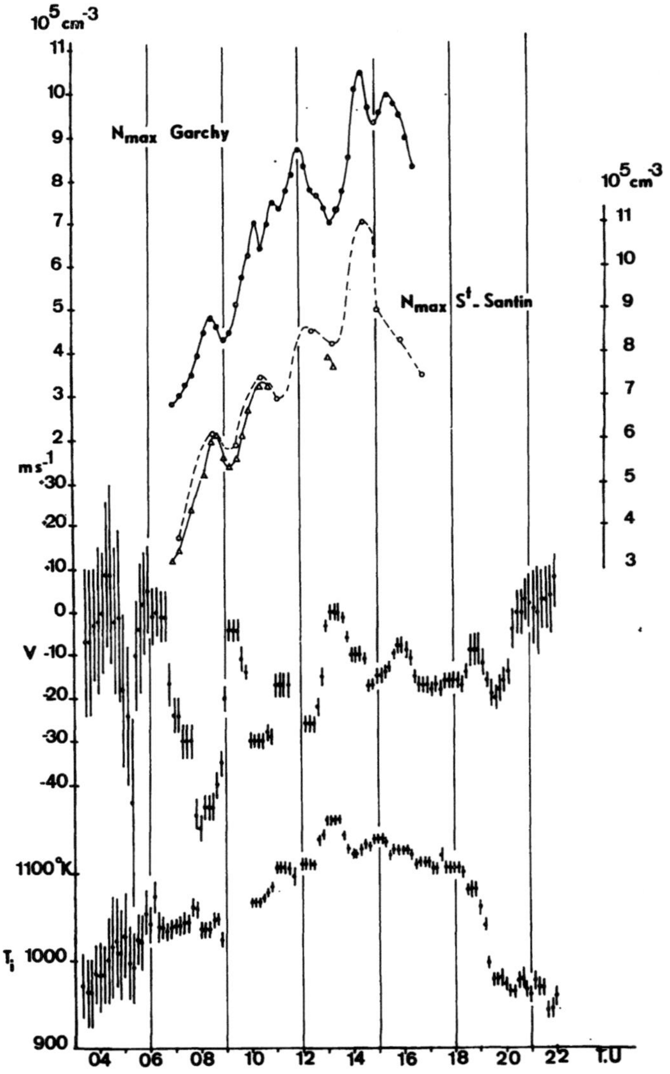

Fig. 6. Large quasi-periodic perturbations observed simultaneously in ion temperature T_i, ion velocity V and peak electron density at St-Santin and Garchy, 16 March 1968 (from J. Testud).

With one exception it gives no information on the horizontal propagation; the horizontal velocity however can be easily deduced from simultaneous observations using classical reflection techniques.

3. Discussion and Synthesis

In the previous section we have presented a fairly extensive cross section of the vast observational information on waves and TIDs in the ionosphere.

We will now present a discussion of the observations by first emphasizing certain aspects of the observations. The following brief paragraph is reproduced from [31]:

"Range–time records from each azimuth showed relatively brief periods during which the backscatter seemed rather smooth without outstanding irregularities and without evidence of motion. The proportion of time that this occurred during the period of regular F2-propagated backscatter averaged about *five per cent*" (underlined by the authors).

Though it is based on one month's sampling of the ionosphere at one location the above statement emphasizes the constant presence of these mesoscale perturbations in the ionosphere, especially at middle and high latitudes. Most other observations support this viewpoint. Yet another important aspect of the observations is that almost all of them – with the exception of the few satellite measurements of neutral concentrations – pertain to perturbations in the plasma component of the upper atmosphere. The interpretation of the observations in terms of similar perturbations in the neutral medium has been very useful in the past for a qualitative understanding of the perturbations. But the recent theoretical work of Hooke [74–78] and others [43, 65, 79–80] has shown that the neutral wave perturbations produce widely different plasma perturbations depending upon the altitude and the orientation of the wave vector relative to the magnetic field and relative to the solar zenith angle. A further observational selection is often imposed by the nature and location of the observing equipment [54]; for example, the horizontal phase velocity (or wavelength) can be greatly in error when measurements are made with inadequate height resolution on nearly horizontal wave fronts [43].

Apart from the above common features, the observed disturbances can be classified into three groups suggested by Georges [43, 44]: large scale disturbances, medium scale disturbances, very short period oscillations. We now discuss each of them in turn.

3.1. Large Scale Disturbances

The large scale disturbances have quasi periods of 30 minutes to a few hours and horizontal dimensions of a few thousands kilometres. They have large amplitude and travel with very high horizontal velocities in the range of 400–700 m s^{-1}. With rare exceptions they move equatorward with little change of shape and amplitude over very large distances. The occurrence of such large scale disturbances is in most cases preceded by geomagnetic activity at auroral latitudes. The gravity wave nature of such disturbances is supported by the observed oscillations of T_i and V_i with incoherent scatter. Recent theoretical work [81–83] has shown that such waves can be generated by the Lorentz force and the Joule heating associated with the intense auroral electrojet currents. Thus most of the large scale disturbances seem to be caused by gravity waves generated in the auroral zone at the times of magnetic disturbances. A few disturbances of this type have, however, been observed even in the absence of magnetic activity and their origin is yet to be understood.

Apart from this important progress about their origin and nature, the propagation mechanism for these waves is still uncertain. Out of the three possible modes of propagation, a freely propagating wave [44], a partly ducted wave [84] or a surface wave [65] we have yet to determine from further observation and theoretical analysis to what extent each one is effective in making these waves propagate over large distances. One possible observational test would be to find out whether the dominant perturbation period is increasing with range from the source (at a fixed altitude) and with decreasing altitude (at a given station). This is the characteristic behaviour of freely propagating waves. A surface wave would exhibit a constancy of phase with altitude, while a ducted mode would exhibit considerable departures from the above characteristics; in addition it would show relatively small attenuation with distance.

Recent incoherent scatter observations [85] have revealed the occurrence of large amplitude, long period ($\sim$10 hours) oscillations in the ion temperature and velocity and in the deduced neutral density, following the onset of an intense magnetic storm. The observed oscillations can only be interpreted as caused by similar oscillations of the neutral parameters. Superposed on this long period oscillation, oscillations of shorter periods ($\sim$2 hours) were also observed. It appears that such oscillations are also generated along with the shorter period ones in the auroral zone but their propagation mechanism and speeds are not known. Thus the perturbations caused by these very long period disturbances can be erroneously interpreted as very large scale TIDs if the observational technique cannot distinguish the period of the perturbation. The storm-associated ionospheric disturbances which were shown in Wright's analysis [86] to progress between Campbell Island and Christchurch (1105 km) with speeds of 400–700 m s^{-1} may be such disturbances of very long period. It is also possible that the moving disturbances (400–800 m s^{-1}) observed by Chan and Villard [41] are of the same kind.

3.2. Medium Scale Disturbances

Medium scale disturbances (with quasi-periods centred mostly around 20–30 min and typical horizontal scales of 100–200 km) are the most persistent ones observed by any technique with a time resolution of a few minutes. The bulk of the extensive observations by Munro [10] as well as Tveten [31] mainly consist of measurements on medium scale disturbances. These disturbances travel with horizontal velocities in the range 100–200 m s^{-1} and have their fronts tilted forward by about 45° from the vertical with the result that the disturbance appears to travel downward at any one station of observation. The results of Munro [10] and Georges [44] show that these medium scale disturbances often change their shape or even disappear over distances of 100–200 km. Munro [10] found such a change of shape to be very frequent even over distances of 40 km, which was the separation distance between his observing stations. In view of the fact that the apparent horizontal scale sizes of these disturbances are typically 100–200 km it is indeed very hard to accept that the disturbance changes its shape within a fraction of its scale length even allowing for dispersive effects. Such a behaviour may arise due to the interference of waves originating from an extended source like the jet stream for example. Further theoretical and experimental studies are necessary for understanding this aspect.

The observations, particularly those of Munro [10] in the southern hemisphere and of Davis and Jones [45] in the northern hemisphere, show rather conclusively that the north–south component of their apparent motion has the same sharp seasonal variation: an equatorward motion in local winter changes to a poleward motion in local summer. Munro finds a seasonally independent eastward component of motion whereas Davies and Jones find a westward motion. This regular feature of seasonal variation has yet to find a satisfactory explanation which may not indeed be forthcoming until the sources of these medium scale disturbances are identified. The seasonal variation in the observed direction of propagation may be the direct result of a seasonal shift in the source locations; alternatively it might result from a seasonal change in propagation conditions such as the seasonal change of middle atmospheric winds which impose a strong directional filtering on the upgoing waves [53, 87]. Very little is known about the sources of these medium scale disturbances though many possible sources are often mentioned in the literature, such as jet streams, lee waves, thunderstorms, non-linear break-up of tidal waves in the upper atmosphere and others. In fact, in future research work on these disturbances most of the effort is to be directed towards gathering observational evidence on and theoretical analysis of the generating mechanisms for these disturbances. The identification of sources is made difficult by the fact that the energy propagates from the source at large angles to the vertical so that the source and the observation point in the ionosphere are usually separated by large horizontal distances. The available observations, however, point to a local lower atmospheric origin rather than to an auroral origin or polar atmospheric origin. The features which support this viewpoint are: (a) the absence of correlation with magnetic activity, (b) the absence of coherence over comparatively short distances like 150 km, and (c) little difference in their amplitudes and other characteristics, with the exception of their direction, as between summer and winter. If these disturbances originate in the winter polar region one would expect the amplitude to be larger in the winter hemisphere.

Still another important area of research on these medium scale disturbances is the investigation (theoretical and experimental) of their effects on the parameters of the thermosphere. Apart from the transitory perturbations of the medium, longer lasting effects like changes in thermospheric temperature and composition etc. are likely through the processes of energy absorption [88] and rectification of the oscillatory perturbations.

The acquisition of further experimental data on the periods, velocities and amplitudes etc. of the medium scale TIDs is not likely advance our understanding of this phenomenon. Only well designed experiments aimed at locating the possible sources or at detecting long lasting effects of the waves on the ionospheric parameters will add significantly to our knowledge in this area.

3.3. Disturbances with Very Short Periods

The major evidence for the presence of wave-like perturbations with periods of a few minutes or less was obtained from hf Doppler records [44] and also (h', t) records [89]. The observed disturbances usually appear as nearly sinusoidal oscillations (in f or h') lasting several hours on some occasions. Wave periods of 3–5 min [44, 47] and also 15–50 s [89] were observed. From the analyses by Georges [44] and by Baker and Davies [48] of several events, such oscillations with a period around 3 minutes were propagating mainly upwards and occurred in limited geographical areas where severe weather effects were present (like clouds with tops above 40000 feet). These observations can be interpreted in terms of acoustic waves with tropospheric origin. The absence of periods exceeding 4 or 5 minutes, which is the acoustic cut-off period of the atmosphere at several levels, is not surprising. The relative absence of periods shorter than 3 minutes is most probably due to filtering and damping effects of the atmosphere [43] rather than due to the source characteristics. Further observations as well as theoretical analyses are needed to understand the generation and propagation processes of this interesting class of waves.

Limited but unambiguous evidence has been obtained [51, 89] for the presence of ionospheric perturbations with infrasonic periods (a few seconds to one minute). Ground-based sensors recorded infrasonic waves from burning rockets at as high an altitude as 180 km [90]. This demonstrates the ability of these waves to propagate in the intervening atmosphere without severe loss though this may not be true under all conditions. The possible generation of these waves by certain large scale auroral forms has been demonstrated by Wilson and Nichparenko [91]. The earlier work of Campbell and Young [92] indicated a similar origin; and in addition, they found a close association between the ground level pressure oscillation and ionospheric current intensity. Maeda and Watanabe [93] found the heating by precipitating particle streams was inadequate for generating these waves while Chimonas' work [94] showed that both Joule heating and the Lorentz force resulting from the intense auroral electrojet currents could give rise to infrasonic waves of appreciable amplitude.

At middle latitudes, the occurrence of infrasonic waves at ground level in association with ionospheric storms was observed in some cases [95], but ionospheric oscillations reported by Shrestha [89] occurred during a quiet period with no evidence of similar pressure oscillations at ground level. Thus the observational knowledge about the sources and the propagation characteristics of these infrasonic waves is rather limited, and much more experimental work is needed in this area.

The possible role of the very short period wave disturbances in modifying the ionospheric processes is yet to be investigated: for example, King [96–97] makes the interesting suggestion that the aurorally generated acoustic gravity waves may cause turbulent mixing of the atmosphere at F region heights and thus enhance the recombination coefficient.

3.4. Ionospheric Disturbances from Known Sources

One important category of observations consists of the ionospheric disturbances originating in known sources. Fairly extensive evidence has been gathered on the ionospheric effects of atmospheric nuclear explosions in the megaton range [98–103]. The observed perturbations are similar in many respects to the naturally occurring TIDs. The most pertinent feature revealed by the observations is that the quasi-period of perturbation increases with increasing range from the source for any given height of observation. This feature as well as many other characteristics were shown to fit into a theoretical description in terms of dispersive internal gravity waves propagating into the upper atmosphere from a source in the lower atmosphere [101, 104].

Ionospheric perturbations originating from the Alaskan earthquake of 1964 were recorded at many places [105–107]. The perturbations can be interpreted in terms of short and long period acoustic gravity waves, in addition to a surface wave mode travelling with the acoustic velocity of the medium. Barry et al. [108] observed the ionospheric perturbation caused by a large ground level explosion of TNT (500 tons) and they interpreted it in terms of acoustic waves propagating upward into the ionosphere.

The above observations taken together provide the most unambiguous evidence to the effect that impulsive disturbances near ground level or in the troposphere can launch pressure waves which propagate to high levels in the atmosphere. The work of Gossard and Munk [109] and Gossard [110] has shown that the internal gravity waves originating in tropospheric weather carry enough energy as they leak upward (from the troposphere) to cause observable perturbations at ionospheric levels.

4. Conclusion

We have reviewed the main points of our observational knowledge on mesoscale wave like disturbances in the ionosphere. No attempt has been made to represent or refer to all the aspects of the reported observations. From the bulk of observations presented here it is clear that the upper atmosphere above 100 km is pervaded by mesoscale wave motions and their associated effects. These mesoscale motions exhibit a wide spectrum of spatial and temporal scales: spatial scales of a few tens of kilometres to a thousand kilometres and periods of a few minutes to 3 hours are commonly observed. This wide spectrum of waves has been successfully described in terms of a unified theory of internal (or acoustic) gravity waves. The waves can however have widely different origins. The Joule heating and Lorentz force associated with the intense electric currents causing auroral substorms are now identified as the source of the long period (1–3 hr), large scale waves which are found to travel large distances equatorward in both hemispheres. There is now increasing evidence that the very short period disturbances (2–5 min) originate in the troposphere and propagate upward. The natural sources of the medium scale disturbances have not been identified so far, though such waves were found to originate from large nuclear explosions and an earthquake. These medium scale disturbances (with typical quasi-periods of 15–40 min and horizontal scale sizes of 100–200 km) are by far the most persistent in their daily occurrence and they exhibit a seasonal reversal in their north–south component of horizontal propagation. This reversal remains unexplained and possibly it is an important clue in identifying the main source of these medium scale disturbances. The constant presence of these mesoscale motions at ionospheric levels, with only their strength and complexity varying, makes it imperative to understand their origins and their varied effects in the upper atmosphere. Future efforts are to be directed mainly towards the identification of the sources, particularly of the medium scale disturbances, and towards the evaluation of the wave disturbance effects (such as the dynamical heating) on the parameters of the neutral and plasma media.

Acknowledgment

We are very grateful to various authors who kindly sent their most recent work concerning the subject of this review.

References

[1] C. O. Hines, Quart. J. Roy. Met. Soc. **89**, 1 (1963).
[2] C. O. Hines, Meteorological Monographs, Amer. Meteor. Soc. **9**, 114 (1968).
[3] R. S. Lindzen and S. Chapman, Space Sci. Rev. **10**, 3 (1969).
[4] B. Haurwitz, W.M.O. Rep. No. 146, T.P. 69 (1964).
[5] H. Kohl, Space Research X, 550 (1970).
[6] C. O. Hines, Canad. J. Phys. **38**, 1441 (1960).
[7] G. H. Munro, Nature, Lond. **162**, 886 (1948).
[8] G. H. Munro, Proc. Roy. Soc. A **202**, 208 (1950).
[9] G. H. Munro, Proc. Roy. Soc. A **219**, 447 (1953).
[10] G. H. Munro, Aust. J. Phys. **11**, 91 (1958).
[11] R. H. Price, in: Physics of the Ionosphere, The Physical Society, London 1955 (p. 18).
[12] K. Toman, J. Geophys. Res. **60**, 57 (1955).
[13] L. Thomas, J. Atmos. Terr. Phys. **14**, 123 (1959).
[14] G. H. Munro and L. H. Heisler, Aust. J. Phys. **9**, 359 (1956).
[15] L. H. Heisler, Aust. J. Phys. **11**, 79 (1958).
[16] L. H. Heisler and J. D. Whitehead, Aust. J. Phys. **14**, 481 (1961).
[17] G. H. Munro and L. H. Heisler, Aust. J. Phys. **9**, 343 (1956).
[18] D. C. Baker and J. A. Gledhill, J. Atmos. Terr. Phys. **27**, 1223 (1965).
[19] F. du Castel and J. M. Faynot, Nature, Lond. **204**, 984 (1964).
[20] G. A. M. King and M. D. Lawden, J. Geophys. Res. **66**, 3962 (1961).
[21] K. H. Bassiliev, Geomagn. i Aeronom. **6**, 773 (1966).
[22] R. W. E. Mc Nicol, H. C. Webster and G. G. Bowman, Aust. J. Phys. **9**, 247 (1956).
[23] L. H. Heisler, Nature, Lond. **183**, 383 (1959).
[24] W. Becker, J. Atmos. Terr. Phys. **22**, 275 (1961).
[25] L. H. Heisler, J. Atmos. Terr. Phys. **25**, 71 (1963).
[26] J. Klostermeyer, J. Atmos. Terr. Phys., **31**, 25 (1969).
[27] G. G. Bowman, Planet. Space Sci. **2**, 133 (1960).

[28] G. G. BOWMAN, J. Atmos. Terr. Phys. **30**, 721 (1968).
[29] N. P. BENKOVA et al., Geomagn. i Aeronom. **9**, 1082 (1968).
[30] J. F. VALVARDE, R.P.L. Stanford University Sci. Rep. No. 1 (1958).
[31] L. H. TVETEN, J. Res. Nat. Bur. Stand. **65** D, 115 (1961).
[32] R. D. HUNSUCKER and L. H. TVETEN, J. Atmos. Terr. Phys. **29**, 909 (1967).
[33] E. N. BRAMLEY and W. ROSS, Proc. Roy. Soc. A **207**, 251 (1951).
[34] E. N. BRAMLEY, Proc. Roy. Soc. A **220**, 39 (1953).
[35] J. C. DE MAISTRE, M. C. CROCHET and D. BARREAU, C. R. Acad. Sci. Paris **272**, 77 (1971).
[36] J. W. FINDLAY, Nature, Lond. **159**, 58 (1947).
[37] C. R. REDDY and B. R. RAO. J. Atmos. Terr. Phys. **33**, 251 (1971).
[38] R. A. VINCENT, Short period gravity waves in the lower ionosphere, private communication (to be published).
[39] J. M. WATTS and K. DAVIES, J. Geophys. Res. **65**, 2295 (1960).
[40] K. L. CHAN and O. G. VILLARD, J. Geophys. Res. **67**, 973 (1962).
[41] K. DAVIES, J. M. WATTS and D. H. ZACHARISEN, J. Geophys. Res. **67**, 601 (1962).
[42] K. DAVIES and D. M. BAKER, Radio Sci. **1**, 545 (1966).
[43] T. M. GEORGES, ESSA Tech. Rep. IER 57-ITSA 54 (1967).
[44] T. M. GEORGES, J. Atmos. Terr. Phys. **30**, 735 (1968).
[45] K. DAVIES and J. E. JONES, J. Atmos. Terr. Phys. **33**, 39 (1961).
[46] T. M. GEORGES, Proc. Symp. on Acoustic Gravity Waves in the Ionosphere, Boulder 1968 (p. 171).
[47] K. DAVIES and J E. JONES, J. Atmos. Sci. **28**, 254 (1971).
[48] D. M. BAKER and K. DAVIES, J. Atmos. Terr. Phys. **31**, 1345 (1969).
[49] T. M. GEORGES, J. Geophys. Res. **72**, 422 (1967).
[50] J. E. TITHERIDGE, J. Geophys. Res. **68**, 3399 (1963).
[51] N. N. RAO, G. F. LYON and J. A. KLOBUCHAR, J. Atmos. Terr. Phys. **31**, 539 (1969).
[52] M. J. DAVIS and A. V. DA ROSA, J. Geophys. Res. **74**, 5721 (1969).
[53] D. H. COWLING, H. D. WEBB and K. C. YEH, Tech. Rep. No. 38, Elect. Engng Res. Lab., Univ. of Illinois (1970).
[54] T. M. GEORGES and W. H. HOOKE, J. Geophys. Res. **75**, 6295 (1970).
[55] J. W. WRIGHT, C. H. MURPHY and J. V. BULL, J. Geophys. Res. **72**, 1443 (1967).
[56] A. KOCHANSKI, J. Geophys. Res. **69**, 3651 (1964).
[57] C. O. HINES, J. Geophys. Res. **70**, 178 (1965).
[58] C. A. REDDY and G. VASSEUR, Space Research XII, 951 (1972).
[59] W. C. KNUDSEN and G. W. SHARP, J. Geophys. Res. **70**, 143 (1965).
[60] K. K. HARRIS, G. W. SHARP and W. C. KNUDSEN, J. Geophys. Res. **74**, 197 (1969).
[61] G. P. NEWTON, D. T. PELZ and H. VOLLAND, Space Research VIII, 980 (1968).
[62] G. P. NEWTON, D. T. PELZ and H. VOLLAND, J. Geophys. Res. **74**, 183 (1969).
[63] D. L. DYSON, G. P. NEWTON and L. H. BRACE, J. Geophys. Res. **75**, 3200 (1970).
[64] G. D. THOME, J. Geophys. Res. **69**, 4047 (1964).
[65] G. D. THOME, J. Geophys. Res. **73**, 6319 (1968).
[66] G. D. THOME and P. B. RAO, Rep. of Raytheon Comp., Spencer Laboratory, Burlington, Mass. (1969).
[67] D. L. STERLING, Aeronomy Rep. No. 19, Univ. of Illinois (1967).
[68] D. L. STERLING, W. H. HOOKE and R. COHEN, J. Geophys. Res. **76**, 3777 (1971).
[69] G. VASSEUR and P. WALDTEUFEL, J. Atmos. Terr. Phys. **31**, 885 (1969).
[70] J. TESTUD and G. VASSEUR, Annls Géophys. **25**, 525 (1969).
[71] J. TESTUD, J. Atmos. Terr. Phys. **32**, 1793 (1970).
[72] J. V. EVANS et al., Radio Sci. **5**, 27 (1970).
[73] L. A. CARPENTER and J. V. EVANS, paper presented at the 1969 Spring URSI Meeting Washington, D.C. (1969).
[74] W. H. HOOKE, J. Atmos. Terr. Phys. **30**, 795 (1968).
[75] W. H. HOOKE, J. Geophys. Res. **75**, 35 (1970).
[76] W. H. HOOKE, J. Geophys. Res. **75**, 7229 (1970).
[77] W. H. HOOKE, J. Geophys. Res. **75**, 7239 (1970).

[78] W. H. Hooke, Planet. Space Sci. **18**, 1793 (1970).
[79] R. A. Nelson, J. Atmos. Terr. Phys. **30**, 795 (1968).
[80] J. Testud and P. François, J. Atmos. Terr. Phys. **33**, 765 (1971).
[81] G. Chimonas and C. O. Hines, Planet. Space Sci. **18**, 210 (1969).
[82] W. Blumen and R. G. Hendl, J. Atmos. Sci. **26**, 210 (1969).
[83] J. Testud, C. R. Acad. Sci. Paris **270**, 241 (1970).
[84] C. A. Reddy, in: Internal Gravity and Acoustic waves — Colloquium, Boulder NCAR Tech. Note No. **43**, 229 (1969).
[85] D. Alcayde et al., in J. Atmos. Terr. Phys. (1972) to be published.
[86] M. D. Wright, Nature, Lond. **190**, 898 (1962).
[87] C. O. Hines and C. A. Reddy, J. Geophys. Res. **72**, 1015 (1967).
[88] C. O. Hines, J. Geophys. Res. **70**, 177 (1965).
[89] K. L. Shrestha, J. Atmos. Terr. Phys. **29**, 107 (1967).
[90] W. L. Donn et al., Science **162**, 1116 (1968).
[91] C. R. Wilson and S. Nichparenko, Nature, Lond. **214**, 1299 (1967).
[92] W. H. Campbell and J. M. Young, J. Geophys. Res. **68**, 5909 (1963).
[93] K. Maeda and T. Watanabe, J. Geophys. Res. **21**, 15 (1964).
[94] G. Chimonas, Planet. Space Sci. **18**, 591 (1970).
[95] G. G. Bowman and K. L. Shrestha, Nature, Lond. **210**, 1034 (1968).
[96] G. A. M. King, J. Atmos. Terr. Phys. **28**, 957 (1966).
[97] G. A. M. King, J. Atmos. Terr. Phys. **29**, 161 (1967).
[98] W. J. G. Beynon and E. Jones, Nature, Lond. **196**, 253 (1962).
[99] G. G. Bowman, Aust. J. Phys. **17**, 409 (1964).
[100] T. Obayashi, Planet. Space Sci. **10**, 47 (1963).
[101] R. V. Row, J. Geophys. Res. **72**, 1599 (1967).
[102] H. Kohl, in: Electron Density Distribution in Ionosphere and Exosphere, North Holland Publ. Co., Amsterdam 1964 (p. 160).
[103] D. P. Kanellakos, J. Geophys. Res. **72**, 4559 (1967).
[104] C. O. Hines, J. Geophys. Res. **72**, 1877 (1967).
[105] R. S. Leonard and R. A. Barnes, J. Geophys. Res. **70**, 1250 (1965).
[106] R. V. Row, J. Geophys. Res. **71**, 343 (1966).
[107] K. Davies and D. M. Baker, J. Geophys. Res. **70**, 2251 (1965).
[108] G. H. Barry, L. J. Griffiths and J. C. Taenzer, J. Geophys. Res. **71**, 4173 (1966).
[109] E. E. Gossard and W. Munk, J. Met. **4**, 259 (1954).
[110] E. E. Gossard, J. Geophys. Res. **67**, 745 (1962).

Space Research XII — Akademie-Verlag, Berlin 1972

F REGION TRAVELLING DISTURBANCES OVER JAMAICA

J. MacDougall

Dept. of Physics, University of the West Indies, Kingston, Jamaica, W. I.

Studies of travelling disturbances over Jamaica (18° N 77° W) in the F region have been made by combining phase height and angle of arrival data. Except for the occasional strong auroral associated disturbances which travel at higher speeds, there is some doubt that the disturbances observed are waves rather than irregularities. Many of the data yield rather ambiguous results which do not fit clearly into any of the traditional models.

1. Introduction

This study of travelling ionospheric disturbances (TIDs) started in the usual way. It was decided to measure the statistical properties of TIDs over Jamaica (18° N 77° W) and set up suitable equipment for doing so. Since financial and spatial limitations did not allow the usual very wide spaced arrays it was decided to measure simply phase height and angle of arrival and combine these two measurements by methods which will be discussed in § 2 to obtain the properties of the TIDs. While (§§ 3 and 4) the phase height and angle of arrival measurements gave good results, combining the two (§ 5) was most often unsuccessful. Some possible reasons for this will be suggested; however, it is hoped that some model may be found which will explain some of our failures.

2. Combining Phase and Angle of Arrival Data to give the Properties of TIDs

It was decided to measure phase height and angle of arrival since it appeared straightforward to combine these to give velocities etc. The most obvious method is by using the partial differential equation

$$\frac{d\Phi}{dt} = \frac{\partial\Phi}{\partial x}\frac{dx}{dt} + \frac{\partial\Phi}{\partial y}\frac{dy}{dt} + \frac{\partial\Phi}{\partial z}\frac{dz}{dt} + \frac{\partial\Phi}{\partial t}$$

where Φ is phase, $\partial\Phi/\partial x$, $\partial\Phi/\partial y$ may be seen to be actually what we measure as angle of arrival in the x and y directions and dx/dt, dy/dt are the two components (V_x, V_y) of the velocity. The z term is assumed negligible since we are more or less looking at a fixed height region and assume motion to be mostly horizontal, and $\partial\Phi/\partial t$ is assumed zero if the changes are only due to horizontal movement. It is most convenient to use this in an integral form,

$$\Phi(t) = V_x\int\frac{\partial\Phi}{\partial x}\,dt + V_y\int\frac{\partial\Phi}{\partial y}\,dt$$

and we may conveniently substitute our integrated angles of arrival into this and solve for V_x and V_y. In practice this never seemed to work well since one never obtained consistent values for V_x, V_y. Most of our analysis was therefore attempted by assuming the conventional model for the irregularities: that they are ripples with elongated wave fronts and move in the direction of the wave normal. For this model we should see the following as a ripple passes overhead: (i) On the amplitude record we should get an increase in amplitude or a "burst" as the maximum of the wave passes overhead. (ii) On the phase height record, if the ripple is shallow we should get a reproduction of the ripple profile. If deep, we should get multiple reflection points and possibly features on the group heights records. (iii) On the angle of arrival records the two components of the angle should swing together along the direction of the wave normal, moving rapidly in the opposite direction to that in which the ripple is moving at the time of the "burst". Thus the angle of arrival data give the direction of motion and using angle and phase height we may obtain the speed.

In some cases this model gave good results, but more often the variations did not agree well with the model.

3. The Phase Height Measurements

Phase height is measured by a simplified Findlay system as described in appendix A. On most days continuous records may be obtained over most daytime hours, but in late afternoon heavy interference and rapidly increasing phase height make measurements difficult or impossible after about 1600. Very few of the "x" and "z" type events described by Munro [1] are seen here but continuous small scale fluctuations are observed. Fig. 1 shows typical records on which the long term diurnal variation is seen as well as the short period fluctuations. Here we are interested only in the short period fluctuations so will not discuss the diurnal variation. The average rms amplitude of the short period fluctuations is about 1 km, there being considerable variability. There is a pronounced seasonal variation in fluctuation amplitude with minima in equinox. It is of interest to see what the spectra of the variations look like and whether there are dominant periods. The larger amplitude of the long period components tends to overload the spectra so in Fig. 2 we show spectra of time differentiated phase height variations. The differentiation reduces the amplitude of the long period components relative to the short. These spectra cover a number of hours of data. It may be seen that while individual days show dominant components these do not occur each day, so that the mean curve would be fairly smooth; what is of interest is that these appreciable components have period of 5–10 minutes. This should not be possible if they are gravity waves. Visual estimates of average short periods also give values of about 8 minutes, which is below the Brunt–Väisälä period for the F2 region. Note that most data were taken on a frequency of about

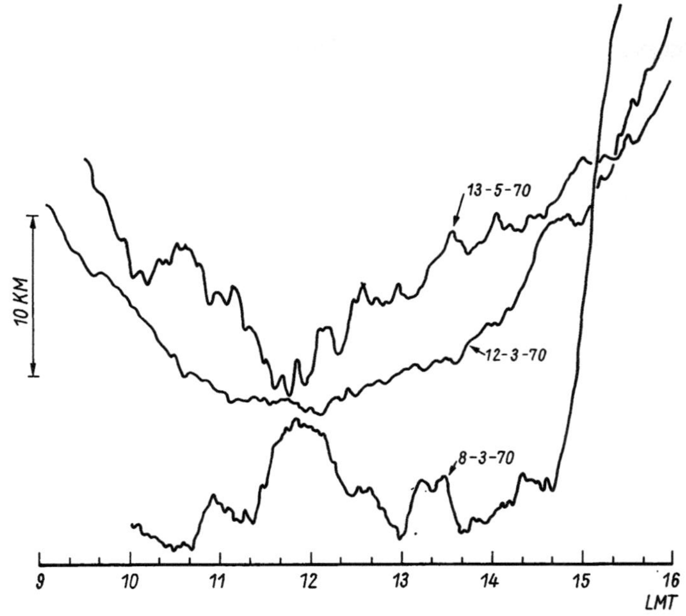

Fig. 1. Typical phase height records showing small (12-3-70), moderate (13-5-70) and large (8-3-70) fluctuations.

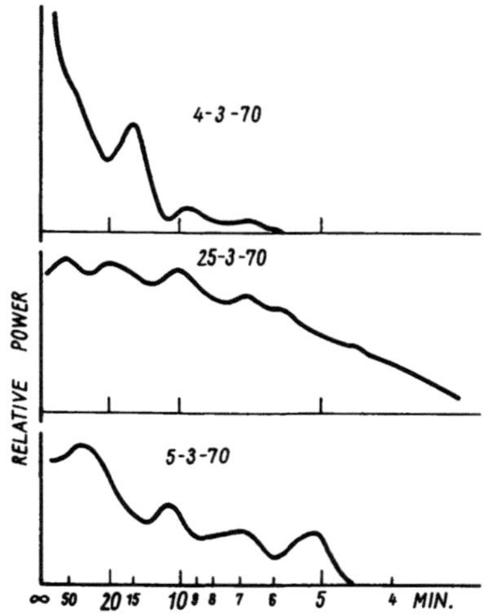

Fig. 2. Typical power spectra of differentiated phase height variations.

6.3 MHz, which was chosen as optimum for F2 region and is usually reflected at a true height of 200–250 km.

A number of possible reasons can be given to explain the periods being shorter than is allowed for gravity waves, but there is no strong evidence to support any one explanation at this time.

4. The Angle of Arrival Measurements

Angle of arrival, which is actually measured as phase difference between two separated antennae, has not as commonly been measured as has phase height. The equipment used in this study is somewhat unique and is described briefly in appendix B. The resolution as the equipment was used for this study was about one fifth of a degree. It had been expected that average rms angular fluctuations would be about 1.7° from Bramley's [2] study. Over Jamaica however they turned out to be only on average about $\frac{1}{2}$° degree. From this one suspects that the irregularities over Jamaica are appreciably weaker than those over England.

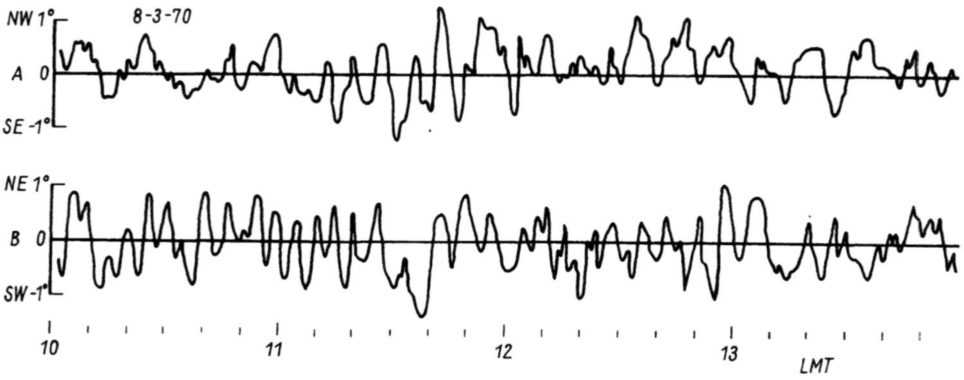

Fig. 3. Typical angle of arrival versus time record.

A typical record of angle (two components) is shown in Fig. 3. Note the appreciable amplitude of the small scale fluctuations. Records are read to only the nearest $\frac{1}{5}$°. Integrating this record to reduce the short period fluctuations relative to the longer gives records such as Fig. 4 and here we may see the diurnal variation. Note that we do not know the exact position of the zero on the vertical scale, so that some of the variation may be due to arbitrary choice of zero. It is of some curiosity that from these curves we may deduce the movement of the overhead reflection point on individual days. A typical example is shown in Fig. 5 and the feature which is most common to these variations is the progressive E to W shift. The N–S shift is different on different days. Using mean diurnal phase height data and these angular variations we find that the east–west speed is closely the speed of the earth's rotation. This is as expected; the daily variation looks like a large irregularity travelling overhead with the speed of the earth's rotation.

Spectral analyses of some of the small scale fluctuations gives results similar to those for the phase height shown in Fig. 2. Surprisingly spectra of the two angular components and phase height (differentiated) for the same day did not agree in detail.

Fig. 4. A typical integrated angle of arrival record. A and B have directions as in Fig. 3. The vertical scale is 3.84 degree minutes per division. Sample from 12 March 1970.

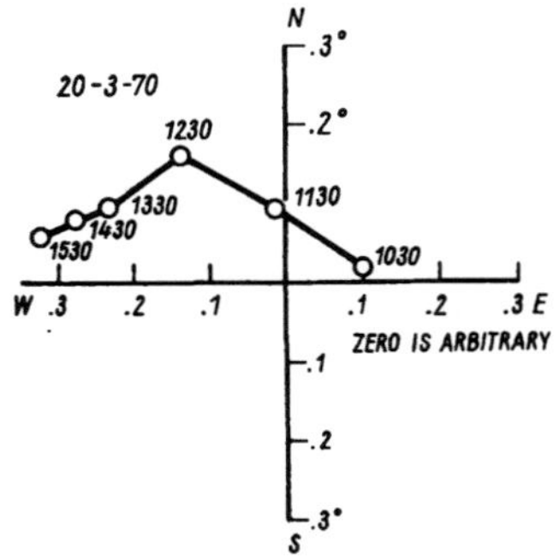

Fig. 5. Movement of the overhead reflection point during the day for 20 March 1970.

5. Combining Phase Height and Angle of Arrival to deduce Properties of the TIDs

The model used for combining phase height and angle of arrival to deduce the motion of TIDs was discussed in § 2; Fig. 6 shows an event which fitted the model well. For those irregularities which fitted well the results agree with other studies, e.g. the mode of speed distribution is about 140 m s^{-1}. The most common problem was that the angular fluctuations did not describe a line as in the model but wandered erratically. Fig. 7 shows a typical example of this. Bramley [2] obtained similar results and was only able to deduce the motions for a small number of TIDs from a large data sample. We were

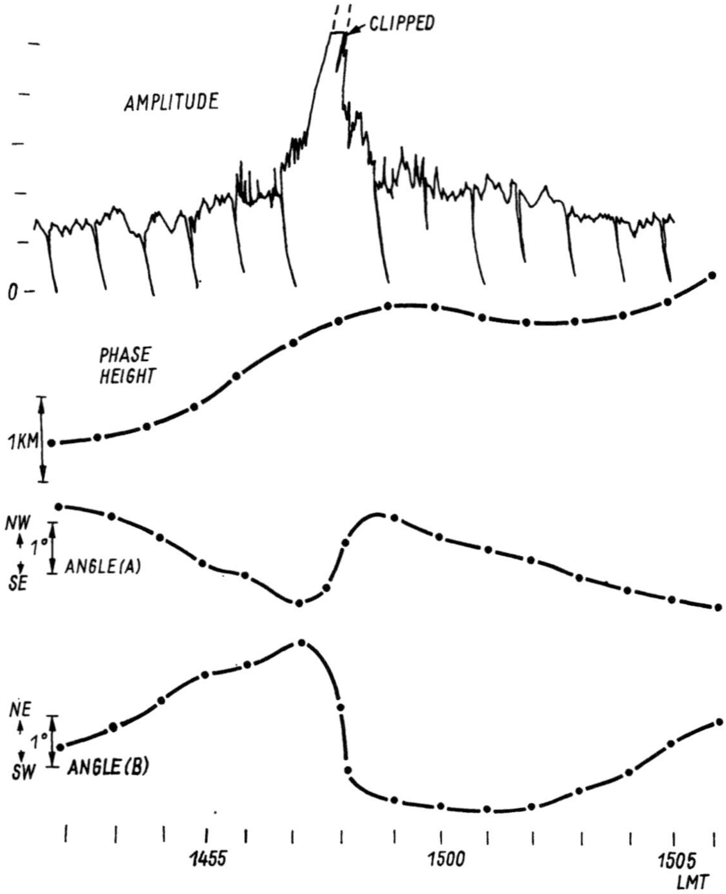

Fig. 6. An event which gave good results using the ripple model. Results are: speed 130 m s^{-1}, wavelength 86 km, direction of motion N 65° E, height fluctuation ± 400 m.

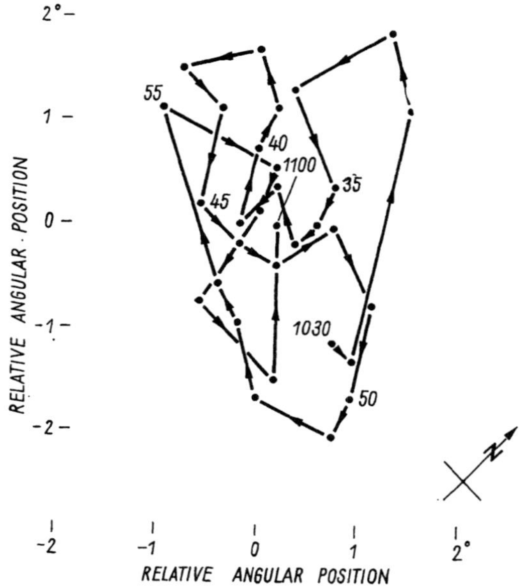

Fig. 7. An example of the confused motion of the reflection point, 29 May 1970.

aware of his problem but thought that with phase height as well as angle of arrival data it should be possible to decipher the data. In some cases this has certainly been true and one can deduce some sort of motion by comparing the two data. In some cases the motion pattern is somewhat involved and Fig. 8 shows motions for parts of two days. Two features are apparent here, the

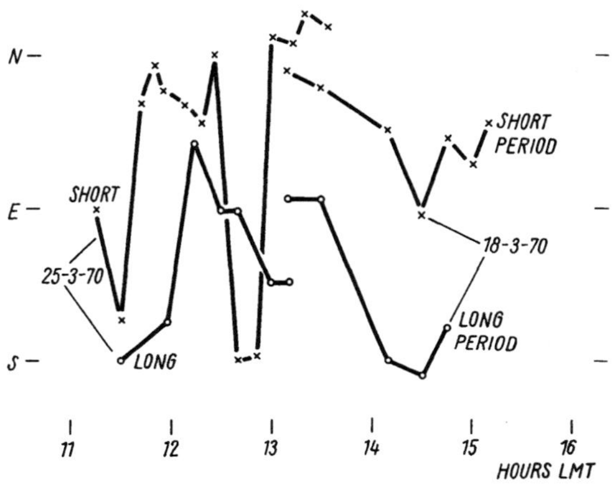

Fig. 8. Directions of motion for TIDs on 18 and 25 March 1970.

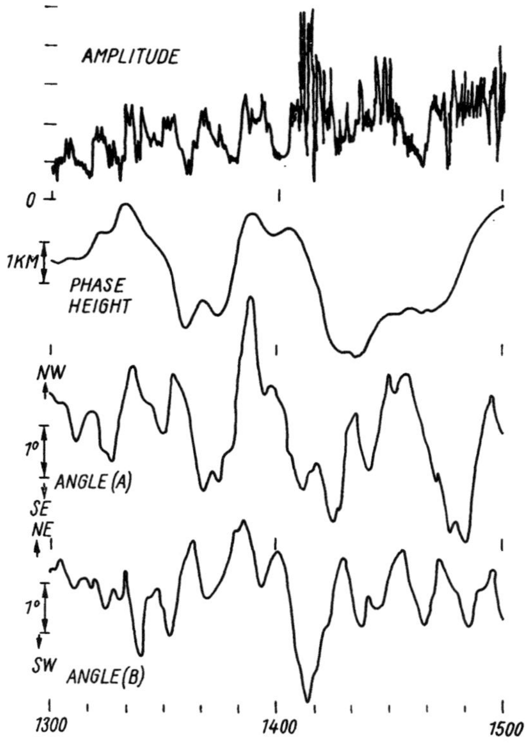

Fig. 9. Measurements from 3 June 1970.

direction of motion is continuously changing and the short period (about 10 min) TIDs are moving in a different direction from the long period (about 20 minutes). There may be some indication that the difference in direction is 90°.

In many cases however even deduction of two independent motions taking place at the same time may not be sufficient to resolve the measurements. Fig. 9 shows a two hour section of measurements from 3 June 1970 and in this case one can see little or no agreement between amplitude, phase height and angles.

One further way of presenting the data is to see whether a bulk plot of angles will indicate any average form for the irregularities. Plotting a long series of angle measurements should give the two-dimensional angular spectrum of the downcoming waves and hence be proportional to the Fourier transform of the average pattern of phase fluctuation. Fig. 10 shows such a plot and it is seen that the general form is relatively circular with perhaps a slight N–S extension. This tends to suggest the average ionospheric pattern is of the same shape but oriented E–W. If, however, one has patterns of ripples moving in random directions one would get much the same pattern so there are two possible interpretations even here.

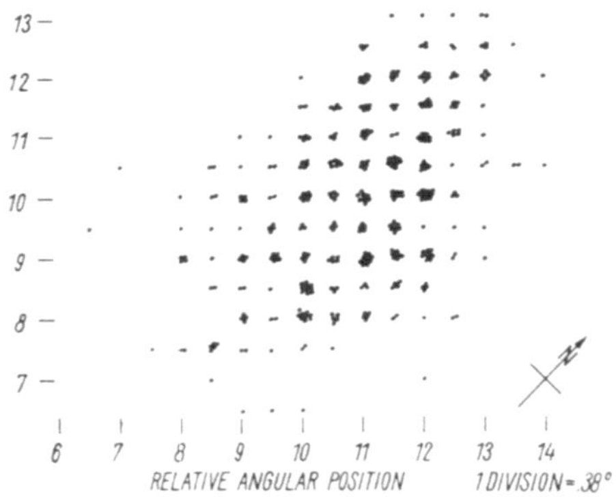

Fig. 10. Angular positions during 6 hours on 25 March 1970.

6. Discussion and Conclusion

We have shown that while the individual measurements of phase height and angle of arrival give good results there are considerable problems in combining the data to give the motions of TIDs. Possibilities are: (i) there are waves moving in several different directions at the same time; (ii) the model is incorrect and the irregularities are more complex than our simple ripples; (iii) some of the structure we are seeing is at lower heights and affects the phases and angles differently. There is no obvious way of distinguishing between these possibilities at present. Studies such as that by Munro [3] wherein a single irregularity can be isolated by some property yield good results whereas studies such as those by Bramley [2], Gusev et al. [4] and the present work, wherein it is more difficult to isolate individual irregularities, have not been quite so obviously successful. In the case of [4] there is the rather odd result of elongated irregularities drifting along their long axis.

In the present study we also find that at times when there is an apparently single set of waves moving from an auroral source as a southerly moving event, there is still no good detailed agreement between phase height and angle using the ripple model.

Appendix A
A Simplified Phase Height Measuring System

The arrangement used is basically that due to Findlay [5], who uses a lock-in oscillator phase locked to the transmitter. It may easily be shown that this is not really necessary and that the arrangement shown in Fig. 11 is quite adequate. In this arrangement the i.f. output from the receiver is mixed with the output from a crystal oscillator whose frequency is 10–20 kHz offset from the i.f. frequency. The resultant 10–20 kHz output from the mixer and filter appear as shown in Fig. 11. The phase of the echo is fixed relative to the

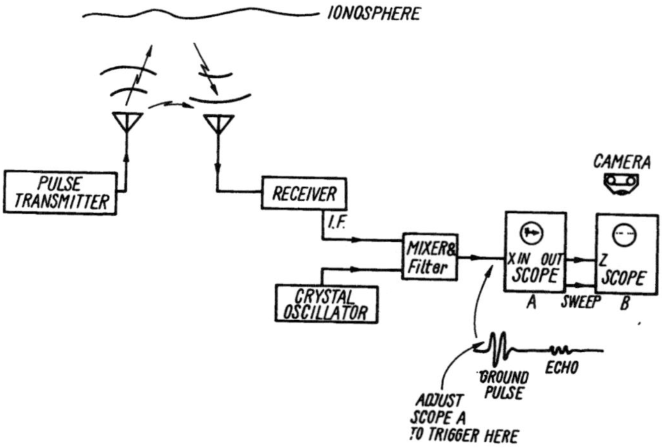

Fig. 11. An equipment arrangement for measuring phase height.

ground pulse if it is not changing in phase height and provided the oscillator in the receiver and the crystal oscillator are reasonably stable. Therefore it is only necessary to choose a phase reference point on the ground pulse, which is done using the oscilloscope trigger, and shifts in echo phase will show relative to this.

This circuit has advantages over Findlay's original circuit apart from the simplicity. It is not necessary to have a connection to the transmitter. All that is required is a strong ground pulse. Also the frequency may be tuned without any adjustments being necessary.

Appendix B
A New Angle of Arrival Measuring Equipment

Almost all standard angle of arrival measurements have involved two identical receivers whose input is from a pair of spaced antennae and the relative phases of the outputs are detected in various ways and give the angle of arrival. This standard arrangement has two disadvantages, the first is the problem of obtaining two phase matched receivers, and the second is that most of the ways of recording the phase are either involved or use large amounts of recording film.

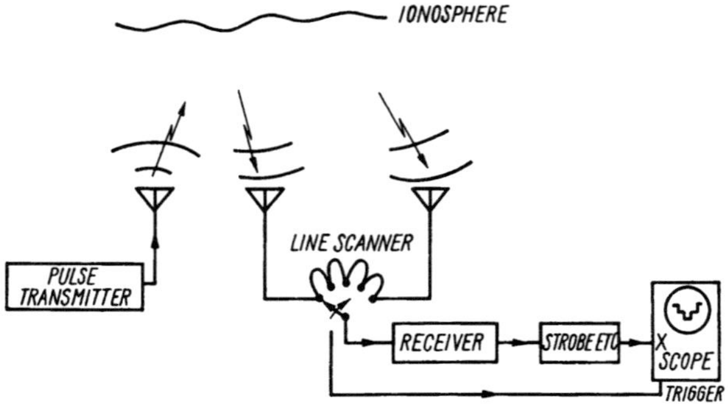

Fig. 12. Equipment arrangement for measuring angle of arrival.

The arrangement shown in Fig. 12 is basically very simple. The two receiving antennae are carefully matched to the transmission lines joining them. Somewhere on the line the two signals will be out of phase (in the middle for an overhead echo if the connection to one antenna is reversed) and by scanning along the line we may find this point, and its position gives us the angle of arrival of the signal. The line scanner is actually a solid state switching network which sequentially samples the line in 16 places, advancing one position with each transmitter pulse. The output from the receiver is strobed and the peak voltage stored between pulses and displayed on the oscilloscope. The null is always easily visible and if the reversed antenna connection arrangement is used its position is independent of frequency. The null is recorded continuously on slowly moving film. If required the system could easily be digitized since the null is in terms of one of the "switch" positions.

This system has been found to work very well. The line scanner, incidentally is a rather interesting way of showing transmission line characteristics. With a steady signal to one input and various loads on the other one can view the standing waves on the line.

References

[1] G. H. Munro, Proc. Roy. Soc. A **202**, 208 (1950).
[2] E. N. Bramley, Proc. Roy. Soc. A **220**, 39 (1953).
[3] G. H. Munro, Aust. J. Phys. **11**, 91 (1958).
[4] V. D. Gusev et al., NASA Tech. Trans. F-20 (1960).
[5] J. W. Findlay, J. Atmos. Terr. Phys. **1**, 353 (1951).

Space Research XII – Akademie-Verlag, Berlin 1972

THE FOCUSING EFFECT ON SATELLITE RADIO OBSERVATIONS DUE TO TRAVELLING IONOSPHERIC DISTURBANCES

G. F. LYON[a] and J. A. D. HOLBROOK[b]

[a] Physics Department and Centre for Radio Science, University of Western Ontario, London, Canada

[b] Satellite Control, Telesat Canada, Ottawa, Canada

Observations of the received signal strength of vhf radio transmissions from orbiting satellites occasionally show signal enhancements too great to be accounted for without invoking some focusing mechanism. Simultaneous Faraday rotation records bear the unmistakable indication of the presence of a travelling ionospheric disturbance (TID). It is shown that the electron density contours resulting from the passage of a typical gravity wave or acoustic wave in the ionosphere can give rise to the focusing effect and that this effect will be most pronounced at certain observation angles. Such observations from a single ground station can provide a measure of the ionosphere 'tilt' due to the TID and the disturbance wavelength.

1. Introduction

It is well known that when electron content measurements are made utilizing orbiting beacon satellites, the periodic content fluctuations that might be expected during the passage of a travelling ionospheric disturbance (TID) are preferentially observed for certain ray paths only. The observational biases involved in such measurements are well documented by Georges and Hooke [1]. Since an acoustic wave or gravity wave propagating through the atmosphere causes periodic electron density variations both vertically and horizontally, a radio wave traversing the perturbed medium will be subject to some angular deviation. This angular deviation will depend upon the direction of the radio wave normal relative to the electron density perturbations. It might be expected then that if observations are made upon an orbiting satellite beacon the angular deviation imposed upon the incoming signal would be a function of the observers "look angle" and may be preferentially observed for certain ray paths, as are the periodic content fluctuations. Furthermore, if the angular deviations are large enough, a focusing and/or defocusing of the received signal could occur.

2. Experimental Results

In 1967 a series of field strength measurements was made on the 20 MHz transmission from the S66 (Explorer 22) orbiting satellite [2]. During these observations several records were obtained showing the characteristic signature of a travelling ionospheric disturbance, that is, marked changes in the Faraday rotation rate. On some of these records a marked enhancement of received signal strength occurred approximately coincident with the most marked change in rotation rate. One such record is shown in Fig. 1. It will be noted

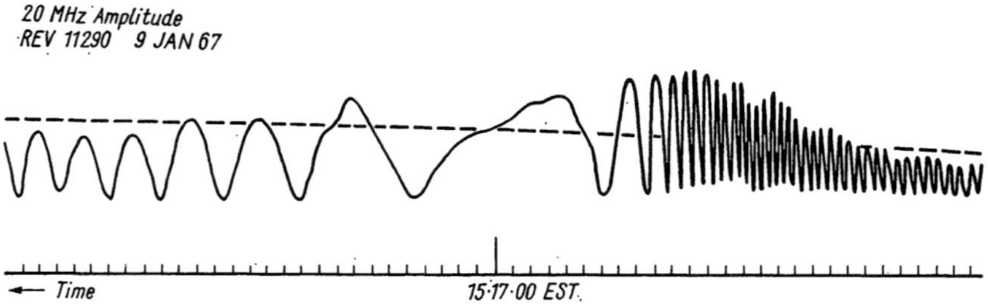

Fig. 1. A portion of the 20 MHz amplitude recording of Explorer 22, rev. 11290, 9 Jan 1967.
Dashed line represents expected amplitude.

that the record is smooth throughout with no sign of scintillation, thus ruling out interference or diffractive effects as an explanation for the enhancement.

3. Analytical Method

The electron density distribution in the ionosphere due to an atmospheric wave may, following Georges and Stephenson [3], be expressed as

$$N = N_0 (1 + \Delta) ,$$

where Δ is the perturbation due to the wave and

$$\Delta = \alpha \cos (\omega t + k_x + k_z) \exp \left[- \left(\frac{z - z_0}{H} \right)^2 \right]$$

where N_0 = unperturbed electron density, k_x, k_z = horizontal and vertical wave numbers of the perturbing wave, ω = angular frequency of perturbing wave, α = amplitude of perturbing wave, z_0 = height of maximum amplitude which falls off to α/e at a distance H above and below z_0.

The angular deviation imposed upon a radio wave traversing the medium can be expressed as [4]:

$$\theta(u) = - \frac{K}{\omega_0^2} \frac{d}{du} \int_0^v N \, dv$$

where u and v are coordinates orthogonal to and along the wave normal, ω_0 is the radio wave angular frequency and $K = 1.6 \times 10^3$ if mks units are used. To apply these expressions to the case of transmission from an orbiting satellite, we assume the wave normal for the atmospheric wave lies in the direction of the sub-satellite track, and transform the expression for Δ to a (u,v) coordinate system with origin at the observer and insert this in the expression for $\theta(u)$ giving for the perturbation part only

$$\theta(u) = \frac{K \sigma N_0}{\omega_0^2} \frac{d}{du} \int_{-\infty}^{+\infty} \cos [\omega t + k_x(v \cos e - u \sin e + z_0 \cot e) + k_z (u \cos e + v \sin e + z_0)] \exp\left[-\left(\frac{u \cos e + v \sin e}{H}\right)^2\right] dv ,$$

where e is the elevation angle of the ray path from observer to satellite.

This yields

$$\theta = -\frac{K \alpha N_0 k_x \pi^{1/2} H}{\omega_0^2 \sin^2 e} \exp\left[-\frac{H^2 (k_x \cot e + k_z)^2}{4}\right] \sin [\omega t + k_z z_0 + k_x z_0 \cot e - k_x (\operatorname{cosec} e) u] .$$

This expression may be evaluated using appropriate values for the parameters of the atmospheric wave, at a value of u equal to zero and any value of ωt between zero and 2π, to give the angular deviation as a function of elevation angle. As will be seen in § 4 the angular deviation changes sign several times as the elevation angle increases from 0 to 180°. There is therefore the possibility of alternate focusing and defocusing as the satellite moves over the observer. An effective focal length may be associated with the electron density variations and it may be shown that

$$\frac{1}{f} = \frac{d}{du} \theta(u) \qquad \text{or}$$

$$\frac{1}{f} = \frac{K \alpha N_0 k_x^2 \pi^{1/2} H}{\omega_0^2 \sin^3 e} \exp\left[-\frac{H^2 (k_x \cot e + k_z)^2}{4}\right] \cos [\omega t + k_z z_0 + k_x z_0 \cot e - k_x (\operatorname{cosec} e) u]$$

which expression may also be evaluated as a function of the elevation angle.

4. Results

The angular deviation, θ, imposed on the ray from the satellite has been evaluated for various atmospheric wave parameters. Fig. 2 shows θ as a function of e for an atmospheric wave with $\lambda_x = 200$ km, $\lambda_z = 100$ km, $z_0 = 200$ km and $H = 50$ km. Only the angular deviations for elevations between 120° and 170° are shown since outside this range the deviations are extremely small. Changing the values of the parameters over the range normally associated with medium scale travelling ionospheric disturbances does not materially affect the shape of the curve. The angular deviations always have a maximum and show the greatest rates of change in the range of elevation angles 150 to 165°. Thus for a northern hemisphere location observing a high inclination orbiting satellite beacon and an atmospheric wave propagating toward the equator, the greatest angular deviations occur when looking to the south at elevation angles in the region of 25–30°. This certainly seems to be borne out by experimental results.

Note that the angular deviation scale in Fig. 2 is in arbitrary units. The value of the deviation will depend of course upon the radio wave frequency

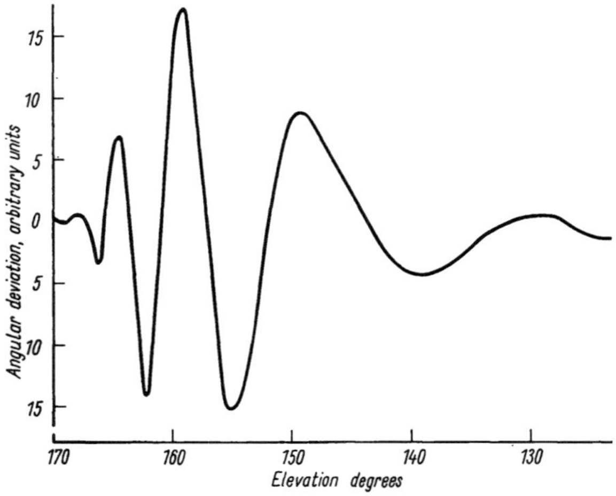

Fig. 2. Angular deviation versus elevation angle for an atmospheric wave with $\lambda_x = 200$ km, $\lambda_z = 100$ km, $z_0 = 200$ km and $H = 50$ km.

and on the values used for the parameters α and N_0. For a frequency of 20 MHz, a disturbance amplitude of 0.1 of ambient and an electron density at the disturbance height of 2×10^{11} m^{-3} the maximum deviation is approximately 2°.

The situation with regard to focusing is not quite so straightforward. As might be expected from the angular deviation changes seen in Fig. 2, the effective focal length changes markedly in magnitude and several times in sign during a satellite passage. Focusing can only occur when the effective focal length is positive and when the object distance (which may be taken as the distance from satellite to height of maximum disturbance) is greater than the focal length. At all other times some defocusing must be present. Effective focal lengths have been computed for an atmospheric wave with $\lambda_x = 100$ km, $\lambda_z = 100$ km, $z_0 = 200$ km and $H = 50$ km. The positive focal lengths only

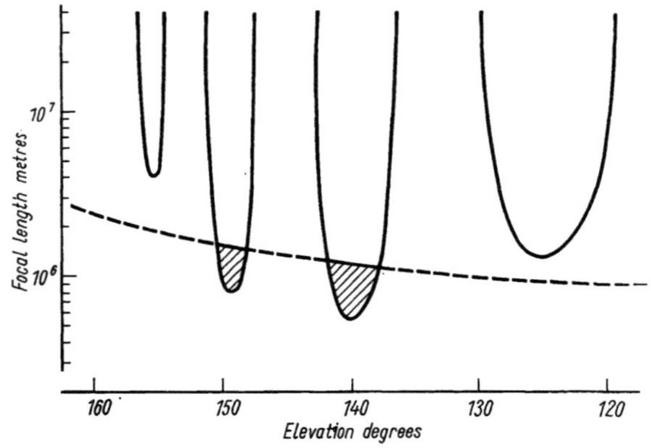

Fig. 3. Positive effective focal length versus elevation angle for an atmospheric wave of $\lambda_x = 100$ km, $\lambda_z = 100$ km, $z_0 = 200$ km and $H = 50$ km. Dashed line represents effective object distance.

are shown in Fig. 3 as a function of elevation angle. Again for elevation angles less than 120° the focal lengths are very large and not effective for the focusing mechanism. Shown also in Fig. 3 is a plot of object distance versus elevation angle. Focusing can occur only in the shaded region where object distance is greater than focal length, that is to say at elevation angles close to 140° or 150°. It is something of a happy coincidence that 140° is just the elevation angle corresponding to the signal enhancement seen in Fig. 1, and that the signal had not been acquired at elevation angle 150°.

It will be appreciated that the focal length curves can be moved up and down relative to the object distances curve simply by a suitable choice of the parameters α and N_0. It can therefore be expected that there would be times when no focusing could take place or times when focusing could occur at three or four elevation angles. Certainly in the experimental records the former can be seen, but in the authors' experience no more than two enhancements have been seen in one satellite passage.

5. Discussion

The analysis of § 3 is in two dimensions only and for a flat earth geometry. Nevertheless it serves to show that the effects arising from the refraction of a radio wave due to the presence of a travelling ionospheric disturbance will be most marked at certain preferential look angles. This phenomenon could possibly be exploited in determining the characteristics of TIDs from a single observing station. A small catalogue of theoretically derived angles of arrival for TIDs of varying parameters could be matched with experimentally determined angles. It serves also to confirm that combining data from different observing stations with different look angles should be approached with caution.

The refraction effect is of course inversely proportional to the square of the radio wave frequency and unfortunately low frequency satellite beacons are at present very rare. Present techniques however permit the measurement of changes in angle of arrival as small as 0.01° at 137 MHz and this is currently being successfully exploited at one of the author's (GFL) institutions.

Acknowledgments

The work described was carried out with the assistance of a grant No. 9511-58 from the Defence Research Board of Canada to one of the authors, GFL.

The authors wish to acknowledge helpful technical discussions with Dr. J. Ehrman and Dr. J. Tunaley of the University of Western Ontario.

References

[1] T. M. Georges and W. H. Hooke, J. Geophys. Res. **75**, 6295 (1970).
[2] J. A. D. Holbrook and G. F. Lyon, J. Atmos. Terr. Phys. **31**, 71 (1969).
[3] T. M. Georges and J. J. Stephenson, Radio Sci. **4**, 679 (1969).
[4] R. M. Turnbull and P. A. Forsyth, Canad. J. Phys. **43**, 800 (1965).

Space Research XII – Akademie-Verlag, Berlin 1972

EVIDENCE FOR WAVES AND WINDS IN THE IONOSPHERIC F REGION

K. Davies and J. E. Jones

National Oceanic and Atmospheric Administration, Enivronmental Research Laboratories, Boulder, Col., USA

A multifrequency spaced transmitter arrangement was used near Oklahoma City in the spring of 1970 to measure ionospheric motions in three dimensions. Characteristic disturbances with periods near 7 min were sometimes observed. These waves would be expected to propagate as acoustic waves in the F region but instead they exhibit the properties of gravity waves. These observations can be reconciled by postulating horizontal winds in the F region (near 180 km) over 70 m s^{-1}.

The objective of this paper is to present evidence for the presence in the F region of neutral winds, obtained from observations of traveling ionospheric disturbances. Based on the work of Hines [1], there is general agreement that many features of traveling ionospheric disturbances can be explained in terms of propagating atmospheric waves. The relative merits of the gravity-wave hypothesis versus the drift hypothesis have been examined by Kent and Wright [2].

The observations reported here were part of a study of acoustic waves generated by severe thunderstorms [3, 4]. Acoustic waves are oscillatory motions of the neutral atmosphere with periods less than the acoustic cut-off period τ_A. Theory shows that the phase speed of acoustic waves exceeds the sound speed, in contrast to the phase speed of gravity waves (with periods longer than the buoyancy or Brunt–Väisälä period τ_B) which is usually considerably less than the speed of sound. The variation of τ_B with height, including effects of temperature gradients, for the 1962 US Standard Atmosphere is shown in Fig. 1.

Hines and Reddy [5] and Booker and Bretherton [6] showed that the intrinsic period τ_i of an atmospheric wave as seen by an observer moving with a horizontal wind component U parallel to the horizontal trace velocity of the wave is related to the period τ observed on the ground by:

$$\tau_i = \frac{V_H}{V_H - U}\tau \tag{1}$$

where V_H is the horizontal trace speed observed from the ground. Thus a wave which appears to be an acoustic wave (i.e. $\tau < \tau_A$) may actually be a gravity wave (i.e. $\tau_i > \tau_B$) in the windy ionosphere. Since V_H and τ can be measured by ground-based radio sounders, the determination of τ_i can yield U or vice versa.

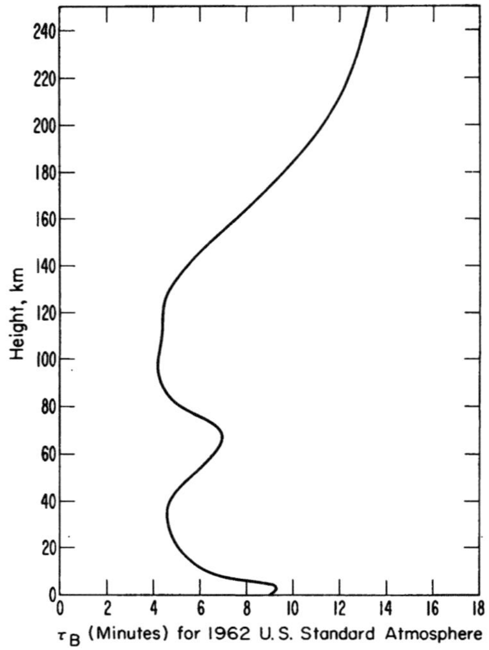

Fig. 1. Height variation of the buoyancy period τ_B for the 1962 US Standard Atmosphere

The measurements to be described in this paper were made with the high-frequency Doppler radio technique (7, Chap. 9). The frequency fluctuations of ionospheric radio echoes were continuously monitored at Norman, Oklahoma,

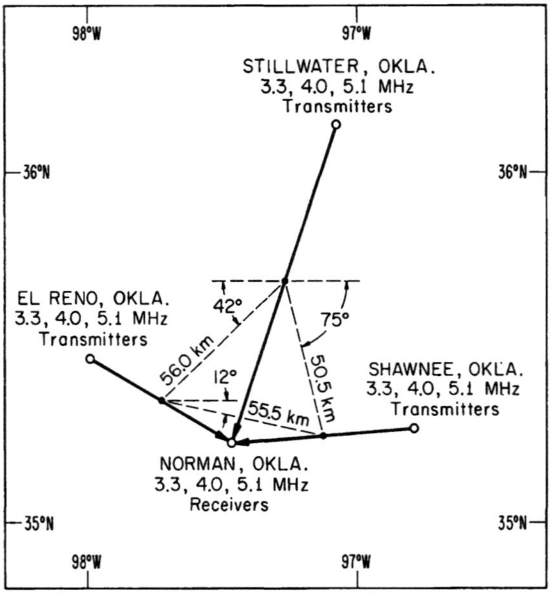

Fig. 2. Location of transmitters, receivers and ionospheric reflection points.

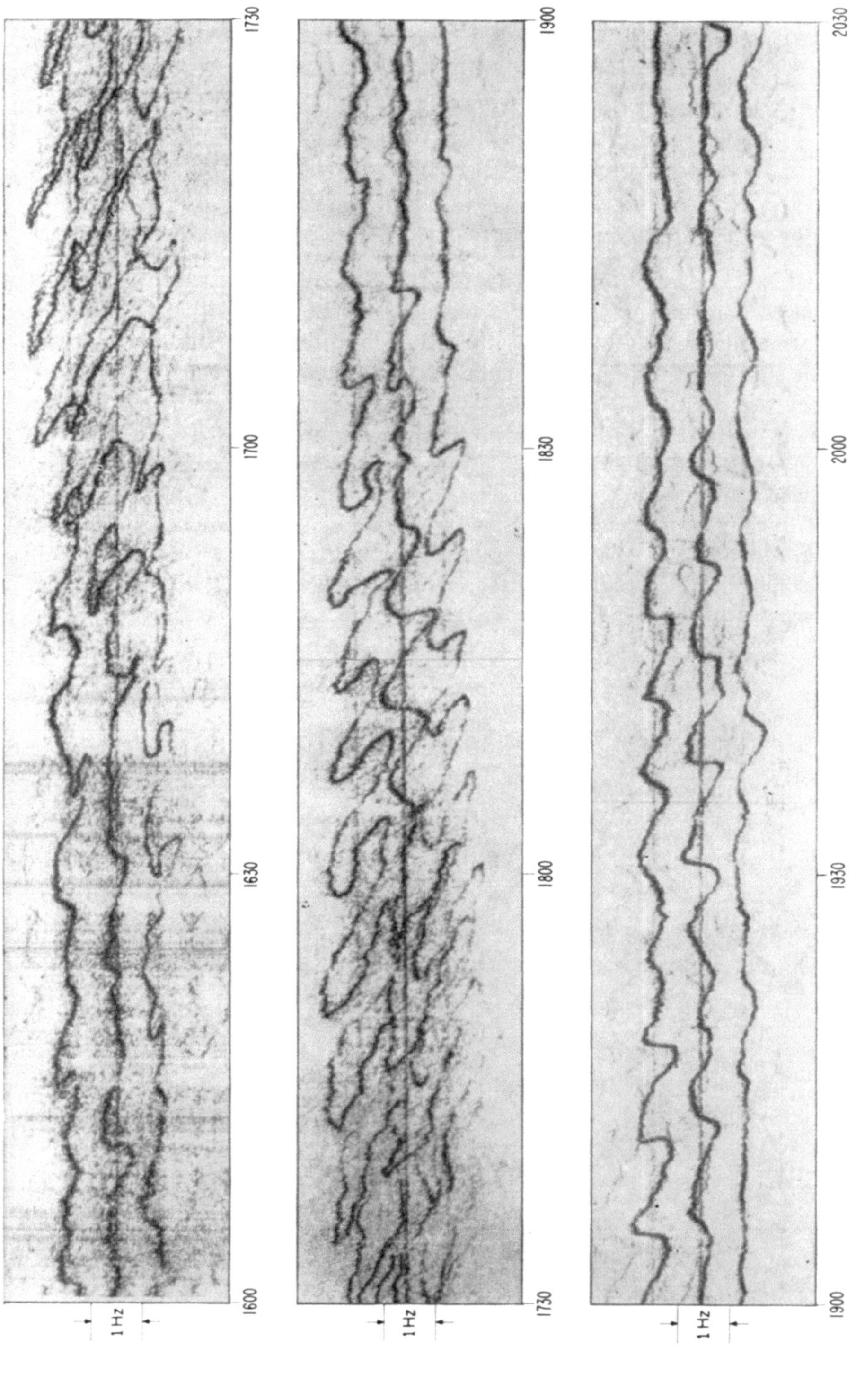

Fig. 3. Doppler records from spaced transmitters showing S-curves, 25 April 1970, 1600–2030 UT. Upper trace, Shawnee, Okla.; middle, El Reno, Okla.; lower, Stillwater, Okla.; all to Norman, Okla. on 5.1 MHz.

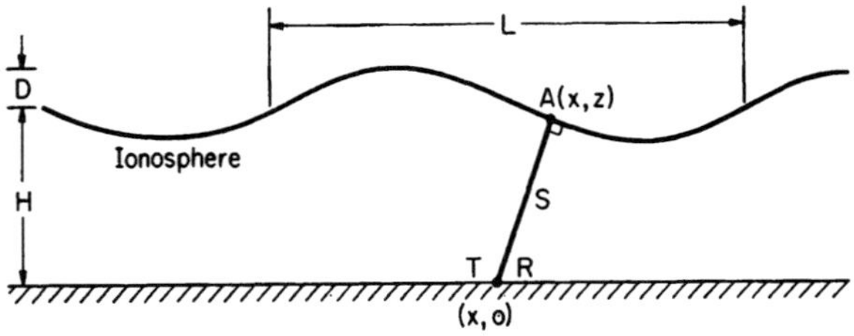

Fig. 4. Ray path geometry for reflection from a horizontally moving sinusoidal reflector.

from spaced transmitters on three frequencies as shown in Fig. 2. The period of the atmospheric wave is found by auto-correlating the Doppler records. From the spacing of the reflection points and the time displacements on a given frequency the horizontal trace velocity can be determined on the assumption of a plane wavefront [8]. Similarly, from a knowledge of the vertical displacements obtained by ionogram analysis, and time displacements on different frequencies, the vertical trace velocity can be determined.

The traveling ionospheric disturbance of interest is shown in Fig. 3. The S-shaped Doppler traces can be explained in terms of mirror-like reflection from a sinusoidally corrugated surface moving with constant horizontal velocity V as depicted in Fig. 4. Provided that the radius of curvature of the

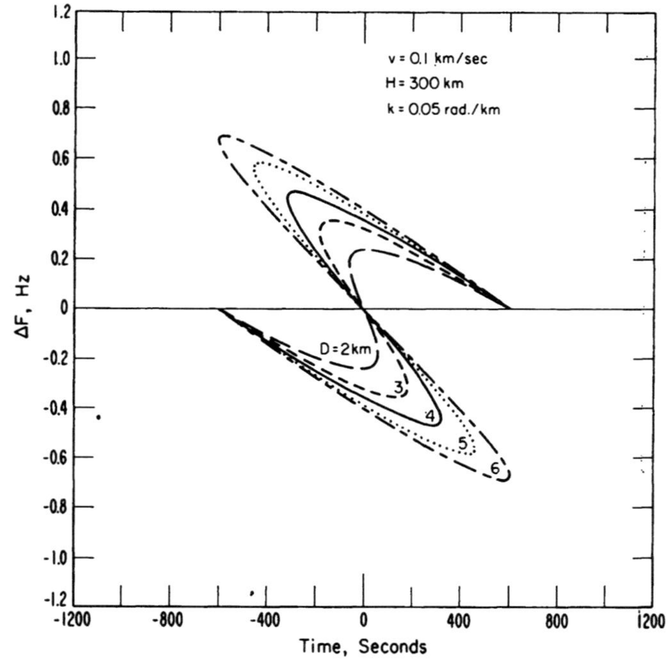

Fig. 5. Calculated radio frequency variation with time for reflection from a horizontally moving sinusoidal reflector. $f = 5$ MHz.

reflector is less than its height, three simultaneous perpendicular reflections can occur [7, p. 250]. If the amplitude of the surface is D and the wave number is k, the radio Doppler shift Δf for a carrier frequency f is given by

$$\Delta f = \frac{2\,f\,D\,k\,V}{c} \frac{\sin\theta}{(1 + k^2\,D^2 \sin^2\theta)^{1/2}} \tag{2}$$

where c is the free space velocity of radio waves and $\theta = k\,x$ is the phase of the reflection point on the corrugated surface. For a part of the cycle there are three values of θ which correspond to the three perpendicular reflections. Some calculated $\Delta f(t)$ curves for sample ionospheric conditions are shown in Fig. 5. In deriving Eq. (2) the transmitter and receiver are assumed to be collocated. This is not true in practice but the differences are immaterial to the present discussion.

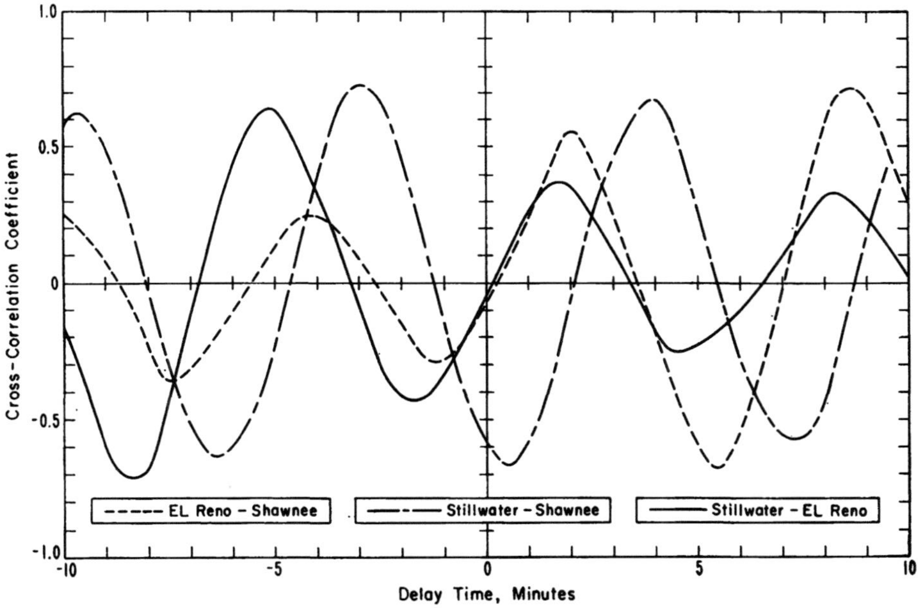

Fig. 6. Cross-correlograms of the "equivalent square-waves" for the S-curves of 25 April 1970, 1755–1835 UT, on 5.1 MHz.

In reality, the ionospheric reflecting surface is not sinusoidal. Both the period and shape vary somewhat from one cycle to the next and this introduces difficulties in measurement of period and time displacements. The zero frequency crossings are essentially independent of the amplitude of the reflecting surface, so we have used these zero crossings to determine the average period and the average time displacements. During the approximate time interval 1800–1845 UT, when the S-curves are most clearly defined, the average period is about 6.8 min. Cross-correlograms of the zero crossings of the signals from El Reno (E), Shawnee (S), and Stillwater (U) are shown in Fig. 6. Because of the difficulty of matching corresponding cycles, we consider the following possible time displacements (in seconds):

$$T_{ES} = +120 \quad \text{or} \quad -250$$
$$T_{EU} = +300 \quad \text{or} \quad -100$$
$$T_{SU} = +180 \quad \text{or} \quad -240$$

Acceptable combinations of time displacements must satisfy the relationship

$$T_{ES} + T_{SU} = T_{EU} \tag{3}$$

to within a reasonable accuracy. This criterion reduces the possibilities to three. Of these three, one corresponds to the highest peaks of the correlograms

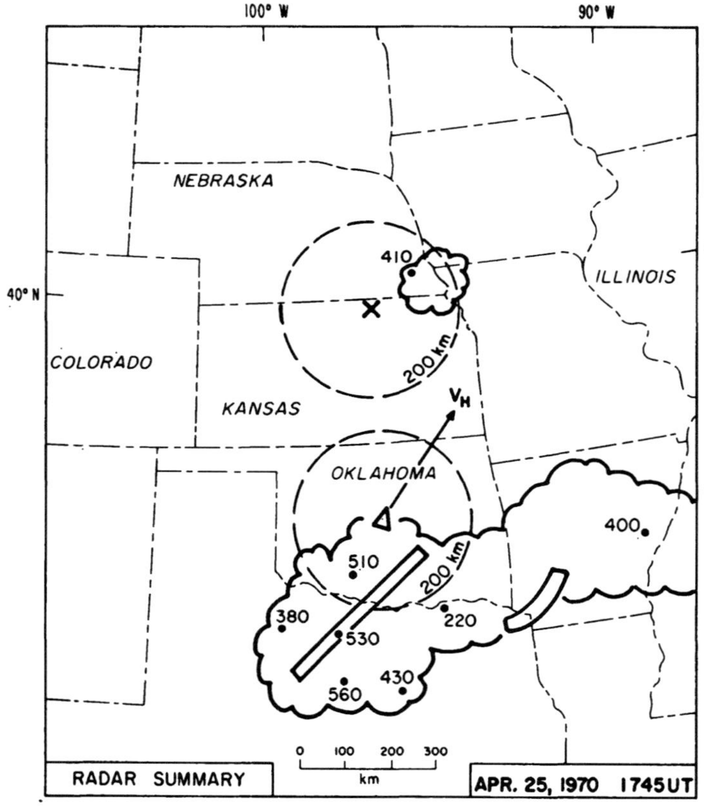

Fig. 7. Radar summary of weather over Oklahoma on 25 April 1970, 1745 UT. The dots represent thunderstorm cells and the numbers represent the heights of cloud tops in hundreds of feet. $V_H = 180$ m s^{-1}, $\sigma = 35°$ at 1800–1845 UT.

and also satisfies (3) best. This one is $T_{ES} = 120$, $T_{SU} = 180$ and $T_{EU} = 300$ s, giving V_H a magnitude of 180 m s^{-1} and a horizontal direction $\sigma = 35°$ E of north. Additional evidence for this value is provided by the weather radar summary shown in Fig. 7. Three of the S-curve disturbances were observed in 1970, all of them when large thunderstorms were present, meeting the criteria for producing ionospheric effects established by Baker and Davies [9] and by Davies and Jones [3]. The direction of V_H strongly suggests that the disturbance of 25 April 1970 originated in the severe thunderstorm shown on the radar summary.

The important feature of these data is that the trace speeds are very much less than the sound speed C in the F region. To compare these data quantitatively with atmospheric wave theory we need the height of reflection of the 5.1 MHz radio wave. Unfortunately, ionograms were unavailable at Oklahoma City so we have analyzed ionograms from White Sands, New Mexico, about 800 km to the west of the area of observation. The heights of reflection at 1800 and 1900 UT were 174 km and 183 km respectively, i.e. an average of 179 km. At this altitude, Fig. 1 shows that the buoyancy period τ_B is 9.5 min. Thus the period of 6.8 min indicates that the disturbance should be an acoustic wave, but the observations indicate otherwise. Three possibilities are suggested: (i) Assuming that $(dT_M/dz)/T_M = 0.0055\ \text{km}^{-1}$, which is the same as for the 1962 Standard Atmosphere at 179 km, the molecular temperature at this height and on this day was less than 500 °K, compared with 1274 °K given by the Standard Atmosphere. (ii) There was a horizontal wind component U opposite to the direction of horizontal phase propagation of an acoustic wave. For an acoustic wave. the phase speed exceeds the sound speed of 720 m s^{-1}, so that the wind speed U must exceed 640 m s^{-1}. This is an unlikely value for a wind. (iii) There was a horizontal wind component U in the direction of horizontal phase propagation which lengthened the intrinsic period τ_i to exceed the buoyancy period τ_B. From Eq. (1) we have, setting $V_H = 180$ m s^{-1} and $\tau = 6.8$ min, $U > 51$ m s^{-1}. The wind speed given in possibility (iii) is certainly a reasonable value and it is worth trying to pin down the value a little further. To do this we estimate the vertical trace speed V_v. From the nature of the Doppler traces it was impossible to determine V_v at exactly the same time as the determination of V_H. During the approximate time interval 1700–1800 UT the time displacements, together with the differences in reflection height obtained from the White Sands ionograms, show two possibilities: $V_v = 158$ m s^{-1} upward and 247 m s^{-1} downward. These eliminate possibility (ii), since they are much less than C. As suggested earlier, the source is probably the thunderstorm (Fig. 7) so that the energy propagated upward, whereas for a gravity wave the phase propagates downward. This points to the value of -247 m s^{-1} for V_v. It is known that the phase of gravity waves propagates near the asymptotic limit Φ_c with the vertical. Putting $\tan \Phi_c = V_H/V_v$ gives $\Phi_c = 126°$. But from gravity wave theory $\sin \Phi_c = \tau_B/\tau_i$ so that $\tau_i \approx 11.8$ min. Hence if the 1962 Standard Atmosphere is correct and $\tau_B = 9.5$ min, then substitution of $\tau_i = 11.8$ min in (1) gives $U = 76$ m s^{-1}. If, on the other hand, we assume no wind, then $\tau_B = \tau \sin \Phi_c = 6.8 \sin 126° \approx 5.5$ min, which corresponds to a molecular temperature of about 280 °K.

Thus we conclude that either (a) the molecular temperature at a height of 179 km is much less than the value given by the Standard Atmosphere, (b) there is a horizontal wind in excess of about 70 m s^{-1}, or possibly (c) a combination of these. We favor (b) because other measurements (unpublished) of horizontal trace speeds of acoustic waves with periods in the 3–5 min range are consistent with the temperature given by the Standard Atmosphere.

References

[1] C. O. HINES, Canad. J. Phys. **38**, 1441 (1960).
[2] G. S. KENT and R. W. H. WRIGHT, J. Atmos. Terr. Phys. **30**, 657 (1968).
[3] K. DAVIES and J. E. JONES, J. Atmos. Sci. **28**, 254 (1971).
[4] T. M. GEORGES, J. Atmos. Terr. Phys. **30**, 735 (1968).
[5] C. O. HINES and C. A. REDDY, J. Geophys. Res. **72**, 1015 (1967).
[6] J. R. BOOKER and F. P. BRETHERTON, J. Fluid Mech. **27**, 513 (1967).
[7] K. DAVIES, Ionospheric Radio Waves, Ginn-Blaisdell, Waltham, Mass. 1970.
[8] K. DAVIES and J. E. JONES, J. Atmos. Terr. Phys. **33**, 39 (1971).
[9] D. M. BAKER and K. DAVIES, J. Atmos. Terr. Phys. **31**, 1345 (1969).

Space Research XII — Akademie-Verlag, Berlin 1972

MOMENTUM DEPOSITION BY ATMOSPHERIC WAVES, AND ITS EFFECTS ON THERMOSPHERIC CIRCULATION

C. O. Hines

Department of Physics, University of Toronto, Toronto, Canada

Representative calculations are made for the amount of momentum deposited by atmospheric waves at thermospheric heights, in various circumstances. It is shown that this source of momentum is likely to be a significant one in the circumstances treated. This conclusion may be of importance to an understanding of the winter-time D region at high latitudes, to an understanding of the so-called 'super-rotation' of the upper atmosphere, and to an understanding of the wind fields and ionization anomalies that occur following magnetic substorms.

Atmospheric waves transport momentum. When they dissipate, they transfer that momentum to the background flow of the atmosphere. Applications of this latter fact have been discussed, to date, primarily with respect to the deposition of momentum at 'critical levels' [1–3], but in fact the process of deposition occurs whenever the waves dissipate, for whatever cause. It is the purpose of this note to draw attention to the possible importance of momentum transfer from waves to the background flow at high altitudes, where turbulence, viscosity, thermal conduction, and ohmic losses (or 'ion drag') act strongly even in the absence of critical layers.

A simplified analysis is appropriate for the purpose. It will be based on the elementary (isothermal) model and results of Hines [4] for plane internal gravity waves of fairly long period and short vertical wavelength. Waves propagating in the $+x$ azimuth of the $x\,z$ plane transfer upwards (i.e. in the direction of increasing z) a flux of x-directed momentum whose time mean per unit horizontal area is given by

$$F_{xz} = \varrho_0 \overline{u\,w} \tag{1}$$

to lowest non-vanishing order in the wave amplitude. Here ϱ_0 is the unperturbed gas density, u and w are the x and z components of the oscillatory velocity, and the overbar indicates a cycle average. The atmosphere is taken to be perturbed from a static state by the wave; more involved relations, that add nothing to the present discussion, must be employed when the background flow is non-zero (as it must become, in circumstances to be discussed).

In the case considered, u and w are 'in phase' if the energy source lies at lower levels, while the amplitude ratio is $u/w = \tau/\tau_g$ where τ is the wave period and τ_g is the (isothermal) Brunt–Väisälä period. Thus,

$$F_{xz} = \varrho_0\, u_0^2\, \tau_g/2\,\tau \tag{2}$$

where u_0 is the amplitude of the sinusoidally varying u.

In the absence of dissipation, $\varrho_0 u_0^2$ is found to be a constant for plane waves, and F_{xz} is therefore independent of height (τ_g having been assumed a constant by virtue of the isothermal assumption). In the presence of dissipation, formulae of much greater complexity come to be relevant, but a crude assessment of the momentum transfer can be made without recourse to them. If, for example, dissipation at some height became so severe that the wave were effectively destroyed over a small height range H, then the horizontal momentum transferred to the atmosphere over that height range, per unit horizontal area and unit time, would be just equal to the incoming flux density F_{xz}. It would lead to an acceleration of the background flow in the x direction at a rate $\dot{U}$ given by

$$(\varrho_0 H)\,\dot{U} = F_{xz} \tag{3}$$

where ϱ_0 and F_{xz} are now to be measured just below the region of dissipation. The combination of (1) and (2) leads immediately to

$$\dot{U} = u_0^2\,\tau_g/2\,\tau\,H\,, \tag{4}$$

with u_0 measured just below the region of dissipation.

The same result applies more generally if H is simply taken as the 'scale height' for the diminution of $\varrho_0 u_0^2$ in the presence of dissipation (without significant reflection) provided one ignores additional terms in the momentum flux, introduced by the dissipative process. If these terms acted to diminish $\dot{U}$ at the height in question, however, they would equally act to increase $\dot{U}$ at some greater height since the total flux of x momentum must ultimately be transferred; and, at the greater height, the resultant $\dot{U}$ would be enhanced because of the diminished gas density there. Thus, (4) provides if anything an underestimate of the accelerations $\dot{U}$ that must be produced at or above the level where dissipation first becomes severe. A convenient assessment can be made at levels where u_0 attains its maximum value — where the usual exponential growth of wave amplitude is just offset by the dissipation process — for there the scale height H that describes the diminution of $\varrho_0 u_0^2$ is just the scale height for the diminution of ϱ_0, namely, the atmospheric (density) scale height.

One rather dramatic application of (4) can be made with the data of Theon, Nordberg and Smith [5], which pertain to temperature variations observed in the mesosphere and lower thermosphere at Pt. Barrow, Alaska, in winter (February 1967). These variations had amplitudes of ± 50 °K, on a mean background temperature of 220 °K, at a height of 85 km or so. They were attributed to internal atmospheric gravity waves with $\tau \simeq 200$ minutes, and would then have been in association with oscillatory quasi-horizontal winds of amplitude $u_0 \simeq 100$ m s^{-1}. With $\tau_g \simeq 5$ minutes, and $H \simeq 6.5$ km, (4) yields an acceleration of 2×10^{-2} m s^{-2} or about 70 m s^{-1} hr^{-1}.

The waves in question might not have reached maximum amplitude in the vertical — the observations terminate in the 90–100 km height range — so this acceleration might constitute a slight overestimate locally, but an underestimate for the acceleration of some higher level. The result must be qualified, too, by virtue of the fact that the relevant waves need not all be propagating in the same azimuthal direction: the momentum deposited by one wave might be partially offset by momentum deposited by another, though the temporal coherence of the temperature variations argues against severe cancellation. And unless the cancellation is severe, the inferred acceleration rate must be recognized as a major contributor to the dynamical balance of the region: it can be offset by the Coriolis force (associated with the earth's rotation) only if the wind it produces is of the order 100 m s^{-1}, for example. Changes in the rate of momentum deposition would likewise produce changes in the general background flow, of a magnitude that could be of great importance to an understanding of anomalous behaviour such as occurs at high latitudes in the wintertime D region.

The waves just discussed probably represent an extreme at those levels. Typical gravity-wave winds are evidently substantially weaker at middle latitudes [6], and much weaker at high latitudes in summer [5]. Moreover, the mid-latitude spectrum appears to be strongly suppressed at heights above 110 km or so [6], very likely in part as a consequence of critical layer interaction with the tides near that level [7] if not simply by non-critical dissipation by turbulence. Regardless of the precise mechanism of loss, the waves will of course deposit their momentum; but a considerable amount of cancellation might be expected in this case, when all azimuths of gravity-wave propagation are taken into account.

On the other hand, the tides themselves are atmospheric waves, the vertically propagating modes being even of the gravity-wave family and having properties closely analogous to those of the plane gravity waves of elementary theory. They carry a great deal of momentum upward, and must deposit it at their levels of dissipation, typically 100–115 km for the principal propagating diurnal mode (or 85–115 km, at tropical latitudes) and somewhat higher for the principal semidiurnal modes. Indeed, the semidiurnal modes tend to retain significant winds of nearly constant amplitude up to the base of the exosphere [8], and may therefore contribute important momentum throughout some 500 km or so of height span. They might very well contribute to the so-called 'superrotation' of the thermosphere [9] and to its variation with height [10], though a proper evaluation of their contribution would require more sophisticated calculations than those attempted here.

To illustrate the possible contribution of the tides, we may take advantage of curves presented by Lindzen [11] for the dominant propagating diurnal mode. In low latitudes at 85 km height, these show a vertical amplitude $w_0 \simeq 0.2$ m s^{-1} almost 'in phase' with the westward component of oscillation, of amplitude $u_0 \simeq 15$ m s^{-1}. These would combine to produce a mean vertical flux density of westward momentum equal to $\varrho_0 \times (1.5\ m^2\ s^{-2})$, and would lead to an acceleration of $1.5\ m^2\ s^{-2}\ H^{-1} \simeq 3 \times 10^{-4}\ m\ s^{-2} \simeq 1\ m\ s^{-1}\ hr^{-1}$ if the amplitude remained constant with increasing height (as it might, due to instability, as discussed by Lindzen [12]). This acceleration is perhaps not so impressive in appearance as that derived previously, but it acts persistently. It would produce an acceleration of 20 m s^{-1} or so per day, day after day.

And at the low latitudes here under discussion, the Coriolis force is not available to offset it.

At higher latitudes, Lindzen's [11] curves show the vertical velocity reversing in phase relative to the westward velocity, and the deposited momentum would therefore be eastward. (The vertical velocity is nearly in phase quadrature with the meridional oscillation, so little mean meridional momentum would be transported.) The upward flux density of eastward momentum at 25°–35° latitude, 85 km altitude, is about equal to that just obtained for westward momentum at low latitude, but in this case dissipation is not severe until heights of 100 km or so are reached. The resultant acceleration at those heights will be correspondingly greater, 10 m s^{-1} hr^{-1} or more.

Apart from the tides, the strongest waves in the atmosphere above 120 km appear to be those launched near that level by auroral currents during magnetic substorms. These are observed to propagate equatorwards, and may be expected to deposit momentum that is similarly directed. The waves described by Testud [13] may be taken for a case study. Their amplitude is relatively independent of height, over the height range 250–425 km, with oscillatory (neutral-gas) winds whose speeds in the horizontal have an amplitude $u_0 \simeq 100\,\text{m s}^{-1}$, with $\tau \simeq 10\,\tau_g$. With $H \simeq 50$ km, Eq. (4) now leads to an acceleration $\dot{U} \simeq 10^{-2}\,\text{m s}^{-2} \simeq 30$–$40\,\text{m s}^{-1}\,\text{hr}^{-1}$. The waves persist for a few hours, which suggests accelerations to a speed well over 100 m s^{-1} in the end, though the Coriolis force would increase in step and gradually deflect the resultant wind into a westward direction. Before such deflection, the induced equatorwards motion would push F region ionization obliquely upwards along the geomagnetic field lines, resulting in a general deformation of the F layer in addition to the oscillatory deformation induced by the oscillatory wind components.

Observations reported by Testud and by Reddy at this symposium include the development of an excess (over normal) equatorwards flow in the F region, with magnitude of the order 50 m s^{-1}, in association with the oscillatory motions of storm-induced waves. They also show the disappearance of that flow after a few hours. It seems reasonable to suggest that the onset of the excess flow results from the deposition of momentum by the wave system as discussed here. The disappearance of the flow might result from the deposition of oppositely directed momentum from a wave arriving from the opposite hemisphere, but is more likely associated with the deflection caused by the Coriolis force (the east–west component of wind being unobservable by the technique employed) or simply by a return to pre-storm conditions.

References

[1] R. S. LINDZEN, Acoustic-Gravity Waves in the Atmosphere, US Government Printing Office, Washington, D.C. 1968 (p. 231).
[2] R. S. LINDZEN and J. R. HOLTON, J. Atmos. Sci. **25**, 1095 (1968).
[3] C. O. HINES, J. Geophys. Res. **75**, 5956 (1970).
[4] C. O. HINES, Canad. J. Phys. **38**, 1441 (1960).

[5] J. S. THEON, W. NORDBERG and W. S. SMITH, Aeronomy Rep. No. 32, Aeronomy Lab., Univ. of Illinois 1969 (p. 18).
[6] A. KOCHANSKI, J. Geophys. Res. **69**, 3651 (1964).
[7] C. O. HINES, J. Atmos. Terr. Phys. **30**, 837 (1968).
[8] R. S. LINDZEN, Geophys. Fluid Dynamics **1**, 303 (1970).
[9] D. G. KING-HELE, Space Research X, 537 (1970).
[10] D. G. KING-HELE, Space Research XII, 847 (1972).
[11] R. S. LINDZEN, Quart. J. Roy. Met. Soc. **93**, 18 (1967).
[12] R. S. LINDZEN, Proc. Roy. Soc. A **303**, 299 (1968).
[13] J. TESTUD, J. Atmos. Terr. Phys. **32**, 1793 (1970).

Space Research XII – Akademie-Verlag, Berlin 1972

INTERACTION BETWEEN GRAVITY WAVES AND IONIZATION IN THE IONOSPHERIC F REGION

J. TESTUD

C.N.E.T. – R.S.R., Issy les Moulineaux, France

During several magnetic storms, large amplitude oscillations of the electronic temperature and density, and of the ionic temperature and velocity, have been observed with the incoherent scatter sounder of Saint-Santin-Nançay. These oscillations have been interpreted as due to gravity waves.

In order to understand more completely these experimental results, this paper presents a theoretical model of the interaction between gravity wave and ionization, in which the diffusion processes and the plasma temperature perturbation are taken into account. A comparison of theoretical results with observations shows that this model permits a coherent interpretation of all the phenomena.

1. Introduction

For a long time, large scale ionospheric disturbances travelling equatorwards with sonic velocity, have been observed to occur following magnetic storms [1]. The role of the neutral atmospheric waves in inducing such large scale disturbances in the ionization seems to be now well accepted.

The incoherent scatter technique has recently emerged as a powerful tool in the study of such disturbances because simultaneous observations in several parameters like electron and ion temperature T_e, T_i, ionic velocity $\boldsymbol{V}_i$, electron density N, allow a complete description of the phenomena. Thus in the F2 region of the ionosphere, several observations made with the incoherent scatter sounder of Saint-Santin-Nançay (France) have revealed the existence, during magnetic storms, of large amplitude oscillations of the four parameters measured: – N, T_e, T_i, V_i which is ion velocity component parallel to the earth's magnetic field $\boldsymbol{B}$. Since two of these parameters, T_i and V_i are very closely coupled respectively to the neutral temperature T_n and to U_n, the meridional component of the neutral velocity, these oscillations have been interpreted as due to atmospheric gravity waves. In fact, this interpretation could be verified in much detail through the study [2] of: (i) The downward phase propagation of V_i and T_i oscillations observed at different heights: an example is shown in Figs. 1, 2. (ii) The influence of the wave on the electron density: in Fig. 3, the oscillation of N_m (electron density at the F2 peak), and V_i are in opposite phase. This has to be compared with the theoretical result of Hooke [3], predicting that the oscillations of U_n and N are in opposite phase near the maximum of the F2 peak. (iii) The horizontal propagation deduced from the

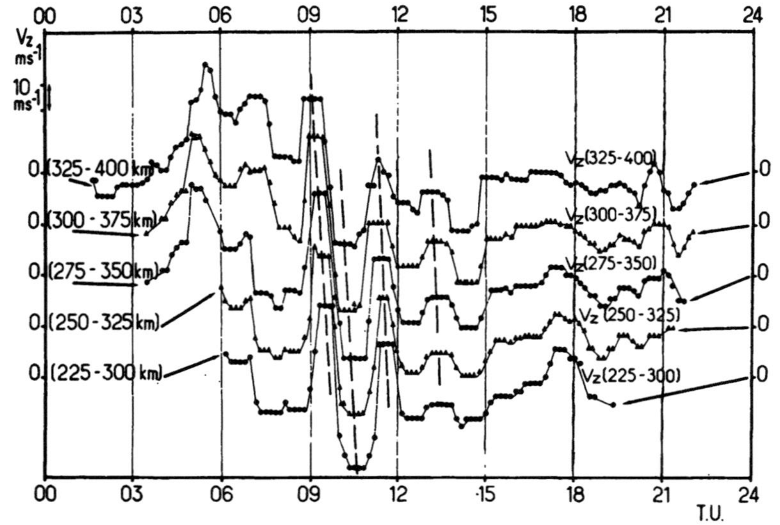

Fig. 1. Diurnal variation of $V_z = V_i \sin I$, for different altitude ranges, as observed at Saint Santin (44.7° N 2.2° E) on 13 September 1967. Two hour period oscillations are clearly visible between 0800 and 1400 UT. Notice the phase shift between upper and lower curves.

ionosonde data of the European network of observatories: an equatorward propagation from the auroral region was found in all cases. However, in the previous works, it was assumed that δT_i was nearly equal to δT_n and δV_i to δU (δU: projection of δU_n over $\boldsymbol{B}$). The aim of this article is to see how far such an assumption is valid.

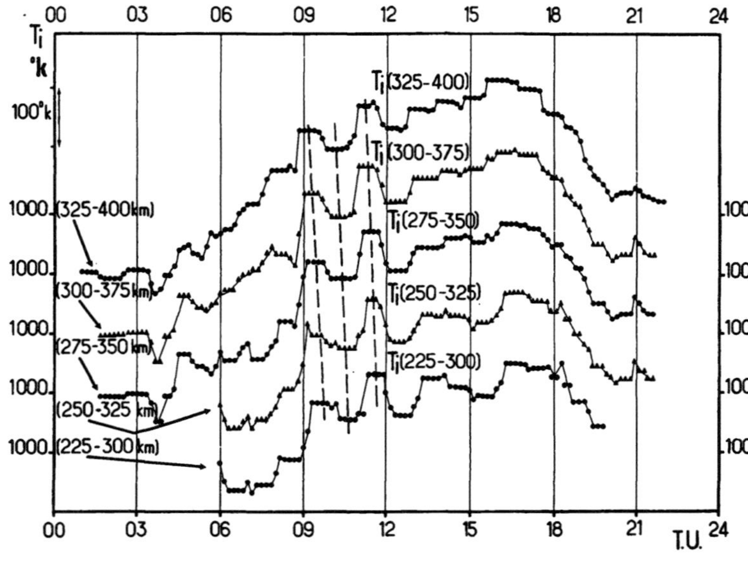

Fig. 2. Same result as in Fig. 1, but for ion temperature T_i.

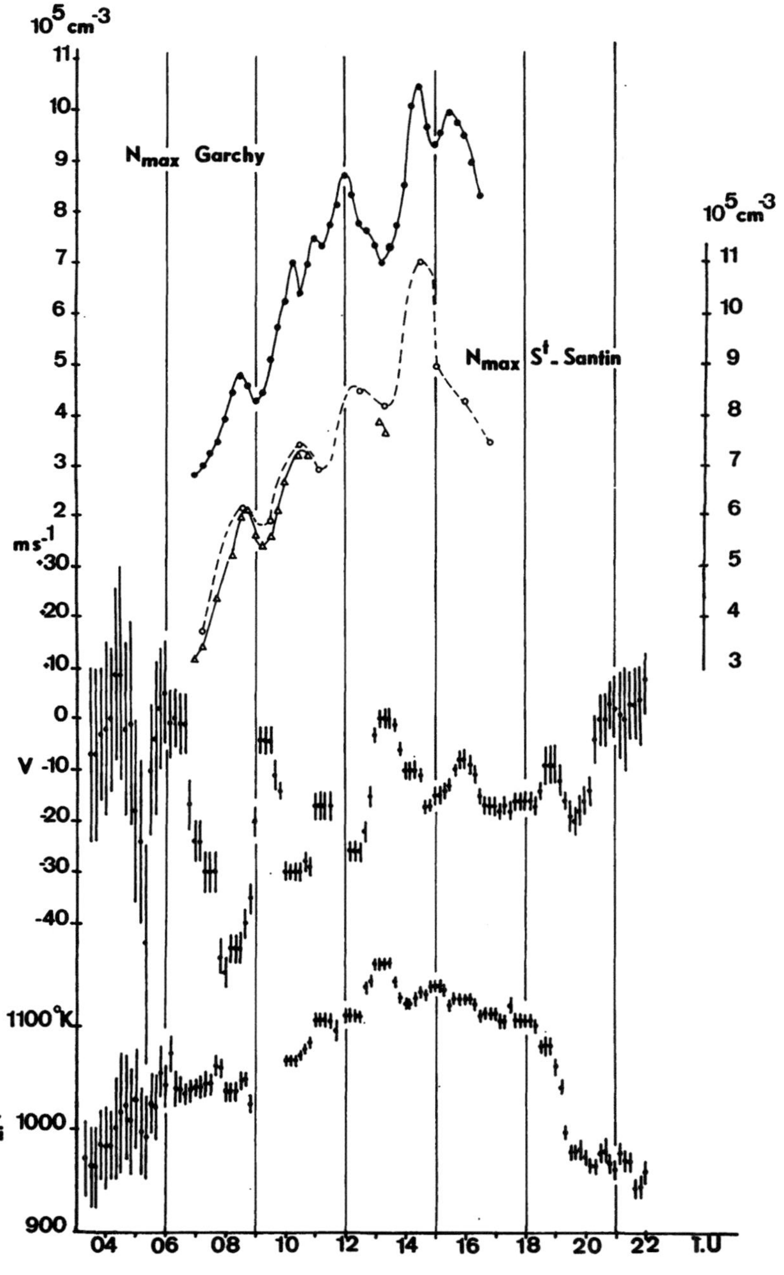

Fig. 3. Diurnal variations observed at Saint-Santin on 3 March 1968 of ionic temperature T_i, ionic velocity $V = V_i \sin I$, electron density at F2 peak deduced from incoherent scatter measurements (circles) and from ionosonde foF2 measurements (triangles). The upper curve corresponds to peak electron density measured from Garchy ionosonde (47.3° N 3.5° E).

More precisely we try to establish a computational model of the interaction between a gravity wave and the F2 layer ionization. In this model, the diffusion processes are taken into account in solving the continuity equation of the ionization. Such a computation is described in [4]. Here we try to improve the previous model by introducing a more realistic influence of the ion and electron temperature perturbations.

2. The Basic Equations

In the F2 region, the continuity equation of the ionization can be written in the following form, neglecting the horizontal gradients:

$$\frac{\delta N}{\delta t} = p - \beta N - \frac{d}{dz} \Phi , \tag{1}$$

where N is electron number density, p production rate, β linear recombination coefficient, Φ vertical ionization flux: $\Phi = N V_{\mathrm{i}z}$, $V_{\mathrm{i}z}$ being the vertical ion velocity.

The vertical velocity $V_{\mathrm{i}z}$ is due to three effects: drag by meridional horizontal wind, ambipolar diffusion along $\boldsymbol{B}$, upward component of the drift perpendicular to $\boldsymbol{B}$ due to an east–west electric field [5]:

$$V_{\mathrm{i}z} = W_{\mathrm{n}} + W_{\mathrm{d}} + W_{\mathrm{E}}$$

$$\left.\begin{aligned} W_{\mathrm{n}} &= U_{\mathrm{n}} \sin I \cos I \\ W_{\mathrm{d}} &= -D\left(\frac{T_{\mathrm{n}}}{2 H_{\mathrm{O}}} + \frac{1}{2N}\frac{d}{dz} N (T_{\mathrm{e}} + T_{\mathrm{i}})\right) \\ W_{\mathrm{E}} &= (E/B) \cos I , \end{aligned}\right\} \tag{2}$$

where I is the inclination of $\boldsymbol{B}$, E algebraic value of east–west electric field, H_{O} scale height of atomic oxygen, $D \cdot T_{\mathrm{i}}$ = ambipolar diffusion coefficient.

The first order perturbation of the electron density caused by the presence of a gravity wave can be obtained by linearizing Eq. (1). One obtains:

$$i \omega \delta N = (\delta p - \delta\beta N_0) - \beta_0 \delta N - \frac{d}{dz} \delta\Phi , \tag{3}$$

where subscript 0 stands for the zero order state and δ for the perturbed one. In the above equation ω is the angular frequency of the gravity wave.

The production term p is proportional to the uv solar flux at a given height z and to the atomic oxygen concentration. In the following we assume that the uv solar flux is not much affected by the wave. This approximation has been fully discussed by Hooke [3], and is certainly valid for the waves with phase plane inclined less than 10° to the horizontal that will be considered. The recombination coefficient β is proportional to the molecular oxygen and nitrogen concentrations.

Then, if $\delta\varrho/\varrho_0$ is the relative perturbation of the neutral density caused by the gravity wave, δp and $\delta\beta$ can be expressed by

$$\frac{\delta p}{p_0} = \frac{\delta\beta}{\beta_0} = \frac{\delta\varrho}{\varrho} .$$

For solving Eq. (3) in δN the main difficulty is now in evaluating the $\delta\Phi$ term:

$$\delta\Phi = \delta N\, V_{iz0} + N_0\,(\delta W_n + \delta W_d)\,,$$

where $\delta W_n = \delta U_n \sin I \cos I$ (δU_n = perturbation of the neutral velocity due to the wave) and δW_d is the perturbation in the diffusion velocity induced by the neutral wave. The contribution due to δW_E is taken as negligible [3]. Neglecting the effect of the δW_d term, Hooke [3] could solve Eq. (3) which then becomes a scalar equation in δN. If we want to take this term into account we must combine Eq. (3) with the following equation deduced from (2) after linearization:

$$\frac{T_{e0} + T_{i0}}{2}\frac{d}{dz}\delta N = \delta N\left(\frac{W_{n0}}{D_0} - \frac{T_{n0}}{2\,H_0} - \frac{d}{dz}\frac{T_{i0} + T_e}{2}\right) + \frac{1}{2}\frac{dN_0}{dz}\delta\,(T_e + T_i) +$$
$$+ \frac{N_0}{2}\frac{d}{dz}\delta\,(T_e + T_i) - \frac{N_0}{D_0}\delta W_n + (\Phi_0 - N_0\,W_{n0})/D_0\frac{\delta\varrho}{\varrho_0}\,. \tag{4}$$

Another difficulty appears now: taking into account the electron and ion temperature perturbation δT_e and δT_i. A simplifying hypothesis such as $\delta T_e = \delta T_i = 0$, or $\delta T_e = \delta T_i = \delta T_n$ (δT_n = neutral temperature perturbation due to the wave) permits us to solve numerically the two differential equations (2) and (4) in δN, $\delta\Phi$. This was done in [4].

A more realistic way of solving (2) and (4) is to consider that δT_e and δT_i are two other unknowns of the problem. Then we have to find two other equations. These can be:

(i) The equation of energy balance of the ions which can be written as:

$$L_{ei} = L_{in}\,, \tag{5}$$

where L_{ei} is the energy loss of the electrons to the ions and L_{in} is the energy loss of the ions to the neutral particles. Linearizing (5) one obtains:

$$(1 + A)\,\delta T_i = (1 - B)\,\delta T_n + (T_i - T_n)_0\,(\delta N/N - \delta\varrho/\varrho + C\,\delta T_e)\,, \tag{6}$$

where A, B, C are coefficients depending on the zero order state model:

$$A = (T_i - T_n)_0\,\left({}^1/_2\,(T_i + T_n)_0 + {}^3/_2\,T_{e0}\right)$$
$$B = (T_i - T_n)_0/2\,(T_i + T_n)_0$$
$$C = 1/(T_e - T_i)_0 - {}^3/_2\,T_{e0}\,.$$

(ii) The empirical law stated by Lejeune and Waldteufel [6] from the incoherent scatter data of Saint-Santin-Nançay, provides a relation as:

$$T_e - T_i = F(z, N) \tag{7}$$

F can be chosen as $F = b(z) - a(z)\,N + c(z)\,N^2$; b, a, c are functions to be determined empirically. Fig. 4 compares the empirical model which has been chosen for our computation with the incoherent scatter data obtained at Saint-Santin-Nançay during the years 1967–1968.

However, it is not obvious that the empirical law still stands when short period perturbations occur in T_e, T_i, N. The triangles plotted in Fig. 4 correspond to a day of observation where large amplitude perturbations with two hour period were present. It appears very clearly that the empirical model also fits these data very well. So it seems reasonable to consider that the empirical law (7) is still applicable during disturbed conditions.

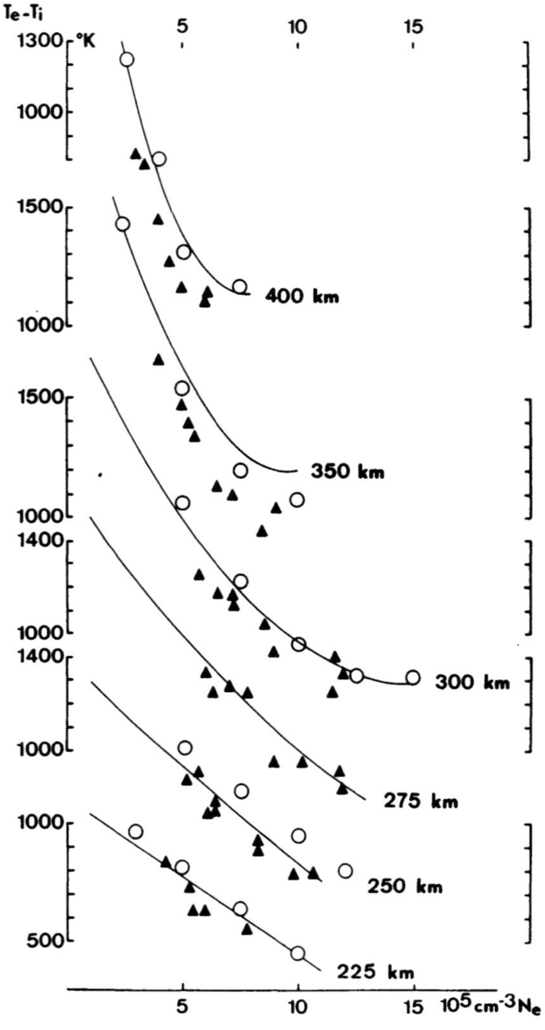

Fig. 4. Illustration of the semi-empirical law $T_e - T_i = F(N_e)$. For different heights the experimental values (circles) gathered from incoherent scatter observations during the years 1967–1968 are compared with the empirical model (continuous line) used in the computations. Dark triangles correspond to the measurements made on 13 September 1967.

Linearizing (7) for perturbations, one obtains

$$\delta T_e - \delta T_i = - [a(z) - 2\, c(z)\, N_0]\, \delta N\,. \tag{8}$$

The problem is now to solve simultaneously the set of equations (3), (4), (6) and (8). As the last two are scalar equations, two variables (for example

δT_e and δT_i), can be eliminated, and the four-equation system can then be reduced to a set of two first order differential equations in δN, $\delta \Phi$, which can be solved with suitable boundary conditions. This is discussed in § 3.

3. Computational Method

Three initial conditions are now to be carefully chosen in order to perform the computation: the zero order state model which is as realistic as possible, the boundary conditions, and the gravity wave model in the thermosphere.

3.1. Zero Order State Model

The zero order state model has been determined by integrating simultaneously equations (1), (2), (5), (7) neglecting the $\partial N/\partial t$ teım in (1). The following assumptions were made: T_{n0} is constant between 180 and 550 km altitude, p_0 and β_0 have height variations like:

$$p_0 = p_{0300} \exp\left[-\frac{z-300}{H_O} - 1.4 \exp\left(-\frac{z-160}{H_O}\right)\right]$$

which corresponds to a zenith solar angle of 45°, and

$$\beta_0 = \beta_{0300} \exp\left(-\frac{2(z-300)}{H_O}\right)$$

z height in km, subscript 300 denotes the 300 km reference altitude; $W_{n0} + W_{E0}$ is constant for heights between 180 and 550 km; D_0 increases with height as the inverse of atomic oxygen concentration:

$$D_0 = D_{0300} \exp\left(+\frac{z-300}{H_O}\right).$$

The numerical values used in the computations are shown in Table 1. These values are typical for the noon time of an equinoctial day. The boundary

Table 1

H_O	T_{n0}	p_{0300}	β_{0300}	$W_{n0} + W_{E0}$	D_{0300}
60 km	1000 °K	333 $cm^{-3}\,s^{-1}$	$1.9 \times 10^{-4}\,s^{-1}$	$-20\,m\,s^{-1}$	1.8×10^{3} mksA

conditions that have been applied are, for the upper boundary condition, that the ionization flux is taken as zero for $z = +\infty$, and for the lower boundary condition the assumption is made that the photochemical processes are predominant below 200 km, which means that $N = p/\beta$ below this height. The zero order state model as computed by this method is shown in Fig. 5.

3.2. Boundary Conditions

For the upper boundary, the most natural condition is that the perturbation of the ionization flux at infinity due to the gravity wave is zero. This condition at infinity can be reduced to a condition at 550 km altitude (the altitude which has been chosen as the upper boundary for the model computation) by the following treatment:

Integrating Eq. (1) between 550 km and infinity, and neglecting the recombination loss process, one obtains:

$$\frac{\delta}{\delta t} C_i = p_{0\,550}\, H_O - \Phi_\infty + \Phi_{550} \,. \qquad (9)$$

C_i is the electron content above 550 km. At altitudes above 550 km, the electron density can be considered at each instant to be in a state of diffusive equilibrium,

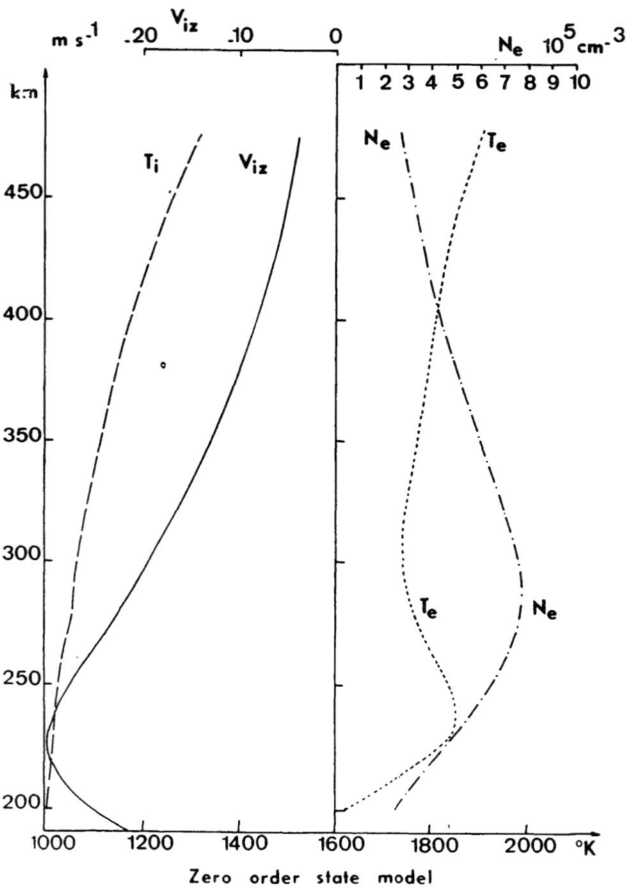

Fig. 5. Zero order state model for T_e, T_i, V_{iz} and N_e, computed by the method described in § 3.1.

if the time variation is not too fast. This approximation can be justified in considering the characteristic time τ of diffusion of the ionization given by $\tau = H_O^2/D\, T_n$ [7]; at 550 km altitude, this time τ is much smaller than the period of the wave considered (2 hours).

Then the electron density can be expressed as:

$$N = N_{550} \exp\left(-\int_{550}^{z} \frac{dz}{H_i}\right),$$

where H_i is the theoretical scale height of ionization:

$$H_i = (T_e + T_i) \Big/ \left(\frac{m\,g}{k} + \frac{d}{dz}(T_e + T_i)\right) \tag{10}$$

and the electron content $C_i = N_{550}\,\bar{H}_i$ where $\bar{H}_i$ is a "mean" value of H_i between 550 km and infinity.

A linearization of (9) leads to

$$i\,\omega\,(\delta N_{550}\,\bar{H}_{i0} + N_{0\,550}\,\delta\bar{H}_i) = p_{0550}\,H_O \frac{\delta\varrho}{\varrho_{0\,550}} + \delta\Phi_{550}\,. \tag{11}$$

($\partial\varrho/\varrho_0$ has been assumed to be constant with height above 550 km, which will be verified by our gravity wave model).

Since most of the ionization present above 550 km is contained in the first scale height above this height, one can expect that $\bar{H}_i$ is close to H_{i550}. With this assumption and linearization of (10), one obtains:

$$i\,\omega\,H_{i550}\left(\delta N_{550} + N_{0\,550}\frac{\delta\,(T_e + T_i)_{550}}{(T_{e0} + T_{i0})_{550}}\right) = p_{0\,550}\;H_O\left(\frac{\delta\varrho}{\varrho}\right)_{550} + \partial\Phi_{550}$$

which constitutes the upper boundary condition at 550 km.

The lower boundary condition is that the diffusion processes become unimportant below a certain height (200 km let us say), which means that below this height the electron density perturbation can be derived from (3) neglecting the term in δW_d:

$$(i\,\omega + \beta_{200})\,\delta N_{200} = (p_0 - \beta_0\,N_0)_{200}\left(\frac{\delta\varrho}{\varrho}\right)_{200} + \left(i\,k_{z\,200} - \frac{dN_0}{dz}\right)\delta W_n\,,$$

where k_z is the wave number of the gravity wave. This constitutes the lower boundary condition at 200 km.

3.3. Gravity Wave Model

In the F region, three processes cause a severe damping of the atmosphere waves: ion drag, thermal conduction, and viscosity; thus, a classical model [8] represented by

$$U_x = A \exp\left(i\,\omega\,t - k_x\,x - k_z\,z + \frac{1}{2\,H}\,z\right)$$

(U_x = perturbation of the horizontal neutral velocity due to the wave) is quite unrealistic. We have therefore chosen a model as:

$$U_x = A \exp\left(i\,\omega\,t - i\,k_x\,x - i\int_0^z k_z(\zeta)\,d\zeta + \int_0^z \alpha(\zeta)\,d\zeta\right).$$

The validity of such a model has been thoroughly discussed by Volland [9]. The model of $k_z(z)$ used in our computations is very close to the one predicted by Volland's formulae taking into account only the heat conduction [10].

For $\alpha(z)$, an evaluation of the damping effect due to ion drag leads us to adopt a model which is slightly different from Volland's [4].

The relative perturbations of oxygen concentration and neutral temperature are defined by:

$$\frac{\delta\varrho}{\varrho_0} = Z_1\, U_x; \qquad \frac{\delta T_n}{T_n} = Z_2\, U_x,$$

where Z_1 and Z_2 are complex quantities which can be obtained from the equations of gravity waves.

4. Results and Discussion

The computations have been carried out for two hour period gravity waves with three different values of the vertical wave number at 300 km: $k_{z\,300} = 0$, $k_{z\,300} = -1 \times 10^{-5}\,\mathrm{m}^{-1}$, $k_{z\,300} = -1.6 \times 10^{-5}\,\mathrm{m}^{-1}$. Fig. 6 shows as a function of height the amplitude of δV_{iz}, the oscillation in ion velocity induced by the δW_n oscillation due to the gravity wave. It appears that δV_{iz} and δW_n amplitudes are very close to each other at 200 km altitude, but differ appreciably

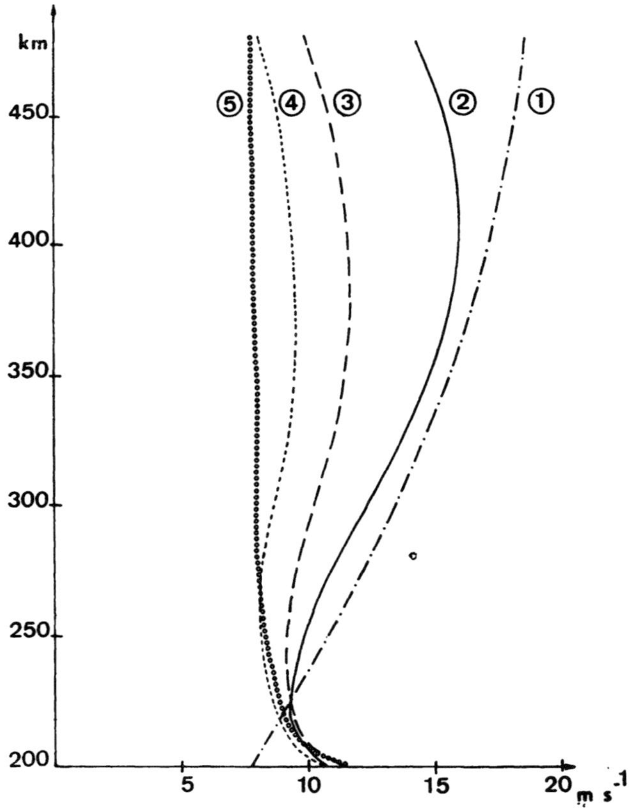

Fig. 6. Computed amplitude of δV_{iz} shown as a function of height for different values of k_{z300}: 2, 0; 3, $-10^{-5}\,\mathrm{m}^{-1}$; 4, $-1.6 \times 10^{-5}\,\mathrm{m}^{-1}$; 5, $-1.6 \times 10^{-5}\,\mathrm{m}^{-1}$, but the influence of T_e and T_i perturbations are neglected. Curve 1 represents the δW_n amplitude model.

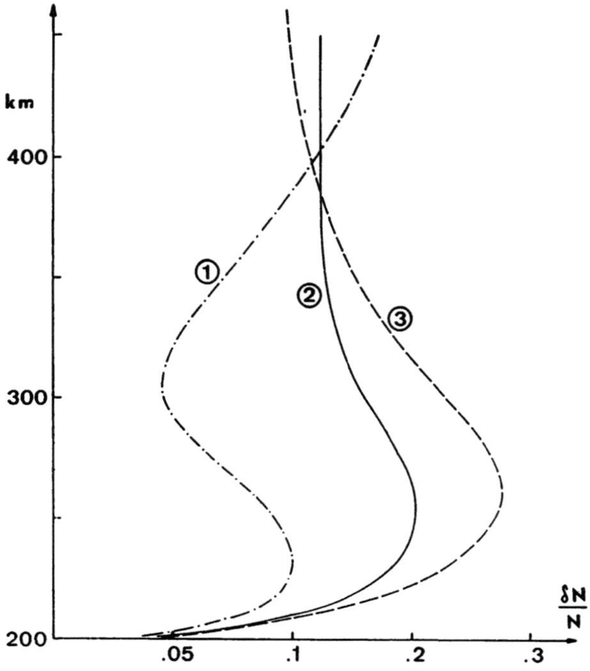

Fig. 7. Computed amplitude of $\delta N/N_0$ shown as a function of height for 1, $k_{z300} = 0$; 2, $k_{z300} = -10^{-5}\,m^{-1}$; 3, $k_{z300} = -1.6 \times 10^{-5}$.
Note. The abscissa scale should be .05, .1, .15, .2.

above 300 km and the larger the absolute value of k_{z300}, the larger is the difference between the two curves. This shows that the diffusion processes play an important role above 300 km altitude, and tend to compensate the effect of the neutral wind perturbation.

The corresponding variation with height of the $\delta N/N_0$ amplitude is exhibited in Fig. 7. A maximum of the amplitude appears near 250 km for the three values of k_{z300} and the larger $|k_{z300}|$ the stronger is this maximum.

Fig. 8 compares $\delta N/N_0$ amplitude results for $k_{z300} = -1.6 \times 10^{-5}\,m^{-1}$ when different assumptions are made: curve 3, the effect of δV_d is neglected; curve 2, the effect of δV_d is taken into account, but with the simplifying assumption of $\delta T_e = \delta T_i = 0$; curve 1, the effect of δV_d is taken into account with the realistic assumptions for δT_e and δT_i described in § 2. Curves 1 and 2 have a quite similar shape and keep relatively close to each other whereas curve 3 diverges appreciably from them above 250 km and corresponds to larger $\delta N/N_0$ amplitude. Thus the effect of the diffusion processes is to reduce the $\delta N/N_0$ amplitude from the value computed with the assumption that $\delta V_d = 0$. Figs. 9 and 10 show the results of the computations concerning δT_e and δT_i, when $k_{z300} = -1.6 \times 10^{-5}\,m^{-1}$. The amplitude of δT_i is very close to that of δT_n (neutral temperature perturbation due to the wave) at 200 km but differs from it more and more with increasing height. This is expected because the coupling between T_i and T_n is very strong at 200 km, but weakens with increasing height.

The amplitude of δT_e is greater by a factor of 5 than that of δT_i. It must be recalled that δT_e is essentially governed by $\delta N/N_0$ through the empirical law (7).

In Fig. 11, the phases of δW_n, δV_{iz}, δT_i, δT_e, $\delta N/N_0$ are compared. It is remarkable that the phases of δV_{iz} and δW_n remain very close to each other in the whole altitude range 200–500 km. The same result holds for the phases of δT_n and δT_i. This means that, when oscillations of $V_{i\parallel}$ and T_i are observed (Figs. 1, 2), the vertical wave number of the gravity wave can be directly deduced from their phase shift with altitude.

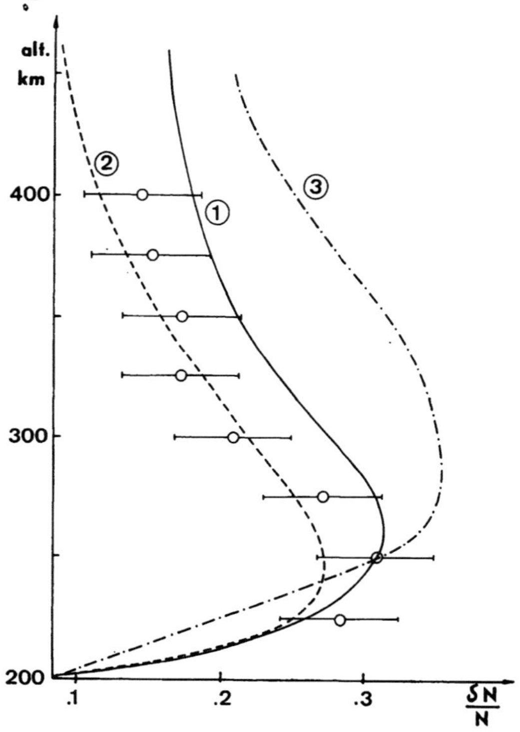

Fig. 8. Computed amplitude of $\delta N/N_0$ under different assumptions: 1, taking into account δV_d, δT_e, δT_i; 2, taking into account δV_d but neglecting δT_e and δT_i; 3, neglecting δV_d. Circles with error bars correspond to data of 13 September 1967. The amplitude of the wave in this figure is 1.7 times greater than in Figs. 6, 7, 9 and 10.

The curves 5 and 4 correspond to the phase variation with height of $\delta N/N_0$ computed under the assumption that $\delta W_d = 0$ (curve 5), and when taking into account the effect of δW_d (curve 4). The difference between the two curves is small. It is to be noticed that curve 4 ($\delta V_d \neq 0$) is smoother than curve 5 and that its shape is quite similar to that corresponding to the phase variation of δW_n.

An attempt to verify this theory has been made by comparing these results with the observations made at Saint-Santin-Nançay on 13 September 1967. The two hour period oscillations of T_i and V_i observed in Fig. 1 and 2 corres-

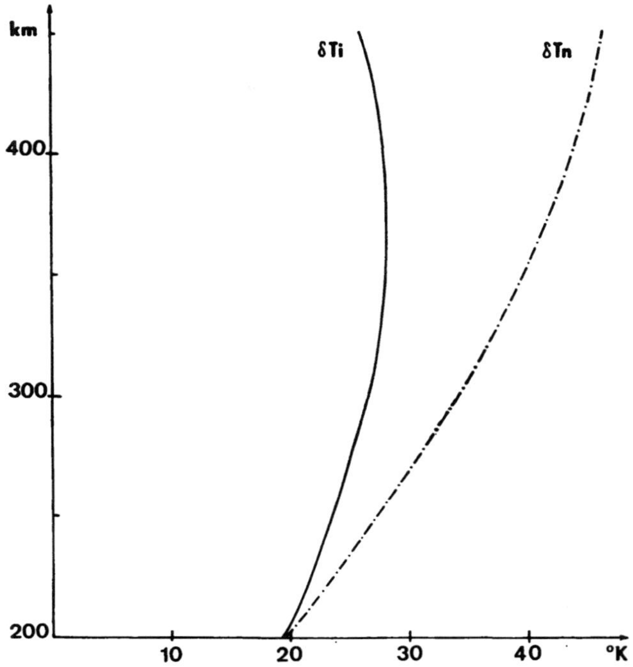

Fig. 9. Computed amplitude of δT_i compared with the model of δT_n amplitude.

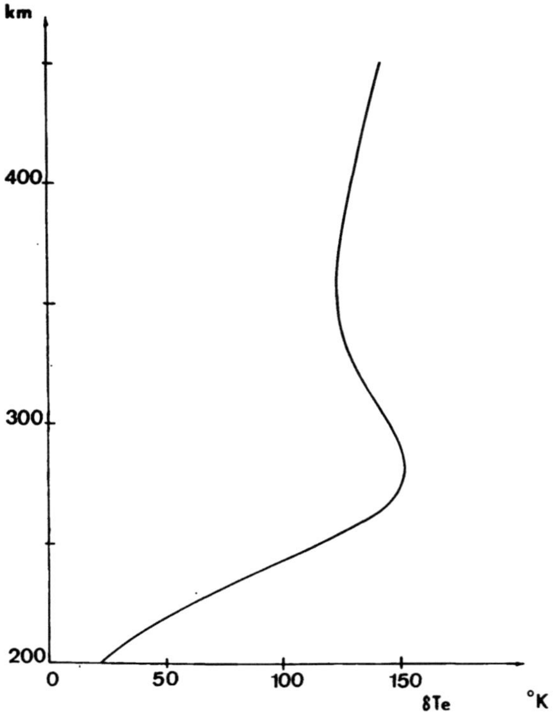

Fig. 10. Computed amplitude of δT_e.

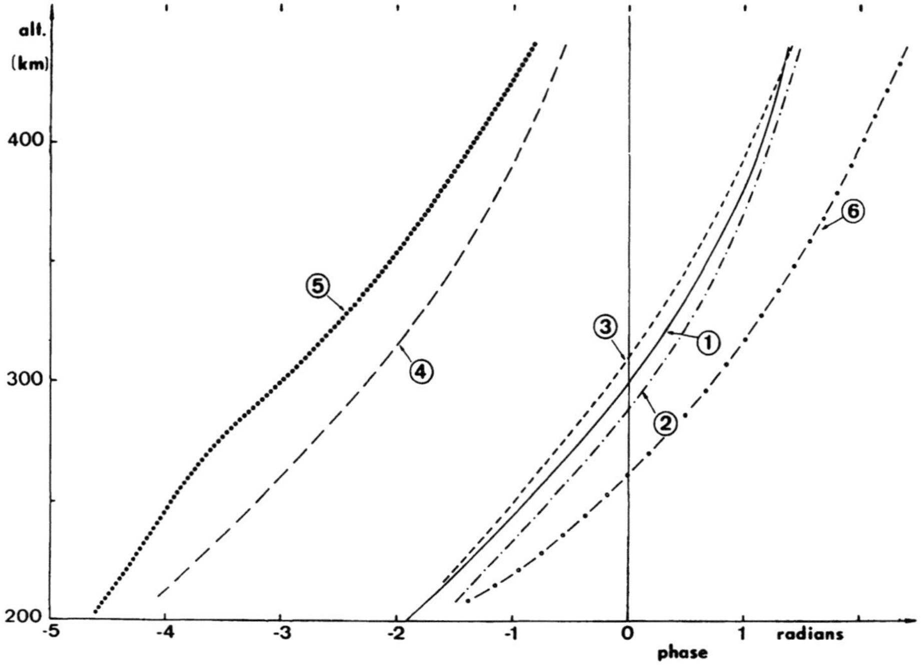

Fig. 11. As a function of height, phases of the model of δT_n and δW_n (curve 1) for $k_{z300} = -1.6 \times 10^{-5}\ m^{-1}$, compared with the computed phase of δV_{iz} (curve 2), δT_i (curve 3), δT_e (curve 6), δN (curve 4), δN when neglecting diffusion effects (curve 5).

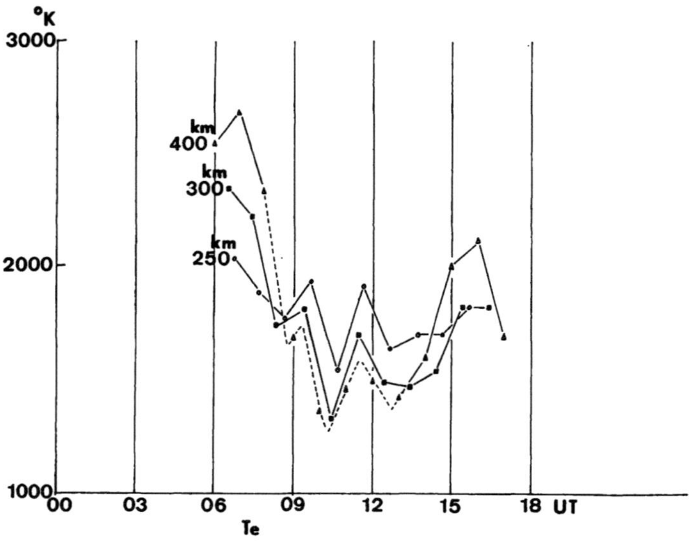

Fig. 12. Diurnal evolution of T_e observed at Saint-Santin on 13 September 1967 for different altitudes.

pond to a mean vertical wave number $k_z = -1.5 \times 10^{-5}\ \mathrm{m}^{-1}$. The amplitudes of δT_i ($\simeq 40$ °K) and δV_{iz} ($\simeq 25\ \mathrm{m\ s}^{-1}$) are approximately constant, which is in agreement with the theoretical results of Figs. 6 and 9. For the same day, the observed amplitude of $\delta N/N_0$ is compared with theoretical curves corresponding to $k_{z300} = -1.6 \times 10^{-5}\ \mathrm{m}^{-1}$ (Fig. 7). The agreement appears to be much better with the curves for which the diffusion effects are taken into account.

The observed oscillations of the electron temperature are shown in Fig. 12. The ratio between of the amplitudes of the observed oscillations of T_e ($=200$ °K) and T_i ($=40$ °K) agrees quite well with the theoretical ratio which is equal to 5.

Thus the described model of interaction between the gravity wave and the ionization is in good agreement with the observations of 13 September 1967. However, the theoretical amplitude of δV_{iz} ($18\ \mathrm{m\ s}^{-1}$), which explains the observed amplitude of $\delta N/N_0$, is slightly different from the observed one ($|\delta V_{iz}| = 25\ \mathrm{m\ s}^{-1}$). This discrepancy is not significant against the uncertainty present in the observed amplitudes of the oscillations due to the discrete sampling of the parameters.

5. Conclusion

We have presented a theoretical model that describes several aspects of the interaction between a gravity wave and the ionization: oscillations of ionic velocity and temperature, influence on the electron density, effect on the electron temperature.

The main results are:

(i) During the interaction, the diffusion processes play an important role and tend to inhibit the oscillation of the ionic velocity and of the electron density.

(ii) The oscillations of V_{iz} and W_n differ appreciably in amplitude, but their phases are almost identical even at 500 km altitude. The same result applies for δT_i and δT_n. This is of great practical importance for interpreting the T_i and V_i measurements obtained at Saint-Santin-Nançay.

(iii) The oscillation of N, V_{iz}, T_i are accompanied by very large amplitude oscillations of the electron temperature T_e. This effect has not been previously mentioned.

A comparison of these theoretical results with experimental data obtained by incoherent scatter, has been made and has led to a consistent interpretation of the oscillations of the four parameters, T_i, T_e, V_i, N observed during the magnetic storms of 13 September 1967.

Acknowledgments

The author would like to thank P. Amayenc, C. A. Reddy, G. Sabelette and C. Taieb for fruitful discussions; G. Vasseur and H. Volland have particularly contributed to this work by their helpful suggestions. He is indebted to M. Lafeuille for the data analysis.

The Saint-Santin-Nançay facility was built by CNET with funds of CNES. It is now supported by CNRS.

References

[1] R. D. HUNSUCKER and L. H. TVETEN, J. Atmos. Terr. Phys. **29**, 909 (1967).
[2] J. TESTUD and G. VASSEUR, Annls Géophys. **25**, 525 (1969).
[3] W. H. HOOKE, J. Atmos. Terr. Phys. **30**, 795 (1968).
[4] J. TESTUD and P. FRANÇOIS, J. Atmos. Terr. Phys. **33**, 765 (1971).
[5] G. VASSEUR, Thèse de Doctorat ès Sciences Physiques, Fac. Sciences Paris (13 Jan. 1971).
[6] G. LEJEUNE and P. WALDTEUFEL, Annls Géophys. **26**, 223 (1970).
[7] H. RISHBETH, Proc. Inst. Elect. Electron. Engrs. **55**, 16 (1967).
[8] C. O. HINES, Canad. J. Phys. **38**, 1441 (1960).
[9] H. VOLLAND, J. Atmos. Terr. Phys. **31**, 491 (1969).
[10] H. VOLLAND, J. Geophys. Res. **74**, 1786 (1969).

Space Research XII — Akademie-Verlag, Berlin 1972

A STUDY OF THE DYNAMICS OF TRAVELING IONOSPHERIC DISTURBANCES

K. C. Yeh

Ionosphere Radio Laboratory, Department of Electrical Engineering, University of Illinois, Urbana, Ill., USA

This paper describes certain experimentally observed properties of traveling ionospheric disturbances and certain deductions on the dynamic parameters of the acoustic-gravity wave using a fairly crude theory. The experiment involves continuous monitoring, at one or three stations, of wave polarization of radio signals transmitted by the geostationary satellite ATS 3. Traveling disturbances occur as quasi-sinusoidal oscillations in the polarimeter output. These events occur nearly daily and can be scaled for studies of dynamic properties. In general, the oscillations are observed to be in the frequency range 0.01 to 0.1 cycle min^{-1}. Experimental results show that these oscillations in the months of February and August tend to have sharper peaks than those of May and possibly also November. By triangulation the three-station data are capable of yielding information on the horizontal wave vector and hence the horizontal phase velocity. The horizontal phase speed is found to have a peak in the range 100–200 m s^{-1}. Using a crude theory, the dynamic variables in the neutral atmosphere associated with these traveling disturbances have been computed.

1. Introduction

Many experimental results have been amassed to indicate that the ionosphere is in constant agitation. One particularly interesting class of agitations is known as traveling disturbances. These traveling ionospheric disturbances (TIDs) have been studied experimentally by many observational techniques. Sufficient evidence has been given that these TIDs are excited by the propagating internal waves of the neutral atmosphere [1, 2]. A strong support to this excitation mechanism comes from the three-station experimental results that the observed wave parameters satisfy the dispersion relation for the internal wave [3]. Under such a premise the experimental data can then be processed to obtain internal wave parameters. This is done here.

The experiment is concerned with continuous monitoring of the wave polarization at 137 MHz. The signals are transmitted by the geostationary satellite ATS 3. For such a fixed ray path, the change in polarization due to Faraday effect is entirely caused by the change in the integrated electron density or electron content from the satellite to the receiver. During most of the experimental period the satellite was located at 47° W. The recording stations were Urbana (40.069° N 88.225° W), Danville (40.058° N 87.564° W) and Brocton (39.717° N 87.908° W), all located in eastern Illinois. More detailed description of the experiment can be found in [3].

2. Experimental Results and Discussion

Fig. 1 shows the spectral occurrence rate of TIDs as a function of wave frequency for four months of 1969. The spectral occurrence rate is defined as the monthly average duration per day in a given frequency range. As seen in Fig. 1, the spectral occurrence rate tended to favor waves within the frequency range 30–40 cycles min^{-1} in February and August while those in May and November tended to spread out more. The higher cutoff at 0.1 cycle min^{-1} corresponds to the 10 min period cutoff and has been reported by Titheridge [4]. As seen in Fig. 1, the higher cutoff occurred at a lower frequency in August than other months. The total area occupied by the histogram is equal to the

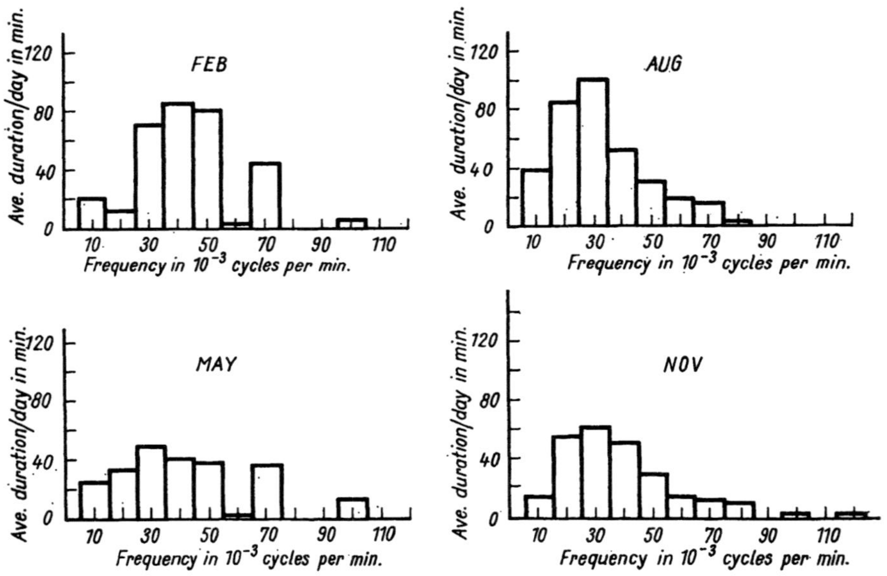

Fig. 1. Average monthly spectral occurrence rate of TIDs for four months of 1969.

average occurrence rate of TIDs of any frequency. On a yearly basis, this average occurrence rate is double humped with a major maximum in July, a secondary maximum in February and two minima in May and November (See Fig. 2). Therefore, TIDs are observable on our polarimeter for one fifth to one third of time on the average. It should be noted that the seasonal variation of the average occurrence rate is only apparent. This is because the electron content perturbation caused by a given internal wave is proportional to the electron content value itself. As the content value goes through a seasonal variation, it tends to modulate the average occurrence rate. An attempt to remove this modulation can be made since the electron content value is measured. After normalization the average occurrence rate shows only a single peak in July which is three times the winter values as shown in Fig. 2. It should be cautioned that because of the experimental biases [5] the values given in both Fig. 1 and Fig. 2 are more likely to be the lower boundary values.

When TIDs are observed at three spaced stations it is possible to deduce from experimental data, in addition to frequency, the horizontal phase velocity [6]. These experimentally obtained parameters on ionization can be used to calculate the associated dynamic parameters in the atmosphere [3] through the use of a theory [5]. It is important to remember that the theory [5] is rather crude as discussed in [7]. In Fig. 3 are shown the computed perturbations in neutral density for fifteen waves observed in the summer of 1968. For these relatively large traveling disturbances chosen for three-station analysis, the percentage change is usually less than 10%. It is also possible to compute other wave-associated perturbations such as velocity and temperature. These computations can be found in [3]. Knowing these perturbations, further computations can be made on viscous dissipation and hydromagnetic dissipation. It

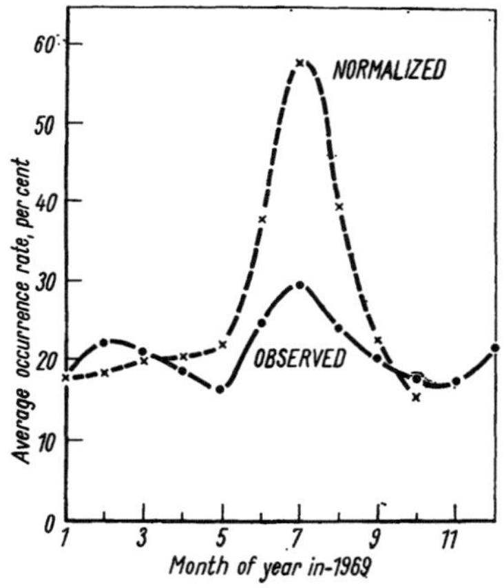

Fig. 2. Average observed and normalized occurrence rate of TIDs.

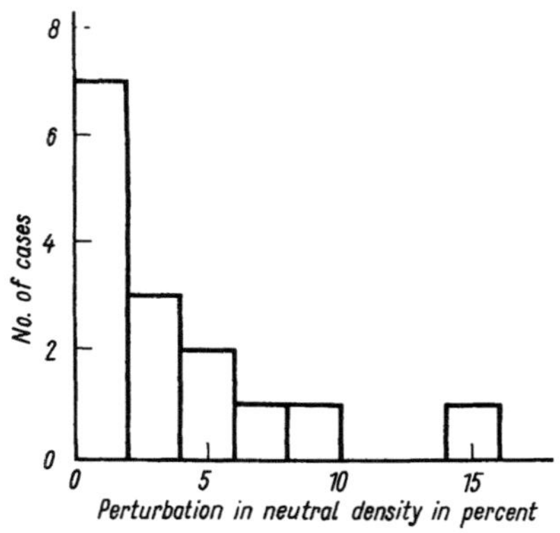

Fig. 3. TID-associated perturbations in neutral density for 15 waves observed in the summer of 1968.

has been found that viscous damping is in general more important than hydromagnetic damping. For the fifteen observed waves the average rate of dissipation has been computed to be 4.2×10^{-6} erg cm^{-3} s^{-1} when normalized with respect to $(\Delta I/I)^2$. Taking the average wave associated electron content perturbation $(\Delta I/I)$ to be 0.4%, the average occurrence rate to be 25%, the average rate of viscous dissipation by the internal wave comes out to be 1.7×10^{-11} erg cm^{-3} s^{-1}. Suppose this heat is dissipated over the height range 200–500 km, then these waves will provide an average heat flux of 5×10^{-4} erg cm^{-2} s^{-1} to the upper atmosphere [8]. This computed heat flux is about 0.1% of the solar heating and therefore is small. But this value must be accepted with the following understanding: (i) The average value of 5×10^{-4} erg cm^{-2} s^{-1} is probably the lower bound because of the experimental biases. The nature of the experiment is such that the semidiurnal tidal component cannot be obtained even though theoretical computations have shown its importance in heating [9]. (ii) On many occasions, $\Delta I/I$ may be as large as 4% or higher. For these cases, the dynamic heating [10] is raised to 10% or more of solar heating. (iii) The dynamic heating due to dissipation by TIDs is present at night when the solar heating is absent.

Acknowledgment

This investigation was carried out under the support of grant NGR 14-005-002 of the National Aeronautics and Space Administration.

References

[1] C. O. HINES, Canad. J. Phys. **38**, 1441 (1960).
[2] J. KLOSTERMEYER, J. Atmos. Terr. Phys. **31**, 25 (1969).
[3] K. C. YEH, J. Geophys. Res. 77, 709 (1971).
[4] J. E. TITHERIDGE, J. Geophys. Res. **73**, 243 (1968).
[5] T. M. GEORGES and W. H. HOOKE, J. Geophys. Res. **75**, 6295 (1970).
[6] D. H. COWLING, H. D. WEBB and K. C. YEH, Tech. Rep. No. 38, Ionosphere Radio Laboratory, Dept. Elect. Engng, Univ. of Illinois 1970.
[7] R. M. CLARK, K. C. YEH and C. H. LIU, J. Atmos. Terr. Phys. **33**, 1567 (1971).
[8] J. TESTUD, J. Atmos. Terr. Phys. **32**, 1793 (1970).
[9] R. S. LINDZEN and D. BLAKE, Geophys. Fluid Dynamics **2**, 31 (1971).
[10] C. O. HINES, J. Geophys. Res. **70**, 177 (1965).

The Ionosphere

Space Research XII – Akademie-Verlag, Berlin 1972

IONOSPHERIC REFRACTION CORRECTIONS IN SATELLITE TRACKING

R. B. BENT[a], S. K. LLEWELLYN[a] and P. E. SCHMID[b]

[a] DBA Systems, Inc., Melbourne, Florida, USA
[b] NASA/Goddard Space Flight Center, Greenbelt, Md, USA

Ionospheric refraction errors in satellite tracking have been studied in detail and a new model developed which provides significantly improved corrections to the corresponding elevation, range and range rate. A number of satellite orbits have been simulated and comparisons made of the relevant ionospheric refraction errors from models using both flat earth assumptions and more precise spherical formulations. The NASA/GSFC ionospheric model was developed by examining many tens of thousands of topside and bottomside profiles over 5° segments of the entire world for the last ten years, thereby covering a large part of the solar cycle. The model gives a complete electron density profile for a particular time, place and solar activity and can incorporate ionospheric data from a site within about 2000 km of the tracking station. Elevation, range and range rate corrections are computed by integrating through this ionospheric profile in the direction of the satellites. Comparisons have also been made to show that range rate corrections derived by taking the derivative of the range are the same as those computed by considering the refraction angles at the satellite. Investigations have been undertaken to examine the improvement brought about by the refraction corrections in the NASA/GSFC orbit determination programs and comparisons made with Faraday rotation experiments and two frequency techniques.

1. Introduction

Ionospheric refraction errors in satellite tracking have been studied in detail by many leading scientists and many methods and formulas exist for computing range, range rate and elevation angle corrections. The main problem however is that in most techniques little thought has been given to the actual electron density profile at the point where the ray passes through the ionosphere. In particular, as the satellite moves one must account for the horizontal electron density gradients incurred along the ray path, which at certain times and places can be quite large. The electron density N versus height h profile is itself a function of latitude, longitude, local time, season and solar radiation. Assuming the critical frequency f_0F2 is known, which itself is a major prediction problem, the thickness and height of the layer can vary considerably during a satellite pass and can result in great errors in satellite tracking corrections. The investigations reported here have attempted to build up a worldwide profile for the ionosphere for use in satellite tracking, where the shape of the topside and bottomside ionosphere, as well as f_0F2 and h_mF2 are computed as a function of latitude, longitude, time and sunspot number.

We briefly describe the approach used and discuss in more detail the refraction effects expected in the ionosphere and indicate some of the results.

2. Technique

It was necessary to investigate data from topside sounding satellites as well as from bottomside soundings over a solar cycle. Over 50000 topside soundings [1] were examined for the years 1961–1966 and comprehensive electron density data from the Ariel 3 satellite [2] linked with ground soundings for the years 1967–1968 gave topside coverage over the solar cycle. For the bottomside ionosphere over 400000 soundings along the North and South American chain [1] were analyzed over the years 1961–1969.

On the basis of earlier work performed at the Radio and Space Research Station in England it was decided to use a profile with an exponential topside layer

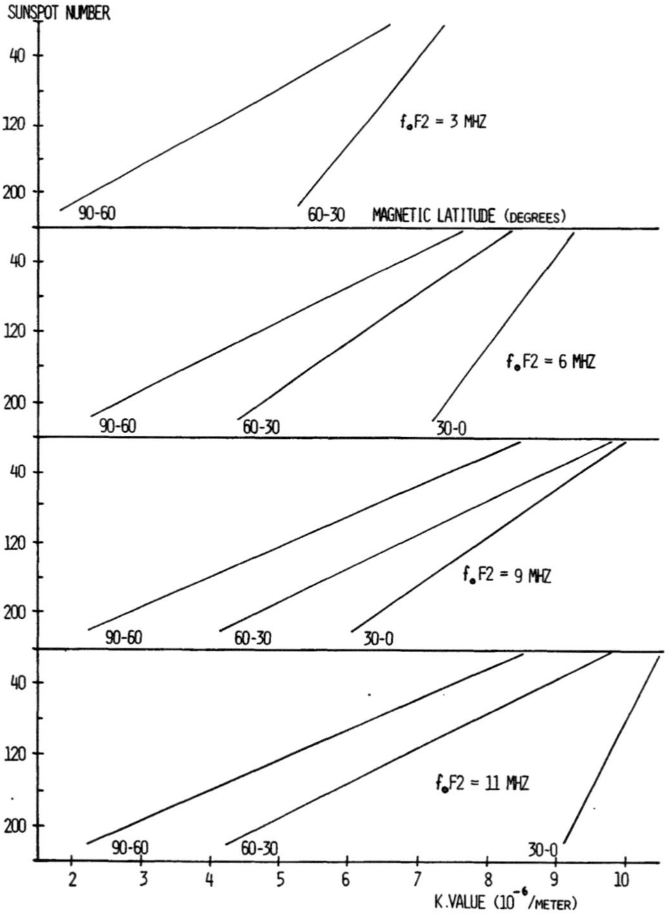

Fig. 1. Variation of decay value K for lower topside profile due to sunspot number, f_0F2 and magnetic latitude.

$N = N_m \exp(-K h)$ and a bi-parabolic bottomside layer $N = N_m (1 - b^2/y_m^2)^2$ although the bi-parabolic formula extended to a distance $y_m/4$ above h_m. (N_m = maximum electron density (at height h_m), h = height above the bi-parabolic layer, b = distance from h_m, K = topside decay constant which is inversely proportional to the plasma scale height, y_m = half thickness of bi-parabolic layer.) In the analysis of the bottomside data, for a particular f_0F2, the value of y_m was computed to give the same total electron content below h_m as was found in the recorded data. The model developed provides N_m, K and y_m. Figs. 1 and 2 indicate the different values of K and y_m found for different

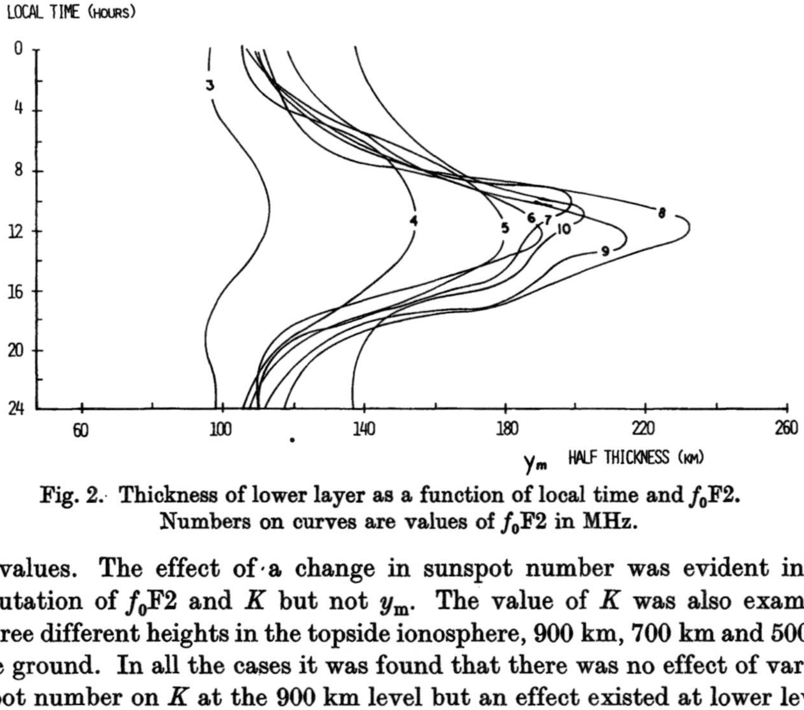

Fig. 2. Thickness of lower layer as a function of local time and f_0F2. Numbers on curves are values of f_0F2 in MHz.

f_0F2 values. The effect of a change in sunspot number was evident in the computation of f_0F2 and K but not y_m. The value of K was also examined for three different heights in the topside ionosphere, 900 km, 700 km and 500 km above ground. In all the cases it was found that there was no effect of varying sunspot number on K at the 900 km level but an effect existed at lower levels; K_{900} was on average 3 or 4 times smaller than K_{500}.

The critical frequency of the ionosphere fluctuates considerably hour by hour as can be seen in Fig. 3. The map illustrated here, and produced by the National Oceanic and Atmospheric Administration (NOAA), Boulder, USA, is for the monthly median value of f_0F2 which, on a day to day basis, may change by $\pm 25\%$. Some of this fluctuation is due to TID's (Travelling Ionospheric Disturbances) and some to solar effects. DBA Systems, Inc. has adopted the NOAA method for predicting f_0F2 and has arranged for this value of f_0F2 to be updated by sunspot number and/or real time data from within 2000 km of the sub-ionospheric point. Such updated information is important as the refraction corrections are approximately proportional to $(f_0\text{F2})^2$. Unless the values of K, y_m, f_0F2 and h_mF2 are known reasonably accurately at each computation point the refraction corrections at satellite tracking frequencies may be in error by at least a factor of three. It is also evident from Fig. 3 that any refraction model that considers the ionosphere vertically above a site to be constant for the whole "visible" sky can lead only to large errors, as rays at lower angles of elevation can have totally different ionospheric characteristics.

The computation of a change in range (ΔR) and range rate ($\Delta \dot{R}$) is a simple matter once the electron density profile is known, as:

$$\Delta R = \frac{40.3\, N_{\mathrm{T}}}{f^2 \left\{1 - \left(\frac{R_{\mathrm{e}}}{R_{\mathrm{e}} + R_{\mathrm{h}}}\right)^2 \cos^2 E\right\}^{1/2}}$$

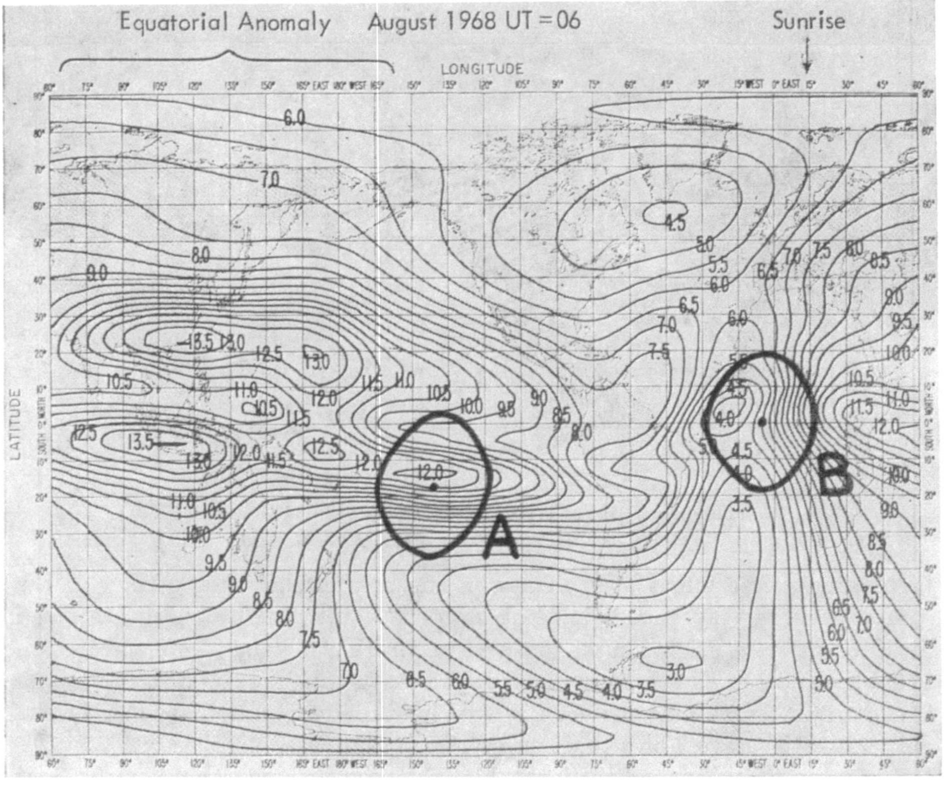

Fig. 3. Predicted world map of f_{x}F2 (MHz) for 06 hours August 1968 showing areas of visibility for two hypothetical tracking stations.

and $\Delta \dot{R}$ is the time derivative of this equation [N_{T} = total electron content, f = transmitting frequency, R_{e} = radius of earth, R_{h} = height of bottom of ionosphere, E = elevation angle]. For $\Delta \dot{R}$ it is necessary to know $\dot{h}_{\mathrm{s}}$ and $\dot{E}$ [h_{s} = height of satellite]. The elevation angle correction ΔE can be computed if we know the angular separation α of the true ray path above and below the ionosphere. In this model we used a proven method developed by Maliphant [3] which shows:

$$\alpha \text{ (radians)} = \frac{1}{2}\left(\frac{f_0\, F2}{f}\right)^2 \xi \frac{\tan \varphi_0 \sec^2 \varphi_0}{\varrho_0} \frac{N_{\mathrm{T}}}{N_{\mathrm{m}}},$$

where $\varrho_0 = \varrho_m \left(1 = 0.5333 \frac{y_m}{\varrho_m}\right)$, $\varrho_m = R_e + h_m$ F2, $\sin \varphi_0 = \frac{R_e}{\varrho_0} \cos E$, ξ is a function of $\left(\frac{f_0 F2}{f}\right)^2 \sec^2 \varphi_m$ and $\sin \varphi_m = \frac{R_e}{\varrho_m} \cos E$.

3. Results

Referring again to Fig. 3 we can consider two tracking sites A and B; the ionosphere seen by the ray from tracking site to satellite from horizon to horizon through the zenith is marked on the figure. Point A sees severe f_0F2 gradients due to the equatorial anomaly and point B due to the sunrise effect. Obviously a satellite travelling north to south will need much different refraction corrections

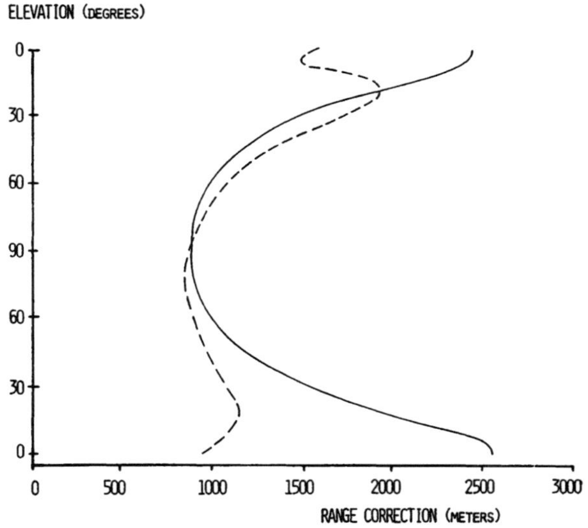

Fig. 4. Expected differences in satellite range corrections for equatorial and polar orbits due to severe gradients in f_0F2. Station India, 28° N 82° E, 10 h UT, frequency 130 MHz. Equatorial orbit ———, Polar orbit ---------.

from one travelling east to west. In fact, Fig. 4 illustrates the typical corrections to range at 130 MHz for two such passes; differences of range correction by a factor of 3 at the same elevation angle are evident. When computing elevation angle corrections it is necessary to know the approximate height of maximum electron density in the ionosphere otherwise errors of a factor of two may develop. Such an example is shown in the graph of Fig. 5.

Many simple ionospheric refraction models assume a flat earth and flat ionosphere and obvious difficulties arise with these assumptions. One of the least obvious effects however is that with a flat ionosphere the refraction angle correction approaches zero as the satellite height increases. The graph of Fig. 6 illustrates this pattern in more detail and compares results from a flat and curved ionosphere under average conditions.

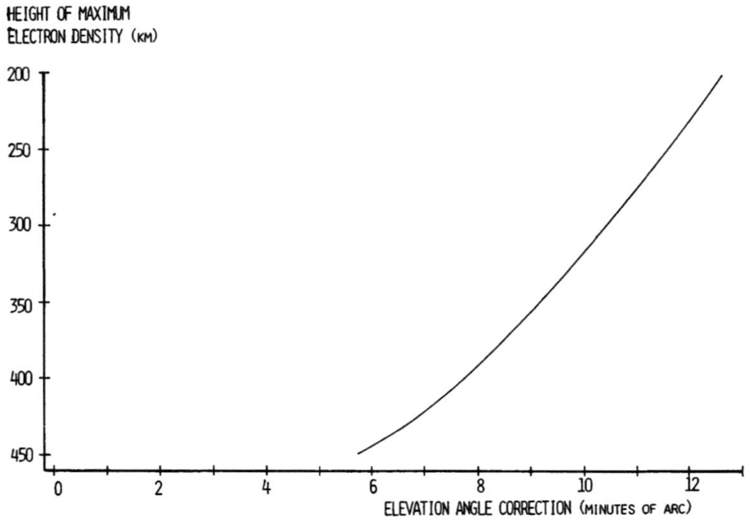

Fig. 5. Evidence of variation in elevation angle correction with the height of the ionosphere. $f = 140$ MHz, $f_0F2 = 15$ MHz, elevation 30°, half thickness 140 km, height of satellite 500 km.

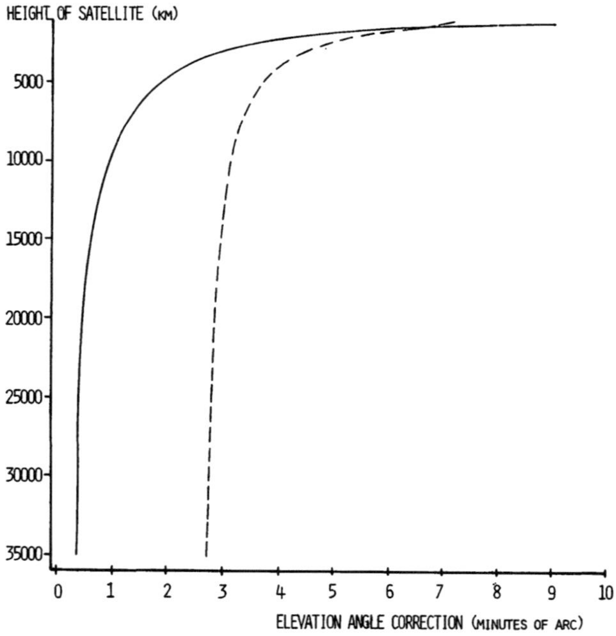

Fig. 6. Indication of error in elevation angle when ionospheric corrections are applied with a horizontal ionosphere. $h_m = 340$ km, $f = 140$ MHz, $f_0F2 = 15$ MHz, elevation 30°, ---- DBA model, ——— model with horizontal ionosphere.

It is possible to check the validity of the models for refraction errors to range and range-rate computations as both these are functions of total electron content and the total electron content can be compared with some other measurements to a satellite such as Faraday rotation. Many such comparisons have been made with the DBA model and Fig. 7 illustrates such a situation for a 6-month period at the height of the solar cycle. Results of Faraday rotation to an ATS satellite taken at Stanford, California, give total electron content. The envelopes for the 30 days of data in each month over a 6-month period are

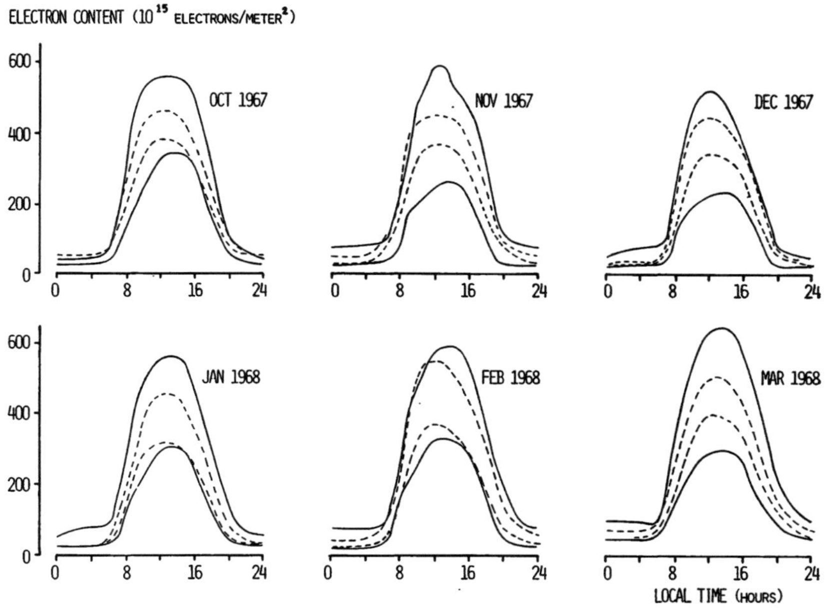

Fig. 7. Envelopes of monthly total electron content obtained from Stanford by Faraday rotation from ATS satellite and DBA ionospheric model including solar effect updating. ——— Faraday, --------- DBA.

plotted in Fig. 7. Also shown in this figure are the envelopes of the DBA model for the same period updated for solar events. Extremely good agreement is reached, the discrepancies possibly being due to travelling ionospheric or other disturbances. Similar comparisons have been made at different parts of the globe and thus it is reasonable to assume that range and range-rate refraction corrections using this model are, on average, quite good. Although the method for computing elevation angle corrections has already been verified experimentally by Maliphant [3] an attempt was made to verify it theoretically by computing range rate for a number of satellites by two methods: (a) differentiating the range; (b) considering the elevation angle of the spacecraft and the angle between the line of sight and the received ray at the spacecraft [4]. Because (a) was computed from total electron content which has been shown to be reasonably accurate and because (b) is computed from the refraction angle formula, then if the two were in close agreement the deviation angle method would be considered quite good. The tests were carried out with many different simulated satellite passes from highly eccentric to circular orbits with many different characteristics. Fig. 8 shows a typical result from an eccentric orbit satellite and in all cases close agreement was found.

The refraction program has also been tested in orbit determination work performed at NASA/GSFC. Three months of minitrack (140 MHz) interferometer data of the satellite BE-C were investigated by an operational long arc orbit determination computer program, and the program run both with and

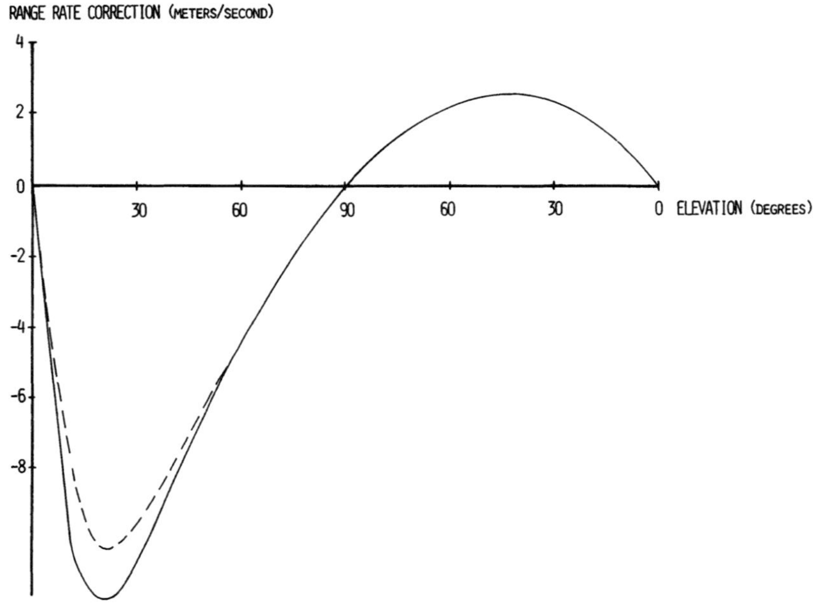

Fig. 8. Comparison of range-rate corrections for elliptical overhead orbit. – – – $\Delta \dot{R} = f$ (angles at spacecraft), ——— $\Delta \dot{R} = d\Delta R/dt$.

without the ionospheric model. The results of this investigation are shown in Fig. 9 where the weighted rms value of the direction cosines α and β are improved by 30%. The modulation of these values seems to bear some relationship to the solar flux, but perhaps the most striking result is the effect of the ionosphere on the inclination angle of the spacecraft which changes by up to 4 seconds of arc. Finally a test was performed using 400 MHz Geoceiver satellite data which has a two-frequency method of eliminating ionospheric effects. The results from nine such passes with elevation angles of from 15 to 83° were compared; the error model used extracted all but the ionospheric refraction from the Geoceiver measurements, leaving the measured ionospheric range correction to be compared with the computed ionospheric range correction in a least squares adjustment. The overall rms residual of the measured and computed range corrections was 1.2 m at 400 MHz. These results are illustrated in Table 1.

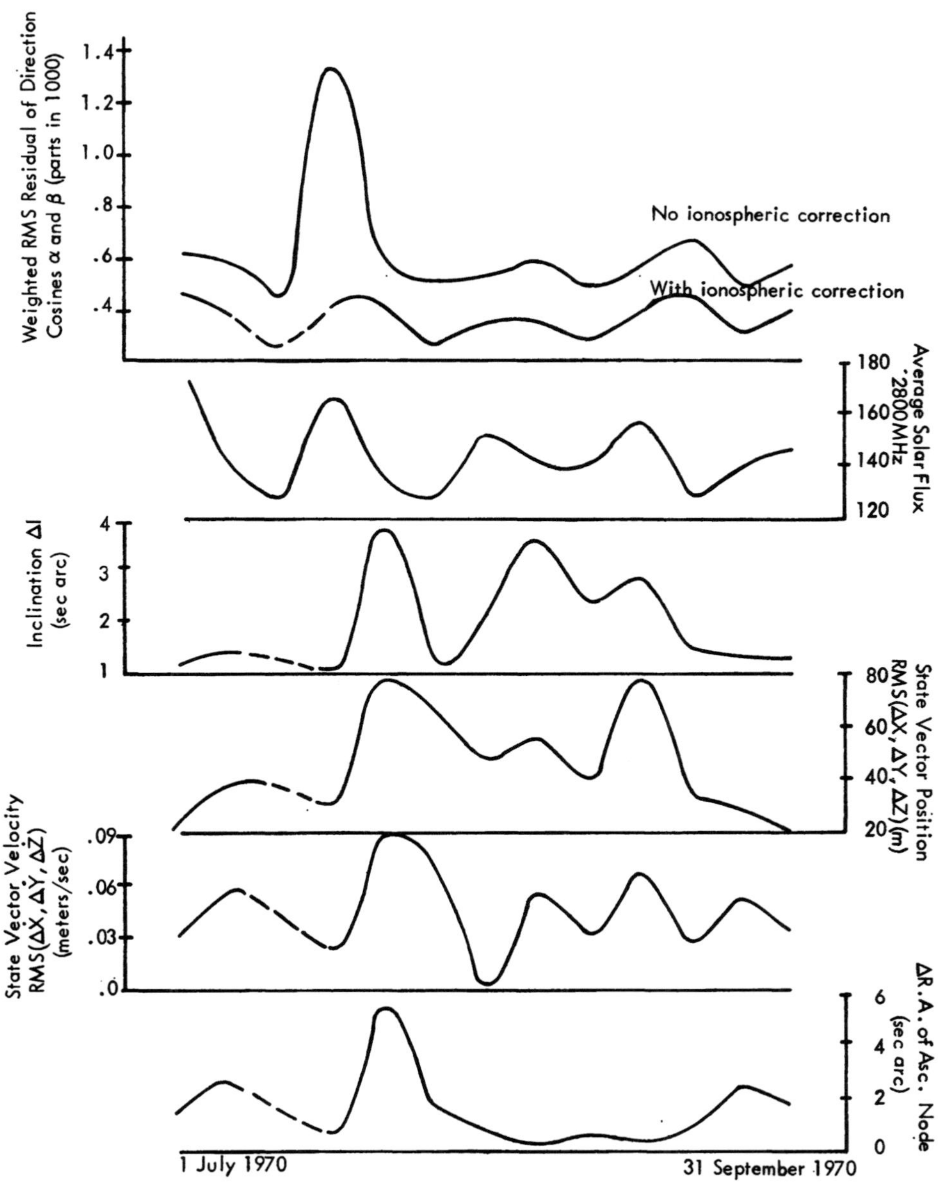

Fig. 9. Results of orbit determination program on three months minitrack data from the BE-C satellite. The top curve shows the improvement in the residuals when an ionospheric model is used and the lower curves indicate the difference in certain orbit parameters due to the ionosphere.

Table 1
Comparison of ionospheric refraction obtained from DBA model and from Geoceiver two frequency results for several passes of Navy navigation satellites

Pass	1 (A)	2 (D)	3 (B)	4 (L)	5 (S)	6 (E)	7 (C)	8 (H)	9 (M)
E_{max} (°)	31.5	57.5	40.0	81.5	33.4	22.7	68.4	37.5	83.3
ΔR_{min} (m)	−13.5	− 75.5	−18.5	− 95.5	−21.5	−15.5	−10.5	−13.0	− 9.5
ΔR_{max} (m)	−23.5	−189.0	−31.0	−246.5	−29.5	−21.0	−23.0	−21.5	−23.0
Point	$\Delta\Delta R$ (m)								
1	0.05	−0.95	−1.30	−2.10	−0.55	0.40	−0.20	−0.30	−0.15
2	0.02	−2.60	0.45	0.50	0.40	−0.25	−0.20	0.80	−0.20
3	0.05	2.70	1.30	2.00	0.55	−0.45	−0.05	−0.50	0.40
4	−0.30	2.70	0.70	1.85	0.02	−0.15	0.70	−0.10	0.50
5	−0.40	1.75	0.30	0.90	−0.20	0.30	0.80	−0.05	−0.90
6	0.45	−0.40	−0.50	−2.40	−0.15	−0.05	−0.25	−0.05	−0.20
7	0.65	−2.45	−0.80	−1.45	−0.25	0.75	−0.60	0.65	−0.20
8	−0.10	−0.05	−0.75	0.40	−0.05	−0.55	−0.60	−0.40	0.40
9	−0.20	−1.90	−0.55	0.40	0.25		−0.30	−0.40	0.15
10		−0.95	1.20	−0.85			0.70	0.20	−0.25
11		2.65		1.15			0.05		0.15
rms (m)	0.40	2.5	1.10	1.80	0.40	0.60	0.60	0.55	0.50

$[\Sigma(\Delta\Delta R)^2/(n-4)]^{1/2}$ (m) rms = 1.20

ΔR = ionospheric refraction computed from predicted ionospheric model for $f = 400$ MHz

ΔR_G = ionospheric refraction from integrated Geoceiver two frequency data measured at $f = 400$ MHz

$\Delta\Delta R$ (m) = residual ionospheric refraction after least squares fit of error model $= (\Delta R_G - \Delta R) - (a_0 + a_1 t + a_2 t^2 + a_3 f(E))$.
The error model takes the influences of a frequency offset $a_1 t$, of a frequency drift $a_2 t^2$, of the troposphere $a_3 f(E)$, and of a constant bias a_0 out of the measured ΔR_G.

4. Conclusions

These results illustrate the success of the DBA ionospheric model. The success is almost totally due to the development of realistic $N(h)$ profiles with provision for updating dependent on solar activity, which show the severe horizontal gradients in the ionosphere. The model's success also is due to the fact that no flat earth assumptions are made and that the computations are repeated at regular intervals along the ray path.

Acknowledgment

This work has been performed for NASA/GSFC under Contract NAS5-11730.

References

[1] World Data Center, private communication.
[2] R. W. PIGGOTT, private communication.
[3] R. G. MALIPHANT, DRTE Report No. 1091, Ottawa.
[4] S. K. LLEWELLYN and R. B. BENT, DBA Systems Report, Melbourne, Florida, USA, Dec. 1970.

Space Research XII — Akademie-Verlag, Berlin 1972

TROPOSPHERIC AND IONOSPHERIC REFRACTION ERRORS IN SATELLITE TRACKING OVER THE INDIAN SUB-CONTINENT

T. R. Tyagi, A. B. Ghosh, A. P. Mitra and Y. V. Somayajulu

Radio Science Division, National Physical Laboratory, New Delhi, India

In this paper an attempt is made to specify inputs of ionospheric and tropospheric parameters in the tropical regions, especially over the Indian sub-continent, in order to evaluate the errors in satellite ranging and position fixing due to refraction effects. Sample calculations showing refraction errors due to ionosphere and troposphere are presented for several frequencies and zenith angles.

1. Introduction

The measured range and elevation angle of a satellite using radio transmissions differ from the actual range and angle due to the refraction effects of the ionosphere and troposphere. These refraction errors in satellite tracking or otherwise have been studied in detail by many scientists. In this paper we have made an attempt to specify inputs of ionospheric and tropospheric parameters in the tropical regions, especially over the Indian sub-continent, in order to evaluate the errors in ranging and position fixing due to refraction.

The ionospheric input parameters are electron content, N_mF2 and h_mF2. Electron content measurements are available in Delhi since 1963 from the Cosmos 5, BE-B and BE-C satellites and cover periods of both low and high solar activity. In addition, observations have been made for several years in Ahmedabad and Calcutta, and over a restricted period of time in Hyderabad, Thumba and Kurukshetra. Vertical incidence ionospheric measurements are available in eight stations from Thumba, located at the geomagnetic equator, to Delhi, with a dip value of 42.4° N. Contours of these parameters can be prepared for the Indian tropical zone. The tropospheric parameters are based on radiosonde observations at eight stations: Jodhpur, Delhi, Ahmedabad, Calcutta, Bombay, Nagpur, Madras and Trivandrum, and surface refractivity measurements at more than 200 stations [1].

Ionospheric refractive indexes based on the electron density profiles built from the above-mentioned input parameters and tropospheric refractivity from the analysis of Venkiteshwaran et al. [1] have been used to estimate the errors in ranging and elevation angle measurements due to ionospheric and tropospheric refraction for a number of frequencies and for different elevation angles of the satellite.

2. Propagation Errors

The apparent range between the spacecraft and aircraft or the ground station is modified by the intervening ionization in two different ways; (i) there is a reduction in the speed of the signal propagation compared with its speed in vacuum; (ii) there is a bending of the ray path due to refraction with a consequent lengthening of the path. Both of these effects combine to increase the time elapsed between transmission and reception of the signal and, therefore, alter the apparent range. We shall examine these propagation effects for the error in the measurement of the range and angle of elevation.

The time delay of a signal propagating through the ionosphere is

$$t = \int_{\mathrm{p}} \frac{ds}{v_g} = \int_{\mathrm{p}} \frac{ds}{c\,\mu} \tag{1}$$

where ds is element of the ray along the optical ray path, c is velocity of light in vacuum, μ is refractive index of the ionosphere at the element ds and p denotes that the integration is performed along the optical path. If we assume a straight line propagation, then

$$\int_{\mathrm{g}} ds = R \tag{2}$$

where R is the true range and g denotes that integration is performed along the geometric path. An apparent range R' can be defined as the distance resulting from multiplying the observed time delay by the speed of light, i.e.

$$R' = C\,t\,. \tag{3}$$

The error made by assuming that the apparent range is actually equal to true range, i.e. by disregarding the ionospheric effects, is

$$\delta R = R' - R = \int_{\mathrm{p}} \frac{ds}{\mu} - \int_{\mathrm{g}} ds\,. \tag{4}$$

We have expressed the ionospheric refractive index as a power series in terms of angular frequency (ω) of the radio wave and have assumed quasi-longitudinal approximation to be valid. Thus μ is given by

$$\mu = 1 - \frac{\omega_{\mathrm{N}}^2}{2\omega^2} \pm \frac{\omega_{\mathrm{L}}\,\omega_{\mathrm{N}}^2}{2\,\omega^3} - \frac{\omega_{\mathrm{N}}^4}{8\,\omega^4} + \cdots \tag{5}$$

where ω_{N} is the angular plasma frequency and ω_{L} is angular gyrofrequency due to the longitudinal component of the geomagnetic field.

We can then express the range error in kilometres by the equation

$$\delta R = \int_{\mathrm{p}} \frac{ds}{(1 - \omega_{\mathrm{N}}^2/2\,\omega^2 - \omega_{\mathrm{N}}^4/8\,\omega^4)} - \int_{\mathrm{g}} ds\,. \tag{6}$$

The vertical electron distribution was represented in the computer programme by a number of spherical shells dividing the ionosphere into layers of uniform electron concentration. Layers were chosen having 20 km thickness from 100 to 1000 km above the earth's surface. The index of refraction was calculated using (5) with the second-order term proportional to $1/\omega^3$ omitted. An advantage was gained by neglecting this term because this allowed precise determination of the amount of error introduced by the transmission path inequality [2]. The ray-tracing programme used Bouguer's law at the layer boundaries to determine the ray path. Each ray path was confined to a plane that contained the satellite, ground station and the centre of the earth. After ray tracing the difference between ray path and straight line path, i.e. the range error, can be calculated.

Below 100 km the electron density is negligible and so the refractive index due to the ionosphere may not differ much from unity. However, the troposphere plays an important role at the lower heights. Relatively thin layers containing abnormally large refractive index gradients are common in the troposphere and are known to be important as a trans-horizon propagation mechanism [3, 4]. If the refractive index of the troposphere at various heights is known the range error can be computed using the above method.

If one starts tracing the ray path from the ground station to satellite using Bouguer's law and a stratified ionosphere, the coordinates of each point along the ray path would be known; and thus the coordinates of the satellite would be known. Therefore the true elevation angle of the satellite would be known. Subtracting this from the initial elevation angle at which the ray is transmitted in order to reach the satellite, one can know the error in the determination of elevation angle.

3. Data Input to calculate the Range and Elevation Angle Error

3.1. Ionospheric Refractive Index

The refractive index for the ionosphere is calculated using Eq. (5). If one assumes that the angular plasma frequency as well as the gyrofrequency is much smaller than the angular frequency, then to a first order approximation Eq. (5) reduces to

$$\mu = 1 - \frac{\omega_N^2}{2\,\omega^2} = 1 - \frac{40.3}{f^2} \int N\, ds \tag{7}$$

and thus the range error is given by

$$\delta R = \frac{40.3}{f^2} \int N\, ds = \frac{40.3\, \langle \sec \Phi \rangle}{f^2} \int N\, dh \tag{8}$$

provided the integration is performed along the geometric path. The quantity $\int N\, dh$ is the columnar electron content and $\langle \sec \Phi \rangle$ is the satellite zenith angle at mean field height. Thus the electron content measurements can be directly used for evaluating the range error caused by group velocity reduction, using Eq. (8) [5]. However, this does not allow correction to the error caused by group path lengthening. To make both these corrections the integral $\int N\, ds$ should be performed along the ray path. If f_0F2 measurements are also available from C-4 ionograms, model electron density profiles can be prepared so as to match f_0F2 and $\int N\, dh$ values.

Electron density profiles during low solar activity are built following a method given by Somayajulu et al. [6]. In essence the method consists of obtaining the profiles below the F2 peak from C-4 ionograms. This is extended up to +1.5 scale heights using Seddon's model [7]. Diffusive equilibrium for the ions

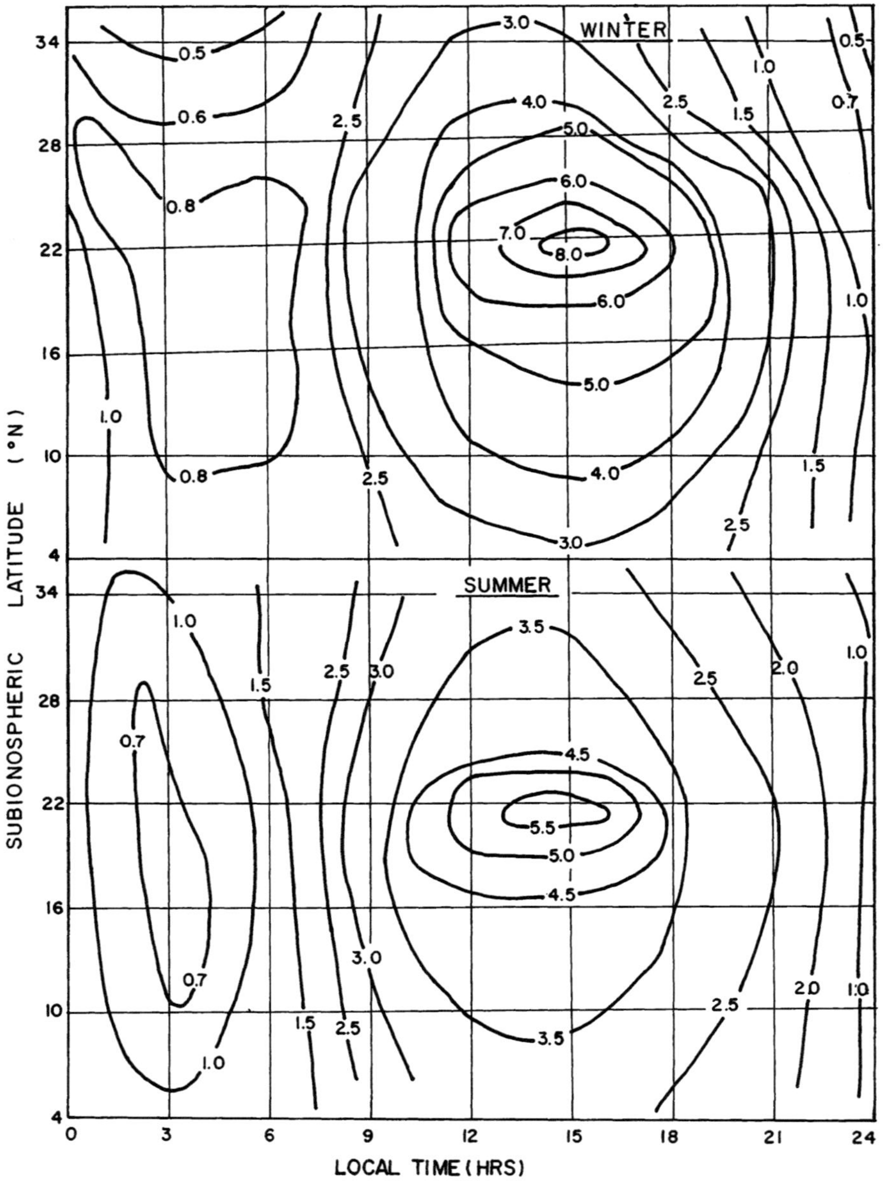

Fig. 1. Electron content contours (10^{17} m^{-2}) over the Indian sub-continent for low solar activity period.

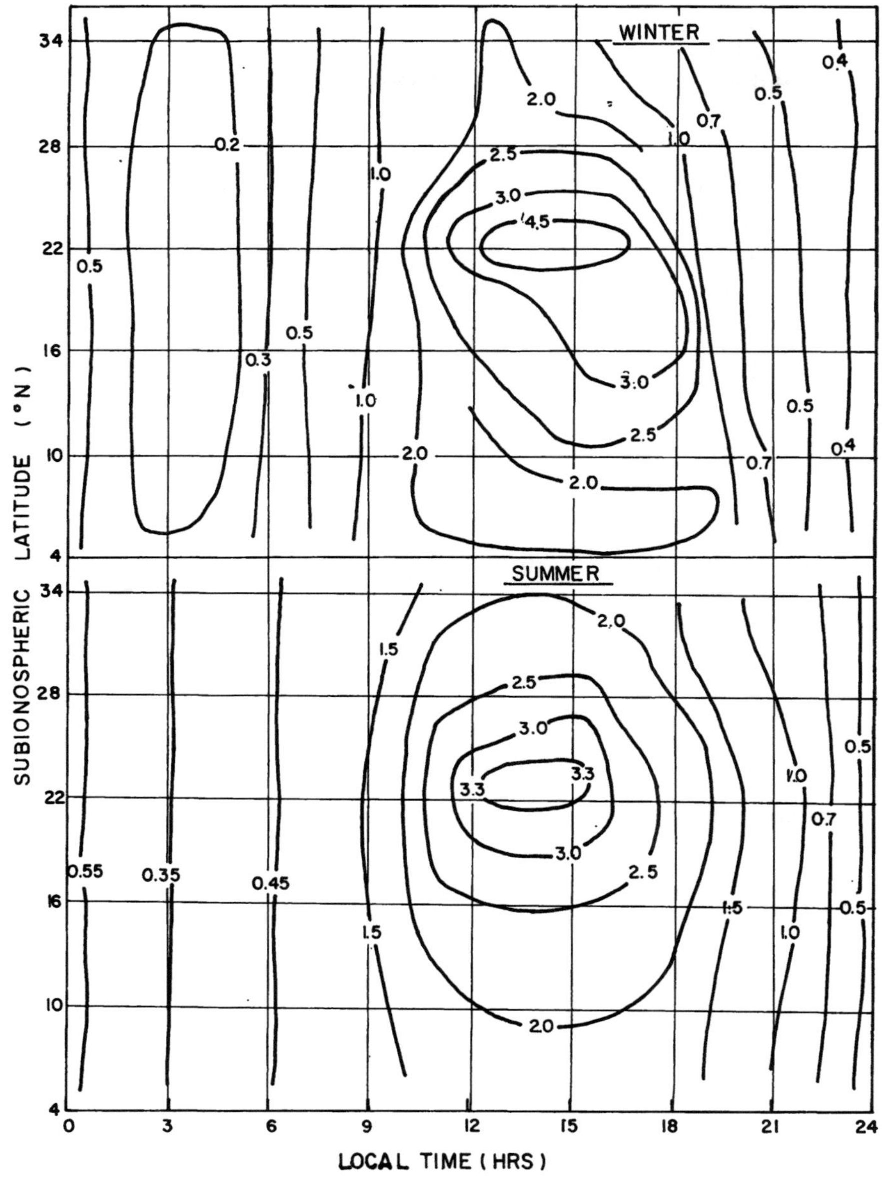

Fig. 2. Electron content contours (10^{17} m^{-2}) over the Indian sub-continent for high solar activity period.

is taken to be established from this height and above, and the profile above this is built using a constant slope equal to the ionic scale height of atomic oxygen. During daytime the transition from O^+ to He^+ ions occurs around 600 km, while at night the transition takes place from O^+ to H^+ at 550 km. During high solar activity the electron density profiles are much different from this model, especially during winter when most of the ionization lies in the topside [8]. Work is in progress to give a high solar activity model for (N, h) profiles and will be reported elsewhere. Knowing the electron density profile the ray path tracing can be done using Bouguer's law and thus

$$\sum_{l=1}^{l=k} \frac{dh \sec \Phi_l}{\mu_l}$$

can replace the integral $\int ds/\mu$ where the ionosphere is assumed to consist of k layers of equal thickness and l denotes the lth layer. Similarly second and third order corrections for the range error can be estimated by calculating μ up to the second and third orders. Figs. 1 and 2 show the electron content contours over the Indian sub-continent for solar minimum and solar maximum. The contours for f_0F2 can be easily prepared from the published values from all the stations in India and in the neighbourhood.

3.2. Tropospheric Refraction

The value of the refractive index μ of the atmosphere is determined from the expression for refractivity

$$n = (\mu - 1)\times 10^6 = 77.6\frac{P}{T} + 3.73\times 10^5 \frac{e}{T^2}. \tag{9}$$

Eq. (9) may be evaluated from the standard weather observations, viz. the atmospheric pressure P in millibars, the temperature T in degrees Kelvin and the vapour pressure e in millibars. It can be seen from this equation that the value of refractivity consists of a dry component D given by

$$D = 77.6\frac{P}{T} \tag{10}$$

and a wet component given by

$$W = 3.73\times 10^5 \frac{e}{T^2}. \tag{11}$$

The dry term which comprises about 70% of the total, contains the vapour pressure, which is dependent on the air mass. Thus it is seen from above that the dry term predominates. However, in India the wet term undergoes large changes from winter to monsoon. Venkiteshwaran et al. [1] have presented the contours of refractivity at sea level, n_0, for different seasons at 0300 and

Table 1

Tropospheric refractivity

Indian mean			$n_0 = n_s \exp(-0.09592\, h_s)$
Average scale height			10.5 km
Average NACA			$H =$ 9.46 km
Seasonal Ranges			
Dehradun July			$H =$ 10.65 km
Dehradun Jan.			$H =$ 8.824 km
Cochin July			$H =$ 10.09 km
Cochin Jan			$H =$ 10.58 km
Dry and wet components, Delhi			
	D_s	W_s	n_s
Jan.	263	41	304
July	246	117	363

1200 GMT, utilizing surface meteorological data extending over large periods and covering nearly 230 stations published by Indian Meteorological Department in 1960. Some data regarding the troposphere are given in Table 1 for reference. When the surface refractivity n_s is known the values of sea-level refractivity n_0 are derived using the relation for Indian stations

$$n_0 = n_s \exp(-0.09592\, h_s) \tag{12}$$

where h_s is the height of the station. Eq. (12) can also be used to calculate the tropospheric refractivity for different heights using n_0 contours, and thus the refraction errors can be evaluated.

4. Model Calculations

Calculations have been made by computer to estimate errors in the measurement of range and angle of elevation. The tropospheric refractive index up to 40 km was evaluated from published refractivity contours. The ionospheric refractive index was evaluated from the (N, h) profiles based on ionosonde data

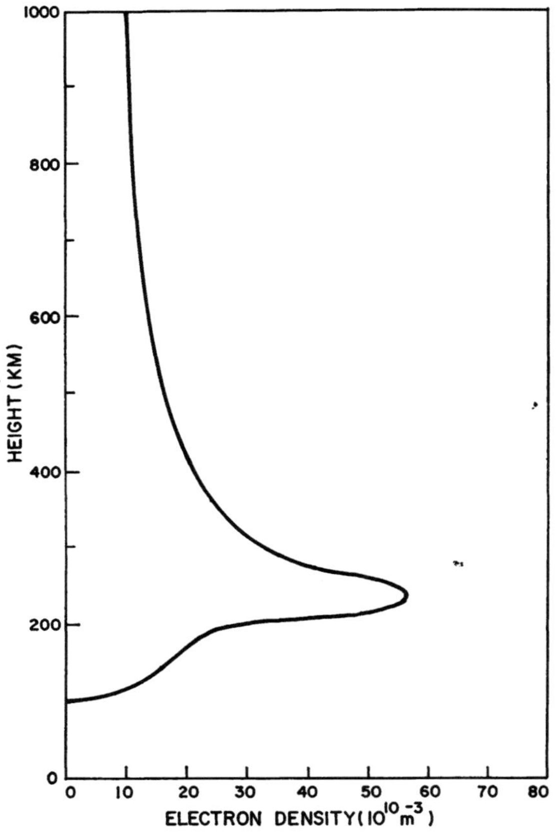

Fig. 3. Sample (N, h) profile valid for low solar activity daytime maximum.

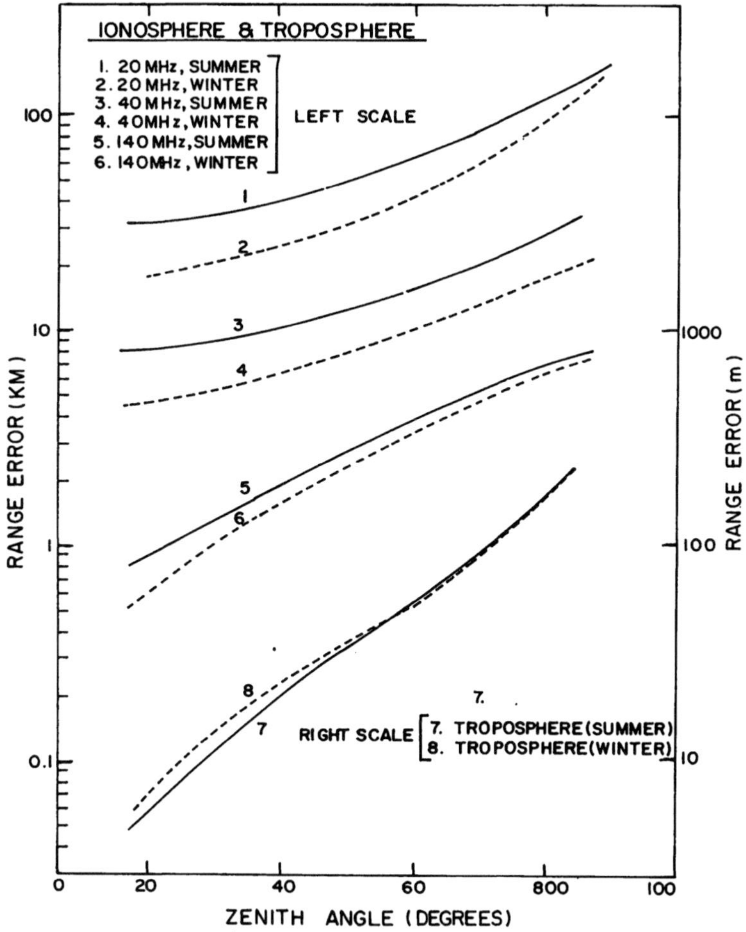

Fig. 4. Ionospheric and tropospheric range error as a function of zenith angle.

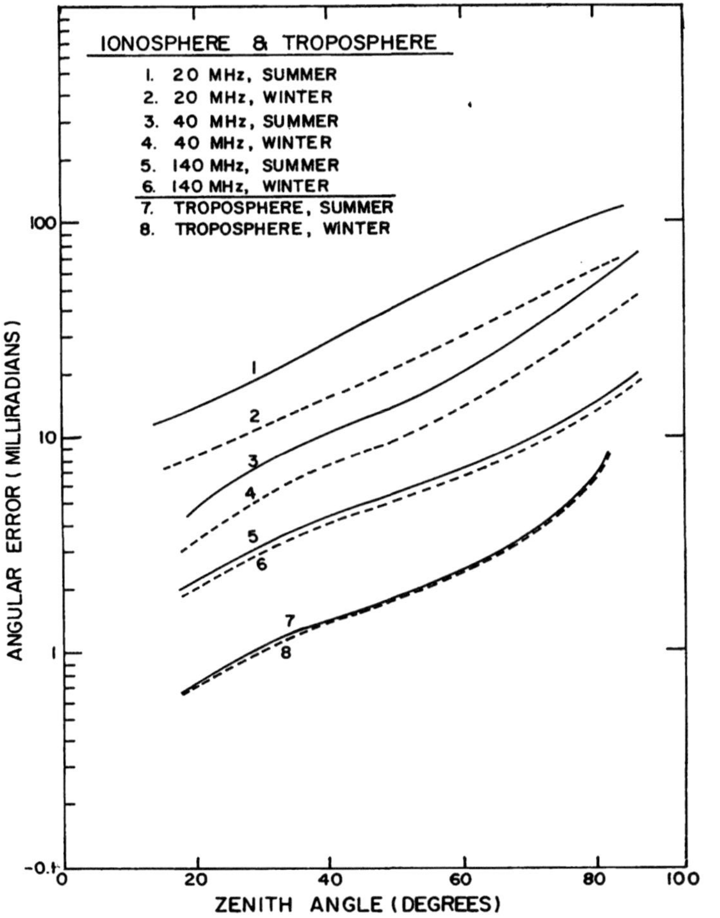

Fig. 5. Ionospheric and tropospheric angle of elevation error as a function of zenith angle.

and electron content measurements. A sample (N, h) profile valid for low solar activity winter daytime is shown in Fig. 3. Calculations have been made for winter and summer daytime during low solar activity for frequencies 20, 40 and 140 MHz and for satellite zenith angles 20, 40, 60 and 80°. Range and angle of elevation errors are plotted in Figs. 4 and 5 as a function of zenith angle.

References

[1] S. P. Venkiteshwaran, V. S. Narayan and H. S. S. Sinha, R.T.R.C. Monograph No. 1, National Physical Laboratory, New Delhi (1970).
[2] J. F. Willman and A. J. Tucker, J. Geophys. Res. **73**, 385 (1968).
[3] F. Eklund and S. Wickerts, Radio Sci. **3**, 1066 (1968).
[5] M. P. M. Hall, Proc. Inst. Elect. Electron. Engrs. **115**, 1595 (1968).
[5] A. V. Da Rosa, Inst. Elect. Electron. Engrs Trans. **AP-17**, 628 (1969).
[6] Y. V. Somayajulu, T. R. Tyagi and V. P. Bhatnagar, Space Research, V 641 (1965).
[7] J. C. Seddon, NASA TN Rep. D-1670 (1963).
[8] T. R. Tyagi and K. K. Mittal, Ind. J. Pure Appl. Phys. 9, 504 (1971).

Space Research XII — Akademie-Verlag, Berlin 1972

A CORRECTION METHOD FOR TRACKING ERRORS DUE TO ATMOSPHERIC REFRACTION

P. EDENHOFER[a], D. GLESNER[a], E. HARNISCHMACHER[b] and V. STEIN[a]

[a] Deutsche Forschungs- und Versuchsanstalt für Luft- und Raumfahrt E. V. (DFVLR), Institut für Flugfunk und Mikrowellen, FRG
[b] Ionosphäreninstitut, Breisach, FRG

System analysis applied to radio tracking of satellites shows that errors due to propagation of vhf waves in the atmosphere are dominant compared with errors due to technical effects. Applying geometrical optics, angle, range and range rate errors can be computed, if the atmospheric refractive index profile between ground and satellite is known. With values of pressure, temperature and humidity measured in the troposphere, and with critical frequencies measured in the ionosphere, model profiles have been constructed. After applying a standard correction to the troposphere, it turns out that the tropospheric residual error is considerably smaller than the ionospheric one. A consistent set of ionospheric model profiles has been derived, using long-period values of ionospheric soundings at Breisach and electron content data from Faraday observations of satellites. Predicted profiles are given hour by hour for periods of ten days.

Itseems that correction based upon predicted profiles reduces the systematic error to about 40% of its original value. Smaller errors can be obtained with measurements of the electron content during a satellite passage, either with the satellite to be tracked or with a beacon satellite. In the first case the accuracy with the computed corrections should be quite high, in the second case an error reduction of 15% seems to be possible.

1. Introduction

When considering errors of a satellite tracking system, it is useful to distinguish the possible errors as technical errors and propagation errors. Furthermore, usually one can distinguish between systematic and statistical errors. Statistical errors have a correlation time τ, which is much smaller than the time t of the duration of the measurement. The influence of the statistical errors upon the result of computation can be reduced by the factor $(\tau/t)^{1/2}$, taking mean values of the measured results. Systematic errors characterized by correlation times $\tau > t$ can influence the accuracy of the tracking data.

The following are examples of the various types of errors:

Systematic technical errors: errors in the distance of the phase centre of the antennae (interferometer system), errors due to time delay in cables and receiver.

Statistical technical errors: errors caused by internal noise.

Systematic propagation errors: errors in angle, range or range rate due to media with an index of refraction varying with height.

Statistical propagation errors: fluctuations of the index of refraction of the troposphere and ionosphere.

If any information is available about the magnitude of the systematic error, this can be used to correct the measured values of angle, range or range rate. In case of the systematic propagation error, one tries to obtain information about physical parameters of the atmosphere, from which the resulting errors can be computed or estimated by means of the electromagnetic wave theory. With this additional information a correction can be made to the measured values, except for a remaining error defined as the systematic residual error. It is the purpose of the correction method presented here to reduce the atmospheric residual error to a minimum. This is especially important when the operating frequency of the system (e.g. the vhf range) gives rise to propagation errors which are large compared with the technical errors. It is planned to apply this method to the correction of tracking data of the first German satellite tracking system at Weilheim (project Z-DBS).

For more details see [1].

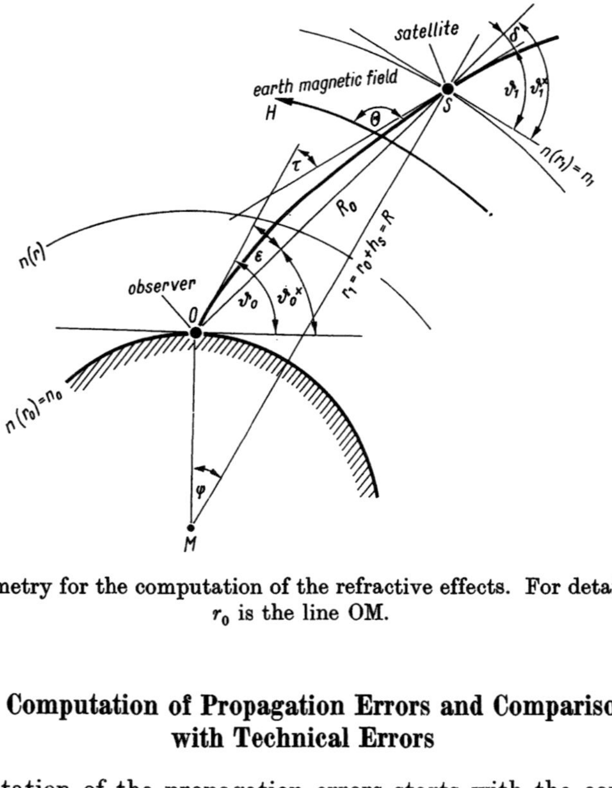

Fig. 1. Geometry for the computation of the refractive effects. For details, see text. r_0 is the line OM.

2. Computation of Propagation Errors and Comparison with Technical Errors

The computation of the propagation errors starts with the construction of the profile of the refractive index of the troposphere and ionosphere. Periodical measurements of the physical parameters of the troposphere are carried out using balloon soundings in the vertical direction. Each day, at noon and midnight, atmospheric pressure p (mb), temperature T (°K) and water vapour pressure e (mb) are measured up to a height of about 12 km above the ground. From these data the refractive index of the troposphere can be derived [2].

Many ionospheric research institutes carry out soundings of the lower ionosphere at time intervals of two minutes, measuring the critical frequency f (MHz) as a function of the apparent height up to the maximum of the F2 layer. From such an ionogramme the actual height h and the electron density N_e can be derived in order to determine the refractive index for the lower ionosphere [3].

Further information about the ionosphere is available, if the total electron content $I = \int_0^{h_s} N_e(h)\, dh$ of the ionosphere is known, by means of Faraday rotation or Doppler frequency measurements (h_s = satellite height). Knowledge of the principal physical processes in the upper ionosphere leads to a theoretical model distribution of electron density versus height. This can be used also for the construction of the refractive index profile. Sometimes the results of satellite soundings for the upper ionosphere are available, e.g. from topside sounders such as Alouette.

The wave equation is solved under the following assumptions:

(i) The wave absorption in the medium is neglected, i.e. the refractive index is a real quantity. The medium is assumed to be linear. In the troposphere this assumption is sufficiently fulfilled for operating frequencies below 3000 MHz. The assumption holds also in the ionospheric region for frequencies above 100 MHz.

(ii) The variation of the index of refraction has to be small within one wavelength λ_0 of the satellite signal. With this assumption the principles of geometrical optics can be applied in solving the wave equation. The assumption results in a limitation of the frequency range towards low frequencies, which is already a consequence of assumption (i). On the other hand, the range of angle of elevation for which the computations are valid is limited to angles of $\theta_0 \gtrsim 20°$. At lower angles of elevation strong scintillations were observed that are thought to be due to ionospheric and tropospheric diffraction effects.

(iii) The refractive medium is assumed to vary only in the radial direction, i.e. horizontal gradients of the refractive index are neglected. This assumption becomes less valid at lower angles of elevation. Consequently, for increasing distances between observation point and sub-satellite point increasing errors arise in the computation of the propagation effects. Strong gradients in the longitudinal direction are present across the sunrise or sunset lines. Horizontal gradients also occur with change in latitude, the total electron content decreasing towards the north. Measurements of several ground stations (with latitudes within 20 and 60° N) showed variations of total electron content from 1:1.5 at 0800 LMT to 1:2 at 1800 LMT.

Fig. 1 illustrates the geometry used for the computation of propagation errors. The plane of the paper is coplanar with the plane through the points M (centre of the earth), O (position of the observer), S (position of the satellite). The equations used for the construction of the refractive index from the values of temperature, atmospheric pressure and water vapour pressure in the troposphere and from the critical frequency in the ionosphere are as follows:

Index of refraction, troposphere:

$$n(h) = N(h) \times 10^{-6} + 1$$

$$N(h) = 77.6\, p(h)/T(h) + 3.73 \times 10^5\, e/T^2(h)$$

N = refractivity, p = atmospheric pressure (mb), T = air temperature (°K), e = water vapour pressure (mb).

Index of refraction, ionosphere:

$$n_{i,e} \approx [1 - X/(1 \pm Y \cos\theta)]^{1/2} \approx (1 - X)^{1/2} \qquad (f_0 > 100 \text{ MHz})$$

$$X = f^2/f_0^2\,, \qquad Y = f_H/f_0$$

f_0 = operating frequency, $f = e(N_e/4\,\pi^2\, m\, \varepsilon_0)^{1/2}$ = plasma frequency or critical frequency, $f_H = \mu_0\, He/(2\,\pi\, m)$ = gyromagnetic frequency.

The formulae for the computation of the propagation errors deduced [4, 5] under the assumptions mentioned above are:

geometrical path: $R_0 = (r_0^2 + r_1^2 - 2\, r_1\, r_0 \cos\psi)^{1/2}$

phase path: $R_p = \int_{r_0}^{r_1} n^2(r)\, r\, [\{n(r)\, r\}^2 - (n_0\, r_0 \cos\theta_0)^2]^{-1/2}\, dr\,,$

group path: $R_g = \int_{r_0}^{r_1} r\, [\{n(r)\, r\}^2 - (n_0\, r_0 \cos\theta_0)^2]^{-1/2}\, dr\,,$

range error: $\Delta R_p = R_p - R_0\,, \quad \Delta R_g = R_g - R_0\,,$

refraction angle: $\tau = \int_{r_0}^{r_1} n_0\, r_0 \cos\theta_0\, n'(r)\, [\{n(r)\, r\}^2 - (n_0\, r_0 \cos\theta_0)^2]^{-1/2} \times \times\, n(r)^{-1} dr$

angle error: $\varepsilon = \arctan\,[(\cos\tau - \sin\tau \tan\theta_1 - n_1/n_0)\,(n_1/n_0 \tan\theta_0 - \sin\tau - - \cos\tau \tan\theta_1)^{-1}]\,,$

kinematic Doppler effect: $\dfrac{\Delta f}{f} = -\dfrac{(r_0 \cos\theta_0^*\, \dot\psi + \sin\theta_1^*\, \dot R)}{c_0}\,,$

atmospheric Doppler effect: $\dfrac{\Delta f_a}{f} = -\dfrac{[r_0 \cos(\theta_0^* + \varepsilon)\, \psi + \sin(\theta_1^* + \delta)\, \dot R]}{c_0}\,,$

Doppler frequency shift error: $\Delta F = \Delta f_a - \Delta f\,.$

Faraday rotation: $\Omega = \pi\, \lambda_0 \int_{r_0}^{r_1} X(r)\; Y(r) \cos\theta\, ds = K\, \overline{M}\, I/f_0^2\,,$

$$K = e^3\, \mu_0 (8\, \pi^2\, \varepsilon_0\, m^2\, c_0)^{-1}\,; \qquad M = H \cos\theta \operatorname{cosec}\theta\,,$$

$$I = \int_0^{h_s} N_e(h)\, dh\,.$$

In the case of an interferometer system (base line 126 m, operating at $f_0 = 136$ MHz) Table 1 compares the systematic technical errors with the systematic propagation errors. Obviously the accuracy of a satellite tracking system at vhf range strongly depends on the efficiency of the method used for the correction of the propagation error.

Table 1

Comparison of the systematic technical errors with the systematic propagation angle error for vhf frequencies

technical errors					
basis: geometrical error: ± 1 mm					0.008 mrad
electrical					0.016 mrad
multipath effect					0.010 mrad
cable					0.020 mrad
receiver: measurement of frequency					0.010 mrad
other phase errors after internal balance					0.020 mrad
propagation errors in the ionosphere and troposphere					
elevation angle θ_0 [°]	20	25	30	40	
angle error ε [mrad]	2.53	1.92	1.49	1.14	

assumptions:

height of satellite: $h_s = 1000$ km

refractive index in troposphere: computed from an exponential profile for refractivity with the values $N_0 = 320$, $H_t = 7.5$ km

refractive index in ionosphere: computed from a Chapman profile for electron density with the values $N_{em} = 1.2 \times 10^6$ cm^{-3} ($f_m = 10$ MHz), $H_i = 60$ km, $h_m = 330$ km.

3. Correction of the Tropospheric Propagation Error

The correction of the tropospheric propagation error was started with the analysis of 80 curves of the modified refractive index built up from balloon soundings of pressure, temperature and humidity in the troposphere. Because there was no remarkable difference between day- and night-time values of the refractivity, the correction profile was constructed as time independent. All results of the 80 measurements lie within the limiting exponential profiles defined by the values $N_0 = 318.5 \pm 23.5$ N-units and $H_t = 8.025 \pm 0.375$ km. These exponential profiles enclose also the mean values of 234 profiles of a Canadian station (78° N 103° W) and the mean values of 274 profiles of a station in the Pacific Ocean (2° S 171° W). From this one can conclude that in the computation of tropospheric errors neglect of horizontal gradients is in good agreement with the real situation, i.e. a correction profile for refractivity would be largely independent of the geographical position of the satellite ground station.

Among the 80 curves there were eight types which showed characteristic and extreme deviations from a theoretical exponential profile, such as refractivity profiles abruptly increasing or decreasing with height. The largest jump in refractivity was reported with $\Delta N = 60$ N-units, a value assumed to be rare. In order to use such realistic profiles in a digital programme for the computation of the resulting propagation errors, these curves were fitted to a series of exponential and straight line functions. The resulting errors lie within the errors attributed to the limiting exponential profiles as mentioned above. Using the mean exponential profile with the parameters $N_0 = 318.5$ and $H_t = 8.025$ km as a correction profile, the tropospheric residual errors can be considered to be small compared with the ionospheric residual errors, see § 4.3.

For an elevation angle of $\theta_0 = 20°$ and for a given orbit of the satellite Geos 2 ($1100 \leqq h_s \leqq 1400$ km), the following values are given: systematic residual angle error: $\sigma = \pm 0.06$ mrad ($= \pm 3.44 \times 10^{-30} = \pm 13.2''$), 6.7% of the systematic angle error; systematic residual range error: $\delta = \pm 1$ m, 16.7% of the systematic range error. From these results we consider the correction of the tropospheric propagation error to be sufficient.

4. Correction of the Ionospheric Propagation Error

4.1. Concept of the Correction Method

Contrary to the tropospheric parameters, the ionospheric parameters show a strong dependence on diurnal, seasonal and solar cycle variations. As a consequence the systematic propagation errors have the same time dependent behaviour. Fig. 2 shows the variation of the angle and the range error from hour to hour in November 1969 for a simulated passage of the satellite Geos 2. The correction method presented here, which was proposed by Prof. Rawer, takes into account the time dependence of the ionospheric medium. It can be divided into two parts: a) the prediction of the electron content and the distribution of the electron density, and b) the correction to these predictions by real-time measurements.

Observations over many years, made by stations investigating the electron content and the distribution of electrons in the ionosphere up to the maximum of the F2 layer, were used to make predictions of the critical frequency and corresponding heights of the E, F1, and F2 layers, when the actual number of

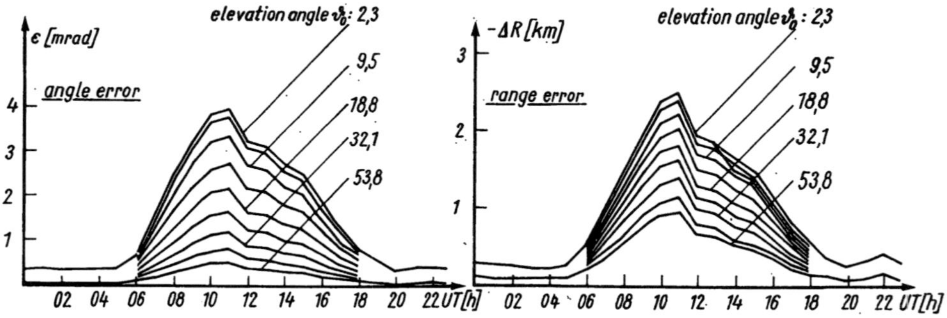

Fig. 2. Variation of the ionospheric angle and range error from hour to hour, 1–11 November 1969 (Satellite Geos 2, $1100 \leqq h_s \leqq 1400$ km).

sunspots is taken into account. The electron density in the lower ionosphere is approximated by four parabolic functions, and the electron density in the upper ionosphere by a cosine and an exponential function. The total set of ten parameters necessary to construct the profile are given for each hour over a ten-day interval. A real-time correction is possible if measurements of the critical frequency and/or the electron content are carried out during the tracking time. Data for the lower ionosphere are continuously reported and are easily available from adjacent ionosonde stations of nearly the same latitude as the satellite tracking station. It is more efficient for a real-time correction to measure the total electron content, using either the satellite to be observed or a beacon satellite.

4.2. Determination of the Prediction Data

According to [5] the height of the base and the thickness of the ionospheric layers up to the height of the corresponding critical frequencies can be determined if the following data are given for each layer: critical frequency, minimum apparent height (apparent height of the base), maximum usable frequency (MUF) determined for a distance of 3000 km (considering only the F1 and F2 layers). These data are constructed for a prediction period of ten days by taking the mean value from the corresponding figures for the previous year and from the corresponding values for a period ten days ahead of the prediction period. By taking now the known parameters for the height of the base (3 parameters), the thickness (3 parameters) and critical frequencies (3 parameters) of the layers, parabolic functions are built up describing the profile of the electron density for the lower ionosphere. The electron content for the lower ionosphere is computed by integrating the parabolic functions from the base of the lowest layer up to the maximum of the F2 layer. In order to predict the electron content of the upper ionosphere and in order to compute the corresponding layer thickness, the total electron content has to be known. A prediction of this value is carried out as follows: From 1966 to 1968 the electron content was measured by means of the satellite Explorer 22 by the ionosonde stations in Breisach, Köln, Lindau (West Germany), Graz (Austria), and Val Joyeux (France). For the period of minimum solar activity (1964–1965), only the electron content data from the station Lindau were used. The data of a station were collected month by month and hour by hour. The mean value $\bar{I}$ is taken from data of the same hour but generally from different days.

The resultant value of the electron content is the mean value of the corresponding smoothed quantities from all stations and is representative of the geographical region 45–55° N 0–15° E. The finally accepted values for the variation of electron content with local time and sunspot number are presented in tabular form. Given the actual sunspot number, the prediction of the total electron content is then possible.

4.3. Correction of the Predicted Data by Real-time Measurements of the Electron Content

Several methods are available for measuring the actual electron content using either the observed satellite itself or a beacon satellite. These methods differ widely in accuracy of measurement, and some require expensive and complicated equipment at the satellite and at the ground station. We consider Faraday measurements a good concept to start with, because this method requires only one satellite frequency. In general, the antenna and receiver at a ground station are not able to determine the direction of the polarization of the incident satellite signal, but this will be possible with the German satellite tracking system at Weilheim. The conditions which have to be met when the Faraday rotation method is applied are discussed in [1].

The expected error of about 15% of the Faraday measurement is mainly caused by taking the mean value of the magnetic-geometric factor M and by neglecting horizontal gradients in the electron density distribution. The differential Doppler method needs no knowledge of the geomagnetic field, therefore

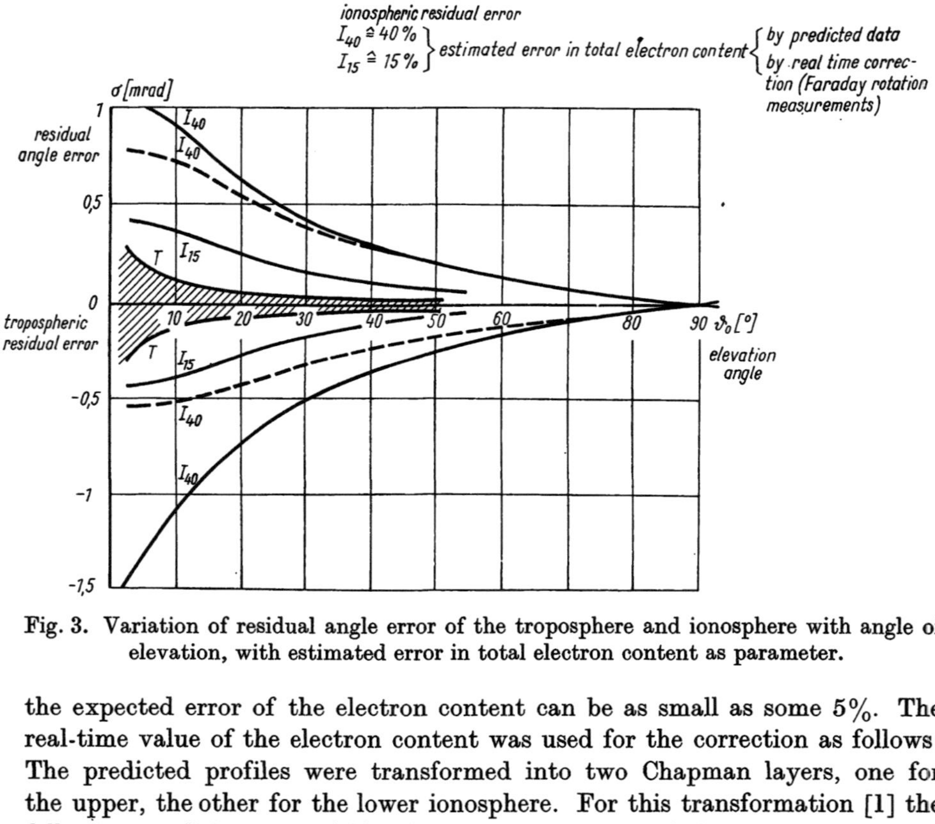

Fig. 3. Variation of residual angle error of the troposphere and ionosphere with angle of elevation, with estimated error in total electron content as parameter.

the expected error of the electron content can be as small as some 5%. The real-time value of the electron content was used for the correction as follows: The predicted profiles were transformed into two Chapman layers, one for the upper, the other for the lower ionosphere. For this transformation [1] the following conditions were fulfilled: (i) the maximum of the critical frequency of the F2 layer and the corresponding height of the Chapman layers are equal to the corresponding values of the predicted profile; (ii) the total electron content remains constant; (iii) the ratio k between the electron content of the upper and lower ionosphere remains constant.

The scale heights H_{i1} and H_{i2} of the two Chapman profiles were computed iteratively. The transformation has the advantage that a fairly simple profile is obtained which can be used easily for real-time corrections. At an elevation angle of $\theta_0 = 20°$ the error due to the transformation is as high as 1.3 and 0.66% for the angle and range error, respectively. The real-time value of the electron content gives rise to a second iteration process generally resulting in two new values H'_{i1} and H'_{i2}. The ratio k again remains constant, when there are no simultaneous measurements of the critical frequency.

Fig. 3 shows the variation of residual angle error with angle of elevation for a given passage of Geos 2. The lines I_{40} enclose the range of the residual angle error, for predicted parameters of the ionosphere (here predicted for the first ten days of May, 1970). The lines I_{15} enclose the range of the residual angle error when the real-time electron content is given by a Faraday rotation measurement. The continuous lines I_{40} are valid for the case when the error in the electron content is affected only by errors in the critical frequency of the F2 layer ("worst case"), the dashed lines I_{40} are valid for the case when the error in the electron content is affected only by errors in the scale height ("best case"). The real residual errors lie between the continuous and the dashed lines. The curves T enclose the tropospheric residual error given by the exponential correction profile of § 3. In order to obtain the same ionospheric residual error (compared with the troposphere) at an angle of elevation of 20°, the electron content has to be known to an accuracy of $\pm 5\%$. Then, however, the error of the transformation and the error due to the neglected geomagnetic field become important.

5. Summary

A correction method for radio tracking errors due to atmospheric propagation effects of vhf waves is reported. The computation of the various errors in phase path, group path, elevation angle, Doppler frequency shift, and Faraday rotation are performed assuming the principles of geometrical optics. Effects of absorption and anisotropy (except Faraday rotation) are neglected and a spherically symmetric distribution of the refractive medium is assumed. These assumptions limit the range of frequency and angle of elevation, where the correction method is applicable. It is found that the correction method can be assumed to be valid for a frequency range $100 < f_0 < 3000$ MHz and for elevation angles $\theta_0 \gtrsim 20°$.

The correction of the tropospheric propagation error turns out to be uncomplicated, because of the relatively time independent behaviour of the tropospheric refractive index. A time independent correction profile can be constructed, by means of which in the vhf range the tropospheric residual error is expected to be much smaller than the ionospheric one.

The correction of the ionospheric error is carried out by using a prediction of the critical frequencies of the ionospheric layers and the total electron content, and by using real-time measurements of the electron content with the satellite to be tracked or with a beacon satellite. The predicted values are valid for each hour and for a period of ten days. A prediction can be uncertain to some $\pm 40\%$ when the actual solar activity is taken into account. The correction of the predicted electron content by a real-time Faraday rotation measurement reduces the ionospheric residual error to about $\pm 15\%$.

The following residual errors in percentage of the corresponding total systematic error are given ($f_0 = 136$ MHz) for a simulated passage of the satellite Geos 2 ($1100 \leqq h_s \leqq 1400$ km) and an angle of elevation $\theta_0 = 20°$ (see Fig. 3):

	Systematic angle error (mrad)	Residual angle error (%)	Systematic range error (km)	Residual range error (%)
Troposphere	0.8	± 6.7	0.006	± 16.7
Ionosphere	1.68	A: ± 40 B: ± 15.5 C: ± 5.2	1.7	A: ± 38 B: ± 15 C: ± 5

A: ionospheric parameters given by prediction
B: real-time correction by Faraday rotation (only one operating frequency necessary)
C: real-time correction by differential Doppler frequency (at least two frequencies)

Acknowledgment

This work was suggested by Dr. W. Fogy, leader of the project Z-DBS. Gratitude is expressed to Prof. K. Rawer for valuable discussions, to Mr. Ch. Münther who made available to us the results of the electron content measurements from the satellite Explorer 22, and to Mr. J. Heisig who contributed the part of the report concerning the prediction data. Gratitude is also expressed to Miss M. Thimm for the evaluation of the numerous ionospheric data from measurements and to Mrs. H. Stark for carrying out the computations and preparing the sub-routines of the digital programme.

References

[1] P. EDENHOFER, D. GLESNER and V. STEIN, DLR-Forschungsbericht 72-, to be published in 1972.
[2] E. K. SMITH and S. WEINTRAUB, Central Radio Propagation Lab., Natl. Bur. of Standards, Rep. No. 1938, September 1952.
[3] K. RAWER, Space Sci. Revs. 3, 407 (1964).
[4] P. EDENHOFER, Proc. Astronaut. Congr., Madrid 1966, 2, 105.
[5] D. GLESNER and P. EDENHOFER, Deutsche Versuchsanstalt für Luft- und Raumfahrt E. V., Mitteilung 68-15, Juli 1968.
[5] C. TAIEB, Annls Géophys. 18, 227 (1962).

Space Research XII — Akademie-Verlag, Berlin 1972

HIGH RESOLUTION MEASUREMENTS OF IONOSPHERIC REFRACTION

T. J. Elkins

Air Force Cambridge Research Laboratories, Bedford, Mass., USA

A long baseline radio interferometer has been constructed and used to measure variable ionospheric refraction by monitoring continuously the apparent angular position of a geostationary satellite. Results are presented for approximately eight months operation during which the interferometer was located at Hamilton, Massachusetts (42.6° N 70.8° W) and the 136 MHz beacon on ATS-3 was used as a source. The interferometer baseline was 10 km long, oriented in the local magnetic meridian plane. The measurements have revealed pronounced diurnal and seasonal variations in the amplitude of the variable refraction component. The diurnal component becomes a maximum near local noon, and appears to be partly controlled by the normal diurnal variation of integrated ionospheric electron content. The refraction perturbations appear to be caused by the propagation of internal gravity waves in the neutral thermospheric gas. The power spectrum and diurnal variation of the mean period of the fluctuations is measured, showing a strong maximum in the range ∼20–40 minutes and very little power below ∼14 minutes.

1. Introduction

The measurement of the variable component of ionospheric refraction, at frequencies in the vhf range, has applications in radio astronomy, satellite tracking, satellite navigation systems and radar systems. An improvement in the state of these arts requires a correspondingly better knowledge of the variable ionospheric refraction phenomenon. Conversely, an improved knowledge of the variable refraction phenomenon can lead to a better understanding of dynamic ionospheric processes, and thereby to a generalization of a limited quantity of experimental data to a variety of geographical locations and observing conditions. This paper describes the results of an experiment designed to measure a component, in the horizontal plane, of ionospheric refraction, using 136 MHz beacon transmissions from the geostationary satellite ATS-3. These measurements were made with a long baseline interferometer at the Sagamore Hill Radio Observatory of AFCRL (geographic coordinates 42.6° N 70.8° W; geomagnetic latitude 57° N).

2. Experimental Details

The interferometer basically consisted of two circularly polarized antennas (gain 20 dB and 16 dB) located 10 km apart along the local magnetic meridian at Sagamore Hill. During the period of approximately eight months in which observations were made, ATS-3 was generally within about 30° of South, at an elevation angle between 35° and 42°. The two antennas were connected to separate receivers whose local oscillators were maintained in phase coherence by means of a synchronizing signal transmitted over a microwave link. The i.f. signals from the two receivers were correlated, again using the microwave link for transmitting the data to the correlator. The dc output of the analog correlator, representing a component of the apparent satellite position, was recorded continuously on magnetic tape. These analog records were later digitized automatically, at a rate of 6 samples per minute, for subsequent computer data processing. The total system sensitivity was such that an rms refraction fluctuation of $\sim$1 second of arc could be detected.

3. Data Description and Data Processing

The records of apparent satellite position obtained as just described consist of three major components:

a. The magnetic north–south component of the true satellite motion. Because the plane of the satellite orbit is inclined to the earth's equatorial plane at a small but non-negligible angle, ATS-3 undergoes an oscillatory motion with respect to the equator, with a period of 24 hours.

b. Large scale, slowly varying ionospheric gradients, due to ionization production and redistribution and to travelling ionospheric disturbances (TID), cause changes in refraction which appear as apparent "motions" of the satellite.

c. Small scale ionospheric irregularities produce a random diffraction pattern on the ground in which the varying phase gradient is measured by the interferometer, appearing as a rapid "flicker" of the apparent satellite position.

At the antenna spacing used in this experiment, the diffraction effects (c) were small compared with the more slowly varying refraction effects. In order to eliminate the undesirable effects of true satellite motion, the data were initially operated upon with a bandpass filter which removed from the data the best least-mean-squared fit of a sinusoid having a 24-hour period. This procedure also removes from the data the true diurnal component of refraction angle variation, but since this study is principally concerned with high-frequency components (i.e. TID), this deficiency is not of serious consequence. The data remaining, after filtering as described, contained substantial components with periods of several hours. Since these components were not of special interest in this study, they were removed by further digital filtering. This second filter had a frequency response flat to within 1.5 dB from 6.6 minutes to 333 minutes period, with a roll-off of 6 dB per octave, at the low-frequency end.

The power spectrum of each individual day's data was then computed, and smoothed by taking a running average over ten successive points. The diurnal variations of fluctuation amplitude and frequency were found by computing, in each hour of the day, the rms amplitude of the deviation from the hourly mean and number of zero crossings of the filtered data.

4. Results

Fig. 1 shows the results of a typical day's data, with the magnetic north–south component of refraction angle, in the horizontal plane, plotted as a function of local time, after filtering. Note the pronounced component with quasi-period ~30–40 minutes, with maximum near local noon. This is a very distinctive

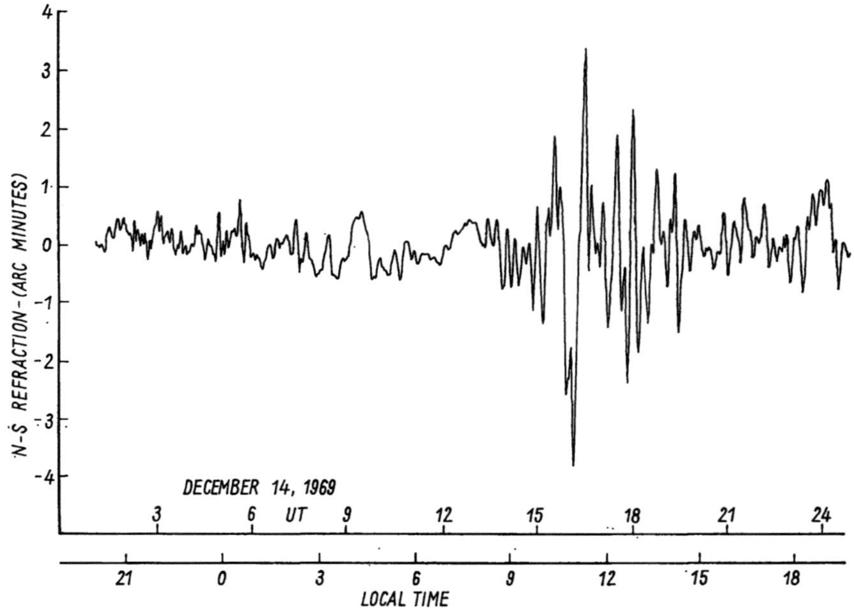

Fig. 1. Typical variation of the north–south component of refraction angle in the horizontal plane for one day, showing the diurnal variation of TID.

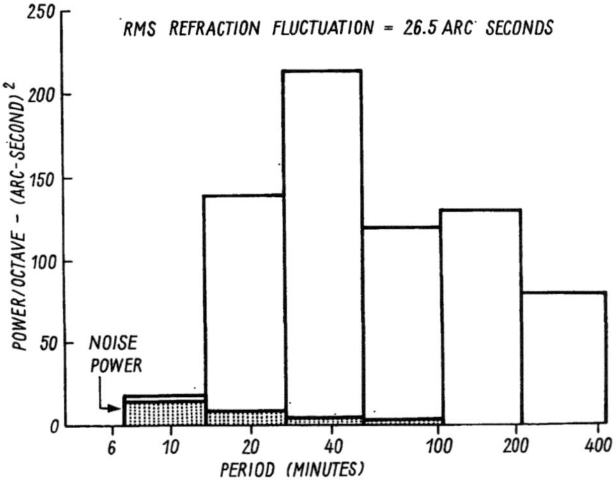

Fig. 2. Power per octave contained in the refraction fluctuations, as a function of period. The figure represents an average over a 14-day interval.

feature of all the data obtained. Fig. 2 shows the power spectrum of data similar to those of Fig. 1, averaged over an interval of 14 successive days. The power per octave is plotted as a function of the period of the fluctuation. There is a pronounced peak of power in the octave centered approximately at 35 minutes period. The shape of the spectrum at the long-period end is modified somewhat by a combination of the effects of the bandpass filter and the spectral smoothing performed. At the short-period end, however, the rapid decrease of power into the noise level is certainly of natural origin. The noise level

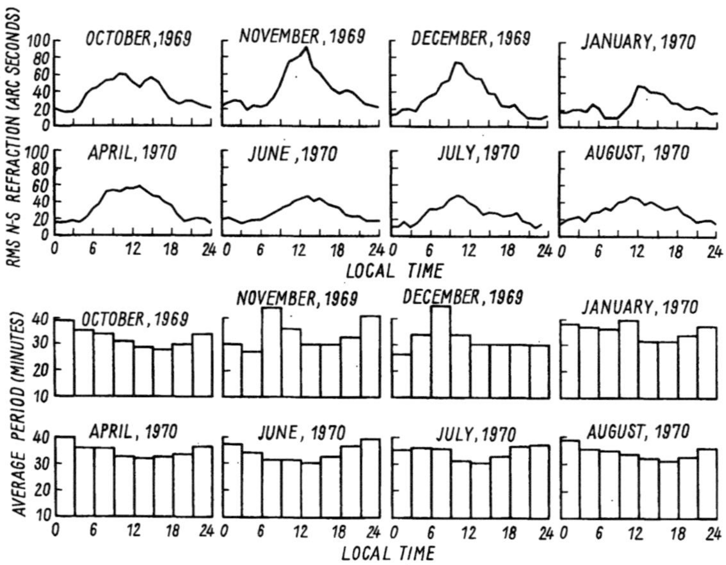

Fig. 3. Diurnal variation of refraction fluctuations, for 8 months' data: above, rms amplitude; below, mean period of fluctuations.

across the spectrum has been computed under the assumption of a white noise spectrum. While this assumption has not yet been well substantiated, the noise levels shown at the highest frequencies are considered to be reliable.

Fig. 3 shows, above, the monthly average of the rms average from the hourly mean of the fluctuations and, below, the mean fluctuation period, for the periods October 1969 to January 1970 and April to August 1970. From these results several general features are apparent: (i) The diurnal variation is such that the maximum rms fluctuation in refraction occurs within an hour or two of local noon. (ii) The seasonal variation has a maximum in November. Note that the periods of recording were such that the coverage near the equinoxes is poor. However, winter values of fluctuation amplitude exceed those for summer. (iii) The rms fluctuation, averaged over 24 hours, is approximately 30 arc seconds, which agrees closely with the estimate obtained from the power spectrum (Fig. 2).

5. Discussion of Results

Under the simplifying assumption of horizontal stratification of the ionosphere, the refractive deviation of a ray traversing the entire ionosphere is proportional to the horizontal gradient of the integrated (total) electron content (TEC) [1]. A preliminary comparison of refraction measurements with simultaneous TEC data along the identical ray path has shown that this relationship is usually approximately satisfied on a time scale of the order of one hour or less. For longer period fluctuations, this simple approximation appears not to be valid.

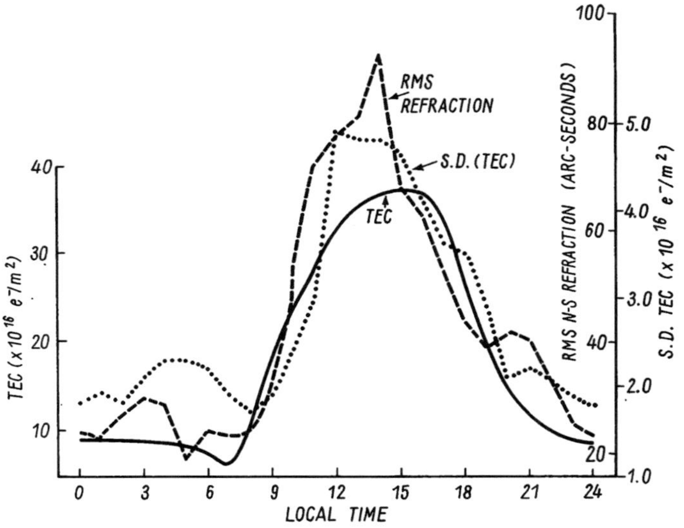

Fig. 4. Diurnal variation of rms refraction fluctuation amplitude, TEC and the standard deviation of TEC variability, for November 1969.

The observed refraction fluctuations, caused by TID, have been observed by a number of different techniques, over a period of many years. They are currently thought to be a manifestation, in the ionospheric plasma, of internal gravity waves propagating in the neutral thermospheric gas. This assumption is compatible with some of the present observations. The sharp decrease of spectral power at the high-frequency end of the spectrum in Fig. 2 is consistent with the theoretically predicted evanescent propagation of gravity waves with period shorter than the Brunt period (~14 minutes in the F region). Simultaneous measurement of TEC variations at each end of the interferometer baseline has consistently indicated a horizontal trace velocity for the waves of ~150 m s^{-1}, which is typical of freely propagating gravity waves with period ~35 minutes.

It has been noted previously, e.g. [2], that the amplitude of TID appears to be an approximately constant percentage, throughout the day, of the TEC. Thus greater fluctuation amplitudes are observed during the daytime, when the TEC is high, than at night, when TEC is low. This effect certainly appears to be responsible for at least part of the diurnal variation of fluctuation amplitude observed. However, closer examination of the data shows that the fluctuation amplitude usually reaches its maximum an hour or two earlier than TEC. This is shown, for the month of November 1969, in Fig. 4. The mean refraction amplitude, the mean TEC, and the standard deviation of the TEC variations are shown as functions of local time. It is highly probable that the TEC variability is predominantly due to TID, but the TEC itself reaches its maximum about two hours later than the fluctuations. This discrepancy may be due to a superposition of a true source effect; i.e. the amplitude of the source of the TID may reach maximum shortly before noon, whereas the TEC reaches its maximum at about 1400 local time.

6. Summary

The results of some measurements of TID, using a long baseline radio interferometer to monitor a 136 MHz geostationary satellite beacon, have been presented. Pronounced diurnal and seasonal variations have been observed in the amplitude of the refraction fluctuations produced by passage of these TID through the ray path from the satellite. The diurnal maximum occurs near local noon and is apparently partly a result of a diurnal variation of experimental sensitivity brought about by the regular daily variation of integrated electron content. There also appears to exist a true source effect which suggests that the intensity of the source of these disturbances has its maximum an hour or two before noon.

The power spectrum and mean period of the refraction fluctuations have also been determined. The dominant period in the spectrum lies in the range ~30–40 minutes. The average rms refraction fluctuation (in the spectral range ~0.25–5 hours period) is ~0.5 minutes of arc. This would indicate the precision with which an astronomical (or extra-ionospheric) source could be located, at a frequency of 136 MHz, in a single day's observation.

References

[1] R. S. LAWRENCE, C. G. LITTLE and H. J. A. CHIVERS, Proc. Inst. Elect. Electron. Engrs. **52**, 4 (1964).
[2] J. E. TITHERIDGE, J. Atmos. Terr. Phys. **30**, 73 (1968).

Space Research XII — Akademie-Verlag, Berlin 1972

BRIEF REVIEW OF SCINTILLATION STUDIES

G. K. Hartmann

Max-Planck-Institut für Aeronomie, Lindau/Harz, FRG

A wealth of experimental data on ionospheric irregularities — ~90% amplitude, ~7% phase, and ~3% angle-of-arrival measurements — is already available and the literature is now very extensive. But there is still no generally accepted theory of the cause of these irregularities. It is important that future experimental observations be planned with a view to deciding between the various theories. More detailed comparisons between theories developed from geophysical aspects, from communication aspects, from tropospheric aspects, and from interplanetary aspects should be made. Mutual effects might then lead to new theoretical and experimental approaches. More detailed theoretical and experimental studies, especially in the high latitude regions, are required in order to resolve discrepancies and ambiguities concerning ionospheric–magnetospheric interactions. Strictly speaking a final and better statement on the correlation between the scintillation boundary and other magnetospheric quantities is still required. Resolving fine structure problems and problems of so-called "patch-like scintillations" requires more sophisticated experiments. Due to recently reported ionospheric–magnetospheric interaction phenomena the high latitude area especially is of specific future interest. A final section briefly summarizes outstanding questions.

1. Introduction

The literature on radio star and satellite scintillations is now very extensive, and a complete review is impossible in the space and time available; instead a few special topics will be selected for discussion. Earlier reviews covering different aspects are given in [1–12].

Not only papers published in scientific journals but also available reports, conference proceedings and lecture series were reviewed in order to demonstrate the wide spread of useful information and to emphasize the problems of obtaining an "overall picture" in this field, e.g. to report correctly about the most important outstanding problems. An additional improved information system is required that enables quick access time to the highest possible percentage of actual and relevant information on the entire world. A wealth of experimental data is already available about the main properties of the irregularities, but there is no generally accepted theory of the cause of the ionospheric irregularities. It is important that future experimental observations be planned with a view to deciding between these theories.

Interplanetary scintillations are treated in [13] and some interesting communication aspects in [14–17]. Communication and navigation engineers require a reliable prediction of the scintillation quantities, which can only be established with the required accuracy when there are more details available about the mechanism of the production of the irregularities.

2. Studies of Scintillation

Propagation of electromagnetic waves through an inhomogeneous ionosphere and troposphere usually results in variations in the amplitude, phase and angle of arrival of the waves received at the ground. These variations are known collectively as scintillations.

2.1. Type of Measurements

Since amplitude measurements require in general only a fairly simple and inexpensive equipment roughly 90% of the scintillation data now available were obtained by amplitude measurements. This is why the expression "scintillation" is often exclusively related to measured amplitude fluctuations. Phase measurements, strictly speaking relative phase measurements, require more sophisticated and much more expensive equipment. Therefore only approximately 7% of the available scintillation data were obtained by phase measurements. Angle-of-arrival measurements require still more complicated equipment, especially a very accurate antenna system. Therefore only 3% of the data were obtained by angle-of-arrival measurements. A brief theoretical treatment is given by Forsyth [18].

A few data are also available on polarization diversity measurements [19]. Frequencies which were used for scintillation studies, emitted by radio stars, low orbiting satellites, geostationary satellites, synchronous satellites and other sources, e.g. quasars [20, 21], are ranged between 20 MHz and 15 GHz.

2.2. Theoretical Considerations

In treating the phenomenon of scintillation we have to deal with the propagation of radio waves through regions of the troposphere and ionosphere in which there exist irregularities of refractive index. The main emphasis will be given to ionospheric irregularities (for reference, some papers related to tropospheric irregularities or resulting scintillations are [22–27]).

There are three possible theoretical approaches [4]:

I. The problem can be treated as one of scattering [6].

II. The second method makes use of diffraction theory [28]. This led to the idea of an equivalent thin phase screen. This is a screen of negligible thickness, which would produce the same phase variations as the actual irregular region. An important parameter in this approach is the rms phase deviation impressed on the wavefront by the phase screen; this is usually denoted by φ_0.

III. The third possible general approach is that of ray optics [21].

It is a mistake to regard the three approaches as being in any way in conflict. Within the range of validity of each, they will give identical results if correctly applied.

In most cases diffraction theory is applied to ionospheric scintillation problems [10, 28–31].

The relation between the quantity which was theoretically derived — under certain assumptions — from diffraction theory and the quantity used experimentally as a measure of scintillation depth was reviewed by Singleton [10]. The approximate expression for the scintillation index, previously considered to be valid for $\varphi_0 < 1$ rad (φ_0 = rms phase deviation) has been shown to have a wider application by taking into account saturation and focusing effects. For geophysical purposes several groups have adopted a special definition of scintillation [32]. Aarons [8, 9] mentioned that this definition can be related to the index S 3 of Briggs and Parkin [29]. See also URSI Information Bulletin No. 168, Sept. 1968. Some further problems in using the scintillation index are mentioned by Oksman [33].

2.3. Frequency Dependence of Ionospheric Scintillations

Following Briggs and Parkin [29] only amplitude fluctuations are considered. They obtain, concerning the scintillation index

$$\frac{S(\lambda_1)}{S(\lambda_2)} = \frac{\lambda_1}{\lambda_2} \text{ (far zone)} \qquad \frac{S(\lambda_1)}{S(\lambda_2)} = \left(\frac{\lambda_1}{\lambda_2}\right)^2 \text{ (near zone)}$$

where λ = wavelength. An inversion effect was reported by Briggs [4]. Golden's [34] findings indicate approximately an inverse square root ($1/\sqrt{f}$) characteristic for equatorial scintillation during the autumnal equinox. There is no satisfactory theoretical explanation yet.

2.4. Height and Correlations of Ionospheric Irregularities

Yeh, Flaherty and Paul [35] give an excellent review on various experimental possibilities to determine the heights of scintillation producing irregularities. The existence of multiple layers of irregularities has been reported by Jespersen and Kamas [36]. Kidd [37] reported heights between 130 and 900 km and Aarons [8] corroborated these results. On several occasions the observed scintillation pattern appeared to have a very regular or sinusoidal nature [38–40]. They determined by spaced receiver techniques the heights of these irregularities in the ionosphere. Hartmann [41, 23] reported that some of the observed regular scintillations are very probably associated with tropospheric inversion layers. A clear distinction between both effects and statements seems to be possible only by using modified experimental techniques.

Other methods for experimental observations are mentioned in [42, 43].

While it is generally agreed that on the average there is a correlation between the occurrence of scintillations and spread-F several authors have noted discrepancies when the occurrence of these two phenomena are compared in detail [44–46].

There is no clear picture of correlation with the sporadic-E layer yet in spite of the fact that now several reports are available [47–53], mainly due to the incompatibility of the various observation techniques.

3. The Three Major Ionospheric Scintillation Areas

A great many ionospheric and geomagnetic parameters display geographical characteristics that suggest the division of the earth into three main latitude zones. For most purposes the geomagnetic latitudes $\pm 20°$ will be taken as defining the boundaries of the equatorial zone, whereas the polar zones lie between $\pm 65°$ and the respective magnetic pole. The intervening zones, between 20° and 65°, will be referred to simply as the medium latitude or temperate zone.

A great variety of physical pictures has been suggested, e.g. for F-region irregularities, in order to account for the observations of various authors. As already pointed out, the apparent discrepancies between these representations are often due to the use of different observing techniques.

Equatorial Region

Equatorial scintillations start abruptly, frequently reaching maximum fading in a few minutes. Once a deep fading is started the level is maintained for several hours. A clear night-time maximum in frequency of occurrence and intensity was reported by various authors. Individual irregularities measure from 200 m to several kilometres in the short dimension. The irregularity is probably aligned along the earth's field, e.g. in the N–S direction, and can be elongated by a factor of 30 to 1 or more. The instantaneous extent of the irregularity region frequently reaches 5000 km [54, 55] with localized control of the irregularities. A maximum scintillation index is observed during the equinoxes with minima during the solstices [34].

The scintillation belt expands towards mid-latitudes with declining solar activity [56, 57]. A negative correlation with K-index has been reported [58, 59] but there are also findings which are in contrast [60].

Apart from Martyn's [61] theory of irregularity enhancement, which was not corroborated by experiments, it seems likely that the irregularity structure exists continuously at the upper levels of the ionosphere where diffusive equilibrium prevails [9]. These irregularities apparently consist of field aligned ducts or sheets of ionization, and are also observed in total electron content measurements during the day [62].

Medium Latitude Region

Mid-latitude amplitude fluctuations of radio signals are noted but neither their depth of fluctuation nor their frequency of occurrence are comparable with equatorial and high latitude fading. Several aeronomic parameters display pronounced activity in the range 50° to 60° geomagnetic latitude. Among others, radio star and satellite scintillation increases sharply northward of 55° [63–66]. A clear night-time maximum in frequency of occurrence and intensity was reported by many authors. The size of the irregularity can reach several hundred kilometres with its major axis in the N–S direction and an ellipticity 1.5–2.0 [67].

The various reports about the seasonal behaviour do not so far reveal a clear picture. However, Hartmann [66] reports that the boundary of the above mentioned "50°–60° belt" is further north in summer than in winter and moves south with increasing solar activity.

A strong positive sunspot correlation was reported by Chivers [68], a positive magnetic activity correlation by Shimazaki [69]. Despite the complexity of this region it seems reasonable to assume that precipitation, from the magnetosphere, of low energy electrons plays an important role for a possible mechanism of generation.

Polar Region

Aarons [8] defined three subelements of the high latitude structure. These are the polar cap, the scintillation oval, and the subauroral region. Observationally, scintillations in the polar region are similar to those observed at low latitudes, only the fading rate is in the mean faster by a factor of 2–10. A clear night-time maximum was reported by many authors. The equatorward boundary of the high latitude region moves to a low invariant latitude of approximately 57° at the period around midnight [65]. Dyson [70] measured the boundary by a Langmuir probe. No clear picture is available concerning patch size and patch alignment and seasonal behaviour.

It seems to be a general rule that the oval-locus of high latitude disturbances moves equatorwards with increasing magnetic disturbance and also, on a much slower scale, with increasing sunspot number [9]. Malko et al. [71] have noted these characteristics for the auroral oval. Muldrew [72] observed that the mid-latitude F-region trough moved towards the equator with increasing magnetic disturbance, and furthermore, exhibited a seasonal latitude movement, being closest to the equator in winter. However, recently there are a few fairly different findings, one by Oksman [73] where he reports: "The boundary seems to be located closest to the pole at the December solstice and closest to the equator at June solstice and it might move towards the pole with approaching sunspot maximum. These movements are suggested to be linked with recently proposed annual and sunspot cycle changes in the size of the plasmasphere".

Very recent studies [74] of electron precipitation with energies $\sim$5 eV show that their boundary seems to match very well the scintillation boundary. However the same seems to be true for precipitating very soft protons. Electrons and protons produce irregularities of different fine structure. Final decisions about the cause of the irregularities require additional and more sophisticated experimental data.

4. Conclusions

1. Theoretical work is needed on the problem of diffraction by a very thick irregular region and by multi-layer structures.

An attempt should be made at a detailed comparison of the theories now existing in the following fields of interest:

a) Ionospheric scintillations treated under geophysical aspects
b) Ionospheric scintillations treated under communication aspects
c) Tropospheric scintillations treated under meteorological and communication aspects
d) Interplanetary scintillations treated under geophysical and space research as well as space communication aspects.

It may then be that mutual effects will lead to new theoretical and experimental approaches to the general scintillation problem.

2. The search for a possible mechanism of scintillation generation must be continued since no general accepted theory is available, though many theories have been proposed.

3. The very complex findings about the movements of the irregularities require further studies.

4. More detailed studies on the fine-structure problems are required.

5. It is important to evaluate in more detail the relationship between the various observing techniques. Perhaps one should try to conceive completely different experimental techniques (one is now in a test-stage in our institute). The fine-structure problems and the problems related to the patch-like structure within the polar cap seem to be strongly dependent on those ideas. More detailed knowledge on ionospheric–magnetospheric coupling can then be expected.

6. Still more detailed investigations are required to support much better the needs of communication and navigation engineers.

Acknowledgment

The experimental scintillation studies, carried out in the Max-Planck-Institut für Aeronomie, were financially sponsored by the German Ministry of Education and Science, grant No. WRK 125.

References

[1] H. G. BOOKER, Proc. Inst. Radio Engrs **46**, 298 (1958).
[2] G. A. HARROWER, in: RASSA*, North-Holland Publ. Co., Amsterdam 1963 (p. 38).
[3] R. S. LAWRENCE, C. G. LITTLE and H. J. A. CHIVERS, Proc. Inst. Elect. Electron. **52**, 4 (1964).
[4] B. H. BRIGGS, Radio Sci. **1**, 1163 (1966).
[5] AGARD, Lecture Series No. **29** (1968).
[6] BOLGIANO, in: AGARD No. **29**, V-1 (1968).
[7] G. G. GETMANTSEV and L. M. EROUKHIMOV, Annals of the IQSY **5**, 229 (1969).

* RASSA: Radio Astronomical and Satellite Studies of the Atmosphere.

[8] J. Aarons, H. E. Whitney and R. S. Allen, in: AGARD No. **41**, 6-1 (1970).
[9] J. Aarons, AFCRL 70-0053, Report No. 106 (1970).
[10] D. G. Singleton, J. Atmos. Terr. Phys. **32**, 187 (1970).
[11] M. Z. v. Krzywoblocki, in: AGARD No. **70**, 35-1 (1971).
[12] G. H. Millman, in: AGARD No. **70**, 4-1 (1971).
[13] V. V. Vitkevich, Report presented at Belgrade Symposium on Solar terrestrial relations (1968).
[14] E. J. Fremouw and A. A. Burns, Radio Sci. **5**, 917 (1970).
[15] J. K. Hargreaves, in: Proc. Symp. on Future Application of Satellite Beacon Experiments, **1**, 16-1, Max-Planck-Institut f. Aeronomie, Lindau 1970.
[16] R. S. Kennedy, Fading Dispersive Communication Channels, J. Wiley & Sons, New York 1969.
[17] M. D. Grossi, in: AGARD No. **41**, 10-1 (1970).
[18] P. A. Forsyth, in: AGARD No. **41** (1970).
[19] T. S. Golden, Radio Sci., **5**, 943 (1970).
[20] A. P. Hewish, F. Scott and D. Wills, Nature, Lond. **203**, 1214 (1964).
[21] E. E. Salpeter, Astrophys. J. **147**, 433 (1967).
[22] J. Aarons, in: RASSA* **65**, North-Holland, Publ. Co., Amsterdam 1963.
[23] G. K. Hartmann, in: AGARD No. **70**, 12-1 (1971).
[24] V. L. Badillo, Radio Sci. **5**, 979 (1970).
[25] J. A. Lane, in: AGARD No. **70**, 15-1 (1971).
[26] L. Fehlhaber, in: AGARD No. **70**, 30-1 (1971).
[27] H. Jeske, in: AGARD NO. **70**, 20-1 (1971).
[28] E. N. Bramley, J. Atmos. Terr. Phys. **29**, 1 (1967).
[29] B. H. Briggs and J. A. Parkin, J. Atmos. Terr. Phys. **25**, 339 (1963).
[30] R. P. Mercier, Proc. Camb. Phil. Soc. **58**, 382 (1962).
[31] S. A. Bowhill, J. Res. Natn. Bur. Stand. **65D**, 275 (1961).
[32] H. E. Whitney, J. Aarons and C. Malik, Planet. Space Sci. **17**, 1069 (1969).
[33] J. Oksman, Space Research X, 882 (1970).
[34] T. S. Golden, NASA/GSFC Preliminary Report No. X-520-70-397 (1970).
[35] K. C. Yeh, B. J. Flaherty and L. M. Paul, Radio Sci. **5**, 967 (1970).
[36] J. L. Jespersen and G. Kamas, J. Atmos. Terr. Phys. **26**, 457 (1964).
[37] W. C. Kidd, Radio Sci. **5**, 975 (1970).
[38] J. Frihagen and L. Liszka, Scientific Report **6**, Kiruna Geophys. Observatory, Sweden (1964).
[39] W. Ireland and G. F. Preddey, J. Atmos. Terr. Phys. **20**, 137 (1967).
[40] T. J. Elkins and F. F. Slack, J. Atmos. Terr. Phys. **31**, 421 (1969).
[41] G. K. Hartmann, J. Atmos. Terr. Phys. **31**, 663 (1969).
[42] W. Calvert and J. M. Warnock, Proc. Inst. Elect. Electron. Engrs **57**, 1019 (1969).
[43] P. F. Checcacci, Space Research VII, 1345 (1967).
[44] D. G. Singleton, J. Geophys. Res., **74**, 1772 (1969).
[45] B. H. Briggs, J. Atmos. Terr. Phys. **12**, 89 (1958).
[46] J. L. Jespersen, in: AGARD Propagation Factors in Space Communications (1967).
[47] M. Anastassiadis and D. Matsoukas, Radio Sci. **5**, 953 (1970).
[48] D. G. Singleton and G. J. E. Lynch, J. Atmos. Terr. Phys. **24**, 353 (1962).
[49] J. Aarons, J. P. Mullen and S. Basu, J. Geophys. Res. **69**, 1785 (1964).
[50] J. Aarons and H. E. Whitney, Planet. Space Sci. **16**, 21 (1968).
[51] Zwi. Houminer, Radio Sci. **5**, 949 (1970).
[52] B. Dueno, Cornell Technical Report No. 37 (1955).
[53] R. B. Dyce et al., J. Geophys. Res. **64**, 1815 (1959).
[54] P. Bandyopadhyay and J. Aarons, Radio Sci. **5**, 931 (1970).
[55] J. Aarons, Radio Sci. **5**, 959 (1970).
[56] J. R. Koster, Report F-3, University Ghana, Accra (1968).
[57] D. G. Singleton, J. Geophys. Res. **73**, 295 (1968).

* RASSA: Radio Astronomical and Satellite Studies of the Atmosphere

[58] N. J. Skinner and R. W. Wright, Proc. Phys. Soc. B **70**, 833 (1957).
[59] K. Davies and A. F. Barghausen, in: AGARDograph **95**, Technavision, London (1966).
[60] S. Rangaswami and K. B. Kapasi, J. Atmos. Terr. Phys. **25**, 721 (1963).
[61] D. F. Martyn, Nature, Lond. **183**, 1382 (1959).
[62] J. Mass, J. Geophys. Res. **72**, 4940 (1967).
[63] A. A. Bol'shoi, Cosmic Research **4**, 502 (1966).
[64] L. M. Eroukhimov et al., Geomagn. i Aeron. **8**, 138 (1968).
[65] J. Aarons, J. P. Mullen and H. E. Whitney, J. Geophys. Res. **74**, 884 (1969).
[66] G. K. Hartmann, Mitteil. Max-Planck-Inst. f. Aeronomie, Nr. 48, Springer Verlag, Berlin 1972 (in press).
[67] V. D. Gusev et al., Dokl. Akad. Nauk SSSR **123**, 817 (1958).
[68] H. J. A. Chivers, J. Atmos. Terr. Phys. **19**, 54 (1960).
[69] T. Shimazaki, J. Radio Research Labs. Japan **6**, 669 (1959).
[70] P. L. Dyson, J. Geophys. Res. **74**, 6291 (1969).
[71] L. N. Malko et al., Geomagn. i Aeron. **6**, 243 (1966).
[72] D. B. Muldrew, J. Geophys. Res. **70**, 2635 (1965).
[73] J. Oksman, J. Atmos. Terr. Phys. **33**, 1727 (1971).
[74] N. B. C. Rao and E. J. Maier, J. Geophys. Res. **75**, 7168 (1970).

Space Research XII — Akademie-Verlag, Berlin 1972

E REGION ELECTRON DENSITY PROFILES

K. Maeda

Kyoto University, Kyoto, Japan

Mid-latitude E region electron density profiles are presented around noon in each season for any solar activity. The profiles have been obtained for summer and winter at low solar activity, based on data of rocket experiments which are relatively abundant.

A theoretical study has been made to adapt a current theory of E region formation to the above experimental results. To determine the solar activity dependence of the principal solar ionizing radiation (1025.7 Å and 977 Å), ionosonde data on f_0 E around noon taken at several stations have been studied, and the Wolf number and cos χ dependence of the peak electron density has been obtained.

Based on these results and using data on solar X radiations (2–8 Å and others) obtained from Solrad satellites, it has been possible to calculate theoretically the electron density profiles for any season and any solar activity. Some further comparisons between theory and experiment are reported for the other cases, in which the data are less abundant. Calculated profiles for representative low and high activity are given, for which no experimental data are available.

Night-time profiles and nocturnal variation of the peak density are also given.

1. Introduction

An attempt has been made by the present author to obtain the electron density profile in the lower ionosphere below 200 km, using the data obtained by rocket experiments. In the first paper [1] all data then available were simply grouped into three classes depending upon the solar zenith angle χ at the time of firing. As the majority of data were limited to solar minimum conditions and all rocket launch sites were located in middle latitudes except one equatorial case, the profile has been obtained for middle latitudes and low solar activity, excluding the data which did not match this limitation [2] (the material obtained in higher latitudes has not been used in these studies). In [2] the data in summer (May–August) for $10° < \chi < 30°$ are gathered in one group and those in other seasons are grouped into two classes depending upon the range of χ, that is, $35° < \chi < 55°$ and $60° < \chi < 80°$. A result of comparative study between the experimentally obtained profiles and a current theory of formation of the E layer as established by Nicolet and Aikin [3], Ohshio [4] and Bourdeau et al. [5] has been briefly described.

An effort to verify the validity of the formation theory of the E layer has been continued further, with a view to predicting the profiles for any season and any solar activity for which no data are available. Some results of this work are presented in [6]. Part of this paper is not adequate, because some uncertainty was involved in the estimation of the solar activity dependence of the solar ionizing radiations, which control the ionization in the E region.

In the present paper the above point has been improved, and some of the models of the neutral atmosphere to be used in theoretical calculations have been corrected, based on the rate of temperature rise near the height of peak electron density N_m (100–120 km), which has been found from cos χ dependence of N_m revealed by the study of the Wolf number dependence of the solar photon flux (1025.7 Å and 977 Å). The atmospheric models above 100 km except for summer are significantly different from the mean model of CIRA 1965. Based on this knowledge on the rate of temperature rise, three seasons in a year have been defined as follows; from April to September as summer, from November to February as winter, and March and October as the intermediate season. But in the atmospheric model representing the summer group the temperature from 100 km down to 80 km is a little higher in April than in other months.

The solar activity dependence of the solar X radiations (2–8 Å and 33.7 Å) which control the ionization below the peak density has been also obtained from observations of the X radiations by Solrad 8 and 9 satellites.

Theoretical calculations have been made for the cases in which the observed data are more or less available. It was found that the agreement between theory and experiment is satisfactory, and the theoretical result has been given for a number of representative solar activities in each season.

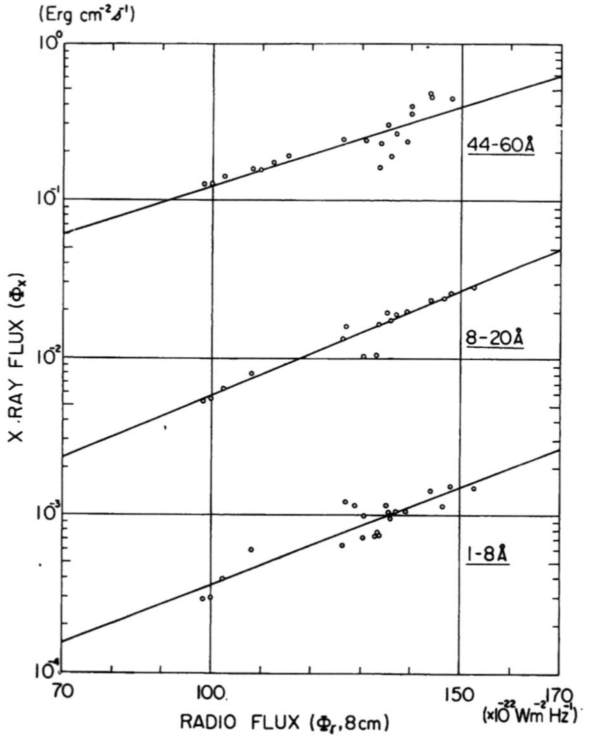

Fig. 1. Solar X ray flux Φ_x versus solar radio flux Φ_r at $\lambda = 8$ cm.

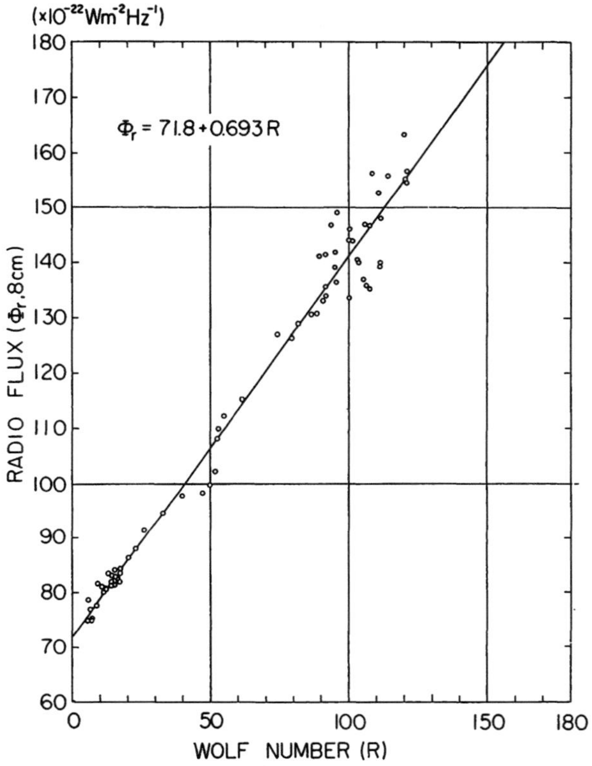

Fig. 2. Solar radio flux (Φ_r, 8 cm) versus Wolf number R.

2. Solar Activity Dependence of Solar Ionizing Radiation

2.1. Solar Activity Dependence of the Solar X Radiations (2–8 Å and 33.7 Å)

The flux densities of 1–8 Å, 8–20 Å and 44–60 Å radiations observed by the US solar X radiation satellites [7, 8] have been given as functions of Wolf number R and solar radio fluxes at 8 cm (Figs. 1 and 2) and 3.19 cm. The straight lines are the relations obtained by the least squares method. From these figures we can obtain Φ_x (1–8 Å) for any solar activity. No observation is available for Φ_x at 33.7 Å, but the relation for 8–20 Å and 44–60 Å can be used as a reference. In later calculations it is assumed that Φ_x (33.7 Å) varies with R in the same way as Φ_x (1–8 Å). It is remarked that R is the three month average of the Wolf number centred at the month of observation of Φ_x, as in other cases in the later sections.

2.2. Solar Activity Dependence of the 1025.7 Å and 977 Å Radiations

The peak electron density N_m in the E region has been measured for many years. The data around noon obtained at four Japanese stations and Washington

D.C. for the period 1944–1969 have been studied to give N_m as a function of R. In this study special attention has been paid to the cos χ dependence of N_m. According to a theory giving the rate of electron production Q generated by a monochromatic radiation when the atmospheric temperature rises linearly with height near the level of N_m (100–120 km), the maximum of $Q(Q_m)$ is given by:

$$\left.\begin{aligned} Q_m &= Q_{m0}\,(1+\varrho\,R)\,(\cos\chi)^{1+\Gamma}\,, \\ Q_{m0} &= \eta\,\Phi_0\,(\sigma\,n_0)^{-\Gamma}\left(\frac{1+\Gamma}{H_0\exp 1}\right)^{1+\Gamma}, \end{aligned}\right\} \qquad (1)$$

where Φ_0 is the unabsorbed solar radiation flux density at $R = 0$ and ϱ the proportional constant. Q_{m0} is a constant depending upon the absorption cross

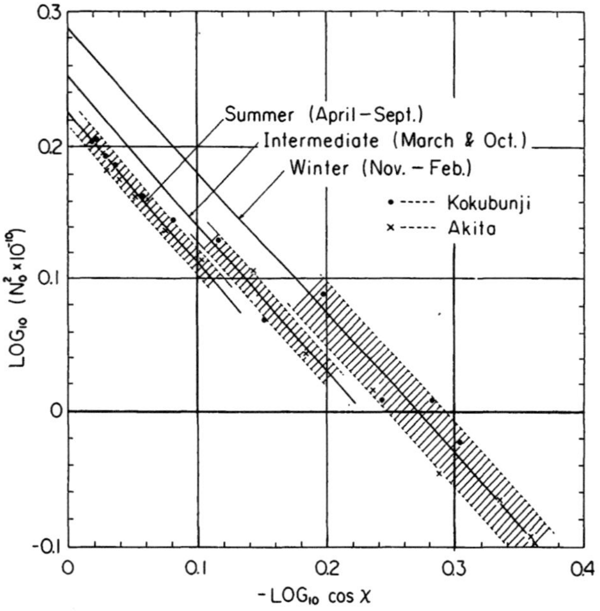

Fig. 3. Relations between N_0^2 and cos χ to obtain N_{00}^2 and Γ (see Eq. (3)).

section σ, the electron yield η, the scale height H_0 and neutral particle density n_0 at a datum level, and Γ is γH_0, in which γ is the rate of temperature rise with height. Here σ, η, the mean molecular weight and acceleration due to gravity are assumed constant.

If we take N_m for noon or near noon (1100–1300 LT) it will be given as follows:

$$N_m^2 = (Q_{m0}/\alpha)\,(\cos\chi)^{1+\Gamma}\,(1+\varrho\,R) = N_0^2\,(1+\varrho R)\,, \qquad (2)$$

$$N_0^2 = (Q_{m0}/\alpha)\,(\cos\chi)^{1+\Gamma} = N_{00}^2\,(\cos\chi)^{1+\Gamma}\,, \qquad (3)$$

where α is the effective recombination coefficient. Taking a particular month for which N_0 can be taken as constant, N_0^2 and ϱ are obtained from observed N_m^2 and R by the least squares method referring to Eq. (2). This was done for twelve months for each station as mentioned above. ϱ was found to range from about 0.0095 to 0.0105. As ϱ should have one fixed value regardless of the month, the average ϱ is obtained and 0.01 is adopted for ϱ. Once ϱ is fixed, the average N_0^2 is calculated based on observed N_m^2 and R, referring to Eq. (2).

Using N_0^2 thus obtained for twelve individual months and their corresponding χ's, Γ and N_{00}^2 $(= Q_{m0}/\alpha)$ are obtained referring to Eq. (3). Fig. 3 shows the plot of $\log_{10} N_0^2\,(\mathrm{cm}^{-6}) \times 10^{-10}$ against $-\log_{10}\cos\chi$ for three seasons defined as above. The points include two Japanese stations (Kokubunji 35.7° N and Akita 39.7° N). Two broken lines in each figure cover all points and a thick straight line represents the average relation. The slopes of these thick lines give $1 + \Gamma$. It is seen that the months from April to September can be grouped in one class, named summer, having $1 + \Gamma = 1.16$, those from November to February in another class, named winter, having $1 + \Gamma = 1.06$, and those of March and October, named the intermediate season, having $1 + \Gamma = 1.10$. Thus $\Gamma = \gamma H_0$ is 0.16 for summer, 0.06 for winter, and 0.1 for the intermediate season. These values of Γ are not sharply determined, but have a certain range.

It should be worth noting that N_{00}^2, which is the value of N_0^2 at $\chi = 0$, is not exactly the same for all seasons and for all stations. This may be caused by difference in α and Q_{m0} owing to the difference of height of N_m and of the season and the place. In the case of Wakkanai (45.4° N) N_{00}^2 is a little lower than in other stations and Γ cannot be clearly obtained because the points are more scattered.

Ignoring the slight difference in N_{00} depending upon the place of interest, the following experimental equations have been obtained.

For summer (from April to September)

$$N_m = 1.31 \times 10^5 (1 + 0.01\,R)^{1/2} (\cos\chi)^{0.58}\ \mathrm{cm}^{-3}\,, \tag{4}$$

for winter (from November to February)

$$N_m = 1.36 \times 10^5 (1 + 0.01\,R)^{1/2} (\cos\chi)^{0.53}\ \mathrm{cm}^{-3}\,, \tag{5}$$

and for intermediate season (March and October)

$$N_m = 1.33 \times 10^5 (1 + 0.01\,R)^{1/2} (\cos\chi)^{0.55}\ \mathrm{cm}^{-3}\,. \tag{6}$$

3. Electron Density Profiles in the E Region

3.1. Electron Density Profiles for Summer and Winter at Low Solar Activity

Among the available data those from eight experiments could be grouped in the summer class and those from eleven experiments in the winter, both for low solar activity. The averages of R and χ are 20 and 60° respectively for the winter group. For the summer group the average χ is 20°, and the average of R is 15 for the part above 100 km (seven data) and 10 for the lower part below 100 km (five data).

Figs. 4 and 5 show the electron density profiles obtained for summer and winter, which are denoted by S-L and W-L. Details of handling the data and of presentation of the result are given in [2]. The arrows indicate the averages of the peak electron densities obtained from ionosonde measurements. The peak electron density in the profile agrees well with the ionosonde value.

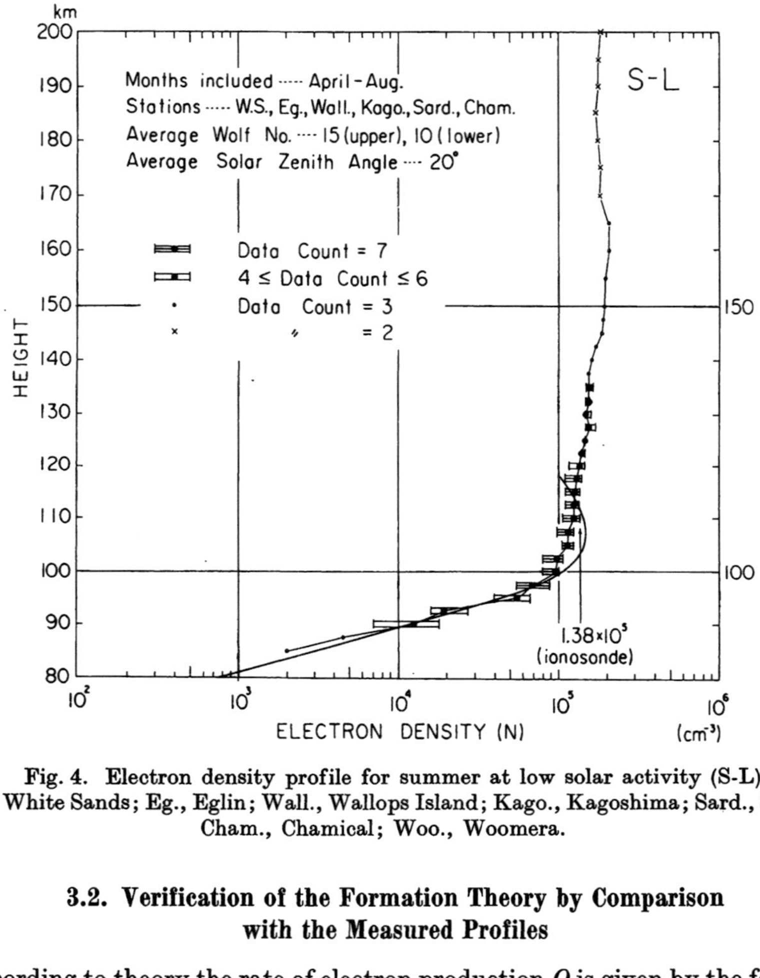

Fig. 4. Electron density profile for summer at low solar activity (S-L).
W. S., White Sands; Eg., Eglin; Wall., Wallops Island; Kago., Kagoshima; Sard., Sardinia; Cham., Chamical; Woo., Woomera.

3.2. Verification of the Formation Theory by Comparison with the Measured Profiles

According to theory the rate of electron production Q is given by the following equations:

$$\left.\begin{aligned} Q &= \sum_{\lambda} \overline{\eta_\lambda \sigma_\lambda}\, \Phi_{0\lambda}\, n \exp(-\tau_\lambda \sec \chi)\,, \\ \tau_\lambda &= \int_z^\infty \overline{\sigma_\lambda}\, n\, dz\,, \end{aligned}\right\} \tag{7}$$

where n is the neutral particle density, λ in suffixes the wavelength and the bars on $\eta_\lambda \sigma_\lambda$ and σ_λ mean the average over particle species, N_2, O_2 and O. To determine $n(z)$ the temperature profiles for all seasons must be given up to about 150 km. For summer and winter the temperature up to 80–90 km given by Nordberg et al. [9] is used and is not changed up to 95 km, and thereafter the summer model is taken as close to the mean model of CIRA 1965. For winter and intermediate season, reference is made to the temperature rise which has been known from our study as described in the former section. The temperatures at 120 km in these seasons are much lower than 355° K given in CIRA 1965. $\Phi_{0\lambda}$'s for the low solar activity were measured by Bourdeau et al.

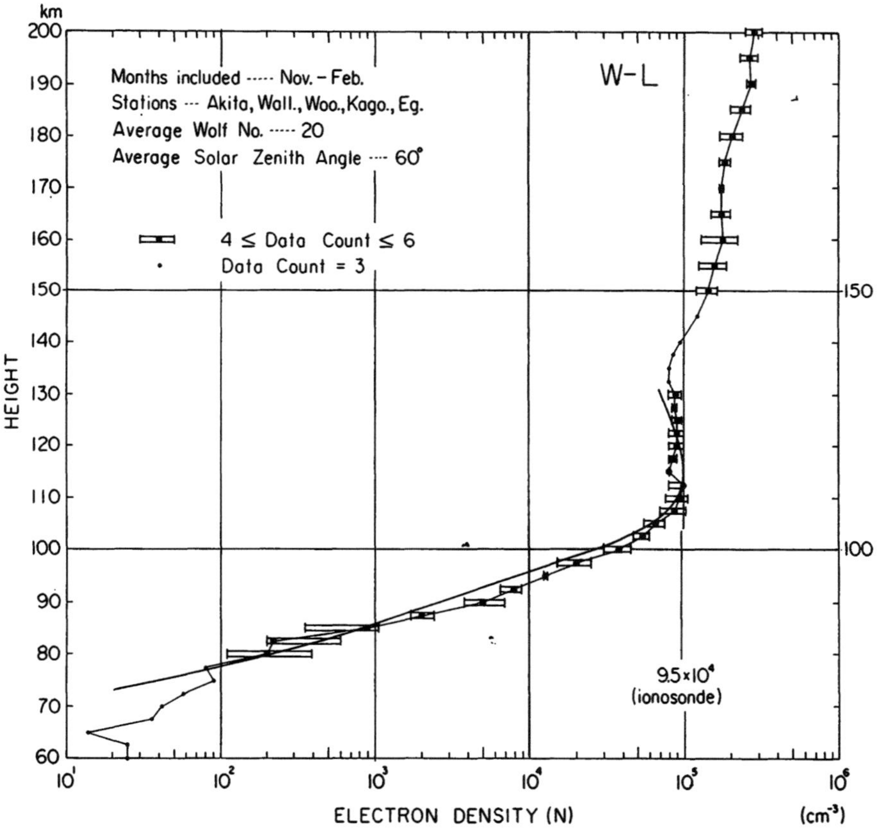

Fig. 5. Electron density profile for winter at low solar activity (W-L) (see Fig. 4).

[5]. R was 9.5 for this experiment. Some minor corrections have been made to give $\Phi_{0\lambda}$'s at $R = 15$ and 10 for summer and at $R = 20$ for winter, based on the R dependence of the $\Phi_{0\lambda}$'s given by Eqs. (4) and (5) and Figs. 1 and 2. $\overline{\sigma_\lambda}$, $\overline{\eta_\lambda \sigma_\lambda}$ and α are functions of height and are given in [2] and [6] together with $\Phi_{0\lambda}$. Mean molecular weight and gravity acceleration are taken from CIRA 1965.

The thin smooth curves in Figs. 4 and 5 are the result of theoretical calculation as described above. Fairly good agreement can be seen between the experiment and the theory.

Some further material of rocket observation is available for medium and high solar activity, though this is not abundant. These data are grouped into summer at high activity ($R = 125$, denoted by S-H) and winter at medium activity ($R = 85$, denoted by W-M), and the intermediate season at low activity ($R = 25$, denoted by MO-L). These are shown in Fig. 6.

Theoretical calculation is made for the above cases. To know $\Phi_{0\lambda}$'s for each case the same corrections as described in the cases of S-L and W-L have been made. The thin smooth curves in Fig. 6 are the result. Here we can see also good agreement between observation and theory.

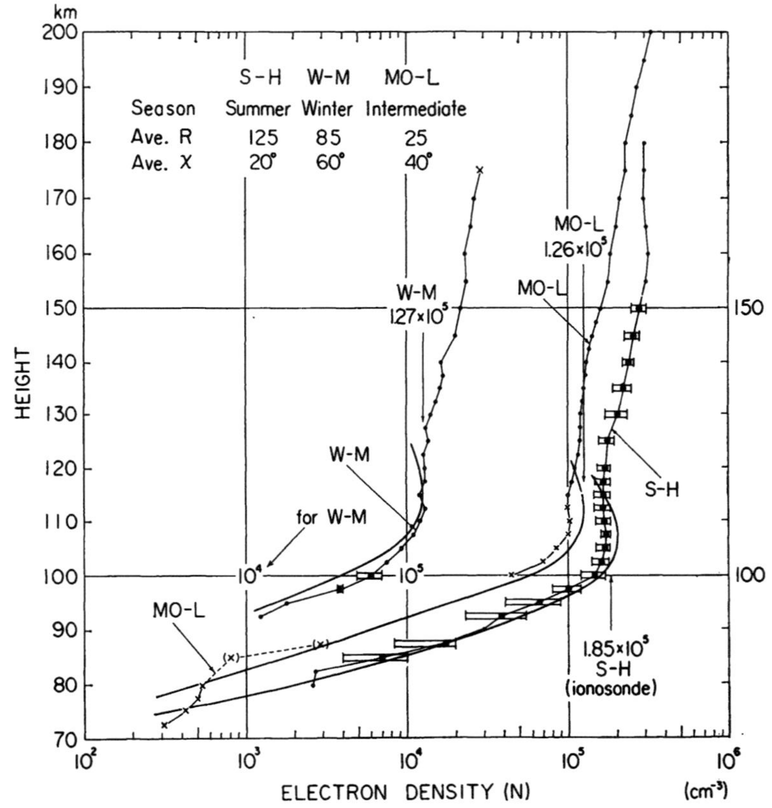

Fig. 6. Electron density profile for summer (high activity, S-H), winter (medium activity, W-M) and intermediate season (low activity, MO-L).

3.3. Calculated Profiles in other Solar Activity

Electron density profiles have been calculated and shown for such particular cases of R and season as S-L ($R = 15$ and 10), S-H ($R = 125$), W-L ($R = 20$), W-M ($R = 85$) and MO-L ($R = 25$), for which experimental profiles are more or less available and comparisons can be made. Further calculations have been made to give the profiles in three seasons for various Wolf numbers represented by $R = 10$, 25, 50, 75, 100, 125 and 150. The result is shown in Fig. 7 for $R = 10$ and 150. The maximum electron densities calculated from

Eqs. (4), (5) and (6) for each corresponding Wolf number are shown by arrows. Generally speaking, the agreement of peak electron densities in the curves with those indicated by arrows is good except for summer. As seen from Figs. 4, 6 and 7, N_m's in summer are generally larger than the arrows. It is supposed that the atmospheric model for summer, which was taken as close to the mean CIRA 1965, would not have been appropriate. The temperature profile above about 100 km should probably be corrected. This point will be studied in the near future.

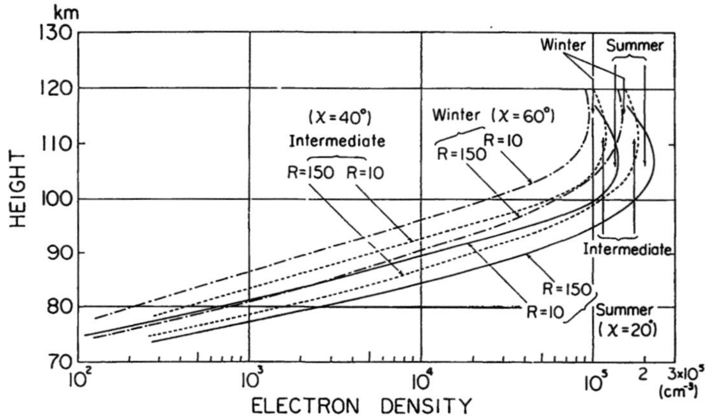

Fig. 7. Calculated electron density profiles for three seasons at $R = 10$ and $R = 150$.

4. Night-time Profiles and Nocturnal Variation of N_m

An extensive study on the mean variation of the night-time E region has been made by Wakai [10, 11], based on low-frequency ionograms taken at Boulder in summer in 1957, 1960 and 1961 and winter in 1958 and 1961. Tohmatsu and Wakai [12] investigated the photoionization of the hydrogen uv geocorona as a possible source of night-time E region formation, with reference to the $N(h)$ profile obtained by a low-frequency ionosonde and a rocket experiment. Here the night-time electron density profiles in the E region obtained from rocket observation are presented and a typical nocturnal variation of N_m is drawn, and an attempt is made to obtain the rate of electron production (Q) by the ionization source which must be consistent with the variation of N_m.

4.1. Night-time Profiles from Rocket Data

In [1] the night-time profiles were presented. Further material based on rocket measurements has since been made available, especially for the local time from 2020 to 2100 (group denoted by N_1), from 2130 to 2320 (N_2) and from 0410 to 0600 (N_5). Newly drawn profiles for the above time intervals (N_1, N_2 and N_5) are presented in Figs. 8 and 9. These involve the data of all seasons and at relatively low solar activity ($R = 55$). The profiles for other times are not presented because the available data are poor.

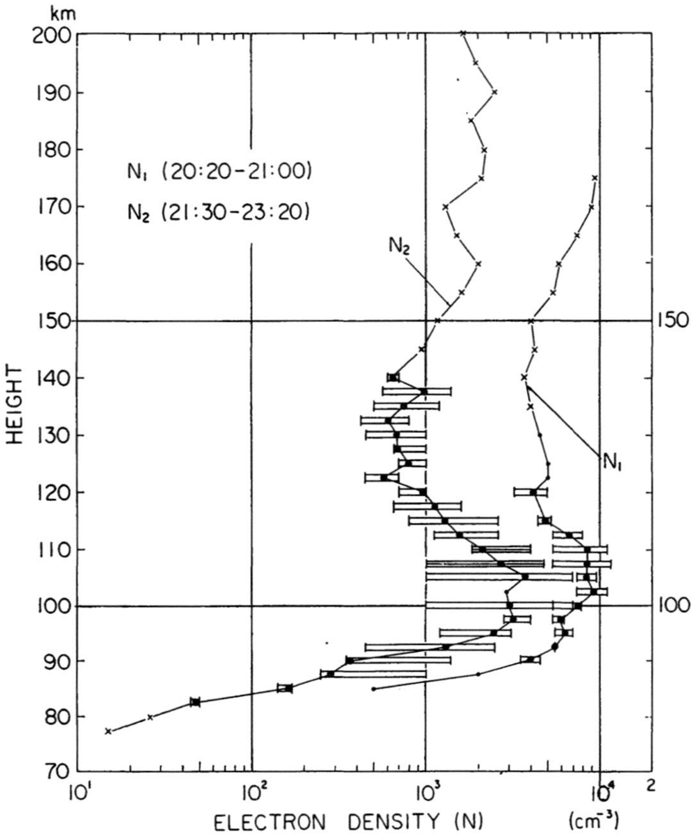

Fig. 8. Night-time electron density profiles: N_1, N_2.

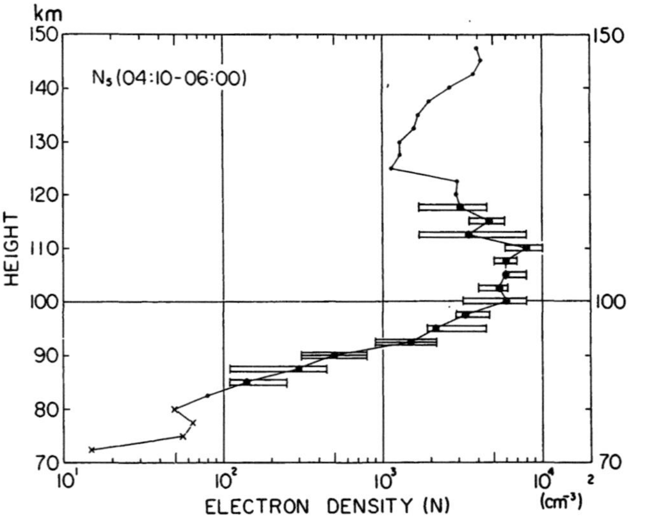

Fig. 9. Night-time electron density profile: N_5.

4.2. Night-time Variation of N_m

The local time variation of N_m during night is shown in Fig. 10. The circled points are taken from the profiles of N_1, N_2 and N_5 shown in Figs. 8 and 9. The two other points are taken from the profiles of N_3 (0010–0050) and N_4 (0300–0350) which are not presented here. The scatter of N_m over about 10 km height range is shown by a vertical bar. The average height is about 105 km. The crosses and connecting curve after midnight are taken from Wakai's paper. A continuous smooth curve by Kimura [13] is an average variation of N_m obtained from broadcast wave experiment. The Wolf number is very large ($R = 187$).

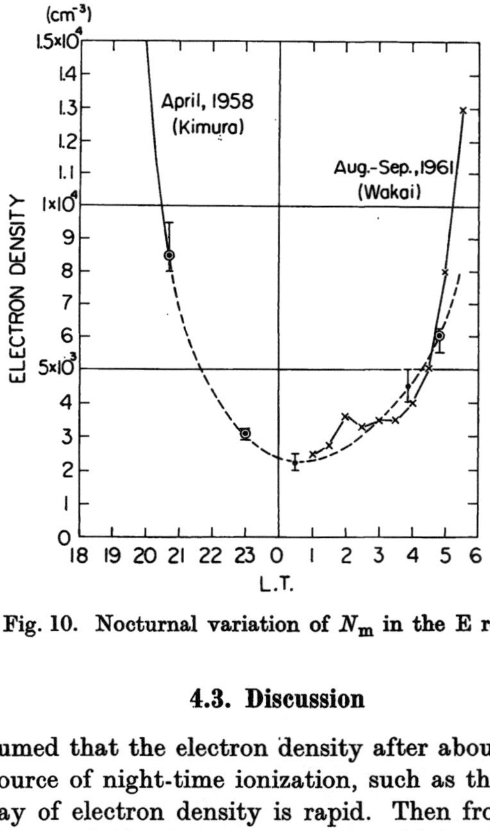

Fig. 10. Nocturnal variation of N_m in the E region.

4.3. Discussion

It can be assumed that the electron density after about 2200 is maintained only by some source of night-time ionization, such as the geocorona, because the rate of decay of electron density is rapid. Then from the values of N_m and $\partial N_m/\partial t$ at 2200 and 0200, α is estimated roughly as 6×10^{-8} cm^3 s^{-1}, if we assume that Q due to the night-time ionization source is symmetrical with respect to midnight. If, for example, we assume

$$Q = Q_0 + \beta t^2 , \tag{8}$$

where β is a positive constant and t is reckoned in hours from midnight, we can determine Q_0 (at midnight) and β. Thus Q_0 is estimated as about 0.3 cm^{-3} s^{-1} and β is 0.074. From this Q_0, α around midnight is calculated. α obtained by this method is 7×10^{-8} cm^3 s^{-1}. Kimura [13] estimated α from his curve after sunset and 5×10^{-8} cm^3 s^{-1} was reported. But if we are allowed to take into account the contribution from the night-time ionization source as described above, the above value of α must be corrected. According to such correction, using the part of the curve from 2030 to 2230, α is obtained as 8×10^{-8}. Based on these studies it can be said that α is from 6×10^{-8} to 8×10^{-8} at the height level of about 105 km. And values of Q_0 and β are approximately consistent with the result obtained by Tohmatsu and Wakai [12].

5. Concluding Remarks

Based on the experience in our study further observations will be made as follows: (i) observations of the daytime $N(h)$ profiles and the solar X radiations (1–8 Å and 33.7 Å) in medium and high solar activity; (ii) observation of the night-time $N(h)$ profiles intensified; (iii) observation of spectral intensity of the geocorona, with particular reference to its dependence on local time, season and solar activity; (iv) observation of the profiles of the atmospheric temperature below 120 km.

Acknowledgments

The author wishes to express his sincere thanks to Prof. H. Tanaka, Research Institute of Atmospheric Research, Nagoya University, who provided the material on solar radio flux, and to Mr. T. Tsujii, who cooperated in handling of the X-ray flux data. He also thanks Mr. S. Fukao, Kyoto University, who undertook the computer calculation.

References

[1] K. MAEDA, J. Geomag. Geoelect. **21**, 557 (1969).
[2] K. MAEDA, J. Geomag. Geoelect. **22**, 551 (1970).
[3] M. NICOLET and A. C. AIKIN, J. Geophys. Res. **65**, 1469 (1960).
[4] M. OHSHIO, J. Radio Res. Labs. **11**, 377 (1964).
[5] R. E. BOURDEAU, A. C. AIKIN and J. L. DONLEY, J. Geophys. Res. **71**, 727 (1966).
[6] K. MAEDA, J. Geomag. Geoelect. **23**, 133 (1971).
[7] R. W. KREPLIN, Annls Géophys. **26**, 567 (1970).
[8] K. TAKAO and T. TSUJII, Rep. Ionosph. Space Res. Japan **24**, 183 (1970).
[9] W. NORDBERG, L. KATCHEN, J. THEON and W. S. SMITH, J. Atmos. Sci. **22**, 611 (1965).
[10] N. WAKAI, Proc. Conf. on Ground-based Radio Wave Propagation Studies of the Lower Ionosphere, Ottawa, Canada **2**, 552 (1966).
[11] N. WAKAI, J. Radio Res. Labs. **15**, 109 (1968).
[12] T. TOHMATSU and N. WAKAI, Annls Géophys. **26**, 209 (1970).
[13] I. KIMURA, Doctorate Thesis, Kyoto University, 125–149, July 1961.

Space Research XII — Akademie-Verlag, Berlin 1972

THE STANDARD PROFILE OF THE MID-LATITUDE F REGION OF THE IONOSPHERE AS DEDUCED FROM BOTTOMSIDE AND TOPSIDE IONOGRAMS

W. Becker

Max-Planck-Institut für Aeronomie, Institut für Ionosphärenphysik, Lindau/Harz, FRG

Day by day noon and midnight ionograms taken at Lindau/Harz (bottomside observations) from 15 December 1957 to 15 July 1964 were converted into electron density–real height profiles. Also more than 500 Alouette 1 ionograms, taken at Solant, Falkland Islands, from 1962 to 1967 were reduced. Exclusively from these data analytical expressions were derived describing the electron density–real height distribution within the F region and its dependence on solar activity. Statistical properties were also investigated.

1. Introduction

Fig. 1 has already been shown in a former study on the seasonal anomaly of the F region [1]. It was this diagram that indicated a successful first order approximation of the noon standard profile of the F region, i.e. a first order quantitative description of the electron density–height distribution in the F region as a function of time, season and solar activity by means of a minimum of parameters.

The physical idea of this profile data presentation was to standardize ionizing solar flux intensity and temperature influences. This was done by normalizing the individual profiles first by the peak density N_cF2 and then by referring all the resulting levels of fixed N/N_cF2 ratios to the respective peak heights h_m and finally by dividing these distances by Y_m. Y_m itself is the half-layer thickness of a parabolic fit to the electron density distribution at the peak height h_m of the F layer. These parameters N_cF2 and Y_m were considered to be the required standardizing parameters for the following reason: The selected interval of observation was centred at about 1300 CET, or 40 minutes after local noon, when quasi equilibrium conditions prevail in the F region, which means that N_cF2 is in fact a representative quantity for the intensity of the ionizing flux and Y_m for the scale height or the temperature of the neutral atmosphere at the peak height h_m. This conclusion was later successfully checked by correlating equinoctial noon Y_m data for years of low and high solar activity with corresponding temperature values deduced from satellite drag measurements [2]. Indeed the four profiles shown in Fig. 1 are practically identical in a height range down to 50% of Y_m below the F2 peak. The deviations below that height range were caused by the composition of, and consequently by scale height changes, in the F1 region.

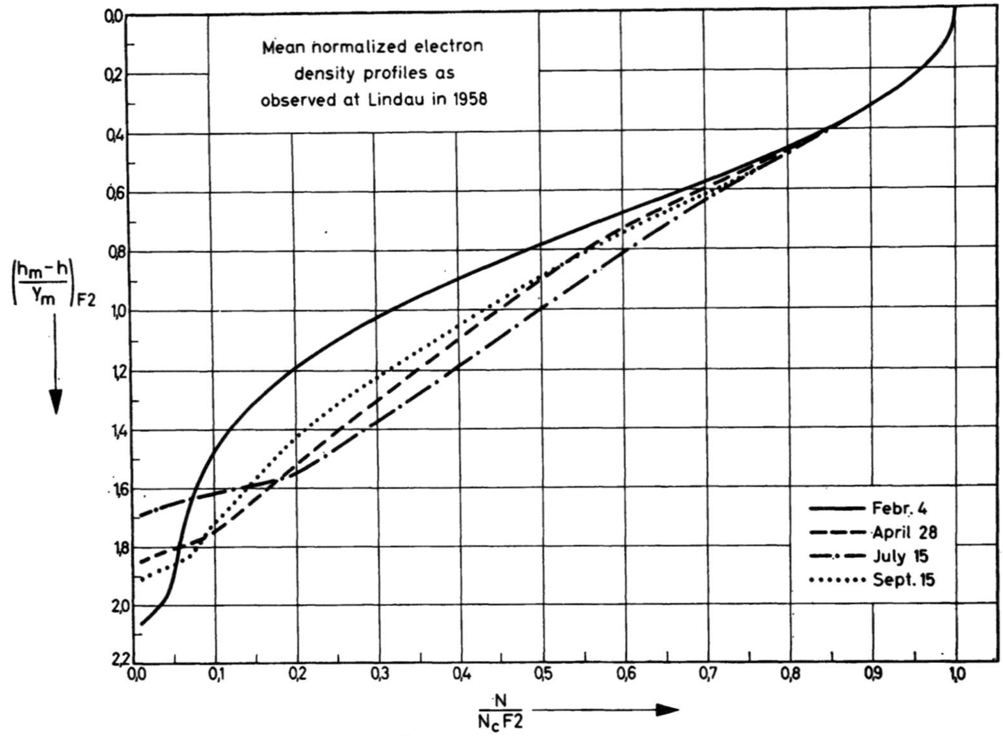

Mean normalized electron
density profiles as
observed at Lindau in 1958
Febr. 4
April 28
July 15
Sept. 15
$\left(\frac{h_m - h}{Y_m}\right)_{F2}$
$\frac{N}{N_c F2}$
0,0
0,2
0,4
0,6
0,8
1,0
1,2
1,4
1,6
1,8
2,0
2,2
0,1
0,3
0,5
0,7
0,9

Fig. 1

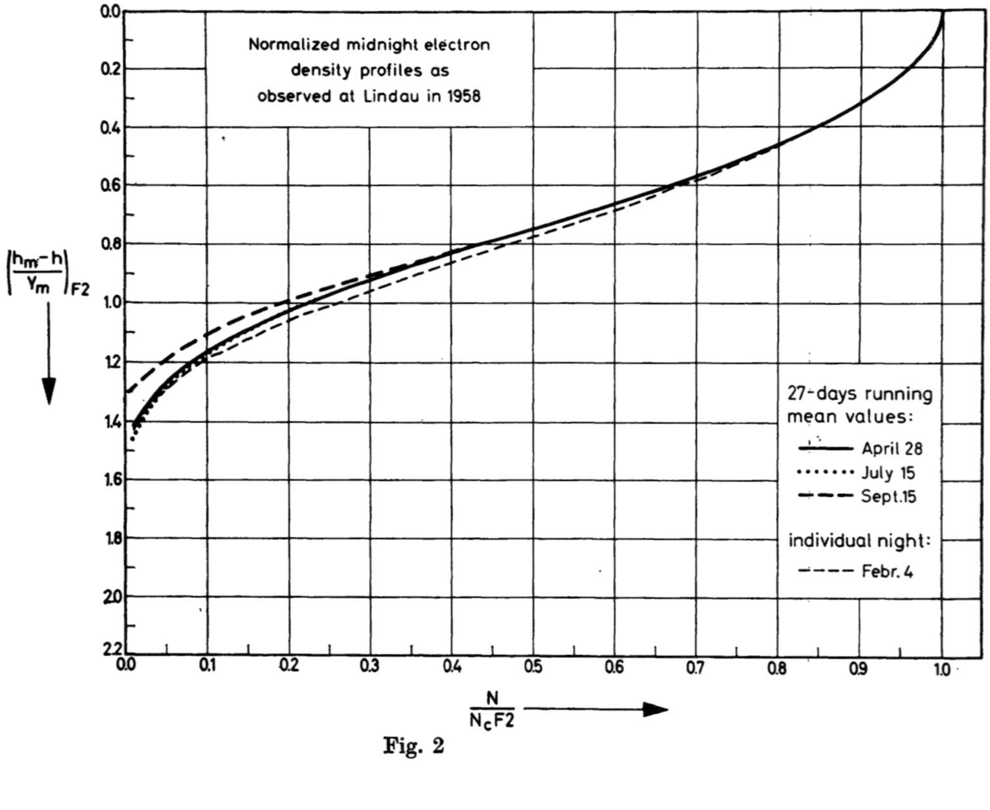

Normalized midnight electron
density profiles as
observed at Lindau in 1958
27-days running
mean values:
April 28
July 15
Sept.15
individual night:
Febr. 4
$\left(\frac{h_m - h}{Y_m}\right)_{F2}$
$\frac{N}{N_c F2}$
0.0
0.2
0.4
0.6
0.8
1.0
1.2
1.4
1.6
1.8
2.0
2.2
0.1
0.3
0.5
0.7
0.9

Fig. 2

Fig. 1. Mean noon electron density profiles normalized with respect to peak electron density N_c F2, height of the peak h_m and half-layer thickness Y_m of a parabolic approximation to the peak for four selected days (from Becker [1]).

One year later a similar study on the standard profile of the midnight F region [3] was even more successful (Fig. 2). Here the height range of seasonal and solar cycle independence extended to at least 90% of Y_m. The reason why the above four special days had been selected for Figs. 1 and 2 can be understood by means of Figs. 3a, b [1]. According to these the four special days represent intervals of extreme noon values of h_m during both sunspot maximum and minimum conditions. They were thought to show also maximum deviations of the individual profiles. The good correlation between Y_m and the 10.7 cm solar flux intensity Φ times $\cos\chi$ (χ is the solar zenith angle) at noon was an extremely welcome result for this attempt to reduce the number of parameters determining the electron density height distribution in the F region. Figs. 4a, b also shown in a former study [3] finally verify the above conclusions drawn from Fig. 3. In addition they also demonstrate the slight spread of the individual heights within the ranges of good correspondence, i.e. between 50 and 90% of Y_m.

2. The Noon and Midnight Standard Profiles of the F Region

In the meantime the preceding study has been completed. The data of interest now cover somewhat more than half a sunspot cycle, more precisely the interval 15 December 1957 to 15 July 1964. Whenever possible four ionograms per day taken at 1230, 1245, 1300 and 1315 during the International Geophysical Years and 1200, 1230, 1300, 1330 during the other years were reduced to electron density true height profiles. The method used has already been described [4, 5]. Also the midnight ionograms considered were taken at Lindau/Harz at 0015, 0030, 0045 and 0100 or at 0000, 0030, 0100 and 0130 during the same years as the noon records. Details of the results will be given elsewhere. This contribution is restricted to the more general features.

2.1. The Bottomside Ionosphere

In Fig. 5 27-days running mean noon values of Y_m are plotted against corresponding values of ($\Phi \cos\chi$), more precisely, each fifth Y_m value. The slightly bent parabola of regression, on the other hand, refers to all the individual noon mean values of the same data set. The correlation coefficient R is 0.853. Only the regression parabolae for noon and midnight intervals of observation are compared with each other in Fig. 6. The analytical expressions corresponding to the two curves are

$$\text{Noon:} \quad Y_m(\text{km}) = 37.6 + 0.783\,(\Phi \cos\chi) - 0.125 \times 10^{-2}\,(\Phi \cos\chi)^2$$
$$\text{Midnight:} \quad Y_m(\text{km}) = 66.3 + 0.835\,(\Phi \cos\chi) - 0.437 \times 10^{-3}\,(\Phi \cos\chi)^2\,.$$

Fig. 2. Mean midnight electron density profiles normalized with respect to peak electron density N_c F2, height of the peak h_m and half-layer thickness Y_m of a parabolic approximation to the peak for four selected days (from Becker [3]).

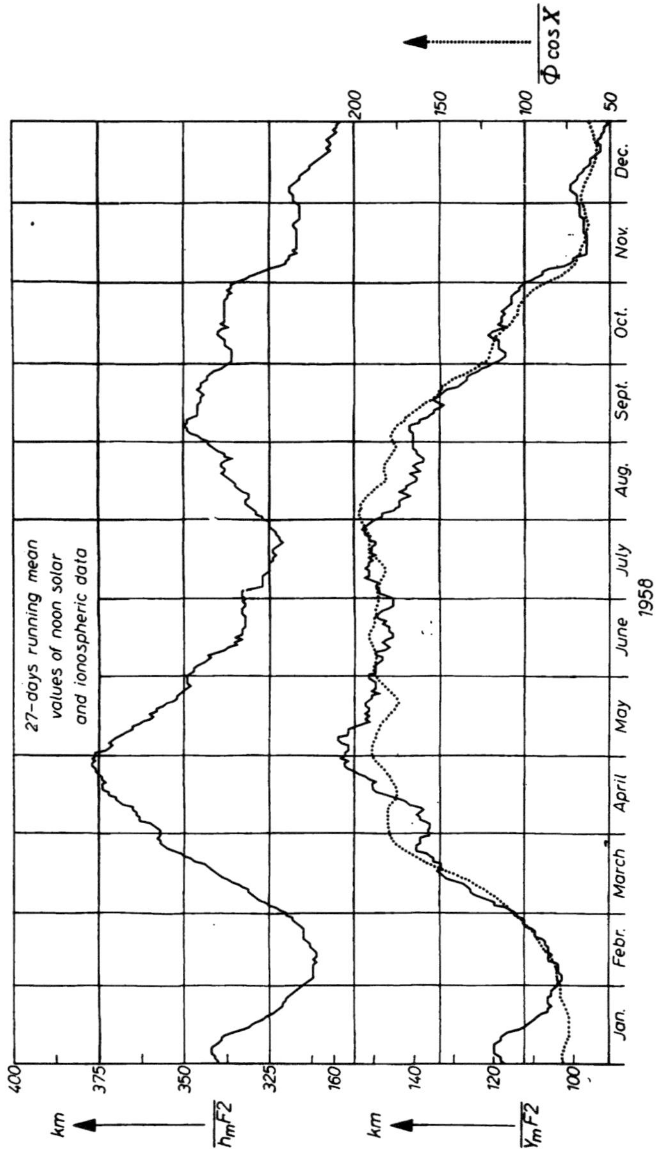
27-days running mean
values of noon solar
and ionospheric data
Jan.
Febr.
March
April
May
June
July
Aug.
Sept.
Oct.
Nov.
Dec.
1958
km
$\overline{h_mF2}$
400
375
350
325
km
$\overline{y_mF2}$
160
140
120
100
$\overline{\Phi \cos X}$
200
150
100
50

Fig. 3a

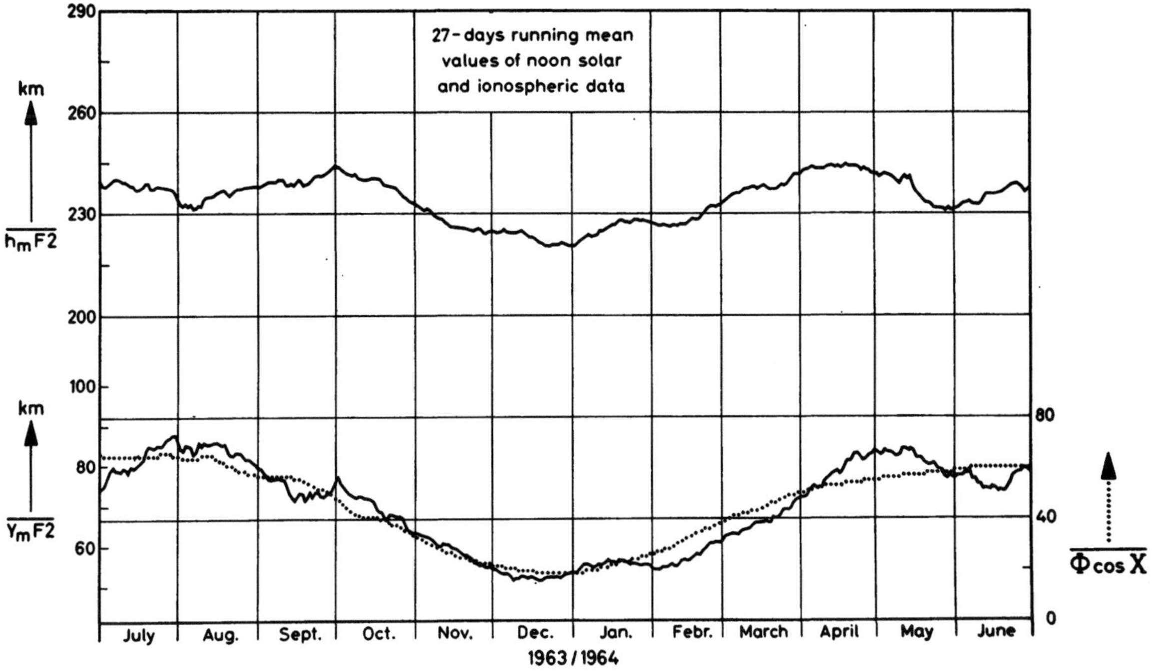

Fig. 3b

Fig. 3. h_m F2 measures the real height of the F2 layer peak. Y_m F2 means the half-layer thickness of a parabolic approximation to the F2 peak. The data were deduced from ionograms taken at Lindau/Harz at 1230, 1245, 1300, 1315 CET (= 15° EST). Φ is the 10.7 cm solar radio flux (10^{-22} W m^{-2} Hz^{-1}) as observed by NRC Ottawa and χ the solar zenith distance (from Becker [1]). a, 1958; b, 1963–1964.

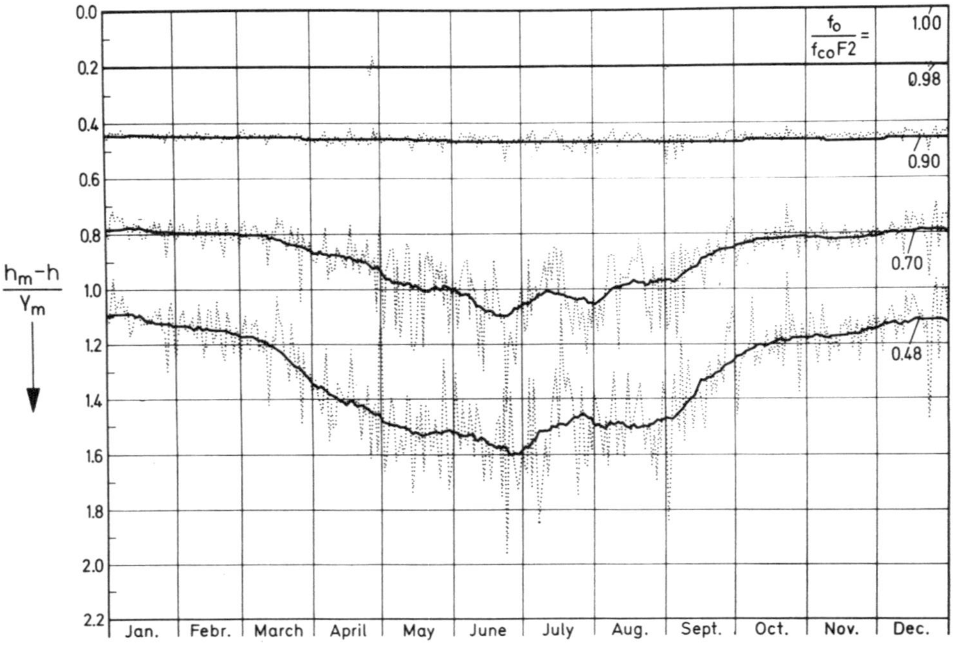

Fig. 4a

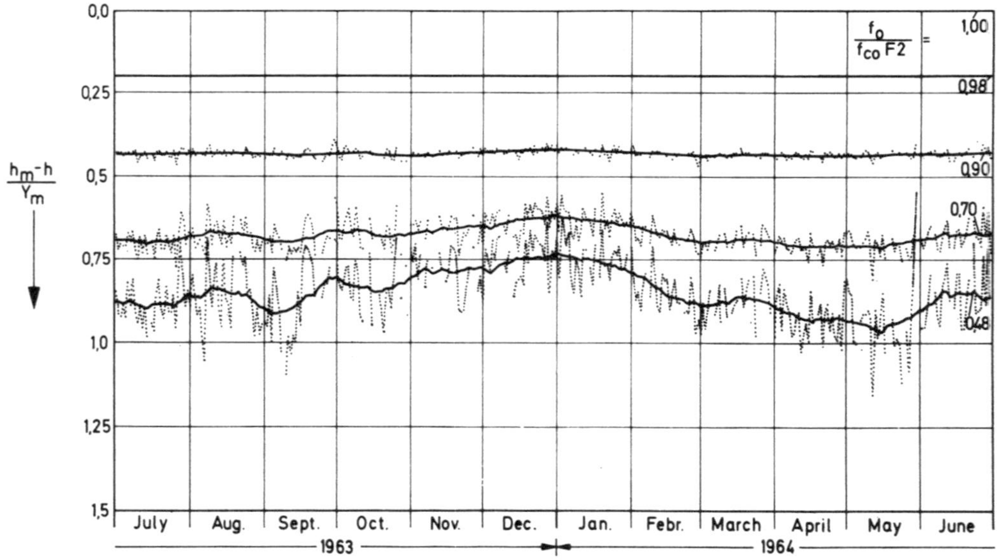

Fig. 4b

Fig. 4. h_m F2 measures the real height of the F layer peak. Y_m means the half-layer thickness of a parabolic approximation of the electron density–height distribution there. f_{co} F2 is the maximum value of the observed plasma frequencies f_o. The daily mean values are connected with each other by dotted lines, the 27-days running mean values by full lines. Further details as in Fig. 3. a, noon 1958; b, midnight 1963–1964.

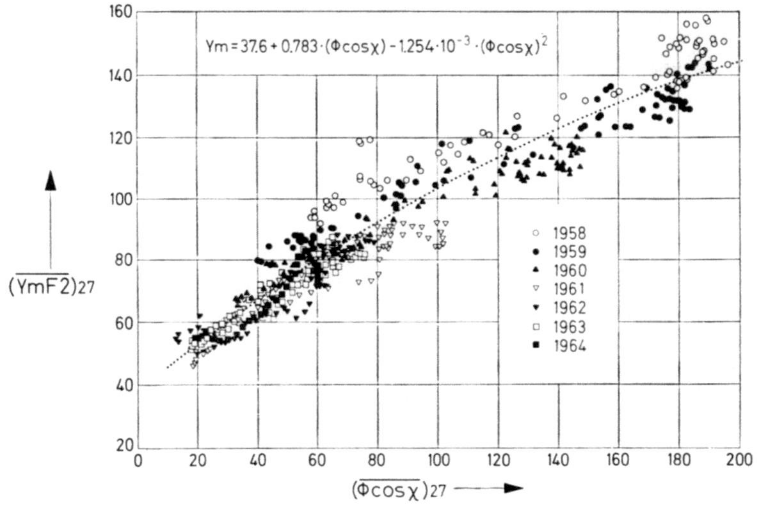

Fig. 5. Y_m F2 means the half-layer thickness of a parabolic approximation to the electron number density–height variation at the F2 layer peak. Φ is the 10.7 cm solar radio flux (10^{-22} W m^{-2} Hz^{-1}) as observed by NRC Ottawa and χ the solar zenith distance. Each 5th pair of 27-days running mean noon values is included in this diagram. The parabola of regression, dotted curve, is deduced from 2083 daily mean noon values.

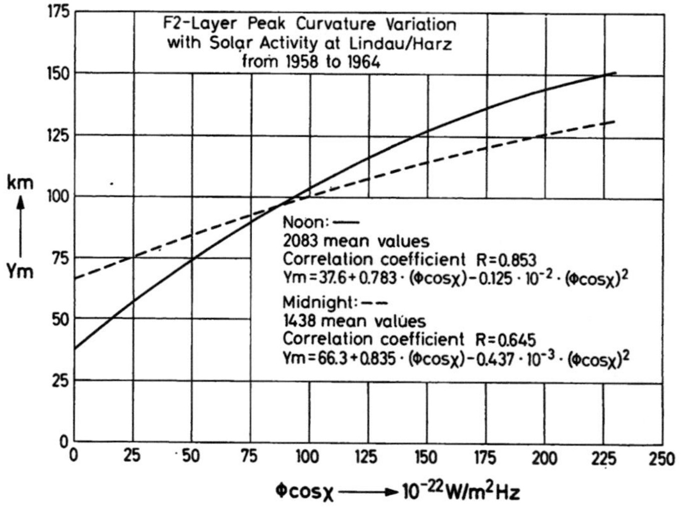

Fig. 6. Parabolae of regression. Y_m means the half-layer thickness of a parabolic approximation to the electron number density–real height variation at the F2 layer peak. Φ is the 10.7 cm solar radio flux as observed by NRC Ottawa and χ the solar zenith distance. These midnight data refer to $\Phi \cos \chi$ values of the preceding days.

It is interesting that the seasonal correlation coefficients were even better than the overall values, for instance, the data for the months March and April from 1958 to 1964 resulted in $R = 0.922$. Equally good statistical correlations could be found for the seasonal peak height variations of the F region (Fig. 7), both for noon and midnight intervals.

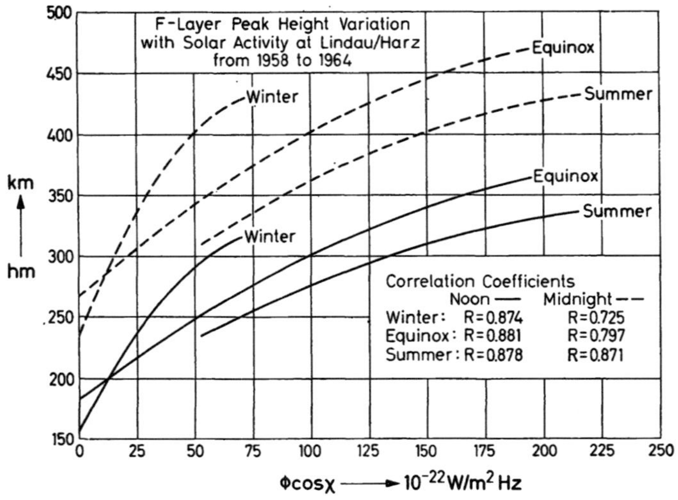

Fig. 7. Parabolae of regression. h_m means the peak height of the F layer. Φ is the 10.7 cm solar radio flux as observed by NRC Ottawa and χ the solar zenith distance. These midnight data refer to $\Phi \cos \chi$ values of the preceding days. Months represented: December and January (Winter); March, April, September and October (Equinox), June and July (Summer).

The analytical expressions for the above parabolae of regression are as follows:

Noon Data

Winter (December, January)

$$R = 0.874$$
$$h_m(\text{km}) = 156.4 + 3.728\,(\Phi \cos \chi) - 2.079 \times 10^{-2}\,(\Phi \cos \chi)^2$$

Equinox (March, April, September, October)

$$R = 0.881$$
$$h_m(\text{km}) = 182.9 + 1.442\,(\Phi \cos \chi) - 2.635 \times 10^{-3}\,(\Phi \cos \chi)^2$$

Summer (June, July)

$$R = 0.878$$
$$h_m(\text{km}) = 176.9 + 1.212\,(\Phi \cos \chi) - 2.190 \times 10^{-3}\,(\Phi \cos \chi)^2$$

Midnight Data

Winter (December, January)

$$R = 0.725$$
$$h_m(\text{km}) = 234.6 + 4.741\,(\Phi \cos \chi) - 2.813 \times 10^{-2}\,(\Phi \cos \chi)^2$$

Equinox (March, April, September, October)

$R = 0.797$

$h_m(\text{km}) = 267.2 + 1.659\ (\Phi \cos \chi) - 3.160 \times 10^{-3}\ (\Phi \cos \chi)^2$

Summer (June, July)

$R = 0.871$

$h_m(\text{km}) = 237.0 + 1.533\ (\Phi \cos \chi) - 2.908 \times 10^{-3}\ (\Phi \cos \chi)^2$

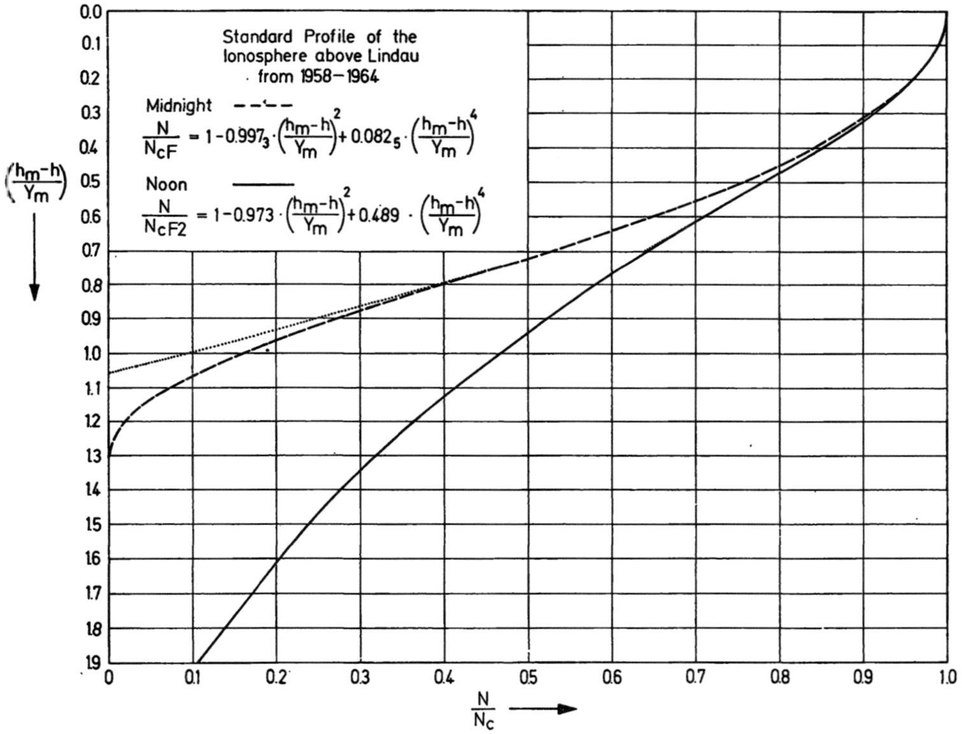

Fig. 8. h_m means the peak height of the F2 layer. Y_m is the half-layer thickness of a parabolic approximation to the electron density–height distribution there. N_c is the peak value of the electron densities N. Lindau 1958–1964.

Because of the seasonal anomaly the overall noon correlation coefficient is 0.786; the corresponding midnight value, additionally affected by scaling difficulties due to frequent spread F conditions and other dynamical effects, was only 0.656. Better coefficients might be obtained by taking account of the total energy input at each preceding day instead of referring to its noon value only.

The slight spread of the individual profile data in Fig. 4 as compared with their 27-days running mean values for the ranges of good correspondence justified the computation of mean standard profiles for both the noon and midnight ionosphere above Lindau. The results are shown in Fig. 8. The dotted lines correspond to the polynomial approximations given on this figure.

In order to deduce an individual profile for a noon or midnight interval and a given day first an estimate of Φ and N_cF2 must be made by means of available prediction data. Then cos χ is calculated for the mid-latitude station of interest. h_m and Y_m can be taken from Figs. 6 and 7. Finally these parameters are inserted into the formulae reproduced by Fig. 8 and the required electron density profile is fully described within the height range of good correspondence, i.e. within 50 and 90% of Y_m below the F layer peak.

2.2. The Topside Ionosphere

Following the same principles topside ionograms taken by Alouette 1 above Solant, Falkland Islands, were also analysed*. The records only occasionally allow the peak electron density, peak height and Y_m to be determined at one time. Out of 1775 ionograms reduced for the years 1962–1967 only 521 examples

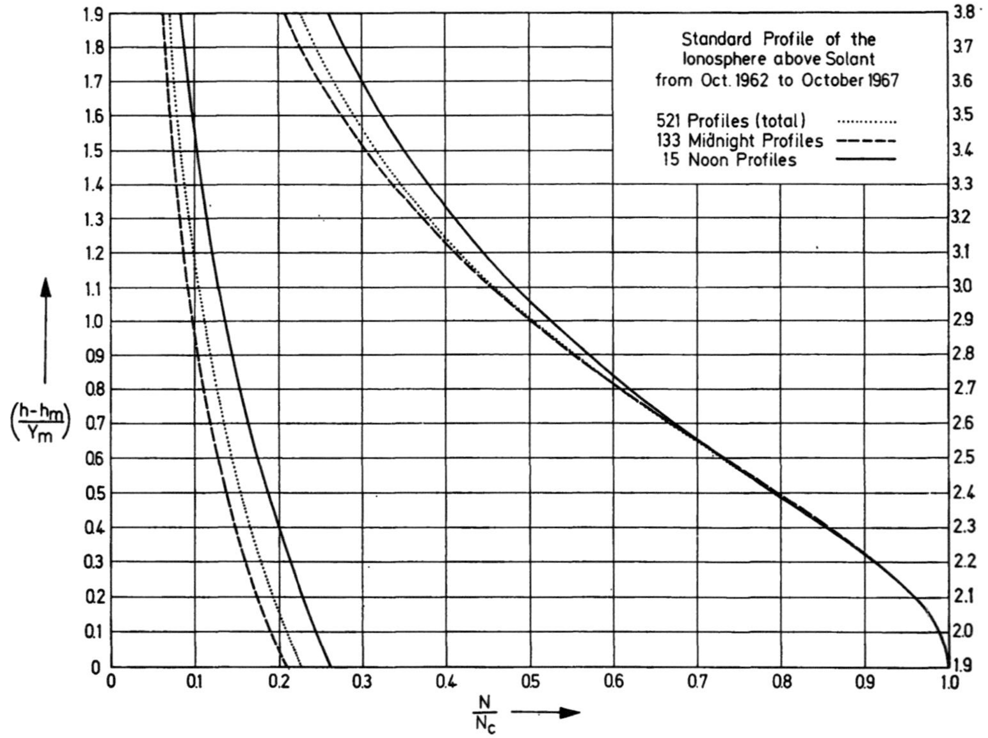

Fig. 9. h_m means the peak height of the F2 layer. Y_m is the half-layer thickness of a parabolic approximation to the electron density–height distribution there. N_c is the peak value of the electron densities N. Solant 1962–1967.

fulfilled the necessary requirements for such an analysis. Therefore, and because of the wide gaps between succeeding passes, running mean values could not be

* They had been provided to us by the World Data Centre at Slough through the courtesy of Dr. J. King and his group, which is highly appreciated.

derived, only mean values. Fig. 9 shows these mean profiles of the outer ionosphere. The dotted curve comprises all the available 521 ionograms. Those for a three-hourly midnight interval (2330–0230 LT) are represented by a dashed curve and the profile for a corresponding noon interval (1130–1430 LT) by a solid curve. As suggested by the bottomside results, only the upward tails

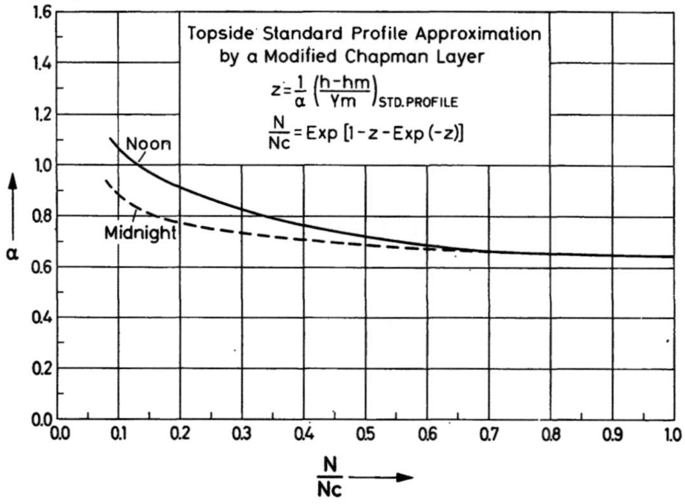

Fig. 10. The normalized real height data of Fig. 9 divided by α are modified z values for a Chapman-layer approximation of the noon and midnight profiles of Fig. 9.

differ slightly from each other. An attempt was made to describe these mean noon and midnight profiles by a modified Chapman layer, Eq. (1):

$$\frac{N}{N_c} = \exp\{1 - z - \exp(-z)\}\,. \tag{1}$$

Table 1

N/N_c	α		N/N_c	α	
	Noon	Midnight		Noon	Midnight
0.08		0.935	0.56	0.707	0.679
0.10	1.067	0.878	0.60	0.694	0.674
0.12	1.019	0.842	0.64	0.681	0.669
0.14	0.986	0.819	0.68	0.670	0.665
0.16	0.958	0.799	0.72	0.662	0.661
0.18	0.933	0.782	0.76	0.656	0.657
0.20	0.912	0.772	0.80	0.652	0.654
0.24	0.874	0.753	0.84	0.649	0.651
0.28	0.839	0.740	0.88	0.647	0.649
0.32	0.810	0.728	0.92	0.646	0.647
0.36	0.788	0.717	0.96	0.646	0.646
0.40	0.767	0.707	1.00	0.646	0.646
0.44	0.751	0.698			
0.48	0.733	0.690			
0.52	0.720	0.684			

These N/N_c values equal those of the above mean profiles if z is modified according to Eq. (2):

$$z = \frac{1}{\alpha} \left\{ \frac{h - h_m}{Y_m} \right\} \text{mean profile}\,, \tag{2}$$

Y_m being the topside thickness parameter. The corresponding noon and midnight α values are about the same near the peak height h_m but differ from each other at greater heights (Fig. 10 or Table 1).

A physical discussion of the above results would be out of the scope of this contribution. Therefore, the author intends to discuss all these aspects in a separate paper which will be published elsewhere [6].

3. Conclusion

Inner and outer F region electron number density–real height profiles for noon and midnight intervals at mid-latitudes were deduced in a self-consistent way, i.e. the corresponding analytical expressions describing their seasonal and solar cycle variations depend on observed parameter values only rather than on incorporated model data. Thus the usefulness is not temporarily limited.

Acknowledgments

The author wants to thank Mesdames W. Donnerstag, A. Jünemann, J. Koch, R. Mittelstädt and Miss I. Falkus for their ionogram scaling efforts, Mr. E. Sarna and H. Benecke for their assistance in the control and processing of the data. and Prof. A. Schülter for allowing the use of the computer IBM 360/91 at his Institut für Plasmaphysik at München-Garching for the above computations.

The financial support for this study granted by the Max-Planck-Gesellschaft and the Minister für Bildung und Wissenschaft is also highly appreciated. This paper is published by permission of the Director of the Max-Planck-Institut für Aeronomie.

References

[1] W. BECKER in: Electron Density Profiles in Ionosphere and Exosphere, North-Holland Publ. Co., Amsterdam 1966 (p. 218).
[2] W. BECKER, J. Geophys. Res. **72**, 2001 (1967).
[3] W. BECKER, Kleinheubacher Ber. **12**, 235 (1967).
[4] W. BECKER, Radio Sci. **2**, 1205 (1967).
[5] W. BECKER, Kleinheubacher Ber. **13**, 37 (1969).
[6] W. BECKER, to be published in Kleinheubacher Ber.

Space Research XII — Akademie-Verlag, Berlin 1972

MODEL ELECTRON DENSITY PROFILES OBTAINED BY EMPIRICAL PROCEDURES

S. Ramakrishnan and K. Rawer

Arbeitsgruppe für physikalische Weltraumforschung, Freiburg, FRG

The CCIR computer programme for monthly median *f*oF2 and programmes developed at ESSA for computing *f*oE, *M* (3000) and *f*oF1 have been applied for the proposed locations for which IRI (International Reference Ionosphere) should finally present tables. For the mid-latitude side complete electron density profiles can be obtained by combining Maeda's model for the E layer and valley, Becker's standard profile for the bottom side of the F region, and a formula for the topside profile up to 1000 km derived from Jackson's topside profiles. Possible use of the new Bent and Llewellyn exponential approximation to the topside is mentioned and proposals for extrapolation from the top to the bottomside are critically discussed.

1. Bottomside Ionosphere

Following the adopted preference schedule parameters of the bottom side electron density profile have been produced from CCIR and ITS (formerly ESSA) computer programmes for three locations of incoherent scatter stations, viz.

Jicamarca (77° W 12° S, near dip equator),
Arecibo (67° W 18° N),
St. Santin (2° E 45° N, mid-latitude).

The programmes are based on the measured monthly median values of many ionospheric sounding stations, over at least one solar cycle. In these programmes solar activity is described by the Zürich sunspot number R and values have been produced for $R = 0$ and $R = 100$. Parameters produced are *f*oF2, *f*oF1, *f*oE and M (3000), a factor closely related to the peak altitude of the F2 layer. Thus the peak electron densities of the main layers are known hour by hour, for three selected seasons and $R = 0$ and 100.

Further, Shimazaki's [1] formula has been used to deduce the corresponding peak altitude from M (3000) values. Recently so-called "average" electron density profiles have been made available by NOAA (formerly ESSA) for the ionospheric stations Huancayo (near Jicamarca), Mexico City (not very far from Arecibo) and Thule (Greenland) as a typical high latitude station. The peak altitude can also be deduced from these profiles though the procedure for obtaining profiles via an "average ionogram" has been criticized. It is intended to derive empirical relations to describe these profiles such that the electron density distribution below the peak height down to the "valley" can be indicated. The valley problem will be resolved by A. K. Paul (private communication) using a summary of his work on the subject carried out at ESSA. For the E-region Maeda's [2] improved profiles are to be used.

For mid-latitudes Becker's [3] standard profiles are available together with his peak altitude and thickness values. These have been obtained by individual profile computation with a very large number of inputs from the ionospheric station at Lindau which is not very far from St. Santin. An attempt has been made to fit these profiles with empirical equations. As night profiles, after reduction for electron peak density and thickness, are independent of season and solar activity, Becker's normalized midnight profile could be described by

$$\frac{N}{N_{\max}} = \frac{\exp\left[-\{(h_m - h)/y_m\}^3\right]}{\cosh\{(h_m - h)/y_m\}}. \tag{1}$$

$N_{\max}$ peak electron density, h_m peak altitude, y_m semi-thickness of a parabolic fit to the nose of the F2 layer. Similar descriptive equations can be used for daytime profiles provided there is no F1. Becker's normalized summer noon profile could, however, not be readily described and may even be not really representative due to the averaging procedure used. It will therefore be necessary to apply a special procedure when F1 is present. Figs. 1 and 2 show the normalized profiles and their approximation for low and high solar activity.

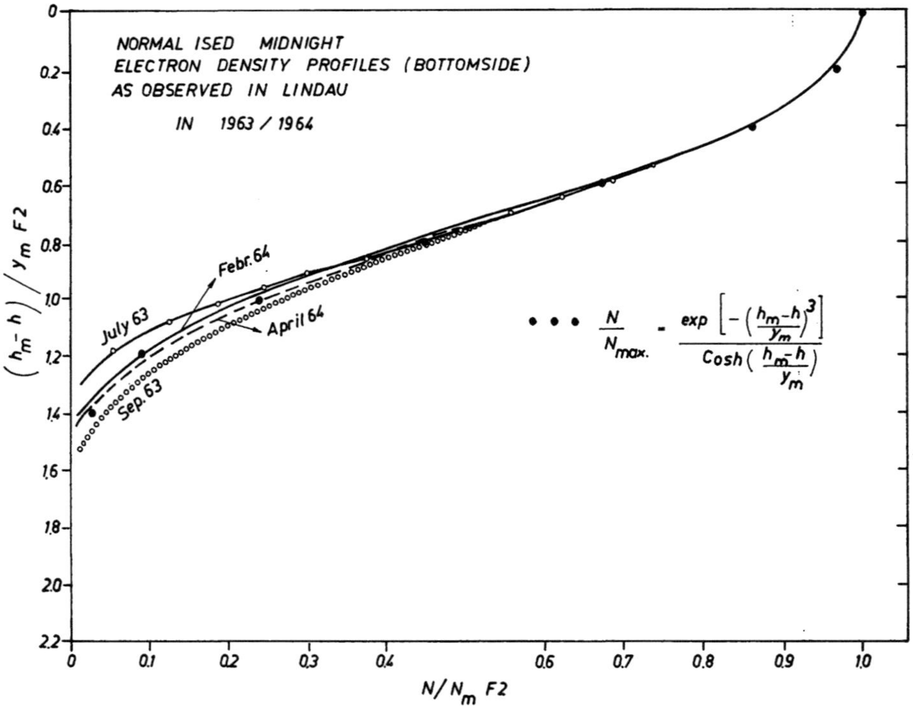

Fig. 1. W. Becker's normalized midnight profiles for low solar activity and our approximation formula (individual dots).

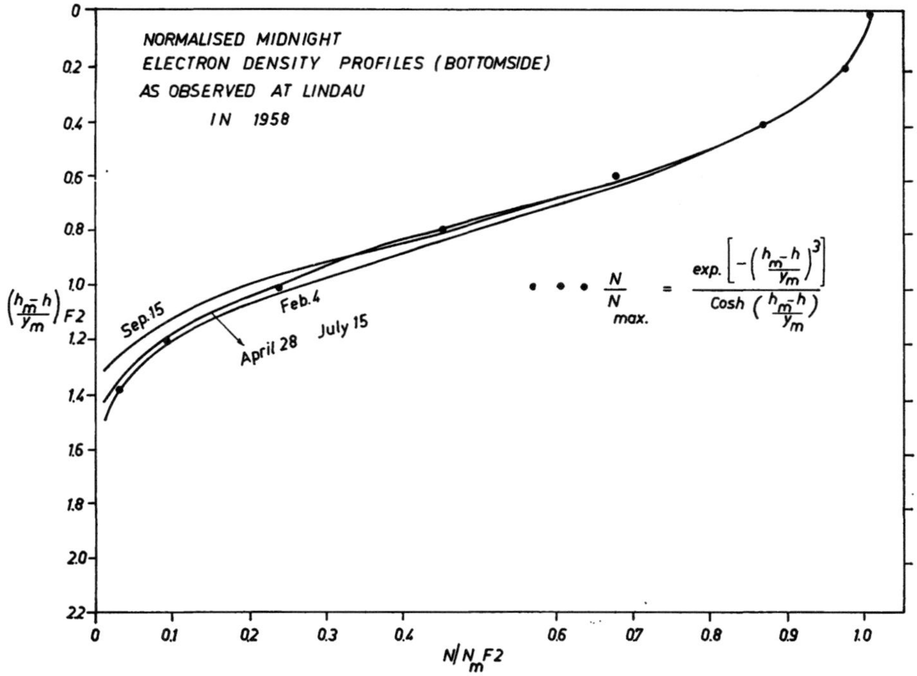

Fig. 2. W. Becker's normalized midnight profiles for high solar activity and our approximation formula (individual dots).

Finally, high latitude profiles from a longer period are now available for Murmansk. These will be compared with the "average" profiles from Thule. It may then be seen whether it is feasible to establish typical high latitude profiles. The "trough" problem certainly needs further consideration, which is under way, in the 'Ionospheric Network Advisory Group' for example.

2. Topside Ionosphere

As to the topside ionosphere it has not been possible to obtain a sufficient number of topside profiles down to the peak. Jackson (private communication) has made available about 30 profiles satisfying this condition. These have been used in order to establish a provisional mid-latitude topside profile first. The electron density values have been normalized with respect to N_{max}, and the heights by introduction of a suitable thickness parameter, Y_m; this is defined by the difference in height of the profile levels corresponding to $^1/_2\, N_{max}$ and N_{max}. The profile so normalized seems to be weakly dependent on season and hour; it varies, however, considerably with the solar activity S (= 10.7 cm solar radio noise). From Jackson's data standard topside profiles could be deduced for 1965/1966 (S between 70 and 100), and for 1968 (S between 100 and 140).

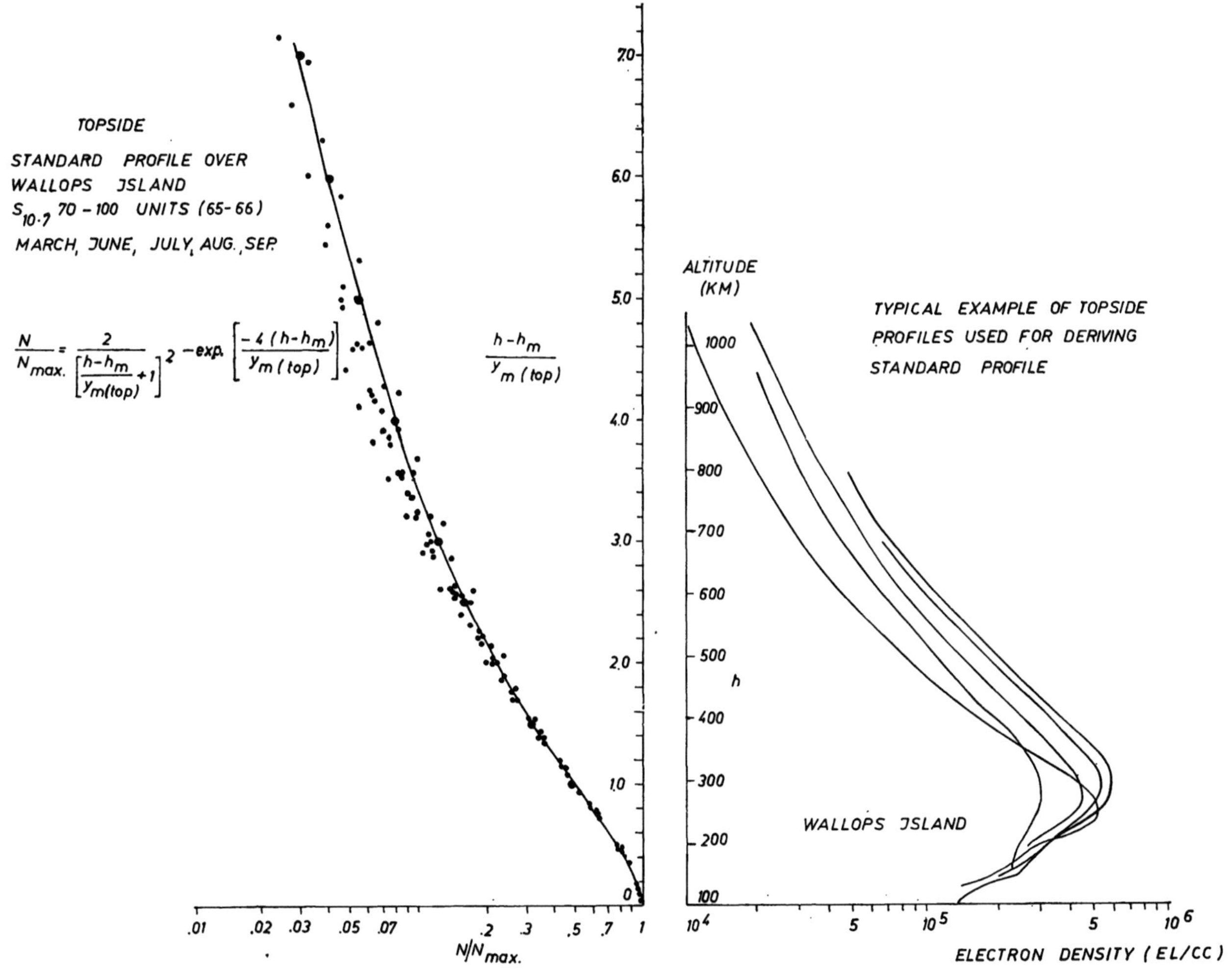

Fig. 3. Left side: standard topside profile deduced from Jackson's data for low solar activity and our approximation formula (full line); right side: typical low activity topside electron density distributions after Jackson.

It was found that the empirical equation

$$\frac{N}{N_{\max}} = \frac{A}{[\{(h - h_{\mathrm{m}})/Y_{\mathrm{m}}\} + 1]^a} - B \exp\left[-\frac{b(h - h_{\mathrm{m}})}{Y_{\mathrm{m}}}\right] \tag{2}$$

fits the observed data reasonably well. The four parameters A, B, a, b depend on solar activity S. Examples for low and medium activity are shown in Figs. 3 and 4. It is intended to use electron content data for determining the topside thickness parameter Y_{m}. Such data for western Europe have recently been summarized by Harnischmacher (private communication) and are available in numerical form. They are based upon measurements of the differential Faraday effect at five stations.

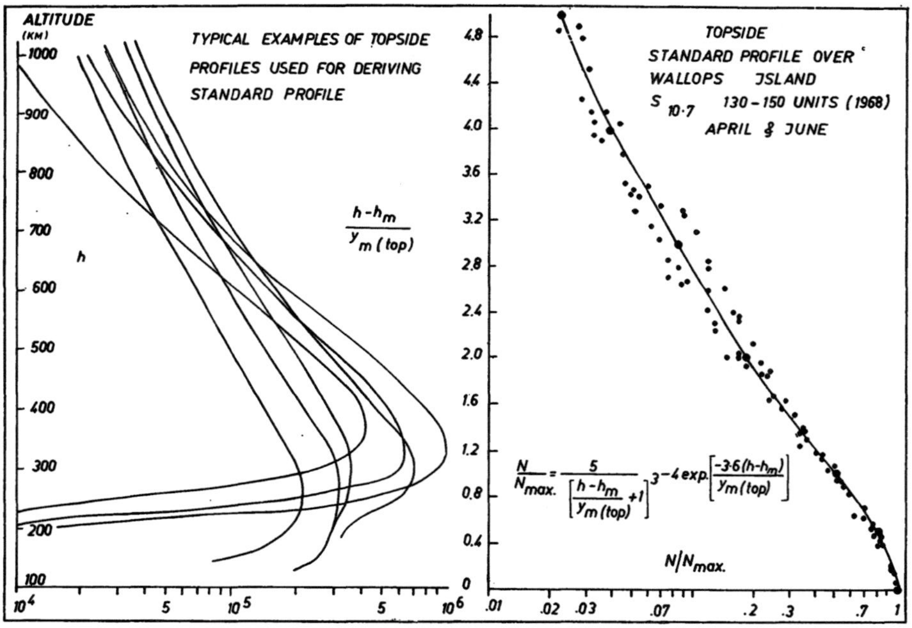

Fig. 4. Left side: typical medium activity topside electron density distributions after Jackson; right side: standard topside profile deduced from Jackson's data for medium solar activity, and our approximation formula (full line).

Quite recently Bent and Llewellyn (private communication) have analysed 60000 topside ionograms in order to establish topside profiles. Due to the now well-known experimental difficulties they were not able to go down to the peak altitude but they analysed the more or less exponential decrease of the electron density beginning at the height level $h_{\mathrm{m}} + {}^1/_2 Z_{\mathrm{m}}$. They apply a biquadratic profile shape

$$\frac{N}{N_{\max}} = \left[1 - \left(\frac{z}{Z_{\mathrm{m}}}\right)^2\right]^2 \tag{3}$$

between $h_{\mathrm{m}} - Z_{\mathrm{m}}$ and $h_{\mathrm{m}} + {}^1/_2 Z_{\mathrm{m}}$. Thus they extrapolate from the bottomside to the lower levels of the topside using the same profile curvature on both sides.

Their data for Z_m and h_m are deduced from the "average" electron density profiles available for stations on the American continent, see above. It seems very promising to use the Bent-Llewellyn profiles for the topside ionosphere up to 1000 km. A detailed report is to appear in a few months. For the peak region a check must be made to see how the data agree with those from other sources.

It is well known that there is another source of information available, namely electron density values deduced from incoherent scatter. Though the absolute value is calibrated with ionosonde data, the shape of the profile is well established.

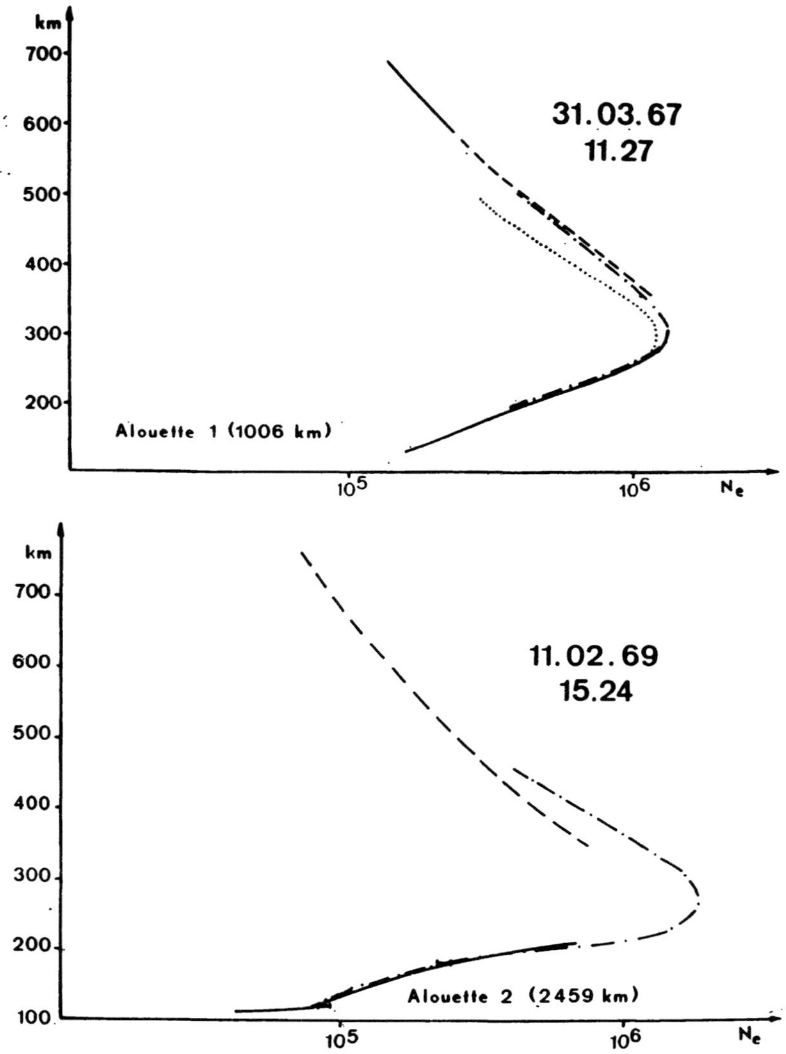

Fig. 5. Two examples of Alouette topside profile — — — and ionosonde bottomside profile ——— compared with incoherent scatter profile —·—·. (This latter was density-calibrated with bottomside ionograms.) Courtesy Bauer and Taieb.

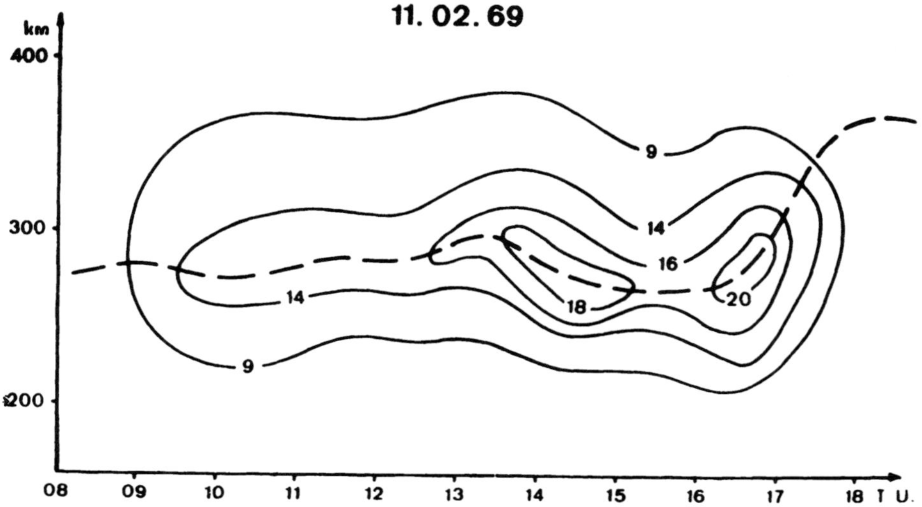

Fig. 6. Contour map of electron density during one day as obtained from incoherent scatter observations at St. Santin. Courtesy: P. Bauer.

Bauer and Taieb recently compared shapes observed at the St. Santin station with bottomside profiles from the nearby sounding station Garchy, and topside profiles from suitable passes of an Alouette topside sounding satellite (private communication). Figs. 5a, b and 6 show typical examples. It should be stated that there are certain differences between these topside profiles and those obtained by incoherent scatter, whereas the agreement is perfect on the bottomside. However, the differences do not seem to be systematic and thus should not be harmful in our context. Such data have been used to derive electron density contours, see Fig. 6. It appears that top- and bottomside profile curvatures are different, the topside thickness being greater. This is a warning against extrapolation from bottom to top without further control.

The considerations above may be helpful for establishing profiles up to a height of about 1000 km. Above that height whistler data are a valuable help, and the geometry becomes more and more field-aligned. On the other hand, in the lowest ionosphere below, say 90 km, the information about the electron density stems mainly from other sources than classical sounding. It seems at present that different methods give different profile shapes. Therefore our considerations are limited to the height range from 90 to 1000 km above ground.

References

[1] T. Shimazaki, J. Radio Res. Japan 2, 85 (1955).
[2] K. Maeda, J. Geomag. Geoelect. 22, No. 4 (1970).
[3] W. Becker, Kleinheubacher Ber. 12, 235 (1965).

Space Research XII — Akademie-Verlag, Berlin 1972

CRITICAL SURVEY OF ELECTRON AND ION TEMPERATURES MEASURED WITH PROBES

W. PFISTER

Air Force Cambridge Research Laboratories, L. G. Hanscom Field, Bedford, Mass., USA

Probes carried on suitable satellites can be a very useful tool for obtaining world-wide patterns of ion and electron temperatures, while rocket probes are needed for details in vertical structure. Difficulties arise in the interpretation of probe data. Probes with an a.c. technique tend to give lower and more likely correct values of temperature, although any type of probe under proper conditions appears to be capable of producing the correct value. The temperature readings can, however, be influenced by effects such as changes in work function. In addition considerable day to day variations in temperature contribute to the spread in measured values, so that it is difficult to arrive at standard values of temperature.

1. Introduction

Probes for electron and ion temperature have been used extensively on rockets and satellites and a substantial amount of data has been accumulated and presented in the literature [1]. Only a limited amount of data has been used so far for morphological studies [2, 3]. This is understandable since not enough confidence exists in the reliability of probe measurements. For instance, discrepancies in temperatures derived from different types of probes or from the incoherent scatter technique have been observed [4, 5]. It is not the purpose of this survey to examine in detail the different probe techniques and their merits. The emphasis is on temperature results in areas which appear most useful at the present time for the development of an International Reference Ionosphere.

All data from high magnetic latitudes generally above 60° are neglected since this region is complicated and deserves a special study. In this survey the rest of the data are divided into electron temperatures measured with satellites, ion temperatures from satellites, and electron temperatures between the E and F2 layer measured with rockets. Not enough rocket measurements of ion temperatures have been reported to be considered as representative.

2. Satellite Data of Electron Temperatures

A comparison of electron temperature measurements between different types of probes on the Explorer 31 satellite is given in Fig. 1 taken from [6]. The electron trap is a skin-mounted planar device with a swept voltage on the

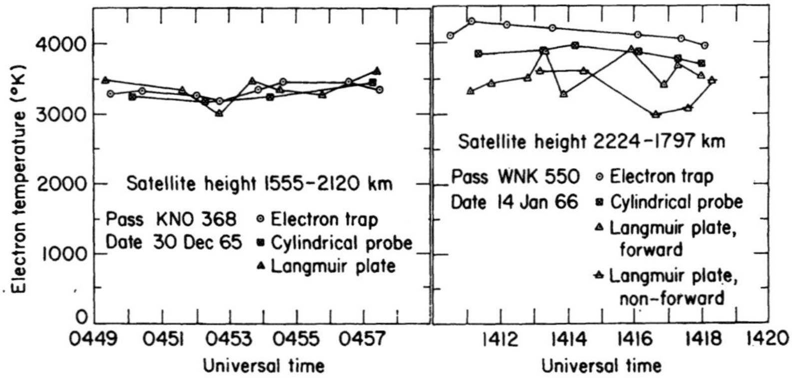

Fig. 1. Data from different types of electron temperature probes on Explorer 31 (from Booker and Smith [4]).

aperture grid [7]. The cylindrical probe is a Langmuir probe* separated from the satellite skin by a coaxially mounted guard electrode [8], a design used by Brace and colleagues extensively for global studies [2, 9–11]. The Langmuir plate is flush mounted with the skin, it is a modified Langmuir probe with an a.c. modulation technique developed by Boyd and colleagues [12]. The three probes in general measure different temperatures depending on satellite orientation and plasma environment. Agreement happens to occur at the magnetic

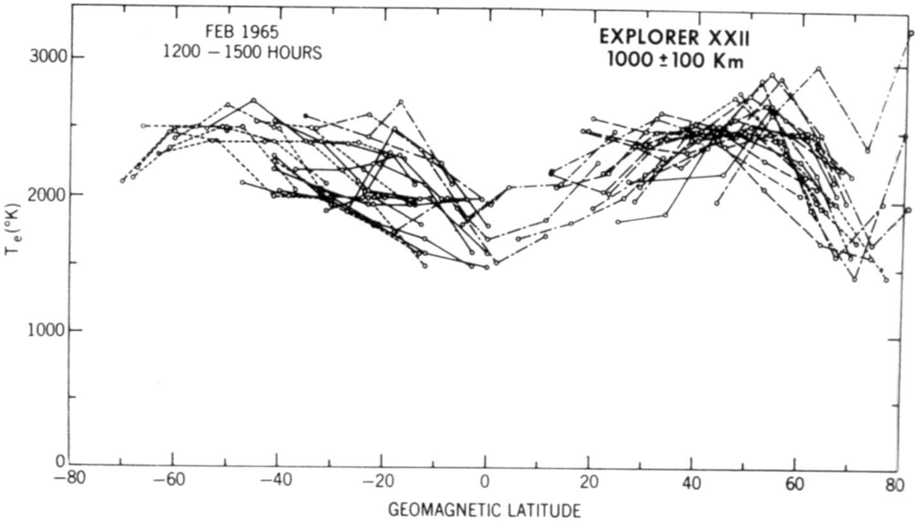

Fig. 2. Actual T_e data points from Explorer 22. The solid lines represent the longitude range 50° E–50° W, the long dashes 100° W–130° W, the short dashes 110° E–150° E, the long and short dashes 50° W–100° W (from Brace et al. [10]).

* Since systematic errors of measurement have the tendency to decrease the gradient of the linear portion of the log current vs. voltage characteristic of the Langmuir probe, the electron temperature derived therefrom will likely be higher than the true temperature.

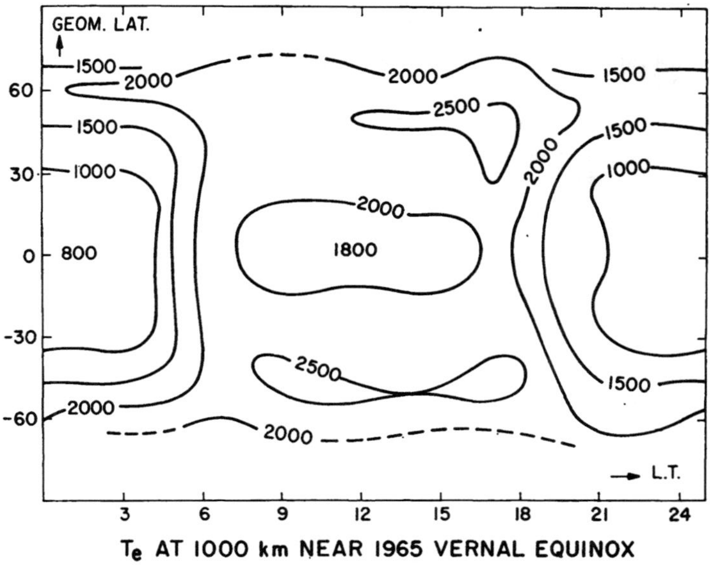

Fig. 3. Contour plots of smoothed T_e data from Explorer 22 (after Brace et al. [10]).

equator. There is a remarkable consistency for each probe from one orbit to the next. We are inclined to accept the lower values measured by the a.c. modulation technique as more correct values of temperature, especially since this technique is more likely to give better agreement with the incoherent scatter technique of radar measurements [5, 13].

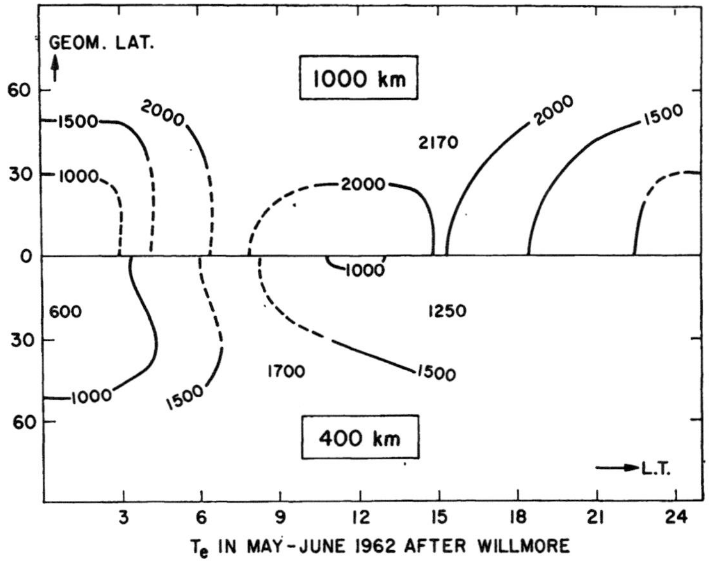

Fig. 4. Contour plots of T_e data from Ariel 1 obtained by regression analysis (after Willmore [3]).

Because of its near circular polar orbit, Explorer 22 is an excellent satellite with which to study the global variations of electron temperature at 1000 km altitude. The instrument on board is the cylindrical Langmuir probe of Brace. Fig. 2 taken from [10] shows the temperatures observed over a 3-week period. Points from a single pass are connected. Different symbols for the connecting lines are used to distinguish passes in different longitude regimes. The scatter of the data can be reduced by looking at these regimes separately. Obviously geomagnetic latitude is not the best choice of abscissa. For a global three dimensional representation, contour plots of smoothed data have been drawn in Fig. 3 as function of local time and geomagnetic latitude. The curves are

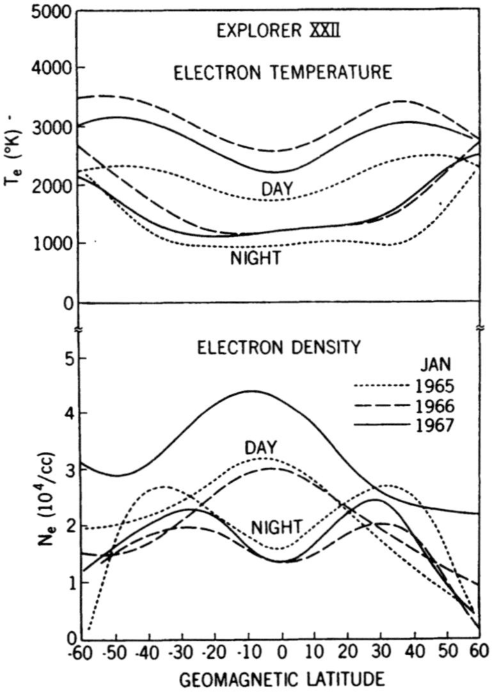

Fig. 5. Latitudinal structure in three consecutive years (from Brace [2]).

for equinox at the minimum of the solar cycle. A few important features are of interest. A temperature minimum of 800 °K occurs at the equator during the night. The temperature rises near sunrise and reaches a peak of about 2200 °K shortly after sunrise. At the equator a broad daytime minimum can be noticed, a second maximum before sunset and a decay into the night. With respect to latitude we find a minimum at the equator, relatively sharp near noon, considerably broader at night and disappearing at the sunrise peak. A maximum near 2000 °K can be seen at 60° latitude in the night and near 2500 °K at 45° latitude in the daytime.

The same pattern can be observed in Fig. 4 on similar contour plots constructed from Ariel 1 data [3]. This satellite was limited to somewhat lower latitudes and due to its elliptical orbit, more data interpolation was necessary to obtain plots for the 1000 km level. This was accomplished by a regression analysis, which has the disadvantage of smoothing out the variations, especially the sharp rise of temperature near sunrise. Otherwise the temperature values agree quite well with those from Fig. 3. However this result might be fortuitous. It could well be that the tendency of the a.c. modulation technique in Ariel 1 to measure lower temperatures is compensated by the influence of the solar cycle, the sunspot numbers in 1962 being higher than in 1965.

An example of some solar cycle dependence is given in Fig. 5 again for data from Explorer 22 [11]. The data from the early phase of the solar cycle show a general rise in temperature with increasing solar activity. Additional data for higher sunspot numbers are available from Explorer 22 and several other satellites. However a compilation of the data in terms of global patterns is still outstanding.

The lower half of Fig. 4 gives contour lines for 400 km altitude, again derived from Ariel 1 data with the regression analysis. There is a gap in coverage for the evening portion, but available evidence [9] from probes on other vehicles indicates a fairly smooth transition to night-time values. The pattern at 400 km is not very different from the pattern at 1000 km, but the temperatures are somewhat lower.

3. Satellite Measurements of Ion Temperatures

Most of the ion temperature measurements available today have been carried out with either one of two types of techniques. One is the spherical ion probe designed by the University College London [12]. The collector sphere is surrounded by a negatively biased grid to exclude the ambient electrons. The voltage on the collector is swept and a Druyvesteyn analysis is obtained by an a.c. modulation technique. Temperature data have been reported for Ariel 1 [14] and Explorer 31 [15]. The other instrument is a planar retarding potential analyzer used mainly by the groups at NASA, Lockheed and the University of Texas. The sweep voltage is applied, either to the retarding grid between the aperture grid at vehicle potential and the collector, or to the collector. Photoelectrons from the collector may be eliminated by a suppressor grid. The instrument is pointed into the direction of flight. A least squares fit to a theoretical expression for the current–voltage curve is the preferred technique of analysis [16]. It yields best estimates for temperature, vehicle potential and concentration of one or more ion species. Nevertheless systematic errors in temperature can be expected due to various sources not considered in the analysis, such as variable grid transparency, sheath distortion, or spacecraft influences. They can be minimized and corrected. The least predictable error is caused by a variation of work function across the retarding electrode. All these systematic errors have the tendency to produce higher temperature readings.

An example of one of the earlier measurements of ion temperature with satellites [17] is presented in Fig. 6. It is interesting first because of the

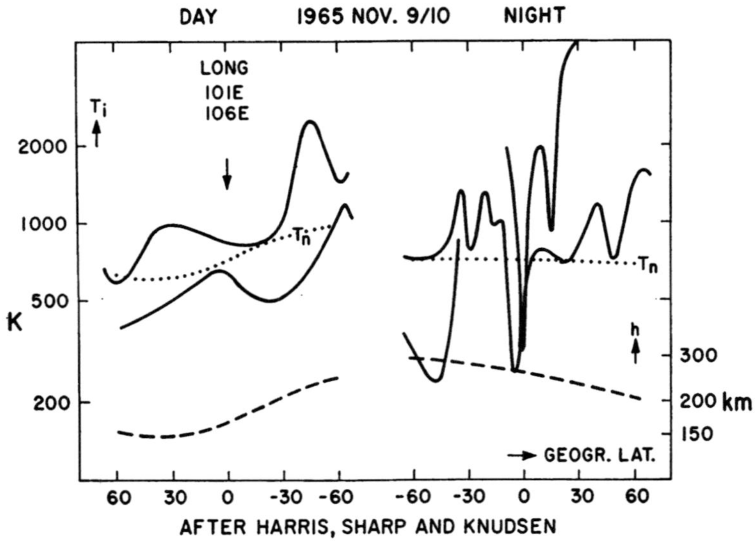

Fig. 6. Orbital profiles of ion temperature from low altitude satellite (after Harris et al. [17]).

extremely low orbit between 150 and 300 km. The temperature is plotted for a full orbit from geographic north to geographic south and back. Latitudes above 60° are omitted. Heights are lower on the dayside and higher at night. The data plotted are two orbits with almost the same longitude at equator

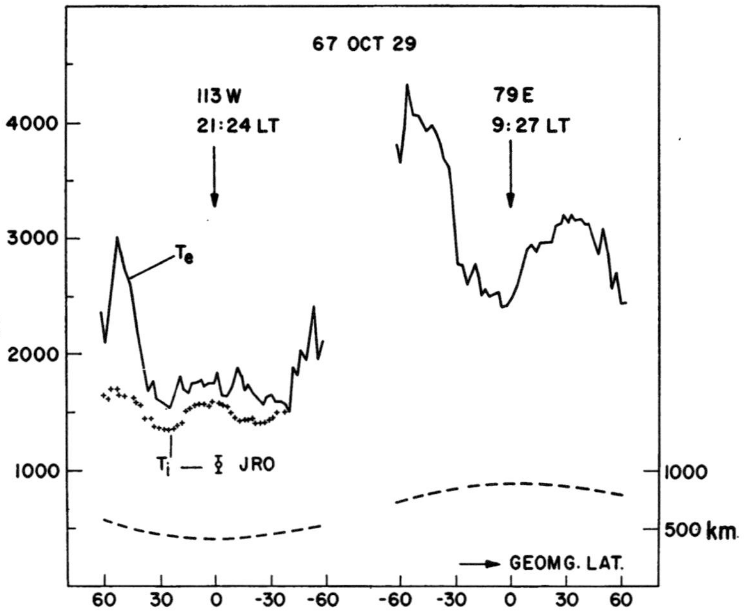

Fig. 7. Orbital profiles of temperatures from OGO 4 (after Chandra et al. [18]).

crossing one day apart. Not shown but of interest is the observation that successive orbits although 23° apart in longitude produce very similar curves, while the day to day variation is considerable. The measured temperatures can hardly be considered as true ion temperatures. On-board data processing was used, lacking the sophistication now used for planar RPA's. The low temperatures measured, which are on the average at the level of the neutral temperature as given by the CIRA 1965 model, are unusual.

Data according to Chandra et al. [18] from OGO 4 with an orbit between 400 and 900 km are shown in Fig. 7. Looking first at the electron temperature we find the values a good 50% higher than in the contour plots of Fig. 3. This is due mainly to the increased solar flux. Otherwise the shape of the curve is not much different. Ion temperatures were measured with a planar ion

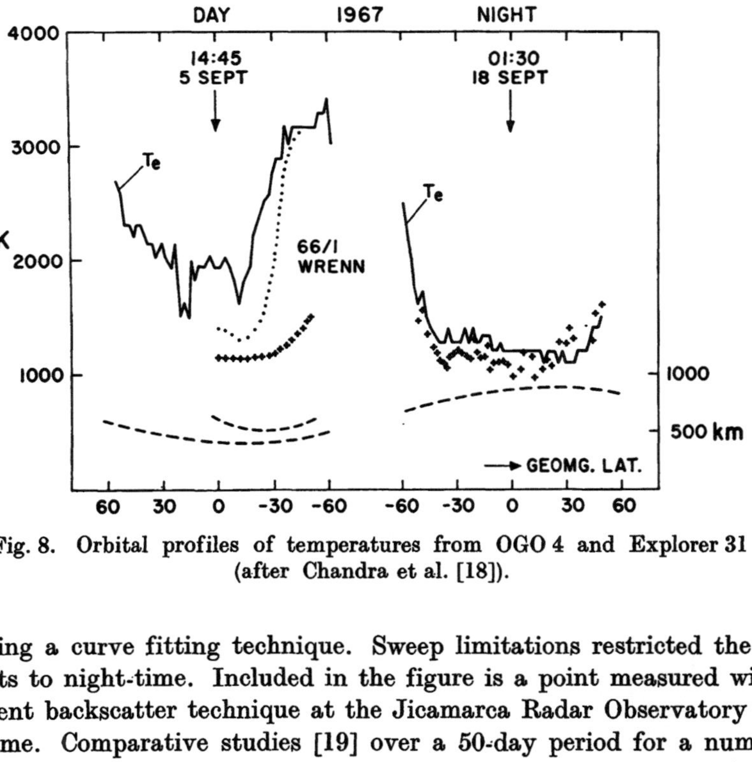

Fig. 8. Orbital profiles of temperatures from OGO 4 and Explorer 31 (after Chandra et al. [18]).

trap using a curve fitting technique. Sweep limitations restricted the measurements to night-time. Included in the figure is a point measured with the incoherent backscatter technique at the Jicamarca Radar Observatory at the same time. Comparative studies [19] over a 50-day period for a number of passes near the Jicamarca location indicate on the average a 40% higher value for this type of ion temperature probe.

Other data from OGO 4, this time with the night-time orbit at apogee are shown in Fig. 8. The ion temperatures are lower and appear to agree well with the radar temperatures at Jicamarca [20]. Near thermal equilibrium is indicated by the electron temperature data. The daytime curve is used to make a comparison, at the same local time but lower sunspot number, with another type of probe flown by Wrenn [15] on Explorer 31 with a highly elliptical orbit. Near perigee the two orbits agree fairly well. Wrenn uses the

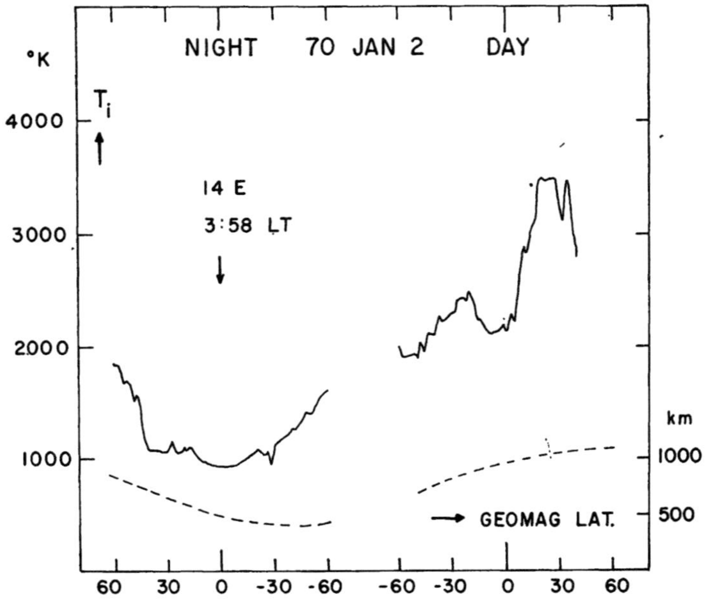

Fig. 9. Orbital profile of ion temperature from OGO 6 (reproduced by courtesy of the University of Texas).

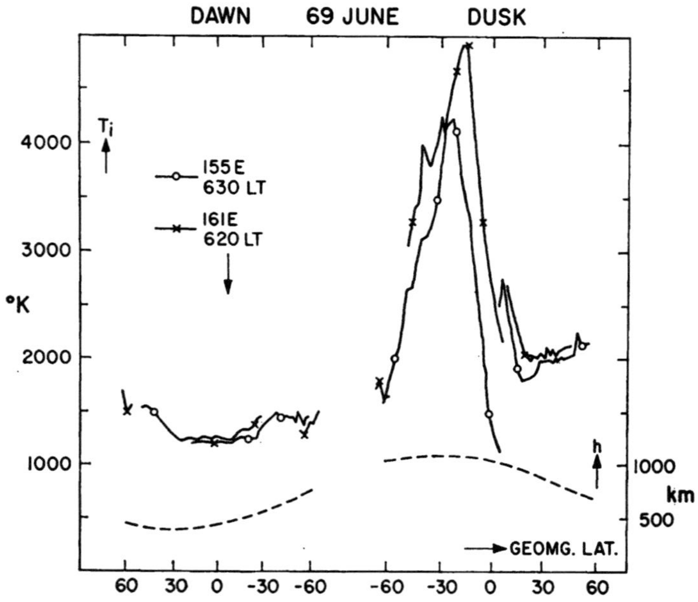

Fig. 10. Orbital profiles of ion temperature from OGO 6 (reproduced by courtesy of the University of Texas).

spherical ion probe with the Druyvesteyn analysis for temperatures. The planar probe on Explorer 31 measures on the average an ion temperature 50% higher than the ion probe with the a.c. analysis [6].

The state-of-the-art with respect to planar RPA data is provided on a fairly routine basis by OGO 6 [21]. A sample of the temperature data kindly made available by the University of Texas is plotted in Fig. 9. The orbit is between 500 and 1100 km. The daytime curve around 1600 LT shows two effects, an increase of temperature with height and with latitude. They add in the northern hemisphere and subtract in the southern hemisphere. The latitude of the temperature peak is always higher at night-time and lower at daytime, as we have seen already in the electron temperatures. Another two-orbit sample with dawn at perigee and dusk at apogee is shown in Fig. 10. The temperature peak is now in the southern hemisphere and unusually high. Whether the South Atlantic anomaly is responsible for the high ion temperatures or for errors due to effects on the work function is an open question.

4. Rocket Measurements of Electron Temperatures

The techniques for measuring electron temperatures with rockets are basically the same as those used with satellites. However some instruments have been reported only in connection with rocket flights, like the nose tip probe [22]. Notable is the ejected package like the dumbbell probe or the thermosphere probe of Brace [23] and Spencer [24]. Both use a Langmuir characteristic, one for a symmetrical spherical geometry, the other one for a cylindrical geometry. Recently results have become available from the Japanese electron temperature probe [25, 26]. It is a planar Langmuir probe. The main feature is to keep the current close to zero by a high external resistor. The Langmuir characteristic is utilized by measuring the shift in d.c. potential caused by different a.c. voltages applied to the electrode. This technique can conveivably avoid some of the problems caused by changes in work function.

An example of difficulties which can occur in probe measurements is illustrated in Fig. 11. The instrument, mounted in a boom, is a spherical electron trap with an aperture grid of variable potential and an electron collector [27]. The data taken at sunset have been collected and reduced by M. Smiddy of AFCRL. They are interesting because they show the effect of rocket illumination. In the complete earth's shadow the measured electron temperature follows the expected neutral temperature profile with exception of the hot layer at 120 km seen on ascent and descent. In total sunlight the measured electron temperature is much higher and a voltage-independent photoelectron current is present. It shows some spin modulation due to the spin dependent illumination of the aperture grid. A similar type of modulation on the current slope used to derive the temperature is not expected. High temperatures are measured as soon as the rocket reaches the penumbra 10 km below the level of full sunlight and before any photoelectron current can be detected. Two possible explanations come to mind. One is an electron gas of higher quasi-

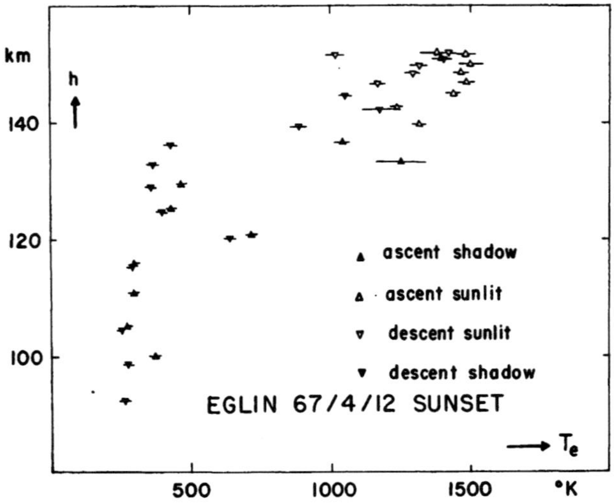

Fig. 11. Effect of illumination on electron temperature measurements.

temperature surrounding rocket and instrument if radiation from the sun is present; or sweep-dependent changes in work function due to irradiation might be blamed for the difficulties.

The variability of electron temperature profiles up to F2 layer heights is shown in Fig. 12 for daytime conditions centering around noon. All data have been taken by Brace and colleagues [24, 28, 29] at Wallops Island with an ejected probe. Each curve is marked by the date, by the *Kp* index and by the 10 cm flux in units of 10^{-22} W m^{-2} Hz^{-1}. The profiles are smoothed with

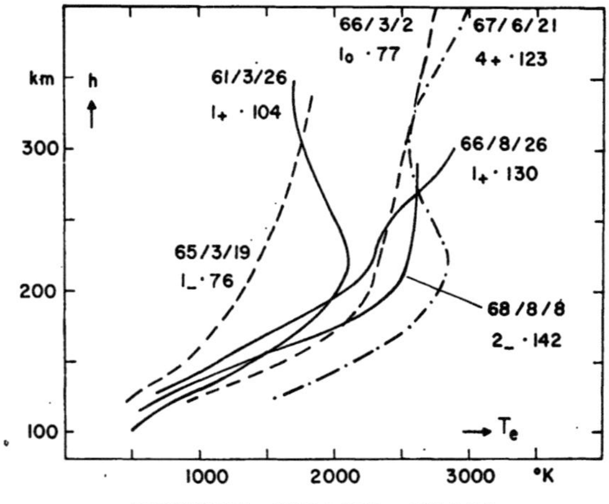

Fig. 12. Electron temperature profiles measured with ejected probes.

respect to small-scale variations, which may or may not have any physical significance. Two types of profiles are evident. One is preferred at higher solar fluxes and displays a pronounced peak value around 220 km. The other type without a peak appears to be characteristic for low fluxes. Large day to day variations are evident. The high F1 layer temperature on the only disturbed day should be noted without drawing premature conclusions.

In general, the electron temperatures tend to be lower at lower latitudes (Fig. 13). The Natal profile and the upper Arecibo profile were again measured with the thermosphere probe, one near the magnetic equator, the other one in comparison with the incoherent backscatter radar of the Arecibo Observatory [30], giving very good agreement. The second profile for Arecibo, available for heights below 180 km, was measured [31] with a gridded spherical electron detector and shows temperatures consistently higher than the radar data by

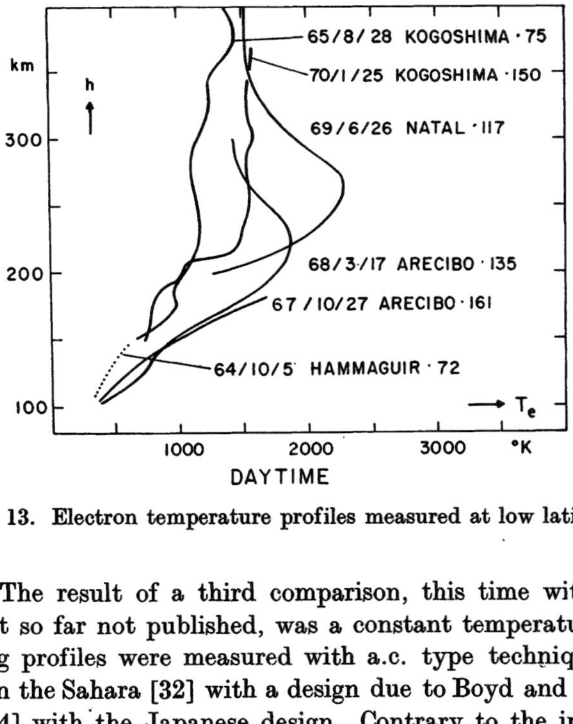

Fig. 13. Electron temperature profiles measured at low latitudes.

about 20%. The result of a third comparison, this time with the nose tip probe [22] but so far not published, was a constant temperature ratio of 1.8. The remaining profiles were measured with a.c. type techniques, the one at Hammaguir in the Sahara [32] with a design due to Boyd and the Kagoshima rockets [33, 34] with the Japanese design. Contrary to the impression given by Fig. 13, the a.c. probes may give as much variation in temperature as other types of probes. For instance, a second flight at Hammaguir two days later measured a temperature twice as high. Additional measurements [35] with the Japanese probe include profiles exceeding those of Fig. 13 in temperature.

A summary of electron temperature results, measured with probes in the E region, was given by L. G. Smith et al. [22]. They found a consistency of the data taken within a few hours even when measured a considerable distance apart, but great variability from day to day. In addition, they noticed a certain

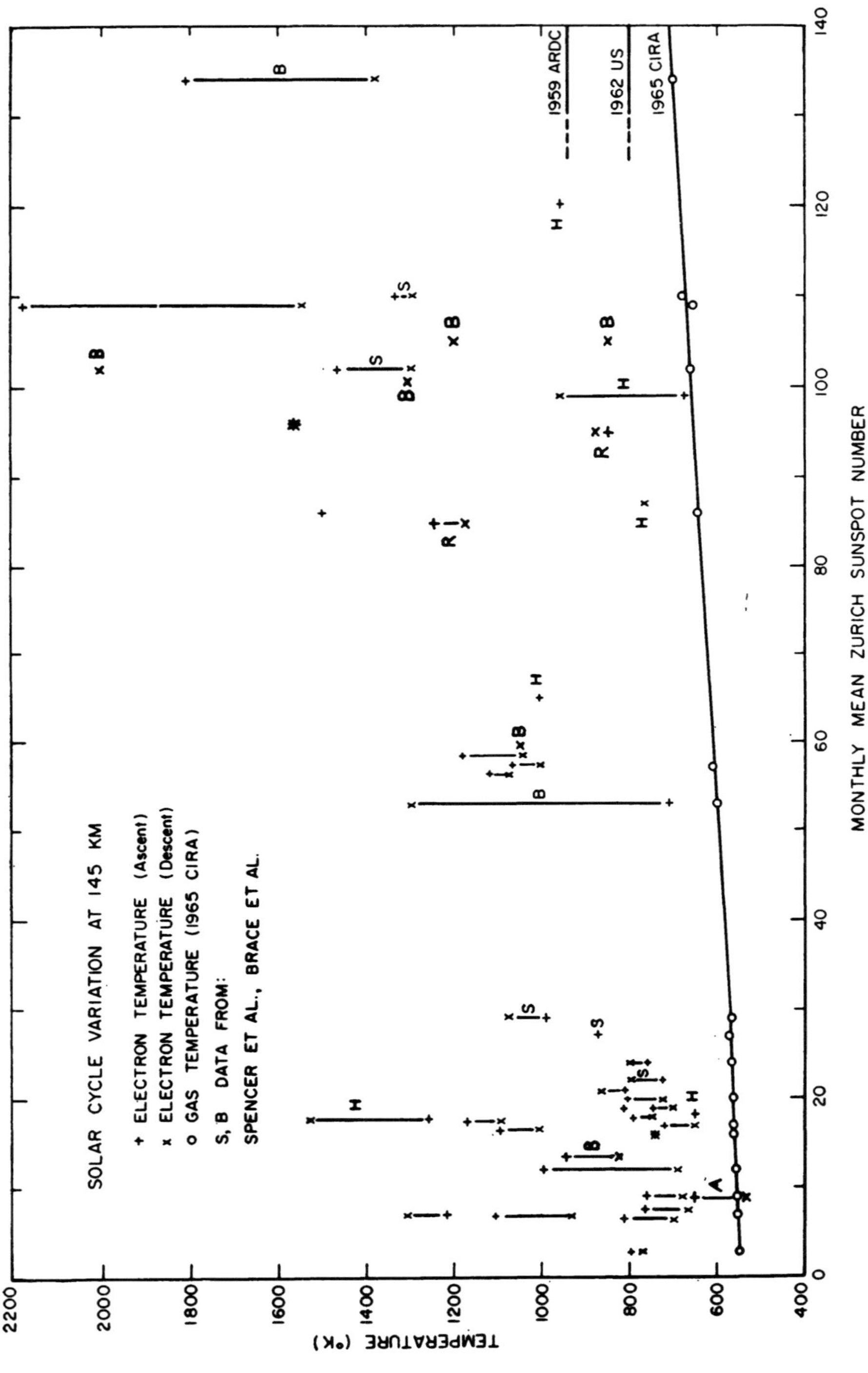

Fig. 14. Daytime electron temperatures from rocket observations plotted against the monthly mean Zürich sunspot number.

trend with solar activity. Their summary figure for daytime temperatures at the 145 km level was used to add 16 additional measurements collected from the literature [28–35] and to construct Fig. 14. Only six measurements fall below the range of the scatter plot established by Smith: four of them are produced either by the British a.c. probe (A) or by the Japanese a.c. probe (H). The remaining two were obtained by Sagalyn (R) and Brace (B) in comparison with the Arecibo Observatory radar. With respect to B there is no difference in instrumental quality compared with most of the other B rockets. It is possible that the design used by Sagalyn on the Arecibo rocket is better for temperature measurements than the earlier design. The knitted wire grid has been replaced by a smooth gold plated spherical copper sheath with holes. One measurement with the Japanese a.c. probe (H) falls above the range of Smith's scatter plot. It is characterized by a highly structured temperature profile [35] which probably indicates some kind of plasma instability.

5. Concluding Remarks

Only tentative conclusions can be drawn from the overall experience with temperature measurements by probes. Certainly considerable day to day variations can be expected in all regions of the ionosphere. Systematic errors of various magnitude have to be faced depending on the technique used and the atmospheric environment encountered. The biggest problem appears to be a delayed dependence of surface work function on probe currents [36]. Good temperature measurements appear to be available from almost any type of technique, but certain techniques like those using a.c. voltages are more likely to produce good results. In spite of some remaining uncertainties, a routine collection of temperature data from satellites of various orbital parameters is very useful for morphological studies. In fact, this collection is pursued presently at various places and it is hoped that a morphological analysis of the accumulated data will be ready for publication soon.

Acknowledgments

The author thanks L. H. Brace of Goddard Space Flight Center, J. P. McClure and S. Sanatani of University of Texas, R. C. Sagalyn and M. Smiddy of Air Force Cambridge Research Laboratories, L. G. Smith of GCA Corporation, and K. Hirao of University of Tokyo for helpful discussions and contributions to this survey.

References

[1] A. P. Willmore, Space Sci. Rev. **11**, 607 (1970).
[2] L. H. Brace, Space Research X, 633 (1970).
[3] A. P. Willmore, Proc. Roy. Soc. A **286**, 537 (1965).
[4] H. G. Booker and E. K. Smith, J. Atmos. Terr. Phys. **32**, 467 (1970).
[5] H. C. Carlson and J. Sayers, J. Geophys. Res. **75**, 4883 (1970).

[6] J. L. Donley et al., Proc. Inst. Elect. Electron. Engrs. **57**, 1078 (1969).
[7] J. L. Donley, Proc. Inst. Elect. Electron. Engrs. **57**, 1061 (1969).
[8] J. A. Findlay and L. H. Brace, Proc. Inst. Elect. Electron. Engrs. **57**, 1054 (1969).
[9] L. H. Brace et al., Planet. Space Sci. **13**, 647 (1965).
[10] L. H. Brace et al., J. Geophys. Res. **72**, 265 (1967).
[11] L. H. Brace et al., J. Geophys. Res. **73**, 1607 (1968).
[12] G. L. Wrenn, Proc. Inst. Elect. Electron. Engrs. **57**, 1072 (1969).
[13] G. N. Taylor and G. L. Wrenn, Planet. Space Sci. **18**, 1663 (1970).
[14] R. L. F. Boyd and W. J. Raitt, Space Research V, 207 (1965).
[15] G. L. Wrenn, Mullard Space Sci. Lab. Internal Rep. ION 6901 (1969).
[16] S. J. Moss and E. Hyman, J. Geophys. Res. **73**, 4315 (1968).
[17] K. K. Harris et al., J. Geophys. Res. **72**, 5939 (1967).
[18] S. Chandra et al., J. Geophys. Res. **75**, 3867 (1970).
[19] J. P. McClure and B. E. Troy, J. Geophys. Res. **76**, 4534 (1971).
[20] J. P. McClure, J. Geophys. Res. **76**, 3106 (1971).
[21] W. B. Hanson et al., J. Geophys. Res. **75**, 5483 (1970).
[22] L. G. Smith et al., J. Atmos. Terr. Phys. **30**, 1301 (1968).
[23] L. H. Brace et al., J. Geophys. Res. **68**, 5397 (1963).
[24] N. W. Spencer et al., J. Geophys. Res. **70**, 2665 (1965).
[25] K. Hirao, in: Electron Density and Temperature Measurements in the Ionosphere, COSPAR Technique Manual Series, 1967 (p. 58).
[26] K. Hirao and K. Oyama, J. Geomag. Geoelect. **22**, 393 (1970).
[27] M. Smiddy and R. C. Sagalyn, Space Research VII, 428 (1967).
[28] L. H. Brace et al., Space Research XI, 1079 (1971).
[29] L. H. Brace et al., J. Geophys. Res. **74**, 257 (1969).
[30] L. H. Brace et al., J. Geophys. Res. **74**, 1883 (1969).
[31] R. C. Sagalyn and R. H. Wand, J. Geophys. Res. **76**, 3783 (1971).
[32] M. Aubry et al., Radio Sci. **1**, 170 (1966).
[33] K. Hirao, J. Geomag. Geoelect. **18**, 333 (1966).
[34] K. Hirao and K. Oyama, paper presented at XIII COSPAR Meeting, Leningrad 1970.
[35] K. Hirao and K. Oyama, Space Research XII, 1335 (1972).
[36] J. Sayers, J. Atmos. Terr. Phys. **32**, 663 (1970).

Space Research XII — Akademie-Verlag, Berlin 1972

OBSERVED SOLAR GEOMAGNETIC CONTROL OF THE IONOSPHERE: IMPLICATIONS FOR REFERENCE IONOSPHERES

H. A. Taylor, Jr.

Goddard Space Flight Center, Greenbelt, Md, USA

The recent availability of nearly continuous, pole to pole profiles of ionospheric composition permits the first detailed view of the significance of solar–geomagnetic processes in determining the distribution of ionospheric components. Bennett rf spectrometer results from the polar orbiting OGO 2, 4 and 6 satellites reveal dominant longitudinal variations superimposed upon the latitudinal distributions of each ion species. The longitudinal variation is important throughout the altitude range of these satellites, being observed in the distributions of molecular ions (N_2^+, NO^+, O_2^+) in the range 400–700 km, and persistent also in the distributions of the light ions (H^+, He^+) observed near apogee in the range 800 to 1100 km. The magnitude of the longitudinal variation may be quite pronounced, with the ambient concentration of O^+ observed at a given altitude and geodetic latitude varying by as much as an order of magnitude between two opposing longitudes, during relatively quiet magnetic conditions ($Kp \leqq 2$). Although the character and magnitude of the longitudinal variations may be different for different ions, a broad and persistent pattern is observed in which the average longitudinal variation appears to be well ordered in terms of solar–geomagnetic geometry. Because of the prominence of the longitudinal variation observed during just one earth rotation, the identification of seasonal, diurnal, and annual variations of the ion composition requires selective analysis of data having a high degree of spatial resolution. Such a study has been conducted with OGO 6 ion results, revealing that (i) seasonal variations in the ion distributions are generally repeatable, and (ii) the annual variation in the topside O^+ and H^+ content and distribution between 1969 and 1970 was minimal. These results would have been largely obscured in the absence of the longitudinal resolution required to provide data at appropriate solar–geomagnetic orientations. This study emphasizes the importance of magnetic field control in regulating the ion distributions, and suggests the need for the use of improved coordinate systems for the correlation of satellite observations and the development of ionospheric models.

1. Introduction

Beginning with the launch of OGO 4 in October 1965, sets of nearly continuous high-resolution direct measurements of thermal ion composition have been obtained for a wide range of locations and geophysical conditions. Together, these results comprise the most complete empirical evidence for the complexity of the distributions of the principal ions which populate the ionosphere within the altitude range of approximately 400–1100 km. An outstanding feature of these results has been the identification of a broad and persistent longitudinal variation observed in the pole to pole profiles of ion composition obtained as the earth rotates beneath the relatively fixed satellite orbit. The pattern of the longitudinal variation has been associated with the wobble of the geomagnetic field relative to the earth–sun orientation, and has been identified in earlier studies as evidence for a "solar–geomagnetic" seasonal variation [1]. Preliminary studies of OGO 4 results have shown that the longitudinal variation is extremely important for global studies of the O^+–H^+ transition level, as well as investigation of the global prominence of ionized helium [2].

Although a number of previous investigators have found some evidence of the longitudinal variation, a definitive identification of this phenomenon has awaited the availability of measurements from the OGO satellites which permit nearly continuous sampling of the longitudinal variation, as the earth rotates beneath the relatively fixed position of the satellite orbit. The fact that ion composition is recorded for *complete* orbits, on a continuing basis, is a major factor in the identification of the ionospheric features described in this paper.

In this study, we examine the longitudinal variation more extensively, using data sets obtained from the OGO 6 ion spectrometer during the period 1969–1970. During this interval, the experiment was operated continuously, and the resultant data are interrupted only occasionally by gaps attributable to spacecraft operations and data tape processing. Accordingly, for most data sets to be studied, longitudinal profiles of the distributions of each of the ion constituents are obtained on a nearly continuous basis, with a separation of approximately 25° of longitude between consecutive orbits, with a period of approximately 105 minutes. In the long sweep mode, the sampling resolution, determined by the spectrometer sweep rate of approximately 35 seconds, is about 3° of latitude between consecutive samples of a given ion along the orbital path. Significant longitudinal variations have also been discovered in data obtained in the short sweep mode which provides a resolution of 0.2° latitude between samples [3].

Although the orbital inclination is nearly polar (82°) the tilt of the dipole axis results in a variable range of latitude coverage between orbits. Wherever possible in the results which follow, an attempt is made to identify the *longitudinal* variation under the most comparable conditions of altitude and latitude, so as to rule out possible confusion produced by changes in these variables. This objective, combined with the precession of apogee–perigee with latitude, season, and local time significantly limits the number of data sets appropriate for the identification of fundamental variations. Within these limits, the results have been chosen to be representative of the diurnal, seasonal, and annual variations observed independently, at particular times and locations.

2. Results

2.1. Ion Distributions in Geodetic Latitude

The longitudinal variation has been observed *throughout* the ion composition, extending from the body of molecular ions populating the lower thermosphere to the light ions of the plasmasphere. To establish the significance of the geomagnetic field in regulating the distributions of these ions, we first examine the longitudinal variability of several of the more prominent ions, as a function of *geodetic* latitude.

In Fig. 1a, profiles of O_2^+ observed near 0630 LT, at 50° E and 154° W respectively, illustrate the extent of the longitudinal variation observed at lower altitudes. As an example, it may be seen that the ambient concentrations of O_2^+ at the equator, near 400 km, differ by as much as a factor of 4 between these contrasting longitudes. At low-southern latitudes, the difference increases to a factor typically near 6, although the zenith angle variation with latitude remains *unchanged* between these two passes. Note also that the altitude variations between the two passes are relatively minor at low latitudes, where significant concentration differences are observed. The magnetic conditions preceding each of these passes were also similar. Defining the maximum value of Kp in the 12-hour interval preceding the pass as Kp_{-12}, $Kp_{-12} = 3^-$ for the 15 March pass and $Kp_{-12} = 2^+$ for the 17 March pass. Accordingly, it is clear that the changes with longitude observed in the O_2^+ distributions, including both the significant change in total concentration as well as the distinct shift in the N–S symmetry of the distributions, are indeed most prominent and are not readily explained when viewed in the geodetic coordinate system.

In Fig. 1b we examine the distributions of O^+ during northern summer, and near 1600 LT. Although at this time apogee occurs near the south pole, and significant altitude variations exist between the northern and southern hemispheres, the altitude–latitude relationship for the two passes compared is very similar and, to a first approximation, altitude effects may be ruled out in considering the longitudinal variation. It may be seen that while the shape and amplitude of the O^+ distributions are similar in the northern summer hemisphere, a distinct variation begins to occur about 30° in the southern winter hemisphere. At 60° S, a difference of an order of magnitude is observed between the O^+ concentrations, although altitude, zenith angle, and prior magnetic activity are closely comparable.

In Fig. 1c, the longitudinal variation in He^+ is shown for equinox, near 0500 LT. In this case, with apogee near the equator, the altitude–latitude relationship is identical for the two passes, and significant differences are observed in the He^+ distributions at the contrasting longitudes. In particular, a pronounced equatorial trough is observed in He^+ on the pass near 160° W, where the He^+ distribution drops by about an order of magnitude near the equator relative to the mid-latitude concentration levels near $\pm 30°$. Note that the trough is broader, although much less pronounced, at the 100° E position. Once again, these data have been chosen for comparable periods of relatively quiet magnetic activity.

In Fig. 1d, contrasting distributions of H^+ obtained during northern summer and near midnight local time exhibit significant differences between longitudes. Both the depth and position of the high latitude proton trough are seen to change significantly with longitude. (Although some difference in the H^+ distributions might be expected as a result of the unavoidable difference in Kp history, this would not be in the direction of reducing the observed longi-

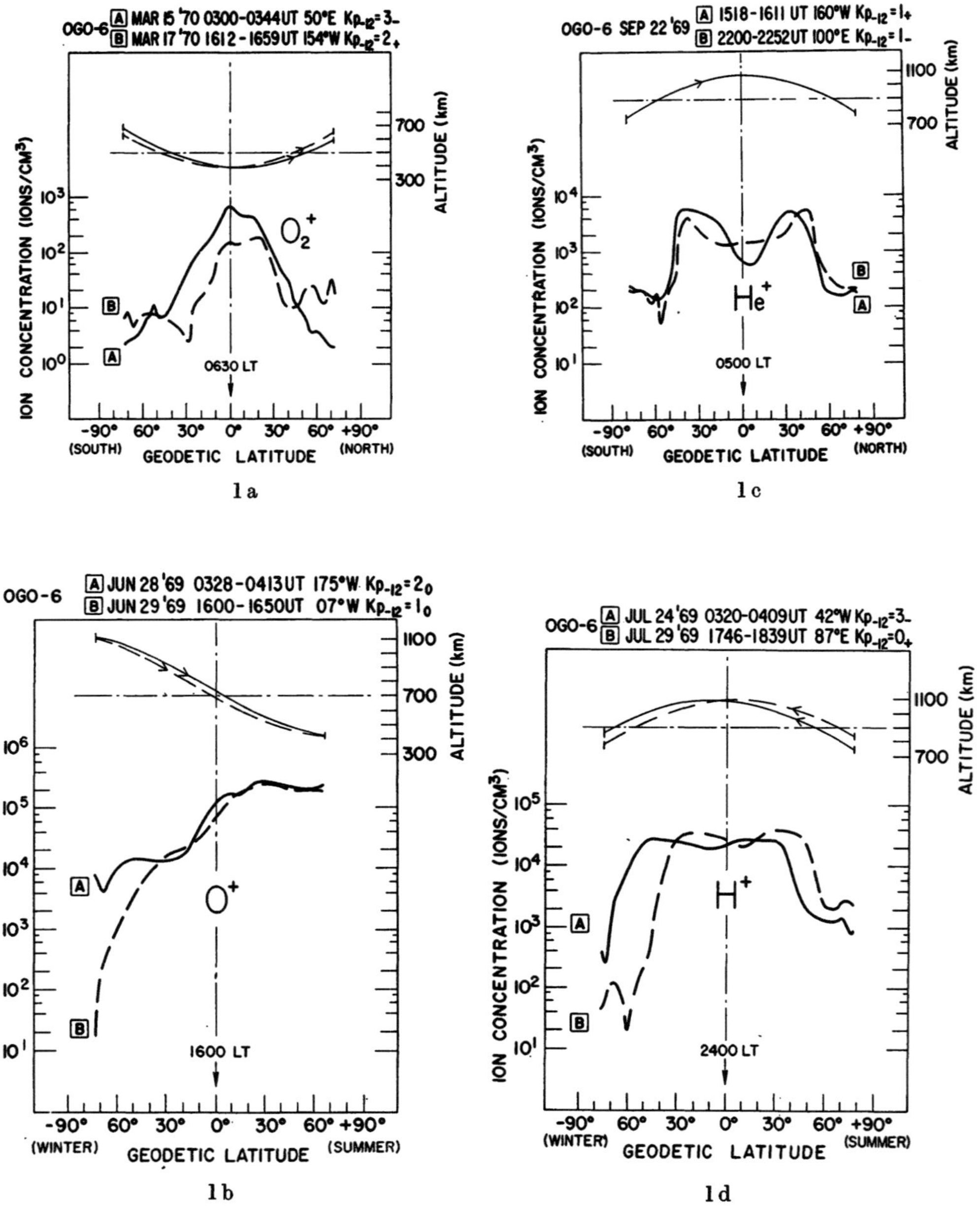

Fig. 1a–d. Examples of pronounced variation in ion distributions observed at positions of contrasting longitude. The notation Kp_{-12} refers to the maximum value of the planetary magnetic index Kp recorded during the twelve hour interval preceding the UT corresponding to the ion profile. The local time specified at the geodetic equator is measured as the orbit crosses the equator. The single time specified in each plot is chosen as representative of the time interval covered by the passes being compared. In panel c, the altitude distribution for the two passes is so nearly identical that the data appear in the form of a single curve.

tudinal variation, since the plasmasphere is observed to *shrink* with increasing *Kp*.). Like the O^+ distributions of Fig. 1b, the H^+ distributions of Fig. 1d exhibit similarity in the summer hemisphere, and yet show prominent differences in the southern winter hemisphere. Like O^+, the H^+ distribution appears to be 'tilted' toward the winter hemisphere, where a winter-time ion depletion appears to be emphasized at certain longitudes.

From the foregoing ion distributions, as well as from the distributions of other ions observed at a variety of local times and seasons, it is clear that pronounced and persistent variations in both the concentration and distribution of ionization are encountered with changing longitude. The persistence of the observed variations strongly suggests that the geomagnetic field must play a dominant role in regulating the ion distribution. For this reason, we next examine the longitudinal variation in terms of *dipole latitude*. For each of the time periods identified in Fig. 1a–d we have selected sets of ion distributions representative of the longitudinal variation encountered throughout a complete earth revolution. In Figs. 2a–d isometric displays of ion distribution sets are given for each of the local times associated with the profiles of Figs. 1a–d respectively.

In Fig. 2a it may seem that the latitudinal distributions of O_2^+ change noticeably as the earth rotates beneath the relatively fixed orbit during the two-day interval of 15–17 March. In this figure the O_2^+ profiles are arranged to illustrate a distribution pattern which appears to be correlated with the changing solar–geomagnetic geometry, indicated by the angle α, which is defined in the noon local time plane as the angle between the earth–sun line and the dipole equator. For the March condition of equinox, the amplitude of α oscillates between $\pm 11°$ (the displacement of the magnetic axis) during the course of one earth rotation. Similarly, the latitudinal distributions of O_2^+ exhibit a cyclic variation, with a tendency for broader and more symmetric distributions at eastern longitudes ($-\alpha$) relative to the western longitudes ($+\alpha$). In addition to this broad pattern, the O_2^+ distributions also exhibit complex features at higher latitudes, beyond about $\pm 45°$.

Further evidence for the link between the persistent longitudinal variation and the changing solar–geomagnetic orientation is given in Fig. 2b. In this case the O^+ distributions obtained near 1600 LT during the 3-day interval 27–29 June reveal a very pronounced correlation between the longitudinal variation in O^+ and the angle α. Because the sun is near the summer solstice position in June, the angle α remains positive throughout the earth rotation, and reaches its maximum amplitude near 7° W. The variation in the O^+ distributions follows a similar pattern, with the southern (winter) hemisphere concentrations falling off dramatically in the range where α is maximum. In the northern (summer) hemisphere, the amplitude and shape of the O^+ distributions is relatively constant. From this pattern it is seen that for solar–geomagnetic winter conditions, the apparent seasonal variation in O^+ is most pronounced, with "maximum winter" (highest α) lower than concentrations observed at "minimum winter" (minimum α) by as much as an order of magnitude. Again, it should be noted that while the altitude–dipole latitude relationship is of course not identical for each ion profile, a direct comparison of pass to pass variations indicates that the range of altitude variability is much less than might be required to explain the observed longitudinal variation in concentrations.

As indicated previously, the longitudinal variation persists in the light ions as well, and in Fig. 2c we see an example of the variability in He^+ observed near 0500 LT during 22–23 September. A significant feature of the longitudinal pattern in He^+ is the changing prominence of the equatorial trough, which is observed to be narrow and deep for the $+\alpha$ condition, in contrast to being broad and relatively shallow for longitudes for which α becomes negative. In addition, although the effect is somewhat more subtle, there is a tendency for He^+ to persist to higher latitudes at magnetic winter (negative α) locations, which is consistent with earlier evidence of a winter bulge in He^+ [2].

The final example of the pattern of longitudinal variability is given in Fig. 2d, in which the distributions of H^+ near 2400 LT are shown for the period 24–30 July. In this case the southern winter hemisphere distributions of H^+ exhibit a pronounced variation with longitude. At the positions of extreme solar–geomagnetic winter (maximum α) the high latitude trough in H^+ becomes much deeper and moves to its highest latitude position. Although it cannot readily be shown in the same figure, the northern summer distributions of H^+ exhibit a comparatively insignificant variation with longitude. As in the case of O^+ (Fig. 2b), the longitudinal variation in H^+ is emphasized in the winter hemisphere. It is emphasized that, for this data set, the longitudinal variation is observed following periods of relatively quiet magnetic activity, with $Kp_{-12} \leqq 3$ for all profiles in the set.

2.2. Identification of Basic Ionospheric Features

Having identified the importance of the longitudinal variation and the qualitative usefulness of the parameter α, it becomes evident that the identification of ionospheric features such as the diurnal or local time variation is directly dependent upon the ability to obtain data under comparable solar–geomagnetic orientations. In the following examples we attempt to identify the repeatability and predictability of certain basic ionospheric features, using the parameter α as a selection guide.

▶

Fig. 2a–d. Isometric displays of sets of ion profiles obtained during time intervals corresponding to the profiles shown in Fig. 1a–d. The profiles are plotted as functions of geodetic longitude, dipole latitude, and ion concentration. The altitude profile indicated on the right hand plot in each family is representative of the altitude–latitude relationship characteristic of the family. The local time specified also represents a typical value for the series of satellite passes.

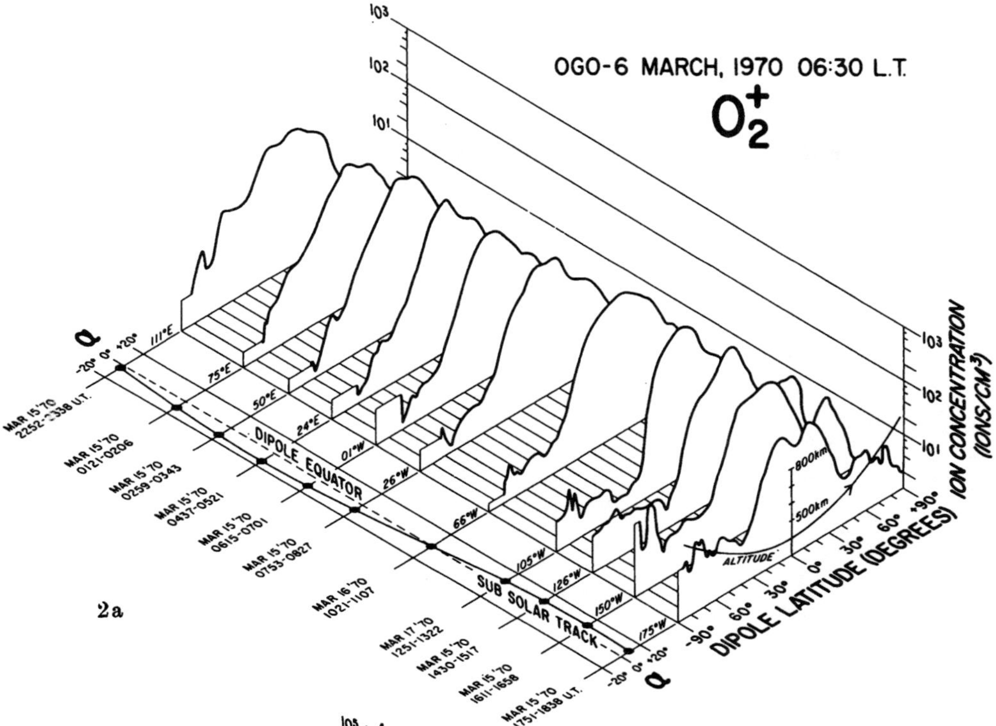
OGO-6 MARCH, 1970 06:30 L.T.
O2+
ION CONCENTRATION (IONS/CM3)
DIPOLE LATITUDE (DEGREES)
-90° 60° 30° 0° 30° 60° +90°
ALTITUDE
500km
800km
DIPOLE EQUATOR
SUB SOLAR TRACK
111°E
75°E
50°E
24°E
01°W
26°W
66°W
105°W
126°W
150°W
175°W
MAR 15 '70 2252-2338 U.T.
MAR 15 '70 0121-0206
MAR 15 '70 0259-0343
MAR 15 '70 0437-0521
MAR 15 '70 0615-0701
MAR 15 '70 0753-0827
MAR 16 '70 1021-1107
MAR 17 '70 1251-1322
MAR 15 '70 1430-1517
MAR 15 '70 1611-1658
MAR 15 '70 1751-1838 U.T.

2a

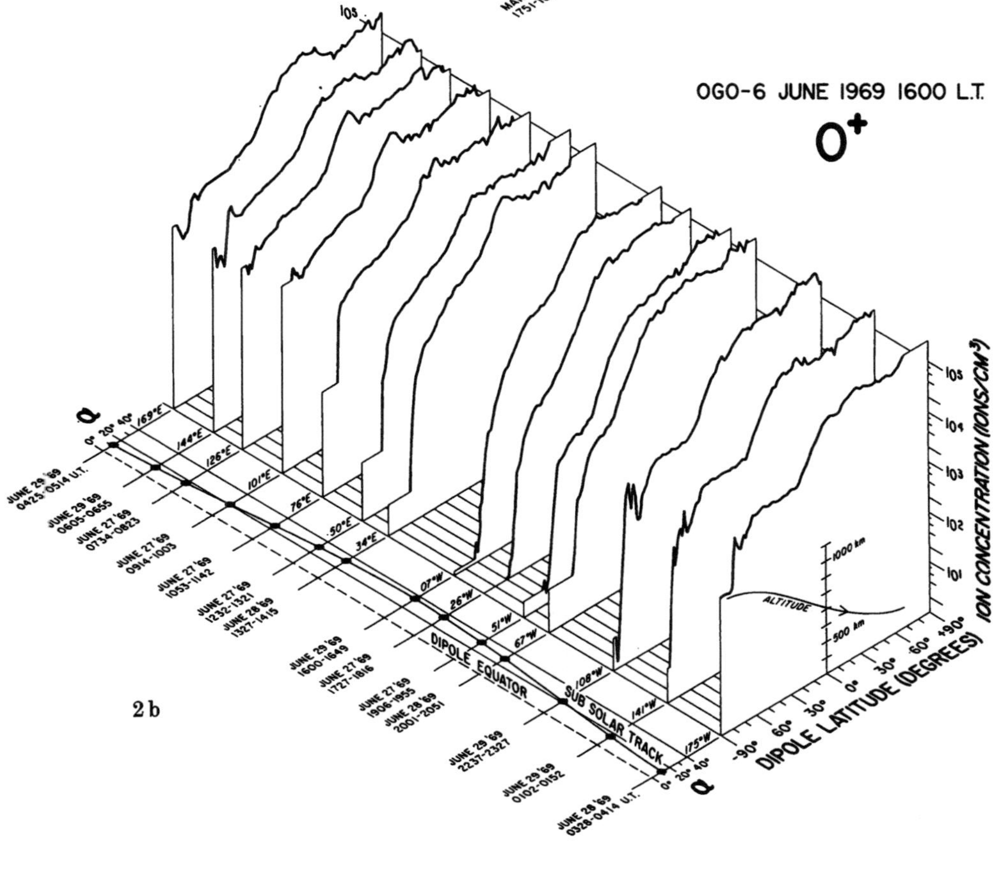
OGO-6 JUNE 1969 1600 L.T.
O+
ION CONCENTRATION (IONS/CM3)
DIPOLE LATITUDE (DEGREES)
-90° 60° 30° 0° 30° 60° +90°
ALTITUDE
500 km
1000 km
DIPOLE EQUATOR
SUB SOLAR TRACK
169°E
144°E
126°E
101°E
76°E
50°E
34°E
07°W
26°W
51°W
67°W
108°W
141°W
175°W
JUNE 29 '69 0425-0514 U.T.
JUNE 29 '69 0605-0655
JUNE 27 '69 0734-0823
JUNE 27 '69 0914-1003
JUNE 27 '69 1053-1142
JUNE 27 '69 1232-1321
JUNE 28 '69 1327-1415
JUNE 29 '69 1600-1649
JUNE 27 '69 1727-1816
JUNE 27 '69 1906-1955
JUNE 28 '69 2001-2051
JUNE 29 '69 2237-2327
JUNE 29 '69 0102-0152
JUNE 28 '69 0328-0414 U.T.

2b

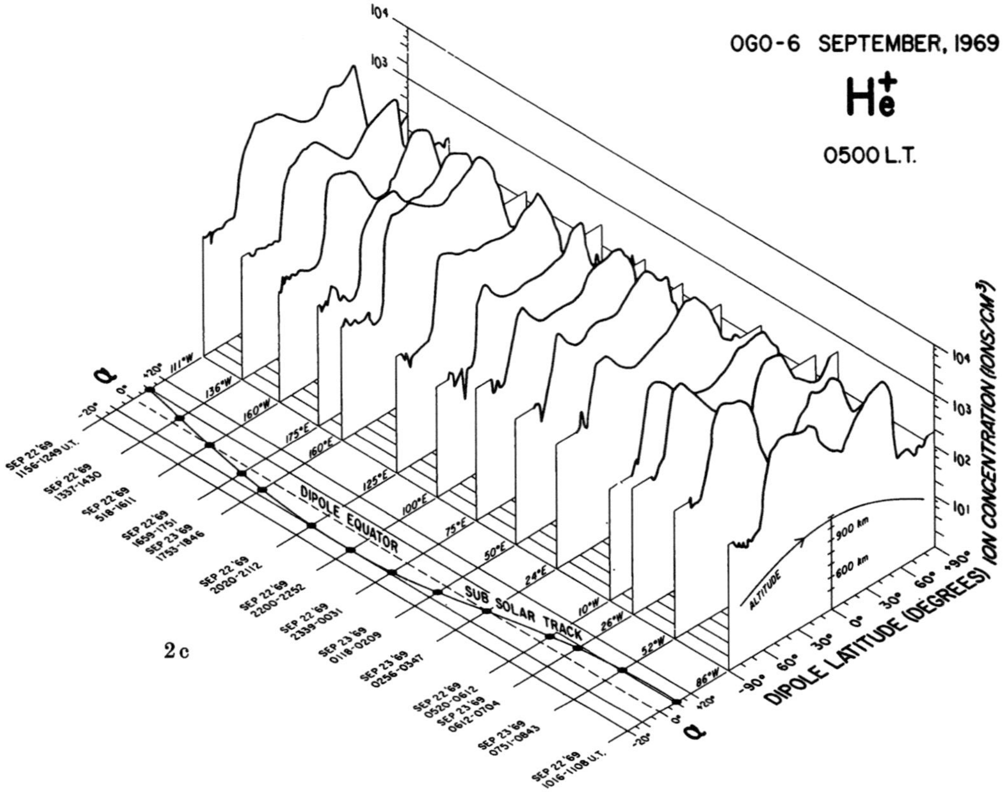
OGO-6 SEPTEMBER, 1969
He+
0500 L.T.
ION CONCENTRATION (IONS/CM³)
DIPOLE LATITUDE (DEGREES)
DIPOLE EQUATOR
SUB SOLAR TRACK
ALTITUDE
900 km
600 km
α

2c

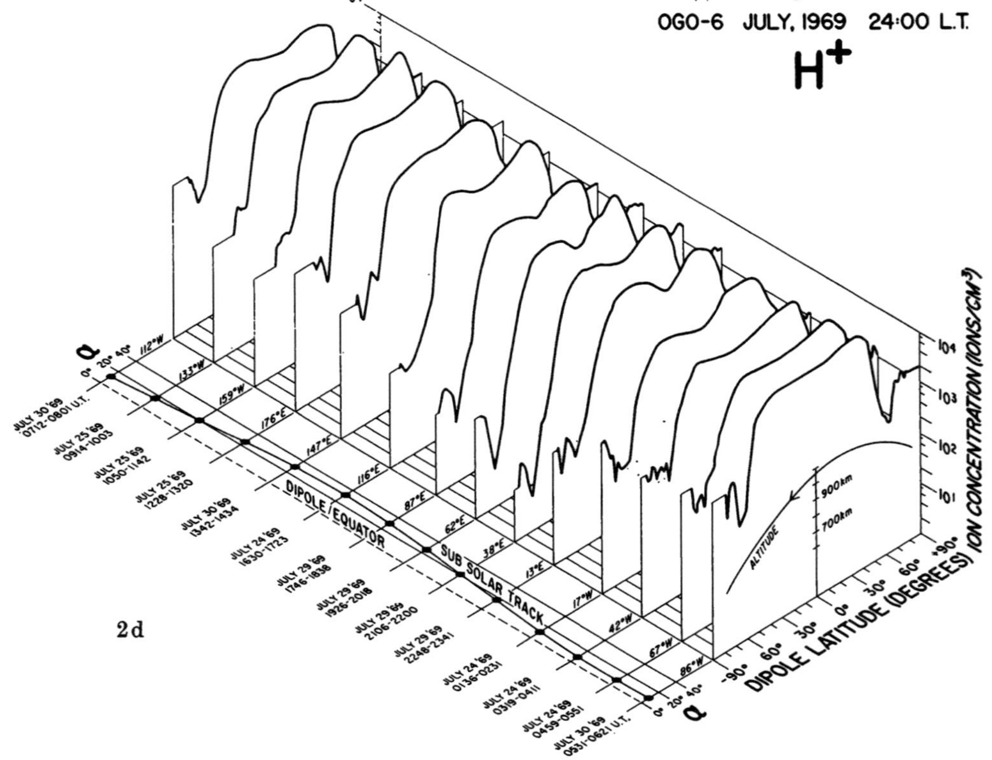
OGO-6 JULY, 1969 24:00 L.T.
H+
ION CONCENTRATION (IONS/CM³)
DIPOLE LATITUDE (DEGREES)
DIPOLE/EQUATOR
SUB SOLAR TRACK
ALTITUDE
900km
700km
α

2d

2.3. The Persistence of Winter–Summer Characteristics of H^+ and O^+ near Dusk

The significance of the longitudinal variation as a parameter for guiding the identification and selection of comparison data is illustrated in Figs. 3 and 4, which show the persistence of winter-summer characteristics in H^+ and O^+, respectively, as observed in both the northern and southern winter hemispheres, near dusk local time. Because of the nature of the OGO 6 orbit, this comparison between data obtained in June and December 1969 results in a local time separation of approximately 2 hours (1600–1800 LT), which adds some evidence

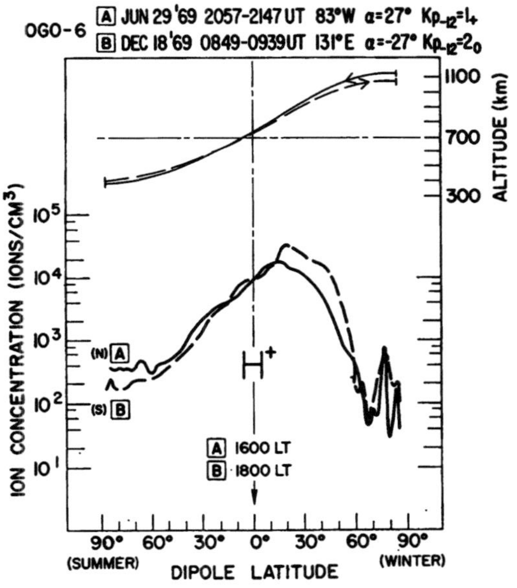

Fig. 3

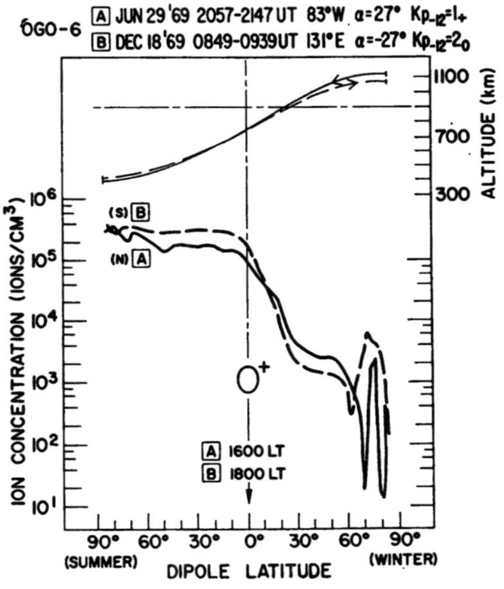

Fig. 4

Figs. 3 and 4. Evidence for the repeatability of summer–winter distributions of H^+ (Fig. 3) and O^+ (Fig. 4) observed between June and December 1969.

of the stability of the ion distributions with respect to local time. In Fig. 3 the profile H^+ obtained during northern winter (December) shows remarkable agreement with the profile obtained during southern winter (June). The longitudes are deliberately chosen to be different, in order to obtain equal and opposite values for α. As before, these passes have been chosen to provide essentially identical altitude–latitude relationships as well as similar magnetic activity histories.

In Fig. 4, O^+ profiles obtained simultaneously with the H^+ profiles of Fig. 3 indicate a similar persistence of the seasonal variation in O^+ between hemispheres. The high latitude, winter hemisphere portions of both the O^+ and H^+ distributions exhibit complex recovery regions which are believed to be associated with auroral zone disturbances. The irregularity of this region confuses the seasonal pattern somewhat, although the general features are fully recognizable. It is emphasized, that for both O^+ and H^+, the seasonal persistence which is illustrated is found only by comparing profiles observed under comparable solar–geomagnetic orientations. It is most significant, that if the comparisons had been performed at the same longitude for both seasons, the close agreement in the ion distributions would have been lost. As selected, however, these profiles reflect nearly identical solar–geomagnetic seasonal conditions and accordingly show good agreement.

2.4. Diurnal Variations in H^+ and O^+

Recognizing the significance of seasonal variations such as identified in Figs. 3 and 4, we have attempted to identify an example of the day–night variation in the ion distributions for equinox conditions. Unfortunately, data sets for equinox and with altitude symmetry about the dipole equator are limited to the dawn–dusk local time sector. Nevertheless, although we cannot yet identify the pure noon–midnight case, it is apparent from the examination of comparison data that the dawn profiles are indeed characteristic of the night-time case, while the dusk profiles are rather typical of the daytime distributions. Accordingly, in Figs. 5 and 6 we examine the contrast in dawn–dusk distributions of H^+ and O^+ respectively. In Fig. 5, profiles of H^+ obtained at dawn

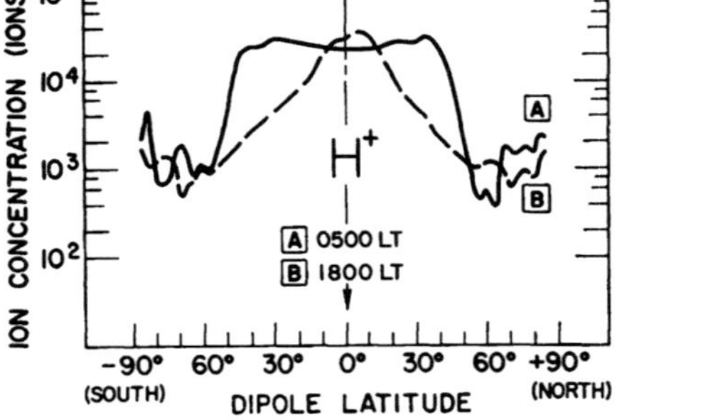

Fig. 5

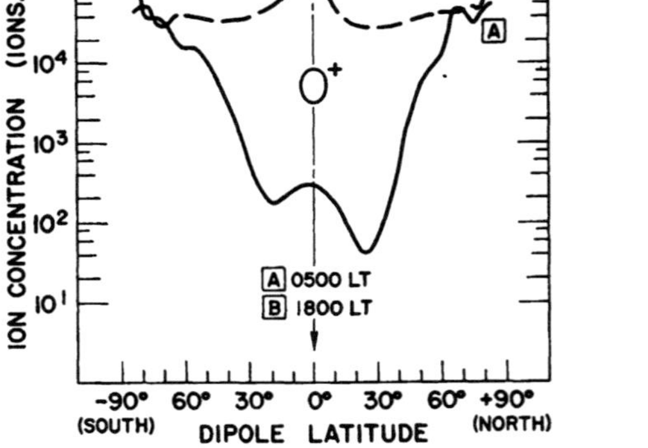

Fig. 6

Figs. 5 and 6. A comparison of H_+ profiles (Fig. 5) and O^+ profiles (Fig. 6) obtained near dawn and dusk local time respectively for near equinox conditions. The 0500 LT profile is representative of night-time distributions, while the 1800 LT profile is more typical of dayside distributions.

(0500 LT) and dusk (1800 LT) provide evidence of a pronounced diurnal variation. At dusk, the mid-latitude concentrations of H^+ are considerably lower than at dawn, with differences as great as an order of magnitude near $\pm 45°$. Consistent with this pattern, the light ion trough is much more pronounced at night with H^+ decreasing by as much as an order of magnitude within 10° of latitude. The diurnal variation in O^+ (Fig. 6) is considerably more pronounced than the variation in H^+. In particular, the extensive depletion of O^+ at low and mid latitudes at night is the dominant feature, with the concentrations of O^+ lower by a factor of more than two orders of magnitude near the equator at night. By comparing Figs. 5 and 6 it is seen that during the day where O^+ is the dominant topside ion, the latitudinal decrease in H^+ toward the poles is relatively gradual, and for quiet, equinox conditions, the light ion trough is not sharply defined. At night, however, where H^+ is the dominant topside ion, a much steeper and more identifiable trough is formed in the proton distributions. Once again, it is emphasized that a selection of α has been used to identify the representative equinox condition. If data were to be selected at the same longitude for both dawn and dusk, the symmetry, and thus the comparability, in the distributions would be significantly degraded.

Several additional features of the day–night ion distributions of Figs. 5 and 6 should also be noted. First, these profiles, in general, exhibit considerable symmetry in the N–S hemispheric distributions, consistent with the fact that these profiles reflect solar–geomagnetic equinox. The lack of seasonal asymmetry is particularly evident in the daytime distributions, which exhibit peaks in the distributions of both O^+ and H^+ at the dipole equator, as well as nearly identical concentrations of the respective ions between the northern and southern hemispheres. The night-time distribution of H^+ also exhibits considerable N–S symmetry during equinox, although the symmetry of the O^+ distribution is somewhat offset at low latitudes by an asymmetrical hump, or equatorial increase. This unexpected increase in O^+ has been observed in other sets of night-time data and is believed to be a characteristic feature, which will be the subject of a separate study.

2.5. Annual Variations in Topside Composition

In order to search for possible long-term variations in the content and distribution of the topside ionosphere, we have selected samples of O^+ and H^+ obtained at times separated by about 1 year. Consistent with the foregoing results, the local time regions selected for this comparison have been deliberately chosen so as to examine the behavior of both O^+ and H^+ under conditions for which each of these ions respectively becomes the dominant ion. Accordingly, in the results that follow, we examine the behavior of O^+ on the *dayside*, with perigee centered about the dipole equator. Conversely for checking the long-term stability in H^+, we examine *nigth-time* distributions obtained with apogee centered about the dipole equator. In order to obtain these conditions, it was necessary to select data obtained during July 1969 and 1970, for northern summertime conditions.

In Fig. 7, we see essentially a lack of annual variation in the distribution of O^+. Both the maximum concentration observed near the dipole equator and the northern and southern hemisphere distributions of O^+ are quite comparable from one year to the next. Although some altitude variation between the two passes being compared was unavoidable, there appears to be no systematic relationship between the altitude differences and the relatively minor concentration differences observed at various latitudes. As noted in Solar Geophysical Data [4] the periods of July 1969 and July 1970 are characterized by similar solar activity levels, representative of conditions near the maximum of Solar Cycle 20, and thus no significant difference in the O^+ content would be expected. In order to identify the similarity in the O^+ distributions observed one year apart, passes were selected with nearly identical α orientations. In this case the α values are 8° and 11° respectively, corresponding to near minimum solar–geomagnetic seasonal conditions, even though the month of June is characteristic of geodetic northern summer. The N–S symmetry reflected in the O^+ profile is consistent with the relatively low values of α, reflecting near solar–geomagnetic equinox, rather than northern summer.

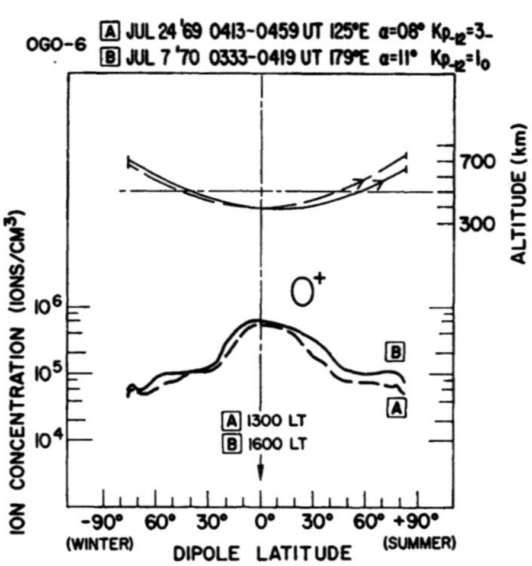

Fig. 7

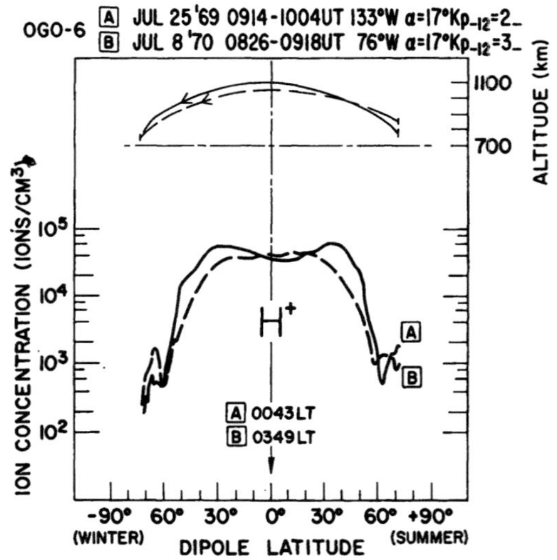

Fig. 8

Fig. 7. Evidence for the lack of annual variation in the distribution of O^+ observed during local afternoon, between July 1969 and July 1970.

Fig. 8. Evidence of the repeatability in the distributions of H^+ observed following midnight between the periods July 1969 and July 1970.

In Fig. 8 we observe a similar long-term repeatability in the distributions of H^+ observed between July 1969 and July 1970, but for higher values of α. Accordingly, in addition to the evidence of a lack of any pronounced annual variation in H^+ at night, there is also superimposed the consistent evidence of a slight asymmetry in the N–S distributions, with the southern or wintertime distributions somewhat lower in concentration for a given dipole latitude. As a result of orbit phasing, comparison of data separated by 1 year also requires a separation of 3 hrs LT, for apogee/perigee centered above the dipole equator (this also applies to the comparison in Fig. 7). Again this LT separation adds some evidence of the range of predictability of the ionosphere. With regard to the relatively small altitude and *Kp* differences involved in the comparison, there is no evidence that changes in either of these parameters significantly influence the overall comparison. In view of these qualitative remarks, we conclude that the repeatability in the O^+ and H^+ distributions over a period of one year is indeed pronounced, and that within several hours variation near midnight and noon, respectively, the ionosphere may be reasonably predictable, near solar maximum.

3. Discussion

3.1. Evidence of Solar–Geomagnetic Control

Overall, the foregoing results contribute evidence for a more pronounced solar–geomagnetic control of topside ionization than has been generally recognized heretofore. Although earlier satellite studies [5–8] provided important clues relative to the significance of the geomagnetic control of ionization, the recognition and documentation of the global longitudinal variation which appears to persist throughout the ion composition has awaited the nearly continuous data coverage made possible by the OGO satellites. As the results indicate, we now see that the longitudinal variation in the latitudinal distributions of a given ion may vary remarkably between two different geographic and or geomagnetic locations during the course of a given moderately quiet day. Accordingly, it has become increasingly clear that correlative studies as well as the identification of certain boundary values for charged particle models must take into account the significance of the longitudinal variation. As implied by the present results, studies of seasonal variations, examination of storm-time variations, and other studies involving a close comparison of spaced observations from satellites will require selection for comparability of solar–geomagnetic coordinates.

The observed dominance of the geomagnetic field in ordering the ion distributions points toward still further complications in the process of selecting satellite observations for correlative and model studies. Since it is clear that the ions are constrained to move along the magnetic field lines, there are several physical processes which may tentatively be suggested as possible mechanisms contributing to the observed patterns of longitudinal variation in the ion distributions.

The first such mechanism is the action of the neutral thermospheric wind in inducing large-scale plasma transport by means of coupled ion drag, directed along magnetic field lines. As shown by the work of Kohl et al. [9] the magnitude and direction of thermospheric winds are sufficient to produce F region anomalies observed within the ionosphere. The likelihood that winds may contribute to longitudinal shifts in the O^+/H^+ transition level has been demonstrated empirically using ion composition data from Explorer 32 [8].

As shown schematically in Fig. 9, the component of the neutral wind acting along magnetic field lines will undergo distinct changes, as the magnetic field rotates relative to the fixed position of the neutral atmosphere pressure bulge. An added mechanism suggested in Fig. 9 is the action of the polar wind [10] indicated by the outgoing arrows at high latitudes. The possible interaction of the *upflowing* polar wind with the *downflowing* component of the thermospheric wind constitutes another example of a dynamic process regulated in part by the wobble of the geomagnetic field in the earth–sun geometry.

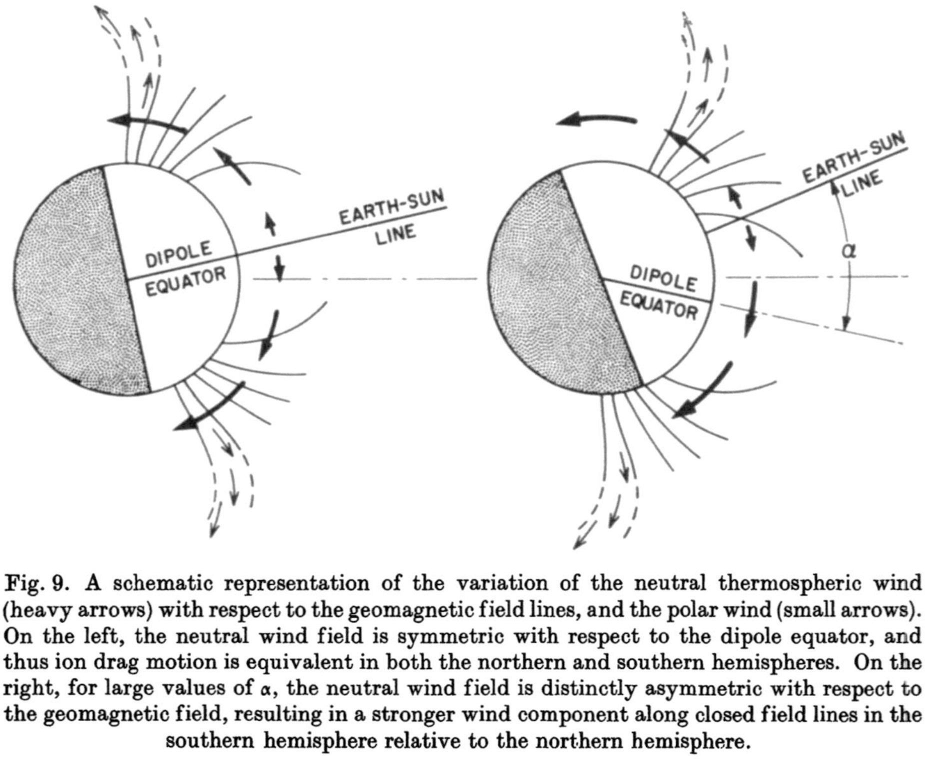

Fig. 9. A schematic representation of the variation of the neutral thermospheric wind (heavy arrows) with respect to the geomagnetic field lines, and the polar wind (small arrows). On the left, the neutral wind field is symmetric with respect to the dipole equator, and thus ion drag motion is equivalent in both the northern and southern hemispheres. On the right, for large values of α, the neutral wind field is distinctly asymmetric with respect to the geomagnetic field, resulting in a stronger wind component along closed field lines in the southern hemisphere relative to the northern hemisphere.

Another possible explanation of certain of the observed longitudinal variations may be that the orientation between the satellite orbit and the geomagnetic field lines varies constantly as the earth rotates. If we assume that the ions are closely constrained to move along magnetic field lines, and that the ionization observed along the orbit results from the upward diffusion of ionization produced at some lower altitude along a given L shell, we may predict the possibility of what may be referred to as an "effective local time" variation in the ion distributions. As shown schematically in Fig. 10, the locus in local time of the feet of the field lines intersecting the OGO 6 orbit may be distinctly different between two widely spaced longitudes. Thus, depending upon the degree of control exercised by the field and upon the importance of diffusion processes in supplying the ionization observed in satellite altitude, the magnetically varying "effective local time" may possibly account for certain observations which appear as 'exaggerated' local time and seasonal variations. The investigation of the possible "effective local time" mechanism is currently under study and will be reported separately.

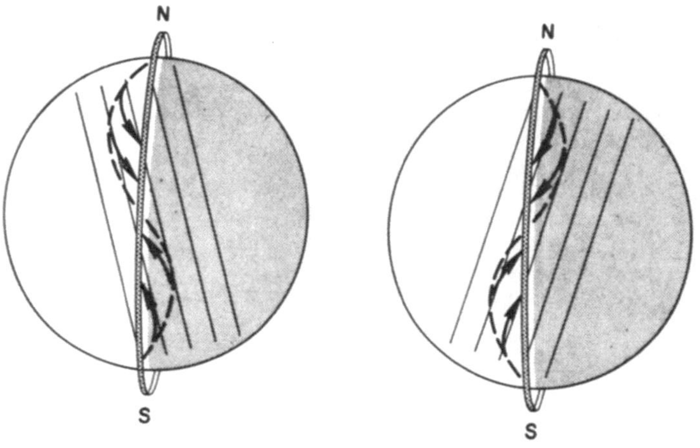

Fig. 10. A schematic representation of the effective local time concept. The earth is viewed from the dusk meridian, with the left panel representing a longitude for which the magnetic field lines are tilted from the dayside in the north to the nightside in the south, while the right-hand panel represents the opposite case. The arrows represent the diffusion of ionization from a reference level (say 100 km) up to the altitude of the orbit. The dashed line represents the locus in local time of the feet of the field lines intercepted by the orbit.

3.2. Evidence of Predictability in Ionospheric Distribution

Recognition of the solar–geomagnetic ordering of the ion distribution provides the tentative result that under certain conditions the topside ion composition may be reasonably predictable, with preliminary evidence that both wintertime and summertime distributions are maintained at the opposite poles, and that annual variations (near solar maximum) are not pronounced. In view of the dominance of the longitudinal variation, and of the relatively small amount of definitive data which may be obtained from a given satellite (due to orbital phasing etc.), however, it is recognized that the present evidence for repeatablity and predictability is preliminary. Nevertheless, the observed day–night variation in O^+ and H^+, as well as the evidence that the distributions of these ions may remain relatively unchanged over ranges of local time of at least three hours, is generally consistent with existing theory, and in good agreement with the ion composition results of Explorer 32 [11].

In summary, to be more conclusive in describing diurnal, seasonal, and temporal variations of the topside ion composition, a more thorough evaluation of the combined effects of changing solar–geomagnetic geometry and varying orbital position must be made. Selective analysis of more 'pure' data sets, free of magnetic storm activity, and descriptive of primary seasonal and local time orientations, and with altitude effects removed, is a prerequisite for developing a clearer empirical model of the ion composition.

References

[1] H. A. TAYLOR, JR., Planet. Space Sci. **19**, 77 (1971).
[2] H. A. TAYLOR, JR., et al., Space Research X, 663 (1970).
[3] H. A. TAYLOR, JR., et al., NASA/Goddard Report No. X-621-71-227 (1971).
[4] Environmental Research Laboratories Solar Geophysical Data Descriptive Text (1971).
[5] P. J. BOWEN et al., Proc. Roy. Soc. A, **281**, 504 (1964).
[6] W. C. KNUDSEN, J. Geophys. Res. **72**, 1941 (1967).
[7] S. CHANDRA et al., J. Atmos. Terr. Phys. **29**, 259 (1967).
[8] H. C. BRINTON et al., Space Research X, 652 (1970).
[9] H. KOHL et al., J. Atmos. Terr. Phys. **29**, 1045 (1967).
[10] P. M. BANKS et al., J. Geophys. Res. **73**, 6846 (1968).
[11] H. C. BRINTON et al., J. Geophys. Res. **74**, 4064 (1969).

Space Research XII – Akademie-Verlag, Berlin 1972

FLARE TIME MODELS OF IONIZATION PROFILES IN THE D REGION

A. P. Mitra and S. D. Deshpande

Radio Science Division, National Physical Laboratory, New Delhi, India

Several electron density profiles have now been obtained with the high power wave interaction facility at Pennsylvania State University, the partial reflection experiment at Ottawa, a series of rocket firings by Somayajulu and Aikin, and multifrequency vlf and riometer experiments. For some of these the time histories as well as the spectral distributions of the ionizing X-ray fluxes are known. These have provided information on nitric oxide concentration at D-region heights, on the changes in the effective recombination rate and on the relative importance of the competing ionizing sources. From these it seems possible to develop models of flare-time D-region electron densities for a wide range of flare conditions in terms of the X-ray flux at 0–8 Å and a parameter such as the flux ratio at 0–3 Å and 0–8 Å to define the flare spectrum.

1. Introduction

Although much work exists on the statistical characteristics of sudden ionospheric disturbances (SIDs), information on ionization profiles is relatively small. This has so far come from ground-based SID observations and from rocket observations. The profiles show a wide variation from event to event. This is expected since to some extent it is a consequence of a wide variation existing in X-ray enhancement as well as in spectral composition. A further complicating factor is the decrease in the effective recombination coefficient observed at heights below 90 km during solar flares [1–3]. The decrease is small or negligible for small X-ray events, but is large (as much as a factor of 5) for large X-ray events, giving an additional electron production rate of 100 to 500 cm^{-3} s^{-1}. In the preparation of flare ionization models, one must, therefore, take account of two factors: (a) the flux and spectral characteristics of the X-ray event, and (b) the nature and magnitude of variation of the effective recombination coefficient expected for the event.

Three profiles, one each for strong, moderate and weak flare conditions are constructed in this work. An examination of the X-ray intensities in the wavelength bands 0–3 Å, 0–8 Å and 8–20 Å, during a large number of flares allows us to set up the appropriate X-ray spectra in the wavelength region below 20 Å for different flare conditions; these spectra, with the three profiles, yield a set of flare time ionization profiles. Absorption effects on 30 MHz cosmic noise signal (SCNA) have been computed for these ionization models. Identification of the ionization profile for any flare would thus be possible with the X-ray observations in 0–3 Å and 0–8 Å bands or with an SCNA effect from these models.

2. Experimental Profiles of D Region Ionization during Solar Flares

In Fig. 1, a number of experimental flare-time ionization profiles are shown. These are obtained from both ground-based and rocket observations. A major flare on 30 January 1968 was recorded on riometers at National Physical

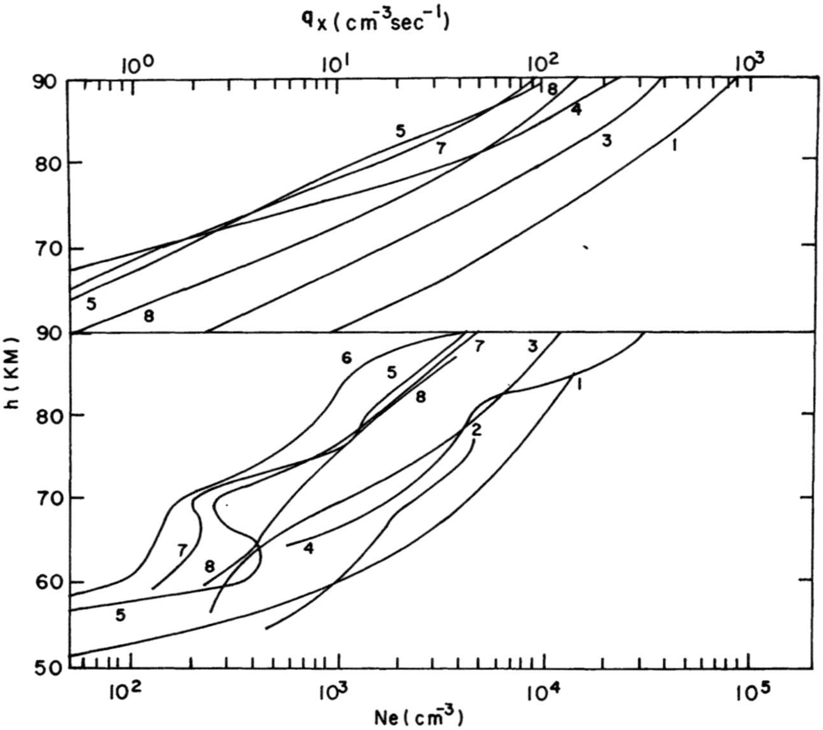

Fig. 1. Production rates and electron density profiles at the peak of selected flare events. 1, 30 Jan. 1968 (A_1/A_2; 2, 1 Mar. 1962 (partial refl.); 3, 21 Oct. 1968 (cross mod.); 4, 21 Aug. 1968 (rocket); 5, 27 Mar. 1969 (cross mod.); 6, 7 Oct. 1948 (lf/vlf); 7, 17 Mar. 1969 (cross mod.); 8, 25 Mar. 1968 (partial refl.).

Laboratory, New Delhi. This flare has been studied by Deshpande et al. [2]. (N_e, h) profiles were deduced from A_2 absorption measurements on 22.4 and 30 MHz at Delhi and on 5.5 MHz at Calcutta. For the flare of 21 August 1968 the ionization profile was measured from a series of three rockets flown by Somayajulu and Aikin [5]; the profile shown is that for the first flight soon after the peak of the X-ray event. The profiles for 21 October 1968, 17 March 1969 and 27 March 1969 were obtained by the Pennsylvania State University group, who used a high-power wave interaction technique and obtained quick-run profiles during the entire course of the flares [1]. The profile for 25 March 1968 was obtained by Belrose and his colleagues [6] using the partial reflection technique. In the events X-ray fluxes in three bands (0–3 Å, 0–8 Å, and 8–20 Å) were also available. The ionization profiles and the production rates for the peak of these selected events are also shown in Fig. 1. In calculating the total electron production rates, the contribution due to the photoionization of NO by Lα is computed on the basis of Meira's recent rocket results.

3. ψ Profiles for Different Flares

There is now a considerable body of evidence that the effective recombination coefficient α decreases during solar flares. In Fig. 2, we present a summary of

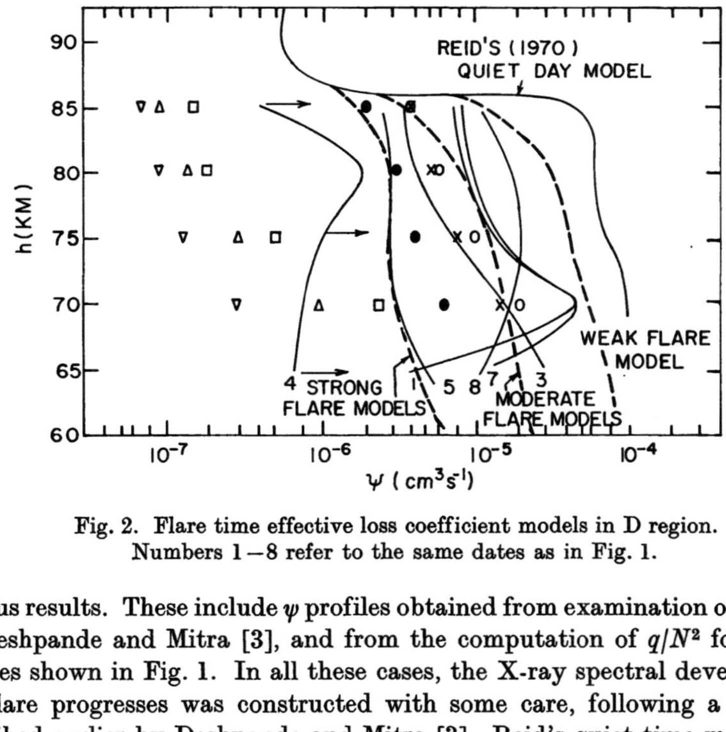

Fig. 2. Flare time effective loss coefficient models in D region. Numbers 1—8 refer to the same dates as in Fig. 1.

various results. These include ψ profiles obtained from examination of 25 SCNAs by Deshpande and Mitra [3], and from the computation of q/N^2 for the flare profiles shown in Fig. 1. In all these cases, the X-ray spectral development as the flare progresses was constructed with some care, following a procedure, described earlier by Deshpande and Mitra [3]. Reid's quiet time model is also shown. All flare ψ profiles are seen to be lower than the quiet time situation and a systematic decrease in ψ profiles with increasing flare intensity can be seen from this figure. The ψ profiles selected to represent the strong, moderate and weak flare conditions are marked.

4. The X-ray Spectra

Four categories of X-ray flares are considered on the basis of the intensity in the 0–8 Å band expressed for a grey-body spectrum at $T = 2 \times 10^6$ °K. For strong flares the 0–8 Å flux level is taken as high as $(2–5) \times 10^{-1}$ erg cm^{-2} s^{-1}, and for moderate flares about 2×10^{-2} erg cm^{-2} s^{-1}. In the threshold case 0–8 Å flux is 2×10^{-3} erg cm^{-2} s^{-1} while for weak flares the flux level taken is

Table 1

Models of X-ray flares and flare spectra

1	2	3	4	5	6	7	8	9
Strong flares	AI	5×10^{-1}	5×10^{-2}	2.5×10^{-2}	2.9×10^{0}	2×10^{7}	8.5×10^{6}	6×10^{6}
	A′II	2×10^{-1}	2×10^{-2}	4.0×10^{-3}	1.4×10^{0}	1.5×10^{7}	8×10^{6}	5×10^{6}
Moderate flares	BI	2×10^{-2}	5×10^{-2}	1.0×10^{-3}	1.2×10^{-1}	2×10^{7}	8.5×10^{6}	6×10^{6}
	BIII	2×10^{-2}	1.5×10^{-2}	3.0×10^{-4}	2.0×10^{-1}	1.5×10^{-7}	7×10^{6}	4×10^{6}
	BIV	2×10^{-2}	5×10^{-2}	1.0×10^{-4}	3.4×10^{-2}	6.5×10^{6}**		3×10^{6}
Threshold case	TIII	2×10^{-3}	1.5×10^{-2}	3.0×10^{-5}	2.1×10^{-3}	1.5×10^{7}	7×10^{6}	4×10^{6}
Weak flares	CII	5×10^{-4}	2×10^{-2}	1.0×10^{-5}	3.5×10^{-3}	1.5×10^{7}	8×10^{6}	5×10^{6}
	CIV	5×10^{-4}	5×10^{-3}	2.5×10^{-6}	9.1×10^{-3}	6.5×10^{6}**		3×10^{6}

1, category of X-ray flare; 2, symbol*; 3, intensity in 0–8 Å ($T = 2\times10^{6}$ °K) (erg cm^{-2} s^{-1}); 4, hardening factor (H. F.); 5, intensity in 0–3 Å ($T = 1\times10^{7}$ °K) (erg cm^{-2} s^{-1}); 6, intensity in 8–20 Å ($T = 2\times10^{6}$ °K) (erg cm^{-2} s^{-1}); 7–9, spectral temperature in the three bands (°K): 7, 0–8 Å; 8, 3–8 Å; 9, 8–20 Å.

* A, A′, B, C and T stand for different 0–8 Å flux levels;
I, II, III and IV for the set of spectral temperatures

** Single equivalent temperature for 0–8 Å band

5×10^{-4} erg cm^{-2} s^{-1}. The large enhancements in the 0–8 Å band during strong flares are not likely to occur without a high degree of spectral hardening. Therefore, neither is the soft spectrum considered for this category of flares nor is the spectral distribution represented by a single equivalent temperature. Two composite spectrum models with separate temperatures below and above 3 Å, based on the ratio of 0–3 Å to 0–8 Å flux levels (to be called the hardening factor, H.F.) as 5×10^{-2} and 2×10^{-2} are selected. Under a situation of moderate inten-

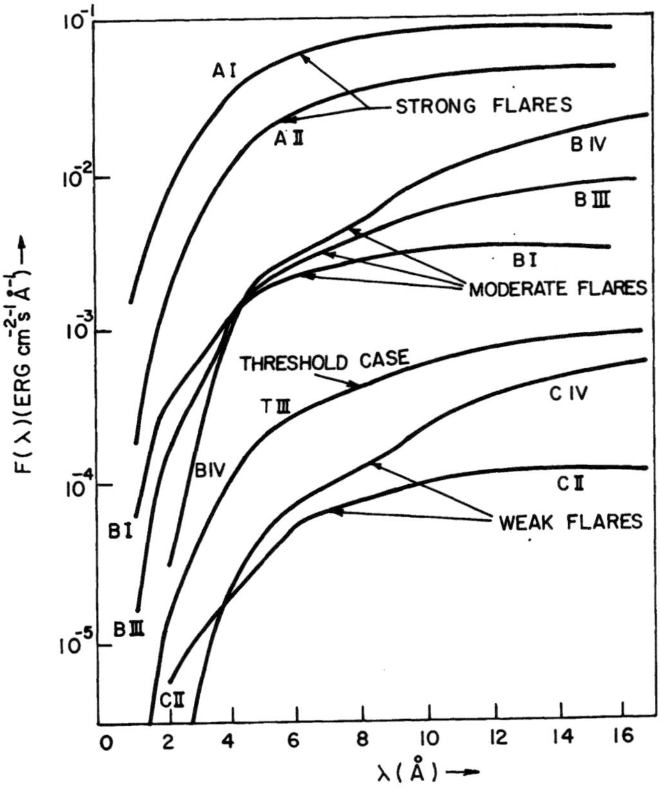

Fig. 3. X-ray spectra representing different flare conditions.

sity, hard as well as soft spectra in large variety are possible. We choose three spectra corresponding to the hardening ratios 5×10^{-2}, 1.5×10^{-2} and 5×10^{-3}; only the soft spectrum (H.F. $= 5 \times 10^{-3}$), is represented by an equivalent single temperature. In weak flares, the spectrum is taken to be soft and corresponding to H.F. $= 5 \times 10^{-3}$. Spectral hardening can occur to some extent during weak enhancements. A hard spectrum corresponding to H.F. $= 2 \times 10^{-2}$ is, therefore, considered for weak flares in addition to the soft one. The threshold spectrum is given by H.F. $= 1.5 \times 10^{-2}$.

It becomes necessary to extend the X-ray spectrum beyond 8 Å, when the spectral distributions are utilized for calculating the production rates. We have considered the 8–20 Å flux levels and equivalent spectral temperatures that are consistent with the spectrum models below 8 Å. The flux levels and spectral temperatures for eight different X-ray spectra belonging to the four broad categories of X-ray flares are given in Table 1, and the spectra are shown in Fig. 3.

5. Production Rate Profiles

The production rates q in the lower ionosphere, mainly in the height range of 60–90 km, were calculated utilizing the X-ray spectra. The usual expression,

$$q = \sum_{\lambda} Q \cdot F(\lambda)$$

where $F(\lambda)$ is flux of radiation at λ in erg cm^{-2} s^{-1} Å^{-1}, and Q is the rate of production at level h, per unit flux (in erg cm^{-2} s^{-1} Å^{-1}) at λ for zenith angle χ [7]. In Fig. 4, are shown the flare time models of production rate profiles for zenith

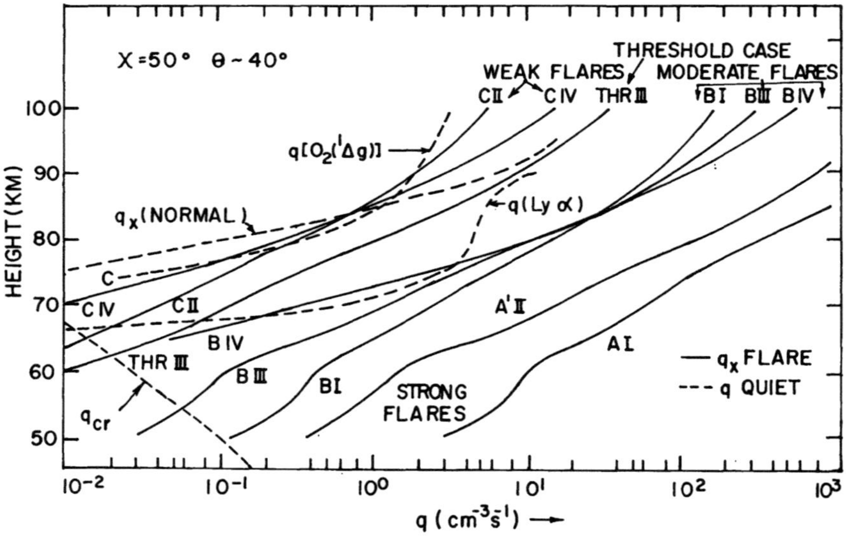

Fig. 4. Models of electron production rates during different flare conditions.

angle of 50°, for the eight spectral models. For comparison, the normal time rates of production due to cosmic rays, Lyman α, non-flare X-rays and $O_2(^1\Delta_g)$ are shown in the same figure. The nitric oxide concentration of Meira [8] were used to obtain $q(\mathrm{Ly}\alpha)$; the values of $q(O_2(^1\Delta_g))$ used are those recently given by Huffman et al. [9].

Flare time production rates exceed the normal time rates of production due to all sources during strong and moderate events; however, for a moderate flare with soft spectrum, the rates of production q_x (flare) are comparable with those due to Lyman α. Under threshold conditions, q_x (flare) values are lower than $q(\mathrm{Ly}\alpha)$, but exceed q_x (normal) at all heights. During weak flares q_x (flare), although much less than $q(\mathrm{Ly}\alpha)$, exceeds the q_x (normal) below 85 km, this being the effect of spectral hardening. Above 85 km flare time rates are lower than q_x (normal) because the latter refer to mid and high solar activity periods with the 0–8 Å, 8–20 Å levels larger than for our weak flare models.

6. Ionization Profile Models

The ionization profiles for strong, moderate and weak flares are shown in Fig. 5. The 0–8 Å X-ray flux, hardening factor and SCNA effect on the 30 MHz signals corresponding to each model has been specified in this figure. Electron

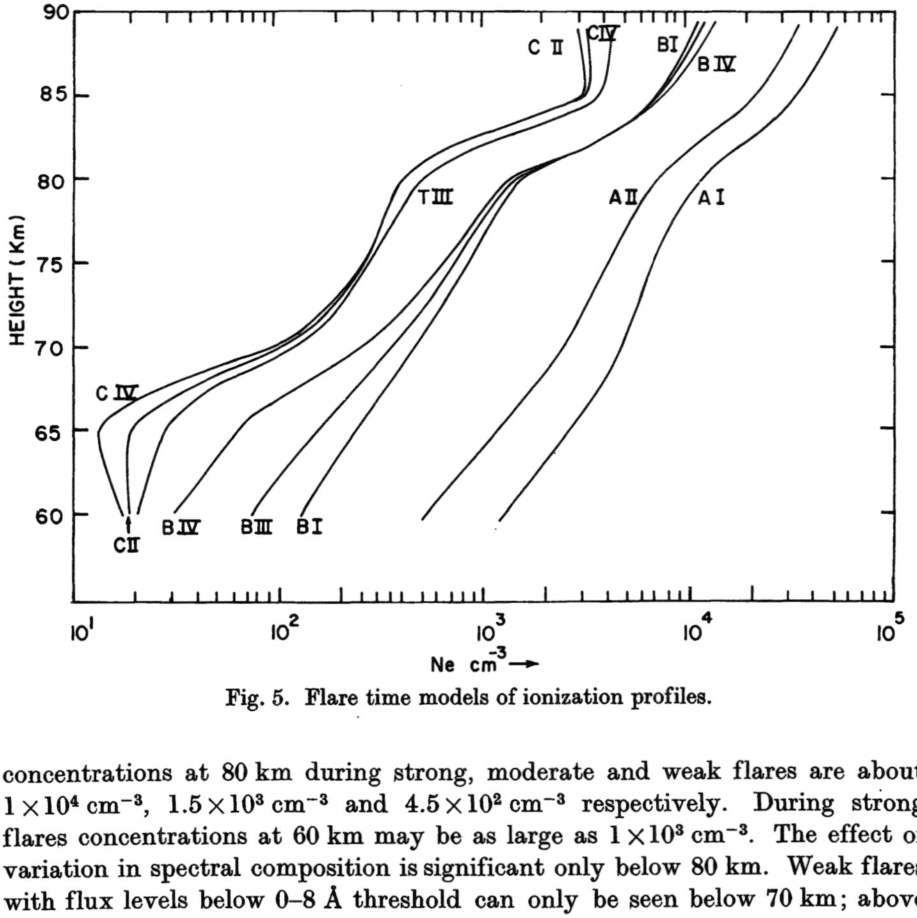

Fig. 5. Flare time models of ionization profiles.

concentrations at 80 km during strong, moderate and weak flares are about 1×10^4 cm^{-3}, 1.5×10^3 cm^{-3} and 4.5×10^2 cm^{-3} respectively. During strong flares concentrations at 60 km may be as large as 1×10^3 cm^{-3}. The effect of variation in spectral composition is significant only below 80 km. Weak flares with flux levels below 0–8 Å threshold can only be seen below 70 km; above this level the ionization is indistinguishable from normal. The SCNA effect on 30 MHz signals during a strong flare is found to be >1 dB; it is between 0.2 and 1 dB for moderate flares but less than 0.1 dB for weak. During strong flares the production rates are large and for corresponding low ψ profiles the electron concentration becomes very high. On the contrary during weak flares, the production rates are low and with high values of ψ low electron concentrations result. A consequence of the variable ψ is then to provide a wider range over which electron concentrations at any level can change with change in flare conditions than is possible for a fixed ψ model.

References

[1] J. N. ROWE et al., J. Atmos. Terr. Phys. **32**, 1609 (1970).
[2] S. D. DESHPANDE, S. GANGULY, V. C. JAIN and A. P. MITRA, National Physical Laboratory, New Delhi, India, Radio Science Division Scientific Report No. 63 (1971).
[3] S. D. DESHPANDE and A. P. MITRA, National Physical Laboratory, New Delhi, India, Radio Science Division Scientific Report No. 61 (1971).
[4] S. D. DESHPANDE and A. P. MITRA, National Physical Laboratory, New Delhi, India Radio Science Division Scientific Report No. 62 (1971).
[5] Y. V. SOMAYAJULU and C. A. AIKIN, paper presented to XIIth COSPAR meeting, Leningrad 1970.
[6] J. S. BELROSE, University of Illinois, Third Aeronomy Conference, Aeronomy Report No. 32, 375 (1969).
[7] M. OHSIO, R. MAEDA and H. SAKAGAMI, Japan Radio Sci. Labs. J., **13**, 245 (1966).
[8] L. G. MEIRA, Ph. D. Thesis, Department of Astro-Geophysics, University of Colorado, Boulder (1970).
[9] R. E. HUFFMAN, D. E. PAULSON, J. C. LARRABEE and R. B. CAIRNS, J. Geophys. Res. **76**, 1028 (1971).

Space Research XII — Akademie-Verlag, Berlin 1972

ION COMPOSITION AND PHOTOCHEMISTRY OF THE E REGION

A. D. Danilov

Institute of Applied Geophysics,
Hydrometeorological Service of the USSR, Moscow, USSR

A survey of the results of ion composition measurements in the ionspheric E-region is presend. It is shown that the ratio of two main positive ions $[NO^+]/[O_2^+]$ increases from day to night and with geomagnetic activity. The behaviour of the ratio $[NO^+][e]/[O_2^+]$ indicates that the reaction with nitric oxide is the dominant process of ion transformation in the E-region and makes possible an estimate of [NO] — values at altitude of 110–130 km.

The behaviour of the ion composition of the E region differs from that of the main part of the ionosphere above 140 km. Above that level the ratio of the concentration of the main ion $[NO^+]/[O_2^+]$ is practically constant with height, does not depend on time of day or solar activity and is equal to 2 ± 1 [1, 2]. But below 130 km sharp increase of $[NO^+]/[O_2^+]$ with decreasing h and strong variations from day to night are observed [3, 4]. The ion composition and photochemistry of the E region are studied in detail in [5] and the behaviour of ion composition at altitudes of 100–130 km in [3, 4, 6–8]. It was shown that it is necessary to invoke ion–molecule reactions with the minor components N and NO for an explanation of high $[NO^+]/[O_2^+]$ values in the E region, and some conclusions were made on the variations of this ratio.

An attempts is made in this work to consider the behaviour of ion concentrations and the photochemistry of the E region on the basis of all the published mass-spectrometric data. We discuss the behaviour of two main ions only and do not touch metallic ions or any other sporadic ion components.

From experiments in which the ion composition at altitudes of 110–130 km was recorded the $[NO^+]/[O_2^+]$ dependences on solar zenith angle for fixed heights 110, 120 and 130 km were derived and are shown in Figs. 1–3. In all the figures crosses denote results of measurements at high latitudes (Fort Churchill, Heiss Island, Andöya) and points represent the results of mid-latitude experiments. The data obtained at high solar activity (the flux of solar radioemission at 10.7 cm) $P_\odot^{10} > 200$ (usual units) are circled by solid rings, and at low solar activity ($P_\odot^{10} < 100$) by dotted rings.

One can see from the lower part of Fig. 1 that most of the experimental data show a well pronounced dependence of the $[NO^+]/[O_2^+]$ ratio on solar zenith angle at 100 km. The ratio systematically increases from noon to night conditions, and the amplitude of the variation is one order of magnitude. There seems to be no tendency for the separation of points observed under different conditions (solar activity, latitude). But the group of points near 90–100° and the point for $Z = 27.5°$ lie far from most of the points. The group was obtained in the Narcisi et al. [15] experiments; the points of this group are marked by triangles. The reason for such a strong deviation of these points from others is unknown. However, for the reasons described below, it is impossible to explain it by an experimental error. The point for $Z = 27.5°$ was obtained by Taylor and Brinton [12]. The results of that flight give anomalous ion composition at higher altitudes. The values of $[O^+]/[NO^+]$ and $[O^+]$ for 150–200 km

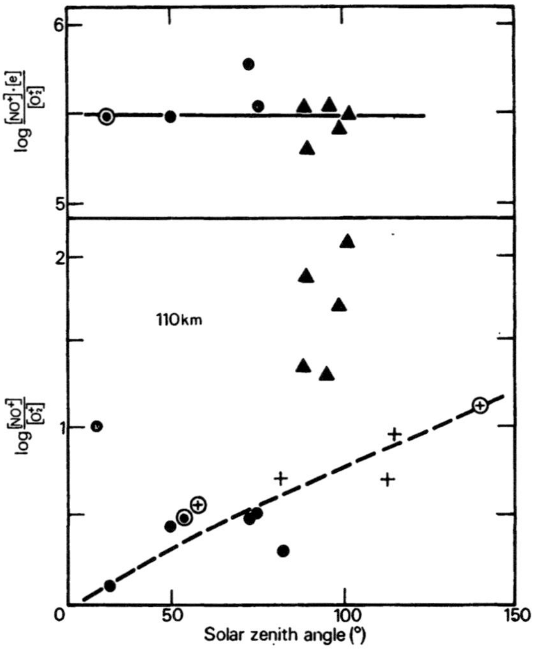

Fig. 1

differ strongly [1, 3] from that obtained in other experiments. In [1, 7] this anomaly of ion composition was explained by the fact that the rocket was launched under disturbed geomagnetic conditions (magnetic storm). The same supposition seems even more sensible for deviation of the flight in the E region because the increase of $[NO]/[O_2^+]$ under disturbed conditions (after solar flare) was detected experimentally [22].

It is accepted now, that the transformation of O_2^+ ions to NO^+ proceeds as a result of ion–molecule reactions:

$$O_2^+ + NO \rightarrow NO^+ + O_2 \tag{1}$$

$$O_2^+ + N \rightarrow NO^+ + O \tag{2}$$

Ions NO^+ recombine then in the dissociative recombination process:

$$NO^+ + e \rightarrow N + O \tag{3}$$

and so the $[NO^+]/[O_2^+]$ ratio is determined by the equation:

$$\frac{[NO^+]}{[O_2^+]} = \frac{[X]\,\gamma_x}{[e]\,\alpha^*_{NO^+}}. \tag{4}$$

Because we still do not know the values of [NO] and [N] well enough to decide which of reactions (1) or (2) play the dominant part in ion balance in the E region, we denote in (4) the concentration of the minor constituent ([NO] or [N]) by [X] and the rate constant of the reaction (γ_1 or γ_2) by γ_x. Some

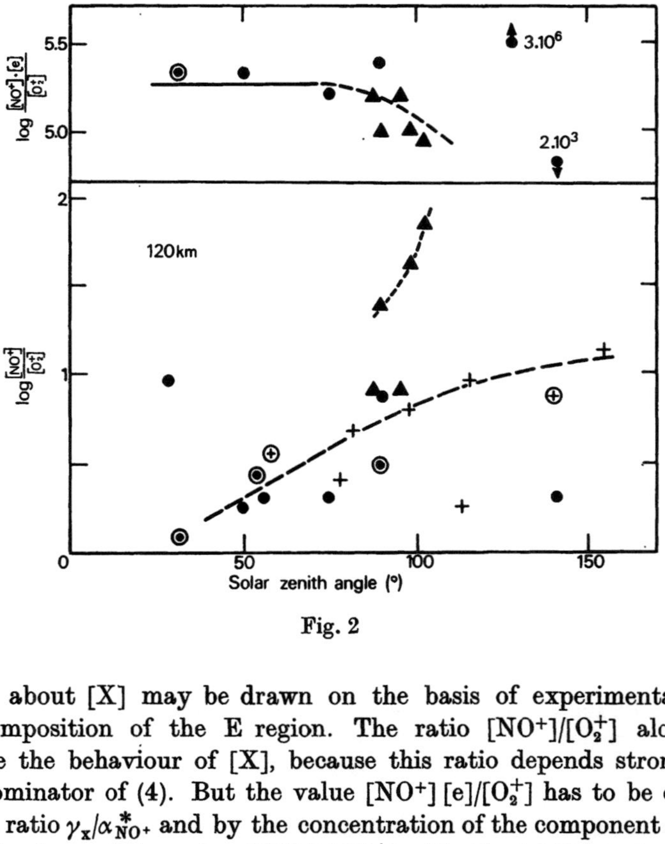

Fig. 2

conclusions about [X] may be drawn on the basis of experimental data on the ion composition of the E region. The ratio $[NO^+]/[O_2^+]$ alone cannot characterize the behaviour of [X], because this ratio depends strongly on [e] in the denominator of (4). But the value $[NO^+]\,[e]/[O_2^+]$ has to be determined only by the ratio $\gamma_x/\alpha^*_{NO^+}$ and by the concentration of the component in question X. The behaviour of the value $[NO]\,[e]/[O_2^+]$ with Z_0 at 100 km for the flights for which the absolute concentrations of ions were published, is shown in the upper part of Fig. 1. The most astonishing fact is that the value $[NO^+]\,[e]/[O_2^+]$ shows relatively small variations during the daytime (less than a factor of 2 from the mean value, shown by a broken line), whereas the $[NO^+]/[O_2^+]$ variations are as high as 2 orders of magnitude. The fact that $[NO^+]\,[e]/[O_2^+]$ values from Narcisi et al. [15] are in good agreement between themselves and with other experiments, shows that there was no mistake and that anômalous $[NO^+]/[O_2^+]$ values are due to low [e] values, observed during Narcisi's experiments.

Fig. 2 is the same as Fig. 1 but refers to an altitude of 120 km. Stronger scattering of data is seen here, but the main peculiarities of the picture are the same. There is a tendency for increase of $[NO^+]/O_2^+]$ from noon to night; the point at $Z_0 = 27.5°$ and points (now only three) obtained by Narcisi differ from most of the data. Constancy of the value $[NO^+][e]/[O_2^+]$ during the daytime and twilight is observed, but there seems to be a tendency for the value to decrease during the night. But the presence only of one point under conditions of deep night ($Z = 141°$), does not allow us to conclude definitely about the variation of [X] from day to night.

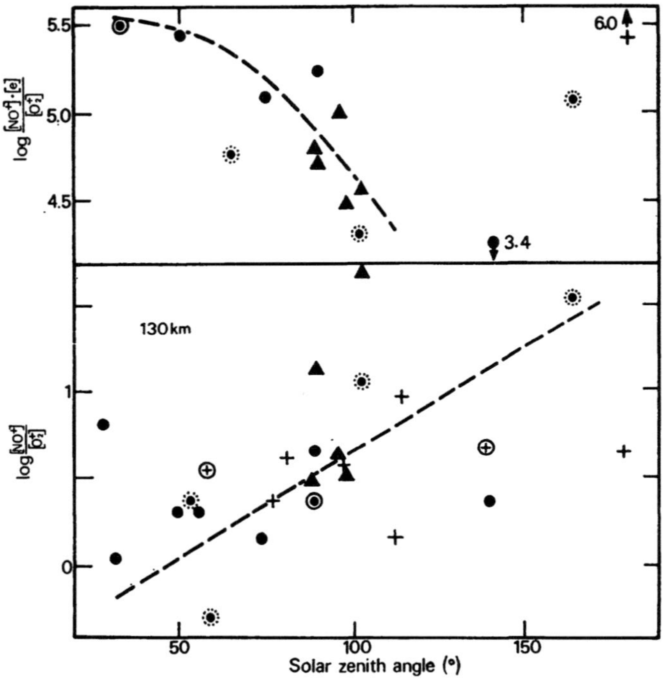

Fig. 3

More recent work suggests that the mean line (dashed line) in the lower part of the figure should curve from the lower point with dotted ring, to below the middle dotted ring, to the upper one.

The same peculiarities of the behaviour of both $[NO^+]/[O_2^+]$ and $[NO^+][e]/[O_2^+]$ can be seen in Fig. 3, which refers to 130 km. The scattering of the data there is much greater than in the previous figures; it shows probably stronger and less regular variations of the electron concentration. However the tendency for increase of the $[NO^+]/[O_2^+]$ ratio from day to night is still observed, and there seems to be no systematic dependence of the ratio on latitude. One can see some difference in the behaviour of $[NO^+]/[O_2^+]$ during the daytime for periods of moderate and low solar activity (two lines in the lower part of the picture). Quite pronounced decrease in the $[NO^+][e]/[O_2^+]$ value from day to night is seen, but again it is difficult to estimate the amplitude of the total day–night variation. It has to be mentioned that part of this variation at 130 km may be produced by electron temperature variation because it may be a little higher during the day than at night and because the $\alpha^*_{NO^+}$ value in the denominator of Eq. (4) is inversely proportional to T_e [25]. But this effect cannot give more than a factor of 1.5–2, so the most part of the variation of $[NO^+]\,[e]/[O_2^+]$ has to be attributed to a real variation of X concentration.

The character of [X] behaviour described leads us probably to conclude which of two components (N or NO) plays the main role in determining the ion composition of the E region. According to the photochemical theory [26] the concentration of atomic nitrogen has to undergo strong (several orders of magnitude) diurnal variations because N formation is strongly dependent on the ionospheric processes involving charged particles. At the same time the variations of NO concentrations have to be relatively small, because neutral particles take part in both the formation and disappearance of NO. Comparing this with the character of the [X] variation on the basis of data on ion composition, one has to conclude that nitric oxide plays a role of X, and the reaction (1) is the reaction which controls the behaviour of ion composition in the E region. This conclusion is in agreement with the Donahue et al. [28] experiment in which strong increase in the $[NO^+]/[O_2^+]$ ratio was observed during the aurora and simultaneously very high concentrations of NO were detected.

We can now determine the values of [NO] on the basis of ion composition data. It seen from Fig. 1 that at 110 km $[NO^+]\,[e]/[O_2^+]$ is practically constant during the day and is equal to 3×10^5 cm^{-3}, which corresponds (with $\gamma_1 = 8 \times 10^{-10}$ cm^3 s^{-1} [27] and $\alpha^*_{NO^+} = 4 \times 10^{-7}$ cm^3 s^{-1} [25] to an NO concentration of 1.5×10^8 cm^{-3}. In the same way at 120 km the $[NO^+]\,[e]/[O_2^+]$ value during the daytime is 2×10^5 cm^{-3}, corresponding to $[NO] = 1 \times 10^8$ cm^{-3}. At 130 km $[NO^+]\,[e]/[O_2^+]$ apparently changes during the day from 3×10^5 cm^{-3} at low Z_0 to about 1×10^5 cm^{-3} during sunset (corresponding values for [NO] are 1.5×10^8 and 0.5×10^8 cm^{-3}). Thus, ion composition data lead us to the conclusion that [NO] has little variation with height in the E region, being equal to $(0.5–1.5) \times 10^8$ cm^{-3} during the daytime.

Summarizing, it is necessary to point out that the increase of $[NO^+]/[O_2^+]$ with increase of Z is observed in the E region and there seems to be no dependence on solar activity and latitude. Apparently $[NO^+]/[O_2^+]$ increases with geomagnetic activity. On the basis of the behaviour of $[NO^+]\,[e]/[O_2^+]$ the conclusions can be formed that the reaction (1) dominates in the E region and that nitric oxide concentration varies slightly during the day and with altitude.

References

[1] A. D. Danilov and G. S. Ivanov-Kholodny, Kosm. Issled. **4**, 439 (1966).
[2] A. D. Danilov, Chemistry of the Ionosphere, Plenum Press, New York 1970.
[3] A. D. Danilov, Space Research X, 725 (1970).
[4] G. S. Ivanov-Kholodny and L. P. Korsunova, Geomagn. i Aeron. **9**, 474 (1969).
[5] W. Swider, J. Geophys. Res. **70**, 4870 (1965).
[6] L. P. Korsunova, Geomagn. i. Aeron. **10**, 15 (1970).

[7] G. S. IVANOV-KHOLODNY and G. M. NICOLSKY, Solntse i Ionosphera, Nauka, Moscow 1969.
[8] A. D. DANILOV, Astrophysics and Space Science Library, Vol. 29: Solar Terrestrial Physics 1970. D. Reidel Publ. Co., Dordrecht 1971 (p. 27).
[9] C. Y. JOHNSON et al., IGY Rocket Report Series NI, 121 (1958).
[10] V. G. ISTOMIN, Iskusstv. Sputniki Zemli **11**, 94 (1961).
[11] V. G. ISTOMIN, Iskusstv. Sputniki Zemli **7**, 64 (1961).
[12] H. A. TAYLOR and H. C. BRINTON, J. Geophys. Res. **66**, 2587 (1961).
[13] J. C. HOLMES et al., Space Research V, 756 (1965).
[14] C. Y. JOHNSON, presented at IGSY/COSPAR Joint Symposium, London 1967.
[15] R. S. NARCISI et al., Aeronomy Report **32**, 358 (1969).
[16] A. D. ZHLOODKO and N. M. KLYUEVA, Space Research X, 746 (1970).
[17] F. ARNOLD et al., Planet. Space Sci. **18**, 171 (1970).
[18] A. D. ZHLOODKO et al., Meteorol. i Gidrol. No. 3, 72 (1970).
[19] T. M. DONAHUE et al., Planet. Space Sci. **18**, 171 (1970).
[20] R. A. GOLDBERG and L. J. BLUME, J. Geophys. Res. **75**, 133 (1970).
[21] A. GIRAUD et al., Space Research XI, 1057 (1971).
[22] A. D. ZHLOODKO, in: Issledovanie Atmospheri i Ionospheri v Period Povishennoi Solnetchnoi Aktivnosty, Gidrometeoizdat, 1970 (p. 44).
[23] R. C. SAGALIN et al., Space Research VII, 448 (1967).
[24] H. OYA et al., Space Research X, 707 (1970).
[25] M. BIONDI, Canad. J. Chem. **47**, 1711 (1969).
[26] A. D. DANILOV and A. N. TRESVIATSKY, report to IUCSTP Symposium, Leningrad 1970.
[27] W. L. FITE, Canad. J. Chem. **47**, 1797 (1969).
[28] T. M. DONAHUE et al., J. Geophys. Res. **76** (1971).

Space Research XII – Akademie-Verlag, Berlin 1972

SIMULTANEOUS ROCKET MEASUREMENTS OF CORPUSCULAR AND SOLAR ELECTROMAGNETIC IONIZING RADIATION AT ALTITUDES UP TO 150–160 KM

V. F. TULINOV and V. M. FEIGIN

Hydrometeorological Service of USSR, Moscow, USSR

The results of simultaneous measurements of corpuscular and solar electromagnetic radiation obtained by meteorological MR-12 rockets during two flights in the middle latitudes over p. Volgograd are discussed.

Measuring instruments contained Geiger counters of various types which provide effective registrations of electrons with energies more than 20 and 35 keV, semiconductor detectors to register protons with energies of 0.5–20 MeV, counters of X-ray radiation in the region of 1–10 Å, and ionization chamber to register radiation in the line Lα.

The relative contributions from corpuscular and electromagnetic sources to ionization of the ionospheric E layer are estimated.

During the last ten years fluxes of charged particles [1, 2] have been detected in the upper atmosphere which together with solar electromagnetic radiation could participate in ionosphere formation. However, because the intensity of so-called "precipitating" particles [3] and short-wave solar radiation [4] are subjected to considerable space–time variations, it is difficult to estimate the relative role they play in ion formation using separate measurements. In this connection it seems important that the principal ionizing radiation Lα, X radiation and corpuscular radiation should be measured simultaneously, so that the contributions of ionizing radiation of every kind to lower ionosphere formation can be estimated under the same space–time and geomagnetic conditions.

For this task to be solved two meteorological MR-12 rockets have been launched to a height of 150–160 km from the mid-latitude site Volgograd ($\varphi = 48°41'$ N) provided with apparatus [5] to register galactic cosmic rays, soft electron fluxes (precipitating particles), soft protons (cosmic rays, albedo or precipitating particles), X-ray solar radiation in the wavelength range $\lambda = 1$–10 Å and solar ultraviolet in the line $\lambda = 1216$ Å. The detectors used and their characteristics are presented in Table 1. Analysis of the data obtained by semiconductor proton detectors leads to the conclusion that the intensity of the soft proton flux (energies 0.5–60 MeV) in the upper atmosphere at mid-latitudes is considerably lower than that of the electron flux. Therefore, the flux of charged particles registered by Geiger counters Zenit and SBT-9 may be attributed to electron and galactic cosmic rays. But when problems of ion formation are considered it is necessary that protons of the energies mentioned above should be taken into account as they are of greater ionizing capability. Measurement results of Lα radiation obtained at different altitudes in one

Table 1

Type of detector	Type of radiation	Energy range	Efficiency area (cm^2)
Geiger counter Zenit	electrons,	$\geqq$25 keV	0.12
	protons,	$\geqq$0.5 MeV	0.12
	cosmic rays (protons)	$>$30 MeV	0.2
Geiger detector SBT-9	electrons,	$\geqq$40 keV	0.18
	protons,	$\geqq$0.5 MeV	0.18
	cosmic rays (protons)	$>$30 MeV	2.5
SBT-9 with magnetic discriminator	electrons,	$\geqq$300 keV	0.18
	protons,	$\geqq$0.5 MeV	0.18
	cosmic rays (protons)	$>$30 MeV	2.5
Semi-conductor detector	protons	0.5–63 MeV	1.25
Geiger detector SPM-9M	X-rays,	$\lambda = 1$–10 Å	7.85×10^{-3}
	cosmic rays (protons)	$>$30 MeV	8
Ionization chamber	ultraviolet radiation	Lα	

of the flights do not contradict this [6], and we shall not consider them in the paper.

Fig. 1 presents some data obtained during two rocket flights launched on 24 October 1969 at 0833 LT (curves 1 and 4) [7] and on 24 December 1969 at 1105 LT (curves 2, 3, 5), the solar zenith angle being $Z = 70°$. Altitude in kilometres is plotted as abscissa; ordinates are counting rate in seconds averaged over 5 s according to the data available from X-ray counters SRM-9 and SBT-9 type and Zenit counters. During the ascent on the second flight in the altitude interval of 75–154 km no information was obtained.

When analysing radiation intensity dependence for all detectors one notes that at heights 15–25 km a maximum was observed and that in the altitude range 35–80 km a plateau occurred. Curves obtained up to about 80 km altitude were in good agreement with the known data on altitude dependence of galactic cosmic ray intensity. It should be mentioned that a similar picture at the same altitudes was observed on the descent rocket trajectory. This is important since agreement between cosmic ray data on the ascending and descending legs of the trajectory indicates correct apparatus performance during the whole flight.

Let us consider the results obtained. Curves 1 and 2 in Fig. 1 show the counting rate (averaged over 5 s) for six SBT-9 counters connected in pairs in parallel (counters were placed along the rocket axis), recounted for one detector. As can be seen in the figure at heights over 80 km a heavy anisotropic additional soft radiation is observed (indicated by a sinusoidal parameter of time intensity change caused by rocket precession). To all appearance, they are soft electrons with small pitch angles (precipitating particles), which, transferring in the so-called "losses cone", can penetrate rather deep in the atmosphere.

In order to account for a possible solar radiation influence on end-window SBT-9 counter readings, the entrance windows of control counters of this type were covered with aluminium (0.075 mg cm^{-2}). Corresponding data are plotted as triangles. A good correlation with the curve 1 shows that scattered solar radiation does not affect counter performance. Curve 3 (Fig. 1) shows counting rate, averaged over 5 s for five Zenit detectors placed in the rocket axis plane on

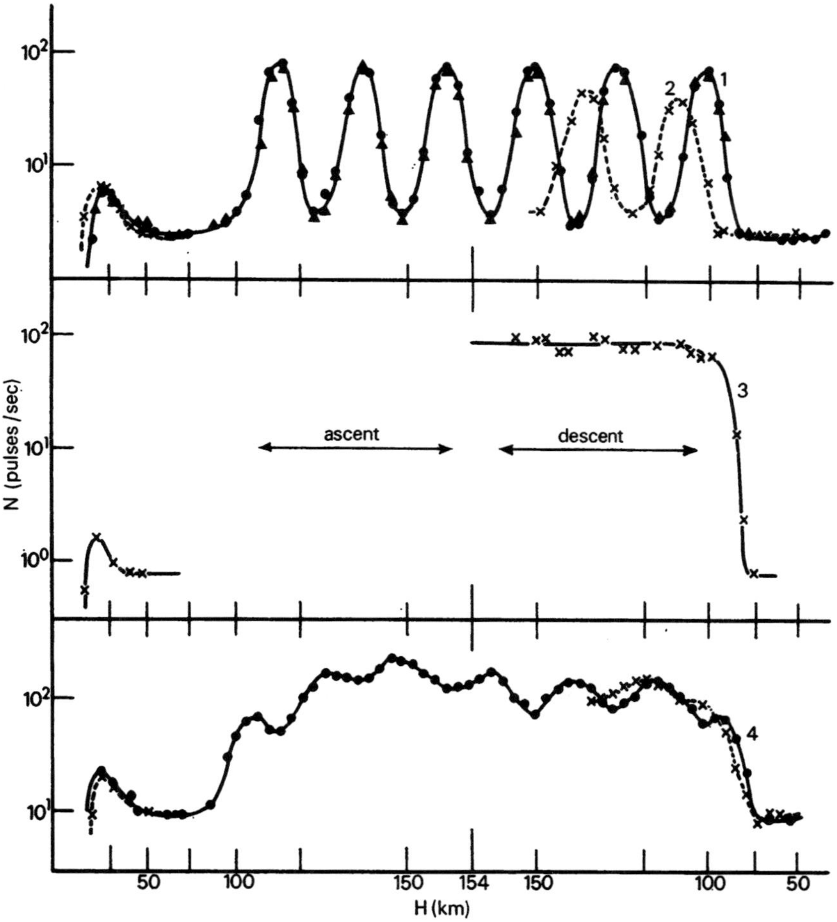

Fig. 1. Radiation intensities observed on 24 October 1969 (curves 1 and 4) and 24 December 1969 (curves 2, 3 and 5)

the arc circle at intervals of 30°. Resulting electron flux intensity is not modulated by rocket precession as diagrams of Zenit counter directivity overlapped.

In Fig. 2 are plotted integral electron flux values for energies $E \geqq 25$ keV (Zenit detector), $E \geqq 40$ keV (SBT-9 detector) and $E \geqq 300$ keV (SBT-9 detector with magnetic discrimination) (circles, launch on 24 October 1969; crosses, launch on 24 December 1969). Having compared the data on integral intensity of electron flux of energies $E \geqq 25$ keV and 40 keV obtained during the flight on 24 December 1969 we found that in the region mentioned the electron flux was distributed in the spectrum $P(E)\,dE \propto E^{-\gamma}\,dE$, where $\gamma = 3$. In the 300 keV energy range index γ somehow decreases and approaches a value of 2. Assuming that the index of the differential spectrum $\gamma = 3$ remains the same up to

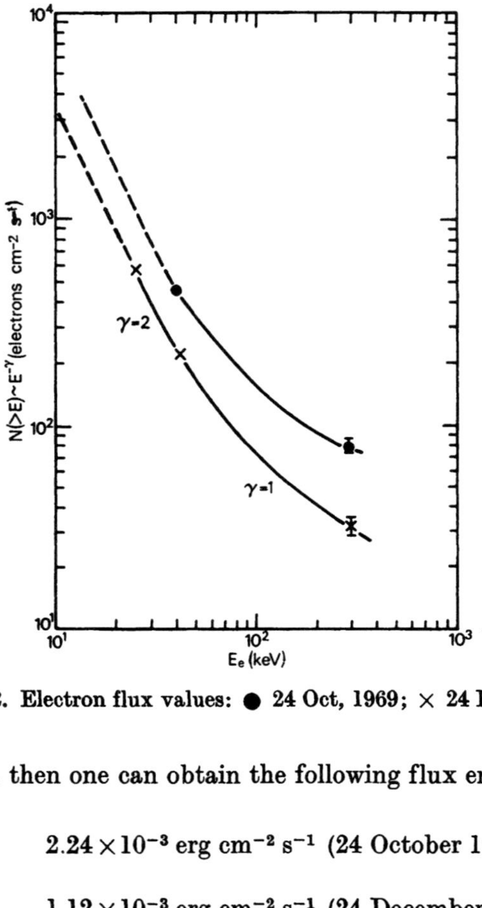

Fig. 2. Electron flux values: ● 24 Oct, 1969; × 24 Dec. 1969.

energies of 1 keV, then one can obtain the following flux energy values carried by electrons:

$$2.24 \times 10^{-3}\ \text{erg cm}^{-2}\ \text{s}^{-1}\ \text{(24 October 1969)}$$

and

$$1.12 \times 10^{-3}\ \text{erg cm}^{-2}\ \text{s}^{-1}\ \text{(24 December 1969)}.$$

Curves 4 and 5 in Fig. 1 present quantum counting rates with wavelength $\lambda = 1$–10 Å, obtained during the first and second flights respectively (exit windows of X-ray sensors were tilted at an angle of 70° to the rocket axis which corresponds to the solar zenith angle). Taking into account that the registration efficiency of X-ray radiation by the counter type used was, approximately, 10% we find the intensity of X-ray solar radiation at an altitude 150 km to be 2.5×10^5 quanta cm^{-2} s^{-1}, which does not contradict (taking into account the known considerable avariations) data available in the literature [8]. This corresponds to the energy flux 5×10^{-4} erg cm^{-2} s^{-1}. X-ray radiation at $\lambda = 1$–10 Å is the principal one to contribute to the ion formation in the D region (lower 100 km). Electrons with $E \geqq 10$ keV energies penetrate into these altitudes. According to rocket data electron fluxes with $E \geqq 10$ keV carry energies

2.16×10^{-4} and 1.08×10^{-4} erg cm^{-2} s^{-1}, respectively, which is only 2.5–5 times lower than the energy flux of X-ray radiation.

Thus, the measurements made showed that even in the day-time electron fluxes appreciably contributed to lower ionosphere formation.

References

[1] L. A. Antonova and G. S. Ivanov-Kholodny, Izv. Akad. Nauk, ser. Geophys., No 5, 756 (1960).
[2] V. F. Tulinov, Kosm. Issled. 5, 240 (1969).
[3] V. F. Tulinov et al., Kosm. Issled. 7, 307 (1970).
[4] R. W. Kreplin, Annls Géophys. 17, 151 (1962).
[5] V. M. Feigin, Centr. Aerolog. Obs. No. 99 (1971).
[6] A. F. Chizhov and I. S. Kim, Izv. Akad. Nauk SSSR, Ser. Fiz. Atmos. Okeana, 6, 7728 (1970).
[7] V. F. Tulinov and V. M. Feigin, Centr. Aerolog. Obs., No. 111 (1972).
[8] R. E. Bourdeau et al., J. Geophys. Res. 71, 727 (1966).

Space Research XII – Akademie-Verlag, Berlin 1972

H_2O^+ IONS AND H_2O MOLECULES IN THE LOWER THERMOSPHERE AND IONOSPHERE

G. M. Martynkevich

Academy of Sciences of USSR, Moscow, USSR

The vertical distribution of ambient H_2O molecules is calculated from the vertical distribution of H_2O^+ ions. Calculated absolute and relative concentrations of neutral H_2O molecules are compared with direct measurements by the author. The agreement of the calculation with the experimental data is noted.

Estimates of the rise time of the equilibrium concentrations of H_2O^+ are made. The minimum value of this time appeared to be 58 hours at an altitude of 240 km. This means that H_2O^+ ions are formed as a result of the reaction $H_2O + O^+ \rightarrow H_2O^+ + O$, in which ambient O^+ ions and H_2O molecules, rather than degassing molecules, take part.

A considerable part of H_2O molecules, or the majority of them, at the levels of $h > 100$ km apparently have an extraterrestrial origin.

H_2O^+ ions are usually considered to be the products of the reaction

$$H_2O + O^+ \rightarrow H_2O^+ + O \tag{1}$$

of desorbed H_2O molecules with O^+ ions [1]. Actually H_2O molecules as well as H_2O^+ ions, measured by mass spectrometers of different types in methodically correct experiments, are of atmospheric origin [2, 3], as can be easily shown.

In accordance with (1)

$$dn_{H_2O^+}/dt = \gamma\, n_{O^+}\, n_{H_2O}\,. \tag{2}$$

Let us first consider desorption to be the only source of H_2O molecules and reaction (1) to be the only source of H_2O^+ ions. In this case $dn_{H_2O^+} = -\,dn_{H_2O}$ resulting in

$$n_{H_2O^+} + n_{H_2O}(t) = n_{H_2O}\,(t = 0) \tag{3}$$

where $n_{H_2O}\,(t = 0) = \mathrm{const}_1$ is the initial (for $t = 0$) number density of desorbed H_2O molecules in the effective zone of the electric field of a mass analyser. Substituting $dn_{H_2O^+}$ in (2) for $-\,dn_{H_2O}$ we have

$$dn_{H_2O^+}/n_{H_2O} = -\,\gamma\, n_{O^+}\, dt\,. \tag{4}$$

Integrating (4) with $n_{O^+} = \mathrm{const}_2$ and $\gamma = \mathrm{const}_3$ we obtain

$$n_{H_2O}(t) = n_{H_2O}\,(t = 0) \times \exp\,(-\,t/\tau) \tag{5}$$

$$\tau = 1/\gamma n_{O^+} \tag{6}$$

$$n_{H_2O^+} = n_{H_2O}\,(t = 0) \times [1 - \exp\,(-\,t/\tau)] \tag{7}$$

Table 1 contains τ values calculated according to (6), Hoffman's et al. [4] and Turner's [5] data and also upper limits of the calculated number density

$n_{H_2O^+}$ due to desorption as well as number density ratio $n_{H_2O^+}$ (desorption)/$n_{H_2O^+}$ (measured). The highest experimental number density of desorbed water molecules n_{H_2O} ($t = 0$) [6, 7] does not exceed $(1\text{–}1.5)\times 10^9$ cm^{-3} at the level of 150 km. t in (5) and (7) is determined by the time during which H_2O desorbed

Table 1

No.	h (km)	n_{O^+} (cm^{-3})	γ (cm^3 s^{-1})	τ (s)	τ (h)	$n_{H_2O\,des}$	$\frac{n_{H_2O\,des}}{n_{H_2O\,meas}}$ (%)
1.	180	3.5×10^4	1.2×10^{-10}	2.4×10^5	66	11	2.7
2.	240	4×10^5	1.2×10^{-10}	2.1×10^4	5.8	120	7.2
3.	300	4×10^5	1.2×10^{-10}	2.8×10^4	7.7	92	4.6
4.	400	1.6×10^5	1.2×10^{-10}	5.2×10^4	14.5	49	2.9
5.	500	9.1×10^4	1.2×10^{-10}	9.3×10^4	25.8	28	1.8
6.	600	5.5×10^4	1.2×10^{-10}	1.5×10^5	42	17	1.5
7.	700	3.3×10^4	1.2×10^{-10}	2.5×10^5	70	10	1

molecules are in the effective mass spectrometer field. The length L of this field is no more than 20–30 cm. Let $L = 10^2$ cm, then $t = L/\bar{v}_{H_2O} \sim 1.7\times 10^{-3}$ s [$\bar{v}_{H_2O} = (8\,k\,T/\pi\,m)^{1/2} \sim 6\times 10^4$ cm s^{-1}, when $T = 320$ °K]. As $\exp(-t/\tau) \approx \approx 1 - t/\tau$, according to (7) one obtains as the upper limit of $n_{H_2O^+}$ due to the desorption the value

$$n_{H_2O^+}(h = 180\text{ km}) \leqq 1.5\times 10^9\text{ cm}^{-3}\times 1.7\times 10^{-3}\text{ s}/2.4\times 10^5\text{ s} \sim 11\text{ cm}^{-3} \quad (8)$$

with measured value $n_{H_2O^+} = 4\times 10^2$ cm^{-3} [4]. Thus the desorption contribution in $n_{H_2O^+}$ at the level of 180 km does not exceed 2.7% for the highest concentration of desorbed water. The most frequently registered number densities n_{H_2O} ($t = 0$) are of the order of 3×10^8 cm^{-3} or lower [2, 3, 7, 8], i.e. the contribution of desorbed water is less than 0.5%. Consequently H_2O^+ ions and H_2O molecules forming these ions according to (1) are of atmospheric origin.

One can calculate the vertical distribution of H_2O molecules from the vertical distribution of H_2O^+ ions measured experimentally. H_2O^+ ions are formed as a result of the reactions

$$H_2O + O^+ \rightarrow H_2O^+ + O \qquad \gamma_9 = 1.2\times 10^{-10}\text{ cm}^3\text{ s}^{-1} \quad (9)$$

$$H_2O + N_2^+ \rightarrow H_2O^+ + N_2 \qquad \gamma_{10} = 3\times 10^{-10}\text{ cm}^3\text{ s}^{-1} \quad (10)$$

$$H_2O + N^+ \rightarrow H_2O^+ + N \qquad \gamma_{11} = 9\times 10^{-11}\text{ cm}^3\text{ s}^{-1} \quad (11)$$

and disappear as a result of the reactions

$$H_2O^+ + O_2 \rightarrow O_2^+ + H_2O \qquad \gamma_{12} = 2\times 10^{-10}\text{ cm}^3\text{ s}^{-1} \quad (12)$$

$$H_2O^+ + e \rightarrow OH^* + H \qquad \alpha' \quad (13)$$

In equilibrium

$$n_{H_2O} = \frac{n_{H_2O^+}(\gamma_{12}\,n_{O_2} + \alpha' n_e)}{\gamma_9\,n_{O^+} + \gamma_{10}\,n_{N_2^+} + \gamma_{11}\,n_{N^+}}\,. \quad (14)$$

The rate constants of reactions (9)–(12) were measured experimentally [5, 9]; the effective recombination coefficient α' was taken from paper [10], and experimentally measured number densities $n_{H_2O^+}$, n_{O^+}, $n_{N_2^+}$, n_{N^+}, n_e, n_{O_2} from papers [4, 11, 12, 13].

Figs. 1–2 present the calculated results of n_{H_2O} and n_{H_2O}/n_{N_2}.** These figures contain also curves obtained in our mass-spectrometer experiments [2, 14] and plotted for comparison. It can be easily seen that the earlier conclusions [2, 14] based on the direct H_2O molecules measurements are entirely confirmed by these calculations: (i) Two maxima of absolute and relative concentrations are distinctly seen at altitudes 95–110 and 120–150 km. (ii) The measured number densities of H_2O molecules may exceed the calculated values by an order of magnitude as a result of hydrogen and oxygen recombination process inside mass analysers. (iii) Both calculated and measured H_2O molecule number

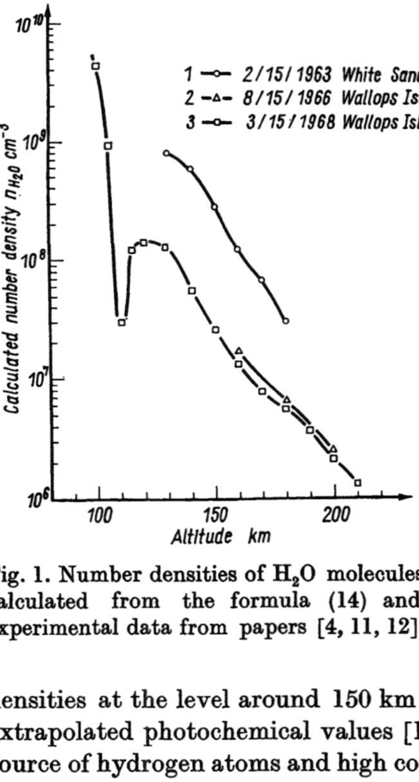

Fig. 1. Number densities of H_2O molecules calculated from the formula (14) and experimental data from papers [4, 11, 12].

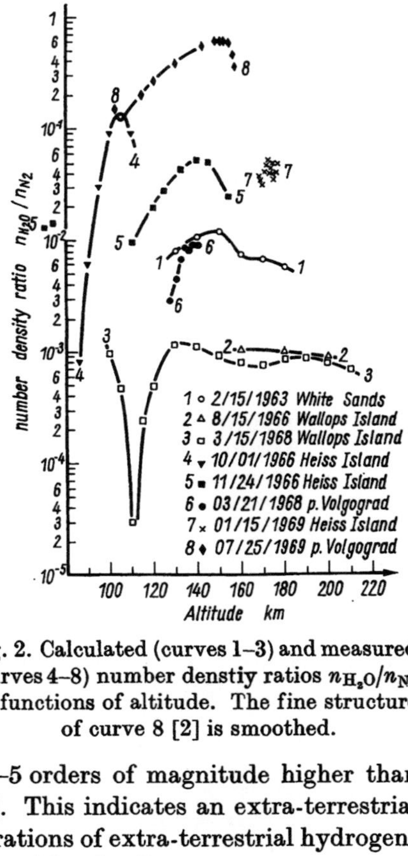

Fig. 2. Calculated (curves 1–3) and measured (curves 4–8) number denstiy ratios n_{H_2O}/n_{N_2} as functions of altitude. The fine structure of curve 8 [2] is smoothed.

densities at the level around 150 km are 3–5 orders of magnitude higher than extrapolated photochemical values [15, 16]. This indicates an extra-terrestrial source of hydrogen atoms and high concentrations of extra-terrestrial hydrogen; the presence of such hydrogen is also confirmed by the direct mass-spectrometer measurements [2, 14].

From the present paper together with that on high atomic nitrogen concentrations [17] the following interpretation of hydroxyl formation and emission can be put forward: High energy hydrogen atoms precipitating into the dense atmospheric layers react with oxygen molecules

$$H_2 + O_2 \rightarrow OH^* + O \qquad (15)$$

forming excited hydroxyl radicals and oxygen atoms. The rate constant of

** See Note added in proof (overleaf).

reaction (15) was measured repeatedly [18]. This reaction, despite endothermic processes, proceeds rather fast because the energy of precipitating hydrogen atoms exceeds significantly its activation energy E ($E \leqq 17.6$ kcal mole^{-1}).

The proposed mechanism of hydroxyl emission excitation is the modified Krassovsky mechanism [19]. Krassovsky assumed that excited O_2 molecules reacted with non-excited hydrogen atoms. The modified mechanism proposes the reverse, i.e. the high energy hydrogen atoms react with non-excited oxygen molecules.

Note added in proof

Recently, measurements of the rate constants for reactions of O^+, N_2^+ and N^+ with H_2O gave values of 2.33, 2.19 and 2.57×10^{-9} cm^3 s^{-1} [20]. Using these new values and taking into account their dependence on temperature in calculations of the vertical distribution of H_2O molecules one obtains following results: (i) absolute values n_{H_2O} decrease by approximately an order of magnitude campared with the number densities calculated using the values 1.2×10^{-10}, 3×10^{-10} and 9×10^{-11} cm^3 s^{-1}; (ii) the rest of the conclusions remain unchanged, i.e. (a) the atmospheric origin of most of the H_2O molecules; (b) the layer structure of its vertical distribution; (c) the considerable excess (by 3–4 orders of magnitude) of H_2O molecule number densities calculated according to the formula (14) over the number densities calculated using photochemical theory [21].

References

[1] A. D. Danilov and A. A. Pokhunkov, Space Research VII, 1167 (1967).
[2] G. M. Martynkevich and E. G. Shvidkovsky, Space Research XI, 865 (1971).
[3] E. J. Schaefer and M. H. Nickols, J. Geophys. Res. **69**, 4649 (1964).
[4] J. H. Hoffman, C. Y. Johnson, J. C. Holmes and J. M. Young, J. Geophys. Res. **74**, 6281 (1969).
[5] B. R. Turner, Aeronomy Report No. 32, 305 (1968).
[6] J. W. Townsend, Jr., E. B. Meadows and E. S. Pressley, in: Rocket Exploration of the Upper Atmosphere, Pergamon Press N.Y., London 1954 (p. 169).
[7] A. A. Pokhunkov, in: Ionosph. Issled. No. 14, 26, Nauka, Moscow 1965.
[8] J. Gross and U. von Zahn, Space Research XI, 875 (1971).
[9] W. L. Fite, Canad. J. Chem. **47**, 1797 (1969).
[10] A. P. Mitra and P. Banerjee, Space Research XI, 1019 (1971).
[11] J. C. Holmes and J. M. Young, Aeronomy Report No. 1 (1963).
[12] R. A. Goldberg and L. J. Blumle, J. Geophys. Res. **75**, 133 (1970).
[13] W. T. Kasprzak, D. Krankowsky and A. O. Nier, J. Geophys. Res. **73**, 6765 (1968).
[14] G. M. Martynkevich, Meteorol. i Gidrol. No. 12, 3 (1969).
[15] M. Nicolet, Aeronomy, Mir, Moscow (1964) (in Russian).
[16] E. Hesstvedt, in: Progress in Astronautics and Aeronautics **22**, 307 (1969).
[17] G. M. Martynkevich and E. D. Byuro, Space Research XII, 691 (1972).
[18] K. Schofield, Planet. Space Sci. **15**, 643 (1967).
[19] V. I. Krassovsky, Iskusstv. Sputniki Zemli No. 17, 3 (1963).
[20] C. J. Howard, H. W. Rundle and F. Kaufman, J. Chem. Phys. **53** (1970).
[21] T. J. Keneshea, S. P. Zimmerman and J. D. George, paper presented at the 14th Plenary Meeting of COSPAR, Seattle, June 1971.

Space Research XII — Akademie-Verlag, Berlin 1972

SOLAR ACTIVITY VARIATION OF $[NO^+]/[O_2^+]$ AND $[NO^+]/[O^+]$ IN E AND F REGIONS

A. P. Mitra and P. Banerjee

Radio Science Division, National Physical Laboratory, New Delhi, India

Studies of solar activity variations of ionospheric ion composition measured principally with rockets at different phases of solar activity reveal that during day and for heights less than 160 km, the ratio $[NO^+]/[O_2^+]$ increases from low to high activity by a factor of 6–2 in the height range 120–150 km, and for heights above 160 km $[NO^+]/[O^+]$ increases by a factor of 1.3–2.4 for the height range 170–200 km as the 10.7 cm flux increases from 70 to 250 units. The increase in $[NO^+]/[O_2^+]$ can be related to an increase in the minor constituents (4 [NO] + [N]) which increases by a factor of 7 at 130 km. The increase in $[NO^+]/[O^+]$ cannot be interpreted entirely on the basis of changes in neutral composition and requires an increase in the value of rate coefficient k_1 for the reaction $O^+ + N_2 \rightarrow NO^+ + O$, thus indicating an increase in vibrational temperature. The estimated vibrational temperatures at 200 km are 1600° and 1675 °K during low and high solar activity as against the gas temperatures of 700° and 1050 °K.

1. Introduction

While substantial increase in the concentration of the ions at heights 100 to 200 km are expected with changes in the incoming xuv flux, the ion ratios $[NO^+]/[O_2^+]$ and $[NO^+]/[O^+]$ should both decrease with the present reaction scheme if the neutral constituents remain unchanged. In earlier work, Mitra and Mitra [1], however, reported an increasing trend of $[NO^+]/[O_2^+]$ with solar activity. In the present paper similar studies for the E and F region are reported, with the addition of recent ion composition measurements for daytime. These confirm the results of Mitra and Mitra, and further show a similar increasing trend in $[NO^+]/[O^+]$, the latter showing at heights above 160 km. This anomaly suggests that there is, in addition to the variation in the concentration of some of the reacting neutral constituents, a variation also in the reaction rates. In this work we examine the physical implications of these anomalies.

2. Observational Results

The ion composition measurements considered in our study are given in Table 1; only daytime measurements are used. The results are summarized in Fig. 1. It is seen that as the solar activity increases $[NO^+]$ increases at the expense of $[O_2^+]$ for heights less than 160 km, and for heights greater than 160 km

Table 1

Daytime ion composition measurements

No.	Date of flight	$F_{10.7}$ (10^{-22} W m^{-2} s^{-1})	χ(°)	Ref.
1.	23 Mar. 1958	268	58	[2]
2.	2 Aug. 1958	254	54	[3]
3.	15 Nov. 1960	183	58	[4]
4.	15 Feb. 1963	70	60	[5]
5.	IQSY	70	60	[6]
6.	3 Mar. 1966	78	49	[7]

$[NO^+]$ increases at the expense of $[O^+]$. It is also seen that for heights less than 160 km the ratio $[NO^+]/[O_2^+]$ increases with solar activity, and for heights greater than 160 km the ratio $[NO^+]/[O^+]$ increases with solar activity.

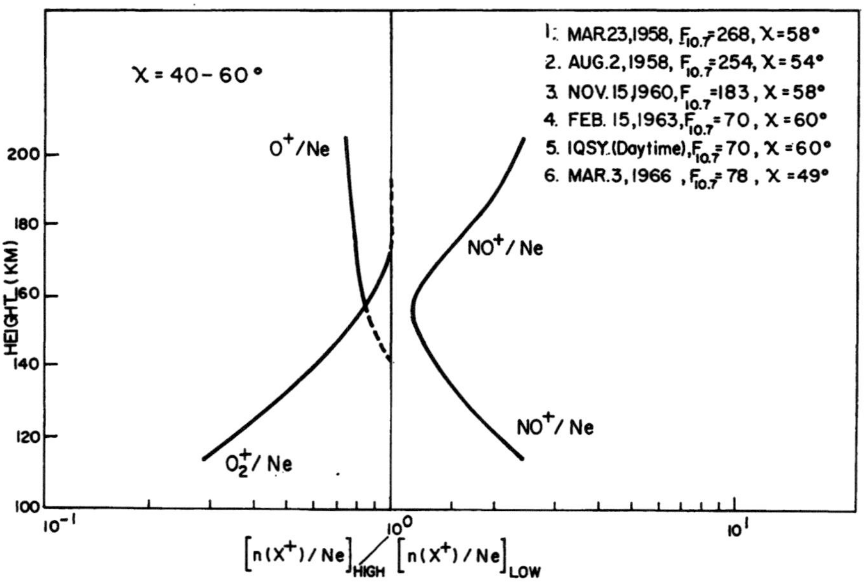

Fig. 1. Variation with height of ratio of the average percentage ion composition for high solar activity to the average percentage ion composition for low solar activity. 1–6 represent the flights considered for evaluation of average model.

3. Theory

Neglecting the transport processes, such as diffusion and atmospheric drifts, the equations governing the concentrations of the ions N_2^+, O^+, O_2^+ and NO^+ are controlled by the reactions listed in Table 2 and are given by

$$[N_2^+] = \frac{q(N_2)}{k_5[O]}, \tag{1}$$

$$[O^+] = \frac{q(O)}{k_1[N_2] + k_2[O_2]}, \tag{2}$$

$$[O_2^+] = \frac{q(O_2) + k_2[O_2][O^+]}{\alpha_1 N_e + S}, \tag{3}$$

$$[NO^+] = \frac{q(N_2) + k_1[N_2][O^+] + S[O_2^+] + k_7[N^+][O_2]}{\alpha_2 N_e}, \tag{4}$$

where S is given by

$$S = 10^{-16}[N_2] + 8\times10^{-10}[NO] + 2\times10^{-10}[N] \tag{5}$$

and q's denote photoionization rates.

Table 2
Relevant ionospheric reactions and their rate coefficients

Reaction	Symbol	Rate (cm³ s⁻¹)
$O^+ + N_2 \rightarrow NO^+ + O$	k_1	1×10^{-12} at 300 °K reduces to 5×10^{-13} at 600 °K
$O^+ + O_2 \rightarrow O_2^+ + O$	k_2	$1\times10^{-9}\,T^{-0.7}$
$O_2^+ + N_2 \rightarrow NO^+ + NO$	k_3	$<10^{-15}$, value used 10^{-16}
$O_2^+ + NO \rightarrow NO^+ + O_2$	k_4	8×10^{-10}
$O_2^+ + N \rightarrow NO^+ + O$	k_5	2×10^{-10}
$N_2^+ + O \rightarrow NO^+ + N$	k_6	2×10^{-10}
$N^+ + O_2 \rightarrow NO^+ + O$	k_7	5×10^{-10}
$O_2^+ + e \rightarrow O' + O''$	α_1	$2.0\times10^{-7}(T/300)^{-1}$
$NO^+ + e \rightarrow N + O$	α_2	$5.0\times10^{-7}(T/300)^{-1}$

We will consider two separate height regions, above and below the level around 150–160 km, such that O+ plays little or no role in the lower region but plays a dominant role in the higher region.

Region 1. $h < 150$ km

Here Eq. (4) becomes

$$\frac{[NO^+]}{[O_2^+]} = \frac{S + q(N_2)/[O_2^+]}{\alpha_2 N_e}. \tag{6}$$

Region 2. 160 km $< h \leqq 200$ km

Here we get

$$\frac{[NO^+]}{[O^+]} = \frac{k_1[N_2] + q(N_2)/[O^+] + k_7[N^+][O_2]/[O^+]}{\alpha_2 N_e}. \tag{7}$$

Considering region 1 first, it is apparent that if S were constant with changing solar activity and since $q(N_2)$ is nearly constant at these heights, then the ratio $[NO^+]/[O_2^+]$ would decrease with solar activity as N_e increases. What we observe is, however, an increase in the ratio. This implies an increase of the parameter S, and hence of the minor constituents parameter [4 NO] + [N]. The values of the parameter are shown as functions of heights separately for low and high solar activity in Fig. 2. We notice that [4 NO] + [N] increases by a factor of 7 at 120 km and about 2 at 150 km. The increase around 120 km is rather large, but it decreases rapidly with height.

For heights greater than 160 km we find a similar anomaly. The ratio $[NO^+]/[O^+]$ increases with solar activity whereas from Eq. (7) this should decrease unless the concentration of N_2 or the value of k_1 increases substantially. The concentration of N_2 in the Jacchia 1970 [8] ionosphere model increases by a factor of 2.4 at 200 km and 2.0 at 180 km. The ratio $[NO^+]/[O^+]$ during the same period increases by a factor of about 2.36 at 200 km and about 2.32 at 180 km. Although the theoretically calculated values of the ratio $[NO^+]/[O^+]$ increase almost by the same amount with $k_1 = 10 \times 10^{-13}$ cm^3 s^{-1}, a discrepancy is observed when the absolute magnitudes of the measured ratio are compared with the theoretical values. The discrepancy appears at heights above 180 km for low solar activity conditions, and above 170 km for high solar activity conditions (Fig. 3). One would thus require for both maximum and minimum conditions somewhat higher concentrations of N_2 or increased values of the

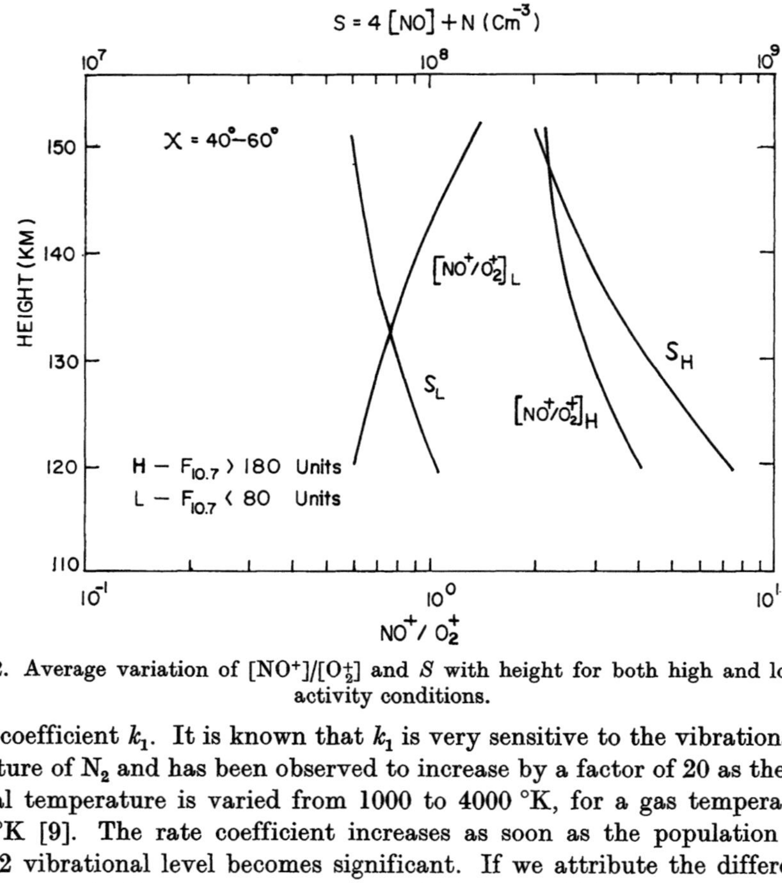

Fig. 2. Average variation of $[NO^+]/[O_2^+]$ and S with height for both high and low solar activity conditions.

rate coefficient k_1. It is known that k_1 is very sensitive to the vibrational temperature of N_2 and has been observed to increase by a factor of 20 as the vibrational temperature is varied from 1000 to 4000 °K, for a gas temperature of 300 °K [9]. The rate coefficient increases as soon as the population of the $v = 2$ vibrational level becomes significant. If we attribute the difference to the increased vibrational temperature of N_2 during high solar activity, which increases k_1, then the values of k_1 required to satisfy the observed values of $[NO^+]/[O^+]$ can be estimated. These are given in Fig. 4. We find, as we had expected, that we require a higher value of k_1; the required value increases rapidly from a value of 5×10^{-13} cm^3 s^{-1} at 160 km to a nearly constant value of about 25×10^{-13} cm^3 s^{-1} at 200 km. The required value at 160 km is the lowest and is, in fact, in complete agreement with Dunkin et al. [10] who found that the coefficient decreases to a constant value of 5×10^{-13} cm^3 s^{-1} at high temperature. The gradual increase of k_1 at greater heights would confirm the

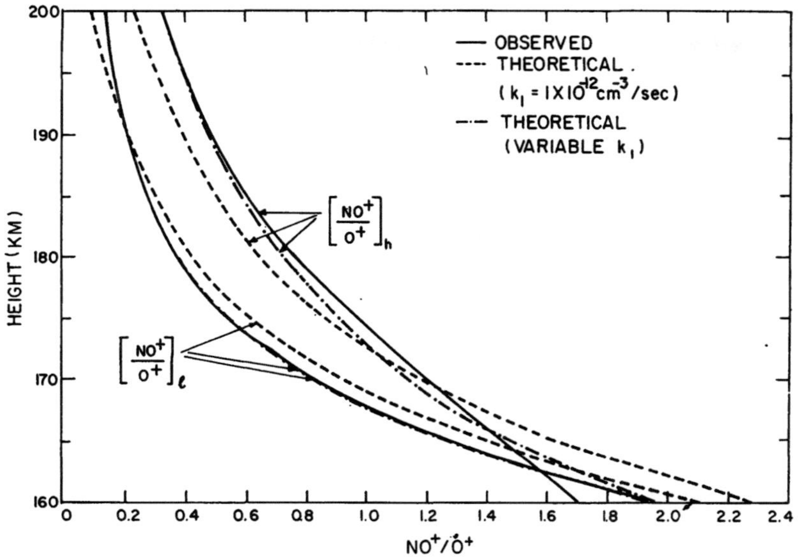

Fig. 3. Average variation of $[NO^+]/[O^+]$ with height for both low and high solar activity conditions.

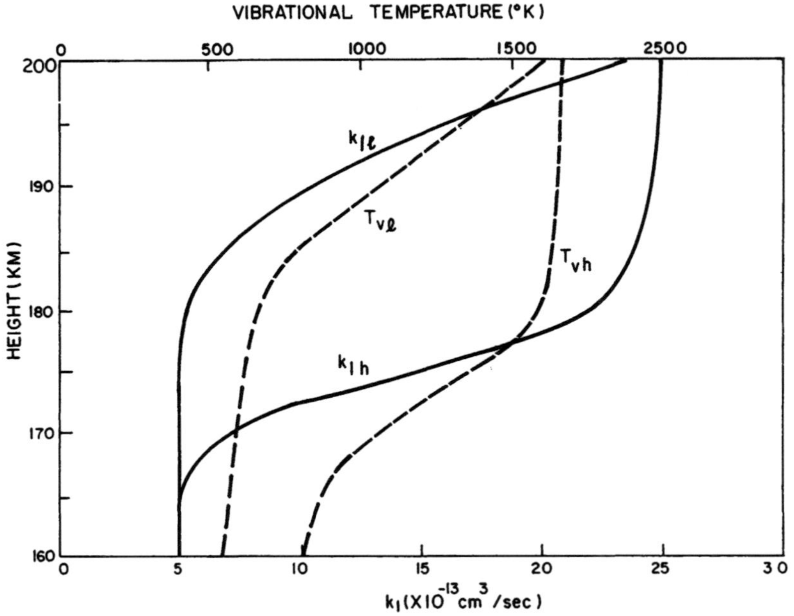

Fig. 4. Variation with height of k_1 and corresponding vibrational temperature required for both low and high solar activity conditions.

concept that k_1 does in fact consist of two parts, one depending on gas temperature and decreasing to a low value around 5×10^{-13} cm^3 s^{-1} at temperatures of 600 °K and above, and the other a function of the vibrational temperature. For low solar activity k_1 continues to have a value of 5×10^{-13} cm^3 s^{-1} up to 180 km, and begins to depart only above this height.

Determination of the appropriate vibrational temperature from these increased values of k_1 is at the moment uncertain. The exact relationships between k_1, T_g and T_v are not known (the observational values of Schmeltekopf et al. [9] referred to earlier were for $T_g = 300$ °K). The nature of the variation of k_1 with T_v for higher values of T_g has not been experimentally determined. Theoretical calculations have, however, been made recently by O'Malley [11] relating k_1 with T_{tr} and T_v; these have subsequently been modified by Van Zandt [12]. From Van Zandt's modification of O'Malley's tables (the two are nearly identical for large T_g and T_v values) the variations in k_1 with height were converted to variations of T_v with height. This was done for both low and high solar activity conditions. The resulting T_v values are given in Fig. 4. For low solar activity conditions, T_v begins to depart from gas temperatures only at 190 km and above; at 190 and 200 km values of T_v are estimated to be about 1200° and 1600 °K while the gas temperatures at these heights are 660° and 680 °K. During high solar activity T_v begins to depart from T_g even at 170 km. Values of T_v at 170, 180, 190 and 200 km are respectively 1050, 1600, 1650 and 1675 °K while the gas temperatures are 920, 970, 1015 and 1060 °K.

It should be emphasized that these quantitative estimates are at present largely uncertain and are to be taken to be indicative only of the range of values that T_v may assume. The important point to note is that with higher solar activity the need for a higher vibrational temperature appears at the relatively lower height of 170 km, but that at 200 km the required vibrational temperatures are not very different during two widely varying solar activity conditions. The actual estimates are, of course, quite sensitive to the type of neutral atmosphere model used; and use of the Jacchia 1971 [13] model will alter these values somewhat, but the general conclusions remain the same.

References

[1] A. P. MITRA and N. P. MITRA, in: Proc. IQSY Symp., Indian National Committee for the IQSY, New Delhi 1969 (p. 362).
[2] C. Y. JOHNSON, Annls. Géophys. **17**, 78 (1961).
[3] V. G. ISTOMIN and A. A. POKHUNKOV, Space Research III, 209 (1963).
[4] H. A. TAYLOR, JR. and H. C. BRINTON, J. Geophys. Res. **66**, 2587 (1961).
[5] J. C. HOLMES, C. Y. JOHNSON and J. M. YOUNG, Space Research V, 756 (1965).
[6] C. Y. JOHNSON, in: Annals of the IQSY, Vol. 5, Solar Terrestrial Physics, Terrestrial Aspects, MIT Press, Cambridge, Mass. 1969 (p. 197).
[7] H. C. BRINTON et al., J. Geophys. Res. **74**, 2941 (1969).
[8] L. G. JACCHIA, Smithson. Astrophys. Obs. Spec. Rep. No. 313 (1970).
[9] A. L. SCHMELTEKOPF et al., Planet. Space Sci. **15**, 401 (1967).
[10] D. B. DUNKIN et al., J. Chem. Phys. **49**, 1365 (1968).
[11] T. F. O'MALLEY, J. Chem. Phys. **52**, 3269 (1970).
[12] T. E. VAN ZANDT, private communication.
[13] L. G. JACCHIA, Smithson. Astrophys. Obs. Spec. Rep. No. 332 (1971).

Space Research XII — Akademie-Verlag, Berlin 1972

VERTICAL WIND PROFILES AND Es FORMATION IN MID-LATITUDES

L. A. ANDREYEVA, L. A. KATASEV, V. P. NESTEROV, D. B. UVAROV, YU. K. CHASOVITIN and G. P. KOMRAKOV

Hydrometeorological Service of the USSR, Moscow, USSR

At the Institute of Experimental Meteorology with the help of rocket MR-12 some experiments were carried out with simultaneous measurements of height wind profiles and electron concentration to check the theory of Es layer formation under the action of wind shear. In the head of the rocket were installed the impedan/ce probe for the measurement of n_e on the ascent and the container for a long luminous trail formation on the descent.

The vertical component of ion velocity and convergence of ion velocity field were calculated according to the equations of wind-shear theory. The calculations showed that a satisfactory correlation between positive extrema of convergence and maxima of electron concentration exists, which is in accordance with the wind-shear theory.

In one of the experiments a thin Es layer was observed with maximum electron concentration approximately one thousand times larger than the background. This layer was recorded at the height predicted by the wind-shear theory, it being known that the convergence was large enough, about $1.3 \times 10^{-2}\ s^{-1}$. In spite of this it seems difficult to explain such high electron concentration in the Es layer on the basis of wind-shear theory only. It is possible that in similar cases some local ionization source is active and in particular a stream of high energy electrons.

In July 1970 three experiments with rocket launchings were carried out at Volgograd; these experiments were aimed at the investigation of the wind regime and height distribution of the electron concentration in the ionospheric E region. Rockets MR-12 were launched at 2102 local time on 1 July, at 2044 on 6 July and at 0240 on 7 July. In the first two launchings the container for artificial cloud formation and probes for determination of electron temperature and concentration (high frequency impedance and Langmuir probes [1, 2]) were installed in the head of a rocket. In these two launchings a mixture containing Na with some Li was used. The electron concentration and electron temperature were measured on the ascent and luminous trails were formed on the descent. A rocket launched on 7 July was equipped only with a container for the long luminous trail formation by means of ejection of a mixture of trimethyl-aluminium and triethyl-aluminium. During all three experiments vertical ionosphere sounding was carried out by means of the automatic ionospheric station.

Clouds were photographed simultaneously from three stations about 40 km apart. Spatial coordinates have been determined according to the technique described in [3]. In calculating the wind velocity it was assumed that the

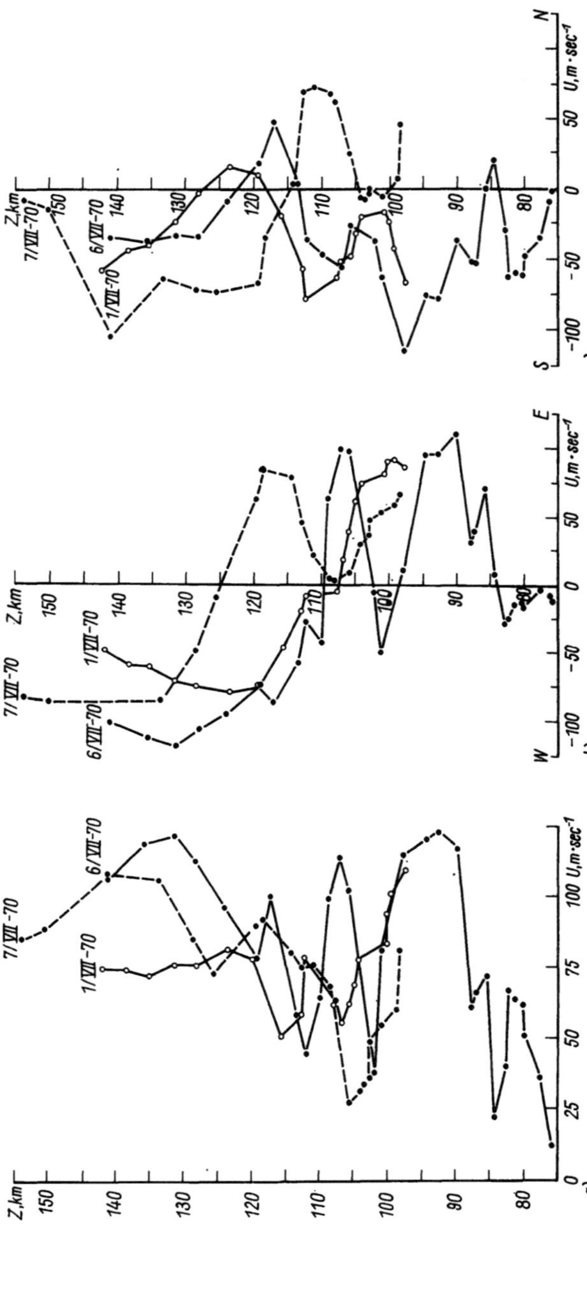

Fig. 1. Profiles of full wind velocity (a), zonal (b) and meridional (c) components. Volgograd 1970, 2102 LMT, 1 July; 2044 LMT, 6 July; 0240 LMT, 7 July.

wind has only horizontal components. The vertical component was not obtained; it seems that its value was less than the standard measurement error, which was about 3–5 m s^{-1}.

The profiles of full wind velocity (a), zonal (b) and meridional (c) components for all three experiments are given in Fig. 1. The wind profiles obtained are characterized by quasi-periodic velocity variations, which are observed up to 130 km. As a rule the maximum values of wind velocity were observed in the 90–102 km height region. Up to 105–108 km the zonal wind is directed mainly to the east and above this level it is directed to the west. However on 7 July this transfer is observed at 124 km. The meridional component is mostly directed to the south. The wind vector in the 90–128 km height region as a rule rotates clockwise in accordance with the tidal-wave theory; some deviations from this observed at 100–110 km may be explained by the simultaneous action of tidal and gravitational waves.

As was pointed out earlier, in two experiments the wind profile and electron concentration have been measured simultaneously. A detailed description by means of n_e measurements is given in [1]. These experimental data and the vertical ionospheric sounding data have been used to check up the mid-latitude sporadic Es-layer formation under the action of wind shear [4–6].

According to this theory the charged particle accumulation (the maximum electron concentration formation) must take place near the heights where the vertical ion velocity component $V_{iz} = 0$, $dV_{iz}/dz < 0$ and also in the height region where the convergence of the ion velocity field $(- \nabla \boldsymbol{V}_i)$ is positive. The vertical ion velocity component and the convergence of the ion velocity field have been calculated for the wind profiles obtained according to the formulae for the stationary case [7, 8] with the help of a computer. The results for the first experiment (1 July 1970) are given in Fig. 2a; in Fig. 2b the simultaneously obtained height profile of electron concentration is presented. From this figure it is seen that the vertical ion velocity component passes zero at 114 km, and the convergence is positive in the 105–108 km and 112–120 km height intervals. In the corresponding profile of the electron concentration the double maxima has been registered in the 108–119 height region; from the ionospheric sounding data at the same time an Es layer with the following parameters was registered: h' Es $= 112$ km, f_b Es $= 2.6$ MHz, f_0 Es $= 3.5$ MHz. This n_e maximum corresponds within the limits of error of measurement to the positive extremum of the convergence at 112–120 km and to the zero value of V_{iz} which is in accordance with the wind-shear theory. In this case the elementary theory is also true [4–6]: the height of the Es layer corresponds roughly to the zero value of the east–west wind component (with the corresponding sign with respect to height). At the same time there is no maximum of electron concentration corresponding to the positive extremum at 105–108 km which may be caused by the imperfection of the wind-shear theory in its present state when detailed physical conditions taking place in the upper atmosphere are not considered.

Two other experiments (6 and 7 July) had been carried out during one night with an interval of 6 hours, therefore it is interesting to consider together the

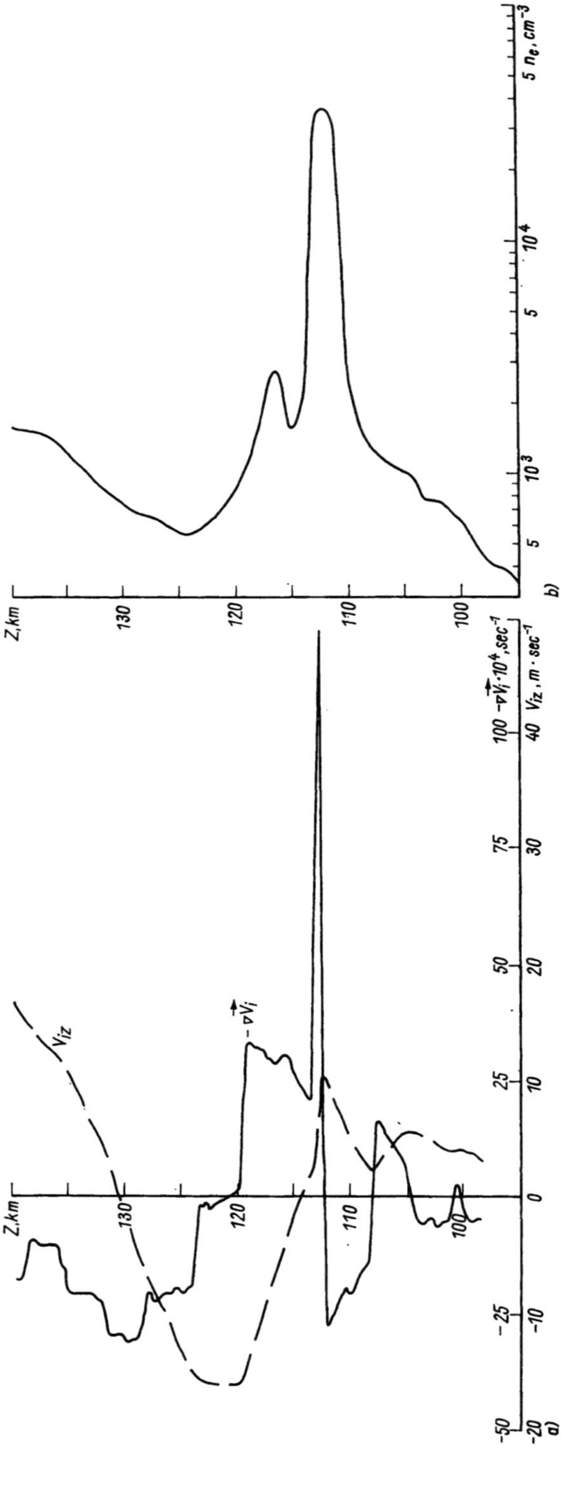

Fig. 2. Profiles of the vertical ion velocity component and the convergence of the ion velocity field (a) and electron concentration (b).

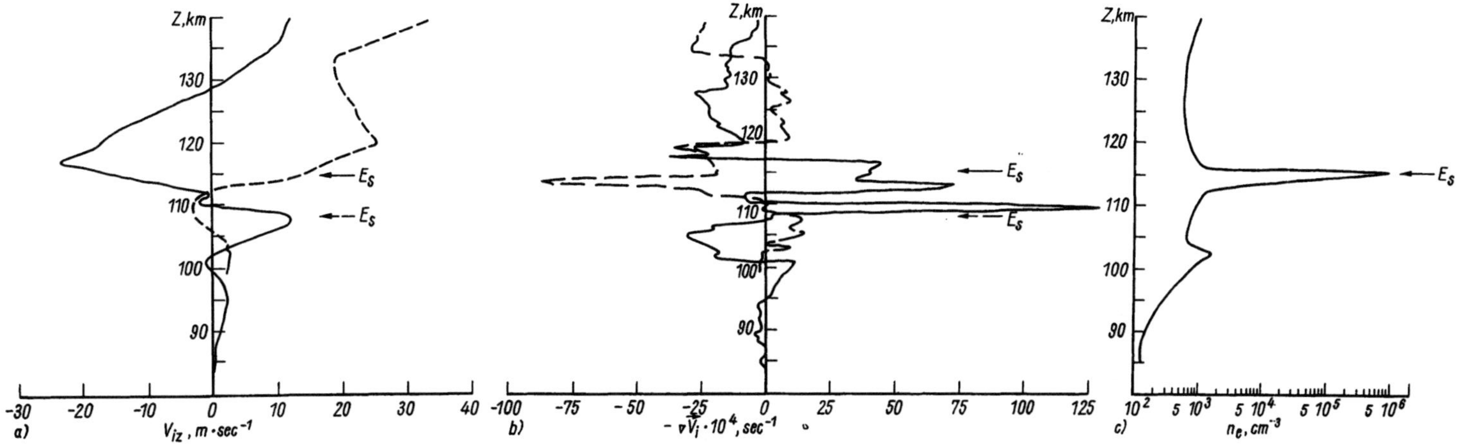

Fig. 3. Profiles of the vertical ion velocity field (a), the convergence of the ion velocity field (b) and electron concentration (c); an improved profile shows also a small maximum around 120 km height. Volgograd, 6 and 7 July 1970. ——— 6 July 1970, 2044 LMT. – – – 7 July 1970, 0240 LMT.

results obtained. Fig. 3 shows the vertical ion velocity component (a), the convergence of the ion velocity field (b) and the electron concentration profile (c) calculated from the experimental data. Here the heights of the Es layer according to the ionospheric data station are shown by arrows. For the experiment on 6 July a very definite positive convergence was obtained at the heights 95–101 km, 107–110 km and 112–117 km. The height where the conditions $V_{iz} = 0$, $dV_{iz}/dz < 0$ are fulfilled lies within the first two intervals. On the electron concentration profile two maxima at 102 and 116 km were registered.

According to the vertical sounding data at this time the Es layer was observed with virtual height 115 km and top frequency 9.3 MHz which corresponds to a n_e maximum at 116 km. As is seen from Fig. 3, the electron concentration maxima satisfactorily correlate with the regions of positive convergence if the two upper regions are treated together as part of the 107–117 km height region. Both n_e maxima also correspond within the limits of accuracy of measurement to the zero values of the east–west wind component. So on the basis of the wind-shear theory in this experiment the levels of the electron concentration maxima may be predicted quite accurately. A lower n_e maximum may be formed only at the cost of the wind-shear mechanism. As elementary calculations show, it is impossible to explain the higher maximum with a concentration one thousand times larger than the background only by the redistribution of charged particles. Such a thin layer with sharp increment of electron concentration may be formed by a local ionization source and, in particular, by a stream of high energy (of the order of 10^5 eV) electrons captured by the earth's magnetic field [9]. In this case the layer formed by the electron streams of high energy is superimposed on the maximum, formed at the cost of the wind shear, and the interpretation becomes more complicated. It is likely that such a situation is observed in a given experiment, and in order to explain the $n_e(z)$ profile obtained it is necessary to carry out detailed calculations taking into account such an ionization source.

In the experiment on 7 July (Fig. 3) positive convergence was registered at 102–109 km and 120–133 km. Within the first interval there is a level where $V_{iz} = 0$, $dV_{iz}/dz < 0$. In this experiment only vertical ionospheric soundings have been carried out, therefore a detailed comparison of the convergence and electron concentration profiles is not possible. According to the ionospheric station data during the experiment the Es layer with parameters h' Es = 108 km, f_b Es = 5.4 MHz, f_0 Es = 5.5 MHz was observed, which is in good agreement with the wind-shear theory.

According to the vertical ionospheric sounding data, which had also been carried out between the launchings, sporadic Es layer occurred during the whole time. Its virtual height, top frequency and blanketing frequency tend to decrease, with some oscillation. Such decrease agrees with the conclusions of [10].

However the absence of the wind structure information in the period during the launchings and information on the $n_e(z)$ profile for the second launching make conclusions about the character of temporal parameter variations in the Es layer impossible.

References

[1] G. P. Komrakov, V. P. Ivanov, I. V. Popkov and V. N. Tyukin, Kosm. Issled. **8**, 278 (1970).
[2] Yu. B. Burakov, V. I. Styashkin and Yu. K. Chasovitin, Proc. Inst. Exper. Meteorol. vol. **16**, Hydrometeoizdat, Moscow 1970 (p. 47).
[3] D. B. Uvarov, Proceedings of the Institute of Applied Geophysics, vol. **15**, Hydrometeoizdat, Moscow, 1969 (p. 189).
[4] W. I. Axford, J. Geophys. Res. **68**, 769 (1963).
[5] J. D. Whitehead, Space Research VII, 89 (1967).
[6] V. P. Nesterov and Yu. K. Chasovitin, Proc. Inst. Exper. Meteorol., Vol. **16**, Hydrometeoizdat, Moscow, 1970 (p. 26).
[7] M. L. MacLeod, J. Atmos. Sci. **23**, 96 (1966).
[8] L. A. Andreyeva et al., Proc. Inst. Exper. Meteorol., Vol. **24**, Hydrometeoizdat, Moscow 1971 (p. 9).
[9] G. S. Ivanov-Kholodny and V. I. Lazarev, Geomagn. i Aeron. **6**, 397 (1966).
[10] G. Chimonas and W. I. Axford, J. Geophys. Res. **73**, 111 (1968).

Space Research XII — Akademie-Verlag, Berlin 1972

WIND SHEAR AND ELECTRON DENSITY MEASUREMENTS BY ROCKET

J. NAKAMURA[a], H. KIMURA[a], T. MATSUOKA[a], T. ASO[b] and S. KATO[b]

[a] College of General Education, University of Tokyo, Tokyo, Japan
[b] Ionosphere Research Laboratory, Kyoto University, Kyoto, Japan

Simultaneous observations of electron density and wind structure in the lower ionosphere region were performed by K-8-16 rocket. Sodium trails were made to determine the wind structure at altitudes from 90 to 120 km. Both the gyro-plasma probe and the Langmuir probe were used to detect the background electron density profile, ionization increase due to chemical release and the existence of a sporadic E layer. The main results are as follows: (i) The wind was turbulent below the altitude of 100 km and had a stratified structure above 106 km. The atmosphere between was transitory. (ii) The gyro-plasma probe measured successfully the absolute value of electron density in the lower E region, particularly sporadic E. (iii) Wind shears were found at an altitude of 101 km which can well explain the sporadic E. (iv) An increased electron density was found by the gyro-plasma probe upon the release of sodium vapour.

1. Introduction

A K-8-16 rocket was launched from Kagoshima Space Centre (31°15′ N, 131°04′ E) at 1914 JST on 2 September 1970. At the launching time the angle of depression of the sun was eight degrees. The sodium trails were made by a thermite combustion type vaporizer with intermittent valve which enabled us to deduce the wind velocity and direction very precisely because of their constricted structure [1]. These trails were made at altitudes from 90 to 125 km in the ascent, and the wind structure was extensively revealed. The barium clouds and the sodium cloud were made at altitudes from 170 to 160 km in descent and served to locate the rocket precisely.

Both the gyro-plasma probe and the electron current probe served to detect the background electron density profile, ionization increase due to chemical release and the existence of sporadic E layer.

2. Wind Direction and Velocity

The corresponding points of the cloud on the photographs taken at each station were identified by those straight lines which are produced as the intersection between the photographic planes and assumed planes with reference to the observed stars including a pair of the stations. These straight lines cut the cloud on the photographs. The angular resolving power in this analysis is expected to be about twenty seconds of arc. The positions of the sodium cloud in ascent are shown in Figs. 1a and 1b.

Wind velocities and directions were calculated from the position of the cloud are shown in Figs. 2a and b. The cloud configuration shows that the atmosphere below the altitude of 100 km has a turbulent structure, above 106 km a diffusive structure, and in between a transient structure. The width of the clouds was about 500 ± 100 m below 100 km, 200 ± 100 m between 100

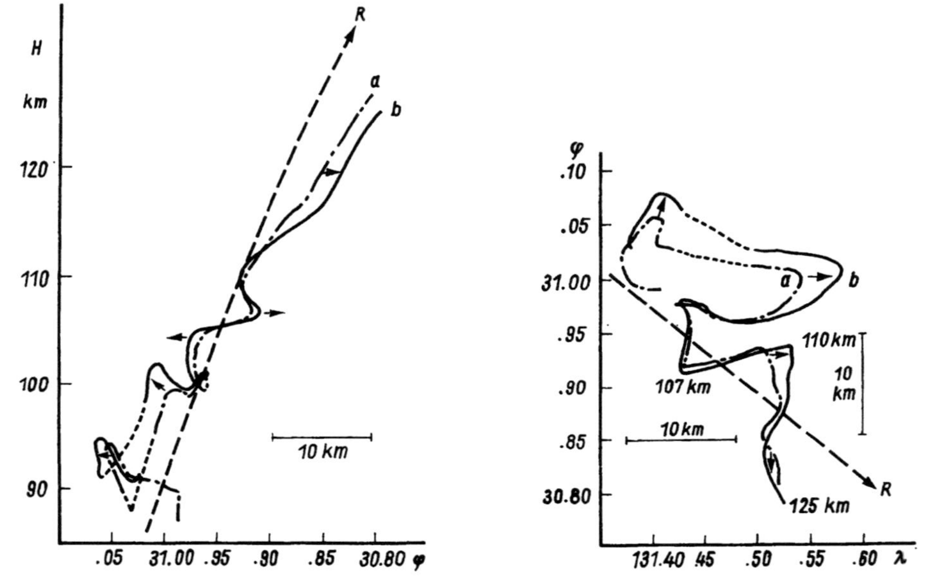

1 a 1 b

Fig. 1. Configuration of sodium cloud: a, 195–200 s; b, 225–230 s after launching. (a) Projection on meridional plane; 'R' indicates the rocket. (b) Projection on horizontal plane; φ = north latitude; λ = east longitude.

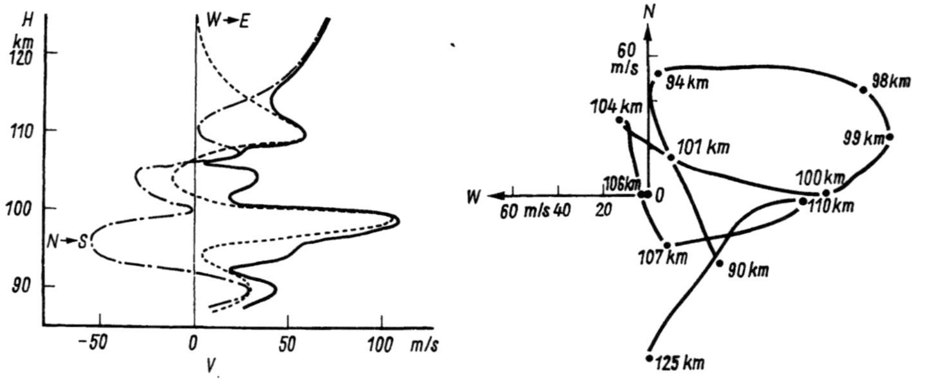

2 a 2 b

Fig. 2. Wind profile: (a) Velocity versus height; – – – W → E; — · — · N → S; ——— magnitude of total velocity. (b) Hodograph of wind velocity.

and 106 km and $1 \sim 3 \pm 0{,}5$ km above 106 km. Fig. 2a shows a vertical component of wind velocity of the order of several, or several tens, of m s^{-1} below 101 km. It seems, however, that this is a small scale motion of the atmosphere. Above 101 km, Fig. 2a shows that the wind velocity has almost no vertical component (smaller than 5 m s^{-1}). In these altitudes the upper atmosphere seems to have a layer structure and moves horizontally.

Fig. 2b shows strong wind shears at 101 km and 106 km. The change of wind velocities and wind directions per kilometre of altitude are 59 m s^{-1} km^{-1} and 180° km^{-1} at the 106 km shear. At 106 km, the wind direction changes

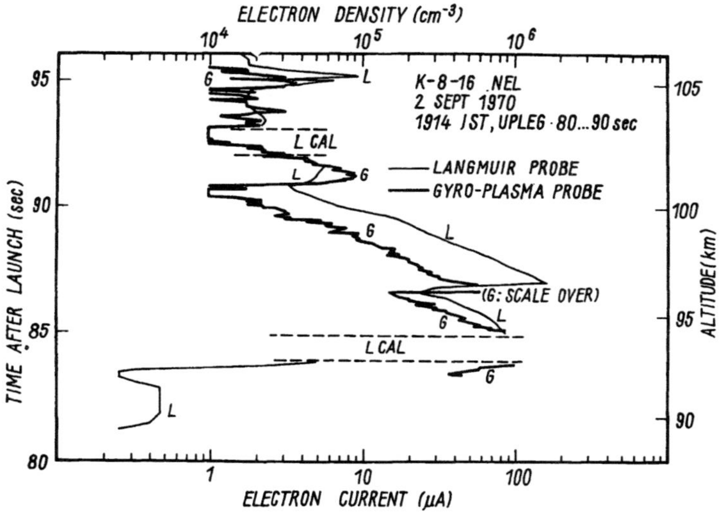

Fig. 3. Electron density increase upon chemical release. Measurements were performed both by the gyro-plasma probe and the Langmuir probe in the ascending path.

from 340° to 160° and the minimum wind velocity is only 3 m s^{-1} which makes a very sharp wind shear.

The vapour release in the descent was not successful. With the Ba cloud the ionization was not sufficient to detect the cloud successfully. The altitude of the sodium cloud release was too high to be discussed here in relation to the wind-shear theory of sporadic E.

3. Electron Density

The rocket appeared in the photographs which were taken for the purpose of the vapour cloud observation; this enabled us to locate the rocket with a high accuracy and was an unexpected advantage of the unsuccessful cloud making in the descent.

Figs. 3 and 4 show the electron density profiles determined by the gyro-plasma probe and the electron current probe. These results show that the maximum density occurs at 101.8 and 105.4 km in the ascent (Fig. 3), 105.8 and 101.4 km in the descent (Fig. 4). Drift measurements by the close-spaced antenna method at Yamagawa Observatory near the launching station gave a result consistent with ours: the altitude of the Es layer was 100 ± 5 km; the drift direction was $60° \pm 20°$ and its velocity was 40 ± 10 m s^{-1}, while from Figs. 2a and b at 101 km, the wind direction was 40°, and wind velocity 18 m s^{-1}. Electron density measurements also detected an increase of electron density by 10^6 cm^{-3} upon chemical release.

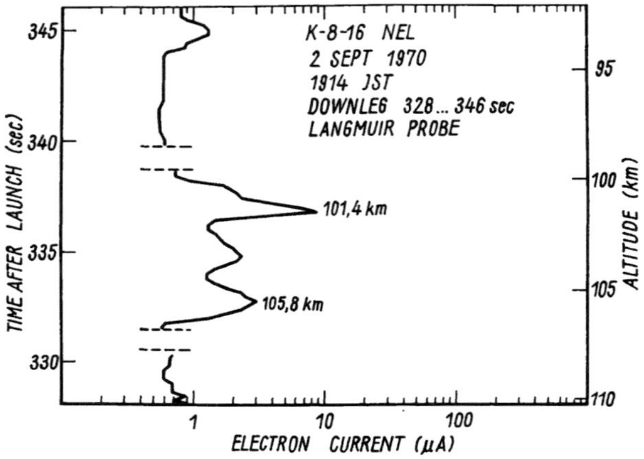

Fig. 4. Micro-structure of sporadic E layer revealed by the Langmuir probe in the descending path.

4. Discussion

The present result is similar to those of Bedinger and Knaflich in 1966 on wind [2] and of Smith in 1966 on electron density [3]. However, we were successful in measuring the electron density in the ascent by probes of two types: a Langmuir probe as in Smith's case and a gyro-plasma probe more suitable for finding the absolute value. Good coincidence of the two results above 100 km confirms correct calibration which applies equally to the measurement in descent where only the Langmuir probe was operative. The gyro-plasma probe was damaged by unexpected heating due to decreasing thermal conduction from the instrument to outside. No damage of the probe by the chemical release is expected for an ac probe such as a gyro-plasma probe. Actually we successfully detected an increased electron density upon release. A discrepancy of the measured result between the two probes below 100 km and before 92 s was due to the different location of these probes.

The relation between the observed wind profile and the electron density distribution is not as simple as expected by a simple wind-shear theory of sporadic E (e.g. [4]). The wind above and below 101 km tends to converge he ionization and this can well explain the peaks around 101 km detected both in the ascent and the descent. However, it is an open question if other smaller peaks could also be similarly explained; the large peak which appeared in the ascent below 100 km is probably due to chemical release (91 km in Fig. 3).

References

[1] J. NAKAMURA and H. KIMURA, Small Rocket Inst. Tech., North-Holland Publ. Co., Amsterdam 1969 (p. 130).
[2] J. F. BEDINGER and H. B. KNAFLICH, Radio Sci. 1, 156 (1966).
[3] L. G. SMITH, Radio Sci. 1, 178 (1966).
[4] J. D. WHITEHEAD, J. Atmos. Terr. Phys. 20, 49 (1961).

Space Research XII — Akademie-Verlag, Berlin 1972

PROFILES OF ELECTRON TEMPERATURE IN THE IONOSPHERE OBSERVED WITH ELECTRON TEMPERATURE PROBE ON A ROCKET

K. HIRAO and K. OYAMA

Institute of Space and Aeronautical Science, University of Tokyo, Tokyo, Japan

Nine electron temperature profiles have been obtained at the Kagoshima Space Centre by the electron temperature probe which is a modification of the resonance probe. These electron temperature distributions are compared with the neutral temperature distribution in CIRA 1965 in the present paper. At a height of 340 km, the electron temperature is slightly higher than the neutral temperature and the greater temperature excess seems to correlate with the geomagnetic activity. This probe can measure the electron temperature even in the bottom side of E layer, and the measured temperature agrees well with that predicted by the model atmosphere.

It is suggested that the neutral temperature in the ionosphere between 100 and 200 km should be carefully re-examined.

A high-temperature layer has been found at a height of around 110 km in measurements at around 11 o'clock.

1. Introduction

Profiles of electron temperature have been observed for a long time by rocket as well as by backscatter of radio waves from the ground. A theoretical explanation of the electron temperature profiles has also been developed by many investigators which explains successfully some typical distributions of electron temperature [1–3].

However, some discrepancies between the *in situ* probe measurements and the backscatter results have recently been discussed, and the backscatter method has been considered to give more reliable data than the probe method. This discrepancy was investigated carefully and is considered to be caused at least partially by the contamination of the probe.

The electron temperature probe, a modification of the resonance probe, is, however, considered to be fairly free from the contamination problem. In this report, the results of nine successful flights of this probe are shown to express some particular profiles of electron temperature in the ionosphere.

2. Observed Data

Nine observations were carried out under various conditions of local time, solar activity, geomagnetic *Kp* index and geomagnetic diurnal variations. The rockets were fired from the Kagoshima Space Centre, Japan, where the

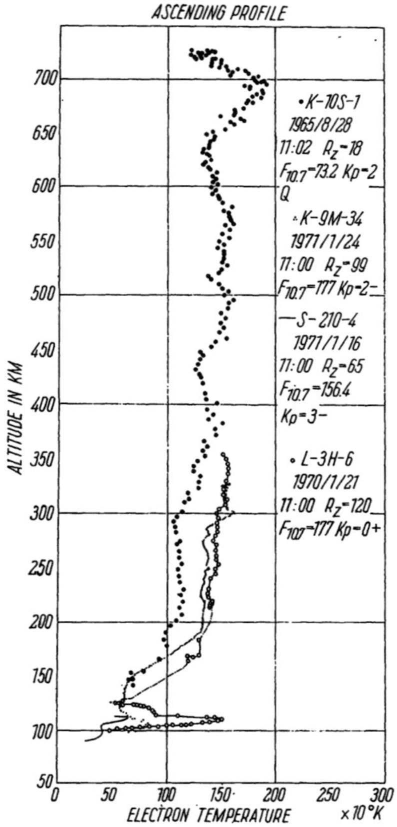

Fig. 1a

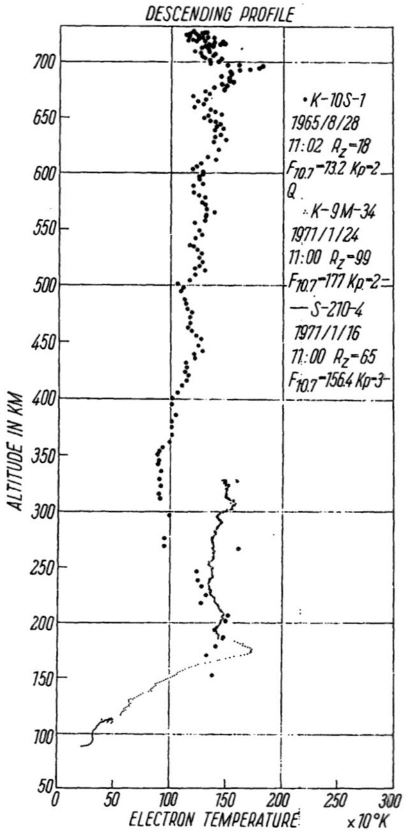

Fig. 1b

Fig. 1. Electron temperature profiles classified according to local time: a, ascending 11 h; b, descending 11 h; c, ascending 14 h; d, descending 14 h; e, ascending 16 h; f, descending 16 h; g, ascending 19 h.

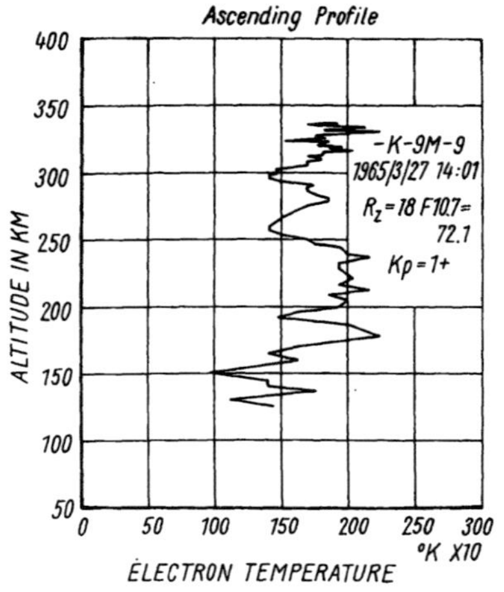

Fig. 1c

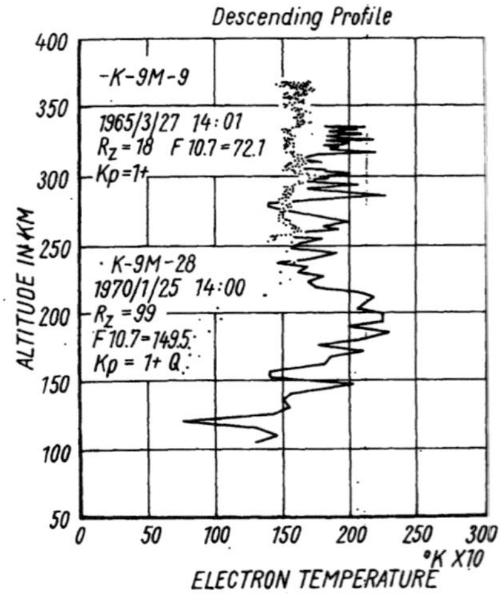

Fig. 1d

centre of the ionospheric dynamo-current for the geomagnetic diurnal variation passes overhead, deviating northwards or southwards. These conditions are shown in Table 1. Following local time, these profiles are classified into four groups, around 11 h, 14 h, 16 h, and 19 h and shown in Fig. 1a–1g.

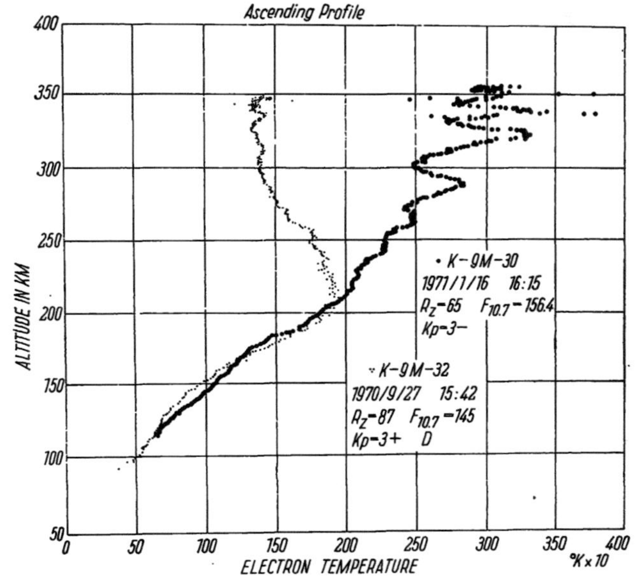

Fig. 1e

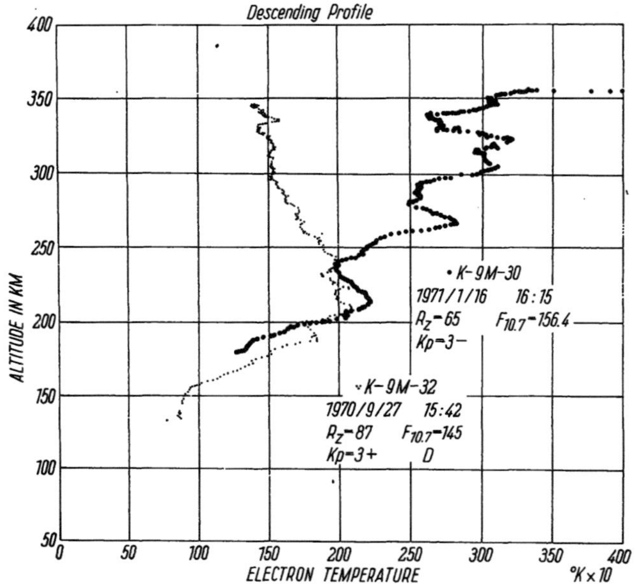

Fig. 1f

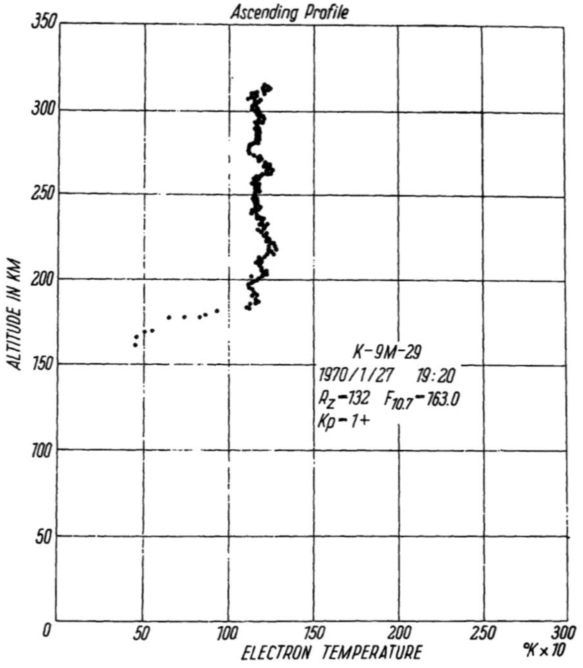

Fig. 1g

Table 1
Environmental conditions of observations

Rocket	Date	Time	Zürich R_Z	Ottawa $F_{10.7}$	Kp	Remarks
K-9M-9	27 Mar. 1965	1401	18	72.1	1+	
K-10S-1	28 Aug. 1965	1102	18	73.2	2	Q
L-3H-6	21 Jan. 1970	1100	120	177	0+	
K-9M-28	25 Jan. 1970	1400	99	149.5	1+	Q
K-9M-29	27 Jan. 1970	1920	132	163.0	1+	
K-9M-32	27 Sept. 1970	1542	87	145	3+	D
S-210-4	16 Jan. 1971	1100	65	156.4	3−	
K-9M-30	16 Jan. 1971	1615	65	156.4	3−	
K-9M-34	24 Jan. 1971	1100	99	177	2−	

3. Discussions

Of these groups of data, two were obtained on internationally quiet days and one was on an internationally disturbed day. These observed profiles of electron temperature show qualitatively reasonable shapes which are expected from theoretical estimation, except in two cases where the profiles show undulations of large amplitude.

Using these data, the electron temperatures at the height of 340 km were compared with the neutral temperatures estimated from the CIRA 1965, corresponding to the local time and solar radio flux of the observing period.

The result shows that the electron temperatures are generally higher than the neutral temperatures, and the excess temperature seems to have a positive correlation with the geomagnetic *Kp* index, except in the two cases mentioned above.

In the lower ionosphere around 100 km, especially in the case of noon observation, a high-temperature layer were always observed, with a thickness of the order of kilometres and a maximum temperature increase of about 1000 °K. The height of the temperature increase is almost coincident with that of the ionospheric dynamo-current. The correlation between the increase of temperature and the geomagnetic characteristics is not so clear, but seems to have positive correlation with geomagnetic activity.

Kato [4] has suggested the possibility of electron temperature increase due to Joule heating in the ionosphere. However, the present observed data are not satisfactorily explained by this theory.

With the S-210-4 rocket, the electron temperature could be obtained above 87 km, with a minimum temperature of about 220 °K at the bottom, which agrees quite well with the neutral temperature of the CIRA 1965 model.

The K-9M-29 rocket was fired on a geomagnetically calm day at 1920 LT which is about 100 minutes after sunset. The electron temperature profile shows a sharp decrease below the height of 180 km and the temperature is 500 °K at the height of 170 km. This electron temperoure is far lower than the neutral temperature which is expected from CIRA 1965, but it is similar to the temperature which was obtained by Blamont et al. [5] from the broadening of Na and K lines at Wallops Island. As it is expected that electron and neutral particles will have the same temperature at this local time on a calm day, the model atmosphere should be re-examined carefully in the range below 200 km.

4. Conclusion

The rocket probe is superior to the backscatter method in detecting the fine structure of the ionosphere, such as a high temperature layer in the lower ionosphere. Also it can measure the electron temperature in the low density parts of the ionosphere such as the night-time E layer or the ionosphere below 100 km.

In conclusion, the electron temperature profile should be more systematically examined by rocket probe especially in the ionosphere below 200 km in order to obtain a comprehensive knowledge of the electron temperature distribution combined with the data obtained by the backscatter method.

References

[1] W. B. Hanson and F. S. Johnson, Mem. Roy. Soc. Liège **4**, 390 (1961).
[2] J. E. Geisler and S. A. Bowhill, J. Atmos. Terr. Phys. **27**, 457 (1965).
[3] A. F. Nagy, E. C. Fontheim, R. S. Stolorski and A. E. Beutler, J. Geophys. Res. **74**, 4667 (1969).
[4] S. Kato, Planet. Space Sci. **11**, 1297 (1963).
[5] J. E. Blamont and M. L. Lory, Nature, Lond. **201**, 593 (1964).

Polar Ionosphere and Precipitation of Low Energy Charged Particles

Space Research XII – Akademie-Verlag, Berlin 1972

THE MORPHOLOGY OF AURORAL PARTICLE PRECIPITATION

W. J. Heikkila

The University of Texas at Dallas, Dallas, Texas, USA

The concepts of the auroral zone and the auroral oval are discussed in the light of satellite observations of auroral particle precipitation. Significant soft electron fluxes are always present around the polar cap. The night-time precipitation patterns reflect the great variability evident in the visual aurora. Highly structured discrete forms are often seen on the high latitude side, with soft electron fluxes at times suggestive of plasma sheet electrons. Narrow stable auroral arcs are due to precipitation of harder electrons, with a peak in the keV range. On the low latitude side there is a broad drizzle region in which trapped electrons are precipitated, primarily in the dawn hemisphere. The day side is characterized by the penetration of magnetosheath plasma to low altitudes in the regions of the so-called neutral points. The term *cleft* is proposed for this magnetospheric feature. Narrow regions of enhanced electron fluxes within this region are not uncommon. Proton fluxes in general tend to occur in coincidence with electron fluxes, but show less structure. In view of these many features the concepts of auroral oval and auroral zone are too simple-minded, and therefore should be modified and elaborated.

1. Auroral Zone and Auroral Oval Concepts

The earliest detailed evidence for auroral particle precipitation came from auroral observations made at the beginning of this century by Störmer in Norway [1, 2]. The auroral zone was defined statistically as the region of most frequent occurrence of auroras; it is a nearly circular ring centered on the geomagnetic pole. During geomagnetically quiet times the auroral zone maximum occurs at about 20° co-latitude, but during disturbances it moves toward lower latitudes.

An extensive body of information on auroral activity has been collected by means of all-sky cameras, particularly since the beginning of the IGY, 1957 to 1958. The locus of the visual auroras at any one instant of time was recognized by Akasofu [3], and also by Feldstein and Starkov [4] whose result is shown in Fig. 1. The visual aurora suggested an oval region of precipitation fixed with respect to the sun, and circumscribing the geomagnetic pole but tipped back toward the night side by about 3°. The locus of the night-time part of the oval, which is the location of the brightest auroral displays, traces out the classical auroral zone as the earth rotates daily on its axis. As the level of activity increases the oval expands both in width and in diameter.

It must be recognized that the zone and oval are concepts which come from a particular type of observation, namely photographic (or visual). Another type of observation, with riometers measuring ionospheric radio wave absorption, provided Hartz and Brice [5, 6] with quite a different pattern (Fig. 2). 'Discrete' absorption events traced out a region resembling the oval, whereas steady or 'diffuse' events traced out what seemed to be the auroral zone at constant geomagnetic latitude.

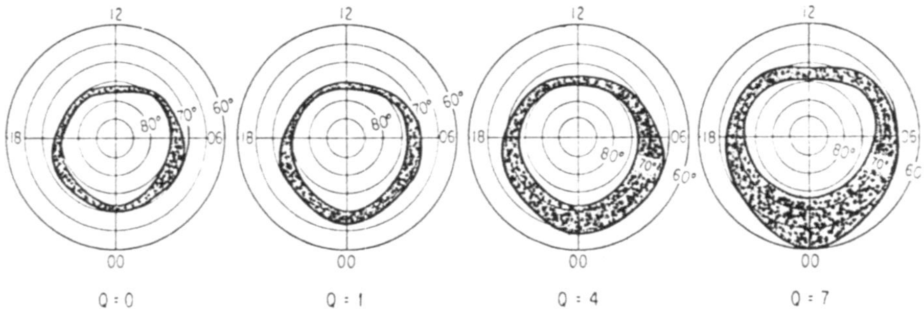

Fig. 1. Auroral oval as a function of geomagnetic activity according to Feldstein and Starkov [4].

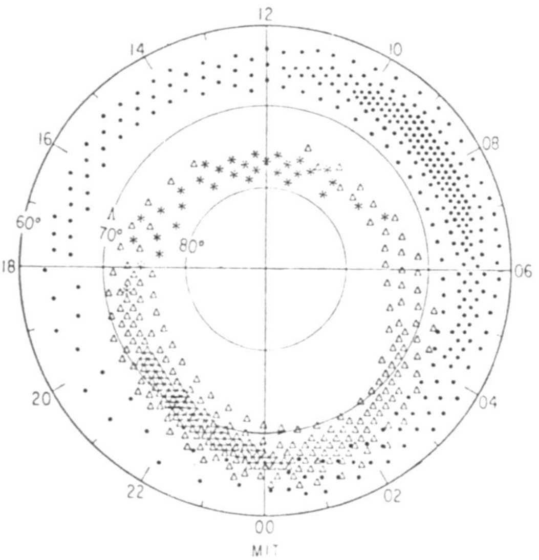

Fig. 2. Patterns of particle precipitation deduced by Hartz and Brice from various observations. Dots show the steady drizzle of more energetic electrons, triangles the softer discrete features, and stars the magnetosheath plasma.

Clearly the differences in the particle precipitation patterns deduced from these various studies are due to observational limitations. Each technique is selective, and emphasizes some feature at the expense of others. For example, the all-sky camera is not particularly sensitive, especially toward the ends of the optical spectrum, and weak auroras may not always be detected with it. Discrete auroras are emphasized, while broad structureless emission can go unrecognized especially in the presence of airglow or twilight. The riometer, on the other hand, is more sensitive to energetic particles above 10 keV than it is to very soft particles. Other techniques, such as ionospheric and geomagnetic, also have their limitations. Perhaps the best indirect observations are the sensitive photometric studies of narrow band or line emissions. These have been reviewed and interpreted by Eather and Mende [7] in the light of available particle observations. However, the final answer on particle precipitation patterns must come from direct observations, especially by satellite instruments. Some observations have already been reported, and these have been reviewed recently by Paulikas [8]. Those earlier particle observations need not be reviewed here again. I will instead present some new data which support, illustrate and extend the deductions based on those observations.

2. Soft Particle Spectrometer Data

These new data were obtained with the Soft Particle Spectrometer (SPS) on the ISIS 1 satellite. The instrument and data presentation have been described by Heikkila et al. [9], and Bunting et al. [10] have provided new data on the detector calibration.

Fig. 3 shows electron and proton data for a complete transpolar pass of the ISIS 1 satellite on 26 July 1969, during moderately quiet conditions ($Kp = 1-$). The record covers both night-time and daytime auroral zones, from left to right. In the spectrograms, in which the counting rate is used for intensity modulation of a presentation of particle energy variation with time, the logarithm of the particle energy is plotted vertically from 10 eV to 10 keV. About eighteen minutes of the pass is shown, beginning over western Canada at geographic latitude 59° N and longitude 132° W, and ending over Europe at 70° N and 37° E. The invariant latitude Λ, given by $\cos^2 \Lambda = 1/L$, is shown at the top as a rough indication of geomagnetic location, but of course that may not be very meaningful near the pole. The spectrogram presentation, with its increasingly dark shades of grey showing higher counting rates, is extremely useful for a qualitative understanding of the data. More quantitative data are plotted below each spectrogram, in the form of number and energy fluxes over the 3-decade range of particle energy.

The spectrograms show spin-modulated fluxes with complex energy and time variations. The spin axis of the satellite at this time was pointed near the north celestial pole (27° co-declination and 100° right ascension) providing a good pitch angle scan at the beginning of the pass, and a somewhat reduced scan at the end. The spin period was 20.4 seconds. This was a pass over the summer pole, and atmospheric photoelectrons escaping from below the satellite [11] were detected throughout most of the pass. The photoelectron spectrum is easy to recognize because of the rather abrupt cut-off above about 60 eV, except where soft auroral or secondary electrons are also present.

As is well known from visual observations the night-time auroral phenomenon is quite variable in its intensity, structure, location, etc., even during undisturbed periods. The direct particle observations reflect this variability. In

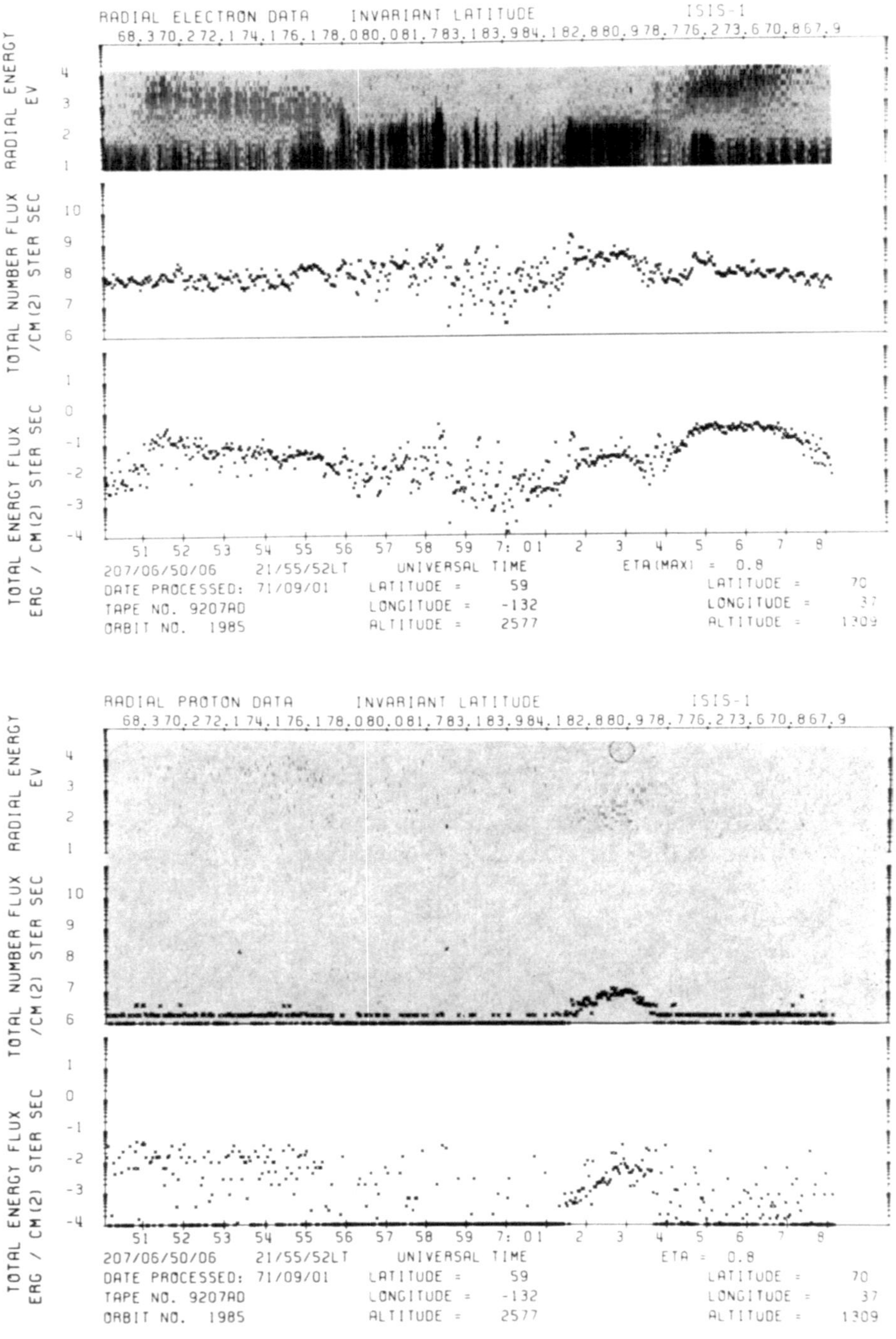

Fig. 3. Data from the soft particle spectrometer on ISIS 1, 26 July 1969 ($Kp = 1-$). The ordinates are $\log_{10}$ of the quantities listed.

Fig. 3 the auroral precipitation is weak, only a few tenths of 1 erg cm^{-2} sr^{-1} s^{-1}. The flux in the region of "discrete" events [5] in this case extended from 78° to over 83° invariant latitude near the pole. It was very soft, a few hundred eV, suggestive of plasma sheet electrons [12]. Fig. 4 shows (on the right-hand side) a narrow night-time antarctic auroral region on 22 June 1969 which was a very quiet day with $\Sigma Kp = 4-$. The electron flux was much harder, in the keV range, and energy deposition reached 1 erg cm^{-2} sr^{-1} s^{-1}. No auroral precipitation was present over the polar cap. A similar form is shown in Fig. 5,

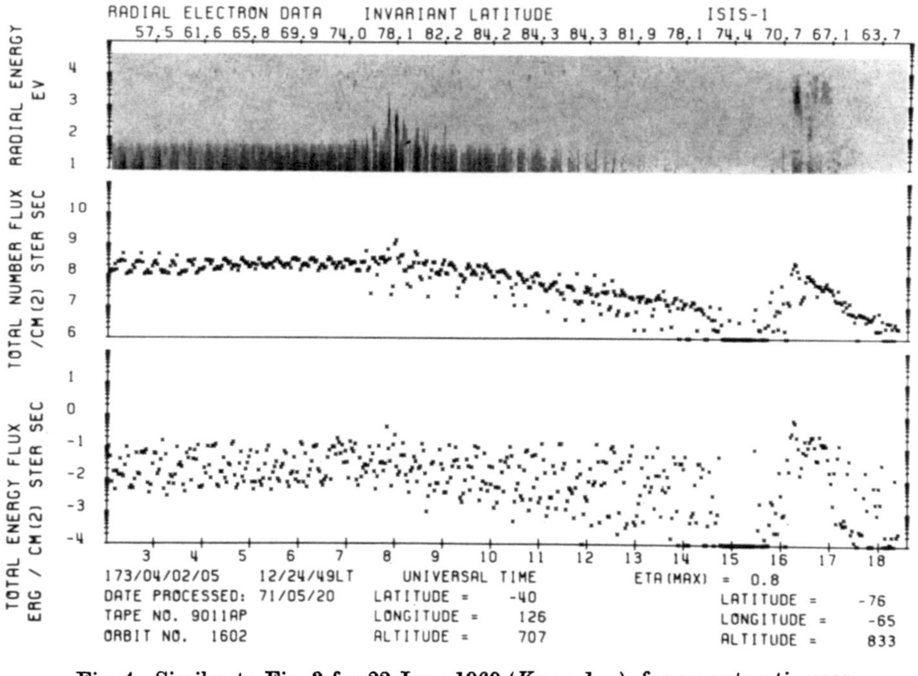

Fig. 4. Similar to Fig. 3 for 22 June 1969 ($Kp = 1-$), for an antarctic pass.

together with an all-sky photograph obtained nearby at Great Whale River. In this case conjugate point photoelectrons were seen immediately to the south of the aurora, demonstrating clearly that field lines just below the auroral form were closed, and "well behaved".

A final example of hard auroral electron precipitation is given in Fig. 6, again in the Antarctic during the quiet period in June 1969. The high latitude boundary on the night side at 75° was extremely sharp; there was no precipitation over the polar cap. The energy flux reached 10 ergs cm^{-2} sr^{-1} s^{-1}, enough to produce an IBC II aurora. A region of hard electron precipitation extended down to $\Lambda \simeq 65°$.

In general we can say that the observed electron precipitation at night follows the patterns deduced from auroral observations. Their detailed understanding, however, requires knowledge of the state of substorm activity. That is an extremely complex subject which is outside the scope of this paper.

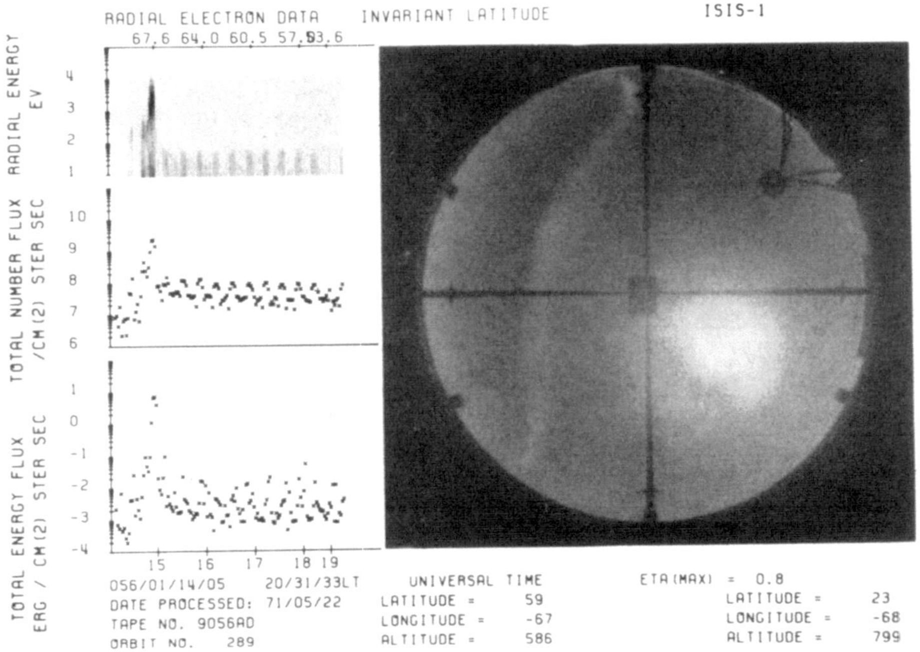

Fig. 5. Similar to Fig. 3 for 25 Feb. 1969 ($Kp = 3$), together with an all-sky photograph taken at Great Whale River showing a quiet arc. An auroral break-up occurred 20 minutes later. The moon is at the lower right hand side.

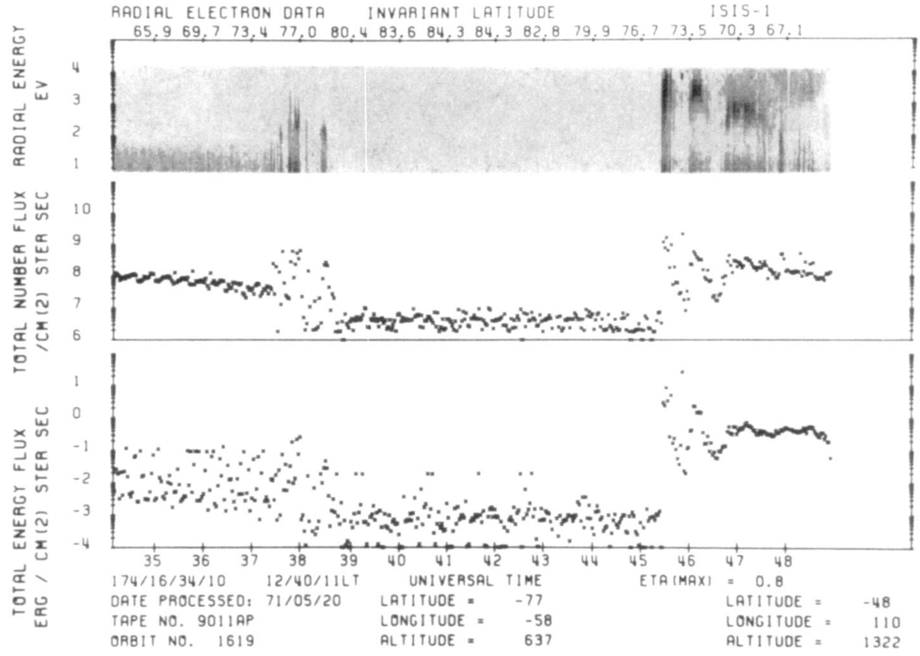

Fig. 6. Similar to Fig. 3 for 23 June 1969 ($Kp = 2-$), early in the growth phase of a substorm.

The more energetic electrons in the keV range and above, seen in Fig. 3 on both the night and day sides, are on closed field lines. Such trapped particles have been studied extensively at higher energies ($E \geqq 40$ keV) [13–15]. Well below the auroral zone most of these particles are trapped, showing two maxima per spin at $\alpha = 90°$ [16]. However, the distribution becomes isotropic over the lower hemisphere (i.e. downgoing particles) at auroral latitudes, and the spectrum softens; this isotropy is evident in Fig. 3. Closer inspection reveals a narrow loss cone for upward going particles, especially on the night side near 0653 UT. These energetic particles are responsible for the steady "drizzle" discussed earlier. These keV electrons produce the mantle aurora [17] and the steady ionospheric radio wave absorption [5]. The energetic electron precipitation on the low-latitude side is missing on the afternoon passes in Figs. 4, 6 and 7. Precipitation may be enhanced in the dawn hemisphere for several reasons. The injection process from the tail occurs near midnight, and electrons and low energy protons are convected toward the dawn side by the magnetospheric electric field [18]. Precipitation is enhanced beyond the terminator because of the build-up in ionospheric electron density which leads to wave-particle interactions and strong pitch angle diffusion [19]. The remaining pseudo-trapped electrons are lost through the magnetopause near the noon meridian [15] due to their normal drift motion, or they may be swept out by enhanced magnetospheric electric fields in the normal storm-time convection [20].

The high latitude boundary of the energetic particle precipitation can be quite sharp, sometimes only a fraction of a degree in extent. More often some structure is present, as in Fig. 3 where a large change in flux occurs at 77°, but some keV electrons precipitate up to 79°. This pattern may be associated with magnetopause structure such as might be caused by surface waves, or by interchange instabilities. The best possibilities for sorting out the particle drift and precipitation patterns would seem to be offered by the occasions when energetic solar particles precipitate over the polar cap.

Above the limit of closed field lines on the day side a strong flux of rather soft electrons and protons is observed on every pass, almost without exception, as in Figs. 3, 4, 6 and 7. In the latter case the SPS was operated in the all-swept mode, with four times higher sampling rate, again on a moderately quiet day ($Kp = 2$). Higher resolution, combined with a good pitch angle scan, brings out the isotropy of the flux; however, there is a well developed loss cone. This flux has been identified as magnetosheath plasma penetrating down to low altitudes through the so-called neutral points in the dayside magnetopause [21, 22, 23]. The geomagnetic field here shows a division between closed mid-latitude lines and open polar cap lines [24]. In cross section the geometry is that of a cusp, referred to as the dayside cusp by Heikkila and Winningham [22], and the polar cusp by Frank [23]. The term *cleft* proposed in Fig. 8 is perhaps more descriptive; according to Webster's *International Dictionary* a cleft is "a V shaped indentation as though made with a splitting wedge". Here the wedge of magnetosheath plasma forces the magnetic field lines apart. Physically this splitting is of course produced by magnetopause surface currents,

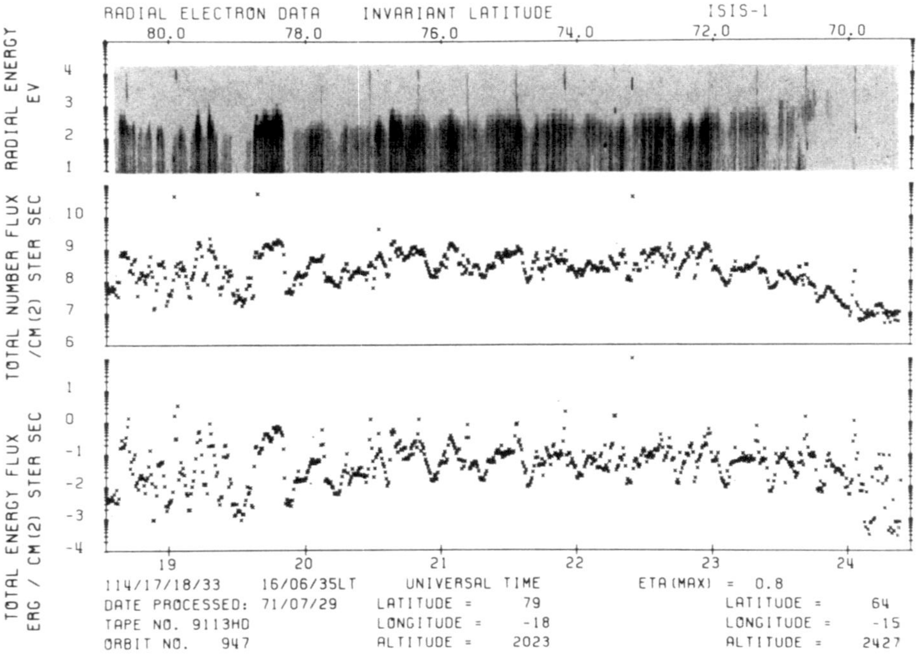

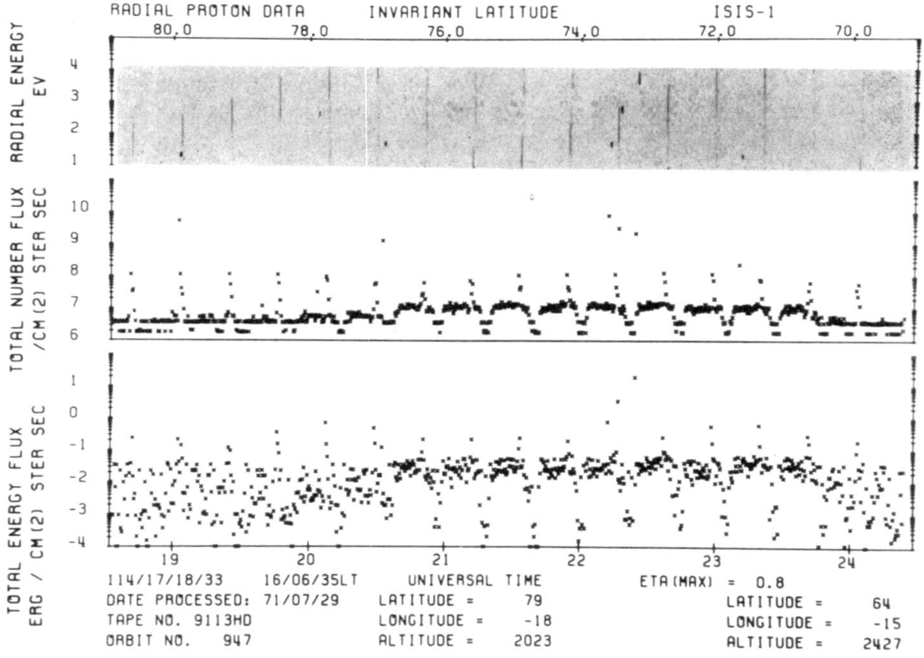

Fig. 7. Similar to Fig. 3 for 24 April 1969 ($Kp = 2$), showing the magnetosheath plasma with four times greater detail.

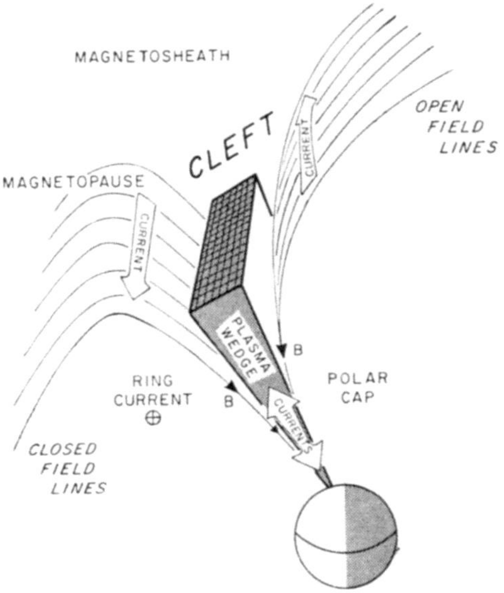
MAGNETOSHEATH
CLEFT
OPEN
FIELD
LINES
MAGNETOPAUSE
CURRENT
CURRENT
PLASMA
WEDGE
B
RING
CURRENT
B
POLAR
CAP
CURRENT
CLOSED
FIELD
LINES

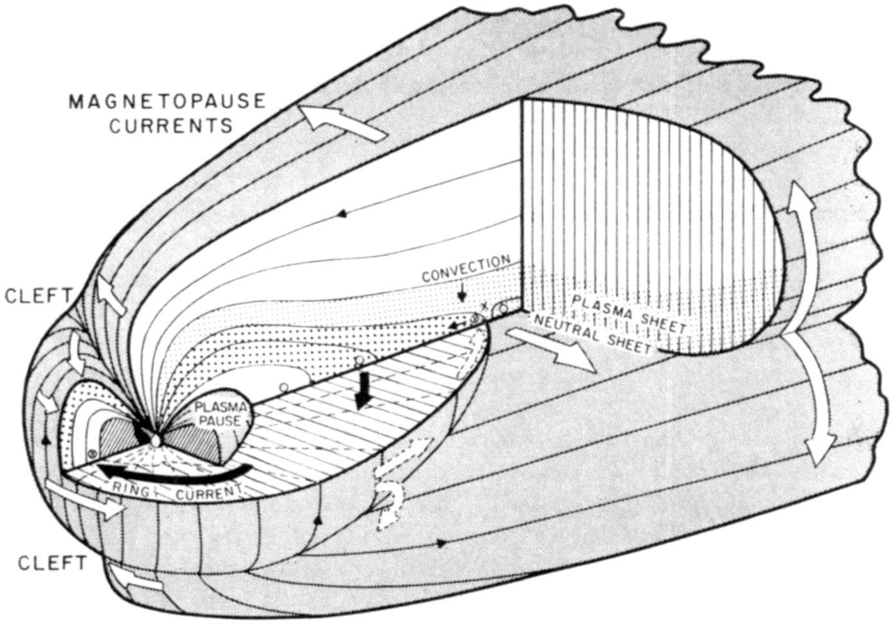
MAGNETOPAUSE
CURRENTS
CONVECTION
CLEFT
PLASMA SHEET
NEUTRAL SHEET
PLASMA
PAUSE
RING CURRENT
CLEFT

or internal magnetospheric currents, directed eastward on the low latitude side, and westward on the high latitude side, as shown in Fig. 8. Note that the low latitude current flows in the opposite direction to the magnetospheric ring current, while the high latitude current is in the same sense as the return for the magnetotail sheet current (see Fig. 9). Field aligned currents, as indicated in Fig. 8 and 9, are also certain to play a major role in the penetration process.

The soft flux extends rather uniformly from 79.5° to 83° in Fig. 3, and from 71° to 77° in Fig. 7. The cleft region may be narrower, as in Figs. 4 and 6, and often shows fine structure. This large extent is difficult to explain on the basis of mapping down any reasonably sized area on the magnetopause. Frank [23, 25] has suggested that the region of magnetosheath plasma penetration extends only 20–30 km in the north–south direction, and that the electron and proton fluxes are displaced in latitude. Fig. 3 shows clearly that this is not the case. The soft ($\simeq$300 eV) proton flux co-exists with the electron flux over the same 3.5° latitude range, about 400 km in extent. Fig. 7 shows a still broader region. This large extent is fully supported by auroral observations [26]. This discrepancy may perhaps be explained by noting the narrow intense electron burst recorded at 0701.40 in Fig. 3, which is like the cusp feature reported by Frank and Ackerson [25]. In this case the burst occurs on the high latitude side of the cleft, whereas Frank reported that it occurs on the low latitude side. In fact, we see such bursts on many passes, and they can occur anywhere within the cleft and even outside it. The protons show no accompanying feature. One remarkable pass reported by Winningham [16] shows three bursts in one otherwise uniform extended region.

It is not unusual for the soft electron flux to extend well into the polar cap region, as in Fig. 3 and 7. However, the proton flux is always more restricted and it usually provides the best identification of the cleft. Auroral photometric observations support this conclusion [26].

Polar cap electron fluxes are generally rather soft and weak, but are by no means uncommon. Two examples are shown in Figs. 3 and 7 (above the region of cleft precipitation, as identified by the proton flux). Interesting variations related to geomagnetic storm activity are often observed. Soft proton precipitation over the polar cap is extremely rare. The subject is new, and will be reported in more detail elsewhere [27].

Positive particle (mainly proton) fluxes generally coexist with electron fluxes throughout the auroral regions, although the proton flux is generally much weaker. The energy flux plot in Fig. 3 perhaps shows the proton precipitation more clearly than the spectrogram itself. The scatter of points is due to the rather short sampling time (11 ms) in the swept mode of instrument operation; better accuracy is obtained in the stepped mode (0.5 s steps) which is normally interleaved with the swept mode; results will be reported elsewhere. This spectrogram does show that the proton spectrum is harder than the electron spectrum below auroral latitudes; in fact we see just the low energy end of the spectrum especially on the dusk side. In the post-midnight sector we often see lower energy protons (several keV) especially after substorms. These proton fluxes below the intense auroral region undoubtedly originate in the ring current, which is formed by particle injection from the magnetotail [18]. This is a distinctly different phenomenon from the soft daytime proton flux from the magnetosheath, penetrating through the cleft.

3. Conclusions

It is time to recognize the many distinctive features in the morphology of auroral particle precipitation. Simplistic concepts such as the auroral zone and the auroral oval are of limited further use, although they have been a great aid in ordering the observations in the past. Elaborations are necessary to take account of the many magnetospheric regions to which the auroras are connected, and the many physical processes which are in operation.

Magnetospheric structure is depicted in Fig. 9. Solar wind plasma, after crossing the bow shock (not shown) flows in the magnetosheath surrounding the magnetosphere [28]. Magnetopause boundary currents driven by the plasma flow, and internal field-aligned and ring currents are indicated by the arrows [29].

Merging of interplanetary and geomagnetic field lines takes place on the front [30], permitting the access of magnetosheath plasma on to the field lines in the cleft. This penetration through the cleft must be a very complicated plasma process; many features such as latitudinal width, rather sharp upper and lower boundaries, current flow and electric field patterns, and narrow electron bursts or jets, remain unexplained. However there is no longer any question as to the reality of the penetration, and of its role in producing the daytime auroras as observed on all-sky photographs.

The extent of the cleft toward the night side is not yet clear. Limited extent is implied by the direct observations of soft particles at low altitudes [16, 21] and by a number of indirect lines of evidence [22]. However, the suggestion has been made that a continuation of the cleft at high altitudes provides a convection path whereby the near-earth plasma sheet is replenished [31, 32]. Otherwise, the plasma sheet would have to be fed via the neutral sheet between the lobes of the magnetotail, or by convection across the polar caps, unless of course a terrestrial origin is considered.

Night-time auroras are generally produced by plasma sheet particles accelerated in the magnetotail [29]. This acceleration involves both longitudinal (Fermi) and transverse (betatron) processes on closed field lines, the energy coming from the general magnetospheric circulation driven by the solar wind [30]. At times the softness of the auroral particle flux, especially at high latitudes, suggests that plasma sheet particles may be precipitated without significant acceleration [33].

Following injection from the tail the particles are convected to different local times and latitudes by magnetospheric electric fields, and by curvature and gradient drifts. These processes depend on particle energy, sign and pitch angle, and are therefore rather complicated. Plasma instabilities involved in the magnetospheric substorm process, and in the penetration process into the ionosphere, produce additional complexities in the precipitation pattern.

The convection carries the particles around to the day side where they produce the broad diffuse auroral phenomena on the low latitude side of the clefts. Such a broad glow would not be readily identifiable on all-sky photographs, especially since this region is generally in twilight if not daylight. Electron precipitation here is strongest before noon, but energetic proton precipitation from the asymmetric ring current may be strongest on the evening side.

In summary, it seems clear that auroral particles have at least two distinct sources. One source is the magnetosheath plasma on the day side penetrating down to low altitudes through the cleft. The other source is the magnetotail, from which particles are injected back toward the earth and energized in the process. There is no clear reason at this time to believe that the two regions are simply connected. The polar caps are connected to the high latitude magnetotail, where the particle fluxes are generally small. Polar cap electron fluxes are by no means uncommon, but their source is a mystery at the present time.

References

[1] C. Störmer, The Polar Aurora, Clarendon Press, Oxford 1955.
[2] J. W. Chamberlain, Physics of the Aurora and Airglow, Academic Press, New York 1961.
[3] S. I. Akasofu, Polar and Magnetospheric Substances, D. Reidel Publ. Co., Dordrecht, 1968.
[4] Ya. I. Feldstein and G. V. Starkov, Planet. Space Sci. **15**, 209 (1967).
[5] T. R. Hartz and N. M. Brice, Planet. Space Sci. **15**, 301 (1967).
[6] T. R. Hartz, in: The Radiating Atmosphere, D. Reidel Publ. Co., Dordrecht 1971 (p. 225).
[7] R. H. Eather and S. B. Mende, NATO Advanced Study Institute, Skeikampen, Norway, 1971, to be published in the proceedings.
[8] G. A. Paulikas, Rev. Geophys. Space Phys. **3**, 659 (1971).
[9] W. J. Heikkila et al., Rev. Scient. Instrum. **41**, 1393 (1970).
[10] W. D. Bunting, Jr., J. Tarstrup and W. J. Heikkila, J. Appl. Phys., April (1972).
[11] W. J. Heikkila, J. Geophys. Res. **75**, 4877 (1970).
[12] E. W. Hones, Jr. et al., J. Geophys. Res. **76**, 63 (1971).
[13] J. B. McDiarmid and J. R. Burrows, Canad. J. Phys. **46**, 49 (1968).
[14] D. J. Williams, in: Astrophysics and Space Science Library, Vol. **29**: Solar Terrestrial Physics 1970 (Proc. IUCSTP Symp., Leningrad), D. Reidel Publ. Co., Dordrecht 1972.
[15] J. G. Roederer, Rev. Geophys. **7**, 77 (1969).
[16] J. D. Winningham, to be published in Earth's Magnetospheric Processes, D. Reidel Publ. Co., Dordrecht 1972.
[17] B. P. Sandford, J. Atmos. Terr. Phys. **30**, 1921 (1968).
[18] S. E. DeForest and C. E. McIlwain, J. Geophys. Res. **76**, 3587 (1971).
[19] N. M. Brice and C. Lucas, J. Geophys. Res. **76**, 900 (1971).
[20] N. M. Brice, J. Geophys. Res. **72**, 5193 (1967).
[21] J. David Winningham, Ph. D. Thesis, Texas A & M University 1970.
[22] W. J. Heikkila and J. D. Winningham, J. Geophys. Res. **76**, 883 (1971).
[23] L. A. Frank, J. Geophys. Res. **76**, 5205 (1971).

[24] D. H. Fairfield and N. F. Ness, J. Geophys. Res. **77**, 611 (1972).
[25] L. A. Frank and K. L. Ackerson, J. Geophys. Res. **76**, 3612 (1971).
[26] W. J. Heikkila et al., submitted to J. Geophys. Res. (1972).
[27] Presented at XV General Assembly of IUGG, Moscow, 1971.
[28] J. R. Spreiter and A. J. Alksne, Rev. Geophys. **7**, 11 (1969).
[29] J. R. Sharber and W. J. Heikkila, submitted to J. Geophys. Res. (1972).
[30] W. I. Axford, Rev. Geophys. **7**, 421 (1969).
[31] L. A. Frank, J. Geophys. Res. **75**, 1263 (1970).
[32] T. W. Hill and A. J. Dessler, submitted to J. Geophys. Res. (1972).
[33] V. M. Vasyliunas, The Polar Ionosphere and Magnetospheric Processes, Gordon and Breach, London 1970.

Space Research XII — Akademie-Verlag, Berlin 1972

ELECTRON DENSITY AND TEMPERATURE PROFILES DURING A MODERATE AURORAL EVENT

J. J. BERTHELIER and R. GODARD

Groupe de Recherches Ionosphériques, St. Maur, France

During a period of moderate auroral activity on 17 April 1969 a Dragon rocket carrying electric field experiments and plasma probes was launched from Andenes (Norway) in the northern auroral zone. Results concerning the electron density and temperature profiles are presented; the profiles show the development and motion of field-aligned irregularities in the F region with anticorrelated variations in density and temperature. The apparent motion of these irregularities corresponds fairly well to a transport of plasma perpendicular to the field lines driven by a perpendicular electric field. Also shown are data concerning two barium clouds released during the descent of the rocket.

1. Introduction

It has become clear in the past few years that one of the major problems to be studied in the field of auroral and magnetospheric physics is the accurate determination of the electric fields in the earth's environment [1]. The effects of these electric fields involve many important aspects of the general problem of the aurora such as the circulation of electric currents, including purely ionospheric currents as well as field aligned currents, the transport of plasma across or along the magnetic field lines, the creation of irregularities due to instabilities in the ionospheric plasma, the heating of the upper atmosphere, etc. In order to obtain precise data on the auroral electric fields and their effects on the ionosphere a cooperative experiment was designed to measure simultaneously on the same rocket dc and low frequency ac electric fields and electron density and temperature. These experiments were respectively made by Laboratoire de Physique Stellaire et Planétaire (dc and ac electric field probes), Max-Planck-Institut für Extraterrestrische Physik (dc electric fields measured by barium clouds technique) and Groupe de Recherches Ionosphériques (plasma probes). The final results from the two first experiments are at present being obtained and this paper only concerns the electron density and temperature measurements.

2. Instrumentation

2.1. Description of the Payload

The experiment was flown on board a two-stage solid propellant Dragon 2 sounding rocket which reached a maximum altitude of approximately 350 km.

A schematic view of the payload is given in Fig. 1. The electric field experiment consisted of two dipoles perpendicular to each other, the sensors being spherical electrodes mounted in the extremity of symmetrical insulated booms 1.50 m long for the upper pair and 2.10 m for the lower pair. The working of such a system and its ability to measure dc and ac electric fields has been previously discussed in detail [2, 3]. The barium canisters were fixed on the two halves of the split nose cone which was jettisoned during the ascent around 70 km

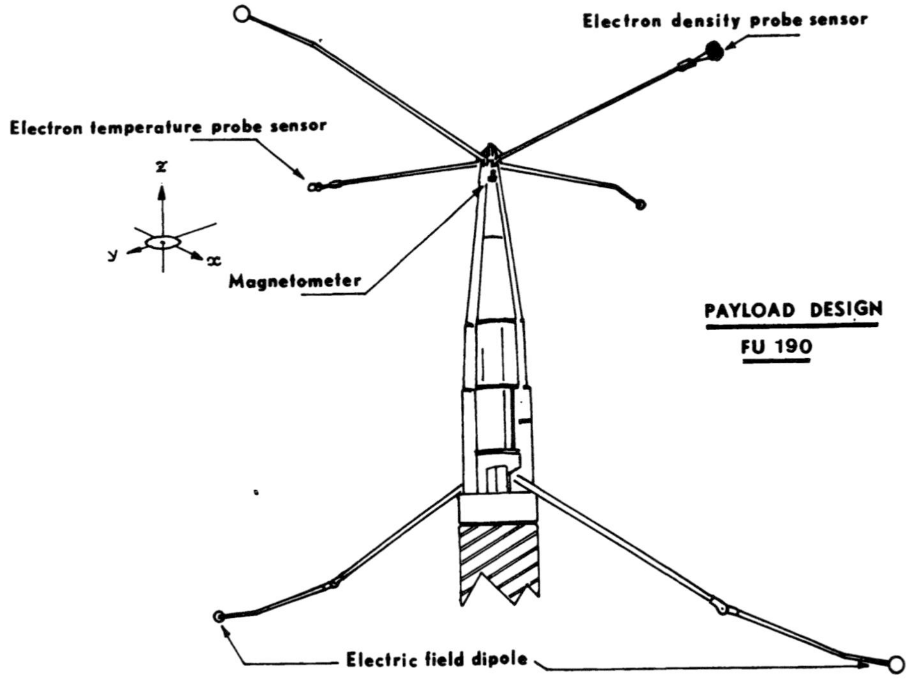

Fig. 1. Schematic view of the payload.

and followed closely the trajectory of the payload itself within 2 or 3 km. The ignitions took place during descent at about 340 and 260 km respectively. The electron density and temperature probes were mounted at the end of two opposite booms, 1.40 m in length, fixed at the top of the payload perpendicularly to the upper electric field dipole and at the same level. The attitude of the rocket in a frame of reference linked to the earth's geomagnetic field was obtained by use of a three-axis magnetometer and a sun sensor. The average accuracy of the restitution of the different Euler angles was typically $\pm 2°$.

2.2. Electron Density Probe

The electron density probe is a radio-frequency capacitance probe made up of two plane parallel grids and, as mentioned above, is positioned at the end of a boom, well away from the ion sheath surrounding the rocket body. A detailed analysis of the probe working has already been published [4]. Briefly the capacitance between the grids is measured using a high frequency of 39 MHz much greater than both the electron collision frequency and gyrofrequency above 70 km so that the change in capacitance from vacuum to plasma conditions is simply proportional to the electron density and the electronics associated with this probe delivers at its output a signal proportional to the electron density.

2.3. Electron Temperature Probe

The electron temperature probe is basically a double Langmuir probe which enables one to measure the slope of the (current, voltage) characteristics in the region where the current is an exponential function of the voltage. The probe sensor consists of two spheres which are biased by means of the sawtooth voltage from −2 volts to +3 volts with respect to the rocket body thus assuring that they are driven beyond space potential. A 6 kHz modulation is also applied to these electrodes and a feedback electronic network automatically adjusts a potential difference ΔV between them in order that the ratio of ac currents is maintained equal to 2:1. It is easy to show that, in these conditions, ΔV is proportional to the electron temperature [5]. This probe, as all of those which are based on current collection, is extremely sensitive to the variations of the contact potential at the surface of the electrodes, either temporal (variations due to change in the potential of the electrode with respect to the plasma) or spatial (non-uniformity of the potential across the surface of the electrode). To reduce these effects we have used spheres made with agglomerated graphite which have proved under laboratory tests to be far better than rhodium or nickel plated electrodes. A very careful and detailed analysis of the actual working of the probe has been made and will be published in the near future [6]; as far as the data reported in this paper are concerned, it is sufficient to say that the overall accuracy of the electron temperature measurement is of the order of 150 °K.

3. Results

The Dragon rocket D.260 was launched from Andenes (Norway) in the northern auroral zone during the late evening of 17 April 1969. The variations of the X component of the magnetic field recorded in Kiruna, not far from the launching site, are shown in Fig. 2 for the days 17 and 18. During the flight the 27.6 MHz riometer at the rocket base indicated a very variable absorption averaging about 1 dB. From ground-based observations it was concluded that the rocket entered an auroral display in the early part of the ascent, and then seemed to remain outside the auroral forms. In Figs. 3 and 4 are shown the electron density and temperature profiles obtained after smoothing of the data to remove the spin modulation effect which produces a decrease in density and an increase in electron temperature in the wake of the rocket.

3.1. E Region

There is a clear difference between the profiles in the E region obtained in the upward and downward legs of the trajectory. The high ionization seen during the ascent between 100 and 150 km reflects the passage through the auroral display and the two secondary maxima observed in density as well as temperature at 105 and 125 km are thought to indicate the location of maximum intensity of the auroral electrojet. The electron temperature is fairly high (~750 °K–900 °K) during this part of the flight but the value of 850 °K found at 150 km is in good agreement with typical values obtained at mid-latitudes by incoherent scatter technique [7]. It seems likely, therefore, that the temperature data are essentially correct and that there is a considerable

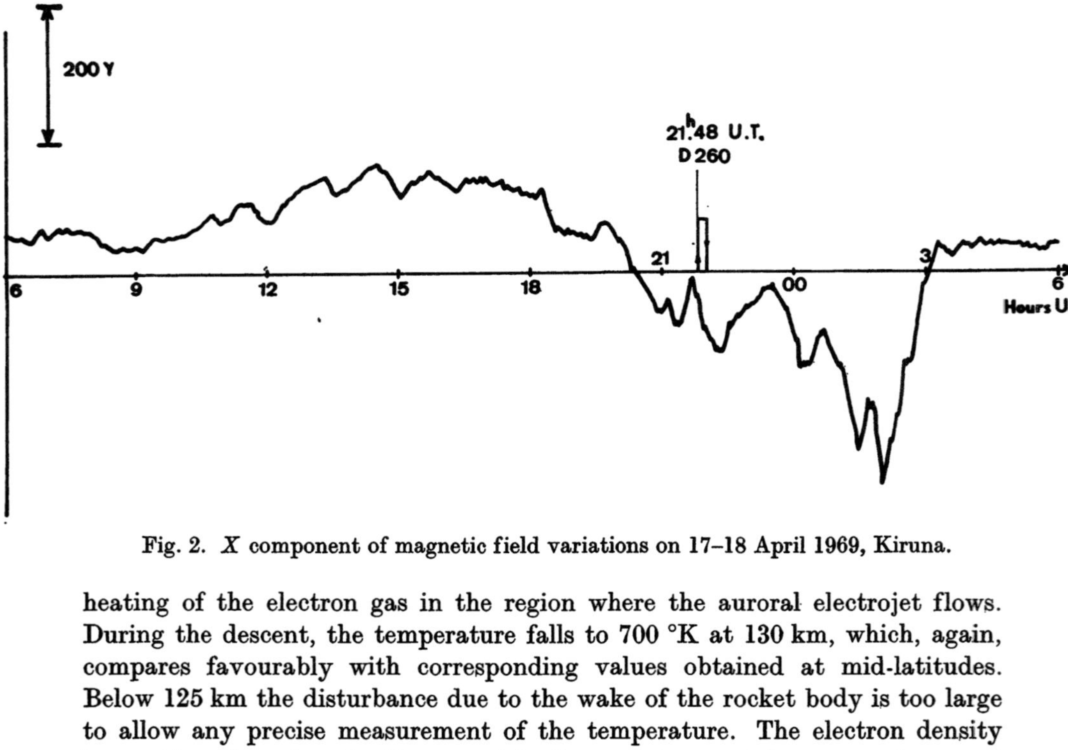

Fig. 2. *X* component of magnetic field variations on 17–18 April 1969, Kiruna.

heating of the electron gas in the region where the auroral electrojet flows. During the descent, the temperature falls to 700 °K at 130 km, which, again, compares favourably with corresponding values obtained at mid-latitudes. Below 125 km the disturbance due to the wake of the rocket body is too large to allow any precise measurement of the temperature. The electron density is considerably reduced with respect to the ascent except for a very sharp and intense sporadic E layer at an altitude of 98 km with a thickness of roughly 2 km.

3.2. F Region: Motion of Irregularities and Static Electric Field

Besides the salient anomalies due to the barium clouds traversal during the descent (Fig. 4), the density and temperature profiles show a very disturbed and turbulent ionosphere above 200 km. Upon the global profile are superimposed numerous irregularities with spatial scale perpendicular to the magnetic field between about 100 m (the best resolution of our equipment) and about 10 km as seen in Figs. 3 and 4. There is a consistent anticorrelation between N_e and T_e, i.e. an increase in density corresponding to a decrease in temperature and vice versa. It is interesting to compare these observations with similar ones made at mid-latitudes [8]. We intend, in the near future, to look for an empirical relationship between N_e and T_e for certain altitude ranges and compare them with those obtained by the aforementioned authors [8].

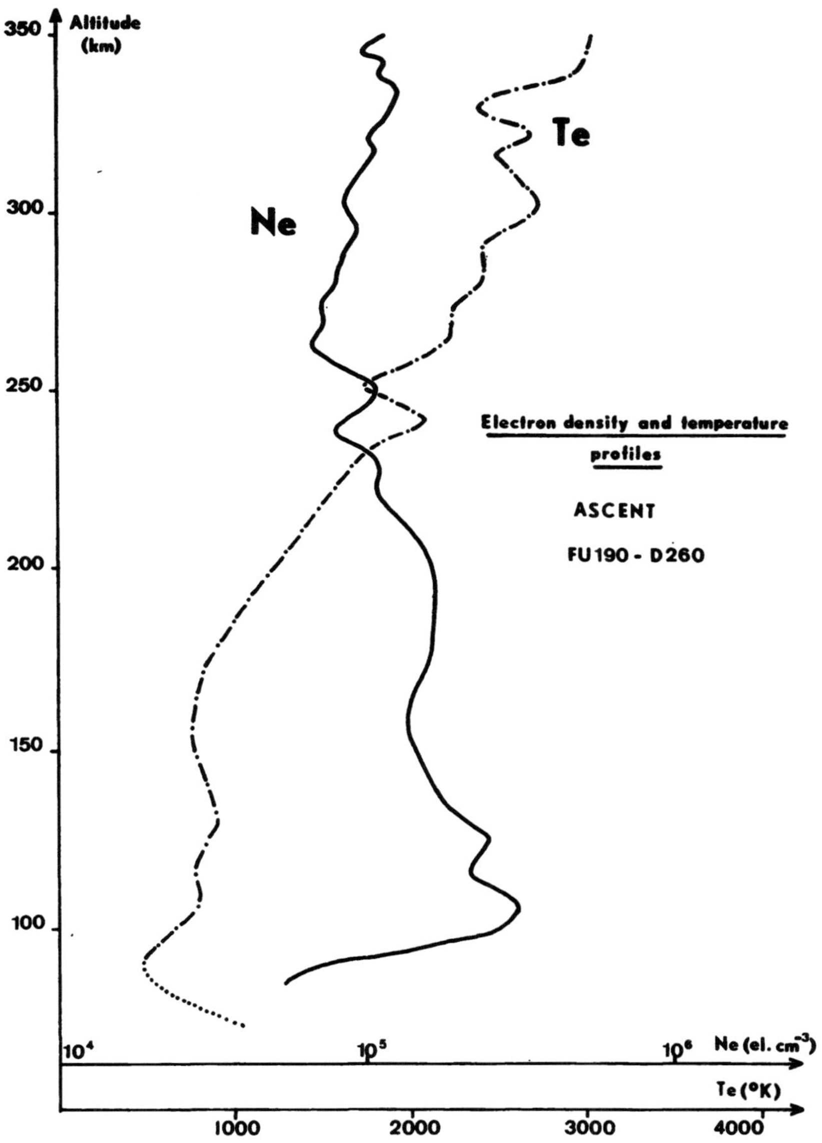

Fig. 3. Electron density and temperature profiles, ascent.

A fact, somewhat more interesting, is the identical aspect of irregularities seen during ascent and descent between 250 km and 320 km (Fig. 5). It is quite certain that irregularities in the high latitude F region are field-aligned and their typical dimensions perpendicular to the magnetic field vary from a few tens of metres to a maximum of the order of a few tens of kilometres [9, 10]. The size of the irregularities corresponding to the maximum in electron density observed at 280, 295 and 320 km is of the order of 10–15 km. As the horizontal distance separating the ascent and descent at 280 km altitude is

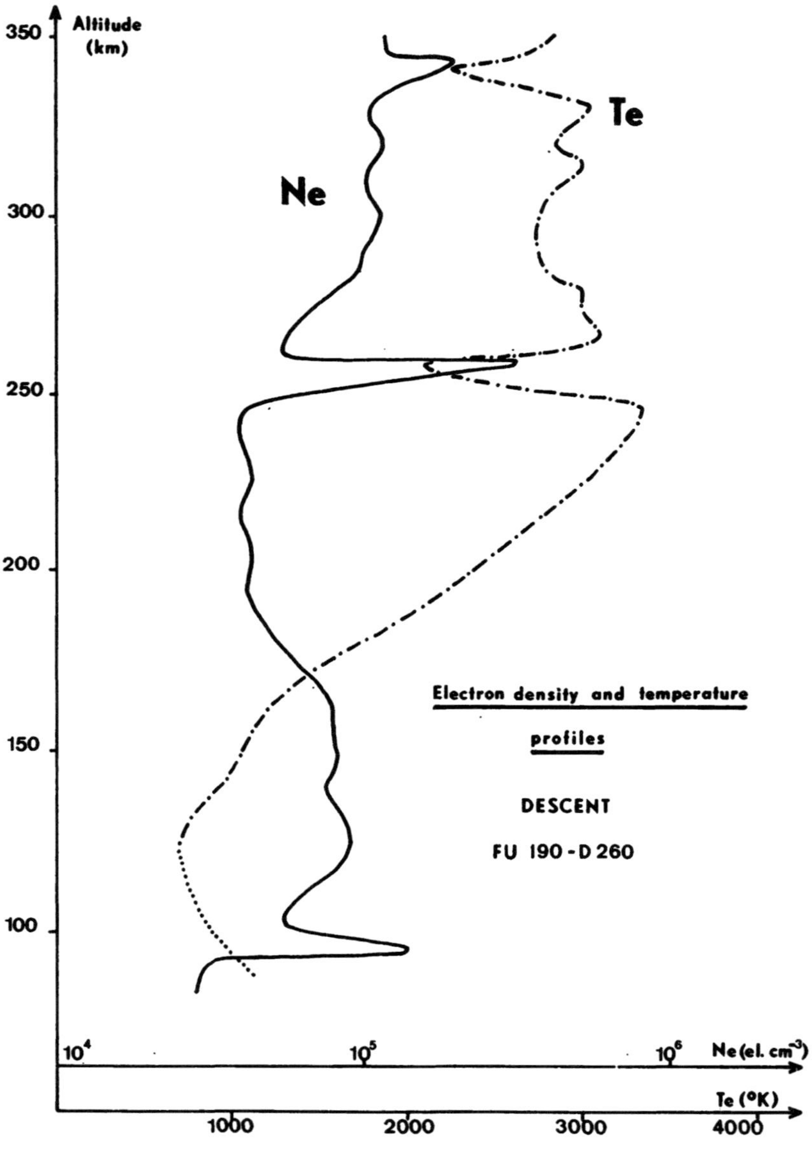

Fig. 4. Electron density and temperature profiles, descent.

equal to 120 km the only possibility of observing the same irregularities at these two different points is that their motion perpendicular to the magnetic field is very close to that of the rocket. In order to check this hypothesis we have tried to derive the drift velocity of the plasma by determining accurately the position of the wake of the rocket as observed by the electron density probe.

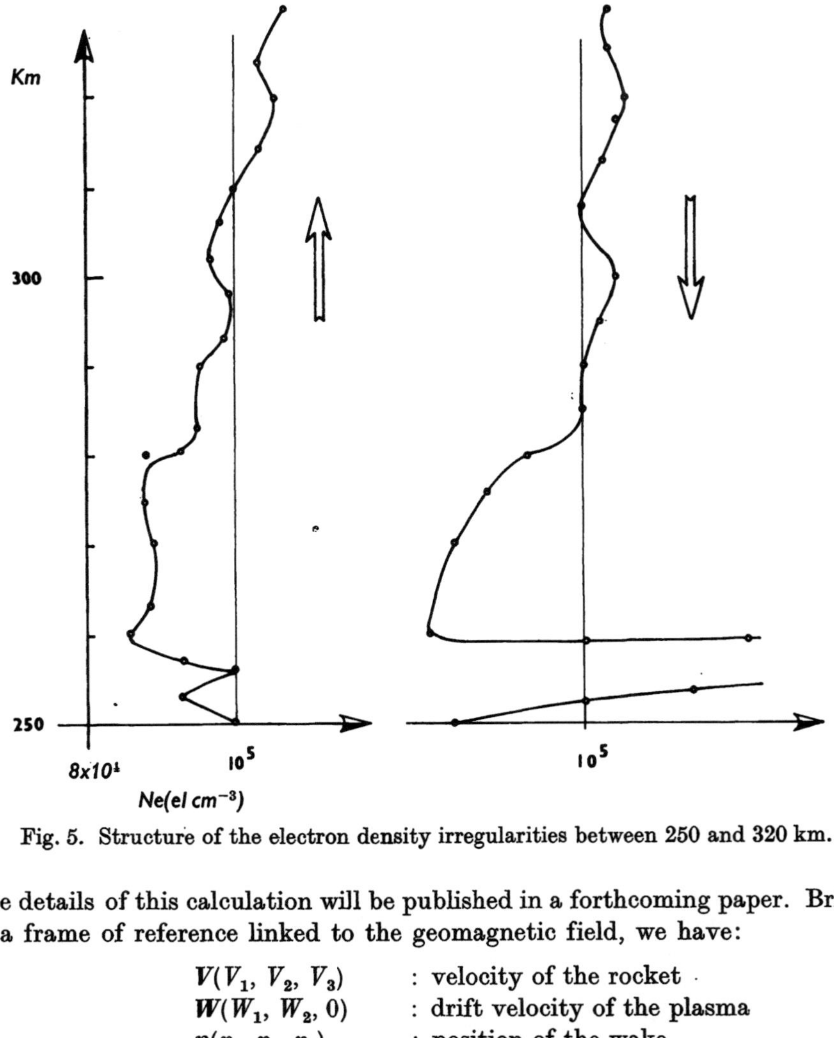

Fig. 5. Structure of the electron density irregularities between 250 and 320 km.

The details of this calculation will be published in a forthcoming paper. Briefly in a frame of reference linked to the geomagnetic field, we have:

$\boldsymbol{V}(V_1, V_2, V_3)$	: velocity of the rocket
$\boldsymbol{W}(W_1, W_2, 0)$	: drift velocity of the plasma
$\boldsymbol{p}(p_1, p_2, p_3)$	: position of the wake
$\boldsymbol{z}(z_1, z_2, z_3)$	: axis of the rocket

When the probe "sees" the wake

$$|\boldsymbol{V} - \boldsymbol{W}, \boldsymbol{p}, \boldsymbol{z}| = 0 \tag{1}$$

or

$$X_1 W_1 + X_2 W_2 + 1 = 0\,, \tag{2}$$

X_1 and X_2 being numerical coefficients depending on V_i, p_i and z_i.

Due to a combination of spin and precession of the rocket we obtain one equation such as (2) for each spin period and all these equations are independent for different spin periods. If the drift velocity of the plasma remains constant over a certain number of spin periods the components W_1 and W_2 of this velocity can be determined by a simple least squares fit method. The assumption that $\boldsymbol{W}$ = constant is easily checked by plotting X_1 and X_2 for each equation (2); this is represented in Fig. 6 where one can see that the drift velocity of the

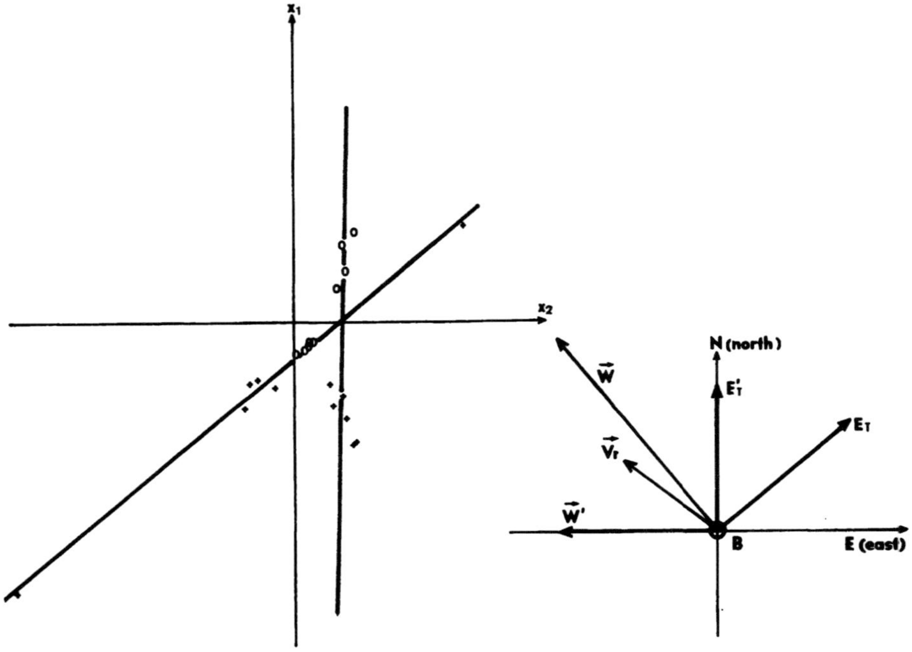

Fig. 6. Drift velocity and perpendicular electric field deduced from wake measurements.

$W_E = -910 \text{ m s}^{-1} \pm 90$	$E_E = 51 \pm 5 \text{ mV m}^{-1}$	
$W_N = +1110 \text{ m s}^{-1} \pm 100$	$E_N = 41 \pm 5 \text{ mV m}^{-1}$	$E_T = 65 \text{ mV m}^{-1}$
$W'_E = -910 \text{ m s}^{-1} \pm 100$	$E'_E = 0$	
$W'_N = 0 \text{ m s}^{-1} \pm 100$	$E'_N = 41 \pm 5 \text{ mV m}^{-1}$	$E_T = 41 \text{ mV m}^{-1}$

plasma is constant for a time corresponding to a north-eastern electric field E_T of $\sim$65 mV m^{-1} and then changes in amplitude and direction, the electric field becoming E'_T northward and of the order of 40 mV m^{-1}.

These results indicate clearly, that, during the time from 270 km in the ascent to 270 km in the descent the plasma drift is in the same general direction as the rocket velocity V_r. As the irregularities seen in the F region are very likely created by instabilities in the E region [11, 12] and as the phase speed of the associated drift waves is very dispersive and lower than the simple drift velocity $E \times B/B^2$ in the F region it is evidently not possible to find a better agreement between $\boldsymbol{V}'_r$ and $\boldsymbol{W}$ but the results presented in Fig. 6 seem to favour the assumption that during descent the rocket crossed the same irregularity as during ascent.

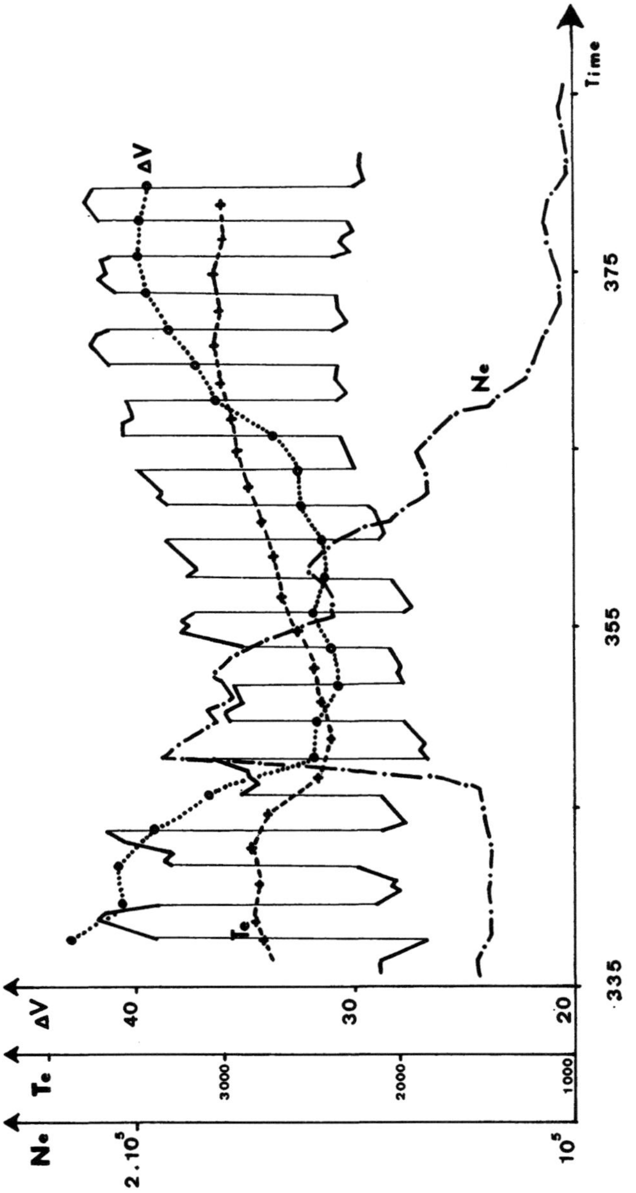

Fig. 7. Electron density and temperature variations during crossing of the first barium cloud.

3.3. Barium Clouds

The barium canisters were exploded at small distances from the rocket and the traversal of the clouds occurred during the first 10–15 s of their life-time; the peak electron density is 1.8×10^5 el cm^{-3} for the first one and 2.3×10^5 el cm^{-3} for the second one. An interesting feature is the decrease in the electron temperature associated with this increase in density: the electron temperature thus decreases from 2800 °K outside the first cloud to 2300 °K inside (Fig. 7) and comparably from 3100 °K outside the second to 2300 °K inside. The neutral density in the clouds is largely higher than the neutral atmosphere density at these altitudes (a factor of $\sim 10^3$), and the temperature of the barium atoms or ions is of the order of 2000 °K (Haerendel, private communication): this could explain the cooling of electrons by collisional thermalization with

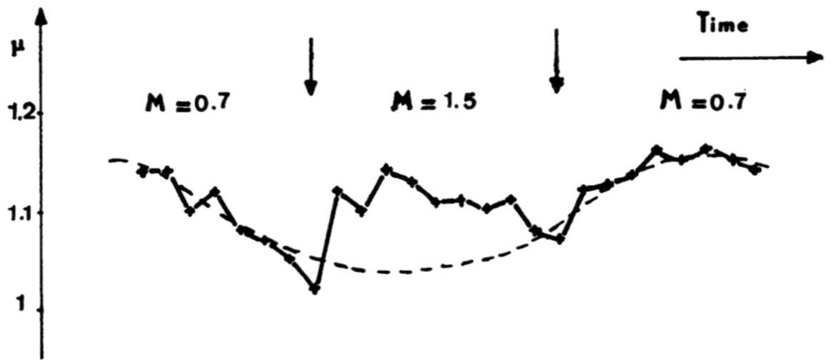

Fig. 8. Effect of the Mach number value on the amplitude of the wake: changes in μ during crossing of first barium cloud. μ = spin modulation depth $N_e(\max)/N_e(\min)$; $\bar{M} = V_r/(2\,k\,T_+/m_+)^{1/2}$

the neutral particles inside the clouds. A more detailed study of this phenomenon is in progress. Another remark concerns the diminution of the apparent contact potential difference between the two spheres during the traversal of the clouds. The cause of such an effect is thought to be due to metallic absorption of the barium ions or neutral particles on the surface of the electrodes leading to a more uniform contact potential on the two spheres [6]. The last point of interest is the change in the amplitude of wake effect when the rocket enters the barium cloud; in the undisturbed ionosphere at 340 km the Mach number ($M = V_{\text{rocket}}/V_{\text{thermal ions}}$) is equal to 0.7 but inside the cloud it increases to 1.5 due to the large mass of barium ions. The consequence of the decrease in density in the wake is clearly seen in Fig. 8 where the modulation depth of the electron density (N_e in front of the rocket/N_e in the wake) is plotted as a function of time.

4. Conclusion

The results reported in this paper, although in a preliminary form, appear to offer valuable data on the high latitude ionosphere during moderate auroral activity. They cover a very wide range of interest: heating of E region under auroral conditions, development and motion of F region irregularities, expansion and structure of the barium clouds. It is hoped that, in the near future, the results from the two other experiments on board the same rocket will provide a good opportunity to proceed further in the analysis of the results.

Acknowledgments

We are indebted to G. Gogly and J. C. Gillon for their efforts during fabrication of the plasma probes, test of the payload and launching operations. This work was supported by the Centre National d'Etudes Spatiales under the auspices of contract No. 66.CNES.035. The data reduction and analysis were made on CNES. IBM 360-65 and DCC 6600 computers.

References

[1] F. S. Mozer, Electric Fields, University of California, Berkeley, Jan. 1971.
[2] F. S. Mozer and P. Bruston, J. Geophys. Res. **72**, 1109 (1967).
[3] M. C. Kelley, F. S. Mozer and U. V. Fahleson, Planet. Space Sci. **18**, 847 (1970).
[4] E. C. McKenzie and J. Sayers, Planet. Space Sci. **14**, 731 (1966).
[5] J. W. Wilson and G. Garside, Planet. Space Sci. **16**, 257 (1968).
[6] R. Godard, Thesis, University of Paris–Orsay, Oct. 1971.
[7] M. Petit, Doctorate thesis, University of Paris (1968).
[8] G. Lejeune and P. Waldteufel, Annls Géophys. **26**, 223 (1970).
[9] P. L. Dyson, J. Geophys. Res. **74**, 6291 (1969).
[10] P. L. Dyson and A. J. Zmuda, J. Geophys. Res. **75**, 1893 (1970).
[11] G. C. Reid, J. Geophys. Res. **73**, 1627 (1968).
[12] G. C. Reid, Planet. Space Sci. **18**, 1105 (1970).

Space Research XII — Akademie-Verlag, Berlin 1972

DETERMINATION OF THE IONOSPHERIC DENSITY AND TEMPERATURE USING A DOUBLE PROBE ELECTRIC FIELD DETECTOR

A. PEDERSEN[a], U. FAHLESON[b] and C.-G. FÄLTHAMMAR[b]

[a] European Space Research and Technology Centre, Noordwijk, Netherlands
[b] Royal Institute of Technology, Division of Plasma Physics, Stockholm, Sweden

Rocket-borne experiments for electric field measurements in the ionosphere have been used also for determining electron density and temperature along the rocket trajectories by intermittently operating the probes in special modes. Density and temperature profiles in the auroral-zone ionosphere above ESRANGE, Kiruna, Sweden, are given for two flights, one during quiet and one during moderately disturbed conditions. The densities obtained are in agreement with ionosonde data. The temperatures found near 200 km were of the order of 1000 °K. Arguments are given why the double probe seems to be a more reliable temperature-measuring device than conventional Langmuir probes.

The probes used were coated with Aquadag. From analysis of shadow effects in the probe signals, the photoyield of that material was found to be $(7 \pm 3) \times 10^{-10}$ A cm^{-2}.

1. Introduction

A number of double probe experiments for measurements of electric fields have been flown on sounding rockets during the last few years. In principle the measurements are made by measuring the potential difference between two symmetrical Langmuir probes. A probe system of this type can also be used for studies of plasma parameters, such as density and temperature, in addition to the electric field measurements [1]. It is the purpose of the present paper to present and discuss temperature and density information obtained from two rocket flights from ESRANGE (65.3° N geomagnetic latitude).

The rockets (Skylarks) were launched during twilight on 14 February 1970 at 1715 CET and 2 April 1970 at 0246 CET. A small magnetic disturbance ($\Delta H = +40$ gamma) was in progress during the first flight and an auroral E layer of a few MHz was observed. No magnetic disturbance was present during the second flight. It is concluded that with very little extra circuitry a double probe on a spinning rocket can be used for separate measurements of electron density and temperature. It is argued that in particular the electron temperature data obtained in this way are more reliable than those obtained by conventional single Langmuir probes.

2. Modes of Operation

2.1. Geometrical and Electrical Configuration

The geometrical configuration of the probe array is shown in Fig. 1. The electric field $\boldsymbol{E}$ measured is that in the frame of reference of the rocket and is related to the field $\boldsymbol{E}_e$ in the earth's frame of reference by the relation:

$$\boldsymbol{E} = \boldsymbol{E}_e + \boldsymbol{v} \times \boldsymbol{B} . \tag{1}$$

The electrical configuration and modes of operation are shown in Fig. 2. The switching sequence was devised to provide cyclically the following three modes: A, high impedance, floating; duration 4 sec; B, high impedance, current bias; duration 5 sec; C, low impedance; duration 1 sec.

In the first flight a technical failure caused the bias to probe 1 to be on continuously, thereby providing the three modes shown in the left-hand column

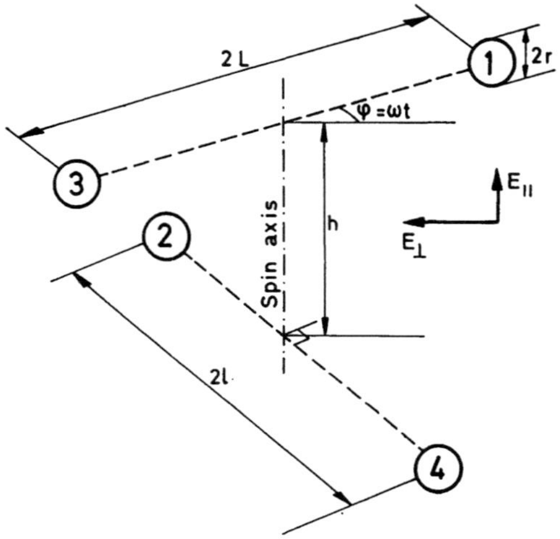

Fig. 1. Geometric configuration of the probe array. The dimensions are: $L = 1.85$ m, $l = 1.95$ m, $h = 1.20$ m, $r = 2.5$ cm. The spin frequency $\omega/2\pi$ was 2.8 rev s^{-1} in the first flight and 2.4 rev s^{-1} in the second flight. Notice that the electric field $\boldsymbol{E}$ (in the frame of reference of the rocket) is decomposed into components parallel and perpendicular to the spin axis, *not* to the magnetic field. Both $E_\perp$ and $E_\parallel$ are non-vanishing, although $E_\parallel$ is the smaller of the two.

of Fig. 2, two of which (A1 and C1) were asymmetric. We shall see that in particular the mode C1 is well suited for electron temperature measurements.

2.2. Symmetric High-Impedance Modes (A2, B in Fig. 2)

The unshunted modes, both floating and biased, are conventional electric field measuring modes, which are well understood [2], and the voltage-difference signals from opposite probes allow for a straightforward determination of the strength and direction of the electric field component perpendicular to the rocket axis, E_1, from amplitude and phase of the observed signal.

2.3. Low Impedance Mode without Bias (C2 in Fig. 2)

Signals from opposite probes are used as a check on the plasma properties, and not for electric field measurements. Current continuity must hold for a probe pair as a whole, and a set of equations can be set up which includes the following parameters: plasma density n_e, electron temperature T_e, ion temperature T_+, ion mass m_+, rocket velocity v_R. At any given altitude v_R is known as well as the approximate ionic composition from which m_+ can be calculated. The ion current, which appears in the current continuity equations, is a function of v_R and can be calculated for a spherical probe when the current is orbital motion limited [3]. This is a good approximation when the Debye shielding distance is comparable with, or larger than, the probe radius [4], as

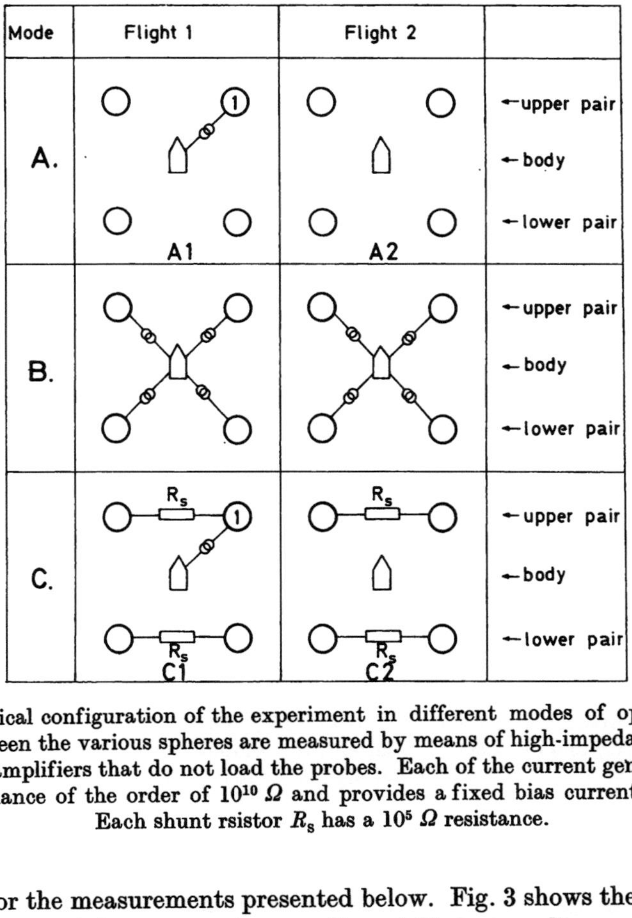

Fig. 2. Electrical configuration of the experiment in different modes of operation. The voltages between the various spheres are measured by means of high-impedance ($2 \cdot 10^9$ Ω) electrometer amplifiers that do not load the probes. Each of the current generators shown has an impedance of the order of 10^{10} Ω and provides a fixed bias current $I_b = 10^{-6}$ A. Each shunt rsistor R_s has a 10^5 Ω resistance.

is the case for the measurements presented below. Fig. 3 shows the relationship between the remaining parameters n_e, T_e and T_+ in two diagrams for typical conditions near 120 km and 200 km respectively, for a flight with apogee near 200 km. It will be shown in the following paragraphs that n_e and T_e can be determined from other modes and this mode (C2) can then be used as a crosscheck or for a crude determination of T_+/T_e.

In the low impedance mode the signal from adjacent probes, for example, $V_1 - V_2$ contains a sinusoidal variation with *twice* the spin frequency. The amplitude of this component, which dominates for a shunt impedance smaller than the plasma impedance, is related in a very simple way to the electron temperature.

Thus it can be shown that when $E\,L \ll R_s\,I_i$ (see Figs. 1 and 2),

$$V_1 - V_2 = h\,E_{||} - (E_{\perp}^2\,L^2/2\,V_e)\cos 2\,\omega\,t \qquad (2)$$

where I_i = ion current and $V_e = k\,T_e/e$.

As $E_{\perp}$ is known independently, the amplitude of the signal $V_1 - V_2$ gives direct information about V_e, i.e. about the electron temperature.

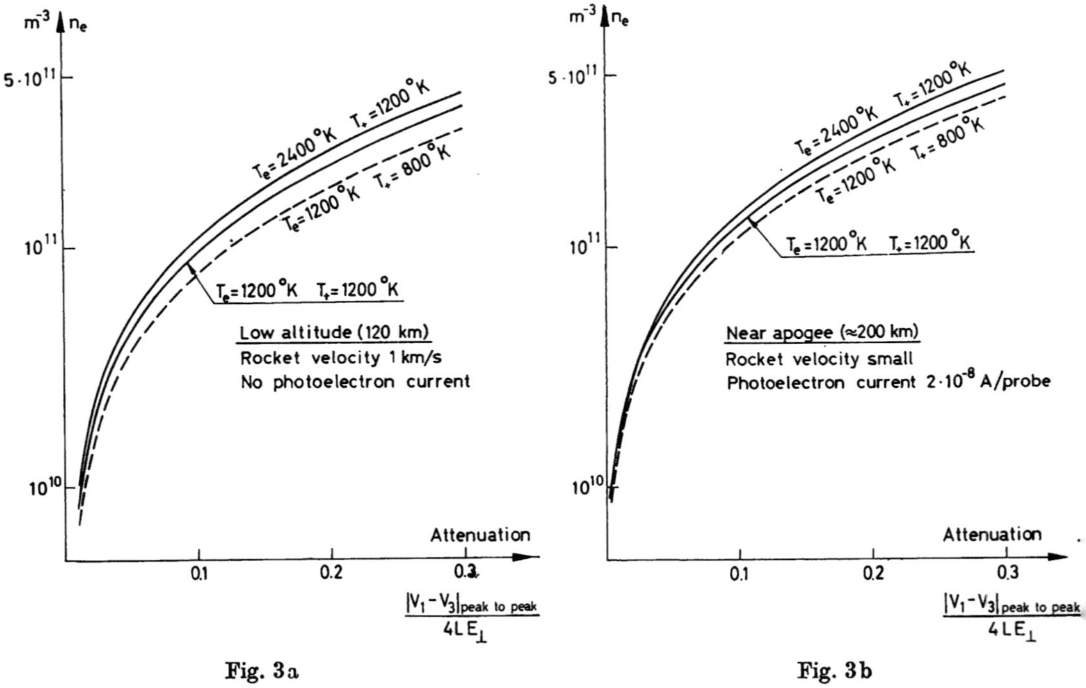

Fig. 3. Plasma density versus attenuation for opposite probes in mode C2. Parameters are electron temperature T_e and ion temperature T_+. Curves are given for two combinations of rocket velocity and photoelectron current, the combinations being typical of low altitude (left) and high altitude (right).

2.4. High Impedance Mode with Asymmetric Bias (A1)

In this mode one probe is at floating potential and the other at a potential that corresponds to a positive bias current, $I_b = 1\ \mu$A. The biased probe will be on the exponential part of the probe characteristic if the plasma density is sufficiently high. For sufficiently low plasma densities the probe will be positive with respect to the plasma. For the flights discussed here the Debye length was larger than the probe radius and an approximately linear relation exists between electron current and a probe voltage at a positive potential [4]. The potential difference between the two probes V_b can in this case be used for determination of plasma density if T_e is known. Fig. 4 gives n_e plotted against

V_b/V_e. It can be seen that this relation is influenced by photocurrent and rocket velocity, both of which are known. However, the relations shown in Fig. 4 are only weak functions of ion mass and temperature.

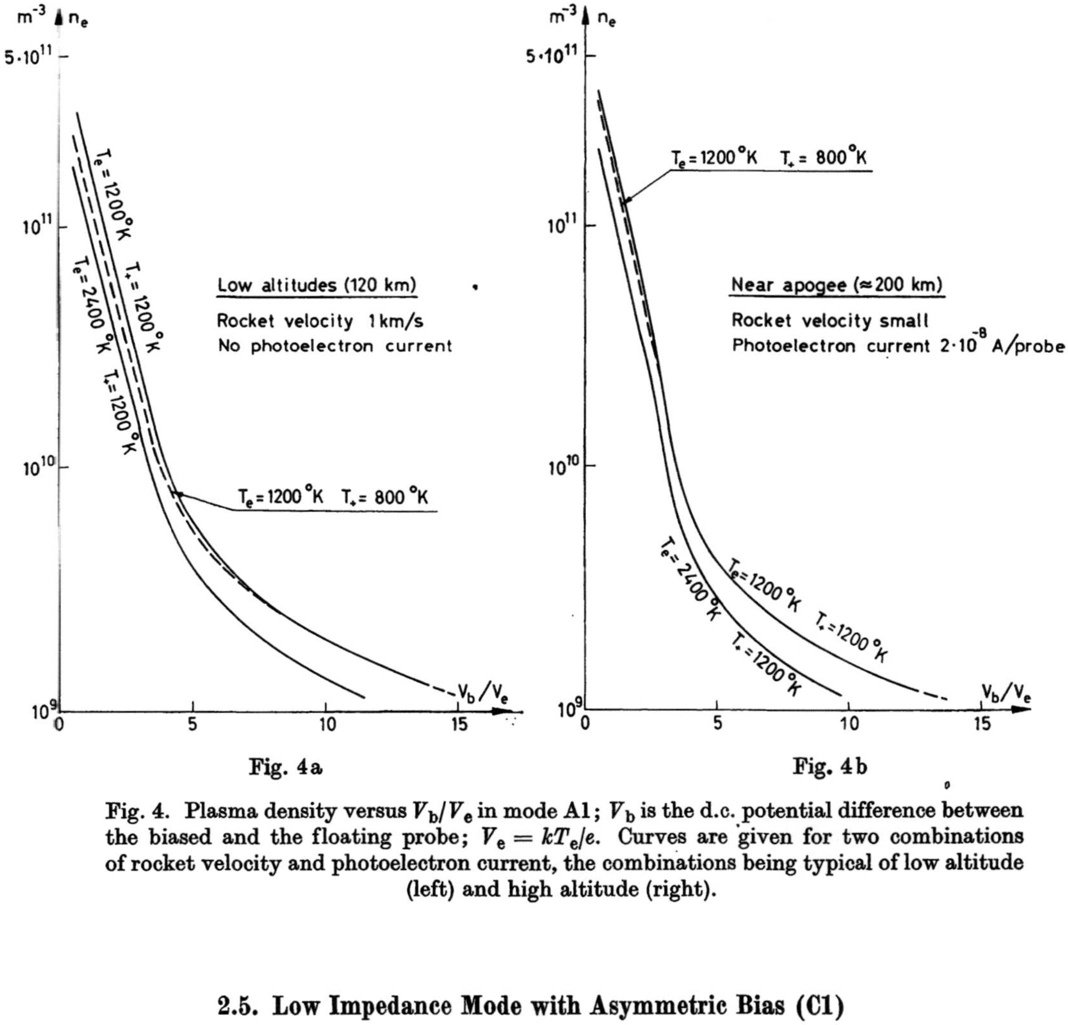

Fig. 4. Plasma density versus V_b/V_e in mode A1; V_b is the d.c. potential difference between the biased and the floating probe; $V_e = kT_e/e$. Curves are given for two combinations of rocket velocity and photoelectron current, the combinations being typical of low altitude (left) and high altitude (right).

2.5. Low Impedance Mode with Asymmetric Bias (C1)

In order to make use of mode C1 for temperature measurements, the bias current must be larger than the ion current but smaller than the random electron current, so that the probes operate on the exponential part of the probe characteristics. $R_s I_b$ was comparable with V_e and $2 E_\perp L$, both of which were of the order 100 mV. The spin of the rocket in the electric field, $\boldsymbol{E}_e + \boldsymbol{v} \times \boldsymbol{B}$, then provided a scanning of the exponential part of the characteristic. The electron temperature can be found as a function of the attenuation $(V_1 - V_3)_{ptp}/4 E_\perp L$ (ptp = peak to peak). This function is shown in Fig. 5.

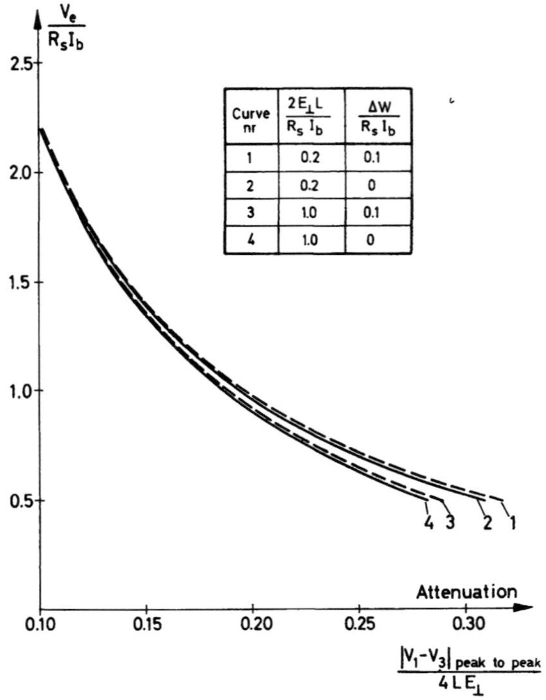

Fig. 5. Electron temperature as a function of attenuation in mode C1. Notice that the curves depend only weakly on the (separately known) parameter $2\,E_\perp\,L/R_s\,I_b$. Notice also that a work function difference $\Delta\,W$ between the probes has only a very small influence. In the present experiment $\Delta\,W/R_s\,I_b$ was smaller than 0.1.

3. Experimental Results

3.1. Electric Field Measurements (A2, B in Fig. 2)

The electric field was also measured with artificial barium cloud experiments ejected from the same rockets, and good agreement was found between these measurements and the double probe [5]. The electric fields measured on the two flights are summarized in Table 1.

Table 1

Comparison of magnitude and direction of electric fields E_e obtained from double probes and barium cloud experiments

	Double probe		Barium	
	Magnitude	Azimuth	Magnitude	Azimuth
Flight 1	22 ± 1 mV/m	350° ± 5°	21 ± 1 mV/m	356° ± 1°
Flight 2	≈ 6 mV/m	180° ± 20°	3 mV/m	180° ± 3°

3.2. Electron Temperature Measurements

The most reliable determination of electron temperature is made from mode C1, which occurred only in the first flight. For the second flight the electron temperature can only be determined from mode C2. The electron temperature profiles derived are shown in Fig. 6. The high electron temperatures marked Ba occurred after barium releases, when the rocket travelled for a short time inside a barium cloud. These results will not be discussed further in the present paper.

A spherical Langmuir probe was included in both flights and gave good data in the second flight (Wrenn, private communication). The plasma densities

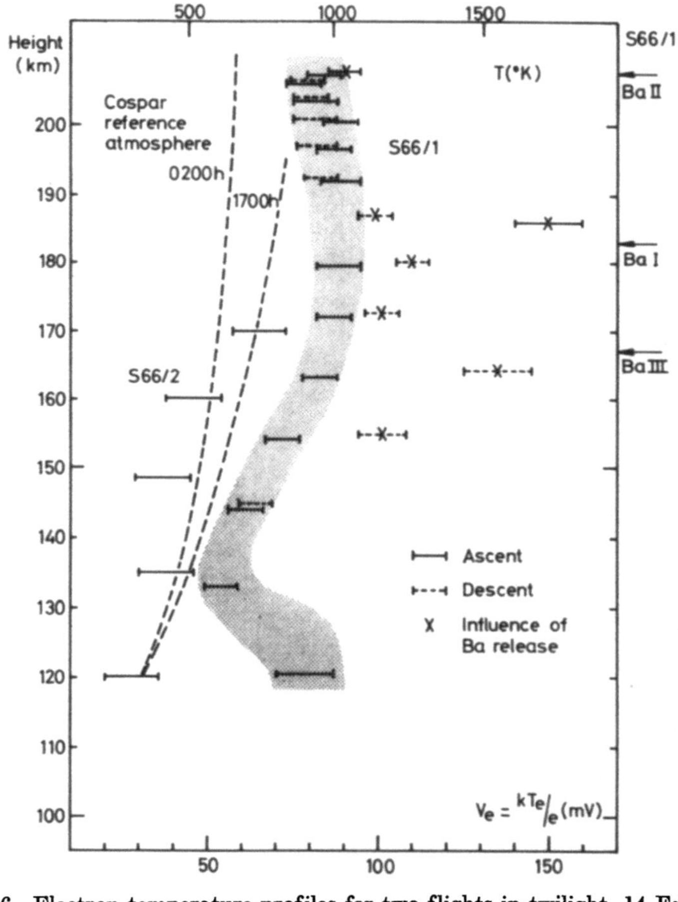

Fig. 6. Electron temperature profiles for two flights in twilight, 14 February and 2 April 1970.

measured with this probe agree with ones obtained by the double probe, whereas the electron temperatures are of the order 1000 °K above the values shown in Fig. 6. The electron temperatures for the second flight (quiet conditions) are derived from mode C2, and are therefore more crude. Only five measurements during ascent are given in Fig. 6 for this flight. At higher altitudes the signal in this mode was obviously not according to Eq. (2), possibly because the plasma density was so low that the floating potential moved near to the plasma potential under the influence of solar radiation. However, the values obtained are in good agreement with theoretical calculations, according to which the electron temperature should be close to the neutral temperature for this time of day [6].

It is interesting to recall that a significant discrepancy has been observed on several occasions between electron temperatures measured with standard Langmuir probes and with other techniques, such as incoherent scatter radars. The Langmuir probes have given higher values than the other techniques. These persistent discrepancies have been discussed extensively in recent papers [7]. The standard Langmuir probe measurements depend on the surface conditions, which in turn are influenced by a voltage bias sweep (or the sweep rate). Bunting and Heikkila [8] have demonstrated that the probe characteristic has a hysteresis effect for a bias voltage varying with time as a sawtooth function. Considerable work has been done in this field, both experimentally and theoretically, but no reliable model exists so far for explaining this change in surface conditions.

The double probe in mode C1 uses a current bias, which keeps the probe on the exponential part of the probe characteristic. The potential variation of each probe (due to spin in the electric field $\boldsymbol{E}_e + \boldsymbol{v} \times \boldsymbol{B}$) is only of the order kT_e/e, which should reduce the hysteresis effect. Furthermore it has been shown that the surface coating used for the double probes, viz. Aquadag, has more favourable surface properties than metals [9]. For all these reasons the present double probe system should represent an improvement over standard Langmuir probes.

3.3. Plasma Density Measurements

With knowledge about $V_e = kT_e/e$ from the previous paragraph, the plasma density can be derived from mode A1 as described in § 2.4. Fig. 7 shows the plasma density versus altitude for the two flights. The data indicate the existence of a density maximum around 130 km in the first flight. This is in qualitative agreement with ionosonde data, which indicated an auroral E layer as shown separately in Fig. 7. Unfortunately no probe measurement was made at the altitude where the maximum density is expected, so a detailed comparison is not possible. Furthermore, as auroral E layers are usually inhomogeneous, a detailed agreement would not be expected. In the case of the second flight V_b (potential difference between a floating probe and a probe biased with 10^{-6} A) was derived from measurements of $V_1 - V$ (rocket body) in the floating and current biased modes respectively. In the density evaluation for the second flight kT_e/e was assumed to be that of the 0200 h COSPAR reference atmosphere ± 10 mV (see Fig. 6). For comparison the densities derived from mode C2 in the first flight are included in Fig. 7. The densities derived from signals in modes A1 and C2 both depend on T_+/T_e, but to a lesser extent in the case of mode A1. The agreement between the two modes is slightly better for the assumption $T_+ = 1/2\ T_e$ than for $T_+ = T_e$; however, these curves are meant more as an llustration of what can be done than as a qualitative result, because the shunt impedance used was not optimized for good measurements in mode C2 with the ionospheric density existing at that time.

3.4. Photoelectron Yield of Aquadag

The rockets were launched in twilight near solar zenith angles when the dominant solar radiation near apogee was the Lyman α line, whereas other wavelengths (10-1000 Å) only penetrated to altitudes well above the apogee of the rockets.

The signal $V_2 - V_4$ on the lowest pair of booms showed a characteristic change when one probe was in the shadow of the rocket due to the absence of

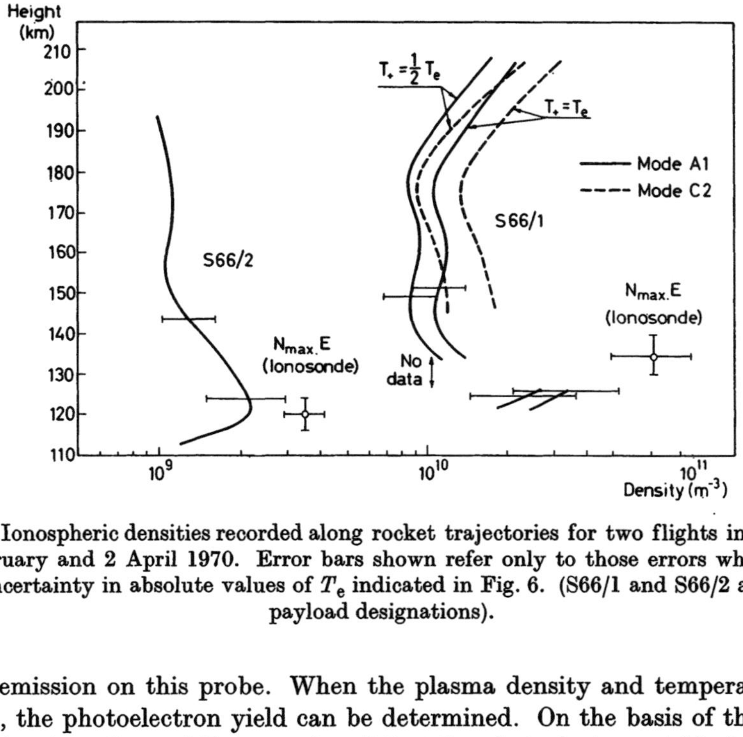

Fig. 7. Ionospheric densities recorded along rocket trajectories for two flights in twilight, 14 February and 2 April 1970. Error bars shown refer only to those errors which result from uncertainty in absolute values of T_e indicated in Fig. 6. (S66/1 and S66/2 are ESRO payload designations).

photo-emission on this probe. When the plasma density and temperature are known, the photoelectron yield can be determined. On the basis of the earlier determination of n_e and T_e it was found that the photoelectron yield of Aquadag for solar radiation at Lyman α and longer wavelengths is $(7 \pm 3) \times 10^{-10}$ A cm^{-2}.

4. Conclusions

It has been shown that spherical double probes for measurement of ionospheric electric fields can be used to obtain temperature and density information by using simple special modes that do not require much extra electronic hardware. The use of a current bias on one of the two probes, and measurement of the potential difference between the two probes in modes with low and high shunt impedances allow determination of electron temperature and density respectively. In particular the electron temperature measurements represent an improvement over conventional single Langmuir probes. As a by-product, the photoyield of Aquadag has been found.

Acknowledgments

We thank Mr. L. Olausson for competent and valuable assistance with the computational work.

This investigation is part of a cooperation between the Department of Plasma Physics, Royal Institute of Technology Stockholm, and the Space Science department of ESTEC. The participation of the Royal Institute group was made possible by financial support from the Swedish Space Research Committee.

References

[1] U. V. FAHLESON, M. C. KELLEY and F. S. MOZER, Planet. Space Sci. **18**, 1551 (1970).
[2] U. V. FAHLESON, Space Sci. Rev. **7**, 238 (1967).
[3] M. KANAL, Science Rept. No. JS-5, Space Physics Research Lab., University of Michigan (1962).
[4] J. G. LAFRAMBOISE, Defense Documentation Center Report AD 634596 (1966).
[5] U. V. FAHLESON et al., Radio Sci. **6**, 233 (19671).
[6] P. STUBBE, J. Atmos. Terr. Phys. **32**, 865 (1970).
[7] H. C. CARLSON and J. SAYERS, J. Geophys. Res. **75**, 4883 (1970).
[8] W. D. BUNTING and W. HEIKKILA, J. Appl. Phys. **41**, 2263 (1970).
[9] E. A. TRENDELENBURG, B. FITTON, D. E. PAGE and A. PEDERSEN, ELDO/ESRO Scient. Techn. Rev. **2**, 1 (1970).

Space Research XII — Akademie-Verlag, Berlin 1972

ELECTRIC FIELDS AND CONDUCTIVITIES DERIVED FROM WAKE MEASUREMENTS ON A ROCKET

K. Knott, A. Pedersen and J. K. E. Tunaley

Space Science Department, European Space Research and Technology Centre, Noordwijk, The Netherlands

Measurements from two rocket flights obtained by means of a combined electron–ion probe are presented. Apart from deriving density profiles, these measurements have been used to study wake characteristics. From the direction of the wake it can be concluded that during one flight the northward component of the electric field in the ionosphere was fairly low (a few mV m^{-1}) while for the second flight the northward component of the electric field was about 20 mV m^{-1}. During this second flight ion and electron wakes did not coincide at altitudes below 100 km. This can be explained by the different mobilities for electrons and ions in a magneto-plasma with collisions. It will be shown that this characteristic feature of ion and electron wakes can yield information about conductivities in the lower ionosphere.

1. Introduction

Space vehicles moving in ionospheric or magnetospheric plasma create a wake by sweeping out the plasma behind them. In the past much experimental and theoretical effort was devoted to the understanding of this phenomenon [1, 2]. However, not until recently have geophysical phenomena, such as ionic composition and drift of the plasma, been derived from wake phenomena [3, 4]. It is the purpose of this paper to show that collisions in the plasma influence the wake formation so that in the presence of an electric field ions and electrons form separate wakes. This splitting of wakes is caused by the different mobilities of ions and electrons and can yield information about ionospheric conductivity.

2. Experimental Technique

A simple ion and electron collector was included in two Centaure rockets launched from ESRANGE near Kiruna in the summers of 1969 and 1970. The aim of these payloads was to study noctilucent clouds with photometric methods, and the launchings therefore took place in twilight. We shall here only discuss the measurements obtained with the probe on these two rockets. The probe had the geometry of a venetian blind mounted in a small frame with the proper bias on the blades to collect charged particles in the electric field between them. Adjacent blades were biased by a voltage of opposite polarity such that every second blade collected ions or electrons, respectively. Each probe was mounted at the side of the rocket, as the purpose of this experiment was the study of the wake effects of both ions and electrons. One can expect to sense wake effects by this set-up as soon as the velocity vector of the spin-stabilized rocket with respect to the plasma deviates from the direction of the spin axis. The rocket, additionally, was equipped with a triaxial fluxgate magnetometer and a solar sensor in order to ensure good attitude reconstitution.

3. Results

One rocket was fired on 11 August 1969 at 0230 local time during quiet magnetic conditions, and reached an apogee of 147 km. The second rocket was launched on 10 August 1970 at 0150 local time when the north–south component of the magnetic field showed a +70 gamma deviation as compared with

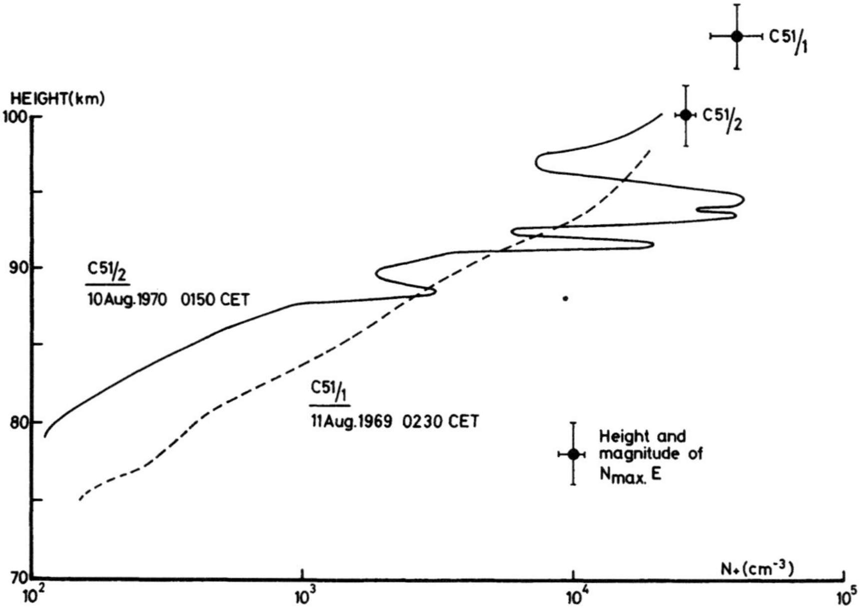

Fig. 1. Ion density profiles obtained during two flights into the auroral ionosphere. Included as dots are the corresponding ionosonde results.

the quiet time value. The peak altitude of this rocket was 154 km. For both flights the electron and ion currents into the venetian blind were recorded over the whole trajectory above 80 km. Both currents are clearly spin-modulated as is expected in the case of wake formation. Above 100 km it was found that the modulation index for the ion current depended on the angle between spin axis and the direction to the sun. The modulation index was smallest when this angle was close to 90°, e.g. when the blades were receiving the lowest possible irradiation. From this we conclude that above 100 km the apparent ion current was dominated by photocurrent: for instance, on the second flight, the solar zenith angle was approximately 92° and the onset of the photoeffect is coincident with the calculated sharp increase in Lyman-α flux near 100 km. For altitudes below 100 km, however, the ion current was not affected by photoemission. From the measured ram current we can therefore calculate the plasma density, and from the phase of the modulation we can derive the ion wake or ram direction.

The electron current collected by the probe did not show any dependence on solar aspect angle at any altitude. It seems, however, to be difficult to convert the electron current into density, as experience shows that rockets in the lower ionosphere adopt a negative equilibrium potential. Nevertheless the phase of the spin modulation in electron current can be used to detect the direction of the electron wake throughout the flight. In summary, we measure the following quantities: plasma density up to 100 km, ion wake direction up to 100 km, electron wake direction throughout the flight.

In Fig. 1 the ion density profiles obtained during both flights below 100 km are presented. For the first flight, which took place under quiet conditions, the density increases smoothly between 70 and 100 km. During the second flight, which took place under slightly disturbed conditions, several sporadic layers have been detected between 88 and 96 km. These profiles have been derived from the maxima in the spin-modulated ion current, which occur whenever the detector is looking into the ram direction.

The phase of the spin-modulated ion and electron current can be used to determine the direction of the electron and ion flow to the rocket. In the case of flight 1 both currents were in phase, which means that both species were streaming to the rocket from the same direction. During flight 2, however, a clear phase shift was found, which indicates different ram and wake directions for ions and electrons at lower altitudes. In other words, the bulk motion of ions and electrons seen from the rocket was not the same. It is worth noting that the high thermal velocity of the electrons does not rule out wake formation, as long as the gyroradius is small compared with the rocket diameter. In this case only the electron drift velocity determines the wake.

Fig. 2 shows how different ram directions for electrons and ions can be explained. The relevant velocities are shown in a plane perpendicular to the magnetic field under the assumption of a northward pointing electric field. The ratio of gyrofrequency to collision frequency for electrons is such that they drift perpendicular to $\boldsymbol{E}$ and $\boldsymbol{B}$ both at 100 and 150 km at a velocity $\boldsymbol{V}_{de}$. Seen from the rocket, which moves at a velocity $\boldsymbol{V}_{R}$, the electrons have a bulk motion shown as $\boldsymbol{V}_{re}$. At the 100 km level the ions perform still enough collisions to have a significant component of $\boldsymbol{V}_{di}$ into the direction of the electric field. Composing the two vectors $\boldsymbol{V}_{R}$ (rocket velocity in the earth's frame) and $\boldsymbol{V}_{di}$ (ion drift velocity in the earth's frame) results in a ram velocity vector $\boldsymbol{V}_{ri}$, which is more or less aligned with the rocket velocity. At 150 km, however, the collision rate has decreased and the ions start having the main drift component perpendicular to both $\boldsymbol{E}$ and $\boldsymbol{B}$. Seen from the rocket, this results in a ram direction for ions which is close to that of the electrons. Still higher up the two drift velocities will become exactly equal and the ram directions will coincide. In summary, in case of a northward pointing electric field and a rocket horizontal velocity roughly toward the north, we expect

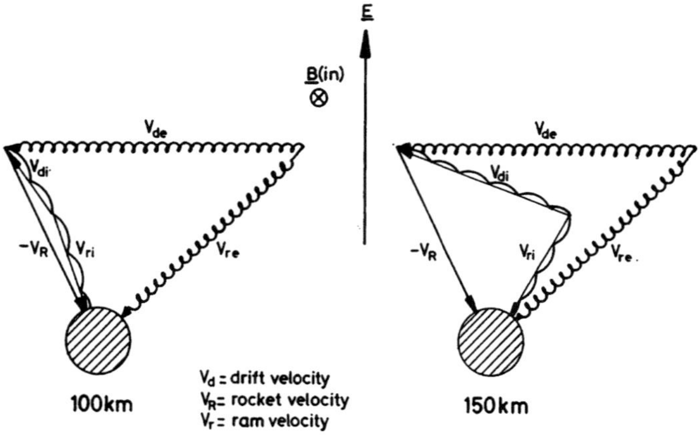

Fig. 2. Ion and electron drift velocities caused by an electric field shown in the frame of reference of the earth (V_{di}, V_{de}) and in the frame of a rocket (V_{ri}, V_{re}). The magnetic field is pointing into the paper.

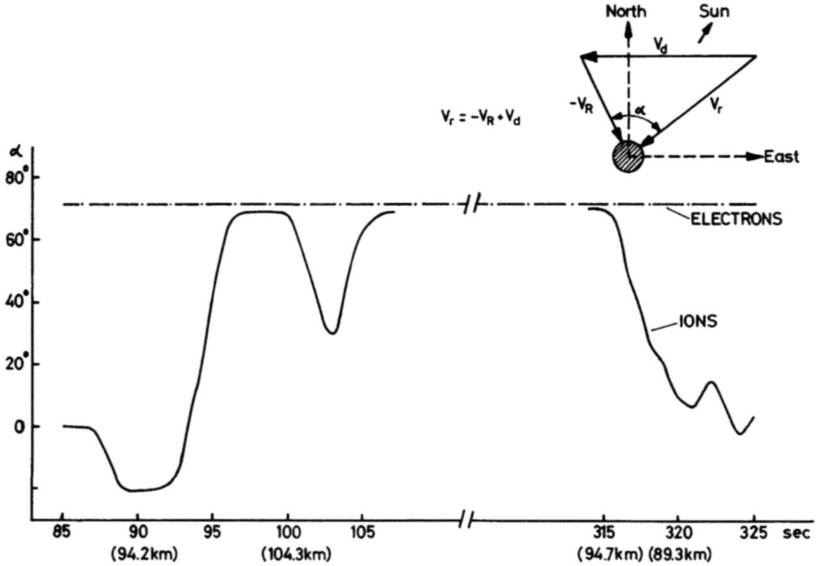

Fig. 3. Observed deviation of ion and electron ram direction from the velocity direction of the rocket in the earth's frame of reference.

at around 100 km a significant difference in ram direction. With increasing altitude the ion ram direction tends toward that of the electrons.

A summary of the observations of electron and ion ram directions for the second flight is shown in Fig. 3. The angle α, which gives the difference in direction between the velocity vector V_R and the observed ram direction, is plotted as a function of time (height). The results are based on ram direction, as mutual electrostatic coupling of the two wakes may influence their orientation. The dotted line shows the results for electrons, and the solid line for ions. The electron ram direction stays constant throughout the flight at an angle of about 70° with respect to the velocity direction. This is an indication of a comparatively strong electric field. In fact, from this angle the magnitude of the electric field can be derived. Below 100 km the ion ram direction coincides with the velocity vector. This is consistent with the ions having a small drift velocity in the lower ionosphere where collisions must be taken into account. The dip at around 94 km on the upward leg may reflect the influence of neutral wind. It should be noted that at this altitude a pronounced sporadic layer has been observed. What cannot be explained by the mobility concept is the fast change in apparent ion ram just below 100 km. According to theory this change should occur much more slowly so that at apogee the two ram directions coincide within the accuracy of the measurement. However, the coincidence of ram and sun direction made it impossible to detect the direction of the ion wake above 100 km. The second dip at 106 km, where α goes to 25°, occurred when the rocket had an attitude with minimum photoeffect on the probe.

For the determination of the electric field strength we make use of the vector equation:

$$\boldsymbol{V}_{\mathrm{re}} = -\boldsymbol{V}_{\mathrm{R}} + \boldsymbol{V}_{\mathrm{de}} \,. \tag{1}$$

In this equation we know the vector $\boldsymbol{V}_{\mathrm{R}}$ from the tracking data. Furthermore we know the direction of $\boldsymbol{V}_{\mathrm{re}}$ from the observed electron ram direction. In order to derive $\boldsymbol{V}_{\mathrm{de}}$, we have to make assumptions on its direction. From previous Ba-cloud observations it seems to be justified to assume, for plasma drifts in the auroral zone, that it is either from east to west or vice versa [5]. With this assumption we can derive $\boldsymbol{V}_{\mathrm{de}}$ from Eq. (1). The electron drift can be expressed in terms of an electric field using the relation:

$$\boldsymbol{E} = \boldsymbol{V}_{\mathrm{de}} \times \boldsymbol{B} \,. \tag{2}$$

During the first flight we observed a very small deviation of the ram and wake direction, which means that the ionospheric plasma was more or less at rest in the frame of the earth. During the second flight, however, a significant drift, corresponding to a field of approximately 20 mV m^{-1}, was found.

4. Conclusions

In spite of photocurrent affecting the ion measurements above 100 km, we can draw the following conclusions from our observations: (i) From the electron ram current, which is not suffering from photoeffect, we can derive the magnitude of northerly electric fields. (ii) It is possible to detect separate wakes for electrons and ions below 100 km in the case of an electric field being present. This is because the electrons have a gyroradius which is small compared with the diameter of the rocket, and the electron drift velocity is comparable with the rocket velocity. (iii) Our measurements below 100 km are in qualitative agreement with ion and electron mobility concepts. We expect that if the disturbing influence of the photoeffect is eliminated on future flights, it will be possible to derive mobility coefficients and estimates on electric conductivity at altitudes between 100 and 150 km from such measurements.

References

[1] V. C. Liu, Space Sci. Revs. **9**, 423 (1969).
[2] U. Samir and A. P. Willmore, Planet. Space Sci. **13**, 285 (1965).
[3] U. Samir and G. L. Wrenn, Planet. Space Sci. **17**, 639 (1969).
[4] K. Knott, G. Brommundt and G. Schumann, Planet. Space Sci. **19**, 345 (1971).
[5] G. Haerendel, paper presented at Int. Symp. on Solar Terrestrial Physics, Leningrad 1970.

Space Research XII — Akademie-Verlag, Berlin 1972

ROCKET MEASUREMENTS OF LOW ENERGY ELECTRONS DURING AURORAL EVENTS

D. J. McEwen

Institute of Space and Atmospheric Studies, University of Saskatchewan, Saskatoon, Canada

Electron spectra with good energy resolution over the energy range 20 keV–20 eV have been obtained from several rocket flights at Ft. Churchill, Canada. Auroral disturbances investigated included proton aurora, auroral break-up and quiet auroras. Some of the observed spectra, absolute fluxes and pitch angle distributions are discussed. Of particular note are secondary electrons observed moving up the field lines during proton aurora. During auroral break-up a sudden marked enhancement of incoming electrons of energy below 1 keV was observed moving down the field lines. It remains to be established whether the electrons were of local origin or injected in the equatorial plane.

1. Introduction

For a few years now investigations of the properties of low energy electrons have been carried on by means of rocket-borne electron spectrometers flown from Ft. Churchill, Canada. One advantage rockets have over satellites for such studies is that particular ionospheric disturbances can be selected for study, and in this paper a number of electron spectra obtained during a variety of auroral disturbances will be described. They include spectra from two rocket flights during normal aurora, one into a proton aurora and one during a major auroral break-up. During the proton aurora appreciable upward fluxes of electrons below 500 eV were observed, evidently secondaries produced by interaction of the incoming proton beam with the atmosphere. Electron spectra obtained during the auroral break-up show for a period of more than a minute a greatly enhanced influx of electrons in the energy range below 1 keV with no change at higher energies.

2. Instrumentation

The electron spectrometer currently in use consists basically of a 14 stage Be–Cu electron multiplier and a pair of cylindrical plates to which are applied controlled voltages initially of +2 kV and −2 kV and decaying exponentially to near zero in 0.5 second. In this way an electron energy range is scanned from about 20 keV to 20 eV. The electron multiplier has a gain of greater than 10^7. Individual pulses at the anode are counted and telemetered either by FM or PCM. Counting rates of up to about 2×10^6 per second can be recorded without difficulty.

The spectrometer aperture is 0.4 cm × 1.0 cm and, with a field of view of 9.8° × 10.8°, the resulting geometric factor is 1.3×10^{-2} cm² sr. The energy resolution $\Delta E/E$ is approximately 0.15.

The spectrometer has been flown in Black Brant II and VB rockets with apogees of the order of 150 and 375 km respectively. It is mounted in the rocket skin looking out tangentially in a direction perpendicular to the rocket axis. Hence as the rocket rotates in flight, normally at a rate of something less than 1 rev s^{-1}, the spectrometer scans through a range of particle pitch angles determined by the orientation of the rocket axis with respect to the magnetic field.

The electron spectra shown in Figs. 1–3 were obtained from an earlier version of this spectrometer with a reduced energy range of 10 keV to 20 eV and lower energy resolution of 0.6.

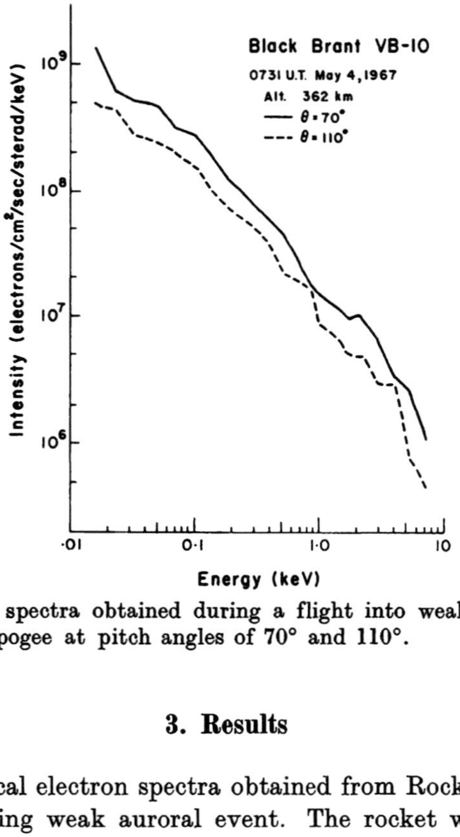

Fig. 1. Electron spectra obtained during a flight into weak aurora around apogee at pitch angles of 70° and 110°.

3. Results

Fig. 1 shows typical electron spectra obtained from Rocket VB-10 launched into an early morning weak auroral event. The rocket was near apogee at 362 km and as the coning angle was small, the range of pitch angles sampled was confined to $\theta = 70°$ to 110°.

The spectrum is relatively soft and fitted by a power law of approximately $E^{-1.25}$. The upward flux at 110° pitch angle is about one half the flux at 70° pitch angle over the whole energy range covered. The rocket at this time was not in the vicinity of discrete auroral forms and the spectrum appears rather typical of precipitation during mildly disturbed ionospheric conditions.

Spectra obtained from the flight of Rocket VB-18 into a quiet diffuse aurora of 5–10 kR brightness around local midnight on 2 August 1968 are shown in Fig. 2. As the aurora was quite stable and extended, the rocket was in or above visual aurora for virtually all of the flight. There were no marked changes of electron flux or spectrum observed during the flight. Both spectra shown were recorded at near apogee with pitch angles about 90°. The lower solid curve is a spectrum typical of much of the flight while the dashed curve represents the maximum intensity observed.

The spectrum is quite flat from 1 to 8 keV with somewhat of a peak at 4.5 keV. The true peak may well be much more pronounced than is shown by this

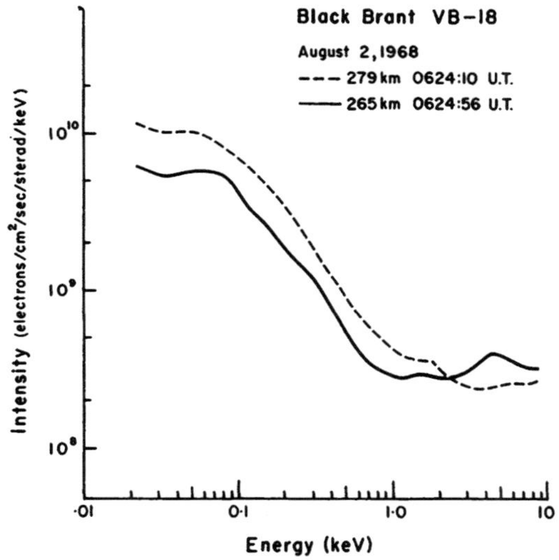

Fig. 2. Electron intensities measured during a rocket flight into diffuse aurora of brightness 5–10 kR at 90° pitch angle. The solid curve is typical of a good part of the flight while the dashed curve indicates the maximum intensity observed.

spectrum obtained with low (0.6) energy resolution. The form is quite similar to that of the peaked spectrum first observed by McIlwain [1] and more recently by Evans [2], Westerlund [3] and others. As Westerlund has pointed out it appears characteristic of electron spectra observed within or above visual auroral forms as distinct from the softer continuum type spectrum observed in general removed from auroral forms and perhaps typified in Fig. 1.

Electron spectra obtained from a similar Rocket, VB-11, launched into a proton aurora at 0340:42 UT on 22 January 1968, are shown in Fig. 3. A Hβ emission intensity of about 50 R was measured from the ground at launch and no visible aurora was observed in the region of proton precipitation, suggesting that the electron influx would not be appreciable. The results from other experiments aboard this rocket have previously been reported by Miller and Shepherd [4] who measured the Hβ hydrogen emission both from the ground and from the rocket with two-channel photometers, and by Whalen and McDiarmid [5] who measured a downward proton flux of approximately 7×10^6

protons $cm^{-2}\, s^{-1}\, sr^{-1}$ above 2 keV energy peaked at 80° pitch angle. The integral upward flux was about one quarter of this value.

The rocket had an abnormally slow spin period of about 50 seconds and was lying nearly horizontal for the middle portion of the flight, so a full range of pitch angles was sampled only once each 25 seconds. The spectra shown in Fig. 3 were recorded over one full rotation of the rocket while the rocket was near apogee and moving slowly through a height region from 360 km to 375 km. The spectra shown over the time interval from $T + 243$ sec to $T + 293$ sec are for pitch angles going from $\theta = 156°$ down to 24° and back up to 156°. There was no appreciable flux change observed at 90° pitch angle at the two

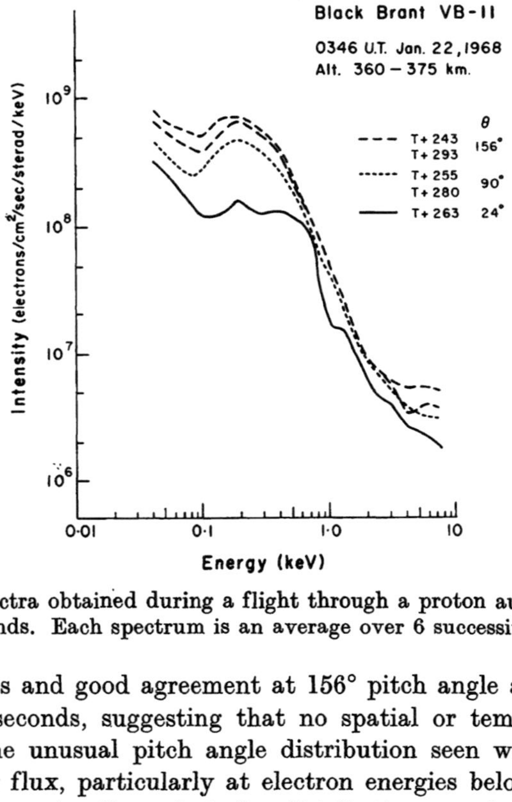

Fig. 3. Electron spectra obtained during a flight through a proton aurora over a period of 50 seconds. Each spectrum is an average over 6 successive scans.

intermediate times and good agreement at 156° pitch angle at the two times separated by 50 seconds, suggesting that no spatial or temporal effect was responsible for the unusual pitch angle distribution seen where there is an upward travelling flux, particularly at electron energies below 500 eV, much greater than the incoming flux. A similar distribution was also observed in the preceding 50 seconds of electron spectral data.

The intensity of incoming electrons is very low, especially in the 1–10 keV region as anticipated from the absence of visible aurora. The observed net flow of electrons upward evidently consists of secondaries produced at lower heights by ionizing collisions of the incoming proton beam. In the energy range of a few keV the dominant flux constituent will be energetic H and the electron production reactions are

$$H + M \rightarrow H^{+} + M^{*} + e\,, \qquad (1)$$

$$H + M \rightarrow H + M^{+} + e\,. \qquad (2)$$

The cross sections for electron production from (1) and (2) are approximately 2×10^{-16} cm^2 and 5×10^{-16} cm^2 respectively as measured by Stier and Barnett [6] and McNeal and Birely [7] for collisions with N_2 and O_2. Both of these values are effective ones averaged over the proton spectrum observed by Whalen and McDiarmid [5].

The electron spectra, such as shown in Fig. 3, thus should allow determination of the energy distribution of the secondary electrons produced by reactions (1) and (2) as well as some indication of the incoming proton flux below 2 keV. The results of this investigation will be reported in detail elsewhere.

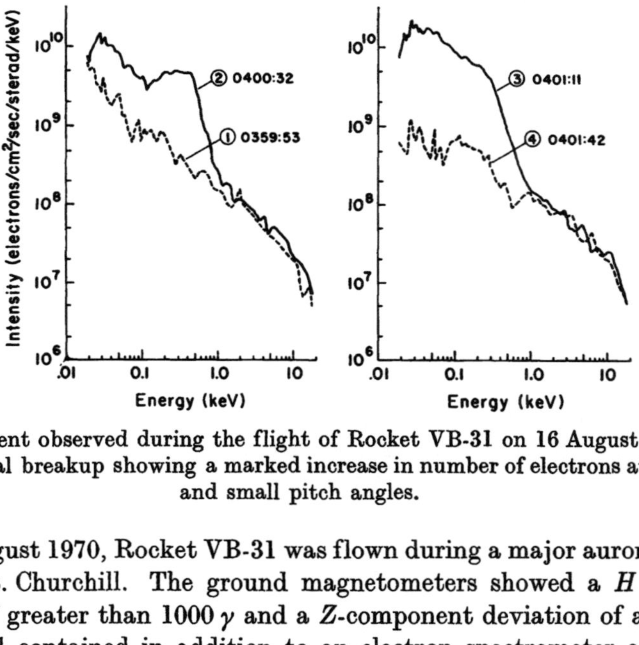

Fig. 4. An event observed during the flight of Rocket VB-31 on 16 August 1970, during a major auroral breakup showing a marked increase in number of electrons at low energies and small pitch angles.

On 16 August 1970, Rocket VB-31 was flown during a major auroral break-up event at Ft. Churchill. The ground magnetometers showed a H component deviation of greater than 1000 γ and a Z-component deviation of about 500 γ. The payload contained in addition to an electron spectrometer a number of electric field and plasma probes.

While the analysis of the data is not yet complete, a number of interesting features were observed in the flight. The most noteworthy event in the electron spectra was a sudden marked enhancement of incoming electrons of energy below 1 keV which appeared as illustrated in Fig. 4. This shows four spectra recorded at small pitch angles during the time interval 0359:53 to 0401:42 UT while the rocket was close to apogee at 265 km. Spectrum (2) shows a factor of 10 increase in intensity of electrons between 250 and 500 eV energy over that of spectrum (1) with no increase at energies greater than 1 keV. The intensity enhancement then gradually extended to energies below 100 eV as illustrated in spectrum (3) and finally decayed to the form shown in spectrum (4). This low energy enhancement was most noticeable at small pitch angles.

This marked increase in low energy electron influx could be explained by a local electric field directed up along the magnetic field causing ambient electrons to be accelerated downward. Preliminary results of Mozer and Kelley

[8] from the electric field probes do show the existence of a parallel field over this time interval but it is in the opposite direction to that which would give the observed enhanced electron influx.

DeForest and McIlwain [9] have reported similar enhanced low energy parallel electron fluxes recorded by the ATS-5 satellite in synchronous orbit for a period of some 4 hours during a magnetic storm on 2 January 1970. They conclude that these are due to injection of plasma into the midnight sector of the magnetosphere. This must be considered a more probable source of the low energy electrons observed during the auroral break-up by Rocket VB-31.

The full analysis of the low energy electron rocket data in conjunction with the electric field and electron density measurements should provide clear evidence as to the source of these electrons, whether they be of local origin or injected at several earth radii. The results of this study will be published in detail in due course.

Acknowledgments

Much of the work on the investigation reported was carried out while the author was at the Communications Research Centre, Ottawa. The assistance of the staff of that laboratory, particularly Drs. I. Paghis and R. Barrington for their support and encouragement, Dr. C. Franklin for consultations on the design of the electronics, and Mr. G. Bird for his conscientious work in constructing and calibrating the several electron spectrometers used, is gratefully acknowledged. The data analysis has been supported by research grants from the National Research Council of Canada.

References

[1] C. E. McILWAIN, J. Geophys. Res. **65**, 2727 (1960).
[2] D. S. EVANS, in: Atmospheric Emissions, Van Nostrand and Reinhold Co., New York 1969 (p. 107).
[3] L. H. WESTERLUND, J. Geophys. Res. **74**, 351 (1969).
[4] J. R. MILLER and G. G. SHEPHERD, Space Research X, 876 (1970).
[5] B. A. WHALEN and I. B. McDIARMID, in: Atmospheric Emissions, Van Nostrand and Reinhold Co., New York 1969 (p. 93).
[6] P. M. STIER and C. F. BARNETT, Phys. Rev. **103**, 896 (1956).
[7] R. J. McNEAL and J. H. BIRELY, in: The Radiating Atmosphere, Reidel Publ. Co. Dordrecht, Holland 1971.
[8] F. S. MOZER and M. KELLEY, private communication (1971).
[9] S. E. DEFOREST and C. E. McILWAIN, J. Geophys. Res. **76**, 3587 (1971).

Space Research XII – Akademie-Verlag, Berlin 1972

EVIDENCE OF A SYSTEMATIC VARIATION OF THE RATIO BETWEEN THE INTENSITIES OF THE AURORAL EMISSIONS AT 5577 Å AND 3914 Å

A. Monfils and J. M. Vreux

Institut d'Astrophysique, Université de Liège, Cointe-Sclessin, Belgium

The processing of intensities recorded by means of photometers aboard a spinning and precessing rocket has included a complete attitude reconstitution of the incident radiation. A selection of measurements made in horizontal planes shows a stronger relative increase of 3914 Å emission near 105 km, the altitude of maximum intensity, compared with the 5577 Å emission profile.

1. Measurements

One of the problems which must be resolved by the mechanisms proposed in order to explain the production of oxygen in the ^1S state during an auroral phenomena concerns the behaviour, as a function of altitude, of both the forbidden line of atomic oxygen at 5577 Å and the (O–O) band of the First Negative system of N_2^+ at 3914 Å.

We have made relevant observations by means of photometers placed aboard rockets; the optical axes were perpendicular to that of the vehicle and the sensitivities were respectively centred on 5577 Å and 3914 Å. These photometers were part of the Centaure payloads C52/1 and C52/2 of ESRO which were launched to an altitude of 136 km at ESRANGE (67° 54′ N 21°06′ E). The C52/1 flight took place on 15 March 1969 at 2331.02 LT. The rocket trajectory was situated between two parallel arcs. Some inhomogeneities, clearly visible to the naked eye, were moving along the arcs. The C52/2 flight took place on 10 October 1969 at 2101.30 LT. The sky was partly cloudy over ESRANGE and the ground-based observations were difficult. Intense and active auroral forms were present all over the sky. Relatively quick motions are not excluded.

The half angles of view of the optical systems were 7°. The divergence of the light beam penetrating the photometers was reproduced when the transmission curves of the interference filters were measured. We have also taken into account the intensity repartition in the (O–O) band of the First Negative system of N_2^+. The absolute calibration of the detectors was achieved by means of a tungsten ribbon lamp of the National Bureau of Standards (estimated precision 20%) [1]. Furthermore, it has been possible to recalibrate the photometers after recovery [2]. Photometric observations, whether made from the ground or from rockets integrate ϱ (the volumetric emission) along the line of sight. It is evident that it is impossible from only one photometer measurement to transform the integrated values without admitting important simplifying hypotheses on the distribution of ϱ_λ (h, θ, φ, t), where h is the distance between the point considered and the ground, measured along the line of force of the magnetic field, λ is the wavelength, θ and φ respectively the geomagnetic latitude and longitude from the point where the line of force passing by the emitting elementary volume intersects the ground.

The analysis of measurements made by means of photometers whose line of sight is parallel to that of the vehicle is generally based on the hypothesis that ϱ_λ (h, θ, φ, t) may be reduced to $\varrho_\lambda(h)$, which is certainly not valid in all cases. As far as the measurements made by means of photometers whose line of sight is perpendicular to that of the rocket spin axis are concerned, the measurement corresponds to the integral

$$\int_{l_1}^{l_2} \varrho_\lambda(h, \theta, \varphi, t)\, dl .$$

It is evident that, in this case, the data handling must be made by selecting the data corresponding to the horizontal optical axis, azimuth by azimuth. This, if hypotheses similar to those used for the first type of measurements are used, permits rational conclusions to be reached. However, we had no direct access to the average value of ϱ. This stage would necessitate a further hypothesis concerning the limits of integration, unless these have been determined experimentally.

2. Results

In order to stress the importance of the method followed for handling of the observations, we shall first present two diagrams corresponding respectively to the C52/1 and C52/2 launchings. They have been obtained by integration of the photometric data for each rotation period of the rapidly spinning Centaure rocket about its axis. As the maxima of the elevation of the spin axis were very close to 90° the average values obtained at these times are representative of the intensities at the corresponding altitude. From these values, unfortunately not very numerous, the average intensity of the aurora as a function of altitude is obtained (Fig. 1a). In the case of the C52/2 launching, the rocket was seriously disturbed at nose-cone ejection. As a result, the precession angle is large ($\pm 20°$) so that the elevation of the spin axis of the rocket is oscillating between 60° and 80°. Fig. 1b has been drawn from the average values for each precession period. It is important to note that for both flights, the ratio I_{5577}/I_{3914} of the average intensities is remarkably independent of altitude, which is in close agreement with preceding measurements.

For the second part of the analysis, we have among our data points retained only those corresponding to the moments when the optical axes of the photometers were parallel to a horizontal plane. We shall present here the results obtained with C52/2 in October 1969 but the same conclusions were reached for the observations made during the flight of C52/1 [2].

During the flight of C52/2, the auroral phenomena studied were presenting different characteristics for different emitting zones. The altitude and the value of the intensity maximum were not constant with azimuth. On the other hand, the ratio of the measured intensities now present important variations with altitude and a different behaviour according to the azimuth. It appears that the gradient of the curve $R(h) = I_{5577}(h)/I_{3914}(h)$ changes sign around the maximum of luminosity. We have summarized these observations for a few azimuths by drawing on the same diagram (Fig. 2) the profile of the intensities

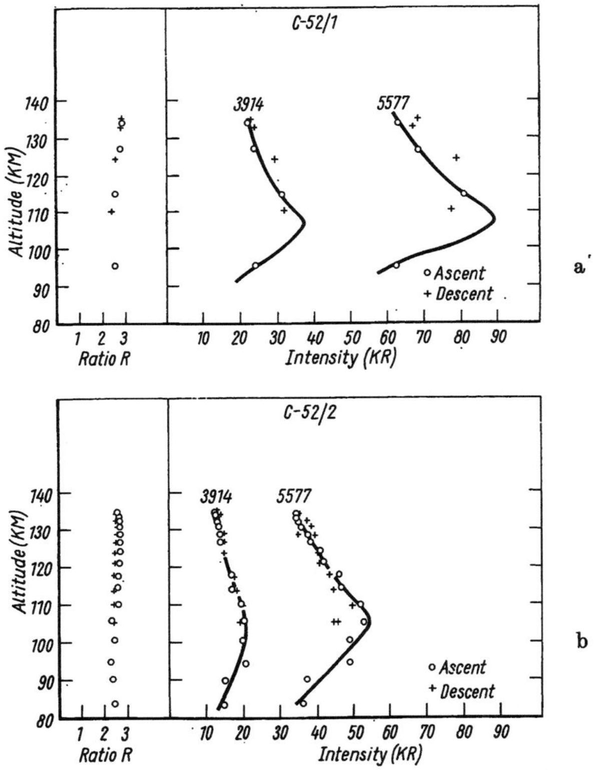

Fig. 1. Integrated intensities of the auroral emissions at 3914 Å and 5577 Å in terms of altitude. The ratio is shown to be roughly constant. a, flight C52/1; b, flight C52/2.

measured in the horizontal plane by the photometer centred on 3914 Å (N_2^+), during the ascent of the flight of C52/2, as well as the corresponding values of the ratio of the intensity of the emissions situated at 5577 Å and 3914 Å. This diagram has been interrupted at 130 km as, beyond this distance, a noticeable dispersion of the measuring points is observed. This is the result of a time effect, due to the fact that at that moment the ascending speed of the rocket is very small.

No feature corresponding to angularly localized phenomena has been presented in this diagram.

The diagram corresponding to the C52/2 flight shows clearly that an analysis of the variation with altitude of the ratio R in a direction corresponding either to the maximum of intensity during each rotation of the rocket about its axis [3] or to the minimum [4, 5] carries important risks. Our observation is only based upon two flights (carried out on intense auroral arcs, whilst many observations in the past corresponded to diffuse forms of the post break-up phase). We consider, as special cases of our description, the only two analyses carried out previously which were likely to lead to the same result, i.e. the work of Romick and Belon [6] and of Duysinx [7]. As far as the latter is

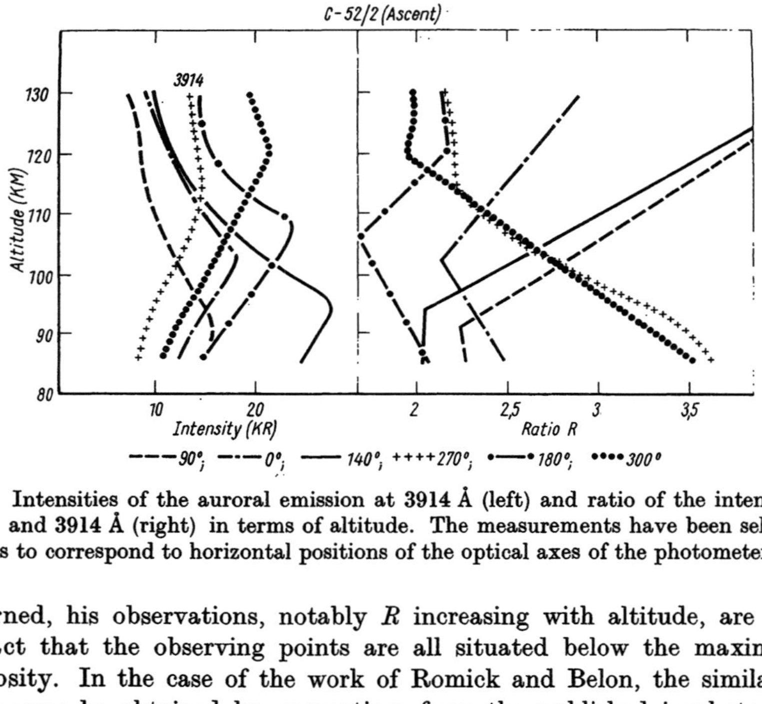

Fig. 2. Intensities of the auroral emission at 3914 Å (left) and ratio of the intensities at 5577 Å and 3914 Å (right) in terms of altitude. The measurements have been selected so as to correspond to horizontal positions of the optical axes of the photometers.

concerned, his observations, notably R increasing with altitude, are due to the fact that the observing points are all situated below the maximum of luminosity. In the case of the work of Romick and Belon, the similarity of results may be obtained by computing, from the published isophotes, what would have been the observation from a rocket equipped with photometers with optical axes perpendicular to the axis of the rocket. The measurements carried out during the flight of C52/2 show that the mechanism which causes the variation of the ratio R is much more dependent on the process of ionization by the flux of primary electrons than on the properties of the atmosphere. The latter would rather lead to an effect depending only on altitude. An apparent decrease of the altitude of the maximum of emission due to an increase of the distance from the emitting zone would imply a decrease of the intensity of the maximum of emission and particularly a flattening of the luminosity curve, which does not seem to be the case (curve corresponding to the azimuth of 140°).

Our observations cannot be interpreted by the superposition of the emission of the night sky on the auroral emission. If this be the case, one would observe an apparent increase of the ratio R in the vicinity of 95–100 km because the contribution of the night sky is negligible at 3914 Å and because this altitude corresponds to the maximum of emission at 5577 Å by the night sky [8, 9]. We observe exactly the reverse, i.e. a minimum of R. The emission of the night sky should rather have a tendency to mask the phenomenon shown here but it may, perhaps, explain partially the relatively high values of the ratio R observed at low altitudes in spite of molecular deactivation by collisions. The influence of the contribution of the night sky may, however, become important in the analysis of launches effected in weaker aurorae. It seems, consequently, reasonable to interpret our observations as the effect of a mechanism that, in the vicinity of the maximum of ionization, would accelerate the greater part of the electrons up to energies of a few tens of eV favouring in this way the emission of the (O–O) band of the First Negative system of N_2^+ with respect to emissions whose sections present a maximum for smaller energies. Such an acceleration mechanism has already been postulated in order to interpret certain measurements showing comparative variations of the emission rate with altitude and of ionic concentration [10, 11] as well as to explain related behaviour with altitude of the (O–O) band of the First Negative system of N_2^+ and some lines of atomic oxygen [12] or even of certain bands of the Lyman–Birge–Hopfield system [13]. Donahue et al. [11] have proposed a parallel electric field of a few tens of mV m^{-1} created by the space charges due to the dumping of primary electrons in the vicinity of the maximum of ionization. This hypothesis seems to us to be compatible with our observations.

References

[1] G. Marette, Comm. No. 111, Inst. Astrophys. Univ. Liège (1971).
[2] J. M. Vreux, Mém. Faculté des Sciences, Univ. Liège (1971).
[3] W. B. Murcray, J. Geophys. Res. **74**, 366 (1969).
[4] K. Hirao, J. Geomagn. Geoelect. **19**, 247 (1967).
[5] K. D. Baker, Annls Géophys. **24**, 285 (1968).
[6] G. J. Romick and A. E. Belon, Planet. Space Sci. **15**, 1695 (1967).
[7] R. Duysinx, Mém. Faculté des Sciences, Univ. Liège (1970).
[8] R. G. H. Greer and G. T. Best, Planet. Space Sci. **15**, 1857 (1967).
[9] I. S. Gulledge et al., J. Geophys. Res. **73**, 5535 (1968).
[10] J. C. Ulwick, W. P. Reidy and K. D. Baker, Space Research VII, 656 (1966).
[11] T. M. Donahue, E. C. Zipf and T. D. Parkinson, Planet. Space Sci. **18**, 171 (1970).
[12] D. J. Strickland, Ph. D. Thesis, Univ. Pittsburgh 1968.
[13] C. B. Opal, H. W. Moos and W. G. Fastie, J. Geophys. Res. **75**, 788 (1970).

Space Research XII — Akademie-Verlag, Berlin 1972

HIGH-LATITUDE PRECIPITATION OF LOW-ENERGY PARTICLES AS OBSERVED BY ESRO I A

W. RIEDLER[a] and H. BORG[b]

[a] Technical University Graz, Department of Communications and Wave Propagation, Graz, Austria

[b] Kiruna Geophysical Observatory, Kiruna, Sweden

Average differential flux values at ~1, 6, 13 keV (electrons) and ~6 keV (protons) at ~10° pitch-angles from the polar-orbit satellite ESRO 1 A (Aurorae) are discussed. The data, which have been accumulated over a period of 20 months, are presented in the form of colour-coded intensity contours as functions of magnetic local time and invariant latitude for four different levels of magnetic activity.

1. Introduction

Many properties of the morphology of low-energy particle precipitation have become known during the past few years. A review has been given by Sharp and Johnson [1]. The purpose of the present paper is to present average values of differential low-energy electron and proton fluxes, which have been collected during a period of 20 months.

1.1. Particle Spectrometer S 71-B

The Kiruna particle spectrometer S 71-B consisted of electrostatic analysers, channel multipliers and the necessary electronics. The sensors were arranged in two boxes of five units each, at angles of 10° and 80° respectively with the satellite axis. In this paper data from the 10° sensors will be used. The energy values selected, together with some other parameters, are given in Table 1. The energy bandwidths were in all cases between 10 and 20%, the acceptance angles of the order ±8°. A detailed description of the experiment is given in [2].

1.2. Spacecraft

The satellite ESRO 1 A (Aurorae) was launched on 3 October 1968 at 1349 PDT (local time) into a near-polar orbit (inclination 93.8°) and decayed on 26 June 1970. The apogee after launch was 1533 km, the perigee 258 km. The spacecraft was magnetically stabilized, with the S 71-B sensors looking upwards over the northern polar region. Redu (Belgium), Tromsø and Spitsbergen (Norway) and Fairbanks (Alaska) have been used as real-time read-out stations for S 71-B. Experiment S 71-B was switched on for the first time on 14 October 1968. It has not been operated on all revolutions, in order to avoid degradation effects of the channel multipliers. In the following study, data collected on approximately 4000 revolutions from 14 October 1968 until 26 June 1970 have been used.

2. Results

2.1. General Remarks

The auroral particle precipitation patterns are presented in (Λ, MLT) coordinates, where Λ indicates invariant latitude and MLT magnetic local (eccentric dipole) time. Values below 54.5° and above 84.5° were not used, mainly because of poor data coverage. The data were grouped according to four classes of Kp: $Kp = 0^0$–1^0, $Kp = 1^+$–3^0, $Kp = 3^+$–5^0, $Kp = 5^+$–9^0. The (Λ, MLT) plane was subdivided into $1° \times 20'$ cells. For each cell and each Kp class, the corresponding differential flux values measured by S 71-B during its operational period over the northern polar region were averaged.

In Figs. 1–4* the results are shown. The average intensities are given by means of a colour code. White colour indicates that no data were available. Yellow colour indicates that no particle count has been obtained, although the satellite passed the area at least once. The flux values were then below the detection thresholds, given in Table 1. Many cells have been sampled several

Table 1

Sensor No.	S1	S2	S4	S6
Particle type	el.	el.	pr.	el.
Centre energy (keV)	5.8	13.3	5.8	1.3
Flux threshold (part. $cm^{-2}\ s^{-1}\ sr^{-1}\ keV^{-1}$)	4.8×10^3	4.9×10^3	3.7×10^3	1.3×10^5

hundred times, especially for higher latitudes. The nominal pitch-angle was 10°. Due to the late switch-on of S 71-B the initial tumbling period of the spacecraft was excluded. Later, the satellite exhibited maximum misalignment angles of only a few degrees.

UT variations of the precipitation patterns have been taken into account only to the extent that they are represented in Kp. Dipole wobbling and seasonal effects have been neglected here, in an attempt to obtain a better statistical coverage of the (Λ, MLT) plane. The units used throughout this paper are: particles $cm^{-2}\ s^{-1}\ sr^{-1}\ keV^{-1}$.

2.2. Protons 5.8 keV

$Kp = 0^0$–1^0 (Fig. 1 a). With the exception of the noon–dusk region practically no measurable fluxes can be seen below ~67°. The increase towards higher values is much steeper on the night side (2 orders of magnitude over 1°) than on the day side. It should be mentioned here that this steep night side slope is

* Between pages 1400 and 1401.

not present in the 80° particle data (not further discussed in this paper). In other words, a pronounced depleted loss-cone anisotropy can be observed at the low-latitude border of the precipitation zone during the night, but not during the day. An exception is the noon–dusk region at low latitudes, where the 80° intensities usually exceed the 10° ones. This fully confirms the earlier findings from individual passes [3], which showed that a depleted loss-cone anisotropy invariably occurs on the night side and also in the noon–dusk region. The main precipitation zone ($\geqq 10^5$) is centred at approximately 68° at midnight and at slightly higher latitudes in the adjacent time regions. A dusk–dawn asymmetry can be observed in the sense that this latitudinal increase is more pronounced on the dawn side. Approximately at 05 h and 19 h marked discontinuities can be observed: on the dusk side the maximum intensity zone is shifted poleward by about 5°; on the dawn side this shift is less pronounced. The maximum day-side zone is centred at ~76°. A "valley" in the average intensity distribution is seen, extending from 09 h to 13 h. There is an indication of a second "valley" around 0130, although statistics are poorer in this region. The width of the precipitation region ($\geqq 10^4$) varies from ~7° at midnight to ~12° on the day side. Due to the noon "valley" mentioned above, it is quite narrow shortly before noon. The polar cap region, if defined by the apparent "boundary" of the precipitation zone, extends down to ~73° at midnight. At noon, this "boundary" is almost 10° higher in latitude. It is almost circular from ~05 h to 19 h MLT, the edges coinciding with the above-mentioned discontinuities.

$Kp = 1^+–3^0$ (Fig. 1b). The precipitation regions show a pronounced broadening compared with $Kp = 0^0–1^0$, whereas the maximum intensities do not change much. The equatorward shift of the southern "boundaries" is ~3°. At the poleward edge a tendency toward disappearance of the noon–midnight asymmetry can be noticed. More yellow colour can be seen on the polar cap indicating a slight diminution of the already low proton fluxes. The discontinuities at ~19 h and, not so clear, at ~05 h are still there. The "valley" has now extended to ~16 h. The fluxes at the low-latitude outskirts in the noon–dusk region have increased somewhat.

$Kp = 3^+–5^0$ (Fig. 1c). The precipitation regions are much broader (~15°), extending equatorward to ~60° at night. The shape of the polar cap region has become more circular and the fluxes have diminished even more. In the noon region, a shift of the polar cap border to lower latitudes by some degrees can be seen, whereas near midnight some higher fluxes are observed up to 80°. The discontinuities are no longer clear, perhaps with the exception of a "tongue" extending to high latitudes around 19 h. The noon "valley" is situated predominantly on the dusk side and the low-latitude fluxes have increased.

$Kp = 5^+–9^0$ (Fig. 1d). The precipitation zones are very broad, extending to $<55°$ in the dusk sector and also in the dawn–noon region. The apparent polar cap area has now merged with the "valley" in the sense that down to 70° no fluxes are observed in the noon–dusk region. In sharp contrast, the average intensities at other local times are now the highest observed (up to 10^7), especially at low latitudes.

2.3. Electrons 1.3 keV

$Kp = 0^0–1^0$ (Fig. 2a). The equatorward boundary of the precipitation region is situated at ~64° for almost all local times. With some minor exceptions, the highest average intensities observed are between 10^8 and 3×10^8. They occur in an almost circular zone centred around 68°–71° on the night side. At ~05 h an abrupt shift towards higher latitudes by about 4° can be seen. On the day side, intensities are lower. From ~12 h to 0130 a "valley" like the one discussed in § 2.2 can be seen, although it is not too well developed in the present case. The polar cap is covered by strong, but very structured precipitation.

$Kp = 1^+–3^0$ (Fig. 2b). An equatorward shift of the boundary by about 2° is noticed. The intensities have increased somewhat, leading to a more extended region of $\geqq 10^8$. The latter appears as two quasi-circular zones, centred on the night side around ~68°–70°, on the day side around ~73°. The shift in latitude takes place at about 05 h and before 19 h. In the noon–dusk region the "valley" is very pronounced for latitudes $<$75°. The observed fluxes over the polar cap have decreased slightly.

$Kp = 3^+–5^0$ (Fig. 2c). The precipitation region extends down to 60° from dusk to late morning. Between noon and dusk the "valley" is strongly developed. The maximum intensities have increased, with peak values $\geqq 10^9$. The two quasi-circular zones have merged into one broad region. In wide areas of the polar cap the intensities have dropped below the detection threshold.

$Kp = 5^+–9^0$ (Fig. 2d). Maximum intensities of $\geqq 10^9$ are concentrated to the morning and especially to the late afternoon regions. Down to $<$55° strong precipitation can be observed. The "valley" after noon is very broad and deep. On the day side of the polar cap region the fluxes are very often below detection threshold. On the night side, however, quite high non-structured intensities can be observed to $>$84°.

2.4. Electrons 5.8 keV

$Kp = 0^0–1^0$ (Fig. 3a). The precipitation zone is almost ringshaped, with practically no observed intensities in the polar cap region. It extends from ~65° to ~77° on the day side, to ~80° on the night side. The division between day and night side high-latitude boundary could be taken around 19 h and 05 h. A "valley" around 02 h is certainly not caused by the poorer statistics in this area.

$Kp = 1^+–3^0$ (Fig. 3b). The ring has broadened considerably, extending equatorward to ~62°. In the polar cap day side region higher fluxes are

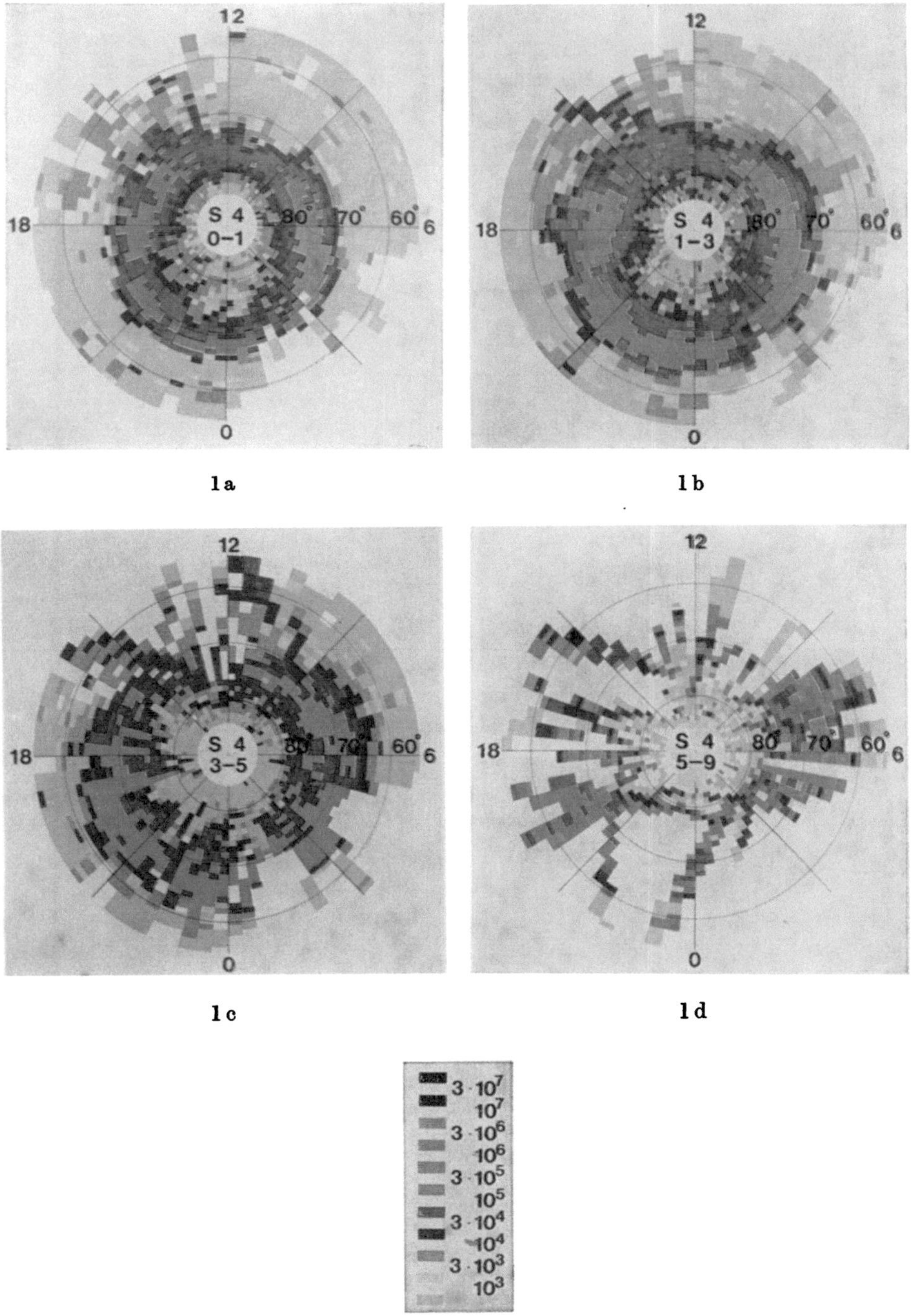

Fig. 1. Average differential fluxes of 5.8 keV protons in (Λ, MLT) coordinates. Units are protons cm^{-2} s^{-1} sr^{-1} keV^{-1}.
a, $Kp = 0^0$–1^0; b, $Kp = 1^+$–3^0; c, $Kp = 3^+$–5^0; d, $Kp = 5^+$–9^0.

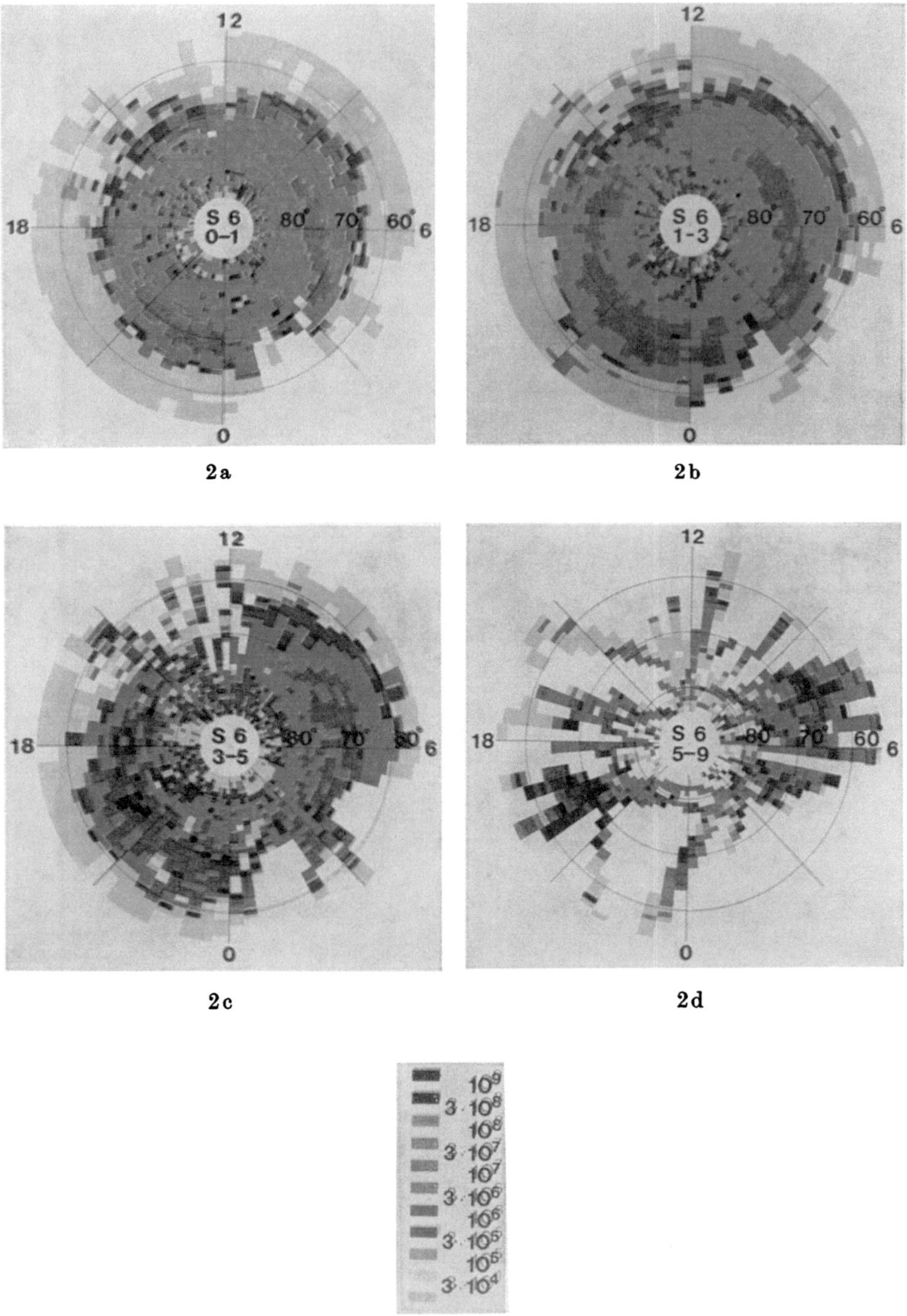

Fig. 2. Average differential fluxes of 1.3 keV electrons in (Λ, MLT) coordinates. Units are electrons $cm^{-2}\,s^{-1}\,sr^{-1}\,keV^{-1}$.
a, $Kp = 0^0$–1^0; b, $Kp = 1^+$–3^0; c, $Kp = 3^+$–5^0; d, $Kp = 5^+$–9^0.

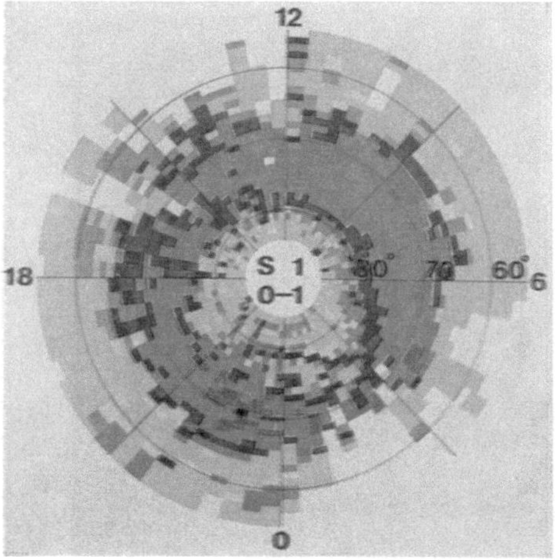

3a

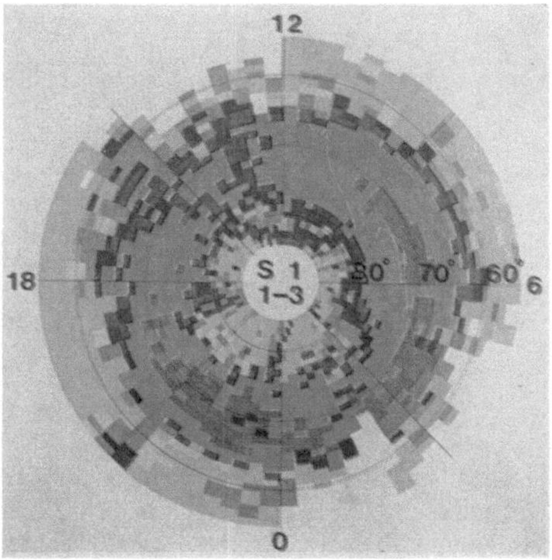

3b

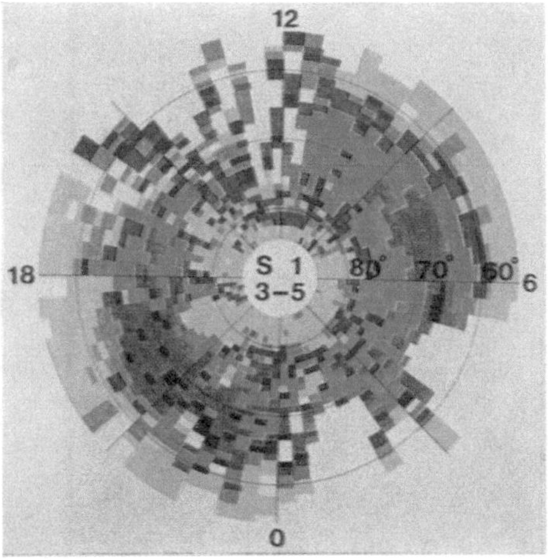

3c

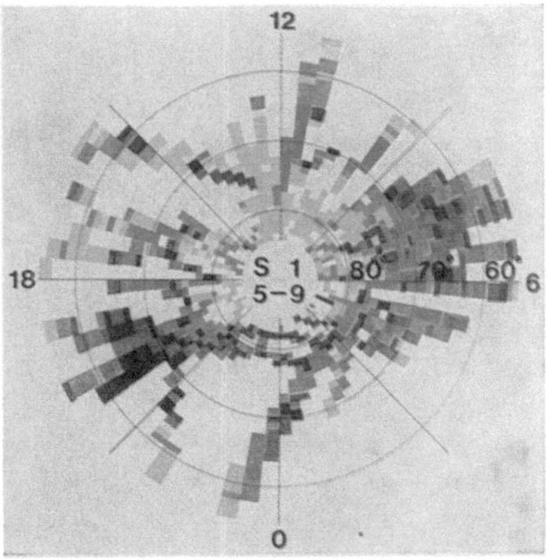

3d

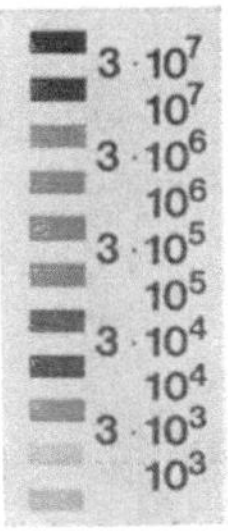

Fig. 3. Average differential fluxes of 5.8 keV electrons in (Λ, MLT) coordinates. Units are electrons $cm^{-2}\, s^{-1}\, sr^{-1}\, keV^{-1}$.
a, $Kp = 0^0$–1^0; b, $Kp = 1^+$–3^0; c, $Kp = 3^+$–5^0; d, $Kp = 5^+$–9^0.

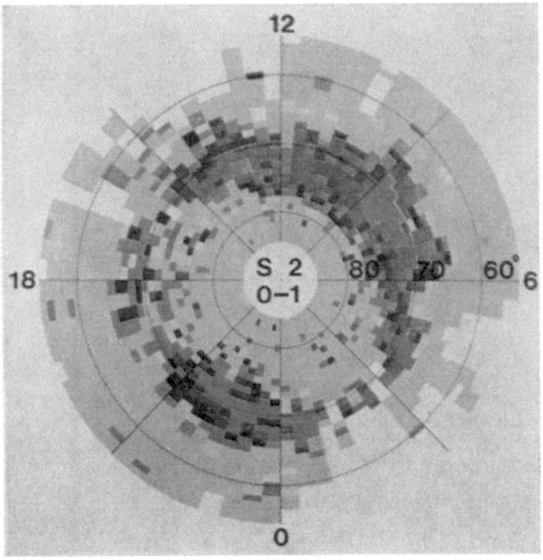

4 a

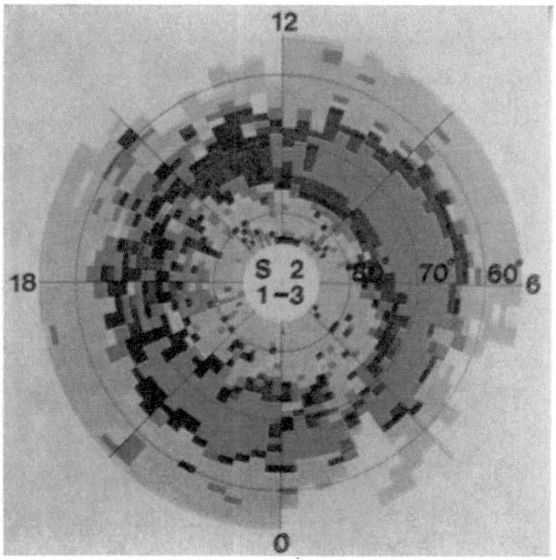

4 b

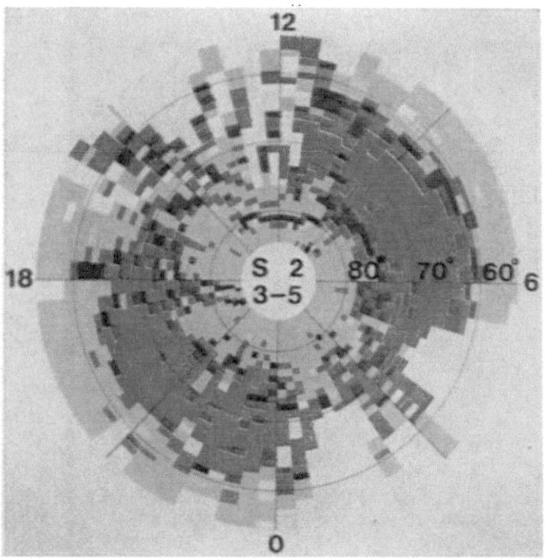

4 c

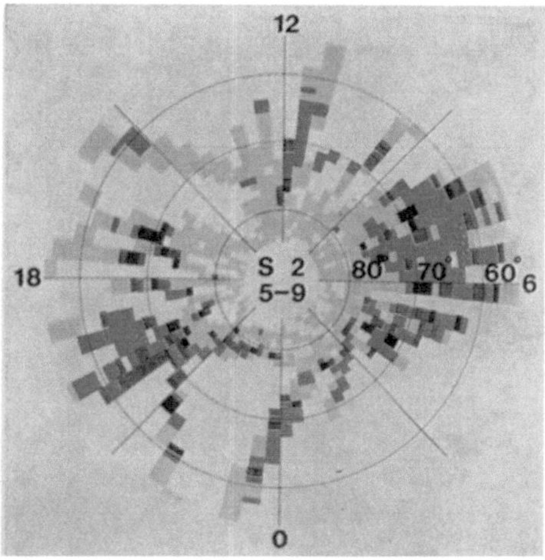

4 d

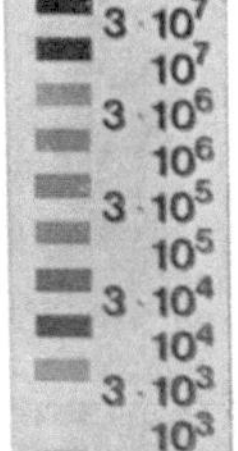

Fig. 4. Average differential fluxes of 13.3 keV electrons in (Λ, MLT) coordinates. Units are electrons $cm^{-2}\ s^{-1}\ sr^{-1}\ keV^{-1}$.
a, $Kp = 0^0$–1^0; b, $Kp = 1^+$–3^0; c, $Kp = 3^+$–5^0; d, $Kp = 5^+$–9^0.

observed from ~05 h to ~19 h. The intensities generally have increased. A "valley" from noon to 15 h is observed, whereas in the 02 h region somewhat higher fluxes are present, compared with Fig. 3a.

$\underline{Kp = 3^{+}\text{–}5^{0}}$ (Fig. 3c). Intensities have further increased, with night side average peak values of about 3×10^{7}. The southern boundary has again shifted equatorward by ~2° and is situated around 60°. The particle flux over the polar cap is less than before and the noon–dusk "valley" more pronounced.

$\underline{Kp = 5^{+}\text{–}9^{0}}$ (Fig. 3d). The "valley" has become considerably broader and deeper. In large areas around noon no fluxes are detectable. At the same time, the maximum intensities around 20 h reach the highest values observed ($\geqq 3 \times 10^{7}$). Also, in the dawn–noon region high fluxes are present. The polar cap night side exhibits some medium–high intensities up to >84°.

2.5. Electrons 13.3 keV

$\underline{Kp = 0^{0}\text{–}1^{0}}$ (Fig. 4a). The highest fluxes are observed shortly before midnight, but the precipitation zone is broadest in the dawn–noon region. A deep "valley" is seen, centred around 02 h. Practically no observable fluxes are recorded in the polar cap area.

$\underline{Kp = 1^{+}\text{–}3^{0}}$ (Fig. 4b). The intensities have increased considerably and the equatorward boundary, where applicable, has shifted by about 2°. The 02 h valley is less pronounced, whereas the noon–dusk "valley" has developed. Some discontinuities can be seen around 05 h and 19 h. Some polar cap fluxes are observed.

$\underline{Kp = 3^{+}\text{–}5^{0}}$ (Fig. 4c). The intensities have further increased and tend to concentrate at dawn–noon and dusk–midnight. Maximum values are around 3×10^{6}. Again, the equatorward shift is about 2°. The noon–dusk "valley" has further steepened and broadened, whereas the 02 h valley remained basically the same. The polar-cap situation is almost unchanged.

$\underline{Kp = 5^{+}\text{–}9^{0}}$ (Fig. 4d). The maximum intensities (in some cases $\geqq 10^{7}$) are clearly concentrated in the dawn–noon and dusk–midnight regions. The precipitation zone extends in these areas down to 56°. The noon–dusk "valley" is very broad and steep. From 12 h to 16 h practically all sampled fluxes are below the detection threshold.

3. Summary and Discussion

From the results of the preceding sections, some of the main points are summarized here:

(i) For both protons and electrons and all energies an equatorward shift of the southern boundaries of the precipitation regions is seen with increasing *Kp*, also leading to a broadening of these regions.

(ii) The northern boundaries do not follow a similar clear-cut pattern. They depend more on the particle type and energy and the local time.

(iii) The average fluxes over the polar cap, which are strongest in the lowest energy ranges, generally decrease with increasing Kp.

(iv) For highly disturbed conditions increasing fluxes are observed at the night side of the polar cap ($\Lambda \geqq 84°$).

(v) A "valley" in the intensity distributions is seen on the day side. Width and depth generally increase with Kp. For 5.8 keV protons it is located around noon for quiet conditions and shifted towards dusk with increasing magnetic activity. For electrons the position is in the noon–dusk region for all activity levels.

(vi) A second "valley" around 02 h MLT is observed, especially for the higher energies. It becomes narrower for increasing Kp and it does not seem to change much above $Kp \sim 3$.

(vii) Discontinuities in the precipitation patterns are observed around 05 h and 19 h MLT. In general the poleward boundaries of the precipitation zones are shifted towards higher latitudes on the day side. This shift is most pronounced for protons (5.8 keV, ~5°) and can also be seen for lower energy electrons.

These observations fit into the general picture developed in the past few years and provide some more details. It is known [4], that the auroral electron precipitation takes place in two more or less distinct zones, a lower latitude circular zone and an oval zone at higher latitudes. Various features of these two zones have been discussed by a number of authors [2, 5–7]. Also, the existence of a soft particle population over the polar caps has been demonstrated. Obviously, it is this population which causes the structured average intensities over the polar cap, reported in the present paper. The observed diminution of these intensities with increasing Kp is in accordance with ground-based observations. The polar-cap aurora is known to be anticorrelated with Kp. Also, crescent-like structures near the night border of the polar cap have been reported. This is supported by the particle observations, discussed in § 2.

In the lower-latitude zone, the particles are probably injected into the midnight region and drift along paths determined by their charge, energy and the external influences. In principle, gradient drift, rotational electric field drift and convection electric field drift influence their motion. In addition, some precipitation mechanism must be acting along the drift path, possibly pitch-angle scattering [8]. The observed "valley" in the noon–dusk region should then mean either that the precipitation mechanism is much less effective shortly after noon, or that a considerable fraction of the particle population is "swept out" by the acting drift mechanisms. In the latter case it is not clear, however, how the new increase around 14–18 h MLT (for electrons) could be explained. The westward drifting protons exhibit the pronounced depleted loss -cone anisotropy distribution, mentioned in § 2.1. The lack of this anisotropy in the noon–dawn sector might be connected with the plasma influx through the day side polar cusps [9, 10]. This direct access of magnetosheath plasma to low altitudes has been postulated for a long time by several authors on the basis of observed discontinuities of different phenomena in the auroral oval (e.g. [11]). A pronounced discontinuity in the morphology of the visual aurora for low Kp has also been reported [12]. For higher Kp values the observed two horseshoe shaped zones merge into one apparent oval. This general behaviour can be traced in the particle data, described in § 2. The discontinuities observed around 05 h and 19 h MLT could then mean that the direct access particle population extends all around the day side to these longitudes. A broad longitudinal region has also been reported by Heikkila [13], who postulates direct access of magnetosheath plasma through large "clefts" in the magnetospheric structure, rather than through narrow neutral-point regions. Thus, our data are in agreement with the picture of two horseshoe shaped precipitation zones of different origin, merging into one zone for higher Kp values.

Acknowledgments

The research reported here has been sponsored by the Swedish Natural Science Research Council, the Swedish Board for Technical Development and by the European Space Research Organisation. We wish to express our thanks to the staff members of the computation centre at the University of Umeå for their support and interest.

References

[1] R. D. Sharp and R. G. Johnson, paper presented at the Advanced Study Institute Aurora and Airglow, 1970, Queens University, Kingston, Ontario, Canada.
[2] W. Riedler, B. Hultqvist and S. Olsen, Arkiv f. Geofys. **5**, 619 (1971).
[3] W. Riedler, B. Hultqvist and H. Borg, Space Research XI, 1163 (1971).
[4] T. R. Hartz and N. Brice, Planet. Space Sci. **15**, 301 (1967).
[5] R. D. Sharp and R. G. Johnson, in: Earth's Particles and Fields, Reinhold, New York 1968 (p. 113).
[6] R. A. Hoffman, J. Geophys. Res. **74**, 2425 (1969).
[7] J. L. Burch, J. Geophys. Res. **73**, 35 (1968).
[8] C. F. Kennel and H. E. Petschek, J. Geophys. Res. **71**, 1 (1966).
[9] W. J. Heikkila and J. D. Winningham, J. Geophys. Res. **76**, 883 (1971).
[10] L. A. Frank, J. Geophys. Res. **76**, 5202 (1971).
[11] V. M. Mishin, T. I. Saifudinova and I. A. Zhulin, J. Geophys. Res. **75**, 797 (1970).
[12] K. Lassen, Phys. Norv. **4**, 171 (1970).
[13] W. J. Heikkila, Space Research XII, 1343 (1972).

Space Research XII — Akademie-Verlag, Berlin 1972

COMPARISON OF 4278 Å N_2^+ EMISSION AND LOW ENERGY ELECTRONS OBSERVED FROM THE SATELLITE AURORAE

G. Gustafsson[a], A. Egeland[b] and C. S. Deehr[b, c]

[a] Kiruna Geophysical Observatory, Kiruna, Sweden
[b] Institute for Cosmic Physics, Blindern, Norway
[c] On leave from University College Alaska, USA

The auroral N_2^+ emission at λ 4278 Å and the precipitation of electrons in the energy range 1–13 keV were measured with a photometer and particle detectors on board the polar orbiting satellite Aurorae. Nineteen passes over the northern auroral zone in the midnight sector have been studied to find a relation between the auroral emission and the precipitating electrons. It has been found that a relatively narrow region of the electron energy spectrum of only a few keV contributes to the main part of the observed N_2^+ emission. Using Rees' method for the calculation of the light emission this is found to be near 2–3 keV. However, it is shown that the energies below approximately 5 keV are given too much weight by this method and it is likely that the electron energy region is near 4–6 keV.

1. Introduction

In this study the measurements from two instruments on board the satellite Aurorae have been compared; namely low-energy electrons in the 1–13 keV region and auroral emission at λ 4278 Å. The satellite had a polar orbit and the results reported here were obtained in October and November 1968, when the satellite had a height of 350–500 km as it crossed the auroral oval. All passes which have been analysed in detail (a total of 19) were obtained within one hour of local magnetic midnight. A description of the particle experiment has been given by Riedler et al. [1] and of the photometer experiment by Deehr et al. [2].

The first section of this report deals with a comparison between calculated intensity of λ 4278 Å from the measured flux of electrons at a pitch angle of 10° and 80° in the energy range 1–13 keV and the measured intensity at the same wavelength. As the calculated and measured intensities are based on absolute values some of the results might be due to calibration errors. Therefore, in the following two sections, an analysis was carried out using relative values which are considered reliable because of in-flight calibrations. It will be shown that the absolute and the relative methods give essentially the same results and support the absolute calibration of the instruments.

2. A Comparison between Calculated and Measured λ 4278 Å Emission

The emission of λ 4278 Å has been calculated according to a method given by Rees [3] assuming that the emission is due to electrons in the range 1–13 keV.

The volume emission rate $\eta(\lambda\,4278)$ (in quanta $cm^{-3}\,s^{-1}$) per unit electron flux F (electrons $cm^{-2}\,s^{-1}$) is

$$\frac{\eta\,(\lambda\,4278)}{F} = \frac{0.013\,(\varepsilon_0/r_0)\,\lambda(Z/R)\,n(N_2)_Z}{\Delta\varepsilon_{ion}\,n(M)_R}$$

where ε_0 (in eV) is the initial energy of the electron; $\Delta\varepsilon_{ion}$ (in eV) is the mean energy loss per ion pair formed; $n(M)_R$ (in cm^{-3}) is the number densities of ionizable atoms and molecules at atmospheric depth R (in g cm^{-2}); $n(N_2)_Z$ is the number density of molecular nitrogen at atmospheric depth Z. The range r_0 (in atm cm) is defined as $r_0 = R/\varrho$ where R is the penetration depth found by Grün [4] to be related to the initial energy by the power law $(\varepsilon_0/1000)^{1.75}$; ϱ is the mass density (in g cm^{-3}); $\lambda(Z/R)$ is the normalized energy dissipation distribution function. The constant 0.013 is the ratio between the emission of λ 4278 Å and the total ionization according to Omholt [5] (that is one quantum per 75 ion pairs formed). The parameters for the ionosphere used here are those given by Rees [3].

The function λ is dependent on the pitch-angle distribution of the electron. In this case information has been available only from the directions 10° and 80°. Judging from these measurements the fluxes are usually isotropic and for the few cases of anisotropy which have been found only minor changes have resulted in the ratio between calculated and measured emission if isotropy is assumed. Therefore, it may be concluded that the discrepancies between calculated and measured emission which will be described below are not due to anisotropic fluxes of electrons.

The photometer has a field of view of 10° which means that the particle data have to be averaged over a time interval corresponding to that taken for the field of view to pass over the aurora created by the particles observed at the satellite. During the passes which are reported here the satellite was at a height of 350–500 km which corresponds to 6–10 s assuming the aurora to be at a height of 100 km. To obtain a continuous spectrum from 1 to 13 keV a linear interpolation between the logarithm of the measured fluxes at 1.3, 2.9, 5.8 and 13.3 keV was carried out. A few examples of the relation between computed and measured emission are shown in Fig. 1. In general it has been found that the calculated emission is at least half the measured emission except in a few rare cases of which orbit 226 in Fig. 1 is an example. Some cases have been found where the calculated emission is twice the measured one (see orbit 294). On the average, for the whole material studied, it has been found that the calculated intensity is slightly higher than the measured one with the method and constants mentioned earlier.

It can be shown that a relatively narrow part of the particle spectrum contributes to the auroral emission. In fact, the result for all the passes investigated is that half the calculated emission is due to particles within only 1 keV of the electron energy spectrum, and this "maximum efficiency region" is located between 2 and 4 keV in 13 of the passes studied here. In the remaining passes the region was at higher energies but always below 8 keV. 80% of the emission is usually between 2 and 7 keV and only in one case did it reach the upper limit of the spectrum. In 5% of the time it was found that 10% of the calculated emission was below 1 keV. It may be concluded that particle measurements in the 1–13 keV region are sufficient to calculate the auroral emission at λ 4278 Å. However, due to the narrowness of the "maximum efficiency region" it is probable that during short periods of variable and peaked spectra the sampling at only four points in the spectrum is not enough to cover all the details.

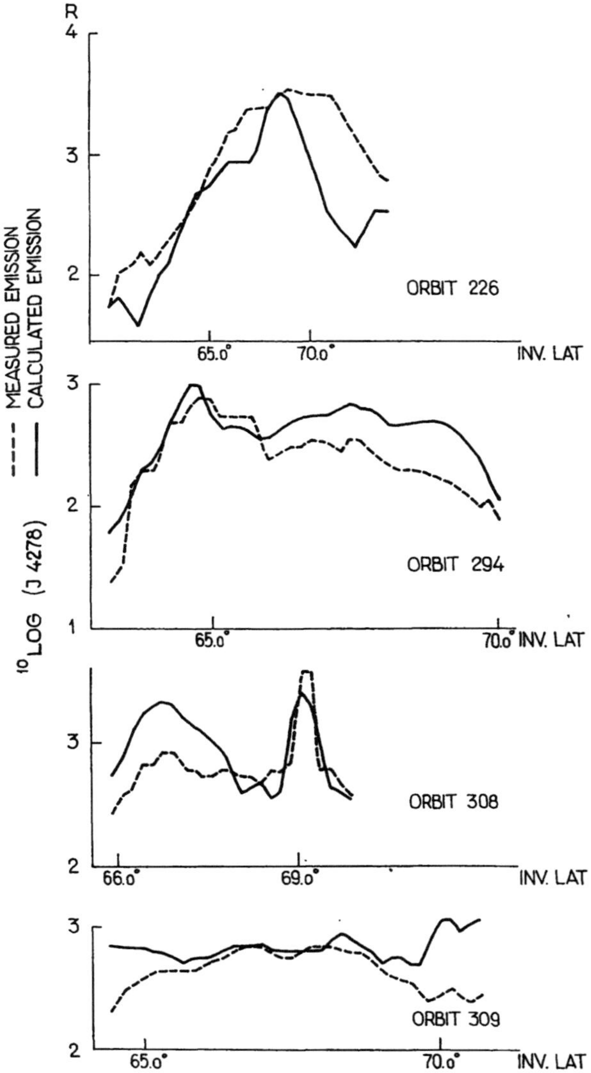

Fig. 1. Examples of the relation between measured and calculated emission at λ 4278 Å for 4 passes. The orbits are from the following dates and times: No. 226, 19 Oct. 1968 at 2353–2357 UT; No. 294, 24 Oct. 1968 at 2016–2018 UT; No. 308, 25 Oct. 1968 at 2014–2015 UT; No. 309, 25 Oct. 1968 at 2156–2157 UT.

A pronounced effect has been found in this study, namely that in the calculation of the emission too much weight is given to the 1–3 keV particle flux. This is demonstrated in Figs. 1 and 2 for orbit 309 where it is seen that both at the beginning and the end of the recording there is an increase in the 1 and 3 keV flux of electrons and in the calculated emission while there is a decrease in the 6 and 13 keV flux as well as in the measured light intensity. Several examples have been found where there is a peak in the spectrum at 3 keV giving a peak in the calculated emission but not in the measured emission. This effect may be responsible for the high calculated intensity in orbit 294

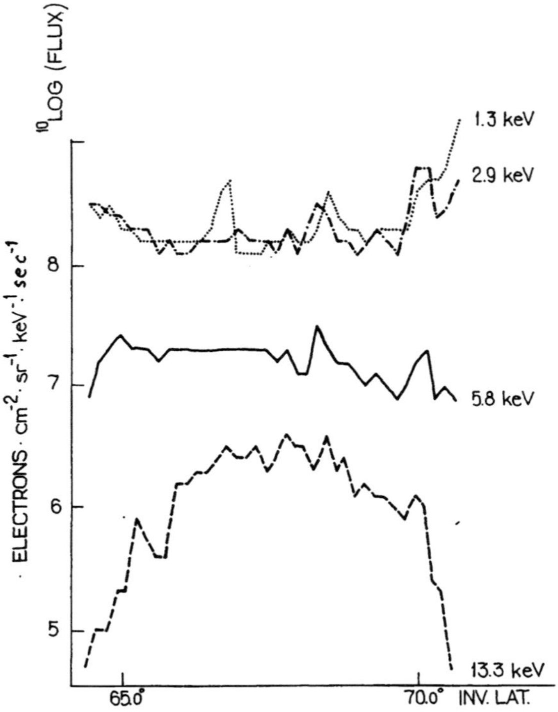

Fig. 2. The spectrum of precipitated electrons corresponding to orbit 309 in Fig. 1.

(Fig. 1) because the spectrum was unusually soft during this pass. The opposite effect could be the explanation for the discrepancy towards the end of pass 226 (Fig. 1). In this case the spectrum was unusually hard which means that a small increase in the weight of the spectrum above 6 keV when calculating the emission will give a large increase in the calculated flux. However, this effect has been noted in only two cases.

3. Cross Correlation between Precipitated Electrons and Emission

The cross correlation between the four different electron channels and the emission of λ 4278 Å was calculated for a number of passes over the auroral oval. Two pronounced features were found: firstly the 6 and 13 keV channels are closely related to each other and to the emission at λ 4278 Å and secondly the 1 and 3 keV precipitations are neither related to each other nor to the λ 4278 Å emission for about half the time. Fig. 3 shows a typical example where the λ 4278 Å-6 keV coefficient is 0.88, 4278-13 is 0.84, 4278-3 is 0.42 and

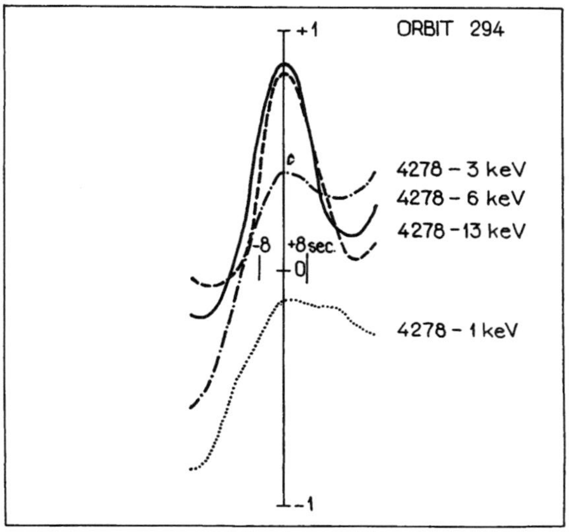

Fig. 3. A typical example of the cross correlation between the emission at λ 4278 Å and different channels of precipitated electrons.

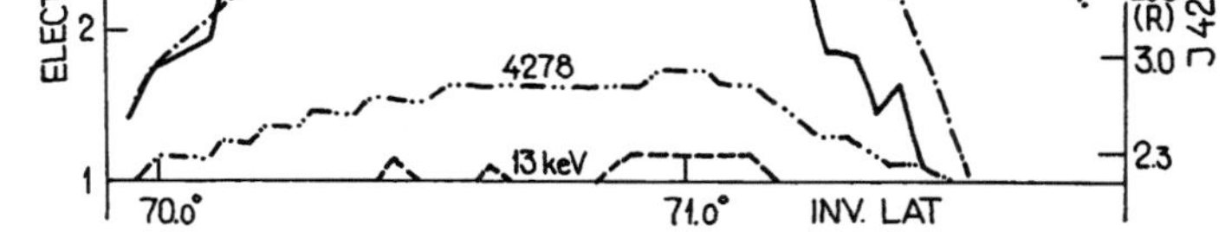

Fig. 4. An example of a satellite pass where 6 keV precipitation electrons are correlated with the emission at λ 4278 Å and 1 and 3 keV precipitation electrons are anticorrelated.

4278-1 is — 0.1 for zero lag (the lag due to the position of the satellite has been corrected for). In particular, it was found that the correlation between 6 keV and λ 4278 Å was always better than correlation between 3 keV and λ 4278 Å at zero lag. It also seems to be quite common that the 1 and 3 keV fluxes are anticorrelated with λ 4278 Å during intense fluxes of 6 keV electrons (see Fig. 4). In this case (orbit 268) at the maximum of the 6 keV precipitation it was found that half the calculated emission came from the spectrum region 3.5–4.5 keV which seems to be too low in the spectrum as judged from Fig. 4. The fact that the 6 keV is more closely correlated with λ 4278 Å than 3 keV indicates also that the part of the spectrum which contributes the main part of the emission is closer to 6 keV than to 3 keV.

4. The Ratio between Particle Precipitation and Emission at λ 4278 Å for different Intensity Levels during a Satellite Pass

If a linear relation exists between precipitated flux in a narrow region of the energy spectrum and emission at λ 4278 Å, it is meaningful to compare the count rate of electrons in the four channels with the emission at λ 4278 Å. In order to minimize the errors, the ratio of the peak to the background intensity was taken from passes for which the peak intensity was greater than twice the background. Comparison of these ratios for electron intensity and λ 4278 Å emission showed that the electron flux in the energy range 2–7 keV fits the emission ratio best. It may be concluded that the observed electron intensity ratio is in an energy range 2–3 keV higher than that which contributes more than half of the calculated emission. This is again an indication that the lowest energies are weighted too high in the calculation of the auroral emission. In all the cases which have been studied here it has been found that the ratio (high intensity/low intensity) at 13 keV is always much greater than the intensity ratio of λ 4278 Å which indicates that higher energies of the spectrum are relatively more important at high intensities of λ 4278 Å. Using the same particle measurements as here and ground-based photometers it was found in an earlier study by Gustafsson [5] that the ratio which best corresponded to the ratio of the light emission was close to 6 keV, which is in agreement with what has been found here.

5. Summary

It has been found in this study that the auroral emission at λ 4278 Å is caused by precipitated electrons with initial energies in the range 1–10 keV and that only in exceptional cases is the energy outside this region of importance. The emission calculated according to the method given by Rees [3] seems to be too high, but as this is critically dependent on the absolute calibration of the instruments on board the satellite and the choice of atmospheric parameters, this result may not be significant.

A detailed study of the relation between calculated and measured emission and the time variation of the particle spectrum has indicated that too much weight has been given to the energies below about 5 keV. In some cases it appears that giving a higher weight to energies above 6 keV should give a closer correlation between calculated and measured emission. No case has been found where the opposite weighting of the spectrum should give better correlation.

This result has been supported by methods of analysis which are not dependent on absolute values. It may also be noted that if the main part of the emission were caused by 2–3 keV electrons the height profile of the emission should have a maximum at about 130 km, which is higher than has generally been found.

References

[1] W. RIEDLER, B. HULTQVIST and S. OLSEN, Ark. Geophys. **5**, 619 (1971).
[2] C. S. DEEHR et al., ESRO–ELDO Bulletin Scientific Supplement. SP-62 November 1970.
[3] M. H. REES, Planet. Space Sci. **11**, 1209 (1963).
[4] A. E. GRÜN, Z. Naturf. **12a**, 89 (1957).
[5] A. OMHOLT, The Optical Aurora, Springer Verlag, Berlin 1971.
[6] G. GUSTAFSSON, Physica Norvegica **4**, 113 (1970).

Space Research XII — Akademie-Verlag, Berlin 1972

EFFECT OF ESCAPING PHOTOELECTRONS IN A POLAR EXOSPHERIC MODEL

J. Lemaire

Institute for Space Aeronomy, Brussels, Belgium

The effect of a photoelectron escape flux on a new polar wind model is discussed. It is shown that the additional electric drag force due to the escaping photoelectrons can accelerate the ions to higher velocities if the photoelectron flux is larger than the escape flux of the thermal electrons. Furthermore, it is shown that, even if the number density of the photoelectrons remains small compared with the number density of the thermal electrons, their kinetic pressure becomes predominant at very high altitudes. As a consequence a positive gradient in the electron "temperature" can be expected in the sunlit polar cap upper region.

1. Introduction

The polar wind concept has been introduced by Axford [1], who suggested that a flux of 2×10^8 cm^{-2} s^{-1} of escaping photoelectrons is required to drag out of the polar ionosphere all the He^+ ions produced in the atmosphere. The effects of such photoelectron fluxes on the density, bulk velocity and pressure distributions of a polar wind model are investigated in the following paragraphs.

2. The Model

The electron and ion temperatures $T_{e,i}$, are assumed to be equal to 3000 °K in the models. The neutral particles (O, H, He) are distributed according to Nicolet–Kockarts atmospheric model ($T_N = 750$ °K); (private communication).

The boundary conditions are taken from the OGO 2 ion-composition measurements [2]: $n_{O^+} = 7 \times 10^3$ cm^{-3}, $n_{H^+} = 320$ cm^{-3} are typical values observed in the sunlit polar cap region at an altitude of 950 km. Using these values and $n_{th.e^-} = 7.32 \times 10^3$ cm^{-3} as boundary conditions, we integrated the hydrodynamical equations (mass and momentum equations for each ionic species) described by Banks and others [3–6]. Each of the hydrodynamical solutions can be characterized by the value of the bulk velocity (or by the upward diffusion flux) at the reference level of 950 km. From the calculated bulk velocity and density distribution one can estimate the deflection mean free path (mfp) of the He^+ ions, and define the baropause as the altitude where the mfp becomes equal to the electron or O^+ density scale height [7]. Above this altitude a kinetic calculation is adopted to determine the structure of the ion exosphere [8, 9]. The requirement that the thermal escape flux of the ions must be equal to the diffusion flux in the collision-dominated region is a new criterion to select a unique hydrodynamical solution. This solution is not necessarily the "critical solution" adopted in completely hydrodynamical polar wind models. Both of these solutions however converge rapidly in the lower altitude range.

The photoelectrons are considered as a separate kind of collisionless particle with a truncated * (no trapped, nor incoming orbits being populated) Maxwellian velocity distribution at the baropause, and a mean energy of 10 eV ($T_{ph.e^-} = 1.16 \times 10^5$ °K).

In the models Nos. 1, 2 and 3, the number densities of the photoelectrons at the baropause ($h_0 = 1250$ km) are respectively 0, 0.2 and 0.4. The corresponding photoelectron escape fluxes $F_{ph.e}$ are 0, 9×10^6 and 1.8×10^7 cm^{-2} s^{-1} at an altitude of 3000 km. The flux of H^+ at the same altitude is calculated to be 1.9×10^7 cm^{-2} s^{-1}; this flux, which is the same for all three models considered here, is determined by the boundary conditions at the baropause.

3. Results and Discussion

The density distributions of O^+, H^+ and of the thermal and photoelectrons are shown in Fig. 1 for these three models. The dashed lines (model No. 4) correspond to a special case where there would be as many photoelectrons

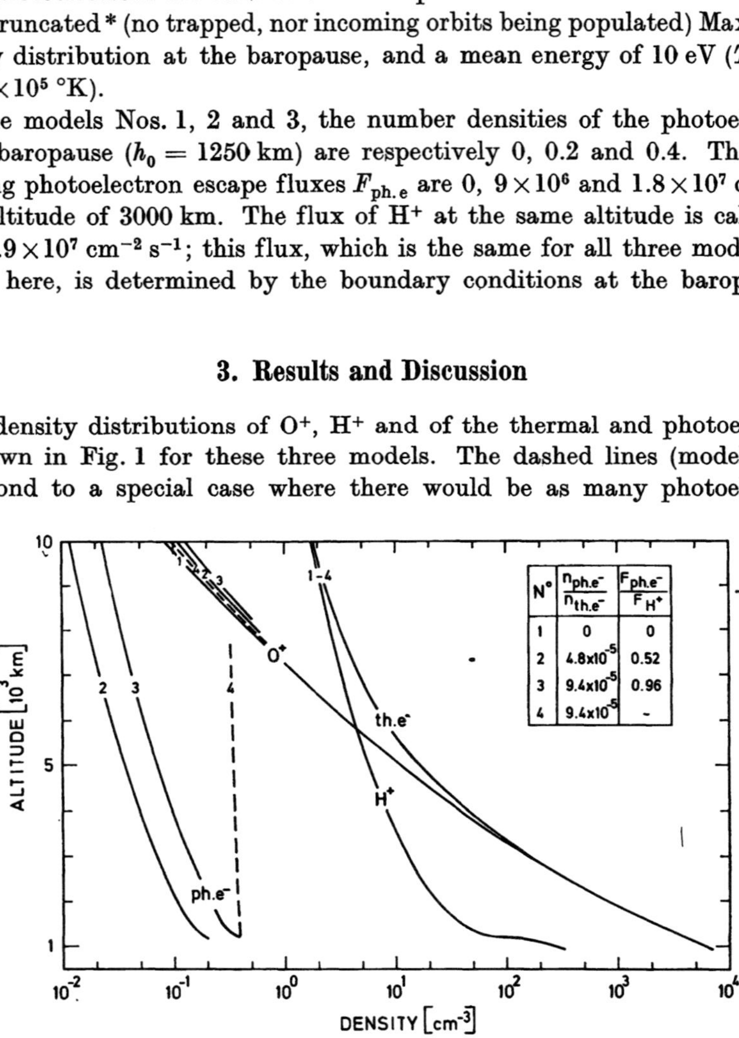

Fig. 1. Density distribution of the oxygen ions, protons, thermal and photoelectrons in four different polar wind models, with the effect of increasing photoelectron escape fluxes shown.

---

* It is assumed that the trapped photoelectrons with two magnetic mirror and/or gravitational reflection points above the baropause are missing in the velocity distribution. The particles coming from infinity and travelling towards the baropause are also absent in the models Nos. 1, 2 and 3.

flowing in as flowing out of the ionosphere. It can be seen that the O^+ distribution is influenced by an outward flux of photoelectrons. This effect becomes most important when $F_{ph.e^-}$ is larger than the thermal electron escape flux, $F_{th.e^-}$. In this case (e.g. model No. 3) the polarization electric field is significantly enhanced compared with that of model No. 1 where no photoelectrons are present. This additional electric drag due to the escaping photoelectrons is most efficient at very high altitudes where the relative abundance of the photoelectrons reaches its maximum value. The larger O^+ density scale height of model No. 3 is a consequence of this effect. Furthermore, the larger electric field intensity in a model with escaping photoelectrons increases strongly the subsonic bulk velocity and escape flux of the O^+ ions. Although the supersonic proton velocity is nearly unchanged below 10000 km altitude, it reaches, at very high altitudes, a laɪger asymptotic value:

$$w_{H^+}(h \to \infty) = \frac{n_{ph.e^-}}{n_{th.e^-}} \frac{1 + F_{th.e^-}/F_{ph.e^-}}{1 + n_{ph.e^-}/n_{th.e^-}} w_{ph.e^-}$$

which depends drastically on the relative abundance of the photoelectrons at large distances ($h \to \infty$); e.g., in model No. 3, $F_{th.e^-}/F_{ph.e^-} = 2.6 \times 10^{-2}$, $n_{ph.e^-}/n_{th.e^-} \to 10^{-2}$, and $w_{H^+}/w_{ph.e^-} \to 10^{-2}$. As $^1/_2\, m_e\, w^2_{ph.e^-} = 10$ eV, it can be concluded that $^1/_2\, m_H\, w^2_{H^+} \to 1.8$ eV, and $w_{H^+} \to 21$ km s^{-1}.

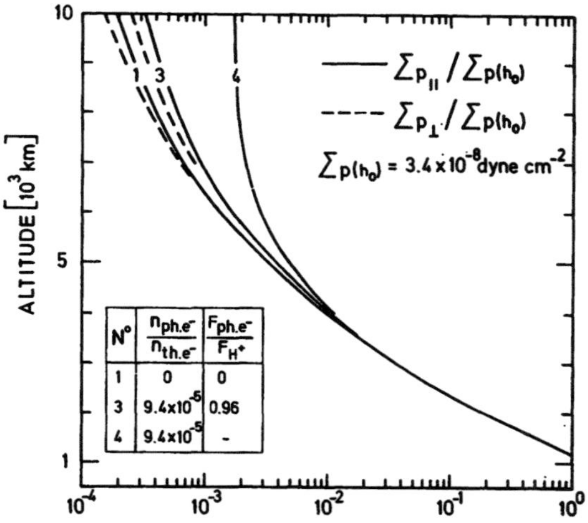

Fig. 2. Normalized total pressure components (parallel and perpendicular to the magnetic field) in four different polar wind models, with the effect of increasing photoelectron escape fluxes shown.

Fig. 2 shows the normalized total kinetic pressure components versus altitude in the models 1, 2, 3 and 4. At heights of 5000 km and above, the electron gas makes the largest contribution to the total pressure. The anisotropy of the kinetic pressure becomes significant above 7000 km. Although the photoelectrons remain a minor constituent in model No. 3, it can be seen from Fig.2 that these particles, however, contribute significantly to increase the kinetic pressure and its anisotropy. As a consequence, when "suprathermal" electrons are present in the polar wind, any "temperature" deduced from the total kinetic pressure (or energy density) of the electrons will increase with altitude. In the lower region of the polar ion-exosphere where the kinetic pressure is mainly due to the thermal ionospheric particles, the energy spectrum of the electrons has a large peak at 0.3 or 0.6 eV.

At very high altitudes it will have a peak near 10 eV, if photoelectrons are present. In the intermediate height range (4000–10000 km) the spectrum would have two peaks whose relative importance depends on the location along the magnetic field lines. As such a doubly peaked velocity distribution generally leads to plasma instability, one can expect that turbulent wave-particle interactions will play a significant role and give rise to energy re-distribution and electrostatic or electromagnetic emissions at high altitudes above the sunlit polar cap region.

Model No. 4 illustrates a case in which all the escaping photoelectrons would be backscattered at a high altitude level. From Fig. 2 it can be seen that the total kinetic pressure would then remain isotropic, and that it would be greatly enhanced. The upward and downward photoelectron flux (that an *in situ* experimental device would measure) at 3000 km, would be equal to 1.8×10^7 ph $e^- \, cm^{-2} \, s^{-1}$.

Acknowledgments

We wish to thank Dr. M. J. Rycroft and COSPAR Secretariat for their help in preparing this manuscript.

References

[1] W. I. AXFORD, J. Geophys. Res. **73**, 6855 (1968).
[2] H. A. TAYLOR, JR., et al., J. Geophys. Res. **73**, 5521 (1968).
[3] P. M. BANKS and T. E. HOLZER, J. Geophys. Res. **73**, 6846 (1968).
[4] P. M. BANKS and T. E. HOLZER, J. Geophys. Res. **74**, 6317 (1969).
[5] J. E. HOLZER, J. O. FEDDER and P. M. BANKS, J. Geophys. Res. **76**, 2453 (1971).
[6] K. MARUBASHI, Rep. Ionosph. Space Res. Japan **24**, 322 (1970).
[7] J. LEMAIRE, Conf. on Theoretical Ionospheric Models, Pennsylvania State University, June 1971 (to be published in J. Atmos. Terr. Phys. 1972).
[8] J. LEMAIRE and M. SCHERER, Planet. Space Sci. **18**, 103 (1970).
[9] J. LEMAIRE and M. SCHERER, Phys. Fluids, **14**, 1683 (1971).

Solar Terrestrial Relationships

Space Research XII — Akademie-Verlag, Berlin 1972

THE INTERNATIONAL MAGNETOSPHERIC STUDY 1975–1977: SCIENTIFIC FUNDAMENTS AND OBJECTIVES

J. G. Roederer

Department of Physics, University of Denver, Denver, Col., USA

The International Magnetospheric Study (IMS) is proposed as an international cooperative enterprise during the limited period 1975–1977, having as its principal objective the achievement of a comprehensive, quantitative understanding of the dynamical processes operating in the earth's plasma and field environment.

In this article a brief review will be given on the present state of knowledge of principal magnetospheric processes; some of the most pressing questions and problems will be analyzed and a summary of the First Report on Spacecraft Missions recently issued by the COSPAR–IUCSTP Study Group for the IMS will be presented.

1. The Earth's Magnetosphere

We may define the magnetosphere as the region of space threaded by magnetic field lines linked to the earth, and filled with ionized gas whose dynamic behaviour is influenced or controlled by this magnetic field.

Since the advent of the space age the study of the magnetosphere has undergone a highly successful stage of discovery and exploration. Through coordinated efforts of ground-based, balloon and rocket research since the International Geophysical Year, and especially as the result of extensive satellite programs supported by individual national space research organizations, we have obtained a morphological description of the magnetospheric field, the particle population embedded in it, and its interface with the solar wind, and we have identified and are beginning to understand many of the physical processes involved.

1.1. The Magnetic Field Configuration

Fig. 1 depicts in comprehensive form the configuration of the magnetosphere. The earth's internal magnetic field presents itself as an obstacle to the flow of solar wind plasma. Since this flow is supersonic, information on this obstacle cannot propagate upstream, and solar wind particles learn about the earth's field presence only very suddenly when they traverse the stationary *bow shock* (Fig. 1). At this shock the solar wind plasma is compressed, heated and slowed down to subsonic speed; it emerges in a state of considerable turbulence and flows along the flanks of the magnetosphere. This layer of plasma, called the *magnetosheath* (Fig. 1), acts like an elastic medium transmitting the kinetic pressure of the solar wind to the earth's magnetic field. A very sharp boundary, the *magnetopause* (Fig. 1), separates the magnetosheath from the magnetospheric field. The latter is confined to a cavity of finite size on the day-side, and is stretched out into a comet-tail-like configuration on the night-side, the *geomagnetic tail* (Fig. 1).

A good quantitative description of the average magnetic field is now available for the regions near the earth's surface, and at low geomagnetic latitudes out to distances of about 7–8 earth radii (R_e). Only scarce magnetic field observations are available in the distant magnetosphere, particularly at high latitudes over the polar regions and far out in the geomagnetic tail.

Notice in Fig. 1 two regions of magnetic field discontinuities: the magnetopause and the *neutral sheet*. The latter is oriented roughly perpendicular to the plane containing the sun–earth line and the earth's magnetic dipole axis (i.e. the plane of the paper in Fig. 1), and extends from the dawn to the dusk

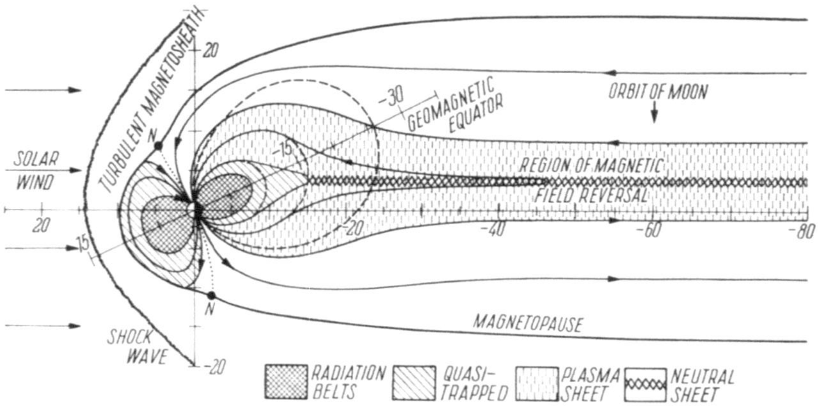

Fig. 1. Magnetic field and particle configuration of the earth's magnetosphere (coordinates shown are in units of earth radii). Courtesy of Dr. N. F. Ness.

wings of the geomagnetic tail. These two-dimensional regions of magnetic field discontinuity are maintained by steady-state electric currents that somehow must be provided by an appropriate flow of low energy charged particles along the corresponding surfaces. Notice also another pair of peculiar field regions, the *neutral points*, located on the day-side magnetopause. Actually, there is a pair of *demarcation lines* (not shown in Fig. 1) extending from the neutral points to the dawn–dusk meridians, which separate open from closed field lines on the boundary. Each neutral point is a vortex of the magnetopause current system.

There are other currents flowing in the magnetosphere, causing measurable effects on the magnetic field, especially during disturbances. They can be divided into several categories. Some currents are of "diamagnetic" type, caused by a density gradient, i.e. an inhomogeneous distribution, of charged particles gyrating in the magnetic field. The quiet-time *equatorial ring current* around the earth, centered at a distance of about 6 R_e, is an example. Then there are conduction currents flowing at the base of the magnetosphere, responsible for magnetic field variations at the surface of the earth that have been known for more than hundred years. These *ionospheric currents* are caused either by the dynamo effect (ionization driven across the magnetic field by neutral winds) or, at high latitudes, by the magnetospheric electric field (whose potential maps into the polar ionosphere along magnetic field lines as if the latter were conducting wires). Strongly coupled to the ionospheric currents in the polar regions are *field-aligned currents* flowing out from, or into, the ionosphere, representing an important physical link between ionosphere and magnetosphere. Finally, electric currents may also arise as the result of positive and negative particles drifting into mutually opposite directions. Their effect is important whenever large fluxes of particles are injected from the tail into the inner magnetosphere. The *asymmetric ring current*, that appears during severe magnetospheric perturbations, is an example.

The magnetic field configuration in the magnetosphere is not at all static. Most of the currents mentioned in the previous paragraph are indeed time-dependent, either in association with the diurnal variation of the solar ionization of the upper atmosphere, or with perturbations triggered by plasma instabilities in the magnetosphere. Variations in solar wind dynamic pressure can cause appreciable changes in the day-side compression of the magnetosphere. Severe compressions following the passage of interplanetary shock waves emitted by solar flares may displace the boundary inwards to as close as 5–6 R_e (compare with Fig. 1). At the same time the tail field intensity increases substantially indicating that also the tail boundary has been severely compressed.

A particularly important sequence of perturbations that affects the magnetosphere as a whole is known under the generic term *substorm* — a sort of "lightning discharge" between the geomagnetic tail and the ionosphere, in which magnetic energy stored in the tail is suddenly released and converted into kinetic energy of the ambient plasma. A severe compression of the magnetospheric boundary is followed by a whole series of substorms in rapid succession — their integral effect constitutes the *magnetospheric storm.*

There are short-term periodic variations of the magnetic field associated with different categories of hydromagnetic waves travelling through the magnetospheric cavity. They correspond to different oscillation modes of the "elastic" system of plasma and magnetic field in the magnetosphere. These periodic variations are detected on the earth's surface in the form of *geomagnetic micropulsations*, with periods between 0.2 second and 10 minutes. These surface variations are important messengers of the state of the distant magnetosphere and may be used as a convenient (and cheap) means to obtain a permanent "diagnostic" of the state of the magnetosphere.

The average magnetospheric field configuration and its typical time variations are fairly well established. However, we still know very little about the electric currents, both steady-state and transient, that are responsible for this configuration and its typical perturbations. Another important shortcoming is the lack of systematic information on the magnetic field in the vicinity of the neutral points. As we shall see later, these points and the corresponding demarcation lines seem to play a key role in magnetospheric dynamics. Also not yet fully established is the detailed magnetic field behaviour across the neutral sheet, particularly near its inner edge (Fig. 1). This is believed to be the region where the substorm "discharge" starts. In summary, neutral points and neutral sheet — the most "sensitive" areas of the magnetosphere from the dynamical point of view — are at the same time among the least explored regions of the magnetospheric field. Finally, although there is evidence that some of the "open" field lines of the magnetosphere may actually leave the magnetosphere and be linked to the interplanetary magnetic field, a systematic experimental attack on this important problem of magnetosphere–solar wind interconnection is still a much-needed program of the future.

1.2. The Particle Distribution

The region of closed field lines of the magnetosphere (Fig. 1) contains a population of energetic electrons, protons and heavier ions in the energy range of a few keV to hundreds of MeV (for protons), that are trapped in the magnetic field — the *Van Allen radiation belts*. On the other hand, the distribution of very low energy particles of the ionosphere plasma extends out to about 4–6 R_e (equatorially). This particle population, the *plasmasphere* (not shown in Fig. 1), is found to terminate rather abruptly on a surface made up of magnetic field lines, called the *plasmapause*. Finally, in the tail there is another plasma "reservoir", the *plasma sheet* (Fig. 1), several earth radii thick and engulfing the neutral sheet, of considerably higher "temperature" than that of the plasmasphere. Between plasma sheet and plasmasphere there is a region of very low particle density called the plasma *trough* (not shown in Fig. 1).

We now have a good description of the morphology of the Van Allen radiation belts, their spectra, fluxes and typical time variations. Many of the dynamical processes that keep the radiation belts going are basically known. Low energy particles from the plasma sheet are convected towards the earth during substorms, accelerated and injected into stably trapped orbits. Small scale, stochastic electric and magnetic field fluctuations cause *radial diffusion* of these particles; since the net diffusive flow is directed radially toward the earth, i.e. toward higher magnetic field intensities, the particles are further accelerated via the betatron mechanism. Simultaneously, resonances between hydromagnetic or very low frequency electromagnetic waves and the particles' periodic bounce or cyclotron motions cause diffusion along magnetic field lines, with a net flow directed from the equator toward the atmosphere. This is a *pitch angle diffusion* mechanism. When a trapped particle precipitates into the atmosphere, it quickly loses its energy by ionization of the medium and is removed from the Van Allen population. It thus is obvious that pitch angle diffusion controls the loss mechanism of these particles. Less obvious is the established fact that particles and waves interact so efficiently in the outer magnetosphere that they are able mutually to control each other's flux or intensity. For instance, when the particle flux increases beyond a certain limit, the waves are enhanced in such a way as to cause the right increase of pitch angle diffusion needed to precipitate away the extra particle flux. Experimental results indeed indicate that particle fluxes in the outer magnetosphere tend to level off at, and never surpass, the maximum values permitted by this wave–particle interaction.

At least a major part of the very stable high energy proton population of the inner Van Allen belt is of a quite different origin. Nuclei of atmospheric atoms are continuously subjected to a bombardment by ultra-high energy cosmic rays; in these collisions energetic neutrons are emitted into all directions. Those travelling away from the earth (*albedo neutrons*) will have a chance to decay into protons and electrons (and neutrinos). If this decay happens to occur in the appropriate region of the magnetosphere, the decay proton will remain stably trapped in the magnetic field (the electron too, but not so stably). The slow secular decrease of the earth's dipole moment seems to be responsible for the gradual displacement of these trapped high energy protons toward lower altitudes (the above-mentioned radial and pitch angle diffusion mechanisms would have a negligible effect on inner belt protons).

Most of the above processes are reasonably well understood. There remain, however, several unsolved problems. The exact proportion of heavy ions in the radiation belts is still a largely unsettled question. Another problem is related to the origin of the particles in the high energy end of the Van Allen spectrum. What proportion of these particles is the result of cosmic ray albedo neutron decay? Is there also a contribution from "regular" Van Allen particles that have undergone various cycles of inward radial diffusion with betatron acceleration and outward diffusion with energy conservation (*bimodal diffusion*)? Other areas where more research is needed are the above-mentioned wave–particle interaction and radial diffusion mechanisms, particularly in regions close to the earth where the plasmasphere should considerably affect pitch-angle diffusion and where field fluctuations should be very inefficient (yet radial diffusion subsists). A most exciting focus of current research is the study of the *low energy end* of the magnetospheric particle spectrum, especially in the range of a few tenths of keV up to, say, 40–50 keV. This represents a transient, highly variable population, that is undergoing the very first stage of acceleration from plasma to Van Allen radiation. This population, which is found mainly in the night-side region of *quasi-trapping* (Fig. 1), but sometimes may penetrate deeply into the radiation belts, appears greatly enhanced during magnetospheric substorms, and is heavily controlled by the magnetospheric *electric field*. It constitutes the fundamental source of particles for the night-side *polar aurora*. Interestingly enough, this population also is the particle source for some of the magnetospheric currents mentioned in the previous section.

We now turn to the plasma sheet, the huge low energy particle reservoir in the geomagnetic tail (Fig. 1). Again, we do have a reasonable description of morphology, particle distribution and flow, and typical time variations. Much less, however, is known about the source of plasma sheet particles. Recent satellite measurements at high latitudes on field lines that pass near the neutral points (Fig. 1) and associated demarcation lines, have convincingly shown that magnetosheath plasma is able to penetrate deeply into the magnetosphere through a "cleft" or *day-side cusp* formed at the neutral points. These solar wind particles should then be convected by the magnetospheric electric field across the polar field lines from the day-side to the night-side. During this convection, the particles should gradually increase the component of their motion parallel to the open field lines; as a combined result of convection and parallel motion, they would finally end up in the tail reservoir. This injection process, although not a true acceleration, would be the "zeroth stage" in the chain of mechanisms to which magnetospheric particles of solar wind origin are subjected during their lifetime. The particular shape of the plasma sheet probably is a direct result of this primary injection mechanism. It is quite obvious that the physical understanding of the plasma sheet is directly linked to the problem of understanding the electric field in the high latitude magnetosphere and in the geomagnetic tail.

One should not discount the possibility of solar wind penetration through the tail boundary along the intersection of the latter with the neutral sheet. Another, secondary, source of plasma sheet particles could be the *polar wind*, a permanent evaporation of ionospheric ions and electrons along the open field lines into the tail.

The plasmasphere can be explored indirectly from ground through radiowave propagation studies (whistlers), and *in situ* with rockets (at its base) and satellites. The plasmasphere, which co-rotates with the earth, is strongly influenced by the electric field in the magnetosphere. Its boundary, the plasmapause, is seen to move inwards during substorms; this process, rather than representing a compression of the plasmasphere, is more correctly described as an "erosion" of its outer layers which are "convected away" as the result of an increase of the dawn-to-dusk component of the electric field. In spite of its name, the configuration of the plasmasphere is not symmetric around the earth; it is bounded by magnetic field lines (at the plasmapause) and bulges out considerably in the dusk–midnight sector of the magnetosphere. It is believed that the plasmasphere is dynamically governed by a threefold interplay between the co-rotational radial electric field close to the earth, the convective dawn-dusk electric field in the outer magnetosphere and the day–night asymmetry of ionospheric ionization (conductivity). In particular, the plasmapause is believed to mark the limit between closed and open equipotential surfaces of the magnetospheric electric field. Obviously, understanding the dynamics of the plasmasphere is directly linked to the problem of understanding the electric field in the magnetosphere at low and intermediate latitudes.

One final word on this magnetospheric electric field. One now has identified three types of electric fields: (i) the *co-rotation field*, associated with the co-rotation of the low and intermediate latitude ionosphere, which on the geomagnetic equatorial plane is directed radially toward the earth; (ii) the *convection field*, directed from dawn to dusk, that seems to be the result of an induction effect by the north–south component of the interplanetary magnetic field flowing past the magnetosphere; and (iii) a transient *substorm field*, probably highly localized near the midnight sector of the magnetosphere, related to the substorm discharge (see next section).

1.3. Magnetospheric Dynamics

We come to the conclusion that the *clue* to a comprehensive understanding of magnetospheric dynamics and associated cause-and-effect relationships must be sought in the *low energy end* of the particle spectrum, comprising both the plasma and the "intermediate energy" population.

Why took it so long to come to this conclusion? The reason is fairly simple: only in recent years has this energy range become accessible to systematic observation. Many difficulties had to be overcome: (i) The lower the energy of a particle, the more difficult it is accurately to measure its parameters (more sophisticated requirements for the detector; perturbing effects of electrostatic potential and motion of the spacecraft.). (ii) It is very difficult to measure and systematically survey the electric field in the magnetosphere. (iii) To study the collective behaviour of plasma particles it is necessary to distinguish experimentally between spatial and temporal variations; this requires *simultaneous measurements* at different well specified positions.

Fig. 2 is an attempt to summarize our knowledge of magnetospheric particle processes in the form of a comprehensive flow chart. On top we show the two primary particle sources, the solar wind and the ionosphere, as well as the main source for high-energy protons, cosmic ray induced albedo neutrons. Possible artificial sources — accelerators flown on rockets, high altitude nuclear injections — are also listed.

Next in Fig. 2 we show the two plasma reservoirs with an indication of the transfer processes: entry through the magnetospheric boundary, polar wind and extension of the ionosphere into the plasmasphere. The dynamics of these plasma reservoirs is the key to the dynamics of the whole magnetosphere. The geomagnetic substorm is the single event, the single macroscopic instability, that shakes up these particle reservoirs and makes them "spill over" their contents, causing aurorae and feeding the radiation belts. This and other transient acceleration processes are listed next in Fig. 2. The first two mainly operate during a magnetospheric substorm. The first one, in particular, is responsible for the recently established earthward convection along the midnight meridian, and acceleration, of plasma sheet particles.

This particle inflow probably represents the most important primary effect of a magnetospheric substorm, caused by an instability that develops across the neutral sheet, in which excess magnetic energy is suddenly released. There are firm indications that this excess energy is being accumulated during periods of time (1–2 hours long) in which the interplanetary magnetic field has a persistently southward direction. During these periods there should be an increased convectional flow of particles and magnetic energy from the high-latitude day-side magnetosphere, over the poles, into the tail caused by the persistently enhanced dawn–dusk electric field. It is possible to picture this process intuitively by imagining the high latitude field lines being convected from the day-side over the poles to the night-side, accumulating along the neutral sheet. The magnetic energy release can then be pictured as the collapse and mutual annihilation of oppositely directed field lines across the neutral sheet. This energy is converted into kinetic energy of particles that will flow along the neutral sheet away from the collapse region, toward the earth (and probably into the distant tail, at the other end). This instability seems to be identical to the one responsible for solar flares and thus may be of a quite universal astrophysical nature.

The particle stream that during a substorm shoots toward the earth along the midnight meridian will be accelerated; as a result, protons and electrons will

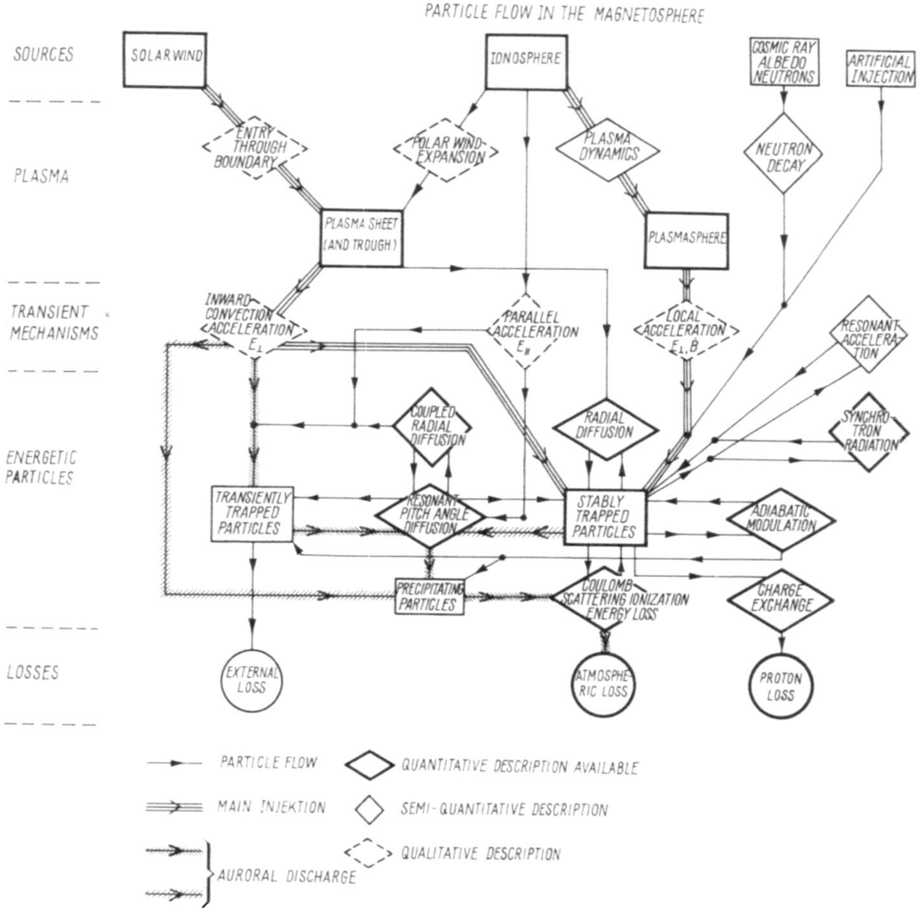

Fig. 2. Flow chart depicting the history of magnetospheric particles.

start following different paths. The "cold" plasma of the ionosphere seems able to provide enough electrons to cancel the effect of this charge separation. These electrons must flow out of the ionosphere (and into it) along field lines, giving rise to field-aligned currents and controlling the electric field directed along the magnetic field lines. This process is labeled "parallel acceleration" in Fig. 2.

The next stages shown in Fig. 2 refer to "energetic particles". They comprise the transient, quasi-trapped population of intermediate energy particles, a fraction of which survives pitch-angle diffusion ending up in stably trapped orbits as Van Allen particles. The rest is precipitated into the atmosphere giving rise to auroral displays. The other processes shown in Fig. 2 are essentially "modulation" processes of radiation belt particles, such as radial diffusion,

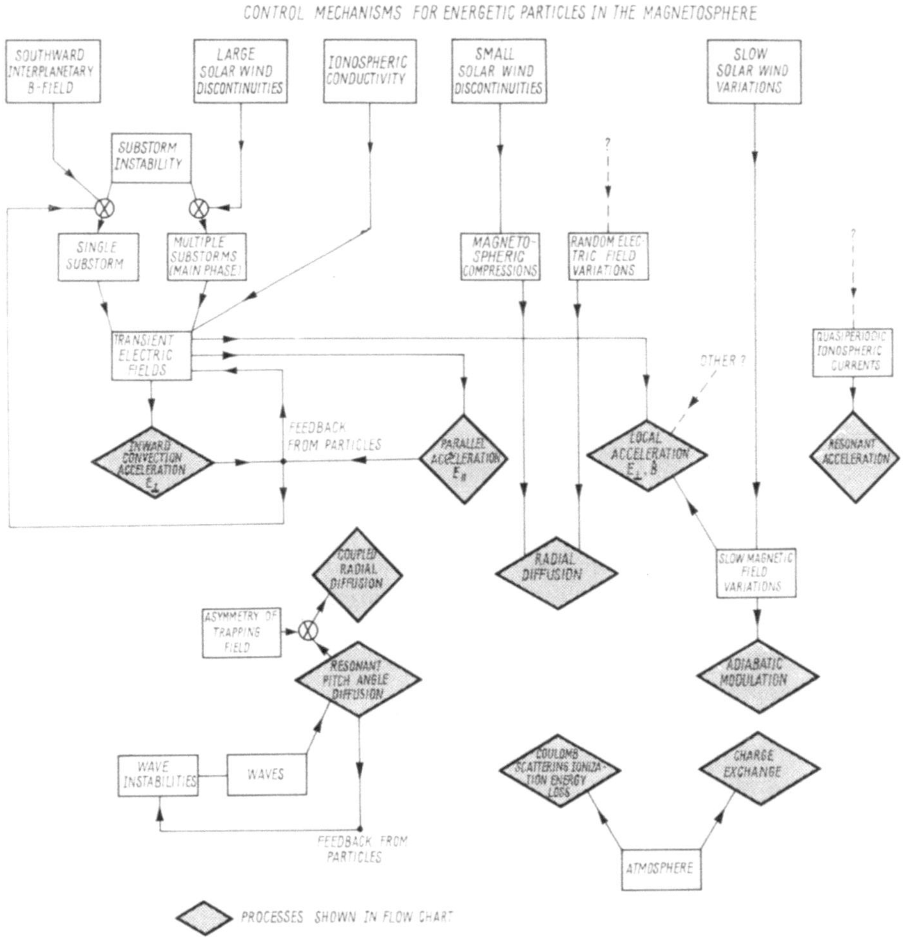

Fig. 3. Control mechanism for energetic particles in the magnetosphere.

coupled radial diffusion (an energy-conserving process associated with pitch-angle diffusion in an asymmetric field), adiabatic modulation, charge exchange (for protons), atmospheric scattering, etc.

Fig. 3 is an almost self-explanatory chart depicting the control mechanisms that regulate or influence the processes shown in Fig. 2.

2. The International Magnetospheric Study 1975–1977

From the preceding section one may conclude that a major quantitative understanding of the fundamental processes governing the magnetosphere is in sight. Yet many crucial questions remain to be answered. Fortunately we are now able to clearly identify the fundamental problems to be attacked, and we are able to pinpoint the regions in space that need further study and to identify the instrumentation, orbits, time correlation of launches and of ground-based, balloon and rocket observations, to be recommended.

The planning for the IMS comes at a time when, on one hand, national space agencies of many countries are turning their attention to the challenge of other planets and the distant solar system, while on the other hand the tax-paying public is focusing its attention ever more outspokenly on the very earthly problems of pollution, social injustice and the perils of war. The magnetosphere is caught right in the middle of these diverging viewpoints, receiving no particular enthusiasm from either. This is one more powerful reason why in magnetospheric research it is necessary more than ever to pull efforts together.

In what follows, excerpts from the First Report of the COSPAR–IUCSTP Special Study Group* on the IMS are reproduced.**

2.1. Introduction to the First Report of the IMS Study Group

The earth's magnetosphere reveals itself as a unique experimental region — a "space laboratory" — for watching, *in situ*, fundamental plasma processes in action, such as collisionless shocks, hydromagnetic waves, instabilities, particle acceleration, trapping and diffusion and, most notably, the structure of the interfaces between collisionless plasmas in different states. These processes are repeated on a large scale elsewhere in the universe, governing the behaviour of stellar atmospheres, the radiation environment of pulsars, supernovae, and the interstellar gas. They are found to occur on a roughly similar scale on the planet Jupiter and on a smaller scale at the interfaces between the solar wind and unmagnetized planetary bodies. It would be practically impossible to scale these collisionless plasma processes down to equivalent standards for detailed experimentation in a terrestrial laboratory.

On the other hand, the magnetosphere represents the outer limit of man's immediate environment. The study of its interaction with the atmosphere — however weak this interaction may be — provides a necessary input to the understanding of processes governing the temperature, composition and circulation of the upper atmosphere, the dissipation of foreign constituents, the origin

* Drs. R. Gendrin, G. Haerendel, J. W. King, N. F. Ness, J. G. Roederer (convener), Z. Švestka, S. N. Vernov and I. A. Zhulin.

** Copies of the complete report can be obtained from the IUCSTP Secretary, Dr. E. R. Dyer, National Academy of Sciences, 2101 Constitution Ave., Washington, D.C. 20418, USA.

and propagation of infrasonic waves, etc. Finally, the role of the magnetosphere in maintaining the polar ionosphere and in controlling energetic particle precipitation and, hence, the ionization at lower altitudes during magnetospheric perturbations, links these studies directly with the field of radio communications, in particular, with the problem of radio blackouts in polar regions during substorms.

Magnetospheric physics is now ripe for a transition from the exploratory stage to one in which missions are planned with the specific objective of achieving a comprehensive quantitative understanding of the cause-and-effect relationships among the dynamical processes involved, such as the solar wind–magnetosphere–ionosphere coupling, the associated energy input and transfer processes, the responses to steady and perturbed regimes of the solar wind, etc. This, in turn, will provide important quantitative information for the physical understanding of some of the processes in astrophysical plasmas and the upper atmosphere mentioned above.

The complexity and the large spatial scale of the magnetospheric phenomena under scrutiny demands a clear separation of spatial and temporal effects in the experimental observations. In particular, it is necessary (i) to determine the actual spatial and temporal extension of the phenomena, (ii) to distinguish between propagating and non-propagating phenomena, and (iii) to determine velocities or flow gradients in propagating phenomena. This necessarily requires simultaneous measurements with instrumentation covering similar dynamic ranges, conducted at spatially different positions, both in space and on or near the earth's surface. Recent isolated efforts to correlate data from different spacecraft that by mere chance happened to be at convenient spatial positions at a given time, have proved to be indeed a most powerful tool for quantitative studies of dynamical processes in the magnetosphere.

Under practical constraints imposed by (i) the existence of satellite missions already approved, with little flexibility for allowing substantial changes at this late stage of development, and (ii) the limited sources available for new missions, a realistic approach to establishing an international coordination of magnetospheric research can be achieved by the following measures:

(a) Definition of program concepts to help participating countries to make decisions about new missions that would represent key contributions serving to fill existing gaps in spatial, temporal, and instrumental coverage.

(b) Recommendation of small adjustments, wherever feasible, of the launch dates, orbital parameters, and instrumental characteristics of the missions already planned in order to optimize the possibilities for the coordination of results.

(c) A coordinated, rational exploitation of data from spacecraft which are, or have been, operating simultaneously in appropriate regions of the magnetosphere during appropriate intervals of time.

(d) Monitoring of solar em, ir, uv, plasma and particle emission in order to maintain continuous knowledge of the energy input to the magnetosphere.

(e) Extensive coordination with ground-based observations and rocket and balloon experiments.

The First Report of the IMS Study Group recommends a modest program during the limited period 1975–1977 towards achieving a maximum possible scientific return per invested effort — compatible with the expected constraints — in the quantitative study of the dynamics of the magnetosphere. It focuses mainly on satellite and space probe missions which require the longest lead time; it is expected that a more comprehensive document, including detailed plans for ground-based, balloon and rocket experiments, will be issued later.

2.2. Physical Problems to be Attacked

In the exploratory stage of studying the solar wind–magnetosphere system we have acquired a general understanding of its macroscopic features, and many of the internal processes have been explored experimentally as well as theoretically. However, most of the dominant processes are still poorly understood, in particular those involved in collective plasma effects.

With respect to the gross structure and dynamics of the magnetosphere some of the unanswered questions are:

(a) Is there a steady state of the magnetosphere and, if so, what is the reference configuration?

(b) What is the total momentum, energy, and particle transfer from the solar wind to the magnetosphere, and how is it controlled?

(c) How does the magnetosphere respond to changes in solar wind parameters?

(d) What is the cause-and-effect sequence of events constituting a substorm?

(e) What are the mechanisms giving rise to auroral forms?

(f) How are the different components of electric currents in the magnetosphere and ionosphere linked with each other?

(g) To what extent are magnetospheric processes determined by the ionosphere or affected by feedback from the latter?

(h) What is the composition and the temperature of exospheric ions and neutrals, and what are their relative contributions to geocoronal photoemissions?

(i) How does the helium in the solar wind affect the helium budget of the earth, and do these mechanisms apply to other chemical elements in the solar wind?

(j) What is the relative importance of the neutral atmosphere, and the polar wind, particle precipitation, and plasma convection in determining the behaviour of the polar ionosphere and the trough region beyond the plasmapause?

(k) What are the relative contributions from the various source and loss mechanisms to the radiation belts, and what is the resulting trapped particle composition?

It should be pointed out that several of these questions are also of interest to solid-earth and fluid-earth physics. For instance, an understanding of the reference configuration of the magnetosphere could add to a more detailed model of the origin of the earth's main field; simultaneous satellite and ground measurements of micropulsations will yield more information on the electrical properties of the earth's crust, and knowledge of the external processes affecting the helium budget will contribute to theories of the evolution of the earth's atmosphere.

In order to find the answers to the above questions, detailed experimental study and theoretical understanding of a considerable number of plasma processes are required. Some of these processes and their roles in magnetospheric phenomena are listed in the following paragraphs:

(1) *The maintenance of regions of abrupt reversals of the magnetic field.* The magnetopause and the neutral sheet of the tail are the most prominent examples. Major problems are the associated transfer of momentum, energy, and mass across the magnetopause, and the occurrence of field-line merging and associated acceleration processes in the neutral sheet.

(2) *The maintenance of magnetospheric plasma convection.* This involves the transport of momentum and energy, both transverse and parallel to the magnetic field, the ohmic losses to the ionosphere, and the reflection properties of regions where the plasma parameters undergo rather sudden changes, as for instance happens at the plasmapause.

(3) *Wave-particle interactions.* These interactions lead to the excitation and absorption of plasma waves, and to the interchange of energy and momentum among the various particle constituents. Some of the geophysical phenomena arising from this are the thermalization of the directed solar wind energy in the bow shock, the acceleration, diffusion and precipitation of particles, and the emission of a wide variety of plasma waves such as vlf, ulf, and electrostatic waves.

(4) *Propagation of waves in an inhomogeneous anisotropic medium.* A wealth of interesting vlf and ulf wave effects is caused by the anisotropy and inhomogeneity of the index of refraction. The flow of wave energy may have a crucial effect on the overall interaction with the particle populations and may strongly influence the equilibrium distribution of the latter.

(5) *Macroscopic plasma instabilities.* The magnetospheric substorm may be considered a consequence of an instability in which, among other processes, magnetically stored energy is suddenly released and converted into particle energy. Neither the pre-substorm configuration of the outer magnetosphere nor the causal sequence of events after onset of the instability is clearly understood.

(6) *Instabilities driven by electric currents.* Strong electric currents may give rise to instabilities leading to plasma turbulence and anomalous electrical resistivity. As a result, substantial electric potential differences may be set up along field lines; this may cause the violation of the "frozen-in" state of plasma and may result in the acceleration of certain particle components.

(7) *Stochastic transport processes.* Perturbations of the magnetic and electric fields and interactions with atmospheric ions and atoms cause the diffusion of charged particles across and along magnetic field lines. The associated diffusion coefficients have to be determined experimentally. Knowledge of their values, particularly their spatial and temporal dependence, is necessary for a quantitative understanding of magnetospheric particle dynamics.

2.3. Approved Spacecraft Missions and Recommendations for Filling Existing Gaps

The First IMS Report reviews those spacecraft missions which will contribute scientifically to the IMS and which have been approved or are being considered for approval for flight during the appropriate time interval. An "approved program" generally refers to one for which the experiments have been selected and a major start in the financial support of the mission is under way. "Proposed programs" include those being formally studied by the appropriate administrative bodies and/or the appropriate scientific study groups.

In order to summarize the separate efforts in a coherent manner the individual missions have been categorized into five orbit classes, which is implicitly a classification of the primary scientific objectives of the IMS. These are:

1. Geostationary–geosynchronous orbit
2. Low altitude (mainly polar) orbit
3. Elliptical orbit
4. Solar wind environment
5. Multiple satellite systems

A summary of the missions discussed in the Study Group Report and identified according to the above classification is presented in Fig. 4. A more detailed description of each mission is given in the original document.

As can be noted in Fig. 4 and concluded on the basis of a more detailed analysis of the First IMS Report, certain critical gaps would exist in the study of the magnetosphere during the period 1975–1977, if only the presently *approved* missions were to be flown. On the other hand, Fig. 4 and the discussion in the First Report reveal that if all *proposed* satellite programs for the IMS period were actually approved, these gaps would be eliminated to a large extent and a reasonably satisfactory program of coordinated magnetospheric research could be put into operation.

The specific recommendations are as follows:

(1) The distribution in longitude of synchronous orbit satellites should be optimized and the payloads of such satellites should allow for reasonable overlap of the observations (for example, in the energy range of on-board particle detectors) wherever possible. In view of the advanced planning stage of the ESRO synchronous satellite GEOS, this should be used as a "reference" spacecraft. Particular points which could possibly be considered by the relevant agencies include: (i) The particle detector thresholds on the US National Oceanic and Atmospheric Administration's synchronous meteorological satellites SMS-B and C; it would be valuable if these were lowered to energy values that would encompass plasma and low energy trapped particles. The longitudes of the SMS satellites could usefully be coordinated with those of other IMS missions and available ground-based facilities. (ii) The synchronous ATS-G satellite (USA) could also make a significant contribution to the IMS project if its payload was coordinated with those of other geostationary satellites. (iii) It is hoped that if there is a follow-up geostationary SIRIO (Italy) satellite, it will be in operation during the IMS.

(2) A serious gap is expected during 1975 in the coverage provided by low-altitude polar-orbiting satellites. It is recommended that new missions with appropriate particle, field, and wave instrumentation be planned for the IMS in close coordination with, and complementary to, AE-C, D, E (USA), and Cosmos and Intercosmos series (USSR), and EXOS-A and ISS (Japan). In

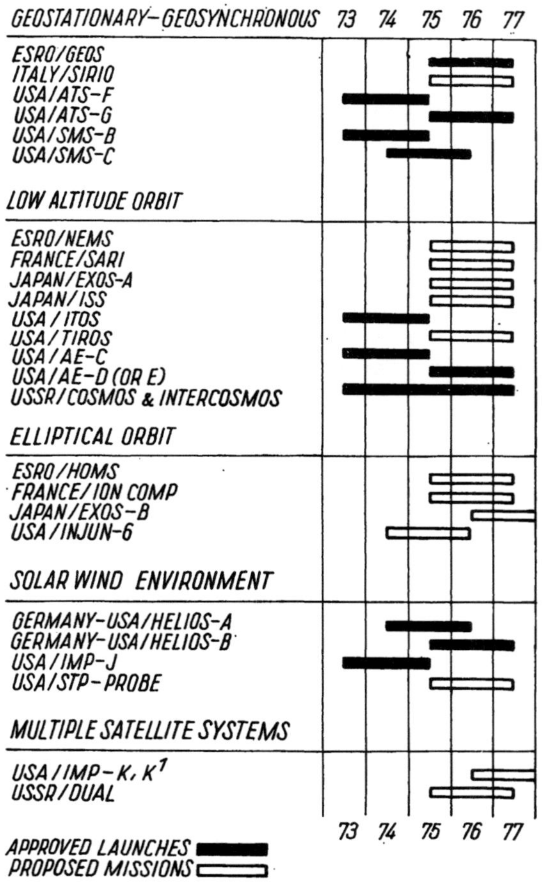

Fig. 4. Approved and proposed missions relevant to the IMS. For further details see text and First Report on Spacecraft Missions issued by the COSPAR–IUCSTP Study Group. 6th item in 2nd group *should read* USA/TIROS-N.

this connection, the SARI satellite proposed within the French National Program and the near-earth magnetospheric satellite (NEMS) being considered by ESRO would make major contributions. Furthermore, particle experiments on the meteorological ITOS (and TIROS-N) satellites (USA) could play an important role if the detector thresholds are suitably chosen.

(3) Another major weakness of the currently planned programs is the absence of definite missions in eccentric polar and equatorial orbits. There is a particular

need for (i) observations near the neutral points at high latitudes and (ii) observations obtained by means of a satellite in polar orbit having an apogee of about 10 R_e on the equator. It would be extremely valuable to have these types of observations carried out simultaneously. The eccentric polar-orbiting magnetospheric satellite (HOMS) being considered by ESRO and the Injun 6 satellite proposed by the USA would all provide important additions to research in this area. The same applies to the planned Japanese EXOS-B satellite, expected to cover the region of the plasmasphere and beyond, and to the French ion-composition satellite (Ion Comp).

(4) Although several proposed space probes, such as Helios and Pioneer G, will be monitoring the solar wind during the IMS, there is a great need for a solar wind probe in a heliocentric orbit near the earth, but outside the influence of the magnetosphere. The Solar Terrestrial Physics probe (STP) being considered by NASA would fill this need.

(5) The separation of space–time effects and the determination of spatial gradients are amongst the most crucial tasks in magnetospheric research. Therefore, mother–daughter and/or cluster satellite missions will make vital contributions to the IMS. The proposed IMP K-K′ pair of satellites (USA), having orbits ranging out to about 15 R_e, would therefore be most important. Furthermore, a pair of scientific satellites in 24-hour orbits (one eccentric 3.2–10 R_e, and the other nearly circular with coincident apogee and perigee longitudes) would provide a wealth of information about important magnetospheric processes. Particular importance is assigned to the dual or triple spacecraft mission (to the neutral point regions and the neutral sheet) that has been discussed by the USSR scientists.

2.4. Ground-based, Balloon and Rocket Observations

Many processes occurring in the magnetosphere correspond to perturbations, which, in the form of either waves or particles, propagate along the field lines giving rise to effects measurable on or near the earth's surface. Electric currents flowing in the magnetosphere and ionosphere affect the surface magnetic field and earth currents. Permanent and standardized geomagnetic, ionospheric, vlf, and particle precipitation observations at fixed locations thus provide a continuous long-term two-dimensional view or "diagnosis" of the state of the magnetosphere, complementary to space observations.

Balloon and sounding rocket measurements, on the other hand, allow us to explore, in detail, many short-term effects at the lower boundary of the magnetosphere, especially during intervals of disturbed conditions. Coordinated on a world-wide scale, these measurements provide a key link between ground-based and space observations.

Ground-based, balloon, and rocket programs give the opportunity to many countries that normally do not carry out satellite space research, to participate effectively in an international study of the magnetosphere, and to make significant contributions.

A continuous patrol of the sun and of several geophysical parameters, and the rapid dissemination of resulting data, has always been of greatest importance to research in Solar–Terrestrial Physics. It will be necessary to count on the continuation of such services during the period of the IMS. Particular importance is assigned to the re-initiation of a rapid magnetogram digitization service, at least covering limited periods of special interest.

Several special programs have been singled out as of greatest importance for the IMS, in complement to the operation of the regular network of solar and geophysical stations:

(1) A wide-scale international effort to study the polar regions.

(2) Comprehensive long-term and short-term observations clustered in the areas near the points at which field lines through geostationary satellites intersect the earth.

(3) Simultaneous measurements at magnetically conjugate areas.

(4) Observations of the position of the plasmapause by all available techniques, particularly whistlers and incoherent scatter.

(5) Rocket observations of precipitated particle flux with simultaneous observations of the X-ray flux from balloons flying at different longitudes along the auroral oval.

(6) Artificial injection of waves from the ground, and of plasma and energetic particles from rockets.

Details of the proposed ground-based, balloon and rocket program will be worked out by the IMS Special Study Group at future meetings with representatives from the various disciplines.

2.5. Data Acquisition and Utilization

The objectives of the IMS can be achieved only if related observations are made simultaneously in particular parts of the magnetosphere. In order to do this efficiently it will be necessary to identify in advance those periods during which particular pairs or groups of satellites are favorably located. This means that the orbits of all IMS spacecraft must be kept under constant review so that advance warnings can be issued about periods of time during which particular combinations of satellite, rocket, and ground-based experiments should be used to acquire data at the maximum possible rate. This operation requires special manpower, computer time, and current satellite tracking information. To that effect it is recommended that an appropriate IMS Satellite and Data Center be established. Further discussions will be conducted to determine whether one or several of the WDC's could take on the responsibility for this Center during the IMS period.

In short, the responsibilities of this Center should be:

(i) In coordination with agencies responsible for actual satellite operation (orbit prediction, data acquisition, etc.) it would determine up-coming intervals of time in which two or more satellites would be in a spatial configuration suitable for the study of some particular problem, during which increased satellite data transmission and the coordination of appropriate ground-based, balloon, and rocket observations would be highly desirable.

(ii) It would then designate these intervals as "special intervals" (or some other name) and notify the interested members of the STP community as far in advance as feasible; but short notice (two weeks) would often be inevitable. It may be necessary to designate regional centers which would undertake the coordination of specific observational programs in their areas during these special intervals.

The circulation of special newsletters and the organization of specialized symposia in connection with the IMS, held in conjunction with the regular meetings of the International Scientific Unions, will greatly contribute to coordination of data analysis and the exchange of information on observational and theoretical results.

Updated reference models of the magnetospheric field, to be used during the IMS for particle flux and field mapping, should be recommended in consultation with the appropriate commissions of IAGA.

Note added in proof

Due to delays in expected launch dates of some of the missions shown in Fig. 4, the proposed dates for the IMS have been shifted to 1976–1978. The Second IMS Report on Ground-based, Balloon and Rocket Programs mentioned in Section 2.4 is now ready for distribution. A "Satellite Situation Center", similar to the center proposed in Section 2.5, will be operated for the IMS at the National Space Science Data Center (USA).

Space Research XII — Akademie-Verlag, Berlin 1972

MEASUREMENTS OF ENERGETIC PARTICLE FLUXES DURING A SLOWLY VARYING ABSORPTION EVENT BY TWO CO-ORDINATED ROCKET FLIGHTS

K. WILHELM[a], G. KREMSER[b], J. MÜNCH[b], M. SCHNELL[b], J. P. LEGRAND[c], N. PETROU[d] and W. RIEDLER[e]

[a] European Space Research Organisation, Neuilly-sur-Seine, France
[b] Max-Planck-Institut für Aeronomie, Lindau, FGR
[c] Groupe de Recherches Ionosphériques, St.-Maur-des-Fossés, France
[d] Centre d'Etudes Nucléaires de Saclay, Service d'Electronique Physique, Gif-sur-Yvette, France
[e] Technische Hochschule Graz, Graz, Austria and Kiruna Geophysical Observatory, Kiruna, Sweden

In order to study the particle flux related to a slowly varying absorption event (SVA), two rockets were launched from ESRANGE, Kiruna, into the same event, the first at 0529 UT and the second at 0631 UT, on 27 November 1969. The SVA event, with maximum absorption of 2.1 dB at 0550 UT, occurred during a magnetospheric substorm. The two payloads, equipped with identical experiments, both reached an apogee height of about 160 km. Electrons were measured in the energy range from 1 to more than 90 keV and protons from 6 to 500 keV. The electron flux ($E > 30$ keV) showed, during both flights, a burst-like structure on the time scale of less than 1 second. During periods of peak precipitation the flux became more isotropic. Low-energy electron fluxes appeared to be relatively stable and near isotropic. In general, the electron energy spectra were harder during peak precipitation and softened considerably between the two flights. The results are interpreted in terms of weak and strong pitch angle diffusion of drifting electrons. Even though the high-energy proton measurements were partly contaminated by sunlight pulses, it was possible to establish that protons did not contribute substantially to the precipitated energy.

1. Introduction

Slowly varying absorption events (SVA events), observed by a relative ionospheric opacity meter (riometer) as an increase of the ionospheric absorption of the cosmic noise in the radio-frequency band, are a characteristic feature of the absorption pattern in the auroral zone during the morning hours. Detailed investigations by means of synoptic riometer studies and balloon-borne X-ray detectors have revealed that a SVA event only occurs when a magnetospheric substorm is in progress [1–5]. Optical as well as rocket and satellite observations have, in addition, enlarged our knowledge of this phenomenon [6–8]. This information was sufficient to demonstrate that the major part of the enhanced absorption is caused by electron precipitating into the atmosphere. Discrepancies between the absorption and X-ray intensity variations led, however, to the suggestion that during the later phases of such an event proton precipitation might substantially contribute to the cosmic noise absorption [3].

The limited data available on the electron energy spectrum and its temporal variations during the SVA event, furthermore, made it difficult to draw reliable conclusions on the origin of the energetic particles; thus, local acceleration, inward movement of a precipitation mechanism in the magnetosphere and magnetic gradient and curvature drift have been discussed as an immediate cause of a SVA event [5, 8–10]. Because of inherent limitations in the proton detection efficiency and in the electron energy resolution for both the cosmic noise absorption and the X-ray observation method, direct measurements of the primary particles were carried out aboard two sounding rockets (ESRO payload Nos. C62/1 and C62/2) which were launched successfully into a single SVA event with a time lag of 1 hour on 27 November 1969 from ESRANGE, Kiruna.

2. Description of the Experiment

In this paper, data will be used from four particle detectors which were mounted at angles of 19° and 60° with respect to the rocket axis. These angles were chosen so as to provide optimum information on the pitch-angle distribution of the particles. Two payloads built to the same specifications were carried by two rockets to apogee heights of 160 km and 164 km respectively. The nominal characteristics of the detectors are compiled in Table 1 giving

Table 1
Nominal characteristics of the particle detectors

Type of particles	Energy (keV)	Geometric factor (sr cm²)	Oopening angle	Viewing angle
e^-	0.7 3 7 11	$\sim 10^{-4}$	$5° \times 10°$	19° and 60°
e^-	>30 >45 >60 >90	0.15	20°	19° and 60°
p	6 22	$\sim 10^{-4}$ $\sim 3 \cdot 10^{-5}$	$12° \times 10°$ $3.5° \times 10°$	19°
p	>150 >250 >350 >500	0.15	20°	19° and 60°

the energy ranges for the electron and proton detectors together with the geometric factor and the viewing angle. The low-energy measurements were performed in differential channels with an energy bandwidth of approximately 20%. The electron detectors [11] utilized magnetic deflection for energy separation whereas the proton detectors [12] were equipped with electrostatic energy analysers. Continuous channel electron multipliers (CEM) with subsequent pulse amplifiers performed the particle detection.

The medium-energy particle detectors [13] used plastic scintillation counters with pulse discriminators for integral measurements. In order to avoid contamination by light, an aluminium cover of 120 μg cm^{-2} for electrons and

100 μg cm^{-2} for protons was evaporated on to the scintillators. A broom magnet was incorporated in the collimator system of the proton detector which effectively removed electrons of less than 500 keV. A PCM telemetry system was included in the experiment which allowed the transmission of information in 24 channels every 50 ms. In addition, low-energy proton data were transmitted in fm channels.

3. Geophysical Conditions

The absorption event into which both rockets were launched occurred during a period of moderate geomagnetic activity ($Ap = 22$). Fig. 1 presents magneto-

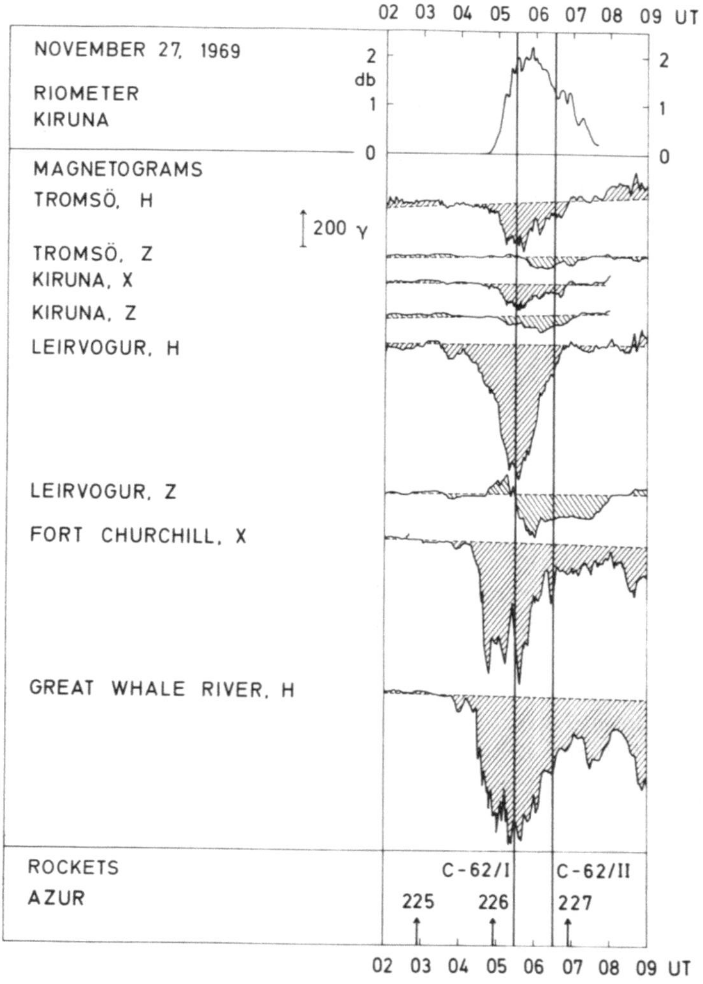

Fig. 1. Ionospheric absorption at Kiruna and magnetometer traces from several auroral stations for the period 02 h to 09 h UT on 27 November 1969.

grams from stations in the northern auroral region as well as the Kiruna riometer record for the time from 02 h to 09 h UT on 27 November 1969. The relevant geomagnetic coordinates and times are given in Table 2. It can be

Table 2

Geomagnetic coordinates and times of stations used in Fig. 1

Station	Corrected geomagnetic Longitude	Corrected geomagnetic Latitude	Eccentric dipole time at 05 UT
Tromsö	105°	67° N	7.4 h
Kiruna	104°	65° N	7.2 h
Leirvogur	69°	68° N	4.5 h
Great Whale	354°	68° N	23.8 h
Ft. Churchill	326°	70° N	22.4 h

seen from this diagram that the enhancement of the absorption starting at Kiruna near 0444 UT followed a break-up phase of a magnetic substorm at Canadian stations. The magnetic activity at Kiruna was small and, if it could be attributed to an electric line current, this electrojet was located always north of Kiruna and, for the latter part of the event, also north of Tromsø. The pattern of this event thus closely resembles those described in earlier studies on SVA events [1–3, 5] and it can, therefore, be concluded that the investigated event can indeed be classified as a SVA event and, in particular, that it occurred during a magnetospheric substorm. The first rocket was fired at 0529 UT just before the maximum absorption at Kiruna was reached and at a time when a rapid northward movement of the electrojet occurred over Leirvogur. The second launch took place at 0631 UT during the recovery phase of the event.

4. Electron Measurements

4.1. General Characteristics

Both rocket flights provided approximately 4 minutes of useful data from the scintillation counters. The apogee heights were not sufficient to assure complete insignificance of possible atmospheric effects on the energetic particles; it is, however, felt that a comparison between the two flights will be less affected by possible absolute errors. The channeltron experiments were only switched on at 127 and 129 km altitude, respectively, and thus no information is available from these detectors for the first part of both flights. During the first flight the 19°-detector axes covered a pitch angle range of maximal 0°–36°, whereas the 60°-detectors pointed between 43° and 77°. The corresponding figures for the second flight are 0°–52° and 27°–93°.

All data from the scintillation counters during the second flight averaged over 1 second are compiled in Fig. 2 together with the pointing directions of the 19° and 60° units. A strong modulation of the electron flux intensity prevailed throughout the flight and was particularly pronounced at lower pitch angles. The peak intensities did not vary appreciably and reached about

2×10^6 cm^{-2} s^{-1} sr^{-1} for electrons with $E > 30$ keV. The relative amplitude of the intensity variations decreased with decreasing energy. Fig. 3, giving a summary of the low-energy electron measurements during the second flight, demonstrates this tendency even for electrons in the keV energy range, though the low counting rates in the channeltron detectors limit a detailed comparison.

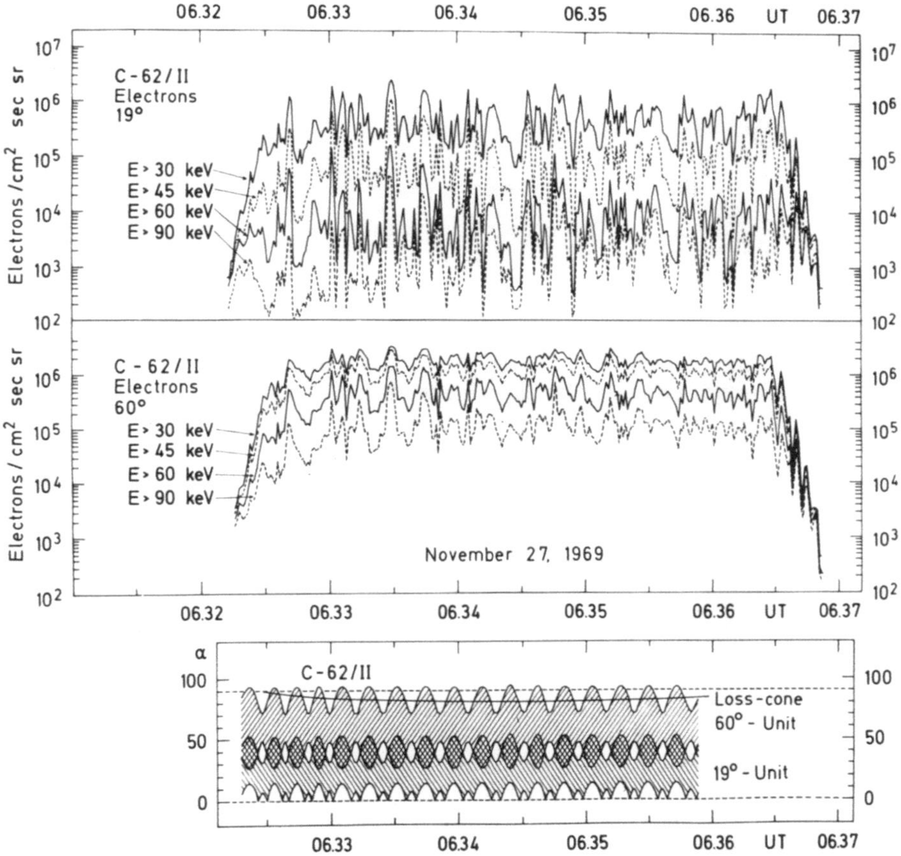

Fig. 2. Electron flux intensities as measured by the scintillation detectors averaged over 1 second during the second flight on 27 November 1969 from 0631 to 0637 UT. The lower part of the figure illustrates the scanning ranges of the 19° and 60° units in pitch angle space as well as the theoretical loss cone.

It can also be seen from the low-energy electron measurements that these particles were almost isotropically distributed in the loss cone.

The observations during the major part of the first flight as far as the gross features are concerned were very similar to those of the second flight. During the initial phase of the first flight until 0531.30 UT, however, intense electron precipitation prevailed which possibly caused saturation in nearly all channels of the scintillation counters. No attempt will be made in this paper to analyse this part of the flight.

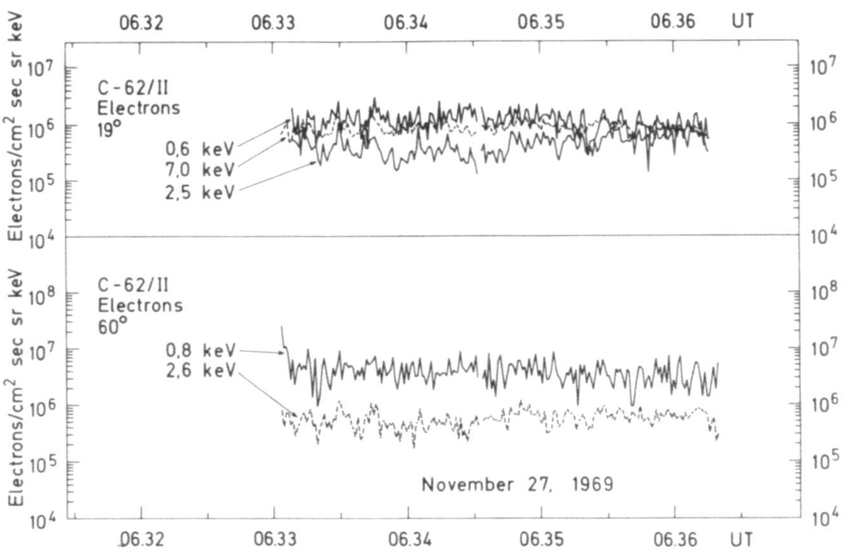

Fig. 3. Low-energy electron fluxes during the second flight averaged over 1 second.

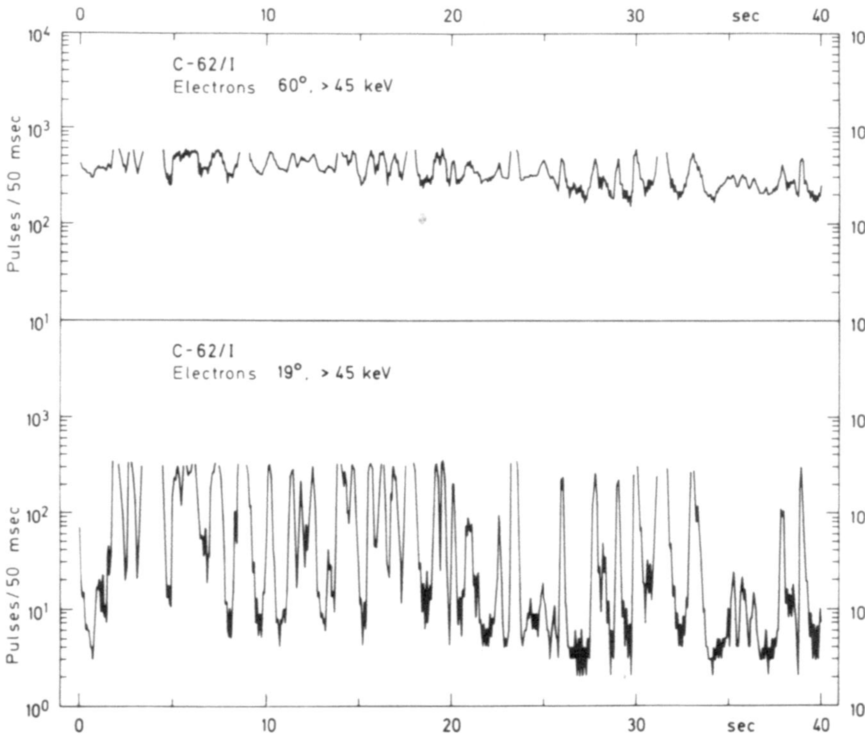

Fig. 4. High time resolution plot of the >45 keV integral channel showing an interval of 40 seconds during the first flight. In some peaks the curves are interrupted as saturation effects might have influenced the high counting rates.

4.2. Time Structure of Intensive Modulations

In Fig. 4, original data are plotted in order to give the structure with the best available time resolution. The interval was taken from the first flight between 0533.20.5 and 0534.00.6 UT. Rapid variations with time scales of less than 1 second, mainly in the low pitch-angle range, were a predominant feature of the latter part of flight 1 and the whole of flight 2. Doubtless, these bursts are the cause of long known X-ray microbursts [5, 7, 14]].

4.3. Electron Energy Spectra

The measurements of the electron energy spectra were of primary importance during these investigations. The analysis so far performed has shown that the spectral parameters depend strongly on the pitch angle range and the flux

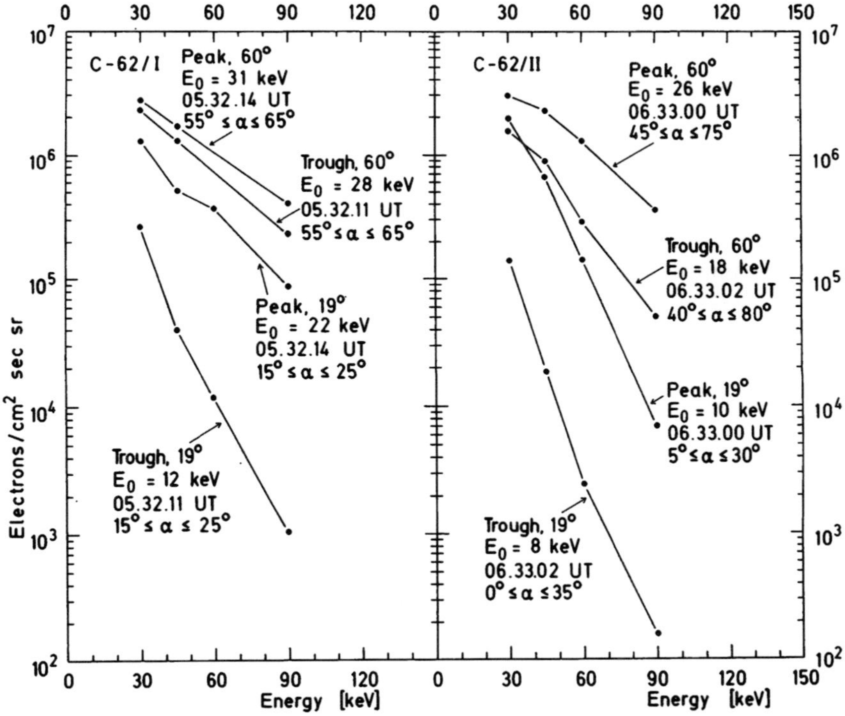

Fig. 5. Integral energy spectra of electrons for two pitch-angle ranges at peak and minimum flux levels during both flights.

intensity, but that there are no persistent variations during each flight. Therefore, a presentation of the energy spectra for particular time intervals as given in Fig. 5 contains most of the relevant information. For each flight the integral spectra measured by the scintillation detectors were evaluated on the basis of 1-second averages for periods of peak and minimum precipitation, both for the high and low pitch-angle range. The results of this analysis are compiled

Table 3

Electron energy spectral parameter E_0 (keV)

Pitch angle range	Flight 1 Peak	Flight 1 Trough	Flight 2 Peak	Flight 2 Trough
40°–80°	31	28	26	18
0°–35°	22	12	10	8

in Table 3 by giving the spectral parameter E_0 under the assumption that the energy spectra could be approximated by functions of the form $\exp(-E/E_0)$.

The low-energy electron measurements have yielded differential spectra and an unambiguous combination with the integral spectra of the scintillation counters requires detailed analysis. However, rough calculations indicate that there is no major incompatibility between the measurements in the different energy ranges. It thus appears as if the energy spectra given in Fig. 5 are representative down to energies of some keV.

5. Proton Measurements

One of the goals of the co-ordinated rocket flights was to clarify whether the amount of energy deposited by proton precipitation into the atmosphere changes considerably during the course of the event.

The flight data from the first launch gives no indication for proton flux ($E > 150$ keV) in excess of $5 \times 10^2\ s^{-1}\ cm^{-2}\ sr^{-1}$ except for those periods when very high electron fluxes were observed which may have contaminated the proton channels. The analysis of the data from the second flight encounters

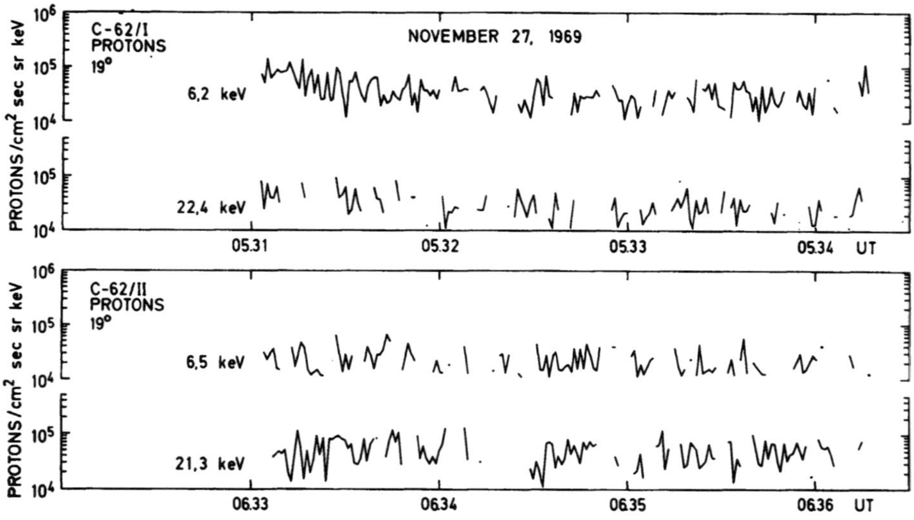

Fig. 6. Low-energy proton fluxes during both flights averaged over 1 second.

another difficulty in that the aluminium cover proved to be not sufficiently light-tight for the prevailing sunlight conditions. The counting rates were thus strongly spin-modulated, but, since no pulses were observed when the 60° detector was looking away from the sun, it was nevertheless possible to conclude that again the proton flux ($E > 150$ keV) did not exceed 5×10^2 s^{-1} cm^{-2} sr^{-1}, at least for the higher pitch-angle range. The low-energy proton flux measured in the 6 and 22 keV energy channels of the 19° detector reached a few times 10^4 s^{-1} cm^{-2} sr^{-1} keV^{-1} as can be seen from Fig. 6 depicting the proton observations during both flights.

6. Discussion and Conclusions

The electron observations have shown that rapid variations of the measured flux on the time scale of less than 1 second are a prominent characteristic of this SVA event. These variations, with almost unchanged features, were present both on the upleg and the downleg of the trajectory despite the fact that the rockets during both flights crossed the magnetic field lines with a mean velocity of 300 m s^{-1} before reaching the apogee and 100 m s^{-1} thereafter. For this reason, it was concluded that the observed variations could be regarded as temporal intensity fluctuations. This conclusion is supported by the similarity of the fluctuations to X-ray microbursts. The variations were much more pronounced at lower pitch angles and the energy spectrum was considerably harder during peak precipitation.

A comparison between the electron energy spectra observed during the two flights has revealed that the slope of the spectrum steepened from the first to the second flight. This steepening occurred both for different pitch-angle ranges and flux intensity levels. This result strongly supports the idea that energetic electrons drifting in the inhomogeneous magnetic field of the earth are precipitated into the atmosphere in the dawn sector, thus being the cause of SVA events. In a recent paper [15] it was suggested that cold plasma flow from the ionosphere might increase the equatorial plasma density on the dayside which would lead to enhanced precipitation of electrons. It seems appropriate to involve the processes of weak and strong pitch angle diffusion [16] for the explanation of the observed time structure of the electron flux above 30 keV and its characteristics of increased E_0 and isotropy during peak intensity as compared with minimum precipitation. Taken together with the distinct features of the low-energy electron flux, the observations, furthermore, support the expectation that the precipitation of low and high energy electrons might be caused by different processes [17].

The steepening of the slope of the electron energy spectrum observed during this SVA event has to be compared with previous observations which have demonstrated a continuous flattening of the spectrum during the morning hours by means of X-ray [18] and optical [6] methods. A flattening of the spectrum was also observed during the later phases of individual SVA events [3]. As this observation seemed to be inconsistent with a drift mechanism it was concluded that an apparent north–south movement of the southern boundary of the precipitation region at the onset of SVA events could not be caused by drift effects. A steepening of the energy spectrum which was observed at the beginning of SVA events was, therefore, attributed to an approach of the precipitation region to the location of the X-ray detector by the same authors. It would appear from the results presented here that a reconsideration of this explanation were necessary to the effect that the initial softening in the X-ray spectrum reflects a true softening in the electron spectrum. This would then allow us to interpret the equatorward motion of the precipitation region by *latitude* dependent drift velocity as was proposed in a study using riometer observations [8].

It is more difficult to reconcile the rocket measurements with the flattening of the energy spectrum as deduced from X-ray observations but a tentative explanation could be found in the following sequence of events: For simplicity, the acceleration mechanisms acting either in the source region of the energetic electrons or on the drift path are assumed to produce electron fluxes with time independent E_0 on the time scale of 1 hour. After an initial period, in which *energy* dependent drift velocities lead to a steepening of the spectrum in the morning sector, the electron energy spectrum at a given location of observation would then be constant in time as far as the generation processes are concerned. The energy parameter E_0 could, however, still be influenced by energy dependent losses. The measurements presented, converted to differential fluxes, have indeed shown that the precipitation rate at 30 keV was nearly constant whereas strong fluctuations occurred at 90 keV. On a time average, more low than high energy electrons were therefore precipitated, yielding a flattening of the energy spectrum in time. Taking into account the finite drift velocity, this is equivalent to a flattening with increasing longitudinal distance from the source region. Preliminary calculations give the result that E_0 would increase by about 5% in 200 seconds. This is too small to be detected in the data of the second rocket flight but compares favourably with the value of 4% in 200 seconds derived from X-ray data published in [3].

Since electron fluxes with a flat energy spectrum normally give rise to much more radio wave absorption than fluxes with a steep spectrum [19] such a change in the energy spectrum might also account for the varying response of the X-ray and the absorption phenomena to the electron precipitation at different stages of SVA events. No need would then arise to postulate substantial proton precipitation together with the electron flux, in accordance with the observations during the rocket flights. We must, therefore, conclude that this experiment did not support the hypothesis that proton precipitation might contribute to the absorption during the later phases of SVA events.

Acknowledgments

We gratefully acknowledge the continued interest of Prof. G. Pfotzer and Dr. A. Bewersdorff. We want to thank Mr. P. Christophersen and Mr. C. Becker for help in the calibration of the low-energy particle detectors and Mr. G. Monecke for assistance in the data processing.

The experiments were supported by the German Bundesministerium für Bildung und Wissenschaft under grant WRK 163, the French Centre National d'Etudes Spatiales and the Swedish Natural Science Research Council. Magnetograms were kindly provided by the IGY WDC, Geomagnetism C1, Charlottenlund, Denmark.

References

[1] R. R. Brown, J. Geophys. Res. **69**, 2315 (1964).
[2] Z. A. Ansari, J. Geophys. Res. **70**, 3117 (1965).
[3] A. Bewersdorff et al., Arkiv Geofys. **5**, 115 (1966).
[4] D. Jelly and N. Brice, J. Geophys. Res. **72**, 5919 (1967).
[5] A. Bewersdorff et al., J. Atmos. Terr. Phys. **30**, 591 (1968).
[6] F. T. Berkey, J. Geophys. Res. **73**, 319 (1968).
[7] M. Lampton, J. Geophys. Res. **72**, 5817 (1967).
[8] L. H. Rosen and J. R. Winckler, J. Geophys. Res. **75**, 5576 (1970).
[9] G. K. Parks and J. R. Winckler, J. Geophys. Res. **73**, 5786 (1968).
[10] K. Wilhelm, J. Geophys. Res. **73**, 7491 (1968).
[11] H. Schütz et al., Mitt. a. d. Max-Planck-Institut für Aeronomie **40**, 71 (1970).
[12] W. Riedler et al., Arkiv Geofys. **10**, 619 (1971).
[13] N. Petrou, Thèse, Conservatoire National des Arts et Métiers (1972).
[14] K. A. Anderson and D. W. Milton, J. Geophys. Res. **69**, 4457 (1964).
[15] N. Brice and C. Lucas, J. Geophys. Res. **76**, 900 (1971).
[16] C. F. Kennel, Rev. Geophys. **7**, 379 (1969).
[17] C. F. Kennel et al., J. Geophys. Res. **75**, 6136 (1970).
[18] A. Bewersdorff et al., Annls Géophys. **22**, 23 (1966).
[19] B. Hultqvist, Planet. Space Sci. **12**, 579 (1964).

Space Research XII — Akademie-Verlag, Berlin 1972

PROTON ENERGY SPECTRA FROM RECENT ROCKET MEASUREMENTS IN THE NIGHT AND MORNING TIME AURORAL ZONE

H. Raethjen

Institut für Reine und Angewandte Kernphysik der Universität Kiel, Kiel, FRG

Three Black Brant V A rockets were launched in February 1970 from Andöya, Norway up to altitudes above 200 km: one of them on 3 February during the break-up of aurora around midnight, the other two in the morning of 14 February during slowly varying absorption of cosmic radio noise. Each of the rockets carried beside other instruments a semiconductor spectrometer able to measure protons in the energy range from 0.13 to 8 MeV. This equipment worked well during the three rocket flights and provided measurements of energy spectra in the range below 1 MeV with maximum integral fluxes from 4×10^2 to 3×10^4 protons $cm^{-2}\,s^{-1}\,sr^{-1}$ for energies >135 keV. The differential energy spectra in the range 200 keV $< E <$ 500 keV can be approximated by exponential law with e-folding energies 75–50 keV. These spectra in comparison with similar measurements in other parts of space suggest that the acceleration of high-energy ($E >$ 200 keV) protons observed in the auroral oval takes place not in the magnetosphere but on the sun.

1. Introduction

Results of particle measurements in the auroral zone by means of rockets were summarized by Pfister [1] and Whalen and McDiarmid [2]. Most of these measurements concentrate on local times around midnight and on the electron component. After the first measurements of protons in the energy range >100 keV by Davis et al. [3] and McIlwain [4] investigations like these remained relatively rare compared with electron measurements. On the other hand the study of high-energy ($E >$ 100 keV) protons associated with auroral phenomena seem still to be of interest at least for two reasons: (i) These protons are able to penetrate the atmosphere down to altitudes <110 km and give rise for instance to ionospheric absorption of cosmic radio noise. This might be a factor in the still unknown process of the SVA event (Slowly Varying Absorption) appearing occasionally in the auroral zone morning sector [5]. (ii) Protons of these energies in the earth's magnetic field have gyration radii large enough to be not affected by local magnetic irregularities as strongly as electrons of equal or even lower energies. Thus protons may reveal more than electrons about the large area magnetospheric dynamics producing polar substorms. A further question still open is the fact that protons of these energies sometimes do and sometimes do not appear to be associated with auroral zone phenomena.

Hence a proton spectrometer was part of three scientific payloads assembled by the cooperation of five German research institutes. Each of these payloads consisted of eight different instruments in order to measure: protons of energies 0.13–8 MeV, electrons of energies 1–1000 keV, optical emissions of different frequencies in the spectral range 920–5577 Å, and the magnetic field parallel and perpendicular to the rocket axis. These payloads were launched from the Norwegian rocket range at Andenes/Andöya by Black Brant V A rockets up to altitudes >200 km at different times and specific geophysical conditions. The coordinates of the launch site are 69.3° N 16.1° E geographic, 67.5° N 113.9° E geomagnetic. Some technical data of the flight are summarized in Table 1.

Table 1

Data of rocket flights

Notation	X1	X2	X3
Date of launch	03 Feb. 1970	14 Feb. 1970	14 Feb. 1970
Time of launch (LT)	0007, 31 s	0426, 19 s	0601, 50 s
Height of apogee (km)	202	210	223
Duration of flight (s)	425	440	453
Distance of impact (km)	124	187	131
Spin frequncy (rev s^{-1})	3.75	4.0	3.87
Coning angle (°)	4	0.5	2.5
Coning period (s)	26	1.75	53

The scientific aim and experimental realization of the entire measuring program is described in detail in [6]. The data received during the flights will be published by the experimenters responsible for the particular measurements. In this article we confine ourselves to the proton measurements. First the instrument will be described briefly, subsequently the measured energy spectra will be presented and discussed.

2. Measuring Instrument

The proton spectrometer used is able to measure protons in the energy range 0.13–8 MeV. A detailed description of the instrument is given in [7]. Fig. 1 shows schematically the principles of operation: The sensor consists of a semiconductor detector (detector 1) as thick as the range of 8 MeV protons; detector 2 is for particles penetrating detector 1. Both detectors are shielded by 8.5 g cm^{-2} of brass leaving open an aperture of 30° full angle above detector 1. A broom magnet removes electrons of energies <1 MeV out of the incoming particle stream. The aperture and detector surfaces together define the geometric factor to 0.063 cm^2 sr. The angle between the axis of this sensor system and the rocket axis was adjusted to 65°. Protons of energies <8 MeV are stopped in detector 1 and give rise to energy-proportional pulses in the following electronic system. These pulses are counted in three integral energy ranges (>0.135, >0.425, >1.5 MeV). Secondly they are analysed by a 16-channel pulse height analyser (PHA). To improve the energy resolution the PHA has

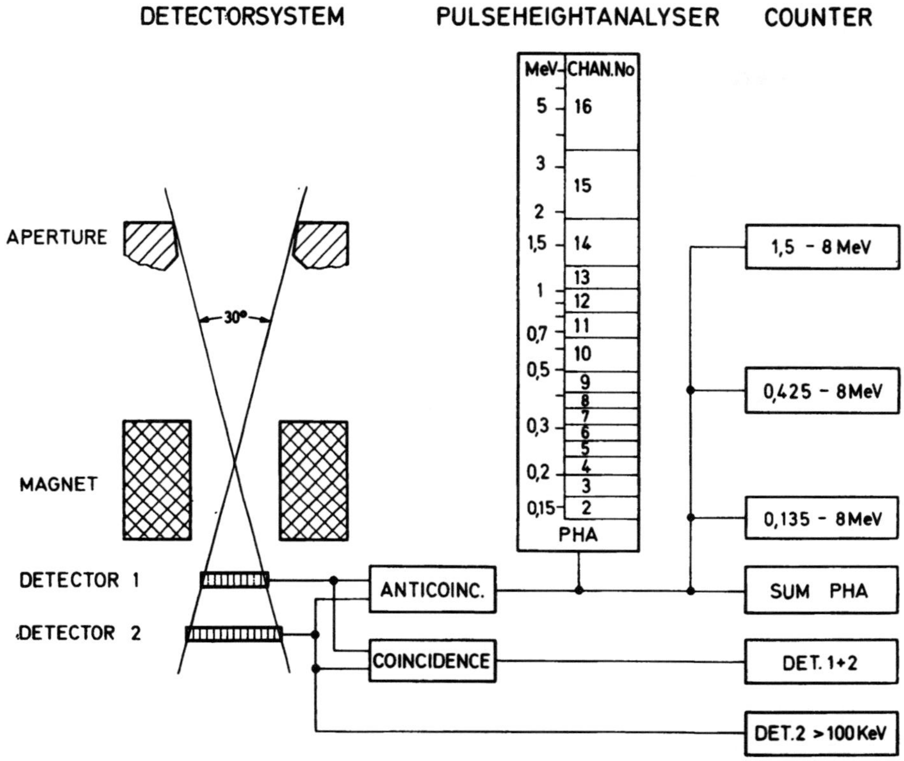

Fig. 1. Operation diagram of semiconductor proton spectrometer with sublinear pulse height analysis.

a characteristic which is nearly logarithmic in the lower range as shown in Fig. 1. Three further counters record all pulses entering the PHA, the coincidences of the detectors $1 + 2$, and all detector 2 pulses >100 keV.

While the counters were in operation continuously over the whole flight time the operation of the PHA was limited to discrete time intervals of 640 ms duration with 1280 ms gaps between. The counting rates and pulse height spectra (if available) were transmitted during the flight every 20 ms by PCM-telemetry [8]. The results subsequently presented are mainly based on the PHA data. The counting rates of the PHA channels are to be multiplied by the factor $834/\Delta E$ to give the differential fluxes in protons $\text{cm}^{-2}\,\text{s}^{-1}\,\text{sr}^{-1}\,\text{keV}^{-1}$. The numerator of this factor contains counting time (20 ms) and geometric factor, the denominator stands for the energy width of the particular PHA channel in keV.

3. Results

3.1. Night Time Measurements

The first payload "X1" was launched at local midnight on 3 February 1970 during the break up phase of a polar substorm. Auroral display was observed through a thin cloud cover. The riometer recorded fluctuating absorption of cosmic radio noise up to 1.5 dB. A bay type magnetic variation of about 250 γ in the horizontal component H was present. Protons of energies up to 800 keV were detected during the flight with maximum integral intensities of 3×10^4

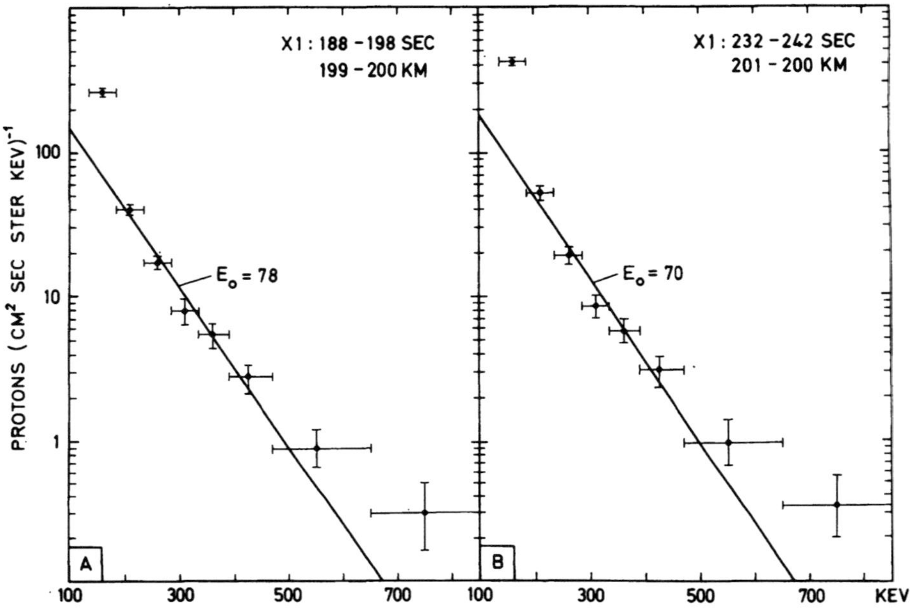

Fig. 2. 10 s averaged differential proton energy spectra measured at selected flight times and altitudes of X1 rocket, Andenes, 3 February 1970.

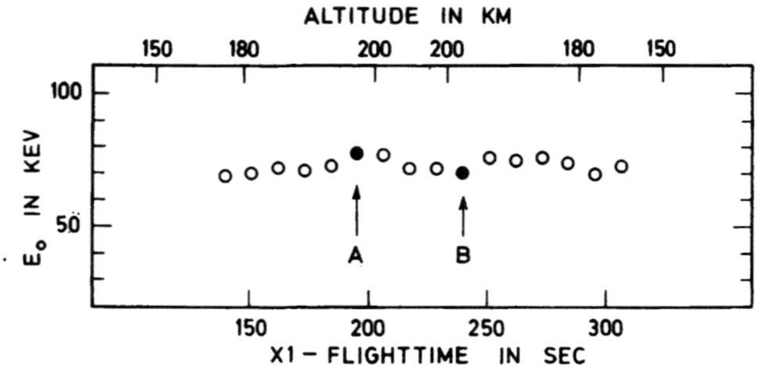

Fig. 3. e-folding energy E_0 of the exponential spectrum approximation in the energy range $200 > E_{\mathrm{p}} > 500$ keV as a function of flight time and altitude of X1 rocket, Andenes, 3 February 1970. A and B indicate the spectra shown in Fig. 2.

protons cm^{-2} s^{-1} sr^{-1} for $E > 135$ keV and 3×10^2 protons cm^{-2} s^{-1} sr^{-1} for $E > 425$ keV.

Fig. 2 shows two typical differential energy spectra measured during the flight of X1 at selected times and altitudes. The differential proton fluxes $d\Phi/dE$ estimated by PHA counting rates as mentioned before are plotted on a logarithmic scale against linear energy scale. The values are time-averaged over 10 seconds (= 37.5 spin periods). Horizontal and vertical bars indicate channel-widths and statistical deviations respectively. The slope of the spectrum flattens considerably with increasing energy. Hence there exists no exponential fit covering the whole energy range picked up by our PHA measurements. However, in order to characterize the spectrum for later comparison we can approximate a middle part of it by the exponential law

$$d\Phi/dE = \Phi_0 \exp(-E/E_0),$$

where $d\Phi/dE$ is the differential flux in protons $cm^{-2}\,s^{-1}\,sr^{-1}\,keV^{-1}$, Φ_0 a constant of the same dimensions as $d\Phi/dE$ the energy in keV and E_0 the so-called e-folding energy, also in keV. This was done for several spectra (like those in Fig. 2) in the energy range 200–500 keV. The e-folding energies found at altitudes above 150 km for this range vary between 70 and 80 keV as the substorm parameters (riometer, magnetometer) fluctuate. This is seen in Fig. 3, where the E_0 values are plotted against flight time and altitude of the X1 rocket.

3.2. Morning Measurements

The second and third payloads X2 and X3 were launched at local morning of 14 February 1970, when slowly varying absorption of cosmic radio noise was recorded by the Andenes riometer*. While the flight of the X2 rocket covered the maximum absorption of 2.6 dB and a subsequent decrease of approximately 1 dB, the X3 rocket took off 1.5 hours later, when the absorption had decreased to 0.3 dB, near the undisturbed level. Therefore the maximum integral intensities measured by X2 (4×10^3 p $cm^{-2}\,s^{-1}\,sr^{-1}$ for $E > 135$ keV and 1.5×10^2 p $cm^{-2}\,s^{-1}\,sr^{-1}$ for $E > 425$ keV) were a factor of 10 higher than those measured by X3. Fig. 4 shows four typical differential energy spectra for this event, time averaged over 10 s each in the same way as was done for the X1-spectra. Three of them come from X2 (A, at ascent; B, at apogee; C, at descent) while only one spectrum could be derived from X3 measurements (for altitudes above 190 km) according to the statistical significance. These spectra seem also to flatten in slope with increasing energy, but not as strongly as the X1 spectra; in particular, the intensity does not rise as high in the lowest energy channel. Hence these spectra are fitted sufficiently by exponential approximation from 130 keV to nearly 500 keV. The e-folding energy E_0 derived from X2 measurements are plotted against flight time and altitude in Fig. 5. Here E_0 does not fluctuate around a constant mean value as shown for X1 but decreases systematically with time from about 75 to 50 keV. The E_0 value of 56 keV derived from the X3 spectrum is comparable with the second half of the X2 flight.

3.3. Conclusions

Results of the previously described measurements are summarized in Table 2. There is one fact which should be discussed in detail: The slope of the exponential

* The record shows the typical gradual increase of absorption within 0.5 hour and a subsequent decrease through about 1.75 hours without significant fluctuation.

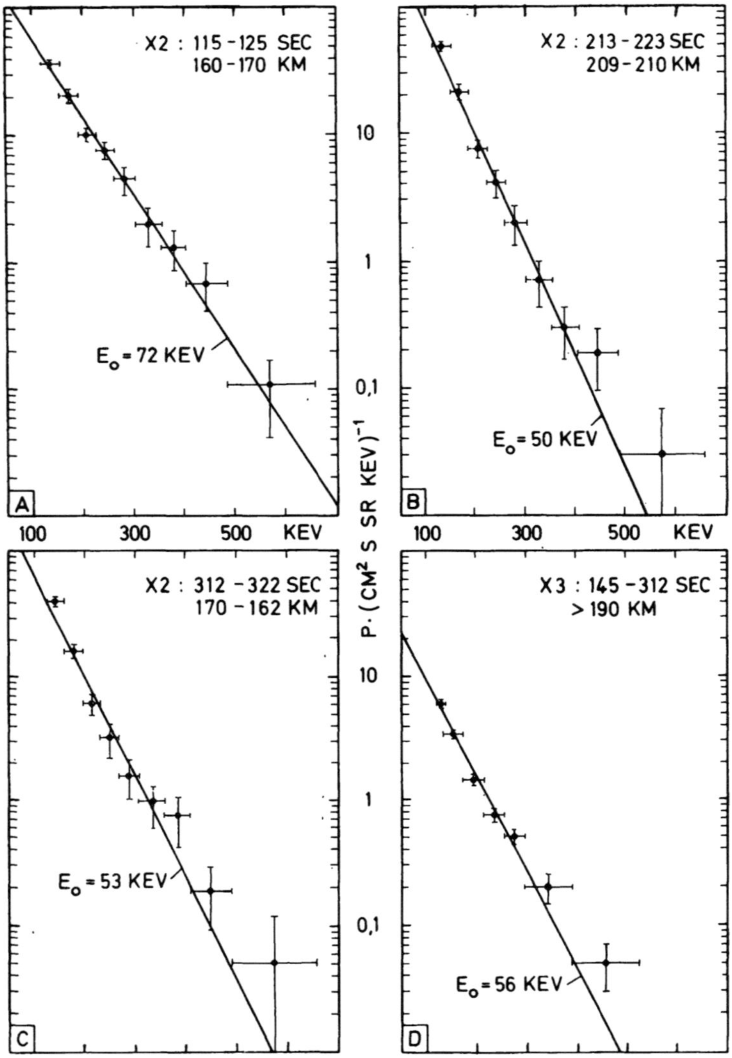

Fig. 4. 10 s averaged differential proton energy spectra: A, X2 at ascent; B, X2 at apogee; C, X2 at descent; D, X3 $>$ 190 km.

Table 2
Results of proton measurements

Notation	X1	X2	X3
Date	03 Feb. 1971	14 Feb. 1971	14 Feb. 1971
Magnetic local time	2110	0630	0810
Auroral zone phenomenon	break-up aurora	SVA max. 2.6→1.7 dB	SVA end 0.3 dB
Integral flux ($E > 135$ keV) (p cm^{-2} s^{-1} sr^{-1})	3×10^4	4×10^3	4×10^2
Differential spectra (200 keV $> E >$ 500 keV)	75 ± 5	$75 \rightarrow 50$	56

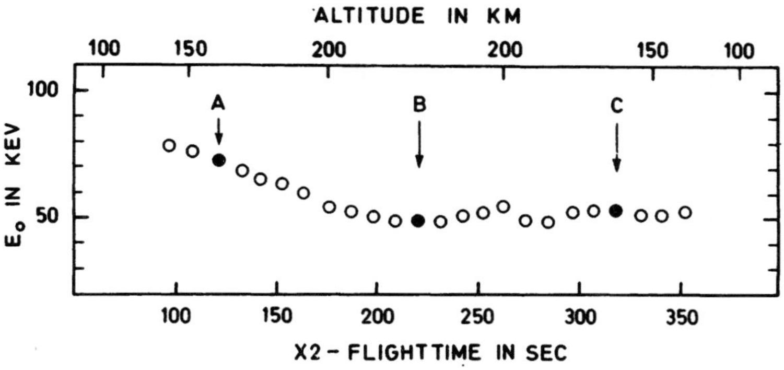

Fig. 5. e-folding energy E_0 of the exponential spectrum approximation in the energy range 130 keV $>$ E $>$ 450 keV versus flight time and altitude of X2 rocket, Andenes, 14 February 1970. A, B and C indicate the X2 spectra shown in Fig. 4.

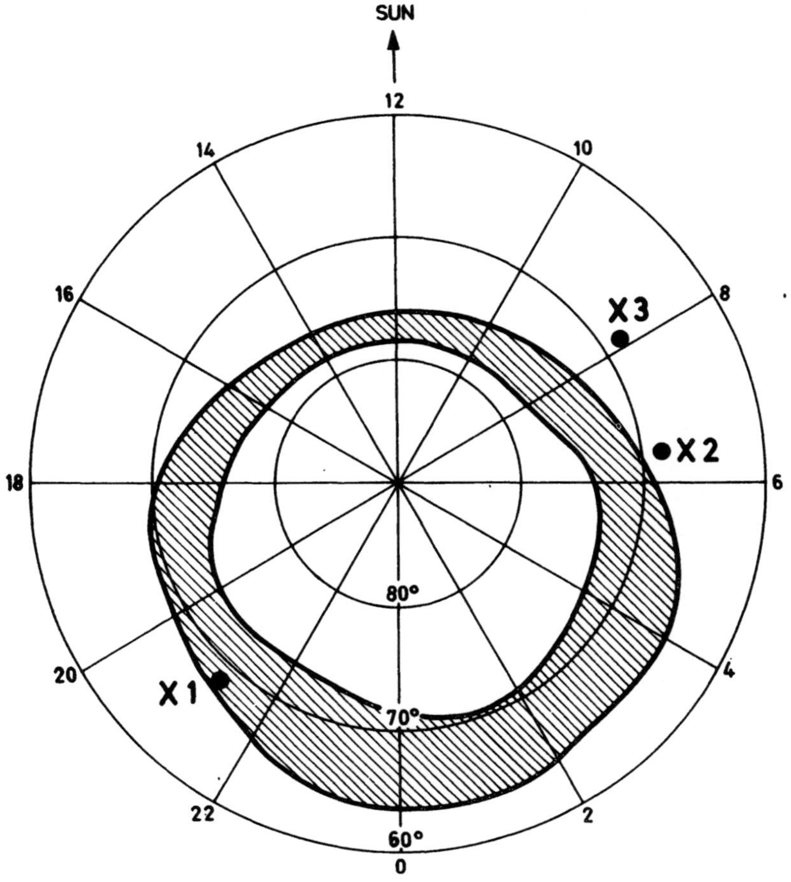

Fig. 6. The mean auroral oval and the positions of rockets X1, X2, X3 in coordinates of magnetic dipole latitude and magnetic local time.

spectrum approximation in the energy range 200 keV < E < 500 keV steepened during the X2 flight in such a way that E_0 decreased from the average value of X1 to a value similar to that of X3. This factor can be looked at in connection with different rocket positions relative to the auroral oval defined by Feldstein [9]. The positions are indicated in Fig. 6: X1 lies in, X2 at the southern edge of, and X3 outside (to the south of) the oval. The oval drawn is an average one; actually it can be different from this in form and position. Considering different patterns of proton precipitation in and outside the auroral oval following Hartz and Brice [10] it is suggested by these results that there are two types of spectra in the energy range 200–500 keV: a flatter one ($E_0 \approx 75$ keV) typical for in-oval precipitation and a steeper one ($E_0 \approx 50$ keV) typical for south-of-oval precipitation.

Both shapes of spectrum were also observed at synchronous altitude* of 6.6 earth radii in the equatorial plane by ATS 1 and some other satellites [11–14]. There the steeper shape is related to geomagnetic trapped radiation during quiet magnetospheric conditions, while the flatter shape is always associated with magnetospheric disturbances. Moreover, always when the spectrum flattens at synchronous altitude, similar fluxes and spectra are found in interplanetary space [11].

Considering further that the geometry of the auroral oval is formed by the boundary region between "closed" and "open" magnetic field lines, there is some indication that the observed high-energy protons ($E > 200$ keV) are not accelerated in the magnetosphere. The protons precipitating south of the oval might come from the reservoir of the radiation belt. The protons precipitating within the oval might come directly from the interplanetary space via "open" field lines. The latter assumption may be confirmed by recent observations made by Fennel [15] of protons ($E > 320$ keV) in the magnetotail and by Krimigis et al. [16] of protons ($E > 300$ keV) in the interplanetary space associated with solar events. Consequently the mechanism for accelerating protons up to the observed energies takes place on the sun, which is the motor for any magnetospheric dynamics. The fact that high-energy protons sometimes do and sometimes do not appear in the auroral zone associated with magnetospheric substorms might be explained by different particular strengths of solar acceleration.

The discussions above are based mainly on the few data of these two single events. Thus the conclusions are somewhat preliminary in character. For better confirmation more measurements from this point of view, if possible simultaneous at different magnetic latitudes, are suggested.

Acknowledgments

The experiment was part of a project in which a number of persons and institutions worked together reliably. I want to thank all who helped to lead

* The altitude at which a satellite rotates synchronously with the earth's rotation (24 hours period).

References

[1] W. Pfister, Space Sci. Rev. **7**, 642 (1967).
[2] B. A. Whalen and I. B. McDiarmid, in: Atmospheric Emissions, van Nostrand Reinhold Corp., New York 1969 (p. 93).
[3] L. R. Davis et al., Space Research I, 721 (1960).
[4] C. E. McIlwain, J. Geophys. Res. **65**, 2727 (1960).
[5] A. Bewersdorff et al., Ark. Geophys. **5**, 115 (1966).
[6] D. Beran et al., Bundesministerium für Bildung und Wissenschaft, Forschungsbericht W 70-59 (1970).
[7] H. Raethjen and R. Ulmann, Atompraxis **16**, 396 (1970).
[8] E. Keppler, Bundesministerium für Bildung und Wissenschaft, Forschungsbericht W 70-69 (1970).
[9] Ya. I. Feldstein, Planet. Space Sci. **14**, 121 (1966).
[10] T. R. Hartz and N. M. Brice, Planet. Space Sci. **15**, 301 (1967).
[11] L. J. Lanzerotti, Phys. Rev. Lett. **21**, 929 (1968).
[12] L. R. Davis and J. M. Williamson, Space Research III, 365 (1963).
[13] T. P. Armstrong and S. M. Krimigis, J. Geophys. Res. **73**, 143 (1968).
[14] S. M. Krimigis and T. P. Armstrong, J. Geophys. Res. **71**, 4641 (1966).
[15] J. F. Fennel, J. Geophys. Res. **75**, 7048 (1970).
[16] S. M. Krimigis et al., Johns Hopkins University, Applied Phys. Lab. Preprint Nov. 1970.

Space Research XII — Akademie-Verlag, Berlin 1972

ROCKET MEASUREMENTS OF CORPUSCULAR FLUX VARIATIONS IN THE POLAR CAP

V. F. TULINOV and V. V. TULYAKOV

USSR Academy of Sciences, Moscow, USSR

The paper presents the results of measurements of corpuscular flux intensities obtained by meteorological rockets M-100 over Heiss Island (81° N) during 1968–1969. In this period over 30 rocket flights, with a maximum altitude up to 95–100 km, were made with SBT type Geiger counters, which enabled effective registration to be made of electron and protons with energies of more than 35 keV and 0.5 MeV respectively.

Data were obtained on the character and magnitude of temporal variations of corpuscular fluxes in the polar zone over the period of the last solar activity maximum.

Since autumn 1968 regular measurements of corpuscular flux intensity have being carried out by meteorological rockets M-100 launched from Heiss Island (Franz Josef Land, 80°31′ N 58°03′ E). In all flights the end-window Geiger counter of type SBT-9 was used as a standard radiation detector which provided effective registration of cosmic rays and of electrons and protons with energies 30–35 keV and 0.5 MeV respectively. The use of unified apparatus is apparently of importance in the study of space–time variations. Semiconductor detectors were also installed aboard some rockets to provide proton registration in the energy range 1–20 MeV. The radiation detectors mentioned above were installed in the head of M-100 rockets in addition to standard apparatus, including temperature and pressure sensors. Thus, one could obtain simultaneous information on the thermodynamic parameters of the upper atmosphere and on the corpuscular fluxes.

It should be also noted that at the Heiss Island observatory a wide complex of ground-based methods of observation were used namely: riometric and ionospheric stations, a station for continuous registering of the magnetic field of the earth, and cosmic ray intensity and aurora observations.

Such unification of ground and direct observations provides fairly complete information about different types of geophysical activity which is essential for determining various casual relationships.

We present here some rocket data on corpuscular flux intensity in the period September 1968–April 1969 and also the first results of comparison of ionizing radiation flux variations with radio-wave absorption variation registered by riometer. Table 1 shows the dates and times of rocket launchings in the period considered.

This paper, being too short to present the primary results for all flights, contains some general remarks about the character of altitude–time–intensity

Table 1

No.	LT	Date	No.	LT	Date
1	1200	25. 09. 68	14	1228	19. 02. 69
2	1216	02. 10. 68	16	1200	05. 03. 69
3	1206	09. 10. 68	17	1200	12. 03. 69
4	1215	13. 11. 68	18	2117	13. 03. 69
5	1202	27. 11. 68	19	0033	14. 03. 69
6	1200	11. 12. 68	20	0030	15. 03. 69
7	1212	18. 12. 68	21	1202	26. 03. 69
8	1211	25. 12. 68	22	2230	01. 04. 69
9	1200	19. 01. 69	23	1132	02. 04. 69
10	1200	25. 01. 69	24	2149	15. 04. 69
11	1243	27. 01. 69	25	0345	24. 04. 69
12	1200	02. 02. 69	26	1100	30. 04. 69
13	1200	07. 02. 69	27	1137	18. 06. 69

dependence. First of all, it should be mentioned that considerable variability, of some orders of magnitude, was observed in the flux intensity from launch to launch at altitudes of approximately 60–95 km. Another essential feature is that the character of particle absorption in the atmosphere differed in different flights. This fact indicates that the energy spectrum of particles and the composition of the radiation are subjected to great changes. For the dependence of counting rate on altitude it is characteristic that the data for all flights up to 40–50 km are in good agreement with the known altitude intensity variation of galactic cosmic rays.

As has been mentioned already large intensity fluctuations occurred at higher altitudes. As an example, Fig. 1 shows an altitude counting rate dependence

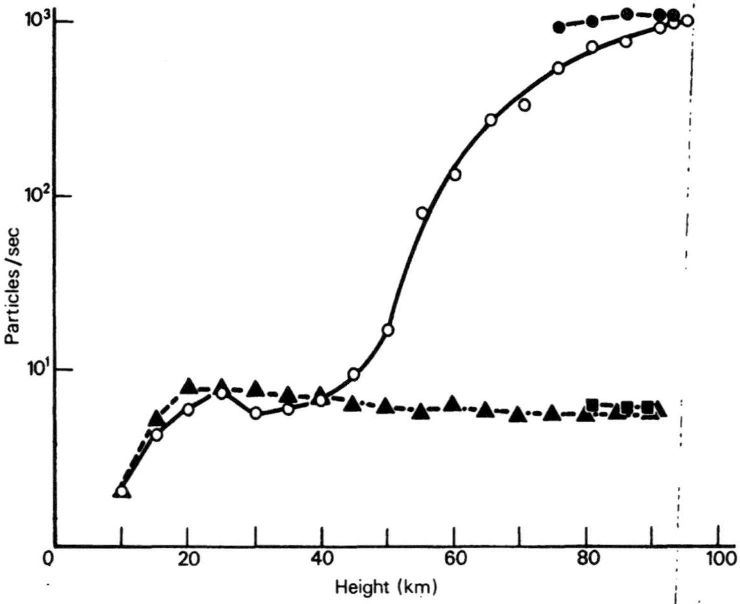

Fig. 1. Dependence of counting rate on altitude in two flights: flight 14: ■ upleg, ▲ downleg; flight 24: ● upleg, ○ downleg.

of particles registered in two rocket flights, when at heights above 50 km minimum and maximum values of radiation intensity, were observed in the whole set of flights. These values are presented in Table 1. In particular, in flight 14 even near the top the absence of additional radiation (to the background of cosmic rays) was detected. In flight 24 a quite strong flux of low-energy particles (counting rate exceeding background by more than two orders of magnitude) was registered together with cosmic rays. These particles are considerably absorbed at high altitudes.

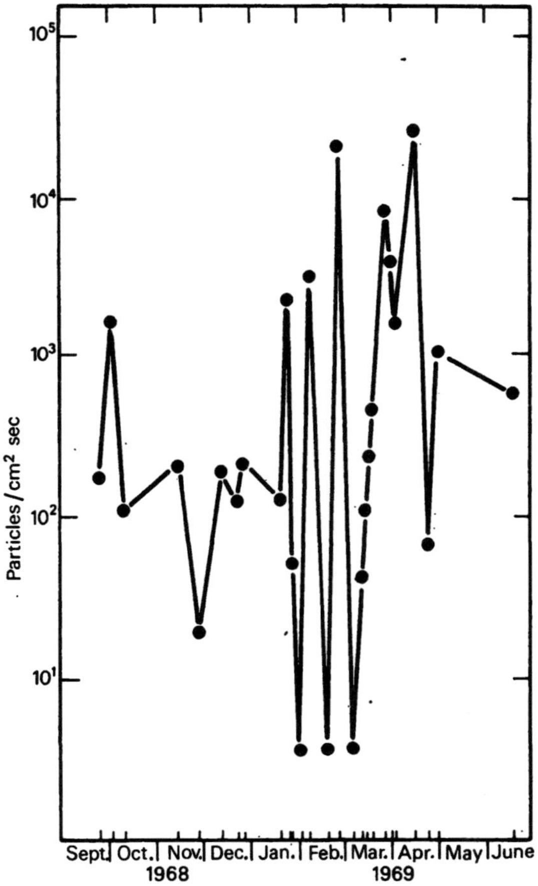

Fig. 2. Intensity variation of corpuscular flux, Sept. 1968–June 1969.

However, while estimating the corpuscular flux actually existing at high altitudes, based on experimental data, one should bear in mind that the registration effectiveness of soft radiation by the SBT-9 counter is essentially lower than that of cosmic rays (due to different effective areas and solid angles). For isotropic particle distribution in the northern hemisphere, as for example in our case, the ratio of registration effectiveness of soft and hard components is equal to 3×10^{-2}.

Flux data from different flights taking account of this circumstance are given in Fig. 2. For convenience of comparison all flux values are adjusted to the same altitude of 90 km. Fig. 2 shows the intensity variations of corpuscular flux over the whole period of observations. As can be seen the intensity changes are of a sporadic nature. It should be noted that considerable fluxes of additional soft corpuscular radiation are present as a rule over Heiss Island; this should be taken into account when analysing the data on geophysical parameters registered in the Heiss Island ground-based observations.

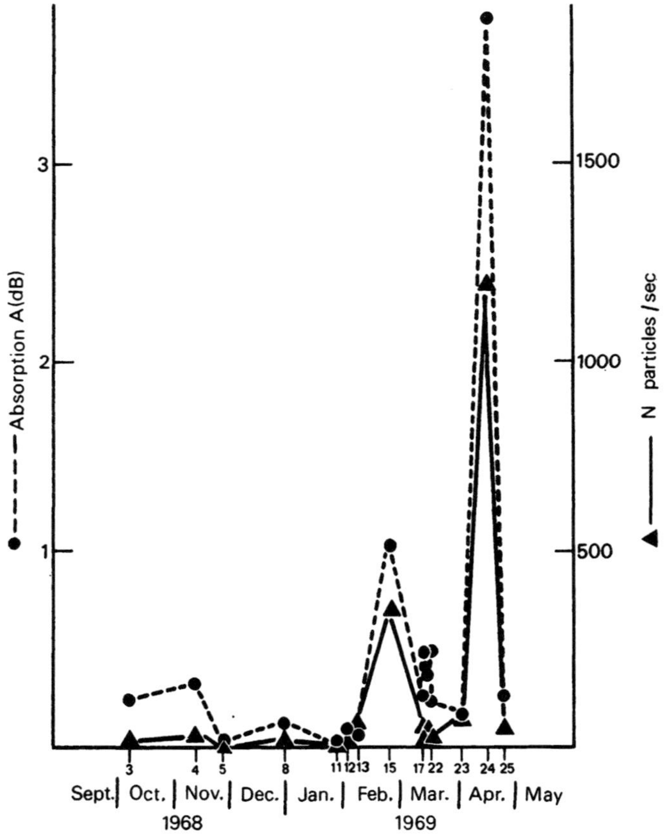

Fig. 3. Comparison of flux intensity and radio-wave absorption.

It is of great interest to compare the direct measurements of corpuscular flux obtained by rockets with simultaneous riometric data. Fig. 3 shows counting rates of particles at an altitude of 90 km (triangle) and radiowave absorption (in dB) at a frequency of 32 MHz (circles) for the same times. As can be seen, the time dependence is approximately the same in both cases. This indicates a close correlation between corpuscular flux intensity and radiowave absorption. In particular, during the strongest two PCA events, occurring in February and April 1969, the correlation between the absorption and the corresponding corpuscular flux increase N can be expressed by the linear function $N = K\,A$. K values are close for both cases.

Space Research XII — Akademie-Verlag, Berlin 1972

PARTICLE PRECIPITATION AND ITS EFFECTS AT HIGH LATITUDES IN THE 2 FEBRUARY 1969 EVENT

G. GUSTAFSSON[a], T. R. LARSEN[b,c], H. PETTERSEN[c], W. RIEDLER[d,a], G. SKOVLI[b], F. SØRAAS[e] and G. R. THOMAS[f]

[a] Kiruna Geophysical Observatory, Kiruna, Sweden
[b] Norwegian Defence Research Establishment, Kjeller, Norway
[c] The Auroral Observatory, Tromsø, Norway
[d] Technical University Graz, Austria
[e] Institute of Physics, University of Bergen, Bergen, Norway
[f] Radio and Space Research Station, Slough, England

A cooperative study has been carried out using data from several particle experiments on board the ESRO 1A satellite and comparing them with ground-based data obtained during the 2 February 1969 event. The data relate to a period near the maximum phase of a particularly intense substorm. High fluxes of electrons and protons were precipitated in the region $L = 2.9$–10. For energies below 40 keV the energy flux due to the electrons was somewhat higher than that for the protons, but above 40 keV the proton energy input was predominant.

1. Introduction

A strong magnetic storm occurred on 2 February 1969 with a sudden commencement at 1502 UT. The maximum of the activity index Kp during this storm was 8_0 in the time interval 18–21 UT when northern Scandinavia was in the 21–24 hour geomagnetic time sector. An extremely strong auroral electrojet was observed at several stations, e.g. the magnetic H component as measured at Abisko at 1908 UT showed a deflection of $-2700\,\gamma$, which is an unusually high value. The maximum of the bay coincided almost exactly in time with the passage of the polar orbiting satellite ESRO 1A/Aurorae over northern Scandinavia. Real time data were acquired by the Tromsø telemetry station in the time interval 1901–1908 UT. Due to the favourable situation of a satellite with several particle detectors covering a large energy range crossing auroral latitudes during a very strong substorm, a cooperative study was undertaken to relate the particle data to various ground-based data.

The particle measurements included in this paper are:

(i) Electrons at 1 and 13 keV and protons at 1 and 6 keV with 10° and 80° pitch angles, channel multipliers flown by Kiruna Geophysical Observatory [1].

(ii) Electrons at >40 keV with 10° and 90° pitch angles, plastic scintillators flown by Radio and Space Research Station, Slough [2].

(iii) Protons at 0.1–6 MeV with 0° and 90° pitch angles, solid state detectors flown by University of Bergen and Danish Space Research Institute [3].

(iv) Protons at 1–30 MeV with 0° pitch angle, solid state detector and scintillator flown by Radio and Space Research Station, Slough [2].

The ground-based measurement include magnetometers, riometers, all-sky camera and vlf phase variations.

2. Presentation of Data

The particle data obtained in the interval 1901–1908 UT are described in §§ 2.1 and 2.2. The satellite was moving northward at magnetic local time 2100, and coverage in L-values was 2.6 to about 13.5 for most of the instruments. The ground based data are presented in §§ 2.3–2.6.

2.1. Electrons

The count rate for the 1 and 13 keV electrons are shown in Fig. 1. For the 1 keV electrons both the 10° and 80° count rates increase at $L = 3.1$ by a factor of at least 200, with isotropic fluxes at higher L values. The fluxes at 13 keV are more variable than those at 1 keV, but detector saturation is sometimes observed in the 1 keV channel (greater than 10^9 cm^{-2} s^{-1} sr^{-1} keV^{-1}). At the southern edge of the precipitation zone a region with relatively soft spectra may be noted, in contrast to the northern boundary at $L = 9.8$ where the spectra do not seem to undergo a systematic change during decay to below count threshold. For the higher energy electron experiment the measured energy thresholds are lower than the nominal values. The 10° (>40 keV) and 90° (>30 keV) count rates (Fig. 1) show a variation with latitude similar to the 13 keV electrons. The 10° and 90° count rates are well correlated; however, it can be seen that the 10° count rate is more variable than the 90°. In general the flux is anisotropic (minimum at 10°), but it approaches isotropy in the peaks of the latitude profile. The peak integral intensity >30 keV observed during the pass is $\simeq 10^8$ cm^{-2} s^{-1} sr^{-1}.

2.2. Protons

The count rates of the 6 keV protons, (Fig. 1), indicate an approximately isotropic flux increasing by about 2 or 3 orders of magnitude with increasing L. This increase and also the subsequent decrease near $L = 9.8$ are not as sharply defined in latitude as for the electrons in the corresponding energy range. However, the proton zone coincides well with the electron zone at the energies discussed. The count rates of the protons in the 115–360 keV range are also shown in Fig. 1. They show approximately the same features noted above for the low-energy protons. The flux is approximately isotropic, increasing by about three orders of magnitude with increasing L. The 0° protons have a marked northern boundary coinciding with that of the higher energy electrons. North of this boundary the pitch-angle distribution of these protons becomes strongly peaked at 90°. The peak intensities observed with both instruments are exceptionally high with 9×10^7 protons cm^{-2} s^{-1} sr^{-1} keV^{-1} at 1 keV 80°, 8×10^6 at 6 keV 10° and 2×10^4 protons cm^{-2} s^{-1} sr^{-1} keV^{-1} in the 115–180 keV energy range. For the higher energy protons the count rate in the 1–6 MeV channel (Fig. 1) increases above the background level at about the same L value as the 115–360 keV protons, but it decreases again at lower L value than the lower

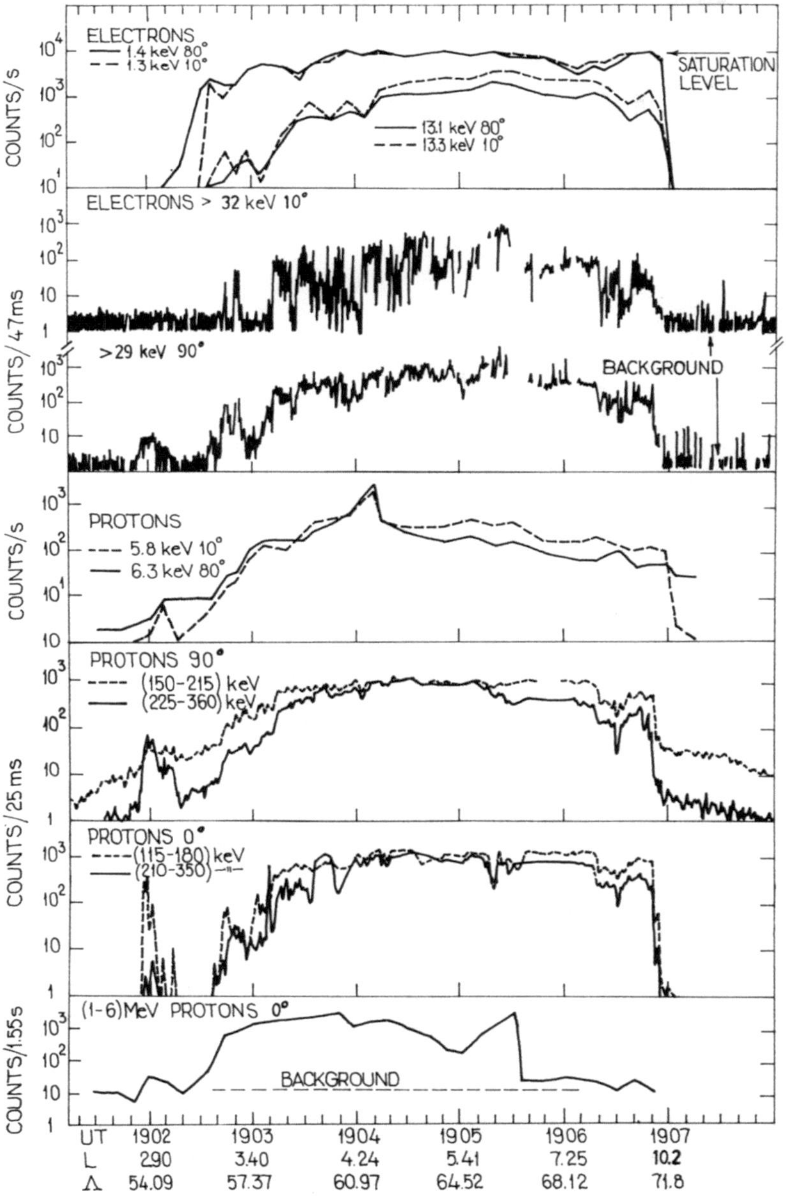

Fig. 1. ESRO 1A particle data for 1901–1908 UT on 2 February 1969, plotted as a function of time with L-values and invariant latitude also indicated.
Note. In the second section, for 32 *read* 40 keV and for 29 *read* 30 keV.

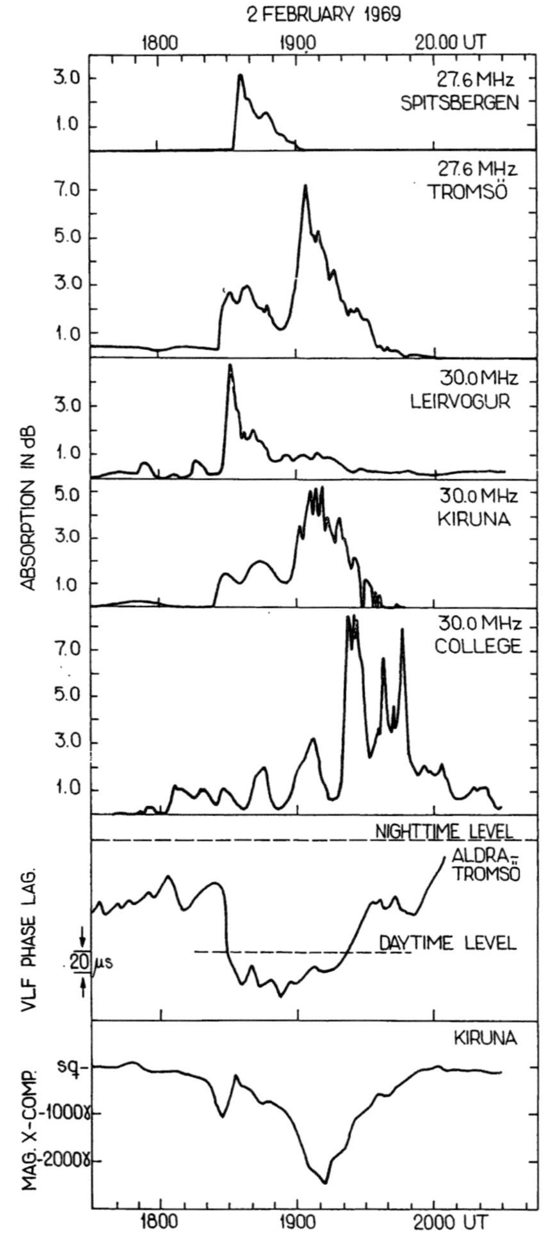

Fig. 2. Riometer absorption at a number of stations, phase recordings of the vlf path Aldra–Tromsø and magnetic X component at Kiruna, plotted as a function of time for 2 February 1969.

energy protons. The peak intensity observed at 1–6 MeV is $\simeq 10^3$ cm^{-2} s^{-1} sr^{-1}, and the intensity at >6 MeV is <5 cm^{-2} s^{-1} sr^{-1}.

2.3. Riometers

The riometer absorption as observed at a number of stations is shown in Fig. 2. It can be seen that the absorption event lasts approximately 90 minutes with little absorption before and after the event for most stations. It may be noted that Spitsbergen ($L = 16$) shows no absorption at all after 1900 in accordance with the observed cut-off of particle precipitation at about $L = 9$–10.

2.4. VLF Phase Measurements

Here some vlf phase measurements on the path Aldra–Tromsø, Norway, will be presented. The path (between $L = 5.2$ and 6.3) is 432 km long. The abnormal component of the sky wave is recorded on a loop antenna adjusted to cancel out the ground wave [4]. Fig. 2 shows the observed phase variations. A large phase advance occurs simultaneously with the onset of magnetic deflection and riometer absorption in the Tromsø/Kiruna area at 1825 UT. Between 1830 and 1920 UT the phase is depressed to below normal daytime phase level. Thus, while the path at this time lies in the night hemisphere, the particle-produced electron densities are so large that the reflection level (triangulation height) is about 30 km lower than during a quiet night. Electron densities of several hundred electrons cm^{-3} must exist at ionospheric altitudes of 60–65 km. This result is upheld by the computed electron density profile $N(h)$ discussed in § 3.

2.5. Magnetometers

The substorm has also been studied using magnetic records from Kiruna, Abisko, Tromsø and Spitsbergen. The X-component measured at Kiruna is shown in Fig. 2. Similar variations were observed for the H component at Kiruna, Abisko and Tromsø. After 1835 UT the magnetic deflection gradually increased until the maximum of the bay at 1908 UT, which was one of the largest bays observed in northern Scandinavia. The H component at Kiruna made an excursion of 2450 γ from the Sq level and at Abisko the corresponding magnitude was 2700 γ. The record from Tromsø was off scale (greater than 1300 γ). Recovery occurred gradually up to about 2000 UT when the Sq level was reached. The other components of the field were always less than 600 γ for the time interval 1500–2000 UT.

The direction of the magnetic disturbance vector gives an indication of the location of the auroral electrojet. During the maximum of the bay at 1905 UT to 1910 UT the Z components at Kiruna and Abisko were less than 200 γ which means that the electrojet had a large latitude extent, and the magnitude of the field shows that the maximum of the current probably was to the north of Kiruna.

The magnetic record from Spitsbergen (78.9° N 11.4° E) showed a different behaviour than the records described above. A negative bay started at 1830 UT with a maximum of 1225 γ at 1842 UT (H component); it had recovered to about 400 γ at 1855 UT and reached the Sq level at about 2000 UT.

2.6. All-sky Camera

All-sky camera pictures taken at Tromsø reveal that a corona-like aurora appeared very suddenly south of the zenith around 1825 UT and quickly developed E–W directed bands. At 1835 UT the aurora became diffuse and covered the whole sky. Inbedded in this a N–S arc developed to the west of Tromsø until 1900 when the complete sky became all covered with a structureless, diffusive aurora which after 1903 slowly faded away.

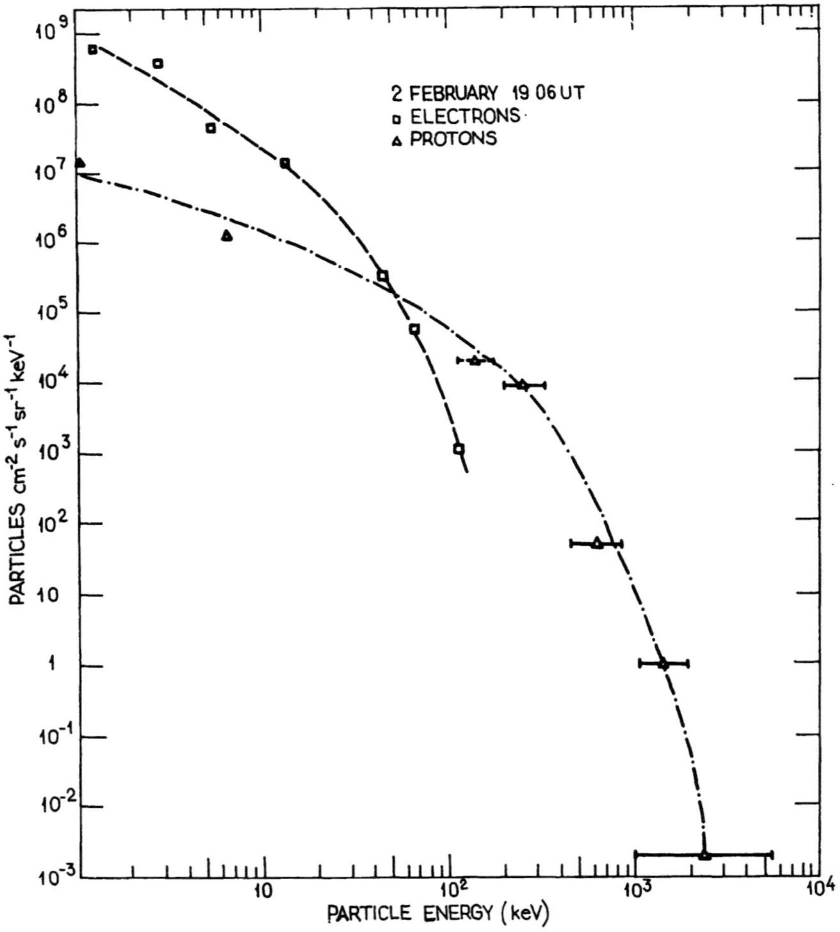

Fig. 3. Electron and proton differential energy spectra at 1906 UT on 2 February 1969.

3. Discussion

In order to discuss the satellite measurements in relation to the ground-based measurements, the differential spectrum of the 0°/10° pitch angle particles is given in Fig. 3 when the satellite was at a latitude similar to that of Tromsø.

Using these spectra the total energy deposited in the atmosphere by these particles can be calculated. The result of the calculation is that the electrons in the energy ranges 1–17, 17–84 and 84–750 keV have an energy flux of 95, 4 and less than 0.1 ergs $cm^{-2}\,s^{-1}$ respectively. For protons in the energy range 1–100 and 100–750 keV the energy flux was 7 and 5 ergs $cm^{-2}\,s^{-1}$ respectively. In the low energy part of the spectrum the energy input is higher for electrons than protons. Fig. 3 shows that the spectra are crossing at about 50 keV.

The ionization production function $q(h)$ resulting from the electron and proton precipitation was computed as follows: the $q(h)$ due to electrons was calculated using a programme based on the formulation by Maeda [5], the $q(h)$ due to protons was obtained from the cross section given by Cook et al. [6] and Sternheimer [7] and using the CIRA (1965) model for the atmosphere. Assuming equilibrium conditions $q(h)$ was converted into electron density $q(h) = \alpha_{eff}(h) \times \times N(h)^2$ where $\alpha_{eff}(h)$ is the effective recombination rate. Above about 90 km where the number density of negative ions is negligible (even at night) the values of α_{eff} are taken from Jespersen et al. [8] who deduced α_{eff} values from measurements of $q(h)$ and $N(h)^2$ (indirectly) during a winter night-time rocket flight from Andenes. Below 90 km we have used values deduced by Folkestad (private communication, to be published) also during a winter, night-time rocket flight from Andenes. The riometer absorption at 27.6 MHz that these profiles produce has been calculated using generalized magnetoionic theory and the results are compared with the observed absorption at Tromsø. The measured absorption is 5.6 dB and the calculated absorption 5.5 dB. In spite of the fact that the proton fluxes observed are unusually high they are not sufficient to produce riometer absorption comparable with that produced by the electrons. Integrating over the ionization profile of the protons, the calculated absorption at 27.6 MHz is about 0.5 dB which is only one tenth of what is observed. It is therefore concluded that the electron-produced ionization predominates in causing the observed riometer absorption.

The riometer data presented in Fig. 2 shows a systematic apparent northward movement (at 1820–1835 UT) in the poleward boundary of absorption. A nearly north–south chain of stations consisting of Kiruna, Tromsø and Spitsbergen, covering 11° in latitude ($\Lambda = 64.5°–75.7°$) was used to calculate the speed of the apparent northward expansion of absorption which was 2.3 km s^{-1}. It was found from the magnetic measurements that the current was widespread in the region Kiruna–Tromsø which corresponds well with the relatively constant particle precipitation observed at similar latitudes. From the precipitation of low-energy electrons it was found that the southern edge of the precipitation zone was at $L = 3.1$. This may be compared with the location of the southern border of aurora in relation to the *Dst* field reported by Akasofu and Chapman [9]. They found that when the aurora appeared as far south as $L = 3.1$ the $Dst(H)$ field was about $-200\,\gamma$. During this event the deflection of the H-component at M'Bour (located at a meridian only 20° W of the satellite pass) was about $-220\,\gamma$ which is in reasonable agreement with the result found by Akasofu and Chapman. However, the electrojet was unsymmetric with less magnitude at other longitudes giving a *Dst* of only $-100\,\gamma$.

Acknowledgment

Drs. T. Berkey and T. Saemundsson have very kindly supplied us with riometer recordings from College and Leirvogur, respectively. Mr. K. Folkestad (NDRE) is thanked for letting us use his unpublished data on D-region effective recombination rates. Mr. L. A. Fletcher at The Omega Systems Project Office has provided part of the vlf equipment in Tromsø.

References

[1] W. RIEDLER, B. HULTQVIST and S. OLSEN, Ark. Geophys. **5**, 619 (1971).
[2] R. DALZIEL et al., J. Atmos. Terr. Phys. **33**, 71 (1971).
[3] F. SØRAAS et al., University of Bergen, Department of Physics, Rep. No. 8 (1969).
[4] T. R. LARSEN, Phys. Norvegica **4**, 177 (1970).
[5] K. MAEDA, J. Geophys. Res. **68**, 185 (1963).
[6] C. J. COOK, E. JONES and T. JOERGENSON, Phys. Rev. **91**, 1417 (1953).
[7] R. M. STERNHEIMER, Phys. Rev. **117**, 485 (1960).
[8] M. JESPERSEN, B. LANDMARK and K. MÅSEIDE, J. Atmos. Terr. Phys. **31**, 1251 (1969).
[9] S. I. AKASOFU and S. CHAPMAN, J. Atmos. Terr. Phys. **25**, 9 (1963).

Space Research XII — Akademie-Verlag, Berlin 1972

DIURNAL, SEASONAL AND *Kp* VARIATIONS IN THE POSITION OF THE HIGH LATITUDE ELECTRON TRAPPING BOUNDARY

D. E. Page and M. L. Shaw

Space Science Department (ESLAB), European Space Research and Technology Centre, Noordwijk, The Netherlands

Data acquired from the ESRO 1/Aurorae polar satellite have been used to determine the variation of latitude of the northern high latitude electron trapping boundary over a period of about 15 months. The position of the boundary has been determined for about four thousand crossings, and an attempt has been made to determine the significance of any seasonal variations in the position of the boundary. The diurnal variation of the boundary position is shown to be well established.

A significant variation with the level of geomagnetic activity has been observed indicating that with increasing *Kp* the boundary moves to higher latitudes during the morning hours and to lower latitudes during the evening. The degree of boundary movement is compared with changes in solar wind velocity and shape of the magnetosphere and the sense of the movement is shown to be in agreement with magnetic field measurements made at the earth. There would appear to be a seasonal variation, with the trapping boundary shifting to higher latitudes at the equinoxes as has been suggested by more limited data used by Feldstein.

1. Introduction

A diurnal variation in the electron trapping boundary discovered by O'Brien [1] has been confirmed by other scientists, e.g. [2–5]. The dependence of the boundary position on Φ, the angle between the earth's dipole axis and the sun–earth direction has been studied in a limited way by Williams and Palmer [6] and more recently by McDiarmid and Wilson [7] who found that the boundary was lowest when $\Phi = 90$ (the equinoxes) whereas Feldstein and Starkov [8] have suggested that the boundary tends to higher values at the equinoxes. It would clearly be of interest to establish how the shape of the magnetosphere changes with time of day, season and magnetic activity, in order to obtain differential quantitative measurements of, for example, magnetospheric structure changes as a function of solar wind changes. In view of this and the non-conclusive evidence so far available a study of the northern 40 keV electron boundary has been carried out on data obtained from the ESRO 1 satellite.

2. Experiment

The detector was a Geiger counter set at 90° to the axis of the magnetically stabilized ESRO 1 satellite, which was launched in October 1968. The orbital inclination was 94°, initial apogee and perigee values were 1530 km and 260 km. The orbital plane precessed at about 0.5° per day. For the first six months data recorded on tape during complete orbits were obtained, but later only real time data received mainly at Fairbanks (Alaska) and Tromsø (Norway) were available.

The Geiger counter, an EON 6222, had a mica end window with a nominal 40% transmission of 40 keV electrons and a proton threshold around 500 keV. It was shielded so that protons not entering through the collimator needed at least 30 MeV to be registered. The geometric factor assuming detection via the collimator was 8.3×10^{-3} sr. The Geiger counter was supplied by the Norwegian Defence Research Establishment, Kjeller, as part of a high speed auroral experiment and ESTEC supplied a ratemeter for slow speed monitoring of the radiation belts. The analogue output saturated at 4.7×10^{3} counts per second, had a time constant of around 3 seconds and was sampled every 12.8 seconds (about 0.7° invariant latitude).

Because of the effects of solar radiation all data taken when the detector axis pointed within 20° of the solar direction were automatically rejected. Data from all altitudes have been combined for the period October 1968 to February 1970, and to facilitate comparison with other results only periods where Kp was $\leqq 4^{+}$ have been used. The boundary position was defined according to McDiarmid and Wilson [7] for locating the background boundary.

3. Results

3.1. Diurnal Variation

Fig. 1 shows the LMT diurnal variation of the 40 keV trapped electron boundary as calculated for 3500 events. This is very similar to the form of the curve obtained by McDiarmid and Wilson [7], and provides some confidence that the definition of boundary position has been very similar in the two cases. The following features should be noted: (i) The dawn–dusk asymmetry is very evident but in the final analysis we see no evidence of a secondary minimum as found at 1200 LMT by Fritz [5] and pointed out by him to be at 1000 LMT in the McDiarmid and Wilson data [7]. At certain intermediate stages in the analysis however, small secondary minima did appear but without obvious explanation. (ii) During the dawn hours the data scatter was greater than in the evening hours. In fact the standard deviations of the two-hourly histograms (Fig. 1) vary with LMT in a manner very similar to the auroral oval variation described by Feldstein [9] (see his Fig. 26). As Feldstein has pointed out, the dawn–dusk asymmetry in the average curve fits the shape of the auroral zone quite well.

3.2. Seasonal Variations

To facilitate comparison we have organized our data again in a form used by McDiarmid and Wilson [7]. The trapped electron boundary Λ_b has been determined for the intervals $75° < \Phi < 105°$ and Φ below 75 or above 105, and the results are shown in Fig. 2. The result that the boundary is lowest in the morning hours when Φ is close to 90° is not repeated. At present no

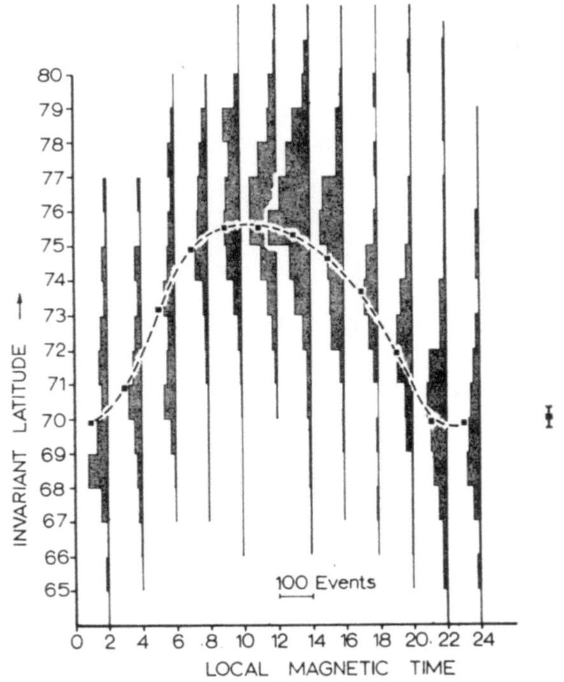

Fig. 1. The average 40 keV trapped electron boundary plotted in invariant latitude against local magnetic time. Northern hemisphere, October 1968–April 1969.

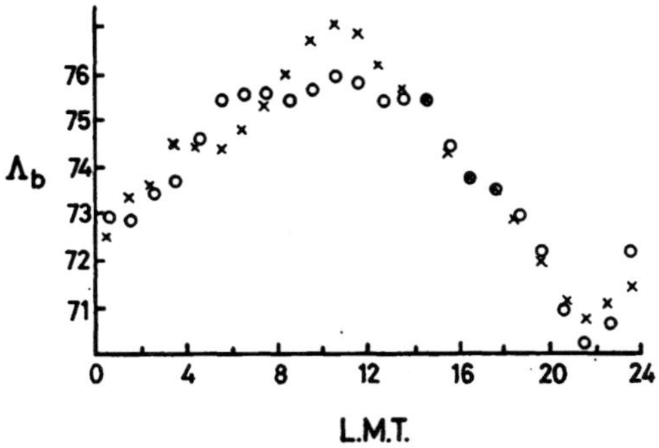

Fig. 2. The average 40 keV trapped electron boundary for: ×, Φ ranges $75 < \Phi < 105$; ○, Φ below 75 or above 105.

positive significance is attached to the relationship plotted in Fig. 2. At intermediate stages in the analysis, i.e. when 1000 or 2000 events had been employed, both curves showed essentially the same diurnal variation but the fine detail was different. This suggests that, because of the physical processes involved, and the fact that these vary with season and time of day, the determination of the accuracy with which the average boundary position is established is not amenable to discussion in terms of purely statistical standard errors of the mean. One other explanation of the inconsistency in the results is that the data coverage was different in uniformity. Both sets of data provide reasonable coverage in terms of Λ and Φ but a critical parameter may be universal time.

The (Λ, LMT) analysis has been divided into intervals of Φ. If indeed we expect the (Λ, LMT) curves to change with Φ this implies that there could

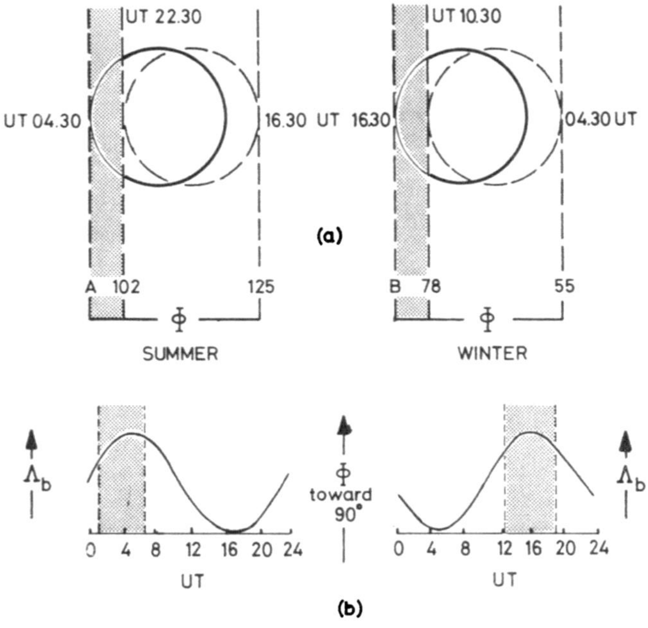

Fig. 3. (a) Illustration of how various universal time periods may be excluded from the analysis by specifying limited ranges of Φ. Part of the Φ range is in the excluded shaded region except precisely at mid-summer or mid-winter. (b) Illustrating exclusion of shaded universal times corresponding to the situations in (a). The (Λ, UT) curves follow Feldstein and Starkov [8].

be a (Λ, UT) variation, since the Φ variation of the dipole axis on any day is a UT phenomenon. If now we consider the winter interval $55 < \Phi < 78$ as illustrated in Fig. 3 it can be seen that these Φ values are attainable for all UT only precisely at mid-winter. As either equinox is approached from the mid-winter position the universal times around 1630 can no longer be included in the analysis because at those times $\Phi > 78°$. Similarly by selecting $102 < \Phi < 125$, universal times around 0430 are missed out except precisely at mid-summer. Thus a bias is automatically introduced into the data if indeed a UT variation exists. Feldstein and Starkov [8] have data which suggest that it does exist and is of the form shown in Fig. 3(b).

3.3. *Kp* Variations

It seems to be generally believed [10] that following strong magnetic activity the trapping boundary moves to lower latitudes. The first experimental indication that the boundary movement may depend on time of day apparently comes from the 1963 38c data [6]. This suggested that during conditions of average disturbance, data at peak magnetic disturbance being excluded, the trapping boundary moved poleward at noon and equatorward near midnight. This was consistent with increasing solar wind distortion of the magnetic cavity. The result of the present analysis is shown in Fig. 4 where it can be

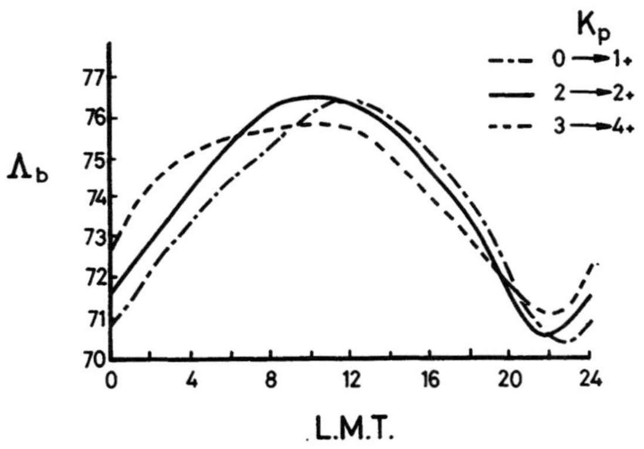

Fig. 4. The average 40 keV trapped boundary for the *Kp* ranges 0–1⁺; 2⁻–2⁺; 3⁻–4⁺.

seen that on average, increasing magnetic activity raises the boundary poleward in the dawn sector but depresses it in the evening sector. Since only *Kp* values up to 4^+ were used probably most 'peak' magnetic disturbance was excluded. For clarity the exact data points have not been drawn but only the smoothed curves. Each *Kp* interval contains roughly 1200 points with the midday and midnight hours having most data.

The similarity of the boundary behaviour to that of other magnetic or ionospheric phenomena is striking. As an example Fig. 5 (reproduced from [11]) shows how the disturbed magnetic field at Bombay on the average increased during morning hours but decreased during evening hours. There are many other examples [12, 13] particularly of deduced ionospheric current regions, which show this reversal just before midday and again before midnight.

McDiarmid et al. [14] suggest that the dawn–dusk asymmetry in the boundary (as in Fig. 1 with all *Kp* up to 4^+ collected together) could be due to an electric field across the magnetosphere, e.g. [12], or due to a distortion of the geomagnetic field. It may be however that both effects are essentially the same arising from a convective flow of solar wind past the rotating magnetosphere [15]. McDiarmid et al. [14] assess numerical values by considering the boundary position for different electron energies. It is suggested that by using one known electron energy (>40 keV) and the amount of boundary shift at each hour of LMT for different *Kp* values, a more quantitative estimate can be made [13] of the 'viscosity' in the interaction between the solar wind and the magnetosphere. More specifically when Kp goes from about 1 to about 4 (Fig. 4) an electric field (or a magnetic field distortion) is produced which is sufficient to move the 40 keV boundary up by about 1.5° at 0400 LMT. This change corresponds to a solar wind velocity change of approximately 350 to 600 km s^{-1} [16]. Similar quantitative calculations could be carried out using the geomagnetic field distortion approach but it is difficult to know the exact

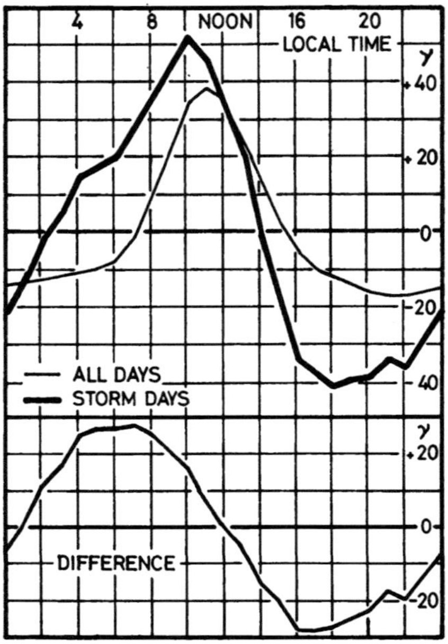

Fig. 5. The solar daily variations of horizontal intensity at Bombay; averages from all days and storm days. *Reproduced from* [11].

form of the field variation and in particular how a change measured at the earth actually appears in the region through which the 40 keV electrons pass. However, comparison of Figs. 4 and 5 indicates that at 0400 LMT a 20 gamma change in magnetic field intensity is accompanied by about a 1° shift to higher Λ in the 40 keV boundary. At least the sense of the variation appears corrrect, electrons having to move further from the earth to conserve invariance in the increased magnetic field. In the light of these results it is attractive to suggest that if the solar wind velocity tended toward zero the dawn–dusk asymmetry would tend to disappear. Extrapolation of Fig. 4 toward 'zero' conditions (lowering the dawn and raising the dusk boundaries) certainly establishes symmetry but is unrealistic because in the limit the whole midday to midnight compression asymmetry must also disappear.

4. Conclusion

The 40 keV electron boundary as measured by ESRO 1 shows an asymmetrical diurnal variation as already found by earlier analysis of more limited data

There is no evidence of a secondary peak. When the analysis as performed by McDiarmid and Wilson [7] is repeated no seasonal effect is established. On average, increasing *Kp* raises the boundary position in the dawn and lowers the boundary in the dusk hours. It is suggested that as the solar wind velocity decreases the dawn–dusk asymmetry decreases. The analysis permits a quantitative relationship between solar wind velocity and boundary movement (or electric field).

Acknowledgments

We gratefully acknowledge continuing encouragement from Dr. E. A. Trendelenburg. We wish to thank our ESRO colleagues engaged on the ESRO 1 project, technicians S. T. Ho and H. Huisman and our data assistant Mrs. A. Minier.

References

[1] B. J. O'Brien, J. Geophys. Res. **68**, 989 (1963).
[2] L. A. Frank, J. A. Van Allen and J. D. Craven, J. Geophys. Res. **69**, 3155 (1964).
[3] I. B. McDiarmid and J. R. Burrows, Canad. J. Phys. **42**, 1135 (1964).
[4] T. A. Fritz, J. Geophys. Res. **73**, 7245 (1968).
[5] T. A. Fritz, J. Geophys. Res. **75**, 5387 (1970).
[6] D. J. Williams and W. F. Palmer, J. Geophys. Res. **70**, 557 (1965).
[7] I. B. McDiarmid and M. D. Wilson, J. Geophys. Res. **73**, 7237 (1968).
[8] Ya. I. Feldstein and G. V. Starkov, Planet. Space Sci. **18**, 501 (1970).
[9] Ya. I. Feldstein, Rev. Geophys. **7**, 179 (1969).
[10] B. Maehlum and B. J. O'Brien, J. Geophys. Res. **68**, 997 (1963).
[11] N. A. F. Moos, in: Colaba Magnetic Data 1846–1905 Bombay 1710 (p. 469).
[12] N. M. Brice, J. Geophys. Res. **72**, 5193 (1967).
[13] W. I. Axford, in: Physics of Geomagnetic Phenomena, Academic Press, New York 1967 (p. 1243).
[14] I. B. McDiarmid, J. R. Burrows and M. D. Wilson, J. Geophys. Res. **74**, 3554 (1969).
[15] W. I. Axford and C. O. Hines, Canad. J. Phys. **39**, 1322 (1961).
[16] C. W. Snyder and M. Neugebauer, Space Research IV, 89 (1964).

Space Research XII — Akademie-Verlag, Berlin 1972

FLUX AND ENERGY SPECTRA OF SOLAR PROTONS OBSERVED ABOARD THE ESRO 2 SATELLITE IN 1968–1969

J. Engelmann

Service d'Electronique Physique, Centre d'Etudes Nucléaires de Saclay, France

From the data registered between May 1968 and March 1969 by a spectrometer aboard the ESRO 2 satellite, the main characteristics of ten proton flares were studied. The range of energies covered is from 27 to 400 MeV. The 18 Nov. 1968 event was studied in more detail. Energy and rigidity spectra at different times were computed and a three-dimensional plot was built, in which the differential flux values are given versus the energy and the time. For each proton flare, the differential energy spectrum integrated over the whole event was computed. These spectra are shown to have the same shape as the source spectra when velocity dependent diffusion appears to be a valid approximation. This was found to be the case for the 29 Sept. 1968 and 25 Feb. 1969 proton flares. In both events, the differential spectra are well fitted by an exponential law in rigidity with e-folding rigidities of 110 and 150 MV respectively.

During the first nine months after launch in May 1968, the spectrometer aboard ESRO 2 registered ten high-energy proton flares. Our apparatus, of which the sensitive part is a silicon detector dE/dx telescope, is designed for measuring the flux and spectrum of solar protons between 27 and 400 MeV. The geometrical factor of the telescope is 0.3 cm^2 sr and the number of energy channels for protons is 75. The characteristics of the instrument did not vary during its three years of orbital life. ESRO 2 being a near earth polar orbiting satellite, the measurements of solar protons can be effective only above the polar caps. More details about the instrument are given in [1].

From the data of each orbit, the differential proton spectrum was computed. Two forms were assumed for these spectra, namely power law in energy, $E^{-\gamma}$, and exponential in rigidity, $\exp(-R/R_0)$. In computing the values of γ and R_0, the standard error of each energy window had to be taken into account. The method of maximum likelihood was used (e.g. [2]); this method enables one to compute also the standard error of the γ and R_0 values. Also, in order to draw the time profile of the event, the channels were grouped into five energy windows (corresponding to mean energies 30, 38, 52, 78 and 160 MeV) and the characteristic curve of each window was obtained from the data of successive orbits. From these curves, the integral fluxes above 30 and 100 MeV integrated over the whole event were computed. The flare characteristics and the integral intensities are summarized in Table 1. As can be seen from the Carrington coordinates, several longitudes were active in 1968. The total intensity over 30 MeV in 1968 is comparable with that observed in the corresponding year (1957) of the preceding solar cycle [3].

Table 1

Flare characteristics and integral intensities

Date	Time (UT)	Type	Heliographic coordinates		Carrington longitude	Integral intensity cm^{-2} $E > 30$ MeV	$E > 100$ MeV	Spectrum (Fig.5)
9. 6. 68	0900	3 B	S 14	W 08	249	3.8×10^6	3.7×10^5	4
7. 7. 68	0805	1 N	N 08	E 64	163	4.1×10^5	6.2×10^4	10
29. 9. 68	1620	2 B	N 16	W 52	246	2.9×10^6	5.6×10^5	6
31. 10. 68	0013	3 B	S 14	W 37	179	1.6×10^6	7.2×10^4	7
1. 11. 68	0844	1 N	S 15	W 47	171	3.2×10^6	9.4×10^4	3
4. 11. 68	0542	1 B	S 15	W 90	176	6.9×10^5	7.0×10^4	9
18. 11. 68	1030	1 B	N 21	W 87	346	8.2×10^7	4.4×10^6	1
2. 12. 68	2202	3 N	N 18	E 80	349	1.2×10^7	5.6×10^5	2
25. 2. 69	0900	2 B	N 14	W 36	72	4.5×10^6	7.8×10^5	5
27. 2. 69	1425	2 B	N 13	W 64	71	1.7×10^6	3.0×10^5	8

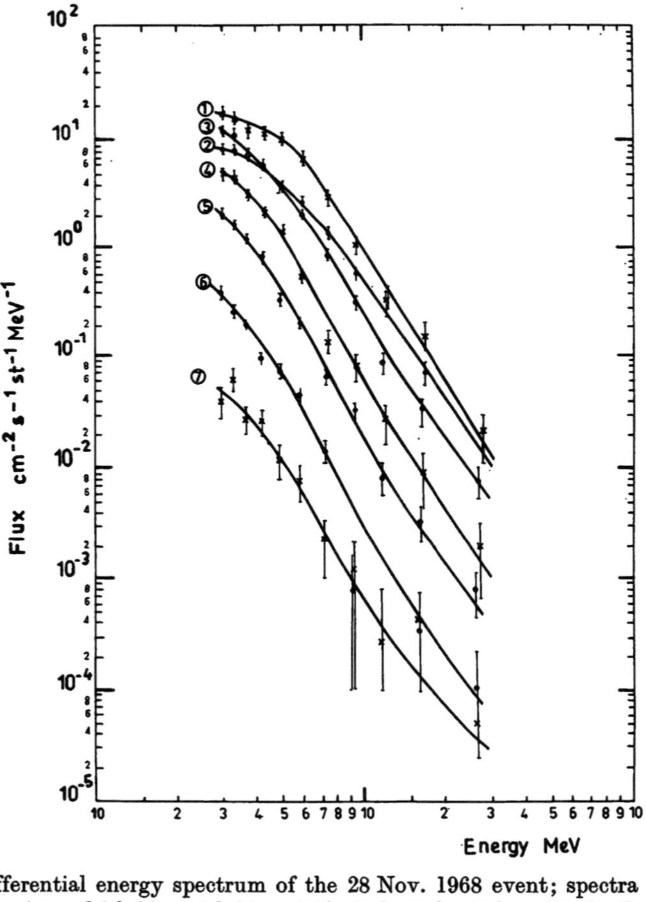

Fig. 1. Differential energy spectrum of the 28 Nov. 1968 event; spectra 1 to 7 were respectively registered 1 h 30 m, 6 h 30 m, 15 h, 28 h, 41 h, 54 h and 77 h after the flare time.

The 9 June 1968 and 25, 27 Feb. 1969 proton events are reported in [4, 5]; we now discuss some characteristics of the 18 Nov. 1968 event, the most important registered. From optical, radio (type IV) and X observations, the acceleration took place between 1026 and 1035 UT [6, 7]. Neutron monitors of cut-off <4.5 GeV registered an increase, and the first particles were detected at 1040 [8]. Data obtained above the south polar cap during 43 orbits after the flare were used; the counting rate was strongly latitude dependent over the north polar cap during the first phase of the event, but above the south polar cap, only a slight enhancement was observed at auroral latitudes [9].

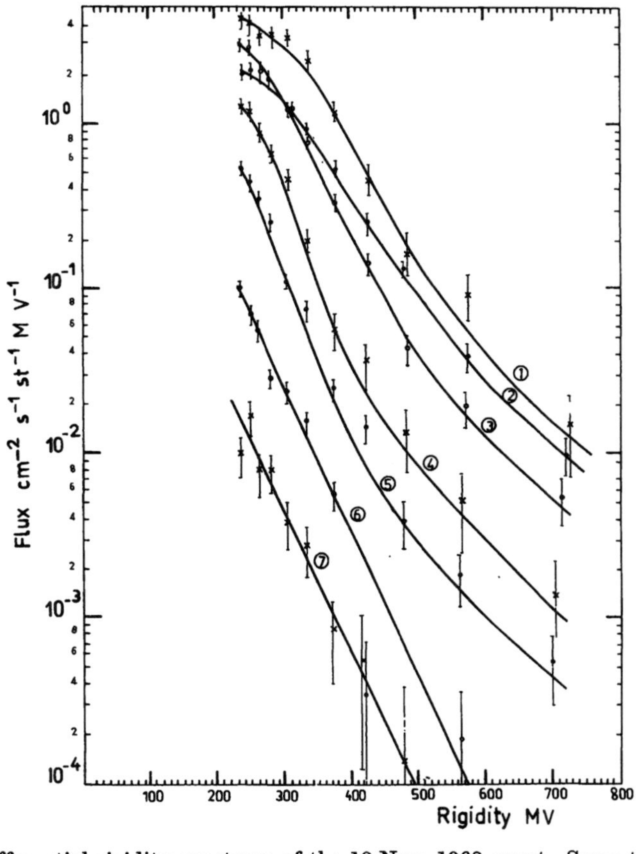

Fig. 2. Differential rigidity spectrum of the 18 Nov. 1968 event. Same times as in Fig. 1.

Examples of energy and rigidity spectra are displayed in Figs. 1 and 2; these spectra were respectively registered 1.5, 6.5, 15, 28, 41, 54 and 77 hours after the flare. The five first spectra are not well fitted either by a power law in energy, or by an exponential in rigidity; but this last form gives a good fit to spectra 6 and 7. Despite the poor fit, the spectral indices γ and R_0 have been computed and are plotted in Fig. 3 against time. As expected, these indices are not constant; the γ index increases during the first 20 hours and tends to 3.2 at later times. A sharp decrease is observed just after the sudden commencement on 20 Nov. at 0904. In order to obtain simultaneously the time profiles of five energy windows and the shapes of the spectra, Fig. 4 was plotted. This is a three-dimensional plot, linear in time, and logarithmic in energy and flux.

The surface generated in this way is characteristic of the whole event. Intersection by a plane parallel to the flux–time plane gives the time profile for one energy window; intersection by a plane parallel to the flux–energy plane gives the energy spectrum at a certain time.

As shown for this event, the γ and R_0 values computed at a certain time cannot be used to characterize an event. We believe that a better choice would

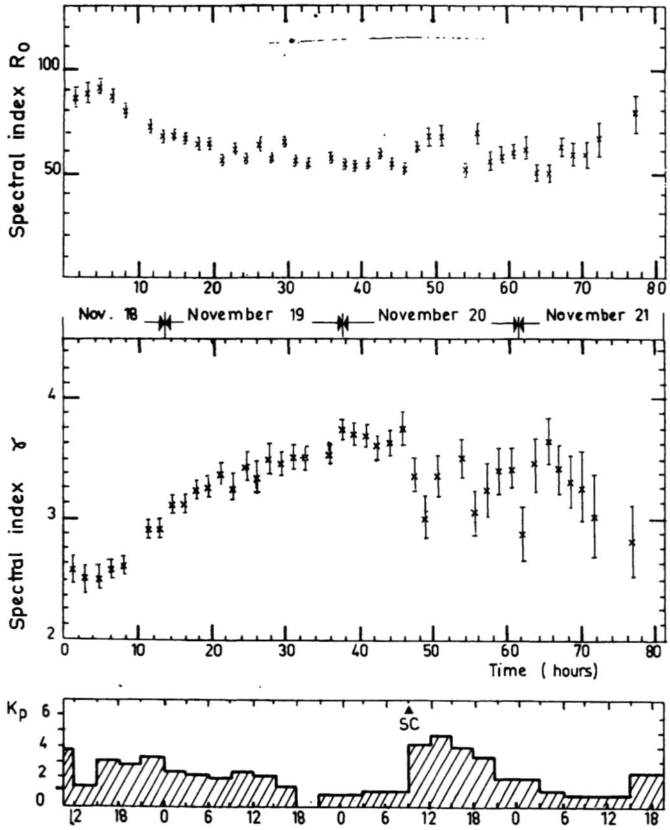

Fig. 3. Spectral indices γ and R_0 versus time for the 18 Nov. 1968 event. The three-hour Kp index is plotted for the same period.

be to use the γ and R_0 values derived from the integrated spectrum, i.e. the spectrum obtained by integrating the flux in each energy window over the whole time of the event. This can be shown in the frame of diffusion theories. In a macroscopic sense, the classical equations still hold if the isotropic diffusion coefficient $\overline{D}$ is replaced by the parallel coefficient $K_{||}$ and if the anisotropy is not great [10]. When integrating over time for instance the equation of the intensity given by Krimigis [11], we find that the intensity is proportional to $(1 + \beta)^{-1} N/\lambda$ (where N is the number of particles within a certain range of energies or rigidities ejected from the sun at the time of the flare, λ is the mean free path of the particles of the energy of interest and β the characteristic coefficient for the radial dependence of the diffusion coefficient generally found to be ~1). That means that for flares in which velocity dependent diffusion is a valid approximation, the integrated spectrum has the same shape as the source spectrum. Such a velocity dependence was observed in particular for the 25 Feb. 1969 event [12].

Although such behaviour is not observed in most of the events, the integrated spectrum may be a better characteristic curve for a flare than, for instance,

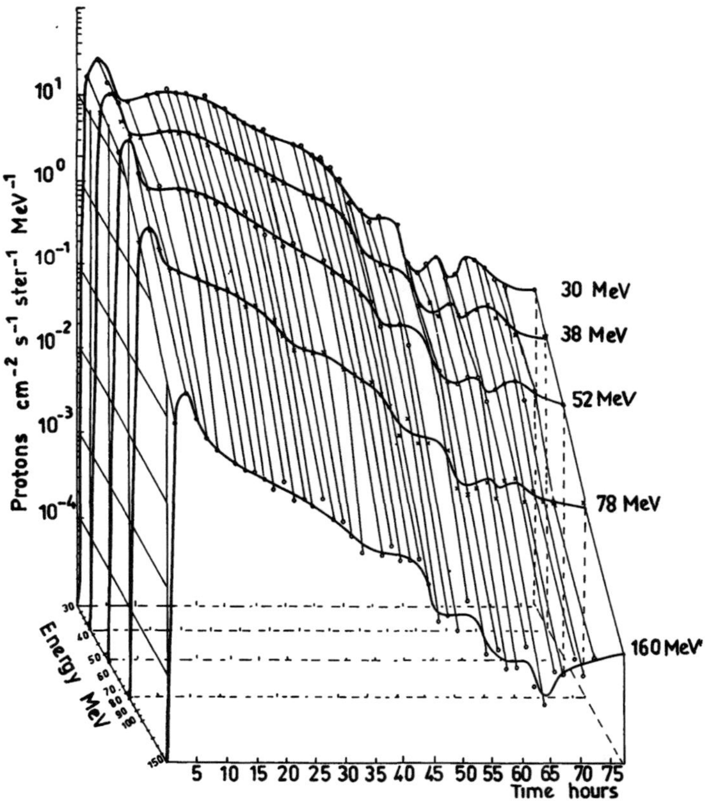

Fig. 4. Flux values in five energy windows versus energy and time (18 Nov. 1968 event).

the spectrum of particles taken at the time of flux maximum. In particular, this type of spectrum can be useful when comparing different species of particles within the same range of energies or rigidities. As an example the proton/α particle ratio at the same rigidity is time dependent (e.g. [13]). Another example is given by the electron/proton flux ratio; in this respect, it is worth pointing out that a good correlation was found to exist between the integrated number of electrons of $E > 7$ MeV and the integrated number of protons of $E > 300$ MeV [14].

The integrated energy spectra of the ten proton flares observed between May 1968 and March 1969 are plotted Fig. 5. As is seen, no event is well fitted by a power law in energy but two of them (29 Sept. 1968 and 25 Feb. 1969) would be well fitted by an exponential law in rigidity. It is remarkable that these events are just those for which velocity dependent diffusion appears to apply. This can be seen by plotting the flux values of different energy windows as a function of the distance travelled by the particles [15]. In both cases the curves for different energy windows have the same shape; the most probable values of the distance travelled are found to be 13 and 8 AU respectively for the 29 Sept. 1968 and 25 Feb. 1969 events. Therefore, for these two solar events,

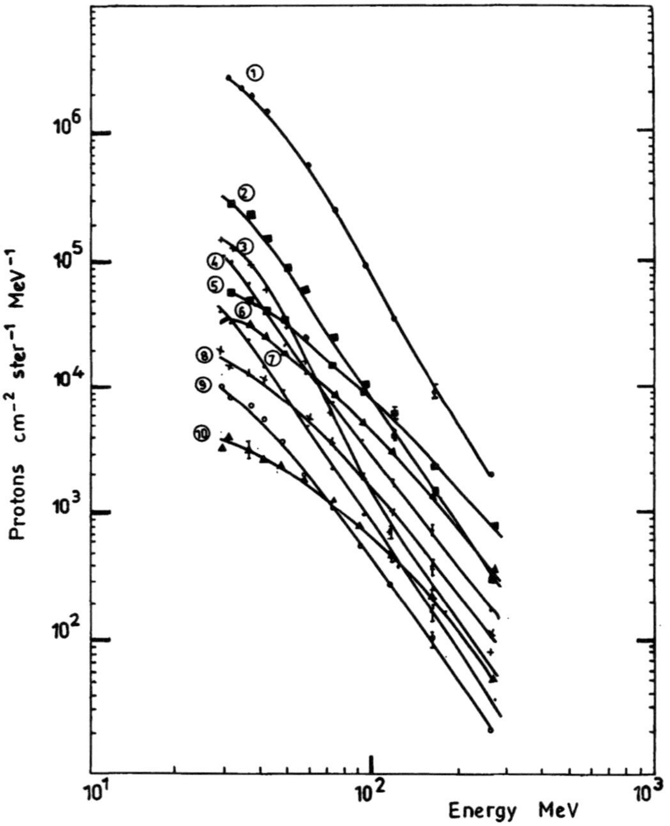

Fig. 5. Integrated differential spectra of ten events: 1, 18 Nov. 1968; 2, 2 Dec. 1968; 3, 1 Nov. 1968; 4, 9 June 1968; 5, 25 Feb. 1969; 6, 29 Sept. 1968; 7, 31 Oct. 1968; 8, 27 Feb. 1969; 9, 4 Nov. 1968; 10, 7 Jul. 1968.

the shape of the source spectrum could be obtained directly from the integrated spectrum. The best fit was given by an exponential law in rigidity, with e-folding rigidities of 110 and 150 MV respectively for the 29 Sept. 1968 and 25 Feb. 1969 events.

Acknowledgments

We gratefully acknowledge fruitful discussions with Dr. Barouch and the assistance of Miss V. Romeo and Mr. Thal in the data processing. We should like to express our thanks to Professor J. Labeyrie and Dr. L. Koch for their constant encouragement and advice during the ESRO 2 project, and to the technical and scientific staff of ESRO.

This work was supported in part by the Centre National d'Etudes Spatiales under contract 70/CNES/320.

References

[1] Y. Amram, F. Axisa, J. Engelmann and L. Koch, Nucl. Inst. and Meth. **75**, 181 (1969).
[2] P. R. Bevington, Data Reduction and Error Analysis, McGraw Hill, New York and London 1969.
[3] L. J. Lanzerotti, Proc. Symp. on Natural and Manmade Radiation in Space 1971.
[4] J. Engelmann, A. Hautdidier and L. Koch, Proc. 11th Int. Conf. on Cosmic Rays, Budapest 1969.
[5] E. Barouch, J. Engelmann, M. Gros, L. Koch and P. Masse, Proc. 11th Int. Conf. of Cosmic Rays, Budapest 1969.
[6] Solar Geophysical Data (1969).
[7] L. Křivský, Proc. 11th Int. Conf. on Cosmic Rays, Budapest 1969.
[8] P. J. Tanskanen, Proc. 11th Int. Conf. on Cosmic Rays, Budapest 1969.
[9] A. Durney, G. Morfill and J. J. Quenby, to be published in J. Geophys. Res. (1972).
[10] K. G. McCracken and U. R. Rao, Space Sci. Rev. **11**, 155 (1970).
[11] S. M. Krimigis, J. Geophys. Res. **70**, 2943 (1965).
[12] E. Barouch et al., in: Intercorrelated Satellite Observations related to Solar Events, D. Reidel Publ. Co, Dordrecht 1970 (p. 448).
[13] G. Wibberenz and M. Witte, in: Intercorrelated Satellite Observations related to Solar Events, D. Reidel Publ. Co, Dordrecht 1970 (p. 499).
[14] J. P. Mercier et al., Proc. Saclay HEOS Colloquium, to be published in ESRO, SP Report.
[15] D. A. Bryant, T. L. Cline, U. D. Desai and F. B. McDonald, Astrophys. J. **141**, 478 (1965).

Space Research XII — Akademie-Verlag, Berlin 1972

MEASUREMENTS OF THE ISOTOPIC COMPOSITION OF PARTICLE FLUXES CARRIED OUT ON SPACECRAFTS SOYUZ, ZOND 8 AND LUNA 16

B. S. Boltenkov, V. N. Gartmanov, G. E. Kocharov, B. A. Mamyrin and V. O. Naidenov

A. F. Joffe Physico-Technical Institute, Academy of Sciences, Leningrad, USSR

The paper provides the results of measurements of elemental and isotope composition of particle fluxes in cosmic space using the method of isotope collecting in special-purpose and natural targets. Special-purpose targets consisted of packets of aluminium foils which permitted us to measure energy spectra of fluxes. The experiments were carried out both inside the magnetosphere of the earth (spacecrafts Soyuz) and outside it (Zond 8). As natural targets were used the lunar probes delivered by Luna 16. The results are compared with those obtained by other authors.

The observations of the elemental and isotopic composition of solar particle fluxes give information about processes taking place in the sun, in the solar atmosphere and in the environment. One of the most convenient methods for solving this task is the method using particles of solar particle fluxes collected in a special-purpose target, which is exposed in space. Natural targets such as meteorites, lunar and other planetary soils etc. can also be used for studying this problem. However, in this case, there are definite difficulties in interpretation of the information because of a complex chemical composition and a poor knowledge of the history of these natural targets. Thus, elemental and isotopic abundances in a particle flux cannot be derived conclusively from analysing natural targets only. Comparison of the results of the analysis of natural and special-purpose targets aids in solving the opposite task, i.e., deriving the natural target history.

The method of collecting particle fluxes in special-purpose targets can be useful to study the following problems:

1. Abundances of elements and their isotopes at the sun. Of special interest is determination of ^3He abundance at the sun. According to Kocharov and Starbunov [1] the energy in the sun can be generated, in particular, by means of reaction ^3He (^3He, 2 p) ^4He.
2. Acceleration processes of solar cosmic rays, and nuclear reactions in regions of solar flares.
3. Physical conditions in the solar atmosphere and solar wind dynamics.
4. Interaction of solar particle fluxes with the magnetic field of the earth.

To understand the degree of participation of each of the processes in creation and change of elemental and isotopic composition of solar particle fluxes, it is necessary to study a number of elements and isotopes in a flux as much as possible in a wide energy range — from energies of the solar wind to solar cosmic ray energies.

Until recently, a small number of experiments has been carried out on the method of collecting particles of fluxes in a target. In the first experiments on Discoverer satellites in 1960–1961 different parts of the satellites have been used as a target [2, 3]. A special-purpose target arrangement has been exposed on spacecrafts Soyuz 4, 5 and 9 in 1969–1970 [4]. All these experiments have been carried out within the magnetosphere of the earth.

There are only three types of experiments which have been carried out outside the magnetosphere of the earth. The first is collecting solar wind ions in a special metal foil, on the moon surface during Apollo 11 and 12 landings in 1969 [5, 6]. The second is the analysis of the trapped solar wind ions in Surveyor 3 material [7] and the third is collecting solar wind particles and more energetic particles in a special-purpose arrangement which has been made on Zond 8 in 1970.

However, very important results have been obtained in these experiments which make it possible to suggest the presence of nuclear reactions in the region of generation of solar cosmic rays during the flare, the occurrence of large fluxes of light elements inside the magnetosphere of the earth which are possibly connected with solar flares, and variations of the isotopic composition of the solar wind. The method thereby appears to be very promising and worth being further developed.

The experimental procedure consisted of two stages: accumulation of ions and nuclei of solar particle fluxes at target exposure, and isotopic analysis of the targets by laboratory equipment on the earth. The laboratory equipment devised especially for these purposes contained a gas extracting device, a mass spectrometer for stable isotopes and a low background device for radioactive isotopes.

The device for extraction and purification of gas enables one to melt samples up to 60 g in weight, to collect into separate ampoules stable isotopes of rare gases and radioactive gases as ^3H and ^{37}A extracted from one sample. In principle the gas extraction technique does not differ from similar devices applied to extract gases from meteorites.

To analyse stable isotopes we have used a specially developed double-beam magnetic resonance mass spectrometer with high resolution, high sensitivity and a line shape with sharply falling wings. The ^3He sensitivity of the mass spectrometer was of the order of 10^6–10^7 atoms.

To analyse ^3H and ^{37}A low background proportional counters have been used. The background counting rate for ^{37}A was *I* imp/day (the counter volume being 0.7 cm^3) and *I* imp/min for ^3H (the counter volume being 20 cm^3).

When producing the targets special attention was paid to the choice of the material with the lowest content of the isotopes measured in the material itself. Tentative analysis showed that in commercial aluminium the content of these isotopes ranges widely, and can reach, for instance, for ^3H and ^3He, 5×10^8 atoms g^{-1}. That is why aluminium used for the targets was additionally purified by vacuum zone melting. As a result ^3H content in aluminium reached $< 2.8 \times 10^6$ atom g^{-1}, ^3He $< 4.4 \times 10^5$ atom g^{-1}, ^4He $< 10^{10}$ atom g^{-1}. These impurities do not impede measuring these isotopes in the solar wind, during high solar flares but they limit the possibility of measuring solar cosmic rays in quiet conditions and cosmic rays in the galaxy.

The technique developed was used by us in the two sets of experiments: inside the magnetosphere of the earth (spacecrafts Soyuz 4, 5 and 9) and outside it (Zond 8).

In the first set of experiments the target consisted of an aluminium film packet. The energy range of different isotopes trapped by the packet was defined by the thickness of the packet (~ 3 g cm^{-2}) and by the shielding substance (~ 0.6 g cm^{-2}) before it. Resolution of measuring the energy spectrum was defined by the thickness of separate films of the packet (0.22 g cm^{-2} for Soyuz 4 and 5, and 0.4 g cm^{-2} for Soyuz 9). In all spacecrafts of the Soyuz type the packets were installed identically.

Table 1

Spacecrafts	Dates	^{3}He			^{3}H		Ref.
		Exp. time (hours)	Energy (MeV)	at cm^{-2}	Energy (MeV)	at cm^{-2}	
Discoverer 17	12–14 Nov. 1960	50	95–180	$4 \cdot 10^8$	90–105	$5 \cdot 10^6$	[2, 3]
Discoverer 18	7–10 Dec. 1960	75	—	—	>90	$9 \cdot 10^5$	[2]
Discoverer 25			140–160	$<3 \cdot 10^7$	—	—	[3]
Discoverer 26	7–9 Jul. 1961	50	—	—	>90	$9 \cdot 10^4$	[2]
Discoverer 29			160–360	$<3 \cdot 10^7$	—	—	[3]
Discoverer 36	12–16 Dec. 1961	100	—	—	>90	$6 \cdot 10^4$	[2]
Soyuz 4	14–17 Jan. 1969	70	75–200	$<3 \cdot 10^6$	35–95	$<1.4 \cdot 10^7$	[4]
Soyuz 5	15–18 Jan. 1969	70	75–85	$2.8 \cdot 10^8$	35–40	$1.4 \cdot 10^9$	[4]
			85–150	$<3 \cdot 10^6$	40–70	$<1.4 \cdot 10^7$	
			150–200	$<3 \cdot 10^6$	70–95	$<1.4 \cdot 10^7$	
Soyuz 9	1–18 Jun. 1970	408	75–200	$<1.8 \cdot 10^5$	35–95	$<1.5 \cdot 10^6$	
Zond 8	21–27 Oct. 1970	151	23–29	$<6.5 \cdot 10^5$	10–13	$<3 \cdot 10^6$	

The experimental results and their comparison with the results published in the literature are given in Table 1. As is seen from the table, the higher content of ^{3}He and ^{3}H was observed on the spacecraft Soyuz 5, and earlier on Discoverer 17, flying during intense solar flares. Enhanced contents of ^{3}H were observed in the absence of high solar flares on the spacecrafts of the Discoverer type and was interpreted by the authors [2] as a flux from the intense flare of 12 November 1960, trapped by the geomagnetic field. A similar situation was probably observed during the flight of vehicles Soyuz 4 and 5 though their orbits were below the radiation belts. The analysis of the solar-geophysical data shows that the period of flight of Soyuz 4 and the joint flight of two vehicles are characterized by relatively quiet conditions. However, weak increases of the solar activity and geomagnetic disturbance were observed after the landing of Soyuz 4 and while the flight of Soyuz 5 was being continued, which may favour the release of particles trapped by the geomagnetic field. The experiment carried out on Soyuz 9 (exposure time 408 hours) produced a result similar to that of Soyuz 4, i.e. the ^{3}He and ^{3}H content in the targets did not exceed the background.

Thus, all the experiments carried out until recently are not sufficient for unambiguous interpretation but indicate the possibility of the existence of ^3He and ^3H short-term fluxes in the magnetosphere of the earth.

Measurements of solar particle fluxes outside the magnetosphere of the earth were carried out during the flight of the spacecraft Zond 8. The aim of the experiment was to detect isotope composition of ions from the solar wind energy range (0.5–1 keV/nucleon) up to 30 MeV/nucleon. For this purpose in the cosmic environment a packet of aluminium targets composed of foils of different thickness was exposed without any shielding. That permitted one to measure the energy spectrum with the interval of 2–3 MeV/nucleon. The area of the target was $\sim$300 cm^2. During the flight of Zond 8 the target was for 140 hours in the sun-oriented state, the angle between the target normal and the sun's direction changing from 0 to 15° on different parts of the trajectory.

The design of the target makes possible a reduction of the thermal load during the exposure time and landing of the spacecraft. Temperature measurements showed that the maximum temperature of the target did not exceed 140 °C.

Table 2

Spacecrafts	Dates of exposure	Exposure time	Isotopic content I/cm^2 ^3He	^4He	$^4He/^3He$	Refs.
Apollo 11	21 Jul. 1969	77 h	$1.27 \cdot 10^7$	$2.33 \cdot 10^{10}$	1860 ± 140	[5, 6]
Apollo 12	19 Nov. 1969	18 h 42 m	$1.77 \cdot 10^8$	$4.47 \cdot 10^{11}$	2450 ± 100	[6]
Surveyor 3	20 Apr. 1967 to 20 Nov. 1969	75 d*	$1.73 \cdot 10^{10}$	$4.8 \cdot 10^{13}$	2770 ± 120	[7]
Zond 8	21–27 Oct. 1970	140 h	$7.3 \cdot 10^8$	$2.0 \cdot 10^{12}$	2870	
Luna 16			$2.36 \cdot 10^{15}$	$6.6 \cdot 10^{18}$	2790 ± 100	

* Integral exposure, equivalent to normal falling of solar wind flux to the target.

Preliminary results of measuring different isotopes in the target of Zond 8 for solar wind energies are presented in Table 2, and for higher energies in Table 1. Table 2 also gives the results of the experiments with foils carried out from Apollo 11 and 12, the analysis of samples taken from Surveyor 3 and lunar probes delivered by Luna 16. As is seen from the table the ratio $^4He/^3He$ in the target of Zond 8 is close to the data of Surveyor 3 and to our data for lunar probes of Luna 16 and differs from those of Apollo 11. Possible causes for this discrepancy are: (i) variations of the $^4He/^3He$ ratio in the solar wind flux; (ii) thermodiffusion of solar wind ions collected by the target. On the basis of the results obtained one cannot make an unambiguous conclusion on the variations of the $^3He/^4He$ ratio existing in the solar wind flux. The solar wind ion losses from the target due to thermal diffusion can result in enrichment of the target by heavy isotopes. Note that temperatures of the target surfaces exposed in the moon on Apollo 11 and 12 did not exceed 90 °C while the temperatures on Zond 8 and Surveyor 3 lie in a higher range (up to 140 °C). At the present time the effect of thermal diffusion on variations in isotope composition of ions accumulated in the target has been studied experimentally. The same experiments specify the trapping probability for the material used for the target of Zond 8. That is why Table 2 gives only the isotope content measured per cm^2 of the target without counting the isotope fluxes in the solar wind.

Besides measuring stable isotopes of rare gases, an attempt was made to measure the content of different radio isotopes (^3H and ^{37}Ar) in the solar wind. ^{37}Ar was not found in the target. On the basis of background measurements one can obtain the upper limit of 2.8×10^3 ^{37}Ar atoms/cm^2. The ^3H content exceeds but little the background and at the present time the upper limit can reach 8×10^7 ^3H atoms/cm^2.

For the analysis of the isotope content of the solar wind we have also studied a natural target — a fine fraction of lunar soil delivered by Luna 16. The particle dimensions in this fraction do not exceed 50 μm. For this analysis samples of ~ 0.3 mg in weight taken from different depths from the lunar surface were used. Within the experimental errors the $^4He/^3He$ ratio does not change with depth and is close to the isotopic ratio detected in the target of Zond 8 and in the samples from Surveyor 3.

In conclusion it should be noted that, in spite of the equipment of the spacecrafts being simple and the exposure method of special-purpose targets being informative for the study of solar particle fluxes, the data obtained by this method are insufficient. This comes from the fact that the number of spacecraft where these experiments could have been carried out is limited and the exposure time is small. The experiment carried out on Zond 8 shows that the exposure can be accomplished completely automatically which widens experimental possibilities and increases the exposure time.

The use of the multilayer targets makes it possible to solve the problem of the composition of solar particle fluxes over a wide energy range.

References

[1] G. E. Kocharov and Yu. N. Starbunov, Pisma v ZHETF **11**, 132 (1970).
[2] D. Tilles, J. DeFelice and E. L. Fireman, Icarus **2**, 258 (1963).
[3] O. A. Schneffer and J. Zähringer, Phys. Rev. Lett. **8**, 389 (1962).
[4] I. A. Alimova, V. O. Naidenov, B.S. Boltenkov and V. N. Gartmanov, Cosmic Studies **9**, 149 (1971).
[5] F. Bühler, P. Eberhardt, J. Geiss and J. Meister, Science **166**, 1502 (1969).
[6] J. Geiss, P. Eberhardt, F. Bühler and J. Meister, J. Geophys. Res. **75**, 5972 (1970).
[7] F. Bühler, P. Eberhardt, J. Geiss and J. Schwarzmüller, Earth Planet. Sci. Lett. **10**, 279 (1971).

Space Research XII — Akademie-Verlag, Berlin 1972

PROTON RADIATION BELT VARIATIONS IN JULY–AUGUST 1970

S. N. Vernov, I. Ya. Kovalskaya, M. I. Panasyuk, A. I. Rubinstein, E. N. Sosnovets, L. V. Tverskaya and O. V. Khorosheva

Moscow State University, Moscow, USSR

The report presents the results of measurements of the radiation belt low-energy protons on Molniya 1 satellite in July–August 1970. The protons were recorded with a junction detector in three differential channels: $130 < E_p < 250$ keV, $250 < E_p < 340$ keV, $340 < E_p < 420$ keV and a single integral channel with $E_p > 420$ keV. The data on the variations in the intensity and spectrum of low-energy protons during strong geomagnetic storms on 25 July and 17 August 1970 are presented. The effect of selective proton acceleration in narrow energy ranges on $L \sim 2.5$ has been found. Irregular proton fluxes of various energies were observed beyond the boundary of the trapped radiation region. The observed effects are compared with the data from the world-wide network of magnetic stations. The results are interpreted in terms of the concept of particle acceleration by the electric fields associated with polar disturbances.

As is known, the structure of the proton radiation belt with energies higher than ≈ 100 keV at low and moderate magnetic activity can be well described by the diffusion theory [1]. During powerful substorms, however, non-adiabatic disturbances of proton intensity occur [2]. It was suggested in [3, 4] that rapid non-adiabatic variations in the intensity of particles with energies of hundreds of keV were due to the drift in strong electric fields appearing during substorms and having characteristic variation times close to the period of longitudinal drift of particles of these energies.

The present report analyses the substorm effect on the dynamics of the proton belt variations. The data used are from a differential proton spectrometer carried on Molniya 1 satellite (apogee 39280 km in northern hemisphere, perigee 470 km in southern hemisphere, period of rotation ~ 12 hours, orbit inclination $\sim 65°$). The protons were detected in three differential channels: $130 < E_p < 250$ keV, $250 < E_p < 340$ keV, $340 < E_p < 420$ keV and a single integral channel with $E_p > 420$ keV. The geometric factor of the detector was $\approx 3 \times 10^{-4}$ cm^2 sr.

Figs. 1 and 2 present the data on the variations in the proton intensity and spectrum in the energy range studied during the storm of 24–25 July 1970. The storm was fairly strong: the Dst variation peaked at $120\,\gamma$ (see Fig. 1). The arrow indicates the moment of Molniya 1 passage through the radiation belts ($L \sim 3$) on 25 July 1970. Fig. 1 presents the ratios of counting rates measured in various channels of the detector during this passage to the counting rates measured when observing the typical quiet profile of the proton belt. The most considerable increase (by ≈ 4 times), with a peak on $L \approx 3.2$, was observed in the data of the $130 < E_p < 250$ keV channel. The high-energy proton intensity decayed in the interval of $L = 3.5$–4.5. A similar effect was observed during the storm on 18 April 1965 [2]. The main contribution to the $E_p > 250$ keV proton intensity decay was probably from the adiabatic variation due to ring current.

An interesting feature is the relative intensity increase in the $250 < E_p < 340$ keV and $340 < E_p < 420$ keV energy ranges on $L = 2.6$ and $L = 2.2$

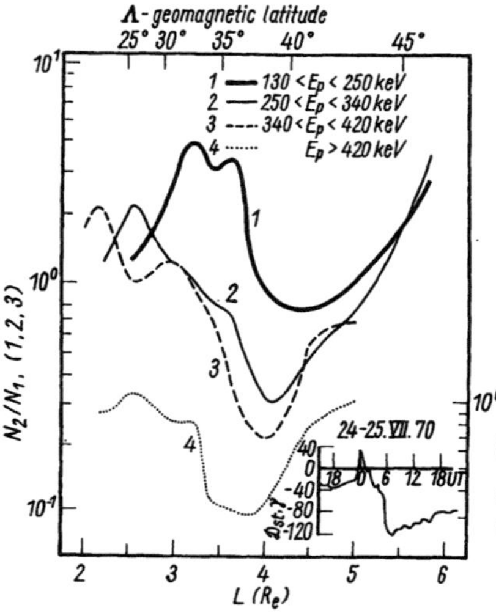

Fig. 1

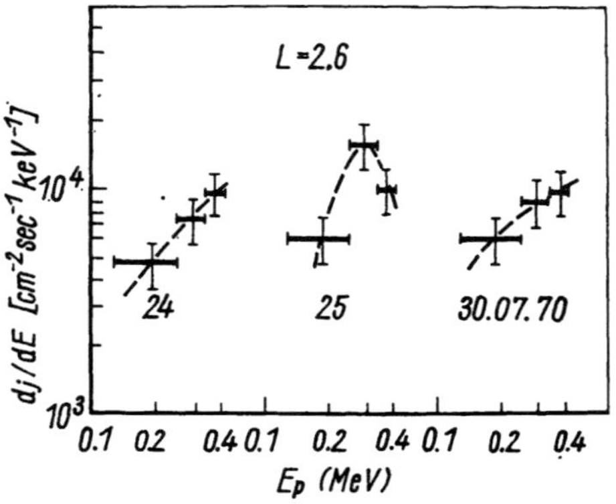

Fig. 2

Fig. 1. Ratio of the counting rates of the proton spectrometer channels (N_1 is the counting rate corresponding to the quiescent state of the proton belt, N_2 is the counting rate during the magnetic storm of 25 July at ~07h UT).

Fig. 2. Variations in the differential proton energy spectrum on $L = 2.6$.

respectively. Such a "selective" proton acceleration in different energy ranges was observed for the first time. Cases of acceleration of such a type were observed in the data on electrons with energies ~1 MeV on $L \approx 1.15$ [5]. Fig. 2 presents the data on the proton spectrum at $L = 2.6$, on 24 July 1970 under relatively quiet conditions, on 25 July 1970 during strong storm (the passage shown in Fig. 1), and on 30 July 1970 five days after the *Dst* variation peak. A transient proton spectrum peak in the $250 < E_p < 340$ keV energy range was observed on 25 July; on 30 July the spectrum was close to the spectrum observed under quiet conditions. As was noted in [5] the most probable cause of such variations is quasi-periodic trains of magnetic and, hence, electric disturbances. In this case those particles on each L-shell whose magnetic drift period is equal to the main period of the train will display the most intense variations. In our case we could simultaneously observe the variations in particles of different energies on different L-shells. It can be easily seen that the magnetic drift period $T = 44/L\,E$ min (E is expressed in MeV) in the peaks of relative variations (average energy for each range is used in the estimates) is the same for all energies and equal to ~1 hour. Fig. 3 presents the data from several magnetic observatories located at various geomagnetic latitudes. It can be seen that the magnetic field variations at the time have a period of ~1 hour which may be related to the DP-2 variation type [6] recorded at all latitudes in the time interval of 02–07 h UT on 25 July 1970. In this case the DP-2 amplitude was several times in excess of typical value. The highest disturbance value was at the pole (up to $\approx 900\,\gamma$). In the auroral zone the disturbance did not exceed $\approx 400\,\gamma$ and only at ≈07h UT a bay of $\approx 800\,\gamma$

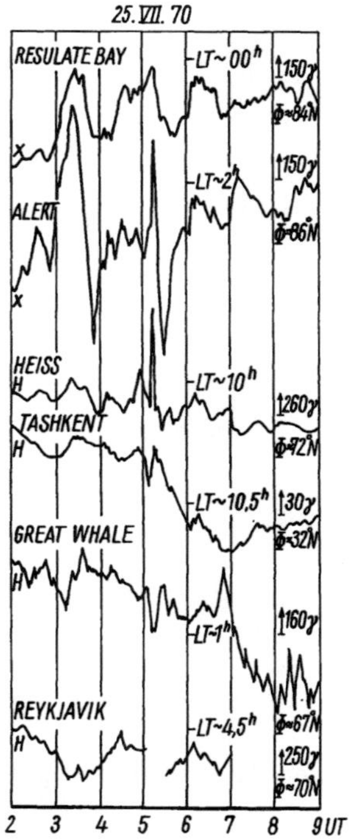

Fig. 3. Magnetic field variations according to the data from some observatories during the interval 02–09h UT on 25 July 1970.

amplitude began to develop which lasted for ~3 hours. Since the main period of magnetic variations corresponds to the abovementioned resonance condition we can conclude that the proton intensity variations are due to this resonance.

It will be noted that the intensity variations on $L = 2.2$ and 2.6 occurred at the inner edge of the belt where the intensity gradient is large. Therefore, the variations are associated with only an inconsiderable redistribution of particles over L. In the case of the protons with $130 < E_p < 250$ keV the variations were observed near the belt maximum and the total particle number increased substantially. In this case the linear approximation used in the resonance theory of particle acceleration in the belts [5] is hardly applicable. The numerical calculation of the particle drift in the magnetic dipole field and transient homogeneous electric field $E = E_0 \sin t/t_0$ [7] shows, however, that in this case too there existed favourable conditions for injection of the $130 < E_p < 250$ keV protons on $L = 3.2$. It follows from the analysis of the drift trajectories that the parameters t_0, E_f (finite particle energy in eV), and L_f (inner boundary of the injected particle belt) are related as follows

$$t_0 \, E_f \, L_f = 6 \times 10^8 \, . \tag{1}$$

The disturbance of ~1 hour duration corresponds to $t_0 \approx 1.2 \times 10^3$ sec. At $E_f \approx 1.8 \times 10^5$ eV, $L_f \approx 3$ the relation (1) is well satisfied. The amplitude of the electric field necessary for such injection is 4×10^{-5} V cm^{-1}. According to modern ideas these are the maximum potential differences within a factor of ≈ 2 attained during the most powerful substorms [8].

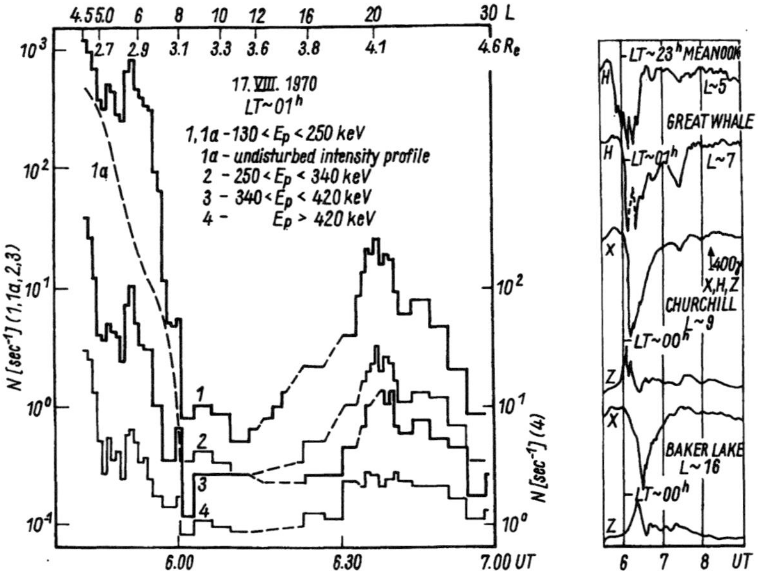

Fig. 4. Intensity variations of protons of various energies during the magnetic storm on 17 August 1970 (0545–0700 UT).

Thus, the results obtained confirm the great importance of large-scale substorm electric fields in the dynamics of proton variations at energies higher than ≈ 100 keV in the magnetosphere.

It is of great interest to compare directly the data on the proton intensity variations with the dynamics of substorm development: Fig. 4 presents one-minute averages of counting rates from different channels of the proton detector in the time interval 0545–0700 UT on 17 August 1970 during a strong magnetic storm ($Kp = 9$). The data from some magnetic stations are also presented. A considerable bay (with an amplitude larger than 1500 γ) was observed in the auroral zone during the time interval studied. At $L \approx 5$ (Meanook station) the bay began to develop at 0545 UT. The analysis of the Z component of the auroral zone stations shows that the polar electrojet centre was displaced: before ~06 h UT the jet was centred south of $L \approx 7$ (Great Whale station), by 0630 it had approached the zenith of Churchill station ($L = 9$). At this moment the disturbance increased abruptly at the high-latitude station Baker

Lake ($L \approx 16$). However, the scanty network of the magnetic stations does not make it possible to conclude unambiguously whether this high-latitude disturbance is independent of or due to the arrival of the electrojet at such high latitudes.

Large variations in the proton belt are associated with the developed bay (see Fig. 4a). The belt profile is strongly jagged; abrupt simultaneous increases in protons of various energies are observed. The intensity decrease at $\approx$06 h UT may be associated with the fact that the satellite, when moving to larger L-values, had left behind the displacing electrojet. It is noteworthy that the intensity of protons of various energies increased abruptly at $L = 18$ at the moment when the disturbance peak was observed at Baker Lake, the station nearest with respect to L ($L \approx 16$). The effects observed may be explained in terms of the hypothesis [9] about the expansion of the auroral oval towards the pole due to magnetic field line reconnection in the magnetotail. The data presented above indicate that the reconnection is accompanied by proton acceleration up to energies of hundreds of keV. It is possible that the change in the character of ionospheric currents during these events is due to an increase of ionospheric conductivity during precipitation of accelerated particles. Reconnection may result in a specific electric current along the corresponding field lines [10].

References

[1] B. A. Tverskoy, Dynamics of the Earth's Radiation Belts, ed. Nauka, Moscow 1968.
[2] F. Soraas, preprint CSFC X-612-69-241 (1969).
[3] S. N. Vernov, T. A. Ivanova, E. N. Sosnovets, L. V. Tverskaya, G. F. Fedorova and O. V. Khorosheva, Izv. Akad. Nauk SSSR **34**, 2270 (1970).
[4] S. N. Vernov, S. N. Kuznetsov, E. N. Sosnovets, L. V. Tverskaya, M. V. Teltsov and O. V. Khorosheva, paper presented at XIII COSPAR Meeting, Leningrad 1970.
[5] J. B. Cladis, J. Geophys. Res. **71**, 5019 (1966).
[6] A. Nishida, J. Geophys. Res. **73**, 1795 (1963).
[7] L. V. Tverskaya, Geomagn. i Aeron. **11**, 530 (1971).
[8] B. A. Tverskoy, Proceedings of Solar Terrestrial Physics Symposium Leningrad 1970.
[9] W. I. Axford, Rev. Geophys. **7**, 421 (1969).
[10] V. M. Vasiliunas, MIT preprint CSR-P-69-24 (1969).

Space Research XII — Akademie-Verlag, Berlin 1972

PRELIMINARY FINDINGS OF A PETREL ROCKET EXPERIMENT TO INVESTIGATE THE VLF EMISSION "CHORUS" IN THE IONOSPHERE

M. J. RYCROFT[a], R. A. USHER[a], S. K. ADJEPONG[a], G. J. DANIELL[a], J. VÁVRA[a*], K. BULLOUGH[b], W. GIBBONS[b] and T. R. KAISER[b]

[a] Department of Physics, University of Southampton, England
[b] Department of Physics, University of Sheffield, England

On 19 November 1970, when the vlf emission known as "chorus" was received by ground-based equipment, a Petrel rocket was launched to an apogee of 130 km from South Uist, Outer Hebrides, Scotland ($L \sim 3.4$). The rocket-borne equipment consisted of a Langmuir probe, a loop aerial, the plane of which contained the spin axis of the rocket, a broad-band amplifier and seven narrow-band AGC amplifiers, the outputs of which were telemetered to the ground. The observations made in five of the narrow-band channels, from 2.2 to 5.4 kHz, and spanning the chorus band, are discussed. The autocorrelation function of the vlf field strength recorded in each of these channels contains a pronounced periodicity, with a period equal to one half the spin period of the rocket. Measurements of the spin modulation ratio are used to test the hypothesis that the signals are propagating in the whistler mode with their wave normals along the geomagnetic field lines. The hypothesis is often, though not always, confirmed. The occurrence of a burst of 9.6 kHz "chorus" is reported. The "chorus" signals become markedly stronger, by ~12 dB, between altitudes 85 and 93 km, above which there is considerable structure in the field strengths observed.

1. Introduction

In order to study the wave characteristics of very low frequency signals (3–30 kHz) in the ionosphere, thus complementing both ground-based and satellite observations of such signals, a research programme was initiated jointly by the Universities of Sheffield and Southampton. Preliminary results from observations of the natural vlf emission "chorus" made during the successful flight of the first Petrel rocket payload of this programme are discussed in this paper. Following a brief review of the experimental equipment, measurements of the signals induced in a single rocket-borne loop aerial are compared (§ 3) with those recorded simultaneously on the ground. The modulation of a sample of the signals, at a period equal to one half the spin period of the rocket, is then discussed quantitatively. The value of the spin modulation ratio is compared with that expected for a signal propagating in the whistler mode, which has its wave normal parallel to a line of force of the geomagnetic field. In § 5 the first observation, at South Uist, of a burst of chorus at the comparatively high frequency of 9.6 kHz is discussed. This seven-second burst was also observed by the rocket-borne equipment. The electron density profile derived from Langmuir probe observations is considered in § 6, as is the structure of vlf field strengths in the ionosphere.

* Now at Katedra elektroniky FJFI CVUT, Prague, Czechoslovakia

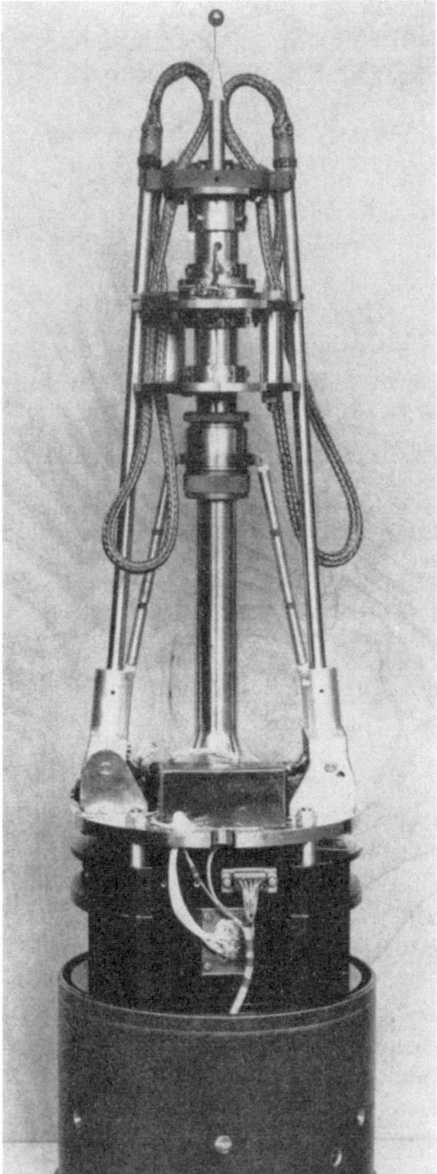

Fig. 1. Photograph of Sheffield–Southampton Langmuir probe–vlf monitor rocket payload. The single component loop aerial is shown in its folded position, as placed under the nose-cone for launch. Below the plate on which the stainless steel column and hinged rods are mounted lies the electronic equipment and, below that, the telemetry unit.

2. Equipment

A photograph of the instrumented payload is presented in Fig. 1. The payload is mounted atop a Petrel rocket, of 7.5 inches diameter. It carries a Langmuir probe, a gold-plated sphere of 0.375 inch diameter mounted on an insulated wire support of 0.04 inch diameter, at the top of a stainless steel column. This column also supports a mechanism for releasing the upper part of the loop aerial following nose-cone ejection. The longer rods hinged on either side of the base of the column then deploy, and a triangular loop is formed, with base 36 inches and height 18 inches. The electrostatically screened loop aerial consists of 60 turns of copper wire (swg 28); its resistance is 21 ohms and its inductance 7 millihenries. The aerial is optimally matched, via a 1:10 step-up transformer, to a low-noise broad-band preamplifier. Observing a field strength of 10^{-12} Wb m^{-2}, at 3 kHz with a bandwidth of 100 Hz, the signal-to-noise ratio is found to be 17.6 dB.

In order to accommodate the large dynamic range ($\sim$60 dB) of chorus signals, the signal is then fed to seven narrow-band AGC (automatic gain control) amplifiers in parallel. The output voltage of each amplifier is linearly proportional to the logarithm of the wave magnetic field strength at the frequency to which it is tuned. Detailed design considerations of these selective channels have been given by Vávra [1]. Five channels, at 2.2, 2.8, 3.5, 4.4 and 5.4 kHz, span the band of chorus observed on South Uist ($L \sim 3.4$). The remaining channels are tuned to 9.6 kHz and, in order to study the wave field in the ionosphere of the signal transmitted by GBR Rugby, to 16 kHz. The rectified and integrated outputs of these narrow-band channels are sampled, using a mechanical commutator placed inside a mu-metal screen, at equal intervals. The sampling rates are 240/second for the 2.8, 3.5 and 4.4 kHz channels, 160/second for 2.2, 5.4 and 16.0 kHz channels and 80/second at 9.6 kHz. The intervals between observations are, therefore, 4.17, 6.25 and 12.5 milliseconds respectively. The sampled analogue outputs are telemetered to ground, where they are displayed on a high-speed uv chart recorder and stored on magnetic tape for subsequent digitization.

3. P47H Observations of Chorus and their Analysis

The first payload of the type described above, designated P47H, was launched at 0515 UT on 19 November 1970. This time was some two hours before sunrise at 100 km in the ionosphere above the range on South Uist, Outer Hebrides, Scotland ($L \sim 3.4$).

At this time, the vlf emission known as "chorus" was received by equipment at a "quiet" site some 10 km from the range. The emissions, of average intensity at this site, are composed of many overlapping rising frequency tones between $\sim$2 and 6 kHz. A Sonagram, i.e. the dynamic spectrum, of a sample of the ground-based recording is presented in the upper part of Fig. 2. Shown as dark bands are some twenty elements of chorus occurring in this three-second

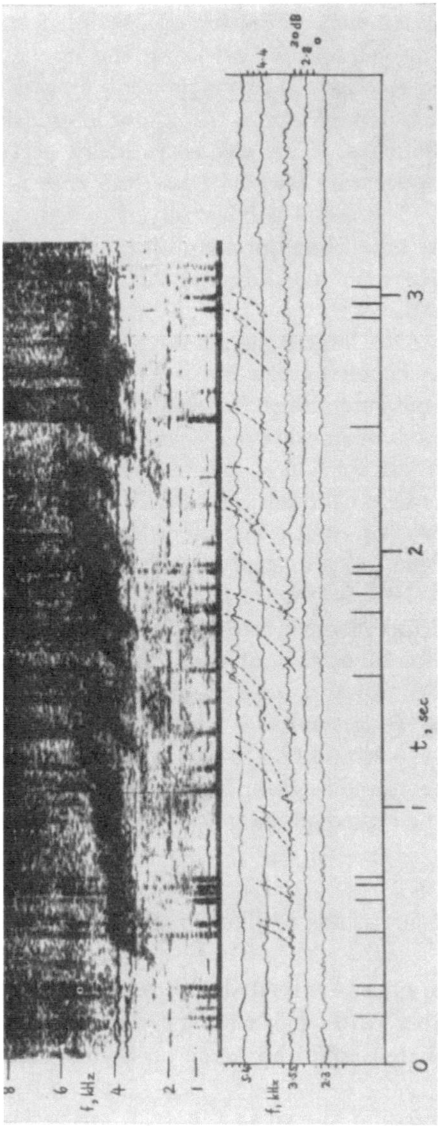

Fig. 2. Upper part: Sonagram of ground-based recording showing chorus. Lower part: Temporal variations of signal intensities at five frequencies within the chorus band. For each channel, the tick marks show outputs corresponding to signal inputs of 0, 10, 20 and 30 dB above receiver noise level.

period. Also apparent are atmospherics, vertical lines in the frequency–time domain, due to electromagnetic radiation from lightning discharges. Unfortunately also present are horizontal lines due to beating between several vlf Morse transmitters. In the lower part of Fig. 2 are presented tracings of the simultaneous uv chart recorder outputs, on a cramped logarithmic scale, of the

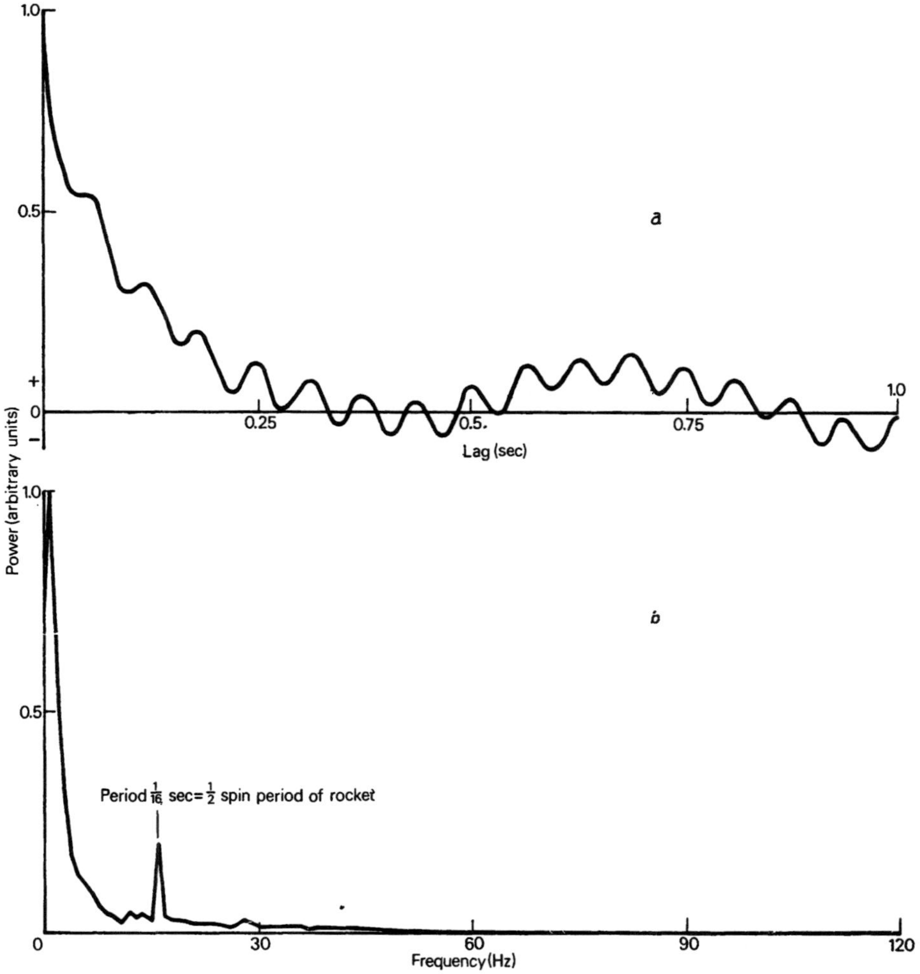

Fig. 3. Upper part: autocorrelation function of 20 seconds of 2.8 kHz data close to apogee, showing a periodicity of $^1/_{16}$ sec associated with rotation of the rocket. The component at $\sim$0.7 sec is probably due to temporal structure of the chorus. Lower part: power spectral density of data showing pronounced component with a period equal to one half the spin period of the rocket.

five lower frequency channels telemetered from the rocket at an altitude of 82 km. Signal enhancements corresponding to the chorus elements are shown as dashed lines. Those corresponding to sixteen atmospherics are shown as solid vertical lines. It is concluded that, amongst those signals received aboard the rocket, are those strong chorus elements and atmospherics recorded at the same time at the ground.

In order to study the rocket observations further, and to search for periodicities therein, the digitized, equi-spaced observations are subjected to an auto-correlation analysis. Fig. 3 is presented as an example of the results of this procedure. The upper part shows the autocorrelation function of 20 seconds of data at 2.8 kHz, taken from 10 seconds before apogee until 10 seconds after. This period was considered to be sufficiently short that changes in the signal strength due to altitudinal changes — only 2 km — were minimized. This is necessary as the time series must be statistically stationary for the analysis to be valid. However, the period is also considered to be sufficiently long for high statistical reliability of all the deductions made. The autocorrelation function has a pronounced sinusoidal component, of which there are 16.0 periods in the maximum lag of 1.00 second. This periodicity is shown in another way, in the lower part of Fig. 3, by the Fourier transform of the autocorrelation function, i.e. the power spectral density [2]. A most pronounced peak exists at the sixteenth point of the power spectrum, which corresponds to a frequency of 16 Hz.

Now, from the modulation of the strength of the telemetry signal, radiated by a dipole, it is found that the spin rate of the rocket about its longitudinal axis was 8.00 Hz. The marked periodicity of the output of the 2.8 kHz channel of the rocket-borne receiver, of $^1/_{16}$ second, is thus equal to one half the spin period of the rocket. Since the observed signals exhibit maximum and minimum intensities for loop orientations 90° apart, this result indicates that the signals are certainly produced by magnetic fields external to the rocket. Although the presence of spin modulation of the signals is shown very conveniently by the computed power spectral density, the method is unsuitable for quantitative measurement of the depth of their spin modulation.

Since the spin frequency of the rocket is known to be 8.00 Hz and since the sampling rate of the 2.8 kHz channel is 240/sec, exactly 30 samples are taken during each complete spin of the rocket. This fact enables another method of analysis to be used: this involves summing the output of, say, the 2.8 kHz channel observed at some time (an arbitrary origin of time, say 10 seconds before apogee) with the outputs observed at intervals of $m/8$ seconds later, where m is an integer taking values between 0 and 160, so that the same 20 seconds of data considered in Fig. 3 are studied again. Division by 160 then gives the average value (over 20 seconds) of the 2.8 kHz output due to chorus at one particular orientation of the loop aerial in space. This value is plotted as a cross at time zero (and also at time $^1/_8$ sec) in Fig. 4. The summing and averaging procedure is then repeated, shifting from $t = 0$ by successively $n/240$ seconds, where n is an integer between 1 and 29, in order to calculate the other values plotted as crosses in Fig. 4. The points are seen to define a sine curve, with a period of $^1/_{16}$ second. The amplitude of this sinusoidal output variation is found by a least squares method to be, in this case, 0.025 volt. The quantitative interpretation of the amplitude of the spin modulation at the output of the receiver in terms of the modulation of the incident vlf signal at twice the spin frequency of the rocket is given in the next section.

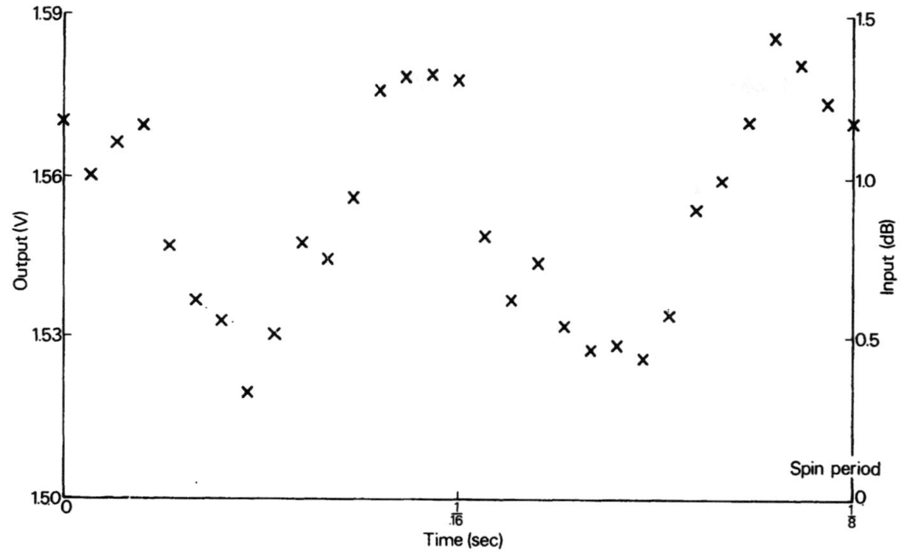

Fig. 4. Average of measurements taken, at 2.8 kHz, at similar orientations of the loop aerial, that is at $^1/_8$ sec intervals, within ± 10 sec of apogee.

4. Theoretical Considerations

Consider a plane wave propagating in the whistler mode with its wave normal $\boldsymbol{k}$ parallel to the geomagnetic field. The wave is thus circularly polarized with its magnetic field $\boldsymbol{H}$ rotating, in a plane normal to $\boldsymbol{k}$, at the angular frequency of the wave ω.

Since the spin rate of the rocket is much less than ω, the electromotive force $\mathscr{E}$ induced in the loop aerial can be calculated by considering the rocket to be stationary. When the plane of the loop lies in the plane containing the spin axis and the wave normal, the magnetic flux cutting the loop and hence $\mathscr{E}$ take on their maximum values. When the loop is perpendicular to this plane, the flux is reduced by a factor $\cos\theta$, where θ is the angle between the wave normal and the spin axis; $\mathscr{E}$ then takes on its minimum value.

Defining the spin modulation ratio R by the relation

$$R = \frac{\mathscr{E}_{\text{max}} - \mathscr{E}_{\text{min}}}{\mathscr{E}_{\text{max}} + \mathscr{E}_{\text{min}}}, \tag{1}$$

it follows that

$$R = \frac{1 - \cos\theta}{1 + \cos\theta} = \tan^2\frac{\theta}{2}. \tag{2}$$

The quantity R is plotted as a function of θ, the angle between the wave normal and the spin axis of the rocket, in Fig. 5.

For the logarithmic AGC system used in the experiment, the output voltage V is directly proportional to the logarithm of the induced emf at the input, $\mathscr{E}$:

$$V = K \ln \mathscr{E} + C, \tag{3}$$

where K and C are constants known from the calibration of the equipment. Differentiating Eq. (3),

$$\frac{dV}{d\mathscr{E}} = \frac{K}{\mathscr{E}}. \tag{4}$$

Combining Eq. (1), (2) and (4), for small values of R,

$$R = \frac{d\mathscr{E}}{\mathscr{E}} = \frac{dV}{K} = \frac{1 - \cos\theta}{1 + \cos\theta}. \tag{5}$$

Referring back to Fig. 4, the amplitude of the spin modulation dV is directly measured and is, in this example, 0.025 volt. Knowing K^{-1} to be 1.92 volt^{-1} for

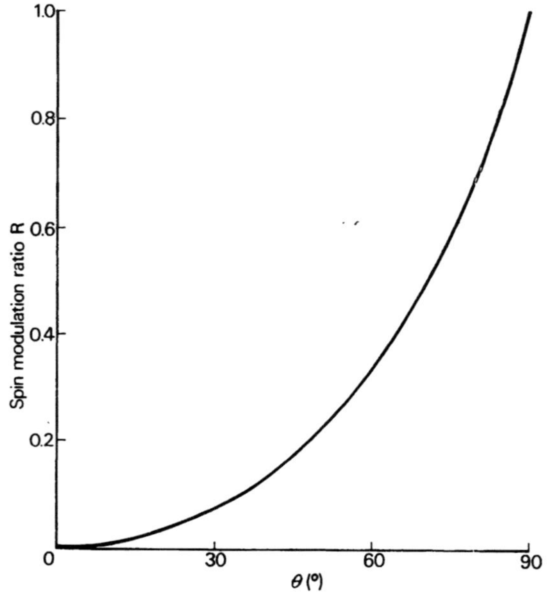

Fig. 5. Graph of spin modulation ratio R (for whistler mode signals with $\boldsymbol{k}$ parallel to the geomagnetic field) against θ, angle between wave normal and rocket spin axis.

the 2.8 kHz channel, it follows from Fig. 5 that a value of $R = 0.048$ corresponds to $\theta = 24.5°$.

Since the payload did not contain a magnetometer, from whose measurements θ could be calculated directly, its value has to be found from estimated values of α and I, where α is the angle of elevation of the rocket spin axis above the horizontal, estimated from the trajectory of the rocket to be 74.5°, and I is the inclination, or dip angle, of the geomagnetic field. From a geomagnetic map [3] this is estimated to be 71.5°. By trigonometry,

$$\cos\theta = \sin\alpha \sin I + \cos\alpha \cos I \cos\varphi, \tag{6}$$

where φ is the geomagnetic bearing of the rocket, 288.5° E. Substituting in the values given, θ is found to be 19.8°. The two values of θ, obtained by different methods, agree reasonably for this part of the data. The spin modulation of

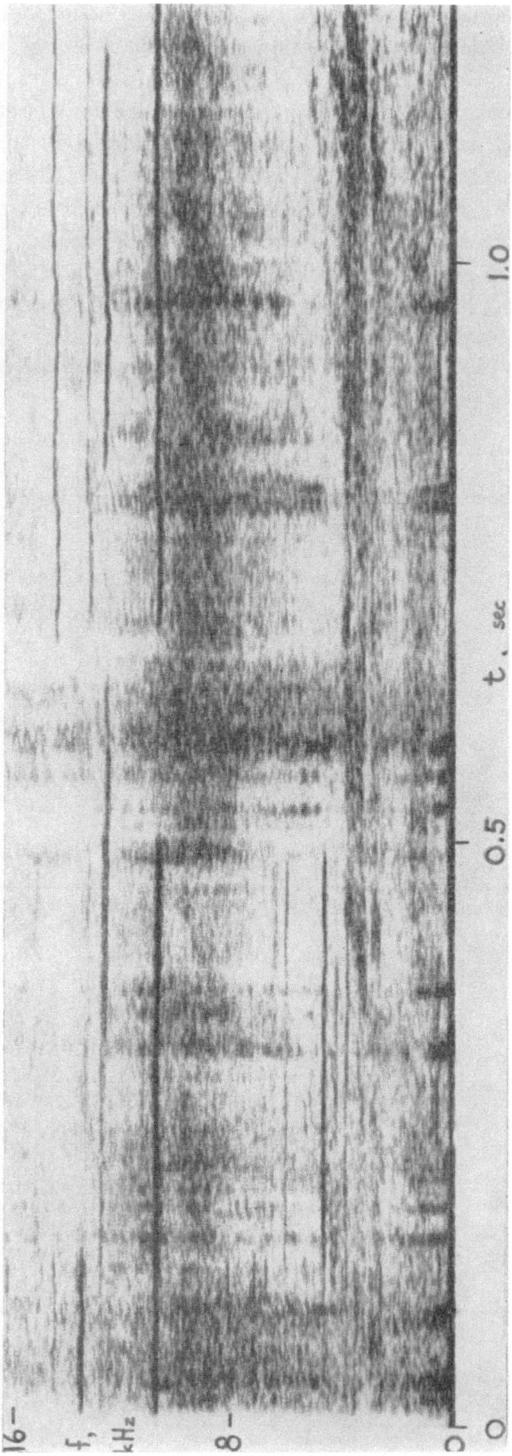

Fig. 6. Sonagram of ground-based recording showing chorus at ~8–11 kHz. Weak chorus is also apparent in the 3–5 kHz band. The horizontal lines between 10 and 16 kHz are due to man-made vlf transmissions.

the received chorus signals is thus consistent with their being a single plane wave propagating in the whistler mode, with wave normals along the geomagnetic field lines.

At other frequencies and at other times during the flight, considerably different values of R are found. It should be pointed out that, particularly on the upleg, R values are often significantly greater than the value $R = 0.030$ expected for $\theta = 19.8°$. Possible reasons for this are currently being investigated. On the downleg, R values both somewhat larger and smaller than this value occur. Since there are several mechanisms whereby the spin modulation would be reduced, R values $\gtrsim 0.030$ are not inconsistent with the belief that the chorus signal in the lower ionosphere can be regarded as a single plane wave propagating in the whistler mode with its wave normal aligned along the geomagnetic field.

5. Chorus at 9.6 kHz

For some seven seconds during the rocket flight, a burst of chorus was recorded by the ground-based equipment in the frequency band ~8–11 kHz.

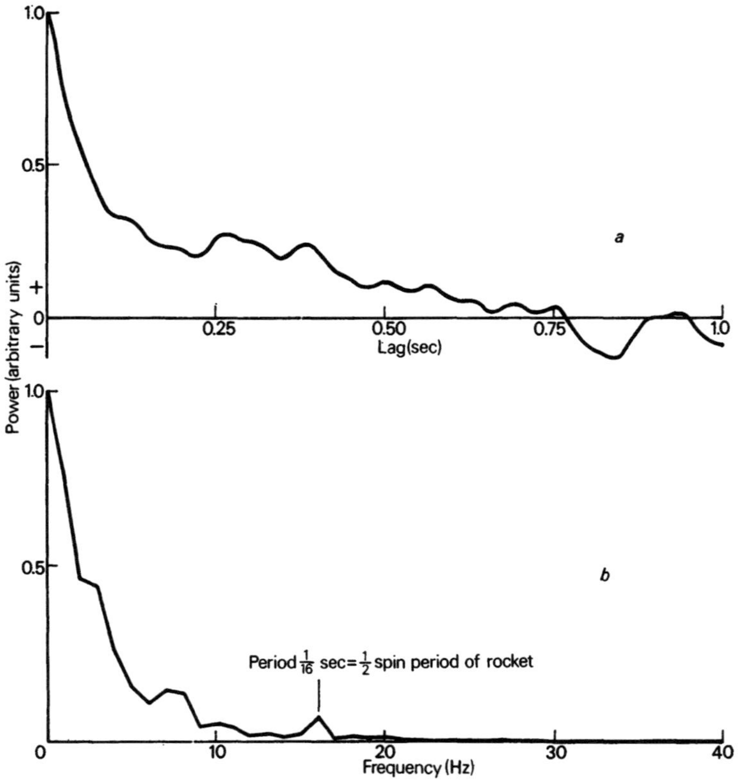

Fig. 7. Upper part: autocorrelation function of $6^2/_3$ sec of 9.6 kHz data recorded on downleg, between 107 and 102 km altitude. Lower part: power spectral density of data, showing a significant peak at 16 Hz, i.e. showing a periodicity of $^1/_{16}$ sec associated with rotation of the rocket.

A Sonagram of part of this burst is presented in Fig. 6. It is the first time that chorus at such a remarkably high frequency has been noted. The discovery is of significance because, assuming that chorus is generated by the transverse resonance instability on the $L \sim 3.4$ field line near the equatorial plane [4], the emission frequency is $\sim$0.4 times the minimum electron gyrofrequency along the field line. This frequency is close to the maximum frequency (0.5 times the minimum electron gyrofrequency) that can be guided through the magnetospheric plasma along field-aligned ducts of enhanced electron density [5], propagate through the ionosphere and be received on the ground. This burst of chorus was also recorded on the 9.6 kHz channel of the rocket-borne equipment. Its strength was $\sim$24 dB above typical signal levels during the rest of the flight. In Fig. 7 are presented the autocorrelation function and power spectral density of a sample, of $6^2/_3$ seconds duration, of these data computed in the manner discussed in § 3. A statistically significant spectral peak with a $^1/_{16}$ second period is clearly seen.

6. Ionospheric Structure

The electron density profile above 95 km altitude, derived from the Langmuir probe observations made on the upleg of the flight, is shown in Fig. 8. The points on the profile are calculated from the current to the probe when a swept potential is applied to it, at 2-second intervals. Using a method discussed by

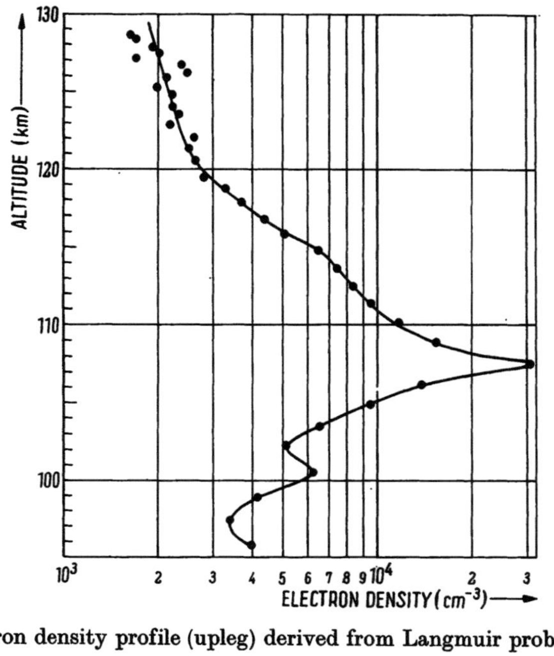

Fig. 8. Electron density profile (upleg) derived from Langmuir probe observations.

Beckingham [6] the current I to the probe, when biased to a potential V above space potential, is given as

$$I = N_e\, e\, S \left(\frac{kT}{2\,\pi\, m}\right)^{1/2} \left(1 + \frac{e\,V}{kT}\right)^n , \qquad (13)$$

where N_e is the electron density, e and m are the electronic charge and mass respectively, S is the surface area of the probe, k is Boltzmann's constant, and T is the kinetic temperature of the electrons, which is assumed here to be 300 °K. The power n is zero for a planar probe, one half for a cylindrical probe and unity for a spherical probe. Since the sphere on a wire support does not behave as a perfect sphere, the best value of n to use in Eq. (13) is found, by minimizing the random scatter of the values of N_e calculated, to be 0.56. The most marked feature of the electron density profile is the 3×10^4 electrons/cm^3 E layer peak at 107.5 km altitude. Only 10 km above this altitude, the electron density is an order of magnitude smaller. The low electron densities found between 120 and 130 km are typical of night-time conditions [7].

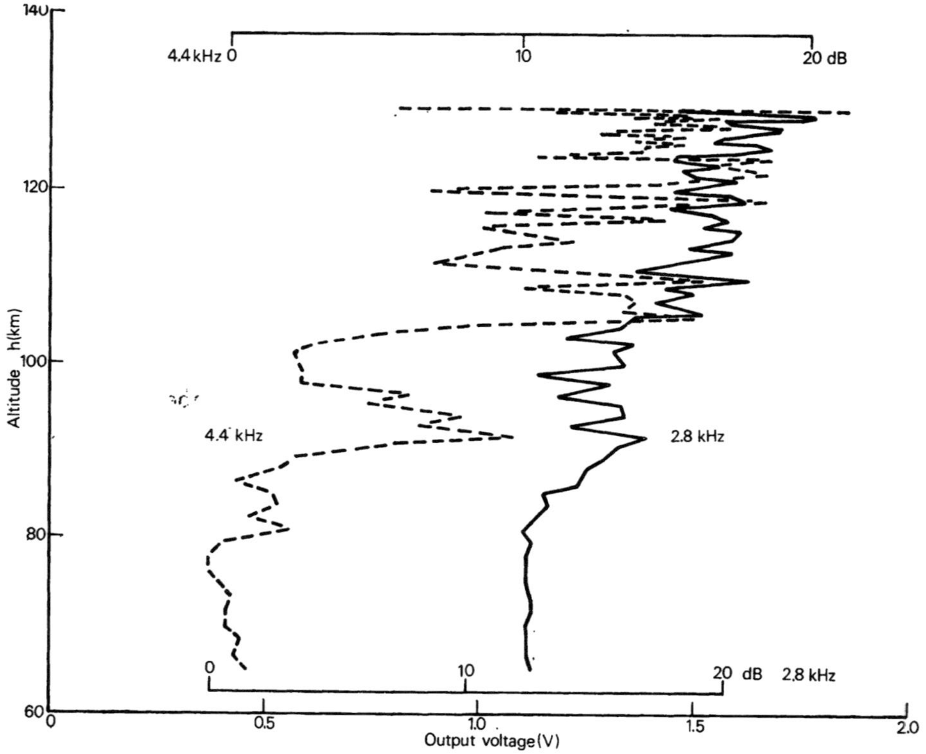

Fig. 9. Height profiles (upleg) of 1.5 sec average signal strengths at 2.8 and 4.4 kHz.

The variation of electron density with height determines the variation of the strength of chorus incident on the ionosphere from above by the processes of refraction, reflection and absorption. The profile of Fig. 8 should therefore be compared with those of Fig. 9, the averages over 1.5 seconds, or 12 complete spins of the rocket, of the observed vlf field strengths at 2.8 and 4.4 kHz, both frequencies being in the chorus band. The signals become markedly stronger, by ~5 dB at 2.8 kHz and ~12 dB at 4.4 kHz, between altitudes of 85 and 93 km, where electron density values are not simply calculable from the probe observations. Above 93 km, there is considerable structure in the field strengths observed. Since there is, in general, no clear correlation, either positive or negative, between the two figures, it is concluded that much of the structure shown in Fig. 9 is due to the temporal variations (~1.5 sec) inherent in signals of the chorus type. Part of the increase of the vlf signal strength with height above 110 km could be attributed to an increase, throughout the flight, of chorus activity and strength as evidenced by ground-based recordings.

Acknowledgments

The authors are grateful to the Science Research Council for financing this project, to D. Hickinson and N. Jenkinson of Dunford Hadfield (Sheffield) Limited for building the electronic equipment, to P. Hughes of the University of Sheffield for technical assistance, to members of the launch team at South Uist, to the Institute of Sound and Vibration Research at the University of Southampton for use of the Myriad computer for data analysis, and to R. Evans and P. Manns for computer programming.

References

[1] J. Vávra, A rocket-borne very low frequency receiver, University of Southampton Report (April 1970).
[2] H. P. F. Swinnerton-Dyer, Computer J. **5**, 16 (1963).
[3] A. J. Dessler, Satellite Environment Handbook, 1961 (p. 119).
[4] N. Brice, J. Geophys. Res. **69**, 4515 (1964).
[5] R. L. Smith, J. Geophys. Res. **66**, 3699 (1961).
[6] P. D. Beckingham, Ph. D. Thesis, University of Sheffield, June 1970.
[7] K. Maeda, Space Research XII, 1229 (1972).

Space Research XII — Akademie-Verlag, Berlin 1972

PLASMASPHERE DYNAMICS INFERRED FROM OGO 5 OBSERVATIONS

C. R. Chappell, K. K. Harris and G. W. Sharp
Lockheed Palo Alto Research Laboratory, Palo Alto, Calif., USA

Data from the light ion mass spectrometer aboard OGO 5 during the period of March 1968 through February 1969 has given a complete local time coverage of the plasmasphere. Inbound and outbound profiles of H^+, He^+ and ($O^+ + N^+$) ions have been measured once every $2^1/_2$ days from an initial perigee height of 300 km to an apogee of about 23 R_e during this year period. Although the ion motion cannot be directly measured by the instrument, the plasmasphere dynamics can be inferred through an examination of the changes in plasmasphere profile between consecutive orbits. These dynamics are discussed and a model of the plasmasphere is developed. Examples of these dynamic processes include dayside filling of the plasmasphere, peeling away of the outer portions of the plasmasphere in the afternoon-dusk sector and inward convective motions of the plasmasphere on the nightside.

1. Introduction

Our objectives in this paper are to develop a picture of plasmasphere motions and to present a time-varying convection electric-field model of the magnetosphere which offers an excellent explanation of the observed plasmasphere dynamics. We will develop our dynamic picture by discussing characteristic H^+ ion density profiles in the different local time regions of the plasmasphere and the changes in these density profiles which are observed during changes in magnetic activity. Data for this study were obtained by the Lockheed light ion mass spectrometer flown on OGO 5. We should emphasize that the OGO 5 instrument *does not measure plasma drift motions*. The dynamics which will be discussed are *inferred* from changes which take place in the plasmasphere profiles between consecutive satellite passes.

2. Magnetospheric Convection Model

The magnetospheric model which will be employed is an extension of the steady-state convection model originally suggested by Axford and Hines [1]. This model explained the observed steep gradient in the plasma density profile, the plasmapause, as the boundary between flux tubes which corotate with the earth, remaining always closed and therefore holding their plasma, and flux tubes which have their motion dominated by a convection electric field probably of solar wind origin. These latter flux tubes are convected to the magnetopause where they lose their plasma.

The basic model is represented by an equatorial flow diagram such as that calculated by Kavanagh et al. [2] and shown in Fig. 1. In this diagram the dotted flow patterns inside the solid line represent the paths of flux tubes which approximately corotate with the earth while the dashed lines show the flow paths of flux tubes whose motion is dominated by convection. The

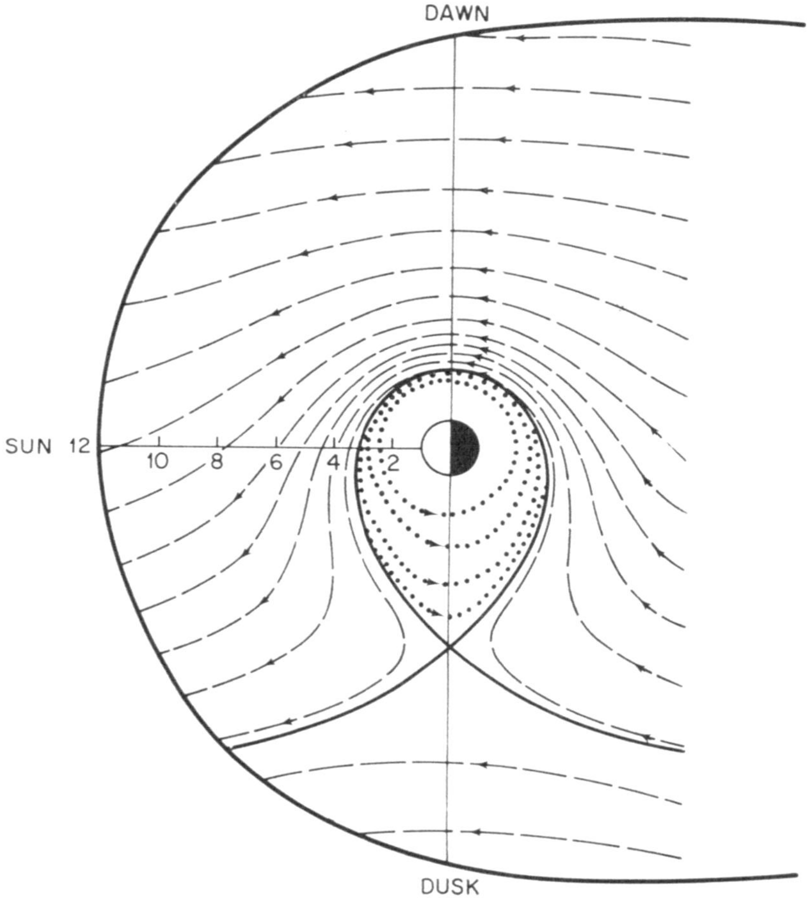

Fig. 1. Convective flow diagram for ambient plasma in the magnetosphere [2]. The closed solid line is the plasmapause position.

closed part of the solid line represents the location of the plasmapause which displays a bulge on the dusk side in general agreement with observation. The shape and location of this boundary and the directions of the flow lines should be considered only in a qualitative manner since the effects of ionospheric conductivity were not included in the model. Note that the flow pattern inside the plasmasphere includes the bulge region as part of the main flow and not as a separate eddy flow.

This diagram represents a steady-state situation where the convection electric field remains uniform. This situation can be thought of as representative of a period of steady magnetic activity. During steady magnetic activity it has been shown [3] that the flow directions inside the plasmapause lead to a $1/R^4$ decrease in density in the bulge region. This density decrease is caused by the expansion of the flux tube volume during drifts to higher L-shells at dusk (see Fig. 1). This $1/R^4$ density profile is indeed found to be characteristic of the afternoon and bulge regions by the OGO 5 observations [3]. The data, then, support this convection model which treats the bulge region as part of the main flow pattern of the plasmasphere.

3. Changes in Plasmapause Profiles due to Non-Uniform Convection

The qualitative flow pattern shown above is for a steady value of the convection electric field. To talk about plasmasphere dynamics, one must adopt the more realistic approach of a non-uniform convection electric field within the magnetosphere — one which on the average increases during magnetically disturbed time and decreases during magnetically quiet time. Then, if the effects of this varying convection field are combined with the effects of filling of the plasmasphere from the ionosphere on the dayside, a more complete model of actual plasmasphere dynamics will be obtained.

First consider the effect that a varying convection electric field has on the equatorial flow pattern of the flux tubes. Fig. 2 illustrates the changes which occur in the flow pattern during different levels of magnetic activity. Part (a) of Fig. 2 shows a sketch of the flow during a steady *active* magnetic period. Region A is the locus of flux tubes which approximately corotate with the earth. Outside this region the dashed lines show the flow paths of the convecting flux tubes which reach the magnetopause. Boundary a represents the plasmapause position during the steady active period. During steady quiet magnetic periods, when the convection field is small, the flow pattern is shown in part (b) of Fig. 2. In this case corotation is dominant out to higher L-values (region B within boundary b). The dotted lines show the flow directions of the convecting flux tubes. In quiet times the plasmapause is represented by boundary b. Region C in parts (a) and (b) represents the normal plasma trough region of the magnetosphere. The flow patterns for steady active and steady quiet magnetic periods are combined in part (c) of Fig. 2 to illustrate the effects of changing magnetic activity. Note that the change from a quiet flow pattern to an active flow pattern reduces the size of the plasmasphere. This increase in magnetic activity causes an inward motion of the plasmapause on the nightside but *not* on the dayside. At the same time a reversal in flow direction occurs in the bulge region.

3.1. Plasmapause Position on Nightside and Dayside

It has been shown [4] that the nightside region of the plasmasphere (2200–0600 LT) reacts very quickly (2–6 hours) and very predictably to changes in magnetic activity with the plasmapause moving to lower L-shells and steepening

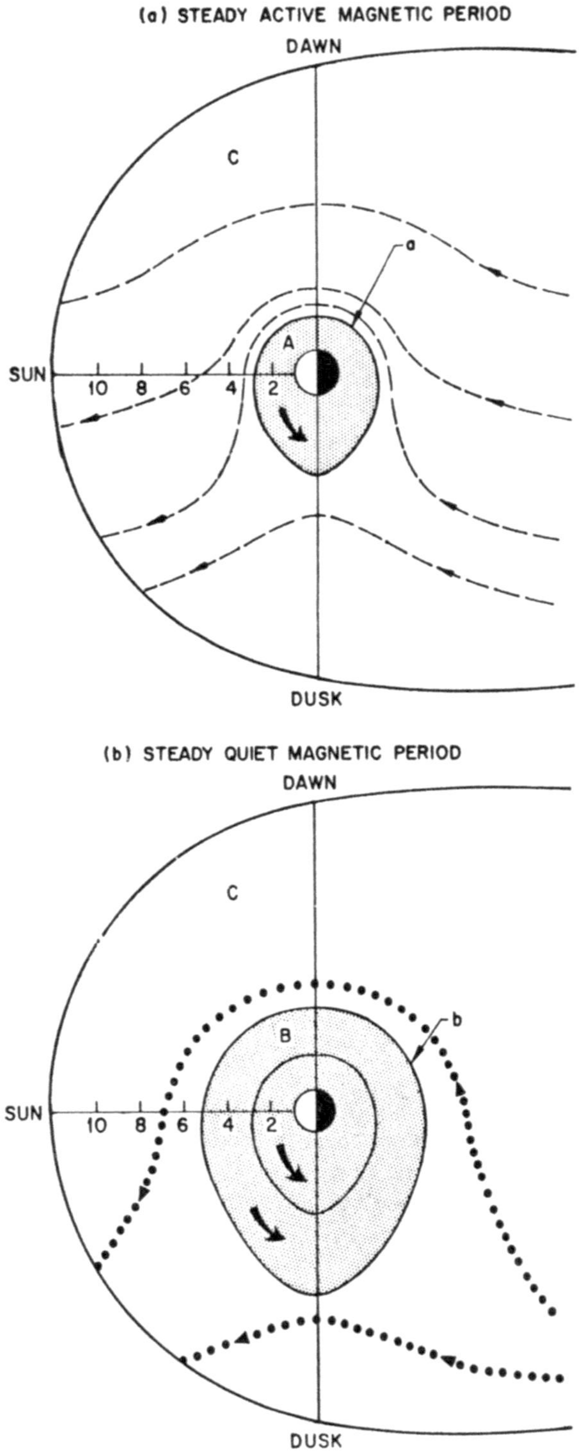

Fig. 2. For caption see opposite

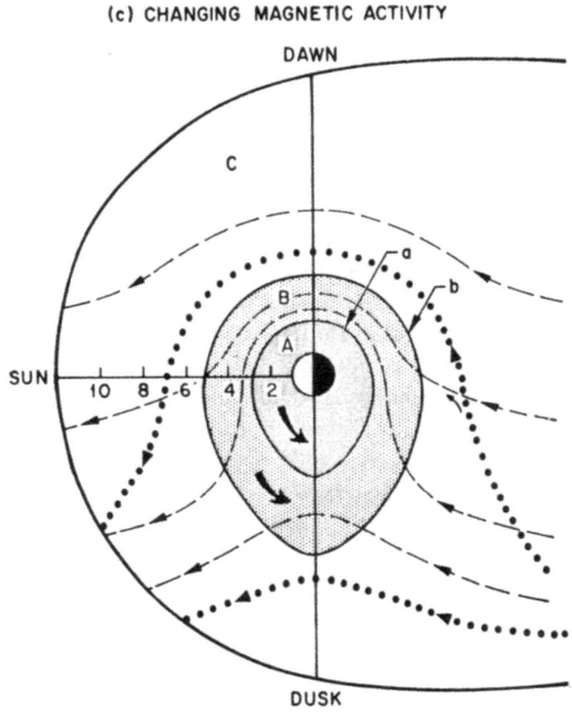

Fig. 2. Three convective flow diagrams showing flow during steady active, steady quiet, and changing magnetic activity periods.

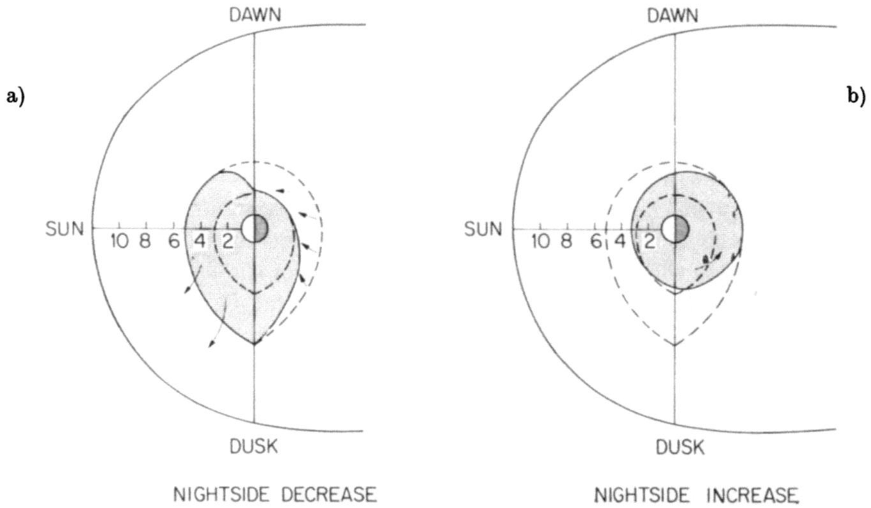

Fig. 3. A sketch of the changes in plasmasphere size and shape during changing magnetic activity. (a) shows a decrease in the nightside plasmapause radius with increasing activity. (b) shows an increase in radius with decreasing activity.

with increasing activity. Unlike the nightside, the location of the dayside (0600–1500 LT) plasmapause position appears to be determined by the level of magnetic activity present during its earlier corotation through the formative nightside region rather than the activity level at the time of the measurement [5]. This sequence of plasmapause position changes can be explained in terms of the varying convection electric field. As the convection field increases, the flux tubes in the midnight-to-dawn sector are driven to lower L-values as shown in part (c) of Fig. 2 (a change from dotted flow lines to dashed flow lines as shown by the small arrow near midnight). This process is illustrated in part (a) of Fig. 3. This inward motion or decrease in L-value cannot occur on the dayside. In fact, the dayside flux tubes begin to be convected toward the magnetopause, tending to peel off the outer edge of the plasmasphere. During times of decreasing activity, the bulge region is allowed to corotate immediately into the nightside region [6] (part (b), Fig. 3) thereby increasing the L-value of the plasmapause in this region. This increase in plasmapause radius cannot immediately take place on the dayside but first occurs in the nightside region and subsequently corotates into the dayside. Hence, it is expected that the nightside sector and not the dayside sector is able to react quickly to changes in magnetic activity and consequently appears to be the formative region for the dayside plasmapause as observed.

3.2. Effects of Dayside Filling on the Plasmasphere Profiles

After the steep gradient or plasmapause position has been determined during corotation through the nightside, the flux tube drifts past dawn where filling of the plasmasphere from the ionosphere begins. During the day this filling occurs both inside and outside the plasmapause position.

Fig. 4 illustrates the effect of filling on the plasmapause profile, assuming a constant upward flow of 3×10^8 ions cm^{-2} s^{-1} at 1000 km [5]. As an example, consider filling following a period of steady quiet to moderate activity. In this situation the region inside of boundary b in the figure is full of plasma. The dawn profile at time T_1 is shown in the small sketch. This shows the plasmapause position at about $L = 5$ with a plasmatrough density of 0.5 ions cm^{-3} which is typical of the densities measured by OGO 5. As the flux tubes drift across the dayside (T_2, T_3, T_4), filling from the ionosphere causes an accumulation of thermal plasma in regions A, B and C. This accumulation is most easily seen in the trough region where background densities are low. Beginning at about 1500–1600 LT, the convective flow direction (dotted line) causes the flux tubes in region C to be peeled away from the main plasmasphere and convected sunward to the magnetopause. This results in the loss of the newly accumulated trough plasma and the maintenance of the plasmapause at boundary b which has expanded to $L \geqq 6$ in the bulge region (profile T_5). The flux tubes inside boundary b retain their plasma (since the flux tubes remain closed) and continue to rotate into the bulge region. After crossing the dusk meridian, part of this plasma is then available to supply the night-

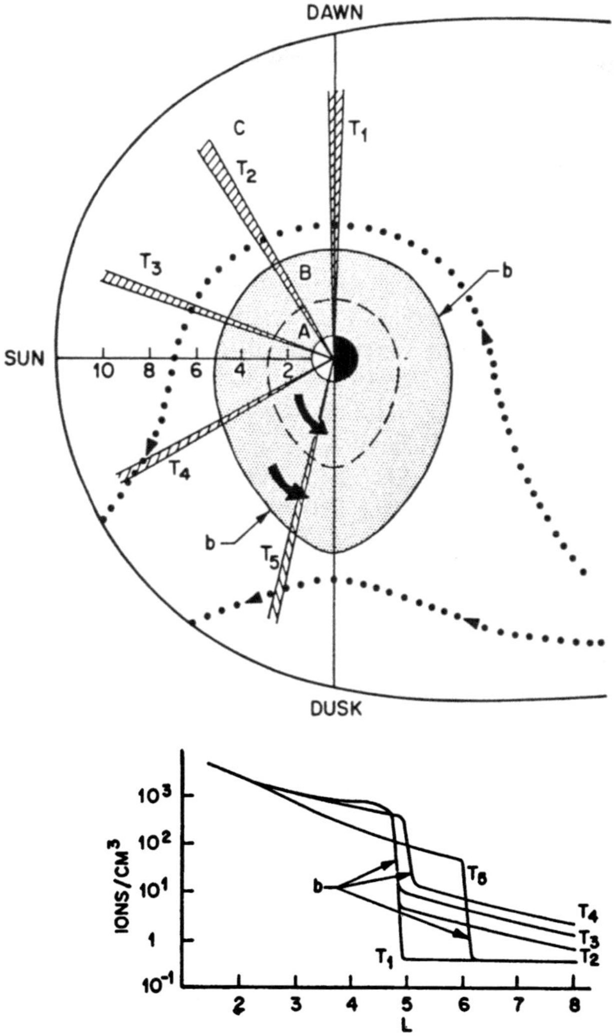

Fig. 4. A sketch showing the filling of the plasmasphere and plasmatrough during the day. The lower portion of the figure shows the changing density profile during corotation across the dayside.

time ionosphere. Filling and loss in this manner should continue as long as there is no change in magnetic activity.

On days of steady high magnetic activity the filling and loss will be similar to the above example except that only the plasma inside boundary a will be retained and the newly accumulated plasma in both regions B and C will be lost through convection to the magnetopause. The predicted effects of dayside filling from the ionosphere have been observed by the OGO 5 spacecraft in the afternoon-dusk sector [5].

This system of dayside filling and peeling away should result in a constant ambient plasma loss near the afternoon magnetopause. In fact, large masses of plasma, which are probably detached from the main body of the plasmasphere, have been observed in the afternoon-dusk sector by OGO 5.

3.3. Regions of Detached Plasma in the Dayside Plasmatrough

Examples of detached plasma regions have been shown in previous work [3, 5]. In each of these cases the H^+ densities in the detached regions exceeded 10 ions cm^{-3}. Detached plasma is found to occur primarily in the afternoon-dusk sector with very few detached regions in the nightside region [5]. This pattern of occurrence is in direct agreement with the prediction of the convection flow diagram. Recall in Fig. 1 that *only* on the dayside are the tubes of plasma carried *away* from the plasmasphere. This peeling away process is most effective in the afternoon-dusk sector where the flow lines turn almost perpendicular to the plasmasphere surface. On the nightside the motion of the tubes of plasma is always inward preventing occurrences of detached plasma regions in agreement with observations.

3.4. Storm Effects on the Plasmasphere Size and Shape

The amount of plasma that is lost through afternoon convection can be drastically increased during magnetic storm periods. At these times when the level of magnetic activity is greatly increased, the sunward convecting plasma region in the afternoon consists both of newly accumulated plasma from the dayside upflow and of plasma which has been peeled off the outer edge of the dayside plasmasphere in the afternoon sector and in the bulge region.

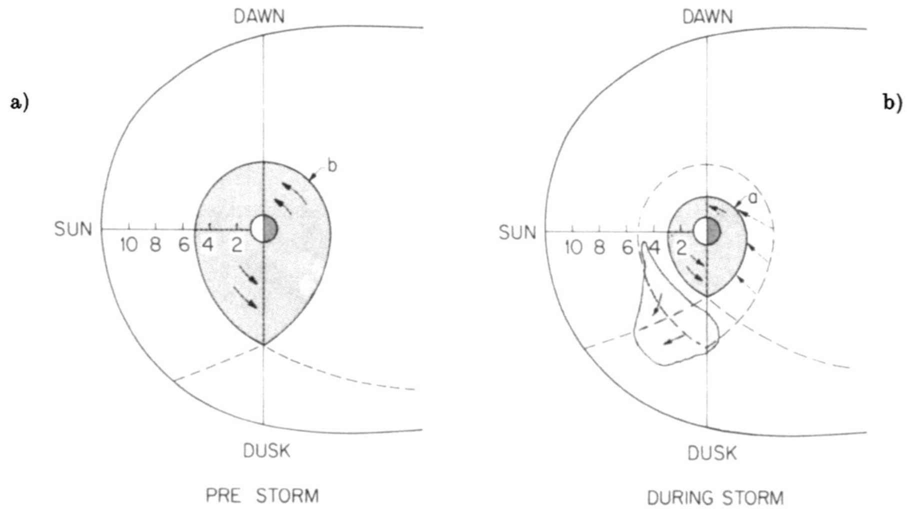

Fig. 5. A sketch showing the effect of a magnetic storm on the size of the plasmasphere. The large prestorm plasmasphere is reduced in size by dayside peeling and bulge detachment and by inward convection on the nightside.

Fig. 5 shows a sketch of the peeling and detachment process which occurs in the plasmasphere during a storm. This sketch is based on the flow direction changes shown earlier in Fig. 2. Part (a) of Fig. 5 shows the plasmasphere during the quiet period before the storm. After the storm onset, part (b) shows how the outer edge of the plasmasphere is separated in the afternoon and bulge sectors and is convected forward. At the same time the plasmapause is pushed inward on the nightside. This sequence of processes predicted by the convection model *can greatly reduce the size and plasma content* of the plasmasphere and at the same time can move regions of very high thermal density through the afternoon-dusk plasmatrough. Both these effects have been observed by OGO 5 [3, 4].

References

[1] W. I. Axford and C. O. Hines, Canad. J. Phys. **39**, 1433 (1961).
[2] L. D. Kavanagh et al., J. Geophys. Res. **73**, 5511 (1968).
[3] C. R. Chappell et al., J. Geophys. Res. **75**, 3848 (1970).
[4] C. R. Chappell et al., J. Geophys. Res. **75**, 50 (1970).
[5] C. R. Chappell et al., J. Geophys. Res. **76**, 7632 (1971).
[6] D. L. Carpenter, J. Geophys. Res. **75**, 3837 (1970).

Space Research XII — Akademie-Verlag, Berlin 1972

ANOMALOUS ORBITAL ACCELERATIONS OF THE PAGEOS SPACECRAFT

D. E. SMITH[a] and K. E. KISSELL[b]

[a] NASA Goddard Space Flight Center, Greenbelt, Md, USA
[b] Aerospace Research Laboratories, AFSC, Wright-Patterson Air Force Base, Ohio, USA

Observational studies of PAGEOS have indicated that the balloon satellite experiences very large periodic accelerations whenever the orbit intersects the earth's shadow. In 1969 the amplitude of these anomalous accelerations was about 10^{-4} revolution per day^2 with a period of 10–15 days and caused along-track deviations of the spacecraft from its expected position of 20–40 km. It has been suggested that these accelerations are abnormal solar radiation pressure perturbations caused by the non-sphericity of the spacecraft and its rotational motion on its axis.

During three months in 1969 the dynamical behavior of Pageos on its axis and the detailed shape of the balloon were determined by the USAF Aerospace Research Laboratories at Wright-Patterson AFB from photometric observations by new methods. These studies showed that Pageos is not spherical but ellipsoidal with major and minor axes approximately 34 m and 27 m respectively and rotating about a minor axis which is itself precessing.

These results on the dynamical behavior of Pageos have been used in a new theory of radiation pressure perturbations developed at Goddard Space Flight Center that includes the effects of spacecraft attitude, rotational behavior and specular and diffuse reflection and have been used to predict the acceleration of Pageos for three months in 1969. The predictions, based on the observed rotational motion, a specular reflectivity of 85% and a spacecraft mass of 55 kg, show that the new theory, except for a possible phase difference, computes the observed anomalous accelerations of Pageos to about 10^{-5} revolution per day^2, or better than 10%. This is believed to be one of the first demonstrations of coupling between the rotational motion of a spacecraft on its axis and the dynamics of its orbit.

1. Introduction

Satellites of large cross-sectional area-to-mass ratio are known to be strongly perturbed by the forces of solar radiation pressure. Such radiation pressure forces indirectly caused the demise of Echo 1 in 1968. Studies [1, 2] of the behavior of two other inflation-deployed spacecraft, the 2.5-m diameter sphere 1963-30D (Dash 2) and the 30-m diameter sphere 1966-56A (PAGEOS), both in high polar orbits, have demonstrated that our present theoretical treatment of radiation pressure indeed goes far in explaining the important secular and long period changes to the orbit. However, these same studies also indicate that both balloon spacecraft undergo accelerations that cannot be explained by the normal treatment of radiation pressure. Neither can these anomalous perturbations be explained by earth-reflected radiation pressure, air drag or luni-solar gravitational effects. Since these satellites were launched for the study of high-altitude air drag and for geodetic position, it is important that their orbital motions be well understood. These anomalous perturbations, recognized in 1969, are evidenced by regular motion of the spacecraft alternately ahead of, and then behind, the position in its orbit to be expected on the basis of the most complete, normal theory. Such accelerations affect, therefore, the mean motion of the spacecraft in its orbit.

This paper reports on the perturbations in mean motion predicted by a new theoretical treatment of radiation-pressure which accounts for non-isotropic scattering of solar radiation from a rotating ellipsoidal body and a comparison of these perturbations with the observed mean motion of the Pageos spacecraft during a time interval when the rotational dynamics had been determined by a new method, namely, photoelectric photometry. The theory necessarily incorporates the orientation for, and precessional movement of, the rotation axis of the satellite. This development is similar to that given by Shapiro [3] for the radiation pressure effects on a cylindrical satellite. We will show that there exists a coupling between the spin and orbital dynamics of the low mass-to-area satellites if their scattering is non-isotropic and that the effect is noticeable in the orbital behavior.

2. Discussion of the Observed Perturbation

A polar orbit at an altitude of about 4000 km intersects the earth's shadow for a period of approximately three months followed by an almost equal period of full sunlight until the orbit again enters the earth's shadow as the earth moves around the sun. This behavior is observed for both Dash 2 and Pageos. During the all-sunlight period, in agreement with theory, virtually no long-period or secular mean-motion perturbations of these spacecraft were detectable in these studies. As soon as the orbits entered the shadow, however, a large perturbation due to solar radiation pressure was observed as expected. Unfortunately, the character of the acceleration of both spacecraft, during every shadow period, differed significantly from what was anticipated. In addition to the normal radiation pressure perturbations, there appeared to be a very large periodic perturbation with an amplitude nearly as large as the normal perturbation. The period of the effect on Dash 2 was many weeks during the first shadow period after launch and steadily decreased so that four years after launch, the period of the anomalous acceleration for Dash 2 had reached two weeks. Three years after launch the period for Pageos was about 11 days.

Since the Pageos launch in 1966, the OPOS Observatory of the USAF Aerospace Research Laboratories at Wright-Patterson AFB, Ohio, has been making photoelectric brightness measurements of the spacecraft [4] as part of a program to explore the capabilities of passive optical photometry for the remote-sensing of spacecraft properties. By utilizing the variations with time of the sunlight scattered from the Pageos aluminized outer surface, it has been possible to deduce, during certain periods, the spin rate, spin-axis orientation and precession, and the shape of the balloon [5]. These studies indicate that the Pageos balloon has gradually become prolate and that it is spinning about a minor axis with a period varying between 190 and 350 seconds. In addition, the spin axis is found to be precessing at varying rates about an axis that always points toward the sun. Fig. 1 shows the location of the rotation axis during the period August–December 1969. It is to be noted that the spin axis precessed very rapidly, at 25 deg day^{-1} during August, increasing to 45 deg day^{-1} at the beginning of November, then dropping again to 25 deg day^{-1} in late November. The angle between the spin and precession axes was about 65°. The physical cause of this dynamical behavior of Pageos on its axis remains unexplained,

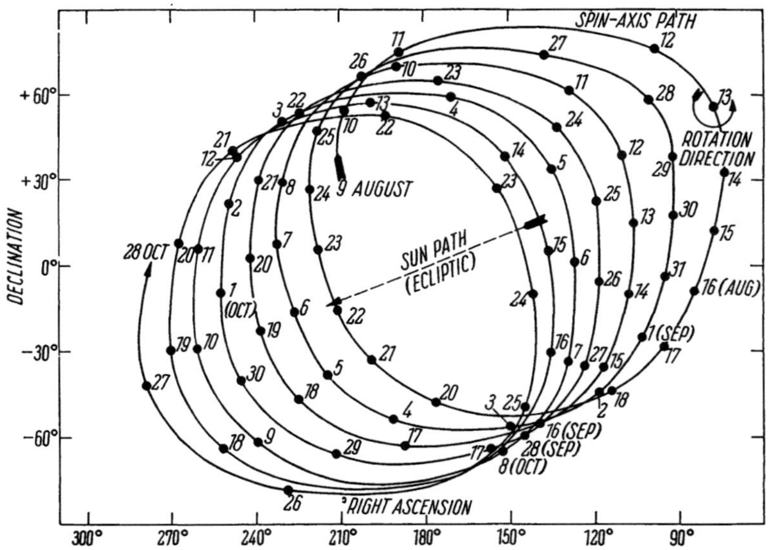

Fig. 1. Smoothed spin-axis positions, 9 August–28 October 1969.

but, because its precession axis appears locked to the sun, the forces of solar radiation pressure might well be the driving mechanism.

3. Theoretical Analysis

The possibility has been examined theoretically [6] that the anomalous orbital accelerations are coupled with the non-sphericity of the balloon and its observed rotational behavior. In particular, a theory has been elaborated for the effect of solar radiation pressure on the orbit of an ellipsoidal-shaped spacecraft, spinning about an axis that is precessing, and including the effects of both diffuse and specular reflection of sunlight from the satellite's surface. The essential modification made to the normal treatment of radiation pressure is that account was taken of the direction in which sunlight is reflected from the satellite. In the normal treatment it is assumed that the reflected radiation is symmetric about the satellite-to-sun line so that the acceleration is always directed away from the sun: this is obviously true for spherical satellites but is not true, in general, for an ellipsoid. Under these circumstances there is a "sideways" component to the radiation pressure force which causes additional perturbations of the orbit, analogous to the lateral accelerations of a tacking sailboat.

4. Application of the Theory

This theory has now been tested numerically against orbital data obtained from the Smithsonian Astrophysical Observatory. The positions of the spin axis of Pageos obtained by the Aerospace Research Laboratories, together with their estimate of the prolate shape of the spacecraft and of its reflective properties were used to compute the acceleration of the spacecraft between

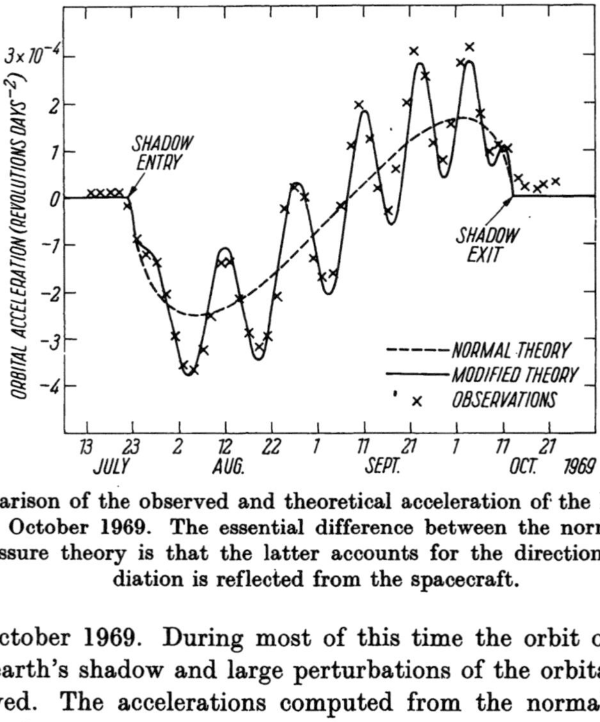

Fig. 2. Comparison of the observed and theoretical acceleration of the Pageos spacecraft from July to October 1969. The essential difference between the normal and modified radiation pressure theory is that the latter accounts for the direction in which the radiation is reflected from the spacecraft.

July and October 1969. During most of this time the orbit of Pageos intersected the earth's shadow and large perturbations of the orbital mean motion were observed. The accelerations computed from the normal and modified theories together with the acceleration observed by the Smithsonian Astrophysical Observatory are shown in Fig. 2. The spacecraft parameters used in the computation of the balloon's behavior were: polar radius 13.0 m; major equatorial radius 17 m; minor equatorial radius 13.5 m; mass 55 kg; specular albedo 0.85, diffuse albedo 0.0.

The agreement between the theory and observations is clearly evident in Fig. 2. It should be noted that the theory does not indicate any perturbation of the acceleration when the orbit is fully in sunlight, a result in agreement with the theory for a spherical satellite. When the orbit is partially shadowed, an oscillation caused by the precession of the spin axis about the sun is clearly evident as is the steadily increasing rate of precession. The differences between the theory and the observations suggest that the root-mean-square of the residuals is about 10% of the amplitude of the perturbation. While the amplitudes agree quite well, there is some indication of a phase difference during the month of September. Except for this phase difference, the residuals appear to be reasonably random and comparable with the slight accelerations observed during periods when the orbit is in complete sunlight.

5. Conclusions

We believe the agreement obtained between the new theory and the observations is one of the first demonstrations of coupling between the rotational dynamics of a spacecraft about its center of mass and the orbital dynamics of the spacecraft about the earth. The perturbing acceleration produces large along-track oscillations of the Pageos spacecraft which, during July to September 1969, had an amplitude of 20–40 km with a period ranging from 10 to 15 days. Such large perturbations have considerable impact on the quality of orbit determination and prediction. For the topocentric observer these were errors of 0.5–1.0 degree in apparent position.

A problem deserving further work is the torquing mechanism of the rotation and precession of these spacecraft. Only a beginning exists in the suggestion by Mar and Vigneron [7] of the Canadian Defense Research Telecommunications Establishment of a thermal-distortion mechanism to allow torquing of the Echo 2 spacecraft.

References

[1] K. Fea, in: Dynamics of Satellites 1969, Springer-Verlag, New York 1970 (p. 295).
[2] K. H. Fea and D. E. Smith, Planet. Space Sci. 18, 1499 (1970).
[3] I. I. Shapiro, in: Dynamics of Satellites 1962, Academic Press, New York 1963 (p. 271).
[4] R. C. Vanderburgh, in: Proc. University of Miami Symp. on Optical Properties of Orbiting Satellites, Aerospace Research Labs Report ARL 69-0078, May 1969 (p. 201).
[5] R. C. Vanderburgh and K. E. Kissell, Planet. Space Sci. 19, 223 (1971).
[6] D. E. Smith and K. H. Fea, NASA Goddard Space Flight Center X-Document 550-70-92 (1970).
[7] J. Mar, and F. R. Vigneron, in: Trajectories of Artificial Celestial Bodies, Springer-Verlag, New York 1966 (p. 151).

Space Research XII — Akademie-Verlag, Berlin 1972

NEAR SUN OBSERVATIONS OF THE SOLAR WIND

P. S. Callahan, P. F. MacDoran and A. I. Zygielbaum

Jet Propulsion Laboratory, California Institute of Technology, Pasadena, Calif., USA

A radio technique which exploits the opposite changes of group and phase velocity in a dynamic plasma has been used to study the solar wind during May, June and July 1970 using the Mariner 6 and 7 spacecraft after their superior conjunctions. The spectrum of the solar wind turbulence in the range 1.6×10^{-4} to 2×10^{-3} Hz has been estimated. The average spectrum falls as $f^{-1.6}$ from 1.6×10^{4} to 6×10^{-4} Hz and then becomes flat. The spectrum shows significant changes from day to day at the low frequencies.

1. Introduction

Following their encounters with Mars in August 1969, both ranging and Doppler transponder signals from Mariners 6 and 7 were received by means of a new, more sensitive ranging system (Fast Acquisition Ranging [1]) at the Goldstone Tracking Station of the Deep Space Network. The improved ranging capability made it possible to obtain range measurements around the Mariners' entire orbits.

The plasma dynamics were measured with a technique known as **D**ifferenced **R**ange versus Integrated **D**oppler (DRVID) which exploits the opposite changes of group and phase velocity [2] as the plasma density changes along the radio raypath. The method is sensitive only to changes in the columnar content, not to the total content.

Between 21 May and 2 July 1970, eleven blocks of data exceeding $4^1/_2$ hours in length and six exceeding $3^1/_2$ hours were obtained. From a preliminary study of these data we have estimated the spectrum of the solar wind turbulence in the frequency range 1.6×10^{-4} to 2.0×10^{-3} Hz.

2. Measurement Technique

The DRVID method is based on the difference in apparent path length as measured by group and phase velocity in a charged medium. In a changing plasma the group and phase velocities change by opposite and roughly equal amounts governed by the change in the electron density [2]. Range code modulation propagates at the group velocity, Doppler at the phase velocity. If the apparent range difference between times t_1 and t_2 is found by differencing range values taken at those times, and by integrating the velocity inferred from the Doppler data over the time interval, the two values will in general not agree. Their difference (hence the name of the technique) is proportional to the change in columnar content in the interval $t_1 - t_2$. As is shown in [2], DRVID(t), as produced by the Deep Space Tracking system, is given in cgs units by

$$\mathrm{DRVID}(t) = 8.06\times10^{7} f^{-2} \int_{t_0}^{t} \int_{\text{raypath}} [d/dt\, N(s, t')]\, ds\, dt' ,$$

where DRVID(t) = two-way range change (cm), f = radio frequency (Hz), $N(s, t')$ = space and time variable electron density (cm^{-3}) and t_0 = arbitrary starting time of observation. For the case of S-band tracking ($f = 2300$ MHz), the change in columnar content, $\Delta I(t)$, referenced to t_0 is given by

$$\Delta I(t) = 6.007\times10^{10}\, \mathrm{DRVID}(t) \qquad (\mathrm{cm}^{-2}) ,$$

where DRVID(t) is in centimeters.

Because of the differencing done to form DRVID the total content information is lost and only the change relative to the unknown content $I(t_0)$ is found. However, there are advantages gained from the differencing: Any effects which affect the group and phase velocities in the same way, such as the earth's troposphere and general relativity, are cancelled out. The relative motion between spacecraft and tracking station is removed. The intrinsic removal of the orbital information from DRVID makes it a far more reliable and easily used indicator of charged particle effects than are Doppler residuals formed from a computed orbit.

The precision of the DRVID measurements depends mainly on the received ranging power. For ranges of about 2.5 AU as at the Mariners' superior conjunctions the received power is −190 dbm. The Fast Acquisition Ranging system, using an integration time of 2 minutes, measures the round trip range with a standard deviation of 5 meters per measurement at such a received power level. This uncertainty is also present in DRVID so that the smallest detectable columnar change under the above conditions is $2\times10^{13}\ \mathrm{cm}^{-2}$. The Doppler uncertainty is negligible by comparison.

For observations within a few degrees of the sun, no data are available to check the DRVID measurements. However, it is possible to check the internal consistency of the data to guard against gross equipment failures. The test involved integrating Doppler (i.e. velocity) residuals from an orbit solution to produce the range change as a function of time. The integrated residuals were differenced with the DRVID data, which is the range change caused by charged particles. The result of such a differencing is to preserve only orbital computation errors if all or most of the plasma effects have been removed. Except in some cases of obvious equipment failure which were discarded, the only effects were diurnal ones caused by errors in the orbit solution. It is presumed that space plasma variations do not have such a diurnal character and will not be confused with orbit errors.

The data used in this discussion were not corrected for the earth's ionosphere. However, the changes of content in the ionosphere are generally much smaller than those observed in the DRVID data. Furthermore, for the data near superior conjunction the measurements were made through the nearly constant noon-time ionosphere, so that changes in columnar content should be small, generally less than 3.5×10^{13} cm^{-2}.

3. Power Spectrum of DRVID Data

The power spectrum of the solar wind turbulence is a subject of much current interest. Results on the spectrum of the magnetic field and velocity fluctuations have been reported by Coleman [3] and others. Recently Intriligator and Wolfe [4] and Goldstein and Siscoe [5] have presented data on the spectrum of proton density fluctuations near 1 AU. Because of the long raypath involved and its nearness to the sun, the Mariner Extended Mission provided a unique opportunity to observe solar wind turbulence in a new domain.

The power spectrum of data may be found by taking the Fourier transform of its autocorrelation function. Because our data are discrete (points every 2 minutes with none missing) and of limited duration, the computation of the power spectrum must be approached with care. The frequencies that can be investigated are limited at the upper end by the Nyquist frequency, 4.17×10^{-3} Hz, and at the lower end by the data span to about 1.6×10^{-4} Hz. The frequency resolution is limited by the number of correlation lags which can be performed while leaving a reasonable number of degrees of freedom (>12). The high frequency spectrum was cut off at one-half the Nyquist rate to avoid problems with aliasing. To reduce the high frequency noise in the data a moving average over two points was performed; it was found to have almost no effect on the resultant power spectrum.

The Fourier transform was performed with formulae given by Bendat and Piersol [6]. The effect of the transform on white noise was investigated and found to be uniform within 10% at all frequencies of interest.

DRVID data measure the columnar content changes and not the local density fluctuations. Therefore the relationship between the power spectrum of columnar content data and that of local fluctuations was investigated. The analysis was simplified by assuming that the spectrum of the turbulence moving with the solar wind is Gaussian. If such a Gaussian spectrum is convected across a raypath which goes near the sun, and if material is distributed about the sun as inverse heliocentric distance squared [7], the spectrum found from the columnar content fluctuations is proportional to the spectrum of the density fluctuations in the medium. Although the observed spectra are not precisely Gaussian, in this preliminary analysis we assume that this proportionality still holds. A more realistic analysis is in preparation.

Fig. 1 shows the composite spectrum from all 17 blocks of data, 21 May to 2 July, which had more than 100 data points. Also shown separately are the composite spectrum from the 11 long blocks of data and from the 6 shorter ones. During the period these data were taken the raypath's points of closest approach to the sun varied from about 12 to 55 solar radii. The composite spectra were formed by simply averaging the points at the given frequency. The two additional points in the spectrum of the long blocks come from spectra done at slightly higher resolution.

Fig. 2 shows four spectra typical of those forming the composite. The individual spectra have about 15 degrees of freedom so that the 90% confidence interval is approximately a range of 0.55–1.50 about any point. The power observed at low frequencies varies markedly from record to record, by as much as a factor of 10^2, while the power of high frequencies is more or less constant. There appears to be no easily identifiable pattern in the low frequency variations as a function of time.

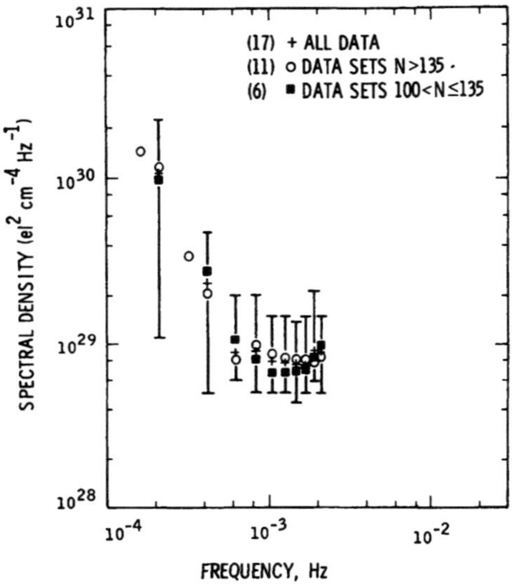

Fig. 1

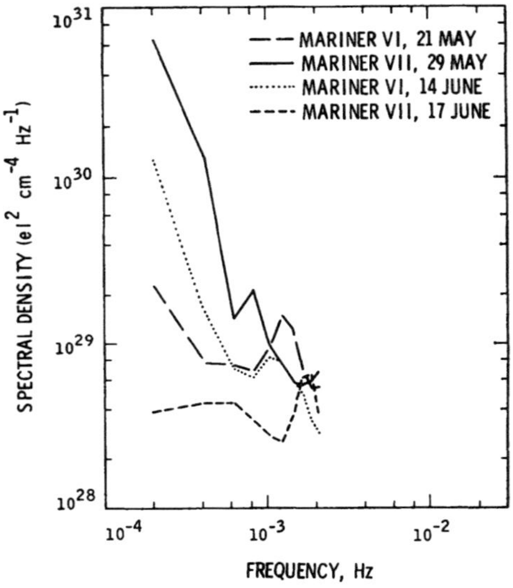

Fig. 2

Fig. 1. Composite power spectrum of DRVID data during May, June and July 1970 ×, all 17 blocks of data; ○, longest 11 blocks; ■, 6 shorter blocks.

Fig. 2. Four examples of individual record spectra during May, June and July 1970. — —, Mariner 3, 21 May; ———, Mariner 7, 29 May; ········, Mariner 6, 14 June; – – – –, Mariner 7, 17 June.

The composite spectrum, on the other hand, is quite well behaved. It falls along a straight line of slope ~ -1.6 from 1.6×10^{-4} Hz to 6×10^{-4} Hz and then rather abruptly becomes flat. The spectral density at which the flattening occurs is more than three times that expected from the tracking system noise. The same shape for the composite spectrum is obtained if straight lines are fitted to the low and high frequency data of each block and these slopes averaged to produce a two-piece linear spectrum. These results agree quite closely with those of Goldstein and Siscoe [5] for the power spectrum of proton density fluctuations, though the flattening at the highest frequencies is much more marked here. From examination of the entire spectrum up to the Nyquist frequency it is believed that aliasing is not the cause of the flattening.

The assignment of errors to the composite spectrum is difficult. The most important cause of uncertainty is the real variations of the power density. The limits shown are such that they encompass 68% of our data. They are quite similar to those derived formally by the procedure of Blackman and Tukey [8].

4. Conclusions

DRVID data offers the possibility of studying many phenomena in the solar wind. The method is especially attractive because it requires no special equipment aboard the spacecraft and the long raypath allows one to probe near the sun at spacecraft superior conjunction.

DRVID data from May to July 1970 shows a power spectrum of solar wind fluctuations falling linearly from 1.6×10^{-4} Hz to 6×10^{-4} Hz and then abruptly flattening to 2×10^{-3} Hz. The spectrum is highly variable at low frequencies but is fairly constant at frequencies above 6×10^{-4} Hz.

5. Acknowledgments

The authors would like to thank Dr. J. R. Jokipii for many helpful discussions and suggestions.

References

[1] P. F. MacDoran and W. L. Martin, Space Programs Summary 37-62, Jet Propulsion Laboratory, Pasadena, Calif. 1970 (Vol. II, p. 34).
[2] P. F. MacDoran, Space Program Summary 37-62, Jet Propulsion Laboratory, Pasadena, Calif. 1970 (Vol II, p. 28).
[3] P. J. Coleman, Astrophys. J. **153**, 371 (1968).
[4] D. Intriligator and J. H. Wolfe, Astrophys. J. Lett. **162**, L187 (1970).
[5] B. Goldstein and G. L. Siscoe, Asilomar Conference on the Solar Wind, March 1971.
[6] J. Bendat and A. Piersol, Measurement and Analysis of Random Data, Wiley, New York 1966.
[7] D. O. Muhleman et al., Proc. CalTech Relativity Conference, November 1970.
[9] R. B. Blackman and J. W. Tukey, The Measurement of Power Spectra, Dover Publications, New York 1959.

Space Research XII — Akademie-Verlag, Berlin 1972

SOLAR COSMIC RAY BURSTS IN NOVEMBER–DECEMBER 1970 ACCORDING TO DATA FROM VENUS 7 SPACE PROBE AND LUNOKHOD 1 STATION

S. N. Vernov, N. N. Kontor, G. P. Lyubimov, N. V. Pereslegina and E. A. Chuchkov

Moscow State University, Moscow, USSR

An increase in low-energy solar cosmic rays began on 5 November 1970 and lasted for ~10 days. Protons with energies of 1–5 MeV, 4–12 MeV and higher than 30 MeV as well as electrons with energies higher than 0.20 MeV were detected. Two intensity peaks on 5 and 7 November 1970 were observed. The character of the change in the intensity, energy spectrum and anisotropy indicate the successive observation of the diffusive and then convective (behind the shock wave front) movement of solar cosmic rays from the chromospheric flare of class 3B which occurred on 5 November 1970 on the eastern part of the solar disk (30° E). The next solar cosmic ray increase began on 11 December 1970 and lasted for more than 10 days. The increase was of complex form and was caused by a series of solar flares which occurred on 10–12 December 1970 on the eastern and central parts of solar disk. At the close of 14 December the cosmic ray intensity decreased abruptly and then decayed exponentially until 22 December 1970. The moment of the abrupt intensity decrease corresponds to the end of the magnetic storm on the earth.

A fairly high proton activity was observed on the sun during the Venus 7 flight [1] the level and intensity variations of which were very close to those of the activity in 1969 [2]. Intensity increases for the 1–5 MeV protons were observed for about 75% of the period. The most considerable intensity and duration of the increases were recorded in August, November and December 1970. Apart from these considerable increases caused by powerful solar flares many smaller increases due to subflares and quasi-stationary fluxes from solar active regions were observed. After the considerable increase in solar cosmic ray intensity in the middle of December 1970 the mean level of the proton activity of the sun began to decrease rapidly as derived from the Lunokhod 1 data [3].

The solar cosmic ray increase in November 1970 was recorded by all the cosmic ray detectors of Venus 7 [1]. All phases were observed: increase, peak and decrease. This event is of interest because of the presence of two distinct peaks for the protons and electrons. Fig. 1 illustrates the event. The figure does not present the intensity curves for the 1–5 MeV and 4–12 MeV protons moving to the sun but shows instead the anisotropy values A_1 and A_2 for the protons with energies of 1–5 MeV and 4–12 MeV respectively. As in the earlier works [1, 4] the anisotropy, A, was determined from the formula

$$A = (n^+ - n^-)/(n^+ + n^-) \cdot (100\%)$$

where n^+ and n^- are the intensities of the protons moving from and to the sun respectively. To analyse the solar cosmic ray intensity increases the power exponents γ of the integral energy spectra of the form $N(E) \sim E^{-\gamma}$ were found: γ_1 for the energy range of the protons moving from the sun higher than 1 MeV and 4 MeV, γ_2 for the same energy range of the protons moving to the sun, γ_3 for the energy range of the protons moving from the sun higher than 4 MeV and for the 30 MeV protons, and γ_4 for the same energy range of the protons moving to the sun. Fig. 1 presents the values of γ_1 and γ_3 only since γ_2 and γ_4 are similarly time dependent and differ in magnitude only, as will be discussed below. The intensity values for the electrons with energies higher than 0.2 MeV presented in Fig. 1 have been obtained on the basis of the method of differences and are shown only to estimate roughly the flux of these

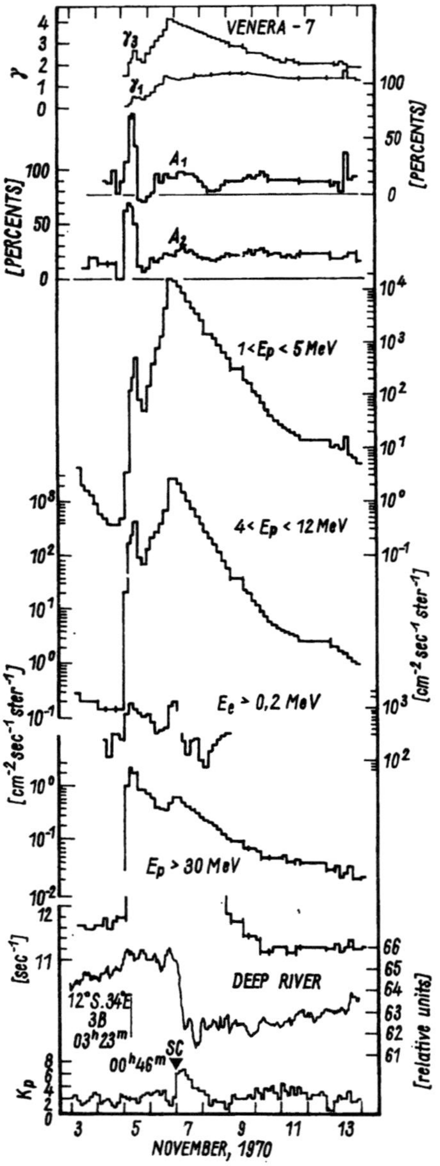

Fig. 1. Solar cosmic ray increase on 5–14 November 1970 according to Venus 7 data:

(1) the values of spectrum exponents γ_1, and γ_3;
(2) the values of anisotropy A_1 and A_2;
(3) intensity of the 1–5 MeV protons moving from the sun;
(4) intensity of the 4–12 MeV protons moving from the sun;
(5) intensity of electrons with energies higher than 0.2 MeV moving from the sun;
(6) intensity of solar protons with energies higher than 30 MeV;
(7) counting rate of the Geiger counter detecting solar and galactic protons with energies higher than 30 MeV;
(8) Deep River neutron monitor data;
(9) data on the Kp index of geomagnetic disturbance labelled with magnetic storm sudden commencements (SC);
(10) moments of solar flares (vertical bars) labelled with the time of peak, coordinates and class.

particles. Two curves are presented for the protons with energies higher than 30 MeV: the counting rate of a Geiger counter detecting both solar and galactic cosmic rays and the intensity of protons of solar origin only This separation was made by deleting the counting rate of the Deep River neutron monitor [5, 6] from the Geiger counting rate normalized over amplitude (relative) and time (including the shock wave delay).

The increase began on 5 November 1970 and lasted at least until 14 November 1970 and was caused by the powerful solar flare of class 3B which occurred on 5 November 1970 at 0323 (peak) on the eastern part of the solar disk (12°S, 34°E) in the active region No. 11019. The flare lasted for about 4 hours and was accompanied by types I, II (class 3), III, and IV radio bursts, and X-ray bursts with peak energy 0.23 erg cm^{-2} s^{-1} in the 1–8 Å range [5]. Ionospheric disturbances of class 3 were simultaneously observed on the earth [5]. At 0046 on 7 November 1970, i.e. 45 hours 23 min after the solar flare peak the sudden commencement (amplitude of 50 gammas) magnetic storm (maximum $Kp = 7 -$) was observed on the earth [5]. The Deep River neutron monitor [5, 6] recorded a Forbush decrease of 6% depth with rapid decay and very slow recovery of the galactic cosmic ray intensity (Fig. 1).

The solar cosmic ray increase on Venus 7 began on 5 November 1970 at 0443 $\pm$ 2 hours. The intensity peak for the protons with energies higher than 30 MeV is delayed by 5 $\pm$ 2 hours with respect to the solar flares and by 9 $\pm$ $\pm$ 2 hours for the protons with energies of 1–5 MeV and 4–12 MeV. The character of the anisotropy (and intensity) indicates that the 4–12 MeV protons arrived somewhat earlier than the 1–5 MeV protons.

Let the diffusion coefficient $Q = r^2/6\,t$, transport path $\lambda = 3\,Q/V$, mean diffusion rate $V_d = r/t$, and anisotropy $A = \lambda/r$ be found on the basis of the time of observation of the intensity peak t, distance to the observation point r, and particle velocity V. Then we shall get for the protons with energies higher than 30 MeV and 1 MeV respectively: $Q = 1.4$ and 0.8×10^{21} cm^2 s^{-1}; $\lambda = 0.6$ and 1.7×10^{12} cm; $V_d = 0.7$ and 0.4×10^9 cm s^{-1} and $A = 5\%$ and 14%. It can be seen that the estimates of the isotropic diffusion parameters obtained are valid for protons with energies higher than 30 MeV only. For protons with energies higher than 1 MeV the anisotropy obtained experimentally is much higher than that calculated. This may be explained by low-energy proton penetration near the sun after the flare (either due to high initial density or through the loop magnetic fields of the corona) into the undisturbed magnetic fields of the neighbouring active regions and their subsequent movement over the long path. It may also be assumed that the interplanetary magnetic field was deformed (extended radially) by the shock wave and plasma from the solar flare. The neutron monitor and the galactic cosmic ray counter indicate a small Forbush decrease in the galactic cosmic rays prior to the commencement of solar cosmic ray increase, which may correspond to the passage through Venus 7 and the earth of a corpuscular stream co-rotating with the sun into a region of magnetic field lines on which the solar cosmic rays from the given flare exist.

On 6 November 1970 at 2036 $\pm$ 2 hours, i.e. 41 $\pm$ 2 hours after the flare on the sun, Venus 7 observed the second peak of the 1–5 MeV proton intensity.

The peak of the protons with energies of 4–12 MeV and higher than 30 MeV was observed on Venus 7 at 0036 on 7 November 1970, i.e. 45 $\pm$ 2 hours after the flare. The values of the peak intensities for protons of these energies moving from the sun are respectively 1.6×10^4 and 2.6×10^3 cm^{-2} s^{-1} sr^{-1}, and 0.7 cm^{-2} s^{-1} sr^{-1} for protons with energies higher than 30 MeV. The second peak anisotropy reaches the values A_1 and A_2 of about 25%. The intensity increases non-uniformly prior to the second peak: a steeper increase was observed from 1437 $\pm$ 2 hours on 6 November 1970. The intensity decrease (after the second peak) can be well approximated by an exponent with time constant 13 hours (for the 1–5 MeV and 4–12 MeV protons); after this (till 13 November 1970) the intensity decrease becomes gradually slower.

The above characteristics of the second cosmic ray intensity peak and almost simultaneous ground observation of the magnetic storm sudden commencement and a Forbush decrease in the galactic cosmic rays indicate that the arrival of solar cosmic rays carried behind the shock wave front was observed from Venus 7 on 6 November 1970. The shock wave should reach the Venus 7 earlier than the earth. If the moment of abrupt steepening of the intensity increase before the second peak of the 1–5 MeV protons is assumed to correspond to the moment of the shock wave arrival at Venus 7 the mean velocity of the shock wave movement between Venus 7 and the sun will be 986 $\pm$ 55 km s^{-1} (with a 35 $\pm$ 2 hour delay). The mean velocity of the shock wave propagation between the sun and the earth which corresponds to the delay of the magnetic storm commencement on the earth found earlier is 910 km s^{-1} and the mean velocity between Venus 7 and the earth is 636^{+160}_{-100} km s^{-1} (with a 10 $\pm$ 2 hour delay). The data on the plasma velocity on 7 November 1970 are given in [5]. The Pioneer 6 space probe located almost opposite the flare detected a 647 km s^{-1} solar wind; at the opposite point, on Pioneer 7 the velocity was 388 km s^{-1} and on Vela 5 it was 576 km s^{-1}. Hence a non-radial plasma stream [7] and inconsiderable shock wave deceleration [8] in the sun–earth direction were observed after the flare. According to the data from Pioneer 6 and Vela 5 the high-velocity plasma stream was detected for several days more. In accordance with this an increased geomagnetic disturbance and slow recovery of the galactic cosmic rays were observed on the earth till about 12 November 1970 (Fig. 1). The long duration of the flare in $H\alpha$ and the above characteristics of the interplanetary medium and cosmic rays may correspond to observation of the shock wave and extended gas piston as shown in [9, Fig. 39].

Thus, during the period of 5–8 November 1970 Venus 7 detected successively the diffusive and convective (behind the shock wave front) propagation of solar cosmic rays from the flare on the eastern part of the solar disk.

It should be noted that another two-peak intensity increase was observed in the electrons with energies higher than 0.2 MeV simultaneously with this event (Fig. 1). The peak intensities on 5 November 1970 at 0842 $\pm$ 2 hours (the first peak) and on 7 November 1970 at 0036 $\pm$ 2 hours (the second peak) were about 1.2 and 1.3×10^3 cm^{-2} s^{-1} sr^{-1} respectively. Thus, both electron peaks were observed simultaneously (within $\pm$2 hours) with the peaks of the protons with energies higher than 30 MeV. The character of the electron intensity change differs substantially from the character of the proton intensity change. This difference is due to the value of the Larmor radius of the particles and to the presence of inhomogeneities of various scales in the shock wave and behind its front.

There is additional information to be considered in the change in the anisotropy and spectrum (Fig. 1). The modulation depth for A_1 ($E_p = 1$–5 MeV) is more considerable than for A_2 ($E_p = 4$–12 MeV). This is associated with a smaller scattering and better localization of the lower-energy protons in the solar plasma inhomogeneities. The short duration of the first peak in the protons with $E_p = 1$–5 MeV and 4–12 MeV, the abrupt commencement of the increase and decrease of the intensity, the impulse character of the anisotropy are indicative of the instantaneous (less than 4 hours) character of the solar cosmic ray injection and subsequent movement of cosmic rays over a long path (of the order of 1 A.U.) [10]. During the whole increase the energy spectrum cannot be represented in the form of a power or exponential function for the whole proton energy range (higher than 1 MeV). Consider therefore the variations in the exponent of the power integral energy spectrum in two energy ranges, γ_1 and γ_3 (Fig. 1). The character of variations in γ_3 repeats well a dependence on the 4–12 MeV proton intensity variations, i.e. the spectrum becomes softer due to a more considerable increase of the 4–12 MeV proton intensity in the first and especially in the second peaks and becomes harder due to a more rapid decrease of these protons during the intensity decay. A similar, but expressed less distinctly, character of the change is observed for γ_1 only before the second peak of the solar cosmic ray intensity. The steep rise can be clearly seen at the shock wave front. After this γ_1 again increased slowly and reached peak values in the region of the plasma piston on 9 November 1970. In this case γ_1 is always less than γ_3. The greatest differences in the values of γ_1 and γ_3 correspond to the peak intensity of the 1–5 MeV and 4–12 MeV protons. We shall call this the "saturation" effect. This effect is expressed as a flatter spectrum in the low-energy range with increasing solar cosmic ray intensity. Using the generalized spectrum of [11, Fig. 2] one can determine the upper limit of possible solar cosmic ray fluxes for various energies. For protons with energies higher than 1 MeV this limit is somewhere in the region of 10^4 cm^{-2} s^{-1} sr^{-1}. The "saturation" effect may be explained in two ways. The first explanation is that an intensity increase in solar cosmic rays causes an increase in their energy density which may exceed the energy density of the magnetic field. In this case the volume occupied by solar cosmic rays will be increased and their intensity will decay. Thus, the balance of the energy densities of solar cosmic rays and magnetic field will serve as a non-linear mechanism limiting the solar cosmic ray flux. The second explanation is that when the solar cosmic rays are accelerated from the plasma their flux cannot exceed the plasma stream.

The next considerable increase in solar cosmic rays which began on 11–12 December 1970 was observed on Venus 7 until 15 December 1970 only (Fig. 2). According to the Lunokhod 1 data it ended about 22 December 1970 (Fig. 3).

Fig. 2 shows the intensity curves for the 1–5 MeV and 4–12 MeV protons moving to the sun, the integral spectrum exponents γ_2 and γ_4, the intensity of solar protons with energies higher than 30 MeV (obtained using the above method), the counting rate of the Geiger counter detecting both solar and galactic cosmic rays, data of the Kp index and from the neutron monitor [5, 6].

The increase is of a complex stepped form. According to the Venus 7 data it began at 0410 $\pm$ 2 hours on 12 December 1970 for the protons with energies of 1–5 MeV, 4–12 MeV, and higher than 30 MeV. The first intensity peak was

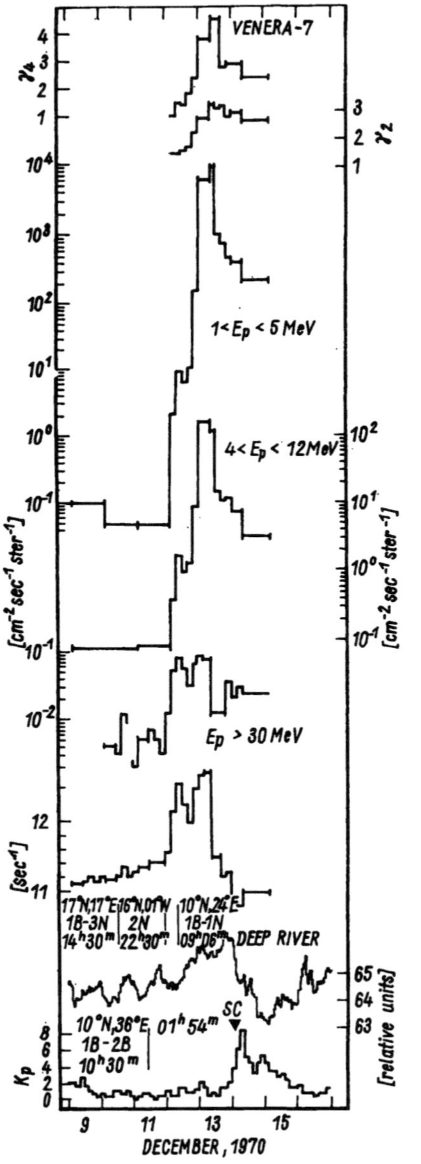

Fig. 2. Solar cosmic ray increase on 11–15 December 1970 according to the Venus 7 data:

(1) the values of spectrum exponents γ_2 and γ_4;
(2) intensity of 1–5 MeV protons moving to the sun;
(3) intensity of the 4–12 MeV protons moving to the sun;
(4) intensity of solar protons with energies higher than 30 MeV;
(5) counting rate of the Geiger counter detecting solar and galactic protons with energies higher than 30 MeV;
(6) Deep River neutron monitor data;
(7) data on the Kp-index of magnetic disturbance labelled with magnetic storm sudden commencements (SC);
(8) moments of solar flares (vertical bars) labelled with the time of peak, coordinates and class.

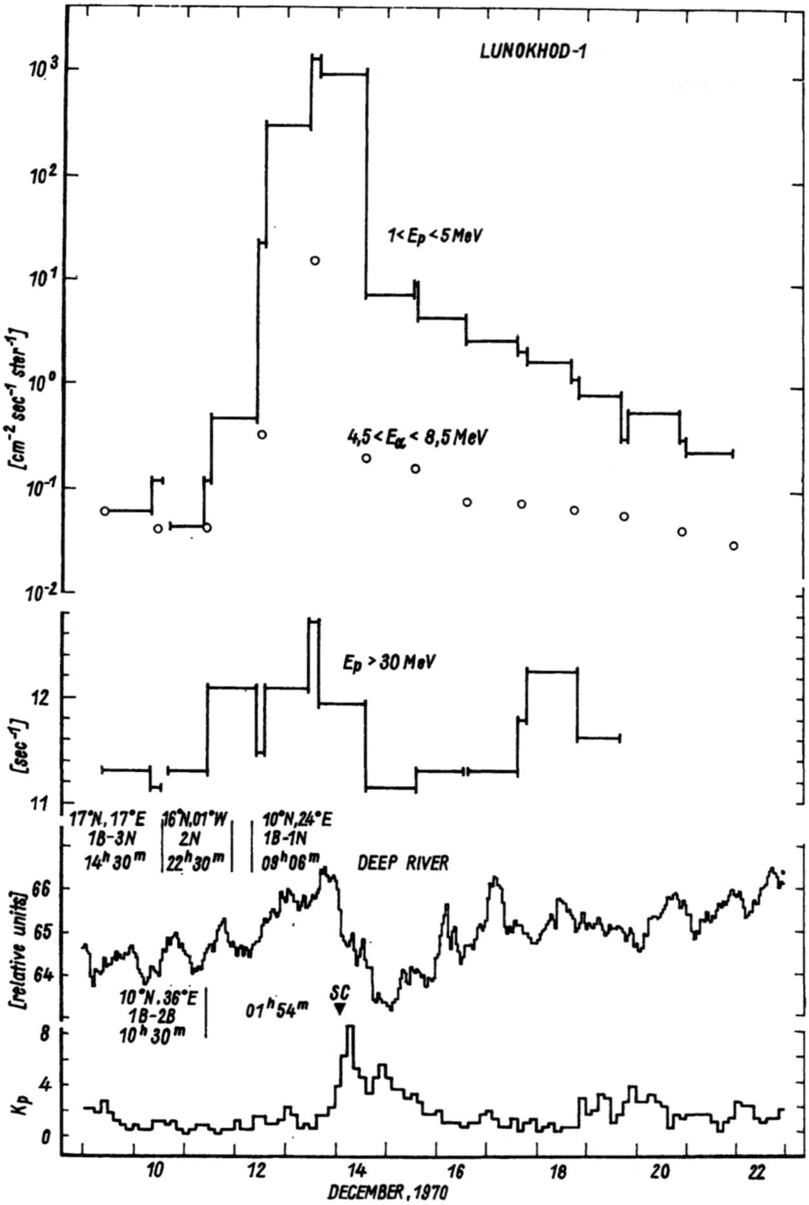

Fig. 3. Solar cosmic ray increase on 11–22 December 1970 according to the Lunokhod 1 data:

(1) 1–5 MeV proton intensity;
(2) 4.5–8.5 MeV α-particle intensity (open circles);
(3) counting rate of the Geiger counter detecting the protons with energies higher than 30 MeV;
(4) Deep River neutron monitor data;
(5) data on the Kp index of magnetic disturbance labelled with magnetic storm sudden commencements (SC);
(6) moments of solar flares (vertical bars) labelled with the time of peak, coordinates and class.

observed on 12 December 1970 (for protons with energies 1–5 MeV), at 1010 ± 2 hours for protons of all energies. The spectrum exponents γ_2 and γ_4 are 1.5, i.e. the spectrum is of power law form over the whole range from 1 to 30 MeV. The next peak was observed for all energies from 00 h to 12 h on 13 December 1970. As can be seen from Fig. 2 the peak is of fine structure which is especially noticeable for the 1–5 MeV protons. The spectrum exponents reach the peak values $\gamma_2 = 3.2$ and $\gamma_4 = 4.5$. Thus, a deviation of the form of the integral energy spectrum from the power law and the "saturation" are observed with increasing intensity. A rapid decrease in the intensity for all particles is observed after the second peak. For protons with energies higher than 30 MeV this decrease reached 12% depth with the minimum on 14 December 1970 at 06 h ± 2 hours.

Assuming similarity to the spectrum behaviour during the previous increase in November and on the basis of the form of the intensity curve for the protons with energies higher than 30 MeV one may assume that a shock wave passed through Venus 7 on 13 December 1970 at 09 h ± 2 hours.

In Fig. 3 the Lunokhod 1 data have been used to show the 1–5 MeV proton intensity, the 4.5–8.5 MeV α-particle intensity and the counting rate of the Geiger counter detecting the protons with energies higher than 30 MeV. Also shown are the values of *Kp* index and the Deep River neutron monitor data [5, 6].

As follows from Fig. 3 the solar cosmic ray increase on Lunokhod 1 began on 11 December 1970 at 15 h ± 2 hours and reached its peak on 13 December at 18 h ± 2 hours. At the end of 14 December 1970 the solar cosmic ray intensity decreased abruptly (by a factor of about 100) and after that, from 15 to 22 December 1970 the 1–5 MeV proton intensity decreased exponentially with a time constant of about 45–50 hours. It should be noted that the α-particle intensity is changed in accordance with the 1–5 MeV proton intensity and is about 1.2% of this intensity.

A magnetic storm sudden commencement (maximum $Kp = 8+$) was detected on the earth on 14 December 1970 at 0154 and the Deep River neutron monitor recorded a Forbush decrease of 5% depth. The time of the Forbush decrease minimum coincides with the time of the second peak of the *Kp* index which is 5+ observed on 14 December 1970 from 21 h to 24 h (Fig. 3). The intensity of the protons with energies higher than 30 MeV on Lunokhod 1 also displays a Forbush decrease on 14 and 15 December 1970. It should be noted that according to the Venus 7 data the electron flux with energies higher than 0.2 MeV did not exceed 100 $cm^{-2}\,s^{-1}\,sr^{-1}$ during this increase and according to the Lunokhod 1 data the electron flux with energies higher than 0.05 MeV was less than 10 $cm^{-2}\,s^{-1}\,sr^{-1}$.

Several considerable and many small flares were observed from 10 to 12 December 1970 on the central and eastern regions of the solar disk [5]. It will also be noted that on 14–15 December 1970 a corpuscular stream from the powerful active regions Nos. 11073 and 11070 in which the flares noted above were observed must have passed through Venus 7 (108×10^6 km from the sun and 21° W of the earth–sun line) and the earth. The most powerful of these flares with respect to the presence of radio-emission of types II and IV and the greatest X-ray burst was the third flare which occurred on 11 December 1970 at 2230.

The first small peak of solar cosmic rays (Fig. 2) may be assumed to correspond to the arrival of the diffusion wave through the corpuscular stream and shock waves from the flares. For the second main peak the intensity increased for more than a day. The moment of observation of this peak corresponded to the arrival of the shock wave through the corpuscular stream. After the second peak the intensity of solar cosmic rays localized behind the shock wave front decreased rapidly by the moment of the arrival of the gas piston [2, 7] formed by either the flares or corpuscular streams from active regions. The subsequent slower decrease in the intensity was due to the reverse diffusion of solar cosmic rays from the localization layer to the sun [4].

Thus, the data from Lunokhod 1 and Venus 7 permit four phases to be isolated in this increase: (1) the diffusion peak, (2) the increase before the shock wave front [12, 13], (3) the solar cosmic rays behind the shock wave front, (4) the solar cosmic rays beyond the plasma piston. However, because of a considerable angular and linear distance between Venus 7 and Lunokhod 1 the events of Venus 7 were observed ~17 hours prior to Lunokhod 1 (except the diffusion peak which has not been determined on Lunokhod 1).

Additional analysis of the maps of the sun [14], the neutron monitor data [6], and the intensity obtained from Pioneer 6 and 8 space probes [5] makes it possible to conclude that the interplanetary medium between the sun, Venus 7, Pioneer 6 and 8 before the flare in November 1970 was sufficiently homogeneous to permit the propagation of solar cosmic rays within a large cone from the flare region to Pioneer 8 and Venus 7, the angular distance between which was more than 60°. The difference observed in the intensity and spectrum can be explained by the east–west asymmetry with respect to the flare and by the distance. The boundary of the powerful corpuscular stream from the active region No. 11073 was located in December 1970 between Venus 7 and Pioneer 6. This resulted in the solar cosmic ray flux being weakened and retarded in the direction of Venus 7 whereas it moved freely to Pioneer 6 and 8 which were almost 70° and 85° to the east of Venus 7.

References

[1] S. N. Vernov et al., in: Space Research (USSR) (in press).
[2] S. N. Vernov and G. P. Lyubimov, preprint NIIYaF MGU (1970).
[3] S. N. Vernov, G. P. Lyubimov, E. A. Chuchkov and N. N. Kontor, in: Lunokhod 1, Izd. Nauka, Moscow 1971 (p. 116).
[4] S. N. Vernov, A. E. Chudakov et al., preprint NIIYaF MGU (1969).
[5] Solar Geophysical Data, ESSA, Boulder, Colorado (1970).
[6] WDC-B2, neutron monitor data for 1970.
[7] S. N. Vernov, E. A. Chuchkov et al., preprint NIIYaF MGU (1970).

[8] G. P. LYUBIMOV, Astron. Tsirk. No. 488, 4 (1968).
[9] A. J. HUNDHAUSEN, Rev. Geophys. **5**, 796 (1970).
[10] S. N. VERNOV, E. V. GORCHAKOV and G. A. TIMOFEEV, Geomagn. i Aeron. **9**, 961 (1969). S. N. VERNOV, E. V. GORCHAKOV and G. A. TIMOFEEV, Izv. Akad. Nauk SSSR, No. 12 (1971).
[11] C. S. WARWICK, J. Geophys. Res. **67**, 1333 (1962).
[12] L. J. LANZEROTTI, J. Geophys. Res. **74**, 2851 (1969).
[13] L. I. DORMAN, N. S. KAMINER and T. V. KOBULADZE, report at Solar-Terrestial Physics Symposium, Leningrad 1970.
[14] Map of the Sun, Fraunhofer Institute, 1970.

Astronomy

Space Research XII — Akademie-Verlag, Berlin 1972

NEUTRON AND GAMMA-RAY EMISSION FROM WHITE-LIGHT FLARES

L. D. DE FEITER[a] and Z. ŠVESTKA[b]*

[a] Astronomical Institute, Utrecht, The Netherlands
[b] Space Science Department, ESTEC, Noordwijk, The Netherlands

Under the assumption that white-light flares are caused by energetic particles penetrating into the photosphere, the known number of protons needed for the white-light emission is used to obtain an estimate of the production of neutrons and gamma-rays occurring at the same time. In the case of the white-light flare of 23 May 1967, the peak flux of neutrons at the earth distance was calculated to exceed 3 neutrons cm^{-2} s^{-1}; thus they should have been detectable in space. This maximum neutron flux would have reached the earth as early as the time of the maximum phase of the flare in the H alpha light. Estimates of gamma rays show that they also should have been detectable at the same time. However, these detectable events only occur when the source particle spectrum is extremely hard, and reasonable estimates show that flares associated with a detectable neutron and gamma-ray flux should be fairly rare phenomena, maybe as rare as the white-light flares.

In past years many experiments were carried out in order to detect neutrons or gamma-rays coming from the sun and from solar flares in particular. So far, all results of these experiments have been negative (see [1] for references). Since, however, the measurements have rarely been made during periods closely following strong solar flares, one cannot exclude the possibility that the flux could exceed the sensitivity limit of the receptors after some particular flares in which the high-energy nuclear interactions were particularly strong.

All the estimates of the neutron or gamma-ray flux made so far have been based on very uncertain assumptions about the number of accelerated particles and their energy spectra in the flare region [1–3]. Recently Švestka [4] and Najita and Orrall [5] have shown that the white-light emission of flares can be explained by penetration of very energetic particles into the upper photospheric layers of the sun and, if one accepts this hypothesis, one can obtain a fairly good estimate of the number of accelerated particles. The white-light emission only occurs when the particle spectrum is very hard, and therefore the white-light flares seem to be the most likely events associated with a detectable flux of neutrons or gamma-rays.

The flare discussed by Švestka and Najita and Orrall occurred on 23 May 1967, and its short-lived white-light emission was photographed and measured by de Mastus and Stover [6]. The appearance of two small white-light areas coincides in time with the hard X-ray and microwave radio bursts, which are

* On leave from Astronomical Institute of the Czechoslovak Academy of Sciences, Ondřejov, Czechoslovakia.

direct manifestations of an acceleration process in the flare region, and the areas were situated in penumbral regions of two big near-by spots of opposite magnetic polarity, i.e. in a position where the acceleration process is to be expected (Fig. 1). Therefore, the supposition that the continuous emission was produced by high-energy particles appears to be reasonably founded. These particles can produce the white-light flare either through heating of the upper photosphere, when mainly protons above 20 MeV would be active, or through synchrotron radiation of relativistic electrons. This second process, however, would need a very high number of relativistic electrons, which would be inconsistent with the simultaneously observed X-ray flux [4, 7], and the number

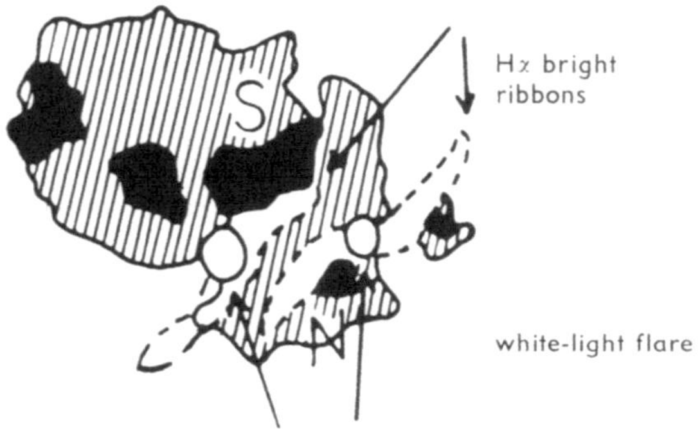

Fig. 1. The white-light flare of 23 May 1967 [6]. S denotes the southern and N the northern, magnetic polarity [4].

of particles needed would be two orders of magnitude higher than in the case of the photospheric heating [6]. Therefore, the hypothesis of photospheric origin of the white-light emission, through a bombardment by high-energy protons, appears most realistic.

One cannot determine the exact number of particles taking part in this process, because several parameters entering the discussion are fairly uncertain (the pitch-angle distribution of the particles, and the fraction of energy reradiated in the continuum, in particular). However, one can easily consider the most unfavourable conditions from the point of view of the neutron and gamma-ray production and thus estimate the minimum resulting flux to be expected at the earth. This consideration leads to the minimum total energy of 2.5×10^{29} erg carried by particles exceeding 20 MeV.

Theoretical considerations on the production of neutrons and gamma-rays [1–3] are based on the number of protons with energy exceeding 30 MeV. This number can be deduced from the required energy assuming different values of the characteristic rigidity P_0, and then the total flux of neutrons and gamma-rays can easily be found from the normalized formulae given by Lingenfelter and Ramaty [1]. Results of these computations are given in Table 1 [7]. Assuming a 5000 second duration of the flux at the earth, the resulting average flux would be $\gg 0.1\ \mathrm{cm^{-2}\,s^{-1}}$ for $P_0 = 200$ MV, thus the peak flux $> 1\ \mathrm{cm^{-2}\,s^{-1}}$,

Table 1

The flux of neutrons and gammas at the distance of the earth in events/cm², integrated over time

P_0	60 MV			100 MV			200 MV		
N (>30 MeV)	8.18×10^{32}			1.49×10^{33}			1.33×10^{33}		
	a	s	total	a	s	total	a	s	total
2.23 MeV	107	147	125	286	760	1046	574	3724	4298
0.51 MeV	34	82	116	48	268	316	170	3192	3362
10–100 MeV	<1	<1	<1	8	45	53	106	1862	1968
4.43 MeV	22	45	67	17	55	72	5	37	42
6.14 MeV	18	41	59	14	49	63	5	33	38
1.63 MeV	10	23	33	7	28	35	3	20	23
7.12 MeV	9	20	29	7	27	34	3	19	22
1.63 & 2.31 MeV	6	13	19	5	16	21	~1	12	13
Neutrons	22	65	87	55	343	398	298	2128	2426

The columns with a and s give the total production of neutrons and gamma's during the acceleration and slowing down phase respectively. For the acceleration phase we have taken $X_1 = 1.6$ g cm^{-2}.

and hence detectable at the earth. Even for lower P_0 values the flux could possibly be detected.

There, however, another factor enters the consideration. If one assumes that the energy spectrum of the accelerated electrons is the same as that for the high-energy protons, one can estimate the total number of electrons with energy above 100 keV. This value can be compared with the number of electrons required to produce the hard X-ray burst. For a power-law spectrum this extrapolation procedure leads only to acceptable values if the differential spectral index $\gamma \simeq 2$; higher values for γ, which are most commonly observed, would lead to excessively high fluxes for the hard X-ray bursts and the microwave bursts. In fact such a hard spectrum was actually observed near the earth [8], in the case of the flare discussed and observations showed an even harder spectrum ($\gamma \simeq 1.5$) early in the event (Fig. 2). This explains why the white-light emission is only observed in flares with extremely hard particle spectra, with $P_0 = 200$ or more, as has been emphasized by Najita and Orrall [5].

Let us consider an example. We will assume that $\gamma = 2$ in the differential energy power-law spectrum in the flare we discussed before, which, with the energy of 2.5×10^{29} erg above 20 MeV, corresponds to 4×10^{35} particles above 100 keV, and 1.3×10^{33} particles above 30 MeV. This last value is also obtained if $P_0 = 200$ MV is considered in an exponential rigidity spectrum. Then theoretical computations [9] lead to the neutron flux near the earth shown in Fig. 3. About 3 neutrons cm^{-2} s^{-1} should be observed in the maximum phase of the neutron flux, which could certainly be detected, and it is important to notice that this maximum flux is coincident with the maximum of the Hα intensity of the flare so that neutrons have to be looked for as soon as the flare brightens.

Now, however, let us consider the same number of 4×10^{35} particles above 100 keV, with $\gamma = 3$, which still is a fairly hard spectrum. The number of protons above 30 MeV would be only 4×10^{30} in this case, i.e. more than two

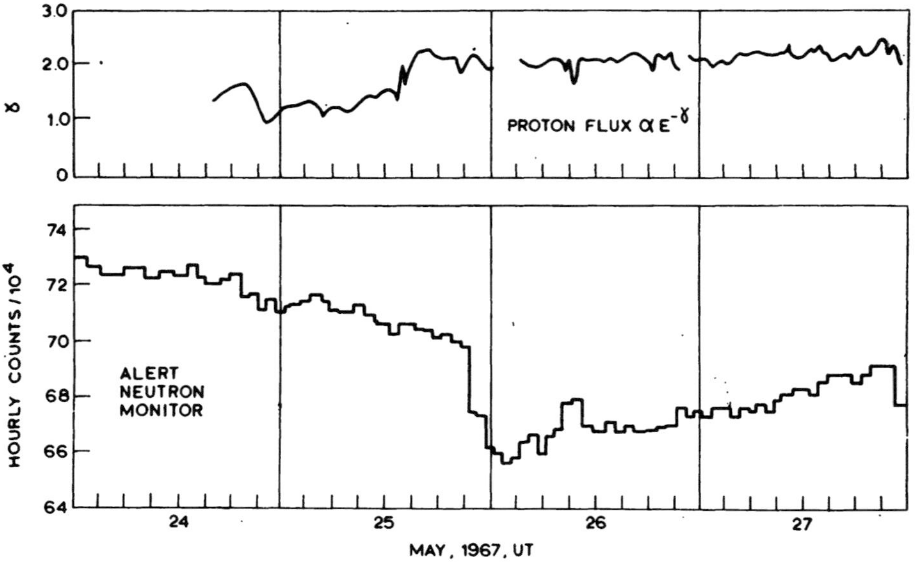

Fig. 2. The differential spectral index γ in a power-law spectrum of protons observed after the flare of 23 May 1967 [8].

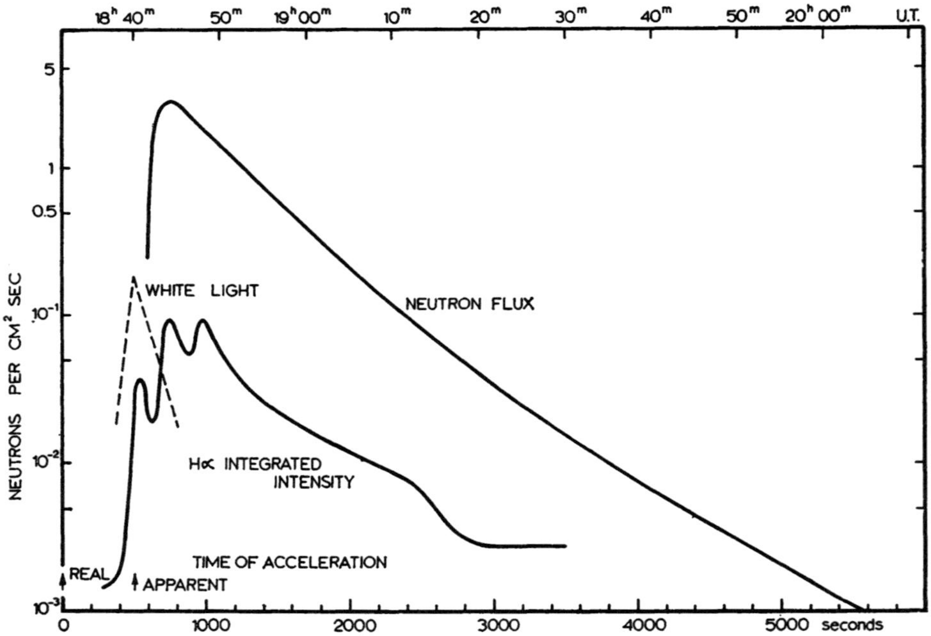

Fig. 3. The minimum expected neutron flux at 1 AU produced by the white-light flare of 23 May 1967. The bottom scale shows time in seconds after the acceleration in the flare region, while the upper scale is the UT time under the assumption that the acceleration occurred at the same time as the white-light flare was observed [9]. A comparison with the Hα development curve (after Tallant, referred to in [5]) demonstrates the early arrival of the neutrons.

orders of magnitude lower, and in consequence of it, the neutron flux at the earth would not exceed 2×10^{-3} neutrons cm^{-2} s^{-1}. There is some evidence that this actually happens. One of the largest flares of this cycle, of 2 September 1966, which produced a PCA of 13 dB, had $\gamma > 4$ in the energy range above 100 MeV [10], or, in exponential spectrum, $P_0 = 60$ MV [3]. It did not produce any white-light emission nor any observable neutron flux, as has been confirmed by negative measurements aboard the Vela satellites (Bame and Asbridge in [3]).

Thus, one can conclude that neutron and gamma-rays should be detected after white-light flares of the type of the flare of 23 May 1967, but this is due to the fact that such flares produce extremely hard particle energy spectra. Since the occurrence of white-light flares is a very rare phenomenon, one can expect that observable fluxes of neutrons and gamma-rays, too, will be very rare phenomena. The more so, however, a search for them is of great importance.

References

[1] R. E. Lingenfelter and R. Ramaty, High Energy Reactions in Astrophysics, W. A. Benjamin, New York 1967 (p. 99).
[2] R. E. Lingenfelter, E. J. Flamm, E. H. Canfield and S. Kellman, J. Geophys. Res. **70**, 4087 (1965).
[3] R. E. Lingenfelter, Solar Phys. **8**, 341 (1969).
[4] Z. Švestka, Solar Phys. **13**, 471 (1970).
[5] K. Najita and F. Q. Orrall, Solar Phys. **15**, 176 (1970).
[6] H. L. de Mastus and R. R. Stover, Publ. Astron. Soc. Pacific **79**, 615 (1967).
[7] L. D. de Feiter, Solar Phys. **19**, 207 (1971).
[8] L. J. Lanzerotti, Report UAG-5, World Data Center A, p. 56.
[9] Z. Švestka, Solar Phys. **19**, 202 (1971).
[10] Z. Švestka and P. Simon, Solar Phys. **10**, 3 (1969).

Space Research XII — Akademie-Verlag, Berlin 1972

OBSERVATIONS OF SOLAR X-RAY EMISSION FROM THE SATELLITE INTERCOSMOS 4 AND THE ROCKET VERTICAL 1*

Yu. I. Grineva, V. I. Karev, V. V. Korneyev, V. V. Krutov, S. L. Mandel'stam, L. A. Vainstein, B. N. Vasilyev and I. A. Ẑitnik

USSR Academy of Sciences, Moscow, USSR

Heliospectrometers with quartz crystals were flown aboard the satellite Intercosmos 4. Spectra of solar X-ray flares in the range 1.8 Å and 8.42 Å with spectral resolution 4×10^{-4} Å and 2×10^{-3} Å respectively were obtained. The electron temperature of the flares was estimated from the intensity of selected lines. The ion temperature was estimated by the Doppler width of the Mg XII ($\lambda = 8.418$ Å) line. The spectrum of the solar corona in the region 8–17 Å was obtained with the use of Bragg spectrometers mounted on the rocket Vertical 1. The lines of FeXVII and hydrogen- and helium-like ions of Mg, Si and Ne were identified. The temperature of the active region was calculated from the ratio of the lines Mg XII ($\lambda = 8.42$ Å) and Mg XI ($\lambda = 9.2$ Å).

Heliospectrometers in the range 1.8 Å and 8.4 Å and an X-ray heliograph with an angular resolution ~20″ were flown on the satellite Intercosmos 4 which was launched on 14 October 1970. The satellite axis was pointed towards the sun and automatically scanned the solar disk within a range of ±1° three

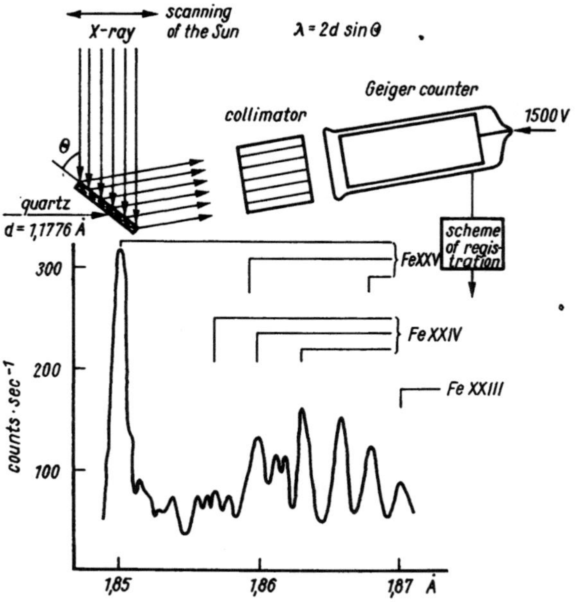

Fig. 1. The schematic design of the heliospectrometer.

* The detailed paper will be published in "Kosmicheskie Issledovaniya" USSR.

times on each revolution. The schematic design of the heliospectrometer is shown in Fig. 1. The quartz crystals with atomic plane spacings $d = 4.246$ Å and $d = 1.1776$ Å were used as dispersing elements. Rocking curve FWHM (full width half maximum) of the crystals was $\sim 10''$. The detectors in heliospectrometers in the range 1.8 Å and 8.42 Å were Geiger counters with Be ($\varrho_x = 160$ mg cm^{-2}) and Al ($d = 10$ μm) windows respectively. The design of the

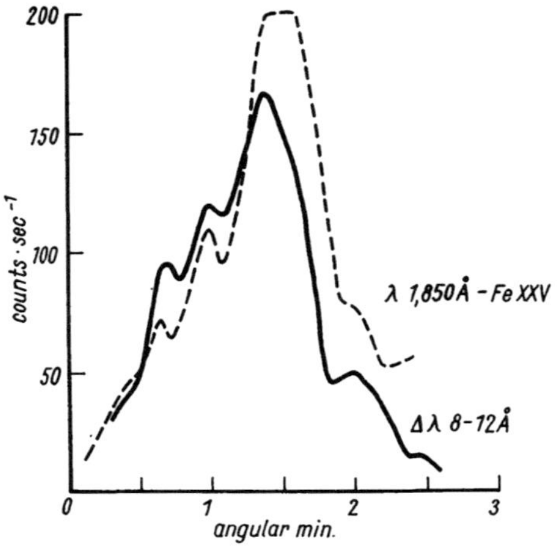

Fig. 2. The spatial structure of the flare region 24 Oct. 1970.

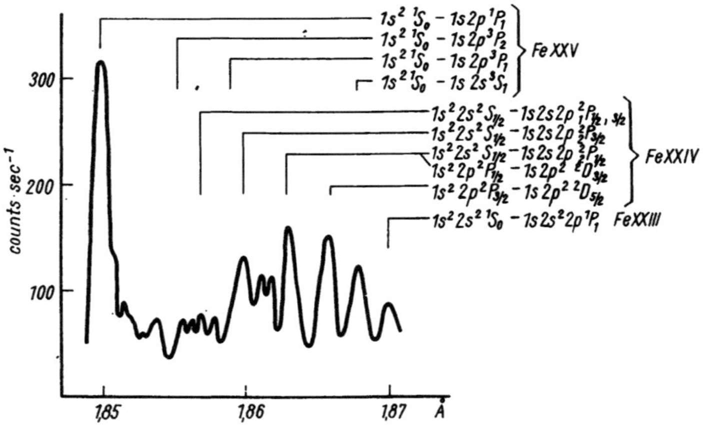

Fig. 3. The flare spectrum 16 Nov. 1970.

X-ray heliograph was similar to that on satellite Cosmos 230 [1]. The detector was a Geiger counter with aluminium window 10 μm ($\lambda = 8$–12 Å). The scans of the flare region (0542 UT, 24 October 1970) obtained by the heliospectrometers ($d = 4.246$ Å) and X-ray heliograph are shown in Fig. 2. It is easy to see that the spatial structures of the flare in the ion line Fe XXV $\lambda = 1.850$ Å and in the range 8–12 Å are very similar. The most interesting spectra were obtained with the heliospectrometer ($d = 1.1776$ Å) which had the highest spectral resolution, ~0.0004 Å at 1.8 Å. The spectrum of this region is characterized by a greater number of lines in comparison with spectra of lighter ions. This can be explained by the fact that the radiative decay of twice excited states for Fe ions ($Z = 26$) is more probable than auto-ionization. The spectrum of a class 2B flare (0104 UT 16 November 1970) is presented in Fig. 3. The wavelengths obtained in two successive scans were in good agreement. This spectral region was previously investigated by Neupert [2] and Friedman [3], but the resolution was much smaller.

To interpret the spectrum we used the calculated values of wavelengths given in Table 1 and relative intensities [4]. The energies of ionic levels were calculated using the perturbation method with hydrogen-like basis functions.

Table 1

Comparison of experimental data and theoretical calculations for the wavelengths of Fe XXV, Fe XXIV and Fe XXIII

Experimental (Intercosmos 4)	Calculated		Data from [2]*)
λ (Å)	λ (Å)	Transitions	
1.850	1.850	FeXXV $1s^{2}\,{}^{1}S_0 - 1s\,2p\,{}^{1}P_1$	1.850
1.854			
	1.855	FeXXV $1s^{2}\,{}^{1}S_0 - 1s\,2p\,{}^{3}P_2$	
1.856			
1.857	1.857	$1s^2\,2s\,{}^{2}S_{1/2} - 1s\,2s\,2p\;{}^{2}_{1}P_{1/2}$	
		FeXXIV $1s^2\,2p\,{}^{2}P_{3/2} - 1s\,2p^2\,{}^{2}S_{1/2}$	
	1.859	FeXXV $1s^2\,{}^{1}S_0 - 1s\,2p\,{}^{3}P_1$	1.855
1.860	1.861	FeXXIV $1s^2\,2s\,{}^{2}S_{1/2} - 1s\,2s\,2p\,{}^{2}_{2}P_{3/2}$	1.860
1.862			
1.863	1.863–1.864	$1s^2\,2s\,{}^{2}S_{1/2} - 1s\,2s\,2p\,{}^{2}_{2}P_{1/2}$	1.875
		FeXXIV $1s^2\,2p\,{}^{2}P_{1/2} - 1s\,2p^2\,{}^{2}D_{3/2}$	1.865 }
1.866	1.866	FeXXIV $1s^2\,2p\,{}^{2}P_{3/2} - 1s\,2p^2\,{}^{2}D_{5/2}$	}
1.868	1.868	FeXXV $1s^2\,{}^{1}S_0 - 1s\,2s\,{}^{3}S_1$?	1.865
1.870	1.870	FeXXIII $1s^2\,2s^2\,{}^{1}S_0 - 1s\,2s\,2p\,{}^{1}P_1$?	1.868 ?

* Obtained by extrapolation of laboratory data.

The Z^{-k} expansion for the energies and ψ-functions includes the second-order terms of the non-relativistic part and the first-order terms of the relativistic part (Breit operator). For $Z = 26$ the accuracy of the method was not less than 10^{-4}. The probabilities of radiative and non-radiative decay were calculated with the same functions.

More detailed analysis of the spectrum will be published later; here we note only the following characteristics: (i) Good agreement was obtained (up to ~0.001 Å) with the calculated wavelength for the majority of the most intense lines. (ii) It follows from the ratio of the line intensities 1.850 Å ($1s^2\,{}^{1}S_0$–$1s\,2p\,{}^{1}P_1$) and 1.866 Å ($1s^2\,2p\,{}^{2}P_{3/2}$–$1s\,2p^2\,{}^{2}D_{5/2}$) that the electron temperature is $T_e = 20 \times 10^6$ °K. It is worth noting that the excitation of both lines comes from the state $1s^2$ Fe XXV (by means of direct excitation and dielectronic recombination respectively). (iii) The transitions ${}^{3}P_{1,2} \rightarrow {}^{1}S$

were found to be unexpectedly weak as compared with the usual intensity of the lines observed for smaller Z by different authors. Our results are however in agreement with the direct theoretical estimation. (iv) Some of the lines have not been identified. They are probably associated with transitions of

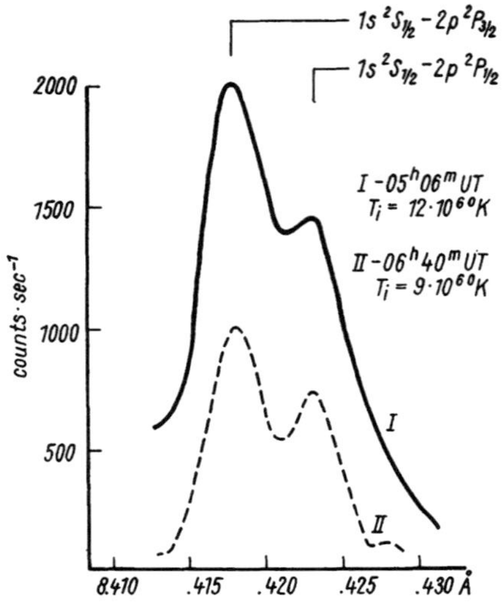

Fig. 4. The line profile for the Mg XII line during the flare of 24 Oct. 1970.

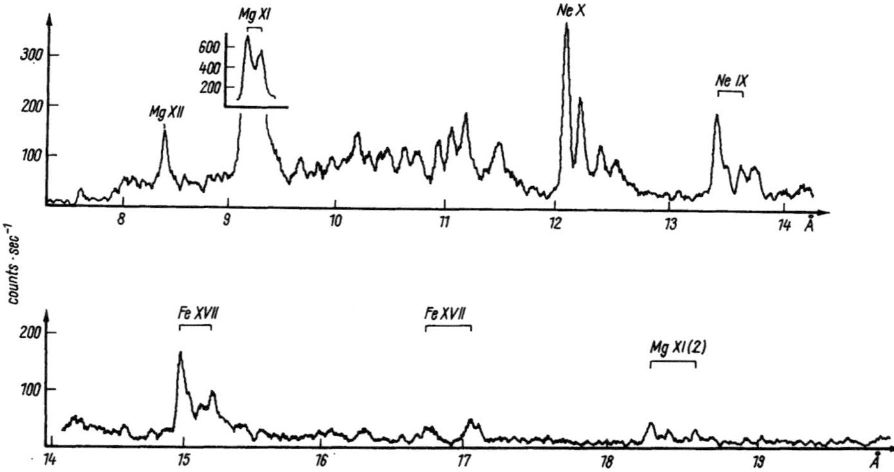

Fig. 5. The solar spectrum obtained on 28 Nov. 1970.

the type 1 s 2 p 3 l–1 s² 3 l, but the theoretical calculations are not yet performed.

The use of two heliospectrometers with different atomic plane spacings made it possible to measure the spectral line width. Tbe line Fe XXV 1.850 Å (scan 0101 UT) was found to have a complex profile with FWHM about 0.002 Å, which may be related to macroscopie motions of flare components with relative velocities about 100 km s^{-1}.

The width of the resonance line Mg XII, 8.42 Å, was measured for a weaker flare (class 2 N, 24 October 1970). Fig. 4 shows scans which belong to different stages of the flare (at 0506 UT, 14 minutes before the optical maximum, and at 0640 UT). It is obvious that both scans shows a clear resolution of fine structure components. Since the angular size of the flare region was practically the same at both times, it is reasonable to account for the observed merging of the doublet components in scan 0506 UT by a greater width of the lines. The flare temperature determined from FWHM was $\sim 12 \times 10^6$ °K for 0506 UT and $\sim 9 \times 10^6$ °K for 0640 UT. The ratio of the doublet component intensities was 1.4, but not 2, as follows from the ratio of their cross sections of excitation.

Within the period of the satellite's activity the rocket Vertical 1 was launched on 28 November 1970 with spectrometers for the range 8–17 Å. KAP crystals were used as a dispersing element. The Geiger counters used had windows from 10 μm Al and 6 μm dacron on which has been evaporated 3000 Å of aluminium. The scanning rate of the crystal was $\sim 5'$/sec. In Fig. 5 one of the scans obtained is shown. Not all the results have been analysed at present, but we can note that the use of appropriate filters in the counters has permitted us to obtain spectra free of scattered radiation and to register more accurately by a greater number of lines in the region $\lambda < 13$ Å. In particular the intensities of the resonance lines Mg XII ($\lambda = 8.42$ Å) and Mg XI ($\lambda = 9.2$ Å) are measured. Their ratio give the temperature of the active region $\sim 5 \times 10^6$ °K.

References

[1] I. L. Beigman et al., Kosm. Issled. **9**, 123 (1971).
[2] W. M. Neupert and M. Swartz, Astrophys. J. **160**, 4189 (1970).
[3] J. F. Meekins et al., Science **162**, 891 (1968).
[4] L. A. Vainstein and Yu. I. Safronova, Astronom. Zh. **48**, 223 (1971).

Space Research XII — Akademie-Verlag, Berlin 1972

OGO 5 DETERMINATION OF THE LOCAL INTERSTELLAR WIND PARAMETERS

J. L. BERTAUX, A. AMMAR and J. E. BLAMONT

Service d'Aéronomie du CNRS, Verrières le Buisson, France

The Lyman-α sky background measured outside the geocorona by two independent photometers on OGO 5 suggested that an "interstellar wind" of neutral hydrogen penetrates into the heliosphere, where solar wind and solar ultraviolet creates an ionization cavity. The shape of this cavity is a function of the interstellar density n_0, the relative velocity V_0, the temperature T and also the solar Lyman-α radiation pressure which may over-compensate the gravitation. For two different values of the solar Lyman-α flux, one which equilibrates the gravitation and one which corresponds to recent measurements and over-compensates the gravitation, the interstellar wind parameters are adjusted to agree with the sky background measurements. Even if a uniform galactic component is present at its maximum allowable value, temperatures are found in the range 10^3–10^4 °K, densities in the range 0.1–0.2 atoms cm^{-3}, and the relative wind velocity between 3 and 9 km s^{-1}.

1. Introduction

Recent measurements [1, 2] have shown that the Lyman-α sky background observed from outside the geocorona was mostly due to the resonant scattering of solar Ly-α on neutral hydrogen atoms coming from the interstellar medium and penetrating into the solar system. The relative motion of the sun and the surrounding interstellar gas allows the hydrogen atoms to penetrate into the heliosphere and to approach the sun before being ionized by charge exchange and extreme ultra violet radiation. The penetration of interstellar gas into the solar system, called the local interstellar wind, can be characterized by a set of three parameters: the velocity vector $\boldsymbol{V}_0$, the density n_0 of the interstellar medium before being ionized by the sun, and the temperature T. The shape of the ionization cavity created around the sun will depend strongly on the values of these parameters. In order to determine these parameters from the experimental evidence (namely the maps of Ly-α emission obtained by both experiments on OGO 5), computations have been pursued on theoretical models of the interstellar wind, and the computed intensity distribution as it should have been seen from OGO 5 was compared with the experimental evidence.

As can be seen from the emission maps of Figs. 1 and 2, obtained during September 1969 (SU-1) and April 1970 (SU-3), the main characteristics of the OGO 5 observations that have to be fitted by comparison with models are: (i) no apparent small-scale structure; (ii) a wide maximum region, centred around a point not far from the ecliptic plane where the intensity reaches 530 rayleighs (1 rayleigh $= 10^6/4\,\pi$ photons $cm^{-2}\,s^{-1}$); (iii) from this maximum region the intensity decreases smoothly towards a wide minimum region located in the opposite direction, where the intensity is 215 rayleighs; (iv) between September 1969 and April 1970 the position of the maximum moved along the ecliptic plane by $45 \pm 5°$, a fact which is explained by the parallax

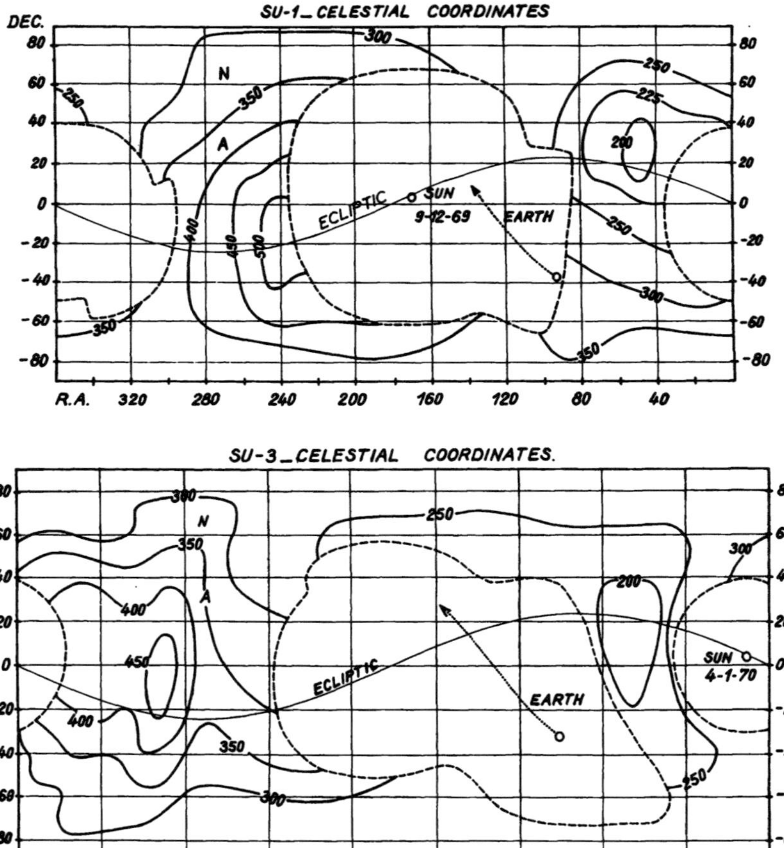

Figs. 1 and 2. Iso-intensity contours of Lyman-α sky background in celestial coordinates as seen from OGO 5 by the Paris experiment during the periods 12–14 September 1969 (SU-1) and 1–3 April 1970 (SU-3). Intensities are measured in rayleighs. The position of the maximum moved by $\simeq 45°$ from SU-1 to SU-3.

effect when the earth moves along its orbit. At the same time the minimum showed only a small motion of a few degrees, if any. The celestial coordinates of the mean position of the maximum between September 1969 and April 1970 are: R. A. $= 263 \pm 5°$, $\delta = -17 \pm 5°$. The intensity measured in the direction of the north ecliptic pole showed in December a 15% decrease with respect to the mea value of September and April.

Calculations on cavity ionization models were conducted in two successive steps, in a manner similar to that used by Blum and Fahr [3]. First the density distribution $n(r, \theta)$ of atomic hydrogen in the solar system is computed (r is the distance to the sun, θ is the angle between the vector $-\boldsymbol{V}_0$ and $\boldsymbol{r}$); then from this model the intensity is computed, by integration of the Ly-α emissivity $\varepsilon(r, \theta) = n(r, \theta) \cdot r^{-2}$ along a line of sight. The physical hypotheses which are included in the computations are the following: (i) Atoms are ionized when approaching the sun under the effect of charge exchange with the protons of the solar wind, and by euv radiation under $\lambda < 912$ Å. The sun is supposed to be a steady and isotropic source of protons, Ly-α and euv; as a consequence both the density distribution $n(r, \theta)$ and the emissivity distribution $\varepsilon(r, \theta)$ have a cylindrical symmetry about $\boldsymbol{V}_0$. (ii) Ly-α radiation pressure and gravitation are the only forces acting on the atoms and determine their Keplerian trajectories in the solar system. (iii) A temperature T characterizes the Maxwellian distribution of the individual velocities of atoms, added to the macroscopic velocity V_0. (iv) The boundary of the heliosphere is transparent to the atoms of hydrogen, which they penetrate without being disturbed. Should this hypothesis not hold, the calculations would not be affected, but the estimated parameters would rather characterize the interstellar medium when inside the heliosphere.

2. Distribution of the Hydrogen in the Solar System

The ionization cavity around the sun is the result of the interaction of the solar radiation and the nearby interstellar medium. As this medium is unknown, the parameters, n_0, $\boldsymbol{V}_0$, T, which characterize it are completely free. However, the density n_0 of the medium may be considered as a normalizing factor which will be determined by comparison with absolute intensity, once the relative shape of the cavity $n(r, \theta)/n_0$ has been adjusted to the observations. Hence, an arbitrary value of $n_0 = 0.1$ atom cm^{-3} was chosen for all the model calculations. There is also a certain degree of freedom concerning the ionizing flux and the Lyman-α flux of the sun, which are not known sufficiently accurately for us to examine only one value of the solar parameters.

2.1. Solar Ionization Flux

We have only considered the charge exchange with protons of the solar wind and ionization by euv radiation ($\lambda < 912$ Å). The ionization by electrons, amounting to approximately 20% of the ionization by protons, was neglected. The total ionization power of the sun may be characterized by the lifetime t_{h} of one atom situated at 1 AU from the sun:

$$1/t_{\mathrm{h}} = \sigma \Phi + J ,$$

where σ is the charge exchange cross section, Φ is the solar proton flux, and J is the total euv photoionization rate per atom. For all the calculations we have chosen the following values:

$$\sigma = 2\times10^{-15}\ \text{cm}^2$$

$$\Phi = 1.5\times10^{8}\ \text{protons cm}^{-2}\ \text{s}^{-1}$$

$$I = 1.6\times10^{-7}\ \text{s}^{-1}\,.$$

This last value, from [3], is twice as large as the value from [4]: the lifetime t_h is then 2.16×10^{6} s for a hydrogen atom at 1 AU from the sun instead of 2.56×10^{6} s as in [4].

2.2. Lyman-α Flux

The Lyman-α flux at the centre of the solar line (considered with a flat profile) has a two-fold effect: the scattered intensity is proportional to it, but more important is the radiation pressure exerted on the hydrogen atoms, which can more than compensate for the gravitational attraction $M\,G\,r^{-2}$. The Lyman-α flux F_0 at 1 AU which exactly compensates the gravitation is obtained from:

$$F_0\, r_0^2\, \sigma_\lambda \frac{h\,\nu}{m\,c} = M\,G$$

where $r_0 = 1$ AU in km, $\sigma_\lambda = (\pi\, e^2/m_e\, c)f = 0.54\ \text{cm}^2\ \text{Å}$, m is the mass of the hydrogen atom, M the solar mass and G the constant of gravitation. The equilibrium value of F_0 is calculated to be 3.32×10^{11} photons $\text{cm}^{-2}\ \text{Å}^{-1}\ \text{s}^{-1}$, giving a "$g$" factor (excitation rate for one hydrogen atom at 1 AU) $g = 1.79\times10^{-3}\ \text{s}^{-1}$.

To take this value for the solar flux greatly simplifies the calculations, since the trajectories of hydrogen atoms are straight lines parallel to their initial velocities. However, recent measurements on OSO 5 [5] of the solar line centre have shown that F_0 is larger than the equilibrium value 3.32×10^{11} photons $\text{cm}^{-2}\ \text{Å}^{-1}\ \text{s}^{-1}$, ranging from 4.2 to 5×10^{11} photons $\text{cm}^{-2}\ \text{Å}^{-1}\ \text{s}^{-1}$, depending on the solar activity. The net action of the sun is then a repulsive force, which deflects the trajectories in hyperbolae with concavity away from the sun. For monokinetic atoms (no temperature) then a forbidden region appears which has the shape of a paraboloid of revolution whose focus is the sun and whose concavity is turned downwind. In our computations we have considered a large number of models with the equilibrium value $F_0 = 3.32\times10^{11}$ photons $\text{cm}^{-2}\ \text{Å}^{-1}\ \text{s}^{-1}$ and a few models with the more realistic value $F_0 = 4.2\times10^{11}$ photons $\text{cm}^{-2}\ \text{Å}^{-1}\ \text{s}^{-1}$, which corresponds to a Lyman-α excitation rate of $g = 2.25\times10^{-3}$ at 1 AU. The numerical treatment of such models is however much more time-consuming than that of the simple straight-line models.

2.3. "Hot" Models and "Cold" Models

To interpret the OGO 5 measurements Thomas [4] studied a certain number of straight-line models, for which it is obvious that, if only monokinetic atoms with velocity V_0 are considered, the ionization cavity will extend indefinitely behind the sun. In this case, the intensity coming from the back part of the cavity would be extremely small, a result contradicted by the observation of 215 rayleighs in the direction of the minimum, if the emission is attributed solely to the scattering of Lyman-α by atoms in the solar system. Thomas considered two different kinds of models: (i) the "cold" model, which depends only on the parameters n_0 and V_0, since the incoming atoms have no thermal velocities; in this case the 215 rayleighs measured in the direction of the minimum would be an isotropic component of galactic Lyman-α emission; (ii) the "hot" model, in which no galactic component is introduced to fit the observations but rather the temperature of the interstellar medium would add to the macroscopic motion peculiar velocities of individual atoms and would "fill" the back of the cavity. Thomas considered an isotropic single velocity distribution characterized by

$$V_T = (2\,kT/m)^{1/2}\,.$$

We preferred to introduce a Maxwellian velocity distribution both for hot and cold models; in this paper the cold model is only cold in the sense that we try to adjust the observations, supposing a 215 rayleigh isotropic galactic background. The use of a Maxwellian velocity distribution is more realistic than the use of a single velocity distribution. However, Thomas' argument [4], that the maximum/minimum ratio Q of scattered intensities is only a function of V_T/V_0, no longer holds for the Maxwellian distribution. We studied separately the influence of V_0 and T on the ratio Q.

3. Intensity Calculations and Comparison with OGO 5 Data

The direction of the incoming flow of neutral hydrogen given by the mean position of the maximum between September 1969 and April 1970 can be inferred directly from the data (R.A. = 263 ± 5°, $\delta = -17 \pm 5°$). As it is very near the ecliptic plane, we placed it right in the ecliptic plane at the longitude $l_e = 263°$ for the intensity calculations; then, the maximum and the minimum, as seen from the earth, are to be found in the ecliptic plane, since the region where the emissivity $\varepsilon(r, \theta)$ is maximum lies in the ecliptic plane, in front of the cavity (Fig. 3). The emissivity is the number of Lyman-α photons emitted per second in 1 cm^3 and is proportional to $n(r, \theta)/r^2$. Fig. 4 represents in a solar ecliptic coordinate system the position of the vector $\boldsymbol{V}_0$, the vector $\boldsymbol{V}_a$ (movement of the sun towards the apex), and the situation of the earth for the SU-1, SU-2 and SU-3 periods.

3.1. Straight-line Models

For a series of straight-line models with temperatures ranging from 10^2 °K to 5×10^4 °K and V_0 ranging from 5 to 30 km s^{-1}, the distribution of intensity in the ecliptic plane as seen from the earth for SU-1 and SU-3 was plotted, and the corresponding parallax was measured graphically. This parallax was then plotted as a function of V_0 and T for all the straight-line models, and the parallax of $45 \pm 7°$ actually observed delimited a zone of allowed parameter values in the (V_0, T) plane of Fig. 5. We can point out that the parallax would not be modified if an isotropic galactic emission I_g were superimposed on the emission of the ionization cavity model; thus the interpretation of the observed parallax effect is independent of the alternative "hot" or "cold" model. This is not the case for the ratio $Q = I_{max}/I_{min}$ of the maximum and minimum intensities observed in the ecliptic plane during SU-1 or SU-3 (the measured ratio was identical for both periods). In fact, for low temperatures the computed ratio becomes extremely large. The measured ratio $Q_m = 2.45 \pm 0.15$, when

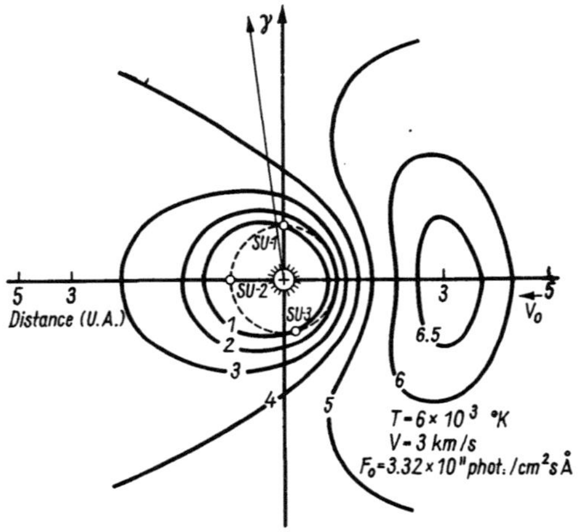

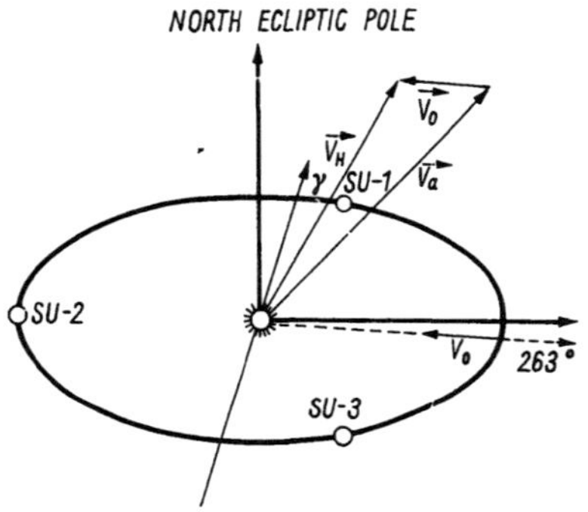

Fig. 3 Fig. 4

Fig. 3. Distribution of Ly-α emissivity in the ecliptic plane (number of photons emitted isotropically per cm^3 s) in arbitrary units for a straight-line model characterized by a temperature $T = 6 \times 10^3$ °K and a macroscopic velocity $V_0 = 3$ km s^{-1}. The Ly-α solar flux $F_0 = 3.32 \times 10^{11}$ phot cm^{-2} Å^{-1} s^{-1} compensates exactly the gravitation, and atoms are crossing the solar system along straight trajectories. The region of maximum emissivity, which explains the parallactic displacement from SU-1 to SU-3, is located at 3 AU.

Fig. 4. The position of the earth during SU-1, SU-2 and SU-3 is represented in the ecliptic coordinate system. The movement of the sun towards the apex is represented by the vector $\boldsymbol{V}_a$; $\boldsymbol{V}_H = \boldsymbol{V}_a + \boldsymbol{V}_0$ is the motion of the neutral hydrogen in the local standard of rest; it is oriented towards a direction between the apex and the direction of the north ecliptic pole for values of V_0 between 3 and 9 km s^{-1}.

interpreted in terms of hot models, cuts in the (V_0, T) plane a second zone of allowed parameters (Fig. 5) which overlaps with the first zone in a region limited by $2 < V_0 < 5$ km s^{-1} and $4 \times 10^3 < T < 9 \times 10^3$ °K. Then a typical hot and straight-line model which fits the data is characterized by $V_0 = 3$ km s^{-1} and $T = 6 \times 10^3$ °K, a set of parameters to which corresponds the emissivity map shown in Fig. 3.

The measured ratio Q_m cannot be used as a criterion to interpret the data in terms of cold models, because the intensity coming from the downwind direction becomes extremely small. Instead one can use the intensities I_N

measured in the direction of the north ecliptic pole at different periods of the year, and particularly the ratio P of the mean value $^1/_2\,(I_N(1) + I_N(3))$ of intensities measured during SU-1 and SU-3 to the intensity $I_N(2)$ measured during SU-2:

$$P = {}^1/_2\,(I_N(1) + I_N(3))/I_N(2)\,.$$

Since the position of the earth during SU-2 is downwind in the ionization cavity, it is expected that this ratio would be >1 which is confirmed by the

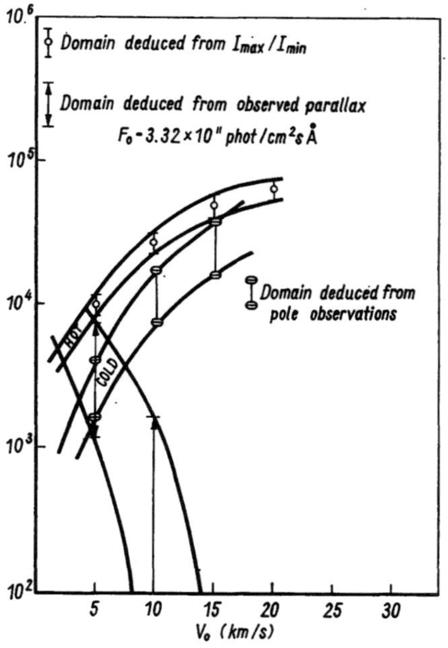

Fig. 5. For straight-line models the domains allowed by the constraints resulting from observations are represented in the (V_0, T) plane. The allowed domain resulting from the observed parallactic displacement is independent of the alternative "hot" or "cold" model. For the "hot" model the domain resulting from the observed ratio of maximum to minimum intensity is represented, for the "cold" model the domain resulting from the observed variation of intensity measured in the direction of the north ecliptic pole over one year is represented.

observations. The accuracy of P_m is less than the accuracy of Q, since we have to compare measurements made at different periods of time; from the Colorado data and the Paris data we obtained for this measured ratio $P_m = 1.15 \pm 0.1$. If we are to interpret this ratio in terms of cold models, we have to subtract from each measurement the intensity I_g of a supposed galactic emission; the maximum possible value of I_g being the minimum intensity I_{min} observed in the downwind direction, the maximum value P_c allowed by observations of the ratio P is:

$$P_c = \frac{^1/_2\,(I_N(1) + I_N(3)) - I_g}{I_N(2) - I_g} = 1.55 \pm 0.1\,.$$

The condition that P must be equal (within the experimental margin of error) to P_m, which is the interpretation in terms of hot models, leads to an allowed domain in the (V_0, T) plane of Fig. 5 which was not represented because it was somehow redundant to the domain allowed by the observed ratio $Q = I_{max}/I_{min}$. On the contrary, the interpretation in terms of cold models (P equal to P_c) leads to an allowed domain in the (V_0, T) plane which overlaps with the domain derived from the observed parallax in a region centred around the point $V_0 = 5.5$ km s^{-1} and $T = 3\times10^3$ °K. This shows that in any case the cold model interpretation leads to a model which is not so very cold.

3.2. Lyman-α Radiation Pressure larger than Gravitation

The same study was made with a solar Lyman-α flux $F_0 = 4.2\times10^{11}$ photon cm^{-2} Å^{-1} s^{-1}, resulting in a deflection of the trajectories of the incoming atoms. The general conclusions of the above paragraph still hold, except that now the allowed region for the hot model lies around the point $V_0 = 3$ km s^{-1} and $T = 1.2\times10^4$ °K, and the allowed region for the cold model lies around the point $V_0 = 7$ km s^{-1} and $T = 7\times10^3$ °K. The effect of an increased solar Lyman-α flux in model calculations is to increase the adequate temperature substantially. Once a set of parameters V_0 and T have been selected, the density n_0 at infinity (outside the ionization cavity) is adjusted to the observed value of the maximum intensity I_{max} for the hot model and $I_{max} - I_g$ for the cold model.

Table 1

Lyman-α solar flux (photons cm^{-2} Å^{-1} s^{-})	Deduced quantity	"Cold" model	"Hot" model
$F_0 = 3.32\times10^{11}$	V_0 (km s^{-1})	5.5 ± 1.5	3.5 ± 1.0
	T (°K)	$3\times10^3 \pm 2\times10^3$	$6\times10^3 \pm 2\times10^3$
	n_0 (cm^{-3})	0.10	0.16
$F_0 = 4.2\times10^{11}$	V_0 (km s^{-1})	7 ± 2.0	3.0 ± 1.0
	T (°K)	$7\times10^3 \pm 3\times10^3$	$1.2\times10^4 \pm 3\times10^3$
	n_0 (cm^{-3})	0.08	0.12

Table 1 gives a summary of the values of n_0, V_0 and T for the different conditions examined in this paper. As absolute calibrations in the uv region are quite difficult, the induced error in the determination of n_0 might amount to a factor of 2. The errors in the determination of V_0 and T are already included in the extent of the allowed domains in the (V_0, T) plane as well as the slight differences between Colorado and Paris data.

4. Conclusion

The present study suggests that to take into account the observed parallactic displacement and the variation of the intensity measured in the direction of the north ecliptic pole, values of V_0 between 2 and 9 km s^{-1} and T between

10^3 and 1.5×10^4 °K have to be selected, whatever might be the solar Lyman-α flux between 3 and 4.2×10^{11} photons cm^{-2} Å^{-1} s^{-1}, and the value of an hypothetic galactic background between 0 and 215 rayleighs. These increased values of T and decreased values of V_0, with respect to previous determinations of the parameters n_0, V_0 and T of the interstellar wind by comparison with Lyman-α measurements [2–4] result in an equivalent value of the density n_0 of the interstellar medium surrounding the solar system. In fact, it seems that the value of this parameter n_0, when adjusted to the data, is not very sensitive to V_0, T, or even F_0, in the limits of their possible values.

References

[1] G. E. Thomas and R. F. Krassa, Astron. Astrophys. **11**, 218 (1971).
[2] J. L. Bertaux and J. E. Blamont, Astron. Astrophys. **11**, 200 (1971).
[3] P. W. Blum and H. J. Fahr, Astron. Astrophys. **4**, 280 (1970).
[4] G. E. Thomas, paper presented at Solar Wind Conference, 21–26 March 1971 Asilomar.
[5] J. E. Blamont and A. Vidal Madjar, J. Geophys. Res. **76**, 4312 (1971).

Space Research XII — Akademie-Verlag, Berlin 1972

NEW INTERPRETATIONS OF EXTRATERRESTRIAL LYMAN-ALPHA OBSERVATIONS

P. W. BLUM[a] and H. J. FAHR[b]

[a] Goddard Space Flight Center, Greenbelt, Md, USA
[b] Institut für Astrophysik der Universität Bonn, Bonn, FRG

The intensity distribution of the extraterrestrial Lyman-α radiation can be sufficiently explained in its structure by the assumption that it is due to the scattering of solar Lyman-α radiation from atomic hydrogen which has entered the interplanetary space from outside the solar system. A large-scale intensity maximum and minimum in the sky background is related to large scale anisotropies of the density distribution of interplanetary hydrogen. However, the exact location of the observed intensity extrema deviate from those directions in which the density extrema are to be expected, and show temporal changes with changing positions of the earth in its orbit to which the observing spacecraft (OGO 5) is bound. Several effects which are quantitatively tested can account for these deviating positions.

The solar Lyman-α radiation pressure affects the orbits and the velocities of the interstellar particles entering the solar system. This leads to enhanced particle losses in the heliosphere, since particles spend a longer time crossing it. This causes a stronger decrease of the density with decreasing distances from the sun than had been calculated without accounting for the radiation pressure. Furthermore, the emission pattern of the solar Lyman-α radiation is anisotropic and rotates with the sun in a 27-day period. This causes a temporal change in the location of the intensity extrema. At the same time it produces hydrogen density anisotropies with extrema deviating in their directions from those which had been calculated without consideration of the radiation pressure.

During recent years various measurements of the extraterrestrial Lyman-α radiation have been carried out by space probes and satellites. At three different times, September 1969, December 1969 and April 1970 the OGO 5 satellite carried out so-called roll manoeuvres during which nearly the total Lyman-α sky background could be registered. The results have been published in several sky maps of Lyman-α isophotes with typical structures which are shown in Fig. 5 [1, 2]. The general features which became apparent from the three sets of sky maps are: (i) no small scale structures of the sky background are apparent. A large intensity maximum, and nearly opposite to it a corresponding intensity minimum, can be found; (ii) the average location of the maximum, seen over the three roll manoeuvres at different times, seems close to the ecliptic at about RA $= 265°$; (iii) the location of the maximum varies by more than $50°$, that of the minimum only by about $15°$ during the period September 1969 to April 1970; (iv) during the same period the absolute intensity of the maximum varies by about 50 rayleighs, whereas the intensity of the minimum is nearly constant. Up to now only points (i) and (ii) could be satisfactorily explained by theory. In the model of the density distribution of interstellar hydrogen in the interplanetary space by Blum and Fahr [3] three important effects had been taken into account; these are the focusing effect of the solar gravitational field, the extinction effects due to charge exchange reactions with solar protons, and the ionization by the solar euv radiation. These effects lead to an anisotropic density distribution with a density maximum in the direction of approach of interstellar matter and a minimum in the opposite direction.

Since the atomic hydrogen in the interplanetary space acts as a scattering agent upon the solar Lyman-α radiation, it is quite obvious that an anisotropic Lyman-α sky background with a large intensity maximum and minimum must occur. The position of this maximum should be determined by the direction of the velocity $\boldsymbol{v}_0$ of approach of interstellar matter. Since this direction is not known, and since deviations from the solar apex direction seem absolutely possible, a location of the maximum near the ecliptic could easily be understood.

On the other hand such a simplified interpretation of the Lyman-α sky background cannot account for any temporal changes, either with respect to the location of the intensity extrema, or with respect to intensity variations at the position of the maximum. The idea that the interstellar matter could have pronounced turbulence structures which would cause a temporal variation of the velocity $\boldsymbol{v}_0$ of approach, in both its absolute value and its direction, while interstellar matter sweeps over the solar system, cannot account for short-time variations with periods of half a year, since the characteristic time scales are of the order of at least $\tau = (n_0\, q\, v_0)^{-1} \approx 10^{11}$ s, where n_0 is the density of interstellar matter, and q is the elastic scattering cross section [4–6]. Therefore the density distribution of atomic hydrogen in the interplanetary space should be given by a system of spherical isodensity shells which are fixed in space. Also the short-time variations of the different solar radiations which determine the extinction of the hydrogen particles entering the solar system from outside cannot explain temporal density variations, since their effects are averaged out along the orbits of the particles [7]. Only periods of the order of or larger than one year can be found in the density structures.

If, however, the spatial structure of the interplanetary hydrogen density is constant in time, what reasons do we have for temporal changes in the Lyman-α sky background? One reason which had already been seen by the two OGO 5 groups from Boulder and from Paris is only virtual by its nature, since it does not account for a true change of the Lyman-α sky background but for a change of the sky as seen from the observer; this effect is called the parallax effect. Due to the orbital motion of the earth around the sun, the three sets of observations of the Lyman-α sky have been carried out at different positions of the earth within the solar system with respect to which the distribution of hydrogen is spatially fixed. Besides this parallax effect there exists another very important fact which has not been taken into account so far but which contributes to a satisfactory explanation of the Lyman-α sky. This is the anisotropic and time-dependent radiation pattern of the solar Lyman-α emission. While spatial and temporal changes in the solar radiation are averaged out with respect to the density distribution of hydrogen, it is instantaneously reflected back from the Lyman-α sky background as far as it concerns the solar Lyman-α emission pattern. From several monitorings of the solar Lyman-α line

[8, 9] it has become obvious that the solar Lyman-α emission cannot be regarded as constant, either in time or in space. On the contrary, intensity variations by more than 30% in amplitude have been found to occur during one solar rotation period, i.e. during 27 days. This proves that the source of the solar Lyman-α emission should be found in both the undisturbed solar disk and the solar activity centres which rotate synchronously with the sun for at least several rotations. From that, however, it must be deduced that not only a longitudinal variation with azimuthal angle φ but also a latitudinal variation with heliographic latitude λ is present in the solar emission pattern. If we admit 30% intensity variations in latitude and longitude of the solar Lyman-α radiation pattern, we can approximately describe both effects, the 27-day effect and the heliographic

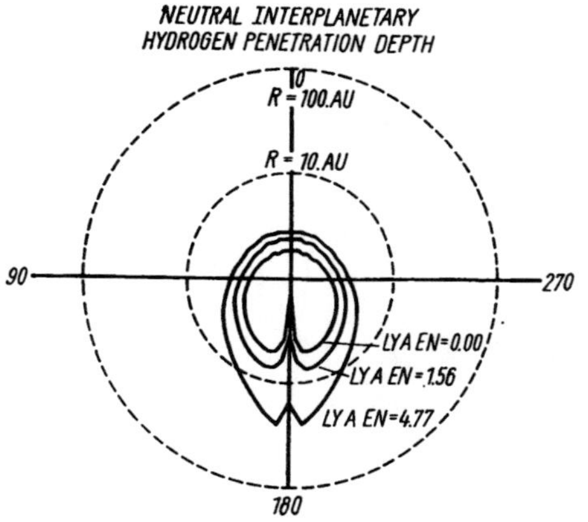

Fig. 1. The penetration depth of interstellar hydrogen is shown by three curves for various conditions of solar radiation pressure. The isodensity lines for the density $n = (1/e)\, n_0$ (n_0 is the density of interstellar hydrogen in the vicinity of the solar system) are shown in a plane which contains the velocity v_0 of approach of interstellar matter. The plane polar coordinates are the distance r from the sun (logarithmic scale) and the angle θ to the velocity vector v_0 ($\theta = 0°$). The three different isodensity lines are calculated for radiation pressures which are due to no solar Lyman-α radiation (innermost curve), and for solar Lyman-α radiations given by Tousey [10] for March 1962 (solar minimum conditions) and July 1959 (solar maximum conditions). ($n_0 = 0.1$ cm^{-3}; $v_0 = 20$ km s^{-1}).

latitude effect, by appropriate sinusoidal functions for an azimuthal phase angle φ and an angle $\bar{\lambda}$ of relative latitude which is the difference $\lambda - \lambda_0$ between the latitude λ of the observational direction in question, and the average latitude λ_0 of the solar activity centres amounting to about 18°. This description fits the observed amplitudes of the variations.

In order to obtain an improved theoretical explanation of the Lyman-α background observations, we will also take into account two further effects that affect the density distribution of interstellar hydrogen in the interplanetary space. These are the velocity dependent Lyman-α radiation pressure and the ionization of hydrogen by solar wind electrons. While the impact ionization by electrons only enlarges slightly the total ionization frequency by about 20% due to the very small cross sections for the electron impact ionization, the solar

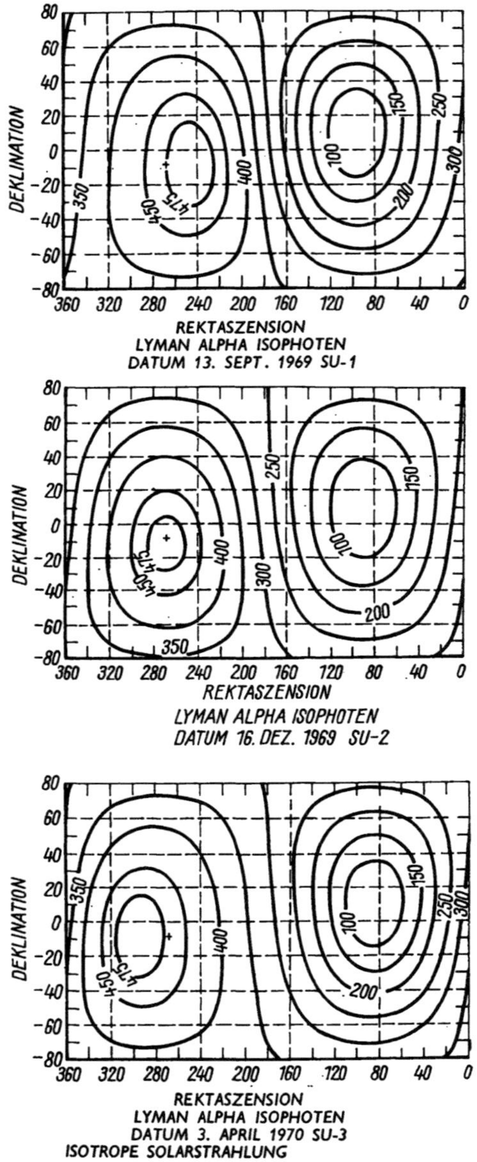

Fig. 2. The Lyman-α sky background for September 1969, December 1969 and April 1970 as calculated from theory is shown by Lyman-α isophotes in celestial coordinates RA and decl. The solar Lyman-α radiation pattern has been taken to be isotropic corresponding to the radiation intensity of the line profile given by Tousey [10] for July 1959 (solar maximum conditions). The apex of the relative velocity v_0 between the solar system and the interstellar matter has been assumed to be at RA $= 265°$, decl. $= -10°$ ($n_0 = 0.1$ cm^{-3}; $v_0 = 20$ km s^{-1}).

Lyman-α radiation pressure which compensates partially or totally the solar gravitation, strongly affects the geometry of the particle orbits and hence the distribution of interstellar hydrogen in the solar system.

Fig. 1 shows three isodensity lines which represent contours of hydrogen density equal to one e-th of its value n_0 outside of the solar system for three different values of solar Lyman-α fluxes. One sees the enlargement of the cavity of ionized hydrogen around the sun with increasing Lyman-α radiation pressure. At the same time a deformation of the original cavity with increasing radiation pressure can be found. The apex–antapex anisotropy which determines the ratio of maximum-to-minimum intensities in the Lyman-α sky background is much more pronounced for high radiation pressures.

We shall now use the improved model of the density distribution of interstellar hydrogen in the solar system and shall calculate to what Lyman-α sky background this model leads. In the next figures we shall demonstrate separately how the three effects mentioned: the parallax effect, the heliographic latitude effect, and the 27-day effect, influence the structures of the Lyman-α sky. In Fig. 2 to 5 we represent the numerical calculations of the Lyman-α sky background as due exclusively to the backscatter of solar Lyman-α radiation from interstellar hydrogen according to our model. Fig. 2 shows the parallax effect under the assumption of isotropic solar emission in theoretical Lyman-α isophotes for those three different times at which the roll manoeuvres SU 1, SU 2 and SU 3 had been carried out by the OGO 5 satellite. The interplanetary hydrogen acting as scattering agent upon the solar Lyman-α radiation, has been calculated for a relative motion of 20 km s^{-1} of the solar system against the surrounding interstellar hydrogen which points into the direction RA $= 265°$, decl. $= -10°$ deviating by about 40° from the direction RA $= 270°$, decl. $=$ $= +27°$ of the solar apex motion relative to the local standard of rest. This direction of the relative velocity $\boldsymbol{v}_0$ has tentatively been chosen in order to obtain better agreement with the observational isophotes. The solar Lyman-α radiation pressure for September 1969 has been calculated from the solar Lyman-α line profile given by Tousey [10] for the time July 1959 (solar maximum conditions). The three sky maps shown in Fig. 2 demonstrate the migration of the intensity maximum and minimum during the period September 1969 to April 1970 due to the parallax effect, i.e. due to the translocation of the earth within the solar system. The maximum with intensities near 500 rayleighs migrates by about 50° in right ascension, whereas the minimum with intensities of about 100 rayleighs only migrates by about 15–20° in right ascension. In this figure the solar Lyman-α emission has been taken to be isotropic in latitude and longitude.

In Fig. 3 we have shown how these sky maps are changed if the 27-day effect, i.e. the longitudinal anisotropy of the solar Lyman-α emission, is taken into account. The upper part of Fig. 3 shows the Lyman-α sky for September 1969 resulting from isotropic solar emission, whereas the lower part of the figure shows the sky if the 27-day effect is taken into account. The phase of the 27-day variation has been assumed to be identical with the phase of the 27-day variation in the 10.7-cm radiation which is continuously monitored. One can see that the longitudinal anisotropy of the solar Lyman-α emission changes appreciably the configuration of the Lyman-α isophotes, not only with respect to translocations of the intensity extrema, but also with respect to the structure of medium-intensity regions. Again it can be found that the migration of the intensity maximum by about 25° in right ascension is more pronounced than that of the intensity minimum.

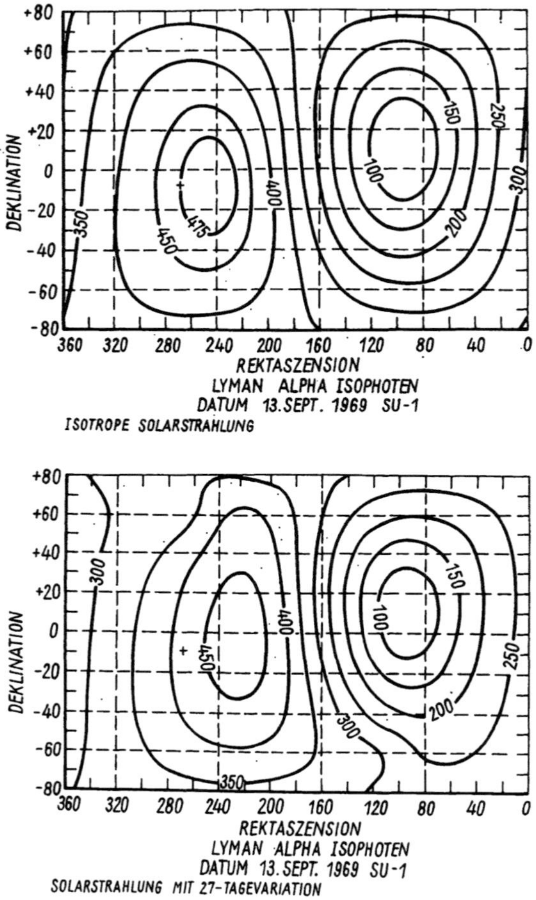

Fig. 3. In addition to the calculations shown in Fig. 2, the 27-day-variation (longitudinal anisotropy) of the solar Lyman-α radiation has been taken into account here. The phase of the 27-day variation has been taken to be identical with that of the 27-day variation in the solar 10.7-cm radio emission which is known for the time in question (13 Sept. 1969). (n_0 = 0.1 cm^{-3}, v_0 = 20 km s^{-1}; apex at RA = 265°, decl. = −10°).

Fig. 4 now takes into account the latitudinal anisotropy of the solar emission pattern, neglecting however the 27-day effect. Again one finds some changes in the structures of the isophotes and in the location of the intensity extrema. While the minimum is nearly stable, the maximum changes its position by about 10° in right ascension.

Including all the above-mentioned effects simultaneously in our calculations, we obtain the Lyman-α sky background as shown in Fig. 5 for September 1969. The upper part of the figure gives the experimentally measured isophotes of Bertaux and Blamont [1], the lower part shows the theoretically calculated sky background. It can be seen that the main features of the theoretically

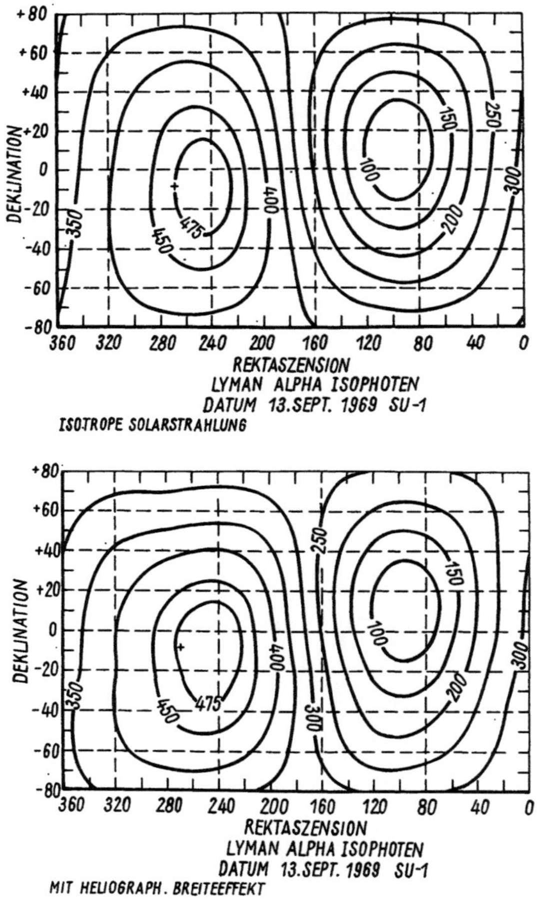

Fig. 4. In addition to the calculations of Fig. 2, here the heliographic latitude effect in the solar emission pattern has been taken into account. ($n_0 = 0.1\ \text{cm}^{-3}$, $v_0 = 20\ \text{km s}^{-1}$; apex at RA = 265°, decl. = −10°).

calculated sky background are in satisfactory agreement with the experimental results. This is true both with respect to the main structures and to the absolute intensities of the sky. If we look at the 300 and 350 rayleigh isophotes which surround the intensity maximum, we find excellent agreement. The location of the maxima nearly agree in right ascension, but differ by about 15° in declination, the location of the minima differ by about 40° in right ascension and 30° in declination. An even better agreement of the location of the extrema and of the configuration of observational and theoretical isophotes can be obtained, if we assume a slightly changed direction of the velocity $\boldsymbol{v}_0$ of approach of interstellar matter: the direction of $\boldsymbol{v}_0$ should be more distant from the ecliptic plane, probably located near RA = 265°, decl. = −20° instead of a declination of −10°. Nevertheless it can be stated that the main features of the observed Lyman-α sky background can be explained by our theoretical calculations which

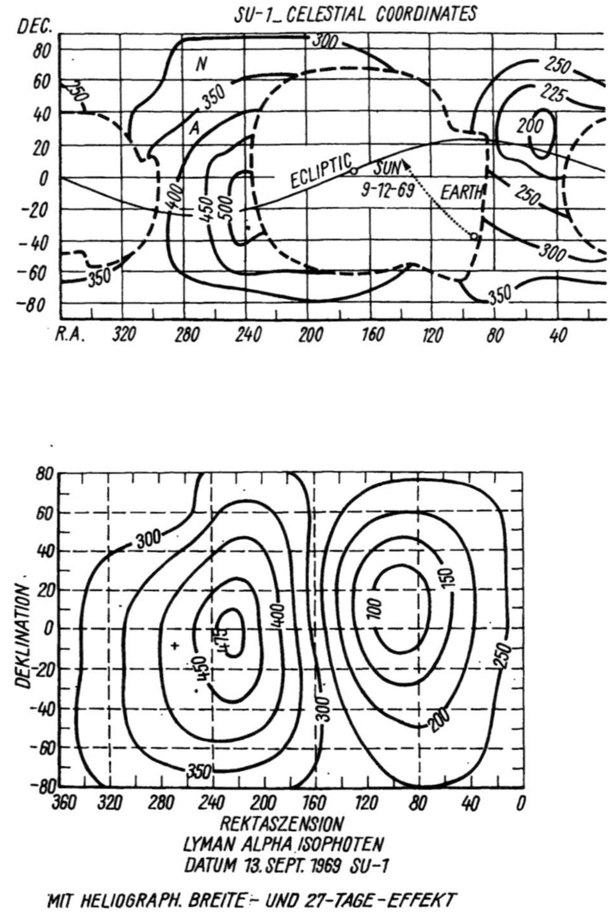

Fig. 5. A comparison of the observational and theoretical isophotes of the Lyman-α sky background. The upper part of the figure shows the Lyman-α sky map for 13 Sept. 1969 given by Bertaux and Blamont [1]. The lower part shows the isophotes for the same date which have been theoretically calculated implying the effects of parallax, the 27-day variation (Fig. 3), and the heliographic latitude variation (Fig. 4) simultaneously. ($n_0 = 0.1$ cm^{-3}, $v_0 = 20$ km s^{-1}; apex at RA = 265°, decl. = −10°).

include the effects of parallax, of the 27-day variation, and of the latitudinal variation in the solar Lyman-α emission pattern. Only the minimum intensities as found by theory seem to be a little too low. With respect to this intensity deficit one could argue that possibly the real sky background is composed of two radiative components, the interplanetary and the galactic.

Remark

During the session where this paper was presented it had been stated by Dr. Bertaux that he obtains a best fit to his Lyman-α measurements with OGO 5 with respect to minimum and maximum intensities by the use of a velocity $v_0 = 5$ km s^{-1} for the approach of interstellar matter in contrast to our value of 20 km s^{-1}. The calculation of hydrogen densities according to our model can also easily account for the observed minimum and maximum intensities in the case of a value $v_0 = 5$ km s^{-1} by simply assuming somewhat higher interstellar hydrogen densities. However, up to now there is no reliable way known to us by which a definite determination of the exact value of the velocity v_0 of approach of interstellar matter can be made. Since one can also account for the observed intensity extrema by other means, we see no urgent reason to assume a velocity value that deviates appreciably from the solar apex motion of 20 km s^{-1}.

In the calculation of maximum and minimum intensities according to our model we have two free parameters at our disposal: the density of interstellar hydrogen outside the solar system and the intensity of the isotropic galactic Lyman-α component. We can always select parameter values to cause our calculated maximum and minimum intensities to coincide with the observed ones at a particular date. A further test of the accuracy of our model would be the plausibility of the assumed values and their constancy, as these parameters do not change within our time scales and should be valid for all times of observations. An investigation dealing with a best fit of the parameters is in progress.

References

[1] J. L. Bertaux and J. E. Blamont, Astron. Astrophys. **11**, 200 (1971).
[2] G. E. Thomas and R. F. Krassa, Astron. Astrophys. **11**, 218 (1971).
[3] P. W. Blum and H. J. Fahr, Astron. Astrophys. **4**, 280 (1970).
[4] W. B. Burton, Bull. Astron. Inst. Neth. **18**, 247 (1966).
[5] P. O. Lindblad, Bull. Astron. Inst. Neth. **19**, 34 (1967).
[6] S. B Pickelner, Ann. Rev. Astron. Astrophys. **6**, 165 (1968).
[7] H. J. Fahr and P. W. Blum, Astron. Astrophys. **8**, 226 (1970).
[8] R. R. Meier, J. Geophys. Res. **74**, 6487 (1969).
[9] L. A. Hall and H. E. Hinteregger, J. Geophys. Res. **75**, 6959 (1970).
[10] R. Tousey, Space Sci. Rev. **2**, 3 (1963).

Space Research XII — Akademie-Verlag, Berlin 1972

INFLUENCE OF INTERSTELLAR HYDROGEN ON THE LOCATION OF THE HELIOSPHERIC SHOCK FRONT

H. J. Fahr

Institut für Astrophysik der Universität Bonn, Bonn, FRG

Parker's hydrodynamical solution of the solar wind yields a supersonic flow with constant outward velocities. When a process of momentum loss of the solar wind plasma due to secondary protons originating from interstellar hydrogen in the heliosphere is taken into account, a decrease of the solar wind velocity towards the outer heliosphere occurs. This velocity decrease influences the location of the shock front and brings it closer to the sun in regions of enhanced hydrogen densities. The numerical calculations show that the geometrical configuration of the shock front is nearly spherical, excluding the downwind region where the focusing effect causes enhanced densities of interstellar hydrogen and helium, and thereby gives rise to a trough in the shock front.

Parker's hydrodynamical solution of the solar wind expansion yields a supersonic plasma flow beyond a critical distance r_c from the sun with an essentially constant expansion velocity. This solution is derived from a hydrodynamical treatment of the coronal expansion that does not take into account any interaction of the solar wind plasma with neutral interstellar matter which enters the solar system. In that case the supersonic solar wind expands with a constant velocity v_p up to a specific distance r_s where the energy density of the solar wind plasma due to the geometrical divergence of the flow has decreased to the value $B_i^2/8\pi$ of the energy density of the interstellar magnetic field. Beyond this distance r_s the interstellar field B_i starts controlling the expansion of the solar wind, which means that the undisturbed hydrodynamical flow of the solar wind plasma in radial directions terminates at this region. At the distance r_s the solar wind is presumed to undergo a shock transition. The location r_s of this shock region which terminates the heliosphere can be estimated by equating the two competing energies to yield:

$$r_s = (2\, v_p/B_i)\, (n_E\, m_p\, \pi)^{1/2} \tag{1}$$

where n_E is the density of solar wind protons at the earth's orbit ($r = r_E = 1$ AU) and m_p is the mass of a proton. Interstellar hydrogen atoms which enter the solar system from interstellar space become ionized while approaching the sun, due to charge transfer reactions with solar wind protons and due to the solar euv radiation. These ionization processes which lead to the generation of secondary protons in the solar wind give rise to the extraction of kinetic energy from the expanding solar wind plasma.

If a solar wind proton of velocity v_p is lost from the plasma ensemble by a charge exchange process with an interstellar hydrogen atom, the solar wind plasma loses momentum $m_p v_p$ and kinetic energy $^1/_2\, m_p v_p^2$. The simultaneous gain of energy due to the creation of a secondary proton can be neglected, since its initial velocity equals that of the former hydrogen atom which is at least about one order of magnitude smaller than v_p. Immediately after its creation the secondary proton is affected by the frozen-in magnetic field which moves with the solar wind. In general secondary protons gain energy by moving in the magnetic field. This energy which is indirectly taken from the dominating kinetic energy of the plasma reappears in a kind of thermal and kinetic energy of these secondary protons. The average motion of secondary protons depends on the direction of the frozen-in magnetic field with respect to the bulk motion $\boldsymbol{v}_p$ of the plasma and is in general not synchronized to that motion. Only that part of the kinetic energy of secondary protons which corresponds to their average velocity component parallel to v_p contributes to the kinetic energy of the solar wind plasma. Due to this fact it is found [1] that the following loss ΔE_{kin} occurs per secondary proton:

$$\Delta E_{kin} = -\,^1/_2\, m_p v_p^2 \,(1 + 2 \sin^2 \alpha - \sin^4 \alpha) \tag{2}$$

where α is the angle between $\boldsymbol{v}_p$ and the frozen-in field $\boldsymbol{B}$, if the secondary proton considered has been created by charge exchange. When the secondary proton is generated by an euv ionization process, the energy loss ΔE_{kin} amounts only to

$$\Delta E_{kin} = -\,^1/_2\, m_p v_p^2 \,(\sin^2 \alpha - \sin^4 \alpha)\,. \tag{3}$$

This indicates that the mean loss of kinetic energy per secondary proton depends both on the velocity v_p of the solar wind and on the angle α between $\boldsymbol{B}$ and $\boldsymbol{v}_p$. Since the lines of the interplanetary magnetic field follow an Archimedean spiral, the angle α is a function of the distance r from the sun. At $\alpha = 0°$ (close to the sun) the energy loss attains a minimum, whereas at $\alpha = 90°$ (far from the sun, i.e. purely azimuthal fields) the energy loss attains a maximum.

In order to calculate numerically the loss of kinetic energy from the solar wind plasma due to the interaction with neutral interstellar matter we shall first assume that the inclination α vanishes in the whole interplanetary region which is of interest for this problem. This case would be realized in space for a non-rotating corona which produces a radially extended interplanetary magnetic field. In that case only the energy loss ΔE_{kin} as given in (2) which is connected with those secondary protons originating from charge exchange collisions has to be taken into account. We shall now calculate the rate of energy loss which would be due to this kind of secondary protons. The density distribution of interstellar hydrogen atoms within the solar system has been calculated by Blum and Fahr [2]. The macroscopic motion $\boldsymbol{v}_0$ of interstellar hydrogen relative to the solar system leads to an anisotropic density distribution within the heliosphere with maximum densities in the direction $\boldsymbol{v}_0$ of approach of interstellar matter and minimum densities in the region opposite to this direction. A solar wind proton which starts from the surface of the sun in a direction θ to $\boldsymbol{v}_0$ has a definite chance $W(r, \theta)$ of reaching a distance r from the sun in a radial flight without having suffered a charge exchange collision. W is given by

$$W(r, \theta) = \exp\left[-\frac{q_{ex}}{v_p}\int_{r_0}^{r} v_{rel}\, n(r, \theta)\, dr\right], \tag{4}$$

where q_{ex} is the charge exchange cross section, v_{rel} is the relative velocity between solar wind protons and neutral interstellar atoms at the point (r, θ), r_0 is the distance of the solar surface from the solar centre, and $n(r, \theta)$ is the hydrogen density at the point (r, θ). That fraction of solar wind protons which has already undergone charge exchange collisions, $1 - W$, has been transformed

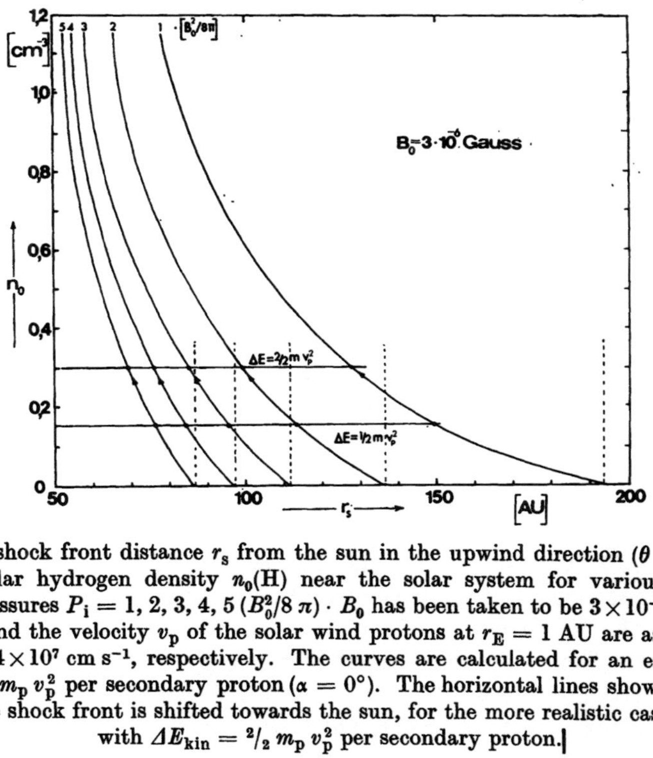

Fig. 1. The shock front distance r_s from the sun in the upwind direction ($\theta = 0°$) versus the interstellar hydrogen density $n_0(H)$ near the solar system for various interstellar magnetic pressures $P_i = 1, 2, 3, 4, 5\ (B_0^2/8\pi) \cdot B_0$ has been taken to be 3×10^{-6} gauss. The density n_E and the velocity v_p of the solar wind protons at $r_E = 1$ AU are assumed to be 5 cm^{-3} and 4×10^7 cm s^{-1}, respectively. The curves are calculated for an energy loss of $\Delta E_{kin} = {}^1/_2\, m_p v_p^2$ per secondary proton ($\alpha = 0°$). The horizontal lines show the amount by which the shock front is shifted towards the sun, for the more realistic case of $\alpha = 90°$ with $\Delta E_{kin} = {}^2/_2\, m_p v_p^2$ per secondary proton.

into fast neutral particles which carry away the energy of the former solar wind protons. According to (2), the kinetic energy density $D(r, \theta)$ which still has remained at the distance r is therefore given by

$$D(r, \theta) = {}^1/_2\, n_E (r_e/r)^2\, m_p\, v_p^2\, W(r, \theta)\,. \tag{5}$$

Owing to the energy loss the density $D(r, \theta)$ of the kinetic energy of the solar wind plasma decreases more rapidly than $1/r^2$. This, however, means that the location of the heliospheric shock front is determined by this kind of energy loss. Instead of Eq. (1) we obtain the following relation:

$$B_i^2/8\pi = {}^1/_2\, n_E\, (r_E/r)^2\, m_p\, v_p^2\, W(r_s, \theta) = D(r_s, \theta)\,. \tag{6}$$

The probability function W can be numerically calculated by the use of the densities $n(r, \theta)$ given in the Blum–Fahr model. The shock front location r_s can then be obtained from Eq. (6) by the determination of the intersection of $D(r, \theta)$ and $B_i^2/8\pi$. The calculation of the location of the shock front according to formula (6) has been carried out for the numerical values of n_E, q_{ex}, v_p, the solar proton fluxes F_E, and euv-fluxes f_E as given in the paper of Fahr [3]. Studies of the Zeeman splitting of interstellar spectral lines by Verschuur [4] and Davies et al. [5] seem to support a value of $B_i = B_0 = 3 \times 10^{-6}$ gauss for the interstellar magnetic field, though larger values cannot be definitely excluded. Fig. 1 shows the result of the computation of r_s in the direction of approach of interstellar matter ($\theta = 0°$, upwind direction). The curves give the location r_s of the shock front versus the density n_0 of interstellar hydrogen outside the solar system for various multiples of the interstellar energy density $B_0^2/8\pi$. It can be clearly seen how r_s is affected by the density of the interstellar hydrogen.

Though the results of Fig. 1 are based on the assumption of $\alpha = 0°$ (radially extended fields) it can easily be seen how the location of the shock front changes, if the real interplanetary magnetic field with $\alpha \neq 0°$ is considered. For $\alpha \neq 0°$ the energy loss ΔE_{kin} in formula (3) due to secondary protons originating from euv ionizations of interstellar hydrogen had principally to be taken into account. Since, however, the energy loss due to such secondary protons at most amounts to one half of that due to protons originating from charge exchange, and since the ionization frequency $\nu_i = q_{euv} f_E$ (q_{euv} is the weighted ionization cross section in the Lyman continuum) is only one fourth of the charge exchange frequency $\nu_{ex} = q_{ex} F_E$ the energy losses due to ionized neutral particles can be neglected to a sufficiently good approximation. In that case we only remain with the energy loss ΔE_{kin} given in (2), which attains its maximum for $\alpha = 90°$ when maximum amounts of kinetic energy are extracted from the solar wind plasma. ΔE_{kin} ($\alpha = 90°$) is exactly twice ΔE_{kin} ($\alpha = 0°$). Therefore the shock front certainly comes closer to the sun compared with the results shown in the curves of Fig. 1, if the real structure of the interplanetary field with $\alpha \neq 0°$ is taken into account. The exact shock front location r_s for the maximum type of energy loss $\Delta E_{kin} = {}^2/_2\, m_p\, v_p^2$ (purely azimuthal fields) can easily be found from Fig. 1 by taking that value of r_s which would correspond to twice the value n_0 that is actually realized in near interstellar space, since the energy loss for n_0, $\alpha = 90°$ is exactly identical with that for $2\, n_0$, $\alpha = 0°$. This is illustrated by the horizontal lines in Fig. 1 that demonstrate the change of r_s for an increase of α from 0° to 90°. It can now easily be argued that the real shock front location r_s should be very close to that value of r_s which is given by Fig. 1 for twice the actual density value n_0. This follows from the fact that the average penetration depth r_{ex} of an interstellar hydrogen atom, which is the distance from the sun where average interstellar hydrogen atoms entering the solar system suffer charge exchange with primary solar wind protons, is between 2 and 10 AU depending on the phase within the solar activity cycle [2]. This, however, means that the energy loss of the solar wind plasma is predominantly determined by the generation of secondary protons outside this region, i.e. for $r > r_{ex}$.

On the other hand the interplanetary magnetic field has an inclination of $\alpha = 45°$ at 1 AU and changes to even larger angles beyond this distance. Therefore the inclinations α of the spirally wound interplanetary field B may not be far from $\alpha = 90°$ at $r = r_{ex}$. Thus the maximum type of energy loss $\Delta E_{kin} = {}^2/_2\, m_p\, v_p^2$ per secondary proton can be assumed to take place all over

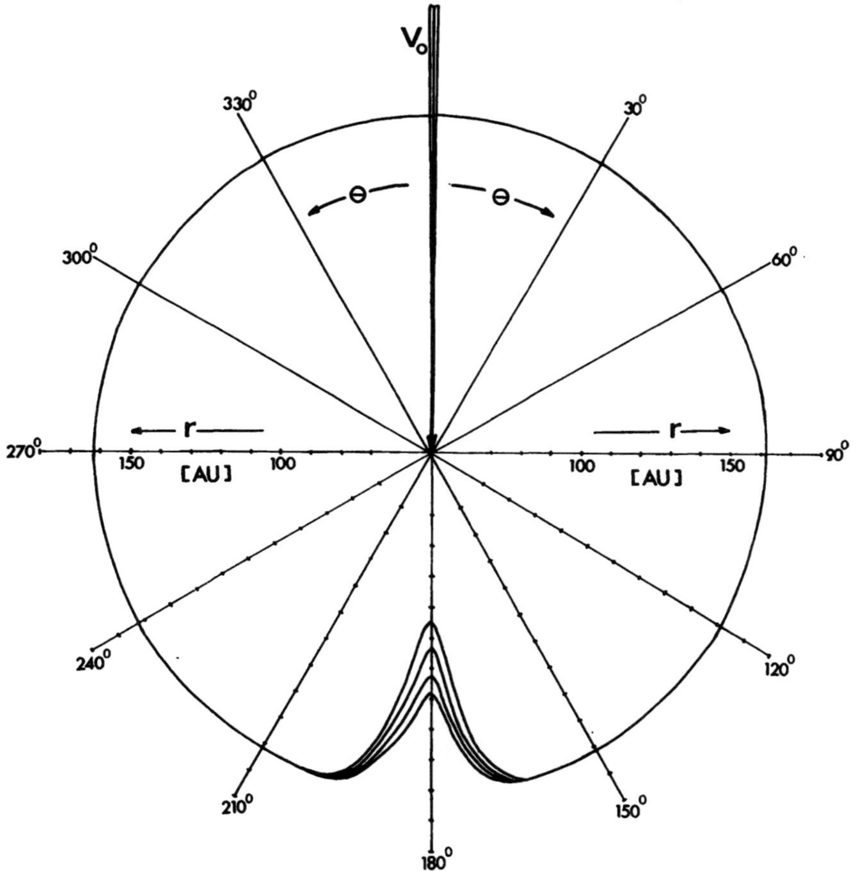

Fig. 2. The geometrical shape of the heliospheric shock front. The polar coordinates are the distance r from the sun (logarithmic scale) and the angle θ with respect to the upwind direction. ($n_E = 5\ \text{cm}^{-3}$, $v_p = 4 \times 10^7\ \text{cm s}^{-1}$, $n_0 = 0.1\ \text{cm}^{-3}$). The energy loss per secondary proton has been assumed to be $\Delta E_{kin} = {}^1/_2\, m_p\, v_p^2$ ($\alpha = 0°$). The shock front cusp in the downwind region is due to the energy extraction of gravitationally focused interstellar hydrogen of temperatures $T = 150\ °K$ (innermost curve), 300 °K, 450 °K and 600 °K (outermost curve).

the interplanetary space. Therefore one must conclude that the location r_s of the shock front for an interstellar density n_0 is very close to that value given in Fig. 1 for the value $2\, n_0$.

The anisotropic spatial distribution of hydrogen in interplanetary space should give rise to a non-spherical shape of the shock front, i.e. r_s should be a function of the angle θ. In spite of the anisotropic distribution, however, the shock front location r_s shows nearly no variation ($\Delta r_s < 2$ AU) with θ within the range $0° \leqq \theta \leqq 170°$; $190° \leqq \theta \leqq 360°$, as is shown in Fig. 2. This is due to the fact that the radial density profiles $n(r, \theta)$ differ only from each other inside the region $r = 5$ AU, whereas the value of r_s is mainly determined by the densities outside $r = 5$ AU. The result of a spherical shape of the shock front for $0° \leqq \theta \leqq 170°$ shown in Fig. 2 contrasts with the suggestions made by Axford et al. [6] and Dessler [7] who argued that the helio-sphere due to the bulk motion of the interstellar matter relative to the solar system should be

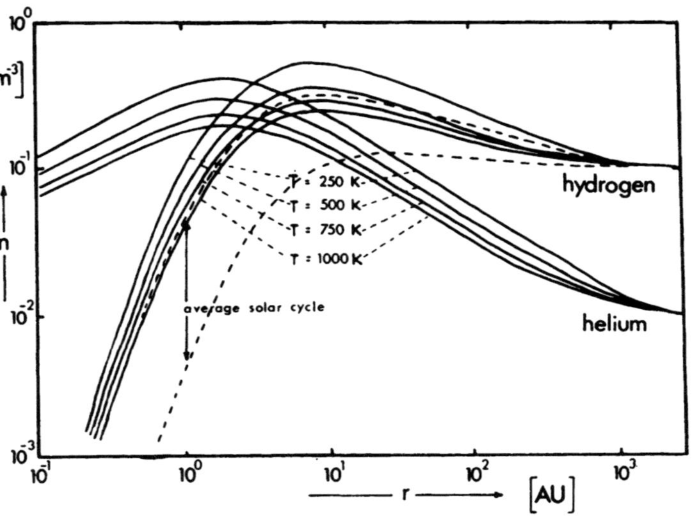

Fig. 3. The densities of interstellar hydrogen and helium are shown versus the distance r from the sun in downwind direction ($\theta = 180°$) for various interstellar temperatures T. Two values of the densities of interstellar gas near the solar system have been taken to be $n_0(\mathrm{H}) = 0.1$ cm^{-3} and 0.01 cm^{-3}. The density reduction of interstellar hydrogen due to the solar-cycle-dependent radiation pressure is shown by the two dashed curves for the $T = 250$ °K profile. (Upper curve, solar minimum; lower curve, solar maximum).

aspherical with minimum values of r_s in the direction of approach of interstellar matter ($\theta = 0°$, upwind direction) and maximum values of r_s in the opposite direction ($\theta = 180°$, downwind direction). A deviation from the spherical shape of the shock front, strictly different from the suggestions of the above-mentioned authors, has to be expected only at angles $170° \leqq \theta \leqq 190°$ where the focusing effect of the solar gravitational field on the inflow of interstellar gas leads to relatively high interstellar densities $n(r, \theta)$ [8]. Fig. 3 shows densities of interstellar hydrogen and helium versus the distance r from the sun along the antapex axis ($\theta = 180°$) for various temperatures T of the interstellar gas. The hydrogen density values have been calculated for a solar wind proton flux of $F_E = 2 \times 10^8$ protons cm^{-2} s^{-1}, a solar euv flux of 10^{11} photons cm^{-2} s^{-1} in connection with a mean weighted ionization cross section of $q_{euv} = 1.2 \times 10^{-18}$ cm^2 that has been determined by averaging the cross section $q_{euv,c}(\nu_c/\nu)^3$ with $q_{euv,c}$ given by Ambarzumian [9] over the solar euv spectrum given by Hinteregger and Hall [10]. The charge exchange cross section between protons and hydrogen atoms has been taken to be 10^{-15} cm^2. The helium values have been calculated for a mean weighted ionization cross section of q_{euv}(He) of 1.6×10^{-18} cm^2 with the critical cross section of $q_{euv,c}$(He) $= 7.3 \times 10^{-18}$ cm^2 at 504 Å taken from [11] in connection with the solar helium continuum radiation below 504 Å given by Hinteregger and Hall [10]. The main charge exchange processes of helium atoms in the solar wind are of the reaction types

$$\text{(a)}\ He^{++} + He \rightarrow 2\,He^{+}; \qquad \text{(b)}\ p^{+} + He \rightarrow H + He^{+}.$$

Reaction type (a) is characterized by a cross section q_{ex}(He) $= 5 \times 10^{-17}$ cm^2 [12], and reaction type (b) by q_{ex}(He) $= 4.8 \times 10^{-17}$ [13]. Owing to the resulting frequencies ν_i for radiative ionization and ν_{ex} for charge exchange, helium can penetrate much deeper than hydrogen into the solar system before it becomes ionized. This becomes obvious in the curves of Fig. 3. It can be seen that the helium densities, though amounting only to one tenth of the density n_0 of interstellar hydrogen at infinity ($r > 10^3$ AU), dominate over the hydrogen densities for distances smaller than $r = 3$ AU. Whereas the solar radiation pressure acting upon the hydrogen atoms compensates between $^1/_3$ and $^2/_3$ of the gravitational force during the 11-year solar cycle [8] the corresponding solar radiation pressure acting upon the helium atoms is negligible compared with the gravitational attraction. This is partly due to the fact that the total energy which is contained in the solar helium-α line at 584 Å (0.03-0.05 erg cm^{-2} s^{-1}) is about two orders of magnitude smaller than that contained in the solar Lyman-α line (3–6 erg cm^{-2} s^{-1}) [10, 14, 15] and partly due to the fact that the Einstein transition coefficient for the helium 584 Å transition $B_{1,2} = 6.54 \times 10^{18}$ cm^3 s^{-2} erg^{-1} is about a factor of 2 smaller than $B_{1,2} = 2.01 \times 10^{19}$ cm^3 s^{-2} erg^{-1} for the Lyman-α transition. This leads to a force K_r due to solar radiation pressure acting on the helium atoms which is about four orders of magnitude smaller than the gravitational force $K_g = 3.96 \times 10^{-24}$ erg cm^{-1}. Therefore the helium profiles do not change during the solar activity cycle, whereas the hydrogen profiles actually do as indicated by the dashed lines in Fig. 3.

Fig. 4 shows how the angular density profile of hydrogen changes during the average solar cycle due to the effect of the solar Lyman-α radiation pressure. The parameter u which gives the compensation of the gravitational force is defined by $u = (K_g - K_r)/K_g$ and varies between $^1/_3$ and $^2/_3$ during the average solar cycle. It is indicated in Fig. 4 that the density differences between the curves belonging to $T = 250$ °K and to $T = 1000$ °K in the antapex direction ($\theta = 180°$) decrease with decreasing values of the parameter u.

Due to the enhanced densities in the region $170° \leq \theta \leq 190°$, the extraction of kinetic energy from the solar wind due to the interaction with neutral interstellar matter is strongly pronounced in this region. This leads to an approach of the shock front to the sun, i.e. the shock front should be expected to have a characteristic trough in the antapex direction; the depth of this cusp depends on several parameters among which are the temperature of the interstellar matter and the effective radiation pressure characterized by the quantity u.

Fig. 2 shows the shock front trough resulting from the energy extraction of interstellar hydrogen atoms with $n_0 = 0.1\ \mathrm{cm}^{-3}$ in the region of the hydrogen cone (antapex direction) for $u = 1$ (solar gravitation uncompensated). For the more real case $u < 1$ realized during the solar cycle, the effect of hydrogen

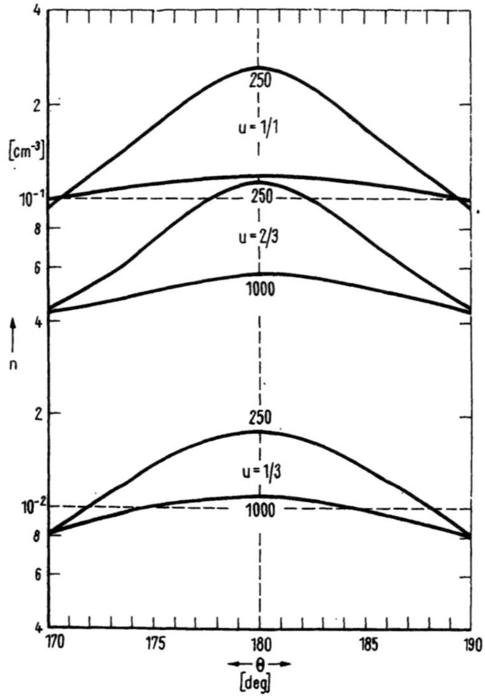

Fig. 4. The interstellar hydrogen density ($T = 250\ °\mathrm{K}$, $T = 1000\ °\mathrm{K}$) at $r = 2$ AU versus the inclination θ with respect to the velocity $\boldsymbol{v}_0$ of approach of interstellar matter for various reductions u of the gravitational force due to the counter-balancing effect of the solar radiation pressure. According to measurements reviewed by Tousey [15] the average solar cycle should be characterized by a value u ranging from $^2/_3$ (solar minimum) to $^1/_3$ (solar maximum) [8]. The density $n_0(\mathrm{H})$ near the solar system has been assumed to be $0.1\ \mathrm{cm}^{-3}$.

on the solar wind energy exhaustion is reduced. On the other hand the extraction of energy due to helium atoms in the solar wind whose densities do not change during the solar cycle has to be regarded additively and contributes well to the antapex energy losses of the solar wind.

References

[1] H. J. Fahr, Preprint 71-10 of Astronomical Institute, University of Bonn (1971).
[2] P. W. Blum and H. J. Fahr, Astron. Astrophys. **4**, 280 (1970).
[3] H. J. Fahr, Planet. Space Sci. **19**, 1121 (1971).
[4] G. L. Verschuur, Phys. Rev. Lett. **21**, 775 (1968).

[5] R. D. Davies, R. S. Booth and A. J. Wilson, Nature, Lond. **220**, 1207 (1968).
[6] W. I. Axford, A. J. Dessler and B. Gottlieb, Astrophys. J. **137**, (1963).
[7] A. J. Dessler, Rev. Geophys. **5**, 1 (1967).
[8] H. J. Fahr, Astron. Astrophys. in press (1971).
[9] V. A. Ambarzumian, in: Theoretical Astrophysics, VEB Dt. Verlag d. Wissenschaften, Berlin 1957 (p. 45).
[10] H. E. Hinteregger and L. A. Hall, Space Research V, 1175 (1970).
[11] Po Lee and G. L. Weissler, Astrophys. J. **115**, 570 (1952).
[12] W. L. Fite, R. F. Stebbings and H. C. Smith, Proc. Roy. Soc. A **268**, 527 (1962).
[13] P. Mason and D. L. Vanderslice, J. chem. Phys. **40**, 3552 (1955).
[14] H. E. Hinteregger and L. A. Hall, J. Geophys. Res. **75**, 6959 (1970).
[15] R. Tousey, Space Sci. Rev. **2**, 3 (1963).

Space Research XII – Akademie-Verlag, Berlin 1972

RECENT RESULTS OF THE GODDARD ROCKET PROGRAM FOR OBSERVING STARS

A. B. Underhill

NASA Goddard Space Flight Center, Greenbelt, Md, USA

Rockets flown by A. M. Smith have yielded spectra of Alpha Vir, Bl V, Zeta Puppis, O 5f, Zeta Ori, O9.5 Ib, Zeta Oph, O9.5 V and Zeta Per, Bl Ib, at about 0.5 Å resolution. Many weak subordinate lines are visible in the range 924 to 1700 Å as well as strong resonance lines, some of which have P Cygni profiles. Sharp interstellar lines from C I, C II, O I and Si II are seen. Levels having term values of at least 63.42 cm^{-1} but less than 158 cm^{-1} are excited which implies a temperature in the range 60 °K $< T <$ 152 °K. Stecher and Smith have confirmed the presence of CO in the interstellar medium from lines in the ultraviolet spectra of Zeta Oph and Zeta Per. Absolute energy calibrations of the ultraviolet energy distribution of Alpha Leo, Alpha CMa, Gamma Ori and Kappa Ori have been derived with accuracy to about $\pm$15% over the range 1350–3100 Å from rocket spectra at 10 Å resolution obtained by D.C. Evans. About half the uncertainty (6–10%) may be attributed to the transfer to primary laboratory energy standards.

1. High Resolution Spectra

A few results from a 1-meter Wadsworth-type spectrograph designed by A. M. Smith for flight on Aerobee rockets are presented. The linear dispersion is close to 4.51 Å mm^{-1} and a resolution of about 0.5 Å has been obtained. The spectra are recorded on Kodak ultraviolet sensitive film (101-01). Studies of the spectra of α Vir, Bl V, ζ Pup, O 5f, and ζ Ori, O 9.5I b, have been published [1–3]. Extracts from spectra of ζ Oph, O 9.5V, and of ζ Per, Bl Ib, are presented here to illustrate the details to be seen on Smith's spectrograms. The spectrograms have been traced with the digitizing microdensitometer at the Goddard Space Flight Center and are presented as tracings of relative intensity plotted against wavelength. The spectral region from 1290 to 1340 Å is shown in Fig. 1. A few lines are identified in this figure to illustrate particular points of interest. Smith has completed a detailed identification study of the spectrum of ζ Oph from 1160 to 1700 Å and is now working on the spectrum of ζ Per. About one-third of the lines cannot be identified from the available laboratory line lists.

Typical lines from the interstellar gas appear in Fig. 1. There is a broad feature due to C I, 2p^2 ^{3}P–2 p^3 ^{3}P^0, at 1329 Å. This multiplet consists of six closely spaced lines and the width of the feature indicates that all may be present. The term values of the ground levels are 0.00, 16.40 and 43.40 cm^{-1}. Better resolution will be required to resolve this multiplet. Two lines from the C II resonance multiplet 2 p ^{2}P^0–2 p^2 ^{2}D at 1334.53, 1335.66 and 1335.71 Å are seen.

Clearly both $2^2P^0_{1/2}$ and $2^2P^0_{3/2}$ have sufficient populations to give interstellar lines. The term values of these levels are 0.00 and 63.42 cm^{-1}. On the other hand, the ground levels of OI have term values 0.00, 158.5 and 226.5 cm^{-1}. Only one line, 1302.17 Å, is seen and this may be due partly to the earth's atmosphere. The lines at 1304.86 and 1306.02 Å are weakly present and probably arise in the earth's atmosphere. Only the interstellar line 1304.37 Å

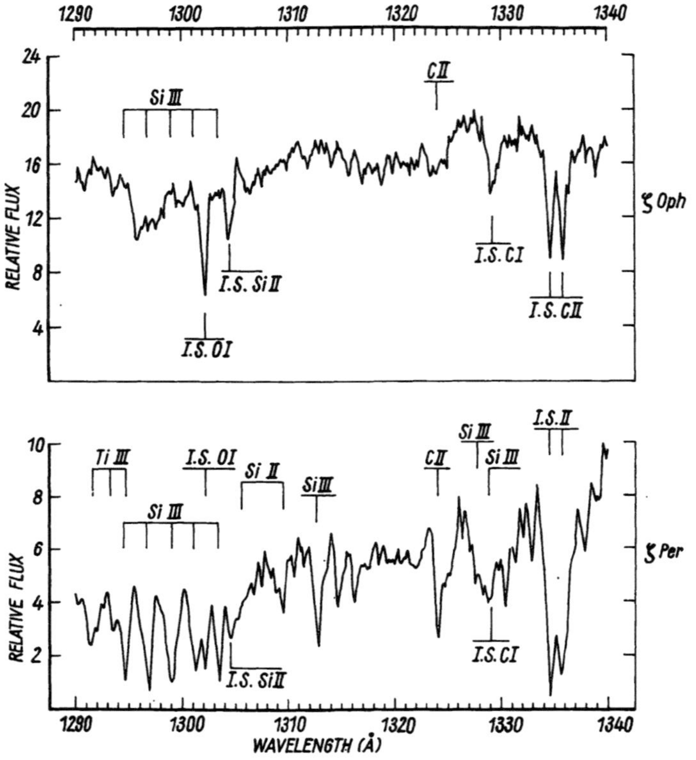

Fig. 1. Intensity tracings of spectra of ζ Oph and ζ Per in the range 1290–1340 Å.

of the resonance multiplet of Si II is present. It arises from the level of zero energy. The line at 1309.27 Å from the level of higher energy does not appear.

If the lowest levels of C I and C II are excited by collisions but only the levels of zero energy of O I and Si II are excited,

$$hc \times 64 < \tfrac{3}{2} kT < hc \times 158 ,$$

which gives the following limits on the temperature of the interstellar gas where the C I and C II lines are formed:

$$60\ °\mathrm{K} < T < 152\ °\mathrm{K} .$$

The complexity of the ultraviolet spectrum of stars is well illustrated by these extracts from the spectra of ζ Ophiuchi and ζ Persei. Lines of Ti III and Si III are much stronger at type Bl Ib than at O 9.5 V. This is in accord with what is known from ground-based spectra.

A longer wavelength section is shown in Fig. 2. Here in addition to the ubiquitous Si III, the resonance lines of Si IV are seen and lines of Al III. Many of the unmarked features seem to be due to the third spectra of the metals and other light elements. The strong feature near 1427 Å is probably C III 3s ^3S–2p3s $^3P^0$. This multiplet has the same lower level as the well-known multiplet 3s ^3S–3p $^3P^0$ at 4647, 50, 51 Å. The resonance lines of Si IV in ζ Per, a Ib supergiant, have an asymmetric shape which suggests outflow motions.

Smith and Stecher [4] have drawn attention to lines of CO in the spectrum of ζ Ophiuchi. Lines from CO appear also in the spectrum of ζ Persei. The line

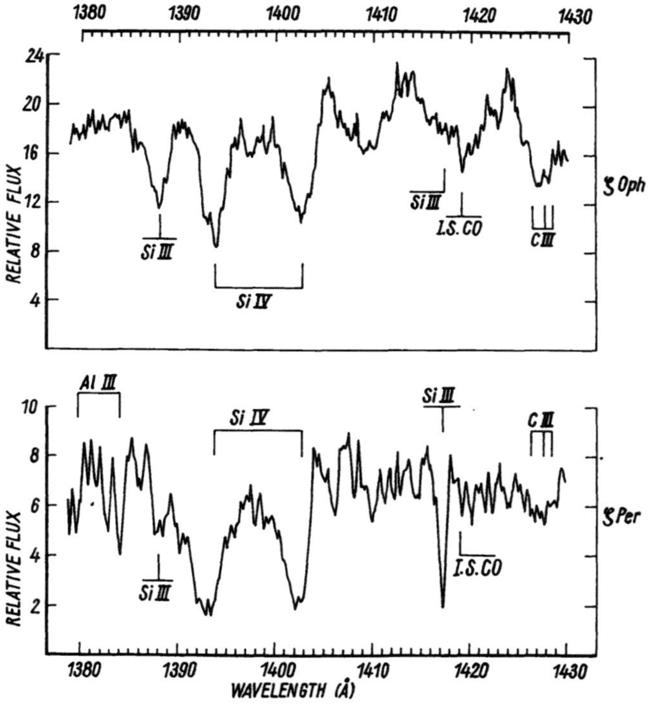

Fig. 2. Intensity tracings of spectra of ζ Oph and ζ Per in the range 1380–1430 Å.

at 1419.0 Å, which represents the 4-0 band of the $A^1\Pi$–$X^1\Sigma^+$ (fourth positive) system, is indicated. Smith and Stecher have argued that the strengths of the CO lines in ζ Ophiuchi indicate a rotational excitation equivalent to a 3 °K radiation field. The definite appearance of *both* C II resonance lines demands a temperature of the order of or greater than 60 °K; the wide appearance of the line attributed to interstellar C I also suggests this.

The immense complexity of ultraviolet stellar spectra which is revealed by a resolution of 0.5 Å makes apparent the great difficulty of deducing any but the crudest physical information from spectral scans at 10 or 20 Å resolution. Future satellite-borne stellar spectrographs having resolutions of the order of 0.1 Å will be needed to resolve the problems touched upon here.

2. Absolute Energy Calibrations

A rocket payload consisting of a 33 cm diameter telescope coupled to a scanning diffraction grating spectrometer has been flown three times and a fourth flight is planned. Two scans of α Leo were obtained on the first flight and on the third flight; in addition one scan of α CMa, two scans of $\varkappa$ Ori and five scans of γ Ori were obtained on the third flight. The second flight yielded no usable data. The spectrum is scanned by three photomultipliers each with a viewing slit 10 Å wide. The total range scanned in three overlapping sections of about 1500 Å each is from 1150 to 4150 Å.

The payload has been calibrated in the Laboratory for Optical Astronomy vacuum calibration chamber by viewing a beam of collimated light simultaneously with the rocket payload and with four reference photomultipliers. The absolute calibration of the reference photomultipliers is determined at 2537 Å using four thermopiles with calibrations traceable to the National Bureau of Standards (NBS). An open window nitrous oxide ion chamber has been used at 1216 Å with limited success. The relative response of the payload was determined longward of 2537 Å by observing tungsten-iodide lamps which had been calibrated by NBS. The range 1150–3400 Å has been referred to sodium salicylate which is assumed to be a constant efficiency quantum detector. The spectral range 2000–2400 Å has been referred to two photodiodes whose absolute quantum efficiency has been determined at NBS. So far all of the techniques of calibration have produced values which range over about 10%, which is well within the goal of the calibration program.

Post-flight recalibration of the two recovered payloads showed no detectable change in the shape of the spectral response and less than a 5% decrease in absolute sensitivity. The search for systematic errors is continuing.

The absolute spectral energy distributions derived for α CMa, α Leo, γ Ori and $\varkappa$ Ori are shown in Fig. 3. There appears to be a systematic difference between the two observations of α Leo shortward of 2600 Å. The first observation is about 10–20% low relative to the second observation which is believed to yield the more reliable spectral shape. The uncertainty in the final data is about $\pm 15\%$ owing to a non-linearity in the response. Corrections have been applied for this non-linearity and continuing analysis may reduce this error range.

There is a distinct dip in the spectrum of α CMa and of α Leo between about 2200 and 2900 Å. This is due to the blending of many strong lines due chiefly to Fe II and Mg II. It is quite noticeable that as the spectral type becomes earlier the point at which the steep decline in the spectrum begins approaches shorter wavelengths. Thus, for α CMa, A1V, the decline begins at about 1700 Å; for α Leo, B7V, it begins at about 1500 Å; for γ Ori, B2III, it begins at about 1200 Å while for $\varkappa$ Ori, B0Ia, it begins at about 1300 Å. The spectrum of the supergiant seems to have a significant dip between 1500 and 1800 Å which may be due to strong lines. None of these stars suffers serious interstellar reddening.

The newly derived intensity distributions are compared in Fig. 4 with the photometric results of others and with the 10 Å resolution scans of Stecher.

The ground based photometry of Mitchell and Johnson has been transferred to the absolute flux system of Oke and Schild. These results are indicated by the subscript MJOS. The present results are drawn as thin lines and can be identified

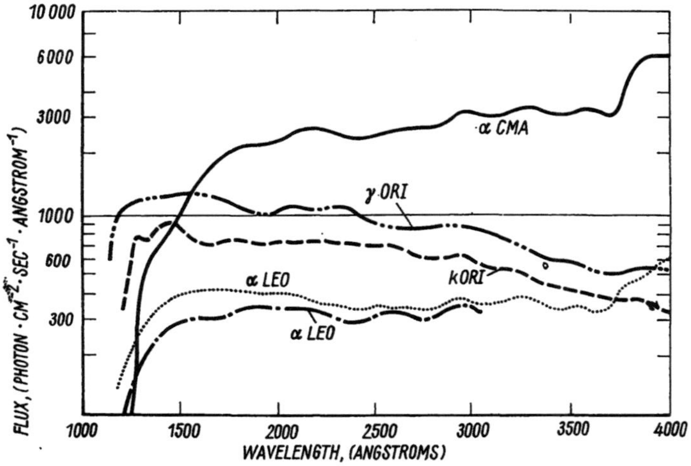

Fig. 3. Absolute energy distributions for α CMa, α Leo, γ Ori and $\varkappa$ Ori determined at 10 Å resolution. The dotted line for α Leo comes from a rocket flight on 14 March 1969. All the other curves come from a flight on 31 October 1970.

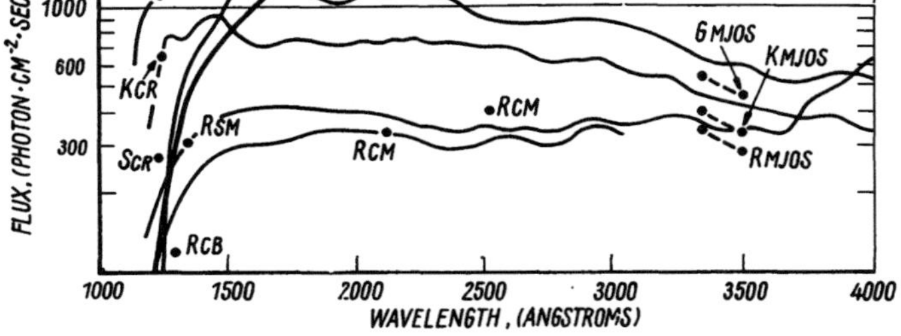

Fig. 4. Comparison of the new absolute energy distributions with the results of others.

from Fig. 3. The zero point of each photometric system has been determined independently. It is clear that the various results agree reasonably well, particularly when the spectrum is fairly flat over the width of the photometric bandpass.

We hope to be able to continue this exacting work in order to establish even more securely known standard energy distributions for a few representative stars. The fluxes are presented as number of photons per cm^2 per sec per Å bandwidth in Figs. 3 and 4.

References

[1] A. M. SMITH, Astrophys. J. **156**, 93 (1969).
[2] A. M. SMITH, Astrophys. J. **160**, 595 (1970).
[3] A. M. SMITH, Astrophys. J. **172**, No. 1 (1972).
[4] A. M. SMITH and T. P. STECHER, Astrophys. J. Lett. **164**, L43 (1971).

Space Research XII — Akademie-Verlag, Berlin 1972

HIGH ANGULAR RESOLUTION ABSOLUTE INTENSITY OF THE SOLAR CONTINUUM FROM 1400 Å TO 1790 Å

G. E. Brueckner and O. K. Moe *

U. S. Naval Research Laboratory, Washington, D.C., USA

Absolute intensities of the solar uv continuum from 1400 Å to 1790 Å have been measured from rocket spectra taken on 13 August 1970. The spectra had an angular resolution of 2 arc sec × 1 arc min and the pointing accuracy of the instrument was ±2 arc sec. This permits us to study the center-to-limb variation of the intensity with a spatial resolution of 2 arc sec. Four positions on the solar disk have been studied corresponding to values of cos θ = 0.12, 0.22, 0.28 and 0.72, where θ is the heliocentric position angle. The measurements have been calibrated using a direct comparison with an absolute calibrated deuterium lamp for wavelengths above 1660 Å. For shorter wavelengths a relative calibration has been made using observations of the relative strengths of the LBH molecular bands from N_2.

The measurements give higher values for the intensity than recent photoelectric measurement, but are in good agreement with the intensities of Widing et al. From the center-to-limb variation above 1550 Å, (T, τ) relations have been deduced giving a temperature minimum of 4450 °K.

1. Observations

Intensity measurements of the solar continuum from 1400 Å to 1790 Å have been made from rocket spectra taken on 13 August 1970. The angular resolution of the instrument is determined by the size of the entrance slit of the spectrograph which is 2 arc sec × 1 arc min projected on the solar disk, and by the pointing jitter of ±2 arc sec. A total number of 18 spectra was taken during the flight with the instrument pointed at four different regions on the sun: at 7, 24, 40 and 300 arc sec distance from the solar limb along a radius near the solar equator and with the slit perpendicular to the radius. This makes it possible to study the center-to-limb variation with a spatial resolution of ±2 arc sec. Exposures were taken using four different exposure times at each position. Of these the two groups with longest exposure times give spectra suitable for continuum measurements. These exposure times were in the ranges 8 to 16 sec and 32 to 64 sec, respectively.

The spectra cover the wavelength range from 1175 Å to 1795 Å, but only for wavelengths above 1400 Å are we presently able to obtain an absolute calibration of our instrument. The second-order spectrum was observed with a 1100 lines/mm grating. A predisperser was used in front of the main grating.

---

* On leave from the University of Oslo.

The spectral resolution of the instrument is 70 mA judging from the half-widths of the sharpest lines and is determined mainly by the width of the spectrograph slit. Fig. 1 shows as an example a small portion of the limb spectrum (7 arc sec) near the Si I ^{3}P edge. A total of about 1100 emission lines are visible in this spectrum, most of them at wavelengths greater than the 1520 Å Si I absorption edge.

The high spectral resolution makes it easy to select good windows for the continuum measurements. The windows have been selected well away from strong emission lines in the limb spectrum and also at safe distance from absorption features in the disk spectrum. Approximately 45 windows have

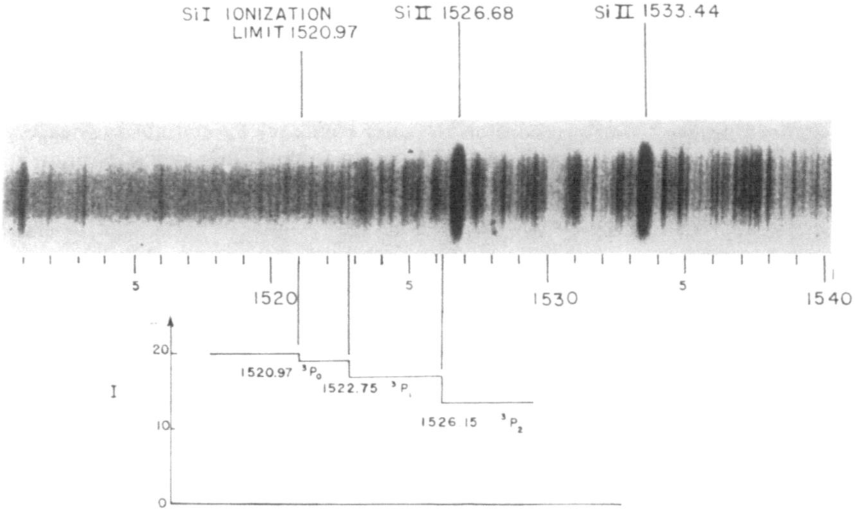

Fig. 1. Part of the limb spectrum (7 arc sec) around 1520 Å.

been picked, the width of the windows ranging from 1 Å to 5 Å. Within each window intensities have been measured at several (usually 5–10) points and the mean value taken as representing the intensity inside the window. The scattering of the intensity values within each window will contain contributions both from measuring inaccuracies and from errors due to undetected weak emission-and-absorption lines. Usually the values scatter by less than 10%.

2. Calibration and Corrections

The instrument was calibrated in two steps. Longward of 1660 Å an absolute calibration was obtained using a deuterium lamp, the intensity of which had been measured against a known synchrotron source by Pitz [1]. Accuracy of the lamp calibration is to $\pm 15\%$. As an independent check the intensity of the lamp was measured using a diode which was calibrated absolutely by the National Bureau of Standards. The resulting intensities differed by less than

10% from the synchrotron calibration values over most of the spectral range of the lamp.

Light from the lamp was made to enter the instrument via a collimator mirror in the same way as sunlight during flight. A direct comparison between the solar intensities and the known lamp intensities is then possible. The simplicity of this method should be noted. At no point is it necessary to know anything about the absolute instrument efficiency which is difficult to measure. Identical calibrations were made before and after flight. No systematic change in instrument sensitivity was detected within the 15% accuracy.

Shortward of 1660 Å we do not have an absolute calibrated continuum light source to compare with. Instead we have used a method for relative calibration proposed by Becker et al. [2] by using a N_2 lamp and observing band intensities in two ($v' = \text{const.}$, v'') molecular band progressions in the Lyman–Birge–Hopfield system and comparing the measured intensities with calculated Franck–Condon factors we could determine the efficiency of film + spectrograph in the wavelength range from 1416 Å to 1752 Å. These relative values were then connected to the absolute calibrated region longward of 1660 Å.

Corrections were necessary for systematic effects of two kinds. Due to scattering at the predisperser a significant amount of first order light (2700 to 3600 Å) was present in the spectrum. While the amount of scattered light is small (less than 1%), there is still a noticeable contribution since the solar intensities in the second order wavelength range is two orders of magnitude weaker than in the 3000 Å region. To determine the correction, film in the flight spectrograph was exposed to ground level sunlight ($\lambda > 3000$ Å) and then on the same film strip a spectrum was taken of the deuterium lamp. By comparing the resulting apparent intensity with the known true intensity of the D_2 lamp a correction method was devised giving the true intensity with an accuracy to $\pm 10\%$ for background film densities of 0.3.

Another necessary correction is for absorption in the earth's atmosphere in the Schumann–Runge continuum of molecular oxygen. The atmospheric absorption was calculated using values for the O_2 density as function of altitude measured by Wildman et al. [3]. These O_2 densities are based on absorption measurements in the Schumann–Runge continuum region. Other workers give different values for the O_2 concentration, generally somewhat higher than the values used by us. However, several recent measurements give values close to Wildman et al. or even lower.

Of the exposures used for the continuum measurements all but one were taken above 140 km. The correction to the intensities at 140 km altitude are less than 12%, which is the value of the maximum correction at 1450 Å. Systematic errors in the correction are probably less than 5%. The only exposure for which the correction is noticeable is the disk exposure at 8 sec exposure time which was taken at 121.5 km altitude. The absorption at 1450 Å is 32% here but may rise to as much as 38% for other reasonable values of the O_2 density. For wavelengths greater than 1550 Å the possible error in the correction is insignificant even for this last exposure, due to the decrease of the O_2 absorption at longer wavelengths.

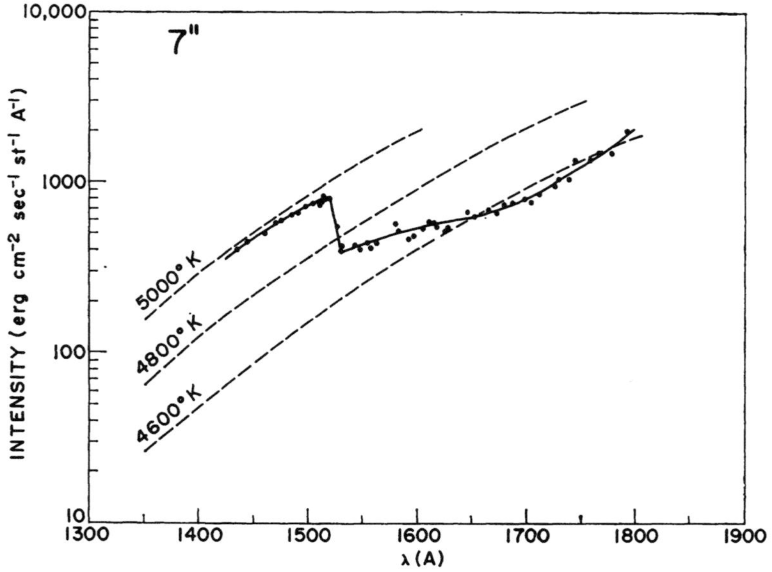

Fig. 2 a

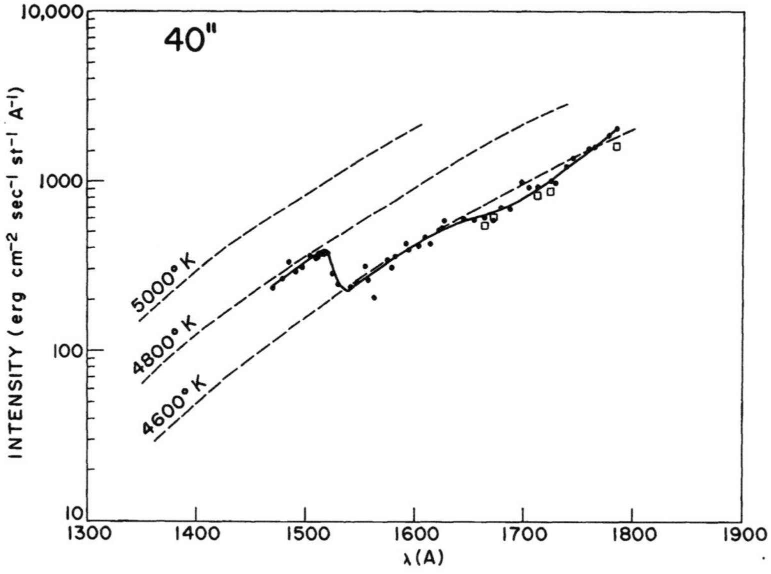

Fig. 2 c

Fig. 2. Solar intensities as function of wavelength in the wavelength range 1400–1790 Å. Intensities from plates with exposure times 8–16 s, ●; 32–64 s, □. Broken lines represent black-body spectra at the given temperatures.

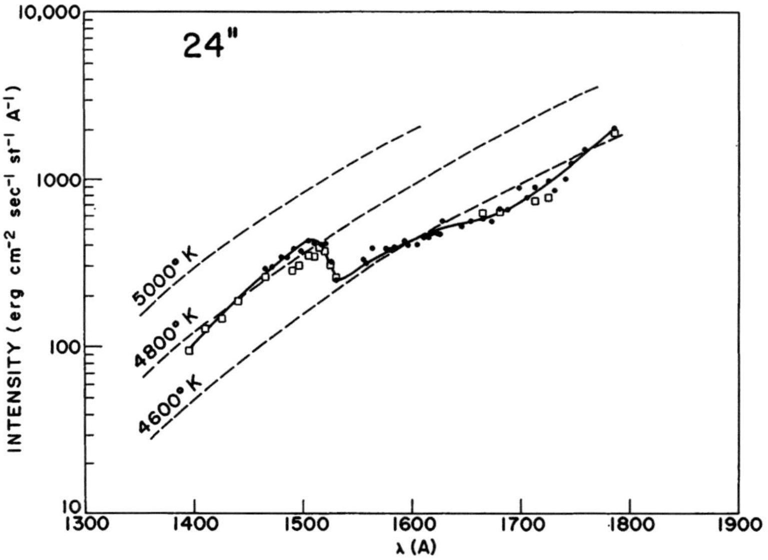
24"
10,000
1000
100
10
INTENSITY (erg cm⁻² sec⁻¹ st⁻¹ A⁻¹)
5000° K
4800° K
4600° K
1300
1400
1500
1600
1700
1800
1900
λ (A)

Fig. 2 b

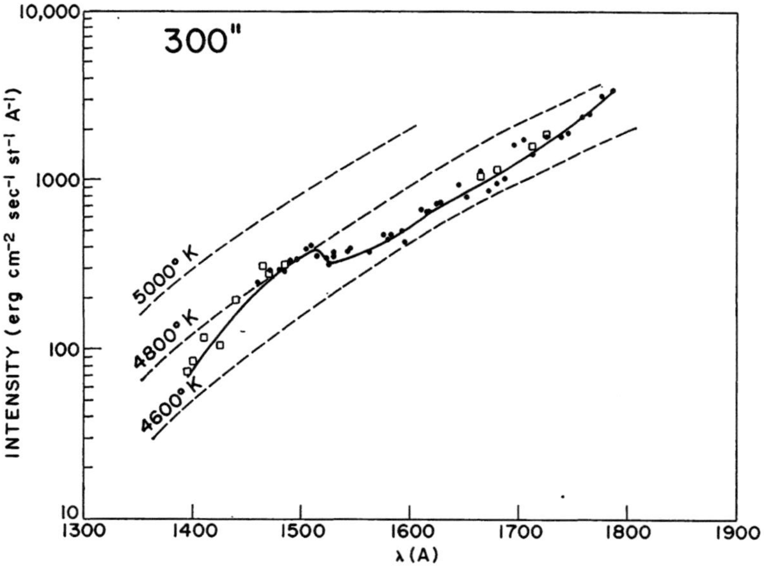
300"
10,000
1000
100
10
INTENSITY (erg cm⁻² sec⁻¹ st⁻¹ A⁻¹)
5000° K
4800° K
4600° K
1300
1400
1500
1600
1700
1800
1900
λ (A)

Fig. 2 d

3. Results

The results of the intensity measurements are shown in Fig. 2. This figure shows the measured intensities as a function of wavelength for the four positions. The dots are values measured from the short exposure time (8–16 s) spectra. Intensities obtained from the long exposure time (32–64 s) spectra are marked with open squares. A solid line representing the mean variation of intensity with wavelength is drawn for each position. Also drawn is the intensity of black-body radiation as a function of wavelength for different temperatures (broken lines).

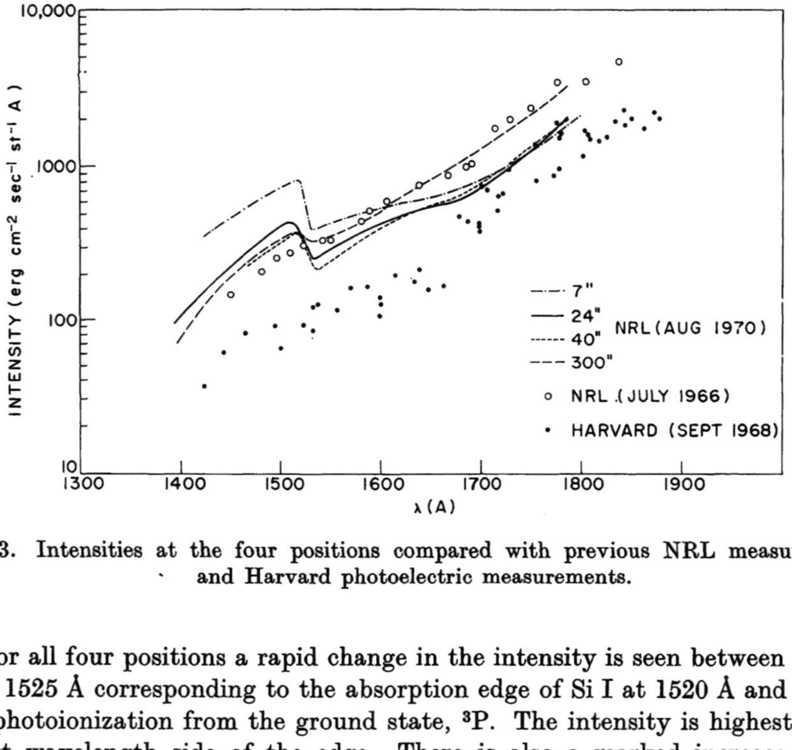

Fig. 3. Intensities at the four positions compared with previous NRL measurements and Harvard photoelectric measurements.

For all four positions a rapid change in the intensity is seen between 1515 Å and 1525 Å corresponding to the absorption edge of Si I at 1520 Å and caused by photoionization from the ground state, ^3P. The intensity is highest to the short wavelength side of the edge. There is also a marked increase in the size of the intensity jump at 1520 Å towards the limb. Above 1530 Å we observe a fairly even increase in intensity with wavelength, the intensity increasing fastest for the disk spectrum. There is no trace of any change in intensity at 1680 Å corresponding to photoionization of Si I from the ^1D level. Any intensity jump larger than 30% would certainly have been detected. The minimum radiation temperature of about 4550 °K is found in the region 1660–1740 Å.

Fig. 3 shows the intensities at the four positions compared with previous measurements from NRL [4] and Harvard's photoelectric measurements [5]. The disk position intensities agree quite well with the older NRL values, while the Harvard intensities lie 150 to 300 °K lower. Another distinct difference from the Harvard measurements is our lack of any intensity change at 1680 Å where the Harvard measurements indicate an intensity jump.

Fig. 4 shows the center-to-limb variation of the continuum intensity at six wavelengths. The curves for the three longer wavelengths reflect the behaviour of the source function in the region around the temperature minimum. The center-to-limb curves for the three shortest wavelengths, all in the emission continuum from the recombination to the ground state in Si I, are remarkably neutral almost to the extreme limb. This could indicate strong non-LTE (local thermal equilibrium) effects in this region. However, the center-to-limb

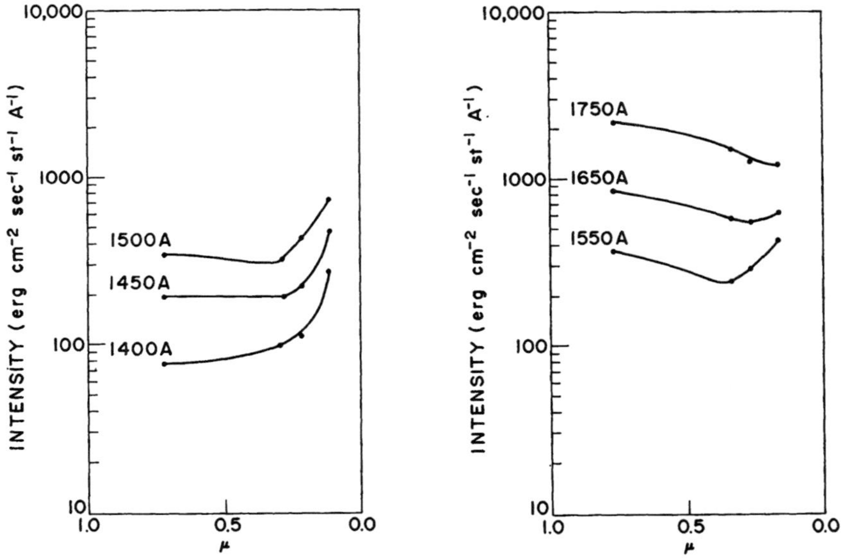

Fig. 4. Center-to-limb variation of the continuum intensity at six wavelengths. The abscissa μ is $\cos\theta$.

curves in this wavelength range are less accurate owing to the uncertainties in the correction for absorption in the earth's atmosphere for the disk spectrum between 1400 Å and 1550 Å.

An attempt has been made to analyze the center-to-limb variation of the continuum intensities to find the continuum source function as a function of optical depth. Only the curves for 1650 Å and 1750 Å have been treated so far, the analysis being very preliminary. By assuming homogeneous atmosphere and LTE a probable value of 4450 °K is reached for the temperature minimum with possible values ranging from 4300 °K to 4550 °K. Using the center-to-limb variation for several other wavelengths and requiring that the temperatures run through the same set of values with depth at all wavelengths a more accurate temperature model for the region around the temperature minimum may be determined.

Half-widths of emission lines have been determined and preliminary results are presented in Fig. 5, which shows the kinetic energy for the ions as measured from the half-widths of the lines, plotted against the electron temperature values corresponding to the maximum abundance of the different ionic species [6]. The largest discrepancy between the electron and ion energies exists in the lower chromosphere, where the values are different by a factor of 50.

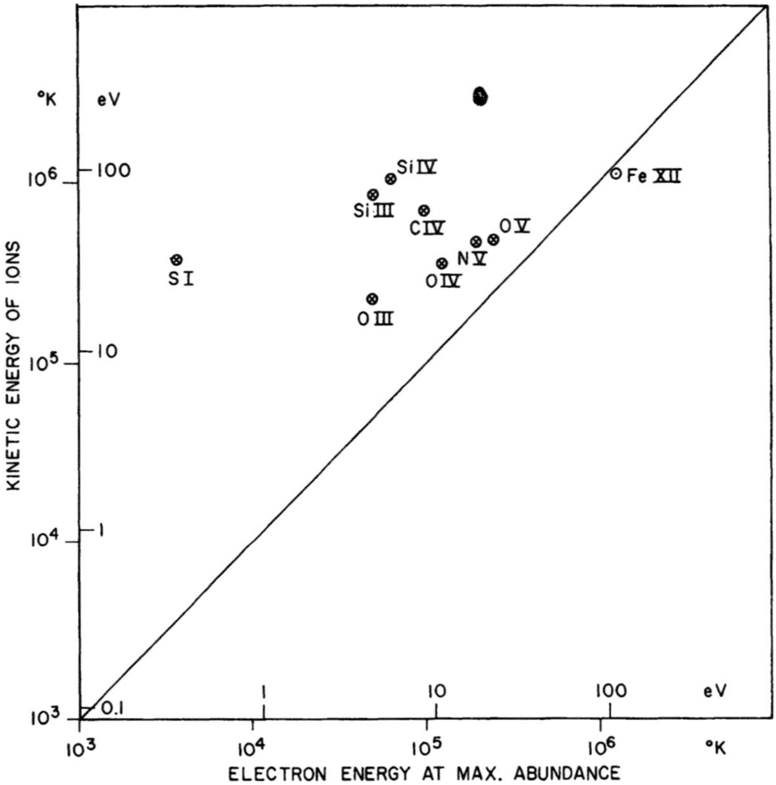

Fig. 5. Kinetic energy of ions against electron energy for ions in the chromosphere and corona.

Towards higher ionization temperatures the discrepancy decreases and disappears in the corona, where the kinetic and ionization temperatures are in remarkable agreement. It should be mentioned that the kinetic energy of ions in the lower chromosphere reaches almost the same value as the kinetic energy of the coronal ions. Only optically thin lines have been used in this investigation.

Acknowledgment

This work was supported by NASA contract DPR S-60404 G.

References

[1] E. E. PITZ, Appl. Optics 8, 255 (1969).
[2] K. H. BECKER, E. H. FINK and A. C. ALLISON, J. Opt. Soc. Am. **61**, 495 (1971).
[3] P. J. L. WILDMAN, M. J. KERLEY and M. S. SHAW, J. Atmos. Terr. Phys. **31**, 951 (1969).
[4] K. G. WIDING, J. D. PURCELL and G. D. SANDLIN, Solar Phys. **12**, 52 (1970).
[5] W. H. PARKINSON and E. M. REEVES, Solar Phys. **10**, 342 (1969).
[6] C. JORDAN, MON. Not. R. Astron. Soc. **128**, 113 (1969).

Space Research XII – Akademie-Verlag, Berlin 1972

LARGE-FIELD STUDY IN THE ULTRAVIOLET (2650 Å) FROM A SOUNDING ROCKET

J. P. Sivan and M. Viton

Laboratoire d'Astronomie Spatiale, Marseille, France

A survey is given of the up-to-date results of a large-field study in the ultraviolet launched from a sounding rocket on 4 April 1967. Systematic errors have been eliminated from the preliminary results and the mean interstellar absorption as well as magnitudes derived for 180 stars ranging from spectral type O7 to G3. A colour spectral-type diagram shows a poor agreement with line-blanketed models around B0–B3 which could be linked to the problem of effective temperature scale, and a large disagreement (0.5 magnitude) in the A–F range. This discrepancy appears also in the UBV colour–colour diagram, and thus is probably due to a line-blanketing problem. A comparison with measurements of different authors for several stars shows that the arbitrary adjustment of the models is satisfactory.

The isophotes of the Milky Way and zodiacal light are given in absolute units and galactic coordinates. The intensities are in good agreement with other measurements.

1. Introduction

A description of this study in the ultraviolet and of the preliminary results has already been given [1]. It involves a large field (5700° sq) photographic experiment sensitive to stellar sources (the magnitude limit reached was $V = 7.5$ for O stars in a 210 s exposure) as well as extended sources, the relative aperture of the instrument being $f/1$. The passband (called U′) was around 1000 Å wide and was centred near 2650 Å. The 210 s exposure photograph is shown in Fig. 1. The U′ magnitude of 201 stars and a colour-spectral type diagram for 100 of these stars were also given in [1]. These results were partial and in particular they were affected by significant systematic errors; the data have therefore been reexamined and corrected, and a rapid glance at the up-to-date results is presented here.

2. Stellar Sources

2.1. Accuracy of Measurements

The systematic errors were mainly caused by the instrument's chromatism and complex corrections were made. At the present time, we think that the systematic error that could still remain in our measurements is a maximum of ± 0.2 magnitude in the relative colour index between an O5 star and a G0 star. This error is very small elsewhere between O5 and A0. The stars' detection

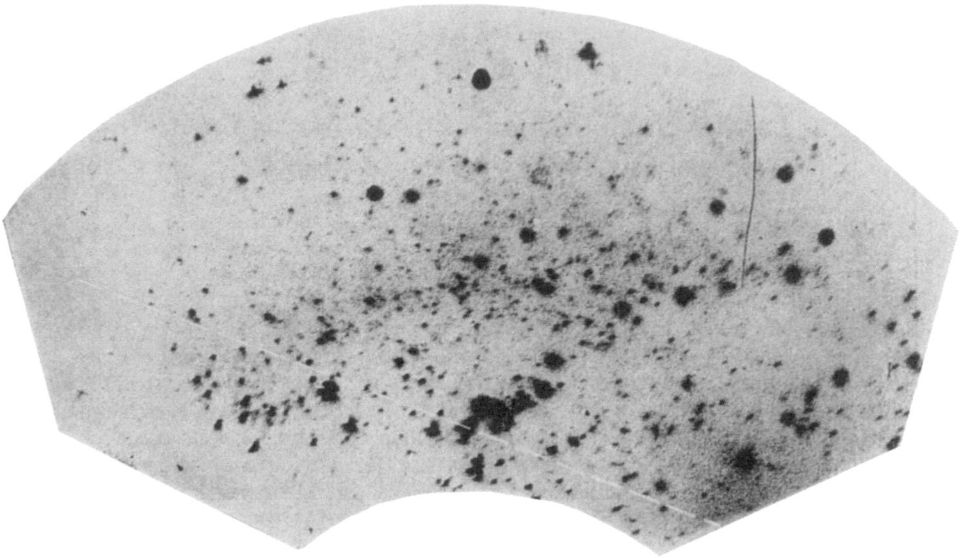

Fig. 1. The 210 s exposure photographs. The Milky Way ranges from 150° (right) to 230° (left) galactic longitude. The zodiacal light (bottom right), Orion (bottom centre) and Auriga (top right) are easy to identify. The white dashed line gives the limit under which stellar and background measurements were not possible because this region was partially obstructed by a mechanical part.

criteria have also been reinforced; in particular, a careful study has been made both of stars contaminated by other weaker stars and of multiple stars. As a result we can now publish the measurements of around 180 stars. The root mean square error of these measurements was evaluated at 0.16 magnitude by comparing stars common to the two pictures taken by the instrument (20 and 210 s exposures).

2.2. Interstellar Absorption

When the preliminary results were published, a ratio of the colour excesses $E_{u'-v}/E_{b-v} = 4$ was adopted in order to calculate the intrinsic $U' - V$ colours. A more precise study enabled us to estimate this ratio at 3.5 ± 0.5 for the effective wavelength of 2650 Å. This figure agrees with the results of Boggess and Borgman [2]. One should note that it was obtained from a certain number of highly reddened stars of spectral types from O5 to B3, which implies *a priori* a possible selection effect. Since the passband was very wide, the reddening correction was calculated by integrating a mean absorption curve obtained from the published results of Stecher [3], Boggess and Borgman [2] and Bless and Savage [4], and by normalizing this mean curve at 3.5 magnitudes at 2650 Å. This enabled us to take into account the maximum absorption "hump" at 2200 Å and thus to avoid errors which may reach 0.15 magnitude for the intrinsic magnitudes of strongly reddened stars.

2.3. Color Spectral-type Diagram

Fig. 2 shows the new colour spectral-type diagram obtained for 146 stars. The solid curve represents the mean distribution of the 52 stars chosen as being pure class V (i. e. excluding peculiar stars), according to the list of spectral types of Johnson et al. [5], Lesh [6] and Cowley et al. [7]. The zero of the U′ — V colours was adjusted for A0 type stars, using this curve. The crosses represent 11 of the theoretical models of Kurucz et al. [8] for effective temper-

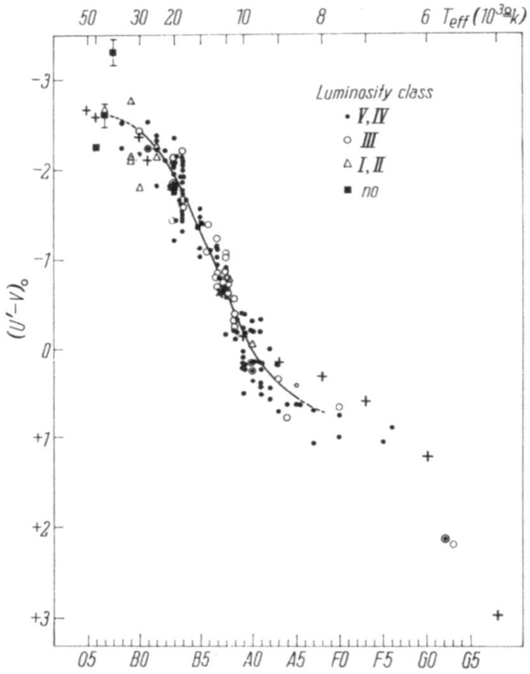

Fig. 2. Colour-spectral type diagram for 146 stars. The solid curve represents the mean distribution of main sequence stars. The large crosses are theoretical models (the model 12000 °K was omitted because the density of stars in the diagram was too high around B8). The star marked ● at about $U' - V = -0{\cdot}5$ gives the position of the Pleiades and ★ at about $U' - V = +2{\cdot}2$ is the sun, integrated over the V and U′ bands.

atures from 5000 °K to 30000 °K, with $\log g = 4$, and the models for 40000 °K and 50000 °K with $\log g = 5$. The choice of these models was made essentially because of the necessity of using homogeneous models covering a sufficiently extended range of effective temperatures. The spectral types corresponding to the models were determined from the scale of effective temperatures published by Morton et al. [9, 10]. The models were adjusted to the curve representing the main sequence in the B5–B9 region since it is the best defined region of the diagram from an experimental and theoretical point of view. Actually for very hot stars two important sources of doubt should be considered. The first is linked to the problem of the scale of effective temperatures of these stars, which could explain the slight disagreement between the models and the measurements in the B0–B3 region. The second is caused by the wide dispersion of the interstellar absorption laws and also by the strong reddening of most of these stars.

The adjustment of the models agrees fairly well with the measured flux published by different authors. This was checked by integrating the measurements of Stecher [11] and Bahner [12] for α CMi and β Tau in the same way as for the models. For β Tau, the point falls less than 0.1 magnitude above our measurement, while for α CMi, the point is too weak by about 0.3 magnitude. The points published by Tousey [13] and Allen [14] for the sun were also integrated and the results agree very well with our results for Capella even though this star is of luminosity class III.

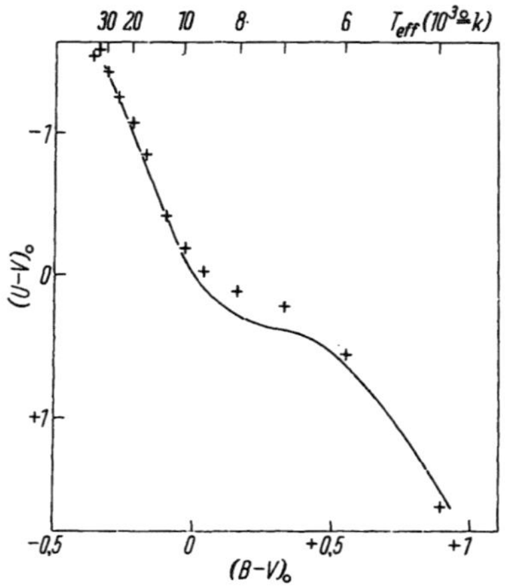

Fig. 3. Colour–colour diagram. The solid curve gives the standard UBV colours and the crosses represent the same models as in Fig. 2.

One notices also marked lack of ultraviolet flux in A and F stars (around 0.5 magnitude near A8). This effect can be also seen in Fig. 3. The solid curve represents the main sequence inferred from standard UBV data. The crosses are the models from [8] which were integrated over the band for U, B and V following the method indicated by Matthews and Sandage [15]. The analogy between the two diagrams is striking in each case and the discrepancy could be due to a blanketing problem. Also, the scatter of class V stars in the B9–A8 region is probably caused by the fact that most of the stars have peculiar spectra. The same can be said for the B2–B3V stars which lie on both sides of the curve, since most of them are emission stars.

3. Extended Sources

Fig. 4 shows, in galactic coordinates, the isophotes obtained for the general stellar background and the zodiacal light from the 210 s exposure. The absolute calibration was made with a calibrated tungsten ribbon lamp and a photomultiplier whose relative spectral response was known. The probable error introduced by this procedure is of the order of 20%. In the neighbourhood of the galactic plane, the measured absolute flux agrees very well with the measurements of Ingham and Sudbury [16] for the same effective wavelength but for 70–130° galactic longitude. A comparison with the measurements of Elsässer and Haug [17] in the V band shows that the mean colour near the galactic plane corresponds to spectral type B6–B8.

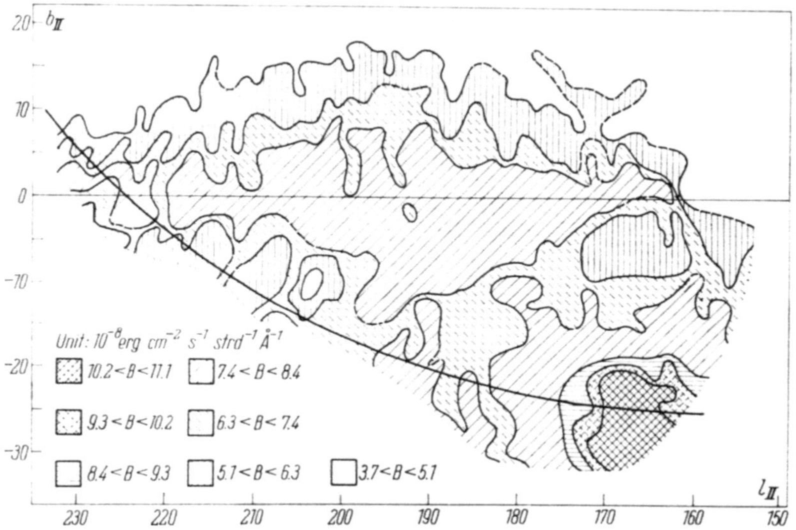

Fig. 4. Isophotes of the stellar background and the zodiacal light (bottom right) in galactic coordinates. Absolute fluxes are in units of 10^{-8} erg cm^{-2} sr^{-1} Å^{-1} at 2650 Å. The solid curve gives the same instrumental limit as in Fig. 1.

4. Conclusion

A revision of the UBV data and spectral classifications of the stars is being undertaken using selected lists which are much more coherent than those previously used. Complementary studies of the distribution of the stars in the diagram, the integration of other models, the influence of stellar rotation and the interpretation of the isophotes of the Milky Way and zodiacal light are in the process of completion.

References

[1] J. P. Sivan and M. Viton, in: Ultraviolet Stellar Spectra and related Ground-Based Observations, D. Reidel Publ. Co., Dordrecht 1970 (p. 120).
[2] A. Boggess and B. Borgman, Astrophys. J. **140**, 1636 (1964).
[3] T. P. Stecher, Astrophys. J. **142**, 1683 (1965).
[4] R. C. Bless and B. D. Savage, in: Ultraviolet Stellar Spectra and related Ground-Based Observations. D. Reidel Publ. Co., Dordrecht 1970 (p. 28).
[5] H. L. Johnson et al., Comm. Lunar and Planet. Lab. **4**, 99 (1966).
[6] J. Lesh, Astrophys. J. Suppl. No. 17 **151**, 371 (1968).
[7] A. Cowley et al., Astron. J. **74**, 375 (1969).

[8] R. KURUCZ, D. F. CARBON and O. GINGERICH, in: Theory and Observation of Normal Stellar Atmospheres, M.I.T. Press, Cambridge, Mass., and London 1969 (p. 401).
[9] D. C. MORTON and T. F. ADAMS, Astrophys. J. **151**, 611 (1968).
[10] D. C. MORTON, Astrophys. J. **158**, 629 (1970).
[11] T. P. STECHER, Astron. J. **74**, 98 (1969).
[12] K. BAHNER, Astrophys. J. **138**, 1314 (1963).
[13] R. TOUSEY, Space Sci. Rev. **2**, 3 (1963).
[14] C. W. ALLEN, Astrophysical Quantities, The Athlone Press, London 1963 (p. 172).
[15] T. A. MATTHEWS and A. R. SANDAGE, Astrophys. J. **138**, 49 (1963).
[16] M. F. INGHAM and G. C. SUDBURY, Nature, Lond. **226**, 526 (1970).
[17] H. ELSÄSSER and U. HAUG, Z. Astrophys. **50**, 121 (1960).

Space Research XII — Akademie-Verlag, Berlin 1972

COMPOSITION OF RELATIVISTIC COSMIC RAYS NEAR THE EARTH AND AT THE SOURCES

M. M. Shapiro, R. Silberberg and C. H. Tsao

Laboratory for Cosmic Ray Physics, Naval Research Laboratory
Washington, D.C., USA

Data on cosmic-ray composition obtained in Gemini XI are combined with other recent measurements on several thousand relativistic nuclei ranging from Li to Ni. Improved charge resolutions has yielded more precise estimates of abundances for elements beyond oxygen. New results reduce the relative flux of Cr by ~3. Our latest distribution function of cosmic-ray path length [exp $(-0.23\ x)$, for $x > 1$ g cm^{-2}] together with newly measured and calculated cross sections yield a revised source composition. Relative to carbon ($C \equiv 1.00$), the *source* fluxes are:
$N = 0.12 \pm 0.03$; $O = 1.02 \pm 0.06$; $Ne = 0.20 \pm 0.03$; $Mg = 0.27 \pm 0.03$; $Si = 0.23 \pm 0.03$; $S = 0.04 \pm 0.02$; $Fe = 0.23 \pm 0.05$. The presence of small traces (~0.02) of Ar and Ca and (~0.01) of Na, Al, and Ni in the sources would improve the overall fit. Estimates of Mg and Si have increased noticeably in the *source* composition since our 1969 Budapest report, while Cr may be altogether absent.

1. Introduction

The composition of the nuclear component of the cosmic radiation has undergone considerable changes between the source regions and the top of the earth's atmosphere due to collisions with the interstellar gas which consists mostly of hydrogen. We have carried out calculations on the transformation of the cosmic ray composition in order to determine the relative abundances of the *primordial* cosmic rays at the source regions. A semi-empirical method of calculating nucleon–nucleus cross sections was designed for this purpose. The elemental abundances of the primordial cosmic rays reflect the chemical composition in the source regions. It may reveal which processes of nucleosynthesis have played a significant role in the evolution of the sources. We have used data obtained in Gemini XI [1], combined with other recent measurements on several thousand relativistic nuclei ranging from Li to Ni and new results on the iron group by Lezniak et al. [2] and Casse et al. [3].

2. Composition of Arriving Cosmic Rays

The primary cosmic radiation arriving at the top of the earth's atmosphere consists mostly of fast hydrogen and helium nuclei in the ratio 7:1, at comparable rigidities above 2 GV. (The term "rigidity" denotes momentum/charge; it is proportional to the radius of curvature of charged particles in a given magnetic field.) The flux of cosmic-ray helium is ten times that of all heavier nuclei combined. Among the latter, carbon and oxygen are the most abundant.

Fig. 1 shows the relative abundances of cosmic rays in the range of hydrogen to nickel. A comparison is made here with the composition of the solar system based on data from solar spectra and meteorites. (For some elements, the meteoritic values are not plotted if they nearly coincide with the solar data.) The solar and cosmic-ray abundances have been normalized with respect to carbon. The outstanding feature of Fig. 1 is the large "excess" of Li, Be, and B in cosmic rays (by about 5 orders of magnitude), compared with that in

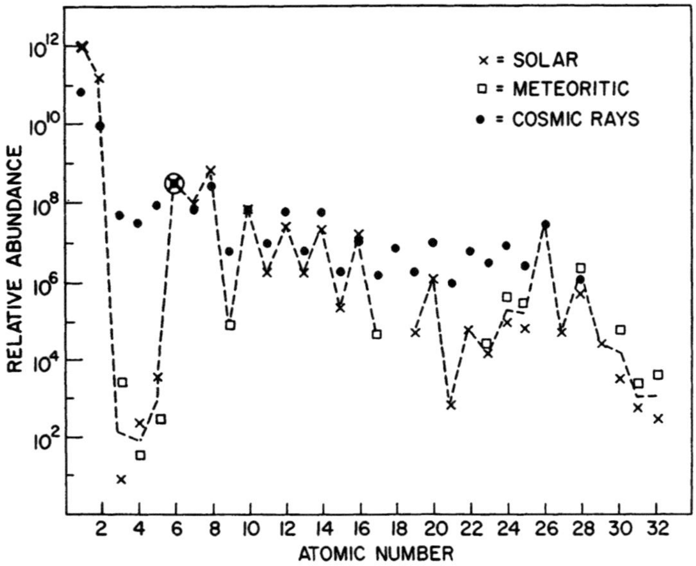

Fig. 1. A comparison of the relative abundances of elements in the cosmic radiation at the top of the earth's atmosphere, and in the solar photosphere, as well as in carbonaceous chondrites. The latter (meteoritic) values are shown only if they deviate appreciably from those of the sun. Solar data are normalized to hydrogen at 10^{12}; cosmic ray data were normalized so that the abundance of carbon coincides with solar carbon.

the solar system. Also, most elements ranging from Cl to Mn (with atomic numbers 17 to 25) are more plentiful in cosmic rays by a factor of about 100.

These overabundances are explained in terms of nuclear breakup of heavier nuclei upon collisions with the interstellar gas; the elements Li, Be and B in cosmic rays [4]* are largely derived from the breakup of carbon and oxygen, and those with $17 \leqq Z \leqq 25$ from the breakup of iron.

3. Transformation of the Primordial Compositions

The primordial composition of cosmic rays can be determined from the relative fluxes of the arriving nuclei including the secondaries generated en route. It is

* However, due to the low altitude of their balloon flight, and uncertainties in correcting for interactions in the atmosphere, these authors concluded that Li, Be and B are absent in cosmic rays above the atmosphere.

necessary, however, to know the partial cross sections for the various modes of breakup. Only a limited number of cross section values have been measured [5]. Most of the pertinent cross sections of nucleon–nucleus interactions have not yet been measured so one must resort to semi-empirical techniques for calculating

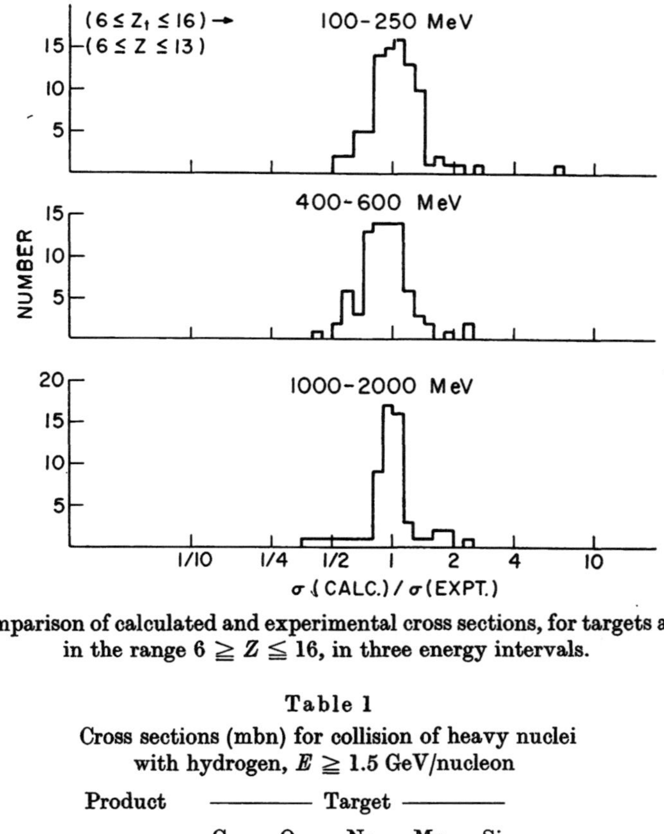

Fig. 2. A comparison of calculated and experimental cross sections, for targets and products in the range $6 \geqq Z \leqq 16$, in three energy intervals.

Table 1

Cross sections (mbn) for collision of heavy nuclei with hydrogen, $E \geqq 1.5$ GeV/nucleon

Product	Target				
	C	O	Ne	Mg	Si
Li	13	28	23	24	24
Be	20	16	15	15	15
B	65	37	27	23	19
C		44	32	23	18
N		76	41	30	23
O			72	53	39
F			45	19	14
Ne				69	55
Na				51	23
Mg					77
Al					52

the unmeasured ones. For this purpose, we have modified and extended [6] the earlier empirical equations of Rudstam [7]. Fig. 2 shows the distribution in the ratios of calculated to experimental cross sections in three energy intervals, for targets ranging from carbon to sulphur. The standard deviation of these ratios from unity is about 20% for energies >1 GeV per nucleon; it increases to 30% at lower energies of the order of 150 MeV per nucleon.

Table 1 shows cross sections for the production of various elements from collision of carbon, oxygen, neon, magnesium and silicon with hydrogen at energies >1.5 GeV per nucleon. These are based on our detailed calculations of partial cross sections, and supplemented by experimental data when available. (It may be remarked that the semi-empirical equations used in the calculations were fitted to the available cross-section measurements.)

No unique value of path length satisfied *both* the abundance ratio of Li-Be-B to C-N-O nuclei, and the ratio $17 \leqq Z \leqq 25$ to iron nuclei. However, an expo-

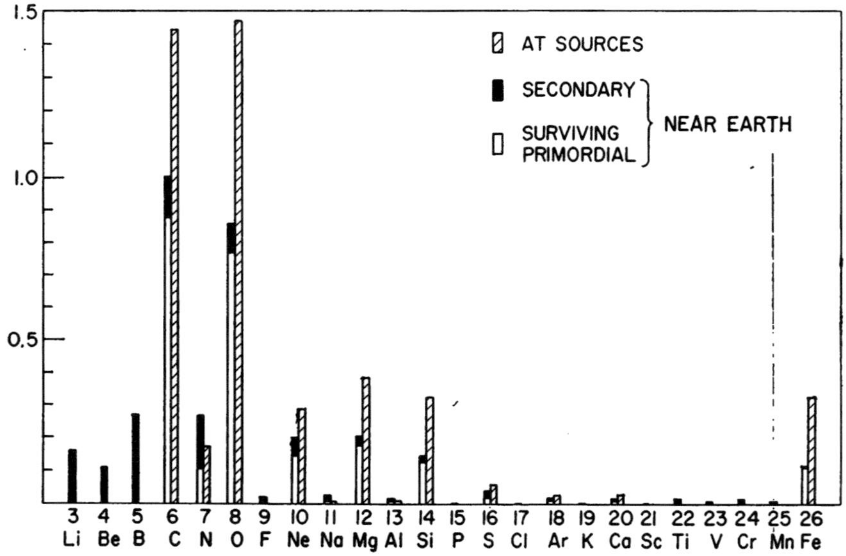

Fig. 3. The cosmic ray abundances at the sources and near earth. The blank portions represent the surviving primordial radiation, the shaded portions the contributions of secondary nuclei that arrive near the earth.

nential distribution in path lengths of the form $e^{-0.23x}$ for $x > 1$ g cm^{-2}, with a gradual rise between 0 and 1 g cm^{-2} satisfies *both* of the above ratios. Such an exponential distribution is to be expected: cosmic rays with longer potential path lengths leak out of the galaxy, and a drop-off at the shortest path lengths is likely if there are no cosmic ray sources relatively close to the solar system.

Next the calculated composition after traversal of an amount of matter (as shown in Fig. 3) was fitted to that observed at the top of the atmosphere. This process was carried out by iteration, adjusting the source composition and a provisional path length distribution function. Fig. 3 shows our newly revised cosmic-ray composition at the sources and near the earth. The arriving carbon is normalized to unity. The blank portions represent the contribution of the surviving primordial cosmic rays and the shaded portions those generated by breakup en route.

4. The Primordial Cosmic-Ray Composition versus the Composition of the Solar System

Table 2 shows the composition of the primordial cosmic rays, normalized to carbon. Hydrogen is the most abundant element—several hundred times more plentiful than carbon. Two values are shown for hydrogen; the smaller value is for a primordial spectrum that follows a power law in rigidity, while

Table 2

Primordial composition*
of the more abundant cosmic rays

H**	2 } $\times 10^4$ 5 }	Ne	20 ± 3
He	$(2.7 \pm 0.5) \times 10^3$	Mg	27 ± 4
C	100*	Si	23 ± 4
N	12 ± 3	Fe	23 ± 5
O	102 ± 6		

* Normalized to carbon

** The two values correspond to $\frac{dJ}{dR} \propto R^{-\gamma}$ and $\frac{dJ}{d\varepsilon} \propto \varepsilon^{-\gamma}$

the other corresponds to a power law in energy/nucleon. Helium is the next most common cosmic-ray element at the source regions, about 20 or 30 times

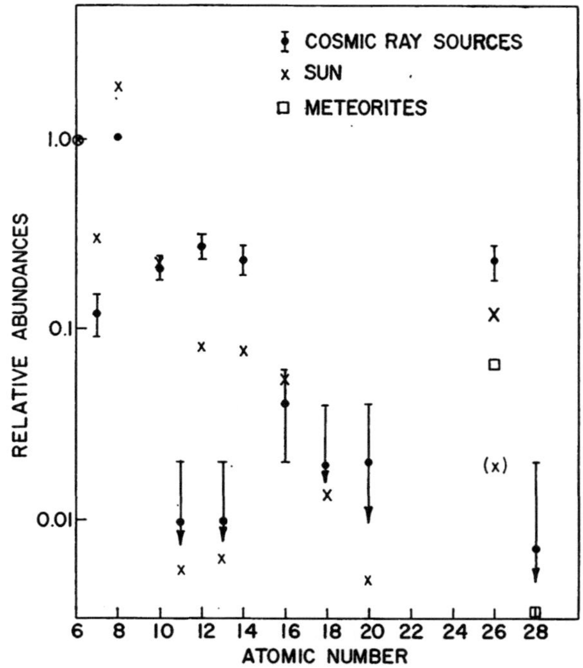

Fig. 4. The relative abundances of elements in the primordial cosmic radiation and in the solar photosphere. The data are normalized with respect to carbon.

as plentiful as carbon. Oxygen is as abundant as carbon. The flux of each of the four elements Ne, Mg, Si and Fe is about one quarter that of carbon. Other elements are scarcer, e.g. nitrogen, about 12% relative to carbon. The presence of small traces (~0.02) of Ar and Ca and (~0.01) of Na, Al and Ni in the sources would also improve the overall fit to the observation at the top of the atmosphere. Furthermore new results [2, 3] which have reduced the relative flux of Cr to one third of the early value [8], no longer suggest its presence at the sources.

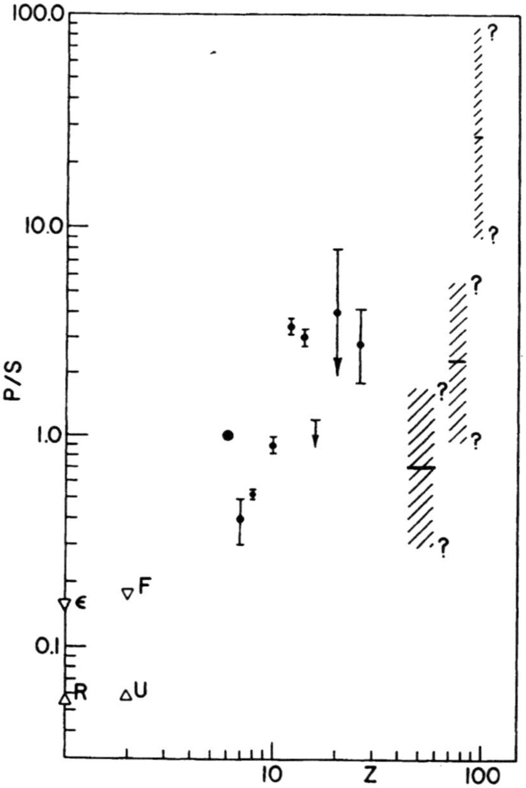

Fig. 5. The ratio of primordial cosmic-ray abundances to those in the solar system, as a function of atomic number Z. The ratios are normalized to unity at carbon. The values ε and R for hydrogen are based on assuming the cosmic ray spectrum to follow a power law in energy/nucleon and rigidity, respectively. The values F and U for helium are based on estimates of solar helium derived from the study of solar particles and solar spectra, respectively. (F denotes Fichtel's group [11]; U, Unsöld [10].) The shaded portions on ultra-heavy nuclei are based on the work of Fowler et al. [15].

In Fig. 4 the primordial abundances of cosmic rays ranging from carbon to nickel are compared with those of the solar photosphere [9–12] and of meteorites [13]. The values are normalized to carbon. Nitrogen and oxygen are relatively somewhat less plentiful in cosmic rays than in the sun. On the other hand, heavier elements like Mg, Si and Fe are notably overabundant in cosmic rays. Also shown in Fig. 4 are the elements Na, Al, Ca and Ni, while our [8, 14] abundance estimates of Mg and Si have been up-dated.

The overabundance of heavier elements in the primordial cosmic rays is even more marked if the comparison is extended all the way from hydrogen to the uranium group. Fig. 5 presents the *ratio* of primordial cosmic-ray abundances to those in the solar system, normalized to unity for carbon. The shaded values for the ultraheavy nuclei are based on the primordial abundance estimates of Fowler et al. [15]. The general trend of larger relative abundances of heavier elements in the primordial cosmic rays is evident. However, some oscillations are superimposed on the trend, e.g. for N and O beyond carbon, for S beyond silicon, and for a set of elements beyond iron.

5. Conclusions

It has been recognized for some time that the comparison with solar composition favours cosmic-ray origin in highly evolved stars, in which the processes of element synthesis have reached an advanced stage, e.g., in supernovae. The latest measurements [2, 3] indicate that the amount of chromium in the primordial cosmic rays (relative to iron) is not anomalously high; special conditions of nucleosynthesis for producing chromium are no longer needed. The preliminary (and statistically limited) data on nuclei of the uranium group ($90 \leqq Z \leqq 96$) indicate that the increase with atomic number in the ratio of primordial cosmic ray abundances to those of the solar system continues beyond iron.

Refernecces

[1] N. Durgaprasad et al., Phys. Rev. D, **1**, 1021 (1970).
[2] J. A. Lezniak, T. T. von Rosenvinge and W. R. Webber, Acta Phys. Hung. **29**, Suppl. 1 (1970); and unpublished work by W. R. Webber et al.
[3] M. Casse et al., Space Research XI, 1397 (1971).
[4] H. L. Bradt and B. Peters, Phys. Rev. **80**, 943 (1950).
[5] P. Fontes et al., Nucl. Phys. **A165**, 405 (1971).
[6] R. Silberberg and C. H. Tsao (to be published).
[7] G. Rudstam, Z. Naturforsch. **21a**, 1027 (1966).
[8] M. M. Shapiro, R. Silberberg and C. H. Tsao, Proc. Int. Conf. Cosmic Rays (Budapest), Acta Phys. Hung. **29**, Suppl. 1, 479 (1970).
[9] K. Kristiansson, Ark. Fys. **25**, 513 (1964).
[10] A. O. J. Unsöld, Science **163**, 1015 (1969).
[11] S. Biswas, C. E. Fichtel and D. E. Guss, J. Geophys. Res. **71**, 4071 (1966).
[12] T. Garz, H. Holweger, M. Kock and J. Richter, Astron. Astrophys. **2**, 446 (1969).
[13] A. G. W. Cameron, in: Origin and Distribution of Elements, edited by L. H. Ahrens, Pergamon Press 1968 (p. 125).
[14] M. M. Shapiro and R. Silberberg, Ann. Rev. Nucl. Sci. **20**, 323 (1970).
[15] P. H. Fowler et al., Proc. Roy. Soc. **A318**, 1 (1970).

Space Research XII — Akademie-Verlag, Berlin 1972

STUDY OF ENERGY SPECTRA OF PRIMARY COSMIC RAYS AT VERY HIGH ENERGIES ON THE PROTON SERIES OF SATELLITES

N. L. Grigorov, V.E. Nesterov, I. D. Rapoport and I. A. Savenko

Academy of Sciences, Moscow, USSR

The results of the Proton 4 measurements of the energy spectrum of all particles in primary cosmic rays in the 10^{11}–4×10^{15} eV energy range are presented. In the 10^{13}–10^{15} eV range the measured spectrum is of power-law form ($\gamma = 1.60 \pm 0.01$) and at energies higher than 2×10^{15} eV the spectrum becomes steeper. The Proton 1, 2, 3 and 4 measurements have shown that the all-particle spectrum has a discontinuity in the 10^{12} eV range where the Proton 1, 2 and 3 measurements have a steeper proton spectrum. The results of the Proton 3 measurements of the primary cosmic ray α-particle energy spectrum in the 10^{10}–10^{12} eV/nucleon energy range are presented.

1. Introduction

Heavy satellite investigations of high and ultrahigh energy cosmic rays beyond the atmosphere began as early as the launches of the Proton 1, 2 and 3 satellites, using the SEZ-14 spectrometer [1] and were continued on the Proton 4 satellite using the IK-15 ionization calorimeter [2]. The results of the Proton 1, 2 and 3 measurements of the energy spectra of protons and all particles in primary cosmic rays are presented in [3]. The present report examines the results of further analysis of the experimental data obtained from Proton 1, 2 and 3 and the preliminary results of Proton 4 measurements of the primary cosmic-ray spectrum.

2. Primary Cosmic Ray α-Particle Energy Spectrum according to Proton 3 Data

The Proton 3 satellite measured the primary cosmic ray α-particle energy spectrum with SEZ-14 directional Cerenkov counters [4]. The detectors record α-particles entering the instrument through its inlet window, the influence of shower particles leaving the calorimeter or the surrounding matter being small. To select α-particles from the flux of primary cosmic-ray particles of different charges each half of the instrument included double proportional counters. To identify α-particles the amplitude of the pulse recorded in each proportional counter must be within 2.7 V_{prob} and 8.0 V_{prob}, where V_{prob} is the most probable pulse amplitude produced by a single-charge relativistic particle in the proportional counter. Knowing the width of the pulse amplitude distribution in the proportional counters from ground muon measurements and considering that for α-particles the distribution width should decrease by a factor of 2, one finds that the probability that the pulses are produced by an α-particle within the above limits is 0.7.

The α-particle energy was measured with an ionization calorimeter.

Electronic circuits selected six-fold coincidence of the signals from the directional detector, from two differential discriminators for the proportional counters as discussed above, from the integral discriminators of energy release

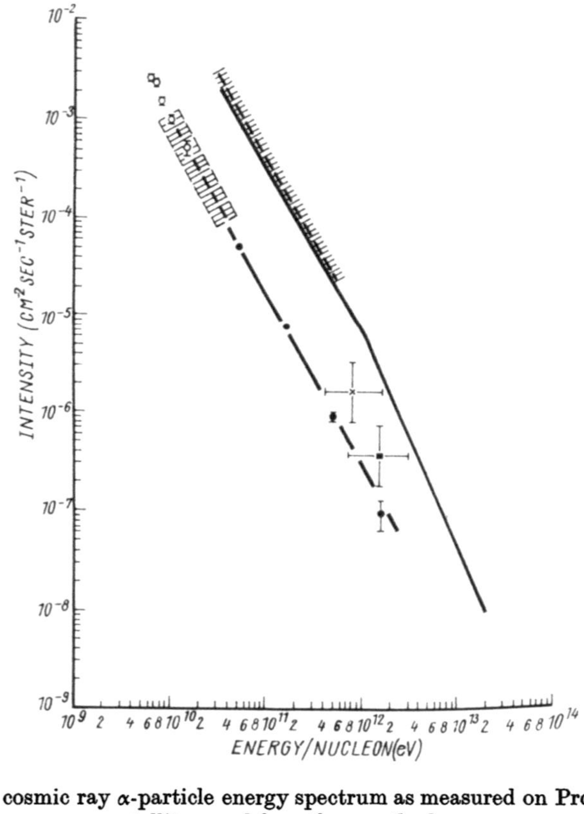

Fig. 1. Primary cosmic ray α-particle energy spectrum as measured on Proton 3 and other satellites, and by other methods.
●, Proton 3 data; |||| data from [6]; ○, data from [5]; ×, data from [7]; ■, data from [8]. upper curve, proton spectrum as measured on Proton 1, 2, 3 satellites [3].

in the ionization calorimeter, and from the scintillation counters arranged above and under the calorimeter. The complex signal (the result of these six-fold coincidences) was supplied to scalers which were interrogated each 8.5 sec by a memory device. The results for each measurement period of half a day were summed.

The intercomparison of the counting rates from these signals for periods when the predominant direction of the instrument axis was pointing to the zenith and to the earth has shown that the intensity ratio for these directions is 3.5:1.

The α-particle to proton flux ratio may be determined by measuring the counting rates from α-particles and protons at the same energy thresholds on the same half of the instrument. For such a determination it is not necessary to know the absolute values of the α-particle and proton fluxes. The measured value of I_α/I_p has to be corrected for the detection efficiency for α-particles (0.7) and protons (0.8) and the fact that the energy release in the calorimeter from α-particles is about 10% in excess of that of protons. With these corrections $I_\alpha/I_p = 0.56 \pm 0.01$.

Fig. 1 presents the α-particle energy spectrum as measured on Proton 3 and the proton spectrum according to Proton 1, 2 and 3 measurements [3]. The particle energy per nucleon is plotted as abscissa, and as ordinate the particle flux in absolute units ($cm^{-2}\ s^{-1}\ sr^{-1}$) obtained by dividing the measured counting rates by 57 cm^2 sr (the geometrical factor of the instrument), by 0.65 (the coefficient including the shielding of the inlet window by the earth) and by 0.7 (the efficiency of α-particle detection with the proportional counter). Fig. 1 presents also the values of α-particle fluxes from [5–8], which are in good agreement with those measured on Proton 3.

The comparison of the α-particle and proton energy spectra in Fig. 1 shows that the spectra are similar up to an energy of 5×10^{11} eV/nucleon, which corresponds to 10^3 GV magnetic rigidity. Because of insufficient statistical accuracy in the α-particle measurements at energies higher than 10^{12} eV/nucleon, it is still impossible to establish whether or not the α-particle spectrum becomes steeper at energies higher than 10^{12} eV/nucleon, as does the proton spectrum.

3. Primary Cosmic Ray Energy Spectrum according to the Proton 4 Data

The Proton 4 satellite was used to extend the measurements of the primary cosmic-ray energy spectrum to much higher energies. The measurements with the IK-15 ionization calorimeter [2] were carried out in two ways: (i) using integral amplitude discriminators dividing the energy range 2×10^{11}–4×10^{15} eV into nine approximately equal intervals (on a logarithmic scale); (ii) using a 64-channel analyser in the 3×10^{13}–10^{16} eV energy range.

In the first case no requirements are imposed as to the kind of particles, the direction of their motion and the point of entry into the calorimeter. The $+$ crosses in Fig. 2 show the results of such spectral measurements for 50 days of the flight. The integral discriminator thresholds were fitted to the detected particle energy by normalizing the intensity to the spectrum as measured from Proton 1, 2 and 3. The $\times$ crosses in Fig. 2 show the results using the pulse height analyser for about 80 days of the flight.

It can be seen from Fig. 2 that the spectrum in the 2×10^{11}–2×10^{15} eV energy range is of a quasi-power form, and agrees fairly well with the spectrum measured on Proton 1, 2 and 3 up to energies of 10^{14} eV. At energies higher than 2×10^{15} eV an abrupt steepening of the measured spectrum of all particles is observed. This cannot be explained as a systematic effect since limiting in the electronic circuits should result in a steepening of the measured spectrum only at energies higher than 10^{16} eV. It can be also seen from Fig. 2 that in the 10^{12}–10^{13} eV energy range the spectrum has a somewhat steeper slope than at energies higher than 10^{13} eV. This discontinuity is studied below.

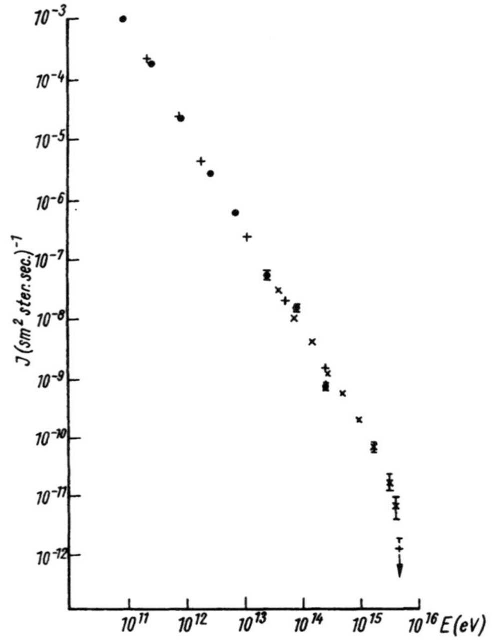

Fig. 2. Primary cosmic-ray (all nuclei) energy spectrum as measured on Proton 4 satellite: +, integral amplitude discriminators; ×, differential amplitude analyser; ●, Proton 1, 2, 3 data.

4. The Discontinuity in the Primary Cosmic Ray Energy Spectrum in the 10^{12} eV Energy Range

The information available from the Proton 1, 2 and 3 satellites was analysed to study the discontinuity in the energy spectrum of all primary cosmic-ray particles. About 50% of the information from these satellites (186 measurement periods broken into 6 series) was used earlier [3] to plot the spectrum. After considering an additional 151 measurement periods (6 series) of Proton 2 and 3 data, the information on the all-particle spectrum was nearly doubled. To find possible deviations of the spectrum from a power law, the measured spectrum was analysed by plotting $\log [I(\geqq E) \cdot E^{\gamma}]$ against $\log (E/10^{11})$ where E is the energy, and I is the particle intensity expressed in arbitrary units. Averaging was carried out for intervals of $\log (E/10^{11}) = 0.3$. For each of these intervals the mean value of $\log [I(\geqq E) \cdot E^{\gamma}]$ and the rms error was determined for some values of γ. It can be seen from Fig. 3, which gives the results for $\gamma = 1.60$, that discontinuity in the spectrum is observed in the 10^{12} eV energy range.

To study quantitatively this discontinuity two approximations were considered and the degree of their agreement with the experimental data was established using the χ^2 test. The value of γ was determined as the weighted mean value of the Proton 1, 2 and 3 data presented in Fig. 3. The values of χ^2 calculated for various values of the integral all-particle spectrum power exponent

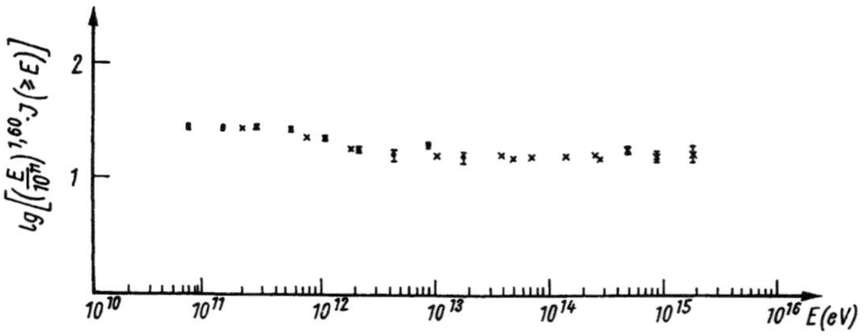

Fig. 3. Energy dependence of log $[I(\geqq E) \cdot E^{\gamma}]$ (at $\gamma = 1.60$) for the average (over all measurement series) values obtained on Proton 1, 2, 3 satellites (●) and for a single series of Proton 4 measurements (×).

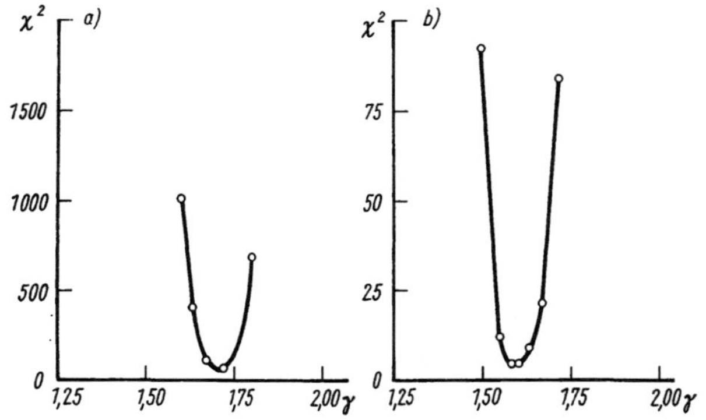

Fig. 4. The values of χ^2 at various values of the integral spectral power exponent γ for the energy dependence of log $[I(\geqq E) \cdot E^{\gamma}]$ shown in Fig. 3 according to Proton 1, 2, 3 data: (a) approximating function, a single constant and a single range; (b) approximating function, two energy ranges with a single value of the exponent γ.

are presented in Fig. 4(a). It can be seen from this figure that with a power law the best agreement with the experimental data is at $\gamma = 1.72$, in agreement with the value $\gamma = 1.73 \pm 0.05$ as given in [3]. The approximation of the spectrum by a power function with a single exponent over the whole energy range is not satisfactory since the minimum value of χ^2 is 68 at 8 degrees of freedom. Approximation of the dependence shown in Fig. 3 over two ranges $0.3 < \log E/10^{11} < 0.9$ and $1.2 < \log E/10^{11} < 2.4$, proved to be more satisfactory. The values of χ^2 for various γ are shown in Fig. 4(b). It can be seen from this figure that the best agreement is observed at $\gamma = 1.60$. At this value of the power exponent χ^2 is 4.5 at 6 degrees of freedom which indicates a fairly good agreement of the proposed approximation with the experimental data. Thus, the primary cosmic-ray energy spectrum can be described best by a function with a discontinuity in the 10^{12} eV range, but with the same $\gamma = 1.60$ at energies below and above this discontinuity.

The irregularity found in the primary cosmic-ray energy spectrum using both sets of measurements favours the actual existence of a discontinuity since the Proton 4 equipment [2] differs drastically from the Proton 1, 2 and 3 equipment [1]. It should be noted that the irregularity is observed in the energy range where the Proton 2 and 3 measurements showed an increase in the steepness of the primary cosmic-ray proton spectrum where the integral spectrum exponent is increased from $\gamma = 1.60$ to $\gamma = 2.3$–2.5 [3].

5. Conclusions

1. The α-particle energy spectrum up to an energy of 10^{12} eV/nucleon has been measured on the Proton 3 satellite using direct methods. The α-particle fluxes detected in the 10^{10}–5×10^{10} eV/nucleon range agree well with the data available elsewhere.

2. The energy spectrum of all particles (protons and heavier nuclei in the 2×10^{11}–4×10^{15} eV energy range) has been measured on the Proton 4 satellite. It has been established using direct methods that, at energies higher than 2×10^{15} eV, the primary cosmic-ray energy spectrum becomes steeper in accordance with the data obtained earlier using indirect methods.

3. A discontinuity found in the spectrum in the 12^{12} eV energy range probably indicates a change in the chemical composition of primary cosmic rays in the 10^{12} eV range. The discontinuity is observed in the energy range where a steeper proton energy spectrum is observed.

Acknowledgment

The authors are indebted to V. V. Akimov, Yu. V. Gubin, N. A. Mamontova and V. L. Prokhin for their help when treating and analysing the experimental data.

References

[1] N. L. GRIGOROV et al., Kosm. Issled. **5**, 383 (1967).
[2] V. V. BUGAKOV et al., Izv. Akad. Nauk SSSR, ser. fiz. **34**, 1818 (1970).
[3] N. L. GRIGOROV et al., Nuclear Physics (USSR) **11**, 1058 (1970).
[4] N. L. GRIGOROV et al., Nuclear Physics (USSR) **11**, **4**, 814 (1970).
[5] K. C. ANAND et al., Proc. 9th Int. Conf. Cosmic Rays, London 1965, **1**, 312 (1966).
[6] R. PINKAU et al., Proc. 11th Int. Conf. Cosmic Rays, Budapest 1969.
[7] P. FOWLER and C. WADDINGTON, Phil. Mag. **1**, 637 (1956).
[8] P. JAIN, C. LOHRMANN and M. TEUCHER, Phys. Rev. **115**, 654 (1959).

Space Research XII — Akademie-Verlag, Berlin 1972

MEASUREMENT OF GENERAL RELATIVISTIC TIME DELAY WITH MARINERS 6 AND 7

J. D. ANDERSON[a], P. B. ESPOSITO[a], W. MARTIN[a] and D. O. MUHLEMAN[b]

[a] Jet Propulsion Laboratory, California Institute of Technology, Pasadena, California, USA
[b] California Institute of Technology, Pasadena, California, USA

Range and Doppler data extending over the interval from August 1969 to December 1970 have been obtained at S-band from Mariners 6 and 7. The range data possess an accuracy of 0.1 μs and constitute the basis of the experimental determination of the relativistic time delay. At superior conjunction, this effect reached a maximum of approximately 200 μs for Mariner 6 and 180 μs for Mariner 7. The prediction of Einstein's theory of general relativity that the propagation of electromagnetic radiation is delayed due to the solar gravitational field has been verified to at least 4%. Non-gravitational forces acting on the spacecraft limit the accuracy of this result. Furthermore, since the solar corona also contributes to a delay in the signal propagation it is important to separate this effect from the relativistic delay. Analysis of the Mariner data indicates a coronal model consistent with those deduced by other methods. Uncertainties in this model as well as the influence of non-modeled forces acting on the spacecraft limit the ultimate precision of the relativity experiment to at best 1%.

1. Mariner Mars '69 General Relativity Experiment

One of the primary objectives of the Mariner Mars '69 Extended Mission is to measure the propagation of electromagnetic radiation in the presence of the solar gravitational field. This is accomplished by achieving a precise knowledge of the Mariner orbits and measuring the round trip distance between the earth and spacecraft.

The effect of the solar gravitational field on the propagation of radio signals has been investigated by Muhleman and Reichley [1] and Shapiro [2]. In addition, a series of related papers [3–5] discuss the measurement and interpretation of the general relativity time delay. In order to express this effect mathematically, we have applied the general, isotropic metric as suggested by Eddington [6] and Robertson [7]. As a consequence of the metric, the one-way relativistic time delay (in isotropic coordinates) [8, 9] is

$$\Delta t = \frac{2\gamma^* m}{c} \ln\left(\frac{r_0 + r_1 + \varrho}{r_0 + r_1 - \varrho}\right)$$

where ϱ, r_0 and r_1 are coordinate distances representing respectively the observer–spacecraft, sun–observer and sun–spacecraft distances. The quantity m (~1.5 km) equals the gravitational radius of the sun and the nominal (Einstein) value of the relativity parameter γ^* is unity. The relativistic time delay is deduced by determining γ^* directly from the data.

2. Data and Experimental Procedure

Phase coherent integrated Doppler and round trip range measurements are available over the interval from Mars encounter (30 July 1969 for Mariner 6, 5 August 1969 for Mariner 7) until December 1970. Most of the Doppler and all the range data were collected with the 64 m Goldstone radio antenna which operated at a transmitted frequency of approximately 2.2 GHz.

For a sun–earth–spacecraft angle of about eleven degrees and greater, the residuals on the Doppler data after the best fit orbit has been obtained indicate an accuracy of 0.01 hertz over a ten minute integration time. In terms of an average range-rate error this is equal to 0.7 mm s^{-1}. The range data have an accuracy of 0.1 μs. However, as the spacecraft approached superior conjunction, the accuracy of both data types was degraded owing to severe noise generated by the solar corona. Thus, range measurements obtained in the superior conjunction interval have been weighted both uniformly and according to a variable scale in order to approximate their accuracy realistically.

Results presented in this paper are based upon a six month arc of tracking data with epochs and superior conjunction information presented in Table 1. The results of an analysis of longer data arcs were presented in [10] and are still undergoing examination.

Table 1

Mariner superior conjunction characteristics

	Mariner 6	Mariner 7
Epoch	4 Feb 1970	3 Feb 1970
Superior conjunction	29 April 1970 17 h	10 May 1970 01 h
Earth–spacecraft distance (km)	402.25×10^6	385.56×10^6
Round trip travel time (min)	44.72	42.87
Sun–earth–spacecraft angle (°)	0.95	1.56
Closest approach of signal to solar surface ($R_\odot$)	2.58	4.90

Analysis of the data to deduce best estimates of various parameters that affect the Mariner orbits is achieved by a differential correction weighted least squares algorithm [11]. The primary parameters and forces affecting the orbit determination include orbital elements defining the spacecraft's orbit, solar radiation pressure, attitude control forces, Mercury and earth–moon masses, solar corona electron density, relativity parameter, earth orbital elements, astronomical unit and station locations. With respect to the solar corona, an electron density model [12] of the form

$$N\,(\text{electrons cm}^{-3}) = \frac{A}{r^6} + \frac{B}{r^{2+\varepsilon}}$$

has been utilized in the orbit determination. A, B and ε represent parameters with nominal values and standard deviations of $(1.30 \pm 0.9)\times10^8$, $(1.15 \pm 0.7)\times10^6$ and 0.3 ± 0.3 respectively. These values were developed from an analysis of classical solar eclipse photometric observations.

It is not feasible to attempt a solution for all of these parameters from the Mariner data alone. However, *a priori* values of certain well known parameters, such as the astronomical unit, can be used with their known standard deviations in order to condition the solution for the parameters under investigation. Thus, the influence of parameters peripheral to the Mariner analysis will be accounted for in the estimate of the standard deviations of parameters directly related to this experiment.

3. Mariner 6 and 7 Data Analysis

Fig. 1 shows the magnitude of the relativistic (Einstein) and nominal solar coronal time delays as well as the quantity of range measurements during this

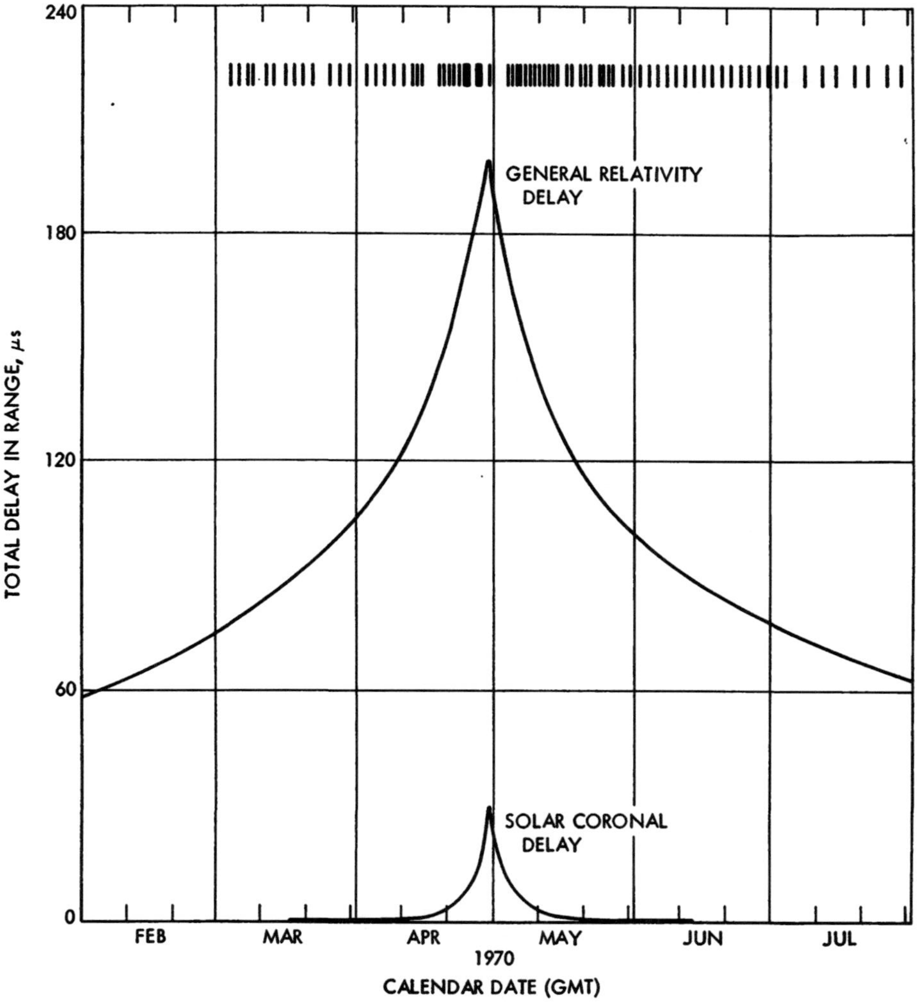

Fig. 1. Mariner 6 propagation delay due to general relativity and the solar corona. The small vertical bars indicate individual range measurements.

six month arc. We have confined our attention to this relatively short arc of data in order to reduce the accumulated effect of random and non-modeled forces acting on the spacecraft as well as to investigate that interval where the relativistic and coronal effects vary most rapidly.

Six months of Doppler and range data were fitted to the Mariner 6 orbit in which a variety of parameter sets were estimated. Range residuals corresponding to a typical fit of data are shown in Fig. 2. A systematic trend in the residuals is evident and as yet has not been positively identified although several possibilities are under investigation.

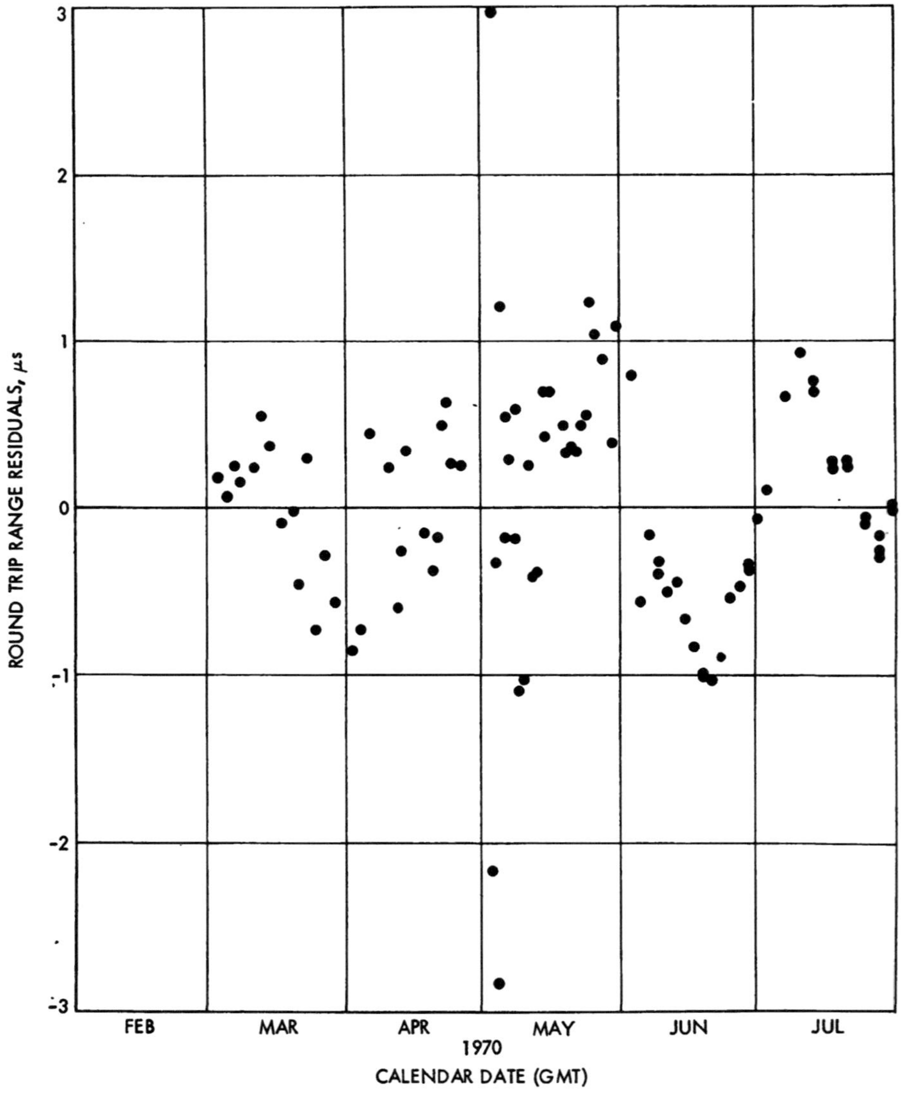

Fig. 2. Mariner 6 range residuals resulting from a typical analysis of a six month data arc consisting of 600 Doppler and 96 range measurements.

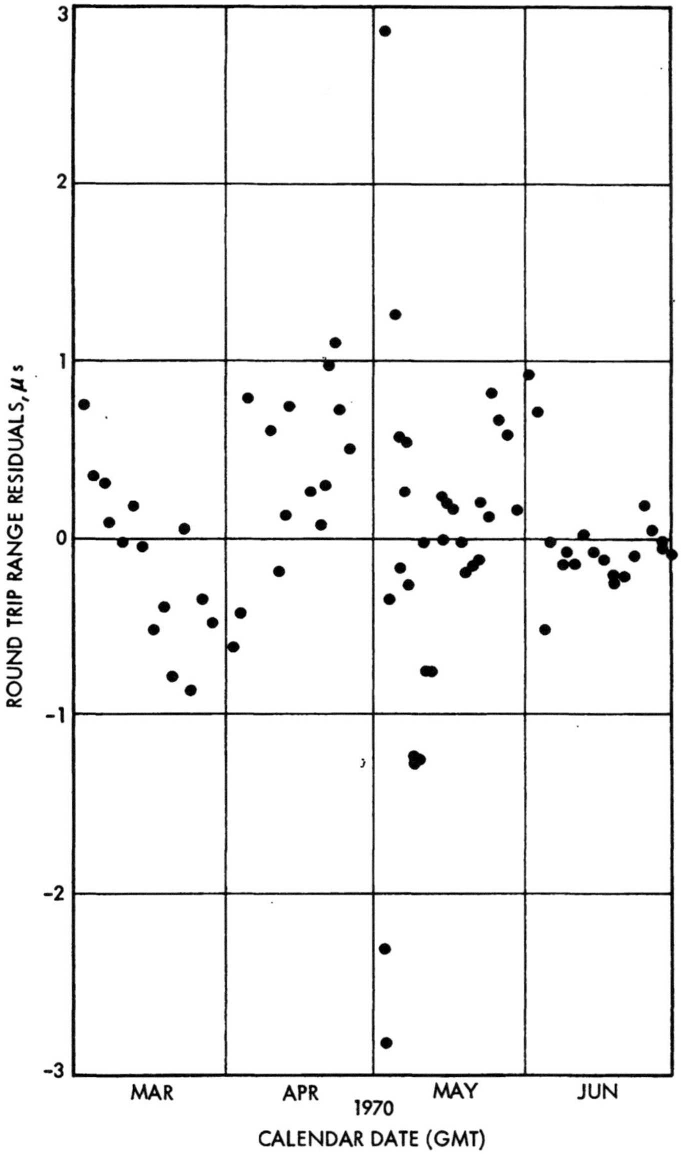

Fig. 3. Mariner 6 range residuals resulting from a typical analysis of a four month data arc.

In order to investigate and reduce the accumulative influence of non-gravitational forces acting on the spacecraft, shorter arcs consisting of four months of data were analyzed. The corresponding solutions, for which a typical set of residuals is shown in Fig. 3, yield results consistent with the longer arcs. Furthermore, the structure of the residuals suggests that we have been at least partially successful in limiting the influence of the source of the systematic trend.

The orbit of Mariner 7 is basically the same as that of Mariner 6 except that at superior conjunction the sun–earth–spacecraft angle is 0.61° larger for Mariner 7. This slight difference in orbits allowed the radio signal to pass farther from the sun and consequently the influence of the solar corona is significantly reduced. In particular, the A term in the corona model contributes less than 1 μs to the total time delay.

Six months of Mariner 7 data were analyzed in manner similar to Mariner 6. Range residuals corresponding to a representative solution (Fig. 4) indicate a small systematic trend of the same magnitude as Mariner 6 as well as some susceptibility to a small attitude control perturbation in mid-June. A four

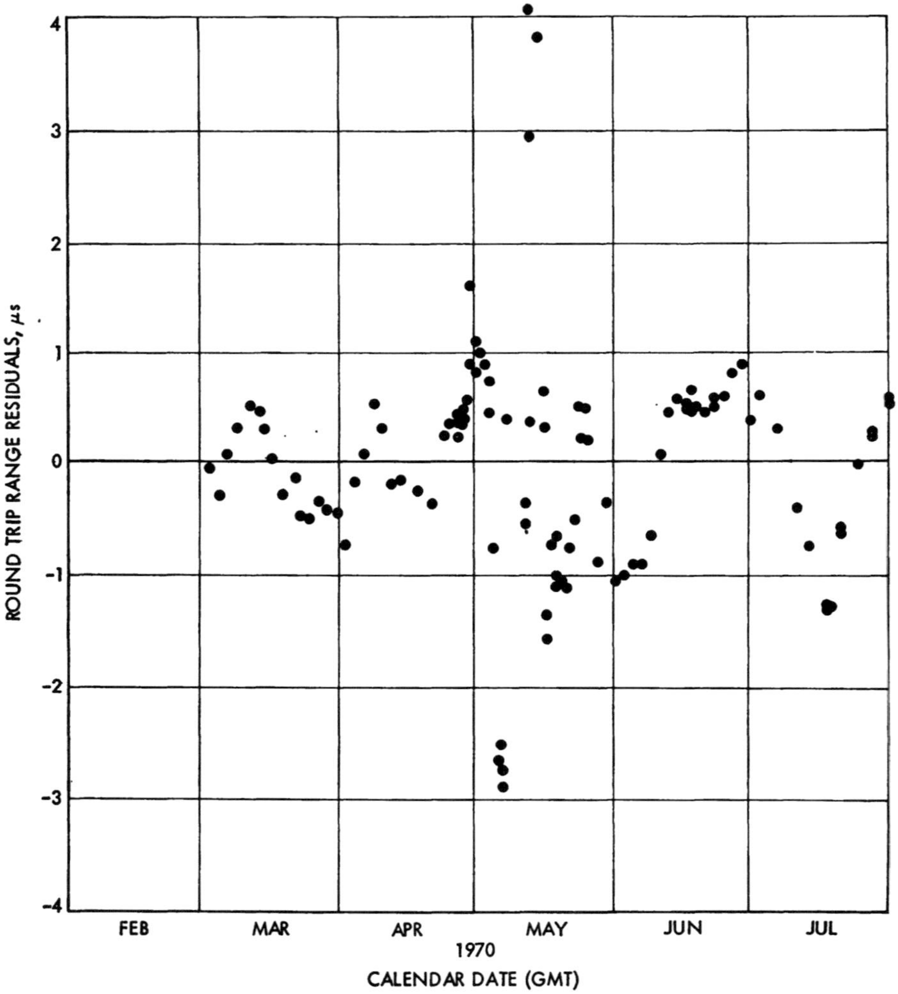

Fig. 4. Mariner 7 range residuals resulting from a typical analysis of a six month data arc consisting of 780 Doppler and 109 range measurements. One data point (12 May 1970) falling outside the border of the figure has a residual of 4.7 μs.

month data arc was also examined and the results indicate slightly larger γ^* solutions compared with the six-month-arc analysis.

The solutions obtained by both spacecraft for γ^* are consistent and in good agreement. By varying both the solve-for parameter and data subsets, one can establish the stability of the solution. Mariner 6 data indicates a more stable solution for γ^* compared with the results deduced from the Mariner 7 data. This is in contrast to the coronal parameters which are much more dependent on "conjunction data" than the relativity parameter. Throughout the analysis, B and ε have an *a priori* correlation coefficient of 0.999 and thus are determined jointly. Analysis of the data from one spacecraft and comparison of the Mariner 6 and 7 results reveal a reasonable consistency between the B, ε solutions. On the other hand, the A corona parameter is extremely dependent on range data within ± 5 days of superior conjunction where, as previously mentioned, the data accuracy has undergone a degradation. Thus, the solution for A must be interpreted cautiously. However, in spite of this situation the relativity solution is stable with respect to the solar corona solution. This is an indication that while the relativistic and coronal delays are certainly correlated their correlation is not crucial and a good separation of both effects is being achieved.

4. Summary of Results and Sources of Error

The effect of non-gravitational and non-modeled forces acting on the spacecraft as well as the degree to which a static coronal model can approximate a dynamic environment constitute the major sources of error. In addition, the possibility exists that the small systematic effect which is present in the data or orbit determination may not be classified in either of the categories mentioned previously. Error sources of lesser importance include uncertainties in the earth's orbit, station locations, time transformations, polar motion and so on.

As a result of a series of error analyses involving simulations of non-modeled forces [13] and earth platform and charged particle induced errors in order to deduce their effect on γ^*, we feel that an *upper bound* to a realistic standard deviation for γ^* is 0.04 in contrast to the formal standard deviation. Thus, the results which summarize the six-month data arc analysis are

	Mariner 6	Mariner 7
γ^*	1.00 ± 0.015	1.01 ± 0.015
B (10^6)	0.20 ± 0.60	0.61 ± 0.59
ε	-0.10 ± 0.26	0.06 ± 0.25
A (10^8)	1.12 ± 0.87	–

All of the uncertainties are formal standard deviations and, at this time, we feel that 0.04 represents a realistic standard deviation for the γ^* solutions.

Analysis of the full arc of data (17 months), a sophisticated simulation of presently non-modeled effects and an investigation of the systematic trends are proceeding. When these investigations are completed, we expect a refined $\sigma(\gamma^*)$.

Acknowledgments

The authors wish to express their appreciation to the Mariner Mars '69 project team at the Jet Propulsion Laboratory for the many hours of time devoted to the extended operations of the Mariner 6 and 7 spacecraft. Special thanks are due the project manager, John R. Casani, and also Victor C. Clarke, Jr., who have shown a very real interest in the general relativistic propagation experiment ever since its inception in early 1966. Also, the support given to the experiment, financial and otherwise, by the Office of Space Science and Applications of the National Aeronautics and Space Administration is sincerely appreciated.

This paper presents the results of one phase of research carried out at the Jet Propulsion Laboratory, California Institute of Technology, under Contract No. NAS 7-100, sponsored by the National Aeronautics and Space Administration.

References

[1] D. O. MUHLEMAN and P. REICHLEY, Jet Propulsion Laboratory Space Program Summary No. 37-29, Vol. IV (1964).
[2] I. I. SHAPIRO, Phys. Rev. Lett. **13**, 789 (1964).
[3] I. I. SHAPIRO, Bull. Astronom. **25** (1965).
[4] D. K. ROSS and L. I. SCHIFF, Phys. Rev. **141**, 1215 (1966).
[5] I. I. SHAPIRO, Phys. Rev. **141**, 1219 (1966).
[6] A. S. EDDINGTON, The Mathematical Theory of Relativity, Cambridge University Press, Cambridge, England 1923.
[7] H. P. ROBERTSON, in: Space Age Astronomy, Academic Press, New York, 1962 (p. 228).
[8] D. B. HOLDRIDGE, Jet Propulsion Laboratory Space Program Summary No. 37-48, Vol. III (1967).
[9] M. J. TAUSNER, Lincoln Lab., MIT Technical Report No. 425 (1966).
[10] J. D. ANDERSON et al., in: Proc. Conf. Experimental Tests of Gravitation Theories, Nov. 1970, Jet Propulsion Lab. Tech. Mem. 33–499 (1471).
[11] T. D. MOYER, Jet Propulsion Lab. Tech. Rep. No. 32-1527 (1971).
[12] D. O. MUHLEMAN et al., in: Proc. Conf. Experimental Tests of Gravitation Theories, Nov. 1970, Jet Propulsion Lab. Tech. Mem. 33–499 (1971).
[13] D. W. CURKENDALL et al., in: Proc. Conf. Experimental Tests of Gravitational Theories, Nov. 1970, Jet Propulsion Lab. Tech. Mem. 33–499 (1971).

High Angular Resolution Astronomical Observations from Space

Space Research XII — Akademie-Verlag, Berlin 1972

HIGH ANGULAR RESOLUTION FROM GROUND-BASED TELESCOPES: A GENERAL OUTLINE OF THE PROBLEM

J. Rösch

Observatoires du Pic-du-Midi et de Toulouse, Bagnères-de-Bigorre, France

Angular resolution is a great advantage in most of the fields of observational astronomy, and a stringent necessity in many of them. For extended objects, the discovery of new structures and hence the understanding of the physical phenomena rely upon the detection and measurement of the smallest possible details; but even for point-like sources, high resolution is extremely rewarding, since a sharp and steady image increases the photometric accuracy and the efficiency of a slit spectrograph and makes the limiting magnitude fainter.

For ground-based telescopes, the present state of the art is such that mechanical or optical defects, or those due to the thermal expansion of the solid parts of the instrument, may be brought well below the thermal effects around the telescope and those of the inhomogeneities of the atmosphere. Improvements have been found for the environment of the telescope, whereas the last effect can be overcome only by "site testing", a problem which cannot be considered as clearly solved yet.

In astronomical observation, angular resolution may be more expressive than modulation transfer function, but more information may be obtained from the wave front reaching the entrance pupil. Observational determination of the structure of the wave front provides a link between the properties of the atmosphere and the image parameters. The smallest details which have been photographed on planets, from the ground, are slightly smaller than 0″2, the closest binary stars photographed have separations of 0″3 to 0″4, and the long exposures on stars give, at best, image-diameters of 0″5. A project is under way of a 200 cm reflector located in a site proved to be good, aiming at 0″1 resolution for short exposures and 0″3 for long exposures. Various solutions which may improve performance are: *compensation in real time* of the atmospheric effects, a solution which is practicable at the moment for some of them only (image motion); *compensation after detection*, i.e. a deconvolution by an atmospheric transmission function which has to be known; *selection after detection*, i.e. select the best of many pictures; *selection in real time*, i.e. detection operates only when resolution is better than a given limit.

Better information could be obtained from the deconvolution of the sharpest images by the sharpest spreading function than from the deconvolution of a "statistic" image by a "statistic" spreading function. Thus, the real atmospheric spreading function should be studied as a function of time, of the size of the objective, of the general and local atmospheric conditions, in order to make more efficient the selection and deconvolution methods; the principles of site-testing would, at the same time, gain very much from such studies.

1. Introduction: Why High Angular Resolution is needed

Because their distances are very large, astronomical objects usually appear as two-dimensional. For each point one should measure the emitted or scattered energy as a function of its two coordinates (x, y), of the wavelength λ of the radiation considered, and of the time t. The most conventional receivers are either dimensionless, as a photomultiplier, or bi-dimensional, as a photographic plate; in that case, one records either an *image* according to x and y for a given time interval and spectral range, or a spectrum according to λ and y for a given time interval and a fixed value of x. In any observation, one should look for

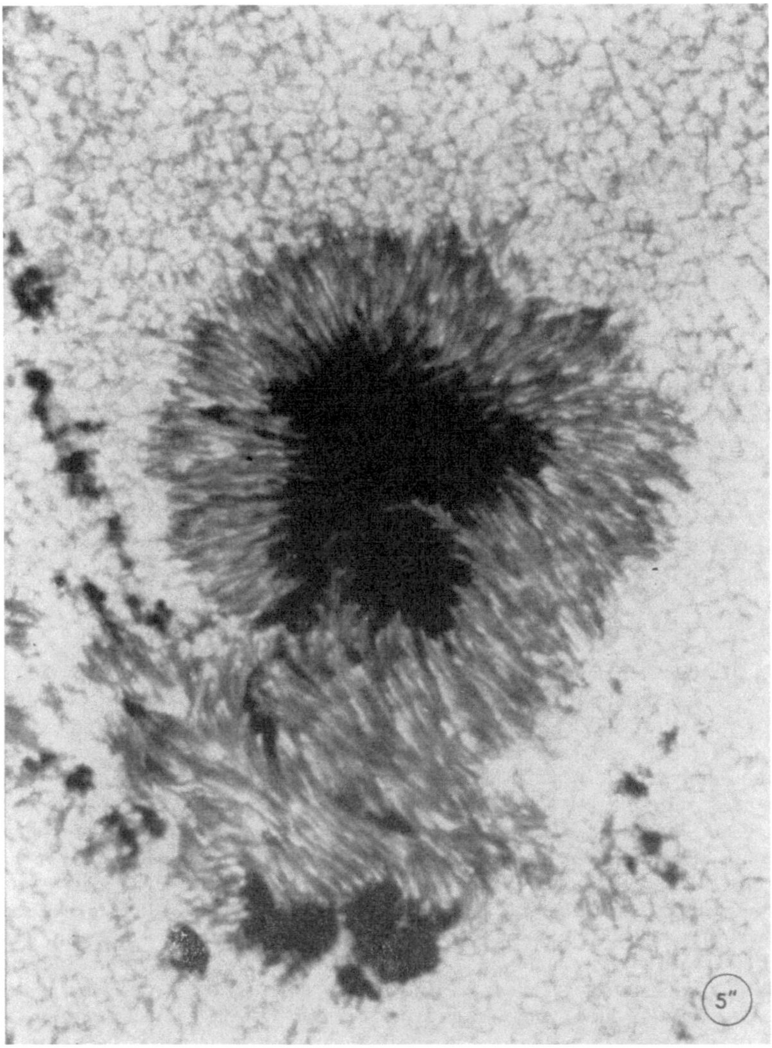

Fig. 1. Fine structure of the penumbra of a sunspot photographed by R. Muller, Pic du Midi, 5 July 1970 (diameter of the objective 38 cm, exposure time 0.05 s). The bright tips of the filaments move towards the umbra. Diameter of the circle 5 arc seconds.

the best resolution over all four variables; but sensitivity limitations force a sacrifice of some of them for the benefit of the others, depending upon the object observed and the information aimed at. The present Symposium deals with the (xy) resolution.

The (xy) analysis (image formation) has a genuine feature which makes it of unique importance: the combination eye + brain has a remarkable ability to integrate the information projected in the form of a two-dimensional retinal image, and, moreover, to understand, if it happens, its time evolution. Thus, such an analysis plays a fundamental role in the *discovery* of structures, and a higher resolution (λy) analysis should take place only *after* the (xy) one, so that one could decide which part of the object should be spectrally analysed, i.e. which value of x should be chosen (Fig. 1).

Now, *high resolution* in (xy) is not only needed for "extended objects": since its limitation results in the spreading and agitation of the image of a fixed point source, it has a direct bearing upon the observation of grouped or isolated stars. For extended objects, the understanding of the structures, mentioned above, obviously requires the best possible disentangling between them; but, quantitatively, the blending of two physically distinct fine structures, or two close point sources, leads to an erroneous estimate of the parameters pertaining to each of them. The extended objects to be considered are: (i) the sun and planets, dealt with in other papers; (ii) galactic and planetary nebulae, for the understanding of which angular resolution will never be too high; (iii) nearby or moderately distant galaxies, where high resolution enables us to identify classes of objects to be related to those of the same class within our own galaxy; (iv) globular clusters, where the drawback of star-blending by lack of resolution is obvious; (v) remote galaxies and quasi-stellar sources for which high resolution would extend valuable diameter measurements, as well as detection, to fainter limiting magnitudes.

Visual binaries (a transition between globular clusters and isolated stars) are a very important class as regards high angular resolution, since, although this is too often overlooked, the whole theory of stellar evolution relies upon an extremely small number of fairly known stellar masses, half of which are derived from absolute orbits of visual binaries; any gain in limiting separation would yield a faster and more accurate determination of orbits, and give access to closer and more numerous binaries, as well as couples having components of unequal brightness and spectral types.

For a point-source, two cases may occur: (i) the general problem of photographic astrometry, the accuracy of which improves with high angular resolution, for the obvious benefit of celestial mechanics, knowledge of galactic structure (through parallaxes and proper motions), and determination of the Ephemeris Time; (ii) an isolated star either observed through a positional astronomy instrument or projected on to the diaphragm of a photometer or the slit of a spectrograph; for photometry high resolution allows for smaller diaphragms, hence a reduction of the background effect, for spectrography it yields a better slit-efficiency and shorter exposures or fainter sources.

Summarizing, high angular resolution is compulsory for a number of problems, and fruitful for most of them, not to speak of the unforeseen discoveries to which it may open the way at any time.

2. Theoretical Aspects of Angular Resolution

For many years, the resolving power of an astronomical instrument with circular aperture of diameter D has been expressed by the angular radius ε of the first dark ring of the theoretical diffraction pattern, i.e. $\varepsilon = 1.22\,\lambda\,D^{-1}$, where λ is the wavelength. For the classical optical range, the new concept of Modulation Transfer Function (MTF) has been worked out (see Appendix). Local path differences occurring somewhere along the light beam bring the function below its theoretical value, thus decreasing the contrast of the image. Whereas MTF is very useful for theoretical problems of information transmission, as well as for practical applications such as routine assessment of optical parts, it is not so useful for astronomers, for astronomical objects never show sinusoidal patterns, but quite often discrete point-sources; thus, the percussional response of the instrument (+ atmosphere) or spreading function, which is equivalent to the diffraction pattern (itself being the square of the modulus of the Fourier transform of the complex amplitudes on the entrance pupil) is directly accessible, rather than the MTF; in any case, convolution of an extended object by the spreading function yields the same result as a treatment through the MTF.

Fourier transform as well as MTF contains only the ratio $\lambda\,D^{-1}$ of the wavelength to a characteristic linear dimension of the entrance pupil. On the other hand, differences in optical path are effective to the extent of the fraction of a wavelength which they equal. Thus: a supposedly perfect instrument has a higher angular resolution the shorter the wavelength used, but it is more difficult to produce a perfect instrument (and the atmosphere itself should be considered worse) the shorter the wavelength at which it is intended to operate. As a consequence, the optimum for a program not depending upon any particular spectral range, e.g. astrometry, would be, in the present state of the art of polishing large mirrors, a 2-meter telescope installed on the moon and operating around 2000 Å, thus leading to $\varepsilon = 0''025$.

Turning back to the spreading function, one should note that it differs from the theoretical one (the Airy pattern for a full circular aperture) because transient or permanent path differences exist along the various rays forming the image. The whole problem of high angular resolution consists in reducing such differences, and may be divided as follows: (i) theoretical description of the consequences of path differences; (ii) description of the effects actually observed; (iii) analysis of the origin of path differences; (iv) attempts to reduce the path differences; (v) attempts to improve the results beyond the residual differences. But a clear-cut distinction should be made, at first, between permanent (or slowly varying) differences arising from mechanical or thermal defects and fast varying differences due to the terrestrial atmosphere and concerning only ground-based instruments.

3. Permanent or Slowly Varying Defects

Permanent and slowly varying defects originate from the fact that one among three conditions is not fulfilled: (i) optical parts should be polished to the best accuracy of the control procedures; (ii) they should not depart from their shape when actually used on the sky; (iii) no other defect should be introduced by the geometrical or mechanical design and construction of the instrument. The first condition (which holds for embarked telescopes too) is within the present reach of optical shops. The second may be jeopardized if the optical parts change their shape either by their own weight (which would no longer be the case in space, but still, to a smaller extent, on the moon) or by temperature changes: A fairly satisfactory compensation of the component of the deformation of a mirror parallel to its axis, when the zenith distance changes, may be obtained by classical procedures [1]; the components normal to the axis are not as easy to compensate, but, although smaller, should nevertheless not been overlooked if the diameter exceeds 2 m and high resolution is aimed at. The problem of thermal deformations has been practically solved by the production of materials having a thermal expansion coefficient hundreds of times smaller than the conventional glass; such small coefficients permit (at least on ground) an increase in the thickness of the mirror, thus decreasing the residual gravity deformations. The third condition relies upon the accuracy and rigidity of the mountings and their motions. The relevant problems may be solved, within a few tenths of a second of arc, for at least 3 m telescopes on equatorial mountings. Beyond that size, one may question whether an altazimuth mounting, making easier the solution of mechanical problems, but relying upon the electronic control for the accuracy required on three non-uniform motions (including field rotation) instead on one basically uniform rotation, is to be preferred. Such a mounting has actually been used for the USSR 6 m telescope.

One should not conclude from what has just been said that the state of the art has brought the permanent and slowly varying defects so well below those due to the atmosphere, that no further improvement is needed. In fact, the two types are additive, so that the smaller the permanent defects, the higher the chances are that the sum falls below a given limit, e.g. the Rayleigh criterion. Thus, any effort to reduce the residual permanent defects should be considered as rewarding.

4. Fast Varying Defects: Theoretical Description

As has already been said, fast varying defects are due to the time, and point-to-point, changes of the refractive index of the terrestrial atmosphere. Thus, they belong to the problem of the propagation of an electromagnetic wave in an inhomogeneous medium, which has been treated by many authors e.g. [2]. However, it is not irrelevant to ask whether or not the wealth of general results already published apply to the astronomical case, or at least may be of some use in it. Indeed, much of the theoretical or experimental work done in the field is intended for application to laser communications, which are to take place in the lower layers of the atmosphere, whereas astronomers observe as high as possible above the horizon, through atmospheric layers of decreasing density, the fluctuations of which are negligible at some tens of kilometres from the observer. Also, the authors, in general, do not consider the effects of the turbulent medium upon the formation of an image as thoroughly as astronomers would like. And finally, the mathematical treatment is necessarily a statistical one, whereas very good astronomical information may be collected in conditions better than average, which would not appear in the statistics.

Therefore, while not neglecting the results mentioned above, we shall look for a description more suitable to the astronomical problems. It may be found by considering the wave front reaching the entrance pupil of the objective, which would be flat for a point-source at infinite distance observed from outside the atmosphere. The wave front is useful because on one hand its shape (or the distribution of the phases in the plane of the entrance pupil, to be completed with the distribution of amplitudes) is a direct consequence of what the light-beam has travelled through; and on the other hand, the Fourier transform of the distribution of complex amplitudes over the pupil yields the distribution of intensities in the focal image of the source, which is precisely what we are looking for; moreover, one may transfer fictitiously on to the entrance pupil the path differences occurring between the entrance and the focal plane.

Since the procedure for deriving the structure of the image from the wave front is known, a very efficient method of analysis would be completed if one could also derive the amplitudes and phases on the entrance pupil from the properties of the atmosphere. But it should first be questioned whether that problem should be treated through the principles of geometrical optics or of wave optics. Tatarski [2] has defined a criterion which he himself considers crude: for an atmospheric cell of size l having a refractive index different from the surrounding medium, at distance L from the pupil and for wavelength λ, geometrical optics is valid provided that $l^2 \gg \lambda L$. Other authors, trying a closer approach, have found, as expected, less restrictive conditions: Taylor [3] gives $20\, l^2 \gg \lambda L$ (20 is not be taken as a strict value); thus, for all quantities in centimetres and $\lambda = 0.5 \times 10^{-4}$, geometrical optics could be used for $4 \times 10^5\, l^2 \gg L$, i.e. as far as some kilometres for a scale of homogeneities of the order of a few centimetres. This already gives a reasonable range for astronomical purposes, and in fact the conclusions drawn from geometrical optics as far as the wave front is concerned agree with the observations; on the other hand, as practical means to obtain the physical parameters all along the beam do not yet exist, one may accept, at the moment, the geometrical solution as a less perfect but reasonably useful one. Furthermore, an experimental test of the Taylor criterion is not impossible. The preceding discussion bears, however, only upon the relation between atmosphere and wave front. Therefore, should one be able to determine the parameters of the wave front, it is not excluded that *empirical* connections, at least, could be found between these parameters and the atmospheric ones, thus completing the link between the atmosphere and the image. It is therefore important on one hand to design instruments for the study of the wave front, and on the other hand to find mathematical representations to allow it to be introduced into equations.

The analytical definition of the wave front may be given, with increasingly closer approximation, by a portion of a plane, the normal to which oscillates with time, a portion of a second-degree surface with curvatures changing with time, a surface in which the height z is a simple sinusoidal function of x and y, abscissa and ordinate, and a surface defined by a two-dimensional Fourier spectrum. Clearly, the validity of these approximations depends, first of all, on the size of the entrance pupil, compared with the reciprocal of the spatial frequencies describing the wave front, or in other words, of the dimensionless number Π (pupil size $\times$ spatial frequency). But two important points should be emphasized:

First, Π is not the only parameter playing a role, since the wavelength of the light defines a linear scale. As a result, data concerning only the directions of the normals to the wave front are not sufficient to define the phenomena, since two surfaces homothetical in the ratio k cannot been distinguished from one another through such data, whereas the results for the images are quite different, unless a wavelength also multiplied by k is used. In particular, the concept of "turbulence atmosphérique" introduced by Danjon [4] as the limiting angular departure of the normal from its average direction does not uniquely characterize a situation; it may be used, at most, to compare with instruments of the same diameter situations which are presumed to be similar, but not to extrapolate the results to what would be observed through instruments of different diameter.

Second, the mathematical treatment of the problem starting from the Fourier spectrum should be undertaken cautiously. Scheffler [5] has found interesting results concerning the spreading function created by stochastic fluctuations of path differences, assuming a Fourier spectrum which, although not unlikely, was chosen more for its mathematical form than for its resemblance to a real one. He thus showed that inhomogeneities mostly affect the wings of the spreading function, but that even when they are important, there remains a central peak not much wider than the central peak of the Airy pattern. Coupinot [6] has resumed the problem by using the MTF and obtained much more direct results than Scheffler, showing, in particular, that the spreading function is obtained by the convolution of the plane-wave spreading function of the same pupil by a function characterizing the wave front; but then it appears that his computations, as well as Scheffler's, are valid only if Π is very large, in other words, there is a gap between the approximation by a second degree surface, valid for Π of the order of 0.2–0.5, and the Fourier spectrum representation, which requires Π to be large. Unfortunately, it turns out that, taking into account the already known properties of the terrestrial atmosphere, telescopes of the order of 1–3 metres, and perhaps more, fall precisely within this gap, which remains to be filled.

Of course, a numerical treatment of experimental data is still possible. A valuable start is therefore to determine, by observation, either the local shape, or the Fourier spectrum, of the wave front, as well as their changes in time, according to meteorological and instrumental conditions, and thence to compute the distribution of the intensities in the image, so as to connect the atmospheric parameters with the information which may be collected. Such a knowledge, defective at the moment, would lead to a better understanding of the relationships between external causes and angular resolution, hence (active) an improvement of such conditions by adequate techniques and (passive) a sounder choice of the most satisfactory places and periods, to a better basis for extrapolation from the results of a small survey instrument to large telescopes planned at the same site, and to the design and operation of methods for improvement of the information.

There are numerous methods for wave front determination, but not all are of easy use with a star as a point source. Rather than using the drift of the wave front before a small objective, it is better to define it at a given moment, over a surface as extended as possible. The Hartmann method can be used over 2 m or more, with a time resolution of a few hundredths of a second, and a sampling in joint elements of a few centimetres, i.e. within the limits of the Taylor criterion.

5. Description of the Effects Observed

One can describe the image of a star (which shows directly the spreading function) or of an extended object (being the addition of the spread images of a set of incoherent sources) in terms of the above descriptions of the wave front. For a small (circular) pupil, covering a portion of the wave front which may be taken as a plane, the image is exactly the Airy pattern, but as the direction of the normal to this plane oscillates in time the centre of the pattern wanders, thus giving the well-known image motion. This is observed almost always with an objective diameter of a few centimetres, quite often with 10 cm, much more seldom with 20 cm; i.e. the "quasi-plane" elements of the wave front usually measure 10–15 cm; smaller elements are not excluded, but their amplitude (in path difference) should be small enough not to alter too much the rings of the Airy pattern. For a larger pupil, one is in the range of application of the criteria given by Danjon [4]; the patterns called by him "à condensations mobiles" and "à anneaux brisés" are easily understood in the approximation of a second degree surface; since the height of such a surface may be split into that of a sphere, plus that of a hyperbolic paraboloid, the defect will look like a sum of defocusing and astigmatism, both continually changing; then the classical pattern resulting from diffraction plus a small residual aberration, with time variations, readily explains the Danjon's patterns.

This image-alteration is no longer described so simply when Π reaches 0.5 or more: for very large values of Π, patterns are of the type described by Scheffler. In fact, one finds, with large instruments, a strong increase of the wings of the profile; the appearance of the central peak is less certain, although some observations by Texereau [7] seem rather significant. For intermediate values of Π, short exposures on a bright star show [8, 9] a "speckle" pattern, something intermediate between simply altered Airy patterns and Scheffler patterns. A long exposure gives a spread image with a centre of symmetry

(Fig. 2). An important point, precisely, is to determine the time scale of the evolution of the granular structure, in order to know the average life-time of a concentrated image, for the choice of an exposure time or of the response time of a selector (see below). Preliminary experiments show that some of

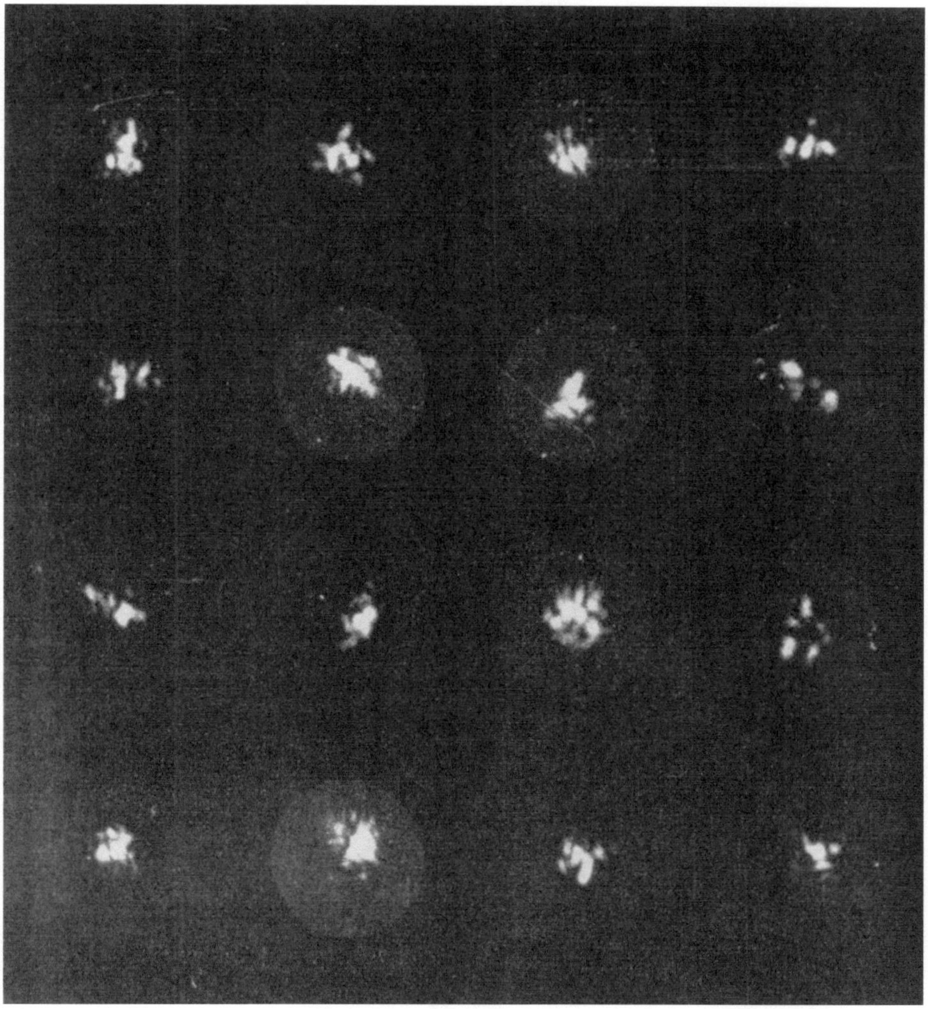

Fig. 2. The focal image of Sirius, exposure time $^1/_{72}$ s. The distance between two exposures, on the print, is about 12 arc seconds.

the elements of the pattern are still present, whereas other have disappeared, after one fiftieth of a second.

Restricting the number of parameters to three nevertheless gives a good estimate of the image quality. The first (image motion) describes the positional changes of the centre of gravity of the distribution of intensities in the focal plane; the second (image blurring or spreading) may define, for instance, the diameter of the circular diaphragm which would admit 50% of the energy; the third (scintillation) depends upon the second derivatives of the path differences on wave fronts at a distance from the instrument; it has not in itself a bearing upon high angular resolution. It is easily seen that, with increasing diameter of the instrument, the blurring (as a ratio to the theoretical image size) also increases rapidly, whereas image motion and scintillation tend to vanish by averaging effect: in fact, they do not exist in any treatment assuming Π very large. Image motion is approximately given by the tilt of the plane which averages the wave front over the pupil, and it follows from this geometrical definition that it is not first-order dependent on the wavelength of the light, whereas the blurring is. Also an isolated inhomogeneous region yields stronger image motion and scintillation the further it is from the instrument.

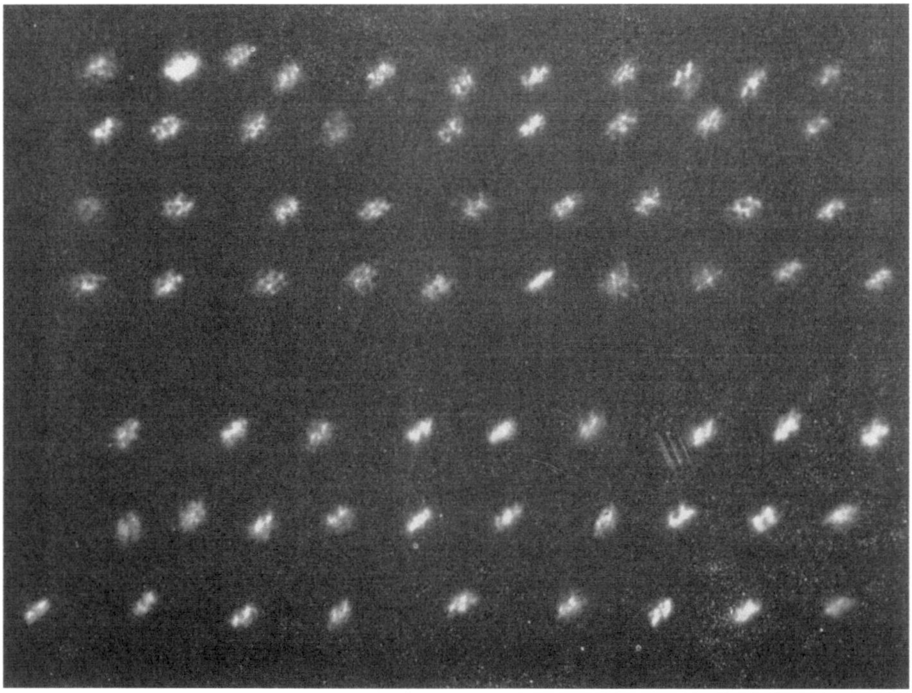

Fig. 3 a

The consequences of the preceding features for various classes of astronomical objects or observations are as follows:

(i) For single-star spectrography with large instruments, the slit-efficiency is reduced by blurring and by motion if it is important (i.e. if the Fourier spectrum of the wave front extends widely towards low frequencies); scintillation does not matter for long exposures.

(ii) For single-star photometry, blurring and motion demand a larger focal diaphragm, otherwise a defect error is introduced by cutting the wings of the spreading function; on the other hand, more energy from the sky background is added, and may be of major importance for faint objects; scintillation plays a direct role and cannot be eliminated except by time averaging.

(iii) Observation of visual binaries is of particular interest: if a 1 m aperture gives a granular image of one of the components, the light beam from the other component at 1″ separation travels through air masses which are 90% the same, so that the image contains the same grains. A short exposure photograph actually shows such "homologous" grains, on which the pointing accuracy is better; as each pair gives an independent measurement of the same quantity, the accuracy is definitely improved; still better, interferometric methods are

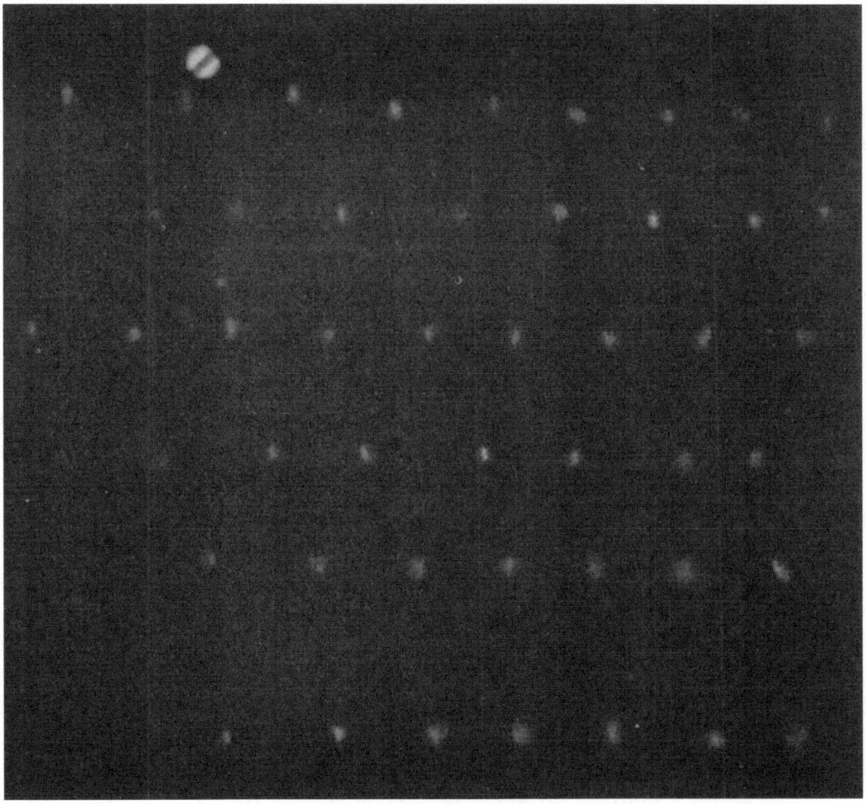

Fig. 3b

Fig. 3. Short exposures ($^1/_{50}$ s) on double stars: (a) ADS 9343 (ζ *Bootis*, magn. 3.5–3.9, separation 1″1), showing homologous structures; (b) ADS 8804 (α *Comae Berenices*, magnitudes 6.0–6.0, separation 0″4).

able to use the whole pair-structure at once, even for very close binaries: in that way Gezari et al. [10] have measured α Aur. (0″057 $\pm$ 0″004) with the 5 m Hale reflector ($1.22\,\lambda\,D^{-1} = 0''027$). Of course, whatever the structure of the diffraction pattern, a reduced overall size will always help in resolving close pairs (Fig. 3).

(iv) When extended objects are considered two new points arise: first, fine structures will not be resolved unless their contrast in the image is sufficient; the contrast depends upon the contrast of the object, the spreading function (particularly its wings), and the response of the receiver. One may revert to MTF, but it is found from experience that the possibility of useful detection of frequencies not far below the cut-off is not to be minimized; secondly, beyond a few seconds of arc apart in the field, the effects of the atmosphere become independent; thus, the amount of blurring changes from one point to another (an explanation of the so-called "réseau photosphérique" described by Janssen) and, still more important, so does image motion, causing a distortion which is very conspicuous, for instance, on series of short exposures obtained on the photosphere. Such a distortion is a severe drawback to accurate angular measurements; long exposures remove it, but at the cost of blurred images. Moreover, these field changes need to be taken into account for correction or compensation of the disturbances (see below).

6. Origin of Atmospheric Defects

Atmospheric defects are all those occurring between the outer space of negligible density and the receiver, including between the various parts of the instrument. One physical parameter only is involved, the refractive index n; and as $n - 1$ is proportional to the density ϱ, the ultimate effect is in the fluctuations of ϱ, whether caused by temperature, pressure or humidity variations, or by two or three of them simultaneously. Three different mechanisms, in principle, are able to produce such variations: adiabatic compressions and depressions upstream or downstream of an obstacle, changes of physical state of water, and mixing of air masses having different densities, either by general or local atmospheric circulation, or by thermal exchange with the ground, obstacles, or parts of the instrument.

The first mechanism is not very efficient, and for wind speed still permitting astronomical observations, the corresponding relative changes of density would not exceed 10^{-4}, with negligible angular deviations of the light rays. The second mechanism occurs in the vicinity of saturation in water vapour; due to the high value of the latent heat, droplets small enough to let the air appear transparent could produce, by evaporation or condensation, localized but not negligible temperature changes; Pagès [11] has observed, pending confirmation, a strong deterioration of the images under conditions of that type. The mere presence of water vapour has, however, an important effect: the refractive index n of a mixture of air having a normal index n_a, with partial pressure $p - e$, and water vapour having a normal index n_e, with partial pressure e, is given by the law of Mascart:

$$n - 1 = (n_a - 1)(1 - e\,p^{-1}) + (n_e - 1)\,e\,p^{-1}\,.$$

As $n_a - 1$ and $n_e - 1$ are different (2.93 and 2.57×10^{-4}), $n - 1$ is a function of e; assuming that the pressure fluctuations are smoothed out very quickly, so that p may be taken as constant, one finds that, around 0 °C, the change in refractive index caused by the presence of water vapour amounting to a fraction r of the saturation equals the change produced by a temperature variation amounting in °C to $0.2\,r$, i.e. 0.1 °C for $r = 0.5$. Thus both phenomena may be of the same order of magnitude, and play the same role in the following paragraphs. Indeed, if there exist somewhere an air mass of different density, it will not produce an optical defect, unless it is mixed with others; therefore, the transport mechanisms play a major role, including when they are caused, precisely, by the presence of air masses of unequal densities.

Whenever a mixture occurs we are only interested in what happens between the first contact of the different masses (since previously there were no optical effects) and a complete homogenization. The time scale in the mixing phenomenon determines the duration of an image of a given quality; the spatial frequency spectrum of the amplitude of the density fluctuations is still more important, since it determines one of the factors of the number Π. Unfortunately, the real air masses are generally re-fed, so that the situation is rather an instable one at small scale, characterized by slowly varying average parameters. A theoretical study is possible [12], but the observation of a star is the most direct source of information, not only for the astronomer interested in a given phenomenon, but even for the atmospheric physicist, who may find there an exploration method with spatial and time resolution difficult to attain by the sensors used *in situ*.

The transport phenomena take place within several dimensional frames: On a planetary scale, the most typical effect is that of jet-streams which, although occurring at elevations where the average density is already much lower than at sea level, may have a non-negligible optical influence because of the strong stirring of different air masses at the level of the tropopause. Such influences have actually been reported by observers, but a complete study remains to be done.

On a synoptic scale, the obvious phenomenon is the alternation of oceanic and continental, or polar and tropical, air masses, the regions of contact of which, along fronts, are of particular importance here. However, it is not demonstrated that any air mass of a given type is always homogeneous in itself. An interesting example is found in Western Europe, where clear sky periods occur when an anticyclonic region extends either from the Atlantic (Azores) or from North-Eastern Europe; images are usually sharp in the first case, and blurred in the second one; the explanation that the cold air masses travelling NE–SW have collected the air heated on the continent during day-time is not conclusive, since the same would hold, to some extent, for atlantic air sweeping over the Iberian peninsula. It is likely that the main difference lies in the inclination towards the pole of the meridional cross section of the polar front: as a result, a northern location of this front gives a better chance that an observatory is higher than the inversion layer which stops the convective ascent of air masses heated on the ground, so that it would have better conditions. On the other hand, the average height of the inversion layer also depends upon more local conditions, e.g. it is lower above a cold ocean (North and South Pacific coast).

On a regional scale, the characteristic factors are the relative situation of oceans and continents, and the relief. The latter produces "mountain waves", the main effect of which is to transform the previously horizontal track of an air parcel into a damped sinuoidal one; there is no harm as long as the flow remains laminar, but a turbulent regime may be created under the crests of the waves, and mix up the air masses arriving from far away with the local ones heated in day-time or cooled in night-time; clearly an observation from a point located downstream of the mountain and at lower elevation will suffer from such effects. The problem is to know the distance from the mountain up to which they still exist; temperature fluctuation measurements tend to indicate several hundreds of kilometres [13]; however, excellent images are observed at Pic-du-Midi, with moderate SW wind, about 30 km downstream of the highest parts of the Pyrenees. As far as the situation relative to an extended water surface is concerned, it has been suggested, apart from the lowering of the inversion layer (see above), that it could provide a solution for day-time high-resolution observations, by taking profit of the "sea-breeze" to remove the thermally unstable ground layer; unfortunately, the air blowing from above the sea is then heated over the land and may be brought back to the shore by an opposite wind at medium altitude [13]. Therefore, one should preferably choose an isolated island, either with summit higher than the inversion layer (which would give better conditions than a isolated peak on land), or quite small (and then, of course, not very high).

On a local scale, since, as we said, the adiabatic effects are negligible, the major role is played by mixing due to thermal exchanges (of opposite signs during day and night) between ambient air on one hand, and natural ground or man-made obstacles on the other. Daily variations of the temperature gradient in the first decametres above ground have been studied from the aerological point of view [14] and from the astronomical one [15]. But these investigations are dealing with what happens on an extended flat ground, and the situation is different on the top of a mountain: in day-time, warm elements may be trapped below the inversion layer and stay within the turbulent region downstream without penetrating into the laminar flow; if the summit is slightly above the inversion layer, it may behave just as an isolated small island in the ocean. Some observations under way at the Pic-du-Midi tend to support this interpretation; the limited duration of good resolution in the morning could be due to the lifting up of the inversion layer, the images deteriorating when it reaches the observatory. At night-time, the situation is easier to understand and, fortunately, more favourable. On a flat area the air which has been cooled at the contact of the ground cannot but stagnate, hence an inverse gradient which, the wind helping, produces inhomogeneous mixtures; on the contrary, from the top of a mountain, cool air flows down and is gently replaced by homogeneous air from above. Conventional temperature records fully confirm these facts: for cloudless sky, from sunrise to sunset, the records show symmetrical arches on a summit, and asymmetrical ones in a valley (even a wide one); from sunset to sunrise, the average temperature is fairly constant on a summit, with quite small variations ($\pm 1°$ or less), and steadily decreasing, with fluctuations. in a valley. Such a stability during the night, (together with

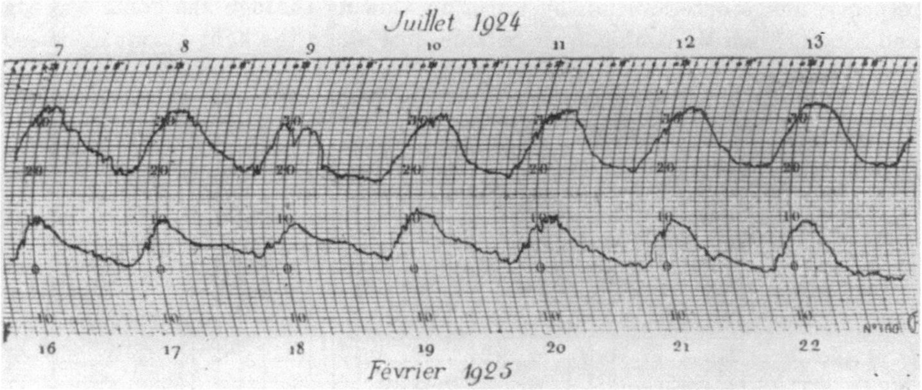

Fig. 4. Temperature records at Forcalquier, alt. ca. 600 m. Durance Valley [4].

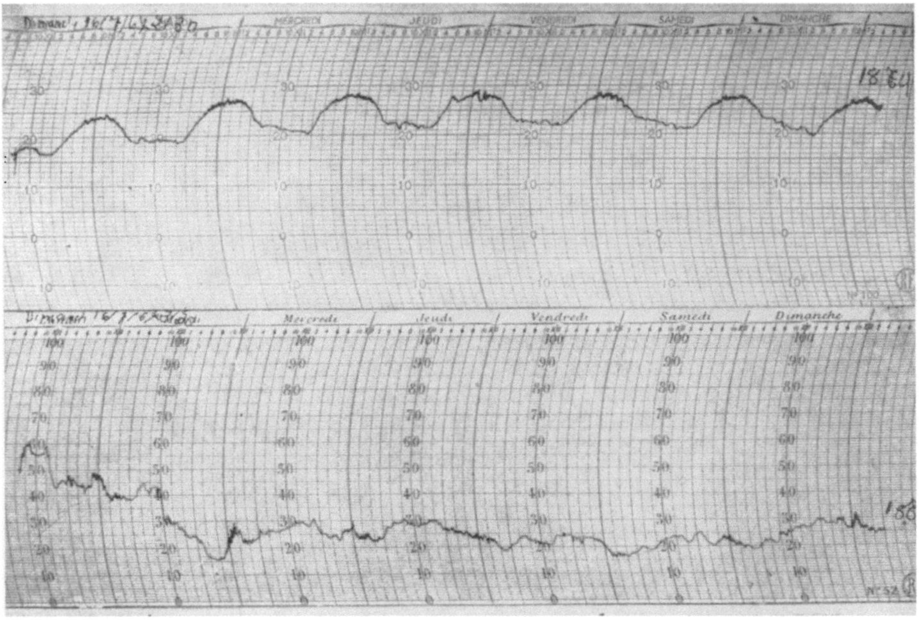

Fig. 5. Temperature and humidity records, Sierra de Gador, 2200 m (Southern Spain).

small daily amplitude) has been observed in Chile [16], at Pic-du-Midi, in Ahaggar [17], and on other mountains (Figs. 4–6).

Finally, on the instrumental scale, although the path lengths are shorter, large temperature differences may cause important disturbances. Such differences may be due either to diurnal heating of the instrument and its surroundings, or to artificial heat sources, and their efficiency depends upon the possibility for the various air masses to mix with one another. The best known examples are: heating of the air in contact with the primary mirror which is warmer than the night-air, (low expansion materials are not different in this respect); image deterioration by warm air flowing through the dome opening and mixing with the ambient air outside, just along the light-beam; increased

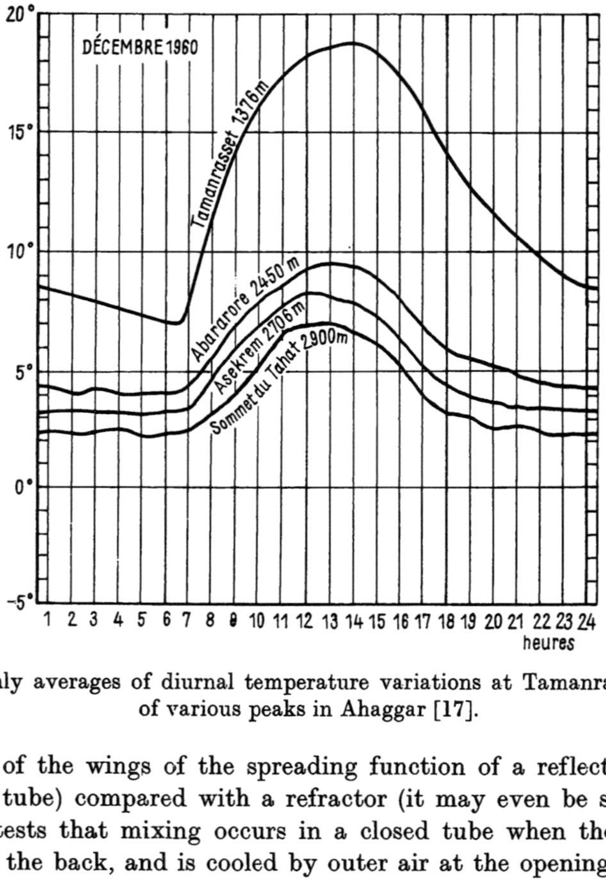

Fig. 6. Monthly averages of diurnal temperature variations at Tamanrasset and on top of various peaks in Ahaggar [17].

importance of the wings of the spreading function of a reflector (open front part of the tube) compared with a refractor (it may even be shown [18] by Hartmann tests that mixing occurs in a closed tube when the front part is higher than the back, and is cooled by outer air at the opening of the dome).

7. Minimization of Atmospheric Defects

Two classes of remedies may be used to minimize atmospheric defects: "passive" ones, taking into account planetary, synoptic, regional and local conditions to choose the best possible location, and "defensive" ones, for the instrumental conditions.

The first class includes many problems which have been very often debated but not conclusively solved. Some general lines begin to emerge, at least for night observations: (i) as good transparency is highly desirable, an elevated location is to be chosen; then, if it is on an isolated mountain, advantage may be taken of the temperature regime mentioned above; (ii) even if the effects of jet streams are not fully known, one should preferably avoid regions over which they usually flow; (iii) the most frequent synoptic situations should be considered, perhaps retaining regions seldom reached by the polar front; (iv) attention should be paid to the frequency of occurrence of an inversion layer, and to its average elevation, above which the observatory should be located; (v) taking into account the direction of prevailing winds, a location downstream of a mountain higher than the observatory should be avoided. Assuming all these conditions to be fulfilled, a direct optical control of image quality cannot be dispensed with. All the methods suggested have two fundamental drawbacks: first, they do not provide a safe extrapolation to the result which could be expected from a large instrument, and secondly, no direct comparison is possible from one to another. The rational solution would be to learn about the Fourier spectrum of the deformation of the wave front, to classify the various types of spectra observed in various meteorological conditions, to determine the minimum number of parameters needed to characterize these types and then to study the filtering of certain spatial frequencies by certain test procedures, so as to correct their results, and to select an association of two or more methods yielding, altogether, a Fourier spectrum allowing for an acceptable extrapolation to a large instrument. Before such an ambitious program reaches its completion, it should be strongly recommended to minimize the risks of the extrapolation by estimating the image quality through a real observation, for instance visual observation of binaries, and photography of the moon and planets, using a telescope some 60 cm in diameter. Application of the preceding meteorological criteria should result in the choice of a very small number of places, all likely to give good results. Experience shows that the cost of any extended site survey is such that the expense of a telescope of that type, on a simplified, light and not cumbersome mounting, would be only a small fraction of the total — and at least the results would be much more significant in the final discussion.

We have considered only night-time observations. For solar observations, no general doctrine seems to be definitely established yet, and it is to be hoped that the surveys under way (e.g. JOSO, Joint Organization for Solar Observation) will cast some light on the problem.

Fortunately there are more resources against near-by or internal disturbances. First, it is possible to measure the average thickness of the turbulent ground layer in order to know the minimum elevation above ground where the telescope should be installed; recording temperature fluctuations along a mast, or with tethered balloons, has been widely used for this purpose, since the fluctuations fade out at the level where laminar flow is reached. Although significant, this method is difficult to apply in many places and for many wind directions. It is, however, possible to work on a model in a wind tunnel; the real Reynolds number cannot be reproduced, but the *shape* of the wake (not what happens inside) is not sensitive to its value; by measuring temperature fluctuations, or by observing smoke puffs, or otherwise, comparison of real and simulated values of the thickness of the wake may be made at one or several points as a check. A study

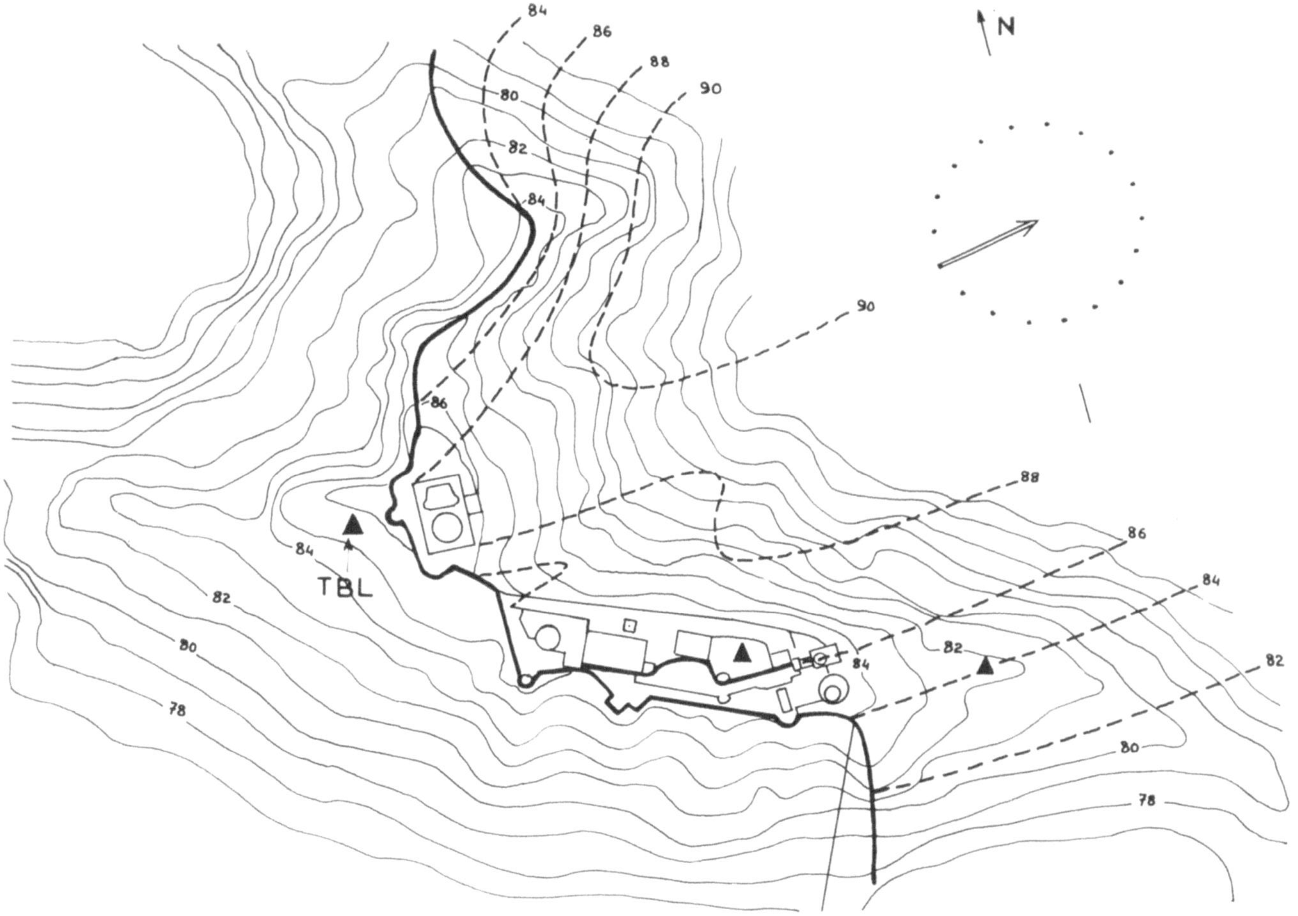

Fig. 7. Example of wind tunnel tests on a Pic-du-Midi model, for the prevailing wind direction (arrow). The wake "takes off" from the ground along the heavy lines; the level lines of the wake are shown superimposed to the ground contour level lines. The TBL (Telescope Bernard Lyot) will be located at elevation 86 (which means 2860 m) as indicated. Other locations, marked by triangular dots, have been considered and then discarded according to the tests [19].

of this type has been conducted [19] for the choice of the exact location of the 2 m reflector on Pic-du-Midi (Fig. 7).

Second, one should try to prevent local air masses of different densities from mixing with one another along the light path, a condition by no means fulfilled in conventional domes and instruments, which, as a consequence, suffer from the disturbances described in the preceding section. Two different concepts may be used to provide against these drawbacks: (i) having the same air-masses all along the path of the light: aiming at that, Couder [20] has designed the 193 cm reflector of Haute Provence Observatory with fans in the dome wall and in the telescope tube just above the mirror so as to fill the tube, and then the dome, with external air; experience seems to show that a non-negligible result is to blow outside the dome the air heated at the contact of the instrumental masses which have kept their day-time temperature, rather than to homogenize the air inside the tube: (ii) partitioning and letting the various air masses find their own equilibrium; then the tube should be completely closed (including at its front end by a parallel plate if the telescope is a reflector) and thermally insulated; the front part of the tube should be connected to the dome by a soft gasket; a solution of this type has been worked out at Pic-du-Midi for the "turret dome" [21]. Three different air masses are therefore separated: outside the instrument and dome, inside the dome, and inside the tube; the second should be kept as near as possible to the external temperature to minimize the residual effects, but it does not need to be perfectly homogeneous; this is a stringent condition for the third one, to which the most careful attention should be paid.

A drastic but expensive and difficult solution is to evacuate the tube, as Dunn [22] has done for the Sacramento Peak Solar Tower. At least, internal heating of the tube should be eliminated: for a solar telescope, the focal plane should be cooled in order to eliminate most of the energy; for a night-observing reflector, the mirror should be cooled in day-time to be, at night, at the same temperature as the air inside the tube; but then, stratification may take place, and it does not seem wise to destroy it by circulating the air in a closed circuit [18]. For a non-evacuated tube, a solution is to fill it with helium, which, thanks to its high thermal conductivity and refractive index very close to unity, may reduce the internal optical effects by a factor twenty or so [23, 24]; however, for application to a large instrument, a number of technological problems would have to be solved.

The first of these problems lies in the production of parallel plates several metres in diameter and homogeneous enough not to introduce permanent defects more severe than the variable ones they eliminate. For the 2 m reflector of the Pic-du-Midi (Telescope Bernard Lyot, TBL) the tube will have closed and insulated walls with gasket connection to the dome, the mirror will be cooled during day-time, and the dome will be ventilated to keep the external temperature. At first, the front part of the tube will be opened, and a parallel plate will be installed when available. To prevent the internal and external air mixing in front of the entrance of the tube has some meaning in itself; but moreover the resulting structures are more streamlined, which reduces the disturbance in the wind flow caused by the dome. Even if the dome is not a closed one, solutions should be preferred which minimize the mixing; for instance, wind-tunnel studies show that the system of one window sliding from the opening backwards around the sphere is better than the classical two windows moving horizontally to the left and right of the opening [25].

8. Means of Improvement

Assuming a large telescope to have been built according to the best methods and located in a place believed to be the best, it will approach its theoretical resolving power less closely and less frequently the larger its diameter. It is possible to make better use of its size (and cost) either by correcting the defects of the image, or by collecting the information only when the deterioration is the least possible.

Correction of the defects may be considered either by introducing adjustable path differences on the beam in order to obtain a real time compensation, or, assuming the spreading function to be known, to make a deconvolution of the recorded image. *Selection* also may be used in two very different ways, pre-detection or post-detection. Whereas the last of these four methods is more empirical, the three first of them have in common that they require a knowledge of the defects themselves or at least of their consequences for a known object, two very severe restrictions. First, we know that these defects vary in time very rapidly, so that the response time of the system which is supposed to detect them and control the compensation or selective recording should be very short; for deconvolution, unless one short exposure is processed by means of the spreading function determined simultaneously, only an average will be obtained. Second, we know that the spreading function, at a given instant, is constant only within a very small field, so that it should be determined on the object itself, if at all possible, or for another object as close to it as possible in the field; otherwise, the real-time methods are meaningless, and deconvolution, again, yields an average. Hence the necessity of knowing more about the behaviour of the defects in space and time before using any of these methods.

Real-time compensation is the ideal solution but, not surprisingly, the most difficult. Babcock [26] has proposed an elegant method based upon the Eidophor television system, but he considers that its working out would be very complicated, and, that, precisely because of the restrictions mentioned above, its field of application would be very limited. However, if the disturbances are divided according to three parameters (§ 5), it appears that the difficulty concerns mostly the compensation of blurring, since scintillation may be readily compensated (but there is no point in doing this except for very special cases) and image motion compensation is nothing but the most refined form of guiding; the many systems which have been used for this purpose are able to provide image motion compensation, provided their response time is short enough; as a consequence, low inertia parts should be used, as for instance a gimbal-mounting parallel plate in front of the image.

Deconvolution meets some mathematical difficulties (uniqueness of the solution, limitation of restored frequencies to the cut-off of the MTF, Gibbs phenomenon, etc.) which we shall not insist upon, since they are not specific to the astronomical problems. Deconvolution should be considered from two different aspects. In some cases, one looks for a quantitative determination of a parameter of the object which is obviously altered by the blurring (even merely by the theoretical diffraction pattern); then a correction is necessary, and could be done by deconvolution of an image by a spreading function assumed to be the one which actually blurred it: an example is the determination of the profile of the solar limb during a partial eclipse using the image of the lunar limb as a spreading test. But if one looks for an improvement of the angular resolution the question may be raised whether the deconvolution of an average image by a spreading function itself a field- and time-average could give better information than the images chosen as the best obtained in a time sequence. It seems clear that one should at least make a deconvolution of the latter by a spreading function the frequency of appearance of which should be of the same order as that for the selected images. Thus, the selection concept already emerges in connection with deconvolution. In fact, image selection has been practised for a long time: the countless drawings of planetary surfaces produced by visual observers are formed with all the details transiently seen (or believed to be seen) on various points of the disk, not necessarily all at once; later on, short-exposure photographs of planets and double stars have been taken by watching for periods of "good images", ignoring whether or not their quality would persist for longer than the reflex of the observer. Of course, this "pre-selection" was completed with what afterwards has been grandiloquently called "post-detection selection", which means simply that a large number of pictures are recorded, and only the best ones are kept.

A selection of that type, although empirical, is nevertheless efficient, but it does not apply to all the problems, and automatic selection should be considered. It could be arranged, in particular, to integrate the information collected only during periods of sharp image, thus extending the selection to the case of long exposures (e.g. to obtain separated spectra of the two components of a close visual binary). But any selection procedure relies upon two conditions (besides the short time-constant needed): the image-quality should be defined by the measured value of a parameter the real value of which, on the object, should be previously known; the role of the selector is to measure it, and trigger the detection whenever it exceeds a given threshold; and such a parameter should be measured on the actual image to be recorded, or at as small an angular distance as possible from it.

The automatic selection techniques are of much easier access, in practice, than the real-time compensation techniques, and several significant attempts have already been performed. Those intended for the photosphere show good examples of the general features mentioned above: the CSIRO group [27], aiming at high resolution pictures of the chromosphere near the limb, uses as quality criterion the slope of the extreme solar limb on the image; for high angular and spectral resolution on the photospheric field (with electron camera)

Rösch [21], followed by Gaujard [28], measures the contrast of the granules, for which a minimum value is known; moreover, as their angular scale is also reasonably known, electrical frequency filtering after scanning at a properly chosen speed eliminates the spurious frequencies due to the atmospheric disturbances. Clearly, each of these selectors has its own field of application and they cannot be interchanged from one region to the other.

Another example has been given by Rösch [8], for the choice of the best image of an isolated star. What is known here is that the angular diameter of the source is practically zero, so that, were it not for the atmosphere, the wave front on the entrance pupil would be a plane; thus, determining the shape of the real wave front gives a criterion. Such a selector would operate as well on small disks, or on a binary with large magnitude difference, in other words, in the same cases as already mentioned by Babcock [26]; but, at the cost of a lower efficiency (i.e. a longer total observing time needed) it could give about as good results as real-time compensation, while more certainly remaining within the reach of the present techniques.

Acknowledgments

This contribution results from the experience gained and the innumerable discussions held over the years, either within the Pic-du-Midi group, or with French or foreign colleagues. Among my nearest co-workers, I would like to thank H. Camichel, F. Chauveau, G. Coupinot, M. Hugon and J. Saissac for the help of their experience in the fields of observation, optics, and meteorology.

Appendix

Spreading Function and Modulation Transfer Function

As the detailed demonstrations of the following theorems are to be found in all textbooks on optics, we shall limit ourselves to showing the links between the various concepts involved.

The image of a *point source* in the focal plane of the telescope is an extended distribution of intensity as a function $D(u\ v)$ of the coordinates u and v in the focal plane. D may be called the *Spreading Function* and it is easily shown that it is equal to the square of the modulus of the Fourier transform of the distribution of complex amplitudes, $F(\xi, \eta)$ over the entrance pupil. In the absence of atmosphere, F would be real and independent of ξ and η.

Let us consider now an extended *incoherent* object, defined by the distribution $S(x\ y)$ of intensities which it would display in the focal plane in the assumption of a perfect and infinitely large objective. The true distribution of intensities in the image through the atmosphere and a finite size objective may be found in two different ways, both of them relying upon the fundamental assumption that $D(u\ v)$ is the same for all points of the focal plane; this is known to be correct, in astronomical practice, only within a field of a very few seconds of arc, and this should be kept in mind.

The first approach is simply to add the intensities produced at each point of the focal plane by all the points of the object, due to the spreading of their image. This leads immediately to the convolution of $S(x\,y)$ by $D(u\,v)$:

$$R(x\,y) = \int_{-\infty}^{+\infty}\!\!\int S\,(x - u, y - v)\, D(u\,v)\, du\, dv \quad \text{or} \quad R = S \otimes D\,.$$

The second approach starts from the convolution of a peculiar object, a chart showing stripes having sinusoidal distributions of intensity, with spatial frequencies μ and ν along the coordinates x and y. The result is that the image still shows the same stripes and same frequencies, but with contrast reduced

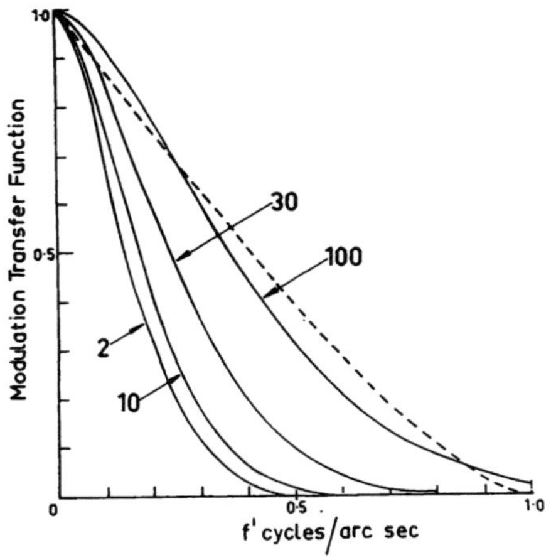

Fig. 8. An example of modulation transfer functions: The full lines show the MTF for an infinitely large aperture, through an assumed atmospheric model, the instrument being located at 2, 10, 30 and 100 m above ground in such an atmosphere. The dotted line is the theoretical MTF for a circular entrance pupil of diameter 11 cm. The MTF for this objective in the atmosphere is obtained by multiplying the ordinates of the dotted line and of the full line corresponding to the elevation of the instrument [15].

by a factor $\Gamma(\mu, \nu)$ which happens to be the Fourier transform of $D(u\,v)$; it may be demonstrated, too, that $\Gamma(\mu, \nu)$ is the autocorrelation function of $F(\xi, \eta)$ when shifted by quantities equal to $\lambda\,\mu$ and $\lambda\,\nu$ along the corresponding coordinates. Γ is the *Modulation Transfer Function*, since it indicates how a modulation, at a given spatial frequency, is filtered by the instrument. If a one-dimensional pattern only is considered, the MTF is expressed by a curve starting from unity for a zero frequency, and reaching zero for a cut-off frequency above which no contrast can be detected (Fig. 8).

Now, if we wish to use the MTF for a complicated object, we should first take the Fourier transform, $\theta(\mu, \nu)$, of the object, then multiply by $\Gamma(\mu, \nu)$ to obtain the inverse Fourier transform of the distribution of intensities in the image. And it is well known that multiplying the Fourier transform of two functions is equivalent to take the Fourier transform of their convolution product. Thus, the same intensity distribution in the image is obtained by both procedures.

References

N. B. — The reference IAU **19** means: Symposium IAU No. 19, Rome 1962, Gauthier-Villars, Paris 1963; also: Bull. Astr. Obs. Paris, **24**, Nos. 2 and 3 (1963).

[1] A. Couder, Bull. Astr. Obs. Paris **7**, 201 and 283 (1931).
[2] V. I. Tatarski, Wave Propagation in a Turbulent Medium, McGraw-Hill, New York 1961 (see also references in IAU **19**, Bibliography).
[3] L. S. Taylor, J. Opt. Soc. Am. **58**, 57 and 705 (1968).
[4] A. Danjon in: Danjon et Couder, Lunettes et télescopes, Edit. Revue d'Optique 1935 (p. 72).
[5] H. Scheffler, Z. Astrophys. **55**, 1 (1962).
[6] G. Coupinot, private communication (1971).
[7] J. Texereau, L'Astronomie **76**, 159 (1962).
[8] J. Rösch, C. R. Acad. Sci. Paris **247**, 422 (1958).
[9] J. Rösch, Symp. on Photoelectronic Image Devices, London 1958, Academic Press, New York 1960 (p. 113).
[10] D. Y. Gezari, A. Labeyrie and R. V. Stachnik, Astrophys. J., April 1972.
[11] J. P. Pages, private communication (1970).
[12] A. M. Obukhov (see references in IAU **19**, Bibliography).
[13] K. O. Kiepenheuer, Joint Organization for Solar Observations (JOSO), Annual Report 1970.
[14] K. Brocks (see references in IAU **19**, Bibliography).
[15] C. E. Coulman, Solar Phys. **7**, 122 (1969).
[16] J. Stock, IAU **19**, 249 (1963).
[17] J. Dubief et al., Boll. Soc. Ital. Geofis. Meteorol. Genova **11**, 3 (1963).
[18] J. Rösch, in: Astronomical Optics and Related Subjects, North Holland Publ. Co., Amsterdam 1956 (p. 311).
[19] J. Saissac, Etude du site de l'OPM sur modèle réduit (Internal note, Observ. du Pic-du-Midi 1969.
[20] A. Couder, Annuaire du Bureau des Longitudes 1961 (p. A2).
[21] J. Rösch, Symp. on Solar Seeing, Rome 1961, Consiglio Nazionale delle Ricerche, Rome 1962 (p. 35).
[22] R. B. Dunn, Appl. Optics **3**, 1353 (1964).
[23] J. Rösch, J. Phys. et Radium **16**, 54 S (1955).
[24] J. Rösch, Appl. Optics **4**, 1672 (1965).
[25] J. Rösch, IAU **19**, 160 (1963).
[26] H. W. Babcock, Publ. Astron. Soc. Pacific (1953).
[27] R. Giovanelli, Symp. on Solar Seeing, Rome 1961, Consiglio Nazionale delle Richerche, Rome 1962 (p. 65).
[28] P. Gaujard, Faculté des Sciences de Paris, Thèse de Docteur Ingénieur (1969).

Space Research XII – Akademie-Verlag, Berlin 1972

HIGH RESOLUTION SOLAR OBSERVATIONS

R. B. Dunn

Sacramento Peak Observatory, AFCRL, Sunspot, New Mexico, USA

Ground-based solar astronomers are slowly accumulating single photographs and occasionally time series with a resolution very nearly equal to that of the telescope aperture. This has encouraged some observers to try even larger apertures, evacuate their telescopes, search for locations with better seeing, and to build automatic seeing monitors. In spite of such efforts a number of problems still persist, for instance, deformation by heating of the vacuum windows, and the residual expansion of the new low-expansion mirror materials. Thus the possibility exists that during those moments of good seeing the telescope is still limiting the resolution.

Recent observations with an Hα Zeiss filter attached to the Sacramento Peak vacuum telescope provide information on the relationships between the network, faculae, spicules, filamentation, and other chromospheric structures. The next step must be to achieve a factor of two in resolution along with velocity and magnetic field maps of equal resolution. The possibilities of obtaining this resolution and a time series are discussed.

1. Introduction

Many solar astronomers, but certainly not all, might agree that the following list represents the state of the art of ground-based solar observatories: (i) Daytime solar seeing is seldom as good as 1 second of arc. On extremely rare occasions it may reach $^1/_2$ second of arc. (ii) To freeze the dancing image, exposures must be short, certainly less than $^1/_{30}$ second. (iii) It is almost impossible to obtain a time series of one hour duration that shows uniformly high resolutions throughout its length. (iv) Most solar telescopes are nearly perfect and are limited in their resolution only by the atmosphere. Turbulence within the telescope is small compared with that in the atmosphere. (v) A telescope with a 12.5 cm aperture has enough resolution to capture all the detail allowed by the atmosphere: in fact it may be a mistake to try a larger aperture since averaging over additional thermal cells in the atmosphere will simply cause more fuzziness in the image. (vi) Refractors provide consistently higher resolution than reflectors. Mirrors deform from solar heating. The coatings become warm and cause local distortions in the telescope optical path. Short optical paths are the most successful. (vii) Mountain tops may not be the best for seeing, although many good pictures, perhaps the best, have been taken from them. Locations on reservoirs appear promising for longer sequences. It is too early to tell about single pictures. Recent developments in telescope construction

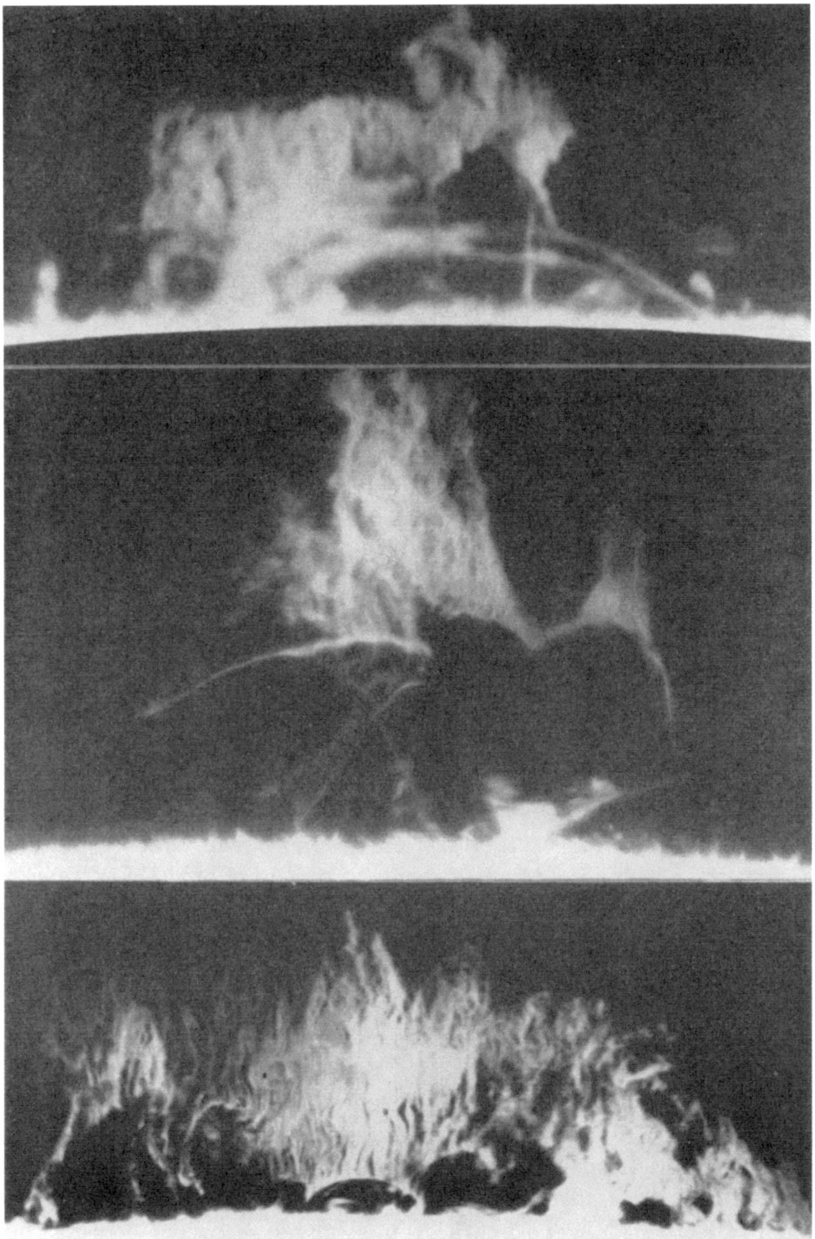

Fig. 1. Prominence pictures: Improvement on the quality of prominence pictures taken with the Schuppman telescope (top), the chromosphere camera (center) and vacuum telescope (bottom) illustrates that the limitations of resolution are imposed by the telescope and not by seeing. The scenes were taken on 17 Dec. 1954, 2 Oct. 1957 and 7 Dec. 1970.

cause us to question this picture. In particular there is doubt as to the validity of statements (i), (iv), (v) and (vi).

Our experience shows that even with new telescopes all the best pictures begin to show about the same top resolution. According to point (iv) this is because the seeing never exceeds a certain quality. However, if a better telescope becomes available at the same location even better pictures are always obtained. This process of improvement has occurred at Sacramento Peak. First there was the Schuppman telescope described by Orrall [1] then the refractor used by Dunn [2] to photograph spicules and now the vacuum solar telescope [3–5]. Fig. 1 shows a steady improvement in the prominence pictures taken with each of these three telescopes. Each time it was thought that the atmosphere limited quality but each time this was wrong. We must now ask if the resolution of the vacuum solar telescope optical system is limited at all times by the earth's atmosphere.

2. Optical Performance of the Vacuum Solar Telescope

The vacuum solar telescope's optical system consists of two 110 cm diameter, 20 cm thick fused-quartz flat mirrors, mounted on top of a concrete tower 41 m high, that reflect sunlight downwards to a 160 cm diameter, 25 cm thick objective mirror 55 m below ground. This mirror is tilted 0.0111 radian to reflect the light back upwards to the spectrographs at ground level. An 86 cm diameter, 10 cm thick, K50 glass window forms the vacuum interface at the top and a fused-quartz window seals the vacuum chamber at ground level. The flats at the top remain at 45° to the incoming light beam independent of elevation and azimuth. The clear aperture of the window is 75 cm so that illumination on the flats is in the form of an ellipse with axes of 75 cm $\times$ 110 cm. Due to the angular diameter of the sun the 75 cm aperture at the top expands to 160 cm when it illuminates the objective mirror down below. This mirror is uniformly illuminated unless the telescope is pointed to the limb of the sun. All the optics except the exit window are in collimated light and must be perfect over their full aperture. The exit window needs to be perfect only over any 25 mm aperture. The exit window is cylindrical to correct a $^1/_4$ second of arc astigmatism introduced by the tilted objective mirror. All other aberrations are negligible. The backs of the mirrors are supported by flexures, the edges are defined by teflon pads. The entrance window is mounted on two O rings. Air pressure may be introduced between the rings to lift the window against the 3600 kg of atmospheric pressure.

2.1. Vacuum Operation

We have concluded that evacuating the inside of this telescope is absolutely essential: with the vacuum it is easy to focus the f/72 system to 5 mm; without the vacuum the image deteriorates completely and one cannot find focus to 5 cm, let alone 5 mm. Apparently the large steel tank is thermally upset despite its being capped at the top by the window and cooled below ground. This upset is also noticeable when one of the manholes is open since air currents are present; the mirrors also become very dusty even with the manholes closed. The improvement of the image due to the vacuum can be seen in Fig. 2, which shows the changes in the images formed by autocollimation from the two sides of the window as the vacuum is applied. The air support is off for these pictures. The path length is 192 m and the images are completely unrecognizable even

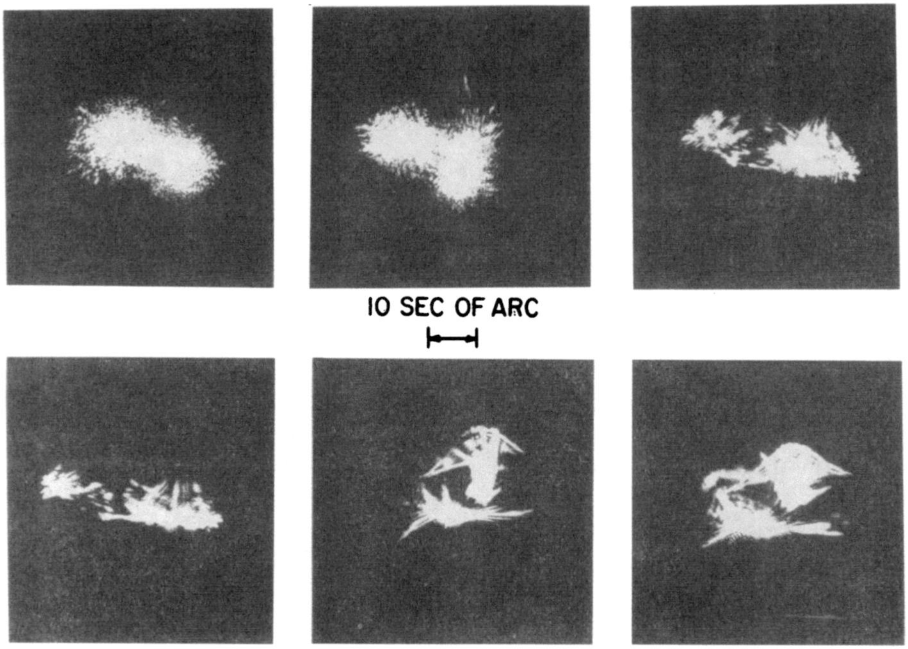

Fig. 2. Vacuum pulldown: Images formed by autocollimation from the two surfaces of the window. Air support is not operative. Top, left to right: no vacuum, 14 cm, 35 cm; bottom, left to right: 40 cm, 49 cm, 53 cm. Full vacuum at Sacramento Peak is 55 cm. The last two images may be rotated.

without sunlight in the system. As the pumps are started, flow patterns are set up and long filaments slowly wave around the edge of the image. By the time the tank is pumped to 10 mm all image motion appears to be stopped. We have concluded that the system will not operate without the vacuum. The system is pumped down to 200 μm in three hours by two 8.3 m^3 min^{-1} pumps. With the pumps off, the tank bleeds up to 300 μm in 5 hours. The pumps cycle for only 15 minutes to bring it back to 200 μm.

2.2. Pressure Effects on the Window

No degradation in optical performance due to pressure on the window has been found. Calculation shows that the bending is only 175 μm and the lens power of the two sides of the window should cancel. The tension in the lower surface is only 17.6 kg cm^2. Most of our image deterioration was first attributed to the uneven support of the window and the air system described previously installed. Fig. 3 shows that this system provides a more uniform support. The image formed by autocollimation from an external optical flat shows no change in the finest detail with or without the air system in operation. Knife-edge tests during vacuum pull-down on images formed by autocollimation from

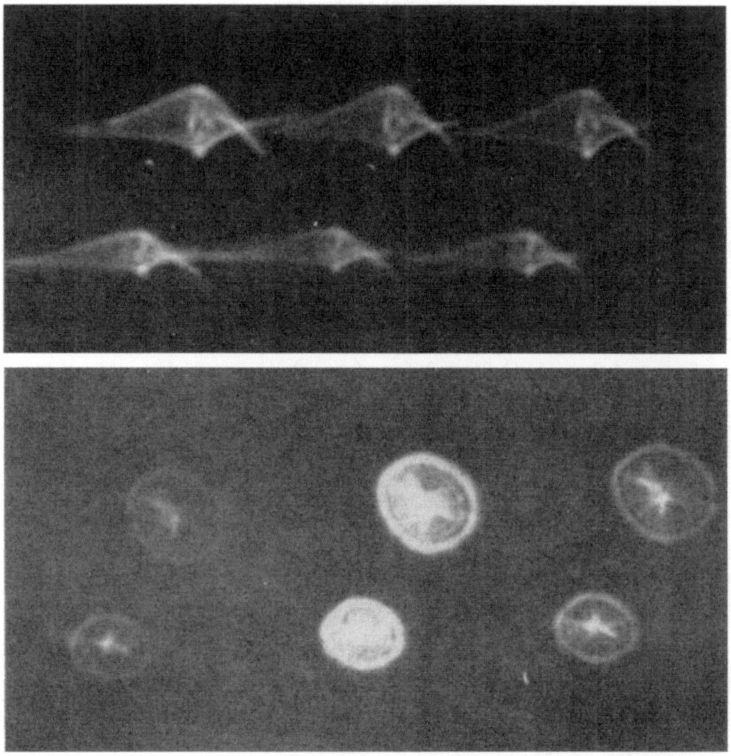

Fig. 3. Air support system: Multiple exposures of images formed by autocollimation from the two surfaces of the window. Top, without air support; bottom, with air support.

an external flat do not show a consistent pattern. Tests using polarizers over the full aperture show slight changes in the stress pattern that disappear when the air system is turned on. The laser interferograms, Fig. 4, formed from the path difference in the window do not show a persistent astigmatism or spherical aberration. We conclude that the pressure effect is small, probably negligible, and completely lost in serious thermal effects (see next section).

2.3. Thermal Effects

We have noticed that the solar image becomes progressively softer as the telescope is exposed during the day to the sun. Knife-edge tests using the edge of the sun show considerable spherical aberration after a day's exposure to sunlight. We know that the edge of the window can be hotter than the center because the window occasionally dews up with condensate and slowly clears from the edge to the center; the image seen by autocollimation from a flat outside the turret is constantly changing, even at night. All these observations point to serious thermal effects.

Table 1

Thermal Distortion

Item	Material	Thickness (cm)	$\alpha \times 10^7$	Waves/°C
Test flat	Cer-Vit	12.5	−2.7	−0.13
Flats	Fused quartz	20	4.3	0.34
Objective	Fused quartz	25	4.3	0.43
Window	K 50*	10	70	0.98
Flats	Cer-Vit (20°)	20	0.1	−0.01
	(−10°)		+1.6	+0.13

* $dn/dt = 25 \times 10^{-7}$

Table 1 shows the change in path expected per degree temperature change for the components in the optical system. The path change in the window is calculated from $\Delta P = L\,\Gamma\,\Delta t$, where $L =$ thickness, $\Gamma = (n-1)\,\alpha + dn/dt$

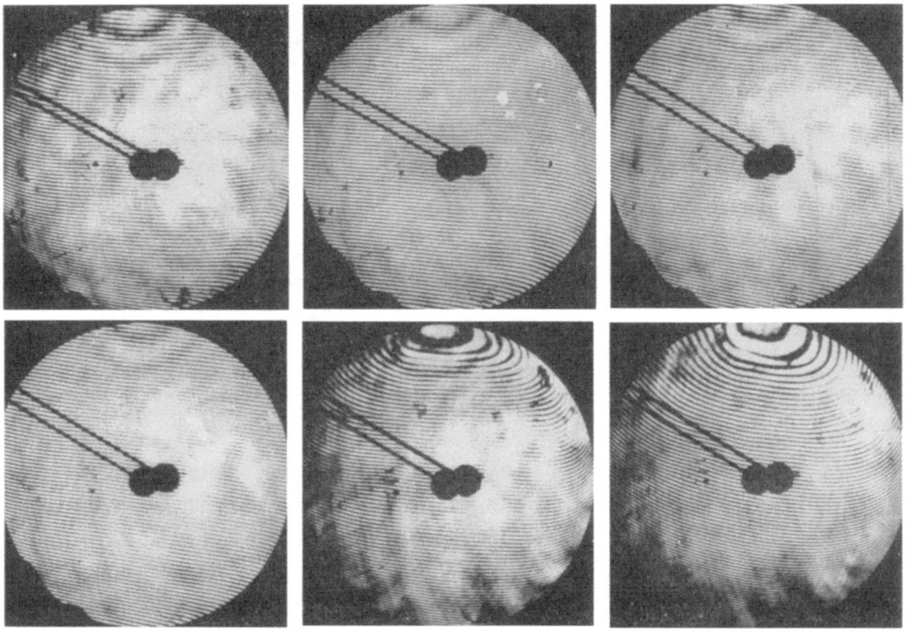

Fig. 4. Interferograms. Image of the window. The pattern is formed by interference between the two surfaces of the window. The window is wedged about 16 μm. The central obscuration is the support for the guider mirror. Compare these patterns with the temperature fluctuations in Fig. 5. All times are UT. For local time subtract six hours. Match the times of the interferograms with those from the center to right hand edge of Fig. 5. Top, left to right, 0004, 0300, 0900; bottom, left to right, 1200, 1611, 2110. Edge of 2110 interferogram is turned up after the day's observing.

with α the coefficient of expansion, dn/dt the change of index of refraction with temperature, and Δt temperature change [6]. Clearly the window is the most sensitive component, although the contribution from the mirrors is not negligible.

Laser interferograms, Fig. 4, show the change of path length within the window itself. These are made by placing a laser with a beam expander at the focus of the telescope and shining it at the window. Each side of the window acts as a flat and reflects the light back through the telescope which focuses an image of the laser in the focal plane. An auxiliary lens forms an image of the window on the focal plane of the camera using both of the reflected images and thus interference bands are seen. The 76 fringes across the interferogram show that the glass is wedged at 16 μm. Each fringe on the interferogram represents one-sixth of a wave-path difference when the window is used in transmission. Therefore the fringes should be straight to $\pm{}^1/_2$ fringe. If we

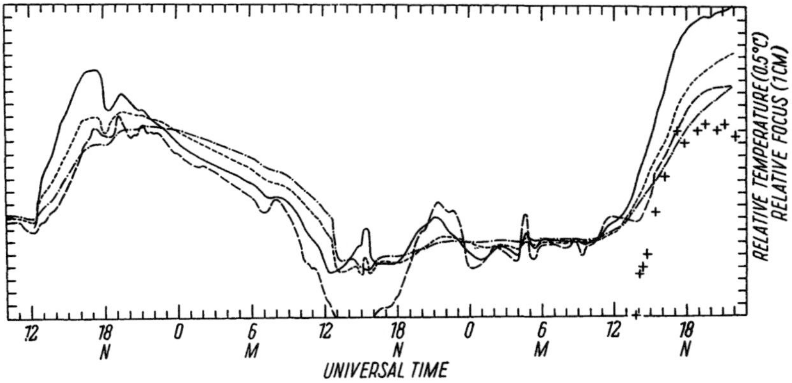

Fig. 5. Temperature measurements and focal change. Relative temperature fluctuations from thermistors attached to the window and its cell. Solid line, average of four thermistors attached to the cell; dotted line, average of four thermistors attached to the edge of the window; dashed line, ambient temperature; dashed-dotted line, temperature measurement of the center of the window; crosses, focal shift measured after opening telescope.

accept a change in focus, they should fit the best circle to within the same tolerance. One fringe also represents a 0.24 °C change in temperature.

All-night ciné photographs of these interferograms have been taken and correlated with temperature measurements and focus changes. These photographs show that the window tends to stabilize at night: first one edge might be turned up with the other straight or turned down; later in the evening the situation might reverse. When the telescope is opened to sunlight heat appears to flow in from the edge with a time constant of approximately an hour causing the edge to be turned up. The temperature measurements, Fig. 5, show that, despite the white paint on the cell and its high conductivity, the steel cell is warming faster than the glass. The flow into the window is not symmetrical since the azimuth turret is on one side and the elevation mirror is on the other. This asymmetry accounts for the unsymmetrical fringe patterns at night. During the day's observing the turned-up edge extends further into the center until the window is greatly distorted. It is seen that the interferogram pattern contains spherical aberration and astigmatism superimposed on a focus change.

Originally a fused-quartz window was used, on the supposition that thermal problems would not exist. Poor telescope performance with this window suggested that the material was not as good as had been thought and that local figuring to correct the inhomogeneous areas was not successful, and a change was made to a glass window. It is now realized that the quartz window is 1.6 times more susceptible than K 50 glass to temperature changes [6]. The coefficient of expansion of quartz is low, but the change of index with temperature is very large. Quartz is not a solution and the glass blank in use must be stabilized. The glass is opaque at a radiation temperature of 20 °C, and the

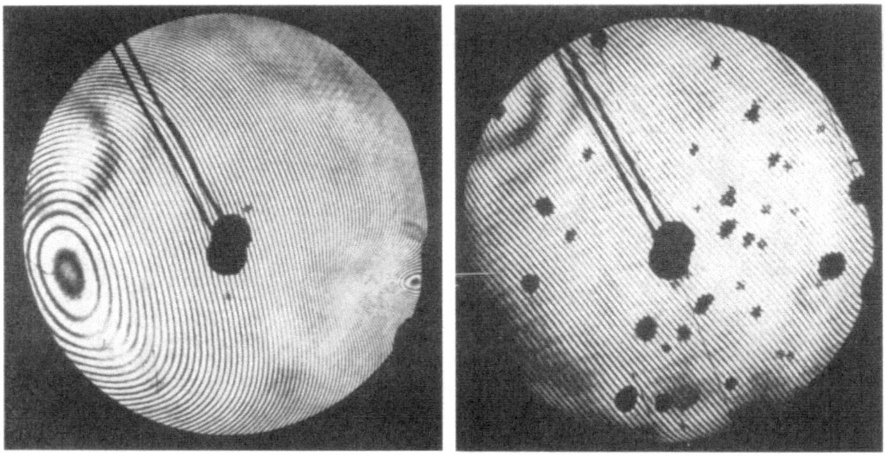

Fig. 6. Interferograms with and without cooling. The left interferogram shows an extreme turned-up edge on the insulated window after a day's observing. Sunlight illuminating a 25 mm square of black tape on the right of the window causes extremely local distortion. White vinyl tape distorts the interferogram by only one half fringe. The right interferogram shows that the active cooling system can hold the fringes reasonably straight even after a day's observation. The spots are from deterioration of the aluminum coating on the main objective mirror.

thermal gradients are small, so that the heat losses may be due to conduction and convection. The thermal time constant varies as the square of the thickness [7]. The vacuum side of the window is completely insulated from heat losses by conduction and convection, so there may be the equivalent of a 20-cm thick blank. Since the stress increases as the square of the reduction in thickness it is not advisable to thin the glass.

An attempt to reduce edge heating by shading the turret with an aluminum shade painted white had no obvious effect; the window was then raised out of the cell by spacing it on an epoxy glass ring $4^1/_2''$ thick; this insulated it from the steel cell and also permitted insulation of the edge with 2″ of urethane foam. Insulation appeared to be a detriment as can be seen in Fig. 6: the steady heat gain seemed higher than before. Water was then circulated around the edge with a helix of rectangular microwave waveguide material to create a heat sink; an uninsulated reservoir was unsatisfactory since it tracked the ambient temperature and an insulated tank became heated from the circulation pump. Finally active cooling with a compressor and freon system was tried; this was promising but the window interferogram still showed a fringe or two of asymmetry which suggested heat was still flowing through the epoxy glass spacer.

At the present time all insulation is being eliminated and the window put back into its steel cell; chilled water is circulated around it to cool both the edge of the window and the cell: Fig. 6 shows this to be a promising solution. For test the system was operated manually, but a control system that cools the water to the temperature of the center of the window will be added.

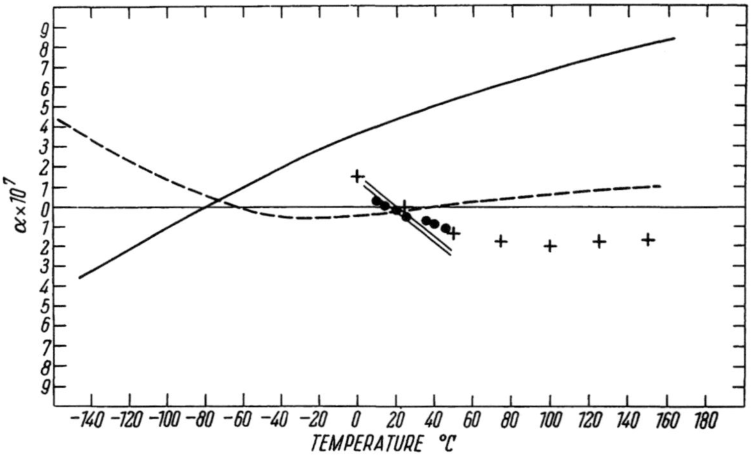

Fig. 7. Expansion of flats as a function of temperature: ———, fused quartz; — — —, Cer-Vit C-101 from Owens-Illinois literature. ═══: measurements of Sacramento Peak blanks by Owens-Illinois. +, measurements by Bradford as extended by Jacobs for a sample of Cer-Vit. ●, preliminary measurements of one Sacramento Peak Cer-Vit blank by Jacobs.

Other solutions might be: To make the edge of the window act like additional glass so that the entire window heats and cools uniformly; the window would then react to thermal inputs like an infinitely wide sheet. A window in the form of a thin shell about 25 mm thick with a radius equal to the clear aperture would have a depth of only 10 cm and would support the vacuum with the stresses equal to the present ones except that the glass would be uniformly in compression throughout the shell, which would improve any residual effects of pressure. The thinness of the glass should reduce the thermal gradients. Such a shell is identical to a Maksutov corrector and would be designed for zero color. It would have to be figured since it would have very large spherical aberration. It might be possible to scan the window in the infrared or scan the interferometer fringe pattern and heat the window locally using infrared radiation at those wavelengths where the window is opaque. Cooling the edge is simple and appears to work and may provide a workable solution to this problem.

The change in focus of the telescope is cause for serious concern. The focal changes from the objective mirror and window can be accommodated by a simple change in focus. However, this is not true for the two flats mounted at 45° in the turret: here the tolerance for departure from flatness is a quarter wave to avoid introducing astigmatism. The astigmatism tends to cancel in the morning or evening when the axes of the two mirrors are crossed; at noon the sense of the astigmatism adds.

The 14.5-cm focal shift observed is made up of: window 4.6 cm, flats Nos. 1 and 2, 4.4 cm each, objective 1.1 cm. The value for the window is calculated by counting fringes in the interferogram and treating the window as a weak lens. Only 10% of the shift comes from the objective mirror, since the solar beam has expanded at this location in the optical system. Calculation shows that the shift in focus attributable to each of the flats comes from a two-wave deformation of the surface which is eight times the quarter wave tolerance for a mirror operating at a 45° angle. A series of tests on a system described earlier [4] that heated the back surfaces of the mirrors to hold them flat required too much heat and too high a temperature gradient.

The obvious solution is to replace the fused-quartz mirrors with one of the new materials such as Zero-Dur, Corning U. L. E. quartz, or Cer-Vit. Installation of Cer-Vit is not expected to reduce the expansion by a factor of eight over the operating temperature range of the telescope (−10 to +30 °C). Recent measurements, Fig. 7, of coefficient of expansion made at Owens-Illinois (see [8] and private communication from S. F. Jacobs) on our blanks and representative blanks of Cer-Vit show that the expansion curves are shifted to higher temperatures and that the slope of the curve in our temperature range is five times greater and in an opposite sense to that advertised. The *average* coefficient between 0 °C and +38 °C is very low; such an average is, of course, meaningless since no representative slope is available. Nevertheless the Cer-Vit should be much better than the fused-quartz mirrors, especially in the region at +14 °C. Of course, in the winter there will be no improvement.

2.4. Photographs with the Vacuum Telescope

From the above discussion it must be concluded that the vacuum solar telescope is far from perfect and that it suffers from a number of serious thermal problems. These problems are probably existent, perhaps to a lesser extent, in all solar telescopes and when overcome will make possible substantially better photographs of solar detail despite the vagaries of the earth's atmosphere. With the vacuum solar telescope diaphragmed to 50 cm it has been possible to photograph in the wing of the Hα line fine "veins" and trails of bright dots at the bottom of clumps of spicules (rosettes). At times this detail appears smaller than the boundaries between granules: these features must be extremely bright to show at all since the granulation visible in the same scene is not

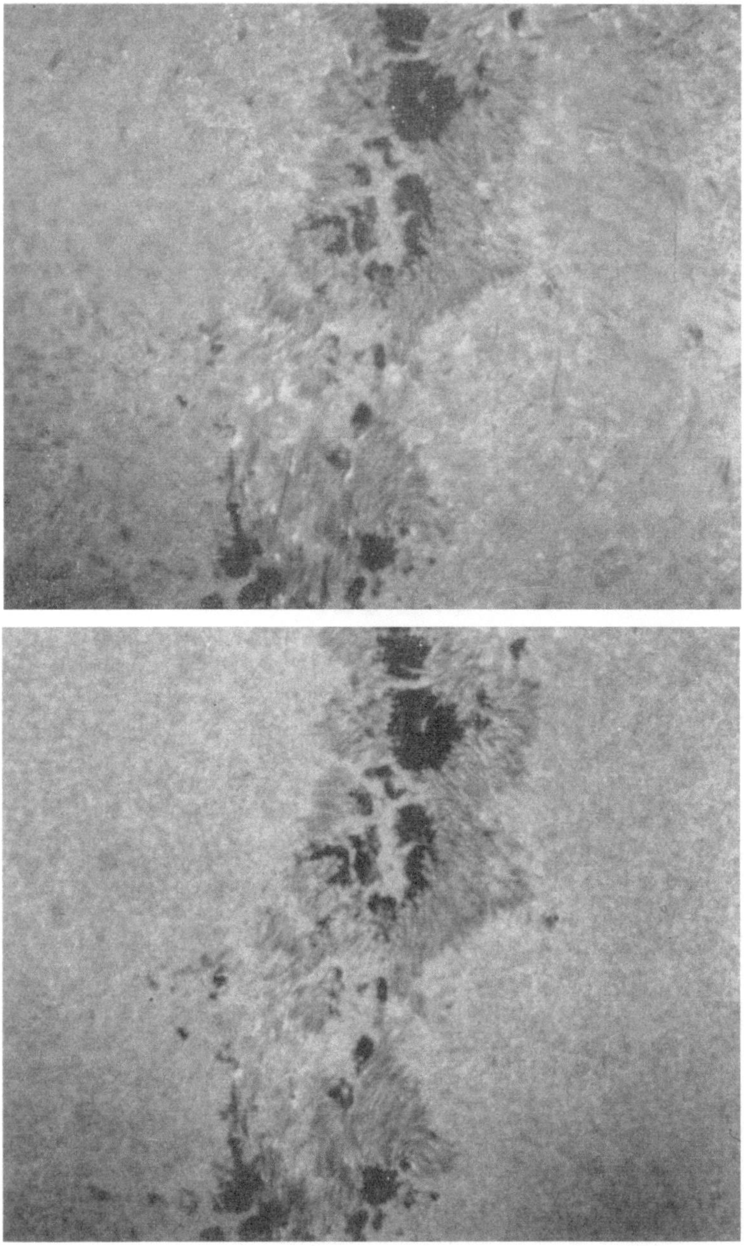

Fig. 8. Bright network in solar chromosphere: The top picture, taken $^7/_8$ Å to the blue of Hα with a $^1/_4$ Å bandwidth Zeiss filter shows a bright, very fine network. This network still shows at 2 Å to the blue of Hα, bottom scene.

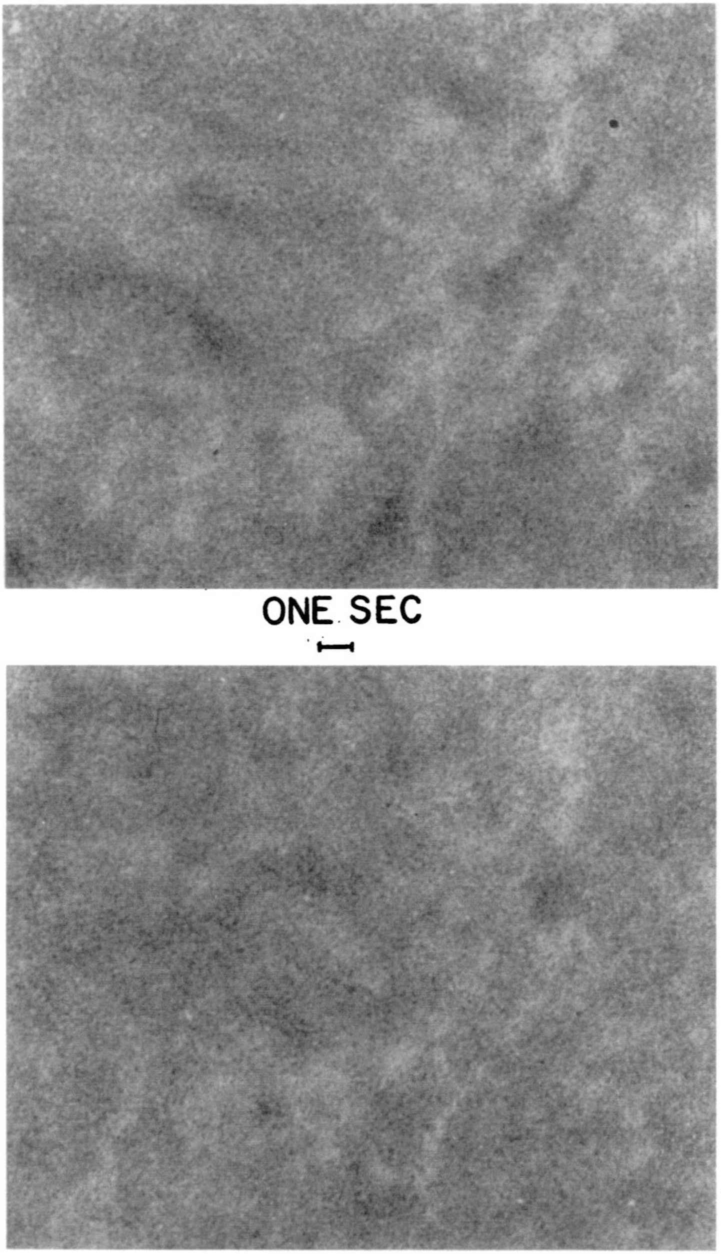

Fig. 9. Vein-like network. An enlargement of the lower left portions of Fig. 8 shows a small vein that is resolved into a zigzag structure in the photograph at the bottom taken 2 Å from the center of Hα. The scene on the top taken $^7/_8$ Å from line center is not as sharp but confirms the reality of the structure.

nearly as well resolved as that photographed by Rösch at Pic-du-Midi. This bright network is visible in Fig. 8, and one of the "veins" is enlarged in Fig. 9.

When the window problem is solved and when the image diameter for the filter system is increased from 22 cm to 55 cm it is expected that the resolving power will be improved by a factor of two.

3. Summary and Conclusions

In the light of these comments certain points in the state of the art as outlined in the introduction may not be quite correct. Consider point (i). Since we do not have a perfect telescope we can not be sure that 0.5 second of arc is the highest resolution that can be obtained for a small area on the sun on a photograph taken with a short exposure. We suspect it is 0.2 or perhaps 0.1 second of arc. Point (iv) is not really settled since each time the telescope is improved we obtain better pictures. A 12.5 cm aperture certainly must be too small, as stated in point (v); a 50 or 75 cm aperture is more realistic. We think we reach the limit of the 50 cm aperture of our tower and certainly Rösch is close to the limit of his 40 cm refractor at Pic-du-Midi. The use of vacuum and the new low-expansion mirror materials suggest that the reflecting telescope may supersede refractors in solar resolution and consequently point (vi) is open to question.

Thus it should be possible to obtain single pictures of solar detail from the ground that have extremely high resolution. The ones we already have encourage us to obtain a time series and spectra of this fine detail which looks fundamental to solar physics. Such observations can only be taken by large aperture telescopes above the earth's atmosphere.

References

[1] F. Q. Orrall, Thesis, Harvard University (1954).
[2] R. B. Dunn, Thesis, Harvard University (1960); also published as Environmental Res. Paper No. 109, AFCRL, Bedford, Mass., June 1965.
[3] R. B. Dunn, Appl. Optics 3, 1353 (1964).
[4] R. B. Dunn, Sky and Telesc. 38, 368 (1969).
[5] R. B. Dunn, IAU Coll. No. 11, Sciences Research Council Publ., Royal Observatory, Edinburgh 1970 (p. 37).
[6] T. Baak et al., Tech. Rep. AFAL-TR-69-19, Wright-Patterson Air Force Base, Ohio (1969).
[7] A. Couder, in: Vistas in Astronomy, Vol. 1, Pergamon Press, New York 1955 (p. 372).
[8] J. N. Bradford, Tech. Rep. No. 48, Optical Sciences Center, Univ. of Arizona (1969).

Space Research XII — Akademie-Verlag, Berlin 1972

HIGH RESOLUTION PLANETARY OBSERVATIONS

W. A. Baum

Planetary Research Center, Lowell Observatory, Flagstaff, Arizona, USA

Some classical examples of good earth-based photographs of planets are presented and estimates of limiting resolution are made. Recent results from the International Planetary Patrol provide a comparison of some favorably situated observatories with one another. Observed resolutions can be interpreted in terms of a modulation transfer function, which provides a basis for deciding how improvements can be achieved. Some current attempts at improving earth-based resolution are described.

Earth-based observations of the planets have, of course, provided the foundation of knowledge to which spacecraft observations are now adding, and they will probably continue to play a major role for a long time to come. During favorable periods, an earth-based telescope network can maintain more complete and continuous coverage than any flyby or orbiter now planned. But there is a big difference in image resolution, and it is the purpose of this review to

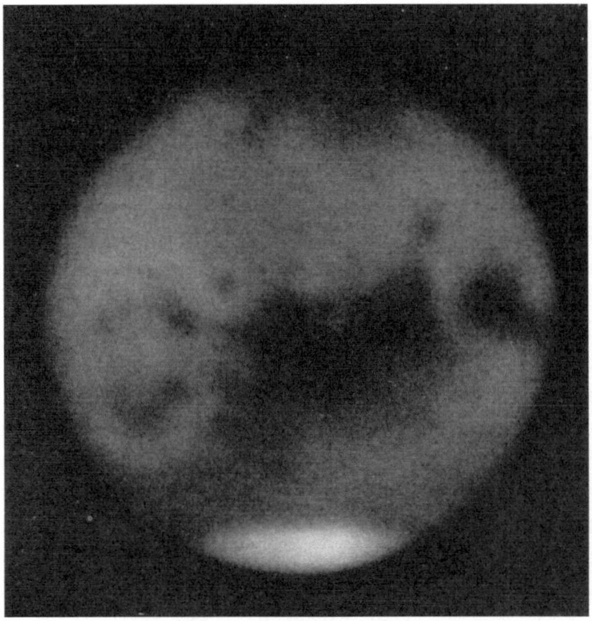

Fig. 1. A very good image of Mars photographed in red light with the 70-cm Lamont–Hussey refractor by E. C. Slipher in 1939.

examine the limitations imposed by resolution on earth-based observations of the planets.

Fig. 1 is an unusually good earth-based photograph of Mars in red light. The bright area at the bottom is the south polar cap, which was tilted somewhat toward the earth at the time this photograph was made. Familiar dark features stand out clearly: the Solis Lacus region (at the left), Aurorae Sinus, Syrtis Major, and Meridiani Sinus (at the right). Some lesser features from which the resolution can be judged in quantitative terms are the following: Near the center of the image are two tiny V-shaped pairs of dark fingers extending upward from the larger dark areas. Other exceptional photographs look much the same, so these small features are evidently real. The right-hand pair of

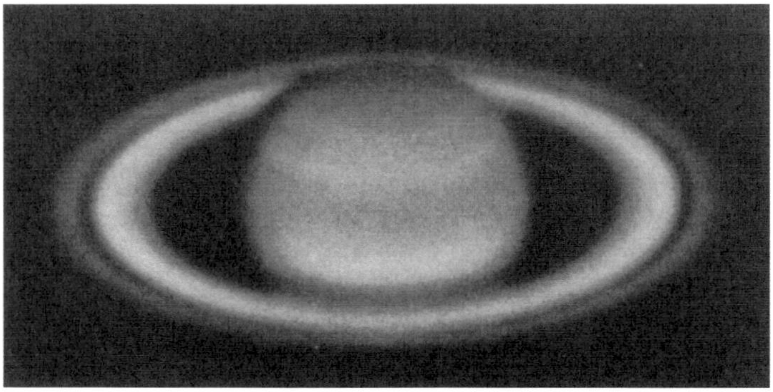

Fig. 2. A very good image of Saturn photographed in green light with the 61-cm Lowell refractor by E. C. Slipher in 1943.

these fingers subtends about 1 arc-second at the upper tip of the V. If, instead, they had a separation only half as great, it is evident that they might not have been recognized as a double marking. Therefore, for a conventional earth-based photographic image of Mars with this scale and granularity, the practical limit is reached at roughly 0.5 arc-second. Typically, the light areas on Mars are only about 1.5 times brighter in red light than the more prominent dark areas, so this limit at 0.5 arc-second pertains to a target of that contrast or less. One finds much the same resolution limit on good earth-based photographs of Jupiter at a similar scale.

A feature of greater intrinsic contrast should be visible down to a smaller size. This is verified by Fig. 2, which shows a very good photograph of Saturn. Cassini's division — the gap between the outer ring and the inner ring — is a high-contrast feature but very narrow. At the sides of the image its maximum width is about 0.7 arc-second, but across the center of the image its projected width narrows to about 0.3 arc-second. Even at its narrowest point, Cassini's division can be detected. However, in photographs with the rings somewhat less tilted, the division cannot be detected all the way across the image. The practical photographic limit for this high-contrast feature falls at about 0.25 arc-second for images of this granularity.

Color photography can be impressive but does not provide any magic superiority in resolution over black-and-white emulsions. Some excellent color photographs of Jupiter, including one kindly made available by Dr. Kuiper of the University of Arizona, were compared with selected black-and-white photographs and were found to resolve no finer detail. Although the visibility of features in black-and-white planetary photographs depends upon the color filters used, the usual bandwidths ($\Delta\lambda/\lambda \sim 0.15$) yield as much discrimination of features as any color emulsion does. Moreover, black-and-white emulsions have the advantage of permitting photometric analysis. Most planet photographs suitable for analysis are therefore on black-and-white emulsions exposed

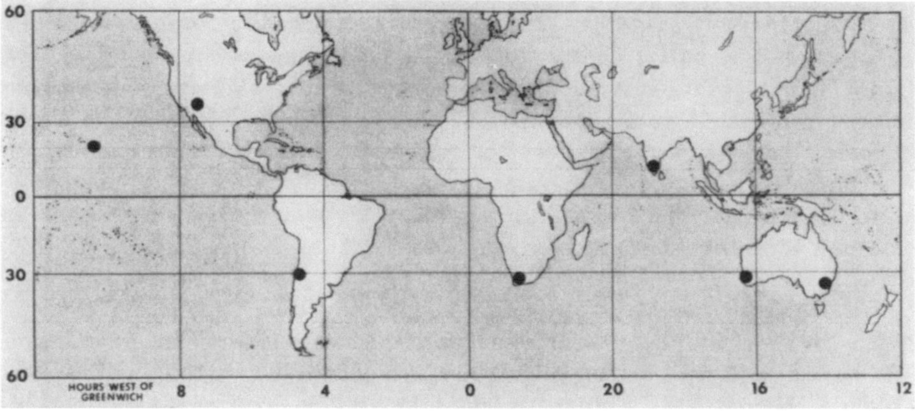

Fig. 3. Locations of the seven observatories now participating in the International Planetary Patrol Program. Identical focal lengths and cameras make possible a statistical comparison of performance, including resolution.

behind color filters, and there is no sacrifice of resolution associated with that choice.

One of the best comparisons of the resolution obtained at various observatories can be had from the International Planetary Patrol Program [1]. At the present time the network includes the stations shown in Fig. 3. Starting from the left, these include the Mauna Kea Observatory in Hawaii, the Lowell Observatory in Arizona, the Cerro Tololo Inter-American Observatory in northern Chile, the Republic Observatory in South Africa, the Kavalur Station of the Indian Institute of Astrophysics in southern India, the Perth Observatory in Western Australia, and the Mount Stromlo Observatory in eastern Australia. The Magdalena Peak Observatory in New Mexico also participated in the network briefly in 1969. The program, which commenced operation in 1969, is coordinated by the Planetary Research Center at Lowell Observatory. In the first two years of operation the patrol has yielded more planet photographs of a quality suitable for analysis than previously obtained at all of the world's observatories since the advent of photography a century ago.

The resolution obtained in planetary images differs considerably from one patrol station to another. This difference shows up statistically in the number of images from the various stations meeting a certain standard of quality, as well as in the number of exceptional images from each. The Kavalur Station in India and the Perth Observatory in Australia have only recently been added to the network, so statistics are not yet available for them; but there is plenty of material from which to judge the relative performance of the others. In 1970 the best station was the Mauna Kea Observatory, which is also the station having the highest elevation — more than 4000 m. Second best was Cerro Tololo at 2000 m elevation in Chile. Owing to the fact that Mars and Jupiter (our main targets) are currently quite far south in the sky, Cerro Tololo (30° S) is more favorably situated in latitude than Mauna Kea (19° N), but the higher elevation of Mauna Kea evidently more than offsets its disadvantage in latitude. The third-best station of the patrol network in 1970 was the Republic Observatory (26° S) in Johannesburg, South Africa. Owing to air pollution around Johannesburg, the total yield of acceptable images from the Republic Observatory was lower than that from the Mount Stromlo Observatory (35° S), but the resolution in the Johannesburg images was consistently higher. The output from Lowell itself (35° N) is now about equal to that from the Republic Observatory, but the relative outputs can be expected to change dramatically when Mars and Jupiter later move north of the equator.

Although the number of superior images may differ markedly from one patrol station to another, one notices how little the best ones differ in resolution. The best from Lowell or Republic look much the same as the best from Cerro Tololo or Mauna Kea. This fact is doubtless due in part to the approaching of limits set by diffraction and emulsion properties. All telescopes of the patrol network have apertures of 60–70 cm, so the Rayleigh limit is about 0.2 arc-second.

It would be a mistake, however, to think that the earth-based imaging of planets should be done routinely with larger telescopes. Even if large diffraction-limited telescopes were widely available, the patrol program experience indicates that their advantage over 60-cm telescopes could only occasionally be detected. This is further confirmed by the almost total lack of better existing planetary images obtained by anyone anywhere.

Suppose that we *do* want to obtain better earth-based observations. What should be our goal and how should we go about it? Let us begin by examining the problem in terms of the *modulation transfer function*, sometimes called the frequency response function. The typical form of this function is sketched in Fig. 4. The broader features of an image are made up of low spatial frequencies, while the smaller details of the image are made up of high spatial frequencies. The downward sweep of the curve represents the way in which the smaller details become diluted or washed-out in comparison with the broader features.

Several factors contribute to this washing out of image detail, including the earth's atmosphere, the telescope, and the photographic emulsion (or other detector). Each can be described by a modulation transfer function. Although these functions differ somewhat from one another, they tend in practice to be downward-sloping curves of the general form illustrated in Fig. 4, differing mainly in the scale along the abscissa. The combined effect of the atmosphere, the telescope, and the detector is also a modulation transfer function. Its Fourier transform is the point spread function, which in practice represents the profile of a recorded star image.

The particular function plotted in Fig. 4 represents a telescope whose ideal performance has been degraded about as little as one can hope for, namely, by an amount equivalent to defocusing until the geometrical confusion disk has a radius equal to the Rayleigh limit. Values labeled on the abscissa correspond to cycles per arc-second for red light (6200 Å) in a telescope of 61-cm aperture, so they are representative of the International Planetary Patrol network under exceptional observing conditions.

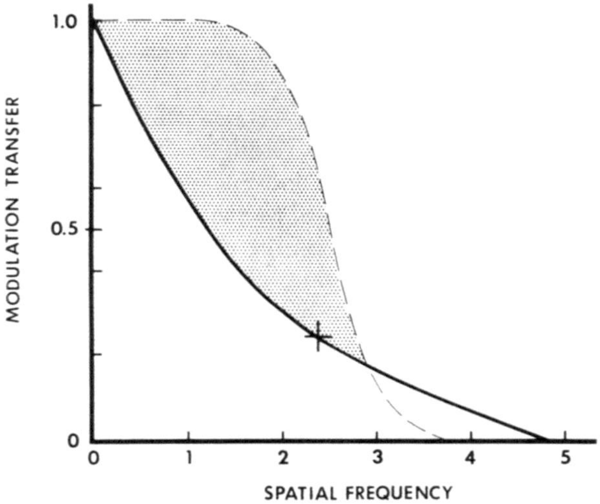

Fig. 4. The typical shape of a modulation transfer function. The cross identifies the neighborhood of the earth-based resolution limit for photographing details on Mars and Jupiter. The abscissa scale varies with observing conditions as described in the text.

Spatial frequencies are supposed to be based on parallel light-and-dark stripes with a sine-wave intensity distribution. Although the classical Rayleigh criterion refers instead to the separation of barely resolved star images, we can arbitrarily choose to talk about stripes that have the same separation. For red light and a 61-cm aperture, the corresponding spatial frequency falls at 3.9 cycles per arc-second, where the modulation is only 8% on the particular curve illustrated. Whether or not the threshold of detection falls above or below that level depends upon the signal-to-noise ratio of the emulsion or detector. For a case in which this criterion is valid, it identifies the spatial frequency that might barely be detected if the striped input image has 100-percent modulation.

But a real planet does *not* have 100% contrast in its features. For red light we have already noted that the light areas on Mars are only about 1.5 times brighter than prominent dark areas, which corresponds to a maximum input modulation of about 33%. A modulation transfer of 24% would reduce this to an output modulation of 8%. Therefore, we cannot easily expect to photograph finer albedo markings on Mars than those associated with the cross in Fig. 4. For a 61-cm telescope and red light, this cross corresponds to 2.4 cycles per arc-second, or a resolution of about 0.42 arc-second. Thus, our earlier estimate of 0.5 arc-second, based on the actual Mars image in Fig. 1, was not very far from optimistic expectation.

Although the resolution of the patrol telescopes may occasionally approach the limit estimated above, it is usually very much poorer than that because of atmospheric "seeing". Typically, "seeing" is the dominating factor in determining the resolution of the system. Although the instantaneous effect of "seeing" on a planetary image is a complicated pattern of distortion and blurring, the time-integrated effect can still be described by a modulation transfer function. In the case of ordinary photography with a typical patrol telescope, if the image size is made larger enough to avoid serious image degradation due to emulsion granularity, the exposure time is long enough (perhaps a few seconds) to introduce a partial time-averaging of the "seeing". Alternatively, a telescope of larger aperture involves less time averaging but introduce a somewhat comparable spatial averaging due to the spanning of a broader column of turbulent air. In either case, differential distortion can be detected among images photographed in rapid succession, but these distortions are found to be small in comparison with the symmetric smearing. For a detector of higher quantum efficiency, and therefore shorter exposure time, this averaging of "seeing" may, of course, be less complete. However, for the purpose of the present discussion, it will be fair enough to assume that "seeing" can be adequately represented by a modulation transfer function. The shape of curve adopted in other studies [2–4] is similar to Fig. 4 except for a less abrupt cut-off at the right-hand end. And, of course, the worse the "seeing", the smaller the number of cycles per arc-second at any point along the abscissa. Since the quality of the "seeing" varies from moment to moment, one can imagine the scale of the abscissa to be fluctuating, expanding and contracting.

Under conditions slightly better than average (but not exceptional), the numbers labeled on the abscissa in Fig. 4 should be divided by about 2 if they are to represent cycles per arc-second for typical earth-based telescopes. Recalling the meaning of the cross, we can say that image detail on Mars or Jupiter will then be approaching threshold around 1 cycle per arc-second. This represents a quality of imaging that occurs often enough for us to follow the things that change, diurnally, secularly, and randomly, on the planets. This is what earth-based planetary image analysis today is aimed at. Although a rare image of exceptional quality is impressive, it is unlikely to tell us something that is not already known. Images of average quality obtained every hour or two are much more valuable. Therefore, if we are looking for useful ways to improve earth-based resolution, we should focus attention on trying to *upgrade the average*, not on trying to excel the best now available. This is made all the more true today by the superior resolution of planetary detail from spacecraft. Instead of trying to compete in the quest for fine resolution, the proper role of earth-based planetary imaging is to provide uninterrupted planet-wide coverage, which is not yet within the capability of space missions now planned.

Three avenues are open to us for improving the resolution of earth-based planetary images, especially those of average quality. One is to increase the signal-to-noise ratio so that the threshold modulation is lower. Another is *post facto* image restoration by applying a selective gain to intermediate spatial frequencies so that the solid line in Fig. 4 is transformed into the dashed line. The third approach is to compensate partially for the effects of "seeing" during the actual exposure at the telescope. We have experimented with all three of these approaches at the Lowell Observatory.

One of the oldest and simplest methods for improving the signal-to-noise ratio is illustrated in Fig. 5. If several nearly identical photographs are super-

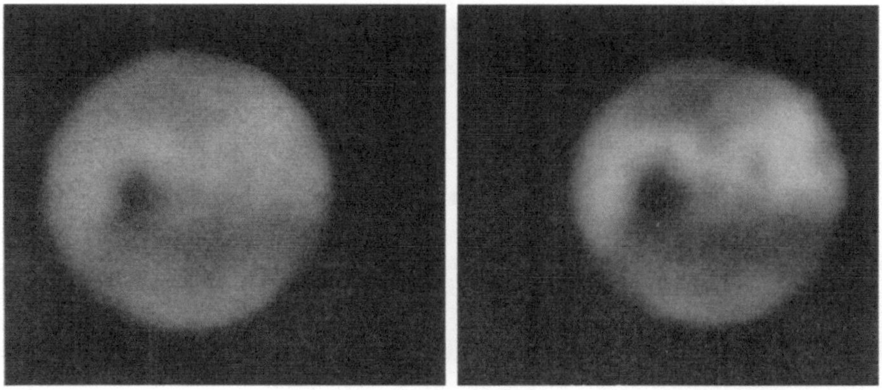

Fig. 5. A single image of Mars (left) compared with a composite of 80 similar images (right). These images were photographed in red light with the Perkins 1.8-m telescope at the Lowell Observatory in 1969. The "seeing" was only average.

imposed by successive projection on to a finer grained emulsion, we obtain a composite photograph that represents the sum of the information content of the originals. The left half of Fig. 5 is a single original image, while the right half is a composite of 80 similar images. The additive copying was done in such a way as to produce about the same average density in the composite image as in the originals. In principle, 80 times as many original grains contribute to this particular composite image, so the grain noise should be lower than that of a single original image by a factor of 9 (the square root of 80). The benefit from this process appears in two ways: (1) the reduced grain noise permits the threshold to fall at a lower modulation level and therefore at a slightly higher spatial frequency, while (ii) the increase in the signal-to-noise ratio provides a proportionally more accurate indication of the relative brightness of larger areas in the image. In other words, there is a small gain in resolution and a larger gain in brightness discrimination. Despite the non-linearity of the photographic process, experiments confirm that the improvement in the signal-to-noise ratio approaches theoretical expectation [5].

What is the advantage of compositing 80 images, instead of using an 80-times longer exposure at the telescope to produce one equally dense original image nine times larger in diameter? The main advantage is that the 80 images can be selected from a greater number so as to favor the better moments during variable "seeing". There is the limitation, of course, that the originals must have been taken within a time interval short enough that the composite image is not significantly smeared by planetary rotation. For Mars this time interval amounts to several minutes, but for Jupiter it is only about one minute.

Another method of improving the signal-to-noise ratio is to use an image receiver having a higher detective quantum efficiency than an unaided photographic emulsion. Electronographic cameras and some types of image intensifier tubes are capable of providing a real gain in this respect, whereas television-

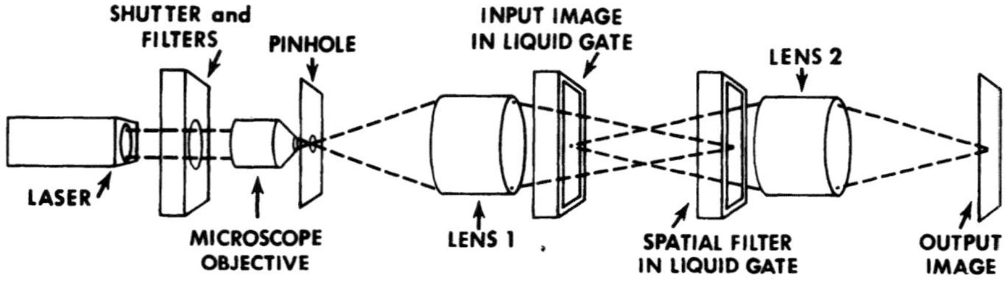

Fig. 6. The optical method of "restoring" images by enhancing a selected range of spatial frequencies, as illustrated by the dashed line in Fig. 4 (from Horner [9]).

type tubes compete poorly with unaided photography. The relative astronomical merits of the various types of image tubes have been reviewed by the writer [6–8].

The *post facto* restoration of images can be undertaken either with an electronic computer or with an optical device. The computer method provides a relatively well-controlled process that has been used successfully in "sharpening" images transmitted by lunar and planetary spacecraft. One of two optical methods is shown in Fig. 6 [9]. It is a quick, low-cost method, but it works less well than the computer. As already remarked, *post facto* restoration is an attempt to compensate for the shape of the modulation transfer function associated with the original photograph. One tries to strengthen low and intermediate spatial frequencies that have been attenuated by the "seeing", the telescope, and the image-recording system. Very high spatial frequencies, at which the original image contains more noise than signal, should be cut off. In other words, one tries to transform the solid curve of Fig. 4 into the dashed curve.

With the cooperation of the Conductron Corporation we have applied the optical method experimentally to some earth-based planetary photographs in the collection of the Planetary Research Center. As shown in Fig. 6, a laser and a microscope objective produce a point source which is re-imaged by lens no. 1 at the plane labeled "spatial filter". When a planetary photograph is inserted at the plane labeled "input image", the features in the photograph produce a diffraction pattern of the point source at the plane of the spatial filter. This filter is a mask that passes part of the diffraction pattern while attenuating or cutting off other parts of it. In so doing, the mask selectivity operates on the modulation transfer function of the original photograph and thereby yields a sharpened photograph in the "output image" plane. In practice, however, it is difficult to avoid problems associated with the imaging of coherent light in the presence of optical imperfections, and it is doubtful that this optical method can be made to work well enough. There is the further complication that photography is a non-linear process.

Another important aspect of image restoration (by either method) is that the signal-to-noise ratio of the input image has to be relatively high. If it is not, the threshold modulation falls too high in Fig. 4, and the dashed curve has to be put too far toward the left, resulting in too small a shaded area. In other words, there is not a broad enough range of low and intermediate frequencies to be usefully enhanced. For example, if one tries to restore the single image in the left half of Fig. 5, any restoration function that would significantly sharpen up boundaries, such as the limb, will also amplify the graininess. Our experiments showed that typical ground-based planetary photographs have too poor a signal-to-noise ratio for effective restoration. A "restored" image looks different, but not better, than the original photograph. There are no planetary features visible in the restored image that cannot also be seen in the original.

One should bear in mind that *post facto* restoration cannot put information into an image that was not there to begin with. It can only rearrange information as an aid to the eye and the brain. The failure to see any improvement in the restoring of existing coarse-grained planet photographs merely tells us that the eye and the brain are already successfully interpreting the full information content of such images. How, then, should we go about obtaining a gain? The relatively fine-grained image at the right in Fig. 5 shows how much the signal-to-noise ratio can be improved by optimizing the original observation for that purpose. As already mentioned, it is a composite of 80 images like the one on the left. Our largest telescope (1.8 m) was used at full aperture, together with a 35-mm camera in a continuous framing mode, in order to blacken as many image grains per unit time as possible. The time interval set by planetary rotation included about 100 consecutive images, out of which the best 80 were selected to be composited.

This sequence of photographs was obtained during average "seeing". We shall therefore know the reality of any planetary features that subsequent image restoration appears to reveal, because there exist some better photographs of this same face of Mars in our collection. In other words, we are testing a method, not looking for unknown Martian features. The eye already sees somewhat more in the unrestored 80-image composite at the right than in the single image at the left, but it probably does not recognize all that is theoretically available from the nine-fold gain in signal-to-noise ratio. It remains to be seen how well image restoration can succeed in revealing the detail that theoretically is locked up in a composite of this kind.

Fig. 7 illustrates the third approach for improving the resolution of earth-based planetary images, namely, compensating partially for the effects of

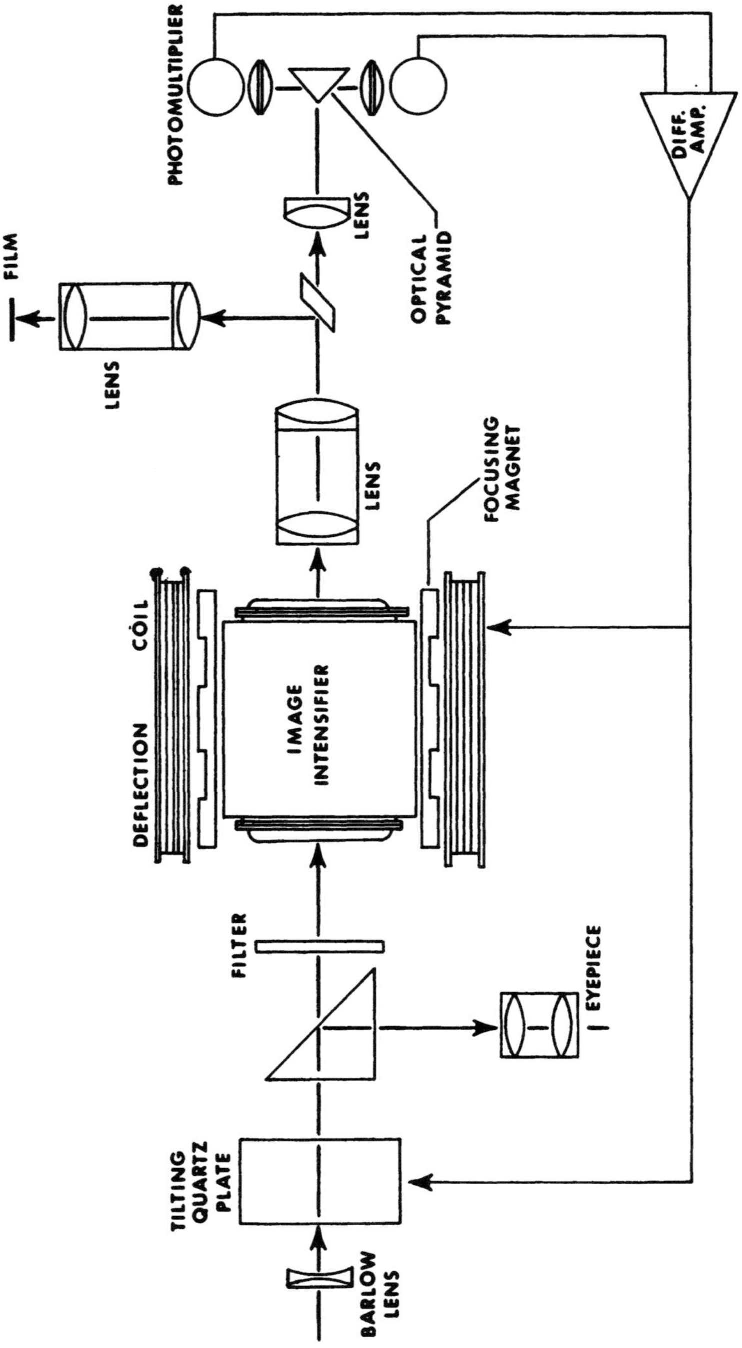

Fig. 7. Schematic diagram of the Lowell "image tranquilizer" for removing image jitter during planetary photography.

"seeing" during actual exposure at the telescope. We call this device an "image tranquilizer". It holds the centroid of a dancing planetary image stationary, but it does not do anything about the differential tremor within the image. In other words, it does not operate on wavefront irregularities smaller than the telescope aperture, as might be undertaken with an Eidophor system or the equivalent [10]. A device similar to ours was developed for other purposes by the Itek Corporation [11], whereas a somewhat different device was applied astronomically at the Dyer Observatory [12].

The image tranquilizer in Fig. 7 is an electronic feedback device that uses photomultipliers to provide an error signal for a fast image deflection system that precedes it. This includes two subsystems: a servo-driven tilting quartz plate to take care of image drift, and deflection coils surrounding an image intensifier tube to take care of image jitter. Following the image intensifier is a beam splitter, so that the stabilized output image can be recorded on photographic film (Eastman 2498). The image intensifier (RCA 33011) is a two-stage magnetically focused tube with a 40-mm field, an S20 photocathode, and a P11 output phosphor. For Mars, we are using an input image scale of 2 arc-seconds per millimeter, a telescope aperture of 23 cm, and a focal ratio of F/450. The exposure time with a red filter is about 20 seconds.

The response of the image tranquilizer is fast enough to provide almost complete image stabilization over the range of jitter frequencies encountered in practice. The effective time-constant of the system is about 5 milliseconds. The response is more dramatic visually than photographically; one can watch the normal dancing of the output image when the feedback circuit is switched off and then see the image suddenly be grabbed and held stationary when the circuit is switched on. As a method for photographing acceptable images during poor "seeing", the image tranquilizer shows promise. The feedback signal is also of interest as a measure of the amplitude and frequency spectrum of "seeing".

The image tranquilizer and the other techniques examined in this review will probably not make major advances over the best earth-based planetary imaging already available. But there appears to be a good chance of raising the quality of the average and of reducing gaps in time coverage, and those are the goals we believe to have scientific value. This is borne out by the nature of recent and current programs of image analysis at the Planetary Research Center [13].

Acknowledgments

This review is based on work supported by NASA Headquarters grants NGR-03-003-001 and NGR-03-003-007.

References

[1] W. A. Baum et al., Icarus **12**, 435 (1970).
[2] J. L. Harris, Scripps Institution of Oceanography, Prog. Rep. 63-10 (1970).

[3] R. E. HUFNAGEL, Woods Holè Summer Study, Vol. 2, p. 19 (1966).
[4] L. O. HENDEBERG and E. WELANDER, Appl. Optics 2, 379 (1963).
[5] H. L. JOHNSON, R. NEVILLE and B. IRIARTE, Lowell Obs. Bull 4, 83 (1958).
[6] W. A. BAUM, Science **154**, 112 (1966).
[7] W. A. BAUM, Physics Bull. (London) **19**, 254 (1968).
[8] W. A. BAUM, in: Advances in Electronics and Electron Physics, Academic Press, London and New York 1969 (p. 753).
[9] J. L. HORNER, in: Evaluation of Motion-Degraded Images, NASA SP-193, 1969 (p. 105).
[10] H. W. BABCOCK, Publ. Astron. Soc. Pac. **65**, 229 (1953).
[11] E. R. ARAZI, J. Opt. Soc. Am. **55**, 599 (1965).
[12] J. H. DEWITT, R. H. HARDIE and C. K. SEYFERT, Sky Telesc. **17**, 8 (1957).
[13] W. A. BAUM, Trans. Int. Astron. Un. **14B**, 136 (1971).

Space Research XII – Akademie-Verlag, Berlin 1972

HIGH RESOLUTION PLANETARY OBSERVATION

G. P. Kuiper

Lunar and Planetary Laboratory, University of Arizona, Arizona, USA

High-resolution planetary observations, visual and photographic, have for several decades seemed to have reached a rather firm limit. The causes for this restriction are briefly reviewed. At the time when space probes make major advances possible, at incomparably higher levels of expenditure, the need arises to re-examine the limiting factors of earth-based planetary resolution. In addition to familiar refinements (of which seven are listed) carried somewhat further, a specific approach, yet insufficiently developed, is the study of the atmospheric boundary layer over the observatory site and the use of an appropriately elevated station.

Planetary observation, both visual and photographic, has always depended on the best image quality (seeing) available. In this respect its main requirement is identical to that of double-star observation. Almost any observing site experiences exquisite seeing at some time. Sites differ in the frequency of such occurrences and in the more readily ascertainable total fractions of clear weather. Tropical sites have a large fraction of good seeing on the nights that are clear, though the frequency of clear nights depends on location and on the monsoon. For instance, at the Bosscha Observatory in Java the seeing is almost always satisfactory whenever the sky is clear, with visual resolution of 0.2 arc sec regularly achievable with the 60-cm visual telescope. In the latitudes 25–35° of both hemispheres, the fraction of clear nights is the largest (with local exceptions) and the seeing is good to excellent a fair fraction of the time, particularly close to the western shores of the various continents. Large continental masses appear to have a disturbing effect on the air parcels so that good seeing is less frequent far away from these shores. This rule has only a statistical validity; exceptional weather conditions may occur anywhere.

With a large telescope of high quality, in excellent adjustment, the limit of *visual resolution* is always near the theoretical diffraction limit. Double stars, of 0.″06 (arc sec) separation, such as 20 Persei, have been repeatedly observed with the 82-inch telescope of the McDonald Observatory and the planet Pluto, the satellites of Saturn, and Triton of Neptune have been measured with the 200-inch Hale teslecope. The measured diameter of Pluto was 0.″23 ± ± 0.″01 [1], and for the Saturn satellites and Triton [2] the following: "5 July 1954, Enceladus (0.″08), Tethys (0.″12), Dione (0.″12), Rhea (0.″24), Titan (0″67), Iapetus (0.″195), all at 9.43 astr. units; and Triton (0.″173) and Neptune (2.″06), both at 30.1 astr. units." (Stars were found to have spurious disks of about 0.″05, a value attributable to the Ross corrector lens at the prime focus, which for safety reasons had to remain in place during measurements, made in the prime-focus cage.)

High *photographic resolution* is not obtained nearly as readily as high visual resolution. This appears due to the fact that high visual resolution results from continued observation over many seconds to several minutes, during which the eye and brain reject all visual records that do not reach to the desired standard, and retain for "storage" only the best of 10^3–10^4 individual impressions. The photographic plate can simulate this process to some extent by the use of multiple exposures. Also, the photography can be monitored, by the observer watching the image and selecting the best moments for the exposures or by some automatic device based on image size. During a program of systematic planetary photography with the 82-inch of the McDonald Observatory in 1948, using both color film and black-and-white, the resolution gap between visual and photographic observation was found to be a factor of about 4. Since then, with improved emulsions and improved thermal control of the air in both telescope and dome, the gap has been narrowed to about 2. Thus, the best photographic resolution obtained with the 61-inch telescope whose Airy disk in visual light is 0.075 arc sec, is about 0.15 arc sec. (Achieved many times in the *Consolidated Lunar Atlas*, 1967, e.g. on the Clavius photograph used as the frontispiece. The finest Jupiter detail recorded on color film has dimensions of 0.″12–0.″15, though 0.″2 is more common.)

Numerous details affecting the *thermal regimes of telescope and dome* must be watched if maximum photographic resolution is to be achieved. Unfortunately, insufficient attention has been given to these factors in some observatories built on excellent sites. Briefly, the requirements are: (i) the entire observatory area must have a good climate in the usual astronomical sense, not too far from the western shore of a continent or on a favorably located island; (ii) the building must be elevated, for the slit opening to be well above the atmospheric boundary layer caused by local topographic features, trees, etc. This level is determined by tests with small balloons and smoke bombs which make the air flow pattern visible. (iii) The building itself should be cylindrical, with a hemispherical dome, having a slit that moves up and over (not horizontally moving doors). This cylindrical observatory should have a diameter not greater than necessary to house the telescope. (iv) The dome should be painted white (TiO_2), which reflects more than $^3/_4$ of the incident sunlight and effectively re-radiates in the infrared the small part absorbed; thus the outer surface remains cool during the day and cools rapidly at night, preventing an updraft of warm air from the building in front of the dome slit. (v) The thermal inertia of the dome should be minimal to achieve equalization of temperatures at night. This is also the cheapest way of building a dome, since a single metal skin over a rib structure of I-beams suffices. (vi) Inside the dome one should have 3 or 4 large exhaust fans spread in azimuth, at about the observing level, capable of running at different speeds, adjusted during operations to cause an inflow of air through the slit at a rate not so rapid as to cause turbulence. This can be checked from the Foucault test. The best arrangement is found to depend on the wind direction and speed.

(vii) Small exhaust fans are also placed around the primary mirror to prevent the formation of air bubbles over it with temperature slightly different from that of the ambient air. These bubbles are well known to all visual observers who have studied the Foucault patterns of bright stars. The average diameter of such bubbles, when permitted to form, appears to be 30–40 cm. (The 82-inch telescope showed usually about 20 of them; the 200-inch, approximately 100). The bubbles grow to full size in 10–30 seconds, and then take off and rise, being replaced by cooler air. The (warm) bubbles act like weak negative lenses and cause the telescopic image to be that of a multiple interferometer. Stellar images are usually confined to an overall diameter of about 0.3 arc sec, containing from 5 to 15 visible components which slowly move over each other with a relaxation time of a few seconds. In double-star observations, one picks the brightest of these images for measurement but planetary photography is, of course, degraded. Experience has shown that these bubbles can indeed be prevented.

Considerable study of photographic materials, image sizes, and development techniques is also essential. In our color photography we have found High-Speed Ektachrome or, more recently, Ektachrome EF, to be the most satisfactory, not used routinely, however. In addition to normal exposures, processed commercially, we expose several rolls at $^1/_4$ of the normal time ($^1/_8$ s vs. $^1/_2$ s for Jupiter at F/75, or F = 115 m) and compensate for this by special processing. Normally, this leads to distinctly improved definition. We have also found that underexposed (dark-looking) color frames can be copied with corrective filters to normal-appearing frames. This requires that some normal exposures are available for reference, though these may have less resolution.

An examination of some 12000 color frames of Jupiter (and many more black-and-white frames) obtained since 8 October 1965, when operations with the 61-inch telescope started, showed that with no more than $^1/_4$ of the telescope time assigned to planetary photography each month, on almost every month a series of exquisite color photographs was obtained, supported by approximately three times the number of black-and-white photographs taken through filters, ultraviolet to infrared. Important supplementary series were acquired each month.

As regards the *use* of planetary photographs, with low-contrast features such as the clouds of Venus, composite prints are the most informative: one may combine from 5 to 10 frames into a single print, increase the contrast, and take out mechanically a linear gradient across the terminator.

The *examination* of the color photographs is best made in binocular vision, using two selected frames. Then, one is able to reject immediately all minor defects, such as scratches, which would be retained, though reduced, in composites. Binocular vision further appreciably enhances visibility of faint contrasts and minute details, giving the impression of a gain in resolution of approximately 1.5 over single frames. Accordingly, this Laboratory is preparing a file of color copies for the planet Jupiter, all 20 mm in diameter, containing all frames from the original set of 12000, that appear to contribute independent information. This number is larger than might be expected since the resolution is not constant over the entire planetary disk, so that many frames must be consulted. The number of frames actually selected for copying is about 1500. A companion file of supporting black-and-white filter photographs will also be produced. Because of the longer wavelengths and higher contrast, the High-Speed IR frames often show somewhat higher resolution.

The *wavelength dependence* of photographic resolution is interesting. On moderately good nights, the resolution may go as $1/\lambda$; on poor nights, no λ effect is noted. This agrees with the visual impression that on poor nights the blurred stellar images do not have a blue fringe. The $1/\lambda$ effect on good nights is probably due to the near-validity of the Rayleigh criterion for good images (deviations less than $\lambda/4$ in the wave front). The criterion may be satisfied from 0.7–0.9 μm (High Speed IR), but not 0.35–0.5 μm making a factor of 2 difference in resolution. Unfortunately, fast films beyond 0.9 μm are not available or they would further improve earth-based resolution in planetary photography with large telescopes, possibly closing the resolution gap referred to. Actually, for low-contrast features there is no gap at present.

Color photography has added a new dimension to photography of the planets. For Jupiter and Saturn the classification of clouds by color almost certainly is a classification by composition; and presumably also one of elevation within the atmosphere. The belts, when observed under the best conditions, usually break up into dozens of individual clouds all of the same color, suggesting a repetitive process of their formation. They appear analogous to terrestrial cloud streams formed by large forest fires or volcanic eruptions. The study of the Jovian planets will be one of the most active scientific endeavors of the 1970's; ground-based astronomy, if carried to the limits of its capabilities, will be able to contribute greatly in two main areas: atmospheric circulation and cloud composition. With still better facilities, resolutions down to 0.10 arc sec should be achievable on a fairly regular basis, on both High-Speed IR and fast color film. Data processing, with allowance for the image blur function, can always yield another factor of about 2 in image definition, as was first demonstrated in 1956 at Johns Hopkins University, and has, for example, been practised in the processing of the Ranger television records by the Jet Propulsion Laboratory in NASA's Ranger Reports (1965, 1966).

Reference must be also made to the exquisite planetary photography with the new 108-cm telescope at the Pic-du-Midi, clearly a superior site that will merit further development, to the important comparative site tests in California by Walker [3], recently extended to other sites in North and South America and Australia [4], and the extensive tests carried out from July 1965 to October 1969 in France, Spain, and Baja, California, by the Paris–Meudon Observatory team (interim reports by INAG). Earlier notable contributions are those by F. G. Pease with the 100-inch telescope of the Mt. Wilson Observatory, (lunar and planetary photography in the 1920's), with the Lick 36-inch telescope by J. H. Moore and F. Chappel (the moon about 1937), and by H. M. Jeffers (Mars 1940–1960), by R. B. Leighton with the Mt. Wilson 60-inch telescope in 1960's, by M. L. Humason with the 200-inch [5] and by G. H. Herbig with the Lick 120-inch telescope, both at the coudé focus, by W. S.

Finsen with the 26-inch telescope of the Union Observatory, (Mars in 1954 and 1956 [5]) and by E. C. Slipher with various telescopes, especially the Lowell Observatory 24-inch and the Lamont-Hussey 27-inch at Bloemfontein, South Africa [6]. The total photographic program represents a tremendous effort (about 1 000 000 planetary and lunar photographs). This may have created the general impression among astronomers that the limits of this technique have been reached. No doubt further progress will be difficult and will require additional refinements and even new approaches if not better observing sites; but it is also true that resolutions of $0\overset{''}{.}1$ must be achievable photographically as they have been achieved visually at all large observatories since the 1890's.

It is probable that the most important advance still within reach is one of seeking the optimum elevations above selected observing sites. Related to this is the determination of the vertical distribution of seeing disturbances within the first few hundred feet above the ground, the region usually assumed to be responsible for the major part of the seeing disturbances, and improved dome-tower design, so as to avoid the creation of new local disturbances. The premium on achieving resolution of about $0\overset{''}{.}1$ regularly, especially during the next decade in the study of Jupiter and Saturn, is so great, in view of the prohibitively expensive alternative of observing changing cloud-covered surface from space probes, that these further efforts in ground-based photography are warranted. It is true that a resolution of $0\overset{''}{.}1$ might be achievable also from a 1–3 m telescope in orbit, but the problems with this approach may have been underestimated. In any case, ground-based planetary photography of highest possible resolution will remain a prime goal of astronomy.

Acknowledgments

Study of photographic materials and development techniques have been carried out at this Laboratory with great competence and persistence by R. le Poole, J. Fountain, S. Larson and R. B. Minton.

References

[1] G. P. KUIPER, Publ. Astr. Soc. Pacific **62**, 133 (1950).
[2] Trans. Int. Astron. Un. **9**, 250 (1955).
[3] M. F. WALKER, Publ. Astr. Soc. Pacific **82**, 672 (1970).
[4] M. F. WALKER, Publ. Astr. Soc. Pacific **83**, 1401 (1971).
[5] Summarized with reproductions in Chapters 16 and 17 of Planets and Satellites: The Solar System, Vol. III, University of Chicago Press 1961.
[6] E. C. SLIPHER, Brighter Planets, National Geographic Society and Lowell Observatory 1964.

Space Research XII — Akademie-Verlag, Berlin 1972

HIGH RESOLUTION OBTAINED BY PHOTOELECTRIC SCANNING TECHNIQUES

J. S. Hall

Lowell Observatory, Flagstaff, Arizona, USA

Several applications of linear scanning of different types of objects are described; examples include double stars, satellites, the Red Spot of Jupiter and a landing site on the moon. This technique allows one to achieve a gain of about an order of magnitude in resolution over conventional photoelectric techniques; it is also effective in providing sufficient data for removing background effects and for the application of deconvolution procedures. Brief consideration is given to two-dimensional scanning, either at the telescope or of electronographic images in the laboratory. It is suggested that some of the techniques described should be given serious consideration for space applications.

1. Introduction

Within the last few years it has been clearly demonstrated that the resolution which is obtained by conventional photoelectric techniques can be greatly improved by scanning a line segment of a focal-plane image. As this is done the intensity is measured by a photomultiplier and recorded by a high-speed integrating system. The data are subsequently processed by a computer. The

Fig. 1. A device used in an early attempt by the author to measure double stars by photoelectric scanning.

scanning mechanism can be mechanical, optical or electronic or some combination of these. First it will be shown how resolution can be effectively improved by the linear scanning technique. Following that, there will be a brief discussion of two-dimensional scanning which represents the ultimate achievement in photoelectric photometry.

About 15 years ago, at the Naval Observatory, an attempt was made by the author to measure double stars with the scanner pictured in Fig. 1. The cylinder, containing evenly spaced slits and surrounding a 1P21 photomultiplier, was driven by a synchronous motor. The orientation was so set that the focal plane images of a double star successively passed through each slit as the cylinder rotated. The output of the 1P21 was displayed on an oscilloscope. This project was abandoned, however, because the persistence of the phosphor screen proved to be inadequate for averaging out the erratic form of individual sweeps. Scanning was tried at the Lowell Observatory by Rakos [1] in 1964. In making photometric measures of Phobos, as it was eclipsed by Mars, he used optical scanning and photographed individual successive oscilloscope sweeps. To derive his results he examined a quarter of a million 35 mm frames. In 1966 the photographic film was replaced by a multichannel analyzer which produced integrated data in a form which could be processed directly by a computer. Rakos' work was followed at Lowell by a number of scanning programs, chiefly an extensive double star study by Franz [2], and polarization and color scans of nebulae, planets and satellites by the author [3] and several others.

2. Scanning in One Dimension

One of the best ways of illustrating the advantages of scanning is to show an example obtained by Franz in scanning double stars. A description of his equipment, together with some preliminary results are given in [2]. In Fig. 2 the ordinate represents light intensity and the abscissa is the position of a slit, 100 μm wide, along its scan line. The star observed is Antares, which has two components separated by 3.2 arc sec. In green light (V) the red component

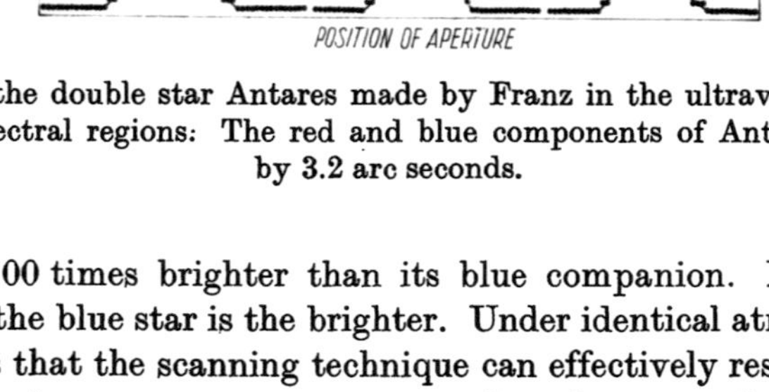

Fig. 2. Scans of the double star Antares made by Franz in the ultraviolet (U), blue (B), and green (V) spectral regions: The red and blue components of Antares are separated by 3.2 arc seconds.

is more than 100 times brighter than its blue companion. However, in the ultraviolet (U) the blue star is the brighter. Under identical atmospheric conditions it appears that the scanning technique can effectively resolve pairs (using deconvolution) whose separations are an order of magnitude less than those measurable by the conventional photoelectric technique of making separate settings and measurements on each star. Certainly the potential of the method

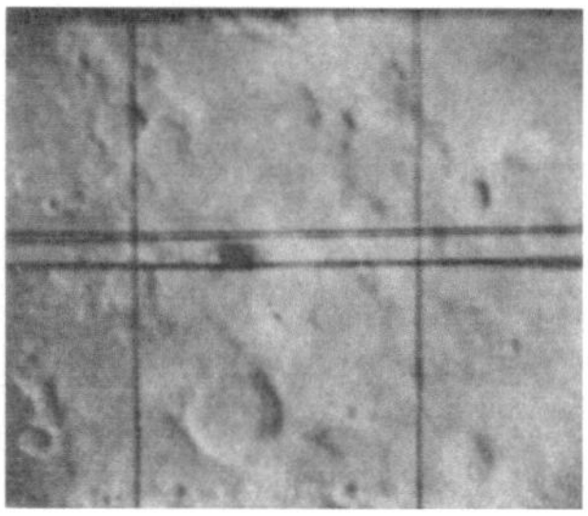

Fig. 3a

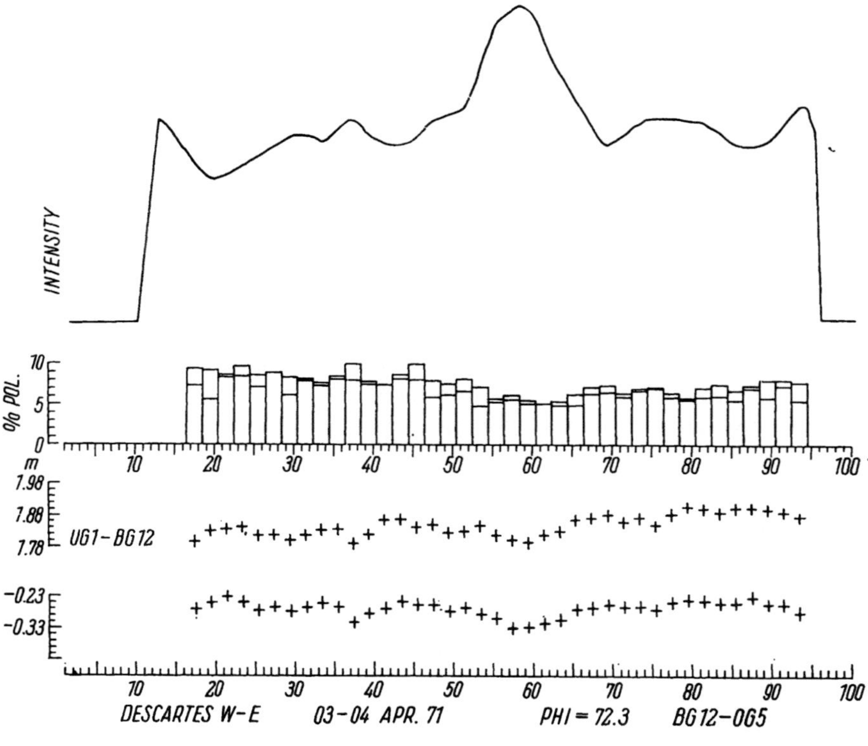

Fig. 3b

Fig. 3. Scans across Descartes by the author and Riley. The circular aperture, 1.2 arc seconds in diameter, is in the center of the black spot which moves repeatedly between the vertical wires. The photograph was obtained as observations were being made. The intensity curve represents the mean of four scans with a depolarizer in the light beam. Two independent polarization measures are represented by the two horizontal caps to each column in the histogram. The color measures are completely independent although a BG12 filter was used in each set. They are expressed in magnitude and are close to the UBV scale but with arbitrary zero points.

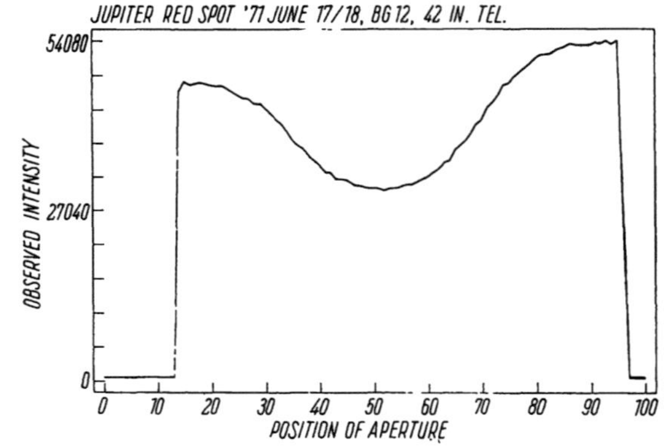

Fig. 4. The integrated intensity obtained from 30 one-second scans of the Red Spot. The scan was parallel to Jupiter's equator and centered on the spot in the orthogonal direction.

should be carefully explored from an orbiting laboratory because it is the very close pairs which are usually of major interest in studies of stellar evolution.

In polarimetry and colorimetry, scanning has yielded large quantities of closely related and accurate data in a short time. Fig. 3 shows some results obtained from scans across Descartes, the proposed Apollo 16 landing site, by the author and Riley. It gives the distribution of intensity, color and polarization at 39 equally spaced points. Here the resolution is limited only by the quality of the seeing and the accuracy with which the scanned surface can be identified. In lunar work we have found it advisable to photograph the region being scanned at the time measures are made.

Recently we have measured the plane polarization of Jupiter's Red Spot and compared it with that of its surrounding regions as the spot moves across the disk. When on Jupiter's central meridian the Red Spot has a maximum size of about 8×4 arc sec. Fig. 4 shows the integrated light from 30 one-second scans. At the time these measurements were made the reddest part of the Red Spot was near its center.

On the night of 6 April 1971, Price of IIT Research Institute with several others at the Lowell Observatory measured the decrease in light from Europa,

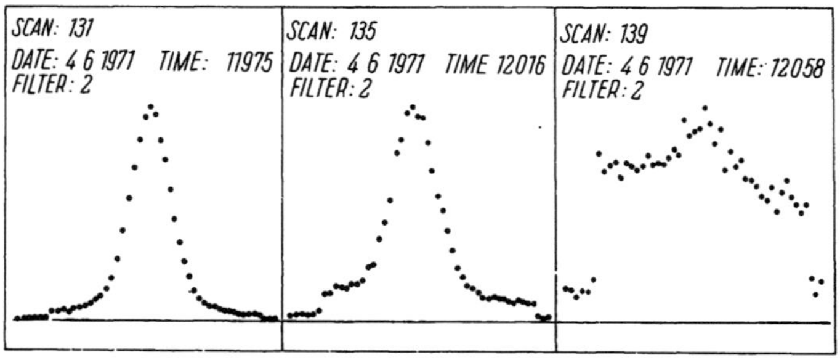

Fig. 5. Three 10 s scans of Europa, 40 s apart, taken with a narrow slit, as Europa was eclipsed by Jupiter.

a satellite of Jupiter, as it went into eclipse. A narrow slit was repeatedly scanned across the satellite, the scan motion covering 9 arc sec. An "instant replay" of three 10 s scans at different stages of the eclipse is shown in Fig. 5. At the left Europa was bright, so no sensible noise is apparent. At mid-eclipse, we see the sky background begin to appear. Near the end of the eclipse the signal from the satellite is much less than that due to the sky. However, the change in the background intensity over 9 arc sec along a radius to Jupiter is readily apparent, so that appropriate corrections for background can be made. Similar results can be obtained, with greater flexibility, by electronic scanning and the use of an image dissector. Hoag at Kitt Peak and Tifft at the University of Arizona are among those who are actively exploring this promising approach.

Although linear scanning has many advantages over more conventional techniques, it is very inefficient. This is because only a small fraction of the available photons pass through the scanning aperture at any one time. McNall et al. [4] at Lick Observatory have used an intensifier system in which the persistence of the output phosphor served as the required memory. This procedure, combined with adequate gain, made it possible for them to achieve the important objective of making efficient use of the photons along the entire spectrum. To accomplish the same purpose, Carruthers [5] at the Naval Research Laboratory used electronography for observation from rockets in the extreme ultraviolet. In the electronographic method the primary electrons from the photosensitive surface are accelerated and focused to an electron image where they produce developable grains in an electron-sensitive emulsion. The extremely fine grain (0.1–0.3 μm) of these emulsions permits very high resolution. High quantum efficiency, linear response, and high resolution make electronography a strong competitor as the ultimate device in photoelectric photometry.

3. Scanning in Two Dimensions

At the telescope it is possible, but rarely practical, to make a series of linear scans regularly separated in the direction perpendicular to the scan and thereby measure a small rectangular area in the form of a raster scan. More efficient methods of two-dimensional scanning involve the use of one or more image tubes. None of these, however, has intrinsically high resolution. The electronographic system such as that now being actively used for astronomical research by Kron et al. [6] at the Naval Observatory's Flagstaff Station closely approaches the ideal objective of achieving simultaneous recording of an area and at the same time providing the extremely high information content which must be achieved if both very high resolution and good area coverage are simultaneously required. Although electronography is not simple, Kron [7], Bied-Charreton et al. [8], and Griboval [9] have greatly improved upon the techniques originally developed by Lallemand. Closed-tube electronography, involving the use of a very thin mica window, has been developed by McGee et al. [10] to the point where apertures of 1 cm × 3 cm have been used at resolutions only a little less than that obtainable with electronographic tubes without a mica window.

The analysis of the very fine-grained emulsions in electronography involves scanning at high resolution in the laboratory and extensive data-handling facilities.

Provided electronographic films can be stored in space and moved to and from the space vehicle, the electronographic approach should produce the best results. If it is not feasible to use films, there are a wide variety of image tubes or combinations of image tubes which should be carefully considered before important decisions are made with regard to space applications. Although image tubes are much easier to use than electronographic techniques, all such tubes require a telescope of very long equivalent focal length to achieve high resolution. This means that the pointing accuracy must be high and the search capabilities must be severely limited. The use of electronic devices to achieve high resolution in astronomy is in a rapid state of flux. All approaches involve scanning, either at the telescope or in the laboratory. All approaches in which many picture elements are simultaneously recorded require extensive data handling. The road ahead is long and sometimes tedious but the objectives are clear and full of promise for ground-based but space-minded astronomers.

Acknowledgments

The observations obtained at the Lowell Observatory which are presented here were made possible by partial support from NASA grant NGR-03-003-001.

References

[1] K. D. RAKOS, Appl. Optics **4**, 1453 (1965); Lowell Obs. Bull. **6**, No. 131, 221 (1965).
[2] O. G. FRANZ, Lowell Obs. Bull. **7**, No. 154, 191 (1970).
[3] J. S. HALL, Lowell Obs. Bull. **7**, No. 143, 61 (1968).
[4] J. McNALL, L. ROBINSON and E. WAMPLER, Publ. Astron. Soc. Pacific **82**, 837 (1970).
[5] G. R. CARRUTHERS, Appl. Optics **8**, 633 (1969).
[6] G. E. KRON, H. D. ABLES and A. V. HEWITT, Advanc. Electronics Electron Phys. **28A**, 1 (1969).
[7] G. E. KRON, NASA SP-256 (1971).
[8] P. BIED-CHARRETON et al., Advanc. Electronics Electron Phys. **28A**, 27 (1969).
[9] P. GRIBOVAL, Bull. Amer. Astron. Soc. **3**, No. 1, Part 1, 3 (1971).
[10] J. G. McGEE et al., Advanc. Electronics Electron Phys. **22**, 11 (1966).

Space Research XII – Akademie-Verlag, Berlin 1972

HIGH RESOLUTION SOLAR INFRARED OBSERVATIONS

P. J. TURON and P. J. LÉNA

Observatoire de Meudon and Université de Paris, France

High resolution images of sunspots and granulation have been obtained at 1.65 μm and 10 μm corresponding to optical depths $\tau_{5000} \simeq 2$ and 0.1 respectively. Infrared imaging techniques are briefly discussed and it is shown that present day technology can provide very high resolution data with "exposure times" much shorter than one second. It is suggested that seeing improvement might allow infrared to give higher spatial resolution than any other wavelength for ground observations.

1. The Properties of the IR Range

Infrared has some special advantages but also serious disadvantages when one tries to achieve high spatial resolution.

1.1. Disadvantages

a. *Diffraction.* At 1.65 μm 1 arc sec resolution requires a 40 cm aperture; at 10 μm it requires a 250 cm aperture, so that infrared observations are likely to be diffraction limited even in the near ir region if small or medium telescopes are used (space observations for instance).

b. *Low contrast* due to the behaviour of the Planck function. A temperature difference of 50 °K in the photosphere gives a brightness contrast of 4% at 0.55 μm but of only 1% at 10 μm.

1.2. Advantages

a. *Continuum opacity* is supposed to be well known and optical depths range from $\tau_{5000} \simeq 2$ at 1.65 μm to $\tau_{5000} \simeq 3 \times 10^{-2}$ at 20 μm or almost two orders of magnitude (Fig. 1). It should be noted that 1.65 μm is really a unique wavelength: nowhere else in the solar spectrum from X or γ rays to radio waves can one see so deeply into the solar atmosphere. It corresponds to the absorption minimum of the H^- ion and although it is not much more transparent than the visible region, there is no better choice (except neutrinos!) if one wants to observe deep layers.

b. *Seeing improvement.* This second advantage has not yet been fully demonstrated. Any observer can tell that "Red is better than Blue" or "Hα better than CaIIK", but, in fact, there are very few quantitative observations of this

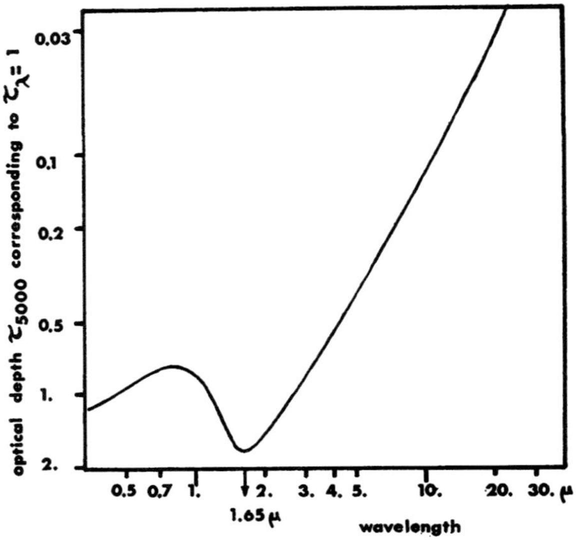

Fig. 1. Monochromatic opacity versus wavelength according to the Harvard Smithsonian Reference Atmosphere [1].

phenomenon. Stein and Woolf [2] report information from the 5 m telescope: diameters of stellar images were only 0.8 arc sec at 10 μm instead of 2–3 arc sec in the visible. Townes and Johnson (private communication from Townes) are building a stellar interferometer using a 10 μm CO_2 laser and a heterodyne technique; they will probably be able to give the answer to this question.

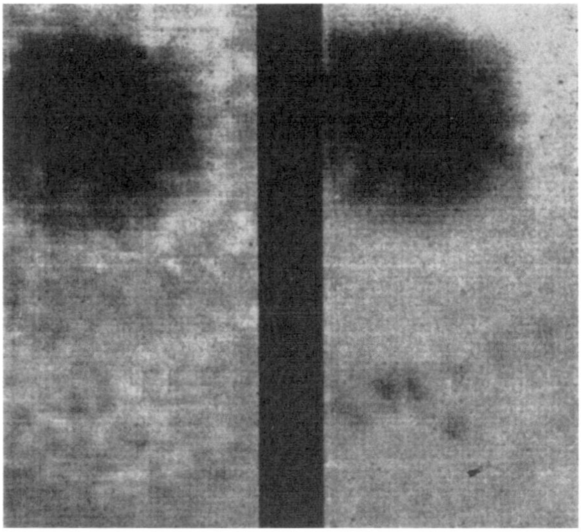

Fig. 2. Sunspot observed simultaneously at 10 μm (left) and 0.55 μm (right); the width of each image is 40 arc sec. Note the bright dots around the spot on the 10 μm picture.

We also observed better images at 10 μm than at 0.55 μm [3] although with very poor image at 0.55 μm due to the use of a slow scanning technique. Resolution at 10 μm was of the order of 2 arc sec which is near the diffraction limit (1.6 arc sec) broadened by the 1 arc sec scanning aperture (see Fig. 2).

Another comparison has been made between 1.65 μm and visible one-dimensional scans of the solar granulation obtained at Meudon in August 1970. We compared normalized power spectra and it appears that 1.65 μm shows more power at high frequencies. However, the visible light shows more power everywhere before normalization, but this is due to the Planck function effect. A statistical study of the "width" of the solar limb also shows a definite advantage of 1.65 μm over 0.55 μm.

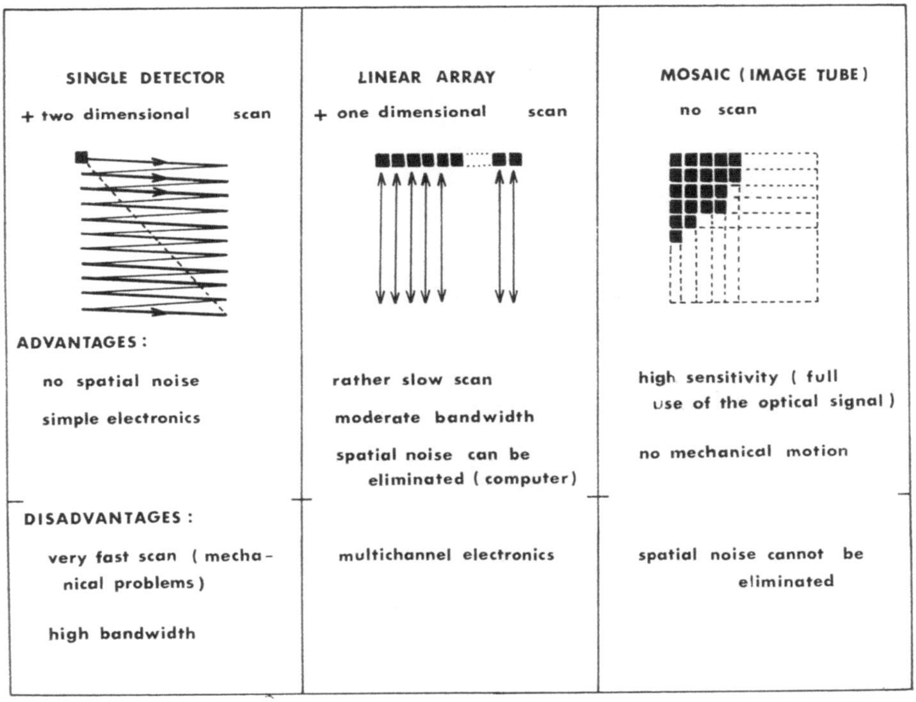

Fig. 3. Three methods in image formation.

Theoretical predictions are also very difficult. One of the best and more recent works [4] predicts a strong wavelength variation of atmospheric blurring. Power laws are found in the literature, with exponents ranging from 1 to 2. A λ or λ^2 law could be valid in special cases depending on the size of atmospheric inhomogeneities compared with the size of the telescope. Other authors give a $\lambda^{4/3}$ law. Of course, a very crude argument can be given: assuming wave front disturbances are independent of wavelength, it is clear that a given geometrical path corresponds to a smaller fraction of a greater wavelength.

Actually the problem is much more complex, but the argument might be at least used for the optical quality of the mirrors which are often far from perfection!

2. Imaging Techniques and First Results

In order to show that seeing improves at longer wavelengths, it is best to demonstrate it with high resolution pictures. Granulation is a good example of the need for higher spatial resolution. Even the best data (Stratoscope)

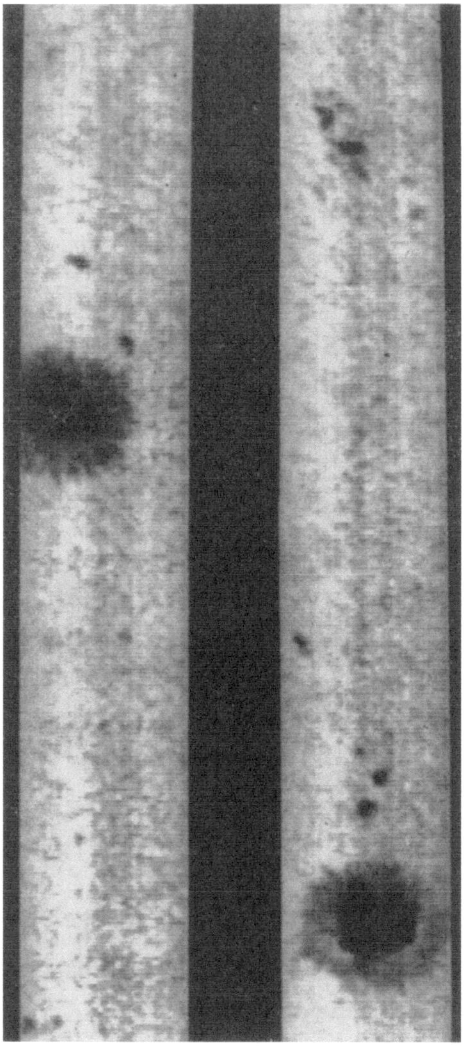

Fig. 4. 1.65 μm images (two different spots observed on 7 June 1971). Penumbral structures are visible but granulation looks rather strange due to the difficulty of adjusting the 64 channels at the same gain (one-dimensional "spatial noise"). This problem can be solved by time averaging each channel during or after the observation. A fast computer is probably the best solution.

do not resolve the narrow intergranular regions. Fig. 3 shows a synoptic diagram of imaging techniques for typical values. The mosaic (image tube) should be the best if one could adjust the sensitivity of each element in order to reduce the so-called "spatial noise". Actually, it is not possible to reach the 1% precision required for granulation work. For other purposes image tubes are certainly better than linear arrays, especially at short wavelengths ($\lambda < 2$ µm). This question has been thoroughly discussed in other contexts, mainly 300 °K thermal imaging [5, 6].

We have built in Meudon a 64-channel photometer using a 64-element PbS array. The scanning rate is assumed to be of the order of 20 arc sec per second of time and this gives 40 arc sec × 40 arc sec every 2 seconds. At 1.65 µm, diffraction limit at Kitt Peak is only 0.25 arc sec. Signal-to-noise ratio is not limited by detector noise and in fact the sensitivity of the array would allow us to scan much faster. The main problem is then the time constant of the Pbs detector. The first results were obtained in June 1971 at Kitt Peak.* They show (Fig. 4) spots and granulation at 1.65 µm with a resolution of the order of 1 arc sec. Centre-to-limb scans, 10 arc sec in width, will be used to obtain temperature fluctuations at optical depth $\tau_{5000} \simeq 2$ as a function of temperature fluctuation at $\tau_{5000} = 1$. This might be particularly useful to check Wilson's theoretical predictions of the behaviour of temperature fluctuations at optical depths greater than unity [7].

Clearly, it should be fairly easy to obtain much higher resolution with much shorter exposure time. There are some problems in matching the PbS time constant and high chopping frequency (or only scanning frequency: it might not be necessary to chop the light) but there are no fundamental limitations. Infrared images 40 arc sec × 40 arc sec can certainly be obtained at a rate of 10–100 per second with a signal-to-noise ratio of 200–1000.

3. Summary of Infrared Observations

1. *Continuum, broad-band observations.* In this case, as we have seen above, there is almost no technical limitation. High quality data can be obtained for deep layers at 1.65 µm or upper photospheric layers at 10 and 20 µm, giving information on: vertical evolution of the granulation from $\tau \simeq 2$ to 0.1; vertical variation of sunspot configuration, umbral diameter, pores, etc.; opacity effects observable near the limb at the sunspot penumbra–umbra boundary; comparison with continuum uv observations and spectroheliograms in the wing of strong lines (Hα + 1 Å . . .).

2. *Line observations, ir spectroheliograms, magnetograms.* The 10830 Å He line has been already used by J. Harvey and D. Hall (spectroheliograms and magnetograms with silicon diodes) and B. Fort (Lallemand camera).

At longer wavelength it is still quite possible to obtain very high resolution data but technical problems become more severe. A spectroheliogram in

* These results will be published elsewhere as a KPNO contribution.

a rotation-vibration transition of CO would probably be very interesting: local thermal equilibrium would allow an easy interpretation and CO is formed in a very narrow range in the solar atmosphere near the temperature minimum region. So, a CO spectroheliogram would give us information on temperature fluctuations in this very interesting layer.

4. Conclusion

1971 technology allows high resolution observation in the whole infrared range up to 20 μm. With existing telescopes, diffraction is a severe limitation above 4 μm but larger infrared telescopes can be reasonably imagined. It has been shown recently [2] that these large ir telescopes cannot be light buckets even for the simple detection of faint sources. Ecconomic arguments were also discussed. The optimum size was found to be much larger than the largest existing solar telescope. If seeing improvement is a real phenomenon this might be the key for higher resolution ground-based observations.

References

[1] O. GINGERICH et al., Solar Phys. **18**, 347 (1971).
[2] W. A. STEIN and W. J. WOOLF, Appl. Optics **10**, 655 (1971).
[3] P. J. TURON and P. J. LÉNA, Solar Phys. **14**, 112 (1970).
[4] C. E. COULMAN, Solar Phys. **7**, 122 (1969).
[5] J. A. HALL, Appl. Optics **10**, 838 (1971).
[6] F. DESVIGNES, Acta Electronica **13**, 9 (1970).
[7] P. WILSON, Solar Phys. **6**, 364 (1969).

Space Research XII — Akademie-Verlag, Berlin 1972

HIGH ANGULAR RESOLUTION SOLAR OBSERVATION FROM BALLOON BORNE INSTRUMENTS

K. O. KIEPENHEUER

Fraunhofer Institute, Freiburg i. Breisgau, FRG

The solar atmosphere is mostly formed and maintained by physical processes (hydro- and magnetohydrodynamics) which occur within regions of dimensions $\lesssim 1$ arcsec (equal to 750 km). Their analysis is seriously impeded by optical disturbances in our own atmosphere. A comparison is given between the angular resolutions hitherto reached or to be obtained in the future on the ground, in the stratosphere (from aeroplanes or from unmanned balloons) and from extraterrestrial equipment. The need to resolve the scale height in the solar atmosphere, which is about 100 km or 0.15 arcsec in pictures and qualified spectra, is stressed. Astrophysical results hitherto obtained by balloon-borne solar telescopes are briefly reviewed. By describing the project Spektrostratoskop, a balloon-borne telescope with high resolution spectrograph, a few technical problems are discussed.

1. Introduction

It is clear today that almost all basic physical processes forming and maintaining the solar atmosphere and performing the energy transfer, occur within very small regions. Most of these processes, their geometry and their time scale are characterized essentially by a scale height or scale length which amounts to about 100 km in the photosphere and in the chromosphere. If one wishes to study, for example, the internal physical structure of a single photospheric granule or a single chromospheric spicule or any other characteristic type of structure and wants to make sound and quantitative statements about the stream velocity, density, pressure or magnetic field within such structures, one would have to resolve details of about 100 km. This corresponds to a telescopic resolution of about 0.15 arc sec. Under vacuum conditions this would require a telescope of about 100 cm aperture (Rayleigh diffraction limit). Such a telescope would reproduce a one-dimensional 0.1 arcsec structure of 100% contrast with 15% contrast in the focus, while a two-dimensional structure will be reproduced with only 2.25% contrast.

The true contrasts on the sun are of course much less than 100% and lie mostly in the range between 5 and 20%. If we therefore ask for 0.15 arcsec resolution with a Modulation Transfer Function (MTF) of 50%, the required telescope will have to have an aperture of 150–200 cm. The situation in the solar case is completely different from the stellar case, where we usually have to do with isolated point sources. A similarly large aperture would be needed to supply, within a reasonable integration time, the necessary flux of photons coming from a region of say 0.1 arc sec × 0.1 arcsec on the sun, in order to guarantee a reasonable photometric accuracy, or to measure a magnetic field of a few gauss.

It is essential that this high angular resolution will be obtained in a *spectrum* of high spectroscopic resolution (c. 10–20 mÅ). Only then reliable and not blurred data will be available for a quantitative physical analysis and interpretation.

We are, of course, still far from this goal of resolving the scale height of the solar atmosphere and it might be appropriate, first of all, to make clear, how far away we are from this ideal in ground-based observation, with balloon-borne instrumentation and with extraterrestrial telescopes.

2. Comparison of Ground, Balloon and Space Observation

2.1. Ground-based Solar Observation

There is much still to do before we can hope to reach an adequate resolution during adequate periods of time, or to resolve the solar scale height on the ground; only a few indications of the optimal results hitherto reached are given here. For short exposure pictures of the photosphere in integrated light (exposure time usually $\lesssim 1/500$ s), quite frequently a resolution of 0.3–0.4 arcsec has been reached with appropriate telescopes. The duration of such excellent periods is mostly a fraction of a second up to a few seconds. At some of the best observatory sites occasionally (at the best perhaps a few times per year) longer periods of such conditions can occur, lasting, with small interruptions, for minutes or even sometimes hours. The quality of pictures obtained during such periods, e.g. in July 1970 on Pic-du-Midi, comes very close to or is practically the same as that obtained by the Schwarzschild group from balloon-borne telescopes.

The situation on the ground is much less favourable when spectra of high spatial spectroscopic resolution are to be obtained which require an exposure time of up to several seconds. Under best observing conditions a spatial resolution of about 1 arcsec has been reached, in a very few cases perhaps 0.8 arcsec. The situation is similar for monochromatic filters (Lyot type) with pass bands of 0.5–0.25 Å or less. Also here 0.8–0.6 arcsec is the present limit. In measuring the magnetic field or recording magnetic maps of the sun, done up to now only with ground-based equipment, the effective angular resolution has never appreciably exceeded 2 arcsec, mostly because of the long integration times needed, in order to exceed the noise level.

It must be clear that all the solar observatory sites at which the above described seeing conditions occur are without exception not really chosen according to meteorological optical criteria. The site testing campaign of the European project JOSO (= Joint Organization for Solar Observation) [1] has shown during the last years quite clearly that in principle air masses exist, for example, over the sea, which are over long periods optically much more homogeneous than air masses which are available over the known observatory sites. The convection in these stable air masses over the sea is known to be 30–100 times smaller than at corresponding altitudes over land. The problem seems to be to obtain access to such an air mass without producing too many technological and logistic troubles. In any case we have good reasons to suspect that duration and frequency of periods with excellent seeing will increase sensibly and that consequently also for longer integration times (spectra, magnetograms, filtergrams) resolutions of better than 0.5 arcsec will be obtained in this way.

Is there a fundamental limit to the angular resolution to be obtained on the ground in day-time under optimal meteorological and topographical conditions? Is there something like a "thermal noise" in the atmosphere defining the optimal angular resolution? I think we are not too far away from answering this important question. From the work of Coulman and Tatarski [2] we know how the Modulation Transfer Function (MTF) of the atmosphere is affected by temperature fluctuations along the line of sight. With 0.04° temperature fluctuation all along the line of sight 0.4 arcsec can be attained. With 0.02° already about 0.15 arcsec has been attained. From measurements above the sea we have found that very often temperature fluctuations of $\lesssim 0.01°$ will occur. So there are good reasons to be quite optimistic about such maritime sites, but to be sure it will be necessary to bring larger telescopes to places strictly chosen according to physical conditions in the atmosphere. I personally have little doubt that we shall come on the ground very close to resolving the solar scale height also in spectra, especially if we learn to handle electronic image amplification without losing angular resolution.

We have to keep in mind also that in planetary and lunar work a resolution of 0.2–0.15 arcsec (personal communication from A. Dollfus and G. P. Kuiper) has already been reached and we have good reasons to assume that this night value can also be achieved in day time, if sites free of convection, without night-to-day changes, can be found. We are certainly not far from this goal, and in a few years more will be known.

2.2. Observations from Aircraft and from Balloon

Efforts have been made to obtain astronomical observations from high flying aircraft, especially during solar eclipses for extending the duration of totality. A telescope of 91 cm aperture will soon be in use (NASA Ames Research Center, Moffett Field, California, C-141 aircraft), especially for the infrared. The mechanical stabilization of such a telescope can be improved obviously to about 1 arcsec. With additional image stabilization inside the telescope even a fraction of 1 arcsec might be reached. For higher requirements the boundary layer around the aircraft will probably be the decisive factor. The independence of ground weather, the possibility of flying rapidly into regions with undisturbed air masses and therefore of high optical quality, and the smaller costs of such flights compared with those of unmanned balloon flights represent an obligation not to ignore this observing tool and to do all that is necessary to find out about its limitations. Also observations from manned balloons have already brought interesting results. Blackwell et al. [3] obtained photographs of the granulation with about 0.5 arcsec resolution with a 29 cm telescope 6000 m above ground 14 years ago. This rather modest resolution is certainly partly due to the fact that these observations have been made below the tropopause without making sure that the tropopause was not disturbed during the observing day.

At a specialized symposium on high angular resolution astronomy one should try to be as realistic and as critical as possible when comparing the various tools of observation. Today, we can no longer say that in solar research the highest angular resolutions have been obtained unconditionally from unmanned balloon-borne telescopes. The resolution of 0.3 arcsec as obtained by Schwarzschild and coworkers with the Stratoscope 1 project from 25 km height with a 30 cm telescope has been reached quite frequently also from the ground, as far as integrated light pictures with very short exposure time are concerned. Even longer time sequences with such a resolution have been achieved on the ground, although very rarely, e.g. on the Pic-du-Midi Observatory. The decisive difference between observation from unmanned balloon and from the ground is that one might have to wait on the ground for such excellent seeing for possibly a year, until the desired meteorological conditions occur, while these conditions can be met in the stratosphere at any time (if we ignore for a moment the long waiting time for a balloon launch, caused by ground weather and by the complexity of such an experiment). This situation will change considerably in favour of balloon astronomy if diffraction limited, high resolution spectra ($\lesssim 20$ mÅ) or magnetograms are achieved, which imply integration times $\gtrsim 1$ sec or 1000 times longer than for integrated light pictures. Such documents with the needed high spectroscopic resolution have *not* been obtained yet from balloons. We return to this problem later.

In comparison with ground-based observation, the technological complexity of balloon-borne unmanned equipment increases very rapidly with improved angular resolution and the exposure time. This holds not only for the construction of a stratosphere-proof faultless optical system which can be manipulated from the ground but also for the complicated procedures of launching and recovering such heavy instruments. Waiting times up to many weeks are often needed until the optimal meteorological conditions for launch and recovery occur, while only a few hours will be available in the stratosphere for astronomical work. With the actual balloon technology it is not possible to launch a balloon telescope at a given moment, which in many cases would be very important, e.g. for the investigation of a large flare or an interesting sunspot etc. On the other hand it must be stated that only with balloon-borne telescopes shall we have the chance to experiment with telescopes able to resolve the scale height in the solar atmosphere reliably and during appreciable intervals of time; this may happen within this decade. Furthermore balloon telescopes of such resolving power will provide the indispensable intermediate step for the development of space telescopes of similar high resolution. They will teach us the appropriate technology and will prepare us to handle this unknown "subtelescopic" world on the sun.

2.3. Space Telescopes for Solar Research

Most of the effort put into the development of space telescopes has been in the extension of the spectrum into the euv and xuv or even in the γ-range, while rather little has been done to obtain an angular resolution significant for solar research. The best angular resolution, 2–3 arcsec, has been reached with rocket-borne spectrographs [4] while from satellites only a resolution of less than 8 arcsec has been obtained. At the moment only the OSO-1 and K projects are aiming at a resolution of about 1 arcsec and may hope to reach it in a few years. An advance into the 0.1–0.2 arcsec regime in the visible and in the euv will probably be reserved for post Apollo technology, for a free-flying solar module maintained and operated through a space shuttle or perhaps somewhat earlier by an improved high resolution version of Sky Lab.

3. Solar Results obtained with Balloon-borne Equipment

Apart from a few successful launches of balloon-borne coronagraphs by Dollfus and Newkirk [5], where the main goal was the reduction of scattered light and not the angular resolution, the basic and important results in solar physics almost all come from Schwarzschild's Stratoscope 1 project flights in the years 1957–1959 [6]. He used a Newtonian reflector of 30 cm aperture and 240 mm focal length. A small fraction of the 23 mm solar image is taken out in the prime focus by means of a small Newtonian mirror and is magnified 25 times. On the photographic negative 1 arcsec corresponds to about 0.3 mm. The exposure time was around 1/1000 sec. The instrument was stabilized by a conventional photoelectric guiding system and was usually brought to an altitude of 25 km. Focusing and pointing was accomplished by means of a television link from the ground. The results obtained mostly concern the solar granulation and are based on a few thousand negatives which at best show an angular resolution of 0.4–0.3 arcsec, which comes close to the diffraction limit of the telescope. In a general way it turned out clearly [6] that the shape of granules is quite irregular, often polygonal. There is a very large number of small granules of only about 300 km diameter. As this value is the diffraction limit of the telescope it is possible that a significant fraction of the granules are appreciably smaller than 300 km. There are also larger granules up to 1800 km in size. From this fluctuating size it follows that granules cannot be convective cells of the Bénard type, but must be a product of non-stationary convection.

The life time of granules [7] is found to be about 6 min, the root mean square relative intensity fluctuation, after correcting for the effects of the instrumental profile is of the order of ± 0.05, corresponding to a root mean square temperature fluctuation in the granulation of ± 92 °K. This value will probably have to be revised if a higher angular resolution is available. Again from Stratoscope 1 negatives Schröter [8] found that granules in the vicinity of sunspots are smaller and have a longer life-time ($\sim$12 min); they have furthermore a somewhat different structure. The dark lines observed within granules in undisturbed parts of the photosphere cannot be seen in the vicinity of sunspots, instead granules seem to be divided there into smaller units.

The width of penumbral filaments [9] seems to be near or even smaller than 300 km (again the diffraction limit of the telescope), their life-time seems to lie between 12 and 90 min, in the sense that after half an hour still a very few filaments can be traced.

Schröter [8] was able, again from Stratoscope 1 time sequences, to obtain some insight into the formation of pores within intergranular lanes. First the intergranular lane widens and at the same time darkens at a certain place along a stretch 2–5 arcsec in length. Then this dark region lengthens forming after 10–20 min a dark ring which surrounds a number of granules. These granules then fade away, become darker and darker, and finally form a pore. He could show furthermore that some bright knots within the penumbra of sunspots showed changes and a systematic motion of 1–2 km s^{-1} away from the spot centre. This corresponds to the motion expected from the spectroscopic Evershed effect. Our information on umbral dots (= umbral granulation) has also been improved by using Stratoscope time sequences [8]: their diameter seems to be $\lesssim 300$ km, their life-time about 0.5 hour.

In the Soviet Union some successful balloon launches have been made in recent years [11] and excellent photographs of granulation with a telescope of 50 cm effective aperture have been obtained [12].

The results of the Stratoscope 1 project might have been valuable to solar research by demonstrating that in the stratosphere an angular resolution of 0.3 arcsec can be achieved by a freely floating telescope, but the astrophysical results obtained have not yet changed our scientific views of the solar photosphere or of sunspots because the angular resolution is practically no better than that obtained on the ground and no spectra of high spectroscopic resolution and appropriate spatial resolution have yet been obtained; only such spectra or filtergrams of an equivalent spectroscopic resolution will supply us with *full* information about the structure and dynamics of the solar atmosphere.

4. Technical Problems of Balloon-borne Solar Instrumentation

We discuss here only a few problems which have not yet been solved completely or which are essential for obtaining the optimal angular resolution. One thing must be stressed: The extreme degree of optical homogeneity of the air mass around and above a telescope in the stratosphere (at altitudes between 25 and 40 km) compared with the ground, and therefore the very high angular resolution which could be obtained there, in principle demands extraordinarily high technical requirements of optical and mechanical performance, stabilization and thermal properties of a balloon-borne instrument. These requirements exceed those needed on the ground by almost an order of magnitude; also the optical adjustment of such balloon-borne instrumentation before flight under the difficult conditions before launch poses indeed very complex problems.

The basic physical information needed for the treatment of the thermal problems arising in a stratospheric telescope have been recognized and discussed quite thoroughly already by Schwarzschild [6]. Two types of disturbance are to be expected: one produced by the thermal deformation of the optical elements, the prime and Newtonian mirrors, the lens systems etc., and the other, which is the more serious, by internal seeing effects generated by the heating of the mirrors and the telescopic structure, and consequently by convective heating of the air along the optical path inside the telescope. A balloon-borne instrument ascending into the stratosphere will be exposed during the first one to two hours to extreme differences of ambient gas temperatures: between ground and tropopause at about 10 km altitude, temperature differences of at least 70° will arise. Then above the tropopause the temperature will rise again, at 25 km to about −45° to −50°, at 35 km to about −20°. In spite of the very small convective heat transfer (only about 10^{-4} W cm^{-2} s^{-1}) the various parts, according to their heat capacity will cool down considerably during their passage through the tropopause. Sun-shadowed parts of metallic structures (e.g. white painted tube structures carrying the telescopic optics) will reach during the ascent of about 65 min duration minimum temperatures around −40° to −50°, and about 30 min after arrival at the ceiling altitude of about 25 km an equilibrium temperature of −20° to −30°, that is about 15–20° warmer than ambient air temperature. This equilibrium is essentially brought about by the sunlight scattered from the ground and from cloud layers, by the infrared radiation coming from the lower atmosphere and by the relatively unimportant convective heat exchange with the surrounding air. The thermal situation of the main mirror is somewhat more difficult. Its weight in relation to its heat capacity, even when not exposed to the sun, will cause a very slow cooling during ascent. A shadowed quartz mirror of 300 mm diameter and 7.5 cm thickness will cool down from +15° on the ground to about −10° or −15°, that is a temperature which still lies 35–40° above air temperature. Schwarzschild found for a similar mirror temperature differences of around 50°. Such a hot mirror would, of course, spoil the internal seeing of a ground telescope completely. In the stratosphere, however, the density of air is only about 1/25 of the ground value and the effect of such a warm cushion of air is not quite so severe. As Schwarzschil [6] has shown, an air layer of 5 cm thickness, 20 °C warmer than the surrounding air will at 25 km altitude cause a change of length in the effective light-path of about 1/10 wavelength, an amount at the limit of tolerance. J. B. Rogerson furthermore showed (see [6]) that in the case of the Stratoscope telescope a nearly laminar flow of air develops above the mirror, which will apparently affect internal seeing rather little, in spite of being almost 50° warmer than ambient air.

If a higher angular resolution by means of larger mirrors is to be obtained, the cooling of the mirror will be still slower and will then really become a serious problem. In such a case, the mirror should be cooled before launch, as has been done with Stratoscope 2, or an active cooling during flight should be envisaged. Another very promising possibility of overcoming these thermal difficulties is of course to take the instrument not to 25 km but to an altitude of 35–40 km, where the ambient air temperature lies already 20–25° higher and where the body equilibrium temperature and air temperature become nearly equal. The reaching of such heights does not in principle imply insoluble difficulties, but the cost of the bigger balloon will be considerably higher and the technology of reliable launching such large, heavy and complex systems is unfortunately still far from being satisfactory.

Finally we consider the deformation of the prime mirror caused by inhomogeneous heating. In the case of Stratoscope 1 the sun-heated front side of the mirror which was thermally isolated from its metal case, was about 6° warmer than the back side. When fused silica or still better Cer-Vit is used, such a temperature gradient, at least for a mirror of 30 cm diameter, will not be critical. Still larger mirrors, in the solar case, should be tempered or cooled artificially.

5. Project Spektrostratoskop

A few more technical problems can be explained by means of the German project Spektrostratoskop [13] which unfortunately, because of a failure during the launch of its first optical flight in 1969, has not yet produced astrophysical results, but which in its technical concept contains a number of innovations. This project was for the first time to supply solar spectra of highest spatial *and* spectroscopic resolution. It is a balloon-borne solar reflector, with aperture 30.5 cm and equivalent focal length including a lens magnifying system of 63 m, which feeds a Littrow type grating spectrograph with a resolution of $\lambda/\Delta\lambda > 400000$. The telescope uses a 30.5 cm f/8 parabolic mirror of Cer-Vit. A small section (about 2 mm × 2 mm) of the primary solar image of 23 mm diameter is reflected by a small Newtonian mirror located near the prime focal plane and is reimaged after 2 times magnification on to the spectrographic slit by a transfer lens system. This image is reflected by the silvered slit jaws of the spectrograph after 12.5 times magnification on to the film plane of the integrated light camera. In this way simultaneously spectra and photospheric images are obtained, the latter showing the position of the spectrographic slit in the photosphere as a dark line with an accuracy down to the resolution limit of the telescope. The dispersion is about 10 mm Å^{-1} and the angular scale 3.3 arcsec mm^{-1} so that 0.1 mm on the film corresponds to spatial and spectral resolution limits of 0.3 arcsec and 12 mÅ respectively.

While the telescope is an open, very rigid scaffold system, the spectrograph is mounted in a double walled Invar system with means of compensation. The spectrograph case is shielded from the outside by a 75 mm styropor cover. To keep the telescopic structure and all instrumental parts below the line of sight at least at the equilibrium sun-shadow temperature, a radiation shield is placed perpendicular to the optical axis of the telescope, thus screening the telescope and all parts below it from the sun. It consists of a group of totally reflecting glass prisms. This system absorbs about 7% of the impinging solar energy and has an emission coefficient close to 1 in the infrared. Under the influence of the earth's radiation it will have, depending to its inclination, temperatures between −40° and −50°.

Much effort has been put into the guiding system, which plays a very important part in obtaining spectra with an exposure time of up to 20 s (for spot umbrae) as compared with white light pictures with an exposure time about 10000 times shorter. The guiding is accomplished in three steps. After the instrument has arrived at altitude, the telescope-supporting frame is oriented at the azimuth of the sun by a simple servo system acting on a torque motor of the upper azimuth bearing, which is a spherical oil pressure bearing. This system continues to work until the azimuth of the gondola is correct within about 1°. A similar servo system orients the telescope then in elevation. The second step of the guiding system uses an auxiliary telescope which senses the solar limb. Error signals from the limb are fed into a servo system controlling the pitch and yaw motors of the general torque system (see below). This intermediate guiding system probably gives already an accuracy of much better than 1 arcsec. Tests in the laboratory indicate that this system can supply a guiding accuracy of the order of 0.1–0.2 arcsec. The third step of image stabilization to about 0.1 arcsec is accomplished by beam stabilization without moving the telescope. The optical image on the spectrographic slit can be deviated by small amounts (±18 arcsec) in two directions by two plane-parallel glass plates, in two axes perpendicular to each other and to the optical axis, in the convergent beam in front of the slit. The plates are driven by galvanometer systems as motors. Similar systems are used as tachometer generators. The error signals for controlling the two plates are obtained from a four-quadrant photocell (Schottky barrier type) placed in the final focus of an auxiliary image, projected through a dichroic beam splitter. The system locks on the centre of gravity of brightness of an arbitrarily chosen intensity maximum in the granulation image within the diaphragm covering the photocathode. It automatically catches a granule as a "guiding star" and holds it within 0.1 arcsec as long as the granule does not suffer too much change during its own evolution. With the average life-time of a granule of about 8 min, we expect about 60% of all 30 s exposures to be perfectly guided. The system has been tested thoroughly by using appropriate prints of Stratoscope 1 granulation negatives and also with artificial granulation. The system is expected to work out to a distance of 0.6 $R_\odot$ from the centre of the disk. A second similar granulation sensor is set at the opposite corner of the white light image and provides an error signal which is fed into the servo system controlling the torque motor about the roll axis of the telescope package.

Owing to large changes of temperature relatively large variations of the position of the focus are to be expected. A fully automatic device for focusing is therefore provided. A small pinhole scanning the granular pattern at the plane of the spectrographic slit supplies a signal proportional to the contrast of the granulation simultaneously before and after the position of the focus and the squares of the contrast signals obtained are compared; the difference zero corresponds then to the best focus position between the two sensors. The focusing is done by a motor-driven shifting lens.

There are two more technical details of importance for high guiding accuracy and long exposures: The bearings carrying the telescope and the actuating system moving the telescope package. The Spektrostratoskop instrument package, containing the telescope, the spectrograph, the cameras etc. is supported by a spherical oil-pressure bearing which in turn is linked to the supporting structure. The viscous friction of this bearing is practically zero in the static case ($\dot{\omega} = 0$) and is especially suited for fine motions. The oil pressure is maintained by a small oil pump. The friction of the bearing is about 100 times smaller than that of the whole instrument with the surrounding air on the ground, and about equal to the air friction at 25 km altitude. This type of bearing has furthermore the advantage of a much larger range of angle compared with flexure bearings. The central spherical bearing is surrounded by a concentric assembly of three torque motors for pitch, yaw and roll control of the package-attitude, which use free-standing air coils of copper wire moving between the poles of powerful permanent magnets. This system of a three-dimensional torque motor does not produce any vibrations which could be dangerous to the performance of the telescope. The torque motors are fed directly by the amplified error signals of the intermediate guiding system.

6. Conclusions

The actual and the future role of balloon astronomy in solar physics can be seen only in a general view, which includes also ground-based and extra-terrestrial observing techniques.

The ultimate observational goal must be the resolution of the scale height in the solar atmosphere ($\sim$100 km = $\sim$ 0.1 arcsec) in pictures and in spectra, in the visible and if possible also in the euv and xuv. There is, unfortunately, little doubt that this goal can not be reached before the end of this decade at the earliest. This will be so on the ground, in the stratosphere and in space. The reasons are fairly obvious: On the ground a solar telescope of suitable thermal properties of 1.50–2.00 m aperture would need to be built and placed on an optimal site, and it might still take 3–4 years to find and to prove this site; the financing of such an expensive instrument and its installation will be very difficult today. On the other hand the prospects of such a project look increasingly promising. In space, there is no hope of achieving an angular resolution of 0.2 or even 0.1 arcsec before 1980, unless a second high resolution version of the Sky Lab A solar telescope is built very early in this decade.

In such circumstances it seems quite reasonable to continue *simultaneously* efforts on the ground, in the stratosphere (using balloons and aircrafts) and in space (rockets, conventional satellites and modules) to approach higher and higher resolution. Only by such a *combined* effort and by continuous competition and cooperation between these three tools will it be possible to achieve such a difficult task as to resolve the scale height in the solar atmosphere during reasonable periods of time.

References

[1] Annual Report of the Joint Organization for Solar Observation 1970/71.
[2] C. E. Coulman and V. I. Tatarski, Solar Phys. **7**, 122 (1969).
[3] D. E. Blackwell, D. W. Dewhirst and A. Dollfus, Observatory **77**, 20 (1957); Mon. Not. Roy. Astron. Soc. **119**, 98 (1969).
[4] R. Tousey, Space Research XII, 1719 (1972).
[5] A. Dollfus, G. Newkirk and J. A. Eddy, Sky and Telesc. **24**, 77 (1962).
[6] M. Schwarzschild, Astrophys. J. **130**, 345 (1959).
[7] J. Bahng and M. Schwarzschild, Astrophys. J. **134**, 337 (1961).
[8] E. H. Schröter, Z. Astrophys. **56**, 183 (1962).
[9] R. E. Danielson, Astrophys. J. **134**, 275 (1961).
[10] R. E. Danielson, Astrophys. J. **139**, 45 (1964).
[11] V. A. Krat et al., Rep. Main Astron. Obs. Pulkovo No. 185, Leningrad 1970.
[12] V. A. Krat, V. N. Karpinsky and V. M. Sobolev, Space Research XII, 1713 (1972).
[13] K. O. Kiepenheuer and J. P. Mehltretter, Appl. Optics **3**, 1359 (1964).

Space Research XII – Akademie-Verlag, Berlin 1972

PRELIMINARY RESULTS OF THE THIRD FLIGHT OF THE SOVIET STRATOSPHERIC SOLAR OBSERVATORY

V. A. KRAT, V. N. KARPINSKY and V. M. SOBOLEV

Pulkovo Observatory, Leningrad, USSR

Preliminary results of observations obtained during the third flight of the Soviet stratospheric solar observatory are reported in brief. The study of fine structure of the solar photosphere (granulation, sunspots) gives evidence of the existence of very small elements (smaller than 200 km). Some conclusions on the life-time of the fine-structure elements in granulation, penumbra and umbra of a large sunspot are derived. The structure of the sunspot umbra is found to consist of a regular network of bright point-like objects and continuous dark background. The spectroscopic observations show the wavy features in Fraunhofer's lines due to the Doppler velocities of granules. The spectrograms show that the chromospheric elements (in $H\alpha$) are at least twice as large as the photospheric granules.

The report deals with information obtained during the third flight of the Soviet solar stratospheric observatory on 30 July 1970. The description of the design and technical data of this observatory have been published [1, 2]. For the third flight the diameter of the main mirror was 50 cm; in the future flights, however, a 1 m mirror will be used. The focal length of the Cassegrain system is 24 m and for direct white light photographs 120 m.

Altogether 882 direct photographs and spectrograms were obtained during the third flight. Each frame has its own photometric calibration. The direct photograph and spectrograms of three sections of the spectrum of the selected regions of the solar surface were obtained simultaneously. During the flight we were able to take photographs of nine sections of the spectrum from 3900 Å to 6600 Å; the width of each part was about 100 Å or 200 Å depending on the dispersion, which varied from 1.25 mm $Å^{-1}$ to 1.90 mm $Å^{-1}$. The time of observations during the third flight was about 7 hours. The following parts of the solar disk were photographed: the quiet photosphere (at angular distance from the centre $\theta = 22°$), a large sunspot No. 359 (numeration in "Solnechnye Dannye"), groups of pores, and the solar limb.

It seems that there are two main criteria for determining the top quality photographs of solar granulation for our instrument: the width of the instrumental profile (point source profile) must not exceed 0.3″, the diffraction limit being about 0.25″, and the observed root mean square intensity fluctuation of granulation must be sufficiently large, say more than 8%. With these criteria we find that 96 photographs can be considered as top quality ones. Another 96 frames were estimated as good and certainly surpass the best ground photographs.

Fig. 1 shows a photograph of the "quiet" photosphere in the central part of the disk. We obtained 20 top quality photographs of this region during twenty minutes. At present we are studying the time sequences of the frames in order to clarify the evolution of various kinds of granules: there are very fine granules of "point-source" size and larger granules of very complicated shape; granules

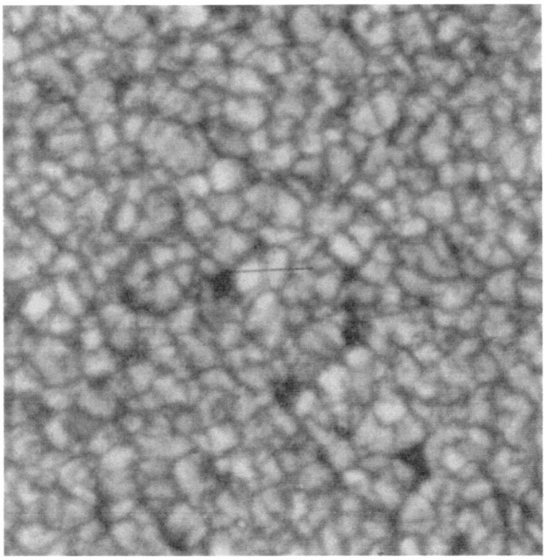

Fig. 1. Fine structure of an undisturbed region of the photosphere. 1'' = 5 mm

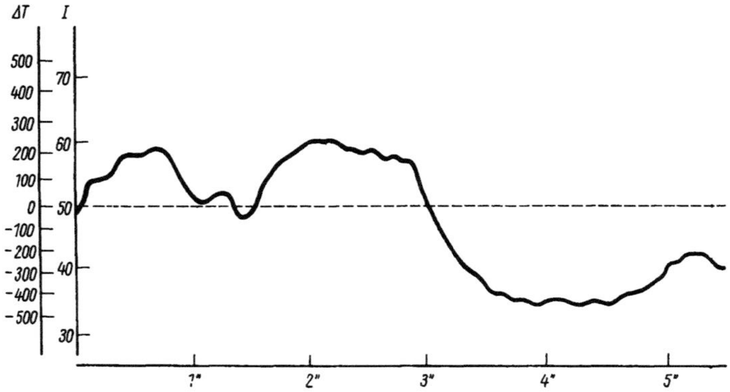

Fig. 2. A microphotometer tracing through a very small granule.

with darkening at the centre are very typical: some look like rings and others have a more irregular shape. Another typical form of granules in some regions is an arc-like shape. Some granules can form stable structures which conserve their configuration for a longer time than the life-time of individual granules. These structures usually have the shape of rings, lines or chains. This stability can be produced by large scale standing gravity waves or by peculiarities of the magnetic field in the upper part of the convective zone. A photometric tracing through a very small granule situated between two larger granules is shown in Fig. 2; the ordinate is the brightness with the mean level value for the photosphere equal to 50. The scale of brightness temperature differences is also given. The horizontal scale shows distance in seconds of arc. The width of the smallest granule on this tracing is 0.25″ and the width of the dark intergranular spaces is only 0.2″. This last value is smaller than Rayleigh's diffraction limit of resolution. However there is no contradiction with the theory because this detail is not isolated. The limiting period of the MTF for our instrument is

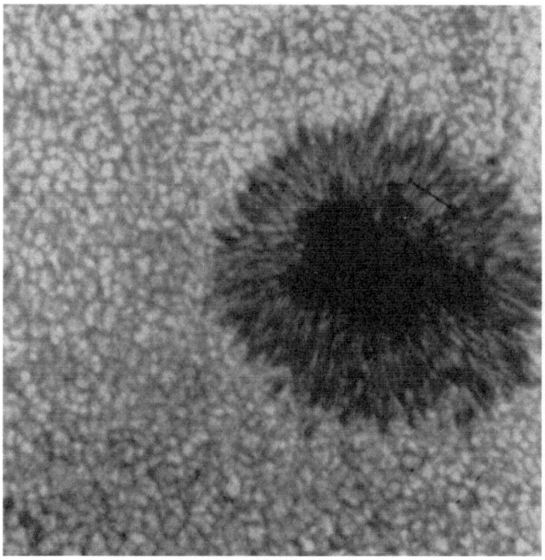

Fig. 3. Fine structure of a large sunspot (granulation and penumbra).
1″ = 2,5 mm

equal to 0.2″, which corresponds to a width of 0.1″. For most granules the contrast between granules and the nearest intergranular space varies from 16% to 35%, which corresponds to a brightness temperature range from 240° to 440°. The true value of the contrast is even higher since instrumental blurring is not taken into account. Our values of temperature differences are noticeably greater than Rösch's estimate which is 250° [3].

We have determined the root mean square intensity fluctuation for the "quiet" photosphere near the centre of the disk. The observed value is equal to 8.8%. This value, corrected for diffraction blurring, is about 12% or 14%, in accordance with Edmonds' data [4]. This corresponds to a temperature difference of 165°. The temperature fluctuation decreases slowly in the vicinity of the sunspot. The brightness in the intergranular space is more homogeneous than the brightness of the granules. The mean width of the granules exceeds by 1.5 times the width of the intergranular spaces.

For the large spot (Fig. 3) observations separated by various time intervals were made over a period of time greater than 3 hours. The essential change in the details of the fine structure of the penumbra is clearly seen after a time interval of about 1 hour. The overall shape of the spot remains the same. The microphotometer tracing of bright penumbra filaments is represented in Fig. 4. The observed width of these filaments is near to the diffraction limit of our telescope, so the actual width is less than 200 km. The true brightness of some

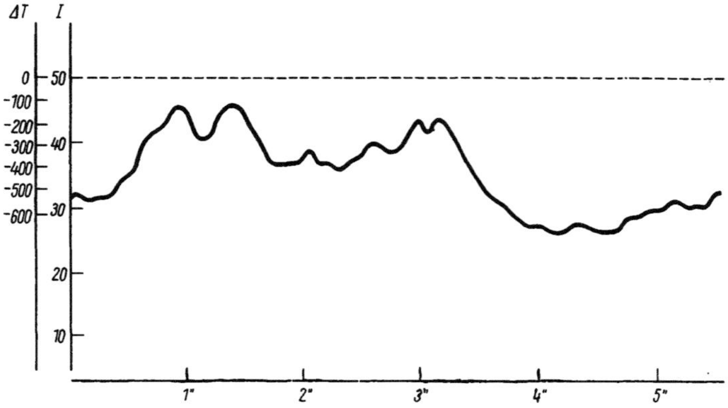

Fig. 4. A microphotometer tracing through the Sunspot penumbra.

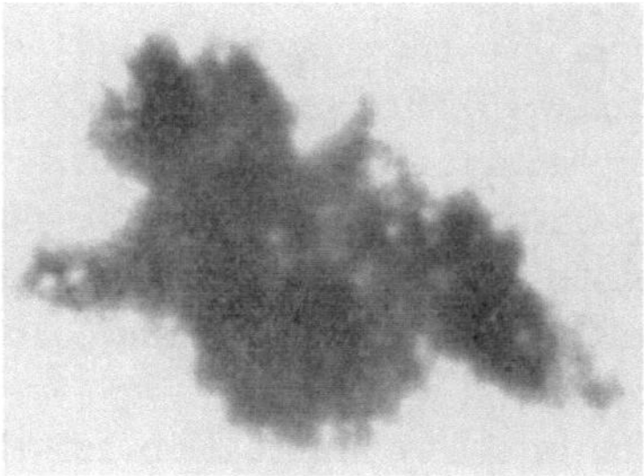

Fig. 5. Fine structure of the sunspot umbra. 1″ = 4.6 mm

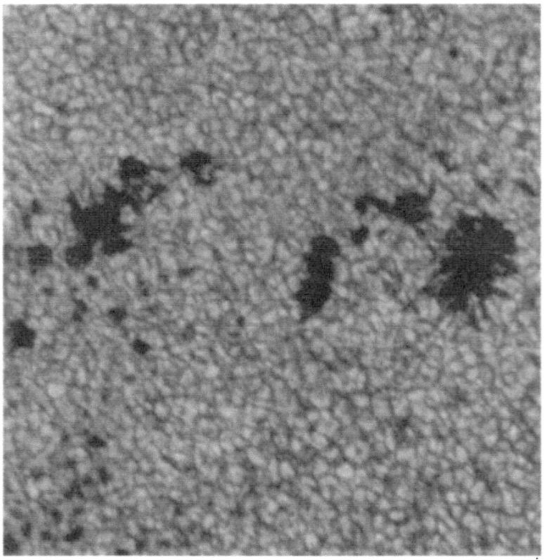

Fig. 6. A group of pores. 1″ = 2.5 mm

bright filaments of the penumbra is equal to or, perhaps, even greater than the mean photospheric brightness.

The sunspot umbra is shown in Fig 5. The exposure time is three times greater than that for the penumbra. A number of bright isolated points and the homogeneity of the darkest parts of the umbra are very remarkable. The width of the finest bright points is near to the diffraction limit and the brightness is near to the mean brightness of the surrounding photosphere. The life-time of these details is about forty minutes.

Fig. 6 shows a beautiful group of pores. One can easily see the sharp border of the pores.

In addition to direct photographs the first balloon-borne spectrograms with high spatial and spectral resolution were obtained. Fig. 7 shows part of the sixth order spectrum centred on the iron line 4376 Å and vanadium 4379 Å. One

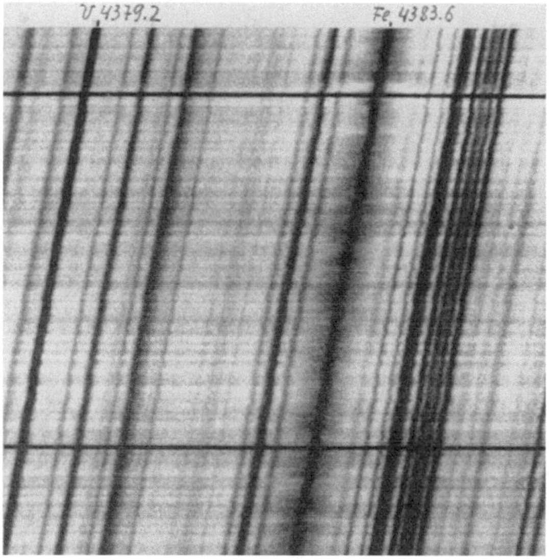

Fig. 7. Sixth order spectrum of the disk centre along the slit.
1″ = 1.4 mm

can see wavy displacements of the lines, the asymmetry of line profiles due to the gradient of the Doppler velocity, and the variations of the physical conditions in the various parts of the photosphere. The size of the finest details is about 0.5″.

It is of interest that the dimensions of the finest elements of the chromospheric network in Hα are twice those of the photospheric granules, approximately 0.8″: this may be the lowest limit of the cells of the chromospheric elements. More detailed analysis of the material will be given in further papers.

The photographs obtained give evidence of the excellent quality of the telescope optics and the absence of vibrational and thermal distortion. The telescope makes possible the solution of a great number of problems of solar physics. Further progress will be made if we improve the spatial resolution, to be able to measure the magnetic fields of the fine photospheric structure simultaneously with observations in white and monochromatic light.

References

[1] V. A. Krat et al., Izv. Pulkovo Observ. No. 185, 124 (1970).
[2] V. A. Krat et al., USSR Astron. Tsirk. No. 597 (1970).
[3] J. Rösch, Trans. Internat. Astron. Un. IIB, 197 (1962).
[4] F. Edmonds, Astrophys. J. Suppl. Ser. 6, 357 (1962).

Space Research XII – Akademie-Verlag, Berlin 1972

HIGH ANGULAR RESOLUTION OBSERVATIONS FROM ROCKETS: SOLAR XUV OBSERVATIONS

R. Tousey

E. O. Hulburt Center for Space Research, US Naval Research Laboratory, Washington, D. C., USA

The current status of high angular resolution xuv solar spectroscopy from sounding rockets wil be reviewed, together with its connection with high spectral resolution. The last few years have seen the commencement of rapid and encouraging progress in vehicles; they will soon no longer limit the achievable angular resolution. Already, monochromatic xuv solar images have been recorded with a few arc seconds resolution. Stabilization of sounding rockets has now reached about two arc seconds, and will soon be even better. New instrumentation must be developed to take full advantage of these improved vehicles, but this will require much research in the laboratory on components and designs. The future for research from stabilized rockets is promising indeed, for they form comparatively inexpensive, highly stable, and relatively simple-to-instrument platforms, suffering only from short life in orbit. But the instrumentation can now be recovered, readjusted or modified, and reflown. It is important that the number of flight opportunities in rockets be increased.

Solar research from rockets, just like solar research from the ground, demands the highest possible angular resolution. Until now rocket systems have set the limitation, the instrumentation has not. Soon, rocket systems may pace the instrumentation, but of this I am not yet certain.

At the outset I should like to raise, and to answer the question, why continue with rockets? Why not use just earth orbiters? For the sun, the answer is simple. Up to now, rockets have made possible the extreme ultraviolet solar observations that have the highest resolution, both angular and spectral. They will continue to do so for a long time, if the development of new rocket systems receives the support it deserves. Their shortcomings, of course, are the short time they provide for observing and the somewhat limited size of instrument they can carry to a sufficient altitude.

Earth orbiters are also required for solar research, especially to observe solar phenomena that vary with time. Skylab A [see following paper by G. K. Oertel] has instrumentation in ATM, the Apollo Telescope Mount, that in angular resolution can equal and perhaps surpass instruments flown in the most advanced rocket systems now available. And of course ATM will provide far more coverage in time, and orders of magnitude more total results than a single rocket flight. But I have the strong conviction that the results from Skylab A and ATM-A will support the need for even more advanced rocket systems, and for new solar experiments that will take advantage of them. The results from ATM-A will also show a need to continue with an ATM-B.

New instruments will be required, but to develop them, continued experimentation with rocket-borne instrumentation is essential.

The most important problem in solar physics, and the one most likely to be fundamental to the understanding of the solar cycle and flares, is the energy transport mechanism. From rocket observations in the xuv made with only moderate angular resolution, it has been shown that the temperature rise from 10^5 °K in the chromosphere to 10^6 °K in the corona takes place in a very short distance, certainly less than 1000–2000 km [1]. But to make more headway toward understanding completely what is happening in this curious region, xuv observations with much more angular resolution than has yet been attained are required.

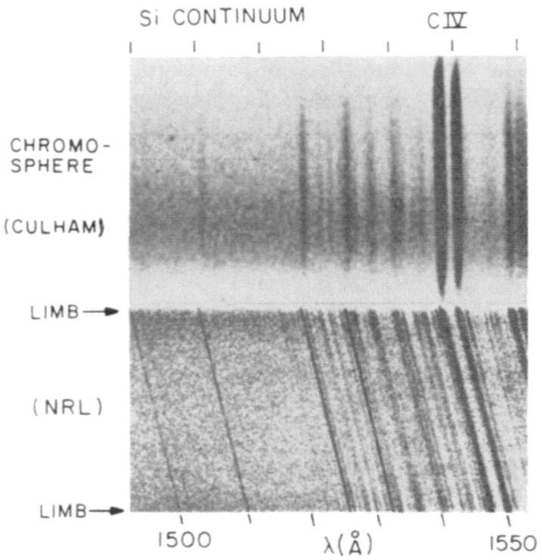

Fig. 1. Sections of a solar spectrum made on 9 April 1965 by the Culham Laboratory with the slit tangential to and 10 arc sec above the limb, and a stigmatic spectrum having about 10 arc sec angular resolution made by NRL on 22 August 1962 with the slit lying along a solar diameter. The off-limb spectrum matches best the ends of the lines in the stigmatic spectrum.

High angular resolution, as defined for this Symposium, is "better than one arc sec". One arc sec probably sets the upper boundary for observations that will really contribute to the complete solution of the energy transfer problem. The goal set by the Astronomy Missions Board [2] is 0.1 arc sec, which is the very most that seems technically feasible in the forseeable future; this may not be possible from rockets.

I shall begin by reviewing the xuv rocket results obtained up to now, with greatest emphasis on high angular resolution. But high spectral resolution, and high time resolution are also of great importance. It is difficult to discuss one alone; all three must be considered when an experiment is designed, and the result is always a compromise.

High angular resolution experiments from rockets can be divided into two categories: measurements across the limb, usually involving resolution in one dimension only, and observation of features on the disk, which requires resolution in two dimensions.

Observations of the limb started about ten years ago, when NRL [3] obtained stigmatic slit spectra in the range 2000 Å to about 500 Å. Not long after, spectra in much of this range were photographed at Culham [4] with a spectrograph having its slit tangential to and just above limb. Fig. 1 shows a small sample of these results. This short spectral range includes SiI ^3P edge and continuum, the C IV chromospheric doublet, chromospheric lines of Si II and Fe II, and lines of Si I that arise from close to the temperature minimum. Because the NRL system was stigmatic, the tip ends of the lines came from the limb. Chromospheric lines like C IV became quite bright at the limb, and were easily distinguished from lines from the temperature minimum that did not. The spatial resolution was 10 arc sec at best, a limit set mainly by instability of the biaxial pointing control system. The Culham spectrum, made with a 4 arc sec slit positioned 10 arc sec above the limb, shows a similar enhancement of C IV, but in a form that is much more suitable for photometry than the tip ends of the NRL lines.

The spectrographic system constructed at Culham [5] was the first solar instrument to be stabilized in all three axes and equipped with a fine-guidance device. The sun was imaged with a mirror on the slit of a single dispersion grating spectrograph. The instrumentation section as a whole was pointed at the sun's limb, set to a pre-selected position in roll, and stabilized with the three-axis Attitude Control Unit system of Elliott Brothers Ltd. But to maintain better than one arc minute or so precision relative to the limb required an additional servo-control device, which was built into the spectrograph at Culham [6]. In 1969 the distance of the slit from the limb was changed during the flight.

The traditional method for obtaining limb spectra of the highest angular resolution is, of course, the observation of the flash spectrum during a total eclipse. The first xuv eclipse measurements were made at NRL [7] in 1958, by flying simple Lyman-α and X-ray detectors into the eclipse umbra using non-pointed, spinning Nike-Asp rockets, launched from a ship. The angular resolution was not high, but emission from plages, the limb, and the disk were separated and distinguished from the corona proper. The next attempt was on 12 November 1966 when Blamont and Malique [8] flew two rockets into the eclipse umbra in South America. They reported that Lyman-α is limb brightened by 30% at the center of a ring 15 arc sec wide, centered 6.4 arc sec inside the ir limb. They achieved an angular resolution of the order of 1–2 arc sec.

The most recent eclipse experiments were carried out on 7 March 1970, under the auspices of NASA. Two Aerobee rockets were launched from Wallops Island and passed into and through the umbra. Each vehicle contained a great deal of powerful instrumentation, designed to photograph the flash and coronal spectra over a wide range of the xuv. Both rockets were oriented precisely and stabilized in all three axes by similar, but not identical, systems developed especially for this purpose. The NRL rocket, an Aerobee-170, was first coarse-pointed by the gyro-controlled Attitude Control System of the Aerojet Corporation; second, precise pointing at the sun's waning crescent was achieved by turning over the control to an eclipse sensor, developed jointly by NRL, the Ames Research Center, and Exotech, Inc.; finally, during totality guidance-control was switched to rate-integrating gyros that held the stability to ± 2.5 arc sec, using a system developed by the Aerojet Corporation and NRL.

The four NRL instruments [9] photographed the ranges 150–450 Å, 700–1060 Å, the Lyman-α profile during the flash phase, and 1400–1960 Å. Failure of the parachute recovery system caused the payload to be lost in 6000 feet of water. On the chance that the film might still be developable, a recovery operation was undertaken. After 15 days the payload was located and retrieved by the Navy, together with NASA. This could not have been accomplished without the accurate radar plot of the trajectory obtained by NASA, and a lot of luck.

Electrolytic effects ruined most of the film. However, a few spectra were salvaged from a camera that had by chance been teflon-coated. These covered the range 1400–1960 Å. The results were reported by Brueckner and Nicolas [10] during the COSPAR Eclipse Symposium. Wavelengths could be determined to 20 mÅ, but the effective resolution was about 60 mÅ, owing to graininess. A spectrum exposed with the moon's limb 4.5–2.2 arc sec inside the sun's limb, showed many lines from the temperature minimum; these lines were absent in a spectrum exposed after second contact, with the moon's limb 0.6–1.4 arc sec above the solar limb. For example, the many high members of the Si I series approaching the ^3P limits disappeared after second contact, whereas lines of Si II and Fe II were still present and the C IV doublet remained very strong. Because of the action of the salt water, only the crudest kind of photometry was possible. Analysis was carried out simply by noting the geometrical conditions when each crescent disappeared into the fog and deriving from this the height above the limb reached by the particular emission line. It was found that Si III, Al III, Si IV, and C IV, actually extend to heights between 4700 and 9600 km.

A model that explains this behaviour qualitatively has been proposed by Brueckner and Nicolas [10]. The model assumes cylindrical spicules extending to 7000 km or more in height, and spicule counts made by Beckers [11]. Each spicule is assumed to have a cool center, sheathed by a transition zone extending to the interspicular corona. It was found that the eclipse results for high chromospheric, optically thin lines were reproduced by this model.

The second eclipse rocket carried two photographic instruments covering the ranges 850–2150 Å and about 2000–3000 Å. Four groups were involved: Culham, The Imperial College, Harvard College Observatory, and York University of Canada [12, 13]. This payload was recovered in perfect condition, and the short wavelength instrument gave excellent results from 977–2150 Å. Several papers [14–17] were presented on these results during the COSPAR Eclipse Symposium. It was noted by Gabriel and Jordan [14], as by NRL [10], that Si II–IV persists to 10000 km during the flash phase, and this can be explained by the spicules. They examined spatial variations around the limb and reported that the electron density and temperature as derived from the relative intensities of various newly discovered and identified coronal lines varied considerably from place to place. A more complete account of the Consortium experiment is to be published by Gabriel et al. [13].

The conclusion from both rocket experiments is that there probably does not yet exist a method that is more powerful than an eclipse for obtaining xuv limb spectra with the very high angular resolution required for deriving models of the solar atmosphere, and that 1 arc sec radial resolution is easily achieved.

But one arc sec resolution is not enough. Dr. Houtgast [18] reported during the Eclipse Symposium that extremely high-speed spectra made over second and third contacts on 7 March 1970 showed that close to the limb great changes take place in line profiles within 100 km, that is, 0.15 arc sec, and that these changes were different from one line to another. His spectra covered the short span 4545–4579 Å with 20 mÅ spectral resolution and fractional-second time resolution. The same must certainly be true in the xuv, at least for the temperature minimum and low chromosphere. The remarkable images from Sacramento Peak and from the Pic-du-Midi, shown earlier in these meetings, also support the need for really high xuv resolution. The goal of 0.1 arc sec seems to be justified.

Eclipses are rare and valuable, but costly. The Wallops Island rocket experiments will not be repeated during the eclipse of 1972. But there is a good chance that some solar rocket experiments may be possible during the 1973 eclipse, which is the last one that is practical for rocket experimentation for many years.

So far I have spoken about photographic experiments only; they are especially suited to rockets because they can be recovered, and because information can be recorded so very much faster by photography than by photoelectric methods. There is, however, one photoelectric experiment that has shown a potential for excellent spatial resolution, and has achieved very high spectral resolution. This is the Ebert monochromator, constructed by Parkinson and Reeves [19] of the Harvard College Observatory, and flown in an Aerobee-150 rocket, stabilized with a Ball Brothers Research Corporation bi-axial pointing control. The slit, 7 arc sec wide and 20 arc min long, was positioned near the center of the disk. Their spectrum covered 1400–1875 Å, with 60 mÅ resolution. Photoelectric recording, of course, has the advantage of greater precision than photographic, generally. But the *accuracy*, that is, the *absolute* value of an intensity, depends on the calibration, hence the one method is no more *accurate* than the other. The HCO spectrum is quite free from long wavelength stray light, and is excellent. By narrowing the slit and reducing spectral coverage, this instrument could be used for limb scanning, if equipped with the proper guidance system.

The best xuv limb spectra secured up to now are the photographic spectra obtained by NRL on 13 August 1970. The instrument was a half-size prototype of the NRL spectrograph to be flown in ATM. It was of the same general type as the Culham instrument, but used double dispersion and secured spectra

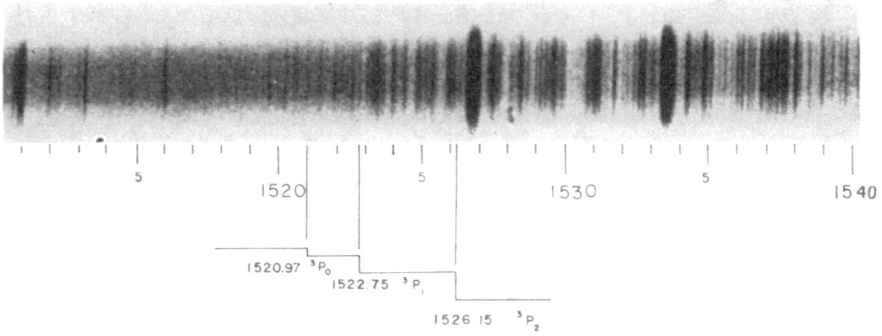

Fig. 2. A portion of the spectrum 7 ± 2 arc sec inside the photospheric limb, photographed by NRL [20] on 13 August 1970 from an Aerobee-150 rocket.

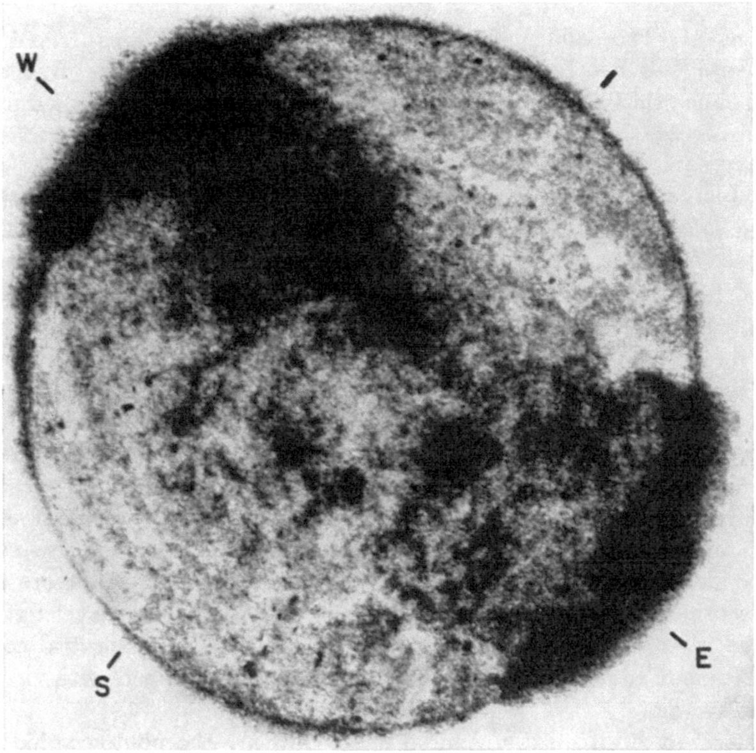

Fig. 3. An xuv heliogram made by NRL on the 13 August 1970 flight using a 25 cm focal length paraboloid and three 1000 Å-thick aluminum filters. The angular resolution is grain-limited to about 10 arc sec.

having 0.07 Å resolution. The slit was 2 arc sec wide and 60 arc sec long, and was tangential to the limb. The angular and spectral resolution values were about the same as for the ATM instrument. The design compromise that made it possible to fly such a powerful instrument in a rocket was that only 20% of the spectral range was covered, and only 18 exposures could be made instead of 6400 in the ATM instrument. The rocket was an Aerobee-150 stabilized in all three axes with the NASA Ames Research Center–Lockheed SPARCS I. In addition, the primary mirror that imaged the sun on the slit was servo-controlled in one axis by a system developed by the Ball Brothers Research Corporation, using an image-dissector tube. Spectra were obtained with about 4 arc sec peak to peak radial resolution at 7, 24, 40 and 300 arc sec inside the limb.

In Fig. 2 is reproduced a short section of an exposure made 7 arc sec inside the limb, covering the range crossing the SiI ^3P continuum edge that was shown in Fig. 1 at lower resolution. These results are reported by Brueckner and Moe [20]. Many emission lines that are absent, or very weak on the disk come into view close to the limb; a great deal of work will be required to identify them. Line widths are unexpectedly great for chromospheric lines that are optically thin, for example, Si III, Si IV, and C IV. This is in agreement with the results from the eclipse spectra.

The principal work on these spectra so far has been the derivation of the continuum intensity and the minimum value of the radiation temperature between the photosphere and chromosphere. The spectra are especially suitable for this purpose because the instrument was carefully calibrated against a standardized deuterium lamp, and because dispersion and resolution were sufficient to expose to view with high radial angular resolution many regions where the continuum is free from lines. The lowest value is at the very limb and is 4550 °K.

I shall now discuss high angular resolution in two dimensions rather than one: first broad band solar xuv images, and then xuv spectroheliograms.

In Fig. 3 one of the best NRL xuv images is reproduced from an exposure made on the 13 August 1970 flight that carried the spectrograph that produced the limb spectra. The instrument consisted of an off-axis paraboloid and a camera, with three unbacked 1000 Å thick aluminum filters in the light path. It photographed the sun in the broad spectral band 170–630 Å approximately, and we call it a heliograph. This is the first heliograph image obtained with complete three-axis stabilization; the resolution is film-limited to about 10 arc sec, since the focal length of the mirror was only 25 cm. In spite of the three filters the image is clean and the contrasts are great. The radiation included the resonance line of Fe VIII through Fe XVI, many other lines from the entire transition region and corona, and also the Lyman series of He II which produces most of the chromospheric background. The quantitative interpretation of these heliograms is not easy, but they are very useful in providing high resolution solar images that show high temperature forms within the regions of activity and also coronal emission above the limb originating from regions on the back side. The range of intensity exceeds the exposure latitude of the film, therefore exposures must be made for various times. With short exposures one can record just the small hot spots in the plages, and with long exposures coronal emission can be followed far above the limb, in one case to 3 R [21]. It would not be difficult to increase the image size by 5 times, and to reach 1 arc sec resolution by using a longer focal length paraboloid.

The NRL spectroheliograph uses a single spherical diffraction grating instead of a mirror, and forms a spectrum of monochromatic solar images. Fig. 4 shows an example; the images overlap, the extent depending on the dispersion, but in spite of this a great deal can be seen and interpreted. A single aluminum filter is used to eliminate stray light of $\lambda > 800$ Å. The image is quite sharp

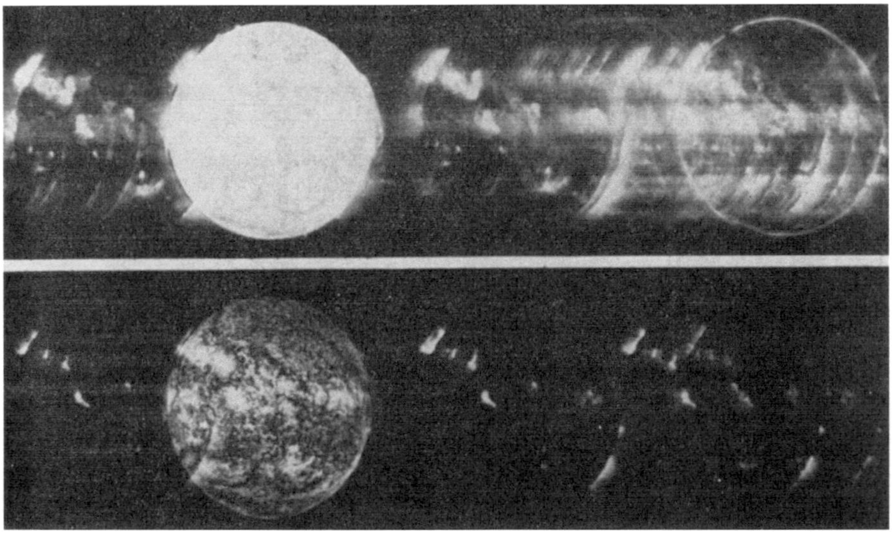

Fig. 4. Sections from two exposures made with the NRL xuv spectroheliograph, flown in an Aerobee-150 rocket on 22 September 1968, the day of the total eclipse in Siberia. The angular resolution is 10–20 arc sec.

on the grating normal, where there is no astigmatism. Spatial resolution of 10 arc sec is easy to achieve with a 1 M radius, 50 cm focal length system, if the rocket pointing system permits. With a 2 M radius grating we have reached about 3 arc sec. The limit with a 4 M grating, which we shall fly in ATM, is about 1–2 arc sec on the normal. Fig. 4 shows two exposures made with a 1 M radius, 3600 line per mm grating on the day of the total eclipse in Siberia. Images in five lines are easily distinguished: In He II 304 Å the entire sun is shown in great detail, especially the chromospheric network. On the right there is an enormous long prominence, and several small prominences are also present. The rest of the emission that is seen above the limb, and in particular most of the emission on the left side is coronal; it is produced not by He II, but by Si XI which lies about 0.5 Å to short wavelengths and is almost blended; also present is emission from Fe XV, 284 Å. There is a large coronal condensation at the upper left, produced by Si XI. At the far right is a prominent ring produced by Mg IX, 368 Å, in the transition region and low corona. In addition to the ring, there is Mg IX emission from active regions and from coronal condensations. The prominences present in He II cause breaks in the Mg IX ring where there is little or no emission. The images of Fe XV, 284 Å, Fe XVI, 335 Å and 361 Å, are easily distinguished, but these coronal lines are recorded only in or above plages, and as coronal emission above the limb produced by plages on the rear side.

The condensation in the upper left of Fig. 4 is of particular interest and an enlargement has been published by Tousey [22]. In He II it is really produced by Si XI, and the 0.5 Å separation makes it seem detached and floating above the limb. In Mg IX the condensation has more structure than in Si XI, and in Fe XV it has less. This increasing diffuseness of condensations and also of plages with increasing temperature is characteristic of increasing altitudes in the solar atmosphere.

Coming out of the center of the He II/Si XI image of the condensation there is a sharp, curved plume; we believe that this is produced by He II, not Si XI. This plume resembles somewhat a feature observed in Lyman-α of H during the 7 March 1970 eclipse, and described by Speer [15]. The condensation itself is very much like one recorded on 7 March 1970 in certain coronal lines [12] and identified and described in [14]. These two condensations look very similar, although the one was recorded in lines of Mg IX, Si XI, and Fe XV that are resonance lines, and the other in lines of Si VIII, Fe XI, Fe XII and S XI that are forbidden transitions.

For the range 1100–2000 Å and greater, an eclipse is required to obtain spectroheliograms like these, because the spectrum is so rich that in the NRL type of xuv spectroheliograph all details would be lost because of the overlapping of images. A conventional Hale-type spectroheliograph could, of course, be flown; but during the limited time of a rocket flight its use would probably be limited to only one or two lines.

From 630 Å to 250 Å, or less, the situation is different, because it is the intense lines that are the most interesting; the overlapping lines are faint, and can be removed by picture processing techniques. However, it is planned to reduce greatly the overlapping of images by limiting the field of view in the direction of dispersion by introducing a grid collimator in front of the grating. If a field of 5 arc min is chosen, the overlap will be reduced by a factor of 6 along the center. The rocket roll position can be chosen to select several of the most interesting solar features present on the particular day, and include them in the 5 arc min wide diametral section imaged in each line.

The most interesting, and certainly the most difficult solar phenomenon to observe in the xuv is the solar flare. High resolution is required, not only angular, but also spectral and in time. To record a flare spectrum with a slit spectrograph is extremely difficult. The NRL spectrograph in ATM will be used for this purpose if there is a suitable flare; pointing is critical, and will be done by the astronaut. Eventually it may be possible to point a spectrograph slit at a flare by ground command, but at present it is not.

Is it possible to obtain an xuv flare spectrum from a rocket? At present, it is not feasible to point the slit of a rocket spectrograph at a flare. Also it is not possible to obtain the first three or four minutes of a flare from a rocket because the rocket requires time to reach altitude after the flare has been detected and a decision to launch is made. But even though the commencement of a flare cannot be recorded, the maximum and later phases can. The NRL xuv spectroheliograph is quite well suited for the purpose because it records the

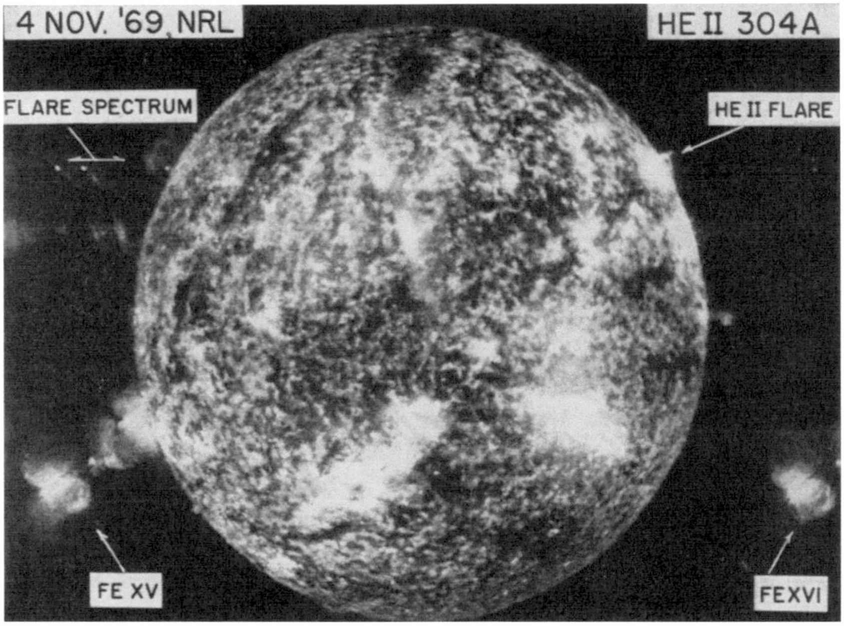

Fig. 5. A spectroheliogram of He II 304 Å, and parts of Fe XV 284 Å and Fe XVI 335 Å obtained by NRL from an Aerobee-150 rocket launched on 4 November 1969 during the course of a flare of importance 2 N. The angular resolution is 5 arc sec. The flare nucleus produces a spectrum of dots.

entire sun and does not require pointing at the flare. In addition the photographic images show the size and shape of the flare in each line. This was, indeed, accomplished [22] by launching the NRL 2M spectroheliograph on 4 November 1969 as soon as possible after a flare was reported by ESSA, Boulder, Colorado. This was the first flight of the 2M instrument, and the images are the best ever obtained. The flare was judged of importance 2N, and the rocket exposures began about four minutes after maximum in Hα.

Fig. 5 was prepared to show He II 304 Å as nearly as possible by itself. In this image, whose original diameter was 9.6 mm, the resolution is about 5 arc sec, about half associated with the instrument, and half with the SPARCS pointing control. In short exposures the resolution approached 3 arc sec. There are present also the strong, overlapping lines Fe XV 284 Å and Fe XVI 335 Å, but in these lines about all that can be seen is the great plage that resembles a butterfly. The flare is very small, and in most lines is just a dot. In He II it is hidden in the limb plage, which is already greatly overexposed. One loop prominence and part of another can be seen above the limb, close to the point where the flare is located.

Fig. 6 shows enlarged sections of the images in He II and Mg IX with the top center located at about 20° E; (a) and (b) contrast the structure in an

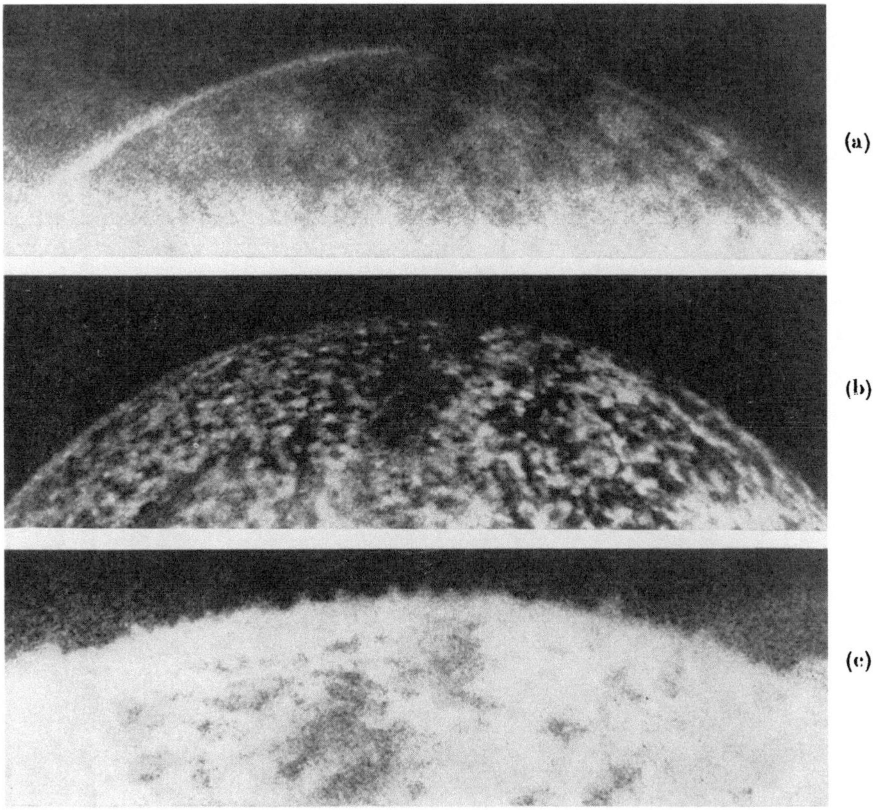

Fig. 6. An enlargement of sections of the 4 November 1969 NRL spectroheliograms showing: (a) Transition region and coronal Mg IX 368 Å. (b) Chromospheric He II 304 Å. (c) Spicules, or clumps of spicules in He II 304 Å, from a long exposure.

optically thick chromospheric line, and in an optically thin line from the transition region and corona. The network, so conspicuous in He II, is absent in Mg IX, but diffuse emission is present in Mg IX that may be associated with the network or activity lower in the atmosphere. The dark region where the emission is quite weak is present in both images. He II has only a faint, narrow bright limb, whereas this feature is exceedingly prominent in Mg IX.

Fig. 6 (c) is a still greater enlargement of Fig. 6 (b) taken from a longer exposure, and shows clumps of spicules about as well as we have been able to

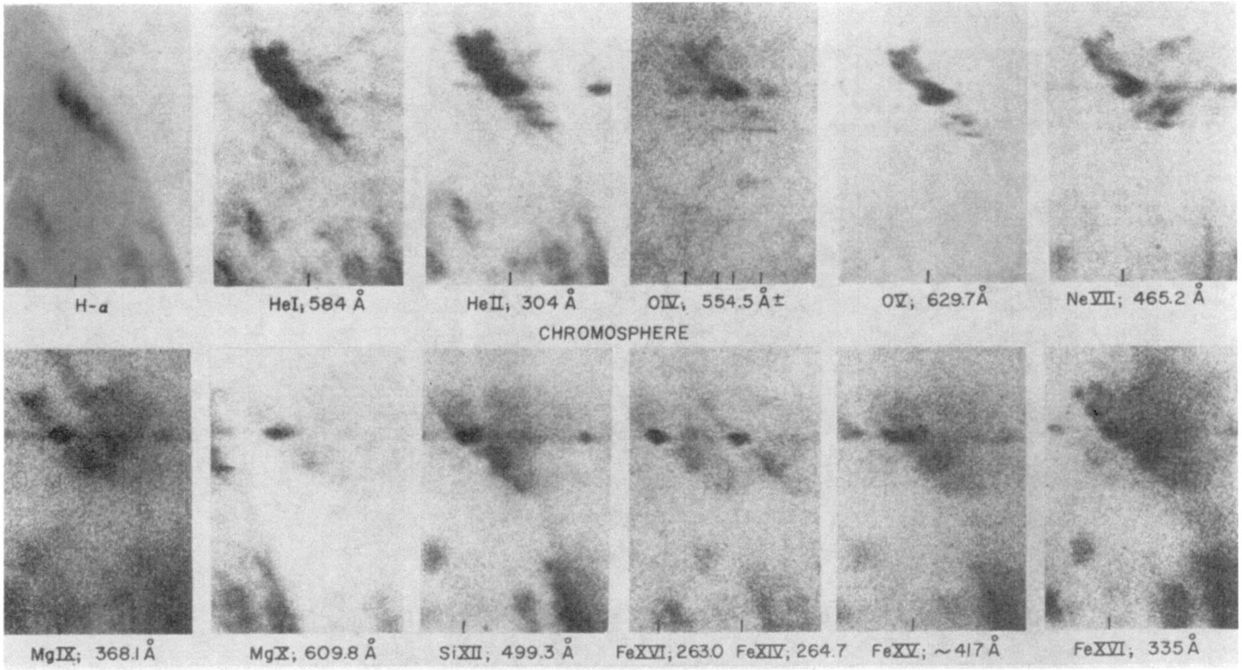

Fig. 7. Images of the 4 November 1969 flare, shown in lines from the chromosphere, through the transition region and into the corona. The enlargements were made from the XUV spectroheliograms of NRL.

resolve them so far. There is no suggestion of the double limb that is characteristic of Hα. It is already within the state of the art, using rockets, to obtain two to three times better spatial resolution than is present in this image.

The flare spectrum was recorded from 171 Å to 630 Å, and shows the form and intensity of the flare in some 430 lines. From this entire spectrum images have been selected in eleven important lines for reproduction in Fig. 7, together with Hα. The arrangement is by increasing height and temperature, the upper set progressing from the chromosphere to the transition region, and the lower from the top of the transition region into the corona. The flare is predominantly ribbon-like in the chromospheric and transition region lines, but included in the structure is the flare nucleus. The ribbon-like emission in the plage becomes blurred by Mg IX and disappears by Mg X, leaving only the nucleus. The approximate apparent size of the nucleus, after correction for pointing instability and the instrucment function, was 6 are sec resolution. One particularly interesting flare image is He II 304 Å, actually from the second order of the grating; alongside to the right is the flare nucleus in Mg X 610 Å. The nucleus in He II was very much overexposed, which resulted in recording on each side an extension to about 0.45 Å along the direction of dispersion. The simple explanation of these extensions is that they are Doppler wings. If so, they correspond to He ions moving at speeds up to 450 km s^{-1} in the line of sight, which was 77° to the sun's vertical at the flare. In the future we plan to obtain flare spectra with greater angular resolution, for flares of different importance, and including flares nearer the center of the disk. We are especially anxious to obtain flare spectra as close as is possible to the commencement.

The final part of my assignment was to say something about the future for very high angular resolution xuv solar research from rockets. I hope I have already convinced you that a vigorous continuing program is essential, not only for its own value, but also as a support for experiments that cannot be conducted from rockets and require some kind of space observatory. The question remaining is, what are the improvements that are required to accomplish more from rockets than I have shown.

First, the rocket system. We still do not have a really satisfactory system. It is a question of time and money. The Aerobee, the principal rocket used in the USA for solar research for about 20 years, has gradually been improved, but not nearly as rapidly or as much as is needed. The present Aerobee-170 has a diameter of only 15 inches, and after installing the guidance, recovery, and telemetry systems, it will carry only about 70–80 pounds of instrumentation to 270 km, the altitude that provides one minute above all significant atmospheric absorption. This payload is not enough. The only rocket now in production that satisfies both the economic and scientific requirements is the Black Brant V. Its diameter is 17 inches, and this extra two inches is extremely useful. It will carry about 250 pounds of instrumentation to about 270 km, which is more reasonable. Other rockets providing similar performance are on the drawing board. The stability of the rocket instrumentation and the ability to point it where one wants are of prime importance. There are various systems and in Table 1 are shown performance data for those that are in general use. Elliott

Table 1

Principal Solar pointing-control systems

Elliott Brothers, UK	
Attitude Control Unit (ACU)	
Lateral stability:	±3 arc sec rms: 10 arc sec peak to peak
Roll stability:	>1 arc min rms; ≈0.25 arc min on limb
Absolute accuracy:	Lateral: 30 arc sec plus 5 arc min/m drift
NASA–ARC–Lockheed	
SPARCS I	
Lateral stability:	2 arc sec peak to peak
Roll stability:	0.15° peak to peak, = 2 arc sec on limb
Absolute accuracy:	Lateral; <15 arc sec initial, plus <15 arc sec total drift (exclusive of instrument)
	Roll: ±5°
SPARCS II	
Lateral stability:	1 arc peak to peak (goal).
Roll stability:	0.15° peak to peak, = 2 arc sec on limb
Absolute accuracy:	Lateral: using SEAS, > ±5 arc sec + >5 arc sec drift including some experiment alignment error
	Roll: ±5°

Brothers, Ltd., made the first. Its current performance permits attaining an angular resolution of considerably better than 10 arc sec, and for short exposures 5 arc sec or better. Absolute pointing and drift are good, but not all that is needed. This is why Culham constructed the auxiliary system mentioned earlier.

The principal system in the USA is the SPARCS, developed by the Ames Research Center and Lockheed. SPARCS I performance permitted attaining 3–4 arc sec angular resolution. Absolute pointing data refer to the sun sensor, not to the instrument, because it is a difficult problem to obtain and hold alignment between the SPARCS sun sensor and the instrument axis. Absolute roll positioning was not as accurate as desired, and was limited by the magnetometer sensing element. SPARCS I is now being replaced by SPARCS II, with lateral stability improved by a factor of two, and there is hope of considerably increased stability at some time in the future. The roll stability has not been improved. Absolute pointing in yaw and pitch can be made accurate to better than ±5 arc sec relative to the instrument axis by using the SEAS solar sensor transfer device, made by Exotech, Inc. Absolute roll, however, has not been improved.

It is obvious that solar pointing systems need more development, and that they can be improved to provide stabilization to tenths of an arc sec. It may be that accurate absolute pointing can best be achieved by command from the ground. This has already been accomplished once, by the Ames Research Center, the Ball Brothers Research Corporation, and NRL during the 13 August 1970 Aerobee-170 flight. NRL and ARC are currently working on a new command system for in-flight control of yaw, pitch and roll, based on an Hα solar slit-plate display that is transmitted to ground by a television system. This would be used in conjunction with SPARCS. An absolutely reliable recovery system for rocket pay-loads is essential. In order to observe flares, at least from the USA, the system must also operate for sea recovery, because only from Wallops Island is there sufficient freedom from launch restrictions to have a reasonable chance of launching during a flare.

My conclusion is that rocket systems should be improved, and can become excellent platforms for xuv solar work in the $^1/_2$–$^1/_4$ arc sec angular resolution range. To obtain these systems soon will require considerably more support than has been available in the past, and this was recommended by last year's Wood's Hole Priorities Study [23].

Finally, is it possible to construct xuv instrumentation that will be able to take advantage of a rocket system that will provide, say, $^1/_4$ arc sec stability? It is, but much work is required in many fields of technology. Much of this effort will contribute very directly to advanced instrumentation for space station observatories. Diffraction is not a limitation, because at 1000 Å, to obtain $^1/_4$ arc sec resolution requires only a 4 inch aperture. But for high angular resolution xuv work it is necessary to use a large mirror to collect enough

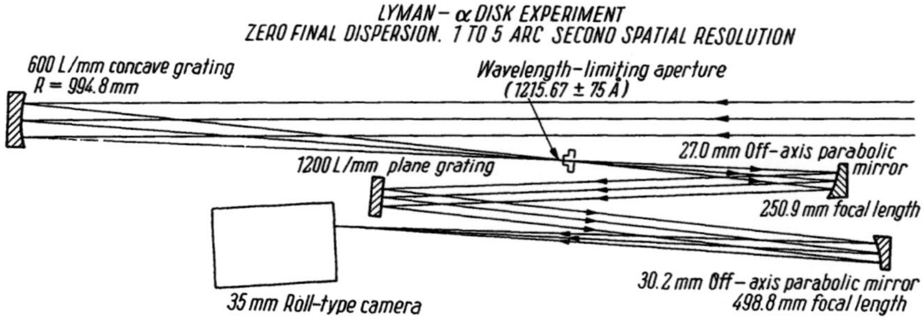

Fig. 8. The high angular resolution H:Lyman-α spectroheliograph of NRL [24]. The angular resolution is 1 arc sec in the central region of the solar disk.

energy. A 12 inch diameter is probably the largest practicable size to use with a 17 inch rocket.

There are several severe technical problems. One is to mount optical elements securely and free from strain. This can certainly be done. Another is the thermal problem; NRL's solution is to use gratings and mirrors of Cer-Vit, and to mount them so that their positions are maintained independent of temperature by means of Invar rods. Improved optical designs are required. NRL has several, in various stages from proposed, to ready-to-fly. Fig. 8 shows a high-resolution spectroheliograph for hydrogen Lyman-α, following a design by Purcell et al. [24]. This instrument has been perfected by Prinz [25] so that it will produce close to 1 arc sec resolution on axis, degrading to some 2–4 arc sec at the sun's limb. A 1 M, 600 line per mm grating is used as predispersor and forms a Lyman-α image on the grating normal. An aperture passes a band about 100 Å wide, centered on Lyman-α. The beam is collimated by an off-axis paraboloid, dispersed again by a plane grating, and reimaged with a second paraboloid at nearly zero dispersion. Results obtained with this instrument in the laboratory

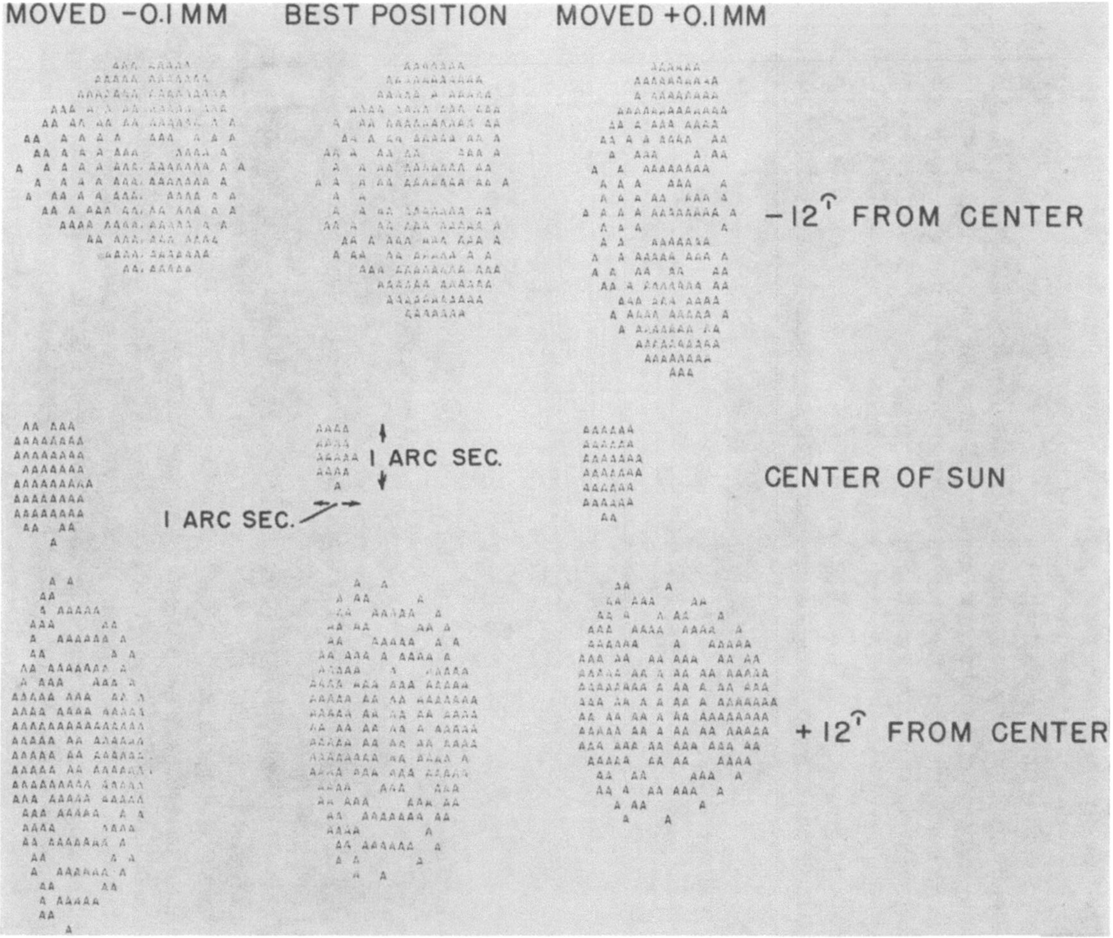

Fig. 9. A ray-tracing printout of the image produced by the Lyman-α spectroheliograph of Fig. 8. The program used 317 parallel rays (from a point source at infinity). Each "A" shows the position where at least one ray intercepted the film. Nine cases are shown, including the on-axis image.

have already resolved a mesh of spacing 5 arc sec between wires of 2 arc sec width, in Lyman-α radiation. Since obtaining these images the resolution has been improved by means of a ray tracing program using 317 rays, developed by Prinz [25]. Fig. 9 shows sample ray tracing results for this instrument. There are shown nine images of a point source for a 4 cm diameter instrument aperture, and with the conditions varied as indicated. Each printed-out A shows where at least one ray intercepted the film. The image at the center is on axis and for the best conditions of focus. The entire image is contained within a 1 arc sec diameter spot. The images to the left and right show the form 0.1 mm in front of and behind the optimum position. The upper and lower sets are for positions 12 arc min from the axis, that is, near the limb. There the image degrades to 2–4 arc sec. In this form of computer printout the density of rays is not shown. The full computation does show the density and gives quite a faithful representation of the image. With this program it is possible

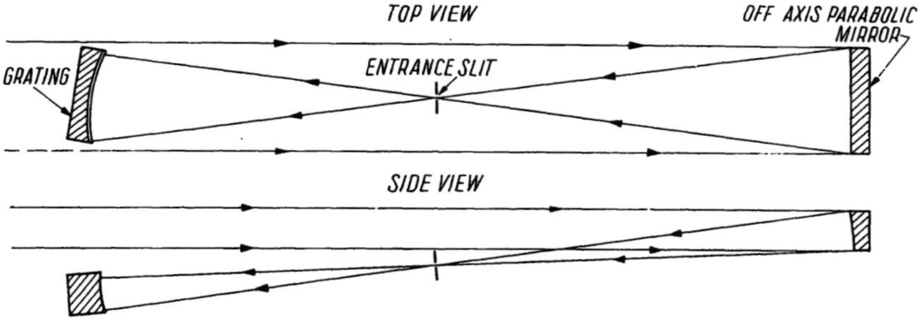

Fig. 10. The optical diagram of the high angular resolution, high spectral resolution photoelectric spectrograph proposed by NRL [26] for OSO-J. The top view is in the plane of the Rowland circle. For OSO-J, channel photomultipliers were placed on either side of the slit and the grating was moved along the Rowland circle by rotation around the circle's center, thus scanning a short section of the spectrum with each photomultiplier.

to vary the various design parameters, and also to adjust the instrument once it is built so as to optimize the performance for a particular observation. For example, one might decide to place a plage 10 arc min in from the limb, or perhaps the limb itself, in best focus, and adjust the instrument to optimize the resolution over a region 10 arc min in diameter.

There are many other optical improvements that can be made in xuv instrumentation, and I mention a few on which NRL is working. Fig. 10 shows a line profile spectrometer having 4 arc sec by 1 arc min angular resolution proposed by Brueckner and Prinz for OSO-J [26]. Solar radiation is focused on a slit with a paraboloid and then passes to a concave grating. It was shown that a gain in performance of a factor of two can be realized by making use of an ellipsoidal grating. The grating produces a spectrum below the slit and with the slit located centrally. In the OSO version, detection is accomplished with photomultipliers placed along the Rowland circle. The instrument could be modified to use photographic recording and flown on a rocket. The point is that the design was carried out by computer. Fig. 11 shows the computer-predicted gain in performance expected for the ellipsoidal grating. The ellipsoidal grating covers with close to a perfect line width about twice the spectral range that can be covered with a spherical grating. To produce as perfect an image, the spherical grating would have to be reduced in width by a factor of two. There remains the question, whether a good ellipsoidal grating can be made. Technological advances may be required. However, Purcell [27] has been able to produce quite satisfactory ellipsoidal, or toroidal gratings by bending, and NRL has flown them in many instruments over the past 12 years.

Another instrument that is just now coming into its own is the echelle spectrograph thanks to great improvements in making echelles that are free from ghosts

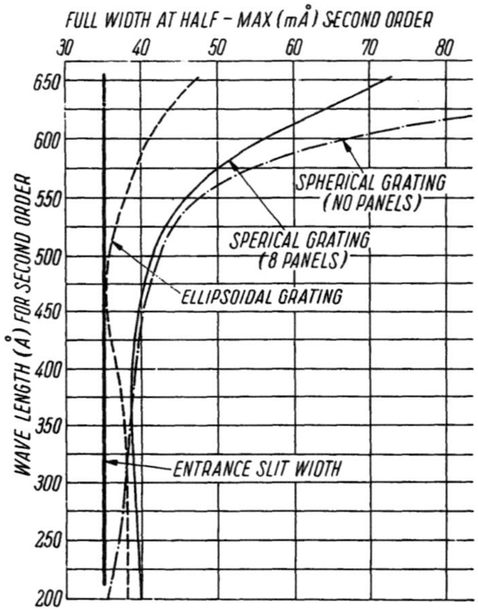

Fig. 11. Curves plotted by ray tracing showing the resolution characteristics for the OSO-J spectrograph of Fig. 10 for three types of diffraction grating. The ellipsoidal grating is far superior to the spherical grating. A small improvement was possible by ruling the spherical grating in a series of panels having slightly changed grating spacing.

and stray light, resulting from work by Harrison at MIT and by Bausch and Lomb. An echelle spectrum obtained recently by the Culham group [28] reached about 1200 Å with a resolving power of the order of 10^5. The angular resolution was 8 arc sec by 3.4 arc min. The slit was held near the sun's center, but flights are planned with the slit near the limb. I am sure that there is much that can be done with echelle grating spectrographs from rockets.

In conclusion, we who are involved in large orbiting observatories realize that their value is great indeed. We do not regret the tremendous effort that is required to instrument them, but this experience has also made us feel very keenly that a strong, progressive rocket program, carried on concurrently, is essential to the entire field of space astronomy, and especially to solar physics.

References

[1] S. R. Pottasch, Space Science Rev. **3**, 316 (1964).
[2] NASA SP-213, NASA, Washington, D.C. (1969).
[3] R. Tousey et al., Space Research IV, 703 (1964).
[4] W. M. Burton and R. Wilson, Nature, Lond. **207**, 61 (1965).
[5] W. M. Burton, A. Ridgeley and R. Wilson, Mon. Not. Roy. Astr. Soc. **135**, 207 (1967).
[6] W. S. Black and D. B. Shenton, The Peaceful Uses of Automation in Outer Space, Plenum Press: New York 1966.
[7] T. A. Chubb et al., Mem. Soc. Roy. Sci. Liège **4**, 228 (1961).
[8] J. E. Blamont and C. Malique, Astron. Astrophys. **3**, 135 (1969).
[9] G. E. Brueckner et al., Nature, Lond. **226**, 1132 (1970).
[10] G. E. Brueckner and K. P. Nicolas, to be published.
[11] J. M. Beckers, Solar Phys. **3**, 367 (1968).
[12] R. J. Speer et al., Nature, Lond. **226**, 249 (1970).
[13] A. H. Gabriel et al., Astrophys. J., **169**, 595 (1971).
[14] A. H. Gabriel and C. Jordan, Presented at the COSPAR Symposium on the Total Eclipse of 7 March 1970, Seattle 1971.
[15] T. L. J. Jones et al. Solar Phys., **21**, 372 (1971).
[16] C. Jordan, Solar Phys., **21**, 381 (1971).
[17] A. H. Gabriel, Solar Phys. **21**, 392 (1971).
[18] J. Houtgast et al., Solar Phys. **21**, 281 (1971).
[19] W. H. Parkinson and E. M. Reeves, Solar Phys., **10**, 342 (1969).
[20] G. E. Brueckner and O. K. Moe, Space Research XII, 1595 (1972).
[21] J. D. Purcell, R. Tousey and M. J. Koomen, Space Research VIII, 450 (1968).
[22] R. Tousey, Phil. Trans. Roy. Soc. Lond. A **270**, 59 (1971).
[23] Priorities for Space Research 1971–1980, Report of a Study conducted by the Space Science Board, Publ. Nat. Acad. Sci. Washington, D.C. (1971).
[24] J. D. Purcell, D. L. Garrett and D. K. Prinz, Naval Research Laboratory, Washington, D.C. (unpublished).
[25] D. K. Prinz, Naval Research Laboratory, Washington, D.C. (unpublished).
[26] G. E. Brueckner and D. K. Prinz, Proposal to NASA, 13 November 1970, Naval Research Laboratory, Washington, D.C. (unpublished).
[27] J. D. Purcell and R. Tousey, J. Opt. Soc. Am. **47**, 1057 (1957).
[28] B. C. Boland, B. B. Jones and S. F. T. Engstrom, Solar Phys. **17**, 333 (1971).

Space Research XII — Akademie-Verlag, Berlin 1972

SOLAR OBSERVATIONS FROM SKYLAB

G. K. Oertel

National Aeronautics and Space Administration, Washington, D.C., USA

Skylab is a manned earth orbiting laboratory (to be launched in April 1973) with a complement of high resolution solar experiments, called the ATM. There will be three manned visits to Skylab with a total duration of about 20 weeks interrupted by additional 12 weeks of unmanned periods. A wide variety of problems in solar physics will be attacked with coordinated observations made by the following instruments on Skylab: a white light coronagraph, a spectrograph for the 950–3900 Å range, a spectrometer–spectroheliometer for the 300–1300 Å range, a spectroheliograph for 150–650 Å, X-ray telescopes, and supporting instrumentation. There will be complementary measurements from the ground and from space with 1″ angular resolution on the sun. The experimenters are currently developing plans for a broad international program of participation in the use of data from ATM. A coordinated program of correlative ground-based observations, sounding rocket launches, and observations from other spacecraft is being developed to provide calibration in orbit and simultaneous observations of solar phenomena over a broad range of the electromagnetic spectrum. The role of the scientist astronaut as an astronomical observer in space will be to recognize and point at targets of opportunity, to diagnose, report on and possibly modify instrument performance, and to retrieve photographic film for transport back to earth.

1. Introduction

The Skylab program has evolved from the Apollo Applications Program (AAP) which consisted of plans for the utilization of backup flight hardware from the Apollo program. The Apollo Telescope Mount (ATM) was originally conceived as an appendage to the Apollo command and service module that would carry a number of solar experiments and be manually pointed at the sun and operated by the crew. The gimballed solar spar was later conceived and adapted to a version of the Lunar Excursion Module in Apollo. It was included in a configuration known as the "wet workshop" (the S IV B stage of the Saturn used as a workshop after the liquid fuels are spent) and later the "dry workshop" (the same but launched without fuel) which is now called the Skylab and combines all active parts of the AAP. One of the primary objectives of the Skylab is to investigate the role of man in a variety of tasks that he can perform in space. The other primary objective is to carry out investigations in the areas of solar physics, earth observations, human physiology, medicine, spacecraft environment, technology, and others.

The Skylab is scheduled for launch on 30 April 1973 (SL 1). Within about one day will follow the first manned visit to the Skylab (SL 2), for a duration of 28 days. After an unmanned period of 60 days, the second visit (SL 3) of 56 days will follow. There will be another unmanned period of 30 days, and a final manned visit (SL 4) of again 56 days duration for a total program length of about 8 months.

2. Description

Most solar experiments on Skylab are located on the ATM. Fig. 1 shows the ATM and its relation to the Skylab cluster. The ATM experiments are

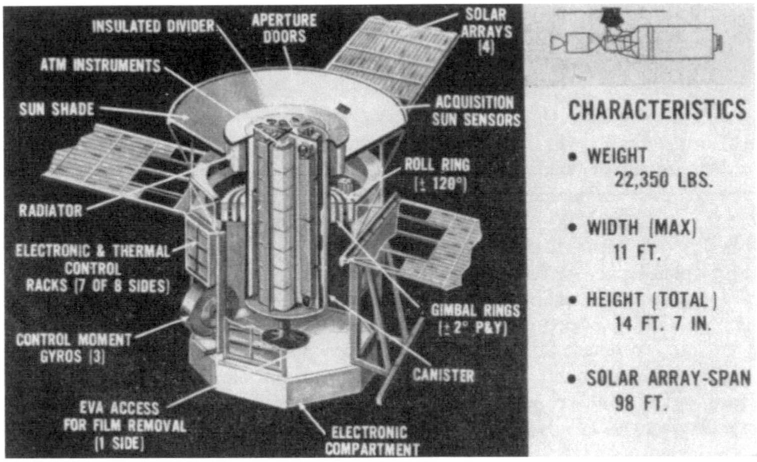

Fig. 1. ATM in relation to Skylab.

co-aligned on a spar and are jointly pointed at the sun during solar observing periods. The spar can be rolled by $\pm 120°$ and controlled in pitch and yaw within $\pm 2°$ each. The pointing accuracy is $\pm 2.5''$ in pitch and yaw, $\pm 10'$ in roll; offset pointing within $\pm 24'$ of sun center can be reached at a slew rate of $80''\ s^{-1}$. The Skylab power system provides an average of 3500 W. The experiments use 300 W, the balance is devoted to thermal control, attitude and pointing control, and other experiment support; Fig. 2 shows the spar layout. The astronauts operate the ATM from a single control and display console in the multiple docking adapter adjacent to ATM which has selector switches, pointing controls, 2 television monitors, and a variety of indicators of experiment status, film usage, and solar conditions.

The investigators and some of the most important parameters of the ATM experiments are given in Table 1. Fig. 3–8 give optical layouts. The following paragraphs describe briefly some of the other important features of the experiments.

2.1. White Light Coronagraph (S-052)

The experiment can operate either in the video mode, for astronaut display or television downlink, or in the photographic mode. The television mode is

Table 1

ATM experiments

Experiment	Investigator	Telescope	Dispersive elements	Detector	Data
S-052	R. MacQueen High Altitude Observatory Boulder, CO	Externally occulted coronagraph	Polarizers	Photographic film and TV	Coronagrams, white light, polarization
S-054	R. Giacconi, AS & E Carleton St. Cambridge, MA	X-ray telescope	Filters, transmission grating	Photographic film and TV	X-ray filtergrams
S-055	L. Goldberg Harvard University Cambridge, MA	Off-axis paraboloid	Normal incidence grating	7 channeltrons	Spectra, up to seven simultaneous spectroheliograms
S-056	J. Milligan MSFC Huntsville, AL	X-ray telescope	Filters	Photographic film	X-ray filtergrams
S-082A	R. Tousey Naval Research Lab. Washington, D.C.	Same as dispersive element	Normal incidence concave grating	Photographic film	Partially overlapping spectroheliograms
S-082B	R. Tousey as above	Off-axis paraboloid	2 predispersers plus normal incidence grating	Photographic	High resolution spectra

Experiment	HAO	AS & E	HCO	NASA	NRL-A	NRL-B
Spectral range (Å)	4000–6000	3–60	300–1400	3–60	150–650	970–3940
Spectral resolution (Å)	White light	up to 0.2	1.2	coarse	up to 0.1	0.08–0.16
Field of view	3.5°	1°	5′×5′ through internal raster	1°	1°	2″×60″
Angular resolution	8″(TV: 15″)	2″–5″	5″×5″	5″	1″–5″	2″×60″
Time resolution (sec)	3–27	1–250	0.04–330	0.3–320	0.25–480	0.15–2000

useful for monitoring, alignment adjustments and diagnostics. Either unpolarized light or one of three orientations of plane polarized light is recorded on film. The three planes of polarization are inclined at 120° with respect to each other.

2.2. X-Ray Telescope (S-054)

A two element paraboloid–hyperboloid grazing incidence mirror assembly consists of three nested mirrors which form a single image on the photographic film. The filter wheel has one open position and five filters (Be(3), CF_2, C_8H_8) for broad-band transmission between 3 and 60 Å. The critical angle for grazing

G. K. OERTEL

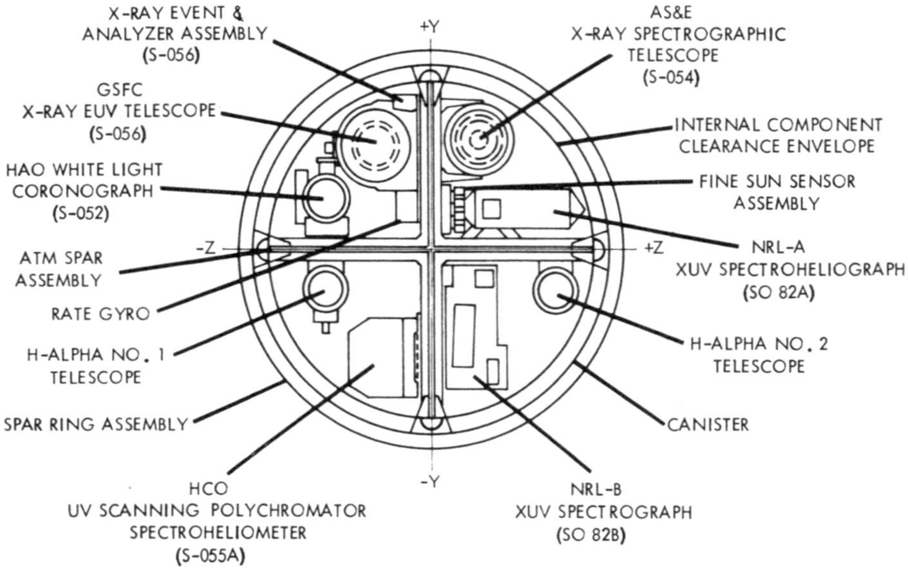

Fig. 2. ATM experiment layout, sun-facing side.

incidence limits transmission to above about 3 Å. A small X-ray imaging system in the center of the main telescope provides a monitor for the console television screen. A transmission grating can be put in the optical path to provide spectroscopy on the brighter X-ray features.

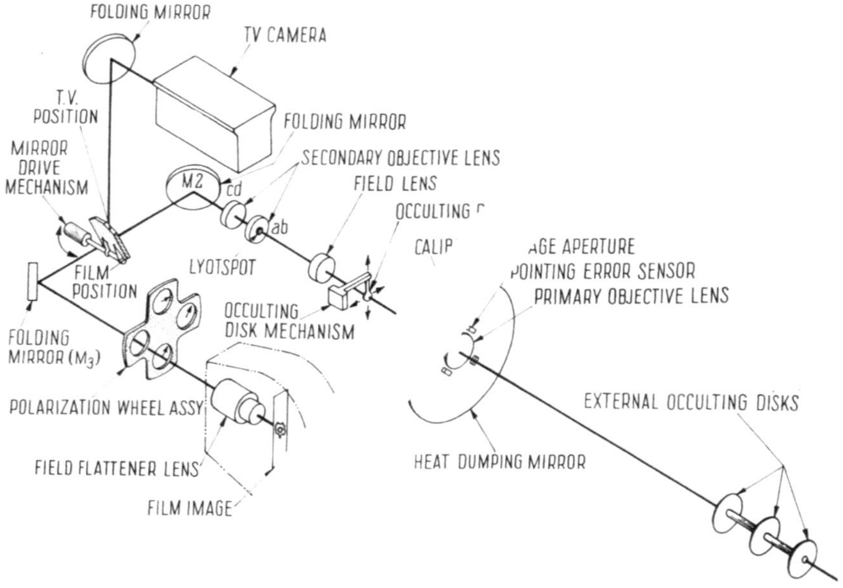

Fig. 3. White light coronagraph (S-052).

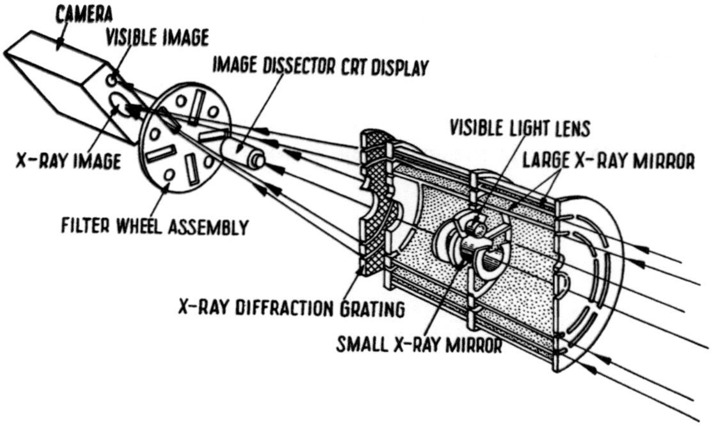

Fig. 4. X-ray telescope (S-054).

2.3. X-Ray Telescope (S-056)

A two-element paraboloid–hyperboloid grazing incidence mirror forms an image of the sun in X-rays on photographic film. A filter wheel with five filters (Be(2), Ti, Al(2)) and one neutral density filter provides broad spectral selection. Proportional counters for the 2.5–20 Å range provide ten-channel spectral information for use by the crew.

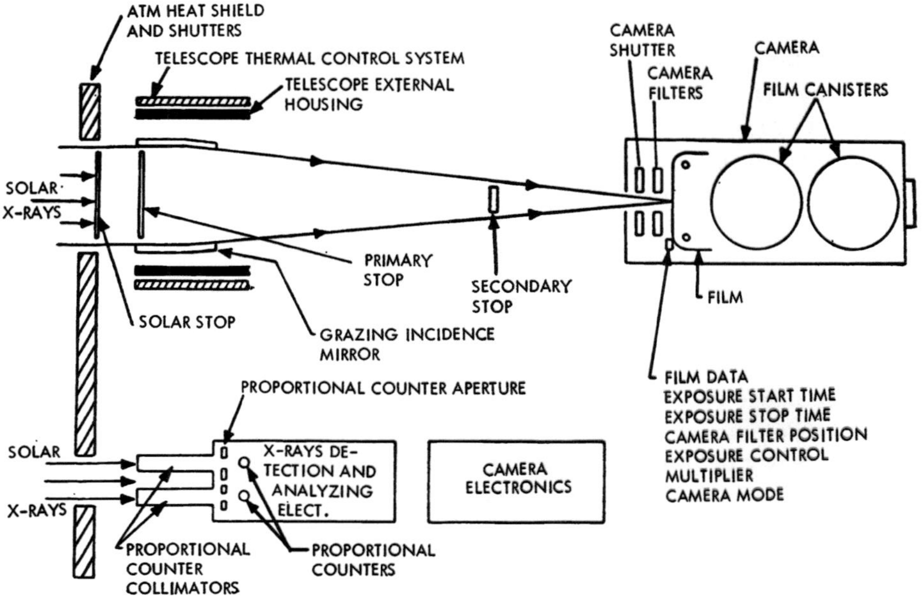

Fig. 5. X-ray telescope (S-056).

2.4. UV Scanning Polychromator/Spectroheliometer (S-055)

An off-axis paraboloid forms a solar image on a 56 μm $\times$ 56 μm entrance slit, corresponding to 5″ $\times$ 5″. A bi-axial drive on the mirror permits raster scanning of a 5′ $\times$ 5′ field of view. A 40 mm wide grating, ruled in gold with 1800 grooves per mm diffracts the light in the spectrometer to seven primary exit slit–channeltron detector assemblies. In the polychromater mode, these view simultaneously the lines of O IV (554.6 Å), Mg X (625.3), the Lyman continuum (896), C III (997.8), O VI (1032), H I (1216), and C II (1336). Alternatively, the grating can be rotated to scan the spectrum in 0.2 Å steps. A zero order detector provides an optical reference within the spectrometer.

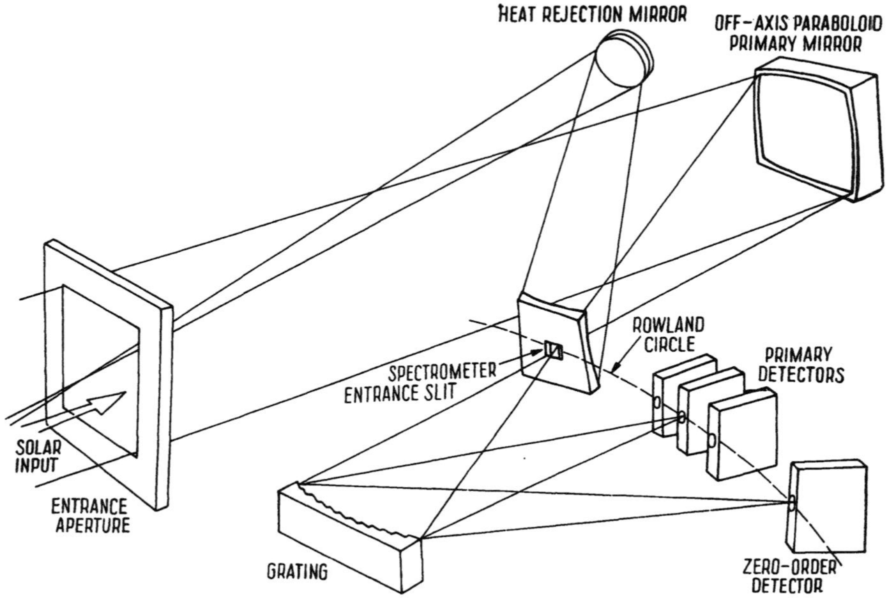

Fig. 6. Polychromator and spectroheliometer (S-055).

2.5. XUV Spectroheliograph (S-082A)

Image formation and dispersion are attained by a single concave Cer-Vit blank ruled in gold with 3600 grooves/mm over a 12 cm $\times$ 12 cm area. The focal length is 2 m. Overlapping solar images (18.6 mm) are formed on photographic film, one image for every solar emission line. The two spectral ranges 150–335 and 321–625 Å are given by two different angular positions of the grating. Stray light is greatly reduced by a free standing thin film Al filter about 3 cm before the film. There are 200 film strips per camera load.

2.6. UV Spectrograph (S-082B)

A 1 m focal length off-axis paraboloid forms a solar image on the entrance slit of the spectrograph. The reflecting slit jaws are viewed by a white light

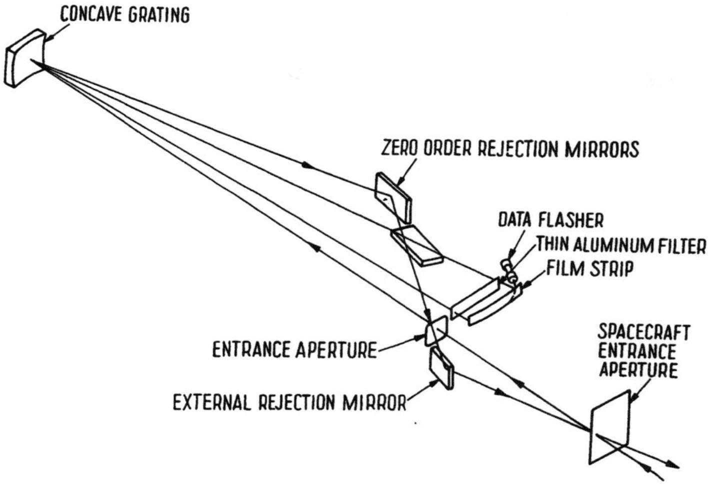

Fig. 7. XUV spectroheliograph (S-082A).

pointing reference television system for positive pointing verification. Inside the spectrograph a predisperser grating assembly with two gratings for the 970–1970 and 1940–3940 Å ranges, respectively, and with 300–150 lines/mm, respectively, serves to disperse stray light out of the film-grating plane. The main grating with a 76 mm × 152 mm ruled area at 600 grooves/mm is acting as the imaging element to form the spectrum on strips of photographic film. It is used in 2nd and 1st orders for the two spectral ranges, respectively, the other order is eliminated by the predisperser grating and main entrance slit. There are eight exposures per film strip and 200 strips per camera.

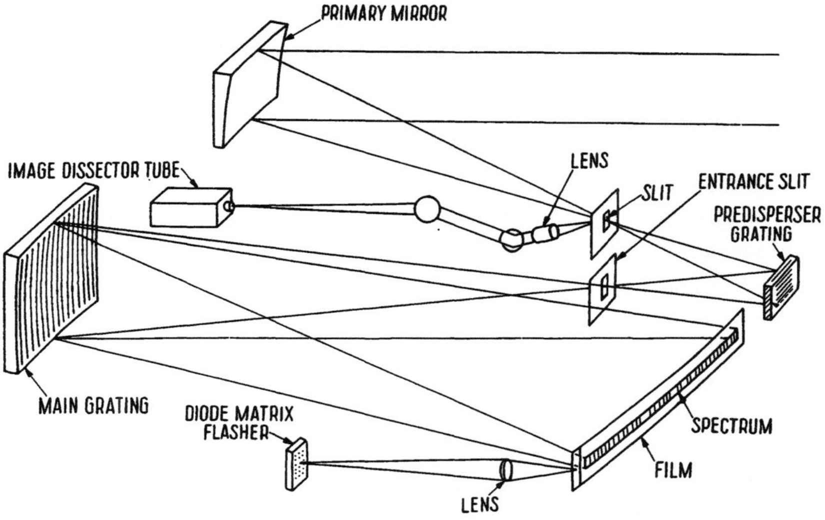

Fig. 8. UV spectrograph (S-082B).

2.7. XUV Monitor

The astronaut can view the sun in xuv with a broadband 170–550 Å telescope with 10–20″ resolution over a field of view of 60′.

2.8. H-alpha

Two H-alpha telescopes with angular resolution 1–5″ and a field of view between 4.4′ and 35′ are provided as an aid to the astronaut and as a record of solar conditions during ATM observing periods. From 1 to 4 pictures/minute can be recorded on photographic film, with telescope No. 1 only. Both telescopes can provide television displays with zoom capability for the astronaut. The telescopes have independently movable mechanical reticles which are to be co-aligned with the S-055 and S-082B experiments, respectively, to provide a visual pointing reference for the field limiting slits of these experiments. The 17 cm aperture f-28 Cassegrain telescope feeds parallel light to a Fabry–Perot filter with interference prefilter. The Fabry–Perot filter is a solid unit, with reflective coatings deposited on opposing surfaces of a single optical flat with precisely made parallel surfaces and thickness. The bandpass is 0.7 Å. The solid Fabry–Perot technology was developed specifically for the ATM telescopes.

3. Operation

The astronaut will use the different displays from H-alpha and the white light corona to xuv and X-rays to recognize targets of opportunity on the sun and to point the ATM at them. He can select and initiate instrument modes and monitor the performance. He can align or correct the performance of some instruments, and he can diagnose and correct problems within the ATM capabilities. He will retrieve the cameras with the exposed film and will replace them with new cameras when appropriate during the mission, and he will return cameras and film to earth with him. The astronauts are undergoing extensive training in solar physics in addition to their normal activities.

The objective of the ATM is the study of solar physics. Accordingly, the observing programs which will be carried out during the Skylab operation do not emphasize techniques or individual experiments, but solar physics problems and a coordinated attack on them. The ATM investigators have identified these major Joint Observing Programs (JOP):

(i) Chromospheric network and supergranulation
(ii) Active regions, morphology and development
(iii) Solar flares
(iv) Prominences and filaments
(v) Center-to-limb studies of the "quiet" sun
(vi) Synoptic observations of longer lived phenomena

In addition, there are JOPs for earth atmospheric absorption, solar coronal transients, solar wind, lunar libration points, solar 300-second oscillations, rocket calibration, velocities in active regions, eclipse observations, and stellar observations. Each JOP consists of the execution of one or several "building blocks", sequences of operations of the individual experiments complete with pointing requirements, data rates, and instrument settings of all kinds. It will remain possible to modify JOPs as new information about the sun becomes available.

As an example, consider the JOP for the chromospheric network in more detail. To understand the physics of this aspect of the chromosphere it will be necessary to determine the three-dimensional structure of the network and the physical conditions — temperature, density, magnetic and velocity fields — as functions of space and, for evolution, of time. The bright points on network boundaries, those seen in X-rays and those in off-band H-alpha, deserve special attention. The ATM experiments will attack this problem by the observation of the network in a variety of spectral emissions formed at different temperatures and heights. Three-dimensional structure can be studied with the X-ray telescopes and the far uv spectroheliograms of S-055 and S-082A. The temperature and density will be derived from uv and xuv spectra from S-082B, S-055, and S-082A and from X-ray filtergrams. Velocities can be measured with the spectrograph (S-082B) and possibly with the spectrometer (S-055) by using slit jaws to divide spectral lines. The X-ray telescopes will show hot spots and will help map the coronal magnetic field. Synoptic observations will show the evolution of the network and its coronal extensions (S-052, S-054, and S-056).

In the JOP, the spectrograph slit is pointed at various positions across a selected prominent supergranular cell, including one pointing along a boundary and one pointing completely inside a cell. The spectrometer (S-055) performs spectral scans at various positions across the cell and the boundary; it also records spectroheliograms across the cell simultaneously in seven wavelengths. The spectroheliograph (S-082A) and the X-ray telescopes have fields of view which cover the sun in its entirety. In studying a bright spot the spectrograph and spectrometer slits are pointed at the spot and the same series of observations is carried out. Finally, synoptic observations over the life of a cell exhibit its evolution.

The ATM investigators, as a group, will speak with a single voice in selecting the ATM observing program during the time slots which will be allocated to ATM operations. A computer will inform them in real time about the options which are open to them after mission constraints are taken into account. The computer will also suggest a particular observing program which is optimized according to prior and current inputs and list alternate, less "optimal", observing programs. After mission constraints are thus taken into account, the final decision will therefore always be made by the ATM investigators.

Unattended or unmanned operation is possible for some ATM experiments, notably the coronagraph (S-052) and the spectrometer (S-055).

4. Supporting Activities

The ATM experiments have undergone considerable testing during their development. In addition to developmental and prototype unit construction, there has been a design verification program with ground-based, balloon, and rocket techniques. For example, half scale models of the S-082 experiments were flown on sounding rockets and have produced spectacular sample results [1, 2]. A verification flight of the coronagraph on a balloon was similarly successful. The Skylab program contributed to the flight of X-ray telescopes on rockets which have produced X-ray filtergrams of typical ATM quality [3].

Absolute photometric calibration and the determination of background will be essential for many interpretations of the ATM data. An extensive calibration program in progress includes laboratory testing and calibration of the prototype and flight instruments, testing at ground-based solar observatories (H-alpha), laboratory research into uv and xuv calibration techniques, plans for intercomparison of ATM experiments during flight, and in-flight calibration with sounding rockets. The sounding rockets and ATM will simultaneously (or nearly so) observe the same area on the sun. To insure comparability, the sounding rocket payloads will have substantially the same optical performance as the corresponding ATM instrumentation and they will be calibrated before and after flight. Current plans call for the calibration of the uv instrumentation once during each manned visit of Skylab, with a total of six sounding rocket flights.

5. Guest Investigators

The problem-oriented approach to solar observations is not limited to ATM. Simultaneous solar observations are desirable over a wide spectral range from the ground, balloons, sounding rockets, and other satellites. As a first step towards initiating a guest investigator program the ATM investigators have defined the JOPs and therefore the solar problems which are most effectively studied with ATM, and many of the specific supporting observations which might be needed. Magnetography, and other radio, infrared, visible, and near uv observations from the ground, and X-ray and gamma-ray spectroscopy from sounding rockets and balloons seem most important. Coordinated observations from space and from the ground will multiply the solar physics return from both activities. The ATM investigators (Table 1) encourage offers of joint investigations with astronomers and physicists in the USA and abroad on the basis of mutual interest. NASA has given exclusive rights for the analysis and interpretation of the ATM data to the ATM investigators for a period of one year; however, joint investigations and international cooperation are encouraged by NASA.

During a partial or possibly a brief total solar eclipse it may become possible to carry out unique ATM experiment operations. The blocking of light from the solar disk leads to a reduction of stray light and the determination of background levels. The earth-lit moon in the field of view of the coronagraph helps to provide absolute photometric calibration. Source confusion in the spectroheliograph (S-082A) is reduced, and occultation can lead to greater angular resolution than the experiments are normally capable of. Most importantly, however, ATM observations near the time of a total solar eclipse will benefit from the unique ground-based solar observations which are made concurrently and may be available for intercomparison.

Recognizing the great importance of simultaneous ground-based measurements, and to help form a nucleus for the guest investigator program which the ATM investigators are initiating, NASA has established the Skylab Ground-based Astronomy Program. It is the primary intent of this program to support improvements in existing ground-based instrumentation the better to support the scientific objectives of ATM. After an announcement of this opportunity, NASA has selected 15 efforts which involve four countries in addition to the USA. For example, several forbidden coronal lines will be observed simultaneously to provide information about the ground-level population of some of the same coronal ions which ATM will observe in excited levels; vector magnetic field and line profile measurements from the ground will support virtually every solar objective of the ATM; collimated X-ray spectroscopy from rockets and balloons will provide important supplementary information for the interpretation of the X-ray filtergrams from ATM, etc.

6. Conclusion

We have had an occasional glimpse at the sun with the instrumental capability of the ATM experiments and know that there is a great deal to be learned. ATM will provide the opportunity to separate the ordinary from the extraordinary, to study the sun with a complement of instruments in a problem oriented way, and to follow the evolution of solar features. We may also expect to discover new aspects of the sun. The opportunity to see the data from the first Skylab observing period before the second and third missions will make it possible to close the cycle of observation, interpretation, and new observation, and to follow up on the unexpected that might be found. It is to be expected that Skylab will be a highlight of activity in solar observations of all kinds, similar to a solar eclipse, and that major advances in our understanding of the sun will result for many years after the mission.

Acknowledgments

Comments and contributions by the ATM investigators, S. Levine of Bellcomm, and by D. L. Forsythe, NASA ATM Program Manager, and scientist astronaut O. K. Garriott are gratefully, acknowledged. Mrs. T. Wise's help in the preparation of the manuscript is especially appreciated.

References

[1] R. Tousey, Space Research XII, 1719 (1972).
[2] G. E. Brueckner, paper presented at COSPAR meeting, Seattle 1971.
[3] G. Vaiana, paper presented at COSPAR meeting, Seattle 1971.

Space Research XII — Akademie-Verlag, Berlin 1972

HIGH RESOLUTION SOLAR OBSERVATIONS FROM SPACE

H. Zirin

Hale Observatories, California Institute of Technology,
Carnegie Institution of Washington, Pasadena, Calif., USA

Present ground-based solar observations are limited by the transmission properties of the terrestrial atmosphere, by the design of instruments and the choice of observing locations. Recent improvements in the latter two have made possible observations with MTF around 7% at $^1/_2$ arc sec and 15% at 1 arc sec. These have revealed remarkable new details of structure, particularly in the chromosphere which lead to new understanding of the magnetic field configurations on the sun. However, the instruments now appear to have been developed as far as the terrestrial atmosphere will permit, and new information will only be determined by space observations. For example, observations from space with a 65 cm telescope should permit a 10% MTF at 0.2 arc sec, and 25% at 0.5 arc sec, a great improvement over the value from the ground. The gain is much greater than one would guess from quoting "limiting resolutions" which give no quantitative value for the modulation transfer function. Larger telescopes would permit corresponding gains.

The problems of construction, alignment and thermal control of a very high resolution solar telescope in space are discussed together with various optical systems. The problems of solar physics in which we would expect new information are also discussed, on the basis of recent high-resolution observations from the ground.

Atmospheric disturbance provides far more serious problems for solar astronomers than for their colleagues in night-time astronomy. The heating effects of the sun on the ground produce serious convection, wind circulation and other difficulties. In addition, variable solar heating effects around domes which are required to reduce wind loading are an additional problem. Although one may obtain $^1/_2$ arc sec seeing over columns as great as 200 inches across at night, even with the smaller apertures used by solar astronomers such seeing is very rare. The possibility of extremely high resolution observations from space holds great promise for the solar astronomer. Furthermore, although extremely good seeing will still not produce disks or shapes for the stellar astronomer in most of his problems, the solar astronomer can always depend on seeing more detail when he goes to finer resolution. In almost every solar phenomenon we observe we wish we could see closer and more clearly the phenomena on that beautiful star that makes our lives possible. The rewards of higher resolution on the sun will be great. We have a beautiful panorama of physical phenomena spread before us every morning — sunspots, granules, plasma instabilities, particle acceleration, relativistic effects, the grand pattern of the spot cycle, the possible oblateness. Every one of these phenomena can be studied, not as inferred from the radiations of a distant point, but in two or three dimensions. In case after case, new physical ideas can best be tested and understood on the sun.

In recent years there have been great strides in the quality of solar observations from the ground. New observatories were located in places where the seeing was better; new telescopes were built which better exploited the seeing in existing sites. Great strides have been made with vacuum telescopes which at least minimize the seeing effects inside the telescope. Each one of these steps has revealed new and fascinating structure, and in many cases helped explain problems which confused us in the past. For example, boundaries between magnetic field regions which were thought to be single dark filaments turned out to be a multitude of fine fibrils at right angles to the dark line. Spicules have been found to be much more complicated phenomena than earlier thought. With rather small improvements we were able to recognize the appearance of emerging flux regions, the first mark of sunspots. The progress has come from higher resolution, longer time sequences of cinematography and spectroscopy and of course the availability of a new range of observations in solar ultraviolet, extreme ultraviolet and X-rays. In every field of astronomy we have found that greater resolution leads to new knowledge, and there is little reason this should be different in the sun.

The true limitations of present ground observations are not generally appreciated. Most observatories are located in poor sites, because the quality of seeing was rarely considered in locating them. Examination of the average patrol picture in *Solar Physics* reveals an abundance of really bad pictures with less than 5 arc sec. The flare patrols operated around the world are mediocre, and all kinds of false deductions on the relation between Hα and other flare emissions have come from bad optical evaluations of flares. Of course these problems could be fixed without launching the flare patrols into orbit, but even for those in good sites, truly high quality seeing is rare. In order to follow the evolution of solar phenomena, it is not only necessary to obtain single pictures of high resolution but to register them very accurately by means of good guiding. Whereas it is easy to obtain single frames of $^1/_2$ arc sec, seeing variations across the disk of the sun often make it impossible to track the sun to an accuracy of better than a few seconds. Thus variations over the field make it extremely difficult to study the evolution of very fine features, and particularly hard to make sequential superpositions of pictures such as is required for some versions of Leighton's method. One would like ideally to obtain a sequence of pictures of an individual region, and then cancel the first picture against the subsequent ones in order to determine the time variation of the autocorrelation coefficient and to find the time behavior of the various phenomena. This is impossible if the successive pictures do not register exactly, and even though the individual frames may be very good, they cannot be superposed reliably.

It is interesting that the development of truly high resolution ground-based systems has not made high resolution space telescopes unnecessary, but, on the contrary, whetted our appetite by showing the marvelous phenomena that high resolution would reveal.

Fig. 1 shows a fine picture from the Big Bear Solar Observatory. Similar pictures have been made with the new tower at Sacramento Peak Observatory. These pictures only look good because so many of their predecessors were bad. But even if we examine such a "state of the art" photograph as Fig. 1 we see that the resolution is not really very good. There are a few dark features which indeed have a width of less than $^1/_2$ arc sec. But one cannot find any two dark features separated by less than 1 arc sec. And there are a large number of areas where the contrast on the sun was relatively low and therefore the resolution is more like 2 arc sec. The reason for this is the role of the modulation transfer

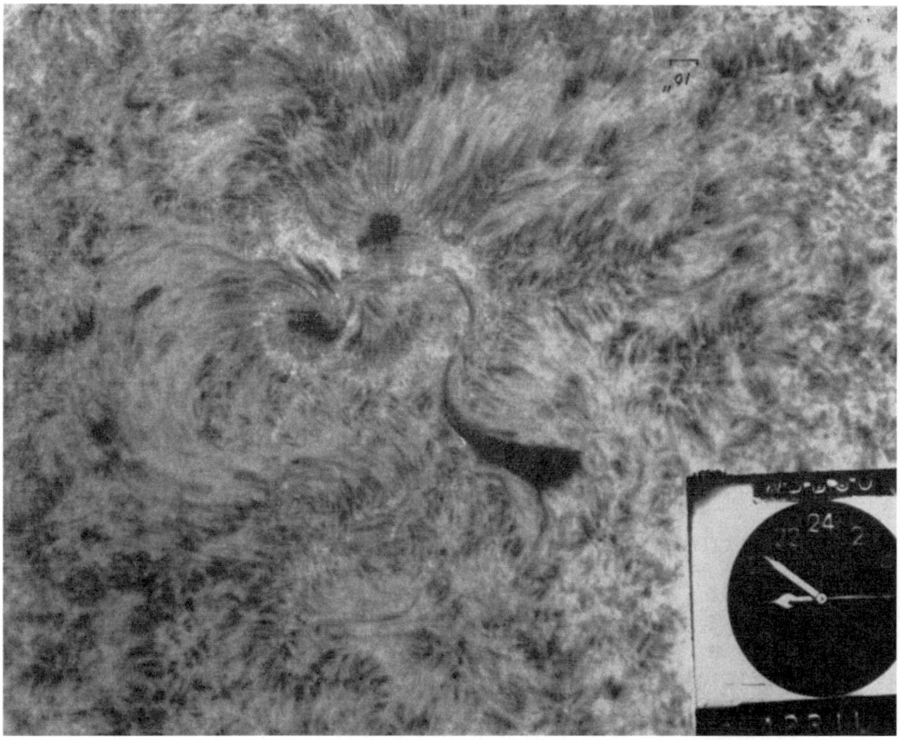

Fig. 1. Hα photo (off line) of a spot group 2 April 1971, at Big Bear Solar Observatory. Ten arc seconds is marked.

function (MTF), a concept well known to optical designers and not well known to astronomers. When we say that a system "resolves" 1 arc sec, we usually mean this for a signal of infinite contrast; more accurately, we mean that if the original signal had 100% modulation at a spatial frequency of 1 arc sec, then the system would produce a 10% modulation at this frequency, which still could be detected by the eye or a detector. Typically, the Rayleigh limit for an optical system corresponds to a modulation of 10%, so that if the contrast in the original image is 20% for spatial frequencies of 1 arc sec, only 2% will be transmitted at the Rayleigh limit, a virtually undetectable difference. Thus those regions of this photograph where the contrast is low have had their modulation reduced to a negligible amount and only areas of high contrast can be resolved in fine detail. Of course the picture shown here represents the product not only of the MTF of the telescope but of the atmosphere and of the film. The film used, Kodak SO-392, is an excellent emulsion for Hα work. Although it is commonly said to resolve 200 line pairs per millimeter (which it does for 100% contrast), it actually produces an MTF of only 65% at 30 line pairs per millimeter. Each one of these factors represents a loss and the loss exists at every spatial frequency, not only the highest.

These matters are well illustrated in Fig. 2. This complicated looking graph was prepared by Ball Brothers Research Corporation in a study of the Caltech 65 cm photoheliograph; the two curves descending from left to right show the modulation produced by a 65 cm diffraction limited system at spatial frequencies

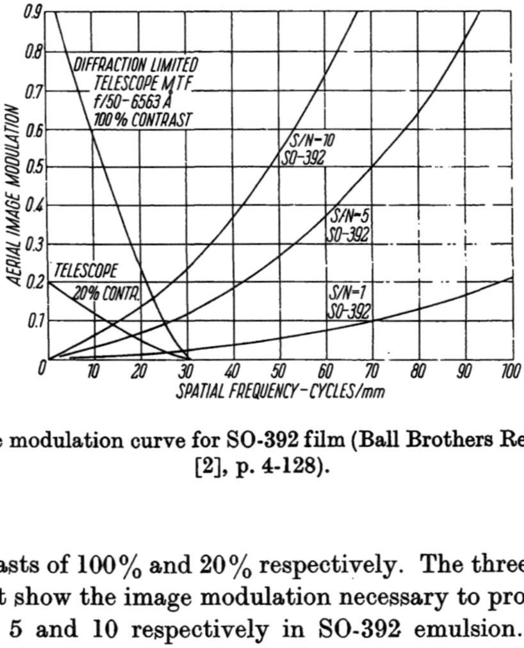

Fig. 2. Aerial image modulation curve for SO-392 film (Ball Brothers Research Corporation [2], p. 4-128).

for original contrasts of 100% and 20% respectively. The three curves descending from the right show the image modulation necessary to produce a signal-to-noise ratio of 1, 5 and 10 respectively in SO-392 emulsion. It is generally considered that a signal-to-noise ratio of 5 is required for reliable photographic detection. We note that this is achievable for frequencies up to 25 cycles per millimeter for objects of 100% contrast (which might be expected for filaments and some other dark solar features) but only permits 18 cycles per millimeter for objects of 20% contrast which seems to correspond more closely to the typical variations on the solar surface. For the telescope focal length of 32.5 m, a peak-to-valley resolution (i.e. a half cycle) of 0.2 arc sec required 15.4 cycles per millimeter or 2.5 full cycles per arc sec. We see from this that all statements on the resolution of a system must specify the contrast of the original signal at a spatial frequency and the signal-to-noise ratio of the detector; simple claims of "resolution of one half arc sec" are no more dependable than the horse power ratings some automobile manufacturers give their product.

The photograph in Fig. 1 may well resolve 0.7 arc sec for 100% contrast, but it only resolves about 1.2 or 1.4 arc sec for 20% contrast. Even if we improve the performance of such a system by a factor of two in the next few years (which may be possible in view of the strides of the last few years) we still achieve a resolution of only 0.7 arc sec at a contrast of 20%. This is a factor three worse than the possible performance of such a system in space and is based only on the best pictures; we indeed have many such pictures, but seeing of this

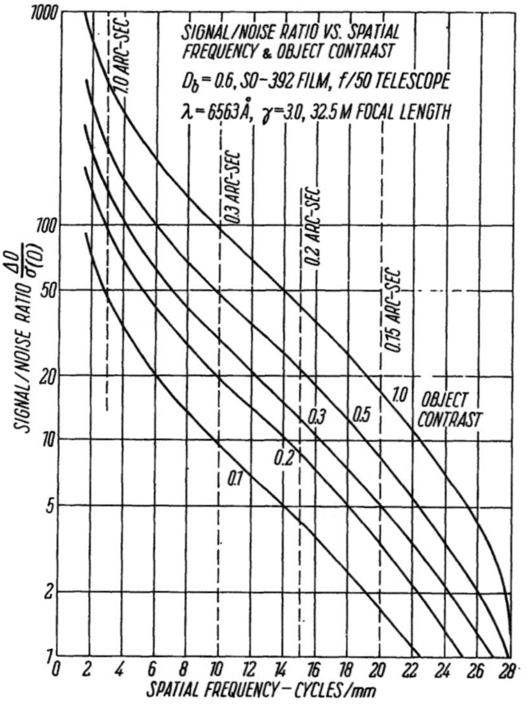

Fig. 3. Signal/noise ratio as a function of frequency and contrast (Ball Brothers Research Corporation [2], p. 4-131).

quality is only present about 30% of daylight hours at our observatory and probably less at others.

What shall we see when high resolution photographs from space are possible, when the new stratoscope will reveal to us the structure of the sun? The scale height in the upper photosphere and low chromosphere are of the order of 100 km, thus one might expect most density variations and structures to be more or less of this size. However, we have found that the magnetic field plays a strong role in ordering the structure of the chromosphere. As one may see from Fig. 1 in regions of enhanced magnetic fields such as plages and sunspot umbrae one finds the finest and smallest structures in the chromosphere, smaller even than the photospheric granulation. We do not have good calculations, but one might guess that the structures in the regions of strong magnetic fields would be even smaller than the 100 km scale height. Although the structures in the quiet chromosphere are somewhat larger, or perhaps of lower contrast (since low contrast features cannot be so well resolved) we must emphasize that at the present we cannot clearly see what these features are. We must again realize that the requirement of "resolution" comes from the days when observers wished to separate double stars. If we replace the single picture element on a point with two (doubling the resolution), we now only get four picture elements in two dimensions. But four data points really tell us very little – you can see that the limit must be 3 or 4 times smaller than the smallest feature. Fig. 3 is another illustration of the signal-to-noise ratio obtained as a function of spatial frequency and object contrast, using SO-392 film. It is seen that for objects of contrast 30% or greater features down to 0.15 arc sec may be detected with a signal-to-noise ratio of five, and objects of 100% contrast, such as filaments and particularly dark fibrils might be detected down to a tenth of an arc second.

In magnetic regions, the aligning force of the magnetic field produces redundant elongated structures which of course are much easier to resolve. The resolution problem is especially critical in solar flares. There tremendous brightness variations in a short period of time place severe requirements on the time resolution and photometric latitude of the system. We have observed many flares, and even though they have a habit of appearing when the seeing is bad, we have seen some with very good seeing. But we cannot say with any accuracy what is going on in these flares, we can only guess by induction from the surrounding large scale structure. Current theories of magnetic field reconnection indicate that they must take place on an extremely small scale in order to occur in the observed time scale. Observations of hard X-ray bursts similarly suggest energy release on a very small scale.

Solar spectroscopists have similarly been focusing on smaller and smaller details on the surface. The spectrum varies from point to point with variations in temperature, pressure, and velocity field. Many of the inconsistencies in earlier models based on the gross characteristics of the atmosphere have been removed by study of the fine scale properties of the spectrum. Most readers will be familiar with the "wiggly line" spectra obtained at the various solar observatories, and the remarkable 250 s oscillation which seems to permeate almost all regions of the atmosphere. The spectroscopic features now being followed are at the very limit of resolution, and even if we did not find any finer structures, there would be an enormous benefit from a better delineation of the features in the 1 arc sec range.

The photospheric granulation has always presented a challenge to high resolution studies. Leighton [1] studied the details of the stratoscope pictures obtained by Schwarzschild and coworkers and found that the profiles of the dark lanes between the granules were all about 0.38 arc sec. This is suspiciously close to the resolution of the system used, and it might well be that the average granule size would be greatly decreased by the discovery of hitherto unobserved finer dark lanes. Although the granule lifetimes have been carefully measured by the Princeton group, the results are still not so definite as one might hope; the pictures dance around, and the whole problem is close to the limit of resolution. The same comments apply to our various studies of the Evershed effect, and other phenomena in sunspots and their penumbrae. In every case we are digging a small signal out of a large noise, and important physical phenomena which we cannot resolve are taking place. Thus the quest for higher resolution will always go on.

A great deal has been learned about the practical problems of the construction of solar space telescopes of high resolution during the Caltech group's studies over the last five or six years of the design of a large aperture solar telescope of 65 cm diameter. Because of the leisurely pace of these studies, we have come to learn a great deal about problems of solar space telescopes which are not obvious at the outset. The principal problems of a solar telescope for use in space are concerned with the maintenance of high optical quality of the system during exposure to one solar constant, and maintaining pointing stability consonant with the resolution; there are the usual additional problems of auxiliary instrument design, exposure control and so forth. Because of the possibility of observation in the ultraviolet, reflecting telescopes are favored. Because the space telescope operates in vacuum, one does not have to worry about long paths in a telescope filled with turbulent air. Because very high optical resolution is required, long focal ratios such as f/50 are favored, because then maximum resolution is obtained from the detecting systems and the effects of optical aberrations are minimized. Ideally, one would like to have a very long off-axis system, where the one reflection would minimize absorption of ultraviolet, and the off-axis design would remove the thermal problems associated with central obscurations. However practical considerations rule out such telescopes, although one might possibly consider systems erected in space. The single mirror system has the additional advantage that all the heat is reflected directly into space, and a small secondary mirror can be used to select whichever part of the sun is desired.

Returning to more practical problems, one must consider systems with secondary magnification to obtain the necessary long focal length. This is possible in three systems, the Cassegrain, Gregorian and Newtonian with secondary magnification. In the Cassegrain system, one has the advantage of a relatively short overall length, because the secondary mirror is ahead of the first focus; however there is a severe difficulty in the fact that the image of the sun falls on the primary mirror, thereby producing a strong source of localized heating on the primary. Analyses of such systems at the Jet Propulsion Laboratory showed that in all possible Cassegrain configurations, the distortion in the primary produced by the solar image was sufficient to exclude this arrangement. One might think that a Cassegrain system could be designed with the image small enough so that it might be passed through the central hole in the mirror for all solar orientations. If we restrict ourselves to looking up to the limb of the sun and no further, the hole must subtend one degree, or $L/57$, if L is the focal length; if we consider further that the secondary obscuration must be less than 20% of the aperture A to prevent loss of resolution we arrive at the result

$$\left.\begin{aligned} 0.2\,A &= L/57 \\ L/A &= \text{f number} = 11.4\,. \end{aligned}\right\} \tag{1}$$

So the effective focal ratio of the two-mirror system must not exceed 11.4. Then the distance d of the secondary from the primary is given by the formula (without field)

$$d/11.4 = 0.2\,A \tag{2}$$

if we still wish to avoid secondary obscuration. This means that the secondary mirror cannot be more than two diameters A distant from the primary, leading us to an f/2 primary. With this short focal length we will need 25 times magnification to produce a focal ratio of f/50, and therefore we will magnify the aberrations of this f/2 primary by 25 times. Thus, we reach a very difficult configuration which essentially rules out the Cassegrain system from consideration. Only the possible development of mirror materials of truly zero expansion coefficient would make it possible to have a Cassegrain system and permit the solar image to fall on the primary mirror. Of course, we have not even considered the heating of the secondary, another difficult problem.

A second possible system is a Newtonian telescope where a small Newtonian flat picks out only the part of the solar image in which we are interested, and reflects everything else. This image is reflected to the side and then magnified by secondary optics. It has the virtue that heat disposal is direct, but the disadvantage that an optical surface is placed right at the primary focus or near to it, resulting in extreme heating problems. It would be very difficult to keep a good surface on such a Newtonian mirror and we have therefore not followed this system further.

Finally we come to the Gregorian system in the configuration adopted by the Göttingen Observatory for its Locarno Station. This is illustrated in Fig. 4 in the configuration adopted for the Caltech photoheliograph. In this system, most of the solar heating is disposed of by a diagonal mirror placed at the first focus. Since this diagonal mirror does not participate in the actual optical train, heating and distortion of it do not bother us. However, the mirror must be cooled to reduce heat transfer to the rest of the secondary structure. The field desired is passed through an aperture in this mirror magnified by an elliptical secondary, and reflected to the side by a flat behind the first mirror. The image may then be dealt with as desired: in our design it is brought back by a flat to the neighborhood of the primary where a number of cameras take pictures in different wavelengths. The Gregorian design has the disadvantage of somewhat greater overall length and some complexity; however it disposes of most of the unwanted solar heat immediately, and utilizes fairly simple optical surfaces. Our telescope has an automatic alignment system which moves the secondary mirror in both tilt and displacement to maintain critical alignment. I am not familiar with the actual success of the Locarno system, although I would suppose it to be limited by the seeing in the Alps. One would expect that the Gregorian would perform well and have minimal scattered

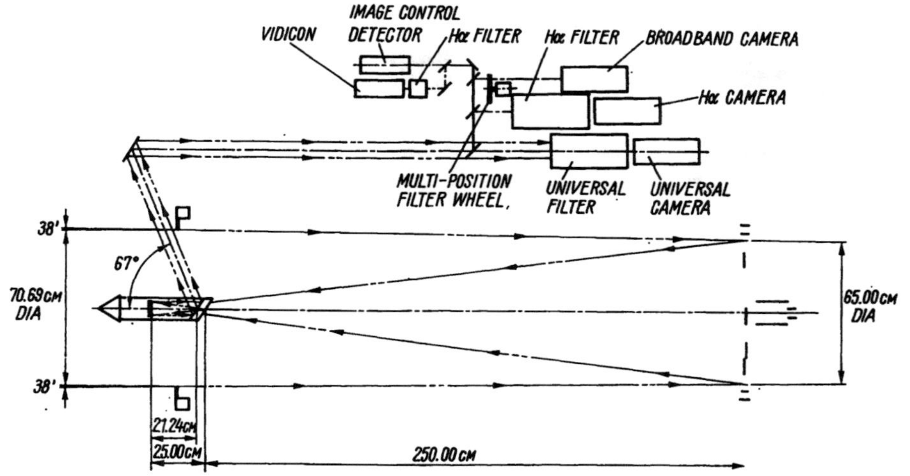

Fig. 4. Folded Gregorian optical concept (Ball Brothers Research Corporation [2], p. 4-3).

light. I have already shown the optical resolution of this system at the center of its field.

The most difficult problem in the construction of the optical system outlined above is the primary mirror. It must perform well in full solar illumination for intermittent intervals. Whereas this is no problem for telescopes limited by terrestrial seeing, it is an extremely difficult condition when one wishes to obtain 0.2 arc sec. This is further aggravated by the fact that the mirror is

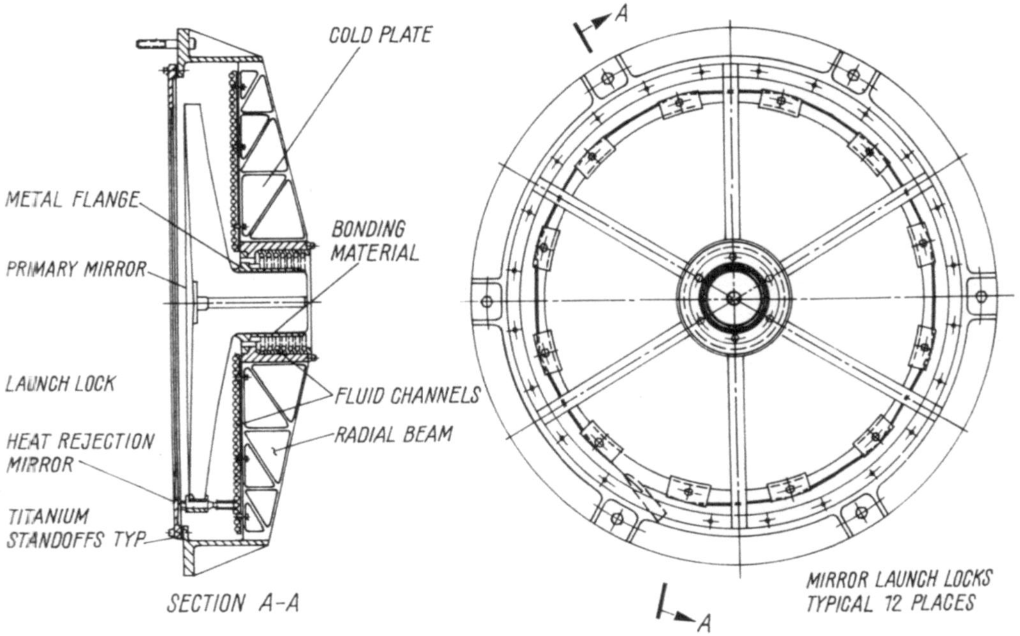

Fig. 5. Primary mirror assembly (Ball Brothers Research Corporation [2], p. 5-7).

situated in a vacuum and can only lose heat by conduction and radiation. We further have the problem that the mirror must be tested under 1 g gravity, and then perform well under 0 g. As the best compromise, JPL adopted the mirror design shown in Fig. 5 where the mirror could be held tightly by a hub without affecting the optical surface, and where the taper to the edge produced a tendency for the mirror to expand into another parabola, rather than a non-parabolic surface, so that simple refocusing could remove some of the changes introduced by thermal expansion and contraction.

Fortunately, two steps may be taken further to alleviate the primary mirror problem: a high reflectance silver coating may be used, and we may use low-expansion glass ceramic material such as Cer-Vit. Even so, it is necessary to surround the primary mirror with a cold shell or at least a cold backplate. Fig. 6 shows the steady-state temperature of the primary for an input of two thirds

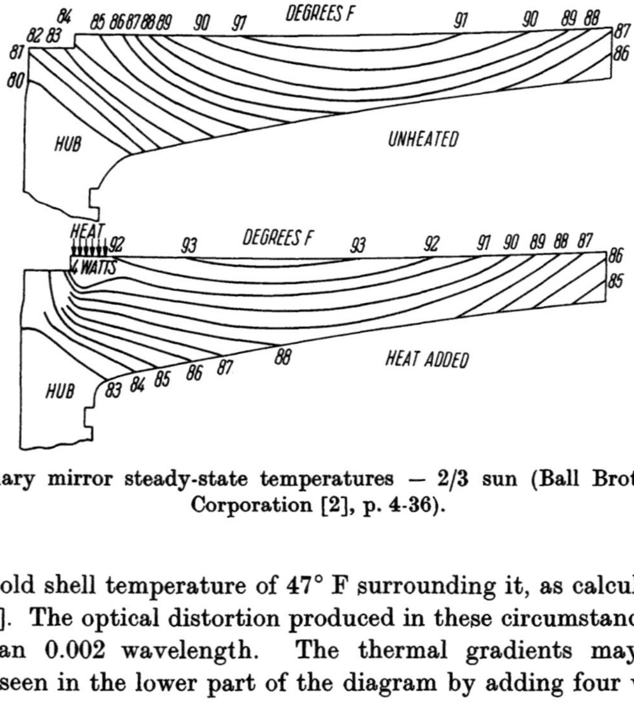

Fig. 6. Primary mirror steady-state temperatures — 2/3 sun (Ball Brothers Research Corporation [2], p. 4-36).

sun and a cold shell temperature of 47° F surrounding it, as calculated by Ball Brothers [2]. The optical distortion produced in these circumstances in Cer-Vit are less than 0.002 wavelength. The thermal gradients may further be reduced as seen in the lower part of the diagram by adding four watts of heat near the shaded center of the mirror. Unfortunately, it is very difficult to specify the exact expansion coefficient of Cer-Vit material before it is delivered, and therefore tests will have to be made on actual systems, in particular on the functional verification unit now being constructed.

Fig. 7 shows a photograph of the functional verification unit for the photoheliograph (minus the primary mirror assembly) which has been constructed at Ball Brothers. The material for the truss structure is invar but the fittings are aluminum. At present no camera package has been specified, although it is envisioned that the telescope will use a universal birefringent filter, an Hα filter and a third camera at which either white light or near ultraviolet photographs will be taken. Extensive consideration has been given to the possible use of video detectors in space, which would make the problems of film retrieval considerably less. Unfortunately, the return beam vidicon systems which are capable of providing the necessary resolution are in a new stage of development,

Fig. 7. The photoheliograph functional verification unit.

and also extremely expensive, so these studies have not gone forward. However, serious consideration should be given to the possibility of the use of such television systems. It would be particularly attractive to use them in subtractive modes, such as a videomagnetograph, for determining fine magnetic field details.

The universal filter under construction by Carl Zeiss in Germany will permit us to make photographs in any spectral line between 4000 and 7000 Å, with band pass ranging from 0.15 to 0.3 Å. It would of course be extremely attractive to make spectroscopic observations with this telescope, particularly in the near ultraviolet range from 3000 to 1500 Å, where it will still provide significant flux. Because the space vehicle is presently undefined, the problem of spectroscopy with the telescope has been left open. However, the stability of the present ATM system, while quite adequate for the short exposures planned in the high resolution photography program, would be quite inadequate for a spectrograph, and some type of internal guiding would have to be provided to keep the image stable over the relatively longer periods of spectroscopic observation. Such devices have been extensively studied by the Fraunhofer Institute.

It is planned to install the completed functional verification unit in a vacuum tank at the Big Bear Solar Observatory within the next year to test it under actual conditions of solar observation.

I think it is clear that with existing hardware, such as ATM, we can look forward to a factor of three or four improvement in resolution over the best we can do at present. If we project forward what such a gain in resolution has meant to solar physics in the past, we see that there is a great deal to be gained, and we are finally coming to the time where ordinary optical solar astronomy can benefit as much from the space program as X-ray and ultraviolet observations have in the past.

Whatever current possibilities may be, further developments of large solar telescopes in space will take place, and it is useful to consider their value and direction; it is self-evident that there is no definite limit beyond which better resolution is not needed: as we pointed out before, a diffraction limited telescope only delivers about 10% modulation at the Rayleigh limit, so that, even if we were not interested in features smaller than that limit, doubling of the aperture would increase the MTF at that resolution by an appreciable factor, and enable us to see the features of lower contrast at that limit. For this reason, most of the things we are now studying with 0.7 arc sec resolution are three or four times larger. Of course it is unlikely that the limiting size will be reached soon, for the magnetic field seems to constrain material to very small structure, and most of the interesting phenomena are linked with fine magnetic fields.

In studying possible large telescope configurations, we should always keep in mind whether the interests of solar physics might not be as well served by smaller instruments with longer observation times, or with more complex auxiliary equipment.

The super-solar space telescope (SSST) will present a great challenge to the ingenuity of the designer. While cinematographic systems can depend on short exposures for reduction of jitter blur, complex analyzing systems will need long term stability to 0.1 arc sec. Fortunately the sun is a very good source for photoelectric guiding, so this may be possible for the whole system, but we should be aware of the possible need for guiding on the solar features themselves. The optical and thermal problems of an SSST in the 1.5 m class are formidable indeed. It would appear that the Gregorian design would still have the advantage. A simple off-axis system at f/50 would be 75 m long; a folded system, on the other hand, would require a substantial off-axis angle to remove the beam before the end of the tube; this would give us the coma of a 3 m system at f/17, and therefore limit the field, unless correctors were to be used. Although the Gregorian system would appear readily adaptable to larger apertures, we should realize that any secondary magnification scheme magnifies the aberrations of the primary. This in turn increases the sensitivity of the system to misalignment, which can of course be reduced by automatic alignment systems. Further, some aplanatic correction would have to be made in the secondary. It would be most valuable to reduce the secondary magnification as much as possible; on the present photoheliograph this is limited by tube length.

The problems of supporting and cooling a large primary are severe, because the mirror thickness is determined by fabrication and launch considerations, and would certainly be much thicker than the 65 cm primary, thus making the problem of removing the heat incident on the front surface more severe. We can expect more progress in zero-expansion materials in the future; further, considerable improvement can be made with selective heating and cooling: there is basically no severe problem in the construction of the SSST that cannot be solved with intelligent design and management.

We should bear in mind that until now no optical solar telescope of any size has been orbited. The Harvard Hα telescope on ATM will be the first; it is of course more of a patrol and finding system, and not designed as a scale prototype. This situation exists because there is no obvious benefit to small optical telescopes in space. Yet if we are agreed that there is a great future for large systems, we should plan an orderly development through smaller systems. In this regard, a small telescope in a sun synchronous orbit that could produce long-term surveillance of the sun with moderate resolution, say 2 arc sec at 30% MTF, would be of enormous value to solar physics: day-to-day data are erratic and variable and even two pictures a day would be a great asset to understanding the development of solar magnetic fields and solar rotation. It is clear that we shall have to go to space for this development for all the effort that has gone into networks has shown that this cannot be done from the ground except for a few months a year.

Acknowledgments

I wish to thank the Ball Brothers Research Corporation and the Marshall Space Flight Center for the use of their various graphs and calculations. This work was supported by NASA contract NGL 05 002 071.

References

[1] R. B. Leighton, Ann. Rev. Astron. Astrophys. 1, 19 (1963).
[2] Ball Brothers Research Corporation Report No. F71-02, Vol. II.

Space Research XII – Akademie-Verlag, Berlin 1972

THE CAPABILITIES OF THE SPIN-SCAN IMAGING TECHNIQUE

T. GEHRELS[a], V. E. SUOMI[b] and R. J. KRAUSS[b]

[a] Lunar and Planetary Laboratory, University of Arizona, Tucson, Ariz., USA
[b] Space Science and Engineering Center, University of Wisconsin, Madison, Wis., USA

A summary description is made of spin-scanning devices for varius space missions. Compared with television, scan-imaging has remarkable advantages and these are listed at the end of the paper.

The first spin-scan imaging camera was conceived by V. E. Suomi for the Applied Technology Satellite ATS 1 and built by the Santa Barbara Research Center. It was launched into geostationary orbit over the Pacific Ocean in December 1966 and has been in continuous operation ever since. An improved instrument, the Multicolor Spin Scan Cloud Camera (MSSCC) was launched in November 1967 on ATS 3. It is situated at 35800 km altitude above the Amazon River and is still being used on a regular daily schedule (Fig. 1).

The two cameras are quite similar in size and design, the primary difference being that the ATS 1 camera is monochromatic, using a single S-11 photomultiplier filtered to give a response of 0.475–0.630 μm. The ATS 3 camera has a 12.5 cm Wynn–Rosin telescope (elliptical primary and spherical secondary mirror) F/3, with three photoetched circular apertures used as field stops in the focal plane, each 38 μm in diameter, giving a resolution of 100 μr or 3.6 km on the surface of the earth at the satellite subpoint.

The field stop apertures admit light to flexible glass fibers which are used to transport light to the three photomultiplier sensors mounted on the camera frame separate from the movable telescope. The photocathodes used are S-20 for the red channel (0.55–0.63 μm) and S-11 for the green and blue channels (0.48–0.58 μm and 0.38–0.48 μm).

The telescope is stepped in latitude by 0.13 mr once each spacecraft revolution. The stepping motion is from north to south, while the spacecraft spin axis is parallel to that of the earth. Thus, at 100 rev min^{-1} nominal spin rate, a full disk image of the earth 18° wide is generated in 2407 lines in approximately 24 minutes. The instantaneous field of view dwell time is 10 μs although horizontal sampling can be done every 3.5 μs to take full advantage of the available communications bandwidth.

The principal interest for this COSPAR Symposium lies in high resolution. The resolution, generally, can be improved by reducing focal plane *aperture size* and/or going to a larger *focal length*. Either solution will require increased *information bandwidth* and correspondingly degraded *signal to noise ratio.* The photomultiplier *reaction time* may also become a problem as the field of view *dwell time* decreases. All of these drawbacks can be counteracted by slowing down the *spin rate* of the spacecraft and/or using a larger *telescope diameter.* The compromises to be made depend on the practical system constraints.

Fig. 1. Spin-scan image obtained from ATS-3.

As an example, the second generation Visible Infrared Spin Scan Radiometer (VISSR) to be flown on the Synchronous Meteorological Satellite (SMS) in October 1972 is constrained to have a 20 minute picture time, 70 kg combined instrument and electronics weight, and 1 km ground resolution with 2.8:1 signal to noise ratio at 0.5% albedo and a visible bandpass of 0.55–0.75 μm. The thermal imaging channels are to have $\leqq$1.4 °K noise equivalent temperature difference for a 200 °K scene. The resulting design is a 40.6 cm diameter F/7 all reflective Ritchey Chrétien (aspheric) telescope with 2.91 m effective focal length. The photomultiplier sensors for the visible channels are operated in parallel behind rectangular glass fibers whose ends form 61 μm × 73 μm field stop apertures in the primary focal plane, yielding resolution of approximately 25 μr with a 2.4 μs dwell time at 100 rev min^{-1}. Information bandwidth and photomultiplier response time are adequate. The thermal channel images in the 10.5–12.6 μm atmospheric window band using radiatively cooled HgCdTe intrinsic infrared detectors with 0.25 mr resolution.

The Multispectral Scanner (MSS) also being built by the Santa Barbara Research Center for the polar orbiting Earth Resources Technology Satellite (ERTS) utilizes a 23 cm aperture F/3.3 Cassegrain telescope with 59 μm square glass fibers as field stops. This yields 77 μr resolution from low orbit (~65 m ground resolution at nadir). Photomultipliers are used for three observation channels in the 0.5–0.8 μm region and photodiodes for a 0.8–1.1 μm band. In addition, HgCdTe detectors are used for thermal imaging in the 11 μm window.

Infrared imaging and radiometry from earth orbiters has also been done with point scanning sensors in the Nimbus weather satellite program (the Medium Resolution and High Resolution Infrared Radiometer experiments) and in the Improved TIROS Operational Satellite (ITOS) scanning radiometer. A picture obtained on ITOS 1 is on the front cover of the 6 August 1971 issue of *Science* [1]. Currently, improved scanners are under development for ITOS. Two of us (Suomi and Krauss) proposed a Line Scan Radiometer (LSR) for the Mariner Venus/Mercury 1973 (MVM) mission. The camera used the slow attitude drift of the three-axis stabilized spacecraft to generate the line stepping function and a small focal plane scan mirror to generate the scan along a line, substituting for the absence of spacecraft spin. The lead time available was unfortunately not sufficient to carry the concept through "breadboard" stage to flight hardware and an essentially pre-existing television system with new optics will be used on the mission instead.

Consequently, the Imaging Photopolarimeter experiment of Gehrels on Pioneer F and G (F to be launched in March 1972) is the only line scan imager now being applied to non-earth oriented missions. As such, it is not a good example for high resolution because the resolution was restricted by spacecraft capabilities. The Pioneer is small and provides for a relatively cheap reconnaissance mission of the asteroid belt and of the Jovian radiation environment, as a precursor to more sophisticated missions such as the "Grand Tour of the Outer Planets", for which launch dates exist in 1977 and 1979. The Pioneer data link handles, when near Jupiter, at best 2048 bits/s. There is no on-board data storage, only a buffer, and the best ground resolution at closest encounter (2 Jupiter radii from the surface) will be 150 km compared with the best earth-based resolution of 600 km.

Considering the 5 rev min^{-1} spin of the spacecraft and the 2048 bits/s data link, 600 μr was all the resolution which could be accommodated. Consequently, a 2.5 cm telescope was large enough, and the instrument itself weighs 5 kg — an optimum weight since the mission could accommodate it without eliminating other experiments. The advantages over a television system were studied at the time and the simplicity of an instrument using photomultipliers was important. Vidicons are considered to be more susceptible to the Jovian radiation belt. The fact that television would have taken a shorter exposure (a fraction of a second, compared with several minutes for a spin scan) would have reduced the image rectification problem, which is considerable for a flyby of Jupiter as we are dealing with a fast trajectory of a rotating spacecraft flying past a rotating planet. The scan lines are far from straight. On the other hand, the geometric reconstruction is straightforward if good timing information and trajectory data are available. Television would be subject to smear limitations and small frame size and has its own peculiar brand of geometric distortions and photometric nonlinearities, especially at the edge of the field.

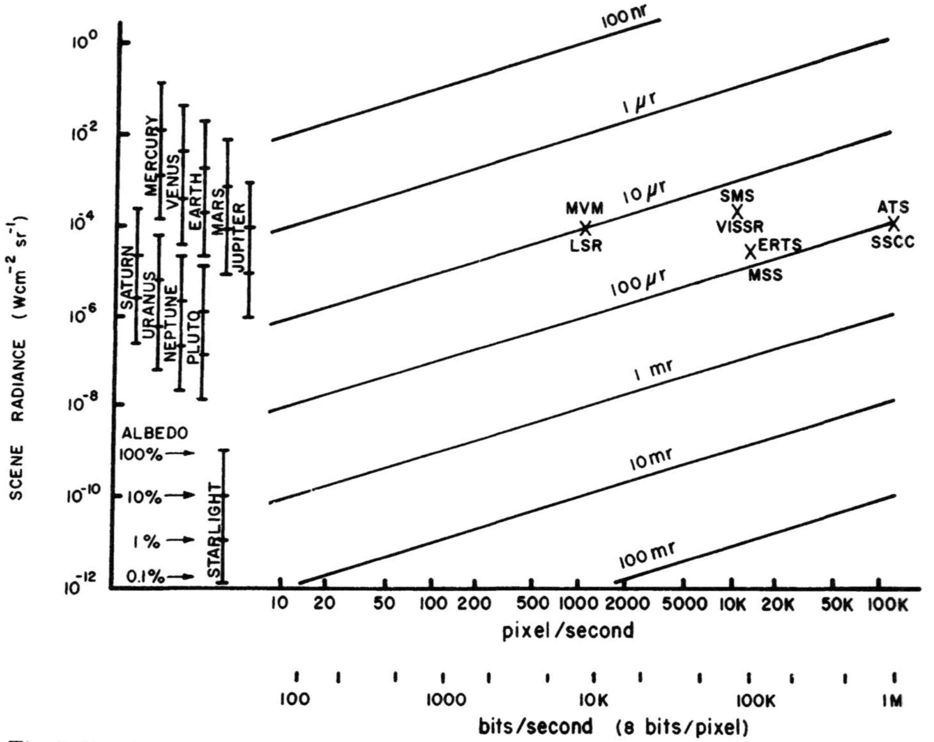

Fig. 2. Resolution per scan line (25-cm optics, 10:1 signal-to-noise ratio at 50% response to 2:1 contrast target). The abbreviations of various missions and instruments are explained in the text.

The imaging of Jupiter will be in two colors: blue (0.390–0.500 μm) and red (0.595–0.720 μm), and the range of visible phase angles will be quite large — much better than from earth. Besides the imaging capability, photopolarimetric data will be obtainable. These modifications are easily added to produce the combination for the Pioneer F and G missions known as the imaging photopolarimeter. The data processing of the imaging is being developed by W. Swindell of the Optical Sciences Center at the University of Arizona. Details of the instrument are given in [2].

A reasonable system for an outer planet spacecraft with a sufficient data link would start with a 25 cm diameter telescope to keep the system weight down. Photoetched field stops can be made in the 30–40 μm range. Smaller sizes tend to imperfections and rough edges. The same is true for glass fibers. While 5 μm fibers can be drawn, the ends cannot be finished smoothly enough to avoid large transmission losses. About 40 μm is the present limit. Fig. 2 shows what a 25 cm optical system can produce in resolution at 10:1 signal to noise ratio as a function of planet brightness (the vertical bars range from 0.1% to 100% albedo) and information bandwidth. A 10 μr resolution appears to be a reasonable limit. A relay lens system should shorten the effective back focal length for ir imaging since smaller detectors are better. With ease, one could add a 10 μr field of view in the visible channel for highest resolution simply by increasing the effective front focal length to 4 meters (F/16) and tolerating the reduction in signal to noise ratio.

In summary, the disadvantages of the spin-scan technique are (i) that it is not as "instantaneous" as is television, (ii) that the instrument requires a spinning spacecraft or a mechanical scan mechanism, and (iii) that it requires duty cycle expansion and a buffer best to utilize the communications link.

We can name seven advantages, over a television system, of the spin-scan technique: (i) practically unlimited format of the field, whereas the 800-line scan of television will image only a part of the planet, and even at that the television has edge distortions; (ii) the detectors are rugged and have wide dynamic range; (iii) multiple detectors allow a wide spectral range; (iv) multiple perfectly registered images in several colors can be made by scanning with several detectors operating simultaneously; (v) the detectors are linear — ideal for combination with a photopolarimeter of $\pm 0.1\%$ precision; (vi) redundancy in detectors is implicit in the design, yielding extremely high reliability; and (vii) the instrument is relatively cheap and lightweight. Even at a heavy duty cycle as on ATS 3, the instrument is proving to be a "workhorse" of long life.

References

[1] P. K. Rao, A. E. Strong and R. Koffler, Science **173**, 529 (1971).
[2] C. E. KenKnight, in: NASA SP-267, NASA, Washington, D.C., 1971.

Space Research XII – Akademie-Verlag, Berlin 1972

THE DATA-HANDLING PROBLEM WITH TELEVISION RECORDING OF SPECTRA

A. B. Underhill and D. A. Klinglesmith

NASA Goddard Space Flight Center, Greenbelt, Md, USA

A television tube is a useful way to record a many-element picture such as is obtained from an echelle spectrograph. The signal on the tube target is scanned and relayed to the data-processing center where it must be reconstituted into the scene that was present in the focal plane of the spectrograph camera. A typical television frame will contain 4×10^6 bits of information which must be processed to remove transmission errors and random noise pulses, to remove periodic noise, to correct for optical and electrical distortion, to remove sensitivity variations across the face of the tube, to calibrate from the photo-electrons back to incident energy and to provide a wavelength scale. Results of a study of these problems is presented. The goal is to provide the equivalent of an intensity tracing of a spectrogram.

1. Introduction

The problem under study is how to convert the telemetry data from an ultra-violet echellogram recorded on a television camera tube in a spacecraft into an intensity–wavelength display similar to that obtained from high resolution spectrograms of stellar spectra obtained with ground-based equipment. A typical

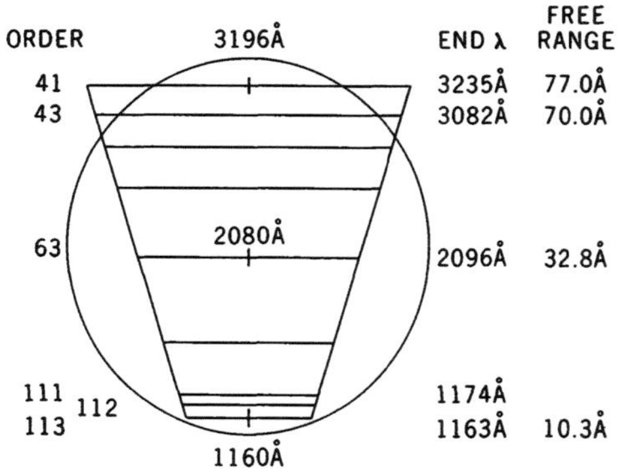

Fig. 1. Schematic of the wavelength distribution resulting from the grating dispersions.

echellogram of the ultraviolet spectrum of a star will appear in the focal plane of the spectrograph on the face of a television camera in the format shown in Fig. 1. The plan is to record the spectrum from about 3200 Å to 1150 Å at about 0.1 Å resolution. If 8 bits per picture element are used to record the light intensity, the total number of bits per picture is about 4×10^6. The television camera tubes have a face plate of 25 mm diameter. Either an ultraviolet-to-visible converter with a cathode sensitive to visible light, or a television camera tube with an ultraviolet transmitting window and an ultraviolet sensitive cathode will be used. The latter combination has several advantages but it is still under development. Nearly all of the desired wavelength range can be accommodated in one display. The echelle grating is used in orders 41–113; the cross dispersion is provided by a grating used in first order. The trapezium shape outlines the region where there is no overlap of wavelengths from one order to the next. In the longest wavelength cycle the available free spectral range is 77 Å; in the shortest wavelength cycle this range reduces to about 10 Å. The short wavelength strips (high orders) are closer together than the long wavelength strips.

Because the positions of the gratings and the face plate of the television camera tube relative to the optical axis of the spectrograph are fixed, a con-

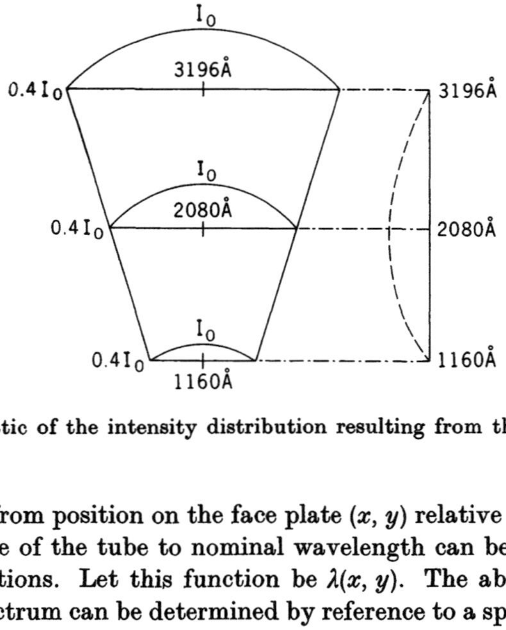

Fig. 2. Schematic of the intensity distribution resulting from the grating blazes.

version formula from position on the face plate (x, y) relative to a set of fiducial marks on the face of the tube to nominal wavelength can be established using the grating equations. Let this function be $\lambda(x, y)$. The absolute wavelength in any stellar spectrum can be determined by reference to a spectrum containing wavelengths known on the absolute scale. A relative wavelength scale within any stellar spectrum can be set up by the observer recognizing a few absorption lines and assigning wavelengths to them.

A constant energy pattern projected through the spectrograph will not be seen in the focal plane of the spectrograph with equal intensity at all wavelengths. Owing to the blaze properties of the gratings each wavelength strip is transmitted with a varying efficiency. This wavelength-dependent efficiency factor is illustrated in Fig. 2. If the intensity is I_0 at the center of a strip, the intensity recorded at the end points, where overlap in wavelength with neighboring orders begins, is about $0.4\, I_0$. In addition, there is a blaze efficiency for the cross-dispersion grating, indicated by the vertical pattern shown on the right side of the diagram. The product of these two efficiency factors gives the echelle ripple, $R(x, y)$. This factor must be determined by pre-flight tests of the optics. It depends solely on the optical properties of the gratings and of the collimator and camera mirrors. A known efficiency correction has to be made dependent on the position (x, y) in the field of view: this position is determined relative to a set of fiducial marks.

No photo-target or photo-emissive cathode is uniformly sensitive over its area. The sensitivity contours vary in shape and may vary with the age of the target. A map of the sensitivity contours must be established before flight and this map should be updated during the operating lifetime of the tube. A typical sensitivity contour map for a cesium telluride ultraviolet sensitive photomultiplier cathode is shown in Fig. 3. This contour map was determined in our laboratories. Measurements are now being made to find typical sensitivity contour maps for the television camera tubes we may use. In the example there is a hole in the cathode, indicated by the cross-hatched area. The cathode departs quite significantly from uniform sensitivity to λ 2537. We believe this

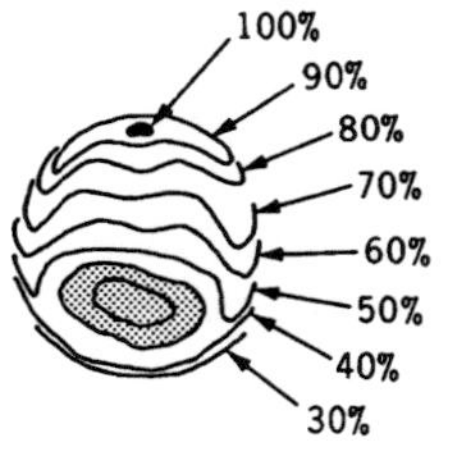

Fig. 3. The sensitivity contours on an ultraviolet sensitive photocathode (Ce–Tl 2537 Å).

cathode to be typical of the average uv-sensitive cathode. The recorded apparent spectral intensities have to be corrected for this factor $S(\lambda, x, y, t)$ which is a function of the wavelength, the coordinates (x, y) on the cathode of each image element, and on time t.

The telemetry data give a set of digital counts for each element of area defined by the size of the reading beam. Such an area is called a pixel. It is usually circular in shape and may have a diameter of 50 μm or less, depending on the geometry of the tube. The positions of the pixels on the target are described by a set of points (ξ, η). The pixels are scanned in succession so that the intensity distribution in the echellogram is contained in a rectangular grid of something like 768×768 pixels. The first transformation to be applied must correct the output for transmission errors and for random noise. Following that, periodic noise must be removed as well as the effect of the sensitivity of the target and the geometric distortion of the tube. In the tubes under consideration the target is close to the cathode, thus a unit transformation brings one to the plane of the spectrograph, that is to the (x, y) system of coordinates. The data can then be corrected for distortions in the focal plane of the spectrograph, known from the optical design, and finally the desired information $I(\lambda)$ may be derived.

2. Corrections applied in the (ξ, η) Plane

A charge that is the result of the photoelectrons which came from the photocathode at point (x, y) will be measured on the target at point (ξ, η). Since the target will not be uniformly sensitive to the photoelectrons, it will be necessary to correct the measured charge $Q(\xi, \eta)$ for the local quantum efficiency in order to determine how many photoelectrons struck the point (ξ, η). This correction is not wavelength-dependent, although it will most probably be time-dependent since parts of the target may become fatigued by overexposure.

A potential problem on the target surface is the positional repeatability of the reading beam. In order to overcome this it is necessary to have fiducial marks on the cathode which will transfer to the target as zero levels of excitation. The shape of the (ξ, η) plane can be determined by the pixel spacing between the fiducial marks and a correction function derived to correct for geometrical distortion in the tube. This correction function will most certainly be a function of the operating conditions and the past history of the camera tube.

Before the charge distribution in the (ξ, η) plane can be accepted as correct, the validity of the raster must be confirmed; that is, the completeness of the television raster and telemetry transmission to the ground must be checked and corrections for line drop-out must be made if needed.

Obvious bright or dark spots may be eliminated by averaging the surrounding points and any electronic ripple that appears corrected out at this point. It may be necessary to use Fourier-transform techniques to remove high-frequency noise.

3. Corrections applied in the (x, y) Plane and the Final Output Format

Once the light intensity per pixel has been corrected for transmission errors and noise and has been transferred to the focal plane of the spectrograph, the wavelength transformation function $\lambda(x, y)$ may be applied to calculate a nominal wavelength of the center of each pixel from the coordinates (x, y) of that central point and the sensitivity correction factor $S(\lambda, x, y, t)$, and the echelle ripple function $R(x, y)$ may be applied to transform the number of electrons per pixel to an intensity value. The next action is to reconstruct the spectrogram. One reconstruction is needed for quick-look purposes in order to monitor the results as regards exposure time and to indicate key wavelengths and order numbers for calculating a precise wavelength. The final reconstruction is to provide data in such a form that the observer may proceed with scientific work.

Fig. 4 illustrates the problem. On the left side a sketch is given of two extreme spectrum strips and the projected shape of the entrance aperture in monochromatic light is indicated allowing for the expected optical aberrations of the proposed optical system. The camera is an off-axis paraboloid and the optical axis is near the lowest strip. The chief aberration is coma which will distort the image of the aperture somewhat as indicated. On the right side sets of pixels are indicated as small squares (really they are inscribed circles) and the projected images of the entrance aperture are indicated.

For quick-look purposes a *picture* of the echellogram may be reproduced by plotting the intensity of light in each pixel on a graphic device which takes account of sufficient shades of grey. In order to obtain an $I(\lambda)$ tracing, using

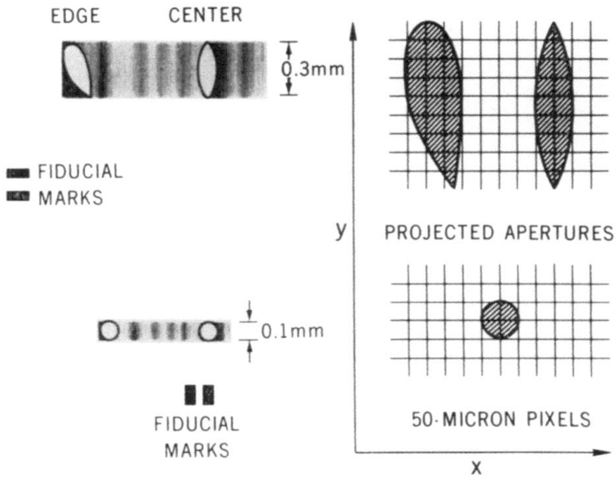

Fig. 4. Reconstruction of the picture.

nominal wavelengths, one must sum the pixels representing each monochromatic image of the entrance aperture and plot these as a function of the wavelength. Precisely how to do this can only be defined when full information about the imaging properties of the spectrograph is available.

Space Research XII — Akademie-Verlag, Berlin 1972

A STELLAR INTERFEROMETER BASED ON COHERENT DETECTION

H. G. VAN BUEREN and H. VAN DE STADT

The Astronomical Institute, University of Utrecht, Utrecht, Netherlands

A stellar interferometer based on the coherent detection (optical superheterodyne) technique, under construction in the Astronomical Institute at Utrecht, is described. Section 1 discusses the principles of the detection method, § 2 the signal-to-noise ratio and sensitivity to be expected when using an optical balanced-mixer system. The latter leads to an increase of sensitivity by a factor between 2 and 4 and to suppression of various laser instability effects. The interferometer construction itself is described in §§ 3 and 4; it consists of two 60 cm all-metal mirrors hydraulically driven, with a single gas laser as local oscillator. The Fourier transform of the brightness distribution of the star is obtained in first order. Stars down to the second magnitude should be easily observable from the ground. Possibly the fourth magnitude can be reached with this system when used in space. In § 5 some astronomical applications of the method and technical details pertaining to space flight application are listed.

1. Principles of Coherent Detection

Two beams of light of neighbouring frequency impinging on a photoelectric device which acts as a quadratic detector produce a radio-frequency (rf) interference signal, provided that the detector is fast enough to follow the difference frequencies involved. The rf signal can be amplified to yield information on the spectral and spatial properties of the light. When one of the beams is a laser beam, consisting of spatially and time coherent monochromatic light with an optically negligible bandwidth, and the other beam is produced by an incoherent and non-monochromatic source (under certain conditions) the frequency spectrum of the rf mixing signal is a faithful reproduction of that of the incoherent source. The relative amplitude and phase distribution of the unknown optical signal have been translated by the radio-frequency domain where they can easily be processed by conventional methods (Fig. 1). If the source has an angular width such that the detector area does not greatly exceed the so-called coherence area of the source as seen by the detector, then the method of coherent detection can be used for spectroscopy of stellar objects, now under development in our laboratory [1].

When the mixing of the coherent and non-coherent light takes place behind two or more receiving telescopes spaced a certain distance from each other, and the same laser or several phase-locked similar lasers are used as the local oscillator for all mixing processes, the resulting rf mixing signals can be mutually correlated (that is, added or multiplied) to yield direct information on the complex degree of coherence of the source, which is the Fourier transform of its brightness distribution. That is, the system acts like a (monochromatic) interferometer, exactly as an optical Michelson stellar interferometer does. By using the coherent detection method, the optical bandwidth is now equal to the electrical bandwidth, so that much greater path-length differences between the two telescopes are allowed. This makes possible much larger spacings between the telescopes without the very stringent mechanical stability required for a Michelson interferometer. The principle of the coherent detection interfero-

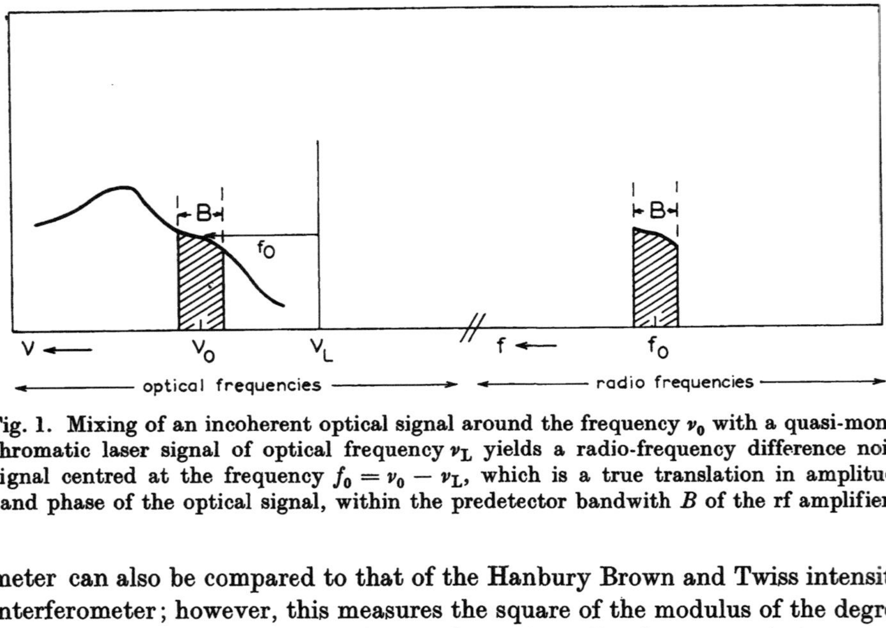

Fig. 1. Mixing of an incoherent optical signal around the frequency ν_0 with a quasi-monochromatic laser signal of optical frequency ν_L yields a radio-frequency difference noise signal centred at the frequency $f_0 = \nu_0 - \nu_L$, which is a true translation in amplitude and phase of the optical signal, within the predetector bandwith B of the rf amplifier.

meter can also be compared to that of the Hanbury Brown and Twiss intensity interferometer; however, this measures the square of the modulus of the degree of coherence, and eliminates phase information. Brightness variations over the source area therefore become more difficult to detect and eventual asymmetry remains unspotted. Before entering into a discussion of the application of the coherent detection method for interferometry, we shall first consider the properties of this method as such, confining our attention to one detection system only.

2. Signal-to-noise Ratio and Sensitivity

It can be shown that if the incoherent source is a point source, the mean square rf mixing current produced by the quadratic detector is given by

$$\overline{i_{rf}^2} = 2\left(\frac{\eta e}{h \nu}\right)^2 \overline{I_L}\, \overline{I_s}\,, \tag{1}$$

where η is the quantum efficiency of the photodetecting device, ν the average optical frequency involved, e and h electronic charge and Planck's constant, and $\overline{I_L}$ and $\overline{I_s}$ the mean laser power and the mean source power on the detector surface, respectively. A bar denotes averaging over a sufficiently long time. However, if the source has a certain angular diameter and the detector, that is, the telescope aperture has also an extended area, the mean square mixing current is reduced to

$$\overline{i_{\mathrm{rf}}^2} = 2\frac{\eta\, e^2}{h\,\nu}\,\overline{I_{\mathrm{L}}}\,\overline{I_{\mathrm{s}}}\,\Delta\,,$$

where Δ, the so called degradation factor or the degree of partial coherence, is a quantity smaller than unity, depending on the relative sizes of source and telescope with respect to distance and wavelength.

For a uniformly bright circular source with angular diameter γ_s and a circular detector area of diameter D one obtains

$$\Delta = \frac{32}{z_s^2}\int_0^{z_s/2} (J_1(z)/z)^2\,dz \tag{2}$$

with $z_s = \pi\,\gamma_s\,D/\lambda$, $z = \pi\,\gamma\,D/\lambda$ (γ being an integration variable $\leqq \gamma_s$; J_1 the Bessel function of first order; λ is the average wavelength of the signals).

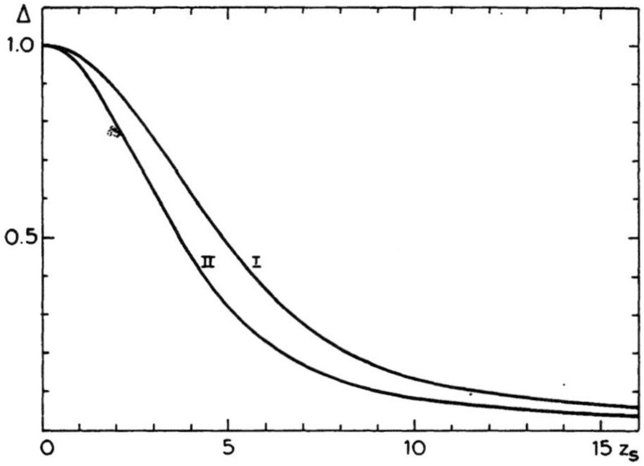

Fig. 2. Degradation factor Δ (or degree of partial coherence) as a function of $z_s = \pi\,\gamma_s\,D/\lambda$: I, for a circular source of angular diameter γ_s and a circular detector aperture of linear diameter D; II, for a square source of angular side γ_s and a square detector aperture with linear side D. λ is the wavelength of the radiation involved.

The integration can be carried out to yield

$$\Delta = \frac{16}{z_s^2}\{1 - \mathrm{J}_0^2(z_s/2) - \mathrm{J}_1^2(z_s/2)\}\,, \tag{2a}$$

which for small z_s can be approximated by the parabolic relation

$$\Delta \approx 1 - z_s^2/2\,. \tag{2b}$$

For other shapes of source and detector, respectively, similar functional relations for the degradation factor are found as illustrated in Fig. 2. Only for values of $z_s \ll 1$ the degradation factor approaches unity and a reasonable mixing signal is obtained. This is often expressed in words, as we have done before ourselves, by saying that the detector area should not exceed much the so-called coherence area of the source.

The signal-to-noise ratio S/N of the system is determined by amplifier noise and laser shot noise. In the ideal case, which can indeed be approached in practice, the laser signal can be made so strong that the shot noise prevails. In that case it can be shown quite generally that

$$S/N = 2\,\gamma\,\overline{N}_s\,\Delta \tag{3}$$

where $\overline{N}_s$ is the mean number of photons received per second on the area of the detector, and per unit bandwidth at the laser frequency. If the source can be considered as a thermal blackbody radiator, $\overline{N}_s$ is determined by the well-known Planck radiation formula. In interferometry one is always interested in sources much smaller than one coherence area of a single telescope, so Δ is practically equal to unity. This implies that for sources of such a size that Δ approaches unity and hotter than about 5000 °K, a visual laser is more favourable than an infrared laser, in contrast to the often read statement that coherent detection always yields a better signal-to-noise ratio in the infrared than in the visual region. Of course it remains true that for longer wavelengths the degradation factor can be more easily made to approach unity: larger telescope apertures can be used and wider sources can be detected with maximum efficiency.

It turns out that in practice for sources of about 5000 °K — using a single telescope of 50 cm diameter and a He–Ne laser emitting 4 mW optical power at 6328 Å — the signal-to-noise ratio per second observing time only amounts to 0.35 for a star of 0^m. In order to increase this value, the received noise signal has to be sampled in the usual way and integrated over a longer time. If the rf bandwidth is B Hz and the integration time is t seconds, a factor $(B\,t)^{1/2}$ can be gained in this way. The bandwidth B determines the spectral resolving power of the system and (for spectroscopic purposes) it cannot be made too large; it has usually a value of 150 MHz corresponding to a resolving power of about 3×10^6 at 6300 Å. With an integration time of 30 minutes the S/N ratio becomes 15 for a 0^m star, that is 1 for a 3^m star. It can further be increased by using longer integration times, a larger telescope aperture or a wider bandwidth, although the latter two quantities cannot be increased indefinitely, for coherence area reasons and because of the rf properties of photodetector diodes. The telescope size applicable for ground-based observations is moreover limited by the disturbing influence of air turbulence, which introduces uncorrelated phase shifts, that is, additional noise. Therefore the practical limit of the sensitivity for coherent detection spectroscopy from ground stations can be put at about the second magnitude. For interferometry from spaceships, to be discussed below, spectral resolving power is of less importance than spatial resolving power, and turbulence is absent, hence the sensitivity might be enhanced so that stars of the fourth magnitude can still be studied*.

* The Hanbury Brown method of intensity interferometry is about equally sensitive (to within a factor 2) with our method. However, turbulence effects are of much less importance there, and larger telescope areas can be used. This is offset by the disadvantages mentioned in § 1.

The previous conclusions and formulae have implicitly been based on the use of an optical as well as an electronic balanced-mixer system. The laser light and the star light fall first from different directions on a beam divider, so that two mixed beams arise in which the signals have undergone different phase-shifts. The two mixed beams are detected by two photodiodes (Fig. 3), which are connected to an electronic circuit in such a way that the two rf mixing currents are subtracted one from another, which is favourable because the phase-shift of $\pi/2$ arising between the transmitted and reflected beams in the beam splitters brings the mixing signals into opposite phase. Moreover, the noise present in the laser beam is completely eliminated in this way, and generally

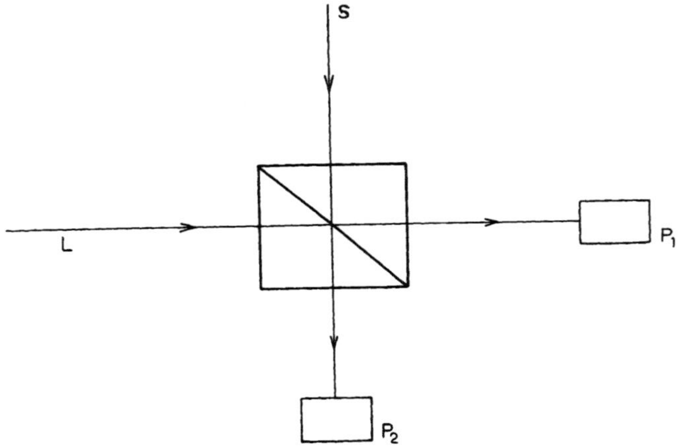

Fig. 3. Principle of the optical balanced mixer. Laser light L and star light S fall from perpendicular directions on a beam-divider cube with 50% reflection and transmission. The two combined signals fall then on two photocells P_1 and P_2, which form part of an electronic subtracting circuit.

the signal-to-noise ratio is enhanced by a factor between 2 and 4, depending on the type of noise involved, over that obtained in a single mixing process. Actually, an even more complicated set-up is employed, in order to be able both to use the polarization directions of the star light and to provide suitable chopping of the beams enabling a synchronous detecting system without loss of light. This is accomplished by applying two polarizing beam splitters, two polarization filters rotating with different frequencies and two optical balancers as described above. For a fuller description we refer to a more detailed paper by van de Stadt [2].

3. The Interferometer

We consider now a coherent detection interferometer system consisting of two telescopes with detector equipment spaced at a distance a. A single laser is used as local oscillator for both. Adding the two rf-mixing signals, each obtained in the way described in the previous section, in a correlating circuit, the resulting mean square rf interference (noise) current is, to a good approximation, given by

$$\overline{(i_{\rm rf}^{(\rm int)})^2} = 4\left(\frac{\eta\, e}{h\,\nu}\right)^2 \bar{I}_{\rm L}\, \bar{I}_{\rm s}\, \Delta \left\{1 + \cos\varphi_{\rm A}(t)\cdot\Gamma\right\}. \tag{4}$$

Here most symbols have the same meaning as before, $\varphi_{\rm A}(t)$ denotes the random phase difference between the beams of star light reaching the two telescopes owing to air turbulence or other uncontrollable causes and Γ is the well-known coherence factor of the extended stellar source as received by the interferometer. For a uniformly bright stellar disk of angular width $\gamma_{\rm s}$ this factor is given by an expression of the form

$$\Gamma = \frac{\sin(\pi\, a\, \gamma_{\rm s}/\lambda)}{\pi\, a\, \gamma_{\rm s}/\lambda}. \tag{5}$$

Proceeding as before in amplifying and detecting the quantity $\overline{(i_{\rm rf}^{(\rm int)})^2}$, a value for Γ is obtained, that is, the diameter of the star can be determined. If the surface brightness distribution of the star is not uniform, or if a double star is studied, the coherence factor yields information on precisely this brightness distribution, since it can be shown that this factor, which in fact is complex and thus characterized by an amplitude as well as a phase, is just the normalized Fourier transform of the brightness distribution $S(\gamma)$ projected along the base-line direction of the interferometer:

$$\Gamma = \frac{\int\limits_0^{\gamma_{\rm s}} S(\gamma)\exp\left(\frac{2\,\pi\, i\, a\,\gamma}{\lambda}\right) d\gamma}{\int\limits_0^{\gamma_{\rm s}} S(\gamma)\, d\gamma}. \tag{6}$$

Measuring Γ at various spacings a of the interferometer telescopes thus enables us to determine the brightness distribution $S(\gamma)$ by applying an inverse Fourier transformation.

Since the coherent detection method yields the complex coherence factor directly, asymmetric brightness distributions can be observed, and such effects as limb darkening can be studied in first order. This stands in contrast to other methods of stellar interferometry where $|\Gamma|^2$ is measured and symmetry effects are suppressed, whereas the limb darkening curve is effectively squared, resulting in much less accurate observational data. The only other way in which asymmetric brightness distributions can at present be directly measured is by the lunar occultation method [3, 4].

4. Practical Construction of the Interferometer

At the moment an interferometer based on the principles described is under test at the Utrecht Observatory. It consists of two mirror mountings (siderostat), as illustrated in Fig. 4, which are temporarily placed in a deserted military fortress somewhere in the Dutch meadows along a base-line which is parallel to the earth's axis, the mirrors viewing the southern parts of the sky. In this way change of effective base-length due to the earth's rotation is prevented. The mirrors are all-metal (aluminium) parabolic mirrors of 50 cm diameter and 2 m focal length, produced on a special type of hydraulic precision lathe devised in the Philips Research Laboratories at Eindhoven, Netherlands [5]. The optical surfaces were shaped by this lathe with an overall tolerance better than 0.2 μm from the required shape, whereas the lathe grooves do not produce

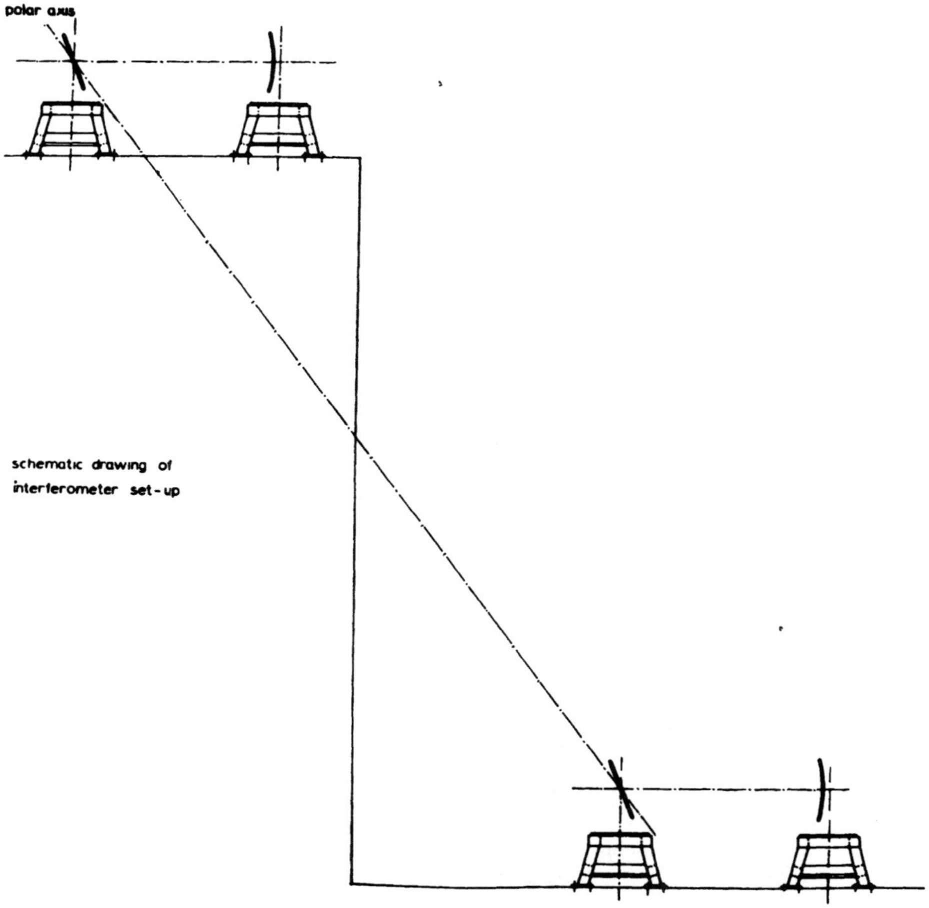

Fig. 4. Sketch of the interferometer installation to be situated on the ramparts of a deserted Dutch fortress. The mirror mountings are of the siderostat type, the base-line is parallel to the earth's axis, the mirrors view the southern part of the sky.

a ripple of more than $^1/_8$ Å. It is a pleasure to acknowledge here the generous and liberal assistance received from the Philips Works for the project. The two star light beams are directed to the detecting equipment by means of flat mirrors, produced in the same way, and combined with the common beam of a He–Ne laser (Spectraphysics Model 122), which is mounted on one of the mirror supports. The precision moving mechanism for the two interferometer mirrors, which have to stay parallel to within about 0″.1 of arc, is hydraulic and again developed by the Philips Research Laboratories. It provides a pointing accuracy of 2″ of arc. Base lengths will be applied ranging from 7 to 30 m, yielding a spatial resolving power of the order of nearly 10^8, corresponding to a few thousands of a second of arc. The observing program comprises stars brighter than the first magnitude among which are Sirius, ($-1\overset{m}{.}4$), Arcturus, ($-0\overset{m}{.}1$), Vega, ($0\overset{m}{.}0$), Capella ($+0\overset{m}{.}1$) and about ten others. No measurements have yet been made, but they are expected to start in the autumn of 1971.

5. Applications of the Coherence Detector Interferometer on the Earth and in Space

The following scientific programmes appear feasible with the aid of an interferometer as described:

(i) Measurement of the angular diameter of stars at different wavelengths. These wavelengths are selected by choosing different types of gas lasers as local oscillators, or even better by using ion lasers that produce at the same time coherent signals at a number of wavelengths.

(ii) Measurement of the limb darkening of stars at different wavelengths.

(iii) Measurement of the brightness distribution of asymmetric sources, in particular of double stars with a separation too small for visual observation.

(iv) Observations of stars where a peculiar brightness distribution is expected, such as fast rotating stars (measured at a wavelength coinciding with a specific absorption line), shell stars, or Wolf–Rayet stars.

The method is limited to the brighter stars, down to at most 4^m, when space techniques are applied. The use of the interferometer in space has two main advantages: (a) The effect of air turbulence, which will certainly be the most disturbing factor for ground-based experiments, is avoided, so that larger interferometer spacings leading to less than 0″.001 separating power can be used. (On the ground, to compensate for turbulence effects, one might have to resort to an interferometer consisting of more than two mirrors, and to applying complicated correlation techniques.) (b) The techniques can be extended to the (near and middle) infrared where a CO_2 or other infrared laser with wavelength in the region 10–50 μm should be used. In this spectral region many interesting bright objects can be observed; moreover, larger telescopes can be applied. According to the authors this is the most promising domain of application, where new and important interferometric data can be expected.

A secondary advantage of the space-borne interferometer is that the effect of the earth's rotation can be avoided, and the base-line direction can be fixed arbitrarily as desired. To start space experiments, the obvious way is to develop at first a balloon-carried interferometer. Preparations for such an experiment are now under consideration. Two telescopes of 50–100 cm aperture are envisaged, a variable base-line is conceived of maximum length 10 m. The two telescopes could be suspended independently from a common point below the parachute train and an extensible mechanical frame would be used to move them apart on ground command. This frame would act as a stabilized platform and house the laser package (Fig. 5). The coarse stabilization could make use of a magnetometer-reaction wheel combination; motor drives should move the telescopes in azimuth and elevation. A coarse-pointing accuracy of about 1° should be sufficient, since the field of view of the telescopes is a few square degrees. The fine-pointing system should be accurate to within a few seconds of arc. Similar balloon systems are now under construction in industry for

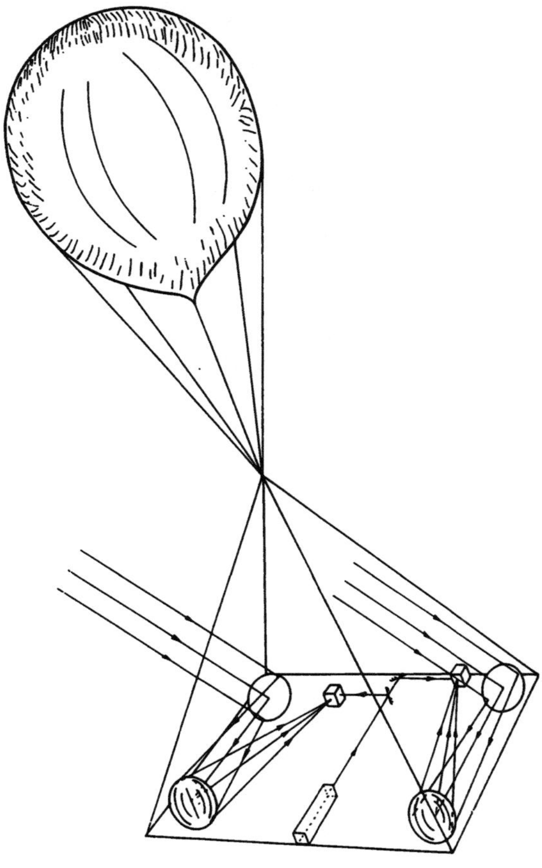

Fig. 5. Sketch of a balloon-borne interferometer.

other balloon-type experiments. Problems will arise in compensating for the pendulum swing of the whole package, which introduces path length and hence phase differences between the two telescope signals. Also, the alignment of the laser beam might require an active system, especially in view of thermal effects.

The correlation of the two rf-mixing signals will be carried out *in situ* and only low-bandwidth signals are transmitted to the ground. The desirable height to be attained by the balloon will depend on the remaining air turbulence and will probably lie between 25 and 30 km.

We foresee a time scheme of about three years from the beginning of the stage preparation of the balloon experiment to its actual flight. If this were successful, a satellite experiment could be planned, perhaps placing the interferometer aboard some type of communication–experiment satellite which already uses stabilized laser equipment, such as the ATS satellites.

Acknowledgment

This work has been made possible by a grant from the Netherlands Foundation for Pure Research Z.W.O.

References

[1] H. NIEUWENHUYZEN, Bull. Astron. Inst. Netherlands **19**, 391 (1968); Bull. Astron. Inst. Netherlands **20**, 300 (1969); Thesis Utrecht University 1970.
[2] H. VAN DE STADT, to be published.
[3] D. S. EVANS, Mont. Not. Roy. Astron. Soc. **115**, 468 (1955).
[4] R. E. NATHER, M. M. MCCUNTZ and D. S. EVANS, Astrophys. J. **160**, L 181 (1970).
[5] H. J. J. KRAAKMAN and J. G. C. DE GAST, Philips Tech. Rev. **30**, 117 (1969).

Space Research XII – Akademie-Verlag, Berlin 1972

TECHNICAL PROBLEMS AND PLANS FOR HIGH ANGULAR RESOLUTION OPTICAL TELESCOPES

M. J. Aucremanne

NASA Headquarters, Washington, D. C., USA

The ability to make measurements above the earth's obscuring atmosphere at wavelengths inaccessible to ground observations has opened up new fields of research. Astronomical instruments located in space as exemplified by OAO 2, can now reach the large remaining region of the electromagnetic spectrum. By coordinating optical telescopes in space with ground-based observatories, the fundamental problems of astronomy can be vigorously attacked.

The complexity and difficulty involved in the development and launch of large optical telescopes with fine pointing and stabilization systems, power systems, and data handling systems, will generate technical problems that demand a strong technology program. This paper summarizes the present status of optical telescope technology. Specifically, it presents some of the technical problems associated with high angular resolution optical space telescopes. It also treats the status of plans for the Large Space Telescope project in the NASA Astronomy Program.

1. Introduction

Man's interest in the stars and planets dates back to the very beginning of man himself. In the early days man thought that these heavenly objects played a distinct role in shaping his life. This keen interest led to the development of the telescope by Galileo around 1610. In the intervening — some 360 — years the telescope has gone from a hand-held instrument which produced a magnification of about 33 times to the present approximately 5 meter or 200-inch telescope at the Hale Observatory on Mt. Palomar which can penetrate far into the universe. The present paper will examine the status of present day technology as it applies to the Large Space Telescope (LST). It will outline some of the problems yet to be solved and will present the NASA plan for the Large Space Telescope and will include the present astronomy program.

Since the start of the space age man has taken his telescopes above the earth's atmosphere. Astronomers have known for many years that the earth's atmosphere, however transparent it may appear, is actually nearly opaque because it absorbs most wavelengths except for the visible windows between 3000 and 7000 Å (Fig. 1). Even the radiation observed by ground-based telescopes is distorted because the atmosphere is not homogeneous but is constantly fluctuating. This distortion limits the sharpness of most astronomical photographs to about 1 arc second. An optical telescope in space can effectively focus photons with wavelengths between 1000 Å and 1 mm, with a sharpness limited in principle only by the ratio of its diameter to the wavelength of light, thus allowing this entire spectrum to be measured with a single instrument.

A 1-m diffraction-limited telescope in space will have significant advantages over any ground-based telescope. Primarily, it would be continuously available and have resolution three times better than that obtained by the 200-inch Hale Observatory telescope. The best resolution of a ground-based telescope can be obtained only for perhaps a few hours a year. Moreover, the space

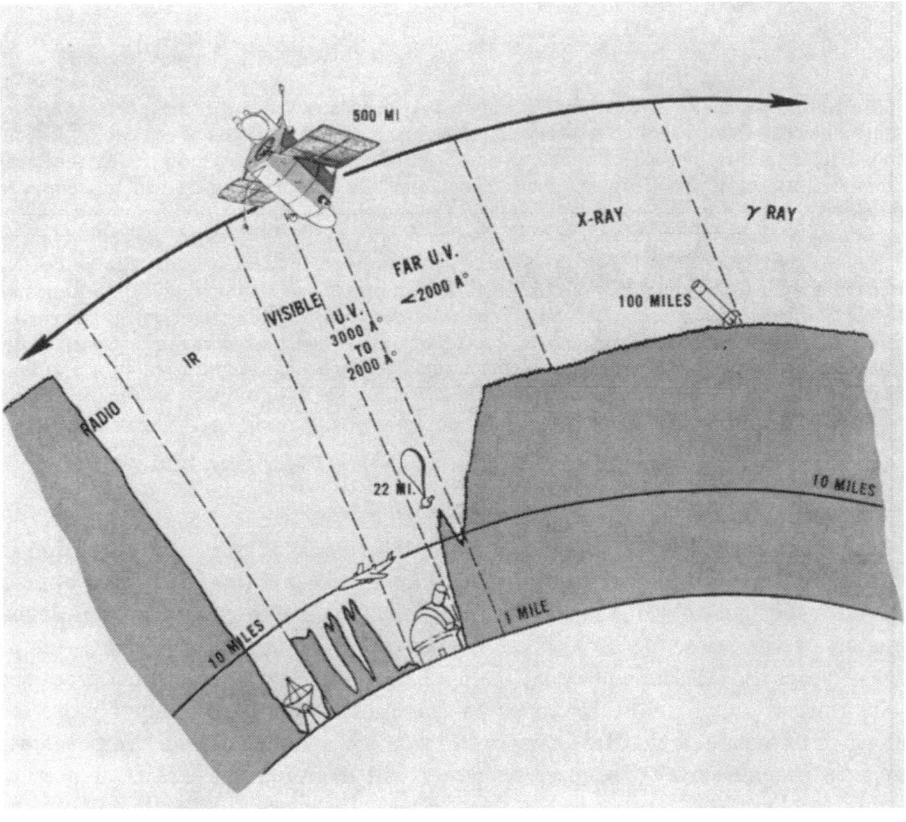

Fig. 1. Atmospheric veiling effects.

telescope can operate in the ultraviolet and the infrared bands which are closed to terrestrial observation.

The LST will provide a significant increase in the knowledge of the spatial structure of astronomical objects in the universe and permit more detailed observations of objects of increased angular resolution. The specific scientific objectives for the LST are as follows:

Scale and curvature of universe.
Evolution of galaxies and quasars.
Density and composition of matter in universe.

Structure of asteroids and cometary nuclei.
Planetary atmospheric and surface structure.
Composition of stars in neighboring galaxies.
Physical nature of pulsars.

To provide the proper perspective for the LST, a brief examination of the current NASA Astronomy Program is in order (Fig. 2). In the solar area four Orbiting Solar Observatories (OSO's) are approved through 1976 with the launch of the Solar Apollo Telescope Mount in 1973. In the ultraviolet region, Orbiting

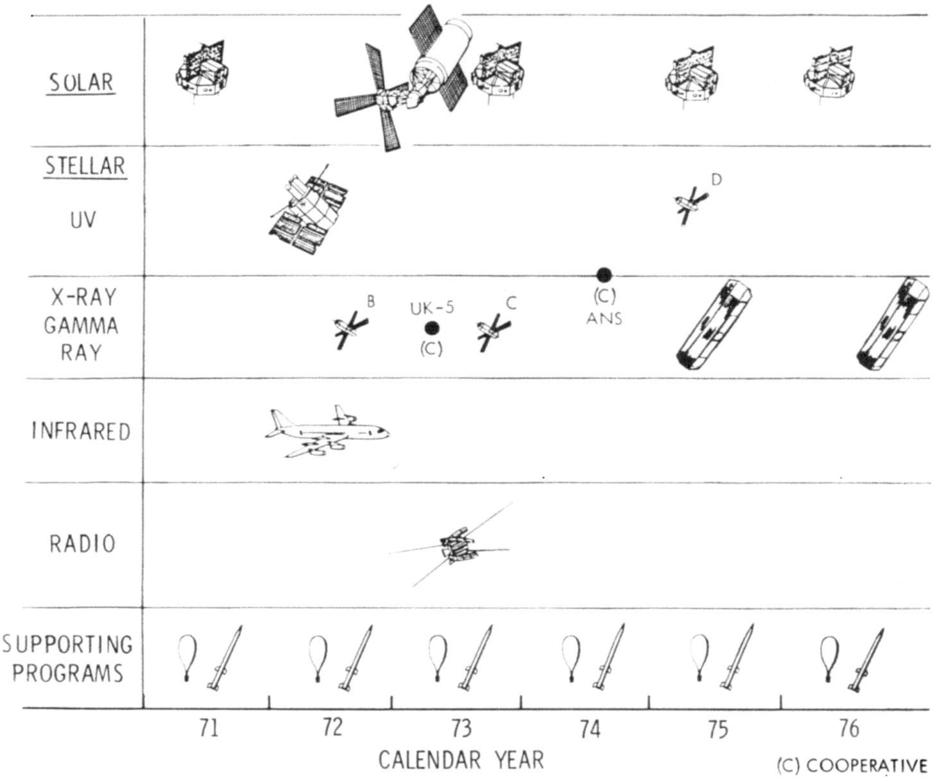

Fig. 2. Major astronomy missions.

Astronomical Observatory-C (OAO-C) containing the Princeton experiment will fly in 1972. The Small Astronomy Satellite-D (SAS-D) is under study as a uv mission with international participation. Two Small Astronomy Satellites for X-ray and gamma-ray astronomy are approved. The SAS-B is being prepared for a launch in 1972 and will measure the spatial and energy distribution of galactic and extragalactic gamma radiation. The SAS-C, scheduled for launch in 1973, will monitor changes in intensity or in the shape of the energy spectrum of X-ray objects whose emission has been found to change with time. Our cooperative projects are with the Netherlands and the United

Kingdom and involve the building of small Explorer satellites with the U.S. furnishing launch vehicles and small experiments for X-ray and ultraviolet astronomy.

In the infrared region it is planned to fly the C-141A jet aircraft. It has been found that jet aircraft can fly above most of the atmospheric constituents that absorb far infrared radiation, permitting observations of celestial objects at these wavelengths. A tremendous increase in capability will be possible with a 0.91 m (36 inch) aperture Cassegrain telescope now being built for use aboard the C-141A aircraft.

In the radio spectrum, the Radio Astronomy Explorer-B (RAE-B) is being prepared for observations beginning in 1973. It will be placed in an orbit which will permit improved measurements to be made, as compared with RAE-A, by using the moon for occultation of strong sources and by removing the satellite from the radio emission generated in the earth's magnetosphere. Finally, under Supporting Programs a continuation of balloons and sounding rockets is shown.

2. Mirror Materials for the LST

The major component of any telescope is the primary mirror; therefore one of the most important elements in the LST will be the primary mirror. Mirror materials have been and will continue to be under investigation for a number of years. Most of the work to date has involved small samples investigated at a variety of temperatures. However, the mirrors to be tested in the future preferably should be of the size that will be flown in the LST.

A study sponsored by the Advanced Research Projects Agency and completed in the first quarter of 1970 established the following goals: (i) a given mirror material must be capable of being fabricated in the sizes and quantities required; (ii) it must be polishable to high surface quality and be a receptive host to one or more types of special purpose optical coatings; (iii) it must be polishable to within $\lambda/50$ rms of its design surface geometry, and remain stable within this requirement over its expected service life and range of environmental conditions. This particular study was undertaken by the Perkin-Elmer Corporation with the Battelle Memorial Institute and the Stanford Research Institute as subcontractors.

The Langley Research Center initiated a contract with the Boeing Company (NASA Contract NAS 1-8851) in 1970 to develop microstrain gauges capable of measurements of the order of 10^{-10} cm cm^{-1}. Torsion, tension, and compression measurements were made on fused quartz, Ultra Low Expansion (ULE) fused silica, Cer-Vit, and iron doped beryllium. Initial tests showed unexpected differences between materials. These differences were accounted for by surface microstresses introduced by the fabrication process. However, when these surface stresses were removed by acid etching, all samples yielded at essentially comparable strains of 3×10^{-10} cm cm^{-1} in torsion and 2×10^{-10} cm cm^{-1} in compression and tension. Further work is being conducted on the same material samples with optically polished surfaces. The Langley Research Center has subsequently determined that the problem of surface stress can be resolved by fabricating primary mirrors of the various materials and of sufficient size so that the data can be extrapolated. Pertinent measurements of the mirrors must be made over an extended period of time to determine if aging affects the mirror materials. In choosing a mirror for the LST, trade-offs must involve the mechanical structure of the telescope, along with the thermal and optical interfaces associated with the total design. The long lifetime, up to 10 years, and other environmental influences dictate a highly stable mirror structure.

Over the past several years, two mirror concepts have been proposed. Perhaps the most basic and simplest approach is the large aperture monolithic mirror

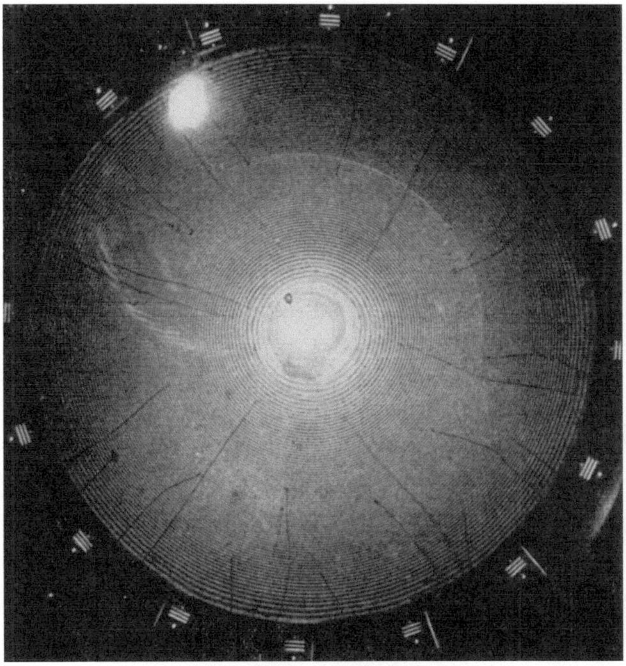

Fig. 3. Stratoscope 2 primary mirror.

that is both thermally and mechanically stable and can function in a passive manner over a long period of time. An example of this is the 1-m fused silica primary mirror fabricated for Stratoscope 2 (Fig. 3). It was figured to better than $\lambda/50$ rms and provided resolution approaching 0.1 second of arc in photographs of Uranus.

The second approach involves a segmented or a thin deformable primary mirror which requires a set of actuators to correct continually the figure of the mirror (Figs. 4 and 5). The segmented approach would not require the actuators to be in service at all times. A major decision between these approaches and associated materials and structural designs must be made. Many mirror materials have been considered as suitable for 2–3 m diameter structures.

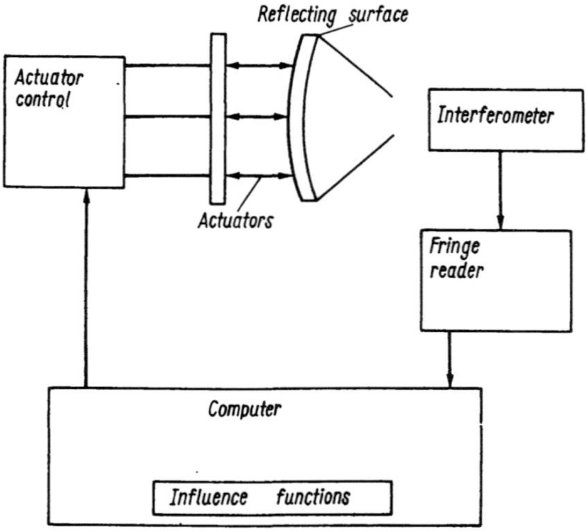

Fig. 4. Active primary mirror.

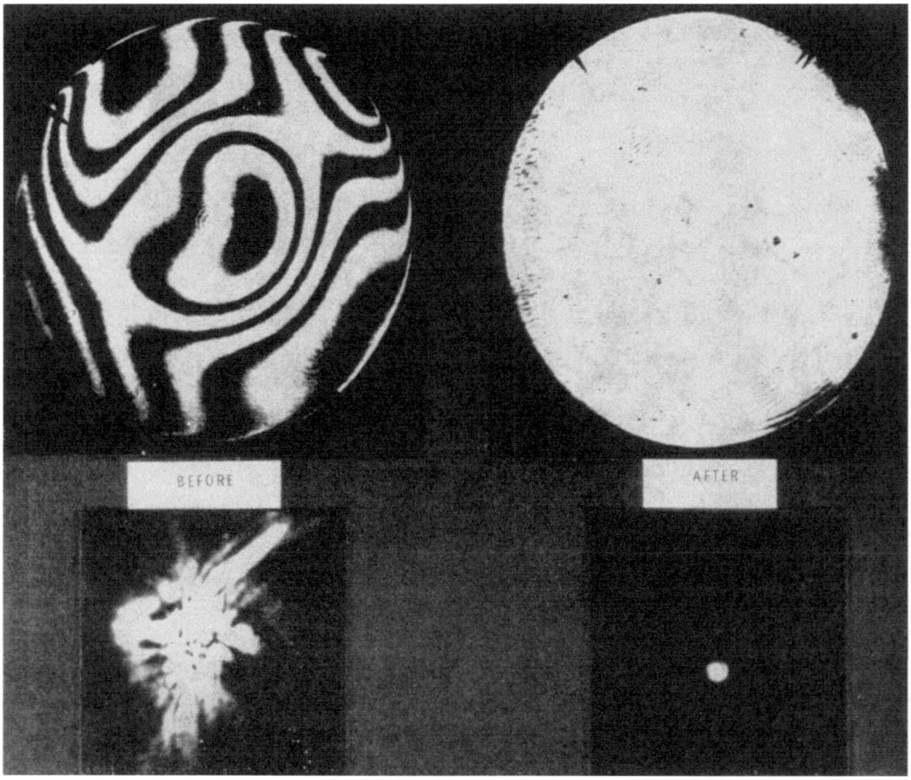

Fig. 5. Interferogram showing operation of active optics system.

Material Property	Owens-Illinois Cer-Vit	Corning ULE	Schott Zerodur	G.E. Type 124	Ohara Oharax Zero
Lightweighting	Yes	Yes	Not at present	Yes	Not at present
Weight saving	Up to 80%	50 to 75%	—	55 to 60%	—
Maximum size built	75 inches	80 inches	—	24 inches	60 inches Solid
Size capability					
Present	165 inches	150 inches	—		
Future	200 to 250 inches	—	—	158 inches built up now	— —
Fabrication type	Monolithic cored	Monolithic fused	Solid		Solid
Minimum dimension					
th	1/32 inch	No minimum	—	1/16 inch	—
αB	1/10 inch	No minimum	—	1/16 inch	—
B&h	Design parameters	Design parameters	—	Design parameters	—
Method of fabrication	Cored out from solid by drilling through rear of blank and removing drilled material	Core formed from flat plates by fusing with heat and pressure Faces fused as sheet to core	— —	Core formed from drawn quartz tubing sealed to face plates with molten glass	—
Core geometry	Any geometric shape	Square and hexagonal cells	—	Any geometric shape	—
Advantage	Test blank before coring Any shape core possible Test for α also	Quality control of α from piece to piece before fusion More uniform deflections (minor)	— —		—
Disadvantage	Question predictability of uniform α of large blank Coring holes in rear		— —	High α α also varies across face sheet core interface	— —

Mirror Blanks Fabrication (Structural Light-weight Capability)

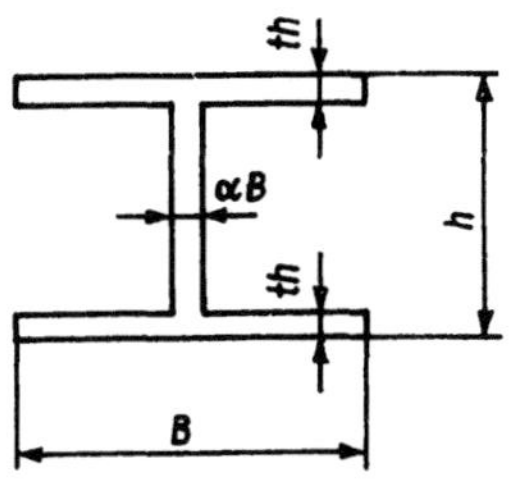

Fig. 6. Mirror blank fabrication data.

These include fused silica, ULE fused silica, Cer-Vit C-101, aluminum, beryllium, molybdenum, silicon, etc. It appears that the only materials available today with proven stability and availability in large sizes are fused silica, ULE, and Cer-Vit (Fig. 6). Of the other materials the Perkin-Elmer report states that: (i) Aluminum cannot be polished directly at this time, therefore electroless nickel coatings are applied which are figured and polished; optical coatings are then applied over this surface. Nickel coatings, however, have poor thermal and temporal stability and as such are considered very risky for prolonged space applications. (ii) Silicon, beryllium, and molybdenum have excellent stability but are size limited at this time (0.5 m for Si, 2 m for Be and 1 m for Mo). Larger sizes are achievable through simple extension of manufacturing techniques or through mosaic techniques by the joining together of smaller pieces by diffusion bonding; however, considerable development in this area is needed. A comparison of the three candidate materials shows that ULE and Cer-Vit have a thermal expansion coefficient lower than fused silica by an order of magnitude or more and have the same high level of stability, and indicates better thermal/optical performance.

Prior to choosing a mirror material and a lightweight concept for the LST mirror, several studies are required. More testing is necessary to choose between a passive monolithic primary or active figure control. Size-effect studies should be conducted on ULE and Cer-Vit to determine influences of lightweight design on mirror mounting, thermal gradients, and temporal stability.

3. Optical Coatings

Vacuum-deposited-aluminum films have the highest normal incidence reflectance of any material studied to date in the spectral region $\lambda > 1000$ Å. Its reflectance, however, is quickly degraded by oxidation to low values in the region $\lambda < 1800$ Å. A solution to this problem is to protect the freshly deposited aluminum with a dielectric film having high transparency in the short wavelength region. Two materials are candidates, lithium fluoride and magnesium fluoride with short-wavelength cut-offs of 1050 Å and 1120 Å, respectively. The reflectance for lithium fluoride and magnesium fluoride protected aluminum is compared with aged aluminum as shown in Fig. 7. Magnesium fluoride protected aluminum ($Al + MgF_2$) provides the experimenter with a stable (in the absence of contamination) optical coating having a reflectance of about 28% or better for $\lambda > 1200$ Å. Experimentation with lithium fluoride protected aluminum (Al + LiF) has not been successful to date. In the 1200 Å $< \lambda <$ 1800 Å region the reflectance is significantly lower than that for $Al + MgF_2$ but better than that for aged aluminum to about 1600 Å. Exposure of Al + LiF to typical laboratory conditions results in reflectance degradation extending to the visible spectral region with the greatest change occurring for $\lambda < 2500$ Å. This is not surprising since LiF is known to be hygroscopic. As expected, the reflectance for Al + LiF can be kept reasonably stable by storage in a dry atmosphere. Since it is not possible to keep a space optical system with Al + LiF optical coatings unexposed, some reflectance degradation must be accepted. The greatest change is expected within the first six months. At this time relatively stable reflectance values ranging from 40 to 50% for 1000 Å $< \lambda <$ 1800 Å can be expected.

Although Al + LiF has been used for both the prototype and flight optics for the OAO-C Princeton Experiment Package at the Goddard Space Flight Center (GSFC), a major problem was the requirement for uniform reflectance and film thickness on the 0.81 m (32 inch) diameter primary mirror. By

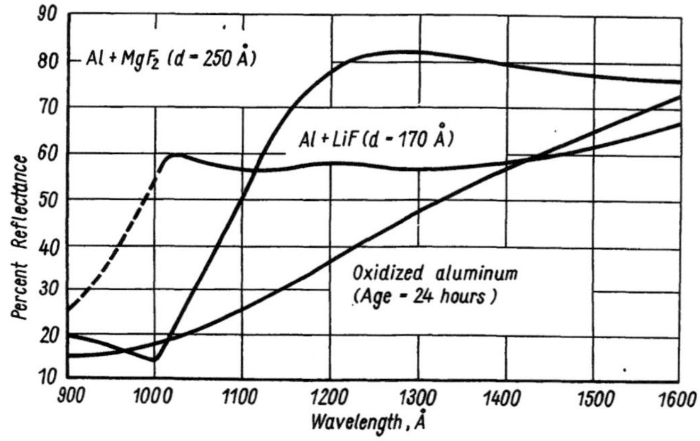

Fig. 7. Ultraviolet reflectance of coated mirrors.

rotating the mirror over a specially designed stationary baffle both requirements were met. Test depositions on a simulated mirror indicated that a radial thickness distribution constant to 5% or less and a radial reflectance distribution constraint to about 2% were possible. For the flight primary mirror a reflectance of 0.632 ± 0.004 rms and 0.643 ± 0.024 rms were obtained at Lβ (1026 Å) and Lα (1216 Å), respectively. Physical thickness measurements were not possible on the samples processed with the flight mirror.

4. Pointing Systems

The Large Space Telescope requires accurate pointing if it is to perform its task of locating faint objects in the sky. The pointing system should be independent of the magnitude, shape, and position of the object. Various pointing stabilities will be called for by the experiments aboard the LST depending upon the experiment, magnification and wavelength. At present the far uv camera with highest magnification is expected to require pointing stabilities in the range of 0.003 and 0.02 arc second. In order to accommodate such requirements the fine error sensor will use a guide star within a reasonable offset angle from the target. According to a GSFC and Kollsman Instrument Corporation study, there are two principal factors which have dictated the use of the offset tracking technique. First, in order to utilize the areas of the sky that have a low stellar density suitable for tracking, it is desirable to have as wide a field of view as possible. The second factor is the tracking accuracy required. A brief review of experimental flight experience will be useful at this point.

4.1. OAO 2

The OAO 2 spacecraft uses a General Electric attitude control system which provides one inertial wheel actuator for each pointing correction in the pitch, yaw and roll axes. Six Kollsman star trackers (three are required at any one time) sense the spacecraft pointing by means of simultaneous tracking of three or more pre-selected stars. The telescope itself is mounted on two gimbals which provide $\pm 45°$ rotation about two orthogonal axes. Each gimbal is positioned by means of a dc torque motor to the commanded angle. A resolver is used for the angle readout, which is the feedback in the closed-loop gimbal operation. Actual predicted tracker accuracy, including gimbal angle readout errors and servo stability, was 20 arc sec for which the predicted OAO spacecraft pointing was 1 arc minute circular error for a duration of 50 minutes during each orbit. Based on spacecraft data the actual pointing accuracy has been maintained at 45 arc sec for two years in orbit.

4.2. Stratoscope 2

In March 1970 Stratoscope 2 carried a 1-m diffraction-limited mirror to an altitude of 24.4 km (80000 feet) and obtained high-resolution photographs of Jupiter, Uranus, and a Seyfert galaxy's nucleus. Pointing of the Stratoscope 2 is accomplished by tracking two stars located within the 50 arc-minute field

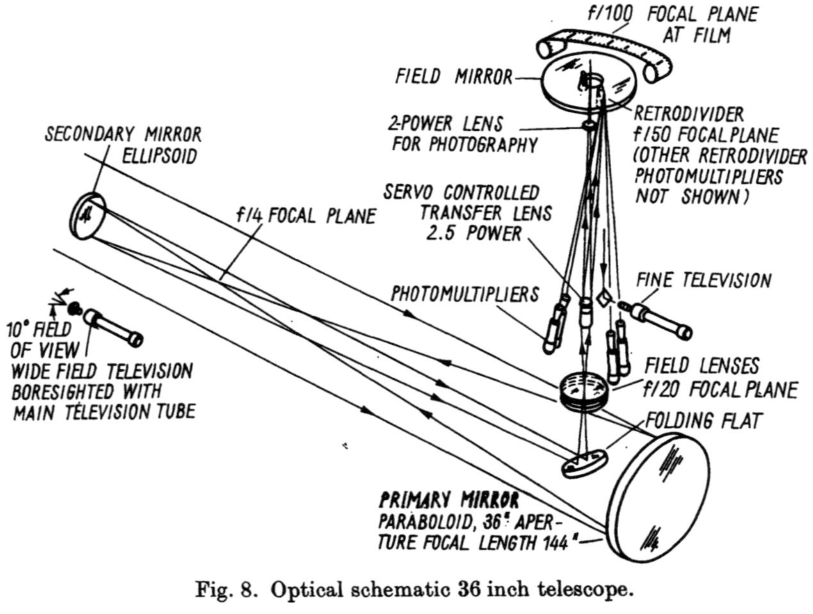

Fig. 8. Optical schematic 36 inch telescope.

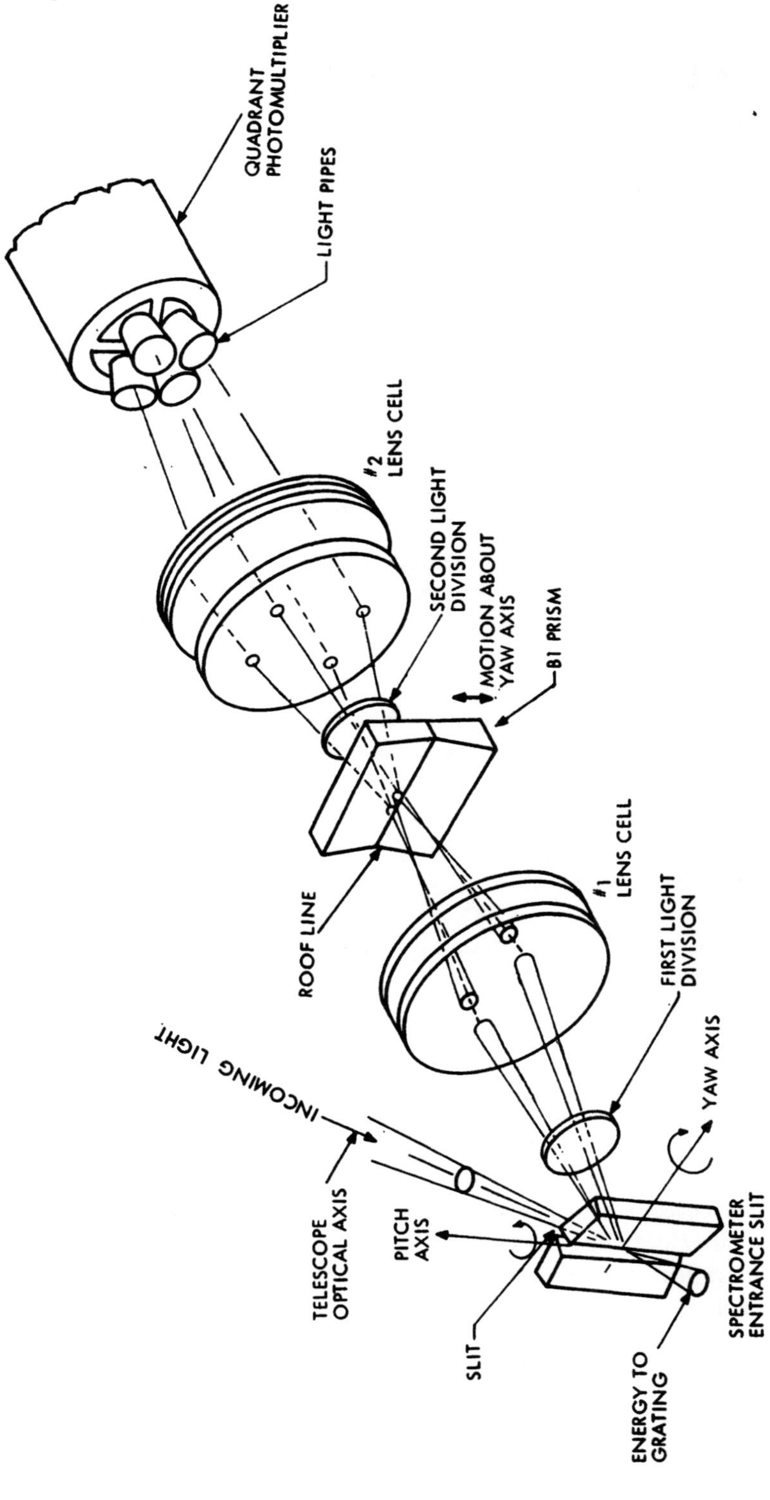

Fig. 9. Orbiting Astronomical Observatory Princeton experiment package fine guidance system.

of view of the telescope while simultaneously holding steady on the object being photographed (Fig. 8). The two star tracking allows one "guide" star to hold the telescope pointing at its object and the second "guide" star to control the roll of the telescope. A gyroscope is often used to control roll during short exposures. The guide stars may vary from 5th to 9th magnitude. Stratoscope 2 employs a two-stage pointing system. Coarse pointing, 15 seconds of arc, is achieved by aiming the telescope at the object and holding it there for as long as required. Within this 15 arc sec pointing, the main telescope transfer lens is positionally adjusted (4–12 Hz bandwidth) so that the image is held fixed in the photographic plane equivalent to an overall pointing accuracy of 0.02 second of arc. The Stratoscope pointing system involves an operator on the ground and through a television system can provide the overall pointing accuracy stated above.

4.3. Apollo Telescope Mount (ATM)

A pointing system is under development for the ATM. The mission for the ATM is high angular resolution measurements of the sun. Instruments aboard the ATM are large and quite heavy compared with those flown on the OSO. The Attitude and Pointing Control System (APCS) consists of two separate but interrelated control subsystems. Its primary portion provides attitude control and stabilization for the entire Skylab assembly. The other, the Experiment Pointing Control (EPC), stabilizes and fine-points the ATM experiment package, further isolating it from disturbance torques to the Skylab assembly. The entire 6000-pound instrument canister rides within two-axis gimbals, located around the canister center of gravity. Pitch and yaw control can be provided within an accuracy of ± 2.5 arc sec for periods up to 15 minutes, utilizing fine-pointing sun sensors for attitude reference. A roll ring allows the experiment package to be manually commanded to any desired roll orientation throughout $\pm 120°$.

5. Fine-Control Error Sensors

The Princeton Experiment Package for OAO-C includes a fine guidance system to meet the 0.1 arc sec specification (Fig. 9). The system, consisting of two field lenses, a roof prism and a quadrant photomultiplier, compares the target star energies falling on the four quadrants of the telescope entrance-slit jaws. The reflective sector of the rotating mirror sequentially reflects the light of each of the four quadrants into the photomultiplier window. The photomultiplier tube output consists of consecutive sets of four pulses whose amplitude, when unequal, is proportional to the error signal and the pitch and yaw axes. Equal amplitude of the pulses indicates a known position. The photomultiplier tube output pulse is routed to the spacecraft stabilization and control subsystem to maintain the 0.1 arc sec pointing accuracy.

6. Telescope Design

A telescope may be improved by careful design, construction, and figuring. Unlike the other subsystems, telescope angular resolution has a physically irreducible limit, viz. its diffraction limit. Regardless of how well the telescope is constructed, or how perfectly it is figured, it will, at best, yield a diffraction-limited image for a point source whose angular spread is equal to $1.22\,\lambda/D$ where λ is the wavelength and D the telescope diameter. Therefore, if all the other subsystems are perfect, a 3 m telescope will limit the overall resolution to 0.04 arc sec when λ equals 5000 Å or 0.02 arc sec when λ is 2500 Å.

Because the system resolution is limited ultimately by the telescope diameter, the purpose of a design should be to come as close as possible to this limit without paying too high a price in terms of complexity, design difficulty, and sacrifice of other system parameters such as field of view.

A NASA study (Contract NASA 3-1925) completed in May 1970 by the Optical Systems Division of Itek Corporation investigated the problems of technology involved in the design, fabrication, and operation of a high quality telescope for space astronomy. The study confirmed the feasibility of a large space telescope up to an aperture of 3 m or more. Specific design concepts will be examined by NASA during the next year so that the Large Space Telescope can be initiated with participation by the scientific community, government and industry.

7. Stray Light Suppression

In almost every program involving star trackers and fine guidance systems the problem of stray light is encountered in some form. The sources are now well understood but in each new design the stray light propagation and its effect on the signal must be examined.

Tests made during the design of the OAO tracker proved that the use of an astrodome or total enclosure of the tracker was not feasible because of noise introduced by impinging sunlight. The resulting open design required special attention to the choice of materials and lubrication of moving parts in vacuum. A stray light shield was successfully developed, which had a one-stage attenuation for earth shine up to 14° and a two-stage attenuation of 10^{12} for sunlight up to 30° from the optical axis. The phototube required special selection and treatment for star tracker duty. Spectral response and tube orientation to the optical center line was found critical to threshold sensitivity. Special shielding was required to eliminate stray light. The stray light environment of the OAO typifies that encountered by most fully gimballed systems. The scientific objectives required the spacecraft to point as close as possible to the sun or the sunlit earth's limb to maximize the availability of guide stars at all times regardless of season or position of the spacecraft in orbit. The stray light environment, the acquisition time, and the tracking accuracy required of star trackers constitute one of the more formidable sets of operational requirements to be met.

8. Image Sensor

The susceptibility of photographic film to radiation in space and the higher quantum efficiency of photoelectric surfaces make integrating television tubes necessary for maximum effectiveness. The primary mission of the sensor is to obtain images and spectra of stellar and planetary objects in conjunction with a large telescope and spectrograph. The Secondary Electron Conduction (SEC) vidicon and image isocon is a variation of the image orthicon. Both the SEC vidicon and image isocon have undergone extensive evaluation by Princeton University. This interest originated with the Stratoscope 1 telescope that used a vidicon camera to guide on sunspots. Since 1963 the use of image orthicons in the astronomical image integration systems has been supported by the National Science Foundation, NASA, and the Office of Naval Research as a part of the Stratoscope 2 program.

In 1964 Princeton University, with NASA sponsorship, undertook a study to investigate the suitability of available integrating televison tubes for use in an orbiting astronomical observatory. A subcontract was awarded to Westinghouse in 1964. This study dealt primarily with the SEC vidicon and opticon. A final report in December 1964 concluded that the SEC vidicon and opticon afforded many advantages for astronomical research.

In the fall of 1966, Princeton contracted with Westinghouse to build an SEC vidicon for use with ground-based telescopes. This contract also called for improving the integration capability of the WX5419-B tube so that 15-minute exposures were possible. This led to coating the interior of the image section with chrome oxide to decrease the field emission and also help block any light generated in the glass wall itself due to high voltage. Camera and tubes were delivered in the summer of 1967. The tubes were evaluated on the 0.91 m (36 inch) telescope at Princeton and on the 1.52 m (60 inch) telescope at Mt. Wilson. The results of these tests were presented at the Fourth Symposium on Photo-Electronic Image Devices held at the Imperial College in London, 16–20 September 1968, and published in *Advances in Electronics*, Vols. 28A, B, 1969, entitled "Integration TV Sensor for Space Astronomy".

In summary, the Mt. Wilson test demonstrated the ability to find and guide on stars that were two stellar magnitude fainter than could be observed by eye and also verified the photometric calibration made at Princeton. The tube was found to have an internal background low enough to allow a one-hour exposure. However, on the negative side, microphonics were excessive and a target storage capability in photoelectrons limited the theoretical signal-to-noise ratio to about 33. Following the test at Mt. Wilson, the results of the evaluation program were discussed with RCA and Westinghouse. With regard to the image isocon, a modest redesign was undertaken to optimize the tube for close scan operation. With regard to the SEC vidicon, it was agreed that the suppressor mesh should be removed in an attempt to reduce the microphonics. One WX5419-B tube without a suppressor mesh has been evaluated in the laboratory with very encouraging results. The resolution was markedly improved. A follow-on study is planned to continue this important work.

9. Data Handling

It would be desirable to observe a celestial scene containing objects covering a broad range of brightnesses. However, for all detectors there is a maximum exposure over which response can be expected to be linear. Additional exposure beyond this point results in a deterioration of the spatial resolution capability of a detector (i.e. a broadening of the image of a point source). For linear operation, the exposure must not be exceeded so that the brightest object does not exceed the saturation point. The dimmest object yielding useful information would then be determined by the amount of noise present in the system and the quality of data being sought. The useful dynamic range of any detector may be extended by taking many exposures and superimposing the results. The brightest object of interest within the field determines the duration of a single exposure and the dimmest determines the total number of exposures required. Practical consideration of experiment time and storage space ultimately limits the sensitivity of the instrument. Were is not for these limitations the signal-to-noise ratio associated with any object, however dim, could be improved indefinitely as a result of an ever larger number of exposures. Where electron image sensors are used, electronic data processing and magnetic storage is usually required.

Photographic film is the most common recording medium for ground-based astronomy because of its information capability, sensitivity, reproducibility, and convenience. Although its quantum efficiency (0.1%) is far below that of typical photocathodes, the possibility of using 35.5 cm (14 inch) square plates goes far to equalize its overall capability with respect to electronic devices. For cases where quantum efficiency is of prime importance (very low energy densities), an electronic image intensifier can be used as a light amplifier for the film, at the sacrifice of a large recording area. In addition, the long-wavelength response of photographic emulsions presents difficulties in its use for ultraviolet image recording. For long-term orbital astronomy, film has two further disadvantages. These are sensitivity to radiation damage, and the difficulty and inconvenience of film resupply and data retrieval. The radiation sensitivity requires heavy storage vaults whose weight grows as a function of storage time. Data retrieval requires frequent visits for on-board processing and electronic transmission.

The television camera tube is a light integrating device. It is this feature which makes it useful for the LST. A multiple exposure scheme can be used to yield broad dynamic range without compromising the resolution capability of the detector. The tube is repeatedly allowed to integrate the light after which the resulting electronic image is scanned and recorded in a memory unit.

10. Thermal Design of Space Telescopes

High angular resolution space telescopes will have rigid requirements for thermal control. Because of the effect of thermal gradients on the primary mirror, the telescope structure, alignment, and the pointing and fine guidance system, it is mandatory that the thermal design be an integral part of the total LST systems design (Fig. 10).

The OAO 2, launched on 7 December 1968, is still being used successfully as a space telescope. Both passive and active temperature controls are used. Passive temperature control is achieved by thermal coatings, super insulation, and conductive-path control. In the active system, louvers and thermostatically controlled electrical heaters are used. The OAO 2 is in a 760 km (470 mile) circular orbit with a period of about 100 minutes subjecting the spacecraft to sunlight from about 65–83% of the time. Thermal control of the electronic equipment is achieved primarily by a passive design approach. Each box is designed to conduct and to radiate all its heat to the heat-sink surface, which is mounted on a honeycomb panel. Heat conducted to the honeycomb panel is radiated to the external skin and subsequently to space. Thermostatically

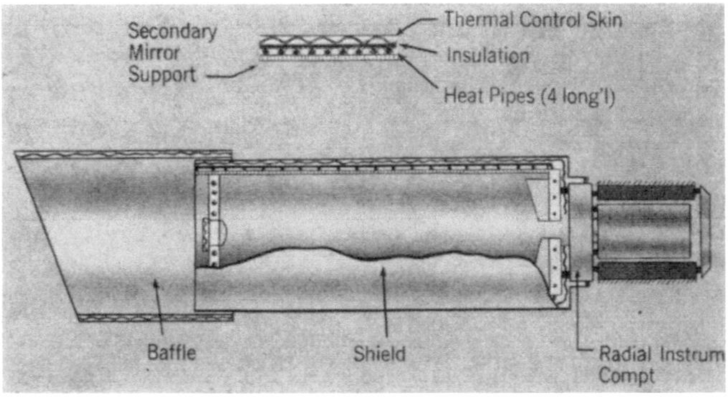

Fig. 10. Observatory thermal control.

controlled heaters mounted on the honeycomb panel are used when minimum design temperatures cannot be maintained passively. To minimize the effect of the heaters on the power supply, additional active control by means of louvers mounted to the honeycomb panel is employed, when significant power savings can be achieved. Louvers are closed during cold conditions, thus reducing the radiator coupling from sink to skin. Open louvers raise the coupling during high temperature conditions.

The OAO telescopes are thermally coupled to the structure and insulated from cold space. If properly insulated, the telescope will run within a few degrees of the mean structure temperature. The spacecraft structure is fabricated of a chemically milled aluminum alloy which has a high coefficient of thermal conductivity. To isolate the structure thermally from heat sinks and sources, stainless steel, nylon, and fiberglass are used as connecting members.

11. Experiment Temperature Control

The experiment package comprises the primary telescope system plus associated electronic equipment. All electronic equipment for experiments not required to be within the telescope envelope is stored in separate compartments and is handled in the same manner as the other spacecraft electronic equipment. Interior surfaces of the telescope and the sun shade are painted black in order to prevent incident earth-albedo reflection from saturating the star image on the detector. To achieve thermal stability, the telescope is insulated from open-end radiation to space and is thermally linked to the spacecraft by conduction through four aluminum lugs and by radiation between the OAO container and the tube. The spacecraft structure is designed to provide the desired mean operating temperature for the telescope. This eliminates temperature differences between the spacecraft and the telescope that would cause misalignment of the star trackers and the telescope. Actual in-flight data shows very good agreement between actual and computer temperatures with 95% of the calculated equipment temperatures being within 10 °F of the flight data.

The LST of the future will have stringent requirements for fine pointing accuracy, reliability, long life, and operational flexibility. Some of the major thermal design considerations that will be required for achieving the LST objectives include tighter temperature control for the telescope, minimum consumption of heater power, and a narrower operating temperature range for equipment with wider variations in the duty cycle and environment of the equipment. New and improved thermal control techniques are being developed to meet these requirements and some of these are discussed below.

The heat pipe which has been under development for a number of years is a device for transferring heat by a closed-cycle evaporation/condensation process. Two highly desirable characteristics inherent in its performance are extremely high effective thermal conductance and near isothermal operations. If used in connection with the thermal control of a spacecraft, heat pipes are ideal for isothermalization and temperature control of equipment with varying duty cycles in a space environment. Attaching heat pipes to the spacecraft structure is one method of producing structural thermal gradients with a relatively low weight penalty. In such heat pipes, sections opposite the hot structure elements will function as evaporators and sections opposite colder structure elements will function as condensers. The OAO-C to be launched in 1972 will carry a network of heat pipes and will provide the data necessary for the design of the thermal control system of the LST.

In summary, the results of a Grumman Aircraft Corporation [13] effort on the OAO spacecraft series and other similar studies have indicated that there are a number of techniques and devices that the designer can use for achieving his objectives:

Isothermal environments achieved with heat pipes.
Cold detectors enabled by coupling to space with heat pipes.
Louvers used for control of absolute temperature in subsystem and instrumentation compartments.
Thermal control surfaces are not exposed to sun.

The trade-offs between the techniques reviewed above are generally effective only when considered for a specific application. As newer techniques are developed, trade-off conclusions may differ. It appears at this time, however, that the solutions to all of the anticipated problems are within the state of the art.

12. Calibration, Simulation, and Test Facilities

A partial step toward a combined thermal vacuum optical facility has been taken by means of the Princeton optical bench. That bench and a special optical dome are mounted on a 3.65 m (12 ft) diameter by 15.24 m (50 ft) high thermal vacuum chamber. Several optical measuring instruments are mounted on an isolated table outside and above the chamber. These instruments consist of a 10.2 cm (4 inch) diameter star simulator, several autocollimators, and theodolites. Electronic levels and a pendulous mirror are used to adjust the lines of sight of these instruments so they are parallel to each other. Hanging from the optical table are three servo-driven columns which penetrate the dome through vacuum-type bellows. Inside the chamber, the actual bench structure is connected to the columns. At the lower end of the bench is a rigid support ring in which the experiment package and a multimode thermal control shroud are mounted. As the thermal gradient on the thermal package is changed through the control shroud to simulate various orbital conditions, the angular relationship among the experiment's fine guidance sensor, telescope axis, spectrometer slit location, and several reference mirrors located on the critical elements can be measured.

Acceptance testing of the OAO Princeton experiment is presently underway. Angle-measuring accuracies of 0.1 arc sec should be demonstrated. With this design, isolation from vibrations and other distortions is sufficiently good to keep the jitter less than 0.1 arc sec.

13. Optical Alignment Facility

To meet the fine alignment and the fine photometric requirements of the OAO 2, the optical alignment facility was designed and built at GSFC. This facility (Fig. 11) incorporates an improved method of optical tooling for the determination of the angular alignment of randomly oriented devices. The optical alignment facility was designed primarily for determining the roll, pitch, and yaw coordinates of star trackers, solar sensors, jets, and other components of the OAO. It is located in a class-10000 clean room. The unique feature of this facility is its inherent capability to measure the pointing direction of any item on the test object, regardless of its location, provided the item can be equipped with a suitable optical test tool.

Calibration chamber concepts have been developed to handle the 3 m diffraction-limited telescope. Because of the logistics involved in handling such a very large stellar observatory, studies should be conducted to test the feasibility of incorporating full-systems simulation and testing capability as well as the calibration features into a single facility. At the present time

Fig. 11. Optical alignment facility.

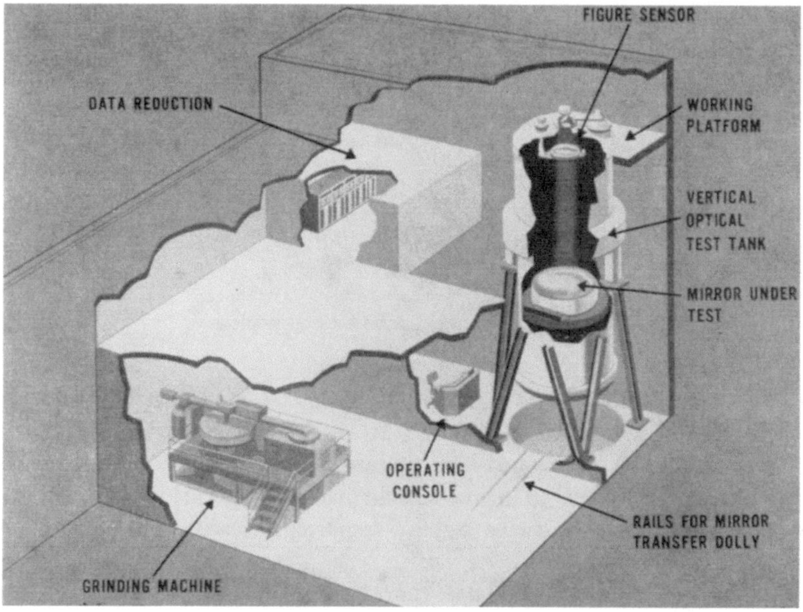

Fig. 12. Optical polishing and test facility.

a vertical optical test tank with a laser unequal path interferometer has been designed by the Itek Corporation. Such a tank is being installed at the GSFC and is expected to be operating within a year. The vertical optical test tank will have the capability of testing the telescope portion of the LST (Fig. 12).

14. NASA Plan for the LST

By following the plan outlined in Fig. 13 one could launch a precursor or intermediate LST in the 1977–1978 time period. This may not be a 3 m diffraction-limited system but it is very likely that a launch of at least a 2 m telescope can be attempted. Certainly by 1981, a 3 m diffraction-limited telescope is realizable.

Spacecraft concepts are being developed which will permit the launch of a large telescope by a Titan 3 launch vehicle. The telescope could operate in an automated mode as a national or international space observatory. Later

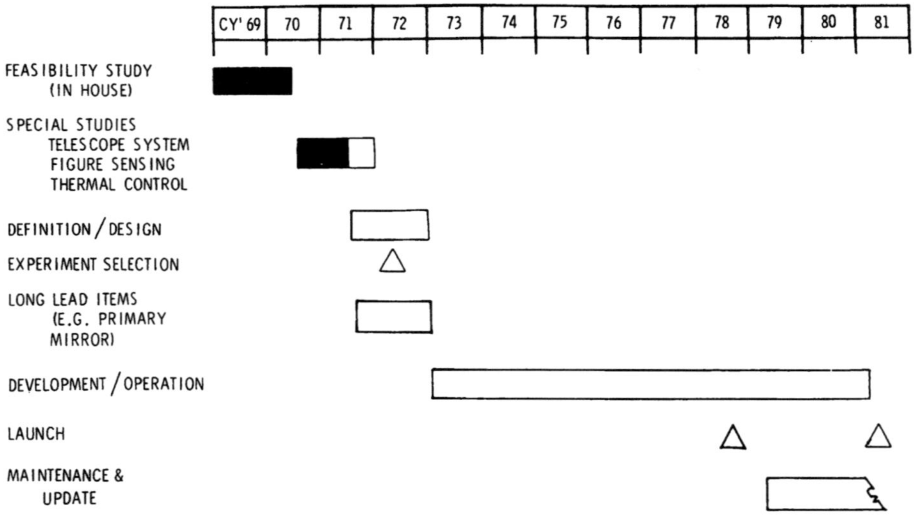

Fig. 13. Large space telescope plan.

when the space shuttle becomes operational, the modularized concepts now being considered would allow the LST to rendezvous with the shuttle for the important functions of instrument update, maintenance, or replacement of malfunctioning modules. This is similar to the operation of ground-based telescopes where once or twice a year some significant update or change of instrumentation occurs.

When the space shuttle becomes operational, it is our intention to use it to launch the 3 m LST as well as to retrieve it from orbit. The mission flexibility provided by the shuttle is quite important in developing a diffraction-limited LST. If it is not possible to test such a large telescope on the ground, the shuttle may provide a means of testing large telescope systems under zero gravity. When operational the space shuttle will permit the launch, test and retrieval of the LST for proper operation, refurbishment, and instrument update. It is our belief that the space shuttle will be a key factor in the LST systems of the 1980's.

In conclusion, the tools and techniques of space flight applied to the Physics and Astronomy Program during the past decade have led to a totally new view of our cosmic environment. We propose to continue the development of new technology and techniques along with proven methods in search of a broadly based but limited set of objectives from the essentially unlimited opportunities that have been brought into our view. Pursuit of these objectives will provide us with a significant understanding of our sun/earth environment and will open to our view as yet unexplored regions of the universe outside our own solar system.

References

[1] Space Research — Directions for the Future, Woods Hole Summer Study, National Academy of Sciences, National Research Council Pub. 1403 (1966).
[2] R. D. Danielson, Interplanet. Science and Technol., pp. 54–56, 58, 60, 62–64, July 1967.
[3] A Long Range Program in Space Astronomy, Astronomy Missions Board Report, NASA SP-213, July 1969.
[4] Scientfic Uses of the Large Space Telescope, National Academy of Sciences 1969.
[5] W. A. Plummer and H. E. Hagy, Appl. Optics 7, 824 (1968).
[6] Optical Materials Study Program, Perkin-Elmer Corporation. Issued by Advanced Research Projects Agency, DoD, Order No. 885, Contract No. DAAH01-69-C-0950, February 1970.
[7] L. R. Canfield, G. Hass and J. E. Waylonis, Appl. Optics 5, 45 (1966).
[8] J. T. Cox, G. Hass and J. E. Waylonis, Appl. Optics 7, 1535 (1968).
[9] A. P. Bradford et al., Appl. Optics 8, 1183 (1969).
[10] Optical Telescope Technology, NASA SP-233, 1970.
[11] Manned Orbital Telescope Study, The Boeing Company, Space Division, NASA Contract NAS 1-3968, D2-84042-1, 1966.
[12] Preliminary Design for a Manned Astronomical Space Telescope, Kollsman Instrument Corporation, Vol. II, NASA Contract NAS 5-11088, July 1969.
[13] Grumman Space Astronomy, Grumman Aerospace Corporation, Vols. I – V, August 1969.
[14] Technology Study for a Large Orbiting Telescope, Itek Corporation, NASA Contract NASw-1925, May 15, 1970.
[15] Instrumentation Package for a Large Space Telescope (LST), Goddard Space Flight Center, X-670-70-480, November 1970.

Papers presented at the XIVth Plenary Meeting of COSPAR and published, but not in Space Research XII

The Moon

J. F. HAYS, Radioactive Heat Sources in the Lunar Interior, Phys. Earth Planet. Interiors, **5**, 77–84 (1972).

Venus and Mars

K. D. PANG, Mariner 9 Ultraviolet Spectrometer Experiment: Photometric Function and Roughness of Mars' Polar Cap Surface, *Icarus* **15**, No. 3, December 1971.

Cosmic Dust

A. P. WAGGONER, N. C. AHLQUIST and R. J. CHARLSON, Measurement of the Aerosol Total-Scatter/Backscatter Ratio, Appl. Optics, (1972), in press.

W. G. MANKIN, R. M. MACQUEEN and R. H. LEE, The Coronal Radiance in the Intermediate Infrared, submitted to *Astrophys. J.* as Letter to Editor.

The Earth observed from Space

B. J. CONRATH, R. A. HAMEL, C. PRABHAKARA, V. G. KUNDE, I. REVAH and V. V. SALOMONSON, Statistical Properties of the Thermal Emission Spectrum of the Earth and Atmosphere obtained with the Nimbus 4 Infrared Interferometer, to be submitted to *J. Geophys. Res.*

K. PETZOLDT, A Comparison between SIRS Radiances and Monthly Mean Maps of the Middle Stratosphere, no reference yet available.

The Earth's Neutral Atmosphere

K. S. W. CHAMPION, The Mean COSPAR International Reference Atmosphere, to appear in CIRA 1972 (Akademie-Verlag, Berlin, in press).

G. V. GROVES, Atmospheric Structure and its Variations in the Region from 25 to 120 km, to appear in CIRA 1972 (Akademie-Verlag, Berlin, in press).

L. G. JACCHIA, Atmospheric Models in the Region from 110 to 2000 km, to appear in CIRA 1972 (Akademie-Verlag, Berlin, in press).

T. TOHMATSU and T. OGAWA, Sounding Rocket Observation of Helium 304 Å and 584 Å glow, *J. Geophys. Res.* **76**, 6136–6145 (1971).

Dynamics of the Thermosphere and Ionosphere above 120 km

T. YONEZAWA, Semi-annual Variations in The Peak Electron Densities of the F 2 and E Layers, *J. Radio Res. Labs. (Japan)* **19**, 1 (1972).

R. LEITINGER, Untersuchungen über die Höhenabhängigkeit der halbtägigen graviationserregten Gezeiten der Atmosphäre, *Z. Geophys.* **37**/5 (1971).

The Ionosphere

F. Arnold, J. Kissel, D. Krankowsky, H. Wieder and J. Zähringer, Negative Ions in the Lower Ionosphere: A Mass-Spectrometric Measurement, *J. Atmos. Terr. Phys.* **33**, 1169–1175 (1971).

F. Arnold and D. Krankowsky, Negative Ions in the Lower Ionosphere: A Comparison of a Model Computation and a Mass-Spectrometric Measurement, *J. Atmos-Terr. Phys.* **33**, 1693–1702 (1971).

B. Betz and D. Krankowsky, Mass Spectrometric Measurement of Positive Ions between 140 and 680 km at a High Latitude, to be submitted to *J. Geophys. Res.*

Polar Ionosphere and Precipitation of Low Energy Charged Particles

T. Hirasawa, T. Nagata, S. Kawaguchi and K. Hirao, Rocket Observations of Electron Density of the Daytime Ionosphere at Syowa Station, Antarctica, *Rep. Ionosph. and Space Res. Japan*, **24**, 237–243 (1970).

R. A. Hoffman and T. Laaspere, A Comparison of VLF Auroral Hiss with Precipitating Low-energy Electrons using Simultaneous Data from two OGO-4 Experiments, *J. Geophys. Res.* **77**, Feb. 1972.

Solar Terrestrial Relationships

I. Kimura, K. Hiraishi and T. Tomimoto, Measurement of Wave Normal Directions of VLF Waves propagating through the Ionosphere by means of the VLF Doppler Technique, to be submitted to *Planet. Space Sci.*

E. Barouch, M. Gros and P. Masse, The Solar Longitude Dependence of Proton Event Delay Time, *Solar Phys.* **19**, 483–493 (1971).

G. A. Doschek, J. F. Meekins, R. W. Kreplin and T. A. Chubb, An Analysis of Soft X Ray Emission Line Behavior during Solar Flares, to be published as a Review Paper presented at 3rd Symposium on UV and X Ray Spectroscopy of Astrophysical and Laboratory Plasmas (IAU Colloquium No. 14, Utrecht 1971), by G. A. Doschek; to appear in *Space Science Reviews.*

INDEX OF AUTHORS

Bei Fragen zur Produktsicherheit wenden Sie sich bitte an:
If you have any questions regarding product safety,
please contact:

Walter de Gruyter GmbH
Genthiner Straße 13
10785 Berlin
productsafety@degruyterbrill.com